Sustainable Development at Universities: New Horizons

Environmental Education, Communication and Sustainability

edited by Walter Leal Filho

Vol. 34

PETER LANG
Frankfurt am Main · Berlin · Bern · Bruxelles · New York · Oxford · Wien

Walter Leal Filho (ed.)

Sustainable Development at Universities: New Horizons

PETER LANG
Internationaler Verlag der Wissenschaften

Bibliographic Information published by the Deutsche Nationalbibliothek
The Deutsche Nationalbibliothek lists this publication in the Deutsche Nationalbibliografie; detailed bibliographic data is available in the internet at http://dnb.d-nb.de.

Copy editing and layout:
Kumpernatz + Bromann.
www.kumpernatz-bromann.de

ISSN 1434-3819
ISBN 978-3-631-62560-6

www.peterlang.de

Table of Contents

II. Implementing Sustainable Development in University Activities

III. Implementing Sustainable Development in Research and in Outreach Programmes

IV. Latin America

Foreword

Forty years after the Stockholm Conference on the Human Environment, and twenty years after the first "Earth Summit" (UN Conference and Environment and Development) held in Rio de Janeiro in June 1992, the world community came together again in Rio to review progress and outline the future steps we need to take towards a more sustainable world. The 2012 UN Conference on Sustainable Development (UNCSD), also known as "Rio+20", was a historic event in the sense that it secured renewed political commitment to sustainable development and addressed new and emerging challenges. The summit was also important in moving forward two specific themes: the green economy in the context of poverty eradication and sustainable development; and the institutional framework for sustainable development.

UNCSD also offered a unique opportunity to discuss the extent to which universities have been providing a meaningful contribution to efforts towards shaping a better world and, perhaps more importantly, to discuss what remains to be done. It was against this background that the World Symposium on Sustainable Development at Universities (WSSD-U-2012), organised by the Research and Transfer Centre "Applications of Life Sciences" of the Hamburg University of Applied Sciences (Germany) and London Metropolitan University (UK), under the auspices of the RCE Hamburg and Region, took place in Rio de Janeiro on 5 and 6 June 2012.

The aims of the World Symposium on Sustainable Development at Universities (WSSD-U-2012) were:

i. to provide universities all round the world with an opportunity to display and present their work (i.e. curriculum innovation, research, activities, practical projects) relating to education for sustainable development at university level;
ii. to foster the exchange of information, ideas and experiences acquired in the execution of projects, from successful initiatives and good practice;
iii. to discuss methodological approaches and projects which aim to integrate the topic of sustainable development in the curriculum of universities;
iv. to network the participants and provide a platform so they can explore possibilities for co-operation.

Last but not least, a further aim of the event was to document and disseminate the wealth of experiences available today. To this purpose, **Sustainable Development at Universities: New Horizons** has been published. Among many interesting chapters, the book contains a description of the Interreg IVB (North Sea) Programme "SKINT", which shows how universities and local authorities may work together in promoting the sustainable use of water resources in northern Europe, with approaches and methods which may be replicable elsewhere in Europe and in other parts of the world. As the title implies, this book is meant to go beyond the current state of knowledge and open the way for new thinking and new horizons in what is a rapidly expanding field, of central interest to both industrialised and developing countries.

I thank the authors for their willingness to share their knowledge, know-how and experiences, as well as Dr Mihaela Sima for her first-class work in supporting WSSD-U 2012 and this book. Enjoy your reading!

Prof. Walter Leal Filho (BSc, PhD, DSc, DPhil, DEd, DL, DLitt)
Autumn 2012

I.
Implementing Sustainable Development in the Curriculum

Chapter 1

Building Sustainability Leaders: A Framework to Prepare Students to Thrive on Complexity and Lead Transformative Changes

Michael Shriberg

Abstract

Leading institutions of higher education are increasingly utilizing the campus as a laboratory to move beyond organizational "greening" into developing the skill set in students to lead deep institutional changes for sustainability in academia and beyond. Unfortunately, the "wickedness" of environmental and interrelated social issues leaves traditional leadership theory, which is still largely focused on industrial efficiency, ill-prepared to provide skills training for the necessary transition to sustainability. Leadership within the environmental movement has moved from a radically decentralized, distributed model to a more professional and hierarchical model yet still lacks basic skills training and assessment models. Therefore, institutions of higher education are left without adequate models upon which to base sustainability leadership training. This paper helps bridge this divide by developing a concept of sustainability leadership that connects the strengths of traditional leadership theory and environmental leadership through skills building in visioning, systems thinking, change management and thriving on complexity. At least 21 institutions have programs focused on building sustainability leadership skills, with varying designs and skills emphases. At the University of Michigan, we are preparing students to thrive in sustainability arenas via an undergraduate sustainability certificate program and a doctoral sustainability fellows program, among other initiatives, with an emphasis on systems thinking, organizational change and personal responsibility in a social context. In the future, skills development in sustainability in academia will define institutional and individual success in this field and must proceed rapidly and effectively to help counteract negative sustainability trends through the empowerment of new leaders.

Introduction

Sustainability leaders are urgently needed. As ecological and social degradation accelerates, wealth disparity widens, and the foundations of our global economy continue to weaken, leadership skills and practices to guide society on a sustainable path are and will continue to be in high demand. The organizations most responsible for shaping these future leaders, and setting the dominant paradigm in which they operate, are institutions of higher education. However, despite the many broad calls to action, and the robust nature of general leadership studies, there is little guidance available for these institutions on empowering students to become sustainability leaders. Dominant leadership theories are not readily adaptable to the unique challenges of sustainability, and the environmental movement has largely failed to systemically address leadership skills building. This paper provides this critical guidance by analyzing leadership theory in the context of sustainability, creating a skills-building framework that provides potential direction for institutions of higher education purporting to build sustainability leadership skills.

Sustainability and leadership theory

Leadership theory and the rise of the consumer society emerged simultaneously, beginning as early as the 1800s, with industrial efficiency as its primary goal. During the early stages, the dominant metaphor for organizations and the people within them was the "machine", with employees acting as cogs (Rost 1997). The leader's role is to make the machine function more efficiently through the use of hierarchical power exerted through positional authoritative relationships (e.g. the standard "command-and-control" model) (Allen et al. 1999; Dugan and Komives 2010; Redekop 2010). The rewards in this system are extrinsic (i.e. money) and the system is geared toward maximizing productivity. This "rational man" model, often termed "transactional" leadership for its focus on keeping transactions moving forward quickly, is elegant in its simplicity: leaders lead, followers follow, and maximizing profit raises all boats (Bass 1985; Burns 1978; Kuhnert and Lewis 1987). Unfortunately, this efficiency-focused model is partly responsible for efficient degradation of the planet's natural and social capital. While this leadership model may work well in times of plenty, and for situations where the management outcome is developed within a social and ecological context, it is inadequate in terms of preparing future sustainability leaders. The dominant training under this model is in corporate efficiency, as exemplified in many MBA programs. Go out and do a better job producing profits and maximizing the use of human capital, and good results will follow. However, in today's globally connected world, where "wicked" problems – defined as where "facts are uncertain, values in conflict, stakes are high and decisions are urgent, and an extended peer community is required for the resolution of the relevant issues" (Gough et al. 1998) – dominate and sustainability is a critical organizing principle, the skill set needed for leaders must reach far beyond power, productivity and good "management".

The struggle to keep up with social context as well as recognize that intrinsic motivation is just as, if not more, important than extrinsic motivatis led to transformational leadership theory (Burns 1978). This model is now viewed as the classic mode of leadership: charismatic leaders at the top of the hierarchy gain followers by building a shared vision and set of goals that "transform" the organization and the individuals within it. The theory chips away at the "rational man" model by recognizing that employees desire inspiration as well as compensation. Typically, the CEO inspires great changes in his (or, less frequently, her) organization through personal powers of motivation as well as persuasion, building trust and loyalty. The outcome is major realignment of organizational goals, mission, culture and values toward a "higher ethical purpose" (Dvir et al. 2002). This "purpose" can be related directly to sustainability, though most often is not, since ecological and social context is not necessarily a precursor or major motivator in organizational actions. The training for being a "transformational leader" involves developing skills around vision and communication, although some argue that transformational leadership potential is largely innate.

While transactional and transformational leadership still dominate the field, emerging leadership models with different foci and more decentralized modes of control are becoming more prevalent. For example, specifically in higher education, the social change model of leadership measures eight core values of change in student leaders: three each at the individual level (consciousness of self, congruence, commitment) and group level (collaboration, common purpose, controversy with civility) in addition to the social value of citizenship and collective value of change for the common good (Dugan and Komives 2010). Used specifically with students and still an outlier in the leadership field, this model approaches the sustainability ideal from a social advocacy and personal change perspective. More directly aligned with sustainability are the emerging fields of ecological leadership (Wielkiewicz and Stelzner 2010), eco-leadership (Western 2010), and deep systems leadership (Satterwhite 2010). These models are based in principles of ecology and systems thinking, using an ecosystem or ecological web as the analogy for an organization. In this context, leadership is relational, context-dependent and deeply embedded in social, ecological and economic systems. Complexity dominates, as individual relationships and webs of interconnection drive organizational action or inaction. Power is dispersed, and positions matter less than situations. This powerful analogy has yet to penetrate deeply into leadership theory, perhaps in part because it is difficult to train people in and operationalize. In fact, leading proponents argue that eco-leadership "cannot be learned through skills training, rather it is a way of being" (Western 2010) and that deep systems leadership is a "lifestyle" more than a skill. This leaves institutions of higher education committed to training future sustainability leaders without a clear skills training pathway.

Environmental and sustainability leadership

Leadership theory and practice in the modern environmental movement has followed a very different path than traditional leadership theory, with attendant differences in leadership skills training. Emerging in local grassroots environmental battles of the 1960s and 1970s over pollution of local waterways, smog, and land protection, environmental leadership was extremely dispersed and non-hierarchical, with no formal training (Snow 1992). Since leaders united over specific issues and arose as a counterbalance to the dominant corporate and governmental forces, the leadership model was radically distributed, not unlike the current ecological leadership models. Many of these individuals and the loose-knit organizations they formed specifically eschewed hierarchy, creating "collaboratives" and other structures that connected individuals and formed partnerships while minimizing hierarchical relationships. One of many such examples is the Ecology Center, an organization founded at the University of Michigan after the first large-scale Earth Day demonstration in 1970. The energy and enthusiasm of the student organizers, and the resources they garnered, led to the creation of a permanent organization which became the premier grassroots advocacy professional organization in the state. The setup was a "collaborative" with group decision-making and no formal lines of authority. Leadership training for organizations like the Ecology Center and early environmental organizers was sparse, in part because the traditional notion of leadership was often rejected.

With local successes growing, but mounting, increasingly complex environmental crises on the horizon, the leadership model in the environmental movement became more professional and hierarchical in the 1980s and 1990s. This change occurred as the dominant modes of pollution in the developed world switched from the visible and local (i.e. point source water pollution, smog) to the global and invisible (i.e. CO_2, toxic chemicals). Concurrently, the advocacy model became more professionalized and even included – in the case of Ecology Center, for example – entrepreneurial endeavours which are structured more like a traditional business. Many organizations (like Ecology Center) adopted a formal leadership structure with the ability to hire and fire, control budgets and set institutional direction, while maintaining aspects of a collaborative decision making structure. These changes mark a potential convergence with the traditional leadership field. However, this transition has occurred with very limited attention to leadership skills building.

Fig. 1: Sustainability leadership bridges environmental and traditional leadership theory and practice

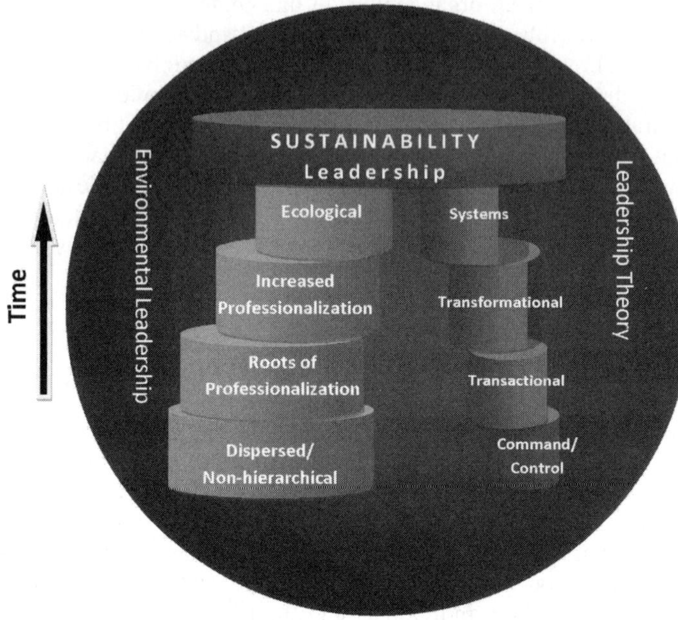

Note: Width of cylinders represents how widely control is held and how mechanistic the thinking is within organizations

Fig. 1 represents the changing and converging nature of environmental leadership and leadership theory. As illustrated, the concept and practice of sustainability has the potential to bridge the leadership divides by merging environmental leadership with standard leadership theory, through its emphasis on systems thinking, long-term relationships and ability to catalyse organizational and individual actions. Colleges and universities have a critical opportunity to advance sustainability leadership in theory and practice by building this bridge through developing student skills that reflect the issues and challenges of the times.

With this theoretical framework, the practical question for institutions and individuals with an interest in building sustainability leaders becomes: how do we adequately prepare students to reach the pinnacle of sustainability leadership? Training students in sustainability leadership poses unique challenges – students need to be trained to embrace complexity, uncertainty and morality. This training involves building specific skills that are broadly and powerfully applicable. For example, visioning is critical – and often underutilized by environmentalists – in building a shared future that is sustainable (Meadows 1996). Without vision, environmental challenges seem dismal, insurmountable and fatalistic. Sustainability leaders must embrace "humility" by seeking multiple voices. This adaptability is in many ways the opposite of what transformational leadership theory teaches, but is critical in developing the adaptive capability of addressing sustainability challenges. But perhaps the most important skill is not only to be unafraid of change, but to embrace and capitalize on change via a systems lens. Dubbed "systems intelligence" by Donella Meadows (2008), this perspective and skill set is vital to all current and future sustainability leaders, and must be built carefully over time.

Review of sustainability leadership programs

As identified through a web search, at least 21 colleges and universities across the US, Europe and Australia now offer programs specifically purporting to build sustainability leadership skills. They face the challenge of cultivating sustainability leadership without established metrics or standards for success. These institutions typically encourage participants to "make a difference" and be leaders in change to a more sustainable future, adopting multiple methodologies and approaches. In these programs, sustainability skills development occurs through:

Foundations in coursework or workshops: While programs vary in structure, the majority are rooted in academic coursework (typically four to six courses) and generally comprised of a core sustainability course, multiple electives, and a capstone course. Programs not based in for-credit coursework generally provide online or in-person workshops.

Interdisciplinarity: Programs highlight the many disciplines, sectors, and aspects of sustainability, often bringing together students or professionals from diverse fields and sectors to learn about these issues. While nearly all programs aim to be "interdisciplinary", curriculums focused on sustainable business are sometimes narrower in scope.

Applied learning: More than half of the reviewed programs involve an applied learning component in the form of a directed project or internship, offering an intensive experience that challenges acquired skills and is an integral part of many sustainability leadership programs. Most often these experiences are a leadership capstone.

Emphasis on Positive Community Impact: Sustainability may be a global issue, but most programs teach practical skills through engagement on a community level. Through capstone projects and internships, programs foster community involvement and civic engagement, using the local community or organization as a microcosm for larger social and ecological changes.

Creations of cohorts and networks: Programs commonly are either cohort-based or strive to create a network of students, professionals, or both. Many programs are explicit in their hope to create a supportive learning community for participants.

While programs are intended for multiple audiences – including students at all levels and in all fields – the majority focus on graduate or professional education. For example, the University of California – Los Angeles (UCLA) Leaders in Sustainability Program aims to shape graduate students of all programs of study into leaders who address sustainability in a multidisciplinary fashion through four required courses, attendance at seminars and conferences, and completion of a hands-on project [1]. The Ithaca College Professional Certificate in Sustainability Leadership provides knowledge of best practices of sustainability to bring back to organizations through six online courses, including a required systems thinking course and five electives related to marketing, communications, management and other organizational elements [2].

The Centre for Sustainability Leadership's program (in Australia) best illustrates the strong workshop model. Geared towards early- to mid-career professionals, the program offers weekly skills development workshops with speakers and discussion [3]. Similarly, Bard College's new Campus to Congress (C2C) Fellowship program brings undergraduates and recent graduates together for intensive weekend training workshops to develop leadership and communication skills [4].

The Sustainability Management Professional Certificate at the University of Colorado Boulder illustrates the experiential, capstone model, with courses on topics such as organizational change and communication supplemented by a 20-hour capstone project [5]. Similarly, the University of Oregon's Environmental Leadership Program consists of a 4-8 credit hands-on course where undergraduate and graduate students work on environmental projects with local businesses, governmental agencies, and non-profits [6]. The program strives to develop students' leadership and professional skills and links them into a local professional network.

Sustainability leadership programs (see Fig. 2) appear to be heavily influenced by transformational leadership model, and often emphasize communication skills. Aspects of social change leadership are also evident in value placed on collaboration, citizenship, the common purpose of sustainability, and change for the common good. There is far less emphasis on transactional leadership in the program descriptions, although the program implementation may revert to this mindset. The conceptual framework for sustainability leadership skills building appears to be emerging through these programs through structures that focus on understanding and applying sustainability and the complexity that accompanies it, often via the "systems lens". Through grounding in the environment (the social aspects of sustainability are absent in most programs) as well as hands-on experience, and a supportive cohort or network, program participants should acquire key change management skills. Less apparent are the visioning and humility skills that help define effective sustainability leadership.

Fig. 2: Sustainability leadership programs

Program Name	Institution	Website	Audience (Undergrad, Graduate, Professional)
Building a Sustainable World Certificate	Saybrook University	http://www.saybrook. edu/phs/academicprograms/ce	Graduate Students
Campus to Congress Fellowship	Bard College	http://www.bard.edu/cep/c2c/	Undergraduate students and recent graduates
Certificate in Socially Responsible Leadership	Queen's School of Business	http://business.queensu. ca/centres/crl/certificate_progr	Undergraduate students
Centre for Sustainability Leadership Fellowship Program	Centre for Sustainability Leadership, Australia	http://www.csl.org.au/? page_id=29	Early to mid-career professionals
Contemplative Leadership and Sustainability (C.L.A.S. P.)	Santa Clara University, Leavey School of Business, California	http://www.scu. edu/business/clasp/index.cfm	Undergraduate students
Certificate in Socially, Environmentally, and Ethically Responsible (SEER) Business Practice for full-time MBA students	Pepperdine University	http://bschool.pepperdine. edu/programs/full-time-mba/seer/overview.htm	Graduate students
Duke Environmental Leadership Program	Duke University	http://www.nicholas.duke. edu/del/	Professionals
Environmental Conflict Resolution and Collaboration Graduate Certificate	George Mason University	http://catalog.gmu. edu/preview_program.php? catoid=5&poid=3796#	Mid-career professionals
Environmental Leadership Program	University of Oregon, Holden Leadership Center	http://leadership.uoregon. edu/get_involved/classes/340	Undergraduate and graduate students
MSU Specialization in Sustainability	Michigan State University	http://www.reg.msu. edu/AcademicPrograms/Progr	Undergraduate students
Sustainability Leadership Graduate Certificate Program	Edgewood College, Madion, Wisconsin	http://www.edgewood. edu/academics/graduate/susta and http://www.edgewood. edu/academics/graduate/susta	Graduate students
Professional Certificate in Sustainability Leadership	Ithaca College	http://www.ithaca. edu/gps/professional/devcert/s	Professionals
Environmental Sustainability: Graduate Certificate	Ohio University	http://www.catalogs.ohio. edu/preview_program.php? catoid=13&poid=2692&bc=1	Graduate students
Leaders in Sustainability Program	UCLA	http://www.environment.ucla. edu/lis/	Graduate students

University of Michigan programs

At the University of Michigan's Graham Environmental Sustainability Institute, we have designed two programs specifically to build sustainability leadership skills to bridge the traditional and environmental leadership gap and directly address "wicked" problems. The Graham Undergraduate Sustainability Scholars Program at the University of Michigan accepts a diverse group of 25 high-performing sophomores each year. The program combines leadership training with organizational change theory and practice, emphasizing interdisciplinary thinking and action. The program complements and adds value to other academic and co-curricular programs, creating engagement across majors and programs and developing a base of social support and a strong network for aspiring sustainability leaders. Scholars engage in a two-year, 11-credit series of coursework, which includes (in approximate chronological order):

Interdisciplinary Thinking for Sustainability: Scholars begin the program by undertaking an interdisciplinary review of core environmental and social problems and solutions through the framework of systems thinking for sustainability.

Sustainability & the Campus: This course integrates knowledge and hones professional skills using the campus as a living-learning laboratory through an intensive team-based project in partnership with University of Michigan operational staff. The focus is on organizational change theory and applying systems thinking.

Field-Based Sustainability Course or Alternative: Scholars are funded to participate in a course or experience that addresses sustainability in a context outside of the campus. Courses include: Sustainable and Fossil Energy: Options and Consequences (Wyoming, USA); Sustainability Challenges & Opportunities in East Africa (Kenya); Sustainable Neighborhoods: Experiential Learning & Active Engagement in Detroit (Michigan, USA); and Toward a New Sustainable Environment in Light of the Changing Face of Rural (and Urban) China. Scholars can petition to substitute an alternative field-based course or experience for these course options, and have conducted experiences in Brazil, Hawaii, Germany, Iceland and other locations.

Interdisciplinary Sustainability Elective: Scholars take at least one elective that approaches sustainability from an interdisciplinary perspective.

Sustainability Leadership Development: Taken in fall of senior year, this capstone seminar explores the concepts and application of sustainability leadership. Through reading seminal works in the field, exploring personal leadership styles and experiences, case studies, and interaction with practitioners, the course develops sustainability leadership skills and prepares Scholars for professional experiences. Specifically addressing the theoretical perspectives described early in this paper, including eco-leadership and deep systems leadership, the course serves as the intellectual and cultural capstone experience.

Beyond coursework, Scholars participate in complementary leadership and cohort-building activities to strengthen connections and foster broad-based learning. For example, Scholars (along with Graham Doctoral Sustainability Fellows, described next) visit the U-M Challenge Program (a ropes course) early in the academic year to build leadership and group process skills, and to begin developing a sense of community. Scholars meet informally at least monthly to discuss topics from a sustainability discussion guide, view documentaries or meet with practitioners, and they engage with speakers and others on campus in an informal setting at least once per semester. Each semester also features social events organized by the faculty director and the students themselves. In addition, Scholars create many other opportunities to engage and connect on their own; for instance, Scholars have self-organized to produce a blog with positive news about sustainability (http://eco-mmendableactions.blogspot.com/). Overall, the Scholars program helps aspiring sustainability leaders harness advanced critical thinking and problem solving skills with an emphasis on systems, values, adaptation, resilience, collaboration and complexity.

The Graham Doctoral Sustainability Fellows Program reaches future sustainability leaders at a critical stage in their intellectual development: while they are writing their dissertations. The goal is to create a lifelong "Community of Scholars" devoted to developing and implementing innovative ideas in pursuit of sustainability. This community is developed using the principles of sustainability leadership, with elements of social change leadership and eco-leadership. Each year, the group is selected to represent a wide range of interests to enhance the interdisciplinary nature of the collaboration, and each recipient receives up to USD50,000 over two years to support their research and intellectual development. Fellows meet monthly in a seminar, during which one student presents his or her research in the context of the larger sustainability questions it raises. After the brief presentation, a Fellow from a different field facilitates a conversation that proceeds online (through a blog) as well as in person (through an informal gathering), culminating in a discussion during the next monthly seminar. The goal is to move beyond the micro-details of doctoral research into the realm of deep sustainability analysis. At the end of each academic year, the Fellows present the results of their discussion to their academic advisers and other key stakeholders at a celebratory event. Moreover, Fellows collaborate on a project, examples of which include a published paper in *Sustainability Science* on the interdisciplinarity of the use of sustainability in different fields, the creation of a place-based sustainability course for Graham Sustainability Scholars and the formation of a new online journal. Through this mix of financial support, intellectual exchange, interdisciplinary collaboration, and social activities, the program seeks to empower future sustainability leaders with the ability to understand and effectively communicate the complex and interdisciplinary nature of sustainability, following the best practices of sustainability leadership skills building.

Conclusion

Institutions of higher education face a profound, difficult and urgent challenge in training future sustainability leaders. Unfortunately, traditional leadership skills training tends to be ineffective when lined up against the "wicked" problem of sustainability since the perspective is often narrow, hierarchical and individual-dependent. Leadership skills training specifically in environmental leadership tends to be collaborative, based in part on the collective models of the early environmental movement, but is virtually non-existent at any scale that is significant enough to meet current and future demand. Therefore, colleges and universities are attempting to meet the pent-up demand for sustainability skills training through curricular, co-curricular and experiential modules. Analysis of 21 of the leading programs shows a strong interdisciplinary grounding and focus on hands-on learning, but a lack of coherent and consistent thought about skill building around vision and humility. At the University of Michigan, sustainability leadership skills development is the focus of two programs – one for undergraduates and one for doctoral students – designed to emphasize visioning, systems thinking and change management.

Sustainability leadership skills training is only beginning to emerge yet holds the key to creating the next generation of sustainability leaders. As our understanding of sustainability challenge evolves and we as a society continue to pilot different approaches to the deep social changes sustainability will require, our approaches to sustainability leadership training is advancing in sophistication. The critical niche of higher education – which is only beginning to be filled – is to not only educate and inspire aspiring sustainability leaders, but to provide a basis in skills training that can be transferred readily and immediately into the professional world and beyond.

References

Allen, K.E., Stelzner, S.P. and Wielkiewicz, R.M. (1999), The ecology of leadership: Adapting to the challenges of a changing world. *Journal of Leadership & Organizational Studies* 5 (2), pp. 62-82.

Bass, B.M. (1985), *Leadership and Performance Beyond Expectations.* Free Press, New York, NY.

Burns, J.M. (1978), *Leadership.* Harper & Row. New York, NY.

Dugan, J.P. and Komives, S.R. (2010), Influence on college students' capacities for socially responsible leadership. *Journal of College Student Development* 51 (5), pp. 525-549.

Dvir, R., Eden, D., Avolio, J.B. and Shamir, B. (2002), Impact of transformational leadership on follower development and performance: A field experiment. *The Academy of Management Journal* 45 (4), pp. 735-744.

Gough, C., Castells, N. and Funtowicz, S. (1998), Integrated Assessment: An emerging methodology for complex issues. *Environmental Modeling and Assessment* 3, pp. 19-29.

Kuhnert, K.W. and Lewis, P. (1987), Transactional and transformational leadership: A constructive/developmental analysis. *The Academy of Management Review* 12 (4), pp. 648-657.

Meadows, D. (2008), *Thinking in Systems: A Primer.* Chelsea Green Publishing, White River Junction, VT.

Meadows, D.H. (1996), Envisioning a Sustainable World. In Costanza, R., Segura, O. and Martinez-Alier, J. (Eds.), *Getting Down to Earth, Practical Applications of Ecological Economics,* pp. 117-26. Island Press, Washington, D.C.

Redekop, B.W. (2010b), Connecting leadership and sustainability. Introduction. In Redekop, B.W. (Ed.), *Leadership for Environmental Sustainability*, pp. 1-16. Routledge, New York, NY.

Rost, J.C. (1997), Moving from individual to relationship: A postindustrial paradigm of leadership. *The Journal of Leadership Studies* 4 (4), pp. 3-16.

Satterwhite, R. (2010), Deep systems leadership: A model for the 21st century. In Redekop, B.W. (Ed.), *Leadership for Environmental Sustainability*, pp. 230-242. Routledge, New York, NY.

Snow, D. (1992), *Inside the Environmental Movement: Meeting the Leadership Challenge*, Island Press, Washington, DC.

Western, S. (2010), Eco-leadership: Towards the development of a new paradigm. In Redekop, B.W. (Ed.), *Leadership for Environmental Sustainability*, pp. 36-54. Routledge, New York, NY.

Wielkiewicz, R.M. and Stelzner, S.P. (2010), An ecological perspective on leadership theory, research, and practice. In Redekop, B.W. (Ed.), *Leadership for Environmental Sustainability*, pp. 17-35. Routledge, New York, NY.

[1] http://www.environment.ucla.edu/lis/

[2] http://www.ithaca.edu/gps/professional/devcert/sustainability/

[3] http://www.csl.org.au/fsl-program/what-is-it/program-overview/

[4] http://www.bard.edu/cep/c2c/

[5] http://ecenter.colorado.edu/sustainable-practices/324-professional-certificate

[6] http://leadership.uoregon.edu/get_involved/classes/340

Chapter 2

Fostering Rapid Transitions to Education for Sustainable Development through a Whole-System Approach to Curriculum and Organizational Change

Cheryl J. Desha and Karlson 'Charlie' Hargroves

Abstract

Despite decades of attempts to embed sustainability within higher education, literature clearly suggests that highly regulated disciplines such as engineering have been relatively slow to incorporate sustainability knowledge and skill areas, and are generally poorly prepared to do so. With current efforts, it is plausible that sustainability could take another two decades to be embedded within the curriculum. Within this context, this paper presents a whole system approach to implement systematic, intentional and timely curriculum renewal that is responsive to emerging challenges and opportunities, encompassing curriculum and organizational change. The paper begins by considering the evolution of curriculum renewal processes, documenting a number of whole system considerations that have been empirically distilled from literature, case studies, pilot trials, and a series of workshops with built environment educators from around the world over the last decade. The paper outlines a whole-of-institution curriculum renewal approach to embedding sustainability knowledge and skills within the DNA of the institutional offerings. The paper concludes with a discussion of research and practice implications for the field of education research, within and beyond higher education.

Introduction

Worldwide, the human species faces challenging times on an unprecedented scale, arising from a range of environmental pressures combined with the required scale and speed of the response needed to reduce such pressures (Brown 2007). 60 years on from Rachel Carson's description of a 'silent spring' (Carson 1962), and 20 years on from the first Rio environment summit (UNEP 1992), issues such as climate change, deforestation, ocean acidification and biodiversity loss provide daily reminders of the urgency for action to restore a number of significant planetary systems.

As discussed by von Weizsäcker et al. (2009), it is imperative that society transforms its relationship with the use of resources in order to do much more with much less and at the same time play a role in restoring degraded systems. Economy-wide efforts will be required to reduce environmental pressures on a meaningful scale (Stern 2006). Further such efforts need to ensure that economic performance is maintained, referred to by Smith et al. (2010) as the challenge to '*decouple economic growth from environmental pressure*', as shown in Fig. 1. Furthermore, positive environmental impacts (reforestation, aquifer recharge, etc) need to be 'recoupled' to economic performance towards sustainable development that ensures the needs of future generations can also be met (WCED 1987).

Fig. 1: The decoupling diagram for transforming the relationship between economic growth and environmental quality

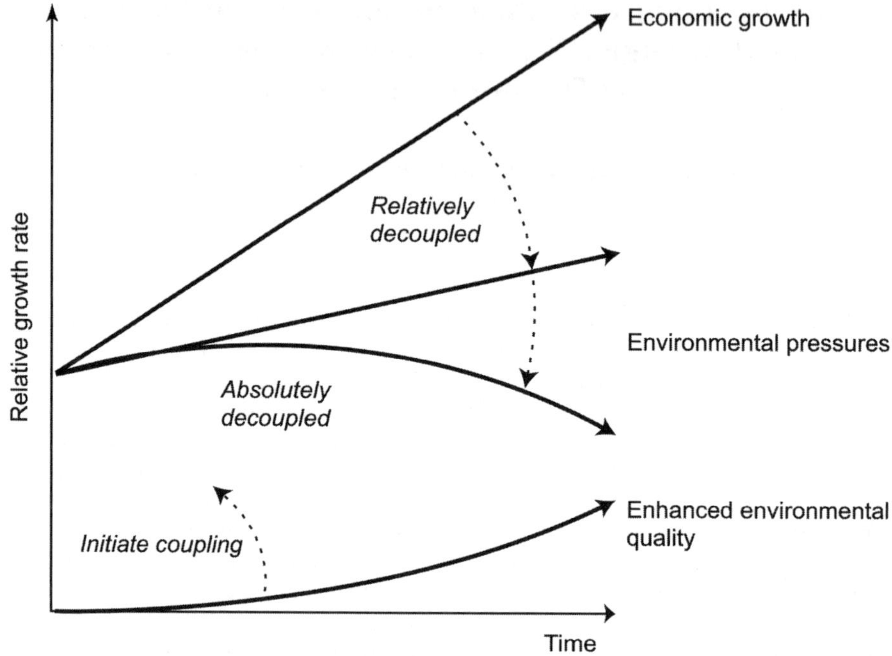

Hargroves et al. in Smith et al. (2010)

In light of this predicament, leaders around the world are now asking, '*How can sustainability knowledge and skills be effectively embedded within society?*', and further, '*What can be done to accelerate the process?*'. These questions stand to dominate much of the agenda in government and capacity building sectors internationally over the coming decades (AASHE 2010; Corcoran and Wals 2008). As highlighted by personalities such as Diamond (2005) and Gore (2006), a significant challenge will be to move from denial and despair about the magnitude and complexity of the challenge, to expedient action. Critical to the transition will be capacity building – education for sustainability (EfS) – across all sectors of society including government and the private sector, in particular for those who are currently in, or who soon will be in decision-making positions. As concluded in one of the early declarations of capacity building in the higher education sector, the Talloires Declaration: '*Universities educate most of the people who develop and manage society's institutions. For this reason, universities bear profound responsibilities to increase the awareness, knowledge, technologies, and tools to create an environmentally sustainable future*' (Shriberg and Tallent 2003).

With this in mind, this paper presents a whole system approach to implement systematic, intentional and timely capacity building within the higher education sector, that is responsive to emerging challenges and opportunities, encompassing curriculum and organizational change. We begin by considering the evolution of curriculum renewal processes, documenting a number of whole system considerations that have been empirically distilled from literature, case studies, pilot trials, and a series of workshops with built environment educators from around the world over the last decade. We then outline a whole-of-institution curriculum renewal approach to embedding sustainability knowledge and skills within the DNA of the institutional offerings, adapted from an existing helix for organizational change and drawing on literature, case studies, pilot trials, and a series of workshops with educators from around the world. Lessons learned from trials to date are highlighted. We conclude with a discussion of research and practice implications for the field of education research, within and beyond higher education.

Capacity building in context

Matching waves of innovation with curriculum renewal

As outlined in '*The Natural Advantage of Nations*' (Hargroves and Smith 2005), following the industrial revolution society has experienced five major waves of technical innovation, with the sixth wave underway providing '*a critical mass of enabling eco-innovations making integrated approaches to sustainable development economically viable*'. With this perspective in mind, each new wave of innovation calls for the world's economies to be 'upgraded' resulting in significant changes across industry, governance, and in due course, the higher education system. For example, the shift to electrification during the third wave called for education and capacity building in many new areas across the professions, such as engineers and designers learning how to generate and distribute electricity and manufacture electrical fittings and appliances; economists and policymakers learning how to set energy prices, and lawyers learning how to consider the new liabilities and risks to infrastructure and equipment from electricity.

If we consider the response to each wave of innovation by the education sector we can see that each wave has called for significant updating of education and capacity building programs, as shown stylistically in Fig. 2. Following each wave the higher education system undertook a curriculum renewal transition to update programs with the new innovations, and applied them to the specific discipline and the needs of the employers of graduates. As the level of innovation in knowledge and skills has progressively increased with each wave this has called for increasingly large-scale efforts to achieve the associated curriculum renewal transition. As the sixth wave has and will continue to create a significant amount of knowledge and skills and considering that there is a serious time imperative to achieve significant reductions in environmental pressures the challenge to update programs in the next two decades will be the greatest challenge the education system has ever faced. As highlighted in a United Nations Environment Program report on working in a low-carbon world, '*... companies in the fledgling green economy are struggling to find workers with the skills needed to perform the work that needs to be done. Indeed, there are signs that shortages of skilled labor could put the brakes on green expansion ... There is thus a need to put appropriate education and training arrangements in place*' (UNEP 2008).

Furthermore, the educators themselves need capacity building. As Walter Erdelen (UNESCO Natural Sciences) reflected in his plenary address to the 2009 World Congress on Higher Education, '*New forces are transforming higher education at a speed that could not have been foreseen 10 years ago ... Higher education institutions play a strategic role in finding solutions to today's leading challenges in the fields of health, science, education, renewable energy, water management, food security and the environment ... We need higher education institutions to train teachers in the conduct of pedagogical research and develop relevant curricula that integrates the values of sustainable development ...*' (Erdelen 2009).

With considerable growth in higher education over the last century, the effectiveness of preparing professionals to contribute to society would appear to be self-evident. The higher education sector has experienced remarkable growth and development, rising to the challenge of times of rapid change and upheaval such as the industrial revolution and the world wars (Perkin 2007). However, despite successes in incorporating the digital wave of innovation into programs over the last two decades, signals clearly suggest that although there are a number of leaders in the sector, generally speaking higher education has been slow to move to incorporate sustainability and is generally poorly prepared to do so (Sterling 2005; Concoran and Wals 2008; Wals 2008; Desha et al. 2009; and Jones et al. 2010).

Fig. 2: A schematic of curriculum renewal transitions, following significant waves of innovation

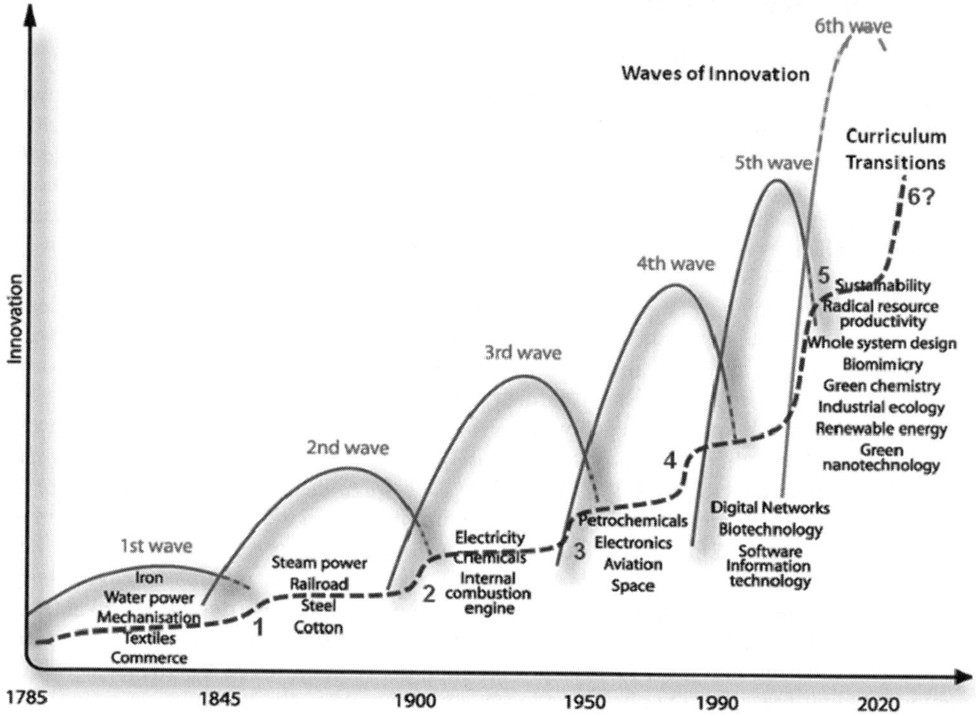

Desha and Hargroves (2011)

Given that a typical (or 'standard') process of undergraduate curriculum renewal for an accredited program – including for example engineering, architecture, planning, law, business or education – may take three to four accreditation cycles (of approximately five-year intervals), the time necessary to fully integrate a substantial new set of knowledge and skills within all year levels of a degree will be in the order of 15-20 years. Further, as the average pathway for a graduate is approximately two to four years from enrolment to graduation, followed by three to five years of on-the-job graduate development, if institutions take the typical approach to fully renew such bachelor programs, this will result in a time lag of around 20-28 years; hence it will be some two to three decades before students graduating from fully integrated programs will be in decision-making positions using current methods. For postgraduate students the time lag will be shorter as students may already be practising in their field and will return to positions of influence; however, accounting for the time to renew programs the time lag is in the order of 8-12 years, depending on the pace and effectiveness of curriculum renewal efforts.

In hindsight, if institutions had acted on the major previous calls for capacity building related to sustainability, such as in *Our Common Future* (WCED 1987), then the standard processes may have been sufficient over the subsequent 20-30 years. However, this window for such a response has well and truly closed. Hence the urgency and complexity of the challenge to reduce environmental pressures, combined with the scope of associated knowledge and skills required to be incorporated into programs, calls for both an improvement in, and acceleration of, the standard approach to curriculum renewal across higher education. Clearly traditional approaches exceed the timeframes needed to significantly reduce a range of environmental pressures as outlined previously (Desha et al. 2009).

Addressing the time-imperative for action

Considering the growing levels of environmental pressures, and that the associated economic impacts are already being felt, it is clear that the current low levels of compliance on environmental performance – in this case capacity building for sustainable development – required by governments will rapidly increase in the near future, shown stylistically to begin at a hypothetical 'Time (*t*)' in Fig. 3. Factors that will influence the timing of 'Time (t)' will include the level of environmental damage and potential collapse of ecological systems such as bee communities required for wide-scale pollination, fish stocks, storm surges and sea-level rises, increased natural disasters, and so on. Following 'Time (t)', the level of environmental performance of an organisation, business, or education institution will dictate the pace at which action is taken to comply with enforcement.

Within the higher education sector, this rapid increase in compliance requirements will be evident in a range of ways, such as: regulatory and policy changes to enforce improved environmental performance in industry and business practices requiring changes to graduate attributes, professional body and accreditation agency requirements for specific graduate attributes to be included in education programs, funding agencies requiring incorporation of related topics in research grant and capital funding applications, and a significant increase in demand from potential employers for graduates with associated graduate attributes.

Fig. 3: Level of commitment to reducing environmental pressure over time

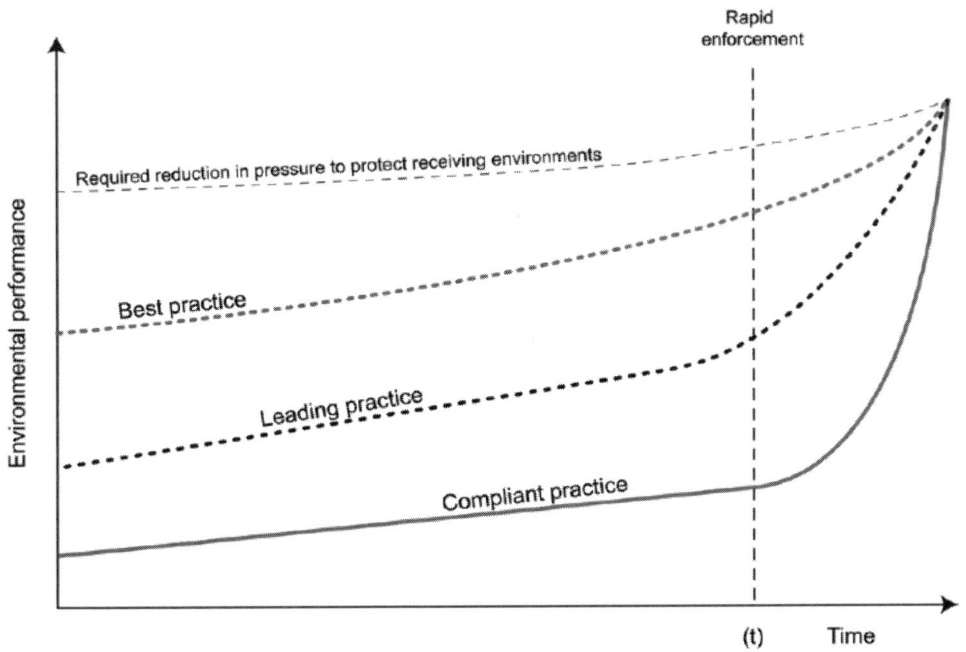

Smith et al. (2010)

Hence the challenge for economies around the world is – in the case of greenhouse gas emissions – to rapidly achieve a low peaking in emissions around 2020, and to then allow a more manageable annual reduction target. The challenge of the higher education sector is that the timeframe to achieve peaking does not allow for the required knowledge and skills, which are largely yet to be incorporated into undergraduate programs, to be developed over the standard curriculum renewal timeline, meaning that it will be largely a postgraduate and professional development challenge. Further, in order to prepare undergraduates to contribute to society achieving gradual sustained reductions after the peaking is achieved, the standard curriculum renewal process will need to be improved and accelerated. In practice this calls for a dual focus, both on engaging with current practitioners and decision-makers on the knowledge and skills required to peak greenhouse gas emissions (such as postgraduate certificates, diplomas and masters programs, along with professional development seminars and short courses), and also focusing on undergraduate programs to develop knowledge and skills required to both maintain the peaking and then achieve gradual sustained reductions balanced across each sector. Such dual-track approaches present a significant challenge to the higher education sector, as graduates and professionals need to be up-skilled in areas that contribute to both agendas.

The authors of this paper have previously investigated approaches to address this urgent dual-track agenda for implementing whole-of-curriculum renewal. From the literature and drawing on case studies, pilot trials, a series of workshops with educators from around the world (summarised in Desha 2010), and working with institutions such as Monash University in Australia (see Rose et al. 2012), it became apparent that key issues associated with embedding sustainability into the curriculum stemmed from the *ad hoc* approach led by champions. This led to curriculum renewal measures that were isolated and vulnerable to institutional changes such as budget, program structure, and human resources. The process was often expensive to implement, and the resultant curriculum was also at risk of duplicated knowledge and skill development. In summary, being an 'aspect' considered in isolation of the main topic (such as engineering, law, business or health) lends to vulnerability, where critical knowledge and skill areas may be deleted or replaced without systemic consideration of learning consequences (Desha and Hargroves 2011).

Addressing the need for systemic curriculum renewal, the authors developed a model for curriculum renewal as shown in Fig. 4. This model is deliberative, including whole-of-curriculum considerations, stakeholder engagement and iterative efforts. It is also dynamic, requiring continual monitoring and evaluation of all aspects including the over-arching graduate attributes.

Fig. 4: The Desha-Hargroves Deliberative and Dynamic Model for Curriculum Renewal

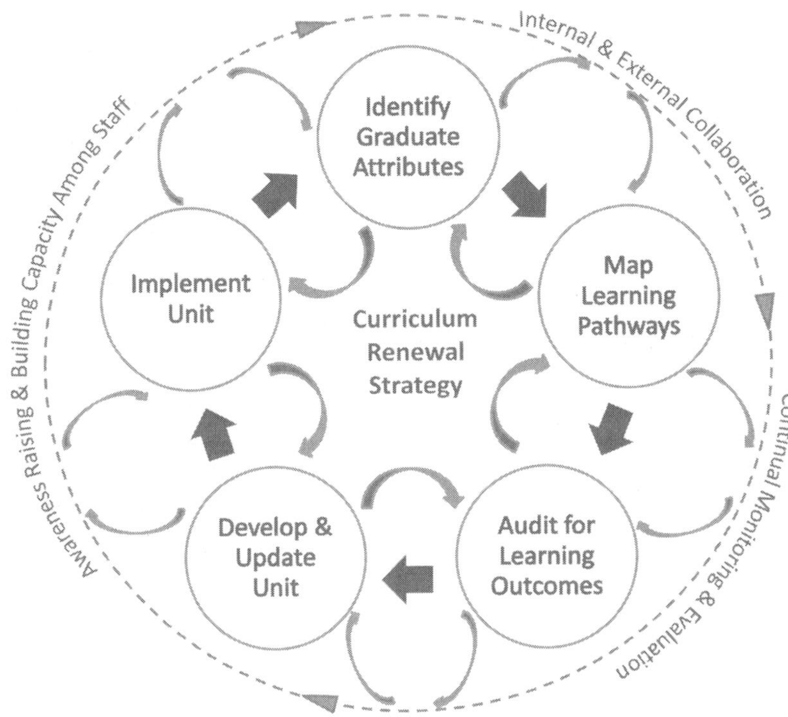

Desha and Hargroves (2011)

However, while the model provides a systemic approach to curriculum renewal, it does not address the speed at which sustainability knowledge and skills need to be embedded within the curriculum. Indeed, the model could be used over any timeframe, running the risk of producing an excellent program that completely misses the time imperatives discussed above. Considering 'time-t', it is difficult to predict when the trigger will occur globally, but as previously discussed the literature suggests that the shift in market and regulatory requirements is likely to occur within the next five to ten years. With this timeframe in mind, for some countries the findings of this study may be slightly pre-emptive of a mainstream shift (i.e. 'pre-t'), when institutions are acting in relative isolation. However, for other countries, significant catalysts are already at work, with institutions already using one or more of the elements in the model.

Clearly another structure is required within the institution to complement this process, embedding timeframes and responsibilities for a curriculum transition in the order of six to eight years rather than 15-20 years. A model for *rapid* curriculum renewal needs to address a number of issues. Firstly, it is important to acknowledge the periodic and highly organic nature of substantial curriculum renewal, which involves many staff within a department completing a variety of tasks. It is also critical to acknowledge the need for a central strategy to drive the process and ensure momentum is sufficient to meet the intended timeframes in a staged transition. Finally, it is critical to acknowledge the role of a number of elements that are prevalent in the literature concerning successful examples of large-scale curriculum renewal, as explained above in the dynamic and deliberative model for curriculum renewal. All of these issues are inter-related and are involved throughout the transition, clearly requiring a whole-of-institution approach to move forward and achieve more than superficial or transient successes in integrating sustainability within the curriculum.

At the level of the institution, there is a need for strategies and structures to incorporate the *EfS* agenda into existing operational frameworks across the many arms of the institution or department – including governance and management, curriculum design and innovation, operations and facilities, marketing, human relations, and stakeholder relations. Departments also need operational capacity to support the transition, including increasing internal professional capacity, and recruiting staff who can deliver the required curriculum. It will also need to promote such opportunities to potential students, and anticipate shifts in student enrolment.

Helix model for a whole of institution approach

When engaging with colleagues around rapid curriculum renewal and EfS, they usually start by discussing their curriculum renewal initiative. In most cases they are doing something around an EfS project, wondering why it is not going as well as they had hoped. When they look at the literature, they see many examples of what can be done, so why are they experiencing problems? The following examples highlight the tenuous nature of new curriculum efforts that are project based, or which rely on individual champions without whole-of-department support:

- We used to run a great course in … but we just didn't get enough enrolments … prospective students just didn't understand what we were offering!
- This initiative has been recognised internationally as a leader in the field … but last year during program restructuring it was removed from the curriculum.
- This is an excellent example of integrating … although the subject is not being run anymore as the convenor left.
- The initiative was very popular with students, but wasn't continued after the project finished.

Indeed, most curriculum renewal examples in the literature today are based around *ad hoc* trials and pilots driven by individuals or small groups in the absence of a strategy for systemic integration. As a result there is a lot of 'reinventing the wheel' between, and within, institutions over time while they wait for program-wide, department-wide and institution-wide strategies. Clearly, attempting holistic change as part of isolated projects is counter-productive, and also counter-intuitive in a world that strives for efficiency! Although attempting such efforts creates experience and builds the capacity of staff involved, it may actually foster an insular approach that involves a minimum of other staff within and outside the university, and can actually hinder systemic efforts.

Our second most common interaction with academics is around frustration about the lack of action in their department. Otherwise energetic individuals express despair about how their colleagues are 'just so protective' about their subjects:

- They won't even consider reviewing their materials, which are 20 years old!
- They are so adverse to any change, it drives me mad!
- They are close to retirement and they just don't care, so I'm going to have to wait for them to go.
- They won't undertake any professional development after being out of direct contact with their field for more than a decade.
- They complain their curriculum is full already, so they can't consider anything else.

- Any curriculum renewal around here is glacial rather than rapid … we work in geological timeframes, with dinosaurs!
- Every year the sustainable development course gets great student evaluations, but the department doesn't see this when they consider program priorities. I always have to fight to keep this course.

From our interactions with colleagues over the last decade, we are convinced that it boils down to action in the absence of an overarching strategic plan across the whole of the university. In 2006, following the development of an electronic-waste education module, the Griffith University Department of Information Services invited us to contribute to their annual planning retreat, which included adding sustainability to several key priority areas. In particular, senior management were keen to consider opportunities for further incorporating a tool developed by TNEP called the management helix for the sustainable organisation, or the 'Sustainability Helix', in their planning for departmental change towards sustainable operations. We subsequently facilitated two workshops to introduce the concepts of sustainability and e-waste, and to consider strategic planning for sustainable operations.

A key lesson from these experiences was that within the university, there were significant groups of staff also considering sustainable operations, which could potentially link in with curriculum renewal initiatives on campus. It also confirmed the benefits of undertaking a highly structured process with a medium-sized group of participants, to quickly arrive at an outcome that management can then take forward (Desha and Hargroves 2006). Another lesson was a revelation with regard to the potential application of an existing model for rapid curriculum renewal. We had previously been involved in developing the Sustainability Helix (Hargroves and Smith 2005) between 2003 and 2006, as the TNEP research team explored organisational change models with a leading researcher and author in the field of sustainable practice, Hunter Lovins (NCI Incorporated). The model is intended for organisations to undertake a concentrated process of significant renewal towards sustainable practice, after which they can return to continuous processes of review and action.

The organisational change model considers six activity streams which were concluded to be critical to success, namely:

- *Governance and management (G)*, including the Dean or Head of the Department or School, and any heads of research and teaching and learning.
- *Operations and facilities (O)*, including any staff employed by the school, or allocated to managing the school's assets, who oversee building operations and maintenance, laboratory and workshop management.
- *Design and process innovation (D)*, including all personnel who are involved in curriculum design and implementation within the school. This spans from program convenors to individual subject lecturers, sessional staff, teaching support staff (who assist with laboratories, workshops, tutorials, marking, etc.) and guest lecturers who may come from elsewhere in the university, or from outside the university.
- *Human resource development and corporate culture (H)*, including any staff employed by the school, or allocated to managing the school's human resources (i.e. annual staff appraisals, contracts, appointments, etc.). It also includes any staff who may be allocated responsibility for the culture of the school.
- *Marketing and communications (M)*, including any staff employed by the school, or allocated to managing the marketing of the schools programs and internal and external communications regarding program content and delivery.
- *Partnerships and stakeholder engagement (P)*, including any staff employed by the school, or allocated to managing interaction with potential employers, industry, schools and the community.

The transition towards sustainable operations involves these activity streams working through several defined stages of achievement. For each of the activities this will produce a series of elements that are required to develop the outcomes required to satisfy the particular stage, forming an 'operational matrix' of elements, activity streams and stages. A strategy is also needed to provide a structure to progress over the defined stages. The strategy essentially becomes the centre-piece of the organisation's transition towards sustainable operations. Focusing on the elements at each stage of the transition across the activities, the aim is for the activities to converge towards the goal of becoming a sustainable organisation, and to align each of the activity streams to achieve this goal.

In reality, the various activity streams are all interconnected in undertaking the tasks that deliver the milestones for each stage, as shown in Fig. 5 through the cross-cutting and oscillating but converging nature of the activity streams. For example, all activity streams will be involved in awareness raising and developing the sustainability strategy, which needs to capture the multiple benefits across the activity streams through their interactions. Furthermore, within each activity stream there is the potential for variability in performance at each stage, through factors such as staff turnover, financial performance variability, change of leadership, office relocations, etc. Essentially this model allows for sometimes random and sometimes fully planned interactions. It also allows for the possibility of overshoot and correction within each of the activity streams as they undertake their tasks in each stage. Although this is shown as a two-dimensional figure, in three-dimensions it depicts a helix structure which has intertwining components. The resulting model involves strategically planning and progressing through several stages towards a sustainable organisation. Using this approach, the model helps organisations to flexibly transition through periods of intense change.

Fig. 5: The Helix Model for a Whole of Institution Transition to Education for Sustainability (EfS)

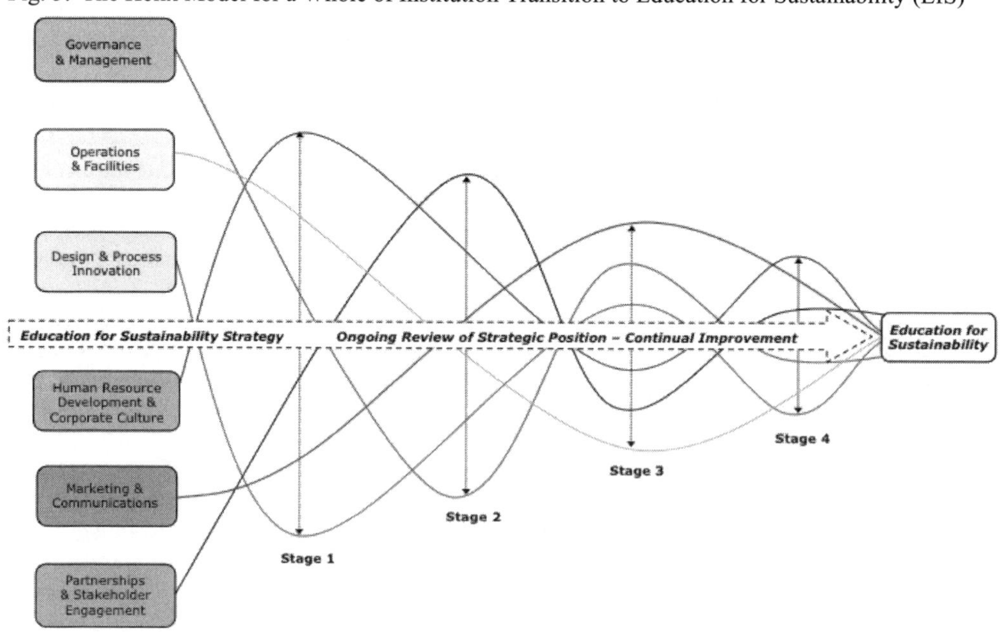

Adapted from Hargroves and Smith (2005)

Entering the model from the left along the central arrow, there is an immediate need for an overarching strategy to be developed, requiring some 'front end' planning regarding aspects such as timing, resources, staffing, budget and risk management before the whole of the department is fully engaged in the process. Moving along the central arrow, there are four stages within the model, each with tasks and milestones that are aligned with the strategy to ensure that a process of rapid curriculum renewal is maintained over the set period. This includes ongoing review of the strategic position to monitor for continual improvement towards EfS. As shown by the boxes and vertical arrows at each stage, there are six elements of curriculum renewal that need to be addressed by each of the six activity streams, at each of the four stages, forming a matrix of tasks to be undertaken throughout the organisation over the set period. The oscillating lines coming from each of the boxes highlight how the elements are considered and progressed in a dynamic manner. Monitoring and evaluation will also need to be built into the EfS strategy, to make sure that progress is documented and reportable. This can include a mixture of qualitative and quantitative measures, from survey data to focus groups and self-reporting on addressing audit recommendations (Desha and Hargroves 2007).

We will now explore this model, discussing three key considerations for rapid curriculum renewal relating to the need for: 1) a strategic, staged approach, 2) strong leadership and support, and 3) a focus on systemic integration.

1. A strategic, staged approach

The process of curriculum renewal needs to be staged, with clear milestones that can be reviewed and reported against. The elements listed in the model form the basis for a strategic approach to curriculum renewal that can be accelerated through focusing on a series of key considerations, namely:

- University/department-wide curriculum renewal requires a strategic approach;
- All key actors within the university/department have a role in curriculum renewal, including management, academics, facilities, human resources, marketing, and stakeholders;
- Efforts are to be staged across preparation, exploration, evaluation and integration;
- Efforts are intended to converge towards a clear goal.

The overarching plan should result in a process that preserves institutional diversity and innovation. This transition includes tasks that are defined with clear staging and an endpoint, whereby the curriculum may be said to have achieved EfS.

The model uses four stages to ensure that the process of rapid curriculum renewal can be planned to fit within the allocated timeframe (Desha 2010). In summary, Stage 1 'Prepare', is about preparing the institution for the intense period of curriculum renewal, through raising awareness amongst staff, investigating the institution's capacity for change, committing to curriculum renewal, and creating strategic documents that can guide the transition. In Stage 2 'Explore', the department continues with awareness raising activities, while exploring existing internal and external initiatives to find examples of curriculum renewal, which could be viable for integrating throughout the curriculum. In Stage 3 'Test and Pilot', the previously *ad hoc* or external EfS initiatives are trialled within the department, to see whether they can be mainstreamed within the program. In Stage 4 'Integrate', the institution completes the embedding of EfS within the curriculum, producing its first graduates who can contribute to genuine progress in economic and social development.

Within these four stages, the challenge is to compress the timeframe for curriculum renewal through the use of progressive tasks and milestones, to ensure that the outcome is rapid (i.e. two accreditation cycles, or six to eight years) rather than standard curriculum renewal which can span up to four accreditation cycles (i.e. two decades). As highlighted by the interwoven strands, different activity streams within the department will be working on tasks together through the transition, each supporting others and being assisted to meet the milestone requirements.

Even the model for rapid curriculum renewal itself does not guarantee a rapid transition to EfS, as there is still the potential for setting long time horizons for each of the stages, in a business-as-usual approach within higher education. Hence, whether or not the model will be useful for rapid curriculum renewal depends on whether there are strong enough catalysts that set clear timeframes for the transition. In engineering for example, although accreditation is currently one of the strongest potential drivers of change, the reality is that accreditation is still relatively weak in expectations, being more likely to incrementally affect the status quo rather than revolutionising it.

It is therefore also important to consider the role of government, industry, the institutions themselves and students in creating the shorter timeframe for the transition:

- In order to make a decision on what action should be taken, the department will need to have a clear understanding about what graduate outcomes (graduate attributes) it is seeking to achieve in order to meet accreditation requirements with regard to sustainability knowledge and skills.
- Although accreditation requirements may be clear for the department, it may also want to ensure that decisions on 'how far' and 'how fast' the curriculum renewal process should proceed are in alignment with market and regulatory requirements. An advisory panel may be useful for reviews, which could be an extension of the role of advisory panels for strategic direction and accreditation reviews.
- If key staff within the department and/or the working party are not familiar with what 'engineering education for sustainable development' means for the department, then awareness raising and capacity building will be important precursors to meetings or workshops about curriculum requirements.

From the preparatory reviews and the developed vision, an EfS strategy can be developed for the department, using the model described in the previous chapter. In summary, the strategy will highlight key tasks and timeframes for each of the four stages of the transition, and key tasks and timeframes across all six elements of curriculum renewal, involving each of the six activity streams. Within the strategy there will also be a work plan for each of the six activity streams, describing how they contribute to the strategy. The authors have determined the following three 'engagement' questions for each:

How can this activity stream contribute to achieving the strategy?
How can this activity stream help others achieve the strategy?
How can others help this activity stream achieve the strategy?

2. Strong leadership and support

The model relies on institutional leadership and support to establish timeframes for curriculum renewal within the institution, and to set the momentum and reduce the many organisational barriers to rapidly embedding new content into curricula. The form of such support will be unique for each institution, including the provision of funding, marketing and resources, and flexibility in rules regarding developing new courses and modifying existing courses.

Management will need to consider a range of incentive mechanisms to gain and maintain such leadership and momentum throughout the rapid curriculum renewal process. Considerations synthesised from the literature are listed in Table 1.

Table 1: Leadership considerations in supporting rapid curriculum renewal

Leadership Action	Description
Strong commit-ment to action	Making a strong commitment from department executive level to the process, including communicating the intent of the activities to reduce staff anxiety about the process.
Internal champion	Securing an internal 'champion' responsible for leading the rapid curriculum renewal process (e.g. to gain executive level commitment and the development of the subsequent plan of action).
Formal requirement for involvement	Including a formal request for staff to participate in the rapid curriculum renewal process, with a clear statement of the timelines and expectations.
Recognition of strengths	Recognising staff who have already embraced EESD and encouraging them to share their curriculum renewal experiences with other staff (e.g. service award, invitation to present, invitation to sit on boards and committees related to EESD).
Flexible workload allocations	Providing research assistance, teaching buy-out, or flexibility in staff appointments (e.g. research/teaching/service proportions) for staff to actively contribute to the EESD curriculum renewal process.
Seed funding provisions	Providing seed funding opportunities (e.g. internal grants) for staff to investigate research opportunities in this area.
Flexible curriculum development	Permitting and encouraging staff to appropriately use existing academically rigorous and cutting edge course materials available online under open-source common attributes licensing arrangements.
Support for professional development	Providing opportunities and financial resources for professional development in the new area, regarding this as a portfolio strengthening activity. This could be linked to specific requirements (e.g. becoming familiar with the topic area, identifying aspects that can be immediately incorporated into existing curriculum, identifying material in demand for post graduate and professional development courses, or attracting regional/ international students faced with sustainable development challenges).

Clearly, departmental leadership needs to be present to action the timeframes set by catalysts such as accreditation, regulation and policy, and employment. This conclusion is supported within engineering education by a 2008 report to the Australian Teaching and Learning Council (formerly the Carrick Institute) on addressing the supply and quality of engineering graduates for the new century, which observed four supporting actions that were common in institutions facilitating significant change, namely vision, leadership, stakeholder engagement, and resources (King 2008).

Those institutions considering the model in the period 'pre-t' may find it more difficult to set substantial short-term goals and allocate budget for curriculum renewal such as graduate attribute mapping and curriculum auditing. Furthermore, these institutions may find it difficult to secure government and industry financial support for innovative renewal initiatives, relying on harnessing momentum from existing institutional projects. Ideally, institutional support would be available at all levels within the institution, from the department through to the university's executive council, encouraging strategic engagement with the elements in the model, and supporting staff who are amenable to engaging with rapid curriculum renewal processes through mechanisms such as workload reallocation, assistance with professional development, and teaching buy-out.

3. Systemic integration

In considering the structure of the transformed curriculum, we have observed a number of strategies being trialled over the last decade, ranging from problem or project-based learning, to inverted curricula and the use of sustainability-related threshold learning concepts (Desha and Hargroves 2012). While there is no 'right' or 'wrong' way of proceeding, it is apparent that an acceptance and embracing of integration and non-linearity is critical to successfully embedding sustainability knowledge and skills within the curriculum. In summary, a rapid transition strategy would ideally incorporate an overarching structure and staging, which provides room for innovation and creativity. At the level of program, three characteristics of curriculum renewal have been observed, each of which have their own challenges and opportunities:

Niche degree versus degree update:

Niche degrees are specialisations existing either within a larger discipline context (for example a Bachelor of Sustainable Energy Systems as a major specialisation within engineering), or as 'hybrid' or 'trans-disciplinary' knowledge areas that cut across traditional boundaries (for example a Bachelor of Environmental Management involving Environmental Science, Engineering, Business and the Arts). At first there may be a tendency to put all of the new learning outcomes in a niche degree to cover the material, rather than integrating the material into current programs; perhaps this may initially seem the easier and faster option. There appears to be inertia from more than half a decade ago, to create new niche degree offers that will attract students (Kronholz 2005; Atkinson 2008); however, audits such as that undertaken by the University of South Australia conclude that academic staff are emphatic that sustainability cannot be addressed in just one (or two) sustainability courses. The authors conclude that, '*no one wants a shorthand reference to "the sustainability class."*' (Koth et al. 2009) In the emergence of successful sustainability bachelor programs, it appears to be important to integrate the core sustainability concepts with a base discipline degree training that is 'job-ready' (for example engineering, science, or law). These programs also seem to contain a number of elective options to cater for the variety of student interests, providing many non-linear pathways for students to articulate with a 'sustainability' qualification.

Using flagship courses:

As with so many other military terms, 'flagship' has crossed over from military meaning into common vocabulary. The original meaning for a 'flagship' is the lead ship in a fleet of vessels used by the commanding officer, which is the largest, fastest, newest, or the most well known. It is used to quickly identify the fleet to others. It now has common derivations such as the 'flagship project' of a government, a 'flagship product' of a manufacturing company or 'flagship store' of a retail chain. Within higher education curriculum development, 'flagship courses' are used frequently within bachelor programs to highlight the institution's commitment to a particular field of knowledge like advanced computer modelling, or project management. A significant benefit of this approach is that, over time, departments can integrate this material into relevant parts of courses across the program to then make way for new flagship courses. Traditionally this process allowed new material on topics, such as ethics and safety, to be vetted by staff, students and alumni to allow demand for the content to be established. The flagship course might be placed anywhere in undergraduate programs, perhaps co-offered as a postgraduate course, and convened by the resident expert or 'champion' within the department. It is often offered as a first-year common course, an elective, or in the form of a PhD or a masters. At an introductory level, for example, a common introductory course is developed to 'kick-start' learning in a certain topic area. At an intermediate or advanced level, flagship courses are developed to cater for learning in a new content area previously not addressed in the program and offered as part of a minor or an elective. More recently, flagship courses have used sustainability-related 'threshold learning concepts' to create a transformational shift in students' approach to the rest of their curriculum (Deha and Hargroves 2012).

Armada courses:

While flagship courses may be an attractive first step, they need to be supported by other courses in the program which reflect the flagship course's key message to provide students with an integrated

> learning experience that develops the required graduate attributes. Without clear assignment of learning outcomes for each existing course and how they reinforce the flagship content, staff who are not directly involved in the flagship course may regard this content as being sufficiently covered and, as a result, they may not seek to incorporate appropriate content from the flagship into their courses, leading to a disjointed overall curriculum. Taking the military analogy a step further, while a 'flagship' flies the fleet's flag, the rest of the fleet, or 'armada', is just as important in making an impression on their target. Hence, when focusing on updating, the various courses that make up the new content areas are integrated across the range of courses as appropriate, and can reinforce or be reinforced by the flagship, depending on the learning path.

However, in adopting an armada approach, the department should consider a number of potential pitfalls. The approach needs commitment from a large number of staff, to amend lectures, tutorials, workshops, assignments, laboratories and site visits etc. There is a risk that, if staff are not committed to such an integrated approach to curriculum renewal, there will be a disconnect between those courses that do address sustainability and those that do not, which could be confusing for students and detract from the program. The approach also needs to be strategic and structured to avoid redundancy and a disorderly revision of content. There is a risk that an unstructured approach could result in possible overlaps or gaps in sustainability knowledge and skills within the curriculum, causing confusion and frustration to staff involved and students undertaking the courses. Whereas staff may teach with a 'course-level focus', students may have a 'program-wide focus' as they are exposed to all courses and can see how they do or do not interact and support each other.

The armada approach also needs to be coordinated to ensure that new content is aligned with regard to language, terms, definitions and overarching message, to provide a clear learning pathway for students. There is also the risk that if staff do not collaborate in the curriculum renewal process, the result will be a fragmented learning experience that does not effectively or fully develop the sustainability-related graduate attributes. It also needs to consider potential interactions with colleagues outside the department, potentially across a number of courses. Curriculum renewal towards EfS requires a multi-disciplinary approach drawing upon the need to acquire knowledge in science, economics and legal disciplines, as well as engineering, so the time needed to involve other departments also needs to be taken into account.

Conclusions

Given that at some time in the next decade there is likely to be an increase in enforcement related to reducing environmental pressures, as with all organisations higher education institutions will have a choice as to whether they move early or wait until enforcement. This decision will affect the level of risk and reward for the institution, with a low commitment to EFS delivering high future risk and low future reward, and a high commitment positioning institutions to capture future rewards and avoid risks. Risks include, for instance, falling student numbers, increasing accreditation difficulties, ineligibility for research grants, and poaching of key staff. Rewards include attracting the best students and staff, staying ahead of accreditation requirements, attracting research funding, and securing key academic appointments and industry funding. Those institutions that are first to have a high commitment will achieve greater rewards before and after enforcement, with those coming after having access to reduced rewards. Further, those who maintain a low commitment will see risks increasing before enforcement as efforts to reduce environmental pressures ramp up, and after enforcement as enforcement efforts become more stringent.

In this paper we have shown how to apply a helix model for organisational change to create a unifying model for rapid curriculum renewal. It is proposed that universities could use this model to rapidly transition curricula to deliver graduates with the knowledge and skills required to underpin a prosperous society in the coming century. Departments seeking to rapidly embed sustainability into engineering curriculum could consider the model and its conceptual areas at the outset, in order to develop a strategy that is suited to its institutional and cultural context.

Drawing on experiences and findings from a decade of engagement with curriculum renewal in engineering education, and building on the Sustainability Helix for organisational change, the model presents a directional but integrated and flexible approach to help organisations to address periods of intense curriculum change. It also highlights the complex and asynchronous nature of accelerated curriculum renewal, where stages allow for sometimes random and sometimes fully planned interactions.

Through the discussion it is evident that the model will not ensure 'rapid' curriculum renewal by itself. Rather, the development of the strategy needs to set the timeframe for rapid curriculum renewal to occur, within a strategic approach to avoid a costly and resource-intensive process that achieves sub-optimum outcomes. Furthermore, institutional leadership and support is also critical to ensure that momentum is maintained through implementation. Finally the structure and strategy need to promote creativity and non-linear actions that cannot be planned but which can be transformational.

Acknowledgements

The authors thank our colleagues from The Natural Edge Project (TNEP) research group for their support and commitment to the team over the years, without whom this dissertation would not have transpired, and Professor John Fien (RMIT), Professor Neil Dempster and Professor David Thiel (Griffith University) for their mentoring and trans-disciplinary PhD supervision. The authors extend gratitude to Monash University (particularly Dr Geoff Rose and Dr Gary Codner), who have been applying the helix model to their curriculum renewal agenda, for access to enquire into their experience. We also thank the more than 150 project and peer review participants from more than 40 countries who have been involved in the international peer review of findings and models described in this paper over the last seven years, through international conferences, symposia, workshops, and online collaborations. Finally, the authors would like to thank all of the undergraduate and postgraduate students at Griffith University and QUT, who have always shown interest and been supportive of developing new and emerging sustainability knowledge and skills within their engineering degree experiences.

References

AASHE (2010), *Sustainability Curriculum in Higher Education: A Call to Action,* Association for the Advancement of Sustainability in Higher Education. Denver, Colorado.

Atkinson, C. (2008), Niche programs branch out into the unexpected. *Canadian Globe and Mail*, 14 February 2008.

Brown, L. (2007), *Plan B 3.0: Mobilising to Save Civilisation.* W.W. Norton & Company, New York.

Carson, R. (1962), *Silent Spring.* Houghton Mifflin, Boston.

Corcoran, P. and Wals, A. (2008), *Higher Education and the Challenge of Sustainability – Problematics, Promise, and Practice.* Kluwer Academic Publishers, Boston.

Desha, C. (2010), Chapter 4: Personal Narrative. In *An investigation into the strategic application and acceleration of curriculum renewal in engineering education for sustainable development* (Doctoral dissertation). Griffith University, Brisbane.

Desha, C. and Hargroves, K. (2006), *Griffith University Senior IT Planning Workshop Sustainability Session: Sustainability Session Report*, 17 November. The Natural Edge Project.

– (2007), Education for Sustainable Development curriculum audit (E4SD audit): a curriculum diagnostic tool for quantifying requirements to embed sustainable development into higher education – demonstrated through a focus on engineering education. *World Transactions on Engineering and Technology Education* 6 (2), pp. 365-371.

– (2011), Informing engineering education for sustainable development using a deliberative dynamic model for curriculum renewal. *Proceedings of the Research in Engineering Education Symposium 2011*, Madrid.

– (2012), Applying Threshold Learning Theory to teach Sustainable Business Practice in Post-Graduate Engineering Education. *Proceedings of the Amercian Society of Engienering Education*, 10-12 June, Texas.

Desha, C., Hargroves, K. and Smith, M. (2009), Addressing the time lag dilemma in curriculum renewal towards engineering education for sustainable development. *International Journal of Sustainability in Higher Education* 10 (2), pp. 184-199.

Diamond, J. (2005), *Collapse: How Societies Choose to Fail or Succeed.* Penguin Books, New York.

Erdelen,W. (2009), Plenary speech – Trends in Global Higher Education. In 2009 World Conference on Higher Education: *The New Dynamics of Higher Education and Research for Societal Change and Development*, Paris, 6-9 July 2009. Kronholz.

Gore, A. (2006), *An Inconvenient Truth.* P.C. a.P. Productions.

Hargroves, K. and Smith, M. (Eds.) (2005), *The Natural Advantage of Nations: Business Opportunities, Innovation and Governance in the 21st Century, The Natural Edge Project*, p. 17. Earthscan, London.

Jones, P., Selby, D. and Sterling, S. (2010), *Sustainability Education: Perspectives and Practice Across Higher Education.* Renouf Publishing, London.

King, R. (2008), *Addressing the Supply and Quality of Engineering Graduates for the New Century*, p. 30. The Carrick Institute for Learning and Teaching in Higher Education Ltd.

Koth, B., Woodward, M. and Iversen Y. (2009), Civil Engineering Education for Sustainability: Faculty Perceptions and Result of an Australian Course Audit. *20th Australasian Association of Engineering Education Conference*, p. 781. Adelaide, 6-9 December.

Kronholz, J. (2005), US Universities Face Turbulent Times Ahead. *College Journal, from The Wall Street Journal Online,* 1 March.

Perkin, H. (2007), History of Universities. *International Handbook of Higher Education*, pp. 159-205. Springer, London.

Rose, G., Codner, G. and Griggs, D. (2012), Integrating Sustainability into Higher Education: Insight from a Case Study of the First Year Engineering Curriculum at Monash University. *Proceedings of the LaTrobe Symposium on Advancing Sustainability in Education and Research*, 20 February, LaTrobe, Australia.

Shriberg, M. and Tallent, H. (2003), Beyond Principles: Implementing the Talloires Declaration". In *Report and Declaration of the Presidents Conference* 1990, Association of University Leaders for a Sustainable Future, Ball State University, Munice.

Smith, M., Hargroves, K. and Desha, C. (2010), *Cents and Sustainability – Securing Our Common Future by Decoupling Economic Growth from Environmental Pressures.* Earthscan, London.

Sterling, S. (2005), Higher education, sustainability, and the role of systemic learning. In Corcoran, P.B. and Wals, A.E.J. (Eds.) (2004), *Higher Education and the Challenge of Sustainability: Problematic, Promise, and Practice.* Kluwer Academic Press, Dordrecht.

Stern, N. (2006), *The Stern Review: The Economics of Climate Change.* Cambridge University Press, Cambridge.

UNEP (1992), *Rio Declaration on Environment and Development*, United Nations Environment Program, Conference on Environment and Development. Rio de Janeiro, 3-14 June.

– (2008), *Green Jobs: Towards Decent Work in a Sustainable.* Low-Carbon World, United Nations Environment Programme.

von Weizsäcker, E., Hargroves, K., Smith, M., Desha, C. and Stasinopoulos, P. (2009), *Factor 5: Transforming the Global Economy through 80% Increase in Resource Productivity*. Earthscan, London.

Wals, A. (2008) (Ed.), *From cosmetic reform to meaningful integration: implementing education for sustainable development in higher education institutes – the state of affairs in six European countries*. DHO, Amsterdam.

WCED (1987), *Our Common Future*, World Commission on Environment and Development. Oxford University Press, Oxford.

Chapter 3

Promoting Education for Sustainable Development through Grassroots Action in Malta: An Example of NGO-University Collaboration

Paul Pace

Abstract

Research in Education for Sustainable Development (ESD) has shown that there is a gap between knowledge and action – a gap that universities involved in ESD have been striving to bridge. The Centre for Environmental Education & Research (CEER) of the University of Malta has attempted to address this issue in schools by supporting Eco-Schools – an NGO-run programme. This paper documents the strategy adopted by CEER in fostering approaches that empower students in decision-making, promoting professional development based on an authentic understanding of ESD and the fostering of a learner-centred pedagogy, and the development of educational material that reflects the realities of schools and their surrounding communities. The results of this strategy have arguably led to the seamless inclusion of ESD as a cross-curricular theme in the National Curriculum Framework (NCF) and greater awareness and commitment to the adoption of sustainable lifestyles and choices at school and community levels.

Introduction

The success of Education for Sustainable Development (ESD), or Environmental Education (EE), in the formal education sector depends on the availability of flexible structures that can adapt to suit the different needs and realities of learners (Centre for Environment Education 2007). However, there is a tendency for formal education systems – particularly those of post-colonial Small Island States, such as Malta[1] – to be built around centralised management structures that tend to institutionalise lack of flexibility in the curriculum in the vain hope of ensuring accountability and quality (Pace 1997; Briguglio and Pace 2003; Mayo et al. 2008; Pace 2009). Such structures entertain the free-market notion that promotes competition as the driving force for innovation – a notion that gradually erodes learner oriented pedagogies into test/exam oriented ones, which in turn redefines what is worthwhile and useful in the curriculum (Ravitch 2010).

Making formal education institutions more effective, efficient and competitive has become the slogan of many political parties and governments and consequently the main driving force of educational reforms. Nevertheless, this stance is developing bureaucratic administrative structures; creating curricular structures that are rigid and monodisciplinary, systematically draining away the enthusiasm and creativity from educators, and engendering a generation of learners who adopt a passive attitude to learning – characteristics that are inherently incongruent with ESD principles (Pace 2010a). This seems to be particularly the case for universities (Pace 2010b), irrespective of the fact that major EE/ESD events (e.g. UNESCO 1980;

1 Malta is an archipelago of islands covering an area of approximately 316 km^2. With a population of just over 417,000, Malta has one of the highest population densities (1,307 persons per km^2).

UNESCO-UNEP 1988; Scoullos 1998; Centre for Environment Education 2007) consider these institutions as important actors in the promotion of sustainability. In fact, Moore (2005a: 543) identifies four barriers that ESD programmes face at university:

- monodisciplinary organisational structures that determine funding, result in territorial conflicts and limit student mobility across disciplines;
- competition among students, between and within, faculties, departments and universities;
- misdirected criteria for evaluating staff and student abilities and achievements; and
- too many priorities, unclear decision-making and hierarchical power structures.

Implicit in this brief analysis is the belief that if universities are to live up to their expectation to be leaders in ESD they need to depart from the rigid structures of their formal settings and move towards more non-formal and informal settings.

The Centre for Environmental Education & Research (CEER)

Officially established in 2004 with the specific task of promoting ESD throughout the nation, CEER is housed within the Faculty of Education (University of Malta) – the only teacher education institution in the country. As is quite typical for small states, where duplication of efforts constitutes an expensive waste of the limited resources available, CEER currently has a monopoly over ESD provision, since all the people qualified to train individuals in ESD are to be found there (Mayo et al. 2008). Consequently CEER is a national reference point, helping the country to develop a coherent vision of ESD and instrumental in the introduction of ESD at policymaking level (Pace 1997; Tanti 2000).

CEER's initial work mainly focused on infusing ESD in schools by preparing future teachers. This involved offering study units as part of initial teacher training programmes and the facilitation of ESD related research. The main themes and the specific research agenda of CEER revolved around the development of curriculum material and investigations into methodologies on how to carry out effective ESD programmes in the formal education sector. Fig. 1 summarises CEER's structure and operation.

However, uptake of ESD principles in schools was very low, primarily because ESD was not mandatory in the national curriculum and consequently teachers (and schools) could opt out from infusing it in their programmes. Consequently, the fate of ESD in schools ultimately depended on the commitment of school administrations and individual teachers and the aspects of the themes explored were predominantly biased towards a romantic view of the environment (with a greater emphasis on the natural environment) in which development is a dirty word and in which harmony with the environment means a 'return to the past' (Pace 2009). Moreover, schools were too busy dealing with the 'more important' aspects of the syllabus, i.e. the examinable subjects, to be bothered with ESD curriculum packs (Pace 1997). The situation was further compounded with the publication of the National Curriculum of 2000, which failed to introduce ESD into the curriculum (Ministry of Education 1999).

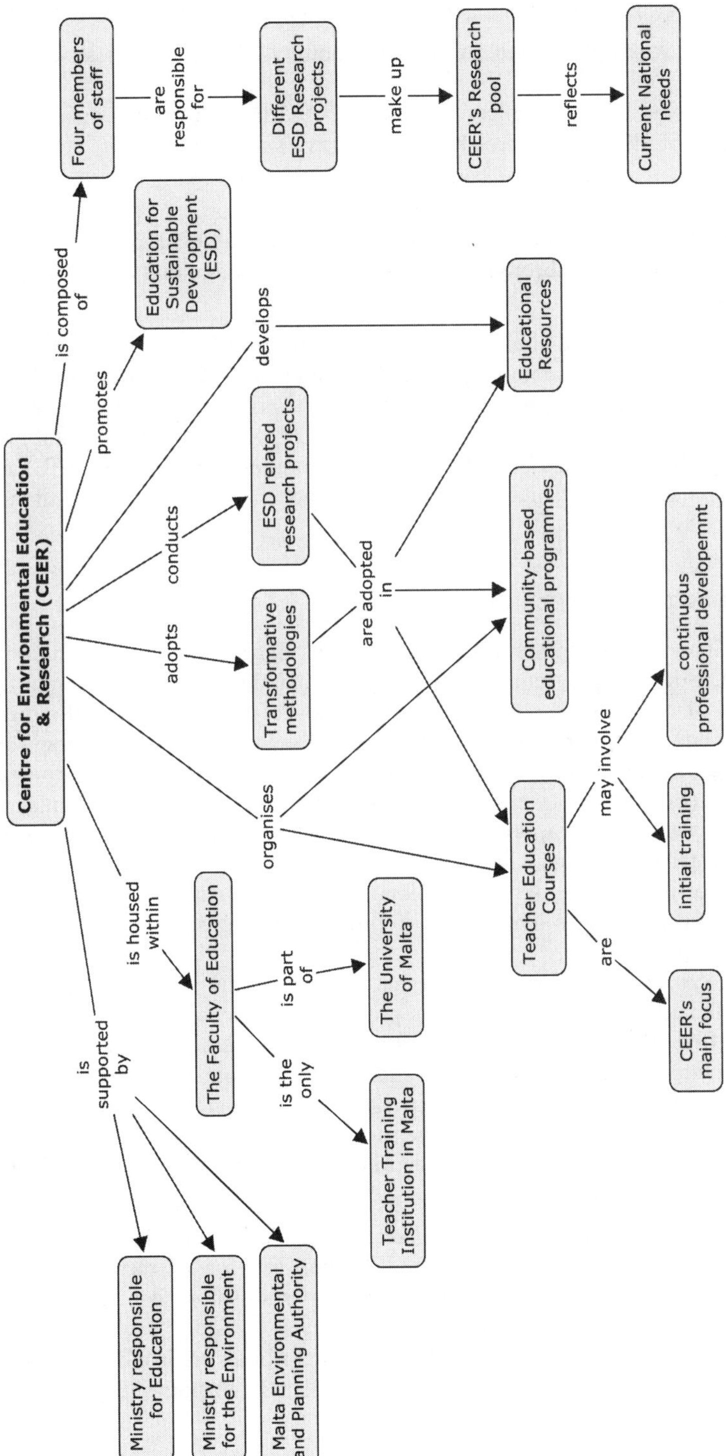

Fig. 1: CEER's structure and operation

CEER's strategy of ESD implementation in schools had to adapt in order to find a way to:

- develop a wider definition of the environment that included sustainable development themes;
- assist teachers in the seamless infusion of ESD themes in the curriculum without adding additional burdens to the already content-heavy subject syllabi;
- actively involve the school community in decision-making for a introducing sustainable development principles in school development plans;
- use schools as catalysts for community-based ESD initiatives; and
- achieve all the above through a non-formal methodology that could by-pass all bureaucratic hurdles.

After considering several alternatives, the implementation of such a strategy seemed achievable through the development of a partnership with Nature Trust (Malta), a local environmental NGO, with a declared commitment towards ESD, which was committed to co-ordinating the Eco-Schools programme in Malta. CEER's director, who was instrumental in the setting up of the initial contacts with the Foundation for Environmental Education (FEE) and the writing of the programme's implementation strategy in schools, was appointed National EkoSkola Co-coordinator by Nature Trust.

The Eco-Schools programme

The Eco-Schools programme, known in Malta as EkoSkola, is one of five international environmental education programmes run by the Foundation for Environmental Education (FEE), a network of NGOs whose aim is the promotion of sustainable development through environmental education (FEE website, n.d.). The EkoSkola programme focuses on student-led action aimed at promoting sustainable development values in the school and surrounding community (Eco-Schools website, n.d.). It seeks to develop environmental responsibility by adopting a whole school approach in the design, implementation and monitoring of a School Environmental Policy that is intimately integrated into the School's Development Plan. The ultimate goal is to infuse, through deliberate choices, sustainable lifestyles into the school's day-to-day functioning so that this environmental ethic gradually becomes an integral part of the school's ethos.

The EkoSkola programme assumes that the participating school is committed to autonomy at all levels because progress along the programme is dependent on the school's readiness to involve different stakeholders (especially students) in decision-making, set and achieve development targets that reflect its needs and available resources, contextualise curriculum work with day-to-day realities, and share experiences and expertise within and without the school. When a school gives evidence that it has achieved these goals, it is awarded the Green Flag – a prestigious, internationally known eco-label. The programme's methodology is based on seven steps (EkoSkola Malta 2011):

1. *Establishing the EkoSkola committee:* usually including the main stakeholders of the school, the committee is the main body responsible for the running of the programme. Its composition needs to ensure that students' participation features as high as possible in Hart's Ladder of Participation (1992: 8).
2. *Conducting an environmental review:* students run an environmental audit of their school identifying issues related to sustainability, their dimension and possible ways addressing them.

3. *Compiling an action plan:* students review and discuss the results of the school audit and develop an action plan that prioritises actions that need to be taken, identifies persons/groups responsible for their implementation and sets deadlines.
4. *Monitoring and evaluating:* students manage the implementation of the action plan ensuring that set targets are met.
5. *Integrating actions within the curriculum:* school concerns about sustainability are used as themes for classroom study and as the basis of school-based actions and policies.
6. *Informing and involving:* outreach is one of the tasks of the EkoSkola committee. The aim is to disseminate the participatory process, initiated through the programme, not just within the school, but also within the surrounding community. Certain EkoSkola schools become a hub for Local Agenda 21 initiatives in their respective communities.
7. *Developing an Eco-Code:* students compile and display a set of principles/actions that define their commitment towards ensuring a good quality of life in their school.

Even a cursory look at the programme shows that EkoSkola provided the Maltese educational system with a tangible opportunity for curriculum development. Curriculum documents (e.g. Ministry of Education 1999; Ministry of Education, Employment and the Family 2011) are usually jam-packed with strategies that purport to foster interdisciplinarity, relevance of learning experiences, active learner participation, critical thinking, assessment for learning, lifelong learning, autonomy, a holistic perspective, and interaction with the community. Nevertheless, these aspirations rarely get beyond the document stage because of bad resourcing or resistance from schools, hence becoming the clichés that characterise these documents. However, these curriculum principles are mirrored in the basic principles underlying the EkoSkola programme – and any ESD programme, for that matter (Bezzina and Pace 2004). Participation in the EkoSkola programme could thus be viewed as a tangible opportunity to implement the national curriculum principles.

Openly supported and financed by Malta's educational and environmental ministries, the management of EkoSkola was entirely in the hands of Nature Trust, whose administration of the programme was in turn monitored by FEE. This ensured maintenance and control of the programme's quality criteria. Launched in six pilot schools in 2002, the EkoSkola programme spread to about 60% of all the schools in Malta (i.e. catering for about 80% of the total student population) in just ten years. Besides its fast rate of dissemination, other factors have been considered as indicators of the programme's success. Anecdotal evidence gathered during school visits from teachers and school senior management teams revealed that the success of the programme was due to three major characteristics of its implementation strategy:

- participation in EkoSkola was completely voluntary and the programme did not invade on school policy, but adapted itself to the individual needs of the participating schools. Consequently school development was intrinsically generated and not externally imposed;
- the programme presented a criterion-referenced model of evaluation, which was a relatively new concept and a welcomed change from the predominant norm-referenced model; and
- the programme was person-oriented rather than task-oriented, putting children and young people at the centre of decision-making in their school and promoting active citizenry.

CEER's role in the EkoSkola programme

Co-ordinating EkoSkola necessitated providing support to schools participating in the programme and creating an infrastructure that would support the initiative in the long term. CEER's three-pronged strategy consisted in:

- fostering approaches that empower students in decision-making;
- providing professional development of pre-service and in-service teachers; and
- developing ESD material that was relevant to the students, the schools and their communities.

Pivotal to the dissemination of the programme was the decision of the ministry responsible for education to appoint five full-time teachers whose sole remit was to support schools in their efforts to implement EkoSkola. As their terms of reference show (Human Resources Development Department 2010), these EkoSkola Teachers were the main vehicles through which CEER could effectively implement its strategy.

Participatory pedagogies

CEER is committed to the promotion of pedagogies that promote participatory learning and higher-order thinking skills, which according to Tal (2005) are necessary for the achievement of the complex and multifaceted objectives of ESD. The argument is further reinforced by Moore (2005b: 337), who claims that educational institutions – particularly higher education institutions – need to create *space for pedagogical transformation* or rather opportunities for learners to engage in reflection, dialogue and action. As opposed to *banking* pedagogies, which view learners as passive *depositories* (Freire 1970), transformative pedagogies facilitate the development of citizens who are informed, engaged and empowered – an essential component of sustainable development.

However, research results conducted on an ESD programme offered by CEER revealed that exposing students to participatory methodologies was not enough to bring about change. Students needed time to acclimatise themselves to their new role as autonomous learners and to examine whether espoused values (considered appropriate) tallied with the actual vales influencing their lifestyles and their choice of teaching style (Pace 2010b).

This acclimatisation period varies from learner to learner, and consequently, courses offered by CEER incorporate the use of peer tutoring, collaborative group work and the use of VLEs (Virtual Learning Environments) that offer personalised support. Another feature of ESD study units offered by CEER is supervised on-site experiences – particularly in EkoSkola schools – followed up by reflection sessions. These experiences help student-teachers develop skills to face the realities and the reluctance to change found in certain schools. True to the saying "*a good example has twice the value of good advice*", CEER's policy is to adopt the accepted ESD principles in the methodology used by lecturers, rather then just preaching about them. Through their reflective journals students commented that this praxis had a positive effect on their personal teaching styles as it provided concrete examples of ways to improve their teaching and their teaching environment (Pace 2010b).

Professional development

The main challenges that professional development in ESD faces include: promoting an authentic understanding of ESD that reorients the mindset that ESD is solely based on environmental issues, and the fostering of a learner-centred pedagogy, which also necessitates a change in predominant classroom practice. Personal development of pro-sustainable development values is the integral feature of study units and courses offered by CEER. Getting teachers on board is essential for the successful implementation of any policy at school (Pace 2009). Considering that in Small States teachers perform multifunctional roles, these courses have other outcomes (e.g. youth and adult education), particularly in the community (Mayo et al. 2008).

CEER considers the preparation of student teachers that are receptive to ESD as a seeding opportunity. As outlined above, the study units offered are based on a flexible methodology that adapts to the needs of the learners and focuses on developing reflective practitioners. Student teachers are also specifically prepared to adopt a whole school approach to ESD – through the EkoSkola programme. However, with some notable exceptions where ESD study units are a compulsory component of their programme, i.e. for students specialising in Primary Education and Science Education for secondary schools, for other students these units are optional. Consequently not all student teachers are exposed to ESD.

CEER also developed in-service teacher training courses as part of the ministry's national summer in-service programme. However, the turnout was not very encouraging because the courses had to be offered on a voluntary basis and most of the teachers were instructed to attend other compulsory courses. What proved very successful were school-based professional development courses that were tailor-made for the specific needs of the teachers and the school. In fact, Bezzina and Pace (2004) claim that professional development at the school site is the key element to initiate and maintain whole school development. These courses were staggered along the scholastic year presenting ESD in small doses that could be digested and absorbed by the school community. Moreover, the sessions were contextualised within the schools' efforts to implement EkoSkola.

Indispensible for the successful running of these school-based professional development sessions was the team of EkoSkola teachers whose professional development CEER took more seriously by providing intensive training on ESD. Besides providing them with regular practical and theoretical sessions on transformative pedagogies, CEER provided the EkoSkola teachers with an opportunity to read for a Master Course in Intercultural and Environmental Management of Schools (3EMI 2009).

Educational material

Countries participating in the EkoSkola programme usually develop curriculum packs about specific environmental issues that are given to participating schools with the intention that they will be implemented. The examination-oriented pedagogy and the crammed syllabi to support it, adopting a similar strategy in Malta would have seen the end of EkoSkola. Initially CEER tried to address this by encouraging student teachers to develop ESD curriculum packs as their dissertation. Considering the large number of students who opted for this option and the high quality of educational packs produced, the idea was successful. However, procuring funding for the publication and dissemination of the projects in schools proved to be an insurmountable obstacle.

The turning point came from the schools themselves. As schools were gradually becoming more and more involved in the EkoSkola methodology, they started holding staff meetings (particularly at the closure of the scholastic year) to identify an environmental theme that was relevant to the Action Plan and that they could develop through their lessons. This initiative proved successful mainly because the teachers themselves wanted it. The EkoSkola teachers tapped into this initiative and through the school-based professional development courses (explained above), they provided support with regard to the resources needed to develop lessons; the organisation of learning; and the integration of classroom practice with the whole school and beyond. This strategy had the added advantage that the class teachers were seen as active participants in ESD, capable of producing educational experiences tailored to the specific needs of their classes, rather than passive technicians always dependent on the availability of externally produced curriculum packs.

Outcomes

From the very start of its involvement with EkoSkola, CEER focused most of its research on gathering information about the impact the programme had on the educational community. In her study, Grech (2010) revealed that being involved in EkoSkola or not did not cause any significant change in the students' level of environmental knowledge. However, EkoSkola caused a marked improvement in the students' and the schools' level of environmental commitment and responsibility. Furthermore, Grech (2010) shows how students from Green Flag holder schools distinguished themselves, in their school and their community, with their insistence on concrete environmental action whenever plans were being discussed. She also claimed that EkoSkola was contributing towards empowering students to take action and become responsible citizens.

The issue of empowerment was specifically dealt with by Zammit (2011). Relying on extensive field observations in two different schools participating in the programme, she reported that the EkoSkola process provided an excellent opportunity for student empowerment. Both Grech (2010) and Zammit (2011) compared participation in the EkoSkola programme to a long educational journey during which students are gradually empowered to assume active and responsible roles at school and in their community.

Educational programmes dealing with authentic real-life environmental issues tend to initiate student and parental environmental communication and hence are conducive to intergenerational influence (Ballantyne et al. 1998; Uzzell 1994). In his study, Sammut (2011) reported that children from EkoSkola schools talk about the environmental activities done at school more than their non-EkoSkola counterparts. However, while conceding that this tendency is a key component for intergenerational influence to take place, Sammut (2011) points out that parental behavioural change was dependent on whether parents felt that the environment was a priority and on spending sufficient time with their children.

From school-based observations, Gauci and Farrell (2009) report that participation in EkoSkola helped students to experience democracy first hand. By being elected onto the EkoSkola Committee and learning to work in a team, students felt that their voice was being heard and appreciated. This feeling of importance led students to be more proactive in school-based and community settings, responsible for their actions and faithful to their commitment.

As highlighted earlier, part of the EkoSkola programme expects students to take an active role in environmental decision-making in their community. There are several cases, which are currently being documented and studied, where students teamed up with local councillors in the design and implementation of sustainable development initiatives in their respective communities. Furthermore, each year students meet Members of Parliament at the EkoSkola Parliament Session, during which they discuss and suggest ways of improving quality of life on a national scale (e.g. MaltaMedia News 2007; Schembri 2011).

Throughout its presence in Maltese schools, EkoSkola has become a brand name for quality education that, based on the research carried out and the feedback obtained from schools, lived up to its claims. Consequently, when the issue of whether to include ESD in the national curriculum cropped up during the workings of the National Curriculum Framework Steering Committee, EkoSkola's reputation served as an example of what ESD could be like in schools. Moreover, from the implementation perspective, schools (and teachers) participating in EkoSkola knew what the ESD provision proposed by the National Curriculum Framework entailed.

Pending issues

Although a lot has been achieved through EkoSkola, old habits die hard and research has outlined the following issues that are still lurking within formal educational institutions:

- a reluctance to depart from monodisciplinary curriculum structures and move on to interdisciplinarity;
- a predominance of issues related to the natural environment in sustainable development discourse;
- a lack of networking between the various educational institutions, which is symptomatic of an innate compulsion of such entities to work in isolation, to compete and 'beat the others at it';
- a preoccupation with the 'products' of education rather than its 'processes';
- token participation alimented by a general belief that students are too young and inexperienced to participate in decision-making; and
- centralised management structures that are still hard to overcome because they have infiltrated and become ingrained in relationships at all levels: educational authorities – schools; headteacher-school; teacher-students; and students-students.

Conclusion

CEER's strategy and its impact on the formal educational system need to be contextualised within Malta's small size. The literature about Small Island States usually attributes most of the problems that such nations have to face to their small size. Nevertheless, there are cases where small can be beautiful. One major advantage is that the people who occupy key administrative roles are few and familiar faces, and hence can be easily identified, targeted and approached. Moreover, the strong community networks that develop in small populations allow Small Island States to rapidly adapt and implement changes (Easterly and Kraay 2000; Laurent 2008). The introduction of ESD in Malta's National Curriculum is a case in point.

Rather than just teaching about flexibility and adaptability, informed by research and a commitment to generate change from the grassroots upwards, CEER let go of its traditional mode of operation. A readiness to try out novel approaches, working closely with an NGO and adapting to the needs of the target audience proved to be a winning combination.

References

3EMI (2009), *Master course I level in intercultural and environmental management of* schools. Retrieved January 22, 2012 from http://www.3emi.eu/?q=en/node/15.

Ballantyne, R., Connell, S. and Fien, J. (1998), Students as catalysts for environmental change: A framework for researching intergenerational influence through environmental education. *Environmental Education Research* 4 (3), pp. 285-298.

Bezzina, C. and Pace, P. (2004), Promoting school development through environmental education. In *Trends: Monograph Series in Education* (1. Malta: Faculty of Education, University of Malta.

Briguglio, L. and Pace, P. (2003), Education for sustainable development in Malta. Case study presented at the SIDS Expert Meeting on "*Capacity development for sustainable development in SIDS: building partnerships for sustainable development through education, public awareness and training*". 3-9 December 2003. University of South Pacific, Fiji.

Centre for Environment Education (2007), *Moving forward from Ahmedabad ... environmental education in the 21st century.* Retrieved January 22, 2012 from http://www.tbilisiplus30.org/Final Recommendations.pdf.

Easterly, W. and Kraay, A. (2000), Small states, small problems? Income, growth, and volatility in small states. *World Development* 28 (11), pp. 2013-2027.

Eco-Schools website (n.d.). *Methodology*. Retrieved January 22, 2012 from http://www.eco-schools.org/page.php?id=48.

EkoSkola Malta (2011), *Getting started*. Retrieved January 22, 2012 from http://www.ekoskola.org.mt/getting-started/.

FEE website (n.d.). [Home page]. Retrieved January 22, 2012 from http://www.fee-international.org/.

Freire, P. (1970), *Pedagogy of the oppressed*. Continuum Publishing Company, New York.

Gauci, A. and Farrell, E. (2009), *Promoting democracy in schools: the EkoSkola experience*. Unpublished Post Graduate Diploma in Education (Administration & Management) long essay. Faculty of Education, University of Malta, Malta.

Grech, S.M. (2010), *Evaluating the impact of the EkoSkola programme on the Maltese education system*. Unpublished BEd (Hons) dissertation. Faculty of Education, University of Malta, Malta.

Hart, R. (1992), *Children's participation: from tokenism to citizenship*. UNICEF Innocenti Research Centre, Florence, Italy.

Human Resources Development Department (2010), Applications for teachers in the peripatetic service to take part in the EkoSkola programme for primary and secondary schools in Malta. *Letter Circular HRD/77/2010*. 27th May 2010. Human Resources Development Department, Directorate for Educational Services, Ministry of Education, Employment and the Family, Malta.

Laurent, E. (2008), *Economic consequences of the size of nations, 50 years on*. Document de travail de l'OFCE: 2008-26. Observatoire Français des Conjonctures Économiques, Paris.

MaltaMedia News (2007), Children tell politicians how to protect environment. *MaltaMedia News*. November 22, 2007. Retrieved January 22, 2012 from http://www.maltamedia.com/artman2/publish/govt_politics/article_4165.shtml

Mayo, P., Pace, P. and Zammit, E. (2008), Adult education in small states: the case of Malta. *Comparative Education* 44 (2), pp. 229-246.

Ministry of Education (1999), *Creating the future together*. Ministry of Education, Malta.

Ministry of Education, Employment and the Family (2011), *Towards a quality education for all: The National Curriculum Framework 2011*. Consultation document 2: Rationale and components. Ministry of Education, Employment and the Family, Malta.

Moore, J. (2005a), Barriers and pathways to creating sustainability education programs: Policy rhetoric and reality. *Environmental Education and Research* 11 (5), pp. 537-555.

– (2005b), Seven recommendations for creating sustainability education at the university level: A guide for change agents. *International Journal of Sustainability in Higher Education* 6 (4), pp. 326-339.

Pace, P. (1997), Environmental education in Malta: trends and challenges. *Environmental Education Research* 3 (1), pp. 69-82.

– (2009), Emerging from limbo: environmental education in Malta. In Taylor, N., Littledyke, M., Eames, C. and Coll, R.K. (Eds.), *Environmental education in context: An international perspective on the development of environmental education*, pp. 73-82. Sense Publishers, Rotterdam.

– (2010a), Education for sustainable development: current fad or renewed commitment to action? *Journal of Baltic Science Education* 9 (4), pp. 315-323.

– (2010b), Self-evaluation as a tool in developing environmental responsibility. *Journal of Teacher Education for Sustainability* 12 (1), pp. 5-26.

Ravitch, D. (2010), *The death and life of the great American school system: How testing and choice are undermining education*. Basic Books, New York.

Sammut, I. (2011), *Intergenerational influence on environmental education*. Unpublished MEd dissertation. Faculty of Education, University of Malta, Malta.

Schembri, D. (2011), Students give MPs an eco-lesson. *The Times* Thursday, June 9, 2011. Retrieved January 22, 2012 from http://www.timesofmalta.com/articles/view/2011 0609/local/Students-give-MPs-an-eco-lesson.369665.

Scoullos, M.J. (1998) (Ed.), *Environment and society: Education and public awareness for sustainability*. Proceedings of the Thessaloniki International Conference, 8-12 December 1997. UNESCO & Government of Greece, Greece.

Tal, T. (2005), Implementing multiple assessment modes in an interdisciplinary environmental education course. *Environmental Education and Research* 11 (5), pp. 575-601.

Tanti, V.M. (2000), *Environmental education in Maltese government schools: A historical review*. Unpublished B.Ed. (Hons.) dissertation. Faculty of Education, University of Malta, Malta.

UNESCO (1980), *Environmental education in the light of the Tbilisi conference*. UNESCO, France.

UNESCO-UNEP (1988), *International strategy for action in the field of environmental education and training for the 1990s*. UNESCO-UNEP, Paris / Nairobi.

Uzzell, D. (1994), *Children as catalysts of environmental change*. European Commission Directorate General for Science Research and Development Joint Research Centre, Brussels.

Zammit, C. (2011), *Journey to empowerment: A case study of the EkoSkola programme at two local primary schools*. Unpublished MEd dissertation. Faculty of Education, University of Malta, Malta.

Chapter 4

Convergence and Confluence: A Systems Thinking Approach to Integrated Sustainability in Higher Education

Jennifer Thompson and Amelia Mayahi

Abstract

Global paradigms such as "one-size-fits-all" have perpetuated unfettered globalization that is linked with social, economic, and environmental degradation. Therefore, critical consciousness is growing within higher education institutions of the urgency of the need to create cultures of sustainability. Systems Thinking Theory is presented as a powerful lens whereby universities might begin the process of realizing these cultures. The University of Memphis is implementing systems thinking theory, not only as a function of thought, but also as the plan of action. Within this paper, focus topics highlight the systems thinking process at the University of Memphis, including the following: outreach, involvement, curriculum design, and national community engagement. Student outreach is achieved by creating a positive and strong first impression; our Freshmen Orientation is the first opportunity for incoming students and their relatives to understand our commitment to sustainability. Involvement is forged by a student-initiated and funded Green Fee. The Green Fee awards this money to various student, faculty and administration led green projects. The curriculum is infused with sustainable learning as it applies to all disciplines of education and interdisciplinary interaction is paramount. In national community engagement, the University aggressively pursued the Tennessee Valley Authority (TVA) and Alliance to Save Energy grant and was awarded funds to hire four student interns to implement our current energy-saving initiatives. The authors selected these initiatives as representative of our understanding of systems thinking within the University structure. The purpose of this paper, therefore, is to take a candid look at global imperatives plaguing sustainable development that are interconnected with higher education institutions, subsequently demonstrating, through practical integration of systems thinking theory, how the University of Memphis is igniting successful innovations in student outreach, involvement, curriculum design, and national community engagement.

Introduction: Critical Awareness

There is critical consciousness growing globally, more specifically in the higher education community, about the lack of sustainability awareness and interrelatedness in terms of social, economic, and environmental issues. The effective integration of campus sustainability development must be understood as inextricably linked fibers within a system. To support this belief, Leith Sharp of Harvard University (2002), defined linked fibers within a system in terms of how individual "green" initiatives can begin to coalesce into a whole university transformation rather than individual sustainability practices. Therefore, creating a culture of sustainability is uniquely derived from the convergence and confluence of sustainable endeavors, hopefully culminating in a spiritual shift of the mind, with deeper meaning embedded in the value of learning. This spiritual shift – acting as a catalyst moment where sustainable development of universities and the humans that inhabit them unite – provides a framework for nurturing axiological principles of sustainability and is the basis for this paper.

The authors purport solutions to obstacles inherent in sustainable development are interdependent within higher education institutions (HEIs). This paper presents how, through practical integration of systems theory and experiential learning, the University of Memphis is igniting successful innovations in student outreach, involvement, curriculum design, and national community engagement. The importance of the systems thinking theory process is its ability to address the complexities inherent within the university structure. This paper is not presented as a prescriptive solution to sustainable development for all universities; rather, it is put forth as an example of what is beginning to work for the University of Memphis. This is in the hope that other universities might benefit from these sustainable development practices. The green initiatives within the University are a continuous work in progress with lessons learned; however, they are potentially moving the Institution into more progressive approaches to sustainable development.

Overview: The Global Imperative

Before examining the role of higher education as an agent for sustainable development, a brief overview of global social, economic, and environmental issues will be discussed. Foremost, there is a paradigm that society has operated within since the industrial revolution indicative of the current global conditions. It is the belief that Universal Design – or "one size fits all" – is the best solution for economic growth and social equality. Moreover, this paradigm is often implemented through "brute force" methodology described by William McDonough and Michael Braungart (2007) in their provocative book, *Cradle to Cradle: Remaking the way we make things*. The detail of "brute force" methodology is too lengthy to discuss in this paper; however, the following synopsis defines some consequences resulting from this societal paradigm.

Social, Economic, and Environmental Global Imperatives

In terms of the environmental degradation, according to the 2010 Gallup Poll, there is a noticeable chasm between how society views current evidence on climate change to that of scientists ("Doomsday Messages" 2010). Two elements identified as contributors to this chasm were the psychological effects of "Doomsday" delivery and society's lack of understanding that 82% of scientists agree that global warming is being accelerated by humans (Doran and Zimmerman 2009). According to Edward Maibach, director of the Center for Climate Change Communication at George Mason University, "Misunderstanding the extent of scientific agreement about climate change is important because it undermines people's certainty that climate change is happening, which in turn reduces their conviction that America should find ways to deal with the problem" (George Mason University 2011, para. 6). It is this mindset that propagates further environmental degradation affecting social and economic health as well, and is discussed in the following.

The degradation of economic and cultural foundations, through unchecked globalization, is perpetuating ecological instability. *The Economics of Happiness* is a documentary[1] highlighting globalization as a primary cause of ecological and social degradation and identifies commodification of personal identity as a byproduct of globalized consumerism[2]. This homogene-

1 *The Economics of Happiness,* "A project of the International Society for Ecology and Culture (ISEC), a non-profit organization whose mission is to protect and revitalize cultural and biological diversity by strengthening local communities and economies worldwide" (www.isec.org.uk).

2 Globalization is defined as the development of an increasingly integrated global economy marked especially by free trade, free flow of capital, and the tapping of cheaper foreign labor markets (http://www.merriam-webster.com/dictionary).

ity feeds the inefficiency of global markets by supporting cheap labor, less regulation on carbon emissions, and acculturation of diverse groups of people.

The aforementioned imperatives are not mutually exclusive, in that each is embedded in the other as part of the problem and the solution. Higher education institutions play a critical role in realizing the solution based upon their role as part of the problem. Dr. William Rees (2003)[3] observed that "... countries whose citizens impose the greatest per capita ecological load on the planet are the countries that have the highest levels of public and private education" (p. 93). This is in part due to the philosophical underpinnings of higher education institutions. Expansionism – the belief that human welfare is equated through continual increasing of economic growth – is believed to be the desired outcome of a fruitful education system. This view diminishes social and environmental stewardship, breeds misconceptions about exaggerated cost for prevention, and emphasizes technology as an answer to all obstacles (Rees 2003) resulting in the commodification of higher education. Therefore, it is imperative that institutions, such as the University of Memphis, evolve current values and belief systems away from this destructive paradigm. The University of Memphis realizes the charge of higher education institutions to change values and beliefs and is implementing systems thinking theory as a solution. This theory is thus augmented by experiential and integrated learning processes within the structure of the campus environment.

Theory into Praxis

Systems Theory

Before demonstrating the integrated praxis of systems thinking within the University of Memphis, a brief overview of this theory is outlined along with a detailed application of systems thinking referred to as Complex Adaptive Systems (CAS). Systems thinking theory – developed over the past fifty years – is a conceptual framework helping humans to understand patterns that exist in life, and embraces the understanding that nature, all humans, and their endeavors "are bound by invisible fabrics of interrelated actions" (Senge 2006: 7). The University of Memphis is investing in this definition of systems thinking and is adapting a form of systems theory called Complex Adaptive Systems. However, this adaptation is not without flexibility. It is fully recognized that systems are parts to a greater whole that are in continuous motion and evolution. The most salient awareness required for successful systems thinking implementation is to see the interrelationships between structure (system) and behavior (humans, environment, etc.). This awareness yields an understanding of how the system works, what makes the system produce poor results, and how to shift the system into better behavior patterns (Meadows and Wright 2008).

Complex Adaptive Systems

Complex Adaptive Systems (CAS) are comprised of three main characteristics: 1) there is self-organization where system components seemingly interact and arrange themselves without design or management; 2) emergence, whereby thousands of micro-interactions lead to co-evolution and evolution finally culminating in an aggregated shift in global behavior, and 3) bottom-up processes whereby leadership is integral at all levels of the organization and members contribute experiences to the whole – this is opposed to unpredictable and uncontrollable top-down systems where organizations become part of the mainstream strategy with

3 Dr. William Rees was educated in Bio-ecology at the University of Toronto and has taught Community and Regional Planning at the University of British Columbia since 1969.

diminished efficiency in systemic shifts (Porter and Cordoba 2009). Complex adaptive systems are comprised of four elements including principles and assumptions, theory, methodology, and strengths and weaknesses (Table 1).

Table 1: Complex Adaptive System Structure

Principles and Assumptions	Densely connected networks of agents, self-organization, and emergence Ongoing learning and bottom up evolution
Theory	Non-linear systems
Methodology	Build and empower learning networks and bottom up processes Provide appropriate incentives Monitor results and make adaptations as needed
Strengths and Weaknesses	Well suited for current turbulent marketplaces Shift to stewardship style of leadership Not applicable in all situations

Note: *Adapted from "Three Views of Systems Theories and their Implications for Sustainability Education," by Porter and Cordoba (2009: 327)*

Within CAS is an adapted system for HEIs specific to sustainability. The complex adaptive systems educational structure (Table 2) demonstrates how the learning networks prepare students, faculty, staff, and administrative personnel to become agents of sustainability within the campus environment. This integrated participation draws together diverse educational and skill levels present within the institution. The network helps to define and achieve goals set forth, thereby building consensus among a larger collective. With a community of intentionality developed, greater success in green activities and events are realized (Porter and Cordoba 2009).

At this juncture, the University augments the CAS system with experiential learning and integrative teaching methods that are embedded in campus activities, outreach, and curriculum. According to Clark and Caffarella (as cited in Baumgartner 2001), the Integrative Approach to teaching takes a holistic view of adult development. This perspective is focused on how the intersection of mind, body, and socio-cultural influences affect development (Integrated Approach section, para. 1). Dirkx (as cited in Baumgartner 2001) discusses the importance of "nurturing the soul" in adult education. Instead of relying exclusively on logic, he invites educators to explore "ways of knowing grounded in a more intuitive and emotional sense of our experiences" (Integrated Approach section, para. 4). Here students transform beyond rational knowledge to extra-rational meaning-making. Additionally, according to McCarthy and McCarthy (as cited in Mullins-Nelson 2008), "Experiential learning puts the student 'face to face with making decisions that are real rather than merely thinking about a situation or case'" (para. 3). Therefore, these methodological approaches to understanding learning as part of mind, body, and socio-cultural influences, along with the action of experiential learning, can prove essential when integrated with sustainable development. Further supporting these approaches the following discussion describes the unique qualities inherent within systems theory and its relation to sustainable development.

Unlike most university approaches to sustainable development whereby the solutions are usually conceptually based (theory) and lack an integrated process (methodology), system thinking is both conceptual and process-based simultaneously. In terms of flexibility, according to Meadows and Wright (2008) systems are non-linear and want to exist without boundaries. However, the inherent nature of humans to organize and establish boundaries works in contradiction to this conceptual framework (Meadows and Wright 2008). Although this might be viewed as a critique of systems thinking, the authors believe this unique framework provides a foundation to help counteract isolated sustainable projects, and ultimately reveal and alter linear thinking patterns within university systems. Moreover, systems thinking can potentially produce mental shifts within the campus community, help universities adapt to continual changes within the system, and act as a methodological organizing principle to sustainable development. Therefore, systems thinking is a mode for creating a paradigm shift in the typical university way of thinking. The following dialogue demonstrates how this shift is occurring within the University of Memphis through specific approaches to sustainable development.

Table 2: Complex Adaptive System Educational Structure

Sustainability	Learning networks create sustainable value in new or existing products and processes Productive innovations from the bottom up
Educational Goals and Skills	Ability to distinguish each system perspective Identify sites of leverage, empower learning and design incentives Win trust and build consensus
Activities and Projects	Greening of Industry Network and Business for Social Responsibility Map systems relevant to campus environmental practices Implement a bottom up sustainability initiative on campus

Note: Adapted from "Three Views of Systems Theories and their Implications for Sustainability Education," by Porter and Cordoba (2009: 329)

The University of Memphis System of Sustainability

Sustainable Systems Thinking Inception

In 2007 the University of Memphis became a signatory of The American College and University President's Climate Commitment (ACUPCC)[4]. Since making that commitment, the University embarked on developing a Strategic Sustainability Plan closely aligned with CAS for HEIs. From the ACUPCC commitment emerged a sustainability committee that operates interdependently and reciprocally.

The University of Memphis realized the sustainability committee must be effective and cohesive and be aware of how the campus is structured. Building upon an existing structure ensures all campus entities are represented, further supporting the CAS structure (Table 2). The process of creating the committee requires that members demonstrate vested interest rather than being nominated to serve. The committee is comprised of appointed and volunteer members (student, faculty, staff, and administrative); however, new incentives for paid positions are being implemented.

The Executive Sustainability Committee meets once a semester and is comprised of individuals in the President's Office. This committee sets the vision and mission of the sustainability plan. The Operations Sustainability Committee is considered the "action" committee and is integrated into the various subcommittees. The Operations Sustainability Committee

4 The American College and University President's Climate Commitment is a network of committed colleges and universities, which believes that higher education contributes in a critical way to reversing the effects of climate change. Its mission is to "accelerate progress to climate neutrality and sustainability by empowering the higher education sector to educate students, create solutions, and provide leadership-by-example for the rest of society" (http://www.presidentsclimatecommitment.org/).

meets once a month and the subcommittees meet bi-weekly or more if necessary. This interplay between committees provides greater communication mitigating potential issues that undermine initiatives. The following describes, in detail, how this system is holistically implemented within student, faculty, staff, and administration-led initiatives such as green funding projects, outreach, curriculum, and national community engagement.

Site Conditions Prior to Green Fee

Site Conditions: Demolition

The University of Memphis: Recycling Zone prototype – Night Rendering Highlighting LED

The University of Memphis: Green Fee Recycling Zone Prototype: 3D rendering

The Green Fee

The Green Fee, started in 2007, is a fund supported and initiated by the students to create an environmentally friendly campus culture. Each year the Green Fee is open to the campus community to fund selected green suggestions or proposals. The Recycling Zone Prototype is an example of a funded project. In 2011 the Department of Architecture submitted a proposal to create a recycling zone and landscape demonstration site on campus. The application parameters specifically address sustainable bottom-up initiatives and accessible learning environments to all campus inhabitants. The purpose is to enhance the physical context of the University by improving current conditions. The recycling zone creates a visible and educational connection by setting an example of the high importance the University places on recycling efforts and sustainable landscape practices. Furthermore, it sends a positive non-verbal message to the campus community by providing an enjoyable exterior space. The project utilized Tiger Blue Goes Green Day[5] to promote awareness. Additional benefits include supporting the University Strategic Sustainability Plan,[6] providing interdisciplinary scholarship and building productive partnerships; creating a sustainable campus of "choice", and supporting retention efforts through enhanced campus environments.

5 Tiger Blue Goes Green is the University of Memphis' annual sustainability expo.

6 University of Memphis Strategic Sustainability Plan strategy for developing an "inviting and sustainable setting" (http://www.memphis.edu/stratplan/index.php).

Outreach

Outreach is a key component to the infiltration of sustainability in the living, learning, and working environment of the University of Memphis. One important outreach strategy is to target the incoming freshmen class. This particular group of students requires a customized outreach focus and presents unique opportunities as follows: 1) there exists multiculturalism in disciplines and backgrounds in the same place at one time; 2) the incoming freshmen's minds are impressionable – therefore this is a perfect time to portray sustainability as part of the University culture; 3) it is an opportunity to embed the seeds to support sustainable leadership in the future. With freshmen orientation grounded in sustainability, followed by four years or more of integrated sustainable outreach events and opportunities, it is the hope of the University that future classes will have positive role models to carry the sustainability message forward.

Freshmen Orientation occurs in various sessions beginning at the end of May and ending in mid-July. Each session holds a new group of incoming freshmen. The event begins with an exhibition hall showcasing the many organizations, initiatives, clubs, and courses. Among these booths are the urban community garden, facilities recycling program, the Green Team, the Sustainability Committee, Tiger Bike Share Program, and sustainability focused courses. Freshmen, along with their parents or guardians, are given a reusable bag to collect free pamphlets and giveaways from the booths. Recycling bins outnumber the trash receptacles in the exhibition hall and are strategically placed near the eating area. The facilities recycling program and the Sustainability Committee provide the freshmen with free reusable water bottles labeled with the "Tiger Blue Goes Green" mascot. They are also given a magnet with the latest recycling information, as well as green websites. Sustainability achievements on the regional, state, and national level are displayed throughout the exhibition hall. University tours include showcasing outdoor recycling zones, the urban community garden, the Tiger Bike Share building, and elements of sustainable design and operation on campus.

Just as students are diverse, so are the university employees. "Greening" employee orientation provides opportunity for employees to understand the role of sustainability in the workplace. It is mandatory that staff employees have fifty hours of professional development training within a year. Tiger Blue Goes Green Classes offers sixteen of those training hours total. The classes include: stormwater awareness, emergency response to spills, recycling fundamentals, urban gardening, hazardous and universal waste, resource management and conservation, and a summary Green Tiger class. The final class is a tour showcasing sustainable areas and practices on campus. Employees who complete all classes in the Tiger Blue Goes Green module are awarded a certificate of completion and a reusable mug and t-shirt with the Tiger Blue Goes Green logo. While these are great incentives, the networking opportunities between departments and acknowledgement of the campus sustainability leaders is most rewarding. The training sessions are job specific as well as holistic, and are great tools for each department to demonstrate how they infuse sustainability practices, specifically at the freshmen orientation event.

In order to have students understand and believe the integration of sustainability in the University system, outreach to faculty and staff is imperative. If faculty and staff do not embrace and practice sustainability in the workplace, then little progress will be reached by committees or clubs. This supports the CAS network under Sustainability and Educational Goals (see Table 2), in that, leverage and incentive is used creatively within an existing process – freshmen orientation.

Curriculum Development

Continuing from outreach endeavors, the University places great emphasis on sustainability within the curriculum. By incorporating the systems thinking approach within the Academia sector, new sustainable methods can be applied to disciplines already in place, thus embracing the CAS approach to sustainability within existing curricula (Table 2). The University of Memphis has applied integrated approaches to curriculum in many disciplines such as: economics, architecture, biology, earth sciences, engineering, anthropology, history, art, K-12 and higher education, nursing, business, and law. Each discipline offers courses centered on sustainable practices that are specific to the discipline. Integration of sustainability in disciplines allows students to understand the important role sustainability can have within their careers and lifestyles. Additionally, students pursuing "green" careers can major in sustainability. The major is in conjunction with several disciplines between the sciences, humanities, and fine arts.

Research and class projects in sustainability courses have a direct relationship and impact on the University. For example, the Department of Architecture Recycling Zone, mentioned earlier in this paper, utilized design and construction resources, both material and human, from the Department of Architecture and other interested student and community organizations, as well as technical expertise of local industry partners. Participating students have the opportunity to take part in the research and implementation of sustainable strategies on campus. Ensuing efforts have created collaborative partnerships with Physical Plant (including the campus landscape architect and the sustainability coordinator), Student Disabilities Services, Residence Life, and the Department of Leadership. Furthering the mission of engaged scholarship, involvement within the greater Memphis community includes the Shelby County Government, the City of Memphis, Memphis College of Art and local businesses (including Memphis Eco-Gardening and Sharp Electronics Corporation).

Another curriculum initiative is the mechanical engineering research project to build a complete bio fuel production machine. A successful partnership between the University dining service, the Physical Plant, and the mechanical engineering students and staff led to the production of bio fuel from the grease and landscape trimmings produced on campus. The bio fuel will be used to power Physical Plant vehicles. These projects prove valuable tools for creating partnerships and facilitate interconnectedness between students, faculty, and staff.

National Community Engagement

The University of Memphis aggressively seeks community engagement beyond the University perimeters. As an HEI, it is imperative that the University serve as the model for quality living and engagement. In June of 2011, the University of Memphis pursued the Tennessee Valley Authority (TVA) and Alliance to Save Energy grant. This grant would financially enable the University to hire four student interns to implement energy efficiency initiatives. The grant is important to the University of Memphis for three reasons: 1) national recognition – TVA is one of the largest energy providers in the United States and well known throughout the Southeast region where the University of Memphis campus is located; 2) the need to reduce energy costs; and 3) the opportunity for students to become Green Campus Interns.

Further elaborating on this position, the Green Campus Interns are selected from various disciplines applying the CAS approach. The interns work directly with University of Memphis and TVA engineers and sustainability leaders. This experience helps students develop management and professional relationship skills while supporting goals to reduce energy consumption. The Green Campus Interns, in addition to energy audits and retrofitting, also host Green Career Panels, create monthly newsletters, and encourage students to sign the Tiger Energy Challenge Pledge. The interns are in charge of their own promotions and managing their budgets, monthly reports, meeting minutes, newsletters, and evaluations. These reports are checked by the campus advisor and then by TVA and Alliance to Save Energy staff. This embedded responsibility and control to the students enables empowerment and critical buy-in among the interns further supporting the CAS model. Currently students from the disciplines of English, Biology, Computer Engineering, and Communication serve as Green Campus Interns at the University of Memphis.

Conclusion

This paper outlines globalized social, economic, and environmental imperatives that threaten the future of society. At the same time the paper defines a sound framework of theory and methodology simultaneously acting as one in order to motivate cultural changes within higher education. The intent of this paper is to demonstrate how the University of Memphis embraces an idea of systems thinking. With passion and hard work, the collective body of students, faculty, staff, and administration are coming together creating incremental, yet invaluable, changes within the campus environment.

In terms of lessons learned, financial support and leadership development are most prevalent. One primary financial issue is failure to include the sustainable development office, coordinator, and initiatives within the University budget. The University is working to incorporate sustainable development into the system through restructuring budgetary processes within the University. Additionally, leadership obstacles arose when sustainable committee positions were appointed and these positions were non-paid. The University is currently working to create paid committee and interdisciplinary positions within all levels of faculty and administration. A lack of transparency within and between the sustainable committees has invited unforeseen issues with communication thereby creating barriers while trying to carry thought into action.

As systems thinking becomes more fully integrated, the University hopes the following prospects will be realized: 1) establishment of a Sustainable Development Department that is funded through the University of Memphis's general funds; 2) creation of paid sustainable coordination positions within each discipline; 3) the fulfilling, through systems thinking, of the goals set within the University of Memphis Sustainability Strategic Plan and the ACUPCC; 4) pre and post systems thinking data collection to document quantitative and qualitative improvements; 5) further publications and presentations of the systems thinking evolution.

The University understands that systems thinking is not a "one size fits all" design methodology. Systems thinking is flexible and malleable depending on the structure of the university. However, to instill a mindset of sustainability that is embedded from inception, rather than applied, it must be presented as an integral concept within all levels of the university structure. The catalyst to the success of systems thinking is interdependent on the soulful investment of all inhabitants of the university community.

References

ACUPCC (American College and University President's Climate Commitment) (2006). Available at: http://www.presidentsclimatecommitment.org (accessed 9 December 2010).

Baumgartner, L. (2001), Four adult development theories and their implications for practice. *Focus on Basics* 5 (Issue B). Available at: http://www.ncsall.net/?id=268 (accessed 19 March 2011).

Doomsday Messages About Global Warming Can Backfire, Study Shows (2010, November). *Science Daily.* Available at: http://www.sciencedaily.com/releases/2010/ 11/101117094248.htm (accessed 9. December 2010).

Doran, P.T. and Zimmerman, M.K. (2009), Examining the scientific consensus on climate change. *Eos,* Vol. 90 (3), pp. 22-23 (doi: 10.1029/2009EOS030002).

George Mason University (2011, November 21), Support for climate policy linked to people's perceptions about scientific agreement regarding global warming. *ScienceDaily.* Available at: http://www.sciencedaily.com/releases/2011/11/1111211 15102.htm (accessed September 2011).

Meadows, H. and Wright, D. (2008), *Thinking in systems: A primer.* Chelsea Green Pub., White River Junction, Vt.

McDonough, W. and Braungart, M. (2002), *Cradle to cradle: Remaking the way we make things.* North Point Press, New York.

Mullins Nelson, B. (2008), Discovering and appreciating diversity through experiential Education. *NSEE Quarterly* Vol. 33 (1.

Norberg-Hodge, H., Gorelick, S., Page, J., Shiva, V., McKibben, B., Korten, D.C., Goldsmith, Z., International Society for Ecology and Culture (2011), *The economics of happiness.* International Society for Ecology and Culture, Totnes, U.K.

Porter, T. and Cordoba, J. (2009), Three views of systems theories and their implications for sustainability education. *Journal of Management Education* 33 (3), pp. 323-347 (doi: 10.1177/1052562908323192).

Rees, W.E. (2003), Impeding sustainability? The ecological footprint of higher education. *Planning for Higher Education,* Vol. 31 (3), pp. 88-98.

Senge, P.M. (2006), *The fifth discipline* (2nd ed.). Random House Business, London.

Sharp, L. (2002), Green campuses: The road from little victories to systemic transformation. *International Journal of Sustainability in Higher Education* 3, pp. 128-145.

Chapter 5

Bridging the Gaps to Develop Graduate Capabilities for Sustainability

Anne Sibbel

Abstract

Considerable research has been undertaken in higher education to promote learning for sustainability. Despite intellectual resources, value attached to knowledge and discovery, and the social responsibilities recognised by universities, progress has been slow. Rather than preparing graduates to adapt to unsustainable conditions, transformative learning must enable them to change the way they work. Transformative learning requires learner-centred and interactive pedagogy. Embedding simulated or real-world experiences is one way to change curricula to promote transformative learning. As part of some ongoing research, this project was concerned with developing the graduate capabilities urgently required for sustainability. Comparing students' pre- and post-intervention responses showed that a novel learning experience had a substantial effect on their capabilities for sustainability. Graduates of the same program with a minimum of one year's work experience were asked to assess the demands of sustainability within their field of work. Reference to their perspectives on real workplace challenges provided a context from which to reflect on the significance of the positive changes in students' capabilities, and so an indication of the effectiveness of the experimental teaching and learning intervention. Based on the findings, it seems that single interventions designed to specifically target sustainability-related attributes, values and motivations may be a useful interim strategy for making urgently needed progress while higher education undertakes the necessary processes of adapting policy and practice to meet these new challenges.

Introduction

In the face of escalating climate change and its pervasive and unpredictable impact on society, it is essential that sustainability becomes a global priority. Universities have a major responsibility to society for education and research, but particularly for transforming students to become future leaders, empowered and motivated to make decisions for sustainability (Adomssent et al. 2007; Anderberg et al. 2009; Barth et al. 2007; Cortese 2003; Cusick 2008; Zilahy 2006). In general terms, this can be achieved by coordinating efforts through research, outreach, campus operations and curriculum design (Lidgren et al. 2006). From the beginning, efforts to promote sustainability through universities were directed at 'greening' the campus, with investment in these initiatives readily rationalised in economic terms (de Ciurana and Filho 2006; Hammond and Churchman 2008; Krizek et al. 2012; Sammalisto and Lindhqvist 2008). There is still a long way to go to fully integrate sustainability as a universal and critical theme across the curriculum in all disciplines (Gilroy 2008; Newport et al. 2003; Savelyeva and McKenna 2011; Sherren 2006; Timmerman and Metcalfe 2009). Certainly demonstrating commitment to sustainability in the ways that resources and waste are managed on campus contributes to a culture oriented to this end, but there is a need to broaden this effort to researching ways to promote teaching and learning for sustainability through the curriculum. Given the embedded intellectual capital, the value attached to the acquisition of knowledge and discovery and the influence of these activities on society, the potential of higher education for promoting sustainability is yet to be fully actualised (McFarlane and Ogazon 2011; Stephens et al. 2008).

Rather than developing knowledge and skills that perpetuate long-established patterns of human endeavour, transformative education presents challenges to personal frameworks of thinking and doing, so can bring about the necessary changes to enable individuals to work towards sustainability (Holdsworth et al. 2008; Podger et al. 2010; Sipos et al., 2008). Transformative education requires a pedagogy that is learner centred, interactive and engages with the real world (Burns 2011; Ferrer-Balas et al., 2008). There are a number of approaches to transformative education, including experiential learning, service learning, work-integrated learning, problem-based or project-based learning (Bacon et al. 2010; Domask 2007; Wals 2009). Essentially, these approaches include the structures and context in which students take responsibility for their own learning by testing new knowledge and skills in a supportive situation. They can assist in orienting students to the changing conditions and expectations of the professions for which they are training. Most importantly, learning through participation in real world tasks offers the scope to prepare graduates for work towards sustainability.

For universities, education for sustainability presents an enormous challenge. Strategies to promote transformative learning for sustainability must take into account the pragmatics of day-to-day teaching and the dynamic processes occurring within university classroom settings. In particular, it is essential to recognise the ethical constraints around teacher-student interactions, including the responsibilities to deliver the prescribed curriculum, assure academic rigour and accommodate the individual differences and the diverse learning needs within a typical student cohort.

One of the barriers to fully integrating sustainability into higher education teaching and learning is the lack of prescribed professional standards which relate to teaching for sustainability (Collier and Armstrong 2010). Directly or indirectly, this probably influences the importance attached to teaching for sustainability in higher education, and the motivation of academics to develop the necessary capabilities for this work.

Associating education for sustainability only with the environmental and ecological disciplines represents a very limited view (Dale and Newman 2005; Reid and Petocz 2005) but may explain why it is not seen as relevant to all disciplines by some academics. Compounding this misconception, general assumptions about the immutability of the 'hard' sciences and mathematics may make the scrutinising of the theory and practice of these disciplines for the sake of sustainability seem quite unnecessary or irrelevant. Such misconceptions mean that there is an urgent need to demonstrate that sustainability can and must be integrated into teaching and learning across all disciplines in higher education.

Another challenge to developing curricula for sustainability is in defining the outcomes, specifically the graduate capabilities or competencies to be developed. This includes a range of variously described personal attributes, values and motivations, discussed widely within the literature (see for example Anderberg et al. 2009; Barth et al. 2007; Blewitt 2010; Dale and Newman 2005; Fenner et al. 2005; Parker 2010; Shephard 2008). Drawing from this literature, they are briefly summarised here.

In their roles as leaders and decision-makers, professionals need to be able to think outside the constraints of the single discipline in which they were trained, so recognising that most problems cross disciplinary boundaries and require solutions made in collaborative multidisciplinary teams. In the context of change associated with global warming, ageing populations, escalating public health problems, continuous technological development, diminishing resources and accumulating wastes, it is unlikely that past university studies will meet future needs in the workplace. Professionals must continue learning, so always reflecting, critically reviewing and updating their knowledge and skills so they can deal with emerging real world problems. There are no precedents to many of these problems, so analysing and resolving them requires creativity and resourcefulness, as well as resilience to persist even when attempts fail. This means that research becomes part of daily professional practice. Above all, responding to the new problems arising at human-environment interface must be a priority, often requiring a change in personal values and an acceptance of broader social responsibilities. This commitment may be reflected in advocating on behalf of others, so the ability to communicate, negotiate and resolve conflict are important political skills required for sustainability. Many of the capabilities outlined here are interrelated, adding to the complexities of researching strategies for teaching and for measuring learning for sustainability.

Methodology

This project investigated the contribution of an intervention to transformative learning, specifically by measuring the change in the students' capabilities for sustainability. Information was also collected about the knowledge, resources and training needs of professionals for work towards sustainability. With no universally accepted strategies for measuring transformative learning, this research also involved designing a data collection tool and validating outcomes.

The participants

Statistics is an example of a field typically seen as rigorous and objective. Choosing to test ideas for teaching and learning in statistics was seen as a way to demonstrate that capabilities for sustainability could be developed within any discipline. Several courses with a major statistics component were selected. These courses are offered to second and third-year students within an undergraduate third-year applied science degree program at RMIT University in Melbourne. Prior to the intervention, it was anticipated that some differences in responses between these two cohorts could be attributed to the length of time that they had been studying at university.

Students' involvement in the research relied on them selecting a sequence of courses over two semesters. As anticipated, the logistics of timetabling and variations in individual students' academic study plans affected participation over the six-month period. Post intervention, it was possible to match the pre- and post-intervention responses of only relatively small cohorts of students ($n = 13$, second-year students; $n = 10$, third-year students). This limited the scope for testing the statistical significance in the trends observed.

Graduates ($n = 52$) from the same program and currently employed for a minimum of one year in a diverse range of workplaces were also recruited. These participants provided information about the opportunities available in professional practice to promote the values, motivations and attributes required for working towards sustainability. This information established the context from which to reflect on the respective roles of higher education and contemporary workplaces in bringing about the necessary changes to achieve sustainability. An important feature of the design of this research, the graduates' responses also provided a reference point, against which the significance of any changes in students' capabilities could be assessed.

The data collection tool

The development of the students' capabilities was assessed by comparing their responses collected using a survey, self-administered before and after the intervention. This survey required them to indicate their level of agreement on a five-point scale with a number of statements relating to attributes, motivations and values, pertinent to sustainability. For the purposes of this research, the statement was considered to indicate the students' capability if they selected 'somewhat agree' (point 4) or 'entirely agree' (point 5) on this scale. While many variables could not be controlled in this setting, it was assumed that the changes were largely due to the teaching and learning intervention.

In the survey, many of the multidimensional capabilities required several different but related statements to describe them. While based on a comprehensive literature review, the statements were somewhat intuitively derived. Without prior testing, it was not possible to predict how the statements would be interpreted. Positive correlations between responses to related statements were interpreted as evidence of consistency in the ways they had been interpreted by the students, and a measure of the validity of the selected indicators of a single, complex attribute, personal values or motivation.

For the graduates, the survey was adapted to be relevant to the workplace. Their responses provided an indication of their professional needs and capabilities, and a measure of how well their undergraduate education had prepared them for promoting sustainability through their daily work.

The teaching and learning intervention

Raw data had been collected from participants in a public event designed to promote awareness of cancer risks and to raise funds for medical research. Rather than setting specific objectives, the purpose of collecting the data had been intuitively understood by the voluntary organising committee. Collected over three years with several variations to the survey in that time, the data described the ways that people had been recruited to participate, their motivations and experiences of the event, and how it may have affected their awareness of the disease. Knowing about any relationships between these variables would help the committee to plan for expanding the reach and scope of this event.

Typically, students learn and practise statistics with prepared sets of data and prescribed correlations and other relationships to test. In contrast, this intervention required the students to consider the needs of an actual client and to deal with the practicalities of incomplete data sets, such as occurs in real world research. Supported by staff through a series of weekly workshops over two semesters, the students had to make decisions about how best to analyse the data to derive the most responsible and useful conclusions for the committee. They had to decide the relationships they should test, and then consider whether the data allowed this. Having completed these tasks, their findings formed the basis for recommendations in a report to the committee about ways to enhance the impact of the event. This intervention also connected to the real world in that the organisation provided feedback on the feasibility of the recommendations and how the report had influenced their planning for the event in the following year. In this way, the work of the students in supporting a potentially important public health initiative could be seen as a service to the community.

Results

There were positive changes in almost all the student capabilities measured in this research, as summarised in Tables 1-10. The significance of the respective changes in sustainability-related attributes, values or motivations are considered and qualified in the following discussion, with reference to the demands of sustainability in real workplaces, as described by graduates and summarised in Table 11.

Knowledge and awareness of the implications of sustainability

Table 1: Self-assessed knowledge and awareness of implications of sustainability

	Second-Year Students			Third-Year Students		
Indicators	**Pre**	**Post**	**Δ**	**Pre**	**Post**	**Δ**
Aware of the environmental impact of decisions	81%	93%	**+12%**	83%	92%	**+9%**
Aware of the economic impact of decisions	76%	78%	**+2%**	89%	92%	**+3%**
Aware of the social impact of decisions	81%	79%	**-2%**	94%	75%	**-19%**
Aware of the bigger (sustainability-related) issues	29%	64%	**+35%**	78%	82%	**+4%**
Recognise more sustainable options	57%	79%	**+22%**	83%	91%	**+8%**

While there was evidence of a positive change in *awareness of the bigger (sustainability related) issues* amongst the third-year cohort (+4%), the effect of the intervention was much greater for the second-year students (+35%) in this respect. The number of students responding affirmatively to statements about their *awareness of the environmental and economic impact of their decisions* increased after the intervention (+12% and +2% for second-year students; +9% and +3% for third-year students, respectively). Interestingly, there were fewer students aware of the *social impact of their decisions* after the intervention (-2% for second-year students; -19% for third-year students). This unexpected trend may be explained in terms of their growing appreciation of the complex and far-reaching social impacts of human activity, often the most difficult to envisage and recognise.

Consistent with these trends, fewer second-year students (57%) initially agreed that they could *recognise sustainable options when making decisions,* as compared to third-year students (83%). As with most of the other indicators of knowledge and awareness, the intervention had a more pervasive effect on the second-year cohort (+22% compared to +8% for the third-year students), according to this indicator.

Political skills

Table 2: Self-assessed political skills

	Second-Year Students			Third-Year Students		
Indicators	**Pre**	**Post**	**Δ**	**Pre**	**Post**	**Δ**
Influence opinions of others	25%	57%	**+32%**	33%	83%	**+50% p=.003**
Encourage others to consider sustainability	48%	36%	**-12%**	72%	82%	**+10%**
Negotiate well	25%	79%	**+54% p=.036**	65%	100%	**+35% p=.037**
Deal well with conflict	40%	71%	**+31%**	50%	75%	**+25%**

Initially, only a small proportion of the students (25% of second years; 33% of third years) were confident that they could *influence the opinions of others* in their class. The development of their confidence in this political skill (+32% and +50%, respectively) suggests that the learning experience had allowed them to realise and develop this attribute. This was a statistically significant change ($p = .003$) for the third year cohort. The extra year of study meant greater exposure to the concept and applications of sustainability so may have made these third-year students more receptive to the intervention in this respect.

Prior to the intervention, many students in both cohorts (48% of second years; 72% of third years) were confident that they could extend their influence to more specifically *encourage others to consider sustainability*. Even more third-year students (+10%) felt this way after the intervention. Coming to terms with the challenges without the extra year of studies in sustainability may explain why there were actually fewer second-year students (-12%) post-intervention who felt able to do this.

Negotiation and conflict management are other important political skills required for work towards sustainability. There were substantial increases in students reporting that they could *negotiate well* after the intervention (+54% for second-year students; +35% for third-year students). The positive changes in this attribute were significant (p = .036 and p = .037, respectively). Consistent with these trends, more students reported a positive change in their self-assessed political capabilities for *dealing with conflict* (+31% and +25%, respectively) post-intervention.

Overall there were some very positive trends in the development of these political skills. However, the small sample size may explain why there was little evidence of real correlations between responses by individual students to the statements referring to capacity to influence opinions of others, to negotiate, to manage conflict or to encourage others to consider sustainability.

Collaborative approach

Table 3: Self-assessed collaborative approach

	Second-Year Students			**Third-Year Students**		
Indicators	**Pre**	**Post**	**Δ**	**Pre**	**Post**	**Δ**
Enjoy working in teams	25%	79%	**+54% p=.002**	61%	75%	**+14%**
Am reliable team member	55%	71%	**+16%**	61%	100%	**+39%**
Expect better solutions with professional collaboration	67%	79%	**+12%**	94%	100%	**+6%**
Ensure everyone's views are represented	30%	79%	**+49% p=.016**	61%	100%	**+39%**

An important indication of this capability was a student's belief that a *collaborative approach would lead to better solutions*. Fewer second-year students agreed with this initially (67% compared to 94% third-year students), providing more scope for change amongst this cohort (+12% compared to +6%, respectively).

It was assumed that, if individuals *enjoyed working in teams,* they would be more likely to adopt a collaborative approach, as required for work towards sustainability. Many more students reported that they enjoyed teamwork after the intervention (+54% for second-year students: +14% for third-year students). This change was statistically significant for the second-year cohort (p = .002). Individuals' preference for working this way might also depend on their perceptions that were *reliable members of the team*. As evidence of the influence of this variable, similar affirmative trends in perceived personal reliability were recorded (+16% for second-year students; +39% for third-year students) post-intervention.

Preparedness to *represent others' different views* as a means to achieving more effective solutions was seen as a useful indicator of a collaborative approach. While the intervention contributed to a positive change in this respect, the impact was greater for second-year students (+49%, p = .016), although a large change was also evident amongst the third-year cohort (+39%).

Optimism

Table 4: Self-assessed optimism

	Second-Year Students			Third-Year Students		
Indicators	**Pre**	**Post**	**Δ**	**Pre**	**Post**	**Δ**
Expect to find solutions to difficult problems faced	24%	69%	**+45%**	56%	69%	**+13% p=.02**
Expect to find sustainable solutions to global human problems	43%	64%	**+21%**	72%	82%	**+10%**
Can make decisions which contribute to sustainability	76%	57%	**-19%**	78%	91%	**+13%**

Students were asked to describe their level of optimism by responding to three statements. Generally, the number of optimistic students increased after the intervention. In some instances, trends were statistically significant. For instance, more third-year students became optimistic that they would *always find a solution to difficult problems* they faced (p = .02). More specific to this research, students' optimism about finding *sustainable solutions to global human problems* increased after the intervention (+21% for second-years, +10% for third-year students). At the same time, fewer second-year students felt optimistic about their ability *to contribute to sustainability* (-19%), in direct contrast to the third-year cohort (+13%) post-intervention. The extra year of studies, in which sustainability was embedded, seems to have enhanced the effects of the intervention on the development of student optimism in terms of this indicator.

Correlations between the responses to some of the optimism statements suggested that students had interpreted them as related to a single capability. However, there were a number of inconsistencies. While a larger sample may have validated the apparent trends in relationships, further research is necessary to explore participants' interpretations of the indicator statements or to identify more reliable co-indicators of optimism.

Resourcefulness

Table 5: Self-assessed resourcefulness

	Second-Year Students			Third-Year Students		
Indicators	**Pre**	**Post**	**Δ**	**Pre**	**Post**	**Δ**
Have access to necessary resources for learning	48%	57%	**+9%**	83%	100%	**+17%**
Prepared to actively search for resources for learning	91%	79%	**-12%**	94%	81%	**-13%**

In the research, being resourceful was indicated by perceptions of *access to resources* and preparedness to *search for resources.* The students' satisfaction with their *access to resources* to complete the intervention tasks (48% for second-year students; 83% for third-year students) increased after the intervention when translated to the resourcing of their studies more generally (+9% for second-year students; +17% for third-year students). Being satisfied that they had access to the necessary resources for learning may explain why both cohorts were less inclined to *actively search* for them post-intervention (-12% and -13% respectively). While this might indicate that the learning context was well supported, these converse trends suggest a need to review the indicators of resourcefulness used here.

Respect for others

Table 6: Self-assessed respect for others

	Second-Year Students			Third-Year Students		
Indicators	**Pre**	**Post**	**Δ**	**Pre**	**Post**	**Δ**
Respect for the views of other professionals	81%	79%	**–2%**	94%	91%	**–3%**
Aware of others' different needs	29%	86%	**+57%p =.004**	56%	83%	**+27%**

By far, the majority of students claimed they were already *respectful of the views of all professionals* (81% for second-year students; 94% for third-year students) before the intervention. It was assumed that respecting others would extend to already recognising that others have different needs. However, this was not the necessarily case. In fact, there were substantial increases in the number of students reporting *awareness of others' different needs* (+57% for second-year students; +27% for third-year students) after the intervention. This was a statistically significant change in this indicator of respect for the second-year students (p = .004).

Values for sustainability

Table 7: Self-assessed values for sustainability

	Second-Year Students			Third-Year Students		
Indicators	**Pre**	**Post**	**Δ**	**Pre**	**Post**	**Δ**
Recognise sustainability as important part of UG studies	62%	79%	**+17%**	79%	82%	**+3%**
Expect sustainability to be important in future work	43%	64%	**+21%**	83%	81%	**-2%**

Assessing affective learning is always a challenge. For this study, changes in values were measured in terms of responses to two indicator statements describing the importance attached to sustainability. With the majority of third-year students (79%) already agreeing that *sustainability was an important part of their undergraduate studies*, there was only a marginal increase (+3%) after the intervention. Apparently they were already convinced. Although many second-year students (62%) had also initially recognised this, considerably more students (+17%) agreed after the intervention.

The effects of the intervention on students' predictions about the *importance they attached to sustainability in future work* varied between second and third-year students (+21% and -2% respectively). As for many other capabilities, substantially fewer second-year students had responded positively before the intervention, suggesting there was greater scope for developing values oriented towards sustainability, and there was evidence to support this. However, assessing affective outcomes, particularly changes in personal values is always problematic, and they could be inferred from responses to many indicator statements used in this study. Rather than drawing conclusions about changes in students' values, this project highlighted the need for further research to understand and measure this important and complex capability which underpins transformative learning for sustainability.

Motivation for sustainability

Table 8: Self-assessed motivation for sustainability

	Second-Year Students			Third-Year Students		
Indicators	**Pre**	**Post**	**Δ**	**Pre**	**Post**	**Δ**
Will actively keep learning about sustainability	48%	64%	**+16%**	61%	82%	**+21%**
Will make decisions to promote sustainability	38%	43%	**+5%**	83%	91%	**+8%**

Although not statistically significant, there was an increase in the number of students who were motivated to *actively keep learning about sustainability* (+16% for second-year students; + 21% for third-year students) after the intervention. Willingness to *make decisions to promote sustainability* was the other behaviour from which motivation was inferred for the purposes of this research, but the changes were less pronounced (+5% and +8%, respectively). While more motivated to continue learning, their growing awareness of the complexity of the decisions to be made could have had a discouraging effect on some students. This might explain why there were no significant correlations between responses to these indicator statements. Like values, this is a complex, multi-dimensional capability which is difficult to measure.

Resilience and adaptability

Table 9: Self-assessed resilience and adaptability

	Second-Year Students			Third-Year Students		
Indicators	**Pre**	**Post**	**Δ**	**Pre**	**Post**	**Δ**
Change approach if mistakes made	38%	71%	**+33%**	50%	75%	**+25%**
Cope better with challenges	24%	64%	**+40% p=.005**	50%	83%	**+33%**

The intervention seemed to build resilience with more students expecting that they would *change their approach, when they made a mistake* (+33% for second-years; +25% for third-year s). There were also substantial increases in the number of students (+40% for second-year students; + 33% for third-year students) who felt they *could cope better* with challenges after the intervention, a statistically significant change (p = .005) for the second-year cohort. Some correlations were evident in responses to statements referring to resilience (*coping better)* and adaptability *(finding a solution, even when it is difficult)*. However, the pattern was not as consistent, as might have been expected.

Creativity

Table 10: Self-assessed creativity

	Second-Year Students			Third-Year Students		
Indicators	**Pre**	**Post**	**Δ**	**Pre**	**Post**	**Δ**
Adopt creative approaches	33%	79%	**+46% p=.011**	39%	92%	**+53%**
Recognise freedom to decide how to approach a task	19%	71%	**+52% p=.016**	56%	83%	**+27%**

Since there are often no obvious or tested solutions, a creative approach to problem solving is important for sustainability. Post intervention, the number of students indicating that they adopted *creative approaches when completing tasks* increased quite considerably (+46% second-year students; + 53% third-year students). This was a significant effect among the second-year students (p = .011).

Recognising opportunities to express creativity in problem solving is an important co-indicator here. Only a small proportion of the second-year students (19%) recognised this prior to the intervention. Having completed the tasks, many more of these students (+52%) realised that they had this freedom, with a statistically higher level of overall agreement (p = .016). Perhaps related to greater opportunities for student-directed learning in the final year of their program, many third-year students (56%) had already recognised that they could make choices in the ways they approached the tasks, so the increase (+27%) after the intervention was not as large for this cohort.

Graduates perspectives on work for sustainability

Table 11: Graduates' self-reported perspectives on work for sustainability

Indicator	Response
Sustainability an important objective in their work	73%
Recognise more sustainable options when making decisions in workplace	73%
Recognise that sustainability was integrated into mission or vision of organisation	54%
Accept responsibility to make workplace decisions for sustainability	58%
Actively trying to continue learning about sustainability	62%
Aware of global issues that affect sustainability of workplace operations	65%
Encourage workplace colleagues to consider sustainability	42%
Able to learn about sustainability through work	48%
Have resources to learn about sustainability through work	54%
Expect to find sustainable solutions to global human problems	60%
Have freedom to choose ways to solve problems in the workplace	75%
Adopt creative approach to solve problems in the workplace	96%
Prepared for work due to UG studies	85%

Most graduates (73%) thought sustainability was an important objective in their work, and could recognise more sustainable options when making decisions in their workplace. However, only half of this group (54%) recognised that sustainability was integrated into the mission or vision of the organisations which employed them. This apparent lack of organisational commitment might explain why only 58% graduates felt it was their responsibility to make those choices. With only 62% of these professionals actively trying to continue learning about sustainability, it is understandable that only 65% of this cohort agreed that they were aware of global issues that affect sustainability of operations in their workplaces. Consistent with these trends, less than half of the graduates (42%) had demonstrated this commitment by encouraging workplace colleagues to consider sustainability.

Other barriers to working towards sustainability were reported. Only approximately half of the graduates reported that had the necessary opportunities (48%) or resources (54%) to learn about sustainability through their work. The lack of ongoing professional development opportunities in this area may have meant many graduates felt overwhelmed by the challenges of sustainability. Predictably, only some (60%) were optimistic that *sustainable solutions to global human problems would be found.* At the same time, most graduates (75%) agreed that they had the *freedom to choose the ways they solved problems in the workplace*, and almost everyone (96%) felt they did so creatively. With most graduates (85%) agreeing that their undergraduate *studies had prepared them well for work*, it seems the workplace was not nurturing and developing their professional capabilities for sustainability.

Conclusion

This research tested a teaching and learning intervention to develop students' capabilities required for sustainability. It demonstrated the scope for transformative learning within a discipline not traditionally seen as being concerned with human values and social responsibilities. The intervention contributed to many positive changes in students' motivation, values and attributes, identified as important for sustainability. In some instances, these self-assessed changes were statistically significant. A larger sample may have confirmed that the positive trends in responses to other indicator statements were also significant.

The tasks included in the intervention had been carefully adapted for the classroom within the existing curriculum framework. As in any teaching and learning context with complex information exchanges and interpersonal interactions, there are many variables which are difficult to control. While assumptions that changes were due to the intervention were acknowledged, they can be challenged on this basis.

Validating the outcomes of any research relying on self-assessment is always problematic. However, there are several reasons why the observed changes are probably a reliable indication of the impact of the intervention. For this research, the positive changes reported were unexpectedly consistent for values, motivation and over a diverse range of attributes. Secondly, considerable time (~ 6 months) had elapsed between students' completion of the surveys. This limited the possibilities for pre-intervention responses to influence the post-intervention responses, so the differences were more likely to reflect actual changes from the students' perspectives.

While awareness about sustainability developed through this intervention, students were also better able to realise the scope of the challenges to be faced in working towards sustainability, so may have become overwhelmed. This may explain a few anomalies in the trends observed, and highlights the important role of higher education in including resilience, adaptability and optimism as learning objectives. The inconsistencies in the correlations between some of the indicators of the sustainability attributes, motivations and values may also explain some of the disparities. On reflection, assessing changes in values was the most problematic because this involves assessment of affective learning. Similarly, motivation could be inferred from responses describing current or predicted behaviours which were used to indicate other attributes. Given the interconnectedness of these capabilities, potential relationships between indicators are difficult to unravel. Overall, the findings here draw attention to the problem of identifying and validating indicators of these complex, multidimensional capabilities to reliably assess transformative learning for sustainability in higher education settings.

The consistently higher numbers of third-year students affirming their capabilities for sustainability prior to the intervention, as compared to the second-year students, could be explained in terms of the probable influence of prior or concurrent studies. Certainly for these third-year students, sustainability was a well-embedded theme in their program. Generally, fewer second-year students initially agreed they had the necessary capabilities for sustainability, and possibly as a consequence, appeared to be much more receptive to the effects of the intervention, as indicated by the changes in their self-assessments. This raises questions about the most effective ways to promote the development of these capabilities through the curriculum. Relying on single, intensive interventions, introduced as additions to existing curricula, may be a less secure investment as compared to a progressive integration of education for sustainability across a whole program. At the same time, many higher education institutions may not yet be prepared to revise policy and practice and commit the resources required for developing and implementing comprehensive sustainability-focused curricula. In the meantime, this research has demonstrated the potential of single interventions for ensuring current cohorts of students leave university with some capabilities in the process of development.

It seems that the students participating in this intervention would finish their studies better prepared for sustainability. However, it is important to know whether changes persist and development continues as students progress through their studies and enter the workforce. This would require a longitudinal study extending from university studies to the workplace, rather than relying on changes taking place in a classroom setting over a relatively short period. The responses of the graduates have highlighted some important issues in this respect. For many graduates, it was the contemporary workplaces which were failing to show a commitment and offer them the opportunities and resources for developing their knowledge and skills for sustainability. It seems that some organisations have yet to understand the importance of striving for sustainability through their day-to-day operations. While there is much to do to develop higher education curricula for sustainability, changes here must be matched by developments in workplaces. Managers need to recognise their roles as mentors to new graduates as they collaborate to balance the economic, environmental and social objectives of the organisation. In this way, commercial, community and government employers can build on these early efforts in developing graduate capabilities for sustainability through transformative higher education.

Acknowledgements

The author would like to acknowledge the invaluable support of Ms Sally Dunn with data collection and analysis, and Mr John Glass for his expert assistance in mentoring students through the intervention.

References

Adomssent, M., Godemann, J. and Michelsen, G. (2007), Transferability of approaches to sustainable development. *International Journal of Sustainability in Higher Education* 8 (4), pp. 385-402.

Anderberg, E., Norden, B. and Hansson, B. (2009), Global learning for sustainable development in higher education: recent trends and a critique. *International Journal of Sustainability in Higher Education* 10 (4), pp. 368-378.

Bacon, C., Mulvaney, D., Ball, T., DuPuis, E.M., Gliessman, S., Lipschultz, R. and Shakouri, A. (2010), The creation of an integrated sustainability curriculum and student praxis projects. *International Journal of Sustainability in Higher Education* 12 (2), pp. 193-208.

Barth, M., Godemann, J., Rieckmann, M. and Stoltenberg, U. (2007), Developing key competencies for sustainable development in higher education. *International Journal of Sustainability in Higher Education* 8 (4), pp. 416-430.

Blewitt, J. (2010), Higher education for a sustainable world. *Education + Training* 52 (6/7), pp. 477-488.

Burns, H. (2011), Teaching for Transformation: (Re)Designing sustainability courses based on ecological principles. *Journal of Sustainability Education* 2 (March), www.jsedimensions.org/wordpress/content/2011 (accessed January 2012).

Collier, G. and Armstrong, P. (2010), *Building our capacity – National professional development initiative for sustainability educators*. Updated edition of paper presented at Enviro 2010 Conference, Melbourne: July 21-23, 2010, http://www.tissues.com.au/profdevel.pdf (accessed February 2012).

Cortese, A.D. (2003), The critical role of higher education in creating a sustainable future. *Planning for Higher Education* 31 (3), pp. 15-22.

Cusick, J. (2008), Operationalising sustainability education at the University of Hawai'i at Manoa. *International Journal of Sustainability in Higher Education* 9 (3), pp. 246-256.

Dale, A. and Newman, L. (2005), Sustainable development, education and literacy. *International Journal of Sustainability in Higher Education* 6 (4), pp. 351-362.

de Ciurana, A. and Leal Filho, W. (2006), Education for sustainability in university studies. *International Journal of Sustainability in Higher Education* 7 (1), pp. 81-93.

Domask, J. (2007), Achieving goals in higher education: An experiential approach to sustainability studies. *International Journal of Sustainability in Higher Education* 8 (1), pp. 53-68.

Fenner, R., Ainger, C., Cruickshank, H. and Guthrie, P. (2005), Embedding sustainable development at Cambridge University Engineering Department. *International Journal of Sustainability in Higher Education* 6 (3), pp. 229-241.

Ferrer-Balas, D., Adachi, J., Banas, S., Davidson, C., Hoshikoshi, A., Mishra, A., Motodoa, Y., Onga, M. and Oswald, M. (2008), An international comparative analysis of sustainability transformation across seven universities. *International Journal of Sustainability in Higher Education* 9 (3), pp. 295-316.

Gilroy, M. (2008), Greening of the American Campus. *The Hispanic Outlook in Higher Education* 19 (6), pp. 40-41.

Hammond, C. and Churchman, D. (2008), Sustaining academic life: A case for applying principles of social sustainability to the academic profession. *International Journal of Sustainability in Higher Education* 9 (3), pp. 235-245.

Holdsworth, S., Wyborn, C., Bekessy, S. and Thomas, I. (2008), Professional development for sustainability. How advanced are Australian universities. *International Journal of Sustainability in Higher Education* 9 (2), pp. 131-146.

Krizek, K., Newport, D., White, J. and Townsend, A. (2012), Higher education's sustainability imperative; How to practically respond? *International Journal of Sustainability in Higher Education* 13 (1), pp. 19-33.

Lidgren, A., Rodhe, H. and Huisingh, D. (2006), A systemic approach to incorporate sustainability into university courses and curricula. *Journal of Cleaner Production* 14, pp. 797-809.

McFarlane, D. and Ogazon, A. (2011), The challenges of sustainability education. *Journal of Multidisciplinary Research* 3 (3), pp. 81-107.

Newport, D., Chesnes, T. and Lindner, A. (2003), The "environmental sustainability" problem: Ensuring that sustainability stands on three legs. *International Journal of Sustainability in Higher Education* 4 (4), pp. 357-363.

Parker, J. (2010), Competencies for interdisciplinarity in higher education. *International Journal of Sustainability in Higher Education* 11 (4), pp. 325-338.

Podger, D., Mustakova-Possadt, E. and Reid, A. (2010), A whole person approach to educating for sustainability. *International Journal of Sustainability in Higher Education* 11 (4), pp. 339-352.

Reid, A. and Petocz, P. (2005), *The UN decade for sustainable development: What does it means for higher education?* HERDSA 2005 Conference Proceedings: Sydney, Australia, http://www.herdsa.org.au/wp-content/uploads/conference/2005/papers/reid.pdf (accessed March 2012).

Sammalisto, K. and Lindhqvist, T. (2008), Integration of sustainability in higher education: A study with international perspectives. *Innovative Higher Education* 32 (4), pp. 221-233.

Savelyeva, T. and McKenna, J. (2011), Campus sustainability: emerging curricula models in higher education. *International Journal of Sustainability in Higher Education* 12 (1), pp. 55-66.

Shephard, K. (2008), Higher education for sustainability: seeking affective learning outcomes. *International Journal of Sustainability in Higher Education* 9 (1), pp. 87-98.

Sherren, K (2006), Core issues: Reflections on sustainability in Australian University coursework programs. *International Journal of Sustainability in Higher Education* 7 (4), pp. 400-413.

Sipos, Y., Battisti, B. and Grimm, K. (2008), Achieving transformative sustainability learning: engaging head, hands and heart. *International Journal of Sustainability in Higher Education* 9 (1), pp. 68-86.

Stephens, J., Hernadez, M., Roman, M., Graham, A. and Scholz, R. (2008), Higher education as a change agent for sustainability in different cultures and contexts. *International Journal of Sustainability in Higher Education* 9 (3), pp. 317-338.

Timmerman, N. and Metcalfe, A. (2009), From policy to pedagogy: The implications of sustainability policy for sustainability pedagogy in higher education. *The Canadian Journal of Higher Education* 39 (1), pp. 45-60.

Wals, A. (2009), Mirroring, Gestaltswitching and transformative social learning: Stepping stones for developing sustainability competence. *International Journal of Sustainability in Higher Education* 11 (4), pp. 380-390.

Zilahy, G. (2006), Toward sustainability: the role of higher education (Editorial) *Clean Technologies and Environmental Policy* 8, pp. 1-2.

Chapter 6

Experiencing Transformation in Sustainability: A Student Perspective

Jennie Winter, Debby Cotton and Vivien Grant

Abstract

Internationally, Education for Sustainability (EfS) is an agenda of growing significance. To date, the focus has been primarily on curriculum content rather than pedagogic processes; however, there is growing recognition that for Higher Education (HE) to contribute to the scale of social change required, it must have transformative capacity so that students can reconsider dominant social framings such as consumerism, and respond to the complex personal and professional challenges that sustainability poses. There is an established literature on transformational learning which advocates a range of participatory pedagogies to promote critical self-reflection; however, in UK HE purpose-designed transformative EfS programmes are rare. More prominent is the trend for Higher Education Institutions (HEIs) to promote sustainability transformation through whole institutional approaches. However, as yet little is known about how students experience transformation in these contexts. The Universities of Bradford and Plymouth in the UK are two of the country's leading sustainability HEIs and therefore provide fruitful settings within which to explore this issue. This paper draws on empirical research which used semi-structured interviews to explore critical moments of transformation and their impact on students' understandings of sustainability. Provisional findings indicate that transformation is mediated by the learner's discipline and the pedagogic approaches utilised therein, as well as wider issues and experiences within their academic and social context. Given the global emphasis on HE as a vehicle for promoting sustainability literacy, the research findings have potential relevance for other national contexts.

Introduction

EfS is an agenda of growing global significance, led at the international level by the ongoing efforts of the United Nations (UN), in particular through the UNDESD (UN 2005-2014). The UK government has also made strong commitments to EfS: for example, the 2005 sustainable development strategy '*Securing the Future*' recommended raising awareness and building a skill base for sustainability through the promotion of 'sustainability literacy' (HM Government 2005). This commitment has been echoed through the Higher Education Funding Council for England's strategic review and action plan which focused specifically on sustainable development in higher education (HEFCE 2008, 2009). Increasingly, universities worldwide are being called upon to contribute to the sustainability imperative (Haigh 2011; Thomas 2009).

> 'Given what academics know about the current ecological condition of the planet, there is an obligation for universities to become leaders in the movement to prevent global ecological collapse' (Moore 2005: 326).

However, whilst curriculum and campus greening efforts have been widespread, examples of significant changes to pedagogic approaches and effective development of sustainability literacy are rare (Jickling and Wals 2008). A contested term, sustainability literacy has been defined as encompassing the perspectives and insights that enable students to 'understand the symbiotic relationships between environmental, social and economic dimensions of sustainable development' (Parkin et al. 2004: 9; Sterling and Thomas 2006). Dawe et al. (2005: 58) describe its key components as declarative knowledge, problem solving skills, holistic thinking, critical evaluation, self-reflection, and informed action. It could be argued that sustainability literacy coalesces around a key point: *transforming* people's perspectives, values and actions, in favour of more sustainable ways of working and living (Moore 2005; Rathzal and Uzzell 2009; Thomas 2009). It follows that a key component of embedding sustainability literacy in higher education is through students experiencing transformation so that they can continue to 'transform [themselves] and society' (UNESCO 2011).

There is an established literature on transformative education that has developed beyond the field of EfS. This work is primarily rooted in Mezirow's (1978; 1997; 2000) work on andragogy and Freire's (1984) notion of conscientisation and critical theory (although other interpretations exist such as Beck and Cowan 1996; Boyd and Meyers 1988: Cranton 1994; Healey and Jenkins 2000; Wenger 1998). This literature suggests that the process of transformation involves revising assumptions about, and perspectives of, the world through processes of engagement and critical reflection (Cranton 1996). Such engagement alters 'frames of reference' the structures and assumptions through which experiences are understood (Mezirow 1997) leading to paradigmatic and behavioural shifts (Taylor 2007). This process is normally undertaken within a purposeful and defined programme of study yet, to date, this is not normal practice in UK universities although some interesting examples of attempts do exist (Tomkinson 2011).

A number of reasons for this omission might be proposed: developing self-reflection, appraising values, and providing opportunities to practise new actions brought about by transformation all require time-intensive pedagogies which are often not appropriate for classroom-based teaching (Taylor 2007) and can be problematic within a confined curriculum. Transformative education is notoriously difficult to implement (Bawden 1997; Sterling and Thomas 2009) and is dependent upon the expertise of the educator, the trust between educator and student, and the openness and responsiveness of the student to the process (SPMC 2002). Moreover, from a curriculum perspective, there are epistemological issues, which mean that transformation is difficult to embed, interpret and measure within outcomes-based curricula (Sterling 2011). Compounding this situation is the dearth of research which considers the student experience of transformation *per se* (Taylor 2007). There remain numerous unanswered questions about the influence and impact of transformative learning on students. For example, Taylor (2007: 187) raises questions about the students' role during the learning process: 'What are their responsibilities? Why do some students openly engage in the process and others refuse to participate? How does the transformation of students affect other students, the educator, the institution, and other significant individuals in the students' lives?' In addition, Moore (2005: 85) asks 'whether students are mentally and emotionally prepared for this type of learning' and whether institutions or staff are able to foster such experiences (de la Harpe and Thomas 2009).

In UK higher education there are very few examples of explicitly transformative educational programmes based upon sustainability. Nevertheless, the drive to foster sustainability literacy in the UK graduate population means there is increasing interest in transformative education in this context, and there are a growing number of UK universities that are attempting to promote sustainability transformation through whole institutional approaches (Hopkinson et al. 2008; Jones et al. 2010). In these institutions, attempts are made to align formal, informal and extra-curricular experiences with a view to providing a transformative context in which students are exposed to 'learning for change' (Sterling 2001: 60). This approach is based upon socio-constructivist perspectives of transformation, in which 'transformation co-emerges in the learner and the setting; that is, the personal and the social in dialectical relationship transform' (Scott 2003: 283). In these scenarios, transformation if it occurs does not take place in response to a purposefully designed programme but instead emerges through a range of experiences and events located in different points in space and time over the course of the students' time at university. This raises questions as to what the student experiences, how they make sense of these experiences, and to what extent the impacts of their experience become manifest in expressed beliefs, values and actions. The research reported here aims to explore such issues in the context of two institutions which are some way along the road towards becoming 'sustainable institutions'. Thus we hope to offer a timely and relevant contribution to this ongoing debate.

Methodology

The findings reported here form part of a wider research project which explored staff and student experiences of transformative learning around sustainability at the Universities of Bradford and Plymouth. These institutions were specifically selected on the basis of their identified successes in developing whole institution sustainability, as evidenced by the UK People and Planet 'Green League' for environmental performance (People and Planet 2011) and other measures (see Hopkinson et al. 2008; Jones et al. 2010). Plymouth University is a 'new university' (one which gained university status in 1992) on a city centre campus. In 2010 it was awarded the number one ranking in the 'Green League', and it has been consistently ranked in the top 10 since the League began in 2005. The university has received other awards, including a 'Green Gown' for continuous institutional improvement around sustainability, and it is home to the Centre for Sustainable Futures (CSF), which provides a focus for research and curriculum development around EfS. The University of Bradford is a 'red-brick', research intensive university which received significant funding to implement a campus-wide sustainability initiative, 'Ecoversity', embedding sustainability practices throughout its campus and curriculum. As well as being a consistently high performer in the Green league, it was recently winner of the International Sustainable Campus Network Awards for Impact and of the Times Higher Education Award 2009 and 2010 for Outstanding Contribution for Sustainable Development. As such, these institutions provide fruitful contexts for exploring students' experiences of institution-wide transformation around sustainability.

The aspect of the research reported here addressed the following key question: 'What are students' perceptions of transformative learning and EfS?' An interpretive approach was considered appropriate to explore personal perceptions of transformative learning about sustainability, before considering divergence and similarities in those perspectives. The researchers used semi-structured interviews to elicit narrative accounts of transformative learning moments around sustainability, using a critical incident approach (Brookfield 1987; Tripp 1993; Fitzgerald 2001). Fitzgerald (2001: 190) describes critical incidents as 'events which involve two or more people; they are neither inherently negative nor positive, they are merely ... occurrences or events which require some attention, action or explanation'. Critical incident analysis is a popular data collection tool in pedagogic research, but there is no evidence of it being used previously to research students' experiences of transformative EfS. However, it offers an innovative approach to prompting reflection which can reveal the complex richness of events and can help to address the methodological limitations of retrospective interviews by focusing on significant events that are more easily recalled.

Students were asked to identify up to three critical incidents and were then prompted to discuss them in greater depth through a series of questions based on Tripp's (1993) critical incident analysis:

1. What happened in the critical incident?
2. How has your thinking changed as a result of this event?
3. What do you do differently as a result of this event? (behaviour)

The sample was purposive, to accommodate different genders, cultural and ethnic backgrounds and to include a range of disciplinary contexts. The disciplines included were Geography, Business and Engineering, encompassing a broad range of approaches to EfS, whilst allowing for easy comparability between institutions. The data collection took place between September and December 2011.

Table 1: Student Sample

	Geography	Business	Engineering	Total
Plymouth	5	5	2	12
Bradford	6	3	4	13
Total	11	8	6	25

Data were analysed using the constant comparative method to draw out cross-cutting themes (Silverman 2005). This involved an iterative process of reading and re-reading data, looking for similarities and differences between accounts, and noting specific references to transformative learning and sustainability. The coding frame was developed through a process of negotiation between the researchers, enhancing inter-rater reliability and the robustness of the findings. Analysis involved consideration of the ways and extent to which students were critically aware of transformative learning moments around sustainability; the extent to which these emerged as a result of their engagement within their specialist fields of study; and the potential for EfS to have transformative capacity beyond their disciplinary context. NVivo software was used as an aid to analysis, and to enhance the ability to sort data by respondents' institution, discipline and personal characteristics. This analysis is still underway, therefore the findings presented here should be considered provisional. In addition, the degree of generalisation which can be made from this kind of research is limited. Whilst statistical generalisation is not intended, the data collected can be used to theorise about the wider applicability of the findings on transformative learning and sustainability using 'theoretical inference' (Hammersley 1998).

Findings

Students' understandings of sustainability were heavily focused on resource depletion and the needs of current and future generations. This was reflected in their accounts of sustainability in their professional and personal lives. These findings present participants' accounts of critical incidents and processes identified as relevant to transformative learning *about*, or *for*, sustainability, whether they occurred within or beyond the formal curriculum. Key themes are discussed below:

(i) Discipline

Discipline of participants was found to be a key mediator in students' understandings of sustainability and transformation. It became clear that the formal curriculum transformed students within their disciplinary context through changing students' perceptions of and interactions with the world to be representative of the particular 'ways of knowing' in their discipline. Transformation, for many students, was viewed as a part of the learning process involved in 'becoming an engineer' (or geographer, etc.). It was evident, therefore, that the more explicit the links between disciplinary content and sustainability were made, the more the student was able to identify with sustainability philosophies and practices as part of their professional identity:

> I think it's our duty as designers to look into the sustainability and the environmental impact of that, and if we don't do that, who is going to? (UB1: 6).

> The sustainability module we did, I guess I just sort of changed the way I thought about it, I suppose. I now try and act sustainably because we're told how important it is. I think it's impossible to go to geography here and not come away more sustainable than you went in (UP1: 18).

This often involved epistemological change, and key disciplinary concepts played an important role in this process:

> Cradle to Cradle is a philosophy .... Originally when we designed products it was from cradle to the dump, but we need to design objects and products that go through their product life and then they're returned into manufacturing, separated into the materials and reused instead of just being thrown away and discarded, because we're running out of materials a lot now (UB1: 6).

(ii) Pedagogy

Knowledge acquisition and the development of new meaning both played an important part within students's beliefs and practices about sustainability and pedagogy emerged as a key underpinning factor in students' accounts. Students identified a wide range of activities including group work, individual tutorials, peer tutorials, involvement in research, professional development portfolio work, work placements, fieldwork and reflective journals, as facilitating opportunities for them to question, experience and test sustainability themes which, in turn, prompted critical incidents within their learning journeys. Experiential learning approaches were frequently mentioned by students in describing critical incidents, and it was clear that first-hand experience had a much stronger and more lasting impact than reading or being told about sustainability activities:

> What really stood out to me and changed my mind, were elements of the design of his home and the thing he had done with his compositing, he had literally composted everything and I actually started doing that at home because it meant I didn't have to put my bins out as much ... that really hit me using the bins so much and he said he only put his bin out once every two months so that was one thing that really stood out (UB8: 8).

Those students who had experience in business and industry either previously or during their degree demonstrated particularly acute awareness of the complexity of sustainability and the disjuncture inherent in applying theoretical sustainability models to real world practice. There was also evidence of an enhanced ability to think holistically and to incorporate systems thinking into problem solving. This is arguably representative of transformation, in professional terms at least:

> If I were to talk about something that is sustainable I would be looking at three aspects it would be social, economic and environmental and it would have to encompass all three ... it has to be a holistic thing because everything is so inter-connected there is no point in working on this if this is completely out of balance … it has to be the whole thing (UP5: 7).

Several advocates of EfS note the role of the university campus as a catalyst for encouraging sustainable living practices and for exemplifying the interconnectedness of social, economic and environmental interests creating a 'sustainability orientated pedagogy of place' (Kagawa 2007: 320). Students who had used the university as a field site were better able to discuss sustainability in the university organisation and apply the principles learnt to other organisational contexts. A side effect of this pedagogic approach was that students critically reflected on their own role as a stakeholder in the university:

> I am doing geography and they use (Institution name) as an example of waste management and how they try and store the waste at source and things like that so I reckon not only is the University continuing its efforts but it makes me feel bad if I say I can't be bothered and they are putting all this money in (UB4: 8).

Where students did not have this experience they saw campus-based sustainability in terms of what (un)sustainable practices were visible to them: double-sided printing, recycling, sustainable procurement, energy inefficiency and use of unsustainable building materials. They reported that the university gave out mixed messages about campus sustainability but rarely reflected upon their own roles and responsibilities.

(iii) Internationalisation

The data suggest that international experiences help to develop students' ability to recognise and articulate causes of, and solutions to, sustainability issues, and to appreciate the role of interconnected global networks on local phenomena. First-hand experiences of poverty, environmental degradation, and responses to sustainability issues often engendered emotive responses in students which encouraged them to be critically reflective, a key feature of transformation:

> From an educational perspective I've changed because I've seen more, I've learnt more. So, I went to Africa so, I mean, I've seen a lot of poverty and I've seen people that aren't as well off as me and people don't have the chance to come here … that's definitely transformed me as a person (UP1: 9).

Internationalisation was also reported as beneficial for developing tolerance for other cultures and an appreciation for multiple perspectives and alternative ways of approaching situations, and this did not have to be as a result of going overseas:

> So if you spend time with someone, you live with someone, you will inherently adapt parts of their culture. I've met a vast number of people from different backgrounds and you meet them and adapt. So if you see something that they do as being better and you like it then you think it's a good idea (UB2: 8).

Both home and overseas students expressed the benefits of exposure to new knowledge, pedagogies and ways of living:

> I think my mind has been changed pretty much since I came to University because in China we got information from the subject books or from the internet or something else but when I came here I found many students and teachers are really active to take part in some activities which involve protecting the environment ... I think it is quite good for students especially in higher education and it forces us to protect our world (UP 12: 2).

Clearly the opportunities in higher education for such inter-cultural mixing are significantly greater that most students' previous experience will have allowed.

iv) Independent living

Students reported that informal aspects of moving to university such as living independently, budgeting, managing social situations and getting involved in new activities and communities were part of an intense cultural shift which contributed to personal development and transformation. Some of these changes were associated with sustainability, such as the need to live on restricted budgets which made cost-effective pro-environmental behaviours such as energy and water efficiency attractive to them. Most students lived in university halls of residence at some point during their studies, and resource efficiency and recycling were an integral part of the choice architecture in these environments which prompted students to reflect on their behaviour:

> To come back on recycling, say you look at it and go why was I throwing it all in the bin and I could have been doing it like this. And really critically evaluating just what you're throwing away and why (UB2: 3).

University was described as a place where significant personal relationships were forged, some of which were identified as impacting significantly on perceptions of and participation in sustainability. Strong relationships such as girlfriends, boyfriends and close friendships exposed students to different ways of life, including participation in outdoor activities, volunteering, recycling and vegetarianism:

> I met a friend and he made me think, "Actually, there's more to life" so, really, it was his influence because he's a very outdoorsy person and that's where he gets his buzz, so, he sort of opened my mind to it. The more time you spend on the moors you think, "Actually, this is really quite impressive and I wouldn't want to destroy it" (UP11: 7).

> That would be my girlfriend. She's very much into looking after her own environment. So I do a relative amount of recycling and am fairly responsive on what I use, what I don't use … If you're with someone of course you end up developing a little bit more into blending approaches of life and that's really helped in terms of understanding and really looking at what you're doing (UB2: 5).

Discussion

The findings of this research exemplify the range of ways in which students experienced transformative learning around sustainability during their time in higher education, and illustrate the importance of disciplinary context as a mediator of student understandings of sustainability. Despite the whole-institution aspirations of the two universities, sustainability remained strongly situated in the discipline in which students are embedded. Through the discipline, sustainability has the potential to act as a new lens for viewing the world, yet there may be some tension between the need to develop students as subject experts, and the more interdisciplinary holistic understandings required for EfS. The students perceived academic and vocational transformation to be common practice at universities but transformation for sustainability appeared heavily dependent upon the expectations of their discipline and tutors. Nonetheless, the formal curriculum, and associated pedagogies, remained highly influential on students' interpretations of sustainability, and on the potential for transformational experiences. Many of the learning contexts mentioned by students incorporated characteristics of experiential and active learning (Kolb 1984; Bonwell and Eison 1991), pedagogic approaches which have previously been identified as being important for EfS (Cotton and Winter 2010).

Nonetheless, there are limitations to the potential for transformation (and behaviour change) within the formal curriculum. Participants reported distinctions between their roles as students and as individuals in everyday life and noted that where learning for sustainability took place within disciplinary boundaries it was often their professional or academic identities which were transformed in favour of sustainability practices; this did not automatically equate to reported transformation outside of this sphere:

> Yeah definitely (I have been changed) as a student, I mean I will definitely take some of the skills from my student life and put them into my work and personal life but it has changed me more as a student than as an actual person (UB4: 6).

This is of interest as the literature on transformative education posits that the students' underpinning epistemological values and worldviews should be changed. These findings suggest that in some cases it is only one dimension of their worldview – one that they are being progressively inculcated into, their professional or academic persona – which has been transformed.

However, aspects of the student experience which arose beyond the discipline and formal curriculum were also influential on students' beliefs and – more importantly – practices. In particular, extra-curricular activities and relationships with peers seem to be crucial in the transformation that occurred during this key developmental period of their lives. The mixing of individuals from different countries, cultures and backgrounds offers a real opportunity for HE to offer EfS experiences in a way that other educational levels may not, and may explain the apparent increase in environmental commitment which has been observed to occur through HE participation (Cotton and Alcock 2012). Sterling (2004: 56), writing about EfS, presents transformative education as one which involves 'the whole person, and affects change in deep levels of values and beliefs through a process of re-perception and re-cognition. It is not then a matter of intellectual and conceptual learning, but learning that engages our emotional and intuitive selves as well' Perhaps this is simply more easily achieved outside of the formal curriculum? Either way, developing trusted relationships with significant others (university faculty, peers, friends, work colleagues) allowed for discursive pedagogies such as extended discussion and debate to act as conduits for evaluating and developing personal and professional perspectives on sustainability. In addition, international experiences were a key indicator of transformation: contexts in which students were exposed to novel experiences and which encouraged them to interact with sustainability knowledge and practice in new ways and in differing social settings were frequently mentioned as being transformative. There are also hints in the findings that the junctures between structured and social learning may be of particular interest in developing transformative learning.

Social identity theory (Tajfel and Turner 1979; 1986) provides a conceptual framework which may help to explain the identified differences between learning in formal and informal contexts. Social identity theory asserts that identification with a reference group or groups is a key component of identity and that this has consequences for behaviour:

> 'Key aspects of our behaviour are motivated by the particular social groups that we belong to. Certain behaviours are more or less ruled in or ruled out for me, simply because I perceive myself as belonging to a particular social group. The roots of these normal behaviours have very little to do with individual rational deliberation. Rather they are heavily influenced by inter-group and intra-group dynamics. And in both cases, my personal influence over prescribed or proscribed action is severely limited' (Jackson 2005: 82).

The signifiers in some disciplines are strongly aligned with sustainability – outside this role this is potentially weakened, depending upon the beliefs and norms of other groups students belong to. However, in disciplines where the sustainability element is less strong, social groupings and extra-curricular roles may strengthen learning about sustainability. In addition, meeting others from very different social groups may act as a catalyst for re-evaluating students' own beliefs and values, an essential pre-cursor to transformation. The challenge for universities is therefore to focus on enabling diverse social learning opportunities, both within and beyond the formal curriculum, and to consider how sustainability values can be embedded as norms amongst the student population. Enhancing the links between the formal and informal curriculum would provide opportunities to align learning in both spheres, and help strengthen sustainability messages for all students.

Conclusions

Whilst this research is still at an early stage, there are indications that higher education may have a particular role to play in EfS. With its combination of transformation associated with the disciplinary context, use of experiential learning approaches and potential for widespread inter-cultural mixing and close working and social relationships with others, the university offers a context with high potential for transformative learning. However, the tensions between the highly-bounded discipline context of the university and the inter-disciplinarity needed for EfS, as well as the increasing marketisation of higher education, may yet prove to be problematic in terms of the potential for encouraging transformative learning. Given that this study focused on two of the leading UK institutions in terms of EfS, it could be argued that the problems encountered in other institutions would be even more significant.

In the light of this research, we offer the following recommendations for enhancing the development of transformative learning around sustainability:

- Support should be put in place to enable faculty to enhance and clarify the links between their discipline and sustainability. The more that a sustainability focus can be integrated into the disciplinary identity, the more likely it is that it will form part of the student transformation process.
- Faculty should be encouraged to utilise experiential and active learning approaches wherever possible. Quite aside from the link suggested here and in other research, between experiential approaches and sustainability, there are sound educational reasons for encouraging this approach which might be used to encourage the less ardent sustainability enthusiasts.
- Developing an internationalised curriculum and student body should be encouraged as a key part of the development of a sustainable university. Since there are clearly some tensions between, for example, international field trips and carbon reduction aspirations, alternative approaches to encouraging inter-cultural mixing should be considered.
- Linking formal and informal learning in higher education offers the potential to make use of some of the extensive learning experiences which students have through interactions with friends and family outside their programme of study. In particular, the potential for social identity to be re-negotiated with a new and varied peer group may promote transformation and greater steps could be taken to build on this through the formal curriculum.

In addition, we would recommend that further research into this issue is conducted. The focus on leading edge institutions can give an indication of what might be possible – but is not necessarily representative of other contexts. We would strongly recommend that national or even international research projects should be undertaken to explore the ways and extent to which these findings are more widely applicable.

Acknowledgement

With thanks to Dr Peter Hopkinson and the University of Bradford for their co-operation and contribution to this research.

References

Bawden, R. (1997), *The Community Challenge: The Learning Response.* Invited Plenary paper, 29th Annual International Meeting of the Community Development Society, Athens, Georgia 27-30 July 1997.

Beck, D.E. and Cowan, C.C. (1996), *Spiral Dynamics: mastering values, leadership, and change.* Blackwell, Malden, MA.

Bonwell, C.C. and Eison, J.A. (1991), *Active learning: Creating excitement in the Classroom.* ASHE-ERIC Higher Education Report, No. 1. The George Washington University, Washington, D.C.

Boyd, R.D. and Meyers, J.G. (1988), Transformative Education. *International Journal of Lifelong Education* Vol. 2, pp. 261-284.

Brookfield, S. (1987), *Developing Critical Thinkers.* OU press. Buckingham.

Cotton, D.R.E. and Alcock, I. (2012), Commitment to environmental sustainability in the UK student population. *Studies in Higher Education* (Pre-publication copy available online at: http://www.tandfonline.com/doi/abs/10.1080/03075079.2011.627423).

Cotton, D.R.E. and Winter, J. (2010), 'It's not just bits of paper and light bulbs': A review of sustainability pedagogies and their potential for use". In Jones, P., Selby, D. and Sterling, S., *Sustainability Education: Perspectives and Practice Across Higher Education,* pp. 39-54. Earthsca, London.

Cranton, P. (1994), *Understanding and promoting transformative learning: A guide for educators and adults.* Jossey-Bass: San Francisco.

– (1996), Types of group learning. *New Directions for Adult and Continuing Education* Vol. 71, pp. 25-32.

Dawe, G., Jucker, R. and Martin, S. (2005), *Sustainable development in higher education: Current practice and future developments.* A report for the Higher Education Academy: York.

de la Harpe, B. and Thomas, I. (2009), Curriculum change in universities: Why education for sustainable development is so tough. *Journal of Education for Sustainable Development* Vol. 3, pp. 75-85.

Fitzgerald, M. (2001), Gaining knowledge of culture during professional education. In Higgs, J. and Titchen, A., *Practice knowledge and expertise in the health professions.* Butterworth Heinemann, Melbourne.

Freire, P. (1984), *Pedagogy of the oppressed.* Seabury Press, New York.

Haigh, M. (2011), Transformative Learning for Global Citizenship: turning to turquoise. *Learning and Teaching in Higher Education* Issue 5. Available at http://resources.glos.ac.uk/tli/lets/journals/lathe/issue5/index.cfm.

Hammersley, M. (1998), *Reading Ethnographic Research: A Critical Guide.* Longman, London.

Healey, M. and Jenkins, A. (2000), Learning cycles and learning styles: the application of Kolb's experiential learning model in higher education. *Journal of Geography* 99, pp. 185-195.

Higher Education Funding Council for England (HEFCE) (2008), *Strategic review of sustainable development in higher education in England.* Available online from: www.hefce.ac.uk/pubs/rdreports/2008/rd03_08/ (accessed 6 March 2009).

– (2009), *Sustainable development in higher education: Update to strategic statement and action plan.* Available online from: http://www.hefce.ac.uk/pubs/hefce/2009/09_03/ #exec (accessed 22 June 2009).

HM Government (2005), *Securing the Future: The UK Government Sustainable Development Strategy.* TSO, London. Available online at: www.defra.gov.uk/sustainable/government/publications/uk-strategy/documents/SecFut_complete.pdf (accessed 6 March 2009).

Hopkinson, P., Hughes, P. and Layer, G. (2008), Sustainable graduates: linking formal, informal and campus curricula to embed education for sustainable development in the student learning experience. *Environmental Education Research* Vol. 14 (4), pp. 435-454.

Jackson, T. (2005), *Motivating Sustainable Consumption,* Sustainable Development Research Network (Briefing One), Guildford.

Jickling, B. and Wals, A.E.J. (2008), Globalization and environmental education: Looking beyond sustainability and sustainable development. *Journal of Curriculum Studies* Vol. 40, pp. 1-21.

Jones, P., Selby, D. and Sterling, S. (2010), *Sustainability education: Perspectives and practice across Higher Education.* Earthscan, London.

Kagawa, F. (2007), Dissonance in students' perceptions of sustainable development and sustainability. *International Journal of Higher Education* Vol. 8 (3), pp. 317-338.

Kolb, D.A. (1984), *Experiential Learning: Experience as the Source of Learning and Development.* Prentice-Hall, Englewood Cliffs.

Mezirow, J. (1978), Perspective Transformation. *Adult Education,* pp. 100-110.

– (1997), Transformative learning: Theory to practice. *New Directions for Adult and Continuing Education* Vol. 74, pp. 5-12.

– (2000), *Learning as transformation: critical perspectives on a theory in progress.* Jossey Bass, San Francisco.

Moore, J. (2005), Is Higher Education Ready for Transformative Learning? A Question Explored in the Study of Sustainability. *Journal of Transformative Education* Vol. 3 (1), pp. 76-91.

Parkin, S., Johnson, A., Buckland, H. and White, E. (2004), *Learning and Skills for Sustainable Development: Developing a Sustainability Literate Society.* HEPS, London.

People and Planet (2011), *Green League.* Available from: http://peopleandplanet.org/ greenleague (access date 30/6/2011).

Räthze, A. and Uzzell, D. (2009), Transformative environmental education: a collective rehearsal for reality. *Environmental Education Research* Vol. 15 (3), pp. 263-277.

Thomas, I. (2009), Problem-Based Learning in Universities: Critical Thinking, Transformative Learning, Sustainable Education and Problem based learning in universities. *Journal of Transformative Education* Vol. 7 (3), pp. 245-264.

Tomkinson, B. (2011), Moving from Vision to Transformation in Engineering and Physical Sciences. *Learning and Teaching in Higher Education* Issue 5. Available at http://resources.glos.ac.uk/tli/lets/journals/lathe/issue5/index.cfm.

Scott, S. (2003), The Social Construction of Transformation. *Journal of Transformative Education* 1, pp. 113-116.

Silverman, D. (2005), *Doing Qualitative Research*, 2nd ed. Sage, London.

SPMC (2002), *Systems Practice for Managing Complexity – Project Philosophy and theoretical basis.* Available frohttp://www.mattnorman.co.uk/spmc/project.ctm (accessed 28 May 2009).

Sterling, S. (2001), *Sustainable education: Re-visioning learning and change.* Green Books, Totnes.

– (2004), Higher Education, Sustainability and the Role of Systemic Learning. In Corcoran, P.B. and Wals, A.E.J., *Higher Education and the Challenge of Sustainability: Contestation, Critique, Practice, and Promise*, pp. 40-70. Kluwer Academic, Dordrecht.

– (2011), Transformative learning in sustainability: sketching the conceptual ground. *Learning and Teaching in Higher Education* Vol. 5, pp. 17-33.

Sterling, S. and Thomas, I. (2006), Education for sustainability: The role of capabilities in guiding university curricula. *International Journal of Innovation and Sustainable Development* Vol. 1, pp. 349-370.

Tajfel, H. and Turner, J.C. (1979), An integrative theory of inter-group conflict. In Austin, W.G. and Worchel, S., *The Social Psychology of Inter-group Relations*, pp. 33-47. Brooks/Cole. Monterey, CA.

Tajfel, H. and Turner, J.C. (1986), The social identity theory of inter-group behaviour. In Worchel, S. and Austin, L.W., *Psychology of Inter-group Relations*, pp. 27-40. Nelson-Hall, Chicago.

Taylor, E. (2007), An update of transformative learning theory: a critical review of the empirical research (1999-2005). *International Journal of Lifelong Education* Vol. 26 (2), pp. 173-191.

Tripp, D. (1993), *Critical Incidents in Teaching.* Routledge, London and New York.

United Nations (UN) (2005-2014), *Decade of Education for Sustainable Development.* Available from http://www.desd.org/ (accessed 21 November 2011).

UNESCO (2011), *Five Pillars of Learning.* Available from: http://www.unesco.org/new/ en/education/ themes/leading-the-international-agenda/education-for-sustainable-development/education-for-sustainable-development/five-pillars-of-learning (access date 03/02/ 2012).

Wenger, E. (1998), *Communities of practice: Learning, meaning and Identity.* Cambridge University Press, Cambridge.

Chapter 7

Enabling University Leaders to Serve as Role Models for Sustainable Development

Chris Shiel

Abstract

In the UK, evidence suggests that the implementation of sustainable development within universities has largely resulted from the work of one or two champions driving change, often from the bottom-up. While such approaches bring success, culture change and the institutionalisation of SD require strategic coordination, the buy-in of senior managers and the support of the University Board (HEFCE 2008). Unfortunately, involvement of the broader group of senior managers has often been lacking; SD is frequently seen as 'low priority' by governing bodies.

This paper will present the outcomes of a small-scale project (funded by the Higher Education Funding Council for England) undertaken to address this perceived gap in SD engagement. The project involved working with Board/Council Members at two distinctly different universities to enable participants to increase their knowledge of SD, explore leadership behaviours to support culture change and secure the commitment of a wider staff base to SD and carbon reduction. Workshops were then cascaded to other institutions.

A description of the project approach is provided with analysis and reflection on outcomes; emerging leadership behaviours for SD are presented. The conclusion suggests that the project is raising the profile of SD and is transferable to other institutions. The paper will be of interest to those who are implementing sustainable development within higher education.

Leadership for sustainability in the UK

In the UK evidence suggests that the implementation of sustainable development across the higher education sector is patchy (Dawe et al. 2005): some universities have made considerable progress; others far less. In those institutions where there has been substantial development it has often been as a result of 'one or two people' (HEFCE 2008: xi) acting as champions to push the agenda forward (Copeland 2008; Jones et al. 2010). Campus greening initiatives (usually but not always led by Environmental Managers within universities) have had some success (Sterling and Scott 2008); however, getting university leaders to accept that education for sustainable development needs to be addressed across the entire curriculum has been a greater challenge. Too often initiatives which have gone beyond 'campus greening' have been perceived as '*just another course or research project, expendable if it does not pay its way*' (Wals and Blewitt 2010: 70). Only a few UK universities have implemented holistic approaches to sustainable development; again, these have largely been driven by a handful of champions advocating change. Evidence that most senior staff within universities are exemplifying effective leadership of this agenda is rarely apparent. Martin and Jucker (2005: 21) suggest that such leaders are not only scarce, but the qualities needed to enable sustainability: '*humility, respect for all forms of life and future generations, precaution and wisdom, the capacity to think systemically and challenge unethical actions,*' is in short supply.

It is frequently the case that sustainable development is seen as a low-priority agenda item for a university's executive team; it rarely appears on the agenda of the Board or Council, as part of the governance process. Such a low priority might be excused in the context of a sector which has been tackling so many other pressing issues, but to continue to regard sustainability as a peripheral activity would be short-sighted. Disappointingly many university leaders fail to see the implications of one of the most significant global issues for the well-being of society; they rarely see the implications of sustainable development for education and their role in particular.

Feedback from Universities that Count (2009) evidences that integrating 'Corporate Responsibility and Environmental Management' into institutions remains a common challenge at sector level (p. 37). Low scoring areas of integration (in the results of their benchmarking survey of universities) include:

- Integration of corporate responsibility and environmental management into strategic decision making
- Building corporate responsibility and environmental management into the development of senior managers (p. 11).

Their report also highlights that the Higher Education sector average, is considerably below the business average in these areas.

Such lack of attention is surprising given the number of clarion calls since *Agenda 21* (UNCED, 1992), for universities to engage with sustainable development. It is particularly surprising in England, where the Higher Education Funding Council for England (HEFCE) has strategically sought to increase engagement (HEFCE 2005; 2008). Further, HEFCE's carbon reduction strategy (HEFCE 2010), which sets challenging carbon reduction targets for the sector, underscores that sustainable development is a key strategic issue for all institutions which requires '*behavioural change and new ways of working*' (p. 15). HEFCE reminds university leaders that SD '*extends beyond the traditional estates function'* to other activities '*including teaching, research and public communications'* (2010: 8). At the same time the assertion is made that '*carbon management is a key strategic issue, so it is a crucial area for governors who should be informed and involved in decision making on the institution's approach to reducing its emissions*'.

The reference to governance is an important point: the Board has to be the main advocate for sustainability if employee engagement programmes are to succeed (Brighter Planet 2010). It is perhaps a lack of advocacy to date on the part of those with governance responsibility which has partly contributed to a lack of strategic engagement. It is undoubted that the institutionalisation of sustainability (which requires culture change and transformational ways of working) has to be strategically coordinated by university leaders, but such coordination also needs the support of Board Members (HEFCE 2008) who offer the potential to ensure change.

Against this background this paper provides an account of a Leadership Governance and Management Funded (LGMF) project which was applied for in 2010, and which aimed to raise the profile of sustainable development with Board members and senior teams. The project aimed to address a perceived internal gap (in terms of leadership development for sustainability) within the two institutions collaborating on the project, before rolling out the approach to other universities. Thus, the secondary aim of the project was to meet a gap in the external context: there is little published work on leadership for sustainable development; there are few initiatives which engage this stakeholder group and enable them to reflect on their contribution to sustainable development.

The overarching aim of the project was to work with the target group to increase awareness of the broad sustainable development agenda, with a specific focus on leadership behaviours to embed sustainable development, and actions to achieve challenging carbon reduction targets.

The objectives were to:

- Provide participants with the opportunity to increase their knowledge;
- Explore the breadth of concerns;
- Identify their potential role in supporting culture change;
- Develop approaches to securing commitment to carbon reduction and sustainable development.

The LGMF project journey

The original project proposal committed the project team to running two workshops at University A and two at University B. Learning would then be cascaded to two further universities. The plan was to target the first workshop at the level of the Board or Council, followed by a second workshop for the senior management team. This was subsequently revised.

Buchanan and Boddy (1992) highlight the need for the change agent to manage three parallel strands of activity: the content agenda, the control agenda and the process agenda.

The development of workshop process and content was undertaken at the same time as 'backstage activities' (Yukl and Falbe 1990), which included inspirational appeals to enhance participation; consideration of the role of the client, a contracting meeting with the Chair, and preparing the ground for change through communication with stakeholders.

In developing the content, consideration began with the literature in relation to leadership and particularly leadership for sustainable development. Consideration was also given to the development/learning model, the influencing process, and how to facilitate change.

There is a vast literature on leadership but very little which considers leadership for sustainability, with the exception of a few texts such as Parkin (2010) and Marshall et al. (2011), neither of whom focuses specifically on university leadership. Whenever the topic of leadership in relation to sustainability is addressed, it is usually in the form of a plea by authors for more effective and strategic leadership of sustainability. A lack of coherent leadership for sustainable development is usually cited, at the same time as acknowledging that most of the drive and innovations that have occurred, have been down to the work of champions.

The 'Sustainability Leadership Relational Model' (Ferdig 2009) enabled the project team to develop ideas around possible leadership behaviours which might result from the project; Yukl's, work on leadership and particularly the 'Essence of Effective Leadership', (pp. 456-457) was also informative. Although Yukl does not explicitly address sustainability, the socially responsible leader he describes should have sufficiently compatible behaviours to lead sustainable development. Reviewing the general leadership theory literature served an important purpose in preparation: it enabled the project team to not only explore ideas and linkages but increased confidence in being able to talk leadership as well as sustainability with the target audience. The decision was made however, not to impose any of the models or leadership behaviours during the workshops, but to let behaviours emerge from participants.

In parallel during preparation, consideration was given to process. It would undoubtedly have been easier to select a 'stand and deliver' method (and some participants might have preferred to be told the solution), but a passive approach to learning was rejected, as was using cognitive dissonance (Festinger et al. 1956), which might have been too negative and challenging in the time available. Outcomes from deliberations were an emphasis on a social learning approach (Bandura 1977) and the potential of action learning (Revans 1982), with flexibility for participants to adapt the approach to suit their own institutional context.

Other considerations in the preparatory stages included:

- Clarifying the role of the client (the University, the chair, the Vice-Chancellor);
- The anticipated outcome (the timeframe prohibited an outcome of SD embedded throughout but developing a process, model and increased advocacy was achievable);
- The need to understand more about how the Board operates;
- World views and the need to surface these;
- The transparency of the learning model (to enable cascade);
- Influencing approaches and ways to 'sell' participation ('thought leadership', for example).

Institutional workshops were preceded by a contracting meeting, which at the first institution involved the VC, the Chair and the Deputy Vice-Chancellor (Estates), the Project Leader and the external facilitator (as credible expert). Potential approaches to project delivery were discussed, as well as time frames. Subsequent contracting meetings (at other institutions) did not always include this combination of participants.

At the first contracting meeting the concepts of 'thought leadership' and 'action learning' were appealing to the Chair. These concepts influenced all further workshops. The decision was also made that workshops should include the Board and wider university team (at both workshops) to increase engagement.

A detailed facilitation plan was developed for each workshop. Table 1 gives a brief outline of the plan, although plans were much more detailed.

Table 1: Outline of workshop activities

Workshop 1	**Workshop 2**
Presentation on drivers for SD and current institutional context	Recap of workshop 1; reminder about SD and institutional carbon emissions
Identification of challenges and opportunities	Reviewed progress on goals and actions; identification of further actions required; identification of leadership behaviours
Individuals ranked top 3 priority areas	Presentation on leadership behaviour
Reflection on which three were most scored; amalgamation of priorities as appropriate	Identified leadership behaviours in relation to strategic management then sustainable development
Consideration of world views	Considered types of behaviour which blocked progress; types to take progress forward.
Individual commitments	Plenary
Action plans developed in small groups: what, why, how and by whom.	Concluding comments from VC/Chair

Each plan was approved by the Chair before the workshops.

A substantial amount of time was taken negotiating participants' availability. The first workshop commenced in January 2011 with a gap of at least three months between workshops.

The workshops were facilitated by the project leader, the external consultant and the two Environment Managers (from each institution) working as a team.

All outputs from the workshops were captured; notes were made of discussions. The delivery team also engaged in post-workshop reflection, gathering summaries of reflective learning.

The next section offers a description of the process, with commentary. Discussion draws from the experience of the process at the first institution, where the process has been completed; the first workshop (with small variations) was then repeated at the partner university and a 'cascade' university.

Workshop One

Carefully scripted briefings were produced for the VC and the Chair; they introduced the project and its significance. This made their roles easier but also ensured that they were 'on message'. The first input was a brief presentation which articulated a persuasive rationale for engagement using an approach to win 'hearts and minds' (rather than evangelism), and emphasising the pay-back. The content included the external context, the drivers (legislative, financial, educational), the institutional context with facts and figures on utilities, carbon emissions, achievements to date, and incentives for engagement.

Participants were then asked to identify the challenges and opportunities of the agenda, as individuals generating a substantial list which was captured on a whiteboard.

Again, as individuals, participants were asked to approach the list and vote for the three items (by marking the board), which they perceived as critical to be taken forward. Votes were scored and through a process of negotiation (and collapsing of some items), the group agreed which were the three to four top-scoring opportunities they wanted to take forward to the next stage.

A 'world views exercise', was then introduced to change the dynamic and increase energy levels. Participants were asked to place themselves on a line which represented a continuum between an Arcadian view (deep green ecological view of our relationship with the planet), and an Imperialist view (man can use resources and is clever enough to fix the problems). They were asked to discuss with those on the line the rationale for their position, relative to others. They were also asked to consider how world views influence leadership and decision making. Engagement with the exercise was fairly enthusiastic (after some initial discomfort from some participants) and participants were interested to see how their colleagues placed themselves. Surprising for the observers were the numbers of the senior team who identified with an Imperialist view. Surprising for some participants was the reflection that world-views impact on all interactions and decisions. As one commented: '*I am a bit surprised to see that my colleagues' views are very different to my own – a bit worrying given that they bring these views to the table.*' No attempt was made at this point to influence world views, or suggest that particular views might be right or wrong; rather the exercise was used to surface the range of perspectives and get participants articulating their personal opinion. The facilitators however noted the challenge of working with so many 'imperialists' and reinforced that world views impact on all decisions. Attention was also drawn to the importance of remaining aware of the diversity of perspectives, the need to be open to new perspectives, and the need to accommodate others as they developed actions and embarked on a journey of change.

Participants were then asked to select groups to develop actions, in relation to one of the 'top challenges.' In these groups, they first fleshed out the issue and with careful facilitation, articulated the what, why, how, by whom, and when. Table 2 offers an example of an output from one worked up action, in relation to the strategic objective of increasing engagement with SD.

Table 2: Developing a Strategic Objective: staff, student and Board engagement

No.	Action (Consult stakeholders research what others do)	By whom	By when
1	Consult stakeholders/ research what others do	E & ET	ASAP
2	Programme of actions (what, why, etc)	ULT	Next board
3	Communications plan (including achievements to date)	M & C and SUBU	Next term
4	VC launch/ make visible through personal examples / champion	VC and others	Next term
5	BIG statement(s) – remove printers – close car parks – carbon sculpture – bicycles	Board/ ULT/ SU	2011-2012
6	Roll out of events	ULT/ SU	2011-2012 (ongoing)
7	Recognition of success/ rewards/ role models	ULT	2011-2012
8	Embedding in new student recruitment campaigns	M & C	2011-12

Participants were asked, at any time during the workshop, to post 'personal commitments' on a board. Participants were generally enthusiastic in relation to the activity; some returned to the board more than once. Examples included:

- I will ask questions about the carbon footprint of new buildings on campus;
- Go paperless;
- Talk about SD more;
- Learn to use an iPad and stop asking for paper copies;
- Turn off PC when not in office;
- Better facilitate options for staff to stop driving to work;
- Carry out research into how students studying in the X building feel about its use;
- Consider SD implications of each Board decision ('SD' impact as a standing item on cover sheet of Board papers).

Articulating personal commitments was seen as a useful part of the process. One participant commented: '*Now I have written it, I'll have to stick to doing it*'. Some commitments (the last, for example) have implications which impact beyond the personal; all are important as a precursor to role-modelling behaviour.

Workshop Two

The second workshop was scheduled for three months later. Again, a detailed facilitation plan was approved; a further warm-up process began with reminders and information distributed. The content included a recap of Workshop One, further slides to enable members to visualise carbon and then group work to review progress, consider learning from actions, and to identify further actions and leadership behaviours. The recap was important, as there were some variations in attendance between the two workshops. Participants needed quite a bit of prompting and were rather slow to engage with reviewing actions. It was evident that some actions had slipped off the radar; some had made little progress. A brief presentation of leadership theory was then included followed by an activity which required participants to focus on those leadership behaviours which maintain progress, those which accelerate progress, and those which block. Finally, they were asked to think about 'infrastructure' needed to move forward; a plenary session captured overall comments on direction and issues.

Evaluation and learning so far

This has been a challenging project. The project's original aim of delivering a workshop for the Board and then a second workshop for the senior management team (which seemed quite easy to complete), was never realised. This was in the end a positive, resulting in increased interaction and more action plans. It also gave the Board a chance to collaborate with a wider group of staff and served to reinforce that this is a collective responsibility. The down-side was that a greater number of participants meant that it was easier for some not to participate, or leave early (absence being less noticeable).

Post workshop reflection concluded that the format of Workshop One worked well. Engagement was enthusiastic but careful facilitation was needed to retain focus. The greatest challenge was maintaining energy levels, given that the workshop took place after a lengthy Board meeting. The project team also noted that facilitating an event which involves participants who are at such a high level (in career terms) is highly stressful, particularly when delivery is within your employing organisation. Using an external facilitator (as part of the team) who had experience of working with government was important in terms of external credibility, but also in providing anecdotes and stories which engaged participants.

All facilitators found Workshop Two, more difficult. It was evident that although some work had been done in between workshops, some participants had done far less, which impacted on engagement. At times, the silences were uncomfortable and it was difficult to avoid stepping in to fill the gaps, however the decision to let participants experience some discomfort, was an important one. As one participant commented: 'a*ctually the leadership behaviour was a negative – 'forgot to take action' – hardly inspiring, will need to do better.*"

An assumption had also been made that participants would have a clear understanding of effective leadership and some understanding of leadership theory. This was not the case. As one participant suggested: '*we make leadership decisions all the time but that does not mean we have ever read anything about leadership. I've learned by doing, as have many others*'. This was countered by another participant, who responded: '*But you haven't been **doing** sustainable development – so you need to do more learning*'. Participants did not find it easy to identify strategic leadership behaviours; frequently the discussion was pitched at the level of middle management behaviour.

The exercise which helped them to think more critically was the one which considered behaviours which maintain, accelerate or block momentum. It generated discussion and yielded positive outcomes (see example, Table 3) which fed into the plenary.

Table 3: Identifying behaviours exercise from Workshop Two

Behaviours which sustain momentum	**Behaviours which accelerate momentum**
• Need a leader! • Passion (sharing), believing in it • Celebrating achievement • Clear sense of mission & well planned • Patience & accepting other people's work • Treat as a 'common sense' action • Rewarding & recognising • Empowering individuals • 'Highly visible' champions (including at Board level) • Being carbon conscious • Case studies/good examples to buy in to	• Positive statement of intent • Impact statement • More frequent/ regular updates/flow of information • Recognition/Incentive schemes/Prizes • Understand why you have won eco-campus award – what next? • Personalised message for individual or team e.g. carbon resp./allowance • Enthusiasm of everyone • Lead by example • Mobilise the almost 20,000 people that make up the institution
Behaviours which block momentum	**What institutional structures need to be put in place to progress activity**
• Making a big issue of it • Negativity • Lethargy/Apathy • Complacency • Cynicism • Someone else's job • Perception there is no show stopper we have to fix/absence of a burning platform • Too many messages • Inconsistent messages from Government & legislation	• No more committees but passionate, visible Board/UET leader and ULT champions – cascading • Three new Board members to be recruited; recruit minimum of one with an interest in sustainability • Meaningful KPI's with investment to back up (measure + do something about it) • 'Stickability' – people, resource, have mandate and skill • Put into a formal project management framework – becoming more disciplined
Behaviours which block momentum	**What institutional structures need to be put in place to progress activity**
• Non-sticking initiative before next • People too busy/ no time • "Individuals will not make a difference" • Unclear communication	• High level performance objectives at board level • Champions at all levels and recruit to that • Sell space for non-used time • It is a good sound business case to include environment as well as VFM in all decisions • Structures are OK • Build into appraisals/part of objectives

The support of the chair and several senior staff (which is in itself a positive behaviour) was undoubted and inspiring, with clear commitment demonstrated to leading change at both workshops. Some Board Members also expressed an interest in further engagement.

A project which involves working with Boards and senior teams is not easy, particularly where the facilitators are internal to the organisation and the underlying message is that there is something they need to learn and do better. As Argyris (1991) suggests, the 'smartest people' are not always pre-disposed to acknowledging that they have anything to learn. Managing the process agenda was therefore a challenge; the skills and emotional resilience needed for this were, as Clarke (1999) suggests, considerable. Working alongside an external 'expert' removed some of the pressure, but also increased work in terms of communication and coordination.

Workshop One felt more successful than Workshop Two. In part this was because participants were confronted with their own inaction in the second workshop, but may also have been because they did not like to reveal their lack of knowledge about leadership behaviour. So far only one institution has completed both workshops; until more have engaged it is difficult to draw conclusions. It did however give rise to a concern that perhaps the 'action' part of action learning had not been fully appreciated. In subsequent contracting meetings and workshops more attention was paid to ensuring that participants understood the approach and to reinforcing ownership of the actions.

The process and materials developed so far (although UK-specific) are transferrable to other institutions, with the proviso that the content is adapted to address institutional context and culture. In relation to the latter, cultural difference (even within UK institutions) can sometimes be huge and has to be considered in terms of preparation (warm up) and during facilitation. Further political processes are different in each institution. Encountering the political has sometimes been bizarre; responding as a *'positive deviant'* (Parkin 2010) has been necessary, as has substantial 'backstage activity' (Buchanan and Boddy 1992).

Other points of note when considering transferability include:

- The immense difficulty of coordinating leadership development with this target group. Board Members find time in full schedules for governance; time beyond that is precious so has to be well spent.
- Access to the Board has to be negotiated through gatekeepers, who play an important role in agreeing dates and communicating information in a timely way. They can slow down or block a project, or contribute wholeheartedly to success. Communication needs to ensure that they are briefed, on side and have the right information to enable them to be seen as knowledgeable and professional.
- The contracting meeting is vital to ensure that those at the top (particularly the Chair) meet the team, own the approach, and can role-model leadership at the start of the process. The contracting meeting has to be followed up with full briefing notes.

The project has already achieved the goal of increasing awareness of sustainable development (over 100 participants have been involved to date), but has it identified the required leadership behaviours? During the workshops the identification of leadership behaviours for sustainable development was sometimes disappointing (and at times like pulling teeth) but behaviours have started to emerge. Thus participants suggest that a leader of sustainable development:

- Demonstrates responsibility for the environment in the personal and professional sphere (exemplifies passion; acts as champion);
- Exemplifies creativity in planning for the future (visioning);
- Develops an SD strategy which embraces all aspects of the business (with key performance indicators);
- Assesses all actions and decisions in relation to SD; facilitates evaluation of the consequences of different actions (better decision-making/more futures-oriented);
- Encourages multiple perspectives, seeking consensus, but learning from diverse views;
- Identifies new ways of working and opportunities to bring in different perspectives;
- Inspires hope, proactively seeks positive SD solutions;
- Ensures SD is addressed on the cover of all papers which come to the Board;
- Ensures Board Membership includes participants who understand and are passionate about SD.

The list will continue to be shaped and refined as the process rolls out and evaluation takes place.

To date, the project has already raised the profile of SD, and resulted in affirmations of commitment by Chairs. A Deputy Vice-Chancellor recently confirmed: '*sustainability is now at the top-table*.' However, as a Chair commented, there is still some way to go to overcome the '*cynicism and scepticism prevalent in some members*.' As the project cascades, the number of governors gaining an understanding of SD is increasing; the number engaged in action will also rise. The Chair of a participating institution commented: '*Yesterday's session was worthwhile – in fact it was much better than I thought it would be and I can see some real upsides for University X. All the governors seem to be very positive about the result so if the objective was to engage the leadership of University X with these issues I think it can be counted as a huge success*.'

Another commented: '*The workshops worked to put us all 'on the same page' in terms of understanding. We have made a start. We are committed to SD and understand what is required of us following the workshops. We need to do more to ensure that SD is a growing and consistent priority and that it is fully embedded in the organisation.*'

Such comments evidence capacity building, but how do you evaluate the success of a project such as this and assess whether leadership behaviour is actually changing? Evaluation of the project will be the subject of a much later paper. An evaluation questionnaire and interviews will explore whether the project has:

- Achieved the goal of broadening leadership support for sustainable development;
- Enabled a shared understanding of sustainable development to be developed within the institution and a greater understanding of leadership behaviours;
- Helped participants to identify behaviours that accelerate the embedding of sustainable development across their institution.

Finally, participants will be asked to reflect on actions and approaches which they identified during the workshops and to identify where they are on the journey, from 'not yet started' to 'fully developed.' They will be asked to identify behaviours which have been particularly successful.

Interviews with Chairs and Vice-Chancellors will also explore their experience of the project and the extent to which they have noted any behaviour change, at the level of governance and executive management. The interviews will not only elicit information but will be used as an opportunity to explore whether further support and changes are necessary.

In the medium term, it should be possible to note an increased reference to sustainable development across all institutional processes and communications; a staff and student survey will test whether the concept is more broadly understood and whether leaders are truly exemplifying the agenda. In the longer term, environmental audits will capture improvements in the very tangible areas of carbon reduction, utilities, waste, etc.

At this point it is evident that the project has broadened participants' understanding of the issues, their role in leading the agenda and the actions required to lead change. It could however be some time before leaders role-model leadership behaviour for sustainable development in their day-to-day activities.

Conclusion

This paper has provided a summary of a project 'in progress' which aims to broaden leadership support for sustainable development. The project has engaged Board Members and senior teams within UK universities in action learning with a view to developing action plans, and considering the behaviours required for SD in taking those actions forward. The action learning approach and workshops have provided participants with the space to reflect on sustainable development and engage with the topic in a way that has not been achieved previously. The approach is transferable to other institutions within the UK and beyond, providing those who lead and govern higher education institutions, are prepared to set aside the time and space to begin engagement. This project has served as a useful vehicle to raise the profile of SD within institutions, broaden understanding of the challenges, and enable some coherent action plans to be developed. More work will be required to follow those actions plans through; substantial work will be required if higher education leaders are to fully incorporate sustainable development into their day-to-day actions.

The project has particularly focused on those responsible for governance and top-level leadership within universities. It is worth remembering however that:

> 'A sustainability leader is anyone who chooses to engage in the process of creating transformative change with others aimed towards a more sustainable future: economically, environmentally and socially.' (Ferdig 2009).

Until such time as those at the top are fully 'walking the talk', others will need to continue championing the agenda.

Acknowledgements

The project described in this paper would not have been possible without the contribution of the wider team. Thanks are given to Ann Finlayson (SEED), Patrick Pica (University of Sussex) and Amanda Williams (Bournemouth University) for their contribution to project design and delivery.

References

Argyris, C. (1991), Teaching smart People how to learn. *Harvard Business Review*, May/Jun 69 Issue 3, p. 99, 11p.

Bandura, A. (1977), *Social Learning Theory*. General Learning Press, New York.

Brighter Planet (2010), *Employee Engagement Survey Report.* http://www.brighter planet.com/research (accessed October 2011).

Buchanan, D. and Boddy, D. (1992), *The Expertise of The Change Agent: Public Performance and Backstage Activity*. Prentice Hall.

Clarke, M. (1999), Management Development as a game of meaningless outcomes. *Human Resource Management Journal* Vol. 9 (2.

Copeland G. (2008), Sustainable development: strategic considerations for senior managers. In Shiel, C. and McKenzie, A. (Ed.), *The Global University: the role of senior managers.* BU/DEA, DEA, London, UK.

Dawe, G., Jucker, R. and Martin, S. (2005), *Sustainable Development in Higher Education: Current Practice and Future Developments*. HEA, York, UK.

Ferdig, M.A. (2009), *Sustainability Leadership.* Sustainability Leadership Institute, http://www.sustain abilityleadershipinstitute.org/atomic.php (accessed September 2011).

Festinger, L., Rieken, H.W. and Scachter, S. (1956), *When Prophecy Fails.* University of Minnesota Press.

HEFCE (2005), *Sustainable Development in Higher Education: Strategic Statement and Action Plan.* HEFCE, Bristol, UK.

– (2008), *Sustainable Development in Higher Education*: update to strategic statement and action plan http://www.hefce.ac.uk/pubs/hefce/2009/09_03.

– (2010), *2010/01 Carbon reduction target and strategy for the sector.* HEFCE, Bristol, UK.

Jones, P., Selby, D. and Sterling, S. (Eds.) (2010), *Sustainability Education: Perspectives and Practice across Higher Education.* Earthscan, London, UK.

Marshall, J., Coleman, G., Reason, P. (Eds.) (2011), *Leadership for Sustainability: An Action Research Approach.* Greenleaf Publishing, Sheffield, UK.

Martin, S. and Jucker, R. (2005), Educating earth-literate leaders. *Journal of Geography in Higher Education* Vol. 29 (1), pp. 19-29.

Parkin, S. (2010), *The Positive Deviant: Sustainability Leadership in a Perverse World.* Earthscan, London, UK.

Revans, R.W. (1982), What is Action Learning. *Journal of Management Development* Vol. 1, Issue 3, pp. 64-75.

Sterling, S. and Scott, W. (2008), Pushing the boundaries: The work of the Higher Education Academy's ESD Project. *Environmental Education Research* 14 (4), pp. 399-412.

UNCED (1992), *Agenda 21: Programme of Action for Sustainable Development.* UN Department of Public Information, New York, NY.

Universities that Count (2009), *Annual Report.* Accessed at http://www.beta.yudu.com/item/details/107038/Universities-that-Count-Annual-Report-2009 in September 2011.

Wals, A.E.J. and Blewitt, J. (2010), Third-wave Sustainability in Higher Education: Some (Inter)-national Trends and Developments. In Jones, P., Selby, D. and Sterling, S. (Eds.) (2010), *Sustainability: Perspectives and Practice Across Higher Education,* pp. 55-74. Earthscan, London.

Yukl, G. (2010), *Leadership in Organisations*, Sixth Edition. Pearson, Prentice Hall, NJ.

Yukl, G. and Falbe, C.M. (1990), Influence tactics and objectives in upward, downward, and lateral influence attempts. *Journal of Applied Psychology* 75, pp. 132-140.

Chapter 8

Making Education for Sustainability Work on Campus: The Proposals of the Roundtable on Environmental Systems and Sustainability

Michael A. Reiter, Will J. Focht, Paul A. Barresi, Susan Gill, Richard C. Smardon, Susan L. Baker, Kimberly D. Reiter, Eric Fitch, Terry Rolfe and Sue Bumpous

Abstract

Sustainability education is increasingly popular, but the content and pedagogical approaches of sustainability-focused programs vary so widely as to make them hard to compare. The Roundtable on Environmental Systems and Sustainability is a collaboration of university faculty and administrators, independent scholars, and environmental practitioners who seek to address this weakness with consensus-based proposals for use in program development and as a potential template for program recognition. The Roundtable proposes "Environmental Systems and Sustainability" (ESS) as the academic domain of relevance, a supradisciplinary field focused on the holistic, adaptive management of the health and resilience of the systems at the human-nature interface necessary to support both stewardship of the natural environment and long-term improvement in the human condition. The Roundtable reached consensus on 12 skills, 15 knowledge areas, and one experience as essential elements of ESS programs that can be embraced in any institutional context, and on complexity as the most important pedagogical touchstone. The Roundtable also proposes that departmental or higher-level placement in the institutional structure offers the best chance for program stability. Roundtable proposals, which represent a continuing discussion open to all interested individuals, are being used or are being considered for use in sustainability-focused programs at several educational institutions.

Introduction

The term "sustainability" has been defined in so many ways and co-opted in so many contexts that it is practically useless. However, the concept that humans must live within the constraints imposed by natural systems and that, therefore, they must understand their relationship to those systems cannot be overstated. Attempts to teach the principles necessary to understand such relationships have largely been undefined and lacking unity, as can be seen in past discussions of the core concepts (or lack thereof) of environmental programs (Caldwell 1983; Orr 1990; Wilke 1995; Soule and Press 1998; Maniates and Whissel 2000). This uncertainty in an otherwise important field of study has led to efforts by both faculty and administrators to understand the nature of interdisciplinary and higher order environmental education, such as the academic discussion groups on interdisciplinary environmental education organized by the Interdisciplinary Environmental Association (IEA) and a comprehensive survey of environmental programs in US colleges and universities sponsored by the Council of Environmental Deans and Directors (CEDD) of the National Council for Science and the Environment (Vincent 2010a, b). While some progress has been made (such as a recent proposed framework for reviewing environmental programs; Smardon 2011), the issues of core competencies, curriculum, administrative structure, and similar topics for interdisciplinary and higher order environmental programs have yet to be addressed as a functional unit.

The Roundtable on Environmental Systems and Sustainability is an ongoing, inclusive collaboration of university faculty and administrators, independent scholars, and environmental practitioners committed to developing and implementing a comprehensive approach to producing graduates of environmental programs with the knowledge, skills, and experiences necessary to respond effectively to the urgent, pervasive sustainability challenges that exist at the human-nature interface. Roundtable participants are committed to seeking a set of consensus-based core principles and recommendations for interdisciplinary and higher-order environmental education that unifies the thus-far piecemeal attempts at disciplinary organization within a framework flexible enough to be applicable to the large number of useful contexts necessary to address today's environmental issues. Roundtable proposals are being used or considered in sustainability-focused programs at several educational institutions (e.g. Reiter et al. 2009), and the Roundtable hopes that its proposals ultimately will serve as a template for recognition of sustainability-focused environmental programs throughout the academy.[1]

The Relevant Academic Domain

The Roundtable proposes "Environmental Systems and Sustainability" (ESS) as the academic domain of relevance to sustainability-focused environmental programs. The goal of this supradisciplinary field is the holistic, adaptive management of the health and resilience of the systems at the human-nature interface necessary to support both stewardship of the natural environment and long-term improvement in the human condition. This goal recognizes the importance and interdependence of human and ecological systems as factors that shape the fate of both. It also implies that neither stewardship of the natural environment nor long-term improvement in the human condition can be achieved without a holistic, integrated synthesis of disciplines aimed at managing the interaction of both human and ecological systems at their interface (Fig. 1).

1 At the time of writing, the Roundtable has convened seven times at locations throughout North America. Reiter et al. (2011) offers a comprehensive account of the Roundtable's work through the fifth meeting in June 2010. This paper summarizes and updates the results of these earlier meetings, and offers a more detailed account of the Roundtable's subsequent work. Parts of this paper have been adapted and summarized from the earlier publication.

Fig. 1: The Roundtable's view of "The Ecosphere" showing the interface of human and ecological systems on which ESS focuses

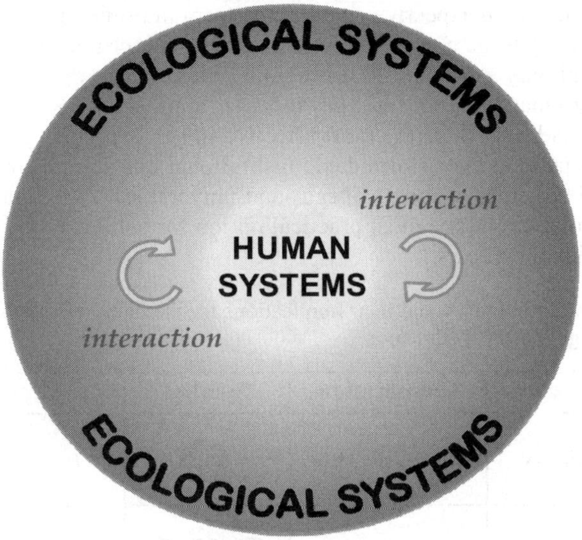

A CEDD-sponsored survey of more than 250 environmental programs in US colleges and universities (Vincent 2010a) revealed three general approaches to environmental education. The *systems science* approach emphasizes the natural sciences as a means of producing scientists capable of diagnosing environmental problems. The *policy and governance* approach emphasizes the social sciences and humanities as tools for cultivating environmental citizens who can engage in collective action. The *adaptive management* approach embraces elements of all three academic domains as vehicles for developing professionals capable of solving practical environmental problems in management contexts. Each of these approaches is consistent with ESS education because each can produce graduates capable of engaging in holistic approaches to understanding and addressing environmental issues. Which approach would be the most appropriate for a given environmental program at a given institution is a function of the mission, traditions, and capacities of that institution and its constituent parts. The effectiveness of any environmental program in achieving the goal of ESS education would depend on the extent to which it were designed and delivered to achieve the disciplinary synthesis that ESS education requires.

Achieving Disciplinary Synthesis

The Roundtable considered the implications of various approaches to disciplinarity for the effectiveness of ESS programs and agreed that the more complete the disciplinary synthesis the more effective the program is likely to be (Fig. 2). Neither the *unidisciplinary* approach (which views sustainability from a single discipline's perspective), nor the *multidisciplinary* approach (which invokes many disciplinary perspectives applied separately), nor the *transdisciplinary* approach (which recognizes points of intersection between otherwise separate disciplines) is likely to be effective in achieving the goal of ESS education because none of them transcend disciplinary boundaries to any significant degree. The *pluridisciplinary* approach (which recognizes some overlap among disciplines but preserves the distinctions among them) is not likely to be much more effective for similar reasons, but could play a valuable transitional role in the evolution of a traditional, discipline-based environmental program to one capable of achieving the goal of ESS education. The *interdisciplinary* approach (which views sustainability from a perspective synthesized from multiple disciplines but applied as a coherent whole) is likely to be much more effective, whether or not some of the courses in an otherwise interdisciplinary environmental program embrace one or more of the lower order approaches to disciplinarity. The *supradisciplinary* approach (which ignores disciplinary boundaries entirely) is likely to be the most effective in achieving the goal of ESS education, but also is likely to face the most formidable institutional obstacles to implementation given the discipline-based structure of most colleges and universities. Accordingly, the interdisciplinary approach is likely to be the most practical option for achieving the goal of ESS education in most institutional contexts.

Fig. 2: Approaches to disciplinarity and their implications for the likely effectiveness of programs in achieving the goal of ESS education

Approach	Connotation	Symbol	Examples
Unidisciplinary	One, Single, Alone		History, Geology
Multidisciplinary	Many, Several		Chemistry + Biology
Transdisciplinary	Across		Historical Archaeology, Geophysics, etc.
Pluridisciplinary	*Belonging to Many*		*Cultural Psychology*
Interdisciplinary	Among		Natural Ecology, American Studies
Supradisciplinary	Above, Over, Beyond		Sustainability

In order to achieve the goal of ESS education, any environmental program would need to achieve learning outcomes that spanned a broad range of substantive domains. The Roundtable reached consensus on 12 skills, 15 knowledge areas, and one integrative experience that would be essential elements of any ESS program (Table 1), although different programs could emphasize these elements to different degrees while still achieving the goal of ESS education.

Table 1: Skills, knowledge areas, and experiences identified by the Roundtable as essential elements of any ESS program

Skills (12)	Knowledge Areas (15)	Experience (1)
Analysis	Administration (Public and Private)	Application of Skills and Knowledge to Environmental Problem-Solving
Critical Thinking	Options Analysis and Selection	
Collaboration	Structure and Function of Geospheric, Hydrospheric, and Atmospheric Systems	
Communication (Written, Oral, Visual)	Environmental and Ecological Economics	
Empirical Reasoning	Environmental Citizenship	
Impact Assessment & Forecasting	Environmental Ethics	
Logical Inference	Environmental Governance	
Problem-Solving	Environmental History	
Quantitative and Qualitative Methods	Health Science	
Research	Human Ecology	
Synthesis	Natural World, Culture, and Belief	
Systems Thinking	Natural Ecology	
	Probability and Statistics	
	Psychology and Behavior	
	Sustainable Design and Technology	

The learning outcomes associated with these 12 skills, 15 knowledge areas, and one integrative experience, the elaboration of which remains on the Roundtable's agenda, could be achieved in any institutional context using any of the three general approaches to environmental education currently in vogue in US colleges and universities, as long as the program as a whole were to adopt the interdisciplinary or supradisciplinary pedagogical approach that ESS education requires.

Program Structure and the Role of Complexity

The approaches to disciplinarity that are most likely to achieve the goal of ESS education could be implemented in the context of any program structure, although some structures are likely to be more conducive to interdisciplinarity or supradisciplinarity than others (Fig. 3). Most existing environmental programs have a "Triangle" or "Inverted Triangle" structure, although the "Diamond", "Hourglass", and "Extensional" structures would be more conducive to the implementation of interdisciplinary or supradisciplinary pedagogical approaches.

As the connotation of the "Extensional" structure suggests, building holistic complexity should be the most important pedagogical touchstone of any ESS program, no matter what its structure. It is within this "Extensional" structure that true holistic, supradisciplinary courses could be established, though this is not an easy task; nor is it common to find them in university curricula. One approach for such a course would focus on an issue or topic starting simply and becoming more complex over time, possibly in a design studio approach allowing for participatory action research. In a supradisciplinary approach, the course would view the whole of the issue at all times but increase the complexity of the view as the course proceeded. In this approach, students would ask and address questions from a holistic perspective, with experts and/or information available to encourage them to move forward. Such a course would emphasize systems thinking, disturbance and resilience theory, holistic thinking, and collaborative skills. For example, one could give the students a series of nested tasks that build complexity within a particular context, such as 1) cooking a sustainable meal, 2) building a sustainable table on which to serve the meal, 3) designing and/or building a sustainable house in which to place the table, and 4) designing a sustainable community in which to locate the house. The benefits of organizing the course in an integrated service-based format would be the increase in student learning within an applied context as well as the benefit to the community from addressing a local issue.

Fig. 3: Structural options for ESS programs and their implications for the implementation of interdisciplinary (ID) or supradisciplinary (SD) pedagogical approaches. Clear: common existing designs. Italics: transitional designs toward a more effective supradisciplinary structure (dark shading)

Structure	Connotation	Symbol	Implications for ESS Programs
"Triangle"	**Broad base, then specialize**	▲	**Least conducive to ID or SD approach**
"Inverted Triangle"	**Disciplinary base, then broaden**	▼	**Least conducive to ID or SD approach**
"Diamond"	*Disciplinary base, broaden, then specialize*	◆	*More conducive to ID or SD approach*
"Hourglass"	*Broad base, narrow, then synthesize*	⧗	*More conducive to ID or SD approach*
"Extensional"	**Simple to complex, but always broad**		**Most conducive to ID or SD**

Administration

The effectiveness of a strong ESS program design can be reduced if placed within a suboptimal administrative structure. Roundtable participants agreed that the optimal administrative structure for an ESS program as described here is an interdisciplinary department (or higher) with its own promotion standards, tenure requirements, and learning outcomes. This structure would provide a budgetary and administrative home for the department, help protect and reward interdisciplinary or higher order faculties needed to make the program function, and help give the program standing equal to more traditional programs. However, other types of structures could also support effective ESS programs if adequate administrative and budgetary support were present. Among those could be designs that included distributed program faculties, split faculty loads, or departmental subunits. Roundtable participants found administrative structures with no discreet university funding, diffuse leadership, and/or no associated degree program likely to be unworkable for ESS programs as defined by the Roundtable.

Of particular importance for creating a successful ESS program is the recognition of the value of supradisciplinary ("horizontal") education by university administrations. Administrators have the ability to insist that ESS faculties should themselves be at least disciplinary integrators if not supradisciplinary holists. Furthermore, administrators can create university units that allow for the hiring and tenure of such faculties; teams of unidisciplinarians are clearly not sufficient because the faculties and concepts placed under multidisciplinary approaches have not been observed to synthesize well.

The Roundtable proposes that appropriate evaluation of ESS programs are more likely to be accomplished using a certification ("truth in advertising") approach as opposed to an accreditation ("value judgment") approach, if for no other reason than there is a wide range of contexts within which an ESS program as described here could be carried out, all of which could be useful in the right place and time. As a result, graduating students with different strengths and weaknesses from different ESS programs (or even the same ESS program) is not a failure of any program but rather a reflection of the scope of ESS and the variety of useful contexts in the field. While any one program cannot do everything, it may do several shades of a few things quite well. Accordingly, evaluating such a program requires a more functional view of the ultimate goal compared to many unidisciplinary programs – in short, producing graduates *functional* in the workforce after their program of study is completed. If desired, it would not be too difficult to move toward true accreditation later with suitable upper-level need and backing.

A framework for certification could be based on a ranking scale with a goal and levels toward that goal. The functional certification minimum for an ESS program would involve all core competencies delivered within some lower-order mode of approach, with higher certification levels based on the "quality" of the delivery (more integrated program designs, higher-order modes of content delivery, integrated training in the faculty, etc.). Since there are several possible structural models being used in existing programs, the framework may require several categories, though the minimum level for certification would remain the delivery of all core competencies within that structural model. Organizations such as the IEA could carry out evaluations on request with a renewal period of five years (to match the accreditation cycle of US colleges and universities), as well as a public registry.

Evaluation criteria for ESS programs would be tied to learning outcomes, but learning outcomes alone are not sufficient because they would be very difficult to measure. Thus additional criteria involving interdisciplinary or higher order curriculum design and faculty qualifications would be useful, as well as the aforementioned administrative structure issues. Program assessments can be carried out by peers (both from within and outside of the institution), employers, and alumni according to ESS criteria. The Roundtable is continuing its development of these recommendations to try to address issues pertaining to poorer versus richer institutions, focusing on the institution versus focusing on the program curriculum and design, and recommending approaches that favor recognition rather than certification.

Conclusions

The work of the Roundtable brings into question the past discussions concerning the inability to determine a common focus or set of core concepts for interdisciplinary environmental programs. Despite the difficulties implied in past studies of environmental pedagogy, the range of types of programs in existence in the US, and the diverse perspectives that have been represented in the Roundtable effort (academics and administrators from both large and small institutions offering liberal arts and applied programs, government agency employees, and practitioners from around the country), Roundtable participants have nonetheless been able to develop a consensus-based vision for a holistic, integrated academic domain called Environmental Systems and Sustainability that participants believe is an effective path toward achieving the academic background needed for creating a sustainable future. This vision includes curriculum guidelines, program designs, core competencies, and administrative structures necessary to formulate a flexible implementation plan, providing a unifying set of recommendations to enhance the effectiveness of sustainability education while still allowing for the flexibility to address the wide range of environmental issues and perspectives that exist today. By implementing these proposals, institutions can more effectively turn out graduates with the common foundation of interdisciplinary or higher order knowledge and skills needed to address complex environmental issues, even as they possess very different contextual knowledge based on the institution's focus and the types of issues for which the students have been trained. The Roundtable is now working toward expanding and defining the curricular aspects of the proposed core competencies with the goal of moving toward a fully implementable modular framework that can be adapted to any environmental program context.

The Roundtable is an open, inclusive, ongoing conversation among academics and practitioners about how best to educate faculty and students for successful sustainability education, scholarship, practice, and informed citizenship. The authors and the Roundtable invite others who share this vision of sustainability to join the conversation and help move the effort forward.

References

Caldwell, L.K. (1983), Environmental studies: Discipline or metadiscipline? *The Environmental Professional* Vol. 5 (3/4), pp. 247-259.

Maniates, M.F. and Whissel, J.C. (2000), Environmental studies: The sky is not falling. *BioScience* 50 (6), pp. 509-517.

Orr, D.W. (1990), Environmental education and ecology literacy. *The Education Digest* 55 (9), pp. 49-53.

Reiter, M.A., Coggins, P.C. and Howse, M.E. (2009), Designing an integrated interdisciplinary environmental science curriculum using an IRMA chart: An example from Bethune-Cookman University. In Leal Filho, W. (Ed.), *Environmental Education, Communication and Sustainability* Vol. 31: *Sustainability at Universities: Opportunities, Challenges and Trends*, pp. 153-162. Peter Lang Scientific, Frankfurt am Main, Germany.

Reiter, M.A., Focht, W.J., Barresi, P.A., Bumpous, S., Smardon, R.C. and Reiter, K.D. (2011), Making education for sustainability work on your campus: The Roundtables on Environmental Systems and Sustainability. In Leal Filho, W. (Ed.), *Umweltbildung, Umweltkommunikation und Nachhaltigkeit [Environmental Education, Communication and Sustainability]:* Vol. 32. *World Trends in Education for Sustainable Development*, pp. 61-76. Peter Lang Scientific, Frankfurt am Main, Germany.

Smardon, R.C. (2011), Developing a national framework for internal/external review of undergraduate environmental studies/environmental science programmes. *Interdisciplinary Environmental Review* 12 (3), pp. 244-257.

Soule, M.E. and Press, D.E. (1998), What is environmental studies? *BioScience* 48 (5), pp. 397-405.

Vincent, S.G. (2010a), *Interdisciplinary Environmental Education: An Exploration of Field Identity and Core Competencies,* National Council for Science and the Environment, Council of Environmental Deans and Directors, Washington, DC. 86 pp.

– (2010b), *A Search for Identity: Exploring Core Competencies for Interdisciplinary Environmental Programs.* Doctoral Dissertation, Oklahoma State University, Stillwater, OK, 187 pp.

Wilke, R. (1995), Environmental literacy and the college curriculum: Colleges and universities have a challenge to meet. *EPA Journal* 21 (2), pp. 28-30.

Chapter 9

Monitoring Changes in the Sustainability Attributes of Higher Education Students in a New Zealand University

Kerry Shephard, John Harraway, Brent Lovelock, Sheila Skeaff, Liz Slooten, Mick Strack and Tim Jowett

Abstract

Higher education institutions are encouraged to develop campus-sustainability initiatives and related curriculum-changes. An expectation is that these initiatives will impact on all of our students (not just those studying 'sustainability-related' programmes), increase the proportion of nations' citizens who embrace sustainability and confirm the status of higher education as a key agent of social change. The challenge for higher education is to better understand what and how students are learning and how this links to what and how universities teach. University teachers have varied views on teaching sustainability outcomes and some have limited engagement with sustainability at a personal level. Few institutions have systematic processes in place to monitor sustainability outcomes. Higher education will find it difficult to meet society's expectations unless it finds appropriate ways to explore and evaluate its impact. Researchers at the University of Otago have been exploring how to benchmark and monitor changes in the sustainability attributes of students and have established a study in several university departments, based on cohort means, multinomial regression modelling and longitudinal analysis, to monitor changes in students' environmental concern, as students experience higher education. The results so far suggest that the environmental concerns of students in this university are highly department-dependent throughout students' stay at university, and that changes, when they occur, are relatively small.

Introduction

Many universities internationally have agreed that it is their role to support the development of citizens who know about sustainability issues, who have relevant skills to put sustainable practices into operation and who, on graduation, may choose to work and live sustainably. Such expectations are explicit, for example, in the Talloires Declaration. Institutions that sign agree to 'Educate for Environmentally Responsible Citizenship' (Association of University Leaders for a Sustainable Future 1994). They are also strongly supported by a number of influential policy statements and agreements. The Brundtland report, for example, emphasises the extensive social changes to which education (including higher education) should contribute (World Commission on Environment and Development 1987). Equivalent values-based educational objectives are sought by some disciplines (see Shephard 2008 for an analysis of values education in medicine) and some institutions have high values-based expectations of their graduates. The University of Sydney, for example, hopes that its graduates will hold personal values and beliefs consistent with their role as responsible members of local, national, international and professional communities (University of Sydney 2004). Implicit within much of this documentation is that these attributes apply to all graduates and not just to those who volunteer to study overtly sustainability-related programmes.

Many higher education institutions have changed to achieve these objectives. Perhaps most obvious is the remarkable range of campus-sustainability initiatives designed to emphasise, particularly to students, that higher education practises sustainability (see for example the Sustainability Tracking, Assessment and Rating System adopted by higher education in the USA (https://stars.aashe.org/). Student involvement in these changes is also notable and exemplified by the student-led People and Planet ranking of UK universities by environmental and ethical performance (http://peopleandplanet.org/green league). It is to be hoped that universities would wish to document the degree to which their graduates were achieving these sustainability objectives. Universities might find room on students' transcripts to record their assessed knowledge about sustainability issues and the skills that they have demonstrated, as examples, in recycling and showing caution towards uncertain outcomes. But what might universities enter in the box marked "Does this graduate display personal values consistent with those expected of the sustainably-minded citizen?" or "Is this graduate likely to continue to act sustainably in the workplace?" These are more complex questions, but not insurmountable objectives. Researchers in higher education have researched values, attitudes, dispositions and behaviours extensively. A wide range of research instruments and tools have been developed to monitor, measure, assess, evaluate or research these attributes (Buissink-Smith et al. 2011). There exists a great deal of contestation about what exactly these instruments measure and how their results should be interpreted, but much is possible. It is possible to monitor students' self-reported behaviours (see, for example, Brody and Ryu 2006), but a long-standing focus has been to explore environmental concern. Of course, citizens may claim high environmental concern, but even so fail to behave sustainably (the attitude behaviour gap [Jay 1990]) but it remains less likely that unconcerned citizens will behave sustainably. The environmental concern of students is worthy of attention.

The research described in this article was conducted to explore what is possible, what is practicable and how it could work in higher education institutions that were minded to monitor their impact on the sustainability attributes of their students. The Revised New Ecological Paradigm Scale (NEP) was used to record levels of students' environmental concern (or ecological worldview) over several years and programmes. The NEP has been extensively used for measuring 'environmental concern', 'ecological worldview' or 'environmental attitude'; and more recently for monitoring how these change (Teisl et al. 2011). The NEP's 15 statements relate to limits to growth, the position of humans in the environment, the fragility of nature and the imminence of ecocrisis. A range of analyses were also conducted to make sense of this data using different statistical approaches to compare changes in how students responded to the NEP.

Methods

Environmental concern was measured using the 15-item NEP (Dunlap 2008) (Table 1), delivered to participants on paper (designed for subsequent optical scanning) during routine classes. Participants were all students enrolled in University of Otago departments (Mathematics and Statistics, Zoology, Surveying, Human Nutrition and Tourism) and they volunteered to be anonymously involved in this research (under University of Otago Category B research ethics approvals). Students were asked to record (by ticking a box) their level of agreement for each statement in the NEP on a five-point Likert-like scale. The participants were also asked to provide, on the research form, some additional socio-demographic information to aid the analysis (year of study, gender and self-reported program affiliations). To allow for follow-up of individuals and at the same time preserving anonymity, participants were asked to write a confidential code word on their survey, that they would remember or could re-calculate, in subsequent years (the code has varied from year to year but most recently has included elements of their birthdate and letters from their name). The coding methodology is described in more detail elsewhere (Harraway et al., in press).

Table 1: The 15-Item Revised NEP Scale

1	We are approaching the limit of the number of people the earth can support.
2	Humans have the right to modify the natural environment to suit their needs.
3	When humans interfere with nature it often produces disastrous consequences.
4	Human ingenuity will ensure that we do not make the earth un-liveable.
5	Humans are severely abusing the environment.
6	The earth has plenty of natural resources if we just learn how to develop them.
7	Plants and animals have as much right as humans to exist.
8	The balance of nature is strong enough to cope with the impacts of modern industrial nations.
9	Despite their special abilities humans are still subject to the laws of nature.
10	The so-called "ecological crisis" facing humankind has been greatly exaggerated.
11	The earth is like a spaceship with very limited room and resources.
12	Humans are meant to rule over the rest of nature.
13	The balance of nature is very delicate and easily upset.
14	Humans will eventually learn enough about how nature works to be able to control it.
15	If things continue on their present course we will soon experience a major ecological catastrophe.

The number of factors that arise from the use of the NEP has been reviewed by Dunlap et al. (2000). Shephard et al. (2009) identified four factors and used their contributory NEP items to describe four consequential implied 'tendencies' (to conserve, to recycle, to be cautious about the future, and to support animal and plant rights). Further research (Harraway et al., in press) has significantly endorsed the utility of the four tendencies model. The NEP, and the data presented here, is also interpretable as a single factor (Dunlap et al. 2000) and analyses of change may be more robust in this form. This approach is adopted in this article.For the research described here, the NEP was initially used at the start and end of a first-year statistics course in 2009 attended by students who self-identified as intending to major in a particular subject area. This data was reported by Harraway et al. (in press). The NEP was used again with many of the same students during their second year (2010) and third year (2011) in subsequent second-and third-year courses in departmental programmes and this data is reported here for the first time. A fourth year is envisaged (2012). The NEP was also re-administered during the first-year statistics paper again in 2010 and 2011, and in subsequent second-year departmental courses and, it is anticipated, in subsequent third-year papers in 2012. Individuals' responses were combined into a summated overall NEP score where higher overall scores indicate stronger pro-environmental concerns (on a scale of 1-5). The data provided in this article is for students majoring in Surveying, Human Nutrition and Zoology, as these groups provided the most complete data sets. Very high percentages (typically >90%) of students who attended each class chose to participate in this research.

Means

At its simplest level, the raw NEP data for individuals is simply combined into a group or cohort mean, with a tacit assumption that the students surveyed within a group are a representative sample of that group.

Longitudinal Analysis using Linear Mixed-Effects Modelling

Longitudinal analysis was applied to cohorts of students responding on multiple occasions over a period of time. A key feature of longitudinal data is that there exists positive correlation between the repeated measurements on individual students at different times (Fitzmaurice et al. 2004). Linear mixed-effects models can measure and provide valid tests on the statistical significance of change over time while taking into account the correlation between observations. This is achieved by the key assumption that there exist two sets of regression parameters described as 'fixed effects' and 'random effects'. The fixed effects parameters incorporate the marginal or population-averaged effects of the different factors in the model which in this case consist of 'Programme of Study' and 'Time' (in weeks from the start of the survey period) and interactions between these. The second set of parameters, which are referred to as 'Random effects', are specific to each individual and are used to allow for the positive correlation between observations.

A key advantage of the linear mixed-effects approach is that it is flexible enough to accommodate issues relating to missing data and differences in the timing of surveys. These are useful characteristics because the attendance at lectures was variable and the differences in the schedule of each course means that it is not practical to administer the NEP surveys at the same time in each department.

The linear mixed-effects model used all of the student data available for the included programmes of study where students had remembered their code from one sample period to another. The model included data from 153 students.

Multinomial Logistic Regression Modelling

An alternative approach to modelling change is to analyse data in a cross-sectional manner, rather than sequentially in relation to time. At any one time, students exist in several cohorts and these can be compared. Teisl et al. (2011) used three logistic regression models to investigate changes in environmental attitudes in response to different courses and teachers. An alternative multinomial logistic regression (MLR) model was developed because of its improved efficiency (only one model is fitted) and because it is easier to interpret the results. The focus of this analysis was to determine whether or not sustainability attitudes change in response to the number of years of study at university and whether this change is different for the different programmes of study. The NEP response of all 521 students sampled in 2010 was analysed. The students in the sample came from first and second-year courses and there were no repeat observations on individual students. The model involves dividing the student group into three equally-sized groups (tertiles). Those with NEP scores in the lowest tertile were categorised as 'brown'; those with NEP scores in the highest tertile were categorised as 'green'; those between were categorised as 'unsure'. The model calculates predicted probabilities of being in each tertile for the different programmes and years of study.

Results

Analysis of Means

Fig. 1 shows the NEP means of the students in the cohort starting in 2009 across the three programmes of study over the years 2009 to 2011. It can be seen that Zoology students have by far the highest mean followed by Human Nutrition and Surveying. Also, the similarity of the means, across the years suggests that the environmental concern of these particular groups of students did not change greatly over time. It is important to note that a comparison of means, to determine the statistical significance of change over time, would not be valid due to the lack of independence between observations made over time on individual students. Independence of data is a strict statistical limitation in simple significance tests.

Fig. 1: NEP means and 95% confidence intervals for the cohort of students starting university in 2009

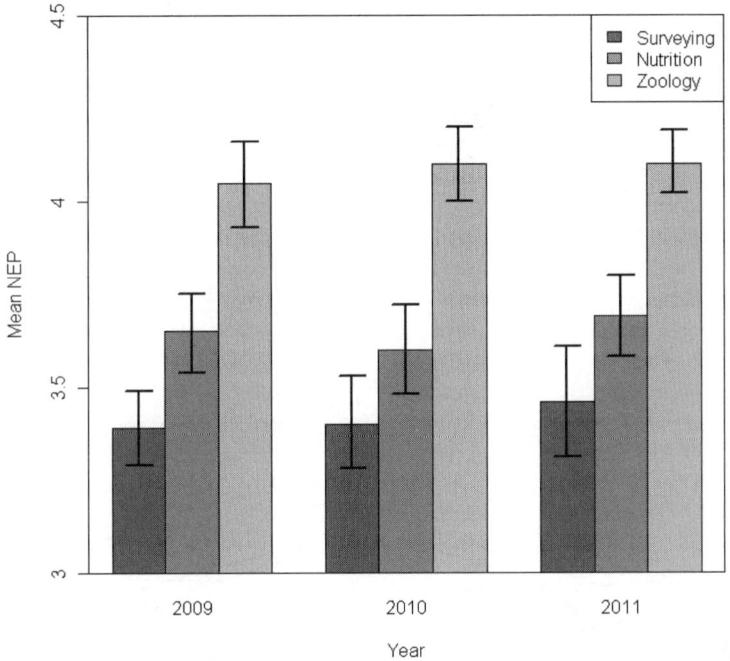

The sample sizes were, in the order they appear in the figure from left to right, 70, 72, 51, 80, 85, 82, 39, 75, and 78.

Longitudinal Analysis Using Linear Mixed-Effects Modelling

A linear mixed-effects model was implemented in the computer programmes SAS and STATA using repeated responses from 153 individual students. Ideally the sample size would have been larger but there were issues with students forgetting their code. The first phase of the analysis was to inspect the data in the form of the response profiles which show how the mean NEP scores of the 153 students change over time (see Fig. 2).

From Fig. 2 it appears that while some students exhibit major change over time, the overall trend is for minimal change, as reflected by commonly observed horizontal to sub-horizontal profiles.

Several candidate linear mixed-effects models were compared including several allowing for interaction between 'Programme of Study' and 'Time'. The best-fitting model, selected using Akaike model-selection criteria, was one that included only 'programme of study' and 'time' but without interaction effects.

The key results from this longitudinal analysis are: Zoology students have the highest mean NEP scores and the mean is significantly higher than both Surveying and Human Nutrition students; the mean NEP score of Human Nutrition students was significantly higher than that of Surveying students; and, for all three programmes of study, there was no evidence to suggest that the mean NEP scores changed over time. This suggests that on average, the environmental concerns of the students remained the same over the three-year period incorporated in this model.

Fig. 2: Response profiles of the 153 students in the Cohort starting in 2009

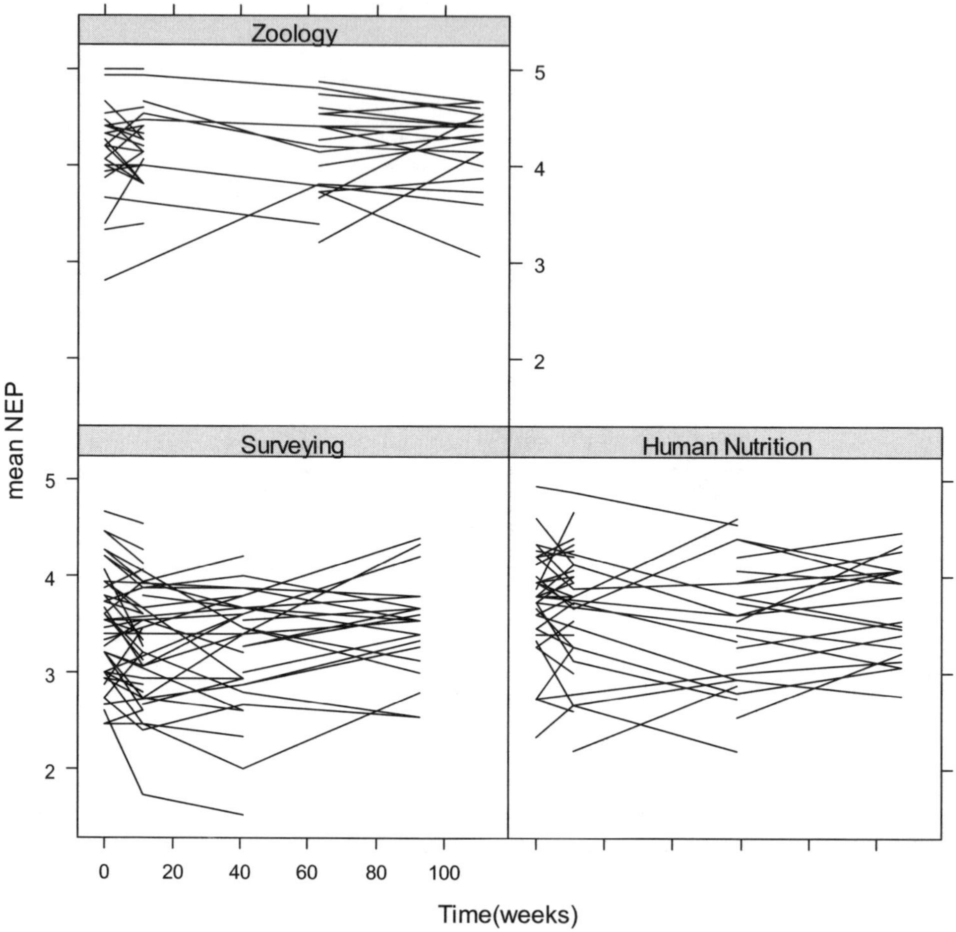

Time is recorded in weeks from the date of the first survey in July 2009.

Multinomial Logistic Regression Modelling

The key results for these programmes from the MLR model are shown in Fig. 3 as changes in probabilities (of being in one of each of the tertiles, Brown, Unsure and Green) between students in their first year and those in subsequent years (Year 2+). The Green probability-difference is positive for all programmes, indicating that the Year 2+ students have a higher probability of being Green compared with Year 1 students. The Brown confidence intervals all include zero indicating that the Brown probabilities are the same for Year 1 and Year 2+ for all programmes. The 'Unsure' confidence intervals for Zoology indicate a reduction in the probability of 'Unsure' in Year 2+. Surveying and Human Nutrition also showed an overall reduction in Year 2+ 'Unsure' probabilities but the difference was not statistically significant.

Fig. 3: 95% confidence intervals for the change in group probability for the three programmes

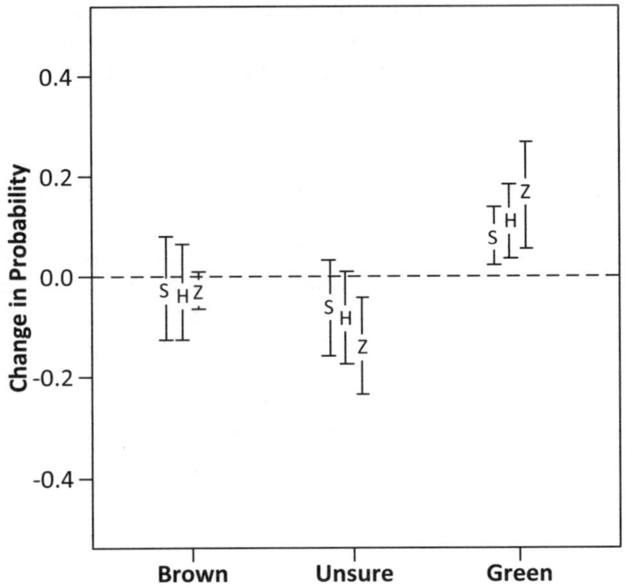

Surveying (S), Human Nutrition (H) and Zoology (Z). The probability point estimates are marked by the letters within each interval.

Multinomial logistic regression modelling suggests that subtle changes in sustainability attributes of these students may occur as they experience higher education. The model suggests that the probability of being in the higher NEP tertile increases with time. At least for Zoology students this is countered by a decrease in the probability of being in the 'unsure' tertile. The model fits the data well and these changes are significant.

Discussion

The need to evaluate change

Higher education institutions take quality advancement and assurance very seriously. Significant efforts are expended by university teachers and administrators every year to evaluate the impact of teaching on student learning, particularly by asking students for their perceptions of the quality of their educational experience. Where evaluation finds problems, institutions respond to improve the quality of their provision. Such evaluation complements formal assessment of the attainment of learning outcomes by students. If the mission of higher education is to extend to preparing graduates to become sustainably-minded citizens, equivalent processes will need to be developed to support the advancement and assurance of quality in this area.

Complex and imperfect approaches and tools

But the processes involved are not simple. In particular, sustainability outcomes are substantially affective (Cotton 2006; Shephard 2008) and not readily assessed (Buissink-Smith et al. 2011). Whether taught explicitly, 'democratically' (Jensen and Schnack 2006) or 'opportunistically' (Cotton et al. 2009), sustainability outcomes may need to be either evaluated, or assessed, or both; eventually for quality assurance processes, but more immediately (and in a quality-advancement context) to enable institutions to progressively develop successful approaches. Whatever approaches are involved, they will, surely, involve more than simply asking students for their opinions about personal change, or ticking a list of learning experiences.

The approach adopted in this research involves the exploration of students' environmental concern and draws much from the work of Dunlap (Dunlap and Jones 2002). To monitor the environmental concern of students may not be a sufficient process to address the aspirations of higher education institutions, but it is a start. Even with this limitation, it is not straightforward to monitor something as complex as environmental concern. Literally hundreds of tools, instruments and research methods have been applied to this quest (Dunlap and Marshall 2007). One tool, the NEP, itself extensively evaluated over many years (Hawcroft and Milfont 2010), has been chosen. It is by no means a perfect tool, but it is helping researchers to make a start on this complex and vital enterprise. Its use as a benchmarking tool has been described (Shephard et al. 2009) but its use to monitor change is work in progress and the substantive subject of this article. Perhaps no other higher education institution is systematically researching how its students' environmental concern is changing as they experience higher education, and researching how best to monitor these changes, but many have signed the Talloires Declaration or otherwise indicated their interest. The University of Otago has recently committed itself to ensure that its graduates develop 'environmental literacy'. The research described in this chapter may help this institution to understand what this will entail and how well it is currently performing in this respect, and may support more extensive and ongoing international engagement with this important higher education issue.

Flaws in this approach

Monitoring change in any group quality is not straightforward. Longitudinal research of population characteristics requires sophisticated statistical underpinning and analysis. It is not sufficient to simply compare the means of population samples from one time-point to another. Even if the process were statistically valid (which in general it is not, due to lack of independence between sample subjects), analyses of means and central variation can miss substantial within-sample changes.

This longitudinal analysis is, statistically speaking, both sophisticated and correct, but its interpretation is dependent on two key characteristics. First, it is essential that students understand that there is no individual link between their own personal level of environmental concern and their standing within the institution. It would be naïve to think that a student's response to the NEP would be an honest reflection of their personal environmental concern if they had any suspicion at all that their graduation, or degree class, might depend on them achieving a 'satisfactory' level of environmental concern. Absolute anonymity within the research is the only assurance that such a link cannot exist. Second, and contingent on the need that students understand that their contribution is anonymous, is the difficulty associated with this requirement. This approach asked students to invent, or calculate, a code word that would allow researchers to match their first and subsequent NEP submissions. Students forget their codes, or miscalculate them. This difficulty does limit the successful application of longitudinal analysis.

Multinomial logistic regression modelling also has limitations. Samples used to estimate the effects of change over time may be biased by extraneous factors. In this study it is possible that the differences between Year 1 and Year 2+ students may be caused by factors relevant only to (in this case) events in 2009 and 2010.

Perhaps relevant to all approaches that explore change in cohorts of university students is a selection bias due to less-academically-minded students dropping out after the first year, leaving subsequent years as essentially different groups, rather than the same group with changed environmental concern. The difficulty in making claims for change is to be emphasised here.

On changes and approaches to monitor them

These analyses suggest that the environmental concern of students does vary in relation to their chosen programme of study, or discipline. Although the first survey described here was conducted some time into the students' first semester, it appears likely that this applies when students first arrive at university – as demonstrated for students at Otago Polytechnic, in a previous study (Shephard et al. 2009). The analyses also suggest that the environmental concerns of University of Otago students (in the programmes and years that we have studied) do not change substantially during their stay at this university. The longitudinal analysis does not, at this stage, suggest any change at all, although the 2012 data may prove to be important. The multinomial logistic regression modelling suggests subtle positive changes, perhaps particularly in students with mid-to high levels of environmental concern. The data does suggest that, on balance, University of Otago students, at all stages of their study, have (relative to other student groups) moderate to high levels of environmental concern (see Hawcroft and Milfont [2010] for comparative data). Surveying students may have NEP scores slightly on the low side of other student groups internationally but perhaps not by a great margin. Absolute levels of environmental concern may not be a pressing issue for higher education internationally, but demonstration of positive change may be all the more difficult with relatively high starting points. It is to be hoped that this research will encourage other university teachers to reflect, with precision, on what changes they seek in their students and how to monitor these changes.

Conclusions and future prospects

This research suggests that exploring changes in the levels of the environmental concern of undergraduate students is a feasible and worthwhile process for universities to undertake. It also confirms that the NEP is a valuable tool if applied in a way that guarantees student anonymity and that it is possible to use this tool and to interpret its outputs in more than one way. The research suggests that institutions may need to use more than one approach to evaluate change to fully understand what changes do or do not occur. It appears that students in one institution may develop stronger environmental concerns as they experience higher education, but the changes that may have occurred in recent years are subtle. If higher education institutions seek more, they will need to address their aims and approaches with clarity and determination.

Acknowledgements

The authors would like to acknowledge the support of University of Otago students in completing the NEP survey, the commitment of involved academic departments to explore how their students change as they experience higher education, and financial support from the University of Otago Committee for the Advancement of Learning and Teaching.

References

Association of University Leaders for a Sustainable Future (1994), *The Talloires Declaration 10 Point Action Plan*, updated version. Retrieved 12th January 2012 from http://www.ulsf.org/pdf/TD.pdf.

Brody, S.D. and Ryu, H. (2006), Measuring the educational impacts of a graduate course on sustainable development. *Environmental Education Research* 12 (2), pp. 179-199 (doi: 10.1080/13504620600688955).

Buissink-Smith, N., Mann, S. and Shephard, K. (2011), How Do We Measure Affective Learning in Higher Education? *Journal of Education for Sustainable Development* 5, pp. 101-114 (doi: 10.1177/097340821000500113).

Cotton, D.R.E. (2006), Implementing curriculum guidance on environmental education: the importance of teachers' beliefs. *Journal of Curriculum Studies* 38, pp. 67-83 (doi: 10.1080/00220270500038644).

Cotton, D., Bailey, I., Warren, M. and Bissell, S. (2009), Revolutions and second-best solutions: education for sustainable development in higher education. *Studies in Higher Education* 34 (7), pp. 719-733 (doi: 10.1080/03075070802641552).

Dunlap, R.E. (2008), The NEP Scale: From marginality to worldwide use. *Journal of Environmental Education* 40 (1), pp. 3-18.

Dunlap, R.E. and Jones, R.E. (2002), Environmental concern: Conceptual and measurement issues. In Dunlap, R.E. and Michelson, W. (Eds.), *Handbook of Environmental Sociology,* pp. 482-524. Greenwood Press, Westport, CT.

Dunlap, R.E. and Marshall, B.K. (2007), Environmental Sociology. In Environmental Sociology Bryant, C.D. and Peck, D.L. (Eds.), *21st Century Sociology: A Reference Handbook* 2, pp. 329-340. Sage, Thousand Oaks, CA.

Dunlap, Riley E., Van Liere, K.D., Mertig, A.G. and Jones, R.E. (2000), New Trends in Measuring Environmental Attitudes: Measuring Endorsement of the New Ecological Paradigm: A Revised NEP Scale. *Journal of Social Issues* 56 (3), pp. 425-442 (doi: 10.1111/0022-4537.00176).

Fitzmaurice, G., Laird, N. and Ware, J. (2004), *Applied Longitudinal Analysis*. Wiley-Interscience, Hoboken N.J.

Harraway, J., Broughton-Ansin, F., Deaker, L., Jowett, T. and Shephard, K. (2012), Exploring the use of the Revised New Ecological Paradigm Scale (NEP) to monitor the development of students' ecological worldviews. *Journal of Environmental Education* (in press).

Hawcroft, L.J. and Milfont, T. (2010), The use (and abuse) of the new environmental paradigm scale over the last 30 years: A meta-analysis. *Journal of Environmental Psychology* 30, pp. 143-158.

Jay, L. (1990), Green about the Tills: Markets Discover the Eco-consumer. *Management Review* 79 (6), pp. 24-28.

Jensen, B.B. and Schnack, K. (2006), The action competence approach in environmental education. *Environmental Education Research* 12, pp. 471-486 (doi: 10.1080/135046 20600943053).

Shephard, K. (2008), Higher education for sustainability: seeking affective learning outcomes. *International Journal of Sustainability in Higher Education* 9 (1), pp. 87-98 (doi: 10.1108/14676370810842201).

Shephard, K., Mann, S., Smith, N. and Deaker, L. (2009), Benchmarking the environmental values and attitudes of students in New Zealand's post-compulsory education. *Environmental Education Research* 15, pp. 571-587 (doi: 10.1080/1350462090305 0523).

Teisl, M., Anderson, M., Noblet, C., Criner, G., Rubin, J. and Dalton, T. (2011), Are Environmental Professors Unbalanced? Evidence From the Field. *The Journal of Environmental Education* 42 (2), pp. 67-83 (doi: 10.1080/00958961003705899).

University of Sydney (2004), Policy Framework. *Policy Framework.* Retrieved 12th January 2012 from http://www.itl.usyd.edu.au/graduateattributes/policy_frame work.pdf.

World Commission on Environment and Development (1987), *Brundtland Report: Our Common Future*. Retrieved 12th January 2012 from http://worldinbalance.net/intagreements/1987-brundtland.php.

Chapter 10

Needs for Tailored Knowledge and Skill-Based Education for Sustainable Development: Balkan Environment Life Leadership Standards Courses

Anđelka Mihajlov

Abstract

The overall purpose of this paper is to link education (at universities) with the strategic goal of adopting the green economy (sustainable growth) concept from the angle of the environmental sector. The research was focused on the Western Balkans region, and more particularly on Serbia (one of the so-called "transition countries", in the process of accession to EU). This paper also tends to point to the rather sad fact that, generally, science in the transition countries has a very limited access to the resources needed for its development and advancement.

Using the reflections of the practices in the Balkans, the methodology applied is desk analysis combined with a list of qualitative questions (and possible answers) about education for sustainable development: how do we teach, what do we teach, who do we teach, who are the teachers, and why do we teach. The findings stress that "university education is not one size fits all" (regions, countries), pointing out that organizing additional educational courses (in addition to mandatory ones) could be a possible solution. One such course, which the author wishes to present in this paper, is Balkan Environment Leadership Standards. The concluding remarks should be taken into account when developing regional/national (sustainable) development strategies, in order to ensure the quality of the knowledge produced and provided. In the context of this paper, education for sustainable development is approached in the context of the specific region of south-east Europe/Balkans. The focus and value of this paper lies in the recommendation for the region to avoid (ongoing) mistakes by copying solutions from other regions, without researching needs and possible performance. The relevant diagnosis can be applied to the majority of transition and developing countries.

Introduction

This paper does not provide an overview of activities in the framework of the Decade of Education for Sustainable Development (2005-2014), declared by the UN General Assembly in 2002. However, it should be mentioned that different overviews exist, some of them focusing on strategy (UNECE 2007, 2009), others on competencies (Parker 2010), curriculum (Hahsmann 2009; Zoller and Scholz 2004), indicators (Reeves 2006), or processes (Scholz et al. 2006; Hansmann et al. 2009).

University education for sustainable development must be internationally focused on the problem, but regionally/nationally tailored. The literature is rich with analysis and regional articles for different regions on development and conflict solution (Mochizuki and Fadeeva 2008; Zilahy and Husingh 2009; Barton and Diouha 2011), but not for South-East Europe/the Balkans. Regional differences influence the meaning/understanding of education for sustainable development. Properly tailored university education for sustainable development is the best prescription for security and peace in the region. In the Balkan/SE Europe region, the question of how to create a wide social and governmental support-base for education for sustainable development is at the same time an issue of the tools necessary to transform a conflict region into a sustainably developed one.

The goal of the research has been already partially presented (Mihajlov 2010, 2011a, 2011b). The rationale for writing this paper came somewhat unexpectedly. After the author's presentation at the Serbian Chamber of Commerce (Mihajlov 2011b) on the results of a preliminary analysis of formally accredited environmental education at Serbian universities, a heated discussion among the industry representatives ensued. The conclusion was almost unanimous – it is mostly up to chemical engineers, who have the necessary knowledge and skills, to organize additional courses in these specific subject areas (pointing out, however, that at present there are no such accredited faculties of environmental studies providing the necessary knowledge and skills, which would be essentially useful to the industrial sector even though they are mostly "copied" from developed countries). With the aid of such an intervention from the industry, the author's research has taken a new course, now focusing on two main issues:

a. formal university education, and
b. possible additional education based on tailored knowledge/skills, and courses on sustainable development.

Moreover, scientists often produce scientific papers only to keep their jobs, not having the courage or initiative to make any substantial scientific progress or reshape the existing scientific disciplines and the already outdated education system, which is, indeed, in desperate need of reshaping. Therefore, green education and sustainable development training courses could act as catalysts for "turning words into action". Investing in the re-skilling of the workforce (new generation leaders) is viewed as a necessary step forward.

It should also be noted that this will contribute to the better preparation of the Western Balkan countries for accession to the EU, keeping in mind that commitment to the Europe 2020 Strategy (more details on ec.europa.eu) particularly implies investment in education and training, together with helping the unemployed to find jobs, boosting research and innovation, allowing cost-effective access to energy, and stepping up energy efficiency policies.

Approach to Education for Sustainable Development at Universities: Western Balkans

Following questions (and possible answers) about the education for sustainable development in the region, building blocks for possible sustainable solutions for education for sustainable development are regionally tailored.

How do we teach: formal and/or vocational education?

In the context of future generations' imperative for education for sustainable development (Mihajlov 2006), i.e. education about integral environmental protection and sustainable development, it is necessary to fully understand the set goals.

University programs (sometimes instructed by the National Accreditation Body) still seem to continue to cling tenaciously to divisions of natural, technical and social science. However, with such clear-cut divisions, it is almost impossible to teach around sustainable development properly in the contemporary context.

In addition, the transition from the centralized political system is affecting the education sector greatly, as state universities are becoming even weaker (mainly due to budgeting problems) while private universities have not yet developed any long-term planning strategy.

What do we teach and what should we teach?

Economists think that *sustainable economy* is sustainable development; environmentalists seem to think that *environmental sustainability* is sustainable development, while social scientists mostly link this to *poverty, happiness and life patterns*. The problem, however, is that there is a lack of general awareness that not enough is actually being done to provide substantial education for sustainable development.

What is mostly taught at Serbian universities are the so-called "pure" disciplines, with inter-sectoral and cross-sectoral approaches rarely adopted. According to recent research (Hill and Knowlton 2009), the environmental, energy, climate (Mihajlov 2011c), environmental health, and waste-related insecurities among the citizens of the Western Balkan region are fairly high (Mihajlov and Stevanovic-Carapina 2010). Thus, taking the cross-sectoral approach is the best way to address these issues. It should furthermore be recognized that many of the social and economic problems cannot be considered separately from the environmental problems (and vice versa), and that instead the social and environmental systems should be studied as one integrated system.

Science educators have the responsibility to ensure that citizens (students, decision-makers and all other education target groups) gain the intellectual tools to be able to engage fully in the ensuing debates and decision-making processes. Inter-governmental organizations, international financial institutions, non-governmental organizations, professional organizations (centers of expertise), the private sector and the international community as a whole, can play a critical role in providing firstly technical and financial assistance for the countries in transition, and secondly the urgently needed education in sustainable development for the future.

Who do we teach and who should we teach?

At Universities, we teach *students*.

However, there are other groups that need to be taught about sustainable development, too. In 2009, a survey related to environmental and sustainable development issues was conducted in the Western Balkans region (specifically in Albania, Bosnia and Herzegovina, Macedonia, Montenegro, Serbia and Kosovo), and at the same time within the following target groups: parlamentarians and state employees, entrepreneurs, independent local government, political parties, NGOs and citizens. According to the data obtained, it is the *decision-makers,* most of all, who need to be further educated about sustainable development and the related issues (Table 1).

Table 1 Opinion (%) as to whether decision-makers' knowledge about sustainable development and environment should be improved

	Overall in Western Balkans
Yes, it absolutely should	47.5
It should	38.3
Yes and no	4.8
It should not	3.8
Absolutely not	0.6
I do not know	5.0
Total	**100**

Note: It was possible to select only one answer

The problem identified above indicates that besides the *students*, *decision-makers* also need teaching. Investing in the re-skilling of the workforce may also be necessary.

Who are the teachers and what should they teach?

"Learning by watching" made the author wonder: Who teaches the 200-250 environmentalists graduating in Serbia every year? Although most professors are acknowledged researchers in particular fields, it could be observed that the knowledge most of them have about sustainable development is basically at a very low level. Furthermore, the transition countries having such a limited access to the basic resources necessary for the growth and development of science, many of these scientists produce scientific papers only to keep their current jobs, not having enough courage to actually look into the future and reshape the already outdated scientific disciplines and educational systems, acting very much in accordance with the "business as usual" principle.

Unfortunately, only a small number of academics are actually willing to accept and face the challenges that come with the new areas of knowledge transfer (Gal and Ptacek 2011). These new areas of knowledge transfer are important for education about sustainable development, because they provide expertise in various fields (consultancy, network-type collaboration); they teach us how to respond to pressure at the time of crisis; they help us utilize and harmonize knowledge about regional and social development to the best effect with the new market demands and policies; and, finally, they teach us how to communicate with different institutions in order to provide not only better funding opportunities, but also a collective learning platform for social interactions.

Why do we teach and why should we teach?

What is the rationale behind teaching: to provide students with diplomas, to provide them with knowledge, or to teach them how to work and keep a job?

Knowledge and skills enhancement programs are needed to prepare the new workforce for a green economy transition (UNEP 2008; Europe 2020: Green Growth and Jobs 2010). Green education and sustainable development training courses could act as catalysts for "turning words into action", which means teaching the leaders of the new generation about new skills and new, greener performance.

However, the question is: are we integrating sustainable development and environmental issues into existing qualifications, and securing the new, emerging skills that are needed on the green job markets? Secondly, how should we address lack of interest in science and engineering leading to a deficit in available technical skills?

Fig. 1: Presentation of a future job marker; education guidance for sustainable development

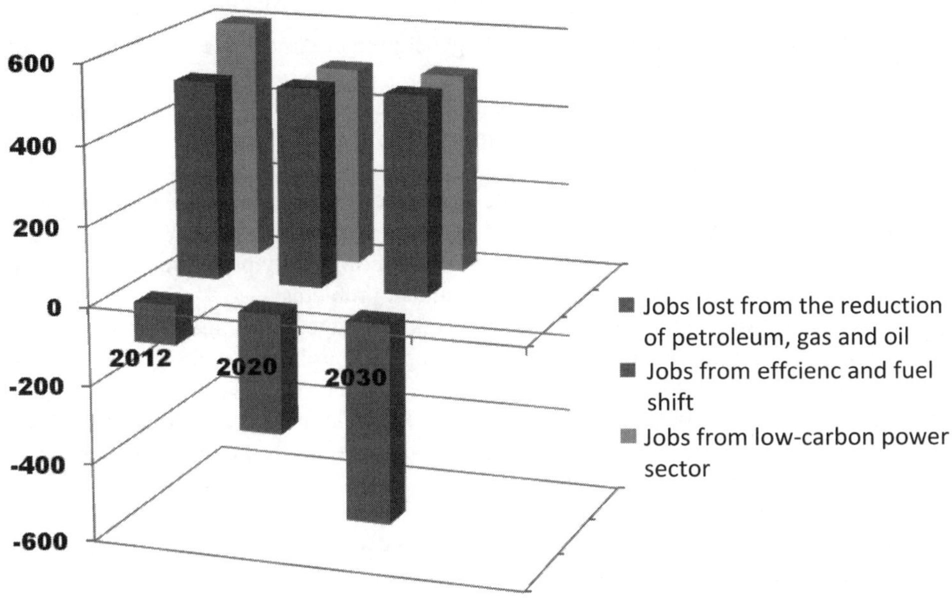

Adapted from Synthesis Report (2010)

Do we still think that the future job markets could "put old wine in new bottles" (see Fig. 1)?

Are we ready to properly educate and train our students for future occupations such as manager in renewable energy, wind-turbine operator, energy auditor, waste managing operator, eco designer, solar energy project designer, Smart Energy expert, Smart Energy manager, carbon trader/broker, environmental procurement manager, etc.?

Are we ready to provide the core (vocational) knowledge for processes, technologies, materials, markets and market dynamics, environment, climate, sustainability, communication, process and planning, and other environmentally driven additional competences needed?

Findings and possible solution: tailored knowledge and skill based education for sustainable development

Regional differences influence the meaning/understanding of education of sustainable development.

The analysis identifies some larger weaknesses in the skills base, mainly in science, technology, engineering and mathematics, as well as in more generic skills such as improving resource efficiency, understanding environmental legislation, and green management and communication, which are highly relevant to the environmental sector. The *teaching circle* forms the basis of the methodology to be applied: needs assessment, course/program design, course/program delivery, and evaluation (and follow-up). The problem with the current education is that it is only the titles that are being changed, while the actual substance of educational programs remains the same.

According to the polls (Hill and Knowlton 2009), the tailored knowledge and skill-based *Balkan Environment Life Leadership Standards Course* could be the possible solution for the Western Balkan countries.

By launching the *Balkan Environment Life Leadership Standards Courses* (Mihajlov, 2011b), as presented in Table 2, a range of goals could be achieved. For example:

- providing a tool for the shaping of the future generation of sustainable development leadership;
- building abilities, relationships and values that would enable decision-makers, civil society organizations, businesses and individuals to improve their performance, skills and professionalism, and achieve their objectives;
- creating new environmentally friendly lifestyles, and fostering environmental ethical responsibility for the future;
- offering the opportunity to those interested in discovering really professional courses dealing with environmental and sustainable development problems;
- assisting target groups to acquire the necessary knowledge, skills and tools for assessing and solving environmental and sustainable development problems, both on a regional level – in their sub-regions and local communities – and on a national level;
- creating higher profile occupations for individuals working in the environmental sector, providing them with relevant certificates, and turning them into acknowledged experts and leaders in the field of sustainable development;
- making target groups more sensitive to environmental issues;
- stressing the importance of incorporating the environmental sector in other sectors, in order to target sustainable development.

The knowledge and skills related to multilateral environmental agreements, which establish the legal and the institutional frameworks for addressing global environmental challenges, can play a significant role.

Table 2: Matrix: Courses as the answer to educational needs in the Western Balkan countries

<table>
<tr><th>Skill- based</th><th>Knowledge-based</th><th>Tool-based</th></tr>
<tr><td colspan="3">Leadership for sustainable development – the new political reality</td></tr>
<tr><td colspan="3">Effective Partnership Development</td></tr>
<tr><td colspan="3">Public participation in environment-related decision making and green communication</td></tr>
<tr><td>Green lobbying and advocacy (for civil society)</td><td>Implementation of environmental legal requirements</td><td>Green economy and budgeting</td></tr>
<tr><td>Environmental policy paper development (for decision makers)</td><td>Basic Security in Climate Change and Energy Security</td><td>Ecological footprint – state's, local, individual</td></tr>
<tr><td>Environmental diplomacy (for decision makers)</td><td colspan="2">Heavy EU directives and responsible investing</td></tr>
<tr><td>Green judges and court experts (for judiciary sector)</td><td colspan="2">Corporate environmental responsibility (for business sector)</td></tr>
<tr><td>Green campaigning (for decision makers)</td><td colspan="2">Synergy in Multilateral Environmental Agreements Governance (for administration)</td></tr>
<tr><td colspan="3">Positive practice in environmental and sustainable development issues (for all target groups, but specifically from the perspective of each particular target group)</td></tr>
</table>

Conclusion

At present, many questions related to sustainable development education at universities remain unanswered, and many issues remain unaddressed, while coordinating activities related to the upgrading of skills and policies still pose a challenge for educational systems in each country.

This paper, on the basis of the research, analysis and experience of the author, attempts to justify the conclusion that "university education is not one size fits all" and "copy-paste" of methodologies and curricula from one region to another does not *a priori* result in the same level of success.

In conclusion, in order to best address the issues related to education for sustainable development in the Western Balkan countries, the following needs to be provided:

- Bachelor-level education at universities (mainly in science, technology, engineering and mathematics), and
- Region-tailored (vocational) courses (such as *Balkan Environment Life Leadership Standards Courses* and the accreditation of additional tailored courses, which are not obligatory at universities within the basic curricula).

Universities (and other registered educational institutions, including CSOs and NGOs) need to develop the tailored knowledge and skill-based education for sustainable development (as the proposed *Balkan Environment Life Leadership Standards Courses* has done), by defining their growth strategy (social visibility, strategic partnerships and engagement), developing a joint transnational knowledge network, and developing the regional and social functions of universities in the region.

Currently, only a few of the proposed courses for the Western Balkans are operational.

Properly tailored (university) education for sustainable development is the best prescription for security and peace in the region. In order to achieve these goals, it is necessary to organize the skill-enhancement programs and interdisciplinary courses, focusing on different issues of sustainable development, and to train and prepare the new workforce for a green economy transition. Universities should shape regional and national sustainable development platforms. Education *for* sustainable development should contain ingredients *of* sustainable development, but not be dominated by the economic mainstream.

As a final observation, it is worth remarking that understanding of problem-solving and critical thinking should be evidenced by leaders in the region, in line with the promotion of synergy with *classical* disciplines.

References

Barton, A., Dlouha J. (Eds.) (2011), *Multi-actor learning for Sustainable Development Regional Development in Europe: A Handbook of Best Practice.* Grosvenor House Publishing Limited (ISBN 978-1-908105-33-2).

*Europe 2020: green growth and jobs (*2010), http://www.euractiv.com/en/priorities/europe-2020-greengrowth-and-jobs-linksdossier-280116.

Gál, Z. and Ptáček, P. (2011), The role of mid-range universities in knowledge transfer: the case of non-metropolitan regions in Central Eastern Europe. In *European Planning Studies*, Vol. 19. Forthcoming.

Hansmann, R. (2009), Linking the Components of a University Program to the Qualification Profile of Graduates: The Case of a Sustainability-Oriented Environmental Science Curriculum. *Journal of Research in Science Teaching* 46 (5), pp. 537-569.

Hansmann, R., Crott, H.W., Mieg, H.A. and Scholz, R.W. (2009), Improving group processes in transdisciplinary case studies for sustainability learning. *International Journal of Sustainability in Higher Education* 10 (1), pp. 33-42.

Hill & Knowlton (2009), *Survey for REA project coordinating by "Environmental Ambassadors"*, www.ambassadors-env.org; www.bellsmovement.org.

http://ec.europa.eu/environment/eussd.

http://ec.europa.eu/europe2020/index_en.htm.

Mihajlov, A. (2006), *Sustainable Development and Environment towards Europe in 95+ steps*, Environmental Ambassadors and Serbian Chamber of Commerce, Monograph, Beograd.

– (2010), Teaching Climate Change: the multiple relationships between climate change related education, climate (un)sustainability, business sector and civil society", 2nd CEE/SEE Regional Colloquium Capacity Building on Global Governance and the UN System: Teaching Climate Change and the United Nations System. http://video lectures.net/andjelka_mihajlov.

– (2011a), *Connecting Academics and UN Practitioners: Natural Resources and Security – Precautionary Dimensions*, ACUNS Vienna Conference, UN Colloquia Session II. http://www.unstudies.org/sites/unstudies.org/files/AnMi%20Connecting%20Academics%20and%20UN%20Practitioners%202011.pdf.

– (2011b), Knowledge for Green Jobs and Green Economy – Analysis from the Perspective of Integral Environmental Protection. In Pavlovic, V. (Ed.), *University and Sustainable Development*, pp. 87-98. Faculty of Political Sciences University of Belgrade, Belgrade, Serbian.

– (2011c), *Climate change and European integration, key note lecture, CCEDEP – 1st Climate Change, Economic Development, Environment and People Conference*, Book of Abstracts, p. 25, Novi Sad.

Mihajlov A. and Stevanovic Carapina, H. (2010), Environmental (In)Security from Citizens Perspective, Chapter 11. In Daniels, J.A. (Ed.), *Advances in Environmental Research* 7, pp. 1-14. Nova Science Publ. Inc., USA.

Mochizuki, Y., Fadeeva, Z. (2008), Regional centers of expertise on education for sustainable development: an overview. *International Journal of Sustainability in Higher Education* 9, pp. 33-42.

Parker, J. (2010), Competencies for interdisciplinarity in higher education. *International Journal of Sustainability in Higher Education* 11 (4), pp. 325-338.

Reeves, T.C. (2006), How do you know they are learning? The importance of alignment in higher education. *International Journal of earning Technology* 2 (4), pp. 294-309.

Scholz, R.W., Lang, D.J., Wiek, A., Walter, A.I. and Stauffacher, M. (2006), Transdisciplinary Case Studies as a Means of Sustainability Learning: Historical Framework and Theory. *International Journal of Sustainability in Higher Education* 7 (3), pp. 226-251

Synthesis Report: Skills for green jobs, European Synthesis Report (2010), Publications Office of the European Union, www.cedefop.europa.eu/download-manager. aspx?id=16440&lang=en&type=publication.

UNECE (2007), *ECE/BELGRADE.CONF/2007/INF/9,* http://www.unece.org/fileadmin/DAM/env/esd/01_Typo3site/ece.belgrade.conf.2007.inf.9.e.pdf.

– (2009), *Learning from Each-Other: the UNECE Strategy for Education for Sustainable Development,* http://www.unece.org/fileadmin/DAM/env/esd/01_Typo3site/LearningFromEachOther.pdf.

UNEP (2008), *Green jobs: towards decent work in a sustainable, low-carbon world*. UNEP, Nairobi.

Zilahy, G. and Huisingh, D. (2009), The roles of academia in regional sustainibility initiatives. *Journal of Cleaner Production* 17 (12), pp. 1057-1066.

Zoller, U. and Scholz, R.W. (2004), The HOCS paradigm shift from disciplinary knowledge (LOCS) – to interdisciplinary evaluative, system thinking (HOCS): what should it take in science-technology-environment-society oriented courses, curricula and assessment? *Water Science & Technology* Vol. 49 (8), pp. 27-36.

Chapter 11

Student Expectations on Content and Pedagogy for Sustainability within Textiles and Apparel Curricula

Anupama Pasricha

Abstract

The purpose of this paper is to inform educators about effective ways to incorporate sustainability in the curriculum and teaching agenda, based on student responses from focus group interviews. This original work is qualitative research that fills a gap in the literature by providing the college student's perspectives on discipline-specific content and teaching/learning pedagogy. Using the grounded theory data-analysis method resulted in themes and subthemes. Thirteen subthemes, which illustrate broader themes of program model, instructional strategies, discipline-specific content, and extracurricular engagement, are identified and explained. This study contributes to the literature by providing a learner's (student's) viewpoints on discipline-specific content and pedagogy for effective learning; in particular as these viewpoints pertain to sustainability and sustainable development. Although focused on one content area, this paper is useful to anyone interested in the Millennial Generation's expectations on content and pedagogy for sustainability in any professional studies curriculum.

Introduction

The issue of sustainability is a colossal and complex challenge facing our society. Although a widely used term, sustainability is still not understood well (Leal Filho 2000). In order to prepare students for understanding and countering this challenge, universities and other educational institutions worldwide are increasing their efforts to incorporate sustainability issues into their curricula. Moreover, the United Nations declared 2005-2014 as the decade of sustainable development and encouraged educational institutions at all levels to transform their curricula (UNESCO 2003).

David Orr, a scholar, educator, and leader in the field of environmental education, ecological design, and climate change, has long been an advocate of sustainability in education. Orr has often stated that the educational system prepares graduates without any broad and integrated awareness or knowledge of the interdependence of life (Orr 2004; 2010). For instance, in the 2010 Worldwatch Institute Report, Orr asks, "What kind of education will enable the rising generation to deal with increasingly complex and portentous global issues?" (Orr 2010: 76). The current educational system nurtures unsustainable paths, and it needs fundamental and systemic change (LeHew and Meyer 2005; Orr 2010; Timpson et al. 2006).

Educators worldwide are engaged in researching and promoting effective ways to incorporate sustainability into the curriculum and to create programs that bring the desired changes in the education system. A number of studies have reported on how to integrate sustainability into the curriculum. Rusinko (2010) developed a generic matrix for integrating sustainability in the curriculum. This matrix ensures a holistic approach that includes social, environmental, and economic dimensions of sustainability. Service learning, problem-based learning, and experiential learning have all been reported as successful pedagogies for sustainability learning (Brundiers et al. 2010; Christensen et al. 2007). Some studies have reported how business schools are including Corporate Social Responsibility and sustainability in their courses (Christensen et al. 2007; Stubbs and Schapper 2011). A majority of these studies focus on the perspectives of educators or institutions. Few studies focus on student perceptions and interpretations of sustainability (Armstrong and Meyer-Brosdahl 2009; Carew and Mitchell 2002; Kagawa 2007; Lundholm 2005).

However, there is a lack of research on learning about student's expectations from their education on sustainability. This research study, which is based on student responses from focus group interviews following the guidelines of qualitative research, attempts to inform educators of effective ways in which to incorporate sustainability into the curriculum and teaching agenda within the field of textiles and apparel. This study contributes to the literature by examining the viewpoint of learners (students) on discipline-specific content and pedagogy for effective learning, particularly as this pertains to sustainability and sustainable development.

Sustainability

The concept of sustainability is not widely understood and may have different meanings for different individuals. Sustainability has been thought of as long-term, durable, sound, and systematic (Leal Filho 2000). Some of the terms used to define sustainability are health, wellness, organicity, environmental consciousness, fair trade, simple living, and buying locally. In 1987, the United Nations World Commission on Environment and Development defined sustainable development or sustainability as "meeting the needs of the present without compromising the ability of future generations to meet their own needs." For the purposes of this study, the definition developed by the United Nations is most applicable.

Sustainable development is grounded on three pillars: social (people), environment (planet), and economic (profit), commonly known as the Triple Bottom Line (TBL) in the corporate world, a term coined by John Elkington (Anderson 2007; Lawrence 2005). According to Kermath (n.d.), sustainability is a condition that encompasses environmental integrity and social justice. Environmental integrity is the "ability of the whole environment to function as natural as possible and to do so without compromising the ability of the ecosphere from supporting all life forms on the planet and to maintain its inherent evolutionary potential" (paragraph 6). Kermath defines social justice as "fair and equitable access to, and distribution of essential resources and power, fairly applied laws and regulations, and the guaranteed opportunity for all individuals and communities to contribute to the pursuit of meeting human needs, improving the human condition, fully realizing human potentials for everyone in safe and clean environments, and to receive an adequate and fair return on their investments of capital, creativity, labor, and time" (paragraph 7). The environmental integrity and social justice issues raised by Kermath fit the UN's definition while addressing the issues of sustainability and social responsibility pertaining to the global textile and apparel industry.

Education, public awareness, and training are the focus of chapter 36 of the Agenda 21 document, a global action plan for delivering sustainable development (United Nations Department of Economic and Social Affairs: Division for Sustainable Development, 1993). Agenda 21 asserts the importance of sustainability and the role of education in preparing individuals to take responsibility for the future of the planet. The documents prepared by the Division of Sustainable Development (DSD) suggest several methods for implementing the sustainable-development goals of the International Work Program of the United Nations (United Nations Department of Economic and Social Affairs: Division for Sustainable Development, n.d.).

Higher Education and the Millennial Generation

The Association for the Advancement of Sustainability in Higher Education (AASHE) is an association of colleges and universities in the US and Canada that is working to create a sustainable future. It was founded in 2006 with a mission to promote sustainability in all sectors of higher education – from governance and operations to curriculum and outreach – through education, communication, research, and professional development (AASHE, n.d., paragraph 1). Several four-year and two-year colleges are members of AASHE and share a common sustainability agenda.

The majority of today's traditional-age college students belong to the Millennial Generation; this generation has also been referred to as Generation Y, the Net Generation, and the iGeneration. The Center of Digital Education (2007) identifies the Millennials as being born between the years of 1977 and 1994. As a group, they tend to be ethnically and racially diverse and politically progressive (PEW Research Center 2009).

Millennials live in a culture called "postmodern," which Morgado (1996) defines as a "body of critical theory about the nature of contemporary society" (p. 41). Contemporary society is a post-industrial society, with a focus on social responsibility and professional commitment rather than exploitation and profit (Morgado 1996). Peter Drucker (1993) refers to this postmodernism as an outcome of transformation:

> Every few hundred years in Western history, there occurs a sharp transformation ... within a few short decades, society rearranges itself – its worldview; its basic values; its social and political structure; its arts; its key institutions .... Fifty years later, there is a new world and the people born then cannot even imagine the world in which their grandparents lived and in which their own parents were born. We are currently living through such a transformation (p. 1).

One private (for-profit) consumer research company, The Hartman Group, Inc. (2008), analyzed the Millennial Generation using techniques that ranged from performing landscape reviews and surveying ethnographical backgrounds to tracking trends, studying social network parties, and creating neurolinguistic maps. According to this research, Millennials were raised in an era of Earth Day celebrations and grew up watching cartoons like *Captain Planet* that portray recycling as "cool" behavior. The company also discovered that Millennials are bored with consuming, have truly global tastes, and view the world through a social construct. For this generation, sustainability is about being community-oriented and supporting small, local businesses that are socially-conscious. Millennials refer to sustainability as "going green," and they are extremely aware of both local and global environments. While they may not understand the definition of organic, they are conscious of socially-responsible companies (Demerrit, president and COO, The Hartman Group, Inc., A Tinderbox Report Power Point, e-mail communication, November 14, 2008). A study on youth and consumption, conducted by the United Nations, surveyed youth from 24 countries; 75% of the respondents agreed that the biggest challenges in the future are reducing environmental pollution, improving human health, and respecting human rights (McGregor 2002; UNESCO 2001).

A study conducted by Carew and Mitchell (2002) on engineering students in Australia revealed that 65% of the participating students held pre- or uni-structural conceptions of sustainability, despite having completed a specific unit on sustainability based on the Structure of Observed Learning Outcome (SOLO) taxonomy. This finding suggests that there may be barriers to understanding the concept of sustainability or that students do not care to learn about it. This study also compiled principles of sustainability from engineering literature, in which sustainability is represented as both an end and a means when applied to decision-making.

Textiles and Apparel

The textiles and apparel industries are considered one of the major product groups within the international trade of manufactured goods (World Trade Organization 2011). Operating on a global basis, the textiles and apparel industries have been growing rapidly, as have the environmental issues associated with them. These industries, considered among the most environmentally damaging worldwide (Anderson, February 2009), are facing problems such as limited natural resources, global warming, and sustainability issues, as well as other social and political challenges (O'Neal 2007). Indeed, the textile industry alone is considered one of the world's largest consumers of water (Anderson, August 2008), and there are several sustainability issues in the industry specifically related to sourcing, production, manufacturing, packaging, marketing, and consumption (Dickson et al. 2006). Other aspects of the industry, such as fast-changing fashion trends and premature product replacement, are also anti-ecological.

With growing awareness, sustainability gained impetus in the textile and apparel industries in 1990s. Today there is an increasing realization that these industries must take responsibility for protecting people and the environment (Dickson et al. 2009; Hethorn and Ulsawicz 2008). To become sustainable, these industries need to make major changes at all levels, including design, production, and marketing. According to Dickson et al. (2006), there are only two choices for individuals when addressing sustainability: 1) either individuals can ignore sustainability issues, or 2) individuals can become the agents of change. Many individuals have chosen to become agents of change, as have such companies as the Adidas group, Eileen Fisher, Levi-Strauss and Co., Nike, Patagonia, and TOMS Shoes. Several corporations have jointly created a Sustainable Apparel Coalition with a vision of creating a positive impact on people and communities without creating unnecessary harm to the environment (Sustainable Apparel Coalition, n.d.).

Conventionally, textiles and apparel education has followed a business model with a focus on innovation, target market, and profit. It is time to change that model to incorporate global citizenship competencies that counter global problems (LeHew and Meyer 2005). These global citizenship competencies ensure the development of informed, empowered, and responsible individuals who understand their relationships within and across communities, as well as understand the consequences of their decisions and actions on themselves, society, and the physical world (Global Citizen Compentency, n.d.). Professional organizations such as the International Textiles and Apparel Association (ITAA) refer to these competencies under the headings of *content and process knowledge* and *professional attitudes and skills* (International Textiles and Apparel Association 2008). Textiles and apparel programs across the US use these competencies as a framework for establishing program goals and outcomes. Utilizing these competencies, today's students and tomorrow's professionals will become able to resolve social, cultural, environmental, economic, and political issues worldwide as they find employment in these industries.

Method

This research study attempts to inform educators of effective ways to incorporate sustainability in the curriculum and teaching agenda within the field of textiles and apparel, based on student responses in focus group interviews, which followed the standard guidelines of qualitative research. In this research study, textiles and apparel were used as a case study for two reasons: (1) the textiles and apparel industries have been considered highly unsustainable, and (2) the researcher, a textiles and apparel educator for more than two decades, wanted to contribute to the systemic change in the education model being advocated by several educators and researchers.

The researcher conducted focus group interviews with college students from two Midwestern universities. The interview protocol included 12 questions that addressed sustainability as part of a larger study. For the purposes of this paper, only the following interview question was analyzed by the researcher: "What would you like to learn in your courses about sustainability, or what do you think is important to include in the curriculum about sustainability? Please give specific examples from your courses."

This qualitative research project included the required pilot-study procedures, development of the interview protocol, approval by the Institutional Review Board, use of focus groups, and use of the grounded theory approach for analyzing the data. Based on the recommendations provided by Krueger (1994), the focus group questions were piloted in steps. Step one was an expert review by four faculty members, two from each university, who have experience with focus groups. Step two involved pre-testing with a small group from the target sample of participants to ensure that the questions were clear; the questions were reviewed by six college students enrolled in the field of textiles and apparel. For their participation, the student reviewers were given a $10 gift certificate redeemable at the university's bookstore, but their comments have not been included in the results. Step three was holding the first focus group interview.

Focus group participants were selected by convenience sampling. Participants were solicited through a class announcement informing them about the objective of the research, eligibility for participation, projected time commitment, proposed focus group interview schedule, compensation for participation, and the voluntary nature of their participation. Interested students were requested to contact the researcher through e-mail. These students were sent an e-mail reminder two days before the scheduled focus groups and informed that snacks and beverages would be provided prior to the focus group. They were told that they would receive extra credit for their participation and a $10 gift certificate redeemable at the university's bookstore at the end of the focus group; they were also cautioned that latecomers would not be allowed to participate because of potential disruption to the interview process. Consent forms and background-information sheets, which the participants were asked to complete, were distributed at the beginning of the focus group sessions. Names of the participants were later given to the instructors. Notwithstanding the two incentives being provided, students may well have chosen to participate because of their interest in sustainability issues.

The seven focus groups included 46 participants, who ranged in age from 18 to 44; of these 46 participants, nine were freshmen, eight sophomores, 12 juniors, 16 seniors, and one unreported, all enrolled in at least one textiles-related or apparel-related course during the Fall 2009 semester. Participants represented a wide variety of majors, including apparel design, fashion merchandising, apparel manufacturing, and product development. A majority of the participants (37) were Caucasian/European Americans. Focus group sessions allowed participants to hear each other's responses as well as make additional responses building on what others had said, creating synergistic effects and providing richer data, and a more rapid collection of that data, in a flexible and informal setting (Krueger, 1994; Morgan, 1988; Stewart et al. 2007). Focus group sessions were transcribed verbatim into a Microsoft Word document. The data were analyzed using a grounded theory approach and a constant comparison method. Themes and subthemes emerged from this data analysis.

Fig. I: Themes and Subthemes representing content and pedagogy for sustainability within textiles and apparel curricula

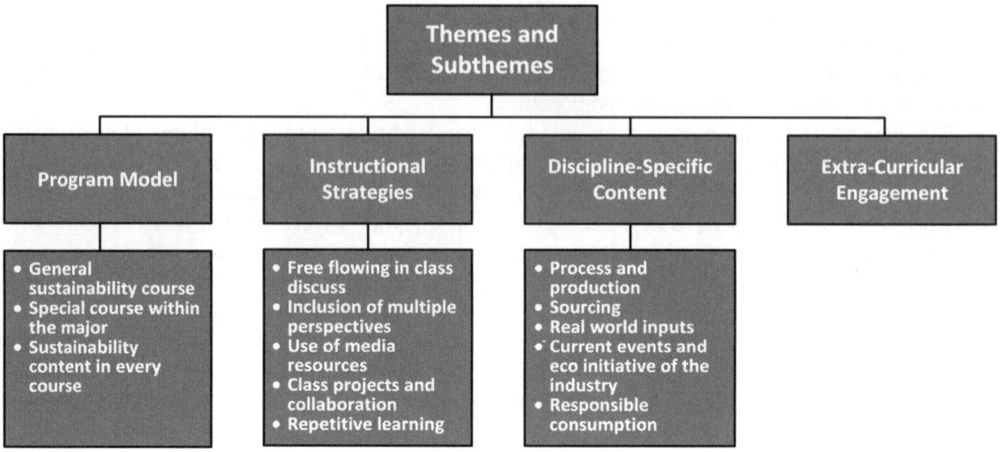

Results

The results provided significant input on the discipline-specific knowledge that students are seeking and the pedagogy that maximizes their learning. Based on the grounded theory analysis of the focus group responses, the following themes were identified: program model, instructional strategies, discipline-specific content, and extracurricular engagement. Each theme is further categorized in subthemes (Fig. 1). The subthemes most frequently mentioned were sustainability course (program model), learning from each other and class projects (instructional strategies), and processing and production (discipline-specific content). Participant quotes are included under several themes; however, names that are used with quotes are fictitious.

Program Model

The theme program model emerged from the focus group discussions on positioning sustainability content within the program. Education for sustainability within the textiles and apparel curriculum is becoming critical. Educators have identified this need, and some colleges and universities have incorporated it either in existing curricula or by designing separate courses or programs focusing on sustainability. The University of Delaware pioneered a certificate program in Socially Responsible Business (http://www.udel.edu/fash/graduate/certificate/). The need for these kinds of programs and courses is expressed by educators and students alike. Several subthemes emerged as participants expressed numerous ideas on how to incorporate sustainability education in textiles and apparel programs.

General Sustainability Course

The participants felt that sustainability is of universal importance across the areas often found in textiles and clothing majors, such as design, merchandising, and production. Required courses dedicated to sustainability that should be required as general or core education courses were mentioned in three focus groups, often being mentioned more than once. Participant Ana stated, "All students in the university take [them] so they know about all the consequences of what they're buying."

Special Course within the Major

Two focus groups agreed that a special course on sustainability is a valuable idea, while participants of another focus group agreed unanimously that a single course may not be as effective as including sustainability within all courses. Participant Julie stated, "I think that you should have a special course that covers everything."

Sustainability Content in Every Course

Within textiles and apparel, sustainability should be included in every course, because if it is only taught in specific courses, those courses may not be required for every major field of study within textiles and apparel. Martens (2006) and Orr (1992) emphasize the significance of making sustainable development a guiding principle throughout the program of study at all educational levels. Participants also felt that it was important and that it should be included in every student's education. As participant Brea stated, "one course on sustainability … probably wouldn't be as effective … because you learn how it affects different parts you know."

Instructional Strategies

Students offered many valuable suggestions for strategies that would facilitate their learning about sustainability. These suggestions included free-flowing in-class discussions, purposeful inclusion of multiple perspectives across coursework, class projects and collaboration, repetitive learning, and use of media.

Free-Flowing In-Class Discussions

In four focus groups, free-flowing in-class discussions, such as focus groups that facilitated their thinking and learning, were identified as an important strategy. Participant Alisha stated, "We have learned so much just by sitting here and talking about it." Another agreed: "I think that getting in groups like this is a good idea because that's what you're going to do at work."

Inclusion of Multiple Perspectives

Students felt it was important to integrate sustainability into every course in order to learn from the multiple perspectives presented by their professors and peers. In other words, the concept of sustainability is best learned and understood through multiple lenses and dialogues. Participant Zoe said, "We are learning different aspects, it is in different professors' point[s] of view, different class peoples' point[s] of view." Another stated, "Assignments where you're doing group discussions and coming out with some deliberations."

Class Projects and Collaboration

Class projects were identified as a common tool for incorporating sustainability. The idea was mentioned at least twice in each focus group. Some popular examples of projects from the focus group discussions were reading about current trends on sustainability; group discussions; research on sustainability and apparel materials; small assignments or small projects; and projects on how to make sustainable products.

Projects allow students to take personal ownership of their learning by examining all aspects of a subject. Student-focused collaborative learning is considered vital to engaging students in their own learning (Harper-Marinick 2001; Kimmons and Spruiell 2005).

Repetitive Learning

Spaced repetition can aid in the efficient retrieval of learned information (deWinstanley and Bjork 2002; Hashimoto et al. 2011). Students emphasized that hearing a message repeated in every class has a lasting impact on their learning. Key concepts can be repeated or referred to in courses that extend across a program. This encourages students to embrace these concepts and apply them to their personal and professional lives.

Use of Media Resources

The Millennial Generation has been given several names, including the iGeneration. This generation, which has been brought up on all the latest digital devices and communication technologies, is highly media-focused, even to the extent of using media as a learning tool. For example, watching a video on a relevant subject becomes a powerful learning tool for Millennials. Participant Alaya stated, “I mean, I learned a lot and I obviously remember the *American Apparel* video.” And participant Maria stated, “That [*China Blue*] video should have been watched by everybody.”

Discipline-Based Content

Student responses and discussions favored learning about the environmental aspects of sustainability. Learning about its social aspects was not mentioned. However, as stated above in the “Instructional Strategies” section of this paper, video resources such as *China Blue* and *American Apparel* were appreciated. Both of these videos focus on social aspects of the apparel industry.

Processing and Production

“Processing and production” refers to learning about the processing and production of raw materials used in the apparel industry. In all of the focus groups, processing and production – and, specifically, processes that materials undergo during production – informed judgments about materials that were either harmful or beneficial to the environment. Participants mentioned that they would like to learn about the composition and source of materials and the beneficial and/or harmful aspects of their use on the environment. Participants also agreed that terms commonly used for environmentally safe materials – organic, for example – need to be better understood. Students also said that learning about the potential effects of these materials on themselves personally will have a greater impact than learning about their effects on the natural environment.

Some focus groups mentioned that their courses only touch upon sustainability, without doing so in any depth or detail. That greater level of detail helps students understand the positive and negative impacts of a material or process. Participants demonstrated a deep interest in learning why various materials and processes are categorized as either beneficial or harmful to the environment, and how certain choices they make will have a positive impact on the environment.

Sourcing

“Sourcing” addresses multiple dimensions, including procuring materials for product manufacture, buying as a professional, and buying as a consumer. Sourcing was mentioned by three groups, which expressed their interest in identifying ecologically friendly companies that could assist consumers in making better choices. One of the focus groups demonstrated a keen desire to make sustainability an important component of buying. Buying is a crucial facet of retail; buyers locate suppliers and purchase products from designers, manufacturers, and others (Diamond and Pintel 2008). Students' awareness of companies that are ecologically friendly will influence their recruitment of suppliers.

Real World Inputs

Two focus groups identified the importance of real world inputs regarding their learning. Guest speakers from companies engaged in sustainable production and practices were considered imperative. In addition, participants encouraged the development of projects that require them to search the marketplace and analyze actual products. Considering the nature of the textiles and apparel major, which attracts the enrollment of many creative students, students want to learn through specific examples that relate directly to their personal lives. Practical examples, in other words, illustrate behavioral changes that students could make themselves as well as provide a process by which to make those changes. As one participant noted, "We're all really artistic people and to see something and go somewhere is a lot more helpful than reading, like, specific examples that relate to our lives."

Current Events and Eco-Initiatives of the Industry

Curriculum can be strengthened by including contemporary examples of developments in industry and initiatives by designers. Sustainability gained momentum in the textiles and apparel industry of the 1990s with the rise of environmentalism and with growing consumer consciousness of new materials – organic cotton, natural colored cotton, and hemp, for example – that are now taking center stage in today's fashion industry (Dickson et al. 2006; Hethorn and Ulasewicz 2008). Students realize that the industry has many initiatives. However, they feel that their knowledge of these initiatives is minimal and that more information should be included in their curriculum.

Responsible Consumption

Awareness of both the positive and negative behaviors relating to sustainability were reflected in five of the focus groups. These behaviors involved sustainable thinking, sustainable practices for textiles and apparel, and buying behavior. Positive behaviors can be promoted by increasing one's knowledge and awareness of materials, practices, and activities that are harmful to the environment and society. A course focusing on developing a responsible consumer is one option.

Consumer education should be included from elementary school or even earlier, as those are the formative years of childhood when behaviors can be molded and changed easily. As textiles and apparel majors, students gain respect and attention for their opinions about products related to their field. If students are well-informed about sustainable products and practices, they will be able to communicate that knowledge and awareness to others.

Extracurricular Engagement

Participation in institutional grass-roots initiatives and student organizations further promotes sustainability. Through the use of campus flyers and other marketing materials, sustainability can be promoted by providing relevant factual data related to energy consumption, water consumption, and waste generation. The enormity of these facts and figures communicates a strong message. As participant Martha stated, "drop shocking facts because that's shocking!"

Highly visible efforts on campus that promote sustainability are important for student engagement (Franz-Balsen and Heinrichs 2007).

Conclusions and implications

Participants realized that it is crucial to incorporate a detailed and in-depth examination of sustainability into every student's education, in addition to student-directed and collaborative learning experiences. These learning experiences could be accomplished through sustainability-focused group discussions, small projects, and research projects involving direct sharing among all participants.

The themes most frequently mentioned were sustainability course (program model) and learning from each other (instructional strategies), followed by processing and production (discipline-based content) and class projects. Students preferred sustainability-related content in every course for two reasons. First, they said that sustainability is applicable to everything, so that including it in every course ensures they are learning about different perspectives in their courses. Second, discussing sustainability in every course ensures that their learning about these issues leaves a lasting impact, eventually becoming an integral part of their personal and professional lives.

The favorite instructional strategies were identified as research projects and group discussions, particularly the rich multiplicity of perspectives that flow from group discussions. The findings from student focus groups suggest that the concept of sustainability is best learned and understood through multiple lenses and dialogues. In addition, using media as a communications channel was preferred.

Participants showed a keen interest in learning why certain materials and processes are categorized as beneficial or harmful to the environment and about how the personal choices they make can improve sustainability. It was encouraging to realize that students become excited when taking an in-depth approach to learning about sustainability.

Educators have a responsibility to energize students with a passion for social and environmental responsibility and to give students the tools to implement these practices in their lives, both as individuals and as professionals. Educational institutions have a compelling responsibility to increase the awareness and knowledge of their students, thereby preparing future professionals for becoming change agents (Calder and Clugston 2003; Dickson et al. 2006). Sustainable textiles and apparel production processes require professionals who are well prepared to work proactively in making sustainable personal and professional decisions (Dickson et al. 2009).

These results suggest the need for course and curricula refinements and changes. Some of the implications of these results are as follows: (1) Reconsider the course content to accommodate extra detail and depth when examining sustainable materials and processes, (2) design new projects with a focus on student-centered and collaborative learning (today's students, given their reputation as the knowledge generation, want ownership of their learning), and (3) develop an entry-level and/or capstone course on sustainability, in addition to incorporating sustainability in every course.

Here is a sampling of ideas that exemplify these implications at the course level:

- Teaching sustainability as a property of textile materials, using tools like Nike's environmental apparel-design tool that allow materials to be selected based on their cumulative sustainability scores (this tool is available at http://www.nikebiz.com/De fault.aspx);
- Group projects and discussion that let students qualitatively and quantitatively assess the sustainability of materials, processes, and brands;

- Collaborative design projects for sustainable design based on materials, adaptable design, longevity through quality, recycling and up-cycling, minimized waste, minimized care, and other sustainability-enhancing features;
- Collaborative research involving students, faculty, and/or industry. Pasricha and Kadolph (2009) reported a brief case study in which sustainability was incorporated as a strategy into a product-development team project.

Some universities have presented case studies on how they have incorporated sustainability in both their teaching and research within the field of textiles and apparel (Langdren and Pasricha 2011; Pontikis et al. 2011).

Suggestions at the program level include an entry-level course that defines sustainability, clarifies associated terms and concepts, and introduces the interdependence of the three tenets of sustainability: people, planet, and profit. The course would provide a foundation for teaching discipline-specific content in each textiles and apparel class. A capstone course that encourages students to apply their knowledge in a term project would facilitate a comprehensive understanding and application of sustainability concepts, engendering a world-view shift in favor of sustainability thinking among graduating students. Adoption of these ideas would be influenced by the specific context of individual programs and institutions. As a result, institutions will need to identify the sustainability goals of their programs, and educators will be required to customize a path to accomplishing these goals.

References

AASHE, n.d. *About AASHE*. Available from: http://www.aashe.org/about (accessed 30 June 2010).

Anderson, K. (2007, April 4), Doing good and doing well: The triple bottom line. *[TC²] Bi-Weekly Technology Communicator*. Available from http://www.tc2.com/newslet ter/2007/040407.html#two (accessed 10 April 2010).

– (2008, August 20), Sustainable apparel products: Defined and flourishing. *[TC²] Bi-Weekly Technology Communicator*. Available from: http://www.tc2.com/newsletter/2008/082008.html#thr (accessed 10 April 2010).

– (2009, February 4), Sustainable consumption: A consumer's perspective. *[TC²] Bi-Weekly Technology Communicator*. Available from: http://www.tc2.com/newsletter/2009/020409.html#alo (accessed 10 April 2010).

Armstrong, C.M. and Meyer-Brosdahl, D.C. (2009), *Surveying knowledge, attitudes, and interest about sustainability among apparel & textiles undergraduates*. Poster presentation and proceedings, International Textile & Apparel Association, annual conference. Bellevue, WA.

Brundiers, K., Wiek, A. and Redman, C.L. (2010), Real-world learning opportunities in sustainability: from classroom into the real world. *International Journal of Sustainability in Higher Education* 11 (4), pp. 308-324.

Calder, W. and Clugston, R.M. (2003), International efforts to promote higher education for sustainable development. *Planning for Higher Education* 30 (3), pp. 30-44. Available from: http://vnweb.hwwilsonweb.com.proxy.lib.iastate.edu:2048/hww/results/external_link_maincontent frame.jhtml?_DARGS=/hww/results/results_common.jhtml.42 (accessed 15 August 2008).

Carew, A.L. and Mitchell, C.A. (2002), Characterizing undergraduate engineering students' understanding of sustainability. *European Journal of Engineering Education* 27 (4), pp. 349-361.

Center for Digital Education (2007), *Teaching the Millennials*. Available from: http://www.center digitaled.com/publications.php?pub_id=87 (accessed 15 August 2008).

Christensen, L.J., Peirce, E., Hartman, L.P., Hoffman, W.M. and Carrier, J. (2007), Ethics, CSR and sustainability education in the Financial Times Top 50 Global Business Schools: baseline data and future research directions. *Journal of Business Ethics* 73 (4), pp. 347-368.

de Winstanley, P. and Bjork, R.A. (2002), Successful Lecturing: Presenting Information in Ways That Engage Effective Processing. *New Directions For Teaching & Learning* 89, p. 19.

Diamond, J. and Pintel, G. (2008), *Retail Buying*. Prentice-Hall, Englewood Cliffs, N.J.

Dickson, M., Rudd, N.A. and Lennon, S.J. (2006), Focused social responsibility: Part 2. *Clothing and Textile Research Journal* 24 (4), pp. 279-281.

Dickson, M.A., Loker, S. and Eckman, M. (2009), *Social Responsibility in the Global Apparel Business*. Fairchild, New York, NY.

Drucker, P. (1993), *Post-capitalist Society*. Harpers Business, New York, NY.

Franz-Balsen, A. and Heinrichs, H. (2007), Managing sustainability communication on campus: Experiences from Lüneburg. *International Journal of Sustainability in Higher Education* 8 (4), pp. 431-445 (doi: 10.1108/14676370710823591).

Global Citizenship Competency-Across-the-Curriculum (n.d.). Available from: www.iup.edu/Work Area/linkit.aspx?LinkIdentifier=idandItemID=83355 (accessed 7 June 2011).

Harper-Marinick, B. (2001, Spring), *Engaging students in problem-based learning*. Forum. Available from: http://www.mcli.dist.Maricopa.edu/forum/spr01/tl1.html (accessed 2 June 2011).

Hashimoto, T., Usui, N., Taira, M. and Kojima, S. (2011), Neural enhancement and attenuation induced by repetitive recall. *Neurobiology Of Learning & Memory* 96 (2), pp. 143-149 (doi: 10.1016/ j.nlm.2011.03.008).

Hethorn, J. and Ulasewicz, C. (2008), *Sustainable Fashion: Why now?* Fairchild Books, Inc., New York, NY.

International Textiles and Apparel Association (2008), *ITAA four-year baccalaureate program meta-goals 2008: International Textiles and Apparel Association*. Available from: http://www.itaaonline. org/displaycommon.cfm?an=1andsubarticlenbr=4 (accessed 7 June 2011).

Kagawa, F. (2007), Dissonance in students perceptions of sustainable development and sustainability. *International Journal of Sustainability in Higher Education* 8 (3), pp. 317-338.

Kermath, B. (n.d.), *Sustainability*. Available from: http://www.uwsp.edu/cnr/GEM/ambassador/ What_is_sustainability.htm (accessed 30 July 2009).

Kimmons, J.V. and Spruiell, P.R. (2005), Using problem-based learning in a multidisciplinary setting. *Clothing and Textile Research Journal* 23 (4), pp. 385-395.

Krueger, R.A. (1994), *Focus Groups: A Practical Guide to Applied Research* (2nd ed.). Sage Publications, Inc., Thousand Oaks, CA.

Landgren, T. and Pasricha, A. (2011), Transforming the fashion and apparel curriculum to incorporate sustainability. *International Journal of Fashion Design, Technology and Education* 4 (3), pp. 187-196.

Lawrence, R.J. (2005), Human ecology and its application for sustainability research. In Leal Filho, W. (Ed.), *Handbook of Sustainability Research,* pp. 121-147. Peter Lang, New York.

Leal Filho, W. (2000), Dealing with misconceptions on the concept of sustainability. *International Journal of Sustainability in Higher Education* 1 (1), pp. 9-19.

LeHew, M.L.A. and Meyer, D.J.C. (2005), Preparing global citizens for leadership in the textile and apparel industry. *Clothing and Textiles Research Journal* 23 (4), pp. 290-297.

Lundholm, C. (2005), Learning about environmental issues: Postgraduate and undergraduate students interpretations of environmental contents in education. *International Journal of Sustainability in Higher Education* 6 (3), pp. 242-253.

Martens, P. (2006), Sustainability: science or fiction? *Sustainability: Science, Practice, and Policy* 2 (1). Available from: http://ejournal.nbii.org/archives/vol2iss1/communityessay.martens.html (accessed 10 December 2008).

McGregor, S. (2002, April), *Consumer citizenship: A pathway to sustainable development?* Keynote at International Conference on Developing Consumer Citizenship. Hamar, Norway.

Morgado, M.A. (1996), Coming to term with postmodern: Theories and concepts of contemporary culture and their implications for apparel scholars. *Clothing and Textiles Research Journal* 14 (1), pp. 41-53.

Morgan, D.L. (1988), *Focus Groups as Qualitative Research*. Sage Publications, California.

O'Neal, G. (2007), Continents, cultures, curriculum: Some thoughts on the future of the profession. *Clothing and Textiles Research Journal* 25 (4), pp. 375-379. Available from: http://ctr.sagepub.com.proxy.lib.iastate.edu:2048/cgi/reprint/25/4/375 (accessed 15 August 2008).

Orr, D.W. (1992), *Ecological Literacy: Education and the Transition to a Postmodern World.* State University of New York Press, Albany, NY.

– (2004), *Earth in Mind.* Island Press, Washington, DC.

– (2010), What is higher education for now? In *2010 State of the World,* pp. 75-82. W.W. Norton and Company, New York.

Pasricha, A. and Kadolph, S.J. (2009), Millennial generation and fashion education: A discussion on agents of change. *International Journal of Fashion Design, Technology and Education* 2 (2), pp. 119-126.

Pew Research Center (2009, December 11), *The millennials.* Available from: http://pewresearch.org/pubs/1437/millennials-profile (accessed 25 December 2011).

Pontikis, K., Martin, A., Cai, Y., Kim, J., Cao, W., Giordano, A. and Torabian-Riasati, S. (2011), Sustainability in Teaching, Research, and Community Practice: The FCS Department at California State University, Northridge. *Journal of Family and Consumer Sciences* 103 (2), pp. 40-46.

Rusinko, C.A. (2010), Integrating sustainability in higher education: A generic matrix. *International Journal of Sustainability in Higher Education* 11 (3), pp. 250-259.

Stewart, D.W., Shamdasani, P.N. and Rook, D.W. (2007), *Focus Groups: Theory and Practice.* Sage Publications, Inc., Thousand Oaks, CA.

Stubbs, W. and Schapper, J. (2011), Two approaches to curriculum development for educating for sustainability and CSR. *International Journal of Sustainability in Higher Education* 12 (3), pp. 259-268.

Sustainable Apparel Coalition (n.d.), *About Us.* Available from: http://www.apparelcoalition.Org/3html. (accessed 14 October 2011).

The Hartman Group (2008), *The Consumer Side of Sustainability.* Available from http://www.hartman-group.com/hartbeat/the-consumer-side-of-sustainability (accessed 18 March 2012).

Timpson, W.M., Dunbar, B., Kimmel, G., Bruyere, B., Newman, P. and Mizia, H. (2006), *147 Practical Tips for Teaching Sustainability.* Atwood Publishing, Madison, WI.

UNESCO (2003), *The United Nations decade for education for sustainable development.* United Nations Educational, Scientific, and Cultural Organization. Available from http://portal.unesco.org/education/en/ev.php-URL_ID=26295&URL_DO=DO_TOPIC&URL_SECTION=201.html (accessed 15 January 2012).

– (2001), *Is the future yours? Research project on youth and sustainable consumption.* Available from: http://www.unesco.org/education/youth_consumption (accessed 12 January 2011).

United Nations Department of Economic and Social Affairs: Division for Sustainable Development (n.d.), *Education and awareness.* Available from: http://www.un.org/esa/sustdev/sdissues/education/edu.htm (accessed 3 August 2008).

– (1993), *Earth Summit: Agenda 21 – The United Nations Programme of Action from Rio.* Available from: http://www.un.org/esa/dsd/agenda21/ (accessed 5 June 2011).

World Trade Organization (2011), *World Trade Report 2011.* Available from: http://www.wto.org/english/res_e/booksp_e/anrep_e/world_trade_report11_e.pdf (accessed 25 December 2011).

Chapter 12

Inclusion of Environmental and Social Aspects of Sustainability in Industrial Design Education

Mariano Ramirez

Abstract

The imperative to teach future generations of industrial designers about the ecological and social sustainability aspects of their practice needs no argument. The profession has generally been blamed for promoting conspicuous consumption and stylistic obsolescence, and designers are considered indirectly responsible for the masses of discarded and short-lived objects in landfill. This paper examines how industrial design education is making up for past errors in design practice. It looks at the undergraduate and postgraduate programs of industrial design universities in various countries around the world, searching for evidences of both ecologically and socially sustainable design in the program descriptions, teaching and learning modules, and galleries of student works appearing on university websites. This research will be useful for industrial design course leaders and academics who are interested in benchmarking the extent to which they cover sustainability in their educational programs, and help them gauge how they fare in educating their students to become more responsible practitioners in the future.

Introduction

The profession of industrial design is credited for the everyday products, packages, and appliances that help make our daily lives and tasks pleasurable; it is equally implicated in the mountains of landfill where most of this "stuff" goes at the end of its useful life (McDonough and Braungart 2002). While negative ecological impacts were hardly a concern in the early days of industrial design, the current state of the environment demands that contemporary designers be made more aware of the consequences of the objects they create and help find a creative solution.

The International Council of Societies of Industrial Design (ICSID) identifies society and environment as important considerations of the industrial design profession. In its definition of design, the council states that the task of design is to enhance global sustainability and environmental protection, give benefits and freedom to the entire human community, and support cultural diversity despite globalization [icsid.org]. Various items in the ICSID Code of Professional Ethics highlight the responsibilities of the industrial designer towards sustainable design. Article II Benefit the User, Article III Protect the Earth's Ecosystem, and Article IV Enrich Cultural Identity require designers to adopt principles of environmental stewardship, to contribute to social and individual well-being, and embody cultural traditions of national societies while incorporating international design principles and standards (ICSID 1987).

The Cumulus International Association of Universities and Colleges of Art, Design and Media acknowledged that "global development and an awareness of the growth of related ecological and social problems are posing new demands and offering new opportunities for design, design education and design research. Design is challenged to redefine itself and designers must assume new roles and commit themselves to developing solutions leading to a sustainable future" (Cumulus 2008). In the Talloires Declaration, an international sustainability and environmental literacy action plan, over 430 university leaders worldwide committed their institutions to "ensuring that all university graduates have the awareness and understanding to be ecologically responsible citizens" (ULSF 1990). The years 2005 to 2014 have been declared the United Nations Decade of Education for Sustainable Development, a challenge and an opportunity for educators at all levels worldwide to encourage changes in behaviour, values and lifestyles such that a more positive future and a transformed and just society could be achieved.

The 1992 Earth Summit in Rio de Janeiro called for a reorientation of education towards sustainable development, recognizing its critical role "for achieving environmental and ethical awareness, values and attitudes, skills and behaviour consistent with sustainable development and for effective public participation in decision-making" (UNCSD 1992). And yet it must be underlined that the pathetic state of our environment is "not the work of ignorant people" but "rather, largely the result of work by people with BAs, BSs, LLBs, MBAs, and PhDs" (Orr 1991).

There is some evidence that industrial design education institutions are starting to seriously consider their broader responsibilities to society and the planet. Questionnaire surveys among undergraduate design educators in Australia and in 221 universities worldwide have generally shown that sustainability issues are gradually being covered in industrial design curricula (Ramirez 2006; 2007). These studies found that the majority of the degree programs studied had either compulsory or elective subjects on sustainable design and that a significant proportion of respondents considered themselves as being adequately informed, interested and even passionate about teaching sustainable product design. Higher education surveys in architecture, interior design, engineering and multiple disciplines have in general shown a relatively low level of sustainability penetration (Fowles et al. 2003; Metropolis 2002; 2003; Nguyen and Pudlowski 1997; Sammalisto and Lindhqvist 2008).

The inclusion of sustainable development concepts in higher education can range from policy statements to integration into courses, curricula, and other activities (Sammalisto and Lindhqvist 2008). Sterling (2004) argues that sustainability integration in education can happen at three consecutive levels: "bolting on", or "education about sustainability"; "building in" or "education for sustainability"; and "transformation" or "education as sustainability".

Aims and Methods of Study

This research investigates whether industrial design institutions present themselves as being concerned about socially and environmentally sustainable design aspects, and if there is evidence in their online materials regarding coverage of these issues.

Firstly, a list of all industrial design degree programs at both undergraduate and postgraduate levels from around the world was compiled. Due to the extremely large number of programs and the difficulty of accurately understanding non-English language material, the study had to be limited to schools in countries with English-speaking majority populations, namely Australia, Canada, Ireland, New Zealand, South Africa, the United Kingdom and the United States.

The study analysed the textual and visual contents of the official websites of industrial design education institutions in the above-named countries. Commercial websites are a key vehicle for communicating an organization's profile (Deloitte 2009); they also provide the most far-reaching medium for promoting an organization's activities, service offerings and capabilities to the world. Website content analysis as a research approach has been used successfully in past studies to investigate the online communication of corporate social responsibility issues (Capriotti and Moreno 2007; Krippendorff 2004). Aside from websites, PDF versions of the university's catalogs, bulletins, prospectuses, degree show viewbooks, and handbooks of undergraduate and postgraduate courses were downloaded and analysed.

This research deliberately considers only information that is available online: it takes the viewpoint of a prospective student who would be searching for a degree program to study and would partly base decisions on the information on curricula, syllabi, course listings, program descriptions, mission-vision statements, academic staff profiles, and galleries of student work that are published on the official university websites. Obviously some institutions diligently post more information on their websites and regularly update while others do not.

The study is only interested in coursework degrees, wherein students undertake a formally structured program of study, with regular classes that should be attended and courses or subjects that must be successfully passed to obtain the degree. Contrastingly, a research degree is self-directed, typically without classes, focused deeply on one's specific interest, and has a substantial written thesis as an outcome. This paper argues that a research degree is flexible enough to be tailored to studying sustainable design if a student proposes it, while in a coursework degree learning sustainable design would only occur if this subject is stipulated in the curriculum as a core unit, as a module, as a major project, or as an elective.

Table 1: Coding system for sustainability content in programs, courses, projects.

	Sustainability statement in program overview	**Sustainability courses**	**Sustainable design projects**	
NNN	No	No	No	0
NNY	No	No	Yes	1
NYN	No	Yes	No	1
YNN	Yes	No	No	1
NYY	No	Yes	Yes	2
YNY	Yes	No	Yes	2
YYN	Yes	Yes	No	2
YYY	Yes	Yes	Yes	3

Courses which claim to cover aspects of sustainable design were loosely categorized into "studio-type" and "lecture-type". Studio classes, which are intrinsic to the art, architecture and design disciplines, are characterized by learning that emerges through investigative and creative action driven by research, exploration, experimentation, critique and reflection; this model emphasizes learning through project-based work, praxis, workshop, and first hand observation (Zehner et al. 2009). In contrast, traditional lecture classes are typically structured with teacher-led classroom activities; these include such courses as design theory, design history, environmental science, ecology, society and culture, and sustainable development.

A coding pattern was devised to analyse the data, whereby it was noted whether or not the program overview mentions aspects of sustainability; whether the curriculum offers modules for learning sustainability or sustainable design; and whether or not the gallery of student works shows evidence of sustainable design projects. Eight possible combinations were found, shown in Table 1. The last column summarizes the strength of the evidence of sustainability coverage; those which have no evidence of sustainability in their program overviews, course descriptions and student projects are rated 0, while those which have evidence in all three indicators are rated 3.

Results and Discussion

All universities in the predominantly English-speaking countries named earlier were included in the study. A total of 402 industrial design degree programs were found: 242 undergraduate and 160 postgraduate. 159 universities are represented in the study. Aside from industrial design and product design, other related offerings were considered, such as design and innovation, design and technology, design and ergonomics, design and society, design futures, design management, furniture design, integrated or transdisciplinary design, product design engineering, product innovation, service design, sustainable design, three dimensional design, and transportation design. Sustainable buildings, interiors, landscapes and spaces are excluded from the study. If a degree can be obtained both as a BA and a BSc, it is then counted as two different entries. Moreover, if an undergraduate qualification is available both as a three-year exit degree and as a four-year honours degree, then these are counted as two separate programs.

Degrees

It is most interesting to note that all the nine British universities mentioned in a previous study as offering undergraduate degrees in ecodesign, sustainable product design, sustainable futures, or design and environment (Ramirez 2007), no longer offer those specialized programs. Email correspondence with some academics in those universities revealed that the sustainable design degrees have been withdrawn due to sustainability aspects now being incorporated into their mainstream product design programs.

In the United Kingdom, the University of Brighton is the only university to offer BA/BSc degrees in Sustainable Product Design. In the United States, the closest undergraduate degree available in sustainable design is the BA/BSc Environmental Studies from the New School University Parsons School of Design. No bachelor's degrees in sustainable design were found in Australia, Canada, Ireland, New Zealand and South Africa.

Most sustainable design specialization is now available only as postgraduate degrees. In the United Kingdom, these can be obtained as MA/MSc Sustainable Product Design from Bournemouth University, the University of Brighton, and the University for the Creative Arts Farnham; MA Design and Environment from Goldsmiths University of London; MA Design Sustainable Futures from Plymouth University; and MA/MSc/ MDes Design and Innovation for Sustainability from University of Derby, Lancaster University, Loughborough University and Cranfield University.

In the United States, MA/MSc Design for Sustainability can be obtained from Philadelphia University and Savannah College of Art and Design. The University of South Australia offers a master's degree in Sustainable Design.

A few new degree offerings address social aspects of sustainability. Maryland Institute College of Art has MA Social Design; the School of Visual Arts in New York has MFA Design for Social Innovation; Glasgow School of Art has MDes Design and Citizenship; and Kingston University in London has MA Design for Development.

Degrees that encourage a radical rethinking of the design process and of the social and ecological impacts of design outcomes are emerging. Studies in Design Futures can be taken at the University of Salford (BA), Goldsmiths University of London (MA), and Griffith University Queensland College of Art (M). Ontario College of Art and Design has MDes Strategic Foresight and Innovation, while Brunel University has MA Design Strategy and Innovation. MDes Design Anthropology and MSc Design Ethnography are awarded at Swinburne University of Technology and the University of Dundee, respectively. New master's programs in the area of Service Design can be found at Ravensbourne, the University of Dundee and Plymouth University; Savannah College of Art and Design offers both BFA and MFA Service Design.

Courses, Subjects, Modules

The data charted in Fig. 1 and 2 show that only 9% of all the programs studied show strong evidence of sustainability coverage in their program overview descriptions, their descriptions of their courses, modules or units of instruction, and in their gallery of student works. 30% of programs show sustainability coverage in two of the above indicators; those which show evidence in just one indicator total 30% as well. Therefore, 69% of the total demonstrates at least one proof of sustainability. Interestingly those programs which do not show any indicators of sustainability coverage also constitute 31% of the total.

The Australian undergraduate degrees are shown to have the highest number and proportion showing sustainability coverage evidence in all three indicators (Fig. 2). While the New Zealand undergraduate degrees seem to come a close second, their actual number is small (n = 3), and there are more in fact in the United Kingdom and in the United States.

Similarly, the New Zealand postgraduate and Ireland and South Africa undergraduate degrees appear to have the highest proportions without any evidence of sustainability coverage, but in reality their numbers are minuscule (NZL n = 2, IRL n = 5, ZAF n = 2). The overwhelming numbers of sustainability-free programs are found in the United Kingdom postgraduate (n = 41) and undergraduate (n = 40).

Of those degrees that offer courses on sustainability or sustainable design, 70% of these were delivered via studio-type classes, and the remaining 30% were lecture-type classes. These findings were markedly different to those of a study five years before (Ramirez 2006), where most sustainability topics were taught via lecture-based courses on design theory, design methodology or generic electives in environmental science.

A higher percentage of undergraduate programs include modules on sustainability or sustainable design (59%) compared to postgraduate (48%).

Fig. 1: Evidence of sustainability inclusion in industrial design courses

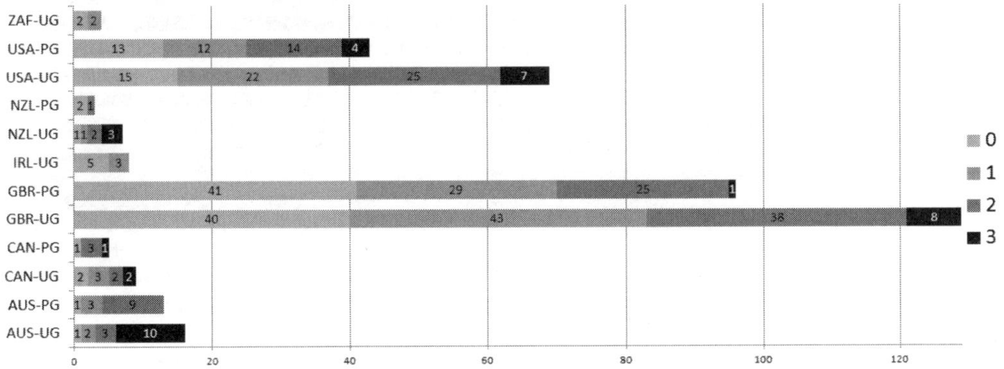

Fig. 2: Proportion of sustainability inclusion in industrial design courses

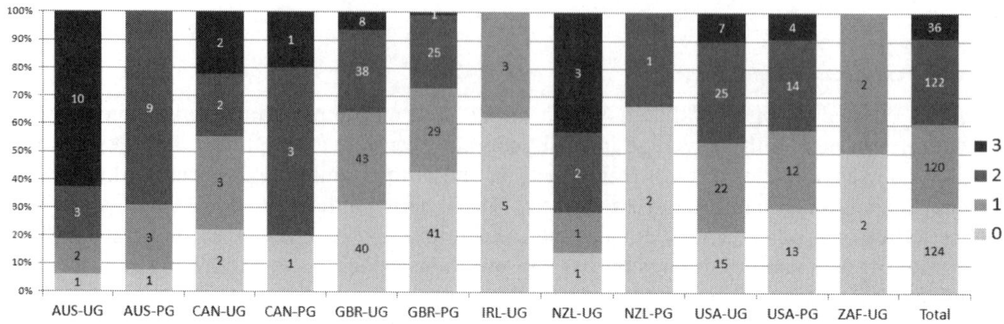

Projects

Examples of student works are more commonly shown on undergraduate websites and in catalogues of end-of-year degree exhibitions than in postgraduate ones. Fig. 3 shows that evidence of sustainable design in student projects can be found in only 16% of programs.

Australian undergraduate programs display the highest proportion with sustainable design student projects (69%), although the number is small (n = 11). The most number of student projects on sustainable design can be found in the undergraduate programs in the United Kingdom (n = 25) and United States (n = 13), although their proportion relative to programs without sustainability projects are small (< 20%).

Fig. 3: Sustainability in online galleries of student works

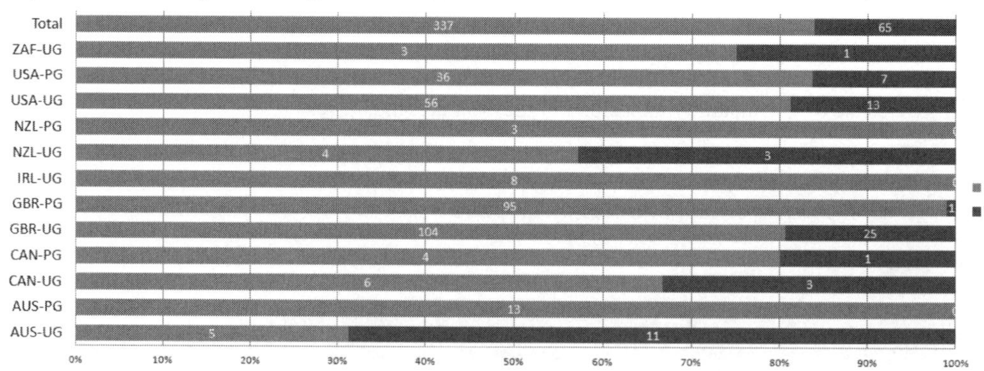

Among design universities in Oceania the most popular categories for responsible design innovations among student projects were "designing for sustainable behaviours" and designing for disabilities, as well as solutions for the elderly, for improving health and safety in workplaces and in public areas, for meaningful educational experiences for children, and for those living in marginal conditions in developing regions of the world.

Self-Description

It is not rare for organizations to "greenwash", or make unsubstantiated claims of adherence to environmental or sustainability ideals (Greer and Bruno 1996). Even universities are known to publicly declare high-level commitments to sustainability without actually demonstrating the fulfilment of these promises (Bekessy et al. 2007). Of the 189 universities in this study, 26 are signatories to the Talloires Declaration (ULSF 1990) which, like many environmental pronouncements, rewards parties before they deliver results and does not hold them accountable for failing to deliver.

Fig. 4: Evidence of sustainability in program overviews, course descriptions, and student projects

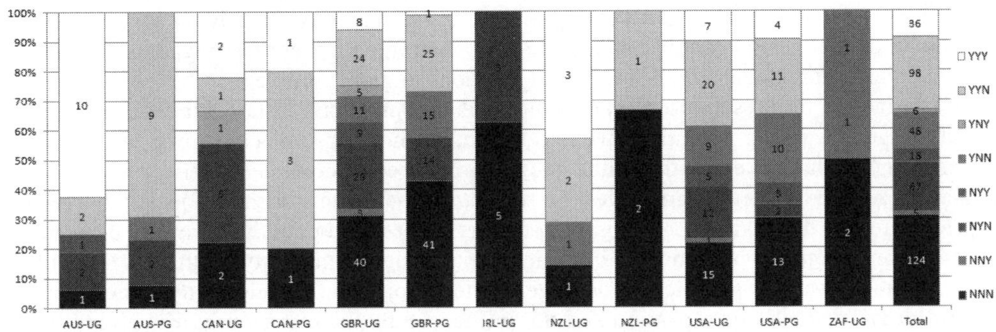

Of the 402 programs in the study, 188 (47%) mention aspects of environmental and social sustainability in the descriptive overview of their programs. There is recurrent reference to the terms "ethical", "societal issues", "ecological", "green", "socio-cultural context", "human-centered", "socially responsible", and "social change". 36 programs (9%) substantiate their commitment by actually having at least one course on sustainability and sustainable design as well as showing examples of student projects that demonstrate understanding of sustainable design principles. A quarter (24%) have one or more sustainability courses but have no student projects to show application of this learning. A tiny fraction (1%) have no course on sustainability and yet some of their student projects display attempts to address social and environmental issues. 48 (12%) express belief in sustainability ideals but do not provide evidence of these in their course listings or in student projects.

It is interesting to note that 90 programs do not mention sustainability in their self-descriptions yet include sustainability courses in their curricula (17%) or show student projects on sustainability (6%).

Australia has the highest proportion and number of undergraduate design programs that actually profess commitment to sustainability and show evidence through inclusion of sustainability in their course descriptions and student projects (Fig. 4).

Discussion and Conclusions

The analyses show that industrial design degrees in universities in Australia, Canada, Ireland, New Zealand, South Africa, the United Kingdom and the United States do cover ecological and social sustainability aspects in various modes and extents. Online descriptions of the programs, the individual courses or study modules, as well as galleries of student works prove this to a reasonable extent. A prospective student who might be concerned about the impacts of designer activities will be pleased to find that 278 of the 402 (69%) degree programs in this study show capacity and commitment to training responsible and ethical designers. However it is ideal that all the remaining design degrees also cross the bridge to include sustainability learning; the many barriers and misconceptions that prevent this easy integration should be addressed in order to pave the way for action (Leal Filho 2000).

It is promising to see that sustainability issues are being increasingly covered in studio courses rather than delegating these to academics in the sociology and environmental science departments. This could suggest that more design educators have become more sensitized to the urgency for design to leave a much smaller ecological footprint as well as recognizing the potentials of design for ushering in long-lasting transformational changes in human society. It also implies that more educators are challenging the current domains of industrial design from being simply product-centric to becoming a more influential agent in driving systemic social change. Industrial design can do much more than churning out more fancy gadgets and obsolescent objects, and indeed contribute to real solutions to global problems.

While it may be presupposed that public declarations about sustainability ideals are attempts at greenwashing potential students, the future-thinking projects shown in the online galleries and graduation exhibition catalogues do provide some credible evidence of application of learned sustainable design principles. To see many graduating students choosing to work on capstone projects that address pressing socio-cultural issues is proof of their growing desire to be more responsible future practitioners in the industrial design community. Although it could be argued that the student's choice to do a graduation project on sustainability issues may be driven by the student's own passion and not influenced by the curriculum, it is highly likely that the student's interest in this topic area might have been grown through their exposure to sustainability in their degree program and interaction with sustainability-focused teachers.

As mentors of the designers of tomorrow, it is our duty that they graduate into good "citizen designers" (Heller and Vienne 2003) with an empathic sense of responsibility and loaded with ethical values to serve all people in need of good design, not just the manufacturers who would pay big money to create more unneeded stuff. Like Papanek (1971) we should advocate that industrial designers go beyond "appearance design", styling, or "design cosmetics", and use their talents to avoid the worsening environmental problems and to solve the pressing needs of disadvantaged minorities in society: the disabled, the elderly, the communities in the developing world, and others often ignored by the design profession.

Most design curricula are flexible enough to allow covert inclusion of sustainability aspects in the projects and assessments undertaken within regular studio courses. There is no need to wait for a wholesale overhaul of the degree program or the creation of specialist degrees, courses or electives before we can make our students aware of their obligation to design with future environmental and social impacts in mind. As responsible design educators we can use our creativity in assimilating sustainability in our teaching and learning activities, even if the program overview does not express commitment to this ideal, or even if the course description does not mention it. As Orr (2007) says, the Hippocratic oath of designers should be "to cause no ugliness, human or ecological, somewhere else or at some later time". It is our duty to guide and mentor our designers-in-training in achieving this noble aim.

Acknowledgments

The author thanks Carl Michelangelo Ramirez and Matthew Constantine Ramirez for their assistance in the data collection from the various university websites.

References

Bekessy, S.A., Samson, K. and Clarkson, R. (2007), The failure of non-binding declarations to achieve university sustainability: a need for accountability. *International Journal of Sustainability in Higher Education* 8 (3), pp. 301-316.

Capriotti, P. and Moreno, Á. (2007), Corporate citizenship and public relations: the importance and interactivity of social responsibility issues on corporate websites. *Public Relations Review* 33 (1), pp. 84-91.

Cumulus (2008), *Kyoto Design Declaration*, Cumulus International Association of Universities and Colleges of Art, Design and Media, Kyoto.

Deloitte (2009), *Sustainable Enterprise Value Map: bridging the gap between shareholder and stakeholder value.* Deloitte Development LLC, New York.

Fowles, B., Corcoran, M., Erdel-Jan, L., Iball, H., Roaf, S. and Stevenson, F. (2003), *Report of the Sustainability Special Interest Group: Architectural Education.* Centre for Education in the Built Environment, London.

Greer, J. and Bruno, K. (1996), *Greenwash: the reality behind corporate environmentalism.* Third World Network, Penang.

Heller, S. and Vienne, V. (Eds.) (2003), *Citizen designer: perspectives on design responsibility.* Allworth Press, New York.

ICSID (1987), *ICSID Model Code of Professional Conduct for Designers.* International Council of Societies of Industrial Design, Essen.

Krippendorff, K. (2004), *Content analysis: an introduction to its methodology.* SAGE Publications, Thousand Oaks.

Leal Filho, W. (2000), Dealing with misconceptions on the concept of sustainability. *International Journal of Sustainability in Higher Education* 1 (1), pp. 9-19.

McDonough, W. and Braungart, M. (2002), *Cradle to cradle: remaking the way we make things.* North Point Press, New York.

Metropolis (2002), Teaching green: making sustainability integral to every designer's education and business. *Metropolis* 22 (3).

– (2003), School survey 2003: taking the pulse of sustainable design education in North America. *Metropolis* 23 (1).

Nguyen, D.Q. and Pudlowski, Z.J. (1997), A comparative study of the perspectives of academics, students and industry on environmental education in engineering courses. *Global Journal of Engineering Education* 1 (3).

Orr, D.W. (1991), What is education for? Six myths about the foundations of modern education, and six new principles to replace them. *The Learning Revolution* 27 (Winter) (52).

– (2007), *The designers' challenge: commencement address.* University of Pennsylvania School of Design, Philadelphia.

Papanek, V.J. (1971), *Design for the real world: human ecology and social change.* Thames and Hudson, London.

Ramirez, M.J.R. (2006), Sustainability in the education of industrial designers: the case for Australia. *International Journal of Sustainability in Higher Education* 7 (2), pp. 189-202.

– (2007), Sustainability integration in industrial design education: a worldwide survey. Paper presented at the ConnectED International Conference on Design Education, Sydney.

Sammalisto, K. and Lindhqvist, T. (2008), Integration of sustainability in higher education: a study with international perspectives. *Innovative Higher Education* 32 (4), pp. 221-233.

Sterling, S. (2004), Higher education, sustainability, and the role of systematic learning. In Corcoran, P.B. and Wals, A.E.J. (Eds.), *Higher education and the challenge of sustainability: problems, promise, and practice.* Kluwer Academic, Dordrecht.

ULSF (1990), *The Talloires Declaration.* University Leaders for a Sustainable Future, Talloires.

UNCSD (1992), Chapter 36: Promoting education, public awareness and training. *Earth Summit: Agenda 21.* United Nations Commission on Sustainable Development, Rio de Janeiro.

Zehner, R., Forsyth, G., Musgrave, E., Neale, D., De La Harpe, B., Peterson, F., Frankham, N., Wilson, S. and Watson, K. (2009), *Curriculum development in studio teaching: Volume 1: Studio Teaching Project final report.* Australian Learning and Teaching Council, Sydney.

Chapter 13

Using Sustainability in the Integrative Training of Engineering Students in Canada

Tatiana Teslenko

Abstract

Incorporation of education for sustainable development (ESD) into the first-year curriculum is a key opportunity for engineering educators. In 2009 the topics of sustainable consumption and design, environmental issues and global engineering were introduced into a first-year engineering communication course in the Faculty of Applied Science at the University of British Columbia in Canada. In 2010 this successful initiative was expanded to include writing research reports on community service learning and sustainable engineering projects in developing countries. This paper presents a pedagogical approach that strives to promote the transition to sustainability through the focus on social learning and global citizenship. Learning objectives include increasing students' awareness of sustainability and providing them with the opportunity to comment on sustainable engineering design. The paper shows how students explore human and social dimensions of sustainability and obtain problem-solving experience by considering the perspectives of multiple stakeholders in their design proposals. An original interdisciplinary approach was employed, with a focus on promoting education for sustainability and global citizenship. The multi-dimensional methodology involves action research based on data drawn from observation, surveys, interviews, and discourse analysis of students' assignments. Learning to engage in a meaningful dialogue on sustainability and global citizenship had a positive impact on students' creativity and their understanding of their future professional identity. Students moved from community service projects to the exploration of global implications of engineering practice. The concept is based on the original idea to use a first-year communication course for integrating ESD with fundamental competencies for engineers. The first-year course is uniquely positioned for promoting ESD because students learn to value sustainability at the very beginning of their university studies. New case studies that are set out in the global context were developed for the course. The described pedagogical approach is applicable in global engineering education.

Introduction

Recent reviews (Allen et al. 2008; Wiggins et al. 2011) suggest that the vast reform that happened in engineering education in the two decades after the Brundtland Report (World Commission on Environment and Development, 1987) has resulted in the widespread inclusion of sustainability topics in engineering programs across North America and the globe. In response to the Report's call to embrace sustainability in higher education, many universities have infused sustainability into their programs (Bielefeldt 2011). In Canada all engineering schools require that students should be familiar with the concept of sustainability. The Canadian Engineering Accreditation Board (CEAB) lists sustainability as one of the key graduate attributes. In particular, Attribute 3.1.9 "Impact of engineering on society and the environment" includes the concepts of sustainable design and development and environmental stewardship (CEAB 2010). Engineering schools face a constant need to improve the teaching of graduate attributes and fundamental competencies (Shulman et al. 2005). Their importance has been extensively discussed in the industry-academe dialogue (Barry et al. 2008; MacLennan 2008; National Academy of Sciences 2002). It has resulted in the adjustment of accreditation criteria for engineering programs by the national engineering accreditation boards (ABET 2009; CEAB 2010). Yet it has been observed that universities do not provide students with systematic training in these competencies (Davis 2010; Reave 2004; MacLennan 2008), and there are warnings that "we're teaching the wrong stuff" (Felder 2008).

This paper presents a pedagogical approach that enables instructors to incorporate a substantial emphasis on global citizenship and sustainability in a first-year course on technical communication. It describes the evolution of students' academic needs, new accreditation criteria, and the institutional move towards outcomes-based assessment. The multi-dimensional methodology involves action research based on data drawn from observation, surveys, interviews, and discourse analysis of students' assignments. Choosing the first-year course as a case-study for education for sustainable development (ESD) is crucial because students learn to value sustainability at the very beginning of their university studies. This approach provides for the introduction of important values, concepts and assignments that are "grounded in processes which can lead to sustainability" (Thomas 2009). This paper also highlights the challenges of teaching in an interdisciplinary space located at the crossroads of education for sustainable development (ESD), composition studies, activity theory and community service learning.

The described pedagogical approach strives to promote the transition to sustainability through social learning (Kates and Clark 2000), i.e. by including community service learning projects that allow students to explore human and social dimensions of sustainability through the perspectives of various stakeholders. Starting with local projects, subsequent assignments involve projects based in the global context. In this way, students move from community service projects to the exploration of global implications of engineering practice. This process helps them to develop an understanding of global citizenship.

Education for sustainable development at UBC

The University of British Columbia (UBC) is the leading university of the Province of British Columbia, and has been consistently ranked among the top five universities in Canada and the top 40 universities in the world. This four-year university serves about 54 thousand students in a broad range of programs, from undergraduate liberal arts and sciences to professional programs, including engineering, law, medical and doctoral programs. Publicly funded, it caters to students from a wide variety of socio-economic levels and shows considerable ethnic diversity among students and professors: "UBC students come from across Canada and around the world, and therefore, are exposed to a wide variety of cultures, languages and customs while studying and preparing to take on their varied roles as global citizens" (Toope, quoted in Harlap 2008). Located in the multi-national metropolis of Vancouver, Canada, UBC sees it as its duty to promote global citizenship in the education of students. UBC's goal is "to prepare students, educators and future leaders for the globally interconnected societies they must serve" (Toope, quoted in Harlap 2008: 1).

The Faculty of Applied Science at UBC has about 6000 students enrolled in 14 engineering programs. The Faculty is "going global", i.e. developing programs and courses that contribute to the global connectivity of engineering education. The following important aspects of its core purpose are listed on the Faculty website: responsibility, professionalism, promoting global citizenship, and enactment of positive change[1] (Engineering at UBC 2011). A lot of attention is devoted to teaching engineering students about sustainability, sustainable development, and sustainable design, especially in the Departments of Civil Engineering, Geological Engineering, Environmental Engineering, Chemical and Biological Engineering. Sustainable engineering is the topic of a growing number of new courses; sustainability-focused modules are also added to existing courses. These initiatives usually take place in senior-level engineering courses (Thomas 2009). Obviously, it is important to embed ESD in senior courses in order to provide students with ample exposure to sustainability topics before they start their professional career. However, there are much fewer attempts to introduce the theme of sustainability in first-year courses (Bielefeldt 2011; Kemppainen et al. 2007), particularly in engineering communication courses.

Communication instruction at the Faculty of Applied Science started with a stand-alone mandatory second-year course, APSC 201: Technical Communication. The topics of this course did not include sustainability. It was first offered in 1993 and employed the same pedagogy that was used in the Department of English at the Faculty of Arts. Initially, the pre-requisite for this course was one of the first-year English courses offered by the Department of English. However, faced with the challenge of introducing fundamental competencies and sustainability into the course, our instructors understood that this pedagogical approach did not fare well in engineering. It led to tensions and conflicts due to the different types of literacy and competencies that engineering students needed. The second-year course gradually evolved during the following 17 years (1993-2010) to accommodate the changing competencies required of our students as per CEAB accreditation requirements. At first the learning objectives were limited to CEAB graduate attributes outlined in Criteria 3.1.7 and 3.1.6: communication skills, as well as individual and teamwork skills. In 2010-11 they were expanded to include three additional attributes: 3.1.8 professionalism, 3.1.9 ethics and equity, and 3.1.12 life-long learning[2] (CEAB 2010). These new objectives resulted in the inclusion of the discourse on sustainability and fundamental competencies for engineers, as well as an in-depth discussion of the above-mentioned integrative skills (a.k.a. "soft" skills).

Pedagogical innovation

Over the last decade the Faculty has encouraged its instructors to speed up the process of integrating fundamental competencies with the students' design experience. In 2001 the Faculty formed a Technical Communication Centre with a mandate to oversee the teaching of integrative skills to engineering students. In 2006 the Technical Communication Centre was transformed into the Centre for Professional Skills Development (CPSD), and several initiatives were launched. In addition to APSC 201, CPSD instructors developed partially-integrated modules in senior capstone courses, training sessions for graduate students, and one fully-integrated second-year program for mechanical engineering students. At the request of the Faculty our communication instructors also developed a first-year course with a sustainability theme, APSC 176: Engineering Communication. Its objectives are to provide students with fundamental competencies and research, critical thinking and advanced communication skills necessary for success in the study and practice of global engineering.

---

1 "Our core purpose is to create and apply knowledge and to educate globally responsible professionals empowered to effect enduring, positive change for society" from http://www.engineering. ubc.ca.

2 See *CEAB 2010 Accreditation Criteria and Procedures Report*, "Accreditation criteria", Part 3, pp. 12-13 (CEAB 2010).

CPSD communication instructors are constantly looking for ways to improve our students' learning experience and to prepare them for effective communication in the global workplace. They believe that a first-year communication course is uniquely positioned for integrating education on sustainability (ESD) with fundamental competencies, particularly the discourse on sustainability and communication. APSC 176: Engineering Communication was developed as a response to the new ethical and democratic imperative to prepare our students for "going global": "Our students must be able to navigate throughout cultural diversity not merely so that business can thrive, but so that our interacting societies can be enriched and our ideas made available across the globe. Our educated students move across many boundaries, and our faculty are enabled to live and work internationally. The ideas we create flow globally. Our service commitments are not limited by state borders" (Toope, quoted in Harlap 2008: 1).

In the expanding global workplace our students will need knowledge of contemporary issues, including sustainability and cultural competence, as well as advanced communication skills (Birch et al. 2009). In addition to analytical reading and interpretive skills, persuasion, ability to adapt messages for complex and varied audiences (MacLennan 2008), the new requirements for getting hired are collaboration within cross-functional teams, management of virtual projects, and intercultural communication. Effective communication is part and parcel of these skills. Based on the context of engineering practice and in order to promote the global nature of engineering work, several international case-studies were developed for APSC 176.

Courses that are intended to help students develop professional attitudes and skills are notoriously difficult to promote and develop. After years of trying to negotiate the tensions and conflicts between different pedagogical needs and approaches inside the classroom, in 2004-2005 it became obvious that these tensions should be negotiated outside the classroom. In 2005-2006 the administration of the Faculty of APSC engaged in extensive discussions and negotiations with the administration of the Faculty of Arts (and its Department of English) in order to develop a new first-year course centered on topics and readings that reflect fundamental competencies for engineers. The purpose was to break down the institutional barrier and create a pre-requisite first-year course in engineering communication.

For transfer credit and articulation purposes, it was important to get multiple stakeholders on board; therefore, course content, learning objectives and outcomes were developed in consultation with the UBC Department of English, the Communication Articulation Committee of British Columbia, and Engineering Communication Program at Simon Fraser University. It was recommended to make this course comparable to English courses offered by our colleagues at the UBC Department of English. However, it could not be a simple adaptation of a first-year English course because our learning objectives and outcomes should incorporate fundamental competencies for engineers. This course would create an opportunity to address our own standards, disciplinary values and pedagogical practices. The goal was to promote a more inclusive use of communication that enables engineering students to draw upon their interest in engineering and their own communicative resources.

The plan was to vertically integrate the two courses, APSC 176 and APSC 201, by moving introductory modules and assignments from the second-year course to the first-year course, thus laying a solid foundation for continuous instruction in communication throughout the four years of studies for the Bachelor of Applied Science degree. APSC 176: Engineering Communication was approved by the UBC Senate in 2005 and was subsequently developed by CPSD instructors at UBC Okanagan (2005) and UBC Vancouver (2008)[3]. It is now offered

3 It was first offered at our satellite campus, UBC Okanagan, in 2005 as part of their Engineering One program. Students' success rate was very high and their feedback on the course was overwhelmingly positive. The course was then offered at UBC Vancouver in 2008.

as a three-credit elective first-year course, for which completion of an English language competency pre-test, LPI (Language Proficiency Index), is a pre-requisite. To be admitted, students have to achieve a high level on the LPI (Level 5).

The discourse on global citizenship and sustainability

The novelty of the course lies in the emphasis on fundamental competencies for engineers, including sustainability and global citizenship. Learning objectives connect the course topics to important themes in engineering education. Initially, the theme of the course was "Can technology solve society's problems?" Assignments were related to "ethical principles that underpin many engineering decisions, whether in the realm of communication, design, or professionalism" (Labun 2007: 3). Students were encouraged to explore and report on recent research initiatives in their intended engineering departments. For their in-class written persuasive proposals, students were assigned to a team and asked to choose an *Engineers Without Borders* project, conduct research and then present the problem and their solution. Students learned to use summary, definition and comparison to structure an argument; to write a research or design proposal; to prepare a collaborative research report and an individual portfolio; and to make a collaborative oral presentation. The course included in-class, take-home and online components, and extensive use of educational technology (multimedia presentations and Web Vista applications).

Sustainability has multiple definitions: it is a concept, an ethical belief, a condition, and a competency, to name but a few. It unites scholarship and practice, local and global perspectives, and a range of disciplines, from natural and social sciences to engineering. An important definition that emphasizes its human and social aspects is provided by the US Environmental Engineering Body of Knowledge. Sustainability is "a condition in which the use of natural resources and cycles in human and industrial systems does not lead to a diminished quality of life due either to losses in future economic opportunities or to adverse impacts on social conditions, human health and the environment" (AAEE 2009).

Human and social dimensions are also emphasized in the definition of sustainable engineering from the American Society of Civil Engineers. Sustainable engineering is "the challenge of meeting human needs for natural resources, industrial products, energy, food, transportation, shelter, and effective waste management while conserving and protecting environmental quality and the natural resource base necessary for future development" (ASCE 2008). An important social dimension is communication, and many documents stress its value for successful practice of sustainable engineering. Engineers must "successfully collaborate with experts from other disciplines and communicate with society and stakeholders in the pursuit of sustainability" (AAEE 2009).

In the interconnected world, global engineering practice should not be limited by regionalism or national boundaries. One of the goals of the course was to implement education in global citizenship into our pedagogical approach. Sustainability is at the heart of the working definition of global citizenship adopted in the course: "A global citizen is someone who feels a duty to respect and protect the Earth, the global community of fellow human beings and all other living creatures. We envision global citizens as individuals who have developed an understanding of the interconnected world and who deeply appreciate and value ecological sustainability and social justice. Global citizens are individuals who are willing and enabled to take action to make the world a fairer place for ourselves and other living creatures" (Harlap 2008: 7).

In 2009, the course theme at UBC Vancouver was changed to incorporate a greater emphasis on sustainability. Course readings for APSC 176 were assigned from a recent textbook "Fundamental Competencies for Engineers" (published by Oxford University Press in 2006), which was written by our engineering and communication professors. To give students a sense of the value and appeal of engineering work, this book uses a conceptual model that is inspired by actual engineering practice. The concept of an international engineering consulting firm, called the Brunel Group, was used in APSC 176. This fictitious company is working on six projects described in the course textbook (Dunwoody et al. 2006). Using this concept helped educators draw the students' attention to global and local perspectives on engineering design. Many engineering companies are active internationally and have formal mentoring programs for new employees. Students were asked to adopt the rhetorical persona of an engineer-in-training who is a member of the mentoring program at the Brunel Group. The following fundamental competencies were selected: sustainable development, ethics, and integrative skills such as teamwork and communication, designing and evaluating projects. These particular competencies are applicable across all disciplines of engineering.

Students began by learning the definition of sustainability and then examined its importance in engineering practice through the following activities:

- Conducting research and writing a formal report about the sustainability of 20th century engineering achievements;
- Investigating the ethical issues involved in sustainable technologies;
- Reflecting on the human and social dimensions of sustainable design;
- Exploring the concept of global citizenship and sustainable engineering practice.

Several themes that interact with sustainability were identified: sustainable consumption, respecting others, communicating effectively, and creating sustainable solutions. Course assignments were designed with respect to the graduate attributes required of graduating engineering students: an ability to formulate or design a system, process, or program to meet desired needs; an ability to function on multidisciplinary teams; an understanding of professional and ethical responsibility; an ability to communicate effectively; the broad education necessary to understand the impact of the solution in a global and societal context; and knowledge of contemporary issues (see CEAB 2010, Criterion 3: Accreditation Criteria).

Students made collaborative oral presentations on two engineering topics throughout the semester: engineering achievements of the 20th century and ethical dimensions of engineering practice. For their collaborative formal report, students researched one of the Greatest Engineering Achievements of the 20th century, as determined by the National Academies of Sciences (Dunwoody et al. 2006: 9). Instead of reporting only on the history of an engineering achievement, students analyzed current and future implications of that achievement. They considered the social, environmental, and economic issues of the technology and commented on sustainability of that achievement for future generations.

Human and social dimensions of sustainability

Recent research (Downey et al. 2006; Mihelcic and Trotz 2010) indicates that the inclusion of human and social dimensions is critical for achieving a balance of societal issues in engineering education. Our students assumed the rhetorical persona of a trainee, were "paired up" with a "mentor" and "assigned" to a sustainable design project in Kyrgyzstan, Peru, Brazil or Canada. The purpose of this assignment was to point out that only technologies appropriate to local culture, skill level and environment of a country would be sustainable. In particular, students were asked to comment on a sustainable design of a biogas generator for a secondary school in rural Kyrgyzstan, the ore slurry pipeline for the new mine in Peru, the new "green" building at a Brazilian university, and the run-of-the-river hydroelectric power plant in the coastal mountains of British Columbia.

Students were asked to evaluate the designs and write a series of assignments. This activity aimed to encourage innovative ideas and solutions. It also allowed us to further explore the social aspects of sustainability, such as attention to the needs of local stakeholders, equity and diversity. It helped students to consider the human dimensions in engineering design, including the important question: "What does bringing local input have to do with sustainable design?" (Dunwoody et al. 2006: 20). Students learned that local stakeholders who are affected by the design must be involved in project evaluation. They were encouraged to conduct further research on the issue, its local and global implications, and determine whether or not the designs were sustainable.

It is important to note that our students are in their first year of the engineering program, and, therefore, they lack the course background necessary to create theoretically sound engineering solutions. For this reason, all calculations and descriptions were provided for them. Similar to results previously reported by C. Labun (2007), our students' design proposals were "entirely conceptual – students are not required to develop a prototype, but rather to work with a team to develop (and subsequently, explain and market) a concept in response to an RFP" (Labun 2007). Students were also encouraged to investigate the economic, environmental, societal and cultural aspects of the country as part of their background research. As a result, solutions they proposed considered these important aspects of sustainable design. Their design proposals explained in what way their designs were economically feasible, socially acceptable and environmentally friendly. Students also expressed an interest in investigating the differences of sustainable technologies in developing and developed countries.

This assignment had two important outcomes. It provided an opportunity for the educators for achieving a balance integrating environmental, human and social dimensions of sustainability. For students it proved to be an important step in furthering their understanding of their professional identity. Critical reflection on human and social dimensions of sustainability and "learning about oneself and others" is important in engineering education because it "informs engineers about the human significance of what they are learning" (Mihelcic and Trotz 2010). Student reflection papers were very creative. They not only responded to a particular case-study, but also demonstrated their understanding of human and social dimensions of sustainable engineering. For example, one first-year student decided to create a strategic plan for the Brunel Group, complete with the Vision and Mission Statements, Goals and Objectives. The following Values Statement was included in this strategic plan: "The Brunel Group relies on determined and innovative engineers who think outside the box while considering how their solutions might affect people and the environment. In particular, we value steadfast observance of the engineering traits of integrity and ethical conduct; a steady profit in order to continue operating; exceptional treatment of our customers, as they are integral to our existence; sustainable development to preserve the world for future generations".

The topics of global citizenship and global development intersect with community service learning. It has been reported in literature that "service learning experience influences students' attitudes towards sustainable engineering" (McCormick et al. 2010). In 2011 the focus of APSC 176 was enhanced by adding topics related to community service learning (CSL). For their research report, students had a choice of investigating either the CSL projects at UBC or the reports of *Engineers Without Borders* (EWB Canada). Working collaboratively, students researched CSL and EWB projects and learned about the work of other charitable organizations, such as *Engineers for a Sustainable World* (Dunwoody et al. 2006: 21). They interviewed CSL administrators, participants and community partners, as well as EWB overseas volunteers, attended presentations and workshops, and developed their own research topic. Their reflection papers indicate that students start to understand the essence of their professional identity.

Most students report that they learnt a lot about themselves as future engineers, about sustainability and community, and about communication in the global context.

Our findings are consistent with those recently reported by engineering educators engaged in ESD (Bielefeldt 2011; Kemppainen et al. 2007; McCormick et al. 2010; Thomas 2009: Wiggins et al. 2011). At the end of the term, 90% of the students were able to write an expanded definition of sustainability and explain the need to protect the environment for future generations. Nearly 95% of students agreed that sustainability was important and were interested in learning more about it. Some students expressed an interest in taking a dedicated course on sustainability and engaging in an extracurricular activity related to sustainability, such as projects of Engineers Without Borders or Engineers for Sustainability. After completing their research projects and evaluating their classmates' presentations, most students were able to identify and explain the sustainability attributes of an engineering project. They demonstrated an increased awareness of sustainability in subsequent course assignments and discussed aspects of sustainability in their research reports and reflection papers.

Conclusion

The theme of sustainability was incorporated into an existing three-credit first-year course, APSC 176: Engineering Communication, in the Faculty of Applied Science at the University of British Columbia, Canada. Our goal was to increase students' awareness of various dimensions of sustainability and give them the opportunity to comment on sustainability issues in engineering design. Dimensions of sustainability were investigated through the use of design proposals, student presentations about engineering achievements, case-studies set in the global context, and research reports about local and global community service learning projects.

Implementing the emphasis on sustainability in a first-year communication course proved to be an important way to embed ESD in the first-year engineering curriculum. As demonstrated by students' comments in their reflection papers, the focus on sustainability in APSC 176 raised students' awareness about the importance of sustainability. This outcome leads to the conclusion that students were not only better prepared for taking upper-level ESD courses, but they were likely to have been enabled to reach greater levels of knowledge in senior courses. Whether this is indeed the case is a question for further research. Another important outcome resulted from the successful use of case-studies set in the global context. The course textbook, "Fundamental Competencies for Engineers" (Dunwoody et al. 2006), was instrumental in enriching our pedagogical approach, and it proved to be applicable for global engineering education. Our first-year students were able to identify sustainability issues in case-studies and to obtain some problem-solving experience through their design proposals. They developed the ability to assess and comment on the sustainability of a technological achievement for future generations. They learned to consider the human, social, environmental, and economic aspects of a technological innovation in a developing country. Engaging in a meaningful dialogue with diverse community partners and volunteers helped them start thinking about their future professional identity and the global engineering work, which is a significant outcome for a first-year course.

References

Accreditation Board for Engineering and Technology (ABET) (2009), *Criteria for accrediting engineering programs. Effective for evaluations during the 2010-2011 accreditation cycle*", ABET Engineering Accreditation Commission. Available at www.abet.org (accessed 12 November 2011).

Allen, D., Allenby, B., Bridges, M., Crittenden, J., Davidson, C., Hendrickson, C., Matthews, S., Murphy, C. and Pijawka, D. (2008), *Benchmarking sustainable engineering education: Final report*. EPA Grant X3-83235101-0.

American Academy of Environmental Engineers (AAEE) (2009), *Environmental engineering body of Knowledge*. AAEE, Annapolis, MD. Available at: http://www.cece. ucf.edu/bok/pdf/EnvE_Body_of_Knowledge_Final.pdf(accessed 11 January 2012).

American Society of Civil Engineers (ASCE) (2008), *Civil engineering body of knowledge for the 21st century: Preparing the civil engineer for the future* (2nd Ed). ASCE, Reston, VA.

Barry, B., Brophy, S., Oakes, W., Banks, M. and Sharvelle, S. (2008), Developing professional competencies through challenge to project experiences. *International Journal of Engineering Education* 24 (6), pp. 1148-1162.

Bielefeldt, A.R. (2011), Incorporating a sustainability module into first-year courses for civil and environmental engineering students. *Journal of Professional Issues in Engineering Education and Practice.* ASCE April 2011, pp. 78-85.

Birch, J., Jaramillo, P., Wosczyna-Birch, K., Adrezin, R. and Richards, B. (2009), Integrating professional skills in the 21st century engineering and technical curriculum. *IMECE 2008: Engineering Education and Professional Development* 9, pp. 293-298.

Canadian Engineering Accreditation Board (CEAB) (2010), *Accreditation criteria and procedures report. Available* at: http://www.ccpe.ca/e/files/Accreditation_Criteria_Procedures_2010.pdf (accessed 10 January 2012).

Davis, M. (2010), Assessing technical communication within engineering contexts. *IEEE Transactions on* Professional Communication 53 (1), pp. 33-45.

Downey, G., Lucena, J., Moskal, B., Bigley, T., Lehr, J. and Nichols-Belo, A. (2006), The globally competent engineer: working effectively with people who define problems differently. *Journal of Engineering Education*, pp. 107-122. Available at http://www. ewb-international.org/pdf/Global%20Competence.pdf(accessed 15 January 2012).

Dunwoody, A.B., Cramond, P.J., Nesbit, S.E., Paterson, C.S. and Teslenko, T. (2006), *Fundamental Competencies for Engineers.* Oxford University Press, Don Mills, ON.

Engineering at UBC (2011). Available at: http://www.engineering.ubc.ca/ (accessed 11 December 2011).

Felder, R.M. (2008), A whole new mind for a flat world. In MacLennan, J. (Ed.), *Readings for Technical Communication*, pp. 11-14. Oxford University Press, Don Mills, ON.

Harlap, Y. (Ed.) (2008), *Road to Global Citizenship: An Educator's Toolbook.* University of British Columbia Press, Vancouver, BC.

Kates, R. and Clark, W.C. (2000), Sustainability Science. *KSG Working Paper*, No. 00-018.

Kemppainen, A.J., Veurink, N.L. and Hein, G.L. (2007), *Sustainability in a common first year engineering program.* Paper presented at the 37th ASEE/IEEE Frontiers of Education Conference, session S2J, 10-13 October, Milwaukee, WI.

Labun, C. (2007), *Teaching engineering communication to first year engineering students.* Available at: cden2007.eng.umanitoba.ca/resources/papers/67.pdf (accessed 11 January 2012).

MacLennan, J. (2008), Why communication matters. In MacLennan, J. (Ed.), *Readings for Technical Communication*, pp. 4-10. Oxford University Press, Don Mills, ON.

McCormick, M., Lawyer, K., Berlin, M., Swan, C., Paterson, K., Bielefeldt, A. and Wiggins, J. (2010), *Evaluation of sustainable engineering education via service learning and community service efforts*. Paper presented at the American Society of Engineering Educators (ASEE) Annual Conference, June 20-24, Louisville, KY.

Mihelcic, J. and Trotz, M. (2010), Sustainability and the environmental engineer: Implications for education, research and practice. *Environmental Engineer: Applied Research and Practice, the Magazine of the American Academy of Environmental Engineers* Vol. 10 (Winter), pp. 27-34.

National Academy of Sciences (2002), *Dialog on the engineers' role in sustainable development – Johannesburg and Beyond.* NAS publications, Washington, DC.

Reave, L. (2004), Technical communication instruction in engineering schools: A survey of top-ranked US and Canadian programs. *Journal of Business and Technical Communication* 18, pp. 452-490.

Shulman, L.J., Besterfield-Sacre, M. and McGourty, J. (2005), The ABET "professional skills" – Can they be taught? Can they be assessed? *Journal of Engineering Education*, pp. 41-55.

Thomas, I. (2009), Critical thinking, transformative learning, sustainable education, and problem-based learning in universities. *Journal of Transformative Education* July 2009 (7), pp. 245-264.

Wiggins, J., McCormick, M., Bielefeldt, A., Swan, C. and Paterson, K. (2011), Students and sustainability: Assessing students' understanding of sustainability from service learning experiences. Paper presented at the 2011 Annual American Society of Engineering Educators (ASEE) Conference and Exposition, 26-29 June 2011, Vancouver, Canada. Available on DVD.

World Commission on Environment and Development (1987), *Our Common Future, Bruntland Report.* Oxford University Press, New York.

Chapter 14

The Value of Authentic Learning Spaces to Facilitate Education for Sustainable Development in Open and Distance Learning

Rudi Pretorius

Abstract

There is agreement that problems dealt with in the field of sustainability are in the category of so-called "wicked problems". Despite this, in many cases students in this field are still confronted with pedagogic approaches rooted in specialization and reproduction. This is not supportive in terms of the development of agents of transformation, which graduates in the field of sustainability should take on. Innovations to address this shortfall have been researched, but not implemented as widely as they should have been. In open and distance learning this is more of a problem than for traditional residential universities. This paper illustrates how the latter problem can be addressed with the introduction of teaching and learning in authentic contexts in open and distance learning courses focusing on sustainability issues. Results are based on a case study of the open and distance learning module "Geography of Tourism", offered by the University of South Africa. The assessment strategy in this module has been transformed to allow students to use their own living environments as authentic learning spaces within which they can experiment with and apply the theory being dealt with. In this way the focus is not on theory as such, but on their unique learning experiences. This fosters a deeper and richer form of learning than is achieved by question and answer, essay or multiple choice formats frequently used in open and distance learning. An action framework for implementation of authenticity in open and distance learning is supplied, which may serve as a useful point of departure for lecturers considering this approach. In this way significant value can be added to education for sustainable development, thus supporting higher education in delivering appropriately capacitated professionals in this field.

Introduction

Approaches to ESD

The problems faced by practitioners in the field of sustainability can be described as ill-structured and "messy" with no single, clear-cut solution. This fits with the characterization by Palmer et al. (2007) of problems in the context of sustainability as "wicked". These problems are not solved within the boundaries of disciplines, and require approaches such as critical systems thinking (Smith 2011), inter- and trans-disciplinary connections (Marinova and McGrath 2004) and scenario and strategy development (Komiyama and Takeuchi 2006). Implementation of these approaches is demanding, requiring more than cosmetic rearrangements portraying education as "a magical wand which could be waved anytime and anyhow to arrive at sustainable development" (Manteaw 2012: 379). In addition, Manteaw (2012) highlights the importance of a focus on mind-sets and practices responsible for the unsustainable present. ESD should aim to unlearn our unsustainable present living patterns, before exploring more sustainable options.

Existing inadequacies

The literature on education for sustainable development (ESD) points towards the inadequacy of many educational offerings to deliver professionals able to deal with the multitude of sustainability issues modern societies are grappling with (Mochizuki and Fadeeva 2010). Factors contributing to this inadequacy include (but are not limited to) the following: a focus on knowledge transfer and reproduction (Reunamo and Pipere 2011), increased specialization in narrowly defined academic disciplines (Bacon et al. 2011), a focus on what should be taught (and not on what should be leant and how) (Sleurs 2008), and limited appropriate real-world learning opportunities (Brundiers et al. 2010). In open and distance learning (ODL) contexts, these issues are more pronounced and need to be addressed more innovatively than in contact teaching. In addition, the idea that with ESD one size fits all needs to be addressed. In order to be effective, teaching and learning interventions need to be adapted to take diverse local and regional contexts into account, as pointed out by Manteaw (2012).

Position statement

This paper connects with the growing body of thinking and experimenting in higher education, as referred to by Mochizuki and Fadeeva (2010), towards more radical innovations in teaching and learning to serve the sustainability agenda more effectively. Over the last few years, increased interest has been shown in competence-based approaches as a means to achieve the transformation to sustainability. This requires a shift in focus from mere acquisition of knowledge and skills to pointing out competencies that are characteristic of subjects (Mogensen and Schnack 2010). According to Blomhøj and Jensen (2003), a competence approach implies some form of action, a sphere of application and consideration of a subjective element as well – in other words, relating it to someone as well as surrounding influences. These competencies need to be acquired by means of personal awareness and an insight and understanding of local situations and their origins, in order to deconstruct local problems and to develop alternatives (Manteaw 2012).

Methodology

This paper is a response to the expressed need for new and alternative approaches to teaching and learning to support the sustainability agenda in a more meaningful way. The point of departure is a review of relevant literature about developments, issues and needs on the interface between ESD, Geography, ODL and teaching and learning (theory and practice). Following this, the notion of exposing students to an authentic learning space in ODL, and thereby adding value to ESD, is introduced and developed. An evaluation is included, based on personal reflection by the lecturer, combined with voluntary student feedback – presented as quotes arranged in a number of categories. This student feedback has been filtered in order to reveal the value of the learning experience in terms of the requirements of the sustainability agenda. Challenges and solutions are shared and an action framework is provided for introducing authenticity in ESD at course level for the ODL context.

Education for sustainable development in context

Pointers such as the vision of the United Nations decade of education for sustainable development (DESD), namely "a world where everybody has the opportunity to benefit from education and learn the values, behavior and lifestyles required for a sustainable future and for positive societal transformation" (UNESCO 2005: 6) provide an indication of how ESD should be characterized. Several questions, however, remain. What should be learned? How should it be learned? What competencies need to be acquired during this learning? Mochizuki and Fadeeva (2010) point out that the required competencies are neither straightforward nor universally applicable. In addition, some educationists are sceptical about competency-oriented approaches, due to their focus on skills development at the expense of the requirement of whole person development. Podger et al. (2010: 345) point out that nothing less than a whole person approach "can hope to cultivate the kind of globally responsible consciousness that can inspire people towards sustainable living".

Recent literature points towards a general opinion that ESD should make provision for not one only, but for a set of capacity-building pathways (Brundiers et al. 2010). This is encapsulated by the framework of transformative sustainability learning with its organizing principle, the so-called "head, hands, and heart" trilogy (Sipos et al. 2007). Associated with this, Brundiers et al. (2010) suggest the following groups of key competencies for ESD:

- strategic – focused on systemic, anticipatory, normative and action-oriented competencies;
- practical – focused on competencies to bridge the gap between knowledge and action;
- collaborative – focused on competencies enabling participative contributions in various contexts.

It follows that in the delivery of ESD, "We have the choice to change our educational mandates to stop teaching for "unsustainability" (i.e. the perpetration and perpetuation of social and ecological crises) and to transform our pedagogical perspective to teach for sustainability (i.e. social and ecological justice). In order to move beyond reproduction of our social ills, teaching for sustainability requires transformation to new ways of approaching education and life" (Sipos et al. 2007).

The value of appropriate teaching and learning interventions

Appropriate pedagogical interventions providing opportunities for students to engage meaningfully with the various issues associated with sustainable development are required. Among other things, a suitable pedagogy will involve learning within real-life contexts through observation, critical analysis and evaluation. The aim is to develop "shaping skills" on the basis of the information that the students have been exposed to, together with their various reflective experiences (Sleurs 2008). Dawe et al. (2005) add to this by emphasizing the importance of:

- Re-connection to reality – the need to involve students in "shaping their world" from the beginning of their studies and to integrate their university experience with personal choices;
- Holistic thinking – the need to move from "reductionist" thinking to facilitation of interdisciplinary and systemic connections between disciplines and critical thinking abilities;
- Personal approach – the need for mutual learning between students and lecturers and for students to take ownership of ESD.

Adoption of contextual learning, as advocated Herrington and Oliver (2000), is one approach from which ESD can benefit. The basic premise of contextual learning is that learning and context should not be separated, or else students experience knowledge as the end product of their learning, and not as one of a number of tools that can be used in problem-solving. This type of approach connects to the plea by Brundiers *et al.* (2010) for real-world learning opportunities in sustainability. An important part of contextual learning is the idea of an apprentice who is observing the community of practice in his or her chosen field. The trick, however, comes with the design and implementation of the contextual learning opportunity, which will eventually be the determining factor if it indeed provides an authentic learning experience.

An appropriate assessment strategy can be regarded as the driving force to determine the extent to which teaching and learning interventions are in congruence with the sustainability agenda. Assessment is regarded as the primary driver of the learning that will be taking place in any course (Swartz and Webb 2002), to the extent that some scholars regard assessment as the "de facto curriculum" (Rowntree 1987). Assessment, however, does not function in isolation with the highest probability of achieving its objectives shown when it is embedded within rich environments for active learning – REALs (Grabinger and Dunlap 1995) – which have the following characteristics:

- Constructivistic learning philosophy;
- Authentic learning contexts;
- Student centred;
- Cooperative learning;
- Generative learning activities;
- Authentic assessment.

Realities of the open and distance learning context

A prime characteristic of open and distance learning (ODL) institutions is that students are not physically present in classes on campuses. They can study at their own pace, at times which suit them. In addition, studying at an ODL institution is usually much cheaper than at residential universities. Taylor (2006) lists the key factors influencing people to turn to the ODL option as: convenience, flexibility, cost, and opportunity. The latest trend in ODL is associated with the prevalence of the Internet, which enables students to access their study materials and interact with fellow students and lecturers wherever they can find a connection. In terms of ESD, the ability of ODL to open up educational opportunities to people who would otherwise not have been able to pursue academic studies is a major advantage.

Despite the potential of ODL as outlined above, the perception exists that ODL study programs are sub-standard in comparison to those provided by residential universities (Barasa, 2011). This is related to the primarily one-sided teaching process followed in the past, where students were treated as passive receivers of study material (i.e. first generation ODL, the so-called "industrial era"). Most ODL institutions have by now realized the value of a learner-centred approach and either have already implemented or are currently implementing steps to move to a more appropriate pedagogic design, also differing from that found at residential universities (University of South Africa 2007). It is accepted now that appropriately designed learning experiences can contribute to involving students in the learning process in a meaningful way, despite the constraints imposed by the ODL context.

In ODL the advantages of appropriately designed learning experiences cannot be emphasized enough, particularly in engaging students with their studies. In addition student motivation is of prime importance, which may differ for the ODL context compared to on-campus situations (Chi-hung 2006). The increased availability and use of information and communication technologies (ICTs) (Lynch et al. 2008) and their implications for ODL need to be addressed as well. Kirkwoord and Price (2008: 14) argue that "important outcomes of learning at a distance can be facilitated and achieved through the use of ICTs, but that the pedagogic design of courses and the associated assessment must reflect this." It cannot be ignored that ICTs offer possibilities for students to grapple with real-world problems, to do collaborative work and to be actively involved in knowledge construction.

The University of South Africa

The module presented as case study in this paper, is offered by the Department of Geography at the University of South Africa (Unisa), which is regarded as a significant example of an ODL institution (Tait 2008). With a head count in 2010 of 293-437 (Unisa Facts and Figures 2010), it is positioned among the largest ODL universities in the world. After mergers between ODL providers in South Africa, a new institution emerged in 2004 that adopted the name "Unisa" and now has the expressed vision: "*Towards* the *African University in the service of humanity"*. The approach to teaching and learning at Unisa since 2007 is geared towards facilitating the transactional process of learning by combining printed study material and appropriate ICT solutions, with a definite move towards e-learning. The aim is to optimize interaction between, among other things, the university and students (University of South Africa 2007).

Case study: "Geography of Tourism" (Department of Geography, Unisa)

Crafting of an authentic learning space

During the past two decades, the Department of Geography at Unisa went through two re-curriculation cycles, associated with a step-wise transformation from a traditional sub-disciplinary focus to an approach using issues in the world, on the African continent and in South Africa as a structuring mechanism (Pretorius 2005). Together with this a structured study program in environmental management was introduced in 2000 and the teaching style in the Department of Geography was transformed to be more student-centred and to involve students more actively with their studies (Pretorius 2004).

The module "Geography of Tourism" deals with the application, relevance and utility of geographical concepts within the field of tourism, with emphasis on the interaction between people (tourists) and the environment (primarily destination areas). The focus is furthermore on the environmental, economic and socio-cultural impacts of tourism, as well as related sustainability and management implications. The tourism management students for whom this module was originally developed find it an "eye-opener" because it requires of them to make a paradigm shift away from profit at all costs to the sustainable use of resources, while maintaining profits.

"Geography of tourism" has recently been redesigned to facilitate a contextual, authentic learning experience. The cover pages of the new study guide are shown in Fig. 1, with the authentic approach evident without even opening it. A constructivist pedagogic (based on REALs) is utilized, requiring students to use their own living environments or to select any other place/destination to make sense of the theory. The assessment strategy "drives" the learning, with a portfolio activity at the end of each study unit. The aim of the activities is that students have to use theory to gain a better understanding of the place/destination selected by them. This is in line with the premise that "… student's localized places of study, work and recreation are the centres of their experiences that help teach them how the world works and how they fit into that world" (Sipos et al. 2007: 70).

Fig. 1: The front and back cover of the study guide for "Geography of Tourism"

Pretorius (2008) – images reprinted with permission of Unisa

At the stage when the module has been completed, students are required to present a portfolio of evidence to showcase how their perspectives on and skills to deal with issues relating to tourist-environment interaction have developed and gained depth. A selection of these portfolio activities are submitted during the semester to ensure that students are on track, to facilitate student-lecturer interaction, to serve as motivation and to assist in preparation for the examination. The semester work contributes 40% and the examination 60% towards their final marks.

Evaluative remarks from the viewpoint of the lecturer

The reason for the revision of "Geography of Tourism" was the realization that the learning experience offered by the previous study package did not increase students' awareness of the issues that confront the world around them; neither did it empower them to do much about it. It was just another module to be ticked off on their "to do list". They also perceived the assessments as "find-and-copy" exercises in the text book and in the examination they expected only to have to give back merely what they had memorized. From the lecturing point of view this was experienced as very unfulfilling. Realizing the untapped potential of the theme being dealt with in this module, especially in terms of the environmental crisis humankind is facing, it was therefore decided to embark on a radical revision of the teaching and learning experience offered by this module.

The point of departure was to create a rich environment for active learning (Grabinger and Dunlap 1995), in which authentic assessments are employed to provide students with the opportunity to engage in real-world situations and to apply and use what they are learning to not only to understand problems better, but also to be able to provide suggestions on how to solve them. The interventions which have been designed and included in the study package to make this possible are as follows:

- Contextualization of the learning experience from the start – first study unit based on an authentic "story" about a tourism practitioner;
- Use of "stories", scenarios, articles, advertisements, newspaper snippets, case studies, etc., to acquaint students with the real world;
- Application of what is learnt in terms of the place where students are staying or any other context of their choice (portfolio activities in the study guide);
- Assignments linked to portfolio activities, which are situated in real-world contexts;
- Activities allowing students to discover the meaning of the learning materials and to develop their own voice on the issues being raised;
- A direct link between the module's aims, outcomes and assessment criteria and the formative and summative assessment.

From the lecturer's point of view, the revision, redesign and redevelopment of the study package for "Geography of Tourism" was a time-consuming but creative experience. It also included an element of risk and uncertainty, not knowing how the final product would be received by Unisa (as an organization) and by the student body – since it differs so vastly in approach not only from the previous study package, but also from most other offerings at Unisa. Now that the new study package has been implemented, it is gratifying to experience how the approach being followed awakens the inner voice of students, and to realize that the effort to change the way they think about the world and their role in it, without forcing any opinions on them, is actually successful. It has to be mentioned, however, that due to the individual and open-ended nature of the student assessment tasks, grading takes an excessively long time. In addition, many more enquiries need to be dealt with because students are not always sure what to do, or need to discuss or share experiences with their lecturer. More flexibility is required because students in the workplace find the assessment tasks time-consuming and cannot always keep to submission cut-off dates.

Evaluative remarks from the viewpoint of students

A variety of responses from students on their experience of "Geography of Tourism" were recorded over the past three years. These included comments submitted as part of their assignments, posts in the discussion forum on the module website, and email messages. Comments were requested in an open-ended format, without any prescription about what exactly should be commented on. Before analysis of the comments, those formulated in terms of exact phrases found in the study guide were discarded and not considered. The comments were then arranged according to a number of themes concerning the value of the learning experiences in terms of the requirements of the sustainability agenda. Analysis of the responses indicates that students who buy into the learning model are positive and feel that they gain a lot, in different ways, by engaging with their study materials. But some students are more critical, especially in terms of the effort and time required to engage meaningfully in all the activities and assignments.

Responses about connecting to the real world and/or thinking out of the box:

- "I really enjoyed the assignment, I like the fact that it was a learning experience and that you have to gather information by yourself. It is the first subject where I can think out of the box and really go out into the business world (in this case tourism) and have a better understanding on what it is all about. I feel good about this module!"

- "Completing the first assignment was not easy. There were so many things I had to think about, before attempting to put pen to paper. I felt the first assignment required a lot of research and information as well as planning. However, with my hard work and dedication to this assignment, I felt this assignment to be interesting, fun and as an environmental management student – I really enjoyed it. This assignment was different to my other modules. It let me be creative and think out of the box, instead of the typical six page essay. However, after completing this assignment I am feeling good and positive about my results I will obtain."
- "This module is very challenging as it has made me to think critically about things that I see everyday …"

Responses about a holistic view and awareness creation:

- "This module made me look deeper at the tourism situation in a broader manner. It is good in my opinion to have tasks such as this for you become so complacent in where you stay that you do not realize what is going on around you has the potential and limitations."
- "I certainly am going to enjoy this module – it opens one's insight into tourism, even beyond the geographical point of view. A person seems to think about things in isolation; yet there is so much that is dependent on interaction between environments, people, resources and nature."
- "I find the topic very interesting – not only to know and understand the natural process … but also to understand the interaction people have (or should have) with the environment."
- "Doing this assignment was incredibly interesting as it encouraged me not only to view my destination in terms of tourism, but also to view it from an environmental point of view."

Responses about the nature of the learning experience:

- "I am glad that this module challenges my creative side as well. It is not like the other modules where I must only answer a long question."
- "I feel I still have a lot to learn but like doing these types of assignments since you can put your own views in and build your own views."
- "Currently I'm excited about this module. This module let students to be involved and participate in their countries tourism places and resources."
- "It's definitely not a case of learning information and repeating it, but of understanding and applying."
- "The module is beyond my expectations – never did I expect the module in Tourism Geography is so exciting and inviting … The module is stunning and amazing. It is exploring a new world of life in career."
- "I feel like a tourism practitioner already because I carried out research, questions and visits all by myself."

Responses about the practical/activity based dimension

- "The completion of this assignment has added some practical dimension which I have found interesting and a bit challenging in terms of resources. I however feel very motivated."
- "…pleasantly surprised that a poster was required to be created instead of just answering a written or a multiple choice assignment."
- "The module is a lot more creative orientated than I expected, not a structured answer but instead to be more objective and look to see all views and outcomes."

Some negative/more critical responses:

- "The assignment was a bit challenging but interesting at the same time."
- "This module is very challenging. It has lots of work and requires too much of time and resources."
- "I am not a great fan of composing scrapbooks (which is what this assignment felt like), however, this is the foundation of one of the skills needed for assessing places with a view of promoting them as tourist places of interest. I am curious to see how skills like these will be tested in an exam."
- "The amount of information available is overwhelming, and it actually made the task of compiling this assignment very difficult."
- "I am staying in a very remote area in South Africa where I can hardly get a library to do my research. May you consider us also …"
- "Next time when such an assignment is given, I think it is better if one is given a chance of answering the questions by writing down the answers rather than making sketches and drawings because one may not be sure whether the questions were answered properly."

Action framework for implementation

This action framework presented in Fig. 2 is partly the result of experience obtained with the design, development and implementation of the revised "Geography of Tourism". This experience was integrated with frameworks for authentic learning environments and authentic assessment (Herrington and Oliver 2000; Gulikers et al. 2004). The requirements for rich environments for active learning (Grabinger and Dunlap 1995) and guidelines for ESD (Dawe et al. 2005) were also considered. The suggested framework should not be perceived as a simplistic sequential, step-by-step process and has therefore not been depicted as such. In other words, at any time during the design, development and implementation process, it might be necessary to revisit and adjust some of the previous assumptions and/or processes. It is therefore a dynamic framework which can be adapted and changed as and when required, and as such it does not have a specific beginning or end.

The framework has a distinct learner-centred focus, with activities and assessments based on the living environments of students (i.e. their study, work and recreation spaces), which are best suited to illustrate to them how the world works and where they fit into it (Gruenewald 2003). Context, including the real-world situation, which ensures authenticity, is embedded right through. To increase meaningfulness, all activities and assessments have to fit within the chosen context/s and be aligned to the study outcomes. Together with this, a learning strategy has to be decided on, which links with a certain theory of learning. A theoretical dimension needs to be added to provide the necessary depth associated with the academic level of the offering. Getting to the finer detail, attention also needs to be given to collaborative opportunities, various perspectives and roles, the design of the activities and assessment tasks and the provision of student support.

Fig. 2: Framework for an authentic learning space, with credit for some aspects to Herrington and Oliver (2000), Gulikers et al. (2004), Grabinger and Dunlap (1995) and Dawe et al. (2005)

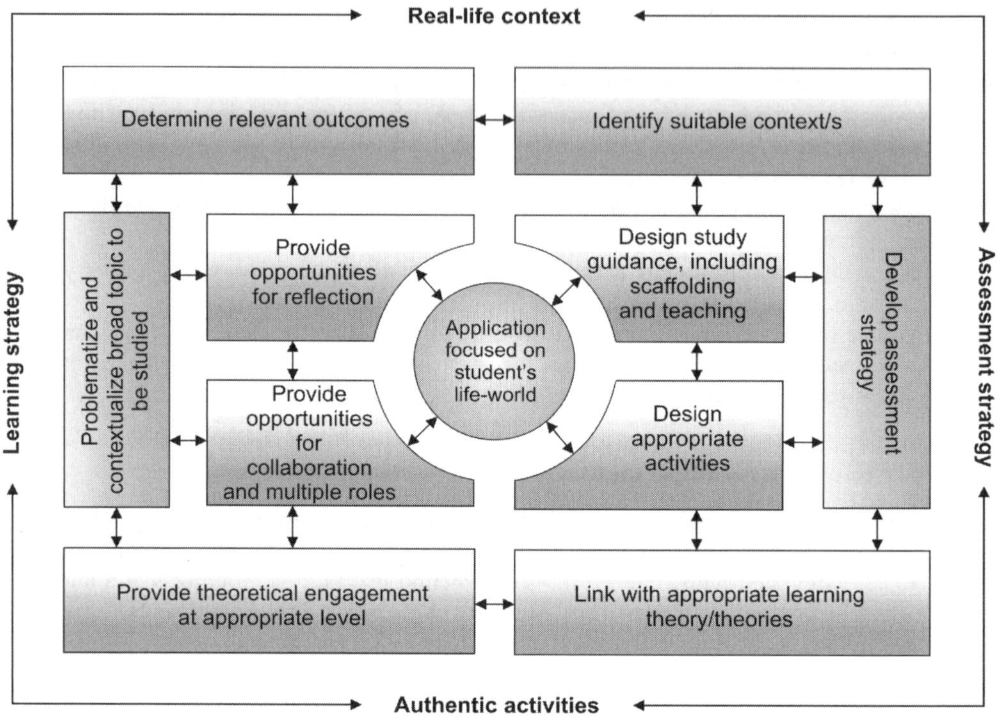

Evaluative discussion and conclusions

Addressing the needs of ESD in an ODL context

ODL institutions have a potentially huge role to play in ESD. This is because of the ability of ODL to make educational opportunities as widely accessible as possible, at a much lower cost than at traditional residential universities. Up to about 2000 the perception that certain fields of study (such as many of the hard sciences) cannot be taught properly via ODL, was quite common. It was even suggested that students who complete all their courses via ODL, might be trained in an inferior way to their counterparts at residential universities. Geography is a case in point, with scepticism voiced in the past about the ability of ODL to do justice to the essence of the subject, namely "... its connection with real, live places" (Gober 1998: 2).

Major strides have been made over the past two decades in terms of improving pedagogy and ensuring quality standards in ODL. Despite this, the challenges involved in offering courses with practical components by means of ODL, is still acknowledged (Scanlon 2011). This is in line with the author's experience with teaching Geography via ODL over 25 years, with a growing frustration that students are not really empowered to perform meaningful activities in job-related contexts after their graduation (Pretorius 2005). If this is indeed the case, these graduates will also not be able to contribute constructively towards a more sustainable future in their respective roles. In terms of meeting the objectives of, for example the environmental management programmes offered by Unisa, this is a concern.

Implementation of a contextual, authentic learning space

It is crucial to determine how to add value to ODL study programmes related to environmental sustainability in order to make sure that the required outcomes are indeed achieved. The author embarked on an experiment to design and implement a contextual, authentic learning space for the module "Tourism Geography", offered via ODL by the Department of Geography at Unisa. The need for this was not related to poor exam pass rates and pressure to improve, but the urge of the lecturer to empower students sufficiently to be able to retrieve and use the abstract academic knowledge they have obtained, to address real-life issues in various contexts. Associated with this was the need to prepare students adequately to make meaningful contributions in terms of the environmental and sustainability crises the Earth is facing.

Evaluative remarks on implementation

Evaluation of the new learning space which has been created for "Geography of Tourism" shows that the design criteria for inclusion of authenticity in teaching and learning (Herrington and Oliver 2000) have been met to a large extent. The ODL context within which this module features, served as a challenge, and where criteria have not yet been fully met, this is mostly related to this aspect. Starting with *context, the real-life opening story* (magazine article format with photos) of a newly appointed tourism practitioner at a resort in the Limpopo Province of South Africa immediately captures the attention of students and provides a clear signal that the approach in this module is different and new. This context is used and referred to again and again right through the module.

The authentic context is enhanced by *supporting activities* based on, among other things, analysis of advertisements of tour packages, newspaper and magazine articles on tourist destinations and issues/problems being experienced and short case studies and/or scenarios. The *assessment tasks* fit into this pattern, with students having to compile a portfolio based on application of the theory in terms of their own location (or else a place of their choice). This portfolio has to be submitted and forms part of the examination for this module. The examination is linked directly to this portfolio, but serves as a validation that the necessary theoretical underpinning has been obtained. *Opportunity for reflection* forms part of the activities being provided at various stages during the study period.

Given the ODL environment, *access to expert performances* is challenging to achieve, although a tailor-made reader, consisting of a variety of readings on the module topic, forms part of the study package. *Multiple perspectives* on the topic are provided by the reader as well as by the ample inclusion of quotes right through the study guide, while students have the opportunity to assume *multiple roles* in the different activities and assessment tasks. Ample *coaching and scaffolding* is supplied throughout the study guide as well as on the website for the module. Concerning the criteria of *collaboration* and *articulation*, the ODL context still represents a constraint, especially since not all students have equal access to ICTs and specifically the module website on the Internet.

The value of authentic learning spaces for ESD in the ODL context

The experience gained with the implementation of the revised study package for "Geography of Tourism" serves as confirmation that "context is king". Generally speaking, students grabbed the opportunity to choose their own context for the development of their portfolios. This is a freedom seldom available to students studying via ODL, and came as a surprise to many of them. The learning taking place during this exercise is much deeper and richer than what is usually achieved by question and answer, essay or multiple choice formats frequently used in ODL. The implementation of this approach shows that, with a bit of creativity and effort, meaningful activity, enquiry and problem-based approaches can be implemented in ODL, thus adding significant value to not only the learning taking place, but also the teaching experience as a whole. The action framework (Fig. 2) which has been developed serves as a useful point of departure for teachers considering implementation of authentic learning in ESD, and is not limited to the ODL context only.

Most satisfying from the teaching perspective is the observation that many students enrolled for "Geography of Tourism" show significant development in their understanding of the world they live in, the environmental issues being faced and the various options available to deal with these issues. Apart from criticism primarily about increased workload, students generally seem to perceive the study approach positively, mentioning "out of the box thinking", awareness creation and the inclusion of a practical component as some of the advantages. From the viewpoint of ODL, the approach followed in this module is challenging because of the difficulties still experienced to facilitate effective student-lecturer and student-student interaction. This is related to the issue of inequalities in access to ICTs and specifically the Internet in the South African context. If this aspect can be improved, it will be possible to pro-actively address many of the incorrect conceptions which students may have in terms of what is expected of them.

Taking authentic learning spaces for ESD in ODL into the future

All indications point at ICTs, specifically the phenomenon of e-learning, becoming more and more important in the facilitation of teaching and learning experiences. It is recognized that the use of ICTs has the potential to add significant value to the teaching and learning experience. This includes Geography, as highlighted by Lynch et al. (2008), as well as the field of sustainability. However, one size does not fit all and use of ICTs has to be approached critically in order to realize where and how specific applications fit into the ICT "space". Some practitioners regard ICTs merely as a mechanism to deliver study materials, whereas others use it to implement new strategies to learn more effectively. Lynch et al. (2008) indicate that ICTs can provide an excellent platform for students to engage with real-world problems, to share ideas and information with each other and to pursue knowledge construction in relevant contexts.

With regard to ICTs and the implementation of authentic learning spaces in ODL, the challenge in the local, South African context in which Unisa operates is narrowed down to access to ICTs. Those without access (or with limited access) should not be marginalized further and mechanisms need to be considered to get them on board before opting for a system of only e-learning. Until that stage, a dual system has to be kept in place to accommodate students with access to ICTs as well as those with limited or almost no access. Such a system is challenging, and unfortunately creates two groups of students who are treated differently. An interesting observation, however, is that students with access to ICTs do not necessarily submit the best work – which might be indicative of the ease of electronic plagiarism and perceptions that it is quick and easy to assemble an assignment electronically overnight.

Acknowledgements

Ms Laura Steyn (Department of Geography, Unisa) for her professional assistance in completing the diagram in Fig. 2.

Ms Marie Nöthling (Directorate: Curriculum and Learning Development, Unisa) for her interest in and support of my development work on alternative approaches to teaching and learning.

Unisa for permission to use the images included in Fig. 1.

References

Bacon, C.M., Mulvaney, D., Ball, T.B., DuPuis, E.M., Gliessman, S.R., Lipschutz, R.D. and Shakouri, A. (2011), The creation of an integrated sustainability curriculum and student praxis projects. *International Journal of Sustainability in Higher Education* 12 (2), pp. 193-208.

Barasa, F. (2011), Promoting open and distance learning in Africa: a critical reflection on rhetoric, real and ideal. Paper presented at the *Conference of Rectors, Vice Chancellors and Presidents of African Universities*, Stellenbosch, South Africa, 27 May – 3 June. Available at http://events.aau.org/userfiles/file/corevip11/papers/fred_barasa_Promot ing_ODL.pdf (accessed 1 February 2012).

Blomhøj, M. and Jensen, T.H. (2003), Developing mathematical modelling competence: Conceptual clarification and educational planning. *Teaching mathematics and its applications* 22 (3), pp. 123-139.

Brundiers, K., Wiek, A. and Redman, C.L. (2010), Real-world learning opportunities in sustainability: from classroom into the real world. *International Journal of Sustainability in Higher Education* 11 (4), pp. 308-324.

Chi-Hung, Ng. (2006), The role of achievement goals in completing a course assignment: examining the effects of performance-approach and multiple goals. *Open Learning: The Journal of Open, Distance and e-Learning* 21 (1), pp. 33-48.

Dawe, G., Jucker, R. and Martin, S. (2005), *Sustainable Development in Higher Education: Current practice and future developments.* A report for The Higher Education Academy. Heslington, York.

Gober, P. (1998), Distance learning and geography's soul. *Association of American Geographers Newsletter* 33 (5), pp. 1-2.

Grabinger, R.S. and Dunlap, J.C. (1995), Rich environments for active learning: a definition. *ALT-J: Research in Learning Technology* 3 (2), pp. 5-34.

Gruenewald, D.A. (2003), Foundations of place: a multidisciplinary framework for place-conscious education. *American Educational Research Journal* 40 (3), pp. 619-54.

Gulikers, J.T.M., Bastiaens, T.J., Kirschner, P.A. (2004), A five-dimensional framework for authentic assessment. *Education, Technology, Research and Development* 52 (3), pp. 67-86.

Herrington, J. and Oliver, R. (2000), An instructional design framework for authentic learning environments. *Education, Training, Research and Development* 48 (3), pp. 23-48.

Kirkwood, A. and Price, L. (2008), Assessment and student learning: a fundamental relationship and the role of information and communication technologies. *Open Learning: The Journal of Open, Distance and e-Learning* 23 (1), pp. 5-16.

Komiyama, H. and Takeuchi, K. (2006), Sustainability science: building a new discipline. *Sustainability Science* 1 (1), pp. 1-6.

Lynch, K., Bednarz, B., Boxall, J., Chalmers, L., France, D. and Kesby, J. (2008), E-learning for Geography's teaching and learning spaces. *Journal of Geography in Higher Education* 32 (1), pp. 135-149.

Manteaw, O.O. (2012), Education for sustainable development in Africa: The search for pedagogical logic. *International Journal of Educational Development* 32, pp. 376-383.

Mogensen, F. and Schnack, K. (2010), The action competence approach and the 'new' discourses of education for sustainable development, competence and quality criteria. *Environmental Education Research* 16 (1), pp. 59-74.

Marinova, D. and McGrath, N. (2004), A transdisciplinary approach to teaching and learning sustainability: A pedagogy for life. In *Seeking Educational Excellence,* Proceedings of the 13th Annual Teaching Learning Forum, Murdoch University, Perth, 9-10 February. Available at http://lsn.curtin.edu.au/tlf/tlf2004/marinova.html (accessed 29 January 2012).

Mochizuki, Y. and Fadeeva, Z. (2010), Competences for sustainable development and sustainability: Significance and challenges for ESD. *International Journal of Sustainability in Higher Education* 11 (4), pp. 391-403.

Palmer, J., Smith, T., Willets, J., Mitchell, C., Sheffield, J. and Fielden, K. (2007), Creativity, ethics and transformation: key factors in a transdisciplinary application of systems methodology to resolving wicked problems in sustainability. Paper presented at the *13th ANZSYS Conference, Systemic Development: Local Solutions in a Global Environment*, Auckland, New Zealand, 2-5 December.

Podger, D.M., Mustakova-Possardt, E. and Reid, A. (2010), A whole-person approach to educating for sustainability. *International Journal of Sustainability in Higher Education* 11 (4), pp. 339-352.

Pretorius, R.W. (2004), An environmental management qualification through distance education. *International Journal of Sustainability in Higher Education* 5 (1), pp. 63-80.

– (2005), A narrative of change: reflecting on two decades of Geography at the University of South Africa. Paper presented at the *Sixth Biennial Conference of the Society of South African Geographers,* University of the Western Cape, Bellville, 5-10 September.

– (2008), *Geography of tourism.* Only study guide for GGH206Y, University of South Africa, Pretoria.

Reunamo, J. and Pipere, A. (2011), Doing research on education for sustainable development. *International Journal of Sustainability in Higher Education* 12 (2), pp. 110-124.

Rowntree, D. (1987), *Assessing students: How should we know them?* (2nd rev. ed.), London: Routledge Farmer.

Scanlon, E. (2011), Open science: Trends in the development of science learning. *Open Learning: The Journal of Open, Distance and e-Learning* 26 (2), pp. 97-112.

Sipos, Y., Battisti, B. and Grimm, K. (2007), Achieving transformative sustainability learning: engaging head, hands and heart. *International Journal of Sustainability in Higher Education* 9 (1), pp. 68-86.

Sleurs, W. (Ed.) (2008), *Competencies for ESD (education for sustainable development) teachers: a framework to integrate ESD in the curriculum of teacher training institutes*, Comenius 2.1 project 118277-CP-1-2004-BE-Comenius-C2.1. Available at: ww.ensi. org/Projects/Teacher_Education/CSCT/ (accessed 29 January 2012).

Smith, T. (2011), Using critical systems thinking to foster an integrated approach to sustainability: a proposal for development practitioners. *Environment, Development and Sustainability* 13 (1), pp. 1-16.

Schwartz, P. and Webb, G. (2002), *Assessment: case studies, experience and practice from higher education.* Kogan Page, London..

Tait, A. (2008), What are open universities for? *Open Learning: The Journal of Open, Distance and e-Learning* 23 (2), pp. 85-93.

Taylor, D. (2006), *Issues, Trends, and Challenges in Distance Education: An International perspective.* Available at https://next.eller.arizona.edu/courses/InternationalManagement/Fall2006/student_papers/final_papers/Distance_Learning.pdf (accessed 31 January 2012).

UNESCO (2005), *United Nations Decade of Education for Sustainable Development (2005-2014): International Implementation Scheme*, October ED/DESD/2005/PI/01, UNESCO, Paris. Available at: http://unesdoc.unesco.org/images/0014/001486/1486 54e.pdf (accessed 30 January 2012).

University of South Africa (2007), *Self-evaluation portfolio for the Commonwealth of Learning trail audit.* University of South Africa, Pretoria.

Unisa Facts and Figures (2010). Available at: http://heda.unisa.ac.za/ (accessed on 1 February 2012).

Chapter 15

Complexity of Campus Sustainability Discourse

Tamara Savelyeva and Jae Park

Abstract

From the outset, the authors note a general lack of a more updated and deep theoretical engagement on the discourses of ecologism and sustainability in universities, and contrast this deficiency with sustainability-related practices. While explaining the interconnected nature of ecologism and sustainability, the authors describe the rise of ecological citizenship and the expected role of universities. Although ecological discourses are comparable across countries and universities, the practical implications of sustainability on campuses vary in degrees of complexity, namely linear, complicated, and complex. To support their arguments with practical understanding, the authors typify the challenges in sustainability practices at the Duke University and the University of Hong Kong by using their earlier works. The findings show similarities in sustainability discourses across campuses characterized by a "double dichotomy" phenomenon in sustainability efforts. The authors propose a metaphor of "ensoulment" to suggest campus-based practices in favor of education for sustainability rather than the dominant greening and public relation efforts.

Introduction

The rapid growth of ecology-related efforts in higher education (HE) reveals the absence of well-articulated theoretical structures to support campus sustainability discourse. The theoretical reflections on the history and intellectual content, pertinent to the topic of sustainability and its relation to a broader notion of ecologism, are sporadic and disconnected. In numerous analytical studies, the key objectives of campus sustainability discourse – analysis, management, and dissemination – have not been pursued as theoretically advanced. As a result, sustainability remains an intellectually non-rigorous field within broader humanistic domains. The lack of theoretical engagement restricts researchers from fully understanding the complex situation of eco-discourses on campuses. How does sustainability discourse fit in within structures and processes of HE?

The case in point is the Western discourses[1] of ecologism and sustainability, which have the most visible applications on university campuses, compared to other ecological movements.[2] Although the origin of concern for nature in the West could be traced back to the Middle Ages in the works of St. Francis (Sessions 1987), the rise of modern eco-discourses reaches the Age of Ecology with Carson's *Silent Spring* (1962) and Meadows's *Limits of Growth* (1972), which, in turn, followed earlier works on "ecological perspective" by Lawrence and Huxley among other writers, and on "ecological conscience" by Leopold (Sessions,

1 The present work follows Gee et al. (1996) to define discourse as "a way of talking, listening, reading, writing, acting, interacting, believing, valuing, and using tools and objects, in particular settings and at specific times to display or to recognize a particular identity."

2 Dryzek (2005) proposes four basic ecological discourses: environmental problem solving, survivalism, sustainability, and green radicalism.

1987). The works of Carson and Meadows are generally regarded as the milestones of Western eco-movements.

The various instances of sustainability discourses going global in HE worldwide can be exemplified by articles of Palma et al. (2011), Fonseca et al. (2011), Naeem and Peach (2011), and Leal Filho (2011). Palma et al. (2011) mapped sustainability courses at federal universities in Brazil. Naeem and Peach (2011) drew a picture of EfS practices in Asia-Pacific region. Fonseca et al. (2011) described the situation of sustainability reports by Canadian HE institutions. A book edited by Leal Filho (2011) linked this globalizing trend with an intrinsic complexity of HE systems and provides comprehensive examples of practical implementations of sustainability discourse in international campuses.

Dewey's (1954) seminal ideas about the public articulate the rise of citizenship and action that underlines the assumptions about the public nature of eco-discourses on campus. According to Dewey, the public is brought into existence by the awareness of "extensive and lasting embroilment and the harm wrought by it" (p. 17). This idea leads to a particular type of public – the ecologically concerned public – that develops a shared ecological awareness in response to environmental damages that inflict potential or real harm on them. This awareness would also be the basis of a discourse: an ecologically concerned public conscientiously organizes and sets itself in motion. In the process of action, the public performs the tasks of defending a common cause and interest (guardians); analyzing problems (interpreters); and taking action (doers or executors):

> Recognition of evil consequences brought about a common interest which required for its maintenance certain measures and rules, together with the selection of certain persons as their guardians, interpreters, and, if need be, their executors (Dewey 1954: 17).

Dewey's typology of public tasks (guardians, interpreters, and doers) can be paralleled with three key objectives of campus sustainability discourse – analysis, management, and dissemination. This parallel allows identifying *ecological citizens* (Dobson 2004) and their roles. With their discourses, ecological citizens shape the ways of apprehending the world with their particular "assumptions, judgments, and contentions that provide the basic terms for analyses, debates, agreements, and disagreements" (Dryzek 2005: 9), and unveil their identity with their actions.

Education system, HE in particular, is one of the key places where such ecological citizens are likely to rise. University is a time-honored global institution that has been addressing "the need for all citizens to have an informed understanding of complex matters" (Barnett 1990: 176), from Christian theology to information and communication technology. Thus, universities are expected to cultivate eco-citizenry by promoting student and faculty involvement with ecological issues in a learned and informed manner. Ecologism and sustainability discourses emerge from this informed and increasingly interconnected international university community. However, the complexity of university structures and processes makes it difficult to articulate and categorize the existing eco-discourses on campus, with a view to finding ways to develop them further or to generate new pathways.

This paper outlines a taxonomy of ecological discourses in HE campuses by making a comparative analysis between sustainability discourse and ecologism in two campuses with different cultural and historical backgrounds, namely Duke University (Duke) and the University of Hong Kong (HKU)[3]. This study follows the earlier work of Savelyeva and McKenna (2011) on the three different forms of campus sustainability – greening, sustainability sciences, and education for sustainability – and intersects these three types of sustainability

3 Duke University sustainability website: http://sustainability.duke.edu; University of Hong Kong sustainability website: www.sustainability.hku.hk.

applications on university campuses with complexity theory's categories of linear, complicated, and complex practices.

Preamble: Ecologism, sustainability, and complexity science

It may be useful to start by differentiating ecology from ecologism. Ecology is the study of ecosystems and is only marginally concerned about sustainability. By contrast, the more recent concept of ecologism consistently juxtaposes sustainability discourse, which adds a clear difference: ecologism is a movement for and with action. Ecologism suggests, demands, and emphasizes the need for a paradigm shift in humanity's relationship with nature, which should bring about fundamental changes in societal and political lives. Ecologism calls for major changes in very assertive or provocative terms, and often assigns priority to the survival of nature over man. Such practical reasoning makes ecologism a moral and ethical discourse. The call for activism, its connection to ethics, and the slowly developing theoretical and ideological roots (Baxter 1999) make ecologism attractive to intellectuals and welcomed by academic communities. Cases in point could be the *Greenpeace* flagship confronting a whaling boat a more action-oriented activism and the "Do not eat blue-fin tuna sushi" campaign from the *World Wide Fund for Nature* (WWF), which is more of a moral education lobby. The "radical green" or "deep ecology movement" (Sessions 1987) brand of ecologism arouses strong attitudes to the point of suggesting, for example, a "landscape without humans," however unfeasible it might be. A milder form of ecologism considers the current environmental predicament as rooted in the "metaphysics of mastery," that is, the dominance of anthropocentrism in Enlightenment humanism that would have caused subordination and exploitation of nature (Bonnett 2007).

Sustainability is the youngest sibling in the ecologism family. It is perhaps the simplest among ecological discourses, and it has become popular since the 1980s. This integrated and integrating discourse appears reasonable and inclusive, as it covers environmental concerns at local and global levels and requires practical movements toward a gradual change in relationships with nature. However, what sustainability means in theory has never been crystal clear within the academic community, regardless of it being the cause of comprehensive debates at all levels of policy and practice. The way something can become sustainable makes it instantly an evolving and longitudinal discourse, highly scalable/negotiable, and, therefore, most relevant to academic institutions and individuals alike.

A theoretical engagement with campus ecologism and sustainability cannot be addressed comprehensively with "either/or" reasoning of classic science. In contrast, complexity science (CS) offers a conjunctive "and" framework for reasoning that allows better handling of some deeply ill-structured and problematic situations in the domain of social sciences (Delorme 2010). CS is a body of multidisciplinary ideas and principles that has been constructed from various domains of knowledge (Page 2011; Waldrop 1992) and offers an analytical tool to understand a wide range of phenomena from ordinary to institutional level events of human life, such as organization learning and corporate management. What CS thinkers have in common is the ability to take a closer look at certain *qualia* of the realities. In relation to sustainability, these realities might include predictability, causality, hierarchy of control, and, where human freedom becomes involved, accountability. These qualities are generally grouped into three degrees: linear (non-complex causal relation), complicated (numerous but predictable rules), and complex (emergence). CS reiterates that "classic" natural and social sciences have long been falling short in their descriptive and explanatory endeavors because they mostly rely on linear and, at most, complicated paradigms. Thus, CS positions itself by differentiation from classic scientific paradigms and chaos theory: "[N]ormal science shows how complex effects can be understood from simple laws; chaos theory demonstrates that simple laws can have complicated, unpredictable consequences; and complexity theory describes how complex causes can produce simple effects" (Anderson 1999: 217). A succinct, yet comprehensive, list of the most frequently visited issues and concepts in CS includes:

- Interconnected and interdependent elements and dimensions;
- Feedback processes promote and inhibit change within systems;
- System characteristics and behaviors emerge from simple rules of interaction;
- Nonlinearity;
- Sensitivity to initial conditions;
- Phase space the "space of the possible";
- Attractors, chaos and the "edge of chaos";
- Adaptive agents;
- Self-organization;
- Co-evolution.

(Ramalingam and Jones 2008: viii)

Method

Two campuses are arbitrarily selected – Duke University and HKU – followed by data collection from the websites created by their sustainability offices, complemented by interviewing sustainability-related staff as well as information from openly available textual narratives and artifacts. The data are manually analyzed employing reflective-interpretive approaches to analyses.

Results: Complexity science and campus sustainability actions

The three degrees of complexity and CS's conceptual paradigms can now be intersected with the framework by Savelyeva and McKenna (2011), which describes three main forms of university-based actions for sustainability (greening, sustainability sciences, and education for sustainability). Related theoretical considerations can be illustrated with two cases of campus sustainability. This study is built on the assumption that HE institutions are multifaceted in their organizational structure and physical sites, with a wide range of enactments of ecological discourses. An easy approach to ecologism is to operate with enforcement of a set of regulations, which is evident in the case of campus greening.

Greening: a linear policy and complicated action

Campus greening is the strongest and the most visible sustainability application in contemporary universities (Savelyeva and McKenna 2011). Greening serves as an assessment yardstick of overall campus sustainability and aims to increase the efficiency in all university operations and inspire sustainable actions. Campus greening policies and initiatives can create an ecologically sensitive academic environment and establish a results-oriented culture of sustainable action. They also exemplify and provide visions of innovation and technological advances that create a positive image for the university. Greening includes a standard list (Keniry 1995) of greening initiatives that are visible on both Duke and HKU campuses:

- Constructing Leadership in Energy and Environmental Design (LEED) certified buildings and retro-fitting old campus facilities;
- Including vegetarian and organic meal options in university catering;

- Updating campus infrastructure with energy and water efficient technologies;
- Using local and organically produced foods in food services;
- Organizing campus recycling initiatives;
- Participating in a sustainability assessment program;
- Including environmental impact specification for cleaning products, laboratory equipment, paper, and building materials.

The linear nature of greening policies allows administrators to explain the multifaceted sustainability action in terms of three well-defined elements: reduce, reuse, and recycle. The goal is predictability (if you recycle one aluminum can, you can save half a gallon of gasoline) and control (if you can control energy consumption, you can control a university budget). Another important characteristic of the linear nature of greening is that it presumes consistency and steady conditions. For example, given the same general conditions for enforcing a green policy on campus, a university usually achieves corresponding results, say, reduction in energy cost, waste production, or water consumption. When conditions are changed (*e.g.* enforcement of new room temperature standards for lecture halls), the outcome (reduction in energy consumption) will settle to a new equilibrium until the next policy takes effect.

Linearity in greening has a number of significant consequences, some intentionally sought, and others unintended. First, it helps universities to materialize sustainability efforts and to measure achievable outcomes provided that proper planning, analysis, and assessment are simultaneously conducted. Second, it narrows the scope of policymaking to a linearly definable scale. Thus greening often allows campus leaders to break down sustainability into piecemeal processes so that even a less-endowed university can make small steps toward its envisioned sustainability. And third, it tacitly makes a "false promise" of absolute control that prevents participants from understanding more complex problems and corresponding solutions required for true sustainability. This is obviously an unintended side-effect of linear greening efforts, which raises the unsolved problems to complicated and complex levels.

However, not all campus-based greening efforts are linear. The large variety of scenarios that draw students and faculty into greening efforts offers a glimpse of the complicated side of campus sustainability. For example, installing recycling bins for sorting waste might not necessarily inspire people to change their habits of dumping all waste in the same garbage bin. To make this happen, universities might need to use media resources to announce and monitor the initiative, educate the public about recycling practices, and promote recycling on their campuses. Analogous actions might be needed for reusing and reducing. It is obvious that greening policies are received by the campus community members with different levels of responsiveness, suggesting the need for a more complicated approach to sustainability, which will be discussed in the following sections.

Sustainability sciences: complicated business

Savelyeva and McKenna (2011) described a second type of campus-based sustainability efforts as "sustainability sciences," with research being its main feature. The idea is that universities as research institutions "could and should develop much of the knowledge base for an ecologically enlightened society" (Mathisen 2006: 111). Campus-based sustainability sciences are complicated (in the taxonomy of CS) and they often transcend disciplinary boundaries. They include either collaborative or single field research projects focused on sustainability and sustainable development (Fig. 1).

Fig. 1: Sustainability research domains on HKU and Duke campuses

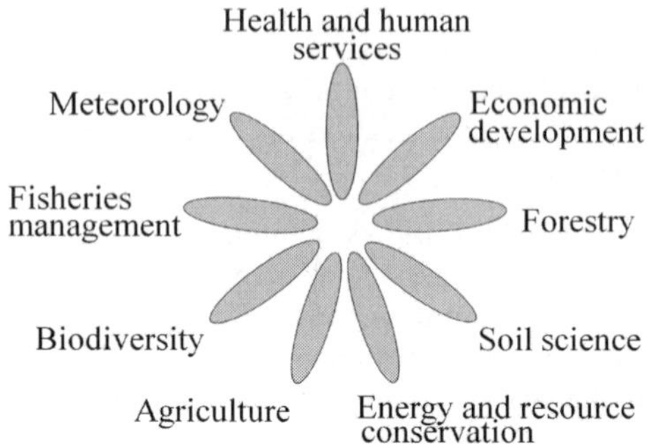

Note: The absence of linking lines among the concentric ovals illustrates the challenge of creating inter- and trans-disciplinary connections.

The complicated nature of sustainability sciences on campuses is attributable to their broad range of disciplines, fields, and topics from the natural sciences to business and public policy. However, the mere presence of sustainability-related research in a university does not make that university sustainable. Existing research projects must be inventoried, and new studies must be launched in coordination without compromising the academic freedom of the faculty and research fellows. A university usually creates a separate research center to identify and monitor sustainability-related research studies, attract governmental and corporate sponsors, and house the new studies on sustainability-related topics.

Education for sustainability: a complex story

In contrast to the divergence of research-centered sustainability sciences, several models of Education for Sustainability (EfS) converge in infusing sustainability into a university's teaching and learning that are intended to engrave stable and lasting dispositions in students and academic staff (Savelyeva and McKenna 2011). At Duke, sustainability-related courses are offered primarily in the Nicholas School of the Environment via its undergraduate, graduate, and professional degrees and programs in biology, environmental sciences, and earth sciences. In its Faculty of Science and School of Biological Sciences HKU also offers majors, minors, and degree programs to undergraduate and graduate students with a focus on issues related to sustainability. Students can take electives on broader sustainability issues (climate change, social responsibility, sustainable building design) in the Faculties of Engineering, Architecture, Business and Economics, Law, and Social Sciences.

The most common ways of delivering EfS to faculty and staff are sustainable-practices training (for administrative staff) and "curricular greening" workshops (for faculty members). These initiatives at the curricular level are typically offered with the support of external grants, and the access to funding generally affects whether a university can offer such workshops. All of these efforts to introduce EfS into teaching and learning, either through adding a sustainability course to the roster or offering a training session to staff and faculty, contribute to the noble goal of capacity building aimed at a particular type of "global citizenship" (Demaine 2004; Pike 2008) by fostering concrete attitudes and values that will produce "ecologically sound" graduates, faculty and administrative staff.

The primary objective of EfS is guided by several United Nations (UN) declarations and initiatives, such as the ongoing 2005-2014 UN Decade of Education for Sustainable Development.

However, almost 40 years after these international declarations and signatories' commitments, EfS remains marginalized and delimited to a particular discipline on campus or a lecture topic within a course. Creative, cross-curriculum, interdisciplinary ways of teaching are rather exceptional due to the disciplinary compartmentalization within university systems themselves (Savelyeva 2012).

The perspective from complexity theory can contribute to a better understanding of the dynamics of classroom relationships regarding sustainability. Taking cues from Miller and Page (2007) it could be argued that EfS should be seen as a "complex adaptive system" rather than simply an "adaptive system." First, multiple agents are involved namely, students, faculty, and administrative staff with various degrees of participation. Second, these agents are not isolated, but interdependent. This tenet has been marked as reflexivity by CS theorists. Third, there is diversity among agents, yet each contributes to their organization with their uniqueness and differences, which provides cohesion in EfS (ref. Sargut and McGrath 2011). For example, students bring in new systemic properties (new agents, ideas, and problems) to every EfS course or training program. These properties are essential for "emergence" in a complex adaptive system, which typically makes "the whole more than the sum of parts." In EfS, such emergence happens during the process of teaching and learning, and they cannot be predicted. For example, an unanticipated yet relevant issue could emerge during a class discussion that might lead to deeper understanding of sustainability. Second, any small difference, called "disturbance" (for example, diverse opinions about climate change within a group) gets amplified through the interaction between faculty and students. This amplification of minor differences holds the potential for short- or longer-term changes in students' and faculties' perceptions of sustainability.

Not all disturbances lead to a major shift in perception and understanding of sustainability, just like not every graduate of the university becomes a global eco-citizen with deeply rooted sustainability ideals. This points to another complex property of the sustainability discourse: unpredictability of outcomes. Accepting the non-linear character of outcomes of EfS helps one to embrace the paradoxical nature of bi-directional relationships within educational processes and to view the lack of absolute control as normal. This acceptance helps ease the perception of personal or pedagogical failure associated with linear implementation sustainability practices. Recognizing and anticipating unpredictability is also a goal of producing a global eco-citizen more realistically. As the focus on measurable outcomes of campus sustainability programs becomes less urgent and somewhat shocking, more attention can be paid to immediate interactions (student-faculty), available resources (existing courses), and defining relationships (agents-nature).

Discussion and conclusion

The findings indicate that ecologism and sustainability discourses of two culturally and geographically distant campuses show significant similarities in:

- efforts to meet eco-standards and rank highly in international ranking systems;
- emphasis on greening over EfS and sustainability research;
- devotion of sustainability efforts to appear as an eco-compliant institution before the general public;
- mostly linear approach to ecologism and sustainability efforts; and
- tacit assumption that complex intricacies of campus sustainability and issues of accountability are not really accessible to the public.

The findings draw a double dichotomy picture of campus sustainability: a dichotomy between campus sustainability discourse and the university's public image, and this in turn with an international ranking of excellence in sustainability. This dichotomy is a result of the HE emphasis on quick and visible sustainability image-building efforts that might be found in other campuses around the world. Although these are reassuring signs of sustainability implementation efforts on campuses, whether university alumni will continue participating and growing as ecological citizens after their graduation is difficult to predict Nor is it certain that academic and non-academic staff will lead their off-campus lives conforming to their informed sustainability ideals. The double dichotomy phenomenon makes sustainability discourse look identical to that found in universities of comparable rank, but the perceptions of those involved can be drastically different. HKU does have a reputation as a green university. However, this perception is countered by a disastrous sustainability situation in the region. This implies that little progress has been made outside the HKU campus. By contrast, Duke enjoys a positive reputation as a highly advanced sustainability-driven university that serves an ecologically concerned community.

The bottom line is that campus sustainability efforts should render more stable habits and lifelong ideals among people in HE. Indeed, the main *raison d'etre* of the university as a social institution is the formation of future citizens, who are expected to be ecologically informed and proactive. This imperative for nurturing a sense of ecological citizenship can be illustrated with an analogy borrowed from Ancient Greek philosophy "ensoulment". In short, ensoulment means the incorporation of a soul into a physical body of living beings.

In the context of sustainability discourse in HE, ensoulment can illustrate the need for higher level of acquired sustainability values, the "ecological soul", for the HE public rather than a merely mechanistic and linear delivery of green knowledge, green rules and green initiatives. But ensoulment is not the same as simply knowing a set of sustainability values. An ecological ensoulment also suggests a pedagogical aim in HE, that is, helping people to embody those values with a firm intellectual commitment and lasting predisposition to sustainability actions. To achieve this state of ecological ensoulment, diverse ways might be considered by HE educators and academics; from classical approaches of moral education scaffolding with a particular focus on the virtue of justice (Kohlberg 1981) and character education (Lickona 1992) to the approach of dual inculcation of "deep ecology movement" and "deep ecology" (Drengson and Inoue 1995; Naess 1973). The real challenge for any of these approaches to ecological ensoulment lies beyond short-term and action-driven campus greening, in a categorically different type of EfS through which ecological citizenship can shine forth from within. Finally, the idea of ensoulment supports a non-mechanistic research perspective, which takes into account "the obvious richness and complexity of human … life" (Johnson 1992: 56). A more conjunctive "and" instead of an "either/or" reasoning (Delorme 2010), as well as multidisciplinary and across-discipline approaches to sustainability research can offer better solutions to deeply ill-structured sustainability problems inside and outside the campus. Without ecological ensoulment, not only will ecological citizenship remain illusory or incomplete, but also a vast amount of invested resources, time, and effort to produce an ecologically sound public would not have been worthwhile.

The main lesson of this study is that both Duke and HKU face the task of campus sustainability in similar ways with comparable greening standards and research orientations. The sustainability discourse as currently articulated and practiced at Duke and HKU, and possibly elsewhere in the global HE system, appears to fall short of the ensoulment of ecological values and sustainability ideals. Not without paradox for these educational institutions, EfS appears as the weakest and most neglected among the three types of campus sustainability efforts (greening, EfS, and sustainability science), when it is absolutely essential for full materialization of ensoulment of sustainability ideals and true ecological citizenship. Perhaps further theoretical and practical developments that take into account the reordered priorities suggested in this paper could open new pathways to success.

References

Anderson, P. (1999), Complexity theory and organization science. *Organization Science* 10 (3), pp. 216-232.

Barnett, R. (1990), *The idea of higher education.* Society for Research into Higher Education and Open University, Bristol, England.

Baxter, B. (1999), *Ecologism: An introduction.* Georgetown University, Washington, D.C.

Bonnett, M. (2007), Environmental education and the issue of nature. *Journal of Curriculum Studies* 39 (6), pp. 707-721.

Carson, R. (1962), *Silent spring.* Houghton Mifflin, Boston.

Delorme, R. (2010), *Deep complexity and the social sciences: Experience, modelling and operationality.* Edward Elgar, Northampton, Mass.

Demaine, J. (2004), *Citizenship and political education today.* Palgrave Macmillan, London.

Dewey, J. (1954), *The public and its problems.* Swallow Press, Athens.

Dobson, A. (2004), Ecological citizenship and global justice: Two paths converging? In Haugestad, A. and Wulfhorst, J. (Eds.), *Future as fairness: Ecological justice and global citizenship*, pp. 1-15. Rodopi, Amsterdam.

Drengson, A.R. and Inoue, Y. (1995), *The Deep ecology movement: An introductory anthology.* North Atlantic Books, Berkeley, Calif.

Dryzek, J.S. (2005), *The politics of the earth: Environmental discourses* (2nd ed.). Oxford University Press, New York.

Fonseca, A., Macdonald, A., Dandy, E. and Valenti, P. (2011), The state of sustainability reporting at Canadian universities. *International Journal for Sustainability in Higher Education* 12 (1), pp. 22-40.

Gee, J.P., Hull, G. and Lankshear, C. (1996), *The new work order: Behind the language of the new capitalism.* Westview, Boulder, Colo.

Johnson, L. (1992), Society, culture, and character development. In Ryan, K. and Lickona, T. (Eds.), *Character development in schools and beyond*, pp. 49-78. CRVP, Washington.

Keniry, J. (1995), *Ecodemia. Campus environmental stewardship at the turn of the 21st century.* National Wildlife Federation, Washington, DC.

Kohlberg, L. (1981), *The philosophy of moral development: Moral stages and the idea of justice.* Harper and Raw, Toronto.

Leal Filho, W. (Ed.) (2011), *World trends in education for sustainable development*, Peter Lang, Berlin.

Lickona, T. (1992), *Educating for character: How our schools can teach respect and responsibility.* Bentam, Toronto.

Mathisen, W.C. (2006), Green utopianism and the greening of science and higher education. *Organization and Environment* 19 (1), pp. 110-125.

Meadows, D.H. and Club of Rome (1972), *The Limits to growth: A report for the Club of Rome's Project on the Predicament of Mankind.* Universe Books, New York.

Miller, J. and Page, S. (2007), *Complex adaptive systems: An introduction to computational models of social life.* Princeton University, Princeton, N.J.

Naeem, M. and Peach, N. (2011), Promotion of sustainability in postgraduate education in the Asia Pacific region. *International Journal for Sustainability in Higher Education* 12 (3), pp. 280-290.

Naess, A. (1973), The shallow and the deep, long-range ecology movement. A summary. *Inquiry: An Interdisciplinary Journal of Philosophy* 16 (1), pp. 95-100.

Palma, L., Oliveira, L. and Viacava, K. (2011), Sustainability in Brazilian federal universities. *International Journal of Sustainability in Higher Education* 12 (3), 250-258.

Page, S. (2011), *Diversity and complexity.* Princeton University, Princeton, N.J.

Pike, G. (2008), Reconstructing the legend: Educating for global citizenship. In Abdi, A. and Shultz, L. (Eds.), *Educating for human rights and global citizenship.* State University of New York, Albany, pp. 223-237.

Ramalingam, B. and Jones, H. (2008), *Exploring the science of complexity: Ideas and implications for development and humanitarian efforts* (2nd ed.). Overseas Development Institute, London.

Sargut, G. and McGrath, R.G. (2011), Learning to live with complexity. *Harvard Business Review* 89 (9), pp. 69-76.

Savelyeva, T. (2012), Escaping the 'structural trap' of sustainability in academia through Global Learning Environment. In Bartels, K. and Parker, K. (Eds.), *Teaching sustainability / Teaching sustainably,* pp. 231-241. Stylus, Sterling, Va.

Savelyeva, T. and McKenna, J. (2011), Campus sustainability: Emerging curricula models in higher education. *International Journal for Sustainability in Higher Education* 12 (1), pp. 55-66.

Sessions, G. (1987), The deep ecology movement: A review. *Environmental Review* 11 (2), pp. 105-125.

Waldrop, M.M. (1992), *Complexity: The emerging science at the edge of order and chaos.* Simon and Schuster, New York.

Chapter 16

Faculty Enrichment Program for Infusing Sustainability Education across the University Curriculum

William M. Stigliani, Catherine Zeman and Gowri Betrabet Gulwadi

Abstract

An essential role of higher education is to prepare students with the knowledge, skills, and habits of mind needed to actively engage them in creating a sustainable future. Key players in achieving this momentous task are the teaching faculty who write course syllabi, develop student learning outcomes, and sit on curriculum committees. For many faculty, however, sustainability is a new concept, about which they lack an organized framework, training and teaching experiences. This paper describes how a regional university in the United States is engaging 26 core faculty leaders strongly committed to sustainability education, and willing to learn the skills required to teach it. The core, representing 19 diverse disciplines, participated in a two-day leadership workshop that immersed them in sustainable thinking and action. Through a focus on systems thinking as a framework for sustainable thinking, carefully selected presentations on topics such as environmental justice and energy for sustainable agriculture, and multidisciplinary discussion-based activities, each participant learned how sustainability could be integrated into their courses, while also broadening and enriching the context of what they teach. This paper outlines the workshop structure, process, activities and discussions, and presents highlights of revised curricula developed by the core faculty for embedding sustainability into their coursework. Post-workshop evaluations and reflections by participants and workshop coordinators demonstrated the invaluable usefulness of the multidisciplinary approach, the importance of a supportive administration, and the provision of a structured but flexible space in which to explore the complex connections of sustainability to their courses. This paper will be of interest to other university teams with similar goals of infusing sustainability in the curriculum.

Introduction

Institutions of higher education have a unique role to play in efforts to create a more sustainable society. They serve as creative incubators for teaching the technologies, policies, and behavior changes needed to make the age of sustainability a reality. With a graduation rate of approximately five million per year from US colleges (NCES 2011), an army of students equipped with this knowledge constitutes the single, greatest means for technology transfer that colleges and universities can offer society. Recognizing the central role of education, UNESCO's "Education for Sustainable Development" (ESD) called for international education initiatives to sustain the four dimensions of sustainability – environment, society, culture and economy (UNESCO 2006). The period 2005-2014 has been designated the Decade of Education for Sustainable Development. Curriculum greening efforts at many universities represent the goals of ESD and recognize a need for more integrative, collaborative and interdisciplinary approaches to learning – ones that involve "training and retraining of educators" (UNESCO 2006: 28) and engage "deep learning" (Warburton 2003). However, this integrated transformative experience cannot happen within disciplinary boundaries and cannot happen overnight. For example, when Abu-Hola and Tareef (2009) surveyed university staff in Jordan, they found that training educators is a necessary component of this transformation.

Pioneering universities in the US such as Northern Arizona and Emory were among the first to create immersive two-day experiences for faculty to reach beyond disciplines (Bardaglio 2007). Through the Ponderosa (Chase and Rowland 2004) and Piedmont projects (Barlett and Chase 2004) there has been an increase in faculty who created or revised courses related to sustainability. Curricular collaborations between educators from diverse disciplines can help students benefit from interdisciplinary dialog, but also point to challenges in releasing students from their disciplinary comfort zones (e.g. Kurland et al. 2010). Successful curricular models have engaged students actively in praxis and real-world problem-solving to communicate the systemic complexity of a sustainable world (e.g. Bacon et al. 2011). Other successful efforts engage faculty and students in a curriculum structured to address learning outcomes related to sustainability (e.g. Allen et al. 2010).

Incorporating education for sustainable development can take place in one of the following ways proposed in a matrix of options (Rusinko 2010): 1) Integrate into existing course(s) minors, majors or programs; 2) integrate into common core requirements; 3) create new discipline-specific sustainability courses, minors, majors or programs; or 4) create new cross-disciplinary sustainability courses, minors, majors, or programs. Integration into existing courses or creating new courses requires interdisciplinary dialog and reflection. A major obstacle to the educational mission is that faculty members teach and conduct research within the confines of their disciplines. They often lack opportunities and support to partake in multidisciplinary enrichment programs that would help them infuse sustainability into their coursework. The faculty leadership program described in this paper is helping to bridge this gap by assembling a team of highly motivated faculty at the University of Northern Iowa dedicated to conversing, learning, and teaching about sustainability from an array of perspectives. Such an experience has broadened their understanding of the multiple dimensions of sustainability, and generated innovative ways of integrating sustainability perspectives into their courses.

The University of Northern Iowa (UNI) made an early commitment to sustainability in 1990 as one of 19 original signatories to the *Talloires Declaration of University Presidents for a Sustainable Future* (Talloires Declaration 1990). UNI has continued that commitment by citing sustainability as a core value in its Strategic Plan, 2010-2015 (UNI 2011), and hiring a Sustainability Coordinator to oversee various sustainability initiatives on campus. In 2011, UNI became one of 23 institutions of higher education in the US to receive a Gold Star ranking from the Association for the Advancement of Sustainability in Higher Education (AASHE) as measured by its *Sustainability Tracking and Assessment and Rating System* (STARS 2012).

Until recently, however, these institutional actions and commitments were not matched by complementary actions among the faculty to promote sustainability in the curriculum, even though, at last count in 2010, there were 45 sustainability-focused and 149 sustainability-related courses at UNI. Moreover, despite having a general education core (liberal arts), there are few interdisciplinary courses in place to engage students in "deep learning" and transformative experiences. About two years ago, Dr. Gloria Gibson, UNI's Executive Vice President and Provost, and UNI's Deans began encouraging interested faculty to explore possibilities for introducing sustainability in the curriculum. The Provost hosted several meetings with faculty members and Deans to discuss options, and last year she provided monetary support for the "UNI Faculty Leadership in Sustainability Education Program." Without this support and encouragement from senior administrators, it is doubtful that this program could have been launched. Twenty-six enthusiastic faculty members representing a wide array of disciplines, are now revising curricula and infusing sustainability into their coursework as a direct result of participation in this program. UNI is the only public university in the state of Iowa to initiate such a faculty support program. The three co-directors of the program are the authors of this paper.

The program began with a two-day workshop in June, 2011. It was guided by the highly successful Ponderosa/Piedmont model (Barlett and Chase 2004) endorsed by AASHE. Barlett and Rappaport (2009) have shown that the model creates enduring changes in teaching, research, interdisciplinary collaboration, and engaged action. AASHE sponsors workshops that train faculty members to apply the model on their own campuses. One of the authors of this paper (W.M. Stigliani) attended the AASHE-sponsored "Sustainability across the Curriculum Leadership Workshop" in June, 2010.

Overview of the UNI Faculty Leadership in Sustainability Education Program

The plan for the program entails the following distinctive steps:

- Recruitment and selection of faculty participants (January-February, 2011);
- Preparation for workshop (posted, beginning of May);
- Structure and content of workshop (June 9-10, 2011);
- Participants revise course syllabi to include a sustainability component (approximately three weeks in summer, 2011);
- Half-day workshop at which the participants submit their revised syllabi (August 19, 2011);
- Participants test their revised courses in classes they will teach (academic year, 2011-2012);
- Participants evaluate the effectiveness of their revised courses in educating their students about sustainability (March 30, 2012 dinner with Provost);
- Program co-directors submit final report to the Provost (June, 2012).

All steps will be covered in this paper – except for the last two, which are works in process.

Recruitment and selection

The degree of interest in sustainability among the 800 UNI faculty was assessed via email polling and meetings with the Deans of the five Colleges for advice about potentially interested faculty. From this inquiry, a list of approximately 80 faculty members was developed, to whom workshop applications were sent. The application form required each applicant to agree to participate in all activities associated with the program, and submit preliminary ideas about how they might include some aspect of sustainability in the courses they would teach in academic year 2011-2012. Five reviewers selected 26 applicants for the program, representing all five Colleges and 19 Departments (see Table 1).

Table 1: UNI faculty members participating in the Faculty Leadership in Education Program indicating their college and departmental affiliations

College	Department (number of participants)
Natural Sciences	Biology (2) Industrial Technology (4) Mathematics (1) Physics (1)
Social and Behavioral Sciences	Geography (2) Graduate Program in Counseling (1) History (1) Psychology (1) Social Work (1) Sociology, Anthropology, and Criminology (2)
Humanities and Fine Arts	Art (1) English Language and Literature (2) Leadership Studies/Communication Studies (1) Philosophy and World Religions (1) Theatre (1)
Education	Curriculum and Instruction (1) Health Promotion and Education (1) Leisure, Youth & Human Services (1)
Business	Management (1)

Number of participants in each department indicated in parentheses

Preparation for workshop

Workshop planning involved selection of speakers, presentation topics, and required readings that included a topic of bioregional significance. Ample time for multidisciplinary sharing of ideas was a key focus. Workshop participants were divided into five discussion groups, each comprising at least five faculty members from a range of disciplines. None of the groups had faculty members from the same department, and all the groups had representation from the sciences, humanities, and fine arts. For example, one group comprised faculty members from the departments of theater, physics, industrial technology, geography, and history; another had representation from the departments of psychology, industrial technology, biology, English language and literature, and leadership studies. The discussion groups met five times during the workshop, offering their reflections on presentations and other topics that were shared at plenary sessions. A project evaluator was recruited, and administered a pre-workshop survey of the participants. A website dedicated specifically for use by the 26 faculty members and program co-directors was created on eLearning, UNI's centrally supported online learning management system. All important information pertaining to the program was posted on the website, thus greatly reducing the consumption of paper.

Structure of the workshop

The UNI Faculty Leadership in Sustainability Education workshop was convened on June 9-10, 2011 from 8:45 am to 5:00 pm at UNI's Center for Energy & Environmental Education (CEEE). A summary of the agenda for the workshop is given in Table 2.

The goals of the workshop were: strengthening intellectual engagement and sophistication; fostering multidisciplinary dialog; and developing pedagogy for teaching sustainability. The workshop participants heard keynote presentations on topics that specifically addressed these goals. Each presentation was followed by breakout sessions in which participants joined their assigned discussion groups, and each group member shared reflections on the relevance and connections of the topic presented to their coursework and discipline. Thus, participants gained new perspectives on the topic from the various disciplines represented. Notes from the group discussions were collected and posted on the workshop's eLearning website.

Table 2: Workshop agenda

Calendar	**Event**
Evening before workshop	Get-together dinner for participants, program co-directors, Deans, and the Provost.
Workshop: Day 1	• Keynote presentations: ➢ Environmental justice ➢ Systems thinking ➢ Energy for sustainable agriculture • Small discussion groups meet in breakout sessions after each keynote presentation to converse on how the presentation is connected with the courses they teach; groups later report their deliberations at plenary sessions. • Other presentations: ➢ Sustainability – What is it? How do we connect? ➢ Eight ways to change a course and connections to the humanities. • Other activities: ➢ Prairie walk ➢ Poetry reading ➢ Reflection exercise/quiet time • Whole group synthesis • Overnight homework assignment: Principles of sustainability and "Big Ideas" in your discipline.
Workshop: Day 2	• Participants report on their homework at plenary session. • Other presentations: ➢ Sustainability at UNI. ➢ Design and its applications to sustainability. ➢ Classroom technology assistance at UNI. ➢ Video presentation "What's the economy for anyway? • Breakout sessions with small group discussions on "What's the economy for anyway?" and "Sustainability, assessment, and learning outcomes." • Other activities: ➢ Display table for browsing books covering various aspects of sustainability. ➢ Flexible time; collecting thoughts for final whole group reflection. • Wrap up: ➢ Whole group reflection and discussion; expectations for summer homework assignment. ➢ Request participants to complete post-workshop survey.

Methodology: Systems Thinking as a Framework for Sustainable Thinking

Systems theory and thinking were developed to study complex systems, and there is a rich historical tradition of their application to numerous areas of study (see Box 1). Because the interactions between the biosphere and human society are inherently complex, systems thinking can be a powerful unifying principle for diverse scholars to examine sustainability through their unique lenses within the common framework of systems analysis (Senge et al. 2008; Klein 2005; Kaufman et al. 2003; Gumport and Snydman 2002; Rover 2002; Kline 1995 and Senge et al. 1994). Despite the capacity for systems thinking to be a foundational component of sustainability education for faculty and students, little specific emphasis has been placed on elevating it to such a status, or testing the potential of its contributions to the pedagogy of sustainability. The adoption of systems thinking at the UNI workshop as a primary organizing sustainability principle is a unique contribution to faculty workshops structured around the Ponderosa/Peidmont model. Moreover, as the UNI faculty apply systems thinking in teaching their revised courses, its efficacy as a pedagogical foundation will be evaluated as the methodology is practiced and tested in the classroom. C. Zeman, who was formally trained in systems thinking history, techniques, and applications to sustainability and leadership problems at the Centers for Disease Control, Environmental Public Health Leadership Institute (EPHLI), will lead the evaluation in collaboration with the other co-directors, the course evaluator, and the participating faculty. The evaluation will be extensively covered in the final report to the Provost.

Box 1 – Evolution of systems theory

While system thinking's origins can be traced back to pre-historical societies and there is a long, rich history of systems thinking development in the arts and sciences, modern systems thinking dates to the efforts of the biologist Ludwig von Bertalanffy and his publication on General Systems Theory (Bertalanffy 1949/1952). Since then, the principles of systems thinking have been shaped by theorists in fields as diverse as philosophy, physics, biology, engineering, social/behavioral fields and business management. Particularly prominent among the theorists and practitioners are Peter Senge, Fritjof Capra, and, in the environmental sciences, Donella Meadows (2008). Meadows describes systems analysis approaches as a means of identifying and studying the parts of complex problems, how they affect one another, how they act together in a system, how that system responds to outside forces and maintains its characteristics over time, and how the system outcomes/results or direction can be changed by changing the relationships between the parts. Systems theorists have determined a number of archetypical systems processes that help to explain how systems function and help develop system changes designed to improve systems outcomes and remediate systems-related problems.

Content of Workshop

Keynote presentations

Presentation 1: Environmental justice. Professor Edwardo Rhodes of Indiana University presented this topic, which entails a common syndrome in the United States where minorities and the poorest citizens bear an unfair share of environmental risk. The topic involves competing economic, social, and environmental concerns that are intertwined in complex ways (Rhodes 2003). They can only be addressed from a multidisciplinary perspective that includes civil rights, political enfranchisement, community healthcare, stricter environmental standards and laws, better education and availability of information, and enhanced job creation, particularly through investment in green industries.

Presentation 2: Systems thinking. Professor Catherine Zeman of UNI presented this topic, which provides an effective pedagogical technique for teaching sustainability. Systems thinking is a formalized way of analyzing relationships of complex systems, and has been applied for decades in the fields of engineering, psychology, and ecology (Meadows 2008). Nature demonstrates that the world sustains itself as a *system*, and human activities that affect it can only be understood by knowing the interactions between the natural and human parts of the system. Dr. Zeman provided participants with the basic tools of systems diagramming and thinking. The presentation stressed that a multidisciplinary, whole-systems framework is implicit for "thinking sustainably," and that it offers great potential value to students' critical thinking about sustainability. It also provides a common "language" and starting point for academics from diverse fields to interact around the theme of sustainability.

Presentation 3: Energy for sustainable agriculture. Dr. Francis Thicke, an organic dairy farmer from Fairfield, Iowa, presented this topic (Thicke 2010). He emphasized the multidisciplinary perspective inherent in a discussion about sustainable agriculture as an alternative to conventional farming. Perhaps even more important, his talk focused on an issue directly relevant to the state of Iowa and much debated among Iowa's citizens. Farms make up about 92% of Iowa's land, and about one-third of the best farmland in the United States is located in Iowa. Most of the state's residents are in some way dependent upon Iowa's fertile soils and crops. Thus, any topic associated with agriculture is not an abstraction to Iowa's citizens, because it is connected to the place where they live. As noted by DeLind and Link (2004), sense of place is important because sustainability, rooted as it is in the earth itself, cannot be taught in a detached, generic way. It requires an understanding of ourselves and our behaviors through affiliation with the space we inhabit. The "pedagogy of place" has been effectively applied to teach sustainability (Barlett 2005). Table 3 shows a sampling of the reflections on the keynote presentations by the small discussion groups during breakout sessions.

Table 3: Sample of ideas of discussion groups on how keynote presentation topics were connected to courses they teach.

Presentation topic	Summary statements of discussion groups
Environmental justice	*Recreation management:* discuss lack of involvement of minorities in leisure activities, and cultural disparities that discourage involvement. US history since 1887: discuss not only inequities involving civil rights, but also unfair exposure to toxic chemicals in minority communities; assign student project analyzing local hazardous waste sites and the communities most affected.
Systems thinking	*Business management:* study corporations as part of larger system that interacts with society and the environment; learn about "triple bottom line" for sustainable companies (equal emphasis on people, planet, and profits). *Manufacturing:* conduct simulation studies on how products are linked to the consumer and society; include impact of globalization on product use and need for regulating products from different parts of the world. *Transportation planning:* consideration of multiple objectives requiring systems thinking with elements (nodes/terminals, links/roads joining places, areas being served) and their interactions.
Energy for sustainable agriculture	*English language and literature:* Iowans' sense of place is often tied to small town culture, and its gradual erosion associated with the rise of industrial agriculture; discussion of these topics could introduce ideas about sustainable agriculture and its potential for revitalizing rural areas. *Industrial technology:* global population growth is closely tied to the land's capacity for food production; what are the advantages and disadvantages of sustainable agriculture in Iowa, and as a model for global agricultural development?

Principles of sustainability and "Big Ideas" in your field

In an overnight assignment after Day 1, participants were asked to think about the most significant ways in which their disciplines connect with sustainability. In the morning of Day 2 they reported back their thoughts in a plenary session. Their numerous ideas can be summarized in a threefold pattern: confronting negatives; advocating positives; and engaging in the cycles of human, natural, and technological relationships. Examples are provided below:

- Economic policy based on consumption is not sustainable, and the pursuit of money and material wealth causes hoarding which leaves less for subsequent generations.
- War has enormous impact on population, land, water, and society in general; unnecessary resources are diverted into military expansion.
- Encourage activities that increase happiness and well-being; think about what is it in life, apart from monetary wealth and collection of "stuff," that makes living worthwhile?
- Creative works, and in particular poems, advance sustainability by insisting on the concrete rather than the abstract, presenting images that appeal to our senses.
- Technological processes can be aligned with natural processes.
- Conserving biodiversity and restoring native ecosystems benefit the health of the natural world.

Sustainability, assessment and learning outcomes

The discussion groups met to reflect on ways to evaluate student comprehension of sustainability concepts, and the types of learning outcomes that could serve as benchmarks for measuring student understanding. There was a focus on both what students should *know*, and what they should *do*. In addition to understanding, they should take active steps such as measuring their carbon footprint, advocating sustainable practices, participating in community projects, and achieving their own "sense of place" in connection with the natural world.

The day ended with a whole group reflection and sharing of ideas and insights participants had experienced during the workshop.

Workshop Evaluation

The goals of the workshop were: strengthening intellectual engagement and sophistication; fostering multidisciplinary dialog among the participants; and developing pedagogy for teaching sustainability. To assess the effectiveness of the workshop in meeting these goals, 10-question surveys were administered online, prior to and after the workshop. Generally, comments about the workshop were quite positive. The workshop was described as "great", "useful", "well run", "a very educational learning experience", with "overall enthusiasm among the participants", and "one of the best I have been involved in". There was also criticism. Some participants felt there was too much material to cover in two days. Others would have modified the organization of the small group discussions to allow for rotation of faculty to form new groups at the breakout sessions, rather than a fixed group membership throughout the workshop. Others hoped to continue the momentum gained at the workshop by sharing their ideas at informal get-togethers throughout the academic year.

Intellectual engagement and sophistication. In the pre-workshop survey, ideas about sustainability ranged from the simple and matter-of-fact, "the ability to maintain an action for the foreseeable future," to the more elaborate and inclusive, "the use of resources in such a way as they continue for future generations and are not depleted or destroyed but rather even enhanced." A majority of participants had included aspects of sustainability in their classes, but had not yet experienced teaching from multidisciplinary perspectives and systems thinking.

The post-workshop survey indicated participants gained new insights from presentations about environmental justice, systems thinking, teaching methods, and sustainable design. They indicated plans to teach sustainability in more specific and comprehensive ways. Three participants hoped to introduce more student engagement in out-of-class projects to enhance sustainability. Almost all participants hoped to continue improving their skills for teaching sustainability in their courses.

Multidisciplinary dialog. The workshop introduced core sustainability issues such as justice, systems thinking, energy, and design, all of which cut across disciplines and connect underlying principles. The agenda also included a poetry reading, a prairie walk, and a farmer's reflections. The discussion groups were specifically organized to connect different academic specialties from around the campus. Participants especially enjoyed what they called the "interactivity" of the workshop with "lots of brainstorming during the group breakout sessions." Eight participants viewed multidisciplinary connection with colleagues as the "single most important thing" they gained from the workshop. Four cited making new associations with four to six faculty members; three entertained the possibility of collaborating with colleagues in teaching sustainability courses; one said, "I made more meaningful connections with faculty from outside my department in two days than I had previously in four years at UNI."

Developing pedagogy for teaching sustainability. The pedagogical goal of the workshop was to "infuse sustainability into the coursework" by "co-discovering innovative ways of integrating sustainability perspectives into existing courses." In the post-workshop survey, participants indicated that they were going to develop a new course, revamp a present course, add readings, incorporate new topics, require a new assignment, and put greater emphasis on sense of place. Four are redesigning existing assignments or inserting more concepts about sustainability in their courses. The majority felt confident they would continue to learn and be better advocates for sustainability education. Several participants had already envisioned new forms of student learning beyond what they learned in the classroom. Participants emphasized that revising a course is an ongoing process rather than a one-time experiment, and that seeking student input about their learning experiences in sustainability would be invaluable for improving the effectiveness of what they teach.

Submission of Revised Syllabi

Building on what they had learned at the workshop, each participant's summer assignment was to integrate an aspect of sustainability into courses they will teach in the upcoming academic year, and to prepare a new syllabus reflecting this change. The revised syllabi were submitted on August 19, when the group reconvened at CEEE for a half-day meeting combining a discussion session with a fieldtrip. The participants presented their revised syllabi in a plenary session. This was followed by a guided tour of a sustainably designed corporate office building. Table 4 shows a sampling from the 26 revised courses that were submitted.

Table 4: Sample of course syllabi that were amended to integrate sustainability

Course	Amendment to reflect sustainability theme
United States History since 1877	Revision will make environmental issues/sustainability an integral part of the course, rather than a module; introduce new readings – primary source excerpts on the environment from 1877-1945 and *US Environmentalism Since 1945: A Brief History with Documents* (Stoll 2006); invite an industrial executive of the steel construction industry present on efforts to reduce its carbon footprint, in contrast to earlier historical practices.
Geographic Information Systems	Developed a lecture and lab exercise for site selection modeling focusing on toxic waste sites, and given environmental and social variables related to sustainability; will prompt students to locate newly proposed waste sites utilizing model to compare different site selections and elaborate on how GIS can be employed for sustainable decision making.
Senior Seminar in Criminology	Sustainability will be introduced as one of four overriding themes in which discussion and consideration will be given to how criminal justice professionals can effectively manage and safeguard natural resources; possible topics of study include dumping trash and old appliances on public lands; proper disposal of hazardous household materials; proper maintenance of vehicle emission systems and noise levels.
Introduction to Electronics	Course component on electrical safety will now include conversation about the long-term environmental impacts of improperly disposed electronic waste products; students will also learn about misleading labeling of electronic devices, and importance of purchasing products with certified labeling indicating reduced hazard risk.
Beginning Poetry Writing	Students will be required to write an "ecopoem" that is rooted in sense of place, shifts focus toward an ecocentric perspective, and evokes an ethical orientation, emphasizing humility and human accountability to the environment.
BioMedical Ethics	Added new section examining the ethics and economics of the US health system from a sustainability perspective; examples given from agriculture, economics, and industry (e.g. how eating food produced by unsustainable agriculture becomes unsustainable healthcare); students were asked to do both presentations and to write a paper that included sustainability as one of five goals of a health care system.
Graphic Design	Course will include a methodology matrix and set of questions for analyzing sustainability issues, and presenting them by visual communication; students will develop E-cards on sustainability issues, and an E-advertisement spot on sustainability.
Teaching Elementary School Science	Course will add a unit on "small animals of Iowa" that will delve more deeply into the natural environment of Iowa to gain more appreciation for Iowa's wildlife; elementary education students will be more prepared to write lessons focusing on behaviors and habitats of Iowa's animal species and their vulnerabilities.
Business Policy and Strategy	Course will undergo major revision to focus on business strategies that create sustainable development; students will read about and discuss recent innovations that make the economic case for adopting sustainability; other topics include the challenge of design; a new framework for managers; sustainable value; the triple bottom line; sustainability's accounts; business reality reshaped.

Conclusions and Future Directions

The workshop provided an effective pedagogical base (systems thinking), illustrated examples from real-world and local applications (e.g. environmental justice, organic dairy farming), and created a friendly environment conducive to brainstorming ideas and helping to link people and ideas in transformative ways. An important strength of this program is its inclusion of faculty from 19 different departments. Students who typically may not engage in discussions about sustainability in their coursework are now being engaged through revised courses in art, theater, social work, mathematics, history, philosophy and religion, English literature, and business management.

In a novel approach, the efforts of participating faculty have been guided through the unifying methodology of systems thinking. The efficacy of this methodology is being tested by evaluating the impact systems thinking has had on the instructor's approach to revising the curriculum, the degree to which it is evident in the organization of the material presented, how well the students have grasped the concept, and whether it has helped to enhance their understanding of sustainability.

The experience with the Leadership in Sustainability Education Program has taught the authors much about the process of sustainability education. After planning and coordinating the activities of this program over the last year, comparing the pre- and post-workshop faculty surveys, and reviewing the revised syllabi, the following components of the process can be highlighted as particularly important for the program's success.

Need for support from senior administrators

Infusing sustainability education across the university curriculum requires the enthusiasm and support of the Deans and senior officials such as the President and Provost. This "leadership from above" provides higher visibility, recognition, and acknowledgement of the importance of sustainability education within the campus community. It also helps to stimulate interest in sustainability among faculty and students, recruit more faculty to take leadership roles in the program, incentivize their efforts, and facilitate harmony and communication between the faculty leadership and the administration.

Need to provide the proper environment for creative work

One rule the workshop co-directors adhered to was to not dictate to the faculty what needed to be changed in their courses. The premise was that faculty members are the experts in their courses, and the role of the workshop was to provide an environment and unifying framework most conducive to evoking their creative thinking about infusing sustainability into their courses. The multidisciplinary group discussions provided the intellectual fusion of disciplines enabling a synergy of new ideas. This approach proved to be successful, as each participant submitted revised syllabi that reflected much thought and ingenuity. In the words of one participant, "I was grateful that the organizers did not preach at us. The information from this workshop will integrate into my courses and my research easier than the previous workshops I attended."

Careful selection of multidisciplinary presentation topics

The presentation topics were specifically chosen to reflect a multidisciplinary perspective rather than any one conventional academic discipline. Categories such as environmental justice, energy for sustainable agriculture, and design for sustainability cut across disciplines and connected underlying principles. The presentation on systems thinking provided basic tools for uncovering those connections. Other chosen topics provided stimulating conversations at the small-group breakout sessions, where faculty representing five different disciplines discussed the topic, viewing it from new, multidisciplinary perspectives.

Importance of multidisciplinary faculty connections

Participants especially enjoyed the contact with colleagues from other disciplines in contrast to previous campus workshops they had experienced. They very much appreciated the interactivity of the workshop and the brainstorming during the breakout sessions. Other events such as the pre-workshop dinner, coffee breaks, lunches and the prairie walk contributed further to the camaraderie that emerged among the faculty. Creating such a diverse group of faculty working together on the common goal of sustainability education contributes indirectly to both the harmony and solidarity of the university, as well as towards pluralistic leadership on campus in sustainability education.

This program, however, is just the beginning of an ongoing process of change at UNI that will require much dedication, patience, hard work, leadership, and vision (see Fig. 1). Ultimately, the contribution of the faculty and a measure of the program's success will be how well the concept and practice of sustainability are taught to a new generation of students, and how well they are motivated to live sustainable lives and strive for a sustainable society. To do so students must be provided with good options including those that extend beyond the classroom. They may further their knowledge of sustainability through academic programs such as a certificate program, or a minor or major in sustainability. They may do community service projects (campus outreach) that will make their local place more sustainable, and help them not only as individuals but also as civic-minded citizens engaging effectively in public discourse and policy debate on matters related to sustainability. As sustainability becomes a way of living for members of the campus community, institutional changes will follow. When that point arrives the university will provide more support and encouragement for the faculty to be engaged in sustainability in their teaching and research, which in turn will draw more students to the study and practice of sustainability. The UNI Faculty Leadership in Sustainability Education Program can be repeated as needed as other faculty are recruited to join the program.

Fig. 1: Projected accomplishments (solid lines) and longer-term changes (dashed lines); role of Faculty Leadership in Sustainability Education Program

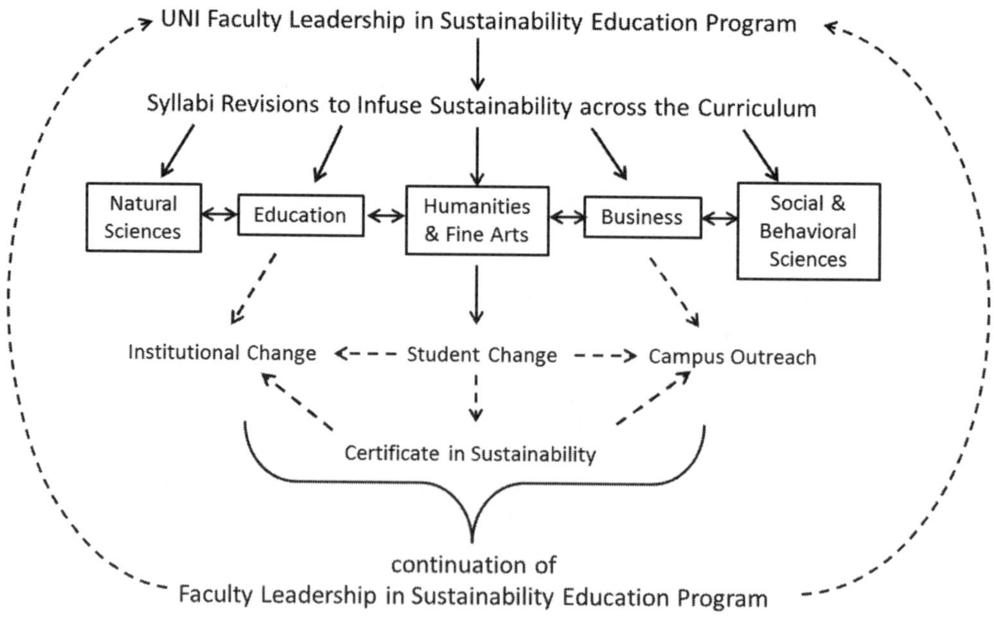

In conclusion, the program described in this paper is only a first step of an ongoing process to elevate the status of sustainability education on campus. Eventually, sustainability needs to be an integral part of every student's education so they are better equipped to meet the challenges of the 21st century and transform our current society into a sustainable one.

References

Abu-Hola, I.R. and Tareef, A.B. (2009), Teaching for sustainable development in higher education institutions: University of Jordan as a case study. *College Student Journal* 43 (4), p. 1287. Retrieved January 29, 2012, from Questia database: http://www.questia. com/PM.qst?a=o&d=5040865773.

Allen, J.H., Gerwing, J.J. and McBride, L.G. (2010), Building capacity for sustainability through curricular and faculty development: a learning outcomes approach. *Metropolitan Universities* 20 (4), pp. 60-74.

Bacon, C.M., Mulvaney, D., Ball, T.B., DuPuis, E.M., Gliessman, S.R., Lipschutz, R.D. and Shakouri, A. (2011), The creation of an integrated sustainability curriculum and student praxis projects. *International Journal of Sustainability in Higher Education* 12 (2), 193-208.

Bardaglio. P.W. (2007), A moment of grace": Integrating sustainability into the undergraduate curriculum. *Planning for Higher Education* 36 (1), pp. 16-22.

Barlett, P. (2005), Reconnecting with place: faculty and the Piedmont Project at Emory University. In Barlett, P. (Ed.) (2005), *Urban Place: Reconnecting with the Natural World.* The MIT Press, Cambridge.

Barlett, P. and Chase, G. (Eds.) (2004), *Sustainability on Campus: Stories and Strategies for Change.* The MIT Press, Cambridge.

Barlett, P. and Rappaport, A. (2009), Long-term impacts of faculty programs: the experience of Teli and Piedmont. *College Teaching* 57 (2), pp. 73-81.

Bertalanffy, L. (1949/1952), *Das biologische Weltbild.* Europäische Rundschau, Bern. English translation: *Problems of Life*: An Evaluation of Modern Biological and Scientific Thought. Harper, New York 1952.

Chase, G.W. and Rowland, P. (2004), The Ponderosa Project: Infusing sustainability in the curriculum. In Barlett, P.F. and Chase, G.W. (Eds.), *Sustainability on Campus: Stories and Strategies for Change*, pp. 91-105. The MIT Press, Cambridge.

DeLind, B. and Link, T. (2004), Place as the nexus of a sustainable future: a course of action for all. In Barlett, P. and Chase, G. (Eds.) (2004), *Sustainability on Campus: Stories and Strategies for Change.* Cambridge: The MIT Press.

Gumport, P.J. and Snydman, S.K. (2002), The formal organization of knowledge: an analysis of academic structure. *Journal of Higher Education, 73*(3), 375-408.

Kaufman, D., Moss, D.M. and Osborn, T.A. (Eds.) (2003), *Beyond the Boundaries: A Transdisciplinary Approach to Learning and Teaching*. Praeger, Westport, CT.

Klein, J.T. (2005), Integrative learning and interdisciplinary studies. *Peer Review* 7 (4), 8-10.

Kline, S.J. (1995), *Conceptual Foundations for Multidisciplinary Thinking.* Stanford University, Press Stanford, CA.

Kurland, N.B., Michaud, K., Best, M., Wohldmann, E.L., Cox, H.M., Pontikis, K. and Vasishth, A. (2010), Overcoming silos: The role of an interdisciplinary course in shaping a sustainability network. *Academy of Management Learning & Education*, 9 (3).

Meadows, D. (2008), *Thinking in Systems: a Primer.* Chelsea Green Publishing, White River Junction, Vermont.

NCES (US National Center for Education Statistics) (2011), *Statistical Abstract of the United States: 2011*, Table 215, http://www.census.gov/compendia/statab/.

Rhodes, E. (2003), *Environmental Justice in America.* Indiana University Press, Bloomington, Indiana.

Rover, D.T. (2002), Interdisciplinary teaching and learning: what, why, and how. *Journal of Engineering Education* 91 (4), pp. 369-370.

Rusinko, C.A. (2010), Integrating sustainability in higher education. *International Journal of Sustainability in Higher Education* 11 (3), pp. 250-259.

Senge, P., Smith, B., Kruschwitz, N., Laur, J. and Schley, S. (2008), *The Necessary Revolution: Working Together to Create a Sustainable World.* Broadway Books, New York.

Senge, P.M., Kleiner, A. and Roberts, C. (1994), *The Fifth Discipline Fieldbook: Strategies and Tools for Building a Learning Organization.* Doubleday, New York.

STARS (Sustainability Tracking, Assessment and Rating System) (2012), Rated STARS institutions, *American Association for Sustainability in Higher Education.* https://stars. aashe.org/institutions /rated.

Stoll, S. (2006) *US Environmentalism Since 1945: A Brief History with Documents.* Bedford/St. Martin's Press, New York.

Talloires Declaration (1990), *The Talloires Declaration: University Presidents for a Sustainable Future,* International Institute for Sustainable Development, Winnipeg, Manitoba. http://www.iisd. org/educate/declarat/talloire.htm.

Thicke, F. (2010), *A New Vision for Iowa Food and Agriculture.* Mulberry Knoll Books, Fairfield, Iowa.

UNESCO (2006), *Framework for the UNDESD International Implementation Scheme.* http://unesdoc. unesco.org/images/0014/001486/148650E.pdf.

UNI (2011), *Strategic Plan 2010-2015 Leadership and Innovation for the Future: Transforming Opportunities into Reality.* University of Northern Iowa, Cedar Falls, Iowa. http://www.uni. edu/stategicplan.

Warburton, K. (2003), Deep learning and education for sustainability, *International Journal of Sustainability in Higher Education* 4 (1), pp. 44-56.

Chapter 17

Climate Justice Partnership Linking Universities and Community Organizations in Toronto, Durban, Maputo and Nairobi

Patricia E. Perkins and Ana Louise Tavares Leary

Abstract

This paper describes a project based at York University in Toronto, funded through the Climate Change Adaptation in Africa program of the International Development Research Centre and the UK Department for International Development (DFID), which is working to increase the participation of marginalized groups, especially women, in urban water governance. Students and faculty members from the University of Nairobi, Kenya; Eduardo Mondlane University in Maputo, Mozambique; and the University of KwaZulu-Natal in Durban, South Africa are working with civil society organizations in the three cities and with York University researchers to show how organizing in local communities can help the vulnerable to deal with climate change.As people in marginalized communities begin to address collectively the impacts of climate change, this summons political attention and allows those with direct experience to influence government policy. Civil society organizations, with support from local and international faculty and students, facilitate and focus this activism. University students help to document the NGOs' work during internships with the NGOs. They also learn community development skills and make contacts. Faculty members publish and disseminate ideas about grassroots climate change adaptation and resulting political responses through presentations, publications and the project's website (www.ccaa.irisyorku.ca).

Introduction

The grave challenges posed by global climate change require creative approaches and contributions from all levels of society in order to develop workable and effective adaptation solutions which address society's most important needs first. This paper reports on the methodology and initial results of an ongoing project which demonstrates the value and potential synergies of university-community partnerships focusing on sustainable development problems.

Partnerships between university researchers/teachers and local community groups exist in many places, but academics generally do not comment on or emphasize the political importance of these relationships. For example, in his excellent recent survey of global democracy-building initiatives, MIT sociologist Xavier de Sousa Briggs mentions the importance of grassroots organizations and their links with government, but does not comment on how universities can facilitate this connection, or indeed on the effects of his own academic research in articulating and building awareness about grassroots knowledge and its policy impacts (de Souza Briggs 2008). Similarly, political ecologist Joan Martinez-Alier highlights the importance of poor people's environmentalism and what it can teach both government and academia, without mentioning the crucial assistance researchers provide when they name and legitimize the knowledge of the poor (Martinez-Alier 2002). In terms of climate change adaptation, community-based approaches have clear political implications, and require appropriate research and training reinforcements, which are the function of universities even if grassroots organizations do not make this connection (Ensor and Berger 2009). As Vandana Shiva has said, "[The] best environmental education is in the minds of those that have suffered environmental destruction. It is from there that the place of learning needs to start" (Shiva 2008).

This paper's explicit focus is the potential of university-based research and training projects to reinforce community-based climate change adaptation and political responses to climate change at the grassroots level. Since universities and poor communities often share physical environments in close proximity, where climate change impacts are felt and available to be studied, university-community collaboration is a creative approach to addressing the challenges of climate change that is both legitimate and promising. The methodology of this paper is to describe one ongoing attempt to build such links, and then to discuss challenges, potential and lessons for the future.

"Strengthening the role of civil society in water sector governance towards climate change adaptation in African cities – Durban, Maputo, Nairobi" is a three-year project based at York University in Toronto, Canada, which links academic researchers and civil society organizations in three African cities with York students and faculty (Climate Justice and Water Management Project 2012). Its goal is to improve watershed governance for climate change adaptation and enhance resilience and adaptive capacity of vulnerable and marginalized groups, especially women. This project, which started in 2010 and runs through the end of 2012, is supported by the Climate Change Adaptation in Africa (CCAA) program – a joint initiative of Canada's International Development Research Centre (IDRC) and the United Kingdom's Department for International Development (DFID). Like the earlier Sister Watersheds Project, also based at York University, this project's methodology includes collaboration between students, NGOs and academics as well as community-based research and environmental education. Project partners based in universities and several NGOs in Nairobi, Maputo and Durban are working together to achieve the following objectives:

- To characterize the institutional framework for urban water governance in the three cities, and explain how different actors within this framework cope with climate change and variability;
- To identify and test viable alternatives for enhancing civil society's role towards adaptation to climate change and variability by vulnerable groups (e.g. by developing education, training and awareness programmes); and
- To share widely the knowledge generated for potential adoption by other cities in Africa.

The project is being implemented by the following community-based NGOs in Africa: Kilimanjaro Initiative (KI) and Kenya Debt Relief Network (KENDREN) in Nairobi; Women, Gender and Development (MuGeDe) and Justiça Ambiental (JA) in Maputo; and Umphilo waManzi (Water for Life) and the South Durban Community Environmental Alliance (SDCEA) in Durban. The University of Nairobi (Nairobi), Eduardo Mondlane University (Maputo), and the Centre for Civil Society at the University of KwaZulu-Natal (Durban) provide academic research coordination and student supervision for this project.

The project focuses on low-income areas of each city, as these tend to be the most severely affected by periodic flooding and other climate change impacts. Residents of low-income areas often lack the ability to protect themselves against the impacts of extreme weather events (Adger et al. 2006; Toulmin 2009; Daley 2008). The capacity-building aspects of this project include training and research sponsorship for students and faculty in the partner universities; support for community-based research, workshops in low-income communities and secondary schools, curriculum and materials development, and skills development within the partner NGOs; training of environmental educators and organizers; contributions to the pool of experienced and qualified community workers in each country; strengthening of all the partner institutions' capabilities to carry out international projects; and contributions to the international literature and professional knowledge concerning water issues, environmental education techniques, and community organizing for improved civil society involvement in governance. The networks being built extend from local and community-based linkages through regional and national-level policy groupings to international academic and policy networks on civil society, watershed management, and governance.

The political process of policy development and implementation depends on the interchange between civil society groups, researchers generating information on current realities, and government. This project attempts to challenge the conventional notion that only educational institutions "produce" knowledge. Understanding community needs and what helps particular civil society groups to see and act to strengthen their role in democratic governance, for example, is something in which community organizations and NGOs have eminent expertise. This collaborative approach, also known as participatory action research (PAR), is broadly defined as "research by, with, and for people affected by a particular problem, which takes place in collaboration with academic researchers. It seeks to democratize knowledge production and foster opportunities for empowerment by those involved" (Kindon et al. 2008).

One objective of the project is to demonstrate how partnerships between academics and non-academics can be very stimulating and effective. This type of partnership encourages and allows the partner NGOs to reflect on and analyze their activities and to document "learning" more systematically than they are often able to do, by bringing student researchers into the NGOs as collaborators/interns. The partnership also encourages universities to be more pragmatic about teaching and research, and to "field-test" approaches towards community organization, equity, and capacity building. Students committed to the project's goals of supporting participatory engagement by local people in municipal water decision-making are given practical opportunities to develop their skills, as a way of hastening each city's climate change preparedness. This project aims to contribute to the integration and meaningful participation of women in formal decision-making processes, as well as to build their adaptive capacity and increase their resilience and ability to cope with climate change.

University-community linkages for improved climate change adaptation

The single most important line of defense against climate change is education. Only with education can local communities, policymakers, finance and insurance institutions, and global agencies flexibly incorporate complex scientific, economic, and socio-political information about a rapidly-changing world into effective and efficient climate change strategies. This is why universities have a crucial role to play, and why processes that extend universities' expertise and information-access to non-traditional communities outside of academia are vital parts of any equity-promoting climate change adaptation strategy. Training students in how to facilitate these extension activities by nurturing their community development, conflict resolution, and environmental education skills is a major contribution which universities can and must make as the world heats up. Faculty members, too, need to progressively focus their research and publication efforts towards helping those first affected by climate change to develop adaptation strategies, and to guide policy development for equitable priority-setting. Civil society organizations which are already working with marginalized communities on environmental issues can play an important bridging role. The ecological knowledge and awareness of local communities is an important resource for governments as they struggle to deal with the increasing numbers and severity of extreme weather events induced by global warming (Ensor and Berger 2009).

Climate change thus challenges universities to develop new ways of heightening their outreach, relevance, and partnerships beyond academia. By creating and sharing new models of how to do this, universities can contribute to climate change adaptation and mitigation strategies which go far beyond a passive scientific understanding of how the Earth is changing.

Between 2002 and 2008, York University, the University of São Paulo (USP) and the Ecoar Institute for Citizenship, an NGO in São Paulo, joined together to work on a project linking academics and activists to strengthen the voice of marginalized people in water management – the Sister Watersheds project. Designed by renowned Brazilian environmental educators Dr. Marcos Sorrentino and Larissa Barbosa da Costa, this project tested out community environmental education in two watersheds containing University of São Paulo campuses those of the Piracicaba and Pirajussara Rivers as well as the Black Creek watershed, where York University is located. The project aimed to increase the potential for marginalized community actors to participate in local water governance. Masters of Environmental Studies students at York University whose theses analyzed aspects of the social-environmental situations of these communities were chosen to take part in Sister Watersheds student exchanges and spend 4-6 months in Brazil. Brazilian students also visited Toronto and researched similar issues there. One particularly productive exchange group, composed of Brazilian graduate students in the fields of hydro-engineering, urban planning, ecological economics, and landscape architecture, contributed a detailed study of how the growth of the York University campus has contributed to increased flooding probabilities in the Black Creek watershed over time. The studies by the USP and York exchange students proved valuable for communities in both countries, and helped their faculty advisors to deepen their understanding of the relation between watershed governance, civil society engagement, and climate change leading to a number of new publications and research contributions.

Similarly, the CCAA project in Africa reinforces the importance of the interdependence of traditional and local scientific knowledge combined with academic work in communities. This in turn encourages NGOs and community organizations to develop their transformative power. This type of work has proved many times to have great political weight. It can also facilitate the spread of important ideas regarding ways to confront not only climate change but also barriers to social and environmental justice (Desai 2002). Such barriers, of course, exist in both the Global North and the Global South. Student exchanges and research partnerships can highlight creative methods of addressing these barriers which are appropriate in both Southern and Northern contexts.

For example, partly inspired by Brazilian examples of green community development and environmental organizing, students and faculty have contributed to activist projects aimed at environmental education and green job creation in the marginalized community near York University in northwest Toronto. One of these, the Green Change project in the low-income Toronto neighborhood of Jane-Finch, is administered through the Jane-Finch Community and Family Centre, a non-profit social services and community development agency. The goal of the project is to engage community residents in discussion about the environment and social justice issues in an accessible way. The project uses a social justice and anti-oppression approach where community members have ownership over the direction and development of the project. A training program for Green Change Agents was developed, which included workshops, tours and visits in the community. The award-winning training program introduced residents to five key areas: energy conservation, waste management and recycling, green active living, social justice and the green economy (Phillips 2010).

York University students have been involved in the project as volunteers, interns, and community-based environmental educators. Three York Graduate Assistants have worked this year on developing new curriculum modules for the next phase of Green Change agents' training. In addition, these students have helped with fundraising, report-writing, and local outreach and environmental education for those living in the social-housing building where a new Centre for Green Change is being built. Faculty members have also been involved; an Environmental Studies professor sits on the Green Change Project's advisory board, and other faculty members have participated in gala fundraising events.

The CCAA project includes the same kinds of student internships locally, followed by international exchange trips which allow students to share their knowledge of climate change challenges and ways of addressing them. Graduate students based in Nairobi have produced an extensive study of the city's physical and institutional infrastructure for water management, worked with the NGO The Kilimanjaro Initiative to develop soccer fields (and resist pressures for expanding low-income housing in flood-prone areas) in the floodplain of the Nairobi River below Nairobi Dam in the Kibera slum, and helped the NGO KENDREN (the Kenya Debt Relief Network) to develop the rationale for considering climate debt relief as a parallel to financial debt relief as a climate justice strategy. In Maputo, students are working with urban farmers producing food on the university campus, as well as residents of several increasingly flood-prone low-income neighbourhoods who are seeking ways to mitigate climate change risks. They are also contributing to curriculum development on climate change for local high schools, working with the NGO Justiça Ambiental. In Durban, students are working with civil society organizations – Umphilo waManzi (Water is Life) and the South Durban Community Environmental Alliance – on a range of activist education and green development projects. By providing skilled interns and volunteers for community organizations' projects, and training university students in how to do community-level work, the project is building local and regional capacity to address climate change.

Research and policy implications of academic-activist partnerships

For both academics and non-governmental or civil society organizations, such partnerships can bring a range of important benefits. These include the following:

Environmental education

Skills needed in community-based organizations and local communities related to evolving green jobs, and for effective public involvement in watershed and other environmental policy processes, can be developed and shared in marginalized communities via organized processes of continuing education (Pinderhughes 2006). These can include environmental education programs for youth and seniors, adults, and in-service training for working professionals such as public health agents/nurses, teachers, and government officials. Local community organizations and NGOs are ideally situated to develop and fine-tune such educational programs, which can lead to increased employment opportunities and/or pay upgrades for local residents.

University students can develop marketable and lifelong skills in community development and workshop facilitation, fundraising, conflict resolution and dispute mediation, management, video production, and many other areas by working with community-based organizations through internships and field experiences. Their writing and research skills can be very helpful to organizations which are stretched and generally unable to document and share their accomplishments and creative solutions to environmental and political problems.

Creative organizing/workshop techniques/strategies and international sharing of ideas

Ideas, designs, and financing proposals can be shared internationally through green/ climate change channels. For example, civil society networking surrounding the United Nations Framework Convention on Climate Change annual COP meetings involves a wide range of community-based environmental NGOs that have participated in the civil society forums accompanying the governmental negotiations. Academic conferences can bring together university-based researchers and local environmental activists for useful discussions on innovative policy and grassroots solutions to the pressing ecological situations people face on the ground.

A conference entitled "How will disenfranchised people adapt to climate change?" at York University in April 2009, organized by the university's Institute for Research and Innovation in Sustainability, focused on local struggles and the importance of traditional ecological knowledge for addressing climate change in Arctic Canada, Brazil, South Africa and India (IRIS 2012). Likewise, Majora Carter's advocacy work highlights the common strategies marginalized areas can use in sparking green community development throughout the Global North and South (Carter 2010).

Role of municipal government and public education

Cities worldwide have begun to develop climate change adaptation plans and to take stock of their new needs – physical, social, economic (Lucon and Goldemberg 2010). When community groups organize to share their expertise and knowledge of challenges, as well as ideas on how to meet them, they may be able to build beneficial partnerships with local governments. Public officials everywhere need training in all aspects of climate change preparedness, and environmental education for the general public is also crucial in a warming world. Here too, networking, university support, and global communication help groups to learn from "best practices" and creative ideas in use elsewhere. University partnerships can facilitate and add "weight" to local organizations' interventions with government officials.

Mobilizing finance

University-based researchers often have access to grants, information and research funding sources which go far beyond what community-based organizations can muster, and which place emphasis on practical applications of the research and teaching involved. For this reason, academic partners can be strong advocates and partners of community-based organizations in their environmental education and political outreach efforts. Academic partners can sit on community advisory boards; edit grant proposals; analyze, publish and disseminate innovative community initiatives; and contribute to local fund-raising efforts in a range of other ways as well.

International networking and information-sharing

How can partnerships between academics and community groups contribute to research priorities, relevance, and success, and to global and local information-sharing which benefits all participants? Because of the power imbalances which are familiar to anyone who has attempted to undertake community-based research, it is important for partners to be very clear about the contributions, goals and benefits of all groups and individuals involved.

Participation in research projects can give community organizations and NGOs unprecedented access to international meetings, networking opportunities, and fundraising possibilities. Partners can develop joint research presentations and panels for international conferences, and jointly explore publication opportunities. Where universities have the means for online networking, Skype conferences, library facilities, and meeting rooms, they can share these and provide access to NGO partners which can be an important benefit for them. The increased dissemination and publicity of community-based activism which flows from academic attention can have important policy benefits. Student-initiated and produced videos, maps, social networking, website content, reports, grant applications, and other products can greatly contribute to NGOs' work.

The benefits also flow both ways, and can strongly influence and benefit university partners as well. For example, York University has begun a tradition of inviting a series of "community scholars" to give seminars, advise students, and contribute in a practical and experiential way to classroom teaching over the course of a term or academic year. At an international level, the Dame Nita Barrow Distinguished Visitorship speaker series at the Ontario Institute for Studies in Education of the University of Toronto brings international women activists from the Global South to lecture and meet with students each year. And of course, access to community-based experts on relevant and pressing research issues is a key benefit for academics of partnerships extending beyond the university.

Research projects tend to lead to opportunities for further research, exploration, and expansion of research partnerships. In this way, they both advance the careers, teaching relevance, and professional development of the faculty members and students involved, and they also contribute to global synergies and advance understanding of complex global problems such as climate change.

Conclusion

The social change which is fundamental to "sustainable development" anywhere – in the South or the North – depends on a synergistic mix of information, public education, leadership, institutional flexibility, and political organizing (as well as serendipity). Research, when it is locally motivated and relevant and reinforces grassroots ecological, social, and political knowledge, is a fundamental pillar of interactive development, generating the current information which underlies public education and organizing around issues of importance for social change (Faber and McCarthy 2001).

The Climate Change Adaptation in Africa (CCAA) climate justice project in African cities is aimed at generating innovative ways of strengthening the ability of civil society to participate in watershed decision-making, within the institutional structures set out by national water laws and other statutes. Graduate students from partner universities visit each other's campuses for short research trips and faculty members collaborate and direct the research, while the integral collaboration of community-based NGOs is essential. The NGOs have long experience with environmental education and organizing in low-income communities, and with presenting information in multimedia ways which are just as accessible to busy politicians and bureaucrats as to barely-literate community residents. The researchers and activists working with the CCAA project are fortunate to share a climate of change-oriented exploration and communication: What water-related issues are the most important to local communities? What are the best, most creative and interesting ways to help get the public involved in and knowledgeable about water issues? When the law creates space at the decision-making table for "civil society", how can that democratic opening be used effectively and how can multiple publics and views be included? These questions are relevant in Canada just as in Mozambique, South Africa and Kenya, and in both urban and rural areas.

There are many examples of synergies between research and development-oriented activism. Master's papers and other research reports, researched and written with project support, include studies of international synergies to address climate change: participatory community organizing, women and water management, gender justice and climate justice, innovations in participatory water governance in times of climate change, politics of climate justice, eco-social commoning of water, conflicts over water management, climate change and water governance, and detailed histories of local green community development and climate justice initiatives.

Without the theoretical insights drawn from academic fields such as action learning and research, participatory planning, the dynamics of institutional change and political organizing, the CCAA project's work would be impoverished. Likewise, the project's grounding in compelling current political and social issues, and its links to communities affected by these issues, are also crucial to its success.

Earlier writers on traditional knowledge and climate change have commented on the desirability of integrating scientific and community knowledge for better understanding of appropriate climate change adaptation processes (e.g. Reid et al. 2009; Berkes and Jolly 2001). This paper broadens that idea to demonstrate that universities and their local communities can fruitfully engage on many levels to build strong sustainable development and climate change adaptation partnerships with great potential. Given the pressing challenges posed by climate change, and the expertise and needs at the community level, university researchers and teachers must build the skills and train students to expand these promising forms of outreach and collaboration.

References

Adger, W., Neil, J., Paavola, S., Huq and M.J. Mace (Eds.) (2006), *Fairness in Adaptation to Climate Change.* MIT Press, Cambridge/London.

Berkes, F. and Jolly, D. (2001), Adapting to Climate Change: Social-Ecological Resilience in a Canadian Western Arctic Community. *Conservation Ecology* 5 (2).

Carter, M. (2010), *Majora Carter: Three stories of local eco-entrepreneurship.* TED Ideas Worth Spreading. Available at: http://www.ted.com/talks/majora_carter_3_stories_of_local_ecoactivism.html (accessed 29 January 2012).

Climate Justice and Water Management Project (2012). Available at: http://www.ccaa. irisyorku.ca (accessed 29 January 2012).

Daley, J. (2008, July 18), *Poor People Can't Worry about Global Warming.* Available at: http://www.telegraph.co.uk/comment/columnists/janetdaley/3560548/Poor-people-cant-worry-about-global-warming.html (accessed 29 January 2012).

de Souza Briggs, X. (2008), *Democracy as Problem-Solving: Civic Capacity in Communities Across the Globe.* MIT Press, Cambridge, MA.

Desai, A. (2002), *We are the Poors: Community Struggles in Post-Apartheid South Africa.* Monthly Review Press, New York.

Ensor, J. and R. Berger (2009), *Understanding Climate Change Adaptation: Lessons from Community-Based Approaches.* Practical Action Publishing, Bourton on Dunsmore, UK.

Faber, D. and McCarthy, D. (2001), The Evolving Structure of the Environmental Justice Movement in the United States: New Models for Democratic Decision-Making. *Social Justice Research* 14 (4).

IRIS (Institute for Research and Innovation in Sustainability), York University (2012). Available at: http://www.irisyorku.ca/ (accessed 29 January 2012).

International Encyclopedia of Human Geography (2008), Participatory Action Research, pp. 90-95. Elsevier, Amsterdam/London.

Lucon, O. and Goldemberg, J. (2010), São Paulo – The 'Other' Brazil: Different Pathways on Climate Change for State and Federal Governments. *The Journal of Environment and Development* 19 (3), pp. 335-357.

Martinez-Alier, J. (2002), *The Environmentalism of the Poor: A Study of Ecological Conflicts and Valuation.* Edward Elgar, Cheltenham, UK/Northampton, MA.

Phillips, M. Ann (2010), *Jane-Finch's Green Change Project Wins Toronto Green Award*. Available at http://yorkwestadvocate.wordpress.com/2010/05/13/jane-finchs-green-change-project-winstoronto-green-award/ (accessed 29 January 2012).

Pinderhughes, R. (2006). Green Collar Jobs: Work Force Opportunities in the Growing Green Economy. *Race, Poverty & the Environment* 13 (1).

Reid, H., Alam, M., Berger, R., Cannon, T. and Milligan, A. (2009), *PLA 60: Community-based adaptation to climate change*. Available at: http://pubs.iied.org/14573II ED.html (accessed 6 May 2012).

Shiva, V. (2008). Quoted in the film, *Somos todos aprendizes: Tratado de Educaçao Ambiental.* Brazilian Ministry of the Environment, Ministry of Education, Ecoar Institute for Sustainability, et al.

Toulmin, C. (2009), *Climate Change in Africa*. Zed Books, London/New York.

Chapter 18

Ecohumanities and Service Learning: New Directions in Pedagogy

Kelly A. Parker

Abstract

Students in a humanities-oriented environmental studies capstone course are immersed in collaborative problem-solving. The primary learning experience is a service-learning project in which interdisciplinary teams of students create and implement programs that promote ecological literacy and sustainability in the broader community. The paper surveys the philosophical context for ecohumanities pedagogy, describes the key learning moments of the course, describes specific outcomes, and offers practical advice for implementation of this pedagogical model.

Introduction: Rethinking Environmental Studies

In 2010, Grand Valley State University introduced a new interdisciplinary undergraduate Environmental Studies minor (Grand Valley State University 2012). Many successful Environmental Studies and Environmental Science programs already exist, and so it was possible to look for a unique educational and social niche that the program would fill. A survey of existing Environmental Studies and Environmental Science programs (Brown University Center for Environmental Studies 2012) made it possible to discern a direction of development in such programs over the past thirty years, so as to assemble a program that responds to particular problematic situations facing environmental education in the early 21st century. Since others had long ago perfected the wheel, our group was free to invent the bicycle. One distinctive feature of the resulting program is that it identifies an *essential* role for the humanities in environmental education. The primary philosophical guidance in identifying that role came from the works of John Dewey (1988) and Hannah Arendt (1958), with adaptation to environmental education as suggested by Plumwood, who coined the term "ecohumanities" (2002), and by Chapman (2007). A second distinctive feature is the reliance on service-learning pedagogy (Corporation for National and Community Service 2008) and the ideals of creating "citizen-scholars" as advocated by the intellectual entrepreneurship movement (Beckman and Cherwitz, 2009), particularly in the capstone course.

Ecohumanities, Reconstruction, and the Human Condition

One of Dewey's core insights is that education produces knowledge via the students' guided reconstruction of the conditions of experience (Dewey 1938: 47). Science itself consists of a certain mode of manipulating the conditions of experience. The scientist renders *experience* as repeatable, quantifiable, highly controlled *experiment* (Dewey 1916: 226-27). Dewey regarded the physical sciences as a paradigm for the activity of knowledge production in general: of all academic disciplines, they have been most attentive to refining and perfecting the methodology of experimental inquiry. But Dewey cautioned that the technical results of highly controlled scientific inquiry are only one among many forms of knowledge; moreover, a purely scientific culture would – if it were even possible – make for a fragmented and impoverished form of collective life. The technical knowledge produced by science is indispensable (and this point is especially relevant in environmental education) but it is not sufficient for the wise reconstruction of the conditions of our shared life. His concern was that the technical sciences themselves cannot tell us which technical problems should take scientific and social priority, nor how the results of such investigation should be used. The general point is familiar: the sciences presume broader social structures that support them, as well as a broader social and political context in which their value and meaning can be determined. For all our successes at increasing both ecological literacy and concern, we still generally lack the kind of experience that fuses artistic and humanistic insight with focused scientific knowledge. As course instructor Kirsten Bartels put it, "Science is a very important component, but let's be honest: Science alone didn't create the environmental crisis we face today, so science alone can't solve the problems we face today" (Moltz 2010). Or, as Bill McKibben quipped, "You don't build movements with bar graphs" (McKibben 2010).

Dewey thus prominently called for new kinds of knowledge and new kinds of social institutions to meet the ever-changing conditions of life. "The prime condition of a democratically organized public is a kind of knowledge and insight which does not yet exist" (Dewey 1988: 339). Most important to the education and inquiry he envisioned is that emerging knowledge must be made compelling to the public through the use of art – including the writer's and orator's art (Dewey 1988: 349). This aim clearly lies within the provenance of the humanities, the disciplines particularly concerned to discern and communicate meaning. The humanities are thus an essential part of effective environmental education. This conception of environmental studies as ecohumanities, an interdisciplinary inquiry that involves the humanities at its core, promises a more unified definition of environmental studies. The environmental studies program requires students to engage scientific, social-political, and humanistic perspectives in their course selection. In the senior-level capstone course they work in project teams to bring all this preparation to bear on a semester-long service-learning project.

Knowledge, Action and Speech

What knowledge and insight should we ask citizen-scholars to develop alongside scientific and policy literacy? Just how can one or a few ordinary individuals – college students – lead the way to discover and implement more environmentally sound institutional, ideational, and practical forms, and make these forms available to others? These are the central questions confronted in teaching the environmental studies capstone course.

Hannah Arendt's discussions of the nature of political engagement provide an appropriate and compelling account of what is needed. In *The Human Condition* Arendt distinguishes among three main modes of human activity. The first is *labor*, "the activity which corresponds to the biological process of the human body, whose spontaneous growth, metabolism, and eventual decay are bound to the vital necessities produced and fed into the life process" (Arendt 1958: 7). "Labor" here may be exemplified by the labor of childbirth, which is a matter of meeting the compelling necessities of biological survival. Arendt identifies the second mode of activity as *work*. This activity transcends the basic exigencies of the life cycle; its origins and ends lie beyond mere physical need. "Work provides an 'artificial' world of things, distinctly different from all natural surroundings" (Arendt 1958: 7). Work is the dominant mode of human activity in modern industrialized societies, where we dedicate our mental and physical capacities to accomplishing tasks, ends, outcomes that are defined primarily by social and cultural structures, rather than by natural need. Because the ends are pre-established, and there are proven and communicable ways to achieve those ends most efficiently, the word "training" may be used to name the process of teaching another to perform either labor or work. Work is necessary and rewarding but it cannot change the status quo.

Arendt calls for us to recover a knowledge and capacity that was once robust but has largely disappeared – for what in *The Human Condition* she calls *action.* Action, along with political *speech*, constitutes a kind of human activity that has been eclipsed by work in the modern world. In her remarkable "Prologue" to *The Human Condition*, Arendt worries that "it could be that we, who are earth-bound creatures and have begun to act as though we were dwellers of the universe, will forever be unable to understand, that is, to think and speak about the things which [owing to the technical discoveries of contemporary science] nevertheless we are able to do" (Arendt 1958: 3). Scientific results are discovered and presented in specialized symbolic languages available only to specialists, who use it to communicate with other specialists *as* specialists. As Dewey said, "All *merely* physical knowledge is technical, couched in a technical vocabulary communicable only to the few" (Dewey 1988: 343). Often, the specialist's knowledge is not even communicable to specialists in other areas; it is even more difficult to convey to the public. This is a fundamental problem, according to Arendt, because "whatever men do or know or experience can make sense only to the extent that it can be spoken about" (Arendt 1958: 4). That is, it can be known, can have public meaning, only so far as it can be spoken about in the public sphere, the political sphere, in speech among fellow citizens. Arendt's plea is that our students begin to learn the lost art of political speech and action.

Ecohumanities draws upon students' existing knowledge of environmental science and policy, then educates them for action, the third of Arendt's modes of human activity. Action is inherently political in that it concerns the formation of new ends, along with new institutions, ideas, and practices by which these ends will be realized. In contrast with work, where the ends are pre-established by external authority, tradition, or custom, the ends of action are *new*, and thus not known with any definiteness until the action has taken place. Action is an open-ended and public process; its results are both unpredictable and irreversible. This open-endedness, the indeterminacy of action that *requires* creative political invention, is the key to addressing novel social problems, as well as to fulfilling the highest aims of humanistic education. "In acting and speaking, men show who they are, reveal actively their unique personal identities and thus make their appearance in the human world" (Arendt 1958: 179). As students interpret what is known about their world, and translate it into meaningful new social forms, they refine and forever change their own social identities. They can become full, free citizens with the positive capacity for speech and action. In this transformation they can thus realize the highest liberating aim of undergraduate liberal education.

A Pedagogy of Civic Action

Typical service-learning strategies introduce a limited practical component into a traditional course structure; the capstone course consists almost *entirely* of the practical service component. The first two weeks of class are devoted to an overview of service-learning principles, with readings from Arendt on the concepts of labor, work, speech, and action, and with a survey of several biographical portraits of successful environmental activists from McDaniel (2005). This is all directed at helping students understand the niche for ecohumanities, and introducing them to the wide field of possibility that is open to them under the concept of "public action." After the first class session students begin to develop project ideas, which involves identifying, contacting and consulting with people and agencies outside the class.

The matter of guiding students toward a workable project is sensitive, both pedagogically and politically. To provide detailed project suggestions or steer them toward my own pet interests would be inappropriate. Pedagogically, since the objective is action rather than work, students need to form their own ends and discover the best ways to achieve them. Politically, since any meaningful project will be perceived as carrying some ideological baggage, students must choose to pick up that baggage of their own free will. They are not there to do the instructor's bidding, but to discover, refine, and pursue their own ends. Students propose a number of project ideas, then decide – via an open-ended democratic discussion – which ones to pursue. Once projects are selected, students form teams and set to work.

Capstone Course Objectives and Outcomes

The course objectives, as stated on the syllabus, indicate that students will:

1. Demonstrate familiarity with the personal and social meaning of works in environmental literature, environmental philosophy, environmental history, and sustainability studies;
2. Conceive, design, and begin to implement an ecological literacy- or sustainability-related project that makes a significant contribution to the community;
3. Demonstrate the ability to collaborate constructively with peers on an ongoing basis;
4. Demonstrate the ability to collaborate successfully with local agencies or other community partners in planning and implementing a major service project;
5. Engage in substantial reflection regarding the role of science, the humanities, politics and civic engagement in modern democratic society. (Parker 2010-12)

Students' success in meeting these course objectives is measured via a variety of activities and instruments.

The once-weekly three-hour *class meetings* allow time for lecture and discussion of assigned readings (objective 1), for collaborative work on group projects (objectives 2 and 3), and for reflective discussion on what they are doing and how it is working (objective 5). Readings come from the instructor as indicated on the syllabus, from student suggestions, and as appropriate from a fifty-item bibliography that is provided along with the syllabus.

The first group task is to assemble a *project area literature review*, an annotated bibliography of peer-reviewed sources relevant to the group's project. Students are encouraged to research within their own areas of expertise (objective 1) and pool the results in a single document (objective 5). The literature review provides an overview of the project, including critical perspectives on similar efforts, as well as detailed technical and historical information that will contribute to the project design and implementation.

Project status updates are key to keeping all students engaged and on track with their contributions to the projects (objectives 2, 3 and 4). Most status updates are posted to the online discussion forum for the course. These are submitted well in advance of the class meeting so everyone can be up to date on their group's progress. On one or two occasions during the semester, students are asked to give impromptu status reports in class. This way they learn to provide a brief account of their work; the impromptu presentations also often reveal where more effort is needed to better focus the project and to prioritize efforts.

The first culminating event of the course is a *public presentation* at the last regular class meeting of the semester. The presentation allows students to present an overview of their project (objective 2), to explain its significance in the context of other environmental and sustainability work (objective 1), to demonstrate constructive collaboration with other group members (objective 3) and with community partners (objective 4). The presentations are publicized to Environmental Studies faculty, to the campus and community, and to university administrators and the community partners involved in the projects, who are specially invited. Students are told to treat the presentation as a professional report to supporting partners and potential funders. This emphasis is not merely a rhetorical matter: since most projects could potentially be continued beyond the semester in which they are developed, it has happened that the presentations have attracted external support.

The second culminating event is a *final paper*, an integrative, first-person report and reflection on the course and the project, due at the designated final exam time for the class. The paper addresses all of the course objectives, but is especially focused on consolidating what students have learned in environmental and sustainability studies (objective 1) and about the role of science, the humanities, politics and civic engagement in modern democratic society (objective 5). This latter aspect of the paper is especially important for graduating seniors. Students should leave the experience of a capstone course conscious of what they have actually done, and in particular, what they have learned about the value of their own undergraduate education and how it relates to public action, civic engagement and service. The final papers indicate that most students do explicitly recognize and appreciate how the service-learning experience differs from, and complements, their other educational experiences.

There are, of course, other outcomes in the course as well, which cannot in principle be stipulated beforehand. Most notably, the particular projects that students develop almost completely determine what environment-related skills and knowledge they will actually acquire during the course (objective 1). Those who worked on the 2008 Recycling Education Project, for example, had to learn a great many pertinent facts about the city's waste management system, including the relation of the municipal recycling program to other disposal methods. A tour of the county's waste-to-energy incineration facility was especially illuminating, as students had the opportunity to see the waste stream firsthand and ask questions of a facility manager. They had to learn the economics of waste collection, sorting, and the market for recycled materials. They met with representatives of the city's Streets and Sanitation Department to learn about participation levels and perceived obstacles to participation, as well as to identify the likeliest population for an outreach and education program. Once that was accomplished, they had to determine the most effective way to contact potential new participants.

To undertake a public action involves much more than having a good idea and being willing to work to make it happen. That the projects are public means there are others in the community who have the authority to facilitate or block them. There are people – city employees, staffers at non-profit agencies, university administrators – whose job it is to look after the proper use of public resources. Students discover that they first of all must persuade these key people to support their efforts. This brings us to a second key area of project implementation: *publicity* may be as crucial as *opportunity* and *permission* to establishing a genuine public action. The role of project publicity had not even been considered in the original planning for the course – it was a pleasant surprise when the Community Garden project was featured on the front page of the school newspaper. It quickly became clear how important the article would be for publicizing the garden and attracting potential participants. The Recycling Education group naturally took notice of this attention and, after some serious in-class discussion about the positive and negative aspects of media coverage, as well as its appropriateness in this educational context, they realized that local news coverage could multiply the effectiveness of their door-to-door outreach efforts many times over. Within a few weeks the Recycling Education Project had been featured in two newspapers and on the evening newscast of the Grand Rapids, Michigan Fox television affiliate, bringing the students' educational message to many thousands of residents. In both projects, students learned how to negotiate the various channels of authority that affect anything in the public realm, successfully working with the relevant officials and the public to bring their projects into being. The course indeed engaged students in genuine action well beyond what had been expected: both the Recycling Education Program and the Community Garden continued to thrive after the course ended.

Landmark Learning Moments

Several turning points in the course may be considered "landmark experiences" for the students. The first is the development of project proposals at the beginning of the semester. All students are invited to propose anything they think will be appropriate, collaboration among students in proposal development is encouraged, and everyone involved understands that the class as a whole will make the final decisions of which to pursue.

Students are required to approach potential community partners in order to establish the genuine need for and feasibility of their projects. Sending enthusiastic students out to contact unsuspecting potential community partners is nerve-wracking for an instructor. Time is therefore devoted to a discussion of the ethics of community involvement, activism, and research. Students explore the ways in which they represent the university to the public, how their projects will affect others, the relevance of professional ethical codes, and the potential need for approval by the university's Institutional Review Board. Likewise, any project that involves a grant application would require pre-approval from our Grants Administration Office. It is important for the instructor in a course like this to be aware of all relevant administrative policies.

Under the instructor's imposed deadline, students decide together which projects to pursue. Ideally, this day becomes a workshop on deliberative democracy. In the end, what matters is not that they have followed a pre-designated procedure, but that everyone chooses to work on a project, and with a group, that they believe in and where they can have a rewarding learning experience.

The second landmark experience might be referred to as "first contact." This is when the students, having chosen their project and established a plan of action, actually go out and confront the "real world" in which they intend to act. The in-class status updates immediately after this point are especially important. Students need a forum in which to report and work through their excitement, frustration, and sense of having bitten off something perhaps bigger than they can chew.

A third landmark experience is the students' encounter with publicity about their efforts. As was mentioned previously, some degree of media coverage can be an effective multiplier of efforts, even comprising an integral part of some kinds of public action project. Any media coverage that the groups receive needs to be discussed in class. It is flattering, gratifying, and somewhat frightening to the students. They are after all more accustomed to presenting their coursework in a private conduit between themselves and a professor, not to describing it on Fox News or in a major daily newspaper. Students need to come to terms with whether media coverage is accurate and useful to their project, and with the factors that influence its making.

A fourth landmark event centers on unplanned educational opportunities that arise during the course of the semester. For instance, we have canceled the regular class meeting at the last minute to attend a campus event and arranged for students to attend a relevant conference. Besides the obvious informational value and networking opportunities that such events provide, students gain an increased sense of how their own efforts fit into the larger field of ongoing public action. Those students who attend the more participatory events also see how they constitute a particularly well-informed and well-placed group of citizens with something worthwhile to offer the public.

The fifth and final landmark moment concerns the final presentation. As indicated previously, it is important that students communicate their results to their instructors, community partners, and other interested parties. It is equally important for the students to go through this process so they can objectively appreciate just what they have done. The Recycling Education project, for example, reported that they had spent sixty man-hours to sign up sixty new households (out of seventy they approached) for the city's recycling program. These households account for a small but real ongoing saving in waste disposal costs to the city; it is reasonable to expect that media coverage of the Recycling Education Project attracted additional participation. The students could see for themselves that, though it was by no means a solution to all the environmental issues we face, their program was a real step forward, worthy of continuing as part of a much larger and more complex social and political effort to address environmental issues. A representative of the mayor's office who attended the presentation emphasized the tangible contribution of their project for the city.

Recommendations

In a pure service-learning approach, the best students will, ideally, seek out research materials and other sound information about their project area. The Project Area Literature Review assignment was introduced when we determined that not all students lived up to this ideal. The new assignment adds a solid academic component to their work, and also helps them understand the role of research for designing and implementing an action or projects.

Given the need for student participation in the class meetings, where they process and reflect on new kinds of learning experience, any class of this sort should be kept to an optimal size for active discussion. Eighteen students enrolled in the course in 2010, with two instructors; fourteen are enrolled in 2012 course, with one. These are probably the maximum workable sizes for such a course: besides the need to allow for meaningful discussion, students need to know that the instructors are able to provide personal support to groups or to individuals who encounter unexpected difficulties. As Dewey noted long ago, student-directed learning often requires more of the teacher's time and ingenuity than does a conventional lecture format (Dewey 1938: 19-20).

Finally, experience with less-successful group projects suggests several specific recommendations. Initial project proposals may be somewhat vague or unfocused, but this is acceptable as long as the proposal includes (1) an established contact with a community partner and (2) a clear civic service component. These external expectations increase students' sense that they need to focus and implement their project quickly. The last observation is that every project group larger than five students has so far run into logistical, personality, or workload-sharing problems that negatively affected the project's outcomes. For projects of this nature, three to four students is optimal; no group larger than five is allowed.

Conclusion

The aim of ecohumanities education is to lead students to develop "a kind of knowledge and insight which does not yet exist" (Dewey 1988: 339). Environmental studies graduates should be effective, imaginative citizen-scholars with a solid knowledge of environmental science and policy; it is equally important that they acquire a deep-seated commitment to the humane values of freedom, integrity, transparency, the public good, and the democratic process. The pursuit of effective public action through collaborative service-learning projects, organized around the idea of ecohumanities and taught in an adaptive, experience-driven style of instruction, compels students to acquire just such knowledge. The environmental studies capstone course uses a well-defined set of fixed activities and instruments to promote such learning, and it encourages students to attend to certain anticipated landmark learning experiences. The ecohumanities/service-learning pedagogy should be readily transferrable to other institutions and settings, though due consideration must be given to ensuring appropriate academic content, class size, and project group size.

Note

This paper is a shortened and updated version of K.A. Parker (2012), Ecohumanities pedagogy: An experiment in environmental education through radical service-learning, Contemporary Pragmatism 9 (1). Used by permission.

References

Arendt, H. (1958), *The Human Condition* (2nd ed). University of Chicago Press, Chicago.

Beckman, G.D. and Cherwitz, R.A. (2009), Intellectual entrepreneurship: An authentic foundation for higher education reform. *Planning for Higher Education* 37 (4), pp. 27-36.

Brown University Center for Environmental Studies (2012), *Environmental studies programs at other universities*. Retrieved from http://envstudies.brown.edu/documents/ ESPrograms.html.

Chapman, R.L. (2007), How to think about environmental studies. *Journal of Philosophy of Education* 41 (1), pp. 59-74.

Corporation for National and Community Service (2008), *What is service-learning?" Learn and Serve America's National Service-Learning Clearinghouse*. Retrieved from http://www.servicelearning.org/what-service-learning.

Dewey, J. (1916), *Democracy and education.* Free Press, New York.

– (1938), *Experience and education.* Touchstone, New York.

– (1988), The Public and Its Problems. In Boydston, J.A. (Ed.), *John Dewey: The Later Works, 1925-1953* (Vol. 2). Southern Illinois University Press, Carbondale.

Grand Valley State University (2012), *Environmental studies*. Retrieved from http://www.gvsu.edu/ens/.

McDaniel, C.N. (2005), *Wisdom for a Livable Planet.* Trinity University Press, San Antonio.

McKibben, B. (2010, July 5), Want to light up a movement? Think art, engage the heart. *Huffington Post.* Retrieved from http://www.huffingtonpost.com/bill-mckibben/want-to-light-up-a-moveme_b_635532.html.

Moltz, D. (2010, January 25), *The new sustainability u. Inside Higher Ed*. Retrieved from http://www.insidehighered.com/news/2010/01/25/gvsu.

Parker, K.A. (2010-2012), ENS 401: Environmental problem solving. *The Agora.* Grand Valley State University. Retrieved from http://agora.phi.gvsu.edu/ens401.

Plumwood, V. (2002), *Environmental Culture: The Ecological Crisis of Reason.* Routledge, New York.

Chapter 19

The Development of a Multidisciplinary Sustainable Business Leadership Degree in a Post-Secondary Institute

Mark Giltrow

Abstract

This paper describes the creation and implementation of a Sustainable Business Leadership Program at an Institute of Technology in British Columbia, Canada. The program is multidisciplinary in nature and has been jointly created by the School of Business and the School of Construction and the Environment. In the creation of the program primary research in the form of Industry, Faculty and Student surveys were conducted to determine demand. Secondary research was also conducted to investigate best practices and the receptiveness by industry for such a multifaceted program. It is concluded that sustainability as a field of study will only be successful if approached from an interdisciplinary pedagogy.

Introduction

Sustainability has become a call to action in our times. It exemplifies the promise of societal development towards a more equitable and wealthy world in which the natural environment and our cultural achievements are preserved for generations to come (Dyllick and Hockerts 2002). Sustainability has been defined as the coupling of environmental, social and business interests. To advance our knowledge researchers studying sustainability will need to integrate knowledge from many different sources; no one discipline could possibly subsume all the knowledge needed (Seager 2008). In fact sustainability is now often described as a "metadiscipline" that surpasses and includes knowledge from many other fields (Mihelcic et al. 2003). Education with the objective of achieving sustainability varies from previous approaches to environmental education in that it focuses more sharply on developing closer links between environmental quality, ecology, business, psychology, socio-economics and the political threads that underlie these disciplines (Tilbury 1995).

In order to advance, sustainability needs to be championed and integrated within post-secondary institutions. Today there is wide consensus that education has an important role to play in motivating and empowering people to participate in changes towards a more sustainable world. Four decades ago Schumacher (1973) described education as the "greatest resource" for achieving a just and ecological society. Since then, major conferences, reports and papers have emphasized the critical role education plays in this cultural evolution. The Brundtland Report, (WCED 1987) states that educators have a crucial role to play in helping to bring about the extensive social changes essential for sustainable development. The 1980 World Conservation Strategy argues that we need a new ethic embracing plants, animals and people if the world as we know it is to survive and flourish. In fact, the report argues that our task as educators is to educate future generations on this new ethic (IUCN UNEP and WWF 1980). Agenda 21 states that education is critical in promoting sustainable development, environmental awareness and the values, knowledge, skills, abilities, and behaviours necessary for effective public participation and decision-making (UNCED 1992). Building from this discourse is a shared vocabulary and direction by educators in mobilizing students and society in the task of achieving sustainability (Agyeman et al. 1996).

Within post-secondary education the business school is beginning to take on a greater role in championing sustainability both on and off the campus. However there are many barriers to integrating sustainability in post-secondary institutes (Leal Filho et al. 2002). For example most business faculty are not currently equipped to meet this challenge. To be effective, sustainable business education must span the established sub-disciplines of business, and integrate topics such as the natural sciences that are not part of the normal curriculum. It will also require faculty to educate students through *transformative* learning processes, where students are challenged and given the opportunity to develop entirely new ways of understanding the role of the business enterprise (Thomas 2005). Currently the more common teaching method is *transmissive* learning, where students are rewarded for adopting and applying relatively narrow models and methods that are inadequate to encompass the meta-issues that are central to sustainability (Thomas 2005).

While a growing number of faculty at colleges and universities are attempting to teach sustainability – there are 60programs in North America offering sustainability-related degrees (http://www.aashe.org) – the obstacles are considerable. Teaching sustainable leadership poses a serious challenge to traditional educators because, as stated above, the topic is fundamentally interdisciplinary. Most colleges and universities, however, are still deeply embedded in the narrow disciplinary paradigm, and faculty members lack the training, resources or cross-disciplinary experience to teach sustainability in their fields. In fact the majority of sustainable leadership programs are taught within the silo of a Business School (http://www.beyondgreypinstripes.org/data-analysis).

Narrow focuses and specialties, traditional within schools of business do not lend themselves to advancing the meta-discipline of sustainability. In fact it has been argued that sustainability rather than being a field in itself, is the locus of interactions between the human and natural systems (Clark and Dickson 2003). As such, by nature sustainability is trans-disciplinary (Seager 2008). While institutes and schools are slow to change, students are seeing the benefit of combining disciplines as they explore sustainability. Many enjoy the cross-pollination found in multidisciplinary teams. Others have been roaming across disciplines in their undergraduate years and cannot imagine doing research any other way (Watanabe 2003). What is needed, then, is for business schools to develop integrated, inter-disciplinary programs that provide transformative learning models to the new meta-discipline of sustainable business leadership.

The Advanced Specialty Certificate in Sustainable Business Leadership (SBL) is offered as a joint credential from the School of Business (SOB-Lead) and the School of Construction and the Environment (SOCE) at the British Columbia Institute of Technology (BCIT). The program has been designed to address the issues outlined above. The program has been built to be transformative, multidisciplinary and integrated with the needs of industry. The purpose of the program is to develop sustainable business leaders who can promote sustainable business approaches and initiatives within organizations ("intrapreneurs") and in new business start-ups (entrepreneurs).

Our research has indicated that graduates from the SBL (http://BCIT.ca/SBL) program have a wide range of employment opportunities, given the technical and scientific skills and sustainable business management training acquired in the program (Rowe 2007). This particular blend of knowledge bases is, and will continue to be, in high demand as companies and organizations shift toward the triple bottom-line and adopt clean technology initiatives. No other Canadian program at this level integrates the knowledge domains of BCIT's SOCE and SOB. This integration is what makes our program unique and prepared to meet the needs of the shifting market, consumer demands, and business climates of the coming decades.

Vancouver's (Canada) economy is one such shifting market, once dominated by natural-resource industries, it is fast becoming primarily knowledge-based (Lewis et al. 2004). The fish canneries and lumber mills are being superseded by globally renowned clean-tech, high-tech, and green enterprises. Green enterprises provide environmentally sustainable economic growth by offering goods and services that meet human needs but reduce our overall ecological footprints. Green businesses rely on renewable energy, use energy and materials more efficiently, and substitute safer substances for toxic chemicals. Examples of green products range from organically produced foods to electric vehicles and solar hot-water heaters. Examples of green services include energy-efficient retrofits and car-sharing enterprises. According to recent estimates, the clean energy sector alone already employs more than two million people worldwide, while 20 million new clean-energy jobs will be added by 2030 (Greenest City Action Team 2010).

In the US, clean energy jobs grew at a rate of 9.1% between 1998 and 2007, while the total number of jobs increased only 3.7%. The City of Vancouver plans to add more than 50,000 new jobs by 2020; of these, 2,000 new green jobs will be added every year, leading to 20,000 new green jobs by 2020 (Greenest City Action Team 2010).

With the growth of this field comes a host of interesting new career opportunities for SBL graduates. As stated above, education in sustainability is becoming increasingly important to the business community as companies grapple with challenges such as how to develop social return on investment (SROI) metrics or understand the impact of corporate carbon footprints on stock prices. In our own industry survey, 40% of respondents felt that the SBL program would increase their employees' opportunities for job advancement. There are also new opportunities for the next generation of managers: managers who are not only trained in traditional business fundamentals but also grounded in an understanding of the multifaceted social and environmental challenges facing 21st century global businesses (Epstein 2004).

The career paths that fall under the broad umbrella of "sustainable business leadership" are as diverse as the students themselves. One student may be interested in social entrepreneurship in the developing world while the next will be seeking advice about clean-tech venture capital careers in Vancouver. A third may be interested in greening the global supply chains. Corporate social responsibility, sustainable product marketing, microfinance, green real estate development, renewable energy, and other interests all fall under the business sustainability umbrella. Our internal BCIT student survey revealed that 29% of the 193 respondents would be interested in completing the SBL program upon completion of their current program in order to pursue some of these emerging careers.

Students recognize the importance of an education in sustainability. The next generation of business leaders see that the future of business will require a new set of skills and responsibilities (Fullan 2005). Between 2003 and 2008, membership of Net Impact (http://netimpact.org), the global organization for business professionals interested in sustainability, increased more than fourfold. By March 2009, over 130 business schools had a Net Impact chapter (http://netimpact.org/).

The SBL program is important to British Columbia and Canada because it recognizes that Canadians are passionate about protecting nature. According to Statistics Canada, nine out of ten Canadians rate the environment as one of their top concerns. However, despite our strong values and extraordinary natural assets, Canada is not achieving its full potential. In an extensive study by the Organization for Economic Co-operation and Development (OECD), Canada finished 28th out of 30 industrialized nations in environmental stewardship (Boyd 2001).

There is a disturbing gap between our values and our poor environmental performance. To close this gap, Canada needs to develop and implement an ambitious and innovative plan for a sustainable 21st-century economy (Boyd 2004). At the provincial level Terry Lake, Minister for the Environment, has publicly stated the importance of such a plan: "British Columbia has become the first province or state in North America to take full responsibility for its emissions, transforming more than 7,000 public buildings into carbon neutral operations in 2010" (Lake 2010). As mentioned, the City of Vancouver has committed to developing 20,000 new jobs in sustainable leadership by 2020: education – and sustainable business education in particular – will support this effort.

Student Supply and Labour Market Demand

Students and industry are very interested in sustainable leadership education and training. Industry needs skilled employees to guide them in integrating sustainability principles into their organizations. The SBL program at BCIT is positioned to meet that need. Moreover, the program is unique in that it delivers a holistic approach to training students by offering courses in environmental and business topics, and integrating these in a capstone project through two capstone courses spanning both terms of the program. This immediate and intensive application of concepts in an industry-focused project will create graduates who are ready to step in and support industry as valued new employees or entrepreneurs ready to launch their own businesses focused on sustainable products or services. As Table 1 indicates, employers from our industry sample see each of the topic areas that the program addresses as important or very important.

Table 1: Industry's Ratings of Course Topic Importance

The following are the topic areas that we are planning to integrate into the program. Please indicate how important you believe each topic is to the program.			
Topics	**Very important**	**Somewhat important**	**Not important**
Ecological Economics	40	46	9
Sustainable Entrepreneurship/ intrapreneurship	49	38	8
Sustainable Ops. Management, Measurement, and System Methods	57	32	5
Sustainable Organizational Change/Leadership and Project Mgmt.	60	28	8
Social Marketing and Consumerism	37	39	20
Principles of Sustainability	57	33	5
Industrial Ecology and Eco-industrial Parks	15	57	20
Climate Change and GHG Management	43	42	8
Social Justice and Sustainable Legal Issues	28	45	22

Further, our industry respondents indicated (Table 2) the most important knowledge- and skill-areas to include in the program are theory, practical knowledge, and presentation skills, with modelling and research skills of secondary importance.

Table 2: Industry's Ratings of the KSAs Needed by SBL Graduates

Please indicate how important it is for graduates of this program to have the following knowledge, skills and abilities.			
Answer Options	**Very important**	**Somewhat important**	**Not important**
Fundamentals/theory	61	29	4
Modeling (climate change, economic scenarios, etc.)	42	42	10
Research skills	52	41	2
Practical knowledge (hands-on experience in research/industry projects)	72	22	1
Presentation skills	62	30	2

In the comment section of the survey, respondents suggested the curriculum would also benefit from courses in accounting/finance, governance/politics and organizational behaviour. These suggestions were integrated in the courses of the SBL.

The coursework and learning outcomes align well with students in the SOB and SOCE. While many SOB and SOCE students will be drawn to the program because of the nature of the courses, the program is marketed to all diploma holders from all schools at BCIT. This marketing effort is important because cohorts with diverse backgrounds are believed to strengthen the interdisciplinary nature of the program and there is institute-wide interest in it (see Table 3). Students with diplomas (or higher-level degrees) from other post-secondary institutions are also welcome to apply to the program, and have done so over the last two years.

Table 3: Students' Interest in the SBL Program by Academic School

Students who stated interest in the SBL program by school.		
School at BCIT	**Count**	**Percent**
Construction and the Environment	32	17.0%
Computing and Academic Studies	29	15.4%
Transportation	3	1.6%
Health Sciences	31	16.5%
Energy	17	9.0%
Business	73	38.8%
Other (please specify)	3	1.6%

As stated, our student survey showed very strong student interest in the program. In the seven-day period from Monday, September 27, 2010 to Sunday, October 3, 2010, 193 students completed the student survey. In this sample, 39% were *very* interested in sustainability issues and 43% were *somewhat* interested in these issues. 30% stated that when applying for their current program they would have considered the SBL program if it were available at that time. Furthermore, 46% saw a need for the program today, and 47% believe the SBL program will increase their opportunities for employment or job advancement in the future. When we looked at the students as potential applicants to the program, we saw that there were 29% who would be interested in attending the program when it was up and running. Of those interested in the program, 28% would like to take it immediately after graduation, and another 60% would want to take it one to three years following graduation from their current program. In our survey of Program Heads and Chief Instructors (20 responses), we found that 81% believed the SBL program would advance the careers of their students and/or make them more employable. Taken together, the industry, student and faculty surveys built a strong case for developing and providing a Sustainable Business Leadership Program in post-secondary education.

The rationale fueling industry's interest in the program is clear: industry needs skilled employees to guide them in integrating sustainability principles into their organizations. The SBL program at BCIT is positioned to meet that need and be a leader in this educational niche. In fact, as Table 4 indicates, almost 40% of the industry sample would consider hiring graduates from the SBL program.

Table 4: Industry's Interest in Hiring SBL Graduates

Would your company/organization consider hiring graduates from the proposed Advanced Specialty Certificate in Sustainable Business Leadership program?		
Answer Options	**Response Percent**	**Response Count**
Yes	39.4%	37
No	17.0%	16
Not sure	43.6%	41
	total responses	94

While industry recognizes the need for trained "green" experts, there is no clear consensus of what makes a job "green" or sustainable (Gulen 2011) and the occupational coding system used by Statistics Canada (National Occupational Classification, or NOC-S) does not have sufficient detail to construct a proper definition of green occupations in Canada. For example, a green occupation such as energy conservation engineer is part of the larger occupational grouping of mechanical engineers, not all of whom would be considered green workers, so we cannot use these statistics to measure the number of green jobs. ECO Canada, in an ECO Canada survey, defines environmental employment as the performance of employment activities that seek to manage the use of, impact on, and enhance the sustainability of the environment. These activities, which could relate to the governance of environmental activities, the supply of environmental products and services, or the development and dissemination of environmental knowledge, may be categorized in any of the following sectors: 1) environmental protection, 2) conservation and preservation of natural resources and 3) environmental sustainability (ECO Canada 2007).

The results of the ECO Canada survey were applied to other data sources such as Statistics Canada's Labour Force Survey and the Census of Canada to come up with counts of green employees. According to the ECO Canada report, in 2006, British Columbia ranked second in the country in terms of environmental employment with 17.6% of all green jobs in Canada. Ontario had the greatest share of the country's environmental employees, at 42.5%, and Quebec ranked just behind British Columbia, at 17.3%.

The industry group with the highest proportion of green employees in British Columbia was the administration and support, waste management, and the remediation sector. There were 18,869 environmental workers in this sector, or 19.8% of all employees in this industry. Education, health, and social assistance had the largest number of green employees of any industry group, at 20,362, which represented 5.3% of workers in these industries. Overall, an estimated 4.3% of the jobs in British Columbia were environmental jobs, a total of 93,462 (BCStats 2010). In 2008, British Columbia's green economy was responsible for approximately 166,000 direct and indirect full-time job equivalents (FTEs) – equal to 7.2% of the province's employment (Globe Foundation 2010).

The 2010 BC Green Economy Report (Globe Foundation, 2010) states that British Columbia's abundant natural resources will continue to form the core of the province's competitive advantage in the new green economy for some time. Low-carbon natural resources such as hydro, wind, wave energy, biomass, and vast reserves of natural gas will be the fuel that drives the province's low-carbon energy future. While British Columbia has strengths in many areas of the green economy, the sectors showing the greatest potential for GDP growth and job creation are: 1) Clean and alternative energy, 2) Energy management and efficiency, 3) Green building and 4) Environmental protection.

GDP from British Columbia's green sectors could grow from CA$15.3 billion in 2008 to between CA$20.1 billion and CA$27.4 billion by 2020. This would represent 10.8% to 14.1% of the total provincial GDP, assuming that total GDP grows at its projected average rate of 2.2% to 2020. In terms of green jobs in the province, an insufficient labour market supply poses the greatest threat to growth. With an average annual green labour force growth rate of 2.6%, direct and indirect FTE jobs in British Columbia's green economy would grow from 166,000 in 2008 to 225,000 by 2020 (Globe Foundation 2010). The labour market demand for leaders in this niche makes the SBL program all the more necessary.

Curriculum

The aim of the SBL program is to develop business leaders who can promote sustainable business approaches and initiatives within organizations (intrapreneurs) and in new business start-ups (entrepreneurs). The goal of the program is to train and education students to:

- critically assess, integrate, and promote the principles of sustainable business strategy, operations, and new product development;
- promote sustainable principles that balance environmental, social, and economic interests;
- manage organizational change by working in multidisciplinary teams focused on a sustainable product or process;
- construct assessments and audits in the field of sustainable business management taking into account relevant scientific, social, and ethical aspects;
- lead the shift toward sustainable business mandates within organizations, and/or the development and adoption of sustainable business products or processes within local industry and communities;
- design sustainable business processes and measure the effectiveness of their implementation.

The SBL program can be completed in two consecutive terms in a cohort-based format. The cohort-based format is important at BCIT as it allows teams to be built and work on projects throughout the tenure of the program, with the hope and possibility of them continuing to work together in a self-created new start-up company. Courses are offered in a flexible schedule including evenings and weekends, and utilizing online delivery approaches as appropriate. The program introduces key business and environmental courses in the first term, followed by more in-depth and specialized courses from these fields in the second term. The Capstone Project spans both terms.

The capstone course integrates all the subjects covered in the SBL program. The capstone course consists of two projects: in the first term students develop a project with an industry partner to help the partner become more sustainable in an area of their business. In the second term students continue to develop their initial project, but now build it into a business proposal so they can pursue the idea in a market as part of an entrepreneurial start-up endeavour upon graduation.

The program uses lectures and labs (tutorials) as the primary transmissive methods to develop the diverse knowledge and applied skills of the learners, supported by flexible learning opportunities. Transformative learning methodologies besides the Capstone Project include case studies, project work, group collaboration, student presentations and field trips. Field trips are incorporated into specific courses to provide additional real world experience and elaborate on material introduced in the classroom. Field trips include visits to organizations that are leaders in sustainable practices, to sites that have been restored, such as brownfields, and to other relevant locations as opportunities arise. There is interest (see Table 5) from industry to support this program directly. Subject experts are willing to give guest lectures, collaborate on research and, to a lesser degree provide actual projects for students in the program.

Table 5: Industry's Interest in Partnering the SBL Program

Would your company partner with BCIT in delivering the Advanced Specialty Certificate in Sustainable Business Leadership program?		
Answer Options	**Response Percent**	**Response Count**
Yes, in the form of guest lectures by qualified staff	17.0%	16
Yes, through support of a practical component of the program by providing access to actual projects	9.6%	9
Yes, through collaboration on applied research projects	14.9%	14
No	23.4%	22
Not sure	30.9%	29
Other (please specify)	4.3%	4
	total responses	**94**

There is no current external certification or accreditation for the SBL. However, the program will be submitted to the Association for the Advancement of Sustainability in Higher Education (AASHE) (http://www.aashe.org/) for their recognition. AASHE is an association of colleges and universities that are working to create a sustainable future. We will also seek recognition from the Disciplinary Association Network for Sustainability (DANS) (http://www2.aashe.org/dans/). DANS is a network of professional associations educating the public about sustainability, developing cross-disciplinary sustainability education projects, and developing curricula and standards to reflect sustainability. Finally, this effort will help in the SOB application for accreditation with the Association to Advance Collegiate Schools of Business (AACSB) (http://www.aacsb. edu). AACSB accreditation represents the highest standard of achievement for business schools worldwide. Institutions that earn this accreditation commit to quality and continuous improvement through a rigorous and comprehensive peer review. Sustainable business practices are considered imperative for education by the AACSB.

Conclusion

There has been a ground-swell of interest in sustainability in post-secondary institutes in North America. This interest has been fuelled by students, faculty and industry. Despite is a growing number of faculty at colleges and universities attempting to teach sustainability, the obstacles are proving considerable. This paper reviews the challenges facing faculty teaching sustainability in a single academic school using transmissive learning methods. It is proposed that teaching this curriculum will require educators to break-down the barriers between schools at their post-secondary institutes and collaborate in the development and delivery of the program. By creating an integrated program, such as the one developed by the author, challenges that arise within an academic institute can better be navigated because institutes will have the full engagement and dedication from the affected faculty and schools. Further because of the nature of the topic, it is argued that the learning method provided to students must fundamentally be transformative with projects that partner with local industry and the community. It is reasoned that a capstone project linking the varied disciplines important to the exploration of sustainable solutions is perhaps an ideal vehicle for developing such transformative learning.

One such program that has pioneered this approach with sustainability in a school of business is the Sustainable Business Leadership Program at the British Columbia Institute of Technology. This multidisciplinary sustainable business leadership program was developed to meet the industry need for trained job applicants and the abundant student interest for careers in sustainability. Our research found that both local students and industry are demanding education in Sustainable Leadership. Industry has identified that there is a need for applicants with an integrated knowledge, skills and abilities in the application of sustainability. Students representing all schools at the Institute are keenly interested in advancing their careers through the SBL program. Whilst this study has demonstrated interest and support for a Sustainable Business Leadership program across important stakeholders, it is recommended that future research demonstrate that such a program leads to valuable outcomes such as alumni and industry satisfaction, graduate placement, and alumni effectiveness as employees.

References

Agyeman, J., Morris, J. and Bishop, J. (1996), Local government's educational role in LA21. In Huckle, J. and Sterling, S. (Eds.), *Education for Sustainability*. Earthscan Publications Ltd., London.

Aspen Institute (2003), *Where Will They Lead? MBA Student Attitudes about Business and Society*. Retrieved from http://www.aspencbe.org/about/library.html#reports.

BCStats (2010), *Measuring Green Color Jobs in British Columbia*. Retrieved from http://www.bcstats.gov.bc.ca/pubs/es/ES2010-1.pdf.

Boyd, D.R. (2001), Canada vs. The OECD: An Environmental Comparison. Victoria: *Eco-Research Chair in Environmental Law and Policy*. Retrieved from www.environ mentalindicators.com.

– (2004), *Sustainability within a Generation: a new vision for Canada*. Retrieved from http://www.davidsuzuki.org/publications/downloads/2004/DSF-GG-En-Final.pdf.

Clark, W.C. and Dickson, N.M. (2003), Sustainability science: the emerging research program. *Proceedings of the National Academy of Sciences* 100 (14), pp. 8059-8061.

Dyllick, T. and Hockerts, K. (2002), Beyond the business case for corporate sustainability. *Business Strategy & the Environment* 11 (2).

ECO Canada (2007), *Profile of Canadian Environment Employment 2007*. Retrieved from http://www.eco.ca/pdf/lmi_industry_final_2007_en.pdf.

Epstein, M.J. (2004), *Making Sustainability Work: Best Practices in Managing and Measuring Corporate Social, Environmental, and Economic Impacts*. Berrett-Koehler Publishers, San Francisco, CA.

Fullan, M. (2005), *Leadership & sustainability: Systems thinkers in action*. Corwin Press, Thousand Oaks, CA.

Globe Foundation (2010), *Building a Strong Low carbon Future*. Retrieved from http://www.bceia.com/media/6607/bcs_green_economy.pdf.

Greenest City Action Team (2010), *Vancouver 2020. A Bright Green Future, An Action Plan for Becoming the Worlds Greenest City by 2020*. Retrieved from http://vancouver.ca/greenestcity/PDF/Vancouver2020-ABrightGreenFuture.pdf.

Gulen, G. (2011), *Defining, Measuring and Predicting Green Jobs*. Retrieved from http://www.copenhagenconsensus.com/Default.aspx?ID=1542.

Huckle, J. and Sterling, S. (Eds.) (1996), *Education for Sustainability*. Earthscan, London.

IUCN, UNEP and WWF (1980), *World Conservation Strategy*. Retrieved from http://data.iucn.org/dbtw-wpd/edocs/WCS-004.pdf.

Lake, T. (2010), *As quoted in the BC Government Publication Carbon Neutral B.C. – Transforming B.C.'s Public Sector*. Retrieved from http://www.livesmartbc.ca/attachments/carbon_neutral_action_reports/CarbonNeutralBC-transformingBCpublic sector.pdf.

Leal Filho, W., Wright, T. and Sharon, A. (2002), Barriers on the path to sustainability: European and Canadian perspectives in higher education. *International Journal of Sustainable Development & World Ecology* 9 (2), p. 179.

Lewis, M., Colussi, M. and Lockhart, A. (2004), *Sustainable Livelihoods*. Retrieved from http://auspace.athabascau.ca/bitstream/2149/999/1/ Background Paper – Sustainable Livelihoods. pdf.

Mihelcic, J.R., Crittenden, J.C., Small, M.J., Shonnard, D.R., Hokanson, D.R., Zhang, Q., Chen, H., Sorby, S.A., James, V.U., Sutherland, J.W. and Schnoor, J.L. (2003), Sustainability science and engineering: the emergence of a new metadiscipline. *Environmental Science and Technology* 37 (23), pp. 5314-5324.

Rowe, D. (2007), *Education for a sustainable future*. Science 317, pp. 323-324.

Seager, T. (2008). The Sustainability Spectrum and the Sciences of Sustainability. Business Strategy and the Environment. *Wiley InterScience* 17, pp. 444-453. Retrieved from http://www.interscience.wiley.com.

Schumacher, E.F. (1973), *Small Is Beautiful: Economics as if People Mattered.* Harper Perennial, New York.

Tilbury, D. (1995), Environmental education for sustainability: Defining the new focus of environmental education in the 1990s. *Environmental Education Research* 1 (2), pp. 195-212.

Thomas, T. (2005), Are Business Students Buying It? A Theoretical Framework for Measuring Attitudes Toward the Legitimacy of Environmental Sustainability. *Business Strategy and the Environment Bus. Strat. Env.* 14, pp. 186-197.

United Nations Conference on Environment and Development (UNCED) (1992), *Agenda 21: Programme of Action for Sustainable Development*. United Nations Publications, New York.

Watanabe, M. (2003), Going multidisciplinary. *Nature* 425 (6957).

World Commission on Environment and Development (WCED) (1987), *Our Common Future: the Report of the World Commission on Environment and Development*. Oxford University Press, Oxford, UK.

Chapter 20

The Sustainable Management and Promotion of Territory: A Strategic Operative Education Plan and Training as a Result of Collaboration between Perugia University, Todi's State Technical Agricultural College and the Local Municipality

Adriano Ciani

Abstract

In 1996, Perugia University's Department of Agricultural Economics, Farm Appraisal and Food Sciences (DSEEA) established a Doctorate in Sustainable Rural Development, Environment and Territory. This educational initiative was followed by several conferences and seminars, organized just after the publication of the famous Brundtland Commission Report of 1987, *Our Common Future*. More recently, in September 2011, the DSEEA promoted an International Summer School (SIS) entitled "The Sustainable Management and Promotion of the Territory" (SMPT) (www.gpst-smpt.com), whose partners were the State Technical Agricultural College *A. Ciuffelli* of Todi, the Alumni Association, and the Local Municipality. The SIS was part of a more extended project called *A.M.A.R. – Associazione Mondiale di Amicizia delle Aree Rurali (World Friendship Association of Rural Areas)* – which we would like to realize in Todi in 2014. In accordance with the A.M.A.R. Project, the first step of the SIS-SMPT meeting (held in Todi from September 4 to 11, 2011) was the drafting of the *Charter of TODI*, which later became the Final Agreement Document, involving participants from 16 countries. The keyword of "Charter of TODI" is "TRADI-OVATION": a combination of the words "tradition" and "innovation". It stands for: "**T**erritory, **R**ural **A**reas, through **D**evelopment, **I**nnovation, **O**rganization, **V**alorization, user-friendly **T**echnology, **I**CT sharing, **O**nline **N**etworking". It can be considered as the "reference pillar" for reflecting on possible concrete actions to be planned by the university system in order to obtain effective holistic sustainability of the projects and their programmes, so as to make them suitable for any territory in the world.

Introduction

The theory of the welfare economy, since its very beginning, has always promoted reflection on growth and development models (Solow 1956; Hirschman 1958). The quantitative importance of development was predominant until the second half of the last century, which meant only minor consideration was given to issues such as environmental equilibrium, intra and inter-generational equity, and sustainability strategy. The principles of sustainable development have been important through several stages, and thanks to this there is currently a widespread heritage among the most important international institutions. It is worth mentioning some of them: Albert Hirschman in 1958; Rachel Carson in 1962 (Carson 1962); the Club of Rome in 1972 (Meadows et al. 1972); the United Nations in 1972 (UN 1972); the WCED in 1987 (WCED 1987); the First World Summit on Sustainable Development in 1992, with the Agenda 21 proposal (UNESCO 1992); the Millennium Goals in 2000 (UNGASS 56 Session 2000); the WSSD of Johannesburg in 2002 (UN 2003); the promotion of the Decade of Sustainable Development by UNESCO (2015-2014) (UNESCO 2005); the UNESCO World Conference on Sustainable Development in Bonn (March 2009) (UNESCO 2009); the G8 University Summit in Turin (May 2009); the UNGASS Resolution for Rio+20 (December 2009); Zero Draft Rio+20 (January 2012) (UNCSD 2012). Along with the Bonn Declaration, other deliberations have also occurred, for example university conferences. The most important are: the Talloires Declaration (1990) (ULSF 1990); the Halifax Declaration (1991); the Earth Summit Agreements (1992); the Swansea Declaration (1993); the Kyoto Declaration (1993); the Copernicus Charter (1993); and the Student Declaration for a Sustainable Future (1995). Numerous scientists, researchers, and universities from all over the world contributed to the evolution of the principles, issues, and methods for the evaluation of Sustainable Development (IUCN-WWF-UNEP 1991; Pearce and Atkinson 1996; Creighton 1996; Leal Filho 2000). Perugia University is located in Umbria, a region in central Italy. Its historical traditions concerning natural resources management can be seen in several famous ancient historical and literary works, such as *Bucolics* by Virgil, *Rerum Natura* by Pliny the Elder, and *Cantico delle Creature* by S. Francesco. The socio-economic context has always been strongly related to the territorial development processes and to the valorization of natural resources. Umbria was also a leading region during the Second World War's most important industrialization period, with its conservation of the land peculiarities and the rural agricultural reality. In 1955 this context was analysed in a noteworthy study by Henry Desplanques, *Campagne Umbre*. It is a significant study, with a penetrating interpretative capacity, concerning the *sui generis* value of the regional area, whose sustainable management is deeply rooted in the awareness of the economic subjects and policymakers (Desplanques 2006). The region is known worldwide as "the Green Heart of Italy and Europe". In this context, Perugia University's main concern is in fact the affirmation of the principles of Sustainable Development Strategy by means of research, didactics, and extension. Since the end of the 1960s, Perugia University's Department of Agricultural Economics, Farm Appraisal and Food Sciences (DSEEA) has developed a tradition of directing crucial research and teaching methods to achieve an innovative model of development. After the publication of *The Limits of Development* at the Club of Rome in 1972, an extensive analysis of the central role played by proper land management was produced: *Earth conquest or re-conquest* (RES 1977). The publication of the famous Brundtland Commission Report *Our Common Future* in 1987 was followed by an intensive sequence of conferences and seminars about the contents and targets of the Sustainable Development Strategy. Its success derived from the positive activities led by the Department in order to acquire a new role within the issues of integrated rural development and sustainable initiatives, at both domestic and international level. Since 1987, an intense network of relationships has been established with fellow researchers and teachers from more than 30 countries in Latin America, Central and Eastern Europe, Asia, and the Mediterranean Basin. In 1994, the Department obtained the funds to organize one of the earliest ALFA Latin America Academic Training projects, *Gestion des Entreprise Agricole et Soutenabilitè du Development (GEASUD)*, which was financed by the EU and in which eight universities were involved, (five Latin American and three European), co-ordinated from Perugia. At the same time, because of increased training demand, sustainable development led to the opening of the Bologna Processes, which were activated first in Agricultural and Environmental Sciences and second in Sustainable Rural Development. One of these is a first-level degree (three years), while the other one is a second-level degree (3+2 years). In 1996, two years later, Perugia University promoted a *PhD in Sustainable Rural Development, Environment and Territory*, attended by Italian and foreign students. This activity increased the visibility and the organizational skills of DSEEA, which in September 2000 also organized the First World Forum of Agro-Tourism and Rural Tourism in Perugia, with the participation of experts and entrepreneurs of more than 65 nationalities. On that occasion, the *International Association of Experts in Rural Tourism and Agro-tourism (IAERT)* was also founded, which was considered an international reference point on those specific topics for several years. Especially after the *Johannesburg WSSD* in 2002, the beginning of the research was dedicated to the young researchers as well as the older academic staff, as long as they focused on the same subjects of the present work during the last decade of the past century. Particular attention was given to the issues related to sustainable development indicators, the Environmental Impact Assessment, the Strategic Environmental Assessment and Support Tools, strictly analysed using the multi-criteria. In 2010 the off-shoot *CARE* (*Centro Ambiente Rafforzamento Economico – Strengthening Economic Environment Center*) was created, and it operated positively, especially in extension services. Recently, in September 2011, the DSEEA promoted an International Summer School (SIS) entitled *Sustainable Management and Promotion of the Territory* (SMPT) (www.gpst-smpt.com) in collaboration with Todi's Agricultural Technical Institute *A. Ciuffelli*, the Alumni Association, and the Local Municipality. The degree, Master's degree and PhD courses are designed to form young technicians and researchers and give them the background knowledge necessary to provide regional niche markets with perspectives in food farming production, handcraft, fashion, and renewable energy, even in a period of international crisis. The input of experts with critical skills in local sustainable development is crucial. The International Summer School was born with the aim of stimulating the mechanisms of benchmarking in the University's activity, with the objectives of keeping the didactics profile high and proposing a repeatable and transferable model to help foreign students realize projects and processes of sustainable development in their own areas (UN 2006). On the other hand, another aim is to provide the local youth (operators, researches, technicians, teachers, policymakers) with continuing attention and the opportunity to transfer their own skills for the consolidation of sustainable development (UNESCO 2007, 2009).

The vision of Perugia University to promote Sustainable Development Strategy

The "future of the world" is the basic theme of the forthcoming Rio+20 Summit, based on the *affirmation of a programme firmly focusing on integrated and sustainable development.* The aim of the above-mentioned programme is to observe the necessity of an integrated amount of support for new synergies, beyond the monothematic approach (often used in development processes and actions) (Ukaga et al. 2010). This innovative yielding approach gives more effective and efficient results. With integration, we intend to reach a model which could help improve the real sectors of economic growth in every sensitive area of the world, by emphasizing also the maximum opportunity to increase the value of these realities (Dulger and Ozdemir 2008). Thanks to their typical specificity, they may – if properly guided – *turn their elements of weakness* to *strength points*, providentially involved in their socio-economic modernization processes. A more endogenous than exogenous development and a bottom-up rather than top-down approach should characterize this course in scientific and technological innovation, so that it has support too. Sustainable development means are not to be considered linked only to utopian ideas (Simai 1995; Selman 1996); rather, by preserving the environment, they lead to a *perspective of intra and inter-generational fairness developed through* at least the following **five key** aspects: ***economic*** *(income adjustment),* ***social*** *(adaptation of the quality of life),* ***environment*** *(productive conservation and sustainable use of natural resources),* ***cultural*** *(management of cultural diversity),* ***management*** *(sustainability management).*

The *formation of the human capital, research development and technological innovation* are essential (Tinbergen et al. 1996). This must be done whilst avoiding the risk of introducing those mechanisms which lead to a break with past and present structures, as well as with the social values hierarchy (Patterson and Theobald 2005) that still marks this criticism and complicates the harsh transition of many countries around the world. The author recommends a smooth introduction of innovations that does not cause any negative impact on the values of hierarchy and social and economic organization in any areas where we will operate. In the present work, the author believes the improved professional formation of the human capital should facilitate the acquisition of the social subjects (Romero 1995), particularly within the weakest areas, using the special flexible ability suitable for the creation of increasingly self-rooted new values shifts, which are combined – when a company engages the processes of modernization – with fairness and sustainability.

Sustainable development: output of political strategy toward the *culture* of all people and nations

For a constant stimulus, actions will take place within society's catalyst areas. In particular, the necessity has been recognized for projects aiming to help them enter higher education, (as far as possible), and, thereafter, the working world. The result would be the acceleration of the process, helping to overcome the current crisis (Smith et al. 2010), of a billion people living below the minimum level of nutrition whilst 20% of the world's population possesses merely 1.4% of global wealth.

In this context, all projects of international cooperation, including mutual consultation, will become operative through a coherent, comprehensive, frequent, clear and penetrating atmosphere, including a modern relationship with the media (ECA 2006, 2007; G8 University Summit 2009). All the above-mentioned serves to enable the gradual establishment and positioning of science as an instrument of peace and solidarity (Summers et al. 2005) through the use of technological innovations, increasing *friendly use* and general *accessibility*. All initiatives should follow the guidelines of the ambitious programme, which point the way to the establishment of a "common house" supplying scientific global-scale research (Delors et al. 1996; Leal Filho 2011a), involving in its wide planning all the potential which each socio-economic reality can bring (Speller 1992), in the areas of education, science and technology, and "spearheading" the required actions to support sustainable development culture (Leal Filho 2010a, 2010b).

Actions will be characterized by progressive work to build a successful scientific framework, based on the values and aiming at a more concrete awareness and knowledge among younger generations (COPERNICUS-CAMPUS 2007; Mitchell 2011): a scientific society within which the University may once again be the *Universitas*, the starting point of a *New Renaissance*, a process related to each country and the world society as a whole.

The multitude of young people alive today require concrete *signs of hope and freedom pointing to a more sustainable life and livelihood* that will lead to a *happier and sweeter society* and overcome the systematic way of working that dominates *like a prison in our modern "easy times"*.

In this sense, science must facilitate the removal of barriers (Rychen and Salganik 2003). The ancestral values of ancient societies are *not to be approached as if they belonged to a museum, rather with the revival and rediscovery spirit* – science should *award property but also humanity* (Allen 2010).

The growing number of university research centres will be characterized both by their spirit of analysis and *even more as a window of consciousness*, as the skills of the autonomous people of the world society are brought together.

The establishment and activation of the International Summer School on Sustainable Management and Promotion of Territory (SMPT)

With the collaboration of the above-mentioned four promoter institutions, together with the support of various local institutions such as the Veralli Cortesi Institute and the La Consolazione Charity Organization of Todi (ETAB), and the sponsorship of the National Order of Agronomy, Surveyors and Land Surveyors, the Province of Perugia organized the activities of the Summer School about "Sustainable Management and Promotion of Territory", which was arranged in Todi at the Agricultural Citadel of the State Technical Agricultural Institute *A. Ciuffelli* from 4 to 11 September 2011.

The idea of the Summer School was born with the objective of providing a solution to the different management requirements of the region. Even though the area is rich in natural resources, it is subject to frequent phenomena of hydrogeological instability, and also to fire hazards in summertime because of the large amount of woods.

In order to limit these risks, the trend is to promote prevention through the learning of GPS and GIS use. The aim is to improve a management method able to safeguard – through constant monitoring – development, good quality of life, and the protection of environmental resources, especially landscape and biodiversity (Leal Filho 2011b).

There is a will to start and consolidate a mechanism that can make "the country talk", by means of new Information and Communication Technology (ICT). For example, the contextual use of websites, webcams, and e-commerce can turn a local firm (handcraft, agricultural, and services) into a globalized entity which is detectable and reachable by any potential client from any place in the world. This procedure leads to a loyalty effect with a high surplus value, since it is linked to the "face" of every operator, and the specific character of the country. This is the direction of processes of growth and innovation aimed at safeguarding internal sustainable development which is repeatable and transferable. From these two attributes it is possible to originate additional opportunities: on the one hand, sustainable development culture for the requirements of the regional reality where Perugia University and SIS operate could be strengthened; on the other, SIS could be promoted with a cognitive system of training and learning through the involvement of foreign students.

The experiences and demands brought by local and foreign people increased the formative activity's value. The several experiences will be compared; they will act as a "fertilizer" and an accelerator, and will enrich the cognitive system. The SIS, with its links to the State Agricultural Technical Institute and to Perugia University, has played the role of and *advanced factory*, which the operators, educators, and single citizens will emulate for the concretization of the sustainability strategy beyond any boundary.

The above-mentioned "TRADI-OVATION activity" about the Territory's Sustainable Management and Promotion, has been conceived as one of the specific requirements of professional training, supplying the best skills for the technicians who work in the area: agronomists, architects, engineers, surveyors and land surveyors. During the workshop week, full immersion projects were carried out involving enhancement of and intervention into some issues in the Todi district: in the castles of Monte Nero and the village of Petroro; the widespread hotel in Massa Martana; and the Arboreal Archaeology Company. All this was done with the addition of a two-day visiting tour. In this framework, the Summer School was attended by young people from 16 countries (Fig. 1), who attended top-level academic lessons and training, with the participation of university teachers from Bern, Budapest CMBS, Iasi (Romania), Chiba (Tokyo), Siena, Tuscia in Viterbo, and Perugia.

Fig. 1: The Group of participants at the educational excursion in Assisi surroundings

The Charter of TODI

Following the careful preparation of the SIS in its educational content and the training within the territory, an extensive discussion during the final session was held, since the majority of the group shared the same basic ideas.

The participants of the International Summer School SIS-SMPT agreed that it is necessary to face the global crisis, which involves manufacturing, financial, social and moral aspects; this is a continuous process that must go beyond the options of sustainable development, green economy and the third industrial revolution in order to give tangibility to the two guidelines of the next World Summit on Sustainable Development (Rio+20, 2012) about an operational implementation of the Sustainable Development strategy to fight against world poverty (UNCSD 2012).

Todi's natural surroundings were declared the most sustainable in the world by the Kentucky University Professor Richard S. Levine. This could potentially be the national and international benchmark to strengthen in order to improve the effective action of guardianship, enhancing and promoting any rural area in the world. The members of the core group decided to draw up the Charter, and agreed to promote the initiative in their home countries, and those of their foreign partners. They also undertook to establish a Local Action Group (LAC), which will help spread this mission all over the world.

The Charter indicated with the keyword **"TRADI-OVATION"** (for short) the acronym of "**T**erritory, **R**ural **A**reas, through **D**evelopment, **I**nnovation, **O**rganization, **V**alorization, user-friendly, **T**echnology, **I**CT sharing, **O**nline **N**etworking); this is the main component with which this innovative process can be structured to give effective credibility, and to erase the increasing disillusionment that usually follows the main international meetings: the author thinks that it is necessary to turn the words into practical actions towards the foundation of the new model of territories' management and promotion. The Charter invites everyone to acknowledge that the *Charter of TODI* and the town's initials, **T.O.D.I.**, are symbolic and of great significance, according to the logic shown by the *words quadrangle* in Table 1.

The last part of the document concerns the aspiration that the national and international Summer School training activities on "Sustainable Management and Promotion of the Territory" become the basic tool for the implementation of the Charter's principles. The Charter of

Todi, through the original representation of the *words quadrangle*, (shown below) is an intriguing frame of reference for the Operational Sustainable Development Strategy. This aspect could be analysed in the future World Friendship Association of Rural Areas (A.M.A.R.) in order to obtain the innovative and integrated sustainable development of a territory anywhere.

Table 1: The meaning of TODI "quadrangle of words"

Territory	**H**olistic	**D**ematerialization	**I**nvent
Tradition	**O**bservation	**D**igitalization	**I**nnovation
Training	**O**rientation	**D**emonstration	**I**CT
Tutoring	**O**rganization	**D**evelopment	**I**nternationalization

The A.M.A.R. Project

The partners, in particular the town of Todi, the *Veralli-Cortesi Institute,* and the *La Consolazione Charity Organization (ETAB)*, are still providing different kinds of assistance. They have always been approachable and have worked, as usual, with intense interest in the issues they are involved with, even in the organization of project stages.

The Summer School is part of a more extended project called A.M.A.R. – *Associazione Mondiale di Amicizia delle Aree Rurali (World Friendship Association of Rural Areas)* which the partnership would like to set up in Todi during 2014, when the State Technical Agricultural College *A. Ciuffelli* will be celebrating its 150th anniversary. The first step of the A.M.A.R. Project was the drawing up of the Todi Charter, before it became a final document of the Summer School activities. The project intends to build up the A.M.A.R. association as a possible international government, with the will and duty to look after Rural Areas' Sustainable Development all over the world (Fig. 2). Many of the most remarkable current "voices" (such as that of Mr Bill Gates) – each within his or her own working and acting field – declared their agreement about the A.M.A.R. project's aims. The project, if well guided towards the development of sustainable management and promotion, can represent the main base to support the *eradication of poverty, to reach a fairer and happier life in the world* (UNCSD 2012).

Fig. 2: A possible logo of A.M.A.R. (World Friendship Association of Rural Areas Project)

Conclusions

During the preparation of the general assembly in Rio+20, it is very important to have a conference designed to discuss and analyse how, when and what has to be done to adhere to the International Sustainable Development Strategy requirement agreed by the University System.

The DSEEA has been a proactive actor since the Stockholm Conference of 1972, which launched initiatives both in education and in research, with positive results. Significant attention was always given to international collaborations.

The planning of the A.M.A.R. Project, the activation of the International Summer School on Sustainable Management and Promotion of Territory and the drafting of the Charter of TODI all showed that responses at international and local levels were very good.

This is a way to support the assertion of the Strategy of Sustainable Development – even if it is probably weak in terms of Cartesian analysis and predictive power. When made concrete it will be fruitful in promoting a culture of territory sustainability. In times of deepening global crisis and a lack of economic governance models, the author maintains that it is sufficient to go out and start working, establishing proper management and promotion of a territory as a great vital resource. A series of mechanisms could enhance wealth and healthy production so as to start defeating the impairing strength of the globalized international speculator lobbies.

The DSEEA promoted its initiative for the diffusion of the principles of sustainable development culture. After a publication about natural resources in the 1970s, DSEEA studied the research, didactics, and extension of the development of issues, and the gradual but inexorable affirmation of the new model of industrial development. The promotion of specific Master's courses and a PhD in Sustainable Rural Development, Environment and Territory from 1996 allowed DSEEA and Perugia University to play a leading role in the professional formation of graduates and technicians working for public administrations who deal with the management of environmental and territorial problems in the Umbria region. This region is extremely interesting from an environmental, productive, historical, and cultural point of view, and a preventive approach to the geological instabilities and fire hazards is required, as well as an innovative effort to valorize and promote the natural resources which represent the basis of the production and wealth of the whole area.

The culture of sustainable development has been promoted over recent years by CARE (Environment Centre and Economic Improvement) to activate a consistent and concrete set of actions coherent to each country's needs, especially concerning the Environmental Impact Assessment and the Environmental Strategic Evaluation.

The foundation of the A.M.A.R. Project represents an opportunity to pursue deep internationalization, solidarity, and international cooperation for the creation of a future world society (African Union 2001, 2006; Couret 2008) that will be more sustainable, fairer, and happier.

The main idea is to create the conditions to give sustainable continuity to a formative action related to the region's future, and to export the formative model of development throughout the entire world (SIDA 2011; OECD 2012).

One future perspective is to improve the formative issues, specifically with the local and international cases of study.

The elaboration and analysis (first of all, the Charter of TODI – "tradi-ovation") by SIS will be gradually transferred through seminars and conferences in local education institutes. The Charter of TODI has been shared by reputed national experts, and is gradually going to be brought into schools and public administrations. It is already present in obligatory elements of the education at the State Technical Agricultural Institute, which is a partner of the project. The next step will be a competition in the schools of the town, with a scholarship for the best essays on the Charter. An additional project aims to integrate the Charter in all the schools of the region. Two municipalities have already approved it as an operative strategy for their government's programme. The second edition is planned for 2012, and its conclusion will be an International Conference on *Territory, Sustainable Development, and Renewable Energy*. The town of Todi have offered a proposal to the executive staff of SIS to prepare the Park Project *Urban Territory and Rural Area Park (T.U.De.R.)* (UNESCO 1995). At present, a preparation phase for an appropriate operative General Agreement between the University and Todi's Municipality is underway.

The International Summer School also represents the basis for a table of discussion on the creation of an off-shoot project on animal welfare. In this area there is a well-established production of broilers with an associational form of management with a large national company. The corporation provides the chicks, and 1-2% of them do not have the required standard features to be sold in the market: these are called "small waste". Normally 10-20 days after being slaughtered they are sent to landfill. For this and other types of farming (pigs, rabbits, pigeons, turkeys, etc.), the "small waste" can be removed from the herd instead of being killed and taken to a new animal welfare company that will produce mostly free-range chickens. The "small waste" turns into the new wealth produced in the territory and promoted as a traditional product. *The pillar of the proposals presented here is an adaptive and visionary approach to establishing a virtuous circle in teaching and learning about sustainable development.*

It is not possible to achieve happiness (considered lawfully indispensible in most countries' constitutions, and one of the most important targets of the next Rio+20) *without big dreams.*

Acknowledgment

The author would like to thank the distinguished unknown reviewers and Prof. Walter Leal Filho for relevant suggestions without which this paper would not have assumed its final form. A special thank you is addressed to all representatives of the Institutional Partnership and members of the Organizing Committee of the International Summer School in Sustainable Management and Promotion of Territory – SMPT for their proactive collaboration in management and for their help in preparing this paper: Prof. Marcello Rinaldi, Head of the

State Technical Agricultural Institute *A.Ciuffelli*, Todi; Prof. Giancarlo Pasqualini, President of the of the State Technical Agricultural College *A. Ciuffelli* – Alumni Association; Dr Avv. Antonino Ruggiano, Mayor of Todi town; Dr Avv. Moreno Primieri, Vice Mayor of Todi; Dr Ezzelino Fulvio Mallozzi, President of Veralli-Cortesi Institute; Dr Bruno Severi, President of La Consolazione – ETAB, Assistance and Charity Organization of Todi; Dr Francesco Moschini, General Secretary of DSEEA, Perugia University; Mr Gilberto Santucci, Secretary of the International Summer School on Sustainable Management and Promotion of Territory – SMPT.

References

Allen, R.E. (2010), *Human Ecology Economics: A New Framework for Global Sustainability*. Routledge, Abingdon, Oxfordshire.

AU-African Union (2001), *New Partnership for Africa's Development Framework Document*. Abuja, Nigeria.

AU African Union (2006), *Second Decade of Education for Africa (2006-2015) – Plan of Action*.

Carson R. (1962), *Silent Spring*, Houghton Mifflin Harcour, Boston's Back Bay, COPERNICUS-CAMPUS – Sustainability Center at the Carl von Ossietzky University Oldenburg & COPERNICUS-CAMPUS University Alliance for Sustainability (2007), *Guidelines for Sustainable Development in the European Higher Education Area*. www.copernicus-campus.org.

Couret, D.G. (2008), *Sustainability in Developing and Developed Countries*. BVSDE, Washington, DC.

Creighton S.H. (1996), *Greening the Ivory Tower. Improving the Environmental Track Record of Universities, Colleges, and Other Institutions*. John Wiley, West Sussex.

Delors J., Al Mufti, I., Amagi, I., Carneiro, R., Chung, F., Geremek, B., Gorham, W., Kornhauser, A., Manley, M., Padrón Quero, M., Savané, M.A., Singh, K., Stavenhagen, R., Won Suhr, M., Nanzhao, Z. (1996*), Learning: The treasure within. Highlights*. Report to UNESCO of the International Commission on Education for the Twenty-first century.

Desplanques H. (2006) (Reprint), *Umbrian countryside, Contribution to the study of rural landscapes of the Central Italy*. Quattroemme, Hardcover, 1404 pages.

Dulger, F. and Ozdemir, Z.A. (2008), Current account sustainability in seven developed countries'. *Journal of Economic and Social Research* 7 (2), pp. 47-80.

Earth Summit Agreements (1992), *Agenda 21 Chapter 36, Promoting Education, Public Awareness and Training*. http://www.iisd.org/educate/declare.htm#Earth.

ECA – United Nation Economic Commission for Africa and AU-Africa Union (2006), *National Strategies for Poverty Reduction and Implementation of the Millennium Development Goals*: An Issues Paper for the African Plenary on National Strategies for Poverty Reduction and Implementation of the Millennium Development Goals March 26-28, 2006, Cairo, Egypt.

ECA – United Nation Economic Commission for Africa (2007), *The Millennium Development Goals in Africa: Progress and Challenges*.

G8 University Summit (2009), *Torino Declaration on Education and Research for Sustainable and Responsible Development (Turin Declaration)*. www.g8university. com.

Hirschman, A.O. (1958), *The Strategy of Economic Development*. Yale University Press, New Haven, Conn. (ISBN 0-300-00559-8).

IUCN – WWF – UNEP (1991), *Caring for the Earth*. Oxford University Press, Oxford.

Leal Filho, W. (2000), Dealing with misconceptions on the concept of sustainability. *International Journal of Sustainability in Higher Education* 1 (1), pp. 9-19.

– (2010a), *Sustainability at Universities: Opportunities, Challenges and Trends*. Peter Lang Scientific Publishers, Frankfurt.

– (2010b),Teaching sustainable development at university level: current trends and future needs, *Journal of Baltic Sea Education* 9 (4), pp. 273-284.
– (2011a), About the Role of Universities and Their Contribution to Sustainable Development. In *Higher Education Policy* 24, pp. 427-438.
– (2011b), Applied sustainable development: a way forward in promoting sustainable development in higher education institutions. In Leal Filho, W. (Ed.), *World Trends on Educationfor Sustainable Development*. Peter Lang Scientific Publishers, Frankfurt.
Meadows, D.H., Meadows, D.L., Randers, J., Behrens III, W.W. (1972) (Club of Rome), *The Limits to Growth*. Universe Books.
Mitchell, R. (2011), Sustaining change on a Canadian campus: preparing Brock University for a sustainability audit. *International Journal of Sustainability in Higher Education* 12 (1), pp. 7-21.
OECD (2012), *Greening Development: Enhancing Capacity for Environmental Management and Governance, Development Co-operation Directorate* (DCD-DAC).
Patterson, A. and Theobald, K. (2005), Emerging contradictions: sustainable development and the new local governance. In Buckingham-Hatfield, S. and Percy, S. (Eds.), *Constructing Local Environmental Agendas,* pp. 155-170. Routledge, London.
Pearce, D.W. and Atkinson, G.D. (1996), *Capital theory and the measurement of sustainable development: an indicator of "weak" sustainability*, Centre for Social and Economic Research on the Global Environment (CSERGE). University College London, Gower Street, London, WC1E 6BT, UK.
RES-Research Centre for Underutilized Resource Recovery (1977), *Land: conquest or reconquest. Tradition, and / or revolution*, Proceedings of the Conference on "Waters and soils : recovering of the underutilized resources". Todi, Italy.
Romero, M.J.R. (1995), The role of the university in sustainable development: challenge and opportunities. *Higher Education Policy* 8 (4), pp. 26-29.
Rychen, D.S. and Salganik, L.H. (2003), *Key Competences for a Successful Life and a Well-Functioning Society*. Hogrefe&Huber, Cambridge (State of Washington) and Göttingen (ISBN: 978-0-88937-272-6).
Summers, M., Childs, A. and Corney, G. (2005), Education for sustainable development in initial teacher training: issues for interdisciplinary collaboration. *Environmental Education Research* 11 (5) (doi: 10.1080/13504620500169841).
Swedish International Development Agency (SIDA) (2011), *SIDA International Training Programme: Education for Sustainable Development in Formal Education,* Source Link: http://www.funds forngos.org/latest-funds-for-ngos/sida-international-training-programme-education-sustainable-development-formal-education/#ixzz1m HpwHDS1.
Selman, P. (1996), *Sustainable Development: Managing and Planning Ecological Sound Places*. Paul Chapman, London.
Simai M. (1995) The Evolving New Global Environment for the Development Process, Tokyo: UN University.
Smith, M.H., Hargroves, K. and Desha, C. (2010), *Cents and Sustainability – Securing Our Common Future by Decoupling Economic Growth from Environmental Pressures*. Earthscan, London.
Solow, R.M. (1956), A Contribution to the Theory of Economic Growth. *The Quarterly Journal of Economics* 70, pp. 65-94.
Speller, P. (1992), Building partnerships for sustainable development: Federal University of Mato Grosso. *Higher Education Policy* 5 (1), pp. 31-32.
Ukaga, O., Maser, C. and Reichenbach, M. (2010), *Sustainable Development: Principles, Frameworks, and Case Studies*. CRC Press, Boca Raton, FL.
Tinbergen, J., Dolman, A.J., Van Ettinger, J. (1996), *Reshaping the international order: a report to the Club of Rome*. Dutton, New York.

ULSF Association (1990), University Leader for Sustainable Future, *Talloires Declaration,* www.ulsf.org.

UN (1972), United Nations Conference on the Human Environment (1972), *Declaration of the United Nations Conference on the Human Environmen*t. www.unep.org.

– (2003), World Summit on Sustainable Development: Johannesburg 2002, *Political Declaration and Plan of Implementation.*

– (2006), Progress report of the Secretary General: *The New Partnership for Africa's Development*: progress in implementation and international support: Fourth consolidated report on progress in implementation and international support.

UNCSD (2012), Zero Draft, *The future we want.* www.uncsd2012.org.

UNESCO – United Nations (1992), *The UN Conference on Environment and Development: A Guide to Agenda 21*. UN Publications Service, Geneva.

– (1995), *The University and Sustainable Urban Development*. Paris.

UNESCO (2005), *Decade of Education for Sustainable Development (2005-2014),* International Launch of the United Nations Decade of Education for Sustainable Development (2005-2014) on March 1 in New York. www.portal.unesco.org/education.

– (2007), *Education for All – Global Monitoring Report: Strong Foundations, Early Childhood Care and Education.*

– (2009), *Bonn Declaration, World Conference on Education for Sustainable Development.*

UNGASS 56 Session (2000), *United Nations Millennium Declaration, Resolution of the General Assembly of the United Nations.* www.un.org/millennium.

UNGASS (2009), *Sixty-fourth General Assembly*, *Plenary 68th Meeting* (PM & Night), *Resolution for the United Nations Conferences on Sustainable Development 2012.* www.un.org.

WCED-World Commission on Environment and Development (1987), *Our Common Future,* Oxford: Oxford University Press.

Chapter 21

International MSc Programme "Environmental Management and Cleaner Production": A Success Story

Jurgis Kazimieras Staniškis

Abstract

To ensure long-term capacity building in the area of sustainable industrial development a new MSc program in environmental management and cleaner production has been developed in the framework of the BALTECH consortium which consists of technical universities in the Baltic Sea region. Sustainable industrial development in the university curriculum is underdeveloped in many countries. The experience of Lithuania and other Baltic countries in implementing this MSc programme in environmental management and cleaner production could be valuable in many different regions around the globe. The paper begins by describing the concept and the structure of the programme, which was initiated by the Institute of Environmental Engineering (APINI) and introduced at Kaunas University of Technology in September 2002. This is a two-year (120 ETCS credits) programme suitable for graduates with qualifications in engineering fields such as chemical engineering, energy and electrical engineering, civil engineering, environmental engineering, and others. The purpose of the programme is to create an example of effective education for sustainable industrial development at the MSc level. Further, the paper describes the skills and knowledge students are provided with and specific conditions of the programme enabling them to understand the interdependence of environmental, technical, economic and social sciences, and to perform interdisciplinary research and development. The programme's contribution to the implementation of sustainable university concept is presented.

Introduction

Main guiding principles of the programme

The focus of environmental work has in recent decades shifted from dealing entirely with emissions and waste from industrial production to include environmental responsibility as a whole and the performance of all types of enterprises, where the environmental properties of processes and products have become more and more important. Environmental issues today affect all types of industrial enterprises. An increasing number of corporations have also come to the conclusion that the sustainable development of any company requires investments in its environmental activities and profile.

Thus, in the framework of the BALTECH consortium, the technical universities of the Baltic Sea region, acting on an initiative of the KTU Institute of Environmental Engineering (APINI), decided to develop and implement a new MSc programme in environmental management and cleaner production based on an integrated industrial ecology approach to current and long-term/strategic environmental issues, and focusing on technologies and concepts in environmental planning and management for a sustainable industrial development. This is a two-year (120 ETCS credits) programme suitable for graduates with qualifications in engineering fields such as chemical engineering, energy and electrical engineering, mechanical engineering, civil engineering, environmental engineering, and others. The programme was launched at Kaunas University of Technology (KTU) in September 2002.

Taking into account the fact that companies applying different preventive environmental measures are growing in number, there is an increasing demand for engineers with high competence in technologies for cleaner production and integrated environmental management. The MSc programme presented here makes it possible to satisfy this demand.

Sustainable industrial development in the university curriculum is still underdeveloped in many universities. The experience of Lithuania and other Baltic countries in implementing a new MSc programme in environmental management and cleaner production could be valuable in many different regions around the globe.

The educational programme is structured according to the following main guiding principles:

Interdisciplinary approach

The environmental field is in its nature multidisciplinary, which is reflected in the structure of the Master's course. Today, engineers need to know how to work alongside economists, biologists, ecologists and many other professionals in order to avoid the negative externalities and/or side effects generated by their economic activities. More sustainable industrial engineering requires the professional formation of industrial engineers with the competence to develop and implement the strategies of incorporating economic, environmental and social dimensions into the practice of their profession (Velazquez et al. 2011). The compulsory part of the course includes areas such as environmental technologies, management, policy, law and economics, and environmental strategies such as cleaner production and eco-design. The optional course programme widens the interdisciplinarity of the programme with areas such as sustainable development, systems analysis, monitoring and modelling, as well as advanced courses in technical areas. To summarise, the programme places a strong emphasis on engineering, management and policy to make sure that graduates have the broad understanding and capability to work with complex issues of sustainable industrial development.

Strong research connection

Most of the optional courses treat advanced topics with close research connections. The subject matter as well as the topics of course assignments and projects are closely linked to the research of the institute giving the course, or to its R&D cooperation with industry, with projects defined by the need for solutions to current environmental or environmental management problems in the industrial company. In particular, the connection with the research is ensured in thesis projects for which active researchers are preferentially engaged as supervisors.

Practical education

Kaunas University of Technology has very strong links to industry and governmental institutions. The Institute of Environmental Engineering has implemented a number of training programmes for industrial enterprises and assisted more than 150 companies in development and implementation of cleaner production projects. MSc students on the programme in environmental management and cleaner production have an opportunity to participate in the projects aimed at improving environmental performance in industry and to acquire valuable practical experience.

Integrated programme

The aim is to provide MSc students with solid theoretical knowledge and hands-on experience in the real world.

The goal of integration can be achieved in various ways: for example, by integrating the principles of environmental communication and informatics into normal business practice in industry. Comparison of the perspectives provided by the multidisciplinary research and systems theory with some of the perspectives of environmental education shows a number of interesting similarities (Leal Filho 1999; 2002). The programme achieves it by a balanced and integrated theoretical and practical education. MSc students use theoretical knowledge in their practical assignments accomplished in industrial companies that result in practically applicable sustainability performance improvement projects.

By integrating systems theory (research) with environmental education and other programmes, students can learn more about the relationships between natural resources, environmental sustainability and human well-being. This knowledge, accompanied by appropriate action, is a prerequisite in maintaining not only our natural resources but also our way of life. Research and environmental education thus require closer cooperation and more involvement in expanding their combined effect. With the support provided by these two areas, industrialists, on the one hand, and educators, on the other, can do more than ever to help people understand and act on environmental concerns (Staniškis and Stasiškienė 2006).

Training in decision-making process

The increased inter-linkage between physico-ecological and socio-economic sectors of the society has placed a variety of pressures on decision-makers in government, society and industry. The decision-making system is a resource allocation mechanism which attempts to rationalise and optimise the system performance to achieve a given set of goals. In this context the decision-maker performs one essential link between the control targets, the control costs and the individuals, groups and components which make up the system being controlled. Additionally, in many cases it is essential to construct a model out of the existing information system to generate the forecasts of future system behaviour. Without a modelling system to allow forecasts, the decision-making process can act only as a feedback control system, which is capable of counteracting deviations from a target only after they have occurred.

The programme provides a system approach as a powerful vehicle for the statement of environmental situations, for reducing the areas of uncertainty in our increasingly complex decision-making situations and different possible control systems, for instance, feedback, feedforward, cascade, adaptive, etc.

Contribution to a sustainable university

The programme provides a great influence on sustainability in relation to whole-university operations. A lot of universities worldwide have actively worked on this issue, because in various ways higher education has a potential impact on sustainable development. If the university succeeds in integrating sustainable development in all or many of its operational activities, the consequence is that exactly one organisation acts in a sustainable way: the university itself. But if sustainable development strategy is integrated successfully in the study programmes, it can expect that a lot of students will, after graduation, eventually reach highly responsible positions in companies, governments, etc. If so, then, in the long term, many organisations will work more and more sustainably – not just one (Roorda 2010).

Many universities have actively worked on sustainability in relation to their infrastructure and management: for example, the so-called "greening campus" projects (Shriberg 2000; Renner 2011), fully functioning environmental management systems (Delakowitz and Hoffman 2000; Sammalisto 2007; Sammalisto and Lindhqvist 2008), etc. Kaunas University of Technology has decided to implement the second model, i.e. to integrate sustainability in all the roles a university has in society.

First, the university acts an educational institution. In this respect, the impact it has on society is, to guide and assist students with their learning processes, and thus deliver educated professionals. Secondly, the university is a research institution and delivers the results of fundamental or applied scientific research to society. Thirdly, a university is an organisation which, as all organisations, e.g. industrial companies, banks, governmental bodies, etc., has all kinds of operational interactions with the outside world, such as procurement and employment of staff members, use of energy and water, and transport of people and goods, as well as the production of waste, atmospheric gases, and other environmental impacts (Roorda 2010).

Besides these three roles, a university has a very important fourth role, which is not always emphasised. This is its direct interaction with society, for instance in the form of joint local Agenda 21 projects, participation in cultural events, strengthening public discussions on sustainable development issues, etc.

The programme has demonstrated that this model of sustainable university facilitates the "snowball effect" that makes the educational role so powerful.

The programme structure

The programme structure is presented in Fig. 1. Compulsory core courses (35-45 ECTS credits) are developed in collaboration with the BALTECH universities. In addition to the compulsory core courses, the candidate shall take a selection of optional courses (45-55 ETCS credits) in two different subject areas (Staniškis and Arbačiauskas 2003; 2011):

Fig.1: Layout of the MSc programme in environmental management and cleaner production

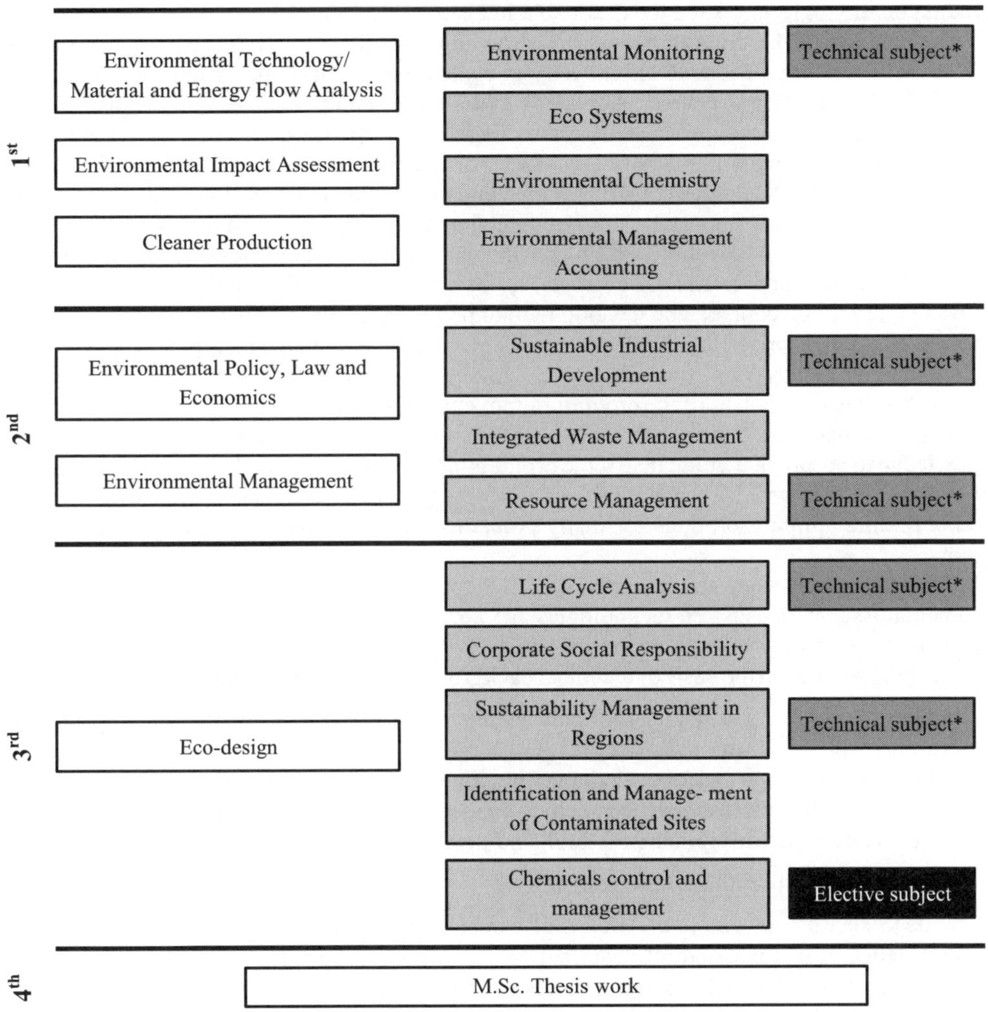

* *Technical subjects in chemical, electrical, mechanical, materials and environmental engineering (five alternative courses in each area).*

- Advanced courses in environmental and related subjects, e.g. industrial ecology, environmental technology, environmental impacts, risk assessment/management, life cycle assessment, environmental impact assessment, etc. (20-30 ECTS credits).
- Advanced courses in one engineering subject, e.g. chemical, process, mechanical or civil engineering (20-30 ECTS credits). The candidate shall in consultation with the director of studies of the MSc programme of his/her university select a number of advanced courses in the subject area of his/her basic engineering programme for acquisition of the necessary engineering competence.

In addition to these courses the candidate may choose free elective courses (approx. 6-8 ECTS credits) in subjects such as language, economics, business management or social sciences.

The last term of the programme is devoted to development of an MSc thesis (30 ETCS credits) on the subjects related to sustainable industrial development.

The contents of the programme are based on an industrial ecology approach, i.e. on industry-environment interactions to aid industry in evaluating and minimising impacts on the environment. The programme courses reflect one of the most important concepts of industrial ecology which, as in the biological system, is rejecting the concept of waste. The programme covers technologies in coping with industrial residues, particularly those aimed at reuse and recycling (Environmental Technologies course); identifying, evaluating and implementing technical and managerial options for improving environmental and economic performance (Environmental Assessment, Cleaner Production and Environmental Management courses); design of industrial processes and products from dual perspectives of product competitiveness and environmental impact (Eco-Design course), and development of policy framework which provides appropriate incentives for enterprises to adopt preventive environmental management practices and to increase their efficiency (Environmental Policy, Law and Economics courses).

The compulsory courses of the programme cover all basic aspects of an industrial ecology approach. Optional courses are used to discuss these issues in more detail and to provide additional knowledge to ensure that graduates of the programme will be able to apply an industrial ecology approach, i.e. they will be capable both conducting systematic analysis of industrial activities and of finding optimal solutions for many problems related to sustainable industrial development.

The programme offers an integrated approach towards current and long-term/strategic environmental issues, focusing on technologies and concepts in environmental planning and management of a sustainable development of industry.

The programme, on the basis of technical background of the student, enables graduates:

- to integrate preventive managerial and technological tools in achieving more sustainable development of industry and society;
- to lead and sustain the process of change in industry, academia and other sectors;
- to understand the interdependence of environmental, technical, economic and social sciences, and to perform interdisciplinary research and development.

This is achieved by providing MSc students with

- skills to identify and assess the effects of human activity on the environment;
- knowledge of national and international environmental policy and legislation and the management of environmental issues in industrial and service systems;
- knowledge of technical systems, strategies and technologies for applying the principles of cleaner production to developing products and production systems;
- experience in implementing preventive environmental measures (Staniškis and Stasiškienė 2006).

The Institute of Environmental Engineering at Kaunas University of Technology has very strong links to industry and governmental institutions. The Institute implemented a number of training programmes for industrial enterprises and assisted more than 150 companies in the development and implementation of cleaner production projects. MSc students of the environmental management and cleaner production programme have an opportunity to participate in the projects aimed at the improvement of environmental performance in industry and to acquire valuable practical experience.

The key issue here is prevention, which is crucial, especially in cases where a product or process is known to cause harm. The preventive principle is to look to changes upstream in a causal network of the system of production and consumption. The preventive nature of cleaner production calls for new approaches to product design, consumer demand, patterns of material consumption and the entire material basis of economic activity. A critical element of efficient production is the accurate and consistent measurement of inputs and outputs. The often-repeated axiom "what gets measured gets managed" has never been more true. Without accurate cost information it is difficult to know what changes are to be made in order to improve profitability in today's highly competitive business climate. Management accounting systems can provide the information required to make these decisions (Staniškis et al. 2008).

Since the beginning of the programme in 2002, students have been involved in more than 50 cleaner production projects based on the preventive approach presented above. The environmental and economic performance of preventive cleaner production innovations is presented in Table 1.

Table 1: Results of the implemented preventive innovations

Number of companies	126
Number of implemented innovations	211
Environmental results (annual):	
El. energy consumption reduced	27 584 000 kWh
Heat energy consumption reduced	60 518 000 kWh
Waste amount reduced	86 700 t
Chemicals consumption reduced	850 t
Air emission reduced	79 500 t
Drinking water consumption reduced	297 500 m^3
Diesel consumption reduced	387 000 l
Environmental results (annual):	
Natural gas consumption reduced	5 883 000 m^3
Fuel consumption reduced	656 800 t
Wastewater amount reduced	622 500 m^3
Industrial water consumption reduced	468 900 m^3
Economic benefit:	
Total investment in CP innovations	16 529 000 EUR
Annual savings from CP innovations	9 605 000 EUR

In recent years a special interest has been devoted to the eco-design of industrial products. With active student participation more than 20 products have been redesigned in order to minimise their environmental impact during the product life cycle. These products range from a refrigerator or corner sofa to very small integrated electronic devices. An electric meter (Fig. 2) was redesigned by applying a life cycle approach, and is now available on the local and international market (Gurauskienė and Varžinskas 2006).

Fig. 2: Electric meter redesigned by applying a life cycle assessment method

Since 2002 more than 100 MSc students have graduated from this programme possessing a very broad perspective on environmental issues and understanding the context of sustainability problems and a wide range of actions. A balanced combination of theoretical and practical education enables graduates to realise that they are capable of accomplishing challenging tasks immediately after graduation, and that they are desirable employees in research organisations, industrial companies, consulting companies and governmental institutions

Programme contribution to engineering education at the technical university level

In advanced engineering education, a key issue is "contextual awareness", i.e. the ability of engineers to view actions, problems and possible solutions in a broad context, in which the consequences may at once be scientific, technical, economic, or social in nature.

To some extent, engineering education is already dealing with the environment, professional ethics, safety and health, and discipline-specific problems related to engineering processes and practice. This content has been often integrated with other subjects, not least engineering design. All too often, engineering students are taught how to react to situations, rather than how to anticipate the consequences of decisions and proceed with them responsibly. This should be regarded as a key point when defining learning objectives and outcomes.

Several areas have attracted considerable attention at the university level; for example, management of the environment, in a broad sense. All engineering processes, projects and products, from the extraction of raw materials right up to the final production of goods and generation of wastes, tend to have an impact on the environment – in most cases a negative one. Energy production and environmental management are fair subjects for sustainability as well. Various actions could be taken towards the limitation of greenhouse gas emissions in the production of energy. Related choices pertain to the use of alternative energy sources, the approach to building design and construction, the production of low-emission vehicles, rational energy use and effective use of material resources.

Sustainable consumption and consumer's protection in a broad sense, including health and safety issues, and the provision of timely and useful public information are other important areas that have attracted attention at university level. Sustainable consumption is defined by the United Nations Environmental Programme as "the use of services and related products, which respond to basic needs and bring a better quality of life while minimising the use of natural resources and toxic materials as well as emissions of waste and pollutants over the life-cycle so as not to jeopardise the needs of future generations".

Corporate social responsibility and ethics in relation to positive or negative consequences and social impact are fundamental to the current debate over sustainable development. It should be reflected in the choices made by authorities in technology, government, business and industry. It may appear that choices are made in a scientific and technical context, even when they have an impact on the economy, the environment, public health and societal conditions.

The MSc programme "Environmental Management and Cleaner Production", as a unit effort became a fertile ground for a larger pilot programme in the whole university. There are various levels of action:

- Development of models consisting of values and corresponding actions, i.e. the university should itself be a model for sustainable operations, research and development.
- Efforts to foster student awareness and vision regarding actual needs, problems, and the potential effect of solutions and actions. This has implications for the choice of teaching/studying approaches and support for student activities beyond the formal curriculum.
- Identification of new skills and occupational profiles that must be translated into education, training, continuing professional education and job requirements. This is a task for leaders and planners at university, department and programme level, in cooperation with employers.
- Development of skills such as crisis communication and management, policy analysis and formulation, interdisciplinary approaches to problem solving; above all, a broader view of problems, their causes, consequences and solutions.

Kaunas University of Technology has undertaken some successful actions at the levels mentioned above and in all four roles (see Fig. 3).The university is a selected pilot institution for the QUESTE-SI project funded by the European Commission under the Lifelong Learning Programme ERASMUS. QUESTE-SI stands for "Quality system of European Scientific and Technical Education for Sustainable Industry". The project is coordinated by EFMD, the Management Education Network, and ENQHEEI, the European Network for Quality of Higher Engineering Education for Industry.

Fig. 3: The main roles of the university

The self-evaluation was performed in four contributing areas for sustainability activity: the university as a whole, education and curriculum, student involvement, research and innovation in cooperation with industry. At university level, the following results have been obtained:

- University authorities view social responsibility and sustainability as integral to the institutional identity and values;
- University management treats sustainability as part of the operational strategy;
- Evidence of good practices: for instance, in energy efficiency, conservation, environmental protection, waste management, sustainable campus etc.;
- At faculty level professional development activities include sustainability as an university issue and as an area with distinct approaches for teaching, learning, and research;
- Research activities include subjects related sustainability.

Findings for the educational dimension are:

- Scientific or technical knowledge and practical skills related to sustainability to be introduced in curricular and student assessment plans;
- Interdisciplinary approaches to problem-solving, life cycle analysis, new approaches to product design and project-based or problem-led learning;
- Acquired familiarity with sustainability-related international standards, the use of exercises that require a broader understanding of problems, solutions and consequences;
- Evidence of discussions with industrial experts, new skills on crisis communication, risk management, policy analysis-formulation-implementation.

In research and innovation in cooperation with industry:

- There is an operational strategy for sustainability-related research, development and innovation projects in partnership with industrial or business firms and, as appropriate, other educational institutions;
- The results of research and cooperation are used to enrich the teaching and learning process;
- There is an adequate research management support and a clear sense of the economic "bottom-line" for sustainability efforts.

Students' involvement:

- There is support for student activities inside the university, for instance, participation in energy saving, waste management campaigns, local and international projects on sustainability;
- The university's departments track the job placement of graduates and seek their feedback on work-related experiences that bear on sustainability.

QUESTE-SI has identified seven topical areas in which sustainability objectives, actions and outcomes are often found:

- Environmental pollution, waste management, remediation;
- Energy conservation, recycling, conversion;
- Eco-building;
- Life-cycle analysis and design;
- Natural and water resources;
- Eco-systems and bio-diversity;
- Health, safety and risk;
- Corporate social responsibility.

All eight topical areas of sustainability are fully covered by the MSc programme and are reflected in the university strategic planning and action.

Conclusions and considerations

Sustainable industrial development in the university curriculum is still underdeveloped in many countries. The experiences of Lithuania and other Baltic countries in implementing a new MSc programme in Environmental Management and Cleaner Production could be valuable all over the world.

The contents of the programme are based on the industrial ecology approach, i.e. on industry-environment interactions to aid industry in evaluating and minimising impacts on the environment. Graduates are capable of conducting systematic analysis of industrial activities and finding optimal solutions for many problems related to sustainable industrial development.

Students' knowledge on technical systems, strategies and technologies for applying the principles of resource-efficient and cleaner production to developing products and production systems are obtained during their participation in the project activities.

The MSc programme has become a fertile basis for a larger pilot project. The programme and staff experience facilitate the development of the university itself as a model for sustainable operations, research and study. Some results of self-assessment are presented in Tables 2 and 3.

Table 2: Institutional recognition of principles of social responsibility and sustainability

Principles of social responsibility and sustainability	**Little or no evidence of recognition & action 1.0**	**Weak recognition & action 2.0**	**Adequate recognition & action 3.0**	**More than adequate recognition & action 4.0**	**Very strong recognition & action 5.0**
1. Social responsibility is demonstrated by the planning, implementation, and realization of sustainability strategies within the institution and its academic units.		2,5			
2. The institution and its academic units recognize social responsibility as a key element of institutional identity and a common thread in matters of sustainability.			3,2		
3. There is a strategic approach to sustainability that involves the institution and its operational practices.				4,1	
4. Specifically, the **engineering division/ school** and its academic units recognize social responsibility as a key element of their identity and a common thread in matters of sustainability.					5,0
5. Faculty professional development activities address sustainability as an area with distinct approaches to teaching, learning, and research			3,5		

6. The institution understands that it is a source and model of emerging values and good practices in matters of sustainability, as well as a base for student personal and professional development.			3,1		

Table 3: Programme "Environmental management and Cleaner Production" evaluation

Programme,Concentrations, Elective Options	**Strategic Indicator**	**Application & Effect**
Title of model program		
Environmental Management and Cleaner Production	1. The programme plan followed by the teaching faculty includes the scientific or technical knowledge, practical skills, and non-technical subjects essential for teaching and learning sustainability in the context of specific topical target areas identified in the plan.	5,0
	2. The programme plan complements the institutional or school-level rationale for the selection of sustainability-related topics, methodologies, techniques and tools, including the use of project-based or problem-led education techniques.	5,0
	3. The program curriculum comprehends the "transverse skills" that graduates will need to recognize and work on sustainability problems.	5,0
	4. The program under review is designed to give graduates the broader "contextual awareness" that they will need to deal with sustainability issues and problems.	5,0
	5. The program faculty, in the various topical target areas, has received training in the formulation and self-assessment of sustainability learning objectives, outcomes or competencies, consistent with the internal quality assurance and/or external review system	4,0

Participation as a pilot university in the QUESTE-SI programme will ensure "contextual awareness" in engineering education, i.e. the ability of engineers to view actions, problems and possible solutions in a broad context in which the consequences may at once be scientific, technical, economic, or social in nature. The pilot experience has proved that

> "sustainability does not simply require add-ons to existing structures and curriculum, but implies a change of fundamental epistemology in our culture and hence also in our educational thinking and practice. Seen in this light, sustainability is not just another issue to be added to an overcrowded curriculum, but a gateway to a different view of curriculum, of pedagogy, of organisational change, of policy and particularly of ethos … We need to see the relationship the other way around – that is, the necessary transformation of higher education towards integrative and more whole state implied by systematic view of sustainability in education and research" (Sterling 2004).

It could be concluded that in order to be effective education for sustainable development should

- ensure the institution itself is a model for sustainable operations, research, and development;

- be addressed in two ways: firstly, through the integration of sustainable development topics across all relevant subjects, programmes and courses and, secondly, through the provision of specific subjects, programmes and courses;
- focus on enabling meaningful learning experiences that foster sustainable behaviour, including within the institution, in the workplace, in families and communities;
- increase cooperation and partnership among members of the educational community and other stakeholders;
- explain environmental problems by means of a life-cycle approach together with economic and social implications;
- use a wide range of educational methods including decision-making processes, scenarios, modelling, case studies, projects, good practice analysis, excursions, etc.

Finally, in a university that is organisationally structured around quality management, the input from the four-role vision (see Fig. 3) has its counterpart in an output evaluation and a system of reporting which, if used as an input to the organisation mission, produces a closed-loop feedback system of continuous improvement.

References

Delakowitz, B. and Hoffman, A. (2000), The Hochschule Zittau/Gohrlitz "Germany first registered environmental management (EMAS) at an institution of higher education". *International Journal of sustainability in higher education* 1 (1).

Gurauskienė, I. and Varžinskas, V. (2006), Eco-design methodology for electrical and electronic equipment industry. *Environmental research, engineering and management,* Technologija, Kaunas, 3 (37), pp. 43-51.

Leal Filho, W. (1999), *Sustainability and university life.* Peter Lang Publishing, Frankfurt.

– (2002), Towards a closer integration of environmental education and industrial ecology. *Int J Environ Sustainable Dev* 1 (1).

Queste Sustainability in Engineering Education (QSI). http://queste.eu.

Roorda, N. (2010), *Sailing on the winds of change. The odyssey to sustainability by the universities of applied sciences in the Netherlands.* PhD Thesis, Maastricht university.

Renner, J. (2011), On-campus innovation for sustainability: evolution of sustainable development on the Joondalup campus, Edith Cowan university (ECU), Western Australia. In Leal Filho, W. (Ed.), *World Trends in Education for Sustainable Development,* pp. 127-149. Peter Lang Publisher, Frankfurt.

Sammalisto, K. (2007), *Environmental management systems: a way towards sustainable development in universities.* PhD Thesis. Lund university, International institute of industrial environmental economics.

Sammalisto, K. and Lindhqvist, T. (2008), Integration of sustainability in higher education. A study with international perspectives. *Innovative higher education* No. 32.

Shriberg, M. (2000), Sustainability management in campus housing – a case study at the university of Michigan. *International Journal of Sustainability in higher education* 1 (2).

Sterling, S. (2004), *High education, sustainability and the role of systematic learning.* In: Corcoran and Walls (Eds).

Staniškis, J.K. and Arbačiauskas, V. (2003), Industrial ecology in university curriculum: new M.Sc. Programme in Environmental Management and Cleaner Production. *Clean Technologies and Environmental Policy* 5 (2), pp. 92-95, Springer-Verlag, Berlin.

– (2011), Integrated Theoretical and Practical M.Sc. Education for Sustainable Industrial Development. In Leal Filho, W. (Ed.), *World Trends in Education for Sustainable development,* pp. 105-115. Peter Lang Publisher, Frankfurt.

Staniškis, J.K. and Stasiškienė, Ž. (2006), An integrated approach to environmental education and research: a case study. *Clean Technology and Environmental Policy* No. 8, pp. 49-58. Springer-Verlag, Berlin.

Staniškis, J.K., Stasiškienė, Ž. and Jasch, Chr. (2008), *Cleaner technologies: environmental management: environmental accounting, investment appraisal and Financing*. Nova Science Publishers, Inc. New York. 345 p.

Velazquez, L., Munguia, N., Esquer, J. (2011), Engineers for more sustainable industrial engineering. In Leal Filho, W. (Ed.), *World trends in education for sustainable development,* pp. 173-184. Peter Lang GmbH, Frankfurt.

Chapter 22

A Model for Enhancing Education for Sustainable Development with Management Systems: Experiences from the Nordic Countries

Tove Holm, Kaisu Sammalisto, Timo Vuorisalo, Thomas S. Grindsted

Abstract

Enhancing education for sustainable development (ESD) in higher education is a global challenge. Most countries have national quality assurance demands for higher education that could benefit ESD. This innovative way to develop ESD has not been investigated before. The authors studied how different higher education institutions enhance ESD, in particular with regard to quality, environmental and integrated management systems. The research was done as an exploratory comparative study. Results from earlier research on ESD curriculum development in recent years were studied. Based on the results, a model of a process for enhancing ESD was developed that can be used in management systems. 27 higher education institutions from the Nordic countries answered a survey and 11 institutions wrote reports of how they are enhancing ESD. The results from the research, survey and institutions were presented at a seminar, where the model was further developed. The authors found that in none of the Nordic countries are indicators for SD included in the quality assurance models of higher education, even though there are demands for enhancing ESD. The majority of the replies stated that the institutions have a clear connection between ESD and quality assurance and it is implemented differently in many parts of the management systems. The findings from the survey and the pilot institutions are restricted to faculty involved in enhancing ESD in higher education in the Nordic countries. The findings can be used by those responsible for developing ESD and/or management systems in different higher education institutions.

Introduction

The objective for the UN Decade of Education for Sustainable Development, DESD, 2005-2014 is to integrate education for sustainable development (ESD) in all education from pre-school to higher education (UN DESD 2011). Higher education has a central role for ESD because future leaders for all labour markets are usually graduates from higher education. ESD in higher education requires cross-disciplinary teaching, because the graduates should be educated to make decisions promoting sustainability after having considered the social, environmental and economic costs and benefits of available alternatives (Sibbel 2009). In Europe UNESCO supports ESD-related research and innovation in higher education, following the strategy for the second half of the UN DESD (UN DESD 2010).

European countries have national demands for quality assurance in higher education, based on European standards and guidelines (Amaral and Rosa 2010). Companies usually start by applying a quality management system according to the standard ISO 9001 into which they later integrate an environmental management system according to the standard ISO 14001 (Bernardo et al. 2009). Researchers have found that one of the benefits of integrated management systems is their contribution to sustainable development (Jørgensen et al. 2006; Kraus and Grosskopf

2008). Several studies examine integrated management systems for quality, environment and/or safety in companies (Bernardo et al. 2009), but as far as is known there is only one that deals with higher education (Holm et al. 2011). Stubbs and Schapper (2011) found that though many higher education institutions support ESD, initiatives on curriculum greening are mainly driven by individuals, and generally in the absence of integrated management systems. A typical explanation for the fact that higher education lags behind companies in developing quality management seems to be that the benefits of the management are not known, or that management is only considered to be relevant to manufacturing (Holm et al. 2011). It seems likely that an integrated management system that encompasses a wider array of aspects than a quality management system, sustainability possibly included, could be considered more relevant by a larger part of the faculty, due to the fact that university faculties represent many different fields. The global aim for quality management in higher education is to secure and improve teaching and learning (Pratasavitskaya and Stensaker 2010), which means that ESD can be enhanced with the management system as far as sustainability aspects are a part of it. An environmental management system requires that personnel with responsibility for tasks that have significant environmental impacts should be competent (Swedish Standard Institute 2004). This means that if the institution identifies ESD as a significant environmental aspect it should offer training in ESD for the entire staff, irrespective of their background.

The aim of this study is to investigate how different higher educational institutions enhance ESD with quality, environmental and integrated management systems – which, as far as the authors know, has not been investigated before. The authors argue that integrated management systems can have great potential for enhancing ESD in higher education, and introduce a model of how this can be further developed. Thus it is argued that enhancing ESD with integrated management systems is an innovative way of promoting SD, because everything that is a part of the quality management system of the institution is evaluated for national demands, and further developed. If the quality management system is further developed to an integrated management system, more aspects, for example ESD, are followed up regularly by being included in the measuring routines of the institution and prioritized by the top management.

Methodology

The objective is to conduct innovative research for ESD and its use in practice. The research was done in a project called "ESD in Academia in the Nordic Countries", in which one objective was to develop a process for including ESD in the curricula in higher education so that it can be used in the quality assurance work.

Published earlier research on curriculum development for ESD from 2009 onwards was collected. This time period was chosen because 2009 was the halfway-point of the DESD and the authors thought that innovative ways to enhance ESD could possibly be found in research from that time. The majority of relevant studies turned out to have been published in the International Journal of Sustainability in Higher Education. All issues of that journal from 2009 until August 2011 were therefore analysed and from them studies dealing with selected keywords were chosen for study. The keywords were: behaviour change, competences, curricula, curriculum, curriculum development, education, environmental management, knowledge creation, learning, learning outcomes, student and teaching. Those papers dealing with sustainability in higher education in a wider area or discipline, which were assumed to include curriculum development, were also chosen. Papers on greening campuses were not included in this study. A model of a pilot process was developed for including ESD in curricula in higher education (Fig. 1) for use in management systems. The literature research was summarized according to what was found in each study according to the elements in the model.

A survey about how ESD is enhanced in higher education institutions in the Nordic countries, especially with the help of their management systems, was carried out by three of the authors (Grindsted, Holm and Sammalisto) and an MSc student. A link to the web-based survey was sent via email to contact persons at higher education institutions in the Nordic countries in June 2011. 27 higher education institutions in Finland, Sweden, Denmark and Norway answered the survey. Only the institutions interested in these matters are assumed to have answered the survey, which means that the results may overestimate the popularity of ESD in higher education in the Nordic countries. The results therefore apparently describe the situation in institutions that have already done something to promote ESD.

Fig. 1: The model for including ESD in curricula in higher education

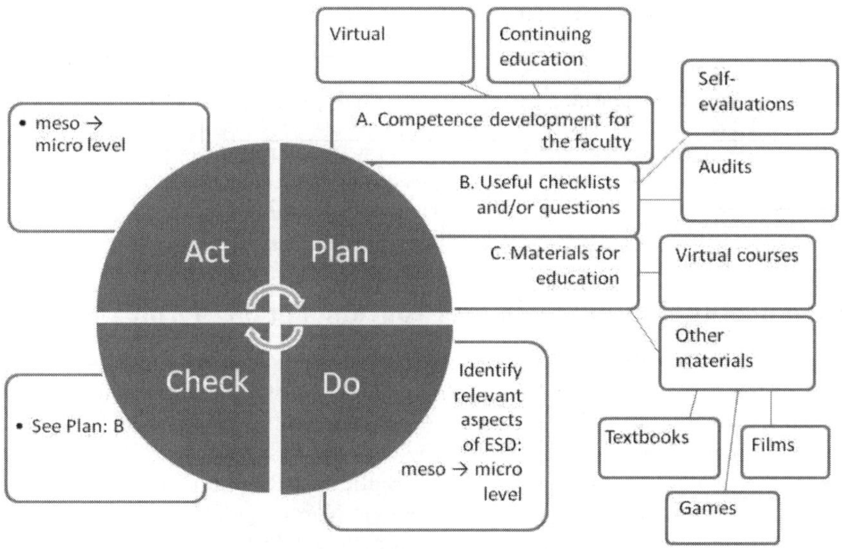

After having categorized the universities according to degree programmes, nine institutions of higher education from Finland, Sweden and Denmark were invited to participate as pilot universities in the project during spring 2011. 11 institutions were selected based on interest expressed in answers by post and/or by conversation, so that a wide range of institutions in the countries would be presented, of different kinds and sizes but also including those institutions at the forefront regarding ESD and/or management systems as well as those who have just begun working in these areas. In September 2011 the pilot universities sent reports on how they were enhancing ESD.

The results from the literature research, survey and work in the pilot universities were presented at the first seminar of the project in October 2011 at the University of Gävle in Sweden, where 27 representatives from 22 different higher education institutions participated. Based on the presentations and audience response the model (Fig. 1) was further developed in a workshop at the seminar. The results of the further development of the model were summarized according to the elements in the model.

Table 1: Demands for ESD and quality assurance in higher education in the Nordic countries

	National strategies for implementing the DESD	Legislation for SD or ESD that relates to higher education	Indicators for SD in quality assurance for higher education
Denmark	Education for Sustainable Development -strategy for the UN Decade 2005-2014, published 2008	A demand for all public buildings to reduce their C02 emissions	–
Finland	Implementation of Baltic 21E programme and Finnish Strategy for the Decade of Education for Sustainable Development (2005-2014), published 2006	–	SD can be an optional subject for the audits in Finland, according to the new model, published 2011
Iceland	Environmental education as part of Iceland's National Strategy for Sustainable Development 2002-2020, published 2002	–	–
Norway	ESD as part of Norway's strategy for sustainable development, published 2007	–	–
Sweden	ESD as part of Sweden's strategy for sustainable development, published 2006	The Swedish Higher Education Act include a statement that universities shall work for SD in all their activities, 2006	–

ESD and quality assurance in higher education in the Nordic countries

The Nordic countries either have national strategies for implementing the DESD or mention it as a part of the strategy for SD (see Table 1) (The Ministry for the Environment in Iceland 2002; Ministry of Education in Finland 2006; Government of Norway 2007; Ministry of Education in Denmark 2008; Lindberg 2010). The regional strategies for ESD in the Nordic countries are the Agenda 21 for Education in the Baltic Sea Region (Baltic 21 E, 2002) and the Nordic strategy for SD in which it is stated that the Nordic countries want to be the leaders in enhancing the UN DESD (Nordic Council of Ministers 2009).

The Swedish Government gave a directive in 1996 to all public authorities in Sweden to implement an environmental management system. In 2006 the Swedish Higher Education Act was revised to include a statement that universities should work for SD in all their activities (Lindberg 2010). None of the other Nordic countries have any legal requirements for higher education to enhance ESD (Nordic Council of Ministers 2011), although in Denmark there is a requirement for all public buildings to reduce their $C0_2$ emissions (Danish Parliament 2005), see Table 1.

Higher education institutions have been required to have national quality assurance demands in the Nordic countries since the mid-1990s (Amaral and Rosa 2010). The formulations of the criteria in the Nordic countries have been developed by national agencies in cooperation with stakeholders (Thune 2001; Franke 2002; Saarinen 2005) and can be seen as broad and soft. Some kind of self-assessment is a part of the accreditation in all countries. So far, the potential benefits of accreditation have been utilized only modestly (Danø and Stensaker 2007).

Connection between ESD and quality assurance

The European standards and guidelines have so far not addressed sustainability (EUA 2011), although it has been discussed in the Nordic Quality Assurance Network in Higher Education (Omar and Liuhanen 2005). Indicators for SD are not yet included in the audits for quality assurance in any Nordic country. However, in Finland they can be an optional subject for the audits in Finland according to the new demands, which will be applied from 2012 (see Table 1) (Universitetet i Bergen 2010; FINHEEC 2011; Magnússon 2011; Swedish National Agency for Higher Education 2011; Danish Accreditation Institution 2011; Danish Evaluation Institute 2011).

In the survey, 63% answered that the institutions have a clear connection between ESD and the quality assurance system. ESD was handled in environmental management systems, some certified according to ISO 14001 and in faculties and departments, where ESD was a part of the vision, policy, objectives, goals and routines. 22% answered that the SD work is ongoing but not yet connected to the management system.

In the reports from the 11 universities one stated that they had a certified integrated management system for quality and environment according to both ISO 9001 and 14001, by which ESD is enhanced. Two other institutions had a certified environmental management system according to ISO 14001, and a third was striving for it. One would include work environment and quality in the environmental management system, and another thought this would be the natural next step. At one university, the responsibility and SD promotion process was described as one of the sub-processes in the quality manual. In one institution ESD was incorporated in the quality assurance policy, and in another the monitoring of the environment and sustainability goals comprised part of quality work. Two answered that there was no connection between ESD and the quality assurance in their institution.

Including ESD in curricula in higher education

The model for including ESD in curricula in higher education consisted of: 1) guidance for competence development for the faculty; 2) guidance for identifying relevant aspects of ESD; 3) useful checklists and questions for self-evaluation and audits; and 4) a list of virtual courses and other related material (Fig. 1).

Previous research indicated that promoting the sustainability competence of teachers requires cooperation within the university, between disciplines, between faculty and students, and with the local communities and society at large, along with competence to integrate multiple ways to look at the world as a multi-lingual and multi-cultural entity (Anderberg et al. 2009; Loranzo-García et al. 2009; Sibbel 2009; Wals 2010; Clark and Button 2011). Brundiers et al. (2010) demonstrated that real world learning is an effective way to introduce interdisciplinary issues and Hansmann et al. (2009) that group processes are essential. Online communication technology can be used to facilitate interdisciplinary research (Dale et al. 2010). There is a need for connectors or coordinating bodies (Bacon et al. 2010). The research of Podger et al. (2010) showed that people are usually more willing to learn when they can relate their own characteristics of identity, morality or work-oriented, to the subject. It is important to take into consideration regional and local values and cultures (Thaman 2010). Savelyeva and McKenna (2011) presented a system for academia to address sustainability through curricula, and Cotgrave and Kokkarinen (2011) a structure for sustainable curriculum design. Loranzo-García *et al.* (2009) demonstrated interconnections for promoting SD.

In the survey, less than half of the respondents stated that they provide uncomplicated education and guidance in SD for the faculty. During further development of the model, a need for annual planning for competence development of SD for the faculty emerged. This should include information for new personnel, annual training for both faculty and staff about ESD and SD, preferably a day with presentations that would include examples for/of ESD and discussions about integrating SD in education. ESD should also be integrated in the curriculum planning process, and SD should be a part of the goals of the courses, which should be monitored. The work should have a budget and the improvement could be based on innovations. Networks of teachers interested in ESD should be established. Information about ESD should form part of regular meetings in which the employees discuss their daily life. Attitudes towards SD should be investigated before inclusion of SD among organization goals.

According to Mochizuki and Fadeeva (2010) academia should first identify competences and then document them in curricula. Rusinko (2010) has developed a general matrix to integrate ESD. In many different studies relevant aspects for ESD in different disciplines have been identified, such as in operations management courses (Fredriksson and Persson 2011), in business administration (Palma et al. 2011; Stubbs and Schapper 2011), and in science, technology, engineering and mathematics (Hopkinson and James 2010; Desha et al. 2009). An integretad sustainability curriculum for environmental studies, engineering, sociology, education and politics department has also been identified (Bacon et al. 2010), and it has been noted that sustainability is a core competence for environmental programs (Vincent and Focht 2009). Competencies to work in teams, in different knowledge communities and to link knowledge and action for SD are important (Brundiers et al. 2010). In the survey most institutions mentioned many relevant aspects for ESD, for example systematic and critical thinking and attitudes. During the further development of the model the participants highlighted that ESD is interdisciplinary, cross-disciplinary and non-disciplinary activity that focuses on long-term holistic thinking. The demands for ESD are discipline-specific. At one institution the Dean's support ensured that the programmes integrate SD.

Many examples of checklists and questions for self-evaluation and audits were found. The STAUNCH model consists of criteria to identify sustainability aspects in the curricula, within the economic, environmental, social and cross-cutting sectors (Glover et al. 2010). Another method is to identify courses with names containing key words for sustainability competences in the curricula (Palma et al. 2011). A compass that can be used for identifying sustainability aspects within curricula is described by Miller et al. (2011). Desha et al. (2009) demonstrated a method for a systematic review of the extent of sustainability content within courses. Cotgrave and Kokkarinen (2011) had asked students to answer a questionnaire about awareness, skills and attitudes around SD before, during and after SD was included in the curricula. Mitchell (2011) showed a method that consists of questions for key stakeholder interviews. The Balance Act is a Nordic campaign that enhances the DESD, and includes a newly developed checklist for identifying different sustainability aspects (Nordic Council of Ministers 2011). In the survey, a third of the repondents answered that they already use checklists, usually as a part of their management systems, which was also mentioned at the seminar. During the further development of the model it was also mentioned that SD should be a part of curricula development and course evaluation, which could be done by the study boards, and as a part of pedagogical training for teachers.

No studies that included results about virtual courses in SD were found, although there are several networks that enhance ESD in the Nordic countries. These include the Baltic University Programme, which is a network of 220 higher education institutions in the region; Idébanken, an organization which publishes a newsletter about SD in the Nordic countries; a website for ESD in both Sweden and Denmark; and in Finland a website for the national SD network for higher education. In the survey half of the institutions answered that they use virtual courses in many different fields.

Table 2 presents examples of enhancing ESD in three of the eleven pilot universities in the project "ESD in Academia in the Nordic Countries". The table is based on reports from September 2011, which were complemented by the authors for this table. The selected universities address sustainability as one strategic area, and one objective is to enhance sustainability in the educational programs. The universities, or parts of them, are based on problem-oriented project work and teaching, sometimes referred as the Aalborg model (Brundiers et al. 2010). Thus the academic culture requires students to have a multidisciplinary and multi-methodological approach, which maintains interaction with ESD (Hansmann et al. 2009). Hence the following methods are very common: problem-based learning, critical thinking, reflective accounts, group discussion, fieldwork, lectures, case studies, etc.

Conclusions

All Nordic countries have requirements for enhancing ESD in higher education, mentioned in both international and national strategies, and in Sweden higher education institutions are required to include SD amongst their objectives. In none of the countries are indicators for SD included in quality assurance of higher education. In the survey, two thirds of 27 respondents answered that their institutions have established a clear connection between ESD and quality assurance. The reports from the eleven institutions showed that one institution had an integrated management system and another is going to integrate quality and work environments into their environmental management systems. ESD is generally enhanced as a component of the vision, policy, processes, objectives, goals, monitoring or routines of a particular institute.

Table 2: ESD promotion in three of the eleven pilot universities in the project ESD in Academia in the Nordic countries, according to the elements in the model

University	**Competence development for the faculty**	**Identified relevant aspects of ESD, or methods for**	**Checklists and questions for self-evaluation and audits**	**Virtual courses and other related material**
Novia University of Applied Sciences (FIN)	Twice a year for the staff. It's voluntary to participate. SD tip of the week on the intranet.	Indicators for ESD: relevant aspects of SD for the profession in the curricula at every degree programme, which is revised every year, and the classification of courses that University of Gävle uses. Questions about how well SD is integrated in surveys to students and personnel. Course descriptions: sustainability aspects. Practice placement: students observe SD aspects in their work environment. Study modules worth 3 or 5 ECTS.	No internal, but the external auditors (of the management system certified according to ISO 9001 and 14001) positive and developing reflections of how ESD is implemented motivates the personnel. See left and University of Gävle	Some degree programmes uses
Roskilde University (DNK)	Training for new personal, voluntary courses	Create a sustainable awareness towards less energy consumption. ¾ of all enrolled students have mandatory courses in sustainability issues.	No, this in being developed as part of the participation at sustainable university partnership.	*Nothing mentionable*
University of Gävle (SWE)	Annual training for new personnel. In regular meetings. Loose network of teachers interested in ESD. Days of inspiration for both students and teachers.	Classification of courses into following categories: A. The main part of the course contains SD B. The course contains some SD C. The course does not yet have moments of SD, but it is possible to include it. D. SD is not relevant for the course E. The course has not been checked for SD content. Also thesis works have been checked in some departments. The aim of the classification is to challenge teachers to think of ESD.	Questionnaire to first and third year students about what they know, how aware they are and how they act regarding ESD in a comparative study with Novia University of Applied Sciences. Survey among personnel every two years about SD including also SD applied in education and research. See column to the left.	*Nothing mentionable*

The results of the study and parallel earlier studies emphasize that ESD is an interdisciplinary or cross-disciplinary activity. Earlier studies that have included guidance for competence development for the faculty show examples of how it can be accomplished, and this study shows that there is a need for such inclusion. In some earlier studies relevant aspects of ESD have been have identified in different disciplines. A lot of different checklists already exist and in the survey a third answered that they already use checklists, usually as a part of their management systems.

The authors think that it is just a matter of time before standards for management systems will be a widespread tool for quality assurance because higher education usually follows the development in society. The authors argue that the presented model of the process for enhancing ESD can be used in management systems, because it is based on earlier research results and practical experience of faculty working with these matters. The model can be used by the personnel responsible for developing ESD and/or quality or integrated management systems in different higher education institutions and is being tested at the eleven pilot universities. Future studies could focus on results from piloting the model for enhancing ESD with management systems in different higher education institutions.

Acknowledgments

The authors would like to thank the Nordic Council of Ministers and the pilot universities in the project for partly financing the project "Education for Sustainable Development in Academia in the Nordic Countries" and Ann-Kristin Hokkanen, Master's degree student at the Department of Biosciences at Åbo Academy University in Finland.

References

Amaral, A. and Rosa, M.J. (2010), Recent trends in quality assurance. *Quality in Higher Education* 16 (1), pp. 59-61.

Anderberg, E., Nordén, B. and Hansson, B. (2009), Global learning for sustainable development in higher education: recent trends and a critique. *International Journal of Sustainability in Higher Education* 10 (4), pp. 368-378.

Bacon, C.M., Mulvaney, D., Ball, B.B., DuPuis, E.M., Gliessman, R.S., Lipschutz, R.D. and Shakouri, A. (2010), The creation of an integrated sustainability curriculum and student praxis projects. *International Journal of Sustainability in Higher Education* 10 (4), pp. 193-208.

Baltic 21E (2002), *Agenda 21 for the Baltic Sea Region Sector Report – Education (Baltic 21E).* Available at: www.hu2.se/Baltic21E.pdf (accessed 21 November 2011).

Bernardo, M., Casadesus, M., Karapetrovic, S. and Heras, I. (2009), How integrated are environmental, quality and other standardized management systems? An empirical study. *Journal of Cleaner Production* 17, pp. 742-750.

Brundiers, K., Wiek, A. and Redman, C. (2010), Real-world learning opportunities in sustainability: from classroom into the real world. *International Journal of Sustainability in Higher Education* 11 (4), pp. 308-324.

Clark, B. and Button, C. (2011), Sustainability transdisciplinary education model: interface of arts, science, and community (STEM). *International Journal of Sustainability in Higher Education* 12 (1), pp. 41-54.

Cotgrave, A.J. and Kokkarinen, N. (2011), Developing a model promoting sustainability literacy through construction curriculum design. *Structural Survey* 28 (4), pp. 266-280.

Dale, A., Newman, L. and Ling, C. (2010), Facilitating transdisciplinary sustainable development research teams through online collaboration. *International Journal of Sustainability in Higher Education* 11 (1), pp. 36-48.

Danø, T. and Stensaker, B. (2007), Still balancing improvement and accountability? Development in external quality assurance in the countries 1996-2006. *Quality in Higher Education* 13 (1), pp. 81-93.

Desha, C.J., Hargroves, K. and Smith, M.H. (2009), Addressing the time lag dilemma in curriculum renewal towards engineering education for sustainable development. *International Journal of Sustainability in Higher Education* 10 (2), pp. 184-199.

EUA (2011), *Mapping the implementation and application of the ESG (MAP-ESG Project). Final report of the project Steering Group*", European Association for Quality Assurance in Higher Education 2011, Brussels, Belgium.

Franke, S. (2002), From audit to assessment: a national perspective on an international issue. *Quality in Higher Education* 8 (1), pp. 23-28.

Fredriksson, P. and Persson, M. (2011), Integrating sustainable development into operations management courses. *International Journal of Sustainability in Higher Education* 12 (3), pp. 236-249.

Glover, A., Peters, C. and Haslett, S.K. (2010), Education for sustainable development and global citizenship: An evaluation of the validity of the STAUNCH auditing tool. *International Journal of Sustainability in Higher Education* 12 (2), pp. 125-144.

Government of Norway (2007), *Norges strategi for bærekraftig utvikling* [*Norway's strategy for sustainable development*]. Available at: http://www.regjeringen.no/en/dep/fin/tema/barekraftig_utvikling/strategi-for-barekraftig-utvikling.html?id=469846 (accessed 31 January 2012).

Hansmann, R., Crott, H.W., Mieg, H.A. and Scholz, R.W. (2009), Improving group processes in transdisciplinary case studies for sustainability learning. *International Journal of Sustainability in Higher Education* 10 (1), pp. 33-42.

Holm, T., Sammalisto, K. and Vuorisalo, T. (2011), *An evolutionary approach to enhancing education for sustainable development with management systems in higher education, a memetic approach.* Working paper, Turku University, Turku, 21 November.

Hopkinson, P. and James, P. (2010), Practical pedagogy for embedding ESD in science, technology, engineering and mathematics curricula. *International Journal of Sustainability in Higher Education* 10 (1), pp. 43-53.

Jørgensen, T.H., Remmen, A.M. and Mellado, D. (2006), Integrated management systems – three different levels of integration. *Journal of Cleaner Production* 14, pp. 713-722.

Kraus, J.L. and Grosskopf, J. (2008), Auditing integrated management systems: considerations and practice tips. *Environmental Quality Management* 18 (2), pp. 7-16.

Lindberg, C. (2010), Structural solutions for ESD in Sweden. In Witthaus, M., McCandless, K. and Lambert, R. (Eds.), *Tomorrow Today,* pp. 92-94. Available at: http://unesdoc. unesco.org/images/0018/001898/189880E.pdf (accessed 21 November 2011).

Loranzo-García, F.J., Huisingh, D., Delgado-Fabián, M. (2009), An interconnected approach to incorporate sustainable development at Tecnológico de Monterre. *International Journal of Sustainability in Higher Education* 10 (4), pp. 318-333.

Magnússon, M.L. (Ed.) (2011), *Quality Enhancement Handbook for Icelandic higher education.* Available at: http://www.rannis.is/files/Handbook_complete_1558767620. pdf (accessed 25 November 2011).

Miller, R.T., Muñoz-Erickson, T. and Redman, C.L. (2011), Transforming knowledge for sustainability: towards adaptive academic institutions. *International Journal of Sustainability in Higher Education* 12 (2), pp. 177-192.

Ministry of Education in Denmark (2008), *Uddannelse for bæredygtig udvikling – strategi for FN's tiår 2005-2014* [*Education for Sustainable Development – strategy for the UN Decade 2005-2014*]". Available at: http://pub.uvm.dk/2008/ubustrategi/ (accessed 23 November 2011).

Ministry of Education in Finland (2006), *Sustainable Development in Education; Implementation of Baltic 21E programme and Finnish Strategy for the Decade of Education for Sustainable Development (2005-2014).* Helsinki University Press, Helsinki.

Mitchell, R.C. (2011), Sustaining change on a Canadian campus: Preparing Brock University for a sustainability audit. *International Journal of Sustainability in Higher Education* 12 (1), pp. 7-21.

Mochizuki, Y. and Fadeeva, Z. (2010), Competences for sustainable development and sustainability: significance and challenges for ESD. *International Journal of Sustainability in Higher Education* 11 (4), pp. 391-403.

Omar, P.-L. and Liuhanen, A.-M. (Eds.) (2005), *A Comparative Analysis of Systematic Quality Work in Nordic Higher Education Institutions*. Helsinki.

Palma, L.C., de Oliveira, L.M. and Viacava, K.R. (2011), Sustainability in Brazilian federal universities. *International Journal of Sustainability in Higher Education* 12 (3), pp. 250-258.

Podger, D.M., Mustakova-Possardt, E. and Reid, A. (2010), A whole-person approach to educating for sustainability. *International Journal of Sustainability in Higher Education* 11 (4), pp. 339-352.

Pratasavitskaya, H. and Stensaker, B. (2010), Quality management in higher education: toward a better understanding of an emerging field. *Quality in Higher Education* 16 (1), pp. 37-50.

Rusinko, C.A. (2010), Integrating sustainability in higher education: a generic matrix. *International Journal of Sustainability in Higher Education* 11 (3), pp. 250-259.

Saarinen, T. (2005), From sickness to cure and further: construction of "quality" in Finnish higher education policy from the 1960s to the era of the Bologna process. *Quality in Higher Education* 11 (1), pp. 3-15.

Savelyeva, T. and McKenna, J.R. (2011), Campus sustainability: emerging curricula models in higher education. *International Journal of Sustainability in Higher Education* 12 (1), pp. 55-66.

Sibbel, A. (2009), Pathways towards sustainability through higher education. *International Journal of Sustainability in Higher Education* 10 (1), pp. 68-82.

Stubbs, W. and Schapper, J. (2011), Two approaches to curriculum development for educating for sustainability and CSR. *International Journal of Sustainability in Higher Education* 12 (3), pp. 259-268.

Swedish National Agency for Higher Education (2011), *The Swedish National Agency for Higher Education's quality evaluation system 2011 – 2014, 2011: 3 R*. Available at: http://www.hsv.se/download/18.328ff76512e968468bc80004249/1103R-quality-evaluation-system-2011-2014.pdf (accessed 25 November 2011).

Swedish Standard Institute (2004), *SS-EN ISO 14001:2004. Environmental management systems-requirements with guidance for use (ISO 14001:2004).* Swedish Standard Institute SIS, Stockholm.

Thaman, K.H. (2010), Teacher capacities for working towards peace and sustainable development. *International Journal of Sustainability in Higher Education* 11 (4), pp. 353-364.

The Danish Accreditation Institution (2011), *Akkrediteringsrådets beretning.* ACE Denmark, Copenhagen.

The Danish Evaluation Institute (2011), *EVA's årsmagasin 2010-2011.* Danmarks Evalueringsinstitut, Copenhagen.

The Danish Parliament (2005). *CIR 27, Cirkulære om energieffektivisering i statens institutioner.* Available at: http://www.google.fi/url?q=http://www.energitjenesten.dk/files/resource_4/Videns kassen/EM%2520af%2520bygninger/CIR%2520nr%252027%2520af%252019..tema.pdf&sa=U&ei=dWLoTvb8N8jQ4QTUtaWQCQ&ved=0CA4QFjAA&usg=AFQjCNEqmR5gWc-6h-_dfmV2rRR Ggmn3XQ (accessed 14 December 2011).

The Finnish Higher Education Evaluation Council (FINHEEC) (2011). *Audit manual for the quality systems of higher education institutions 2011-2017.* Available at: http://www.finheec.fi/files/1335/KKA_0311.pdf (accessed 21 November 2011).

The Ministry for the Environment in Iceland (2002), *Welfare for the Future. Iceland's National Strategy for Sustainable Development 2002-2020.* Available at: http://eng.umhverfisraduneyti.is/media/PDF_skrar/Sjalfbar__roun_enska.pdf (accessed 31 January 2012).

The Nordic Council of Ministers (2009), Sustainable Development – New Bearings for the Nordic Countries. *Revised Edition with Goals and Priorities 2009-2012.* Scanprint a/s, Copenhagen.

– (2011), *Kunnskap for Bærekraftig utvikling – en oversikt over tilbud til voksne i Norden* [*Knowledge about Sustainable development, an overview over the offer to adults in the Nordic countries*]. Available at: http://www.nordvux.net/download/6800/klima_og_ miljo.pdf (accessed 25 November 2011).

Thune, C. (2001), Quality assurance in higher education on Denmark. In Dunkerly, D. and Wong, W.S. (Eds.), *Global Perspectives on Quality in Higher Education.* Ashgate Publishers, Aldershot.

United Nations Decade for Sustainable Development (UN DESD) (2010), *UNESCO Strategy for the second half of the United Nations decade of education for sustainable development.* Available at: http://www.unesco.org/new/en/world-conference-on-ecce/single-view/news/unesco_strategy_for_the_united_nations_decade_of_education_for_ sustainable_development_2005_2014/ (accessed 21 November 2011).

– (2011), *United Nations Decade for Sustainable development 2005-2014.* Available at: http://portal.unesco.org/en/ev.php-URL_ID=47100&URL_DO=DO_TOPIC&URL_SECTION=201.html (accessed 21 November 2011).

Universitetet i Bergen 2010, *Handbook for quality assurance of university education.* Available at: http://www.uib.no/ua/dokumenter/studiekvalitet/kval-sikring_2010_ENG.pdf (accessed 29 November 2011).

Vincent, S. and Focht, W. (2009), US higher education environmental program managers' perspectives on curriculum design and core competencies: implications for sustainability as a guiding framework. International *Journal of Sustainability in Higher Education* 10 (2), pp. 164-183.

Wals, A.E.J. (2010), Mirroring, gestaltswitching and transformative social learning: stepping stones for developing sustainability competence, *International Journal of Sustainability in Higher Education* 11 (4), pp. 380-390.

Chapter 23

Developing People and Transforming the Curriculum: Action Research as a Method to Foster Professional and Curriculum Development in Education for Sustainable Development in Higher Education

Gisela Cebrián, Marcus Grace and Debra Humphris

Abstract

Research is currently being undertaken at the University of Southampton that aims to embed the principles of Education for Sustainable Development (ESD) within its undergraduate curriculum. This methodological paper seeks to discuss the use of an emancipatory action research (EAR) approach within a critical paradigm as an approach that can foster staff professional development and curriculum development in ESD in Higher Education (HE). In this paper the main assumptions underpinning EAR within a critical paradigm are discussed, as well as its methodological principles and some of the criticisms of this approach. A case is made to locate EAR within a wider framework of educational change, namely transformative learning theory, as a theoretical framework to guide the development of EAR projects with academic staff members. One of the consistent problems identified in embedding ESD holistically in HE is the need to promote staff development programmes as a tool to build capacity within academic staff members to rethink their current teaching and professional practice. The use of EAR as a methodology can be a transformative strategy and a research approach to foster professional learning within academics that can lead to curriculum development and to broad organisational learning in ESD in HE.

Introduction

Although the conceptualisation of sustainability remains controversial and different perspectives and definitions exist around the term (Bell and Morse 1999; Dresner 2002; Eichler 1999), there is political agreement on the need to build awareness and develop strategies and action plans to deal with current societal global challenges such as climate change, poverty and environmental degradation. The Declaration of the United Nations Decade of Education for Sustainable Development (abbr. DESD, 2005-2014) has represented a catalyst in the process of integrating the principles of ESD into all levels of education. Education is considered to play a key role in fostering a more sustainable, equitable and socially just society (UNESCO 2005). According to UNESCO (2009: 2) *'ESD is based on values of justice, equity, tolerance, sufficiency and responsibility'*, with respect as its core. The existence of diverse views of sustainability and diverse ways to embed ESD are acknowledged as a positive element to ensure that new developments are culturally and locally relevant but with *'consensus around a range of key principles covering the scope, purpose and practice of ESD'* (Wals 2009: 25). Tilbury (2011a: 7), in a recent expert review, highlights important ESD processes and learning such as *'processes of collaboration and dialogue; processes which engage the 'whole system'; processes which innovate curriculum as well as teaching and learning experiences; and processes of active and participatory learning'*. HE is a principal agent for this positive transformation towards sustainability, because of its key mission of knowledge generation and transfer through research and teaching (UNESCO 2005). There exist numerous projects across the globe that have tried to embed the principles of sustainability in HE, but these are often fragmented and only involve a small part of the university community without connecting strategies and actions and without effecting the required social and cultural transformation of HE (Tilbury 2011b). As she states (p. 19):

> Attempts to mainstream this agenda across higher education have so far failed to have impact. To make sense of this challenge, one needs to appreciate that sustainability is more a journey than a checklist as worldviews pervading thinking and practice need to be questioned.

In England, the Higher Education Academy (HEA) and the Higher Education Funding Council for England (HEFCE) have made significant efforts to support the integration of sustainability in English universities. Despite good intentions and the support given to different research projects that seek to embed ESD into the HE curriculum, further research in the area is still required (HEFCE 2008). By contrast, practically all HE institutions have their own environmental managers and environmental policies, usually more focused on the environmental management of the university itself (Jones et al. 2010a). What could explain this situation is that, for environmental managers, the implementation of sustainable practices is seen to lie within their professional services role, while for academics introducing sustainability into the curriculum it is seen as added work without real relevance to their professional development or academic career (Jones et al. 2010a).

In 2005 HEFCE published its first strategy for Sustainable Development and an action plan, which was updated in 2008. In the update on the strategic statement and action plan in 2008 the vision of HEFCE (2009: 3) is outlined as follows: *'Within the next 10 years, the Higher Education sector in this country will be recognised as a major contributor to society's efforts to achieve sustainability – through the skills and knowledge that its graduates learn and put into practice, its research and exchange of knowledge through business, community and public policy engagement, and through its own strategies and operations*'. HEFCE has taken on the role of facilitator and catalyser of the integration of sustainability. Jones et al. (2010a) view this position of HEFCE has made difficult the real progress towards sustainability in university and curriculum development in this area. This paper seeks to discuss the potentialities of using EAR with academic staff as a research approach to encourage transformative personal, professional and organisational learning towards advancing curriculum development in ESD in HE.

Action research to foster professional development towards curriculum development in ESD in Higher Education

There are numerous authors who make reference to the obstacles to and opportunities for embedding ESD within HE (Corcoran and Wals 2004; Dawe et al. 2005; HEFCE 2008; Leal Filho 2009; Moore 2005). One of the consistent problems identified is the need to promote staff development programmes as a tool to build capacity within academic staff members to rethink their current teaching and professional practice (Dawe et al. 2005). Jones et al. (2008), when interviewing academic staff members on their views concerning ESD in HE, identified limiting factors such as staff interest, lack of awareness, time and capability. These issues are associated with staff worldviews, understandings and beliefs. To progress the agenda of sustainability, members of universities need to be committed and willing to develop their practice, share personal and professional experiences and be open to questioning about their current practice, and at the moment it is predominantly only the champions and enthusiasts who do this (HEFCE 2008). Jones et al. (2010a) identify three inhibitors in curriculum development and improvement towards sustainability: academics tend to see sustainability as an imposed matter and not connected to their discipline; academics think that they do not have enough knowledge and competencies to integrate sustainability issues in their subjects; the ethos of the institutions is seen as an obstacle to embedding sustainability holistically in the curriculum. Sustainability in HE calls for innovative and interdisciplinary practice (Cotton and Winter 2010; Dawe et al. 2005; Jones et al. 2010b; Tilbury 2011b). Many academics in the field of sustainability in Higher Education claim the need for a paradigm shift – an epistemological change towards sustainability (Jones et al. 2010a; Sterling 2004). According to Tilbury (2004: 51) *'a change of epistemology, from reductionism towards holism, from objectivism towards critical subjectivity, and from relativism to relationalism'* is essential for deep learning in all agents that are part of the university. Two of the key texts about sustainability in HE (Corcoran and Wals 2004; Gough and Scott 2007) give special attention in the first chapters to the role and purposes of HE in the current society. There is a need to rethink the existing paradigm of HE and to engage the university community in the debate about the purposes of HE in the current society (Bawden 2004; Gough and Scott 2007). The curriculum, pedagogy, structure, organisation and ethos are shaping dimensions of education and for these reasons Sterling (2004: 50) argues that:

> Sustainability does not simply require an 'add-on' to existing structures and curricula, but implies a change of fundamental epistemology in our culture and hence also in our education thinking and practice.

ESD can be seen as an education for a sustainable social transformation, which stimulates critical reflection and the clarification and reassessment of values, should be innovative and constructive, culturally appropriate and action-orientated (Huckle and Sterling 1996; Sterling 1996; Tilbury 2004; Tilbury and Wortman 2004). Furthermore, it is about encouraging transformative learning, the capacity to challenge existing patterns and worldviews, to construct new knowledge collectively, to challenge and improve practice, and to critique and examine sustainability issues (Sterling 2001). Much of the educational research in the area concerning university staff has focused on identifying their understandings, beliefs, values, attitudes and visions towards ESD (Cotton et al. 2009; 2007; Jones et al. 2008; Reid and Petocz 2006). Little research exists concerning staff professional development aimed at transforming engagement with ESD. Tilbury et al. (2004) carried out a participatory action research project in order to engage lecturers in exploring ways to change the curriculum towards sustainability. This was undertaken using an innovative and interdisciplinary approach to encourage professional development, curriculum and organisational change in Australian universities. Roberts and Roberts (2008) carried out a staff development event to provide a space for exchanging and sharing innovative practice in ESD in the university context. Barth and Rieckmann (2012) illustrate the case of an academic staff development programme implemented in an Ecuadorian university and its resulting impact on individual competence development, innovation in teaching practices and organisational development due to new ideas and competencies acquired by staff through their participation in the programme. These are the few studies that have been found in the literature concerned with professional development of academic staff members in the area of ESD. Holdsworth et al. (2007) identify a lack of continuous professional development (CPD) programmes on ESD and the need to develop CPD programmes that enable the examination of personal understandings and beliefs in order to embed sustainability in the curriculum. There is a need to endorse research in curriculum and staff development within an organisational environment if we are really aiming to implement the principles of ESD within university curricula and generate holistic organisational change (Barth and

Rieckmann 2012; Holdsworth et al. 2007). Action research is a methodology that can contribute to the transformation of professional practice, generate new knowledge and promote the progress of the current state of ESD in HE (Warburton 2003). According to Kemmis (2010) action research can contribute to making a better world through inquiry within the real world, providing new ideas for practice, new ways of thinking, new ways of doing and new relations (to others and to the world).

As Kemmis (2010: 425) states:

> *To find out what needs to be done differently, we need action research – action research that will inform our individual praxis and inform our collective praxis.*

An action research approach can offer opportunities to engage academic staff members in a learning-by-doing process, where critically reflecting on existing worldviews, clarifying values and acting towards sustainability in a practical, but also in a theoretical way are the core elements.

Theoretical assumptions of critical theory

Corcoran et al. (2004) analysed 54 case studies on sustainability in HE and identified the lack of theorising research methodology as an obstacle to better understanding and improvement of sustainability issues in HE. A critical case-study approach within a critical theory paradigm can lead to a better understanding and improvement of situations. The theorisation of research methodology has been frequently misplaced in research concerning sustainability in HE, where the focus has been on presenting examples of good practice or case-studies (Corcoran et al. 2004). There is a need to develop research designs within a critical paradigm that links sustainability with organisational learning and social theory (Fien 2002). For critical theory the main goal of research is to transform reality. According to Guba and Lincoln (2005) and Neuman (1994), the main assumptions relating to critical theory are:

- Society and reality are always evolving, so it is possible to study any given situation at any given moment.
- People are creative and adaptive and can change conditions and transform reality.
- People are limited by cultural and historical conditions but they can transform current beliefs and behaviours, enabling them to develop new understandings that can change present oppressive structures and relations. It brings about empowerment of individuals and allows collective groups to participate actively in society.
- Critical theory is action-orientated in the sense that it critically analyses the present situation and allows individuals and groups to imagine and construct a different future through action plans.
- Research is value-laden. Social research is political and moral. The role of the researcher is based on values that involve the judgment of viewpoints and alternatives, of both participants and themselves.
- The focus is on change, contradictions or conflicts within social systems to support the understanding of reality (critical socio-historical analysis to identify the values and beliefs that underpin the ways of doing and thinking) and to help to transform it. The aim is to expose underlying assumptions and structures.
- Social sciences research is with others and not on others. Participants in the research are seen as co-researchers and co-learners.
- Reality is socially constructed but it is governed for structures that are not visible, theory is critique in the sense that it exposes conditions and contributes to the empowerment of people enabling them to visualise an alternative path to a better society and world.

The conduct of research in ESD within a critical paradigm can foster the epistemological change advocated by ESD experts which is necessary to transform thinking and practice to embed the principles of ESD within higher education institutions.

Framing the different dimensions of action research methodology

There are many types of action research approaches, including participatory action research, emancipatory or critical action research (EAR), collaborative inquiry, active learning and narrative inquiry. These are used in a variety of disciplines such as social and health sciences, education, organisation and administration studies, environmental sciences and management. Action research can be undertaken by a teacher, by a group of teachers working in the same school or a group of teachers from different schools, managers, researchers or experts within a group of practitioners, university departments, students and advisers (Cohen et al. 2000). Somekh (2006: 8) identifies eight methodological principles for action research:

- integrates research and action;
- is conducted by a collaborative partnership of participants and researchers;
- involves the development of knowledge and understanding of a unique kind;
- starts from a vision of social transformation and aspirations for greater social justice for all;
- involves a high level of reflexivity;
- involves exploratory engagement with a wide range of existing knowledge;
- engenders powerful learning for participants;
- locates the inquiry in an understanding of broader historical, political and ideological contexts.

These principles provide an overview of action research and represent an umbrella for the different kinds and approaches of action research. Some of the essential features of action research are: the vision that all the participants in the research have something to contribute and in turn have something to learn; the participants are seen as co-researchers or co-learners, including the researcher; knowledge and theory are inseparable from practice; the main purpose is the improvement of a real situation or problem; reflection and action are two core elements; and the whole learning-by-doing process is what counts. Action research offers opportunities for staff and curriculum development in the field of ESD in HE.

For Noffke (2009) there are three main dimensions of the action research approach: the personal, the professional and the political. The personal dimension makes reference to the practitioner as researcher and the process of self-reflection, planning and introducing changes to improve self-practice. It is related to the main purpose of action research, described by Elliot (1991: 49) as *'to improve practice rather than to produce knowledge'*. McNiff and Whitehead (2002: 15) state that *'action research is an enquiry by the self into the self, undertaken in company with others acting as research participants and critical learning partners'*. The professional dimension corresponds to professional development, usually in education concerning teachers with the aim of enhancing the teaching profession and developing its practices. The political dimension is usually embedded in the previous dimensions, but the purpose is mainly generating democratic processes to empower groups often without a voice, such as lower socio-economic groups and underprivileged communities. It is linked to power and political issues, structures, participation and decision-making in the community, and idealistically leads to educational and social change (Kemmis 2010).

For Noffke (2009) the three dimensions (personal, professional and political) are interconnected, but researchers tend to focus on a single aspect. By contrast Carr and Kemmis (2009) argue that it is not possible to give a different emphasis to an action research project, because it will inevitably involve

- the self-transformation of participants as understanding is developed;
- the professional development of the community of practitioners;
- the empowerment of participants due to a democratic process that can potentially change the social system, the institution and the world.

All these aspects are relevant for educational practice. The emancipatory or critical approach explores consciously the relationship between these three faces (Carr and Kemmis 1986). For this reason EAR, placed in a critical paradigm and discussed in the next section, brings about personal, professional and organisational learning towards curriculum development in ESD.

Emancipatory action research to foster ESD in Higher Education

EAR is a powerful method for cooperative self-inquiry used by individuals in a society to improve wisdom and integrity of their social and/or educational practices. EAR is also about understanding the practices, situations and conditions in which they occur. It relates to the whole structure of social and political constraints that are influencing the social and educational situation (Carr and Kemmis 1986; 2009; Kemmis and McTaggart 2005). According to Carr and Kemmis (2009: 78): *'Emancipatory action research is a form of research that seeks to create the kind of communicative space within which practitioners can participate in making decisions, taking action and collaboratively inquiring into their own practices, their understandings of these practices, and the conditions under which they practice'.*

Carr and Kemmis (1986) argue that one can only understand education as part of the reproduction of the social structure. Two main functions of education can be defined: reproductive of the current social and cultural model and transformative for a better and worthwhile living society. EAR is transformative and political, not in the sense that gives a vision of what a good society is, but in the sense that it recognises the existence of different worldviews and the value and importance of generating reflective processes with regards to different perspectives, with the premise of equal respect to diverse ideas, opinions and understandings. It is about local knowledge, daily life, cultural context and local situations (Somekh 2006). As Kemmis (2010) suggests, the question is what exactly constitutes a good society or a worthwhile living world, and to answer this question it is necessary to think about the assumptions that underpin what a good society is in relation to education, because the current educational model assumes what a good society is. This is a critical idea for progressing ESD at all levels of education. According to Sterling (2001) a paradigm shift is necessary to provide a real education for change, where the main role of education is to offer a deep and transformative learning experience and provide opportunities to develop learning communities for sustainability. EAR is a suitable research approach to assist the creation these learning communities. One of the first debates that we need to address is the purpose of HE and the extent to which the current educational model is reproducing a non-sustainable society (Bawden 2004). EAR is a methodology that aims at offering an open space for communication and participation of individuals in decision-making in a collective and reflective manner, acting in their own practical settings, sharing and acquiring a better understanding of the self and the collective situation and the significant circumstances that influence behaviour. EAR can give opportunities to challenge existing worldviews, beliefs and values, and can empower practitioners to improve their practice and help improve the educational and social situation towards sustainability.

Criticisms of critical theory and emancipatory action research

There are some academics who are critical of critical theory and EAR. Morrison in the nineties argued that there may be no connection between ideology critique and the emancipation of societies or even practitioners; for him there is no scientific evidence in the relationship between critical theory and emancipation (Cohen et al. 2000). Concerns exist around action research that makes it a highly problematic approach. Gibson (1985) stated that the agenda of EAR is extremely particular, prescriptive and problematic. Gibson's criticisms include:

- it is not realistic;
- it tends to control and prescribe;
- it assumes a view of emancipation and action research;
- it favours critical collective reflection instead of individual self-reflection;
- it assumes that power can be distributed simply by empowering people;
- it assigns more importance to groups rather than individuals.

Kemmis and McTaggart (2005) mirror some errors made in previous research due to an over-optimistic and enthusiastic view of EAR, where the ability of action research to transform individuals and contribute to social and educational change is overemphasised. The empowerment of research participants is complicated and the change tends to be technical rather than emancipatory. Another of the myths identified is in respect to the role of the facilitator, who usually acquires the role of consultant or expert. This is seen as controversial to the role of equal participation of participants as co-researchers. However, understanding the facilitator role as unbiased or purely technical discards its duty in assisting the social change implied in EAR (Kemmis and McTaggart 2005). Moreover, it is suggested that action research requires both more theory and more action, with more proactive participation needed from research participants in research and action.

There are a number of challenges associated with the use of EAR. Reflexivity and critical reflection are essential elements of this research methodology to examine successes and failures, as well as the learning process as a whole that results from EAR (Kemmis 2010). A case can be made to locate EAR within a wider framework of educational change, namely transformative learning theory. Transformative learning theory (Mezirow 2009) emphasises critical reflection, dialogue and holism. This learning theory can guide the development of EAR projects with academic staff members to promote emancipatory rather than technical change to embed ESD in the HE curriculum.

Fostering transformative learning in the academic community

Transformative learning theory is successfully used in the field of adult education. Taylor (2009) defines it as a learning process that lets participants challenge existing worldviews, beliefs, feelings, values and assumptions based on past experiences. By a process of critical reflection decisions are taken collectively about new ways of understanding. Four main elements in transformative learning are identified: individual experience, critical reflection, dialogue, and a holistic orientation, referred to as the affective and emotional part. The exploration of emotional issues is considered as fundamental to engage people emotionally in the reconstruction and transformative process (Mezirow 2009). The engagement of academic staff in communicative and critical reflective spaces within the community can lead to the deep and transformative learning required to change the existing paradigms associated with sustainability (Bawden 2004). Mezirow (2009: 167) argues that transformation is *'the process of becoming critically aware of how and why our assumptions have come to constrain the way we per-*

ceive, understand, and feel about our world; changing these structures of habitual expectation to make possible a more inclusive, discriminating, and integrating perspective; and finally, making choices or otherwise acting upon these new understandings'.

Transformative learning in the field of ESD is advocated as a theoretical approach that can contribute to engaging students in deep learning, engaging them in a cognitive, psychomotor and effective sphere (Sipos et al. 2008). Transformative social learning is envisioned as key for the development of sustainability competence amongst students, fostering a new way of understanding and being by reflecting critically on self-assumptions and beliefs (Wals 2010). In the field of ESD, transformative learning has been discussed mainly as a teaching and learning method and as a result of the student learning experience in ESD. In this paper we refer to transformative learning theory as a theoretical framework that can guide EAR with academic staff members to promote staff and curriculum development in ESD. Transformative learning theory is aligned with EAR because they share values and assumptions. In this sense, EAR is a transformative strategy and a research approach that can foster transformative learning in staff engaged in professional development programmes using an EAR approach. As Cranton and King (2003: 36-7) see professional development as *'an opportunity to cultivate transformative learning'*, which *'gives us a new perspective on our goals, what we do in our practice, and how we think about our work'*. There is a growing use of action research in combination with transformative learning theory; however, as Taylor (2007: 177) states, these studies *'provide little insight into the challenges associated with this type of research, particularly how to manage both the highly emotive consequences associated with transformative learning and the learner-centred approach often inherent to action research, within the context of a clearly defined research agenda'*. Gravett (2004) used an action research approach to facilitate transformative learning within university staff, with the purpose of changing their perspectives and professional practice. As she states (p. 270) *'action research has the potential to explore transformative learning in educational settings'* to help academic staff members explore new ways of teaching and engage students in an authentic learning process towards embedding sustainability.

Conclusions

Academic staff members have a significant role to play in moving forward the agenda of sustainability in HE. One of the preventive factors identified to embedding ESD as a whole is the fact that staff need support such as CPD programmes to build up the capacity within the academic community to change current teaching and professional practice towards ESD. It has been argued that the use of EAR as a methodology can foster transformation in this area and can be a research approach that catalyses professional learning within academics to bring about curriculum development and to broad organisational learning in ESD in HE. The participation of the community of practice becomes indispensable in building the community capacity required to deal with sustainability and engaging in an innovative way of knowing, thinking and doing. EAR can be a suitable methodology to conduct research on

- staff beliefs, values and experiences;
- professional development in ESD;
- developing shared visions, understandings and action towards ESD;
- finding out ways to improve university practice towards embedding ESD in the curriculum;
- structural and institutional barriers to implementing curriculum change;

- bottom-up innovation strategies;
- curriculum development from different disciplines;
- generating personal, professional and organisational learning.

At the present time, little research exists that uses EAR within the field of ESD as a methodology to promote staff professional development. Further research using EAR needs to be done to evaluate the potentialities, outcomes and impact of EAR in order to drive the professional development, curriculum development and the claimed epistemological change required to transform ESD practice in HE.

The lack of theorising research methodology in sustainability in HE has been identified in the literature as an obstacle advancing this agenda. The conduct of research in ESD within a critical paradigm can foster the epistemological change advocated by ESD experts which is necessary to transform ways of thinking and practice to embed the principles of ESD within higher education institutions. The involvement of academic staff in critical reflective practices can develop the deep and transformative learning required to change the existing paradigms associated with sustainability. There is a need to conduct research within a critical paradigm that can build a better theoretical understanding and can connect sustainability to organisational theory.

There could be great potential in combining transformative learning theory with an EAR approach, to endorse the transformative personal and professional development that can lead to innovative curriculum initiatives to embed ESD in higher education. Locating EAR within transformative learning theory can help bridge the existing gap identified between staff development and curriculum development in education for sustainable development in higher education. Further research needs to be conducted to assess the possibilities and outcomes in using this approach and in achieving transformative learning in ESD among academic staff members. This combination is promising in developing new understandings and ways of teaching and proceeding towards creating more sustainable curriculums and universities.

References

Barth, M. and Rieckmann, M. (2012), Academic staff development as a catalyst for curriculum change towards education for sustainable development: an output perspective. *Journal of Cleaner Production* 26 (1), pp. 28-36.

Bawden, R. (2004), Sustainability as emergence: The need for engaged discourse. In Corcoran, P.B. and Wals, A.E.J. (Eds.), *Higher Education and the Challenge of Sustainability: Problematics, Promise and Practice*, pp. 21-32. Kluwer Academic Publishers, Dordrecht.

Bell, S. and Morse, S. (1999), *Sustainability Indicators: Measuring the Immeasurable?* Earthscan Publications Limited, London.

Carr, W. and Kemmis, S. (1986), *Becoming critical: education, knowledge and action research.* Deakin University Press, Geelong, Victoria.

– (2009), Educational Action Research: A Critical Approach. In Noffke, S. and Somekh, B. (Eds.), *The SAGE Handbook of Educational Action Research*, pp. 74-84. SAGE Publications Ltd, London.

Cohen, L., Manion, L. and Morrison, K. (2000), *Research Methods in Education.* RoutledgeFalmer, Abingdon.

Corcoran, P.B., Walker, K.E. and Wals, A.E.J. (2004), Case studies, make-your-case studies, and case stories: a critique of case-study methodology in sustainability in higher education. *Environmental Education Research* 10 (1), pp. 7-21.

Corcoran, P.B. and Wals, A.E.J. (Eds.) (2004), *Higher education and the challenge of sustainability: problematics, promise, and practice.* Kluwer Academic Publishers, Dordrecht.

Cotton, D., Bailey, I., Warren, M. and Bissell, S. (2009), Revolutions and second-best solutions: education for sustainable development in higher education. *Studies in Higher Education* 34 (7), pp. 719-733.

Cotton, D., Warrena, M., Maiborodaa, O. and Baileya, I. (2007), Sustainable development, higher education and pedagogy: a study of lecturers' beliefs and attitudes. *Environmental Education Research* 13 (5), pp. 579-597.

Cotton, D. and Winter, J. (2010), It's Not Just Bits of Paper and Light Bulbs': A Review of Sustainability Pedagogies and their Potential for Use in Higher Education. In Jones, P., Selby, D. and Sterling, S. (Eds.), *Sustainability education: perspectives and practice across higher education*, pp. 39-54, Earthscan Publications Limited, London.

Cranton, P. and King, K.P. (2003), Transformative Learning as a Professional Developmental Goal. *New Directions for Adult and Continuing Education* 2003 (98), pp. 31-38.

Dawe, G., Jucker, R. and Martin, S. (2005), Sustainable development in higher education: Current practice and future developments. Report for the Higher Education Academy. Available at: http://www.heacademy.ac.uk/assets/York/documents/ourwork/tla/sustainability/sustdevinHEfinal report.pdf (accessed 12 December 2011).

Dresner, S. (2002), *The Principles of Sustainability*. Earthscan Publications Limited, London.

Eichler, M. (1999), Sustainability from a feminist sociological perspective: a framework for disciplinary reorientation. In Becker, E. and Jahn, T. (Eds.), *Sustainability and the social sciences: a cross-disciplinary approach to integrating environmental considerations into theoretical reorientation*, pp. 182-206, Zed Books, New York, NY.

Elliot, J. (1991), *Action Research for Educational Change*. Open University Press, Buckingham.

Fien, J. (2002), Advancing sustainability in higher education. Issues and opportunities for research. *International Journal of Sustainability in Higher Education* 3 (3), pp. 243-253.

Gibson, R. (1985), Critical times for action research. *Cambridge Journal of Education* 15 (1), pp. 59-64.

Gough, S. and Scott, W. (2007), *Higher Education and Sustainable Development: Paradox and possibility*. Routledge, Abingdon.

Gravett, S. (2004), Action Research and Transformative Learning in Teaching Development. *Educational Action Research* 12 (2), pp. 259-272.

Guba, E.G. and Lincoln, Y.S. (2005), Paradigmatic Controversies, Contradictions, and Emerging Confluences. In Denzin, N.K. and Lincoln, Y.S. (Eds.), *The SAGE Handbook of Qualitative Research*, pp. 191-215. Sage Publications Inc, Thousand Oaks, CA.

Higher Education Funding Council (HEFCE) (2008), *HEFCE strategic review of sustainable development in higher education in England*. Report to the Higher Education Funding Council for England by the Policy Studies Institute, PA Consulting Group and the Centre for Research in Education and the Environment, University of Bath. Available at: http://www.hefce.ac.uk/pubs/rdreports/2008/rd03_08/rd03_08.pdf (accessed 20 January 2011).

– (2009), *Sustainable development in higher education 2008: Update to strategic statement and action plan*. Available at: http://www.hefce.ac.uk/pubs/hefce/2009/09_03/09_03.pdf (accessed 20 January 2011).

Holdsworth, S., Wyborn, C., Bekessy, S. and Thomas, I. (2007), Professional development for education for sustainability: How advanced are Australian universities? *International Journal of Sustainability in Higher Education* 9 (2), pp. 131-146.

Huckle, J. and Sterling, S. (Eds.) (1996), *Education for Sustainability*. Earthscan Publications Limited, London.

Jones, P., Selby, D. and Sterling, S. (2010a), Introduction. In Jones, P., Selby, D. and Sterling, S. (Eds.), *Sustainability education: perspectives and practice across higher education*, pp. 1-16. Earthscan Publications Limited, London.

– (2010b), More than the Sum of their Parts? Interdisciplinarity and Sustainability. In Jones, P., Selby, D. and Sterling, S. (Eds.), *Sustainability education: perspectives and practice across higher education*, pp. 17-37. Earthscan Publications Limited, London.

Jones, P., Trier, C.J. and Richards, J.P. (2008), Embedding Education for Sustainable Development in higher education: A case study examining common challenges and opportunities for undergraduate programmes. *International Journal of Educational Research* 47 (6), pp. 341-350.

Kemmis, S. (2010), What is to be done? The place of action research. *Educational Action Research* 18 (4), pp. 417-427.

Kemmis, S. and McTaggart, R. (2005), Participatory Action Research: Communicative Action and the Public Sphere. In Denzin, N.K. and Lincoln, Y.S. (Eds.), *The SAGE Handbook of Qualitative Research*, pp. 559-603. Sage Publications Inc, Thousand Oaks, CA.

Leal Filho, W. (2009), Sustainability at universities: opportunities, challenges and trends. In Leal Filho, W. (Ed.), *Sustainability at Universities: Opportunities, Challenges and Trends*, pp. 313-319. Peter Lang, Frankfurt.

McNiff, J. and Whitehead, J. (2002), *Action Research: Principles and Practice.* RoutledgeFalmer, London.

Mezirow, J. (2009), Transformative Learning Theory. In Mezirow, J., Taylor, E.W. and Associates (Eds.), *Transformative Learning in Practice: Insights from Community, Workplace and Higher Education*, pp. 18-32. Jossey-Bass, San Francisco, CA.

Moore, J. (2005), Barriers and pathways to creating sustainability education programs: policy, rhetoric and reality. *Environmental Education Research* 11 (5), pp. 537-555.

Neuman, W.L. (1994), Social research methods: qualitative and quantitative approaches, Allyn and Bacon, Massachusetts.

Noffke, S. (2009), Revisiting the Professional, Personal and Political Dimensions of Action Research. In Noffke, S. and Somekh, B. (Eds.), *The SAGE Handbook of Educational Action Research*, pp. 6-24. SAGE Publications Ltd, London.

Reid, A. and Petocz, P. (2006), University lecturers' understanding of sustainability. *Higher Education* 51 (1), pp. 105-123.

Roberts, C. and Roberts, J. (2008), Starting with the staff: how swapshops can develop ESD and empower practitioners. *Environmental Education Research* 14 (4), pp. 423-434.

Sipos, Y., Battisti, B. and Grimm, K. (2008), Achieving Transformative Sustainability Learning: Engaging Head, Hands and Heart. *International Journal of Sustainability in Higher Education* 9 (1), pp. 68-86.

Somekh, B. (2006), *Action research. A methodology for change and development.* Open University Press, Milton Keynes.

Sterling, S. (1996), Education in Change. In Huckle, J. and Sterling, S. (Eds.), *Education for Sustainability*, pp. 18-39. Earthscan Publications Limited, London.

– (2001), *Sustainable Education: Re-Visioning Learning and Change*. Schumacher Society Briefing No. 6, Green Books, Dartington.

– (2004), Higher Education, Sustainability, and the Role of Systemic Learning. In Corcoran, P.B. and Wals, A.E.J. (Eds.), *Higher Education and the Challenge of Sustainability: Problematics, Promise and Practice*, pp. 49-70. Kluwer Academic Publishers, Dordrecht.

Taylor, E.W. (2007), An update of transformative learning theory: a critical review of the empirical research (1999-2005). *International Journal of Lifelong Education* 26 (2), pp. 173-191.

– (2009), Fostering Transformative Learning. In Mezirow, J., Taylor, E.W. and Associates (Eds.), *Transformative Learning in Practice: Insights from Community, Workplace and Higher Education*, pp. 3-17. Jossey-Bass, San Francisco, CA.

Tilbury, D. (2004), Environmental Education for Sustainability: A Force for Change in Higher Education. In Corcoran, P.B. and Wals, A.E.J. (Eds.), *Higher Education and the Challenge of Sustainability: Problematics, Promise and Practice*, pp. 97-112. Kluwer Academic Publishers, Dordrecht.

– (2011a), *Education for Sustainable Development: An Expert Review of Processes and Learning.* UNESCO, Paris.

– (2011b), Higher Education for Sustainability: A Global Overview of Commitment and Progress. In GUNI (Ed.), *Higher Education in the World 4. Higher Education's Commitment to Sustainability: from Understanding to Action*, pp. 18-28. GUNI, Barcelona.

Tilbury, D., Podger, D. and Reid, A. (2004), *Action Research for Change Towards Sustainability: Change in Curricula and Graduate Skills Towards Sustainability*. Final report. Available at: http://aries.mq.edu.au/publications/other/Education/ACTS_Report. pdf (accessed 15 May 2011).

Tilbury, D. and Wortman, D. (2004), *Engaging People in Sustainability*. Commission on Education and Communication, IUCN, Gland.

UNESCO (2005), *United Nations Decade of Education for Sustainable Development (2005-2014): Draft International Implementation Scheme*. Available at: http://portal.unesco.org/education/en/file_download.php/e13265d9b948898339314b001d91fd01draftFinal+IIS.pdf (accessed 10 November 2011).

– (2009), *UNESCO World Conference on Education for Sustainable Development: Bonn Declaration*. Available at: http://www.esd-world-conference-2009.org/fileadmin/download/ESD2009_BonnDeclaration080409.pdf (accessed 15 November 2011).

Wals, A.E.J. (2009), *Review of Contexts and Structures for Education for Sustainable Development 2009*. UNESCO, Paris.

– (2010), Mirroring, Gestaltswitching and transformative social learning. Stepping stones for developing sustainability competence. *International Journal of Sustainability in Higher Education* 11 (4), pp. 380-390.

Warburton, K. (2003), Deep learning and education for sustainability. *International Journal of Sustainability in Higher Education* 4 (1), pp. 44-56.

Chapter 24

Introducing the Methodology Background of the Lifestyle Index of Sustainability

Kira Giesecke

Abstract

Since the United Nations Conference in Rio de Janeiro many indicators and composite indicators have been developed to measure sustainable development. However, with such developments, many problems have arisen which have led to misinterpretation and misunderstandings. Therefore, calculation methods have to be chosen with caution so that composite indicators measure what they are supposed to measure. This paper will present a way of evaluating and comparing different methods, in order to find appropriate methods for a new composite indicator, the Lifestyle Index of Sustainability (LIS). In doing so, different normalization, weighting and aggregation methods have been critically discussed according to predefined quality criteria, in order to establish a solid foundation to calculate this index. The LIS is an instrument which will be developed for students. It should be used as a university experiment, where students can learn how their lifestyles affect the environment, human beings, and societies. The conclusion is drawn that re-scaling as normalization method, the Analytical Hierarchy Process in combination with equal weighting as weighting method, and the linear additive procedure as aggregation method are preferred in order to ensure simplicity, comprehensibility and transparency for the LIS. Students can therefore understand how the numbers have been put together to generate the aggregated LIS. The paper is useful not only for students, but also for anyone who is interested in learning more about different calculation methods of indicator concepts and how to evaluate them.

Introduction

At the 1992 United Nations World Conference on the environment and development in Rio de Janeiro, the first call was made for an indicator concept to assess sustainability, to provide a foundation "for decision-making at all levels and contribute to a self-regulating sustainability of integrated environment and development systems" (United Nations 1992: Agenda 21, Chapter 40: 346). Since then, an extended number of indicator models and composite indicators have been developed to measure sustainable development (Schlossberg and Zimmerman 2003; Singh et al. 2009; Van de Kerk and Manuel 2008; Saisana and Tarantola 2002).

Indicators have become the most important tool for measuring sustainable development (Fahy and Ó'Cinnéide 2008; Bell and Morse 2008; Hanley 2000).

Existing composite indicators like the Gross Domestic Product (GDP), the Human Development Index (HDI) or the Ecological Footprint (EF) follow the objective to support politicians and corporations in their decision-making processes, from local to global level. Furthermore, they show which activities should and should not be implemented in order to lead the society more in the direction of sustainability (Munda and Nardo 2005a; Singh et al. 2009). Even though indices are recognized as a useful tool for policy and decision-makers, "composite indicators can send misleading or non-robust messages if they are poorly constructed or misinterpreted" (Nardo et al. 2005: 6). They have many methodological difficulties (Freudenberg

2003), which can lead to a point where it is important to inform stakeholders about the limitations of methodologies, their use and interpretation of the composite indicators (Gasparatos et al. 2008). Making the right choice in the methodology used to calculate a good composite indicator is very important. Problems can be found in choosing weights for indices, as seen from studies by Munda and Nardo (2005b), Chowdhury and Squire (2006), or Sawada et al. (2004). In addition, the choice for the right aggregation method in combination with weight, is critically discussed in literature from the studies by Munda and Nardo (2005b), Linkov et al. (2004) and Ebert and Welsch (2004) . Even the famous ecological footprint and its methodology are criticized. For example, Fiala (2008) gives evidence concerning the failure of the ecological footprint. There are already studies reviewing different methodologies to develop appropriate composite indicators. Pohekar and Ramachandran (2004) reviewed and discussed over 90 papers to analyse the applicability of various methods. Nardo et al. (2005) also developed a handbook on constructing composite indicators. In this study, the main methodologies can be found, including benefits and drawbacks. Nardo et al. (2005) also give advice on which methodology should be used for certain indicators. Zhou et al. (2010) pointed out that even though composite indicators have been widely accepted, they depend heavily on the underlying construction scheme where data aggregation has been applied. Therefore, it is critically important to develop quality criteria, which supports the development process of finding the right calculation methods. In the literature there are different approaches to defining quality criteria for sustainable composite indicators, as seen in studies of Hak et al. (2011), Grüning (2002), Kromrey (2009), Singh et al. (2009), Lang (2003), Lautso et al. (2002), Mayer (2008), Caeiro et al. (2012), and Schuh (2001).

In this paper, different calculation methods of composite indicators will be critically discussed and compared in terms of their relevance to the Lifestyle Index of Sustainability (LIS). In doing so, different calculation methods will be presented and evaluated with regard to their benefits and drawbacks, as well as to the chosen quality criteria. The focus is on providing a way of finding the right calculation methods for the LIS, which has scientific foundation and is simultaneously simple and easy for students to understand, so it may be used in Higher Education Institutions (HEIs). Subsequently, these points are discussed and a conclusion is drawn.

Concept of the Lifestyle Index of Sustainability (LIS) with the chosen quality criteria

The LIS will be developed as an instrument for students to provide information about selected areas of life and its relationship to environmental, social and economic issues of sustainability. It acts as a tool for students to use to assess individual behaviours and identify where there are saving potentials. The LIS identifies the students' impact on the environment, human beings and society by showing how sustainable their lifestyle is in points: 100 points means the lifestyle is sustainable and 0 points means the lifestyle is unsustainable. The LIS has been structured hierarchically to guarantee that each user can see where their strengths and weaknesses are within their lifestyle.

A basic limitation of the LIS is that it will not consider every effect and pressure on the environment, on human beings, or societies. It also does not provide solutions on how to change certain unsustainable behaviour patterns into sustainable ones. Rather, it acts as an instrument which gives the user an idea of how their lifestyle currently looks with regard to sustainability, and of which behaviours have the greatest impact on the environment, human beings, or societies.

To be able to develop the LIS on a scientific basis, but so that it is simple and easy to understand and can be used in an experiment in HEIs, certain quality criteria have to be chosen to optimize the development process of the LIS. A good composite indicator should possess standard quality criteria. Singh et al. (2009) listed certain characteristics which an indicator set should have. They suggest that a composite indicator should fulfil the following criteria: validity, reliability, comparability, simplicity, and data availability, in order to generate the indicator. Lang (2003) added two further aspects: objectivity and communicability. Meadows (1998) pointed out that "it should always be possible to go down as many levels as necessary to see the numbers that have been put together to make the aggregate indicator" (Meadows 1998: 23). Schuh (2001) used criteria for analysing different decision procedures for sustainable development: completeness, transparency, comprehensibility, objectivity, accuracy and validity, reliability, flexibility, practicability and economy, uniqueness and data availability. These criteria seem to be suitable for the LIS and have been adjusted (Table 1).

Table 1: Quality criteria adjusted to the LIS

Quality criteria	Description
completeness	The method has to include all three dimensions as the LIS prescribes, taking into account qualitative and quantitative indicators
transparency, comprehensibility, and objectivity	As Meadows (1998) said, it is important to guarantee that everybody understands and sees how indicators have been put together to aggregate the LIS. Transparency is very important, in order to comprehend every step. It is also important to minimize subjectivity to give the LIS a plausible basis.
accuracy and validity	Does the method measure what it is supposed to measure? This means, the method should be able to keep the information which a student enters into the system, to reflect their lifestyles in terms of sustainability
reliability	Is the method robust enough when the application is repeated?
flexibility	Is the method adaptable? Is the method adjustable enough to different situations and is it possible to add or to remove indicators?
practicability and economy	Is the effort in proportion to time and cost justifying?

Source: own illustration based on Schuh (2001)

On this basis, the following calculation methods will be evaluated.

Analysis of calculation methods and their relevance to the LIS

This section aims to analyse the calculation methods used to measure the concept of sustainability. Six normalization-, four weighting- and three aggregation methods are presented and discussed according to the benefits and drawbacks with regard to the chosen quality criteria and to the relevance of the LIS.

Normalization methods

Before producing an adequate composite indicator, it is necessary to normalize different units of indicators into a common dimensionless unit, as a first step. This means that indicators can have different measurement units which cannot be aggregated – like trying to add up apples with pears (Nardo et al. 2005; Mayer 2008). Normalization methods are of critical importance because they circumvent the danger of aggregating different measurement units (Freudenberg 2003). It is also necessary to receive identical value ranges. If basic indicators are already expressed in the same unit – similarly to the study about an e-business readiness composite indicator (Nardo 2004) – then no standardization is necessary. Six normalization methods will now be presented and subdivided into ordinal and metrical normalization. The normalization methods for transforming original values into ordinal values are: 1) Ranking, 2) Categorical scales, 3) Normalization into an outranking matrix for the Condorcet-Kemeny-Young-Levenglick ranking procedure. Into metrical values they are: 4) Z-scores, 5) Transforming in highest and lowest value (Re-scaling) and 6) Distance to reference countries.[1]

Ordinal normalization methods

1. Ranking

Ranking is one of the easiest normalization methods (Nardo et al. 2005; OECD 2008). It is used to bring countries into an appropriate scale, ranking their performance in a respective area. When variables are already in logical order, then no ranking is necessary (Singh et al. 2009). It is used by Fagerberg (2000) in the Information and Communication Technology Index.

The main benefits of this method are that it is simple to construct and is independent from outliers (Nardo et al. 2005). In addition, it provides an opportunity to compare country performance over time (OECD 2008). The main drawback of this method is clear. Ranking "bears the problem that the distance between any two adjacent positions" (Grupp and Mogee 2004: 1380) cannot be traced back. They can "be very small, or large, in original value" (Grupp and Mogee 2004: 1380). Thus, information gets lost on the absolute level and it is impossible to recognize any decision about difference in performance (Nardo et al. 2005).

Even though it is a simple normalization procedure and is not sensitive to outliers or extreme values, it does not reflect the original data in its entirety. For the LIS it is necessary to provide transparency by staying with original values, in order to minimize the information loss. As a result, ranking is not an appropriate method for the LIS.

2. Categorical scale

By using categorical scale, values of indicators will be categorized into quantitative or qualitative categories. A categorical scale could be, for example, one, two or three stars as quantitative category or "'fully achieved', 'partly achieved' or 'not achieved'" (Nardo et al. 2005: 49) as qualitative category (OECD 2008; Nardo et al. 2005). Nicoletti et al. (1999) used this method for the indicators of product market regulation. They normalized raw data which had been collected in qualitative information from the survey, or into other databases, and transformed the data into a scale from 0-6: the least restricted (0) and the most restrictive (6) provision in the country (Nicoletti et al. 1999).

The benefit of this method is that it is reliable with regard to outliers and extreme values. Moreover, the predefined scores for the categories will be the same over time, because "any change in the definition of the indicator over time will not affect the transformed variable" (OECD 2008: 28). The disadvantages are that the information between the values gets lost through the building of categories and no small improvements will be recognized (Nardo et al. 2005) and the rank order among the variables also disappears.

When comparing the disadvantages of this method in relation to the criteria for a good composite indicator, this method is not suitable either.

---

1 For more information regarding other normalization methods see Nardo et al. (2005).

3. Normalization into an outranking matrix for the Condorcet-Kemeny-Young-Levenglick ranking procedure (CKYL).

This method normalizes raw data from the 'Impact Matrix' into an 'Outranking Matrix' by pair-wise comparison "according to the whole set of sub-indicators used" (Nardo et al. 2005: 77). The normalization value of country A is the sum of the weights of the subindicators which are performing better than country B. The question to be asked here is which subindicators have a higher value in country A than in country B. The weights for each indicator are used to calculate the score, not the original value (Nardo et al. 2005). An example with two countries and eight indicators is illustrated in the 'Impact Matrix' below;

$$\textit{Impact Matrix:}\begin{bmatrix} \frac{\textit{Countries}}{\textit{indicator}} & 1 & 2 & 3 & 4 & 5 & 6 & 7 & 8 \\ A & 50 & 30 & 60 & 50 & 40 & 30 & 60 & 10 \\ B & 30 & 10 & 20 & 40 & 50 & 40 & 70 & 20 \\ \textit{weights} & \frac{1}{8} & \frac{1}{8} & \frac{1}{8} & \frac{1}{8} & \frac{1}{8} & \frac{1}{8} & \frac{1}{8} & \frac{1}{8} \end{bmatrix}$$

Assuming that country A performs better in the indicators 1,2,3 and 4 of eight than country B, and the score for country A with equal weights of 1/8 is 0.5 by multiplying 4 by 1/8, country B then obtains the reciprocal score which is each 0.5. The results are, subsequently, collected and provided in the outranking matrix:

$$\textit{Outranking matrix} = \begin{bmatrix} \textit{Countries} & A & B \\ A & o & 0{,}5 \\ B & 0{,}5 & o \end{bmatrix}$$

This outranking matrix is further used in a new mathematical aggregation convention called the 'Condorcet-Kemeny-Young-Levenglick ranking procedure' (CKYL) (Munda and Nardo 2003), which is explained in the section 'Aggregation methods'. This method is used by Graymore et al. (2009) in their study about an index of regional sustainability in Australia, as well as by Girardin et al. (2000). They assessed potential impacts of agricultural practices on the environment.

The benefit of this normalization is that missing information can be easily determined by the Impact Matrix. Also, the direction of each indicator can be detected easily, by looking at which country performs better in some indicators or vice versa (Munda and Nardo 2003). It does not take into account the real values, but rather the weights which are attached to each indicator (Munda and Nardo 2003). Therefore, it is not sensitive to outliers or extreme values. However, this normalization no longer reflects the indicator's real information (Nardo et al. 2005). As a result of the drawbacks, this normalization method, like the other two discussed above, is not suitable. In addition, it is only used in a combination with the CKYL method, which is also not appropriate (see the section 'Aggregation' methods).

Metrical normalization method

1. Z-scores

The second normalization method is the use of standard z-scores, which is a very common method in constructing composite indicators (Singh et al. 2009; Nardo et al. 2005). The actual value of an indicator is subtracted from the mean and divided by the standard deviation.

$$Z = \frac{(actual\ value - mean\ value)}{standard\ deviation}$$

It transforms all variables into a common scale and assumes a 'normal' distribution, having an average of zero and a standard deviation of one (Freudenberg 2003). This "avoids introducing aggregation distortions stemming from differences in variable means" (Freudenberg 2003: 11) and is therefore useful if calculations obtain distorted values (Singh et al. 2009). The method is used by Floridi et al. (2011) in their study for constructing a composite indicator for Italian regions, by the Yale Center for constructing the Environmental Sustainability Index (Yale Center for Environmental Law and Policy et al. 2005) and by the European Commission for the Internal Market Index (Tarantola et al. 2002). Another way to calculate the actual value is from the median rather than from the mean, which is less sensitive to outliers (Freudenberg 2003).

It is a good normalization method if there is an intention to reward exceptional behaviour, because this method is extremely sensitive to outliers. If such an intention is not required, the outliers have to be adjusted by different treatments (OECD 2008). The disadvantage is that it does not reflect the scale effect and depends on the mean, which is also sensitive to outliers and extreme values (Nardo et al. 2005). Even though this method stays in line with the information of the original value, it loses the scale effect. Therefore, it is not suitable for the LIS either.

2. Transforming in highest and lowest value (re-scaling)

This method is a very common normalization technique, where variables with different measurement units are transformed into a pre-defined value range, for example like 0 to 100 points. It is also called re-scaling method (Nardo et al. 2005). There are different forms to calculate the units into points.

Freudenberg (2003) illustrates the formula like many others (Nardo et al. 2005; Klugmann 2010; Knutsson and Ostwald 2006; Floridi et al. 2011; Munda 2005) as:

$$100 * \left(\frac{actual\ value\text{-}minimum\ value}{maximumvalue\text{-}minimumvalue} \right)$$

This linear function can be also illustrated as:

$$y = m * x + b$$

The wording 'minimum' and 'maximum' is not well chosen here. More appropriate might be 'highest' and 'worst', the reason being that a maximum value can indicate a 'bad' value, like long-term unemployment, while others can be measured as 'good', such as life expectancy (Cherchye et al. 2003). To avoid misinterpretation, the maximum and minimum values should be rewritten as best and worst values. The value range can be defined using three criteria: 1) existing rules and regulations; 2) experience with previous study projects, historical data or data in the literature; and 3) "minimum and maximum values of the different alternatives with respect to an indicator" (Alarcon et al. 2010: 3). This method is used in the Human Development Index (Klugmann 2010), in construction of composite indicators for Italian regions (Floridi et al. 2011), and in an index for sustainable livelihoods (Knutsson and Ostwald 2006).

If any changes are made concerning the minimum and maximum, the range value must be newly defined. Outliers and extreme values can also be very sensitive to the indicators. On the other hand, if the distance between values is very small, this method increases the effect on the index more (OECD 2008; Nardo et al. 2005). This method sticks to the original scale and reflects the information of the original data without information loss. Even though it is sensitive to outliers, it is a useful normalization method for the LIS, because it keeps the original value. Therefore, transparency, simplicity and comprehensibility are ensured.

3. Distance to a reference

This method calculates the relative position of a given indicator (X_n) to a reference value:

$$I = \frac{X_n}{X_r}$$

whereby the reference value is a fixed value at the initial time (X_r).This method is used by Adriaanse (1993, in Nardo et al. 2005) in the Environmental Policy Performance indicator.
One benefit is that a decision-maker receives a quick overview if the country performs above or below average. In addition, trends can be monitored to show what direction decisions must take (Nardo et al. 2005). A disadvantage is that outliers and extreme values can be sensitive and therefore unreliable (Nardo et al. 2005; OECD 2008). Another objection is that there is no value range, which makes comparison impossible. The reference could be, for example, a benchmarking country or physical unit for carbon dioxide (Nardo et al. 2005; OECD 2008). Another variation of the method is to use, for example, the country itself as a reference at an initial point of time (Nardo et al. 2005) to compare the performance of the chosen country over time. As discussed above, it is less sensitive than the other method but still can distort the data set. This method is not appropriate for the LIS because it is not flexible enough to adopt the requirements of the LIS, such as using the predefined value range, which is 0 to 100 points. Also, this method is not able to calculate with a zero value, which can be very important in terms of sustainability.

Weighting methods

After the normalization process the indicators need to receive weights that can be aggregated in an expressive way. This indicates a decision-making process of which method should be used to identify corresponding weights for categories, indicators, and sub-indicators to aggregate the information (Nardo et al. 2005). It is also necessary to correct information when indicators imply high correlation, because it can lead to distortion (Yigitcanlar and Dur 2010). In the following, five weighting methods will be described with regard to their benefits and drawbacks and how the methods should be applied. The methods are: 1) equal weighting; 2) the Factor Analysis; 3) the participatory technique and experts judgement, as well as 5) the Analytical Hierarchy Process (AHP).

1. Equal weighting

Equal weighting is a method which is often used in many composite indicators (Nardo et al. 2005). It is chosen because of the lack of scientific background to define different weights (Nardo et al. 2005; Van de Kerk and Manuel 2008; Knutsson and Ostwald 2006). It has found application in the Sustainability Society Index (Van de Kerk and Manuel 2008), in the composite indicator for Italian regions (Floridi et al. 2011), in the Commitment to Development Index (Roodman 2005), and in the Environmental Sustainability Index (Yale Center for Environmental Law and Policy et al. 2005). The use of this method is advantageous when developers have no knowledge of whether an indicator is more important than another. However, equal weighting is not free of judgment (Nardo et al. 2005). By weighting categories equally, the problem that can arise is that the different amount of sub-indicators beneath the categories can have different weights. For example, there are three categories with equal weights of 33.33%. The first category has five sub-indicators and the second and third categories have three sub-indicators each. Each sub-indicator in category 1 obtains 6.67%; each sub-indicator in categories 2 and 3 receive 11.11% (Floridi et al. 2011). Another disadvantage with this method is the danger of the double counting effect, which may happen by combining indicators with a high degree of correlation (OECD 2008). To avoid the double counting effect, only indicators with a low degree of correlation should be used, or weights should be adjusted, meaning that highly correlated indicators should receive fewer weights (OECD 2008). The amount of indicators should also be reduced to a basis which guarantees "transparency and parsimony" (OECD 2008: 21).

Another option is to omit weightings when no information is available (Knutsson and Ostwald 2006).

Equal weighting is very easy to conduct. It is simple and replicable, so that students can follow the procedure. In addition, equal weighting enables comparability of countries, because the weight does not reflect a country's own interests and priorities. However, equal weighting is still a kind of weighting and therefore implies a small rate of subjectivity. Nevertheless, it is a useful weighting method for the LIS, in the absence of scientific foundation to support the weighting decisions.

2. Statistical methods

To overcome the double counting effect, statistical methods like the Factor Analysis (FA) can be deployed. This method combines indicators with the highest correlation to create the smallest possible number of factors (Nardo et al. 2005). If there is no correlation between variables, FA does not have the ability to calculate any weights (Nardo et al. 2005). The method is used in the study by Ansoms and McKay (2010), who carried out a quantitative analysis of poverty and livelihood profiles in the case of rural Rwanda. It has also been used in the study by Salvati and Zitti (2009), who were exploring the importance of various components of a synthetic index, according to sustainability and weighting of ecological and economic indicators. It further found application in the Internal Market Scoreboard (European Commission 2001), in the success of software process improvement (Emam et al. 1998), and in the Environmental Performance Index (Esty et al. 2006). The benefits are that it chooses weights on a non-subjective and reliant basis (Yale Center for Environmental Law and Policy et al. 2005; Singh et al. 2009; European Commission 2001), as well as solving the double counting problem (Nardo et al. 2005). Objections are that the method can only be conducted with correlated indicators (Nardo et al. 2005). The transparency is not given, because there is no knowledge of how the weights are actually combined together and "does not reflect priorities of decision-makers or the budget constraints that limit free choice among a range of policy options" (Yale Center for Environmental Law and Policy et al. 2005, p. 66). It is sensitive to outliers, small-sample problems, data shortage and modification of data (Nardo et al. 2005).

It is a useful tool to overcome the problem of subjectivity, which makes the index more trustworthy and comparable among countries. On the other hand, it might not be easy to follow how the mathematical methods chose the weights. This method reduces the dimensions of the indicators into factors which need to be newly defined. Therefore, the factors may not be compatible with the desired three indicator categories which are already predefined as environment, economy and social categories of the LIS. Thus, this method might be not suitable for the predefined hierarchical structure.

3. Participatory technique and Expert judgement

As the name implies, the weights of the participatory technique will be received by a panel of participants, which combine different opinions of experts, citizens or politicians (Nardo et al. 2005). Štreimikienė et al. (2009) concluded in their study that indicators should be weighted by stakeholders' preferences. This method has been used by Parker (1991, in Nardo et al. 2005) in his study about the 'concern about the environmental problems index'.

The benefit is that participants can give their own input into the development process of indices and can contribute to a consensus for policy action (Nardo et al. 2005). This method is only appropriate when a purpose is well-defined (OECD 2008), for example when the public knows the issues on the public agenda (Nardo et al. 2005). On the other hand, weights will be selected that will create inconsistencies and subjectivity depending on local priorities (Nardo et al. 2005) and make a comparison between countries difficult. Also, non-experts might not be not able to define meaningful weights according to the importance of different indicators (P.R. Consultants 2000).

The method 'expert judgement', also called 'budget allocation', is a technique where only experts with special know-how and experience are brought together to find an appropriate weighting system to make the index reliable (Nardo et al. 2005). The experts receive a budget of a number of points which they have to distribute over the subindicators (OECD 2008). This procedure makes it different to the participatory technique. The method is used in the e-business readiness composite indicator by (Nardo 2004), in the Ecoindicator 99 (P.R. Consultants 2000), and by Utne (2008). Utne (2008) deployed this method combined with the Analytical Hierarchy Process, which is described below. She used it in her study investigating whether small fishing vessels were more sustainable than large fishing vessels. The advantage is there is no technical manipulation and the weights are improved by reliability. The procedure basically produces a debate to find an agreement for policy action (Nardo et al. 2005). The method is appropriate when there is a need for experts' opinions, with their extensive know-how and experience (Nardo et al. 2005). A disadvantage is a high rate of subjectivity, as in the participant approach. Decisions are made which depend on "local specific conditions" (Nardo et al. 2005: 67). Therefore, comparison between countries might be difficult, because each expert from another country would set different priorities for indicators that would reflect their level of urgency, or the need for political intervention, and not the importance of each indicator (Nardo et al. 2005). Another disadvantage is that if there are an extensive number of indicators, experts might find it difficult to give meaningful weights and to understand the seriousness of the impact of each indicator (P.R. Consultants 2000).

As a result of the high rate of subjectivity, both methods are unsuitable. The reason is that the LIS should be used internationally; therefore, comparability is necessary, and this is not given by these methods.

4. Analytic Hierarchy Process (AHP)

The AHP was developed by Thomas Saaty around 25 years ago (Utne 2008; Nardo et al. 2005) and is a multiple criteria decision-making tool which is very widely used (Vaidya and Kumar 2006). This method is used to analyse the preferences of a decision-maker within a decision-making process. Basically, the decision-maker determines weights via pairwise comparison between alternatives to a criterion, and the criteria will be weighted with regard to the goal that has to be achieved (Bodin and Gass 2003). This is achieved on a nine-point scale:: "1 = equal, 3 = moderately dominant, 5 = strongly dominant, 7 = very strongly dominant and 9 = extremely dominant" (Saaty 2006: 558). The intermediate figures (2, 4, 6 and 8) are for compromise, and the reciprocals figures (1/2, 1/3, 1/4, 1/5, 1/6, 1/7, 1/8, 1/9) for opposite judgement (Saaty 2006). The question the decision-maker has to ask is, "Is alternative A more important than alternative B and by how much?" (Bodin and Gass 2003: 1489). There are six steps which can be followed to conduct this method:

1) Stating the problem; 2) Identifying objectives of the problem; 3) Defining criteria which affect behaviour; 4) Creating a hierarchy with different levels which include the "goal, the criteria, sub-criteria and alternatives" (Vaidya and Kumar 2006: 2; 5) Comparing each component of each level and bringing them into a numerical scale. Either the results of a pairwise comparison can be equal, which is one, or the component receives one figure such as 8 (on the nine-point scale). In this case, the other component receives the reciprocal score, which is 1/8. It is outlined in a reciprocal matrix (Ishizaka and Labib 2011):

$$\text{reciprocal matrix} = \begin{bmatrix} 1 & a_{12} & \ldots & a_{1n} \\ a_{21} & 1 & \ldots a_{ij} & .. \\ \ldots & a_{ij}=\frac{1}{a_{ij}} & 1 & .. \\ a_{n1} & \ldots & \ldots & 1 \end{bmatrix}$$

6) Testing the inconsistency via calculation of the maximum of Eigenvector, consistency index and the consistency ratio. The consistency ratio is calculated by CR = CI/RI, whereby RI is the random index (see Saaty 1977, in Ishizaka and Labib 2011). The CR should be not more than 10% (Utne 2008; Bodin and Gass 2003). Finally. 7) repeating the procedure until the values are in the desired scale (Vaidya and Kumar 2006). Inconsistency cannot be avoided because it is part of human nature, and it appears during any decision-making or judgments. Therefore, Saaty assigned a scope of having approximately less than 10% inconsistency (Nardo et al. 2005). Bodin and Gass (2003) pointed out that it is better that the weights reflect the view of the decision-maker than have 0% inconsistency, which could lead to a decision potentially being wrong. If the Eigenvector is above 10% it is suggested to repeat the process. Utne (2008) uses the AHP in her study of measuring which vessel is more sustainable. It is also often used in medical and healthcare contexts (Liberatore and Nydick 2008).

The benefit of this method is that it can include qualitative and quantitative data and support the transparency of the index (Nardo et al. 2005; Vaidya and Kumar 2006; Ishizaka and Labib 2011). It creates a hierarchy where a decision-maker is able to better focus on the criteria when assigning the weights (Ishizaka and Labib 2011). However, the method is quite time-consuming and costly, because the decider needs to define the structure of the hierarchy with its goal, criteria and alternatives, and conduct a number of pairwise comparisons (Ishizaka and Labib 2011). The decision-maker cannot choose the weight objectively. Therefore, the outcome depends on who chooses weights and what should be weighted (Nardo et al. 2005). Another problem of the AHP is that different structures lead to different results (Ishizaka and Labib 2011). If one criterion has more sub-criteria than another criterion, then this criterion tends "to receive more weight than when they are less detailed" (Ishizaka and Labib 2011: 3)

(Ishizaka and Labib 2011). Another disadvantage is that the choice of weights can be irrelevant. This means when A is more important than B by two-zero, and B more important than C by thee-zero, it is expected that A is more important than C by six-zero (2*3 = 6) (Ishizaka and Labib 2011).

The AHP implies a logical hierarchical structure which is easy to follow and would reflect the desired structure of the LIS. It also works with chosen criteria to evaluate the indicators to the goal. Therefore this method seems to be appropriate. Subjectivity needs to be considered where this can be minimized using the inconsistency index.

Aggregations methods

The last step, to finalize the development process of composite indicators, is the choice of an adequate aggregation method, which should be in line with the overall goal. In the literature, many aggregation methods can be found (Nardo et al. 2005) and "the choice of an appropriate method depends on the purpose of the composite indicator as well as nature of the subject being measured" (Yale Center for Environmental Law and Policy et al. 2005: Appendix A: 64). In this section the three main aggregation methods – 1) linear additive aggregation; 2) geometric mean; and 3) the Condorcet-Kemeny-Young-Levenglick ranking procedure – are illustrated with their pros and cons and in which index they have been conducted.

1. Linear additive aggregation

The linear additive aggregation is the most widely used aggregation technique found in the literature (Nardo et al. 2005). This method is the "summation of weighted and normalized indicators" (Nardo et al. 2005: 74). It can be expressed as:

CI

whereby $\sum_q w_q = 1 \; and \; 0 < w_q < 1 \; for \; all \; q = 1, \ldots Q$

The requirement of using this method is that all measurement units have to be normalized into a common unit to avoid any distortions (Yale Center for Environmental Law and Policy et al., 2005). The benefit of this method is of course its simplicity, reliability and transparency (Mayer, 2008). Many developers have therefore used this method, such as Van de Kerk and Manuel (2008), for the Sustainability Social Index; Pohl (2006) for her ITT Flygt Sustainability Index; Rees and Wackernagel for the famous 'ecological footprint' (Van den Bergh and Verbruggen 1999); Tarantola et al. (2002) for the Internal Market Index; Utne (2008) to measure the sustainability rate for different vessels; and Pratt et al. (2004) for the Environmental Vulnerability Index. However, this method raises the issue of compensation (Mayer 2008; Nardo et al. 2005; Munda and Nardo 2005a). The weights "have the meaning of substitution rates and do not indicate the importance of the indicator associated" (Nardo et al. 2005: 76). This means poor performance in one indicator can be compensated by good performance in another indicator. It further implies a discrepancy between how weights have been chosen for the indicators and the real meaning when this method is used (OECD 2008); see also Munda and Nardo 2005a). Assuming we have four indicators with values of 21, 1, 1, and 1, and with values 6, 6, 6, and 6 if simple aggregation is used, then the composite of both is 6. From this result, no differences among the indicators are detectable, only that the 21 makes up for the poor performances of the others. Therefore, weights in a linear aggregation method always have a sense of compensation (Munda and Nardo 2005a). As a result of the benefits, in addition to the fact that the method is often used by scientists, it seems to be an appropriate method. It is simple and comprehensible. Therefore, students can follow how the aggregated index has been put together. To reduce the trade-off effect, which is the main disadvantage, Van de Kerk and Manuel (2008) argue that if the highest point score is reached in every category, indicator and subindicator, then no trade-off effect appears in the index itself. Pratt et al. (2004) argue that the simple aggregation method is used for the Environmental Vulnerability Index because there is no complex model which can offer any advantage against the linear aggregation method. Morse and Fraser (2005) explored whether methodological transparency was an adequate solution to the problem posed by highly aggregated indices. They concluded that an index should provide transparency and be left open for other stakeholders. Freudenberg (2003) and Meadows (1998) agree that composite indicators should be as transparent as possible in order to provide the information which is needed to understand the methodology and to avoid misinterpretation. Taking these arguments into account, the linear additive aggregation method is a suitable approach for the LIS.

2. Geometric mean aggregation

The geometric mean aggregation is the second method which can be used to aggregate different indicators to a composite.

The formula is as followed;

$$CI = \prod_{q=1}^{Q} x^{wq}$$

This method can be used when measurement units are not equal and indicator values have positive values. In the Human Development Index (Klugmann 2010) the geometric mean is used. In the Environmental Sustainability Index it is used as a potential alternative (Yale Center for Environmental Law and Policy et al. 2005). Additional to the benefit above, this method has a weaker compensation effect compared to linear aggregation. Using the same example with the four indicators, the result for the first index is 2.14 and 6 for the second. Any increase in the low absolute value has a greater impact on the composite indicator than any increase in upper absolute values (Nardo et al. 2005; OECD 2008). To extend this example, if we assume that the second indicator increases by one, the index with 2.14 scales up to 2.456 and the index with 6 to 6.236. This shows that the first index increases by 0.316 and the second only by 0.236. However, it still implies a kind of compensation effect between indicators, even though it is not as strong as the linear additive aggregation (Nardo et al. 2005).

Compared to the linear aggregation method the mathematical formula deals with square roots and exponents. This might be not easy to comprehend for students and non-professionals. The trade-off effect can be minimized by the same measures as discussed above. However, the geometric mean is not as suitable as the linear aggregation method is for the LIS.

3. Condorcet-Kemeny-Young-Levenglick ranking procedure (CKYL)

If no compensation is desired, a non-compensatory multi-criteria approach must be conducted to guarantee there is no trade-off between indicators (Nardo et al. 2005). Munda and Nardo (2003) use a non-compensatory multi-criteria approach called the 'Condorcet-Kemeny-Young-Levenglick ranking procedure', which is a mathematical aggregation convention. The normalization in the outrank matrix (see the section 'Normalization') belongs to this aggregation method. "According to CKYL the ranking of countries with the highest likelihood is the one supported by the maximum number of sub-indicators for each pair-wise comparison, summed over all pairs of countries considered. The final ranking would be with the highest score" (Munda and Nardo 2003). This method is used by Graymore et al. (2009) in their study about an index of regional sustainability in Australia, and by Girardin et al. (2000). The benefit is that it is robust against outliers and extreme values and keeps only ordinal information. It finds a compromise between two or more purposes and therefore avoids the trade-off effect (OECD 2008). It can combine qualitative and quantitative data (Nardo et al. 2005) and "weights are never combined with intensities of preference" (Munda and Nardo 2003: 14), which ensures that they are important coefficients (Munda and Nardo, 2003). However, it is extremely time-consuming and costly, because the CKYL procedure depends on the amount of indicators which need to be compared pairwise (Nardo et al. 2005). It also raises the problem of irrelevant alternatives, such as the AHP method, which means that country A performs better than country B and country B performs better than country C, but country C performs better than A (Nardo et al. 2005).

If no compensation is desired, this method is an alternative to the linear aggregation method and the geometric mean. It gives the possibility of using weights not as a substitution rates. However, this method is only useful when there are a limited number of circumstances. In the case of the LIS, this method might be not suitable because there will not be a limited amount of students who will use the LIS. Therefore, the significant effort required to conduct this method with so many students is not compensated by the positive effect of non-trade-off.

Discussion and Conclusion

Since the Rio conference in 1992 and the publication of Agenda 21, many indicator concepts and composite indicators have been developed. However, composite indicators have a high rate of methodological difficulties. This paper explored what calculation method was the most appropriate one for the Lifestyle Index of Sustainability. To achieve this purpose, six normalization methods, four weighting methods and three aggregation methods were presented and their benefits and drawbacks were discussed, according to 10 quality criteria. The 10 quality criteria were completeness, transparency, comprehensibility, objectivity, accuracy and validity, reliability, flexibility, practicability and economy. In Table 2 below, a summary can be seen of all methods according to the quality criteria and their relevance to the LIS.

Table 2: Summary of calculation methods with their relevance to the LIS

Quality criteria / Calculation methods	completeness	transparency, comprehensibility, and objectivity	accuracy and validity	reliability	flexibility	practicability and economy	Relevance to the LIS
Normalisation methods							
Ranking	++	++	--	-	+	++	+
Categorical Scale	++	-	--	++	-/+	+	-
Normalisation for the outranking matrix	++	+/-	-	-	-	--	--
z-scores	-	-	+	++	+	-	-
Re-scaling	++	++	++	++	++	+/-	++
Distance to a reference	++	+			--	+	--
Weighting methods							
Equal weighting	++	+	-	+	++	++	+
Factor Analysis	++	-	--	-/+	-	-	--
Participatory technique and expert judgement	++	-/+	-	-	-	--	--
Analytic Hierarchy Process (AHP)	++	+	++	+	+	-	++
Aggregation methods							
Linear additive aggregation	+	++	+	-	+	++	++
Geometric mean	+	-	+	-/+	+	+	-
CKYL ranking procedure	++	--	+	+	-	--	--

Source: own display (++ criterion is fully achieved, + achieved, – mostly not achieved, – not achieved)

The method 'Transformation in best and worst value' (re-scaling) is the most suitable normalization method for the LIS. Although it seems more time-consuming and costly than other methods, it sticks to the information of the data set and to the scale effect, and is therefore valid and accurate. All other methods lose either information, the scale effect, or both. Of course this method is sensitive to outliers, but there are treatments to adjust these, such as the 5% rule, removing or replacing outliers (Nardo et al. 2005). If the LIS is distorted by outliers, then one of these treatments should be applied. In addition, this method is often used by the most famous of the indices presented in this paper, giving this method a scientific basis and reliability.

According to the different weighting approaches, there is not only one method which can be used for the LIS without any limitations. The participatory technique and expert judgment are not suitable, even though Morse and Fraser (2005) concluded that indicators and data selection should be chosen via local stakeholders and international experts. The reason for this decision is justified by the argument of Nardo et al. (2005), who point out that the danger behind this method is the high rate of subjectivity, which is influenced by the interests and priorities of the respective country. The Factor Analysis is also not an appropriate method, because it does not reflect the purpose and structural hierarchy of the LIS. The other two methods, equal weighting and the AHP, are more appropriate but not fully satisfying. Therefore, the method for the LIS needs to be extended to guarantee a lower rate of subjectivity, ensure comparability and still be simple and replicable. One extension is to combine the AHP with results of studies and literature which support the decision for certain weights. AHP will then test the inconsistency. There is no one decision-maker; there are rather scientific studies that decide the weights to the indicators. If no scientific background is found, equal weights might be used as an alternative. In addition, the sensitivity analysis will be conducted to ensure the weighting procedure via AHP is robust enough to reflect the purpose of the LIS and to reduce the rest of the suspected subjectivity.

Compared to the geometric mean and the CKYL procedure, the linear additive aggregation comes into the equation. It is transparent, easy to understand and replicable, which is necessary in order to meet the quality criteria, despite the problem of the compensation effect as Munda and Nardo (2005a) discussed in their paper. Scientists such as Van de Kerk and Manuel (2008), Pratt et al. (2004), Morse and Fraser (2005), Freudenberg (2003), and Meadows (1998) argue that the highest score should be achieved in every category, indicator and subindicator. The model should display transparency, so that everyone can see the score within the model to understand how the aggregated index has been put together. Therefore, an index should not be understood as the final number which only gives orientation and decision-making support. It is only a number at the end of a hierarchy of an indicator system. The LIS will consider these recommendations to minimize the trade-off effect.

The purpose of this paper was to find the most suitable calculation methods for the LIS, in order to provide scientific foundation and simplicity and comprehensibility for students, with the aim of making HEIs able to use this composite indicator as an experiment in their lectures, helping to motivate young people to move towards a sustainable lifestyle.

It can be concluded that the re-scaling method, as a normalization method, is the most suitable one to fulfil the criteria of the LIS. A combination of AHP and equal weighting will be used to define adequate weights for the subindicators, indicator sets and indicator categories. Of the three aggregation methods, the linear additive technique reflects the needs of the LIS best, despite the fact that it suffers from a compensation effect, which can be reduced by transparency and sustainability achievement at each level.

The LIS is already in the development process and will probably be finished in 2013. The instrument will then be ready to test and use in HEIs to enable students to calculate their lifestyle in terms of three dimensions of sustainability and find ways to change certain unsustainable behaviours. This paper is important in understanding how methods are chosen for the LIS, so that students and non-experts are able to follow the mathematical background of the LIS. Consequently, the LIS can be applied and interpreted correctly.

References

Alarcon, B., Aguado, A., Manga, R. and Josa, A. (2010), A Value Function for Assessing Sustainability: Application to Industrial Buildings. *Sustainability* 3 (1), pp. 35-50.

Ansoms, A. and McKay, A. (2010), A quantitative analysis of poverty and livelihood profiles: The case of rural Rwanda. *Food Policy* 35 (6), pp. 584-598.

Bell, S. and Morse, S. (2008), *Sustainability indicators: measuring the immeasurable?* Earthscan, London.

Bodin, L. and Gass, S.I. (2003), On teaching the analytic hierarchy process. *Computers & Operations Research* 30 (10), pp. 1487-1497.

Caeiro, S., Ramos, T.B. and Huisingh, D. (2012), Procedures and criteria to develop and evaluate household sustainable consumption indicators. *Journal of Cleaner Production* 27, pp. 72-91.

Cherchye, L., Moesen, W. and Van Puyenbroeck, T. (2003), *Legitimately Diverse, yet Comparable: On Synthesising Social Inclusion Performance in the EU.* Available at: http://ideas.repec.org/p/wpe/papers/ces0301.html (accessed 21 December 2011).

Chowdhury, S. and Squire, L. (2006), Setting weights for aggregate indices: An application to the commitment to development index and human development index. *Journal of Development studies* 42 (5), pp. 761-771.

Ebert, U. and Welsch, H. (2004), Meaningful environmental indices: a social choice approach. *Journal of Environmental Economics and Management* 47 (2), pp. 270-283.

Emam, K.E., Goldenson, D., McCurley, J. and Herbsleb, J. (1998), Success or Failure? Modeling the Likelihood of Software Process Improvement. *International Software Engineering Research Network technical report ISERN-98-15.*

Esty, D.C., Srebotnjak, T., Kim, C.H., Levy, M.A., de Sherbinin, A. and Anderson, B. (2006), *Pilot 2006 Environmental performance index*. Yale Center for Environmental Law and Policy, New Haven. Available at: http://www.yale.edu/epi/2006EPI_Report_ Full.pdf (accessed 27 December 2011).

European Commission (2001), *Internal Market Scoreboard*. Available at: http://ec.europa.eu/internal_market/score/docs/score18_en.pdf (accessed 5 December 2011).

Fagerberg, J. (2000), *Europe at the Crossroads: The Challenge from Innovation-based Growth*. Available at: http://ideas.repec.org/p/tik/inowpp/20000913.html (accessed 6 February 2011).

Fahy, F. and Ó Cinnéide, M. (2008), Developing and testing an operational framework for assessing quality of life. *Environmental Impact Assessment Review* 28 (6), pp. 366-379.

Fiala, N. (2008), Measuring sustainability: Why the ecological footprint is bad economics and bad environmental science. *Ecological Economics* 67 (4), pp. 519-525.

Floridi, M., Pagni, S., Falcorni, S. and Luzzati, T. (2011), An exercise in composite indicators construction: Assessing the sustainability of Italian regions. *Ecological Economics* 70 (8), pp. 1440-1447.

Freudenberg, M. (2003), *Composite indicators of country performance: a critical assessment*. Working Papers, OECD Science, Technology and Industry.

Gasparatos, A., El-Haram, M. and Horner, M. (2008), A critical review of reductionist approaches for assessing the progress towards sustainability. *Environmental Impact Assessment Review* 28 (4-5), pp. 286-311.

Girardin, P., Bockstaller, C. and Van der Werf, H. (2000), Assessment of potential impacts of agricultural practices on the environment: the AGRO ECO method. *Environmental Impact Assessment Review* 20 (2), pp. 227-239.

Graymore, M.L.M., Wallis, A.M. and Richards, A.J. (2009), An Index of Regional Sustainability: A GIS-based multiple criteria analysis decision support system for progressing sustainability. *Ecological Complexity* 6 (4), pp. 453-462.

Grupp, H. and Mogee, M.E. (2004), Indicators for national science and technology policy: how robust are composite indicators? *Research Policy* 33 (9), pp. 1373-1384.

Grüning, M. (2002), *Performance-Measurement-Systeme: Messung und Steuerung von Unternehmensleistung*. Deutscher Universitäts-Verlag, Dresden.

Hak, T., Kovanda, J. and Weinzettel, J. (2011), A method to assess the relevance of sustainability indicators: Application to the indicator set of the Czech Republic's Sustainable Development Strategy. *Ecological Indicators* 17 (0), pp. 46-57.

Hanley, N. (2000), Macroeconomic Measures of 'Sustainability'. *Journal of Economic Surveys* 14 (1), pp. 1-30.

Ishizaka, A. and Labib, A. (2011), Review of the main developments in the analytic hierarchy process. *Expert Systems with Applications* 38 (11), pp. 14336-14345.

Klugmann, J. (2010), *Human Development Report 2010*, 20th Anniversary Edition, *The Real Wealth of Nations: Pathways to Human Development*. Available at: http://hdr.undp.org/en/media/HDR_2010_EN_Complete_reprint.pdf (accessed 20 May 2011).

Knutsson, P. and Ostwald, M. (2006), *A Process-Oriented Sustainable Livelihoods Approach – A Tool For Increased Understanding of Vulnerability, Adaptation and Resilience*. Mitigation and Adaptation Strategies for Global Change.

Kromrey, H. (2009), *Empirische Sozialforschung*, 12th ed. Lucius & Lucius, Stuttgart.

Lang, A. (2003), *Ist Nachhaltigkeit messbar?* (1st ed.). Ibidem Verlag.

Lautso, K., Spiekermann, K. and Wegener, M. (2002), *Modelling Policies for Urban Sustainability*. Paper presented at the 42nd Congress of the European Regional Science Association (ERSA), 27-31 August 2002, pp. 1-18. Dortmund.

Liberatore, M.J. and Nydick, R.L. (2008), The analytic hierarchy process in medical and health care decision making: A literature review. *European Journal of Operational Research* 189 (1), pp. 194-207.

Linkov, I., Varghese, A., Jamil, S., Seager, T., Kilker, G. and Bridges, T. (2004), Multi-criteria decision analysis: a framework for structuring remedial decisions at contaminated sites. *Comparative Risk Assessment and Environment Decision Making*, pp. 15-54. Kluwer.

Mayer, A.L. (2008), Strengths and weaknesses of common sustainability indices for multidimensional systems. *Environment International* 34 (2), pp. 277-291.

Meadows, D. (1998). *Indicators and Information Systems for Sustainable Development*. The Sustainability Institute, Hartland VT.

Morse, S. and Fraser, E.D.G. (2005), Making "dirty" nations look clean? The nation state and the problem of selecting and weighting indices as tools for measuring progress towards sustainability. *Geoforum* 36 (5), pp. 625-640.

Munda, G. (2005), 'Measuring Sustainability': A Multi-Criterion Framework. *Environment, Development and Sustainability* 7 (1), pp. 117-134.

Munda, G. and Nardo, M. (2003), *On the methodological foundations of composite indicators used for ranking countries*. Joint Research Centre of the European Communities, Ispra.

– (2005a), *Constructing consistent composite indicators: the issue of weights*. Available at: http://www.magtud.sote.hu/constructing-consistent-composite-indicators-the-issue-of-weights.pdf. (accessed 4 August 2011).

– (2005b), *Non-compensatory composite indicators for ranking countries: A defensible setting*. Official Publications of the European Communities, Luxembourg.

Nardo, M. (2004), *The E-Business Readiness Composite Indicator for 2003: a Pilot Study*. European Commission Joint Research Centre.

Nardo, M., Saisana, M., Saltelli, A. and Tarantola, S. (2005), *Tools for Composite Indicators Building*. Available at: http://collection.europarchive.org/dnb/2007070213 2253/http://farmweb.jrc.ec.europa.eu/ci/Document/EUR%2021682%20EN.pdf (accessed 8 August 2011).

Nicoletti, G., Scarpetta, S. and Boylaud, O. (1999), *Summary indicators of product market regulation with an extension to employment protection legislation*. OECD, Paris.

OECD (2008), *Handbook on constructing composite indicators: Methodology and user guide*. Organization for Economic.

P.R. Consultants (2000), *Eco-indicator 99 Manual for designers*. Ministry of Housing, Spatial Planning and the Environment, The Hague.

Pohekar, S.D. and Ramachandran, M. (2004), Application of multi-criteria decision making to sustainable energy planning-A review. *Renewable and Sustainable Energy Reviews* 8 (4), pp. 365-381.

Pohl, E. (2006), *Towards Corporate Sustainable Development: The ITT Flygt Sustainability Index*. Institutionen för Samhällsteknik.

Pratt, C., Kaly, U. and Michell, J. (2004), *Manual: How to Use the Environmental Vulnerability Index*. SOPAC Technical Report 383. Available at: http://www.vulnera bilityindex.net/Files/EVI Manual.pdf (accessed 20 May 2011).

Roodman, D. (2005), *The Commitment to Development Index: 2005 Edition*. Center for Global Development, Washington, DC.

Saaty, T.L. (2006), Rank from comparisons and from ratings in the analytic hierarchy/ network processes. *European Journal of Operational Research* 168 (2), pp. 557-570.

Saisana, M. and Tarantola, S. (2002), *State-of-the-art report on current methodologies and practices for composite indicator development*. EUR 20408 EN Report.

Salvati, L. and Zitti, M. (2009), Substitutability and weighting of ecological and economic indicators: Exploring the importance of various components of a synthetic index. *Ecological Economics* 68 (4), pp. 1093-1099.

Sawada, Y., Kohama, H., Kono, H. and Ikegami, M. (2004), *Commitment to Development Index (CDI): Critical Comments*. Available at: http://dakis.fasid.or.jp/report/pdf/DP_1_E.pdf (accessed 4 September 2011).

Schlossberg, M. and Zimmerman, A. (2003), Developing Statewide Indices of Environmental, Economic, and Social Sustainability: a look at Oregon and the Oregon Benchmarks. *Local Environment: The International Journal of Justice and Sustainability* 8 (6), pp. 641.

Schuh, H. (2001), *Entscheidungsverfahren zur Umsetzung einer nachhaltigen Entwicklung*. Technical University Dresden, Faculty Economic and Business Sciences (45), pp. 1-31.

Singh, R.K., Murty, H.R., Gupta, S.K. and Dikshit, A.K. (2009), An overview of sustainability assessment methodologies. *Ecological Indicators* 9 (2), pp. 189-212.

Štreimikienė, D., Girdzijauskas, S. and Stoškus, L. (2009), Sustainability Assessment Methods and Their Application to Harmonization of Policies and Sustainability Monitoring. *Environmental Research, Engineering and Management* 48 (2), pp. 51-62.

Tarantola, S., Liska, R., Saltelli, A., Leapman, N. and Grant, C. (2002), *The Internal Market Index 2004*. Available at: http://ec.europa.eu/internal_market/score/docs/ relateddocs/ 2004-im-index_en.pdf (accessed 21 December 2011).

United Nation (1992), *AGENDA 21*. Available at: http://www.un.org/esa/sustdev/documents/agenda21/english/Agenda21.pdf (accessed 21 May 2011).

Utne, I.B. (2008), Are the smallest fishing vessels the most sustainable? – trade-off analysis of sustainability attributes. *Marine Policy* 32 (3), pp. 465-474.

Vaidya, O.S. and Kumar, S. (2006), Analytic hierarchy process: An overview of applications. *European Journal of Operational Research* 169 (1), pp. 1-29.

Yale Center for Environmental Law and Policy, Center for International Earth Science Information Network, Columbia University and World Economic Forum Joint Research Centre, European Commission (2005), *2005 Environmental Sustainability Index Benchmarking National Environmental Stewardship*. Available at: http://www.yale.edu/ esi/ (accessed 21 May 2011).

Van de Kerk, G. and Manuel, A.R. (2008), A comprehensive index for a sustainable society: The SSI – the Sustainable Society Index. *Ecological Economics* 66 (2-3), pp. 228-242.

Van den Bergh, J.C.J.M. and Verbruggen, H. (1999), Spatial sustainability, trade and indicators: an evaluation of the ecological footprint. *Ecological Economics* 29 (1), pp. 61-72.

Yigitcanlar, T. and Dur, F. (2010), Developing a Sustainability Assessment Model: The Sustainable Infrastructure, Land-Use, Environment and Transport Model. *Sustainability* 2 (1), pp. 321-340.

Zhou, P., Fan, L.-W. and Zhou, D.-Q. (2010), Data aggregation in constructing composite indicators: A perspective of information loss. *Expert Systems with Applications* 37 (1), pp. 360-365.

Chapter 25

Impacts of Sustainability Courses on the Behavior of Business School Students

Menno de Lind van Wijngaarden and Julia Planko

Abstract

This paper introduces a model about the impacts of sustainability management courses on the beliefs and behavioral intentions of business students.

Business schools have been heavily criticized for delivering graduates who have contributed to unsustainable business practices and the financial crisis. Moreover, there is a growing demand amongst businesses for sustainability literate graduates. Therefore, sustainable management courses are important in the business curriculum.

Based on a literature review, as well as essays and evaluations of former sustainability course students with regard to their values and beliefs concerning sustainability, a model has been developed. It aims to test the impacts of the sustainability courses on students' problem awareness and beliefs, as well as on their behavioral intentions to contribute to sustainable development in the future.

The model can help curriculum developers to incorporate sustainability into their business courses.

Introduction

Business schools have, in the past, been considered notorious for delivering graduates showing unethical behavior in their professional role. After the socially irresponsible and illegal actions of companies like Enron, Tyco and WorldCom, business schools have responded to societal demands by offering courses teaching social and personal responsibility towards the various stakeholders of corporations. Furthermore, business schools have paid no attention to unsustainable business practices in the past, like unrestrained resource use, environmental degradation and pollution. Business schools were seen as the institutes handing students the tools to make profits regardless of moral and environmental consequences. A report by KPMG in 2009 stated that 90% of businesses find ethical education important and that one third of surveyed businesses found the ethical and moral education of new graduates and starters insufficient (KPMG 2009). The increased criticism of the role of business schools in the current financial crisis has also led to reflection and discussion on the curriculum of business schools (Datar et al. 2010)

Many companies, business associations, governments and international organizations are adopting social responsible policies that suggest that corporations can simultaneously make profits and be good citizens (UNRISD 2000). Organizations like CSR Europe have 80 multinational companies as members, and three-quarters of G250 companies have a corporate responsibility strategy that includes clearly defined objectives (KPMG 2008). This shows that business is embracing the concept of sustainability. Business schools have to deliver graduates who can fulfill demands for a workforce which is educated with regard to sustainability. As indicated by Jabbour, there is a need for empirical research on the question of the effect business schools have on the behavior of graduates who go on to become managers (Jabbour

2010). This research is a first attempt to create a model and to build up empirical evidence for this correlation.

The authors' hypothesis is that sustainable management courses (Business Ethics, Corporate Social Responsibility, Sustainable Management) have positive impacts on the specific beliefs and behavior business students will show later as managers and as part of the business world.

To test this hypothesis, we use Ajzen's model of planned behavior to predict the future behavior of students once they work for businesses; we adapt the model to apply to sustainable behavior.

In the following sections, we describe what is important in order to change a student's mindset with regard to sustainability. Next, we explain Ajzen's model of planned behavior and adapt it to behavior which fosters sustainable development.

The need for change in values, beliefs and attitudes to enable sustainable development

People's beliefs and values are very important for sustainable development. Most of the environmental, social and economic problems societies face today arose because of the current value system (Brown 2005). In order to achieve sustainable development, changes in human values, attitudes, and behavior are necessary (Leiserowitz et al. 2006). Values are abstract ideals, which "define or direct us to goals, frame our attitudes and provide standards against which the behavior of individuals in societies can be judged" (Leiserowitz, Kates, and Parris, 2006). Attitudes are the evaluation of a specific behavior as positive or negative, good or bad. "Behavior refers to concrete decisions and actions taken by individuals and groups which are often rooted in the underlying values and attitudes" (Leiserowitz et al. 2006).

Leiserowitz et al. argue that a change in human values and attitudes is a requirement and major tool toward achieving sustainable development. According to them, the main barrier to sustainable behavior consists of values and attitudes which do not favor sustainability. The second barrier to sustainable development is an individual's lack of knowledge, time, money, self-efficacy, access and skills (Leiserowitz et al. 2006). To overcome these barriers, education with regard to sustainable development could play a role.

Educators need to provide students with sufficient knowledge and with alternative models that they can apply in business life. They need to teach them how to make decisions that will contribute to sustainable development (Herremans and Reid 2002). Moreover, without sufficient knowledge with regard to sustainability, students may feel overwhelmed by today's environmental problems. "They may feel that the problems are so large and complex they cannot change them. This may lead to apathy or cynicism" (Rowe 2002).

The focus in the literature is often on "environmental literacy", an awareness of the environment and responsible behavior toward human and nature relationships (Wolfe 2001). It is important that students not only attain environmental literacy, but also that they understand the concepts of social responsibility and sustainability. The authors want to introduce the collective term "sustainability literacy" for these three concepts.

Brunton considers teaching students about sustainability has been successful if they have not only learnt about alternative models of acting and evaluating, but also been inspired to reflect on and possibly change their values, attitudes and behaviors (Brunton 2006).

The theory of planned behavior connected to sustainability-related beliefs and behaviors

In order to understand how beliefs, values and attitudes can impact behavior, Ajzen's "theory of planned behavior" will be reviewed. Ajzen showed that an individual's values, norms and attitudes can be used as antecedents of planned behavior, which he calls "behavioral intention", which again is an antecedent of actual behavior. An important factor in Ajzen's model is what he calls "perceived behavioral control, the perceived ease or difficulty of performing the behavior in question" (Ajzen 1991). Not only an individual's values and attitudes shape his behavior, but also the ability to perform this behavior and the expected outcome the individual thinks his behavior will yield. If he anticipates that his behavior will be unsuccessful or not show the intended outcome, he is less likely to perform it. "Perceived behavioral control" is connected to Bandurra's principle of self-efficacy, "the judgment of how well one can execute courses of action to deal with prospective situations" (Ajzen 2002).

All three predictors (attitudes, norms, perceived control) can make independent contributions to the behavioral intention. Behavioral intention again, together with behavioral control, can be used to predict actual behavior. To translate it to this topic: a certain set of values, and knowledge regarding alternative actions which are more sustainable than conventional ones, will only result in the student taking that 'more sustainable decision' if he is in a situation where he feels his decision/action will have an impact on the outcome. Therefore, it is not only important to trigger changes in students' values and beliefs (cf. Brown 2005; Brunton 2006; Leiserowitz et al. 2006), but also to make them realize that they are in a position to change something.

In their research about environmental specific behavior, De Groot and Steg demonstrate that value orientations (i.e. general beliefs) affect behavior specific beliefs. Behavior specific beliefs are (i) problem awareness and (ii) ascription of responsibility, which means the individual feels responsible for problems. Values, attitudes and specific beliefs are antecedents of behavioral intention (De Groot and Steg 2007).

The authors combined Ajzen's theory of planned behavior with the findings of De Groot and Steg, and applied it to businesses' contribution to sustainable development (see Fig. 1).

The impact of sustainability literacy courses on the specific beliefs and behavioral intentions of business students

At the HU Utrecht Business School, there are a number of courses devoted to sustainable development. The various aspects of sustainable development are integrated in courses such as Business Ethics, Sustainable Management, Corporate Social Responsibility, and Microfinance. These courses have been subject to evaluation by the students through the use of open questions that invite students to contemplate the impact of the course on their views of business practices and their own activities. These evaluations have led to responses such as the following:

> The most important challenge for the business world is to apply corporate social responsibility and become more global sustainable, the business life needs to acknowledge the fact that their activities can harm the world and so are responsible to help the world and the communities they are active in in order to improve the world's current position (Student 1).

> I think I will try to have a role in the conservation of our nature; I will change my standard of living (like using less water and electricity, recycling, etc.) (Student 2).

> … and therefore, I will take the initiative to start to become more environmental friendly myself (Student 3).

> My way to contribute, personally, to this will be trying to choose my job depending on the company, taking care of the regular aspects of a job, but also trying to choose a sustainable company to work in (Student 4).

These anonymous evaluations were analyzed according to the impact on the students' perception of the role of business in society and the personal effects of learning about sustainability. Almost all respondents mentioned that, although they were already quite aware of environmental and social problems, the courses made them more aware of the role that business could play in finding solutions. Many held businesses responsible for problems, but also saw corporate social responsibility as a solution. A smaller group of students mentioned the personal impact the course had had on their life. They expressed the intention to change their behavior, and a few even mentioned that they had started to behave more sustainably.

The courses in the curriculum have been designed according to the competency-based learning principle, and aim to not only impart knowledge to students but also train skills and change attitudes. The sustainability-related courses are knowledge based, contain interactive settings and provide room for reflection. These courses could prepare students for being a change agent in business after graduation. Although the evaluations indicate that there is some effect, the question remains of how effective these courses are in creating change agents.

There has been a noticeable increase in attention to courses in graduate and undergraduate business curricula dealing with the topic of environmental and social issues (World Resources Institute 2001) but little is known about the effects of these courses upon the attitudes, behavioral intentions and awareness of students.

The prerequisite of any behavioral change is that individuals realize there is a problem. De Groot and Steg called this "awareness of problem". Students need to be aware that there is a problem (unsustainable development), in order to contribute to its mitigation. Yet students can be aware that there is a problem, but they may not realize that businesses can help mitigate (or scale back the creation of) the problem. So the second prerequisite is that they understand that businesses are (partially) responsible for the problem. De Groot and Steg call this "ascription of responsibility". They apply this to the individual's feeling of responsibility for the problem. The authors use it in a different way – referring to business responsibility. Students need to understand that businesses are (partly) responsible for the problem, in order to change their (planned) behavior.

Once students understand that there is a problem and that businesses create and contribute to the problem, they might be inclined to try to act as change agents. According to Ajzen's theory of planned behavior and Bandurra's principle of self-efficacy, "perceived behavioral control" plays an important role (Ajzen 2002). An individual is more inclined to show a certain action when he or she feels that they have the ability to achieve results, or that their actions will yield an outcome. Translated to the authors' model, this means managers will only show a certain behavior when they think that the behavior will actually have an impact on the outcome.

Students have to perceive that they have behavioral control to mitigate the problem, in order to change their (planned) behavior.

Perceived behavioral control, attitudes and social norms are all antecedents of behavioral intention, whereas perceived behavioral control often has a more prominent role. Even if attitudes and social norms are positive towards a certain behavior, if the individual does not feel that his or her actions will have any impact on a specific situation, they are not likely to show the behavior. Because attitudes regarding sustainable behavior are usually positive (based on the literature and the authors' qualitative study of students' course evaluations), as the quotes below illustrate, it was chosen not to include it into the model.

> I feel global sustainability is a huge topic in industry ...[that]... we still yet need to explore and understand. Through this course, I learned so much about how there's so much we still need to do to improve the environmental condition of our society. (Student 5).

> I loved this course!! It gave me a great joy to see this is a compulsory subject and that the students were forced into attending lectures, reading about the real life and the actual and future state of our planet, society and so on. (Student 6).

The aggregation of increased awareness, feelings of responsibility and perceived behavioral control leads to changes of (planned) behavior in terms of behavior they plan to show in the job, and in terms of their job choice. There can be, however, an intervening effect (Ajzen 1991). In the timeframe between the sustainability course and the moment where the sustainable behavior can be displayed at the job, there may be changes in intentions or in perceptions of behavioral control, which could lead to a less accurate prediction of behavior.

The authors assume that if the sustainability course has changed the students' behavior specific beliefs and control beliefs, this will have a twofold impact on their behavioral intention regarding their future work life. First of all, they are expected to take different decisions, which contribute more to sustainable development by influencing or even guiding the companies they work for towards a sustainability transition. Second, it is assumed that some might not only show different behavior (or take different decisions or carry out different actions) within the company they work for, but that they already take different decisions before starting to work for a company, namely in their job choice. The authors assume that some of them might find it more important to work for a company that has a good reputation with regard to sustainable practices.

Fig. 1 displays the model developed.

Fig. 1: The impact of sustainability literacy courses on behavioral intentions of business students

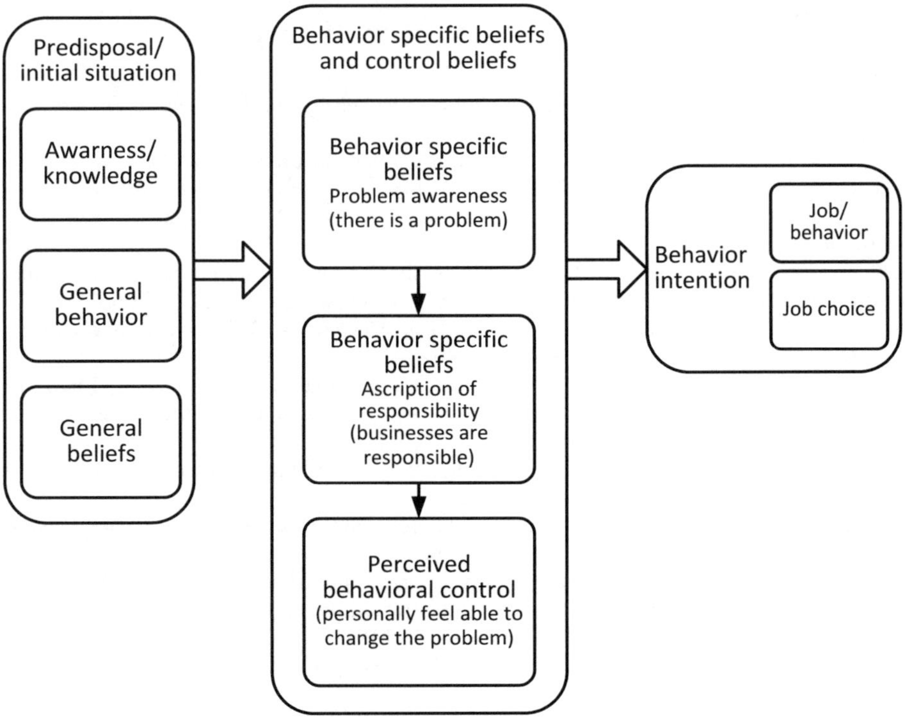

Conclusion

An increasing number of business schools are answering the calls from society to deliver students that contribute actively to sustainable development. This can be done by integrating sustainability literacy courses into the business curriculum.

In order to establish sustainability literacy courses the proposed model can be used. The model will be helpful for educators who teach sustainability courses at business schools or who have to convince curriculum designers of the importance of sustainability courses in the business curriculum. The model can be used for designing sustainability courses, as it illustrates the importance of teaching the student that there is a problem (unsustainable development), that businesses are (partly) responsible for the problem, and that the students will have the power to change the problem by changing their behavior.

The model has been developed based on a literature review and qualitative research. In order to predict the effects of sustainability literacy courses on students' future behavior in the workplace, Ajzen's model of planned behavior was adapted. The proposed model will provide a better insight into the factors that will need to be included in the curriculum in order to effectively change behavior towards sustainability. The sustainability literacy courses will be therefore more influential with regard to the choices these students make later in life when choosing jobs or selecting companies to work for, and act as change agents whenever the opportunity arises.

The developed model brings forward the factors underlying behavioral intentions and helps predict actual behavior of business students after participating in sustainable literacy courses. These factors now need to be quantified by further research. A questionnaire based on the model has already been developed. The survey will be conducted among students before and after they take a sustainability-related course.

References

Ajzen, I. (1991), The Theory of Planned Behavior. *Organizational Behavior & Human Decision Processes* 50 (2), p. 179-122.

– (2002), Perceived Behavioral Control, Self-Efficacy, Locus of Control, and the Theory of Planned Behavior. *Journal of Applied Social Psychology* 32 (4), pp. 665-683.

Brown, B.C. (2005), Theory and Practice of Integral Sustainable Development. *AQAL – Journal of integral theory and practice* 1 (2), pp. 1-70.

Brunton, K. (2006), Education for sustainable development: Principles for curriculum development in business subject areas. *Investigations in University Teaching and Learning* 3 (2), pp. 36-46.

Datar, S.M., Garvin, D.A. and Cullen, P.G. (2010), *Rethinking the MBA: business education at a crossroads.* Harvard Business Press, Boston.

De Groot, J.I.M. and Steg, L. (2007), Value Orientations to Explain Beliefs Related to Environmental Significant Behavior: How to Measure Egoistic, Altruistic, and Biospheric Value Orientations. *Environment and Behavior* 40 (3), pp. 330-354.

Herremans, I.M. and Reid, R.E. (2002), Developing Awareness of the Sustainability Concept. *Journal of Environmental Education* 34 (1), pp. 16-20.

Jabbour, C.J.C. (2010), Greening of business schools: a systemic view. *International Journal of Sustainability in Higher Education* 11 (1), pp. 49-60.

KPMG (2008), *KPMG International Survey of Corporate Responsibility Reporting 2008*. Available at: http://www.kpmg.com/Global/en/IssuesAndInsights/ArticlesPublications/Documents/International-corporate-responsibility-survey-2008.pdf (accessed 27 March 2012).

– (2009), *Bedrijfsleven vindt ethische vorming studenten onder de maat*, Press release, 21 September 2009.

Leiserowitz, A.A., Kates, R.W. and Parris, T.M. (2006), Sustainability values, attitudes, and behaviors: a review of multinational and global trends. *Annual Review of Environment & Resources* 31 (1), pp. 413-444.

Rowe, D. (2002), Environmental literacy and sustainability as core requirements: success stories and models. In Leal Filho, W. (Ed.), *Teaching Sustainability at Universities*. Peter Lang, Frankfurt.

UNRISD (2000), *Business responsibility for sustainable development.* United Nations Research Institute for Social Development, Geneva.

Wolfe, V.L. (2001), A survey of the enviornmental education of students in non-environmental major at four-year institutions in the USA. *International Journal of Sustainability in Higher Education* 2 (4), pp. 301-315.

World Resources Institute (2001), *Beyond grey pinstripes 2001.* World Resources Institute; Aspen Institute Initiative for Social Innovation through Business, Washington, DC, New York, NY.

Chapter 26

Sustainable Development in Higher Education in the Muslim World

Fayyaz Vellani and Naveed Nanjee

Abstract

To date, there is a dearth of research on how higher education institutions in the Muslim world are engaging with notions of sustainable development. Expanding 'sustainable development' beyond its current definition of development which considers the needs of future generations, the authors argue that a much more holistic approach to this concept, which incorporates Islamic values of ethical action, stewardship and the careful management of resources needs to be deployed. In particular, this paper examines the Aga Khan University (AKU) and how it engages with an ethos of sustainable development in a manner which is in keeping with Islamic values. The founding principles of this university – from its very inception – are concerned with being in harmony with its surrounding environments. In addition to the case study of the AKU, these principles have relevance to broader contexts beyond the Muslim world.

Introduction

In 1987, the United Nations' World Commission on Environment and Development (WCED) wrote a report entitled "Our Common Future," proposing long-term strategies for environmental protection in harmony with development. The idea of sustainable development, defined as development which "meets the needs of the present without compromising the ability of future generations to meet their own needs" (World Commission on Economic Development 1987), emerged from the report. Sustainable development had two broad goals: (1) to encourage developed nations to moderate the levels of industrial growth which cause environmental degradation, and (2) to assert the rights of developing nations to create economic growth balanced with environmental protection.

Following the United Nations Conference on Environment and Development in 1992, also known as the Earth Summit, a set of goals known as Agenda 21 was established. The aim of Agenda 21 was for member states to "contribute in a meaningful way to public policy decisions and actions leading to sustainable development" (Strong, in Robinson and Dale 1996: iv). In many nation-states, particularly in the industralised West, efforts were undertaken to "influence key decisionmakers [sic] in government, industry and non-government organizations" (ibid.), including the establishment of postsecondary sustainable development research institutes and efforts to integrate sustainable practices into the operations of universities themselves. Thus, universities, charged with knowledge and discourse production and removed from the partisan electoral demands faced by governments and the profit-seeking imperative faced by corporations, emerged at the forefront of debates and practices around sustainable development. As Cortese (1999: 8) notes:

> If we are to achieve a sustainable future, institutions of higher education must provide the awareness, knowledge, skills and values that equip individuals to pursue life goals in a manner that enhances and sustains human and non-human well-being.

The focus on environmental protection as secondary to economic growth has limited conceptual explorations of sustainable development in the context of Muslim countries. While the governments of some countries in the Muslim world are currently experiencing radical change, universities in these countries can provide necessary forums for discussion and planning around practices which are sustainable in the broadest sense, as evidenced by efforts by various faith communities, including Muslims, to prioritise and "promote co-existence, peace, and sustainability through education and activism" (Interfaith Center for Sustainable Development).

In the current geopolitical climate, universities in the Muslim world are uniquely positioned to contribute to this discourse owing to a number of factors: (1) the dramatic rise of democratic movements in these countries, and the concomitant impact of these on the region; (2) the global dependence on fossil fuels emanating from this region and the necessity of research exploring alternatives; (3) the increasing strategic political importance of areas with large Muslim populations, e.g. the Middle East, West Africa and the Far East; (4) universities' roles in cross-cultural interaction and the global exchange of research and knowledge. For example, New York University established its first four-year campus outside of New York in Abu Dhabi. The role of universities is expanding outside their local contexts. In this regard, universities can direct both local and international environments towards becoming better stewards of the Earth. Indeed, the university selected in this paper as a case study, the Aga Khan University in Pakistan is a leading example of an institution which aspires to ethical relationships with its surrounding communities.

Although universities in the Muslim world are uniquely positioned to contribute to these discourses, there is little scholarship on the relationship between Islam and environmental protection. The existing research in this area highlights the idea that sustainability is fundamentally compatible with the principles and practices of Islam (Khalid and O'Brien 1992). For example, in their paper *Islam on Sustainability* Marinova et al. argue that Islam as a religion and as a set of cultural practices "contains a range of guidelines and prescriptions which encourage its followers to pursue a sustainable life on earth" (2008: 123). Their exegesis of Qur'anic quotes highlights the need for humans to live responsibly on earth and consider the needs of future generations. Specifically, one of their arguments is that sustainability, as interpreted in the Qur'an, is measurable, as highlighted in the following verse:

> We send down water from the sky according to (due) measure, and We cause it to soak in the soil; and We certainly are able to drain it off (with ease) (23:18).

A final point of relevance from their paper has to do with the importance of education to Muslims and the identification of two different kinds of knowledge: that which is generated versus that which is gathered (Renard 2004, in Marinova et al. 2008). Interestingly, the authors argue that an Islamic approach to sustainability ought to include considerations of education which is ethical and which guides everyday living, a point which we later argue in relation to AKU's engagement with its surrounding community. As they note, "this transmission of knowledge and generation of new knowledge within a constantly changing ecological and social world are essential for sustainability education" (Marinova et al. 2008).

Another recent work, *Green Deen: What Islam Teaches About Protecting the Planet* (Abdul-Matin 2010) is about the author's quest to seek environmental ethics through the practice of his religion and as an introspective approach to how communities can engage with Islam and environmentalism. The book therefore highlights the connections between Islamic teachings and environmentalism and serves as a blueprint for communities seeking to implement environmental change at local level. We argue that the AKU exemplifies these connections by highlighting the importance of development which is harmonious with its local environments and which engages community participation.

Those who have contributed to scholarship on Islam and ecology (Nasr 1976; Abdul-Matin 2010), point to the unique contributions which Islam can make to these discourses by way of the intrinsic values it espouses. Given their roles as leaders of knowledge and discourse production, universities in the Muslim world can cite these congruencies between Islam and the environment. Dating back to the establishment of Al-Azhar University in Cairo, the world's first university, higher education institutions in the Muslim world have always viewed it as their responsibility to provide blueprints for ethical societies, rather than existing simply for educational or vocational purposes.

Seyyed Hossein Nasr has alluded to the difficulties humans would face in sustaining life on earth and wrote about it in his seminal work *Man and Nature: The Spiritual Crisis in Modern Man*. Since then, following the growth of the environmental movement, and the evolution of the concept of sustainable development, many interpretations of this term have emerged, as well as critiques of it. Nasr's particular contribution, upon which this paper builds, is that the environmental challenges faced by humans today embody not only problems in the material world, but also a deeper existential crisis indicative of a spiritual gap in conceptualising the relationship between humans and the earth. In his view, unless this deeper crisis is addressed, humans will not find ways of living sustainably on earth. Nasr writes (1976: 17):

> Today, almost everyone living in the urbanized centres of the Western world feels intuitively a lack of something in life. This is due directly to the creation of an artificial environment from which nature has been excluded to the greatest possible extent. Even the religious man in such circumstances has lost the sense of the spiritual significance of nature.

This 'lack of something in life' cannot be remedied by sustainable development according to its current definition; Nasr alludes to a deeper spiritual and existential crisis which can be more fully addressed by the holistic approach contained in teachings of Islam. He continues:

> Furthermore, even this type of secularized and urbanised existence is itself threatened, through the very domination of nature that has made it possible, so that the crisis brought about through the encounter of man and nature and the application of modern sciences of nature to technology has become a matter of common concern (ibid.).

Nasr's argument is that science alone cannot prevent environmental catastrophe. Such prevention requires a deeper examination of the relationship between humans and the earth; an examination which can be fostered through the core religious values of Islam and other faiths. In terms of the application of these values, while Muslim countries were present as part of the World Conference on Environment and Development and the Earth Summit, there are principles in Islam which exemplify and push the concept of sustainable development beyond simply economic progress with environmental protection. Development, if following Islamic practices, should always be enhancing the relationships between humans and the earth.

Methodology

This paper addresses how higher education institutions in Muslim countries engage with notions of sustainable development using the Aga Khan University (AKU) as a case study, employing empirical material obtained from discussions with architects, designers and consultants, and having reviewed planning documents, blueprints and strategies. As indicated, very little research has been conducted on the role of universities in the Muslim world, and the AKU is distinct from other institutions in that it belongs to a wider network of institutions within the Aga Khan Development Network (AKDN), whose work includes architecture, culture, development, microfinance, civil society, education and healthcare. Therefore, notions of sustainability for the AKU have the potential to employ the best practices from other organisations within the broader network, and holistically approach some of these topics. For example, the Aga Khan Rural Support Program (AKRSP) was the first microfinance initiative of its kind, and its outcomes and achievements inform the work of AKU in thinking about how the university engages with the surrounding communities facing poverty. Similarly, the Aga Khan Award for Architecture, the largest architectural award in the world, has looked favourably upon designs which are sensitive to the environment, and designs for new AKU campuses will build on the best practices and aesthetic sensibilities identified by the award's steering committee. The significance of this research is to highlight the role of universities as vehicles for promulgating sustainability in the context of Muslim countries. As sustainable development becomes an increasingly important value, AKU can play a leading role within the broader network of institutions to which it belongs.

Concepts of environmental stewardship in Islam

The Qur'an, which is the holiest scripture in Islam, is believed to be the word of God, as revealed to the Prophet Muhammad in seventh-century Arabia. One of the key themes of the Qur'an is law, governing the ways in which Muslims should aspire to live (Rippin 1990). As a primary text, the Qur'an is the most important point of reference in actualising Islam as a way of life. It is employed here is to highlight values which espouse ethical, responsible human behaviour.

Although the Qur'an was compiled more than 1400 years ago, before the formal language of environmentalism existed, five key Qur'anic concepts can be interpreted in light of their implications for relationships between humans and their environments. Universities in the Muslim world have the capacity to incorporate these values found within the Qur'an to exemplify ethical responsibility, and the Aga Khan University is an example of an institution which seeks to operate in keeping with these values.

The first of these is *tawhīd* (meaning "oneness"), the central principle in Islam. Indeed, the oneness of God is so fundamental to the religion that without monotheistic belief, one cannot be considered Muslim. The concept can be further extended to the oneness of everything – everything has one source. In the context of the environment, *tawhid* can be interpreted to extend to humans and their surroundings: "it is the primordial testimony of the unity of all creation and the interlocking grid of the natural order of which man is intrinsically a part" (Hasnain 2006: 211). As found in the Qur'an: "Thou seest not in the creation of the All-merciful any disorder" (67:3).

The Qur'an reaffirms this single created order in terms of the uniformity of all its elements; God, the Creator is one, and all living things originate from Him; the world is a unified, ordered, single creation. The ontological unity flowing from *tawhīd* translates into the role of humans in maintaining order on earth. This is expressed in the Qur'an as *khalīfah*, which is an understanding that God has given the earth to mankind to protect and take care of it in an ethically suitable manner. This concept is exemplified in the following verse from the Qur'an:

> It is He Who hath made you (His) agents, inheritors of the earth: He hath raised you in ranks, some above others: that He may try you in the gifts He hath given you: For thy Lord is quick in punishment: yet He is indeed Oft-Forgiving, Most Merciful (6:165).

The above quotation mirrors Biblical passages bestowing dominion over the earth to humankind. However, in another Qur'anic quotation, this inheritance is coupled with responsibility for the earth, and not simply ownership:

> In Allah's earth, and let her come to no harm, or ye shall be seized with a grievous punishment. And remember how He made you inheritors after the 'Ad people and gave you habitations in the land: ye

> build for yourselves palaces and castles in open plains, and carve out homes in the mountains; so bring to remembrance the benefits (ye have received) from Allah, and refrain from evil and mischief on the earth (7:74-75).

The case study institution, the AKU, with its humanistic ideals of improving quality of life, aspires to abide by this principle of "no harm". AKU is a steward of the environment in Karachi. The university takes ownership of its role as an agent for good, and works to improve the lives of its students and those it serves. This work is conducted in a manner which is sensitive to the surrounding environment. As the founder and Chancellor of the university, His Highness the Aga Khan has repeatedly stressed the need for individual and institutional actions to be guided by Islamic ethics of compassion, responsible stewardship of resources and planning for the long term impacts of any endeavours on a given environment and its populations. As he notes:

> In Islam, the Holy Qur'an says that man is God's noblest creation to whom he has entrusted the stewardship of all that is on earth. Each generation must leave for its successors an enhanced and sustainable social and physical environment (His Highness the Aga Khan 2008: 72).

Building on the notion of *khalīfah,* the third concept highlighted in the Qur'an is that of *mizan*, defined as balance. This requires an understanding of the balance between the natural environment and also the balance between human beings and the earth. As noted in the Qur'an:

> He has created man: He has taught him speech (and Intelligence). The sun and the moon follow courses (exactly) computed. The sun and the moon follow courses (exactly) computed. And the herbs and the trees Both (alike) bow in adoration. And the firmament has He raised high, and He has set up the Balance (of Justice), in order that ye may not transgress (due) balance. So establish weight with justice and fall not short in the balance (55: 3-9).

This illustrates humanity's place in the context of the elements within nature – to be in harmony with them, rather than 'transgressing' the balance alluded to in the Qur'an. In another passage of the Qur'an it is stated that "acquisitiveness turns you away" (102:4), highlighting the fact that the accumulation of material goods will not bring fulfillment and could cause an imbalance. This Qur'anic teaching reminds Muslims of the enormity of the universe and that they are merely a small part of it. The AKU, by its very presence in the densely urban environment of Karachi – a city not noted for its green space – attempts to create such a balance by virtue of its existence. The use of natural elements throughout the campus indicates the university's adherence to the concept of balance between humans and the environment. Indeed, the fourth concept, *fiṭrah*, defined as natural disposition, identifies the place of humans in God's larger creation.

> So set thou thy face steadily and truly to the Faith: Allah's handiwork according to the pattern on which he has made mankind: no change (let there be) in the work (wrought) by Allah: that is the standard Religion: But most among mankind Understand not. (30:30)

The concept of *fiṭrah* is evident in the mission of the AKU to improve societies through human endeavour, and specifically in the ways in which it has engaged with its surrounding communities. For example, when the university was developed, significant planning went into the contribution of the university in its surrounding areas, known as katchi abadis (slums), and specifically targeted those populations with the improvement of their quality of life (UN Habitat 2003). Throughout its work – whether in providing healthcare to local populations, educational training to improve the lives of students and their surrounding societies, or its efforts to operate in an ethical manner – the AKU is inspired by *fiṭrah*.

As noted in the earlier Qur'anic (55:3-9) quotation "He has set up the Balance (of Justice)", justice, known as *'adl*, is another important ideal to which Muslims aspire. Humans can exemplify this value by providing resources to those who are in need; teaching principles of justice, and serving others. As noted in the Qur'an:

> God commands justice, the doing of good, and liberality to kith and kin, and He forbids all shameful deeds, and injustice and rebellion: He instructs you, that ye may receive admonition (16:90).

Justice is described on two levels – human and divine. The following Qur'anic quote illustrates *'adl* in the context of creation – situating justice in the higher order (*fiṭrah*) of creation: "We created not the heavens, the earth, and all between them merely in (idle) sport. We created them not except for just ends" (44:38-39).

The AKU embodies this Qur'anic ideal of justice in two senses: (1) environmental justice through its conscious efforts to use resources responsibly, and (2) through creating educational resources which will provide long-term improvements to the community. In light of the above, the Qur'an can be used to illustrate values to guide ethical and sustainable practices. For the future betterment of relationships between humans and the environment, universities in Muslim contexts should incorporate these values into the core of their operations. The Aga Khan University embodies these values in its current practices at its original campus in Karachi, and is creating a series of landmark green campuses in East Africa and Pakistan.

The values of sustainable development at AKU:

> "Inspired by Islamic ethics, humanistic ideals and the philosophy of The Aga Khan Development Network, the University is committed to building an environment that fosters intellectual freedom, distinction in scholarship, pluralism, compassion and humanity's collective responsibility for a sustainable physical, social and cultural environment" (Aga Khan University).

According to this mission, humans are responsible for taking care of the earth and this is in keeping with Islamic teachings. These teachings embody the Qur'anic notion that humans are *khalīfah,* or stewards of the earth – inheriting it and providing for its perpetuation for future generations.

The centrality of environmental stewardship to the AKU, as a central responsibility and goal rather than an afterthought, dates back to its inception. Founded in 1983, the university includes sensitive considerations for a broad range of environments – human, social, physical and cultural – from its very inception and this is in harmony with the wider AKDN within which it exists. As Kassam notes, "The AKDN's goal of relieving ignorance, disease and deprivation is profoundly relevant to environmental issues" (Kassam 2003: 493).

The university was designed to blend in organically with its physical surroundings as embodied in the choice of building materials, for example, and to appropriately meet the educational and social development needs of Pakistanis in unobtrusive and intellectually sophisticated ways. One of the campuses located in Karachi is AKU's Institute for Educational Development (Institution for Educational Development). This Institute aspires to be "A Role Model for Teaching Excellence" (Aga Khan University), but more specifically, it aims not just to provide students with Master's degrees but also to ensure that graduates have significant impact on the poverty-ridden areas surrounding the university. For example, graduates of IED contribute to the teaching at Prince Aly Khan Boys School which serves "poor, destitute and orphan children in low-income areas of Karachi" (ibid.). Alumni of IED's Master of Education programme have launched a comprehensive school improvement programme. In seeking to create equality of opportunity for those less able to access quality education, this initiative exemplifies the Qur'anic notion of *'adl.*

In a country where quality education is not prevalent, the IED is engaged in educational sustainability, creating highly educated human resources who can contribute to the development of the educational standards – and therefore quality of life – of the country as a whole. Additionally, the programme's ethos embodies the idea that students should not just improve themselves through the acquisition of a Master's degree, but also view service to others as an integral part of their educational and career development.

Since its founding, the university has continually sought to improve its surrounding environment, which most recently has resulted in plans for major expansion in the form of a new campus to house its Faculty of Arts and Sciences (FAS). FAS is one of two new green AKU campuses being developed, the other being in Arusha, Tanzania at the foot of Mount Kilimanjaro. Education City in Karachi will consist of a host of institutions, including FAS, incorporating all of the ethical notions which AKU already embodies, and its plans will be environmentally sustainable in terms of its energy use, infrastructure and ongoing maintenance. For example, the buildings on the campus will be powered by various renewable forms of energy such as solar, wind and geothermal power. The proposals for the new campuses indicate the AKU's ongoing commitment to developing sustainably and, beyond simply employing the term 'sustainable development' in plans and materials, this notion is deeply embedded within the ethical framework of the university and its wider network.

Conclusions

The case study of AKU illustrates three main ways in which universities can work towards sustainability; through their values, operations and curricula. The values, in Muslim contexts, can be inspired by the Qur'an, but more broadly, universities can embody notions of engagement with their surrounding communities. Furthermore, institutions need to incorporate sustainable practices into all areas of operation, including maintenance of buildings, development of new campuses and stewardship of resources including the ethical investment of university funds. Finally, for the purpose of educating future generations about the importance of sustainability, universities should create curricula which reach across all disciplines so that every academic endeavour considers the relationship between humans and the natural environment.

Having examined the AKU, an argument can be made that employing the concept of sustainable development in a manner which focuses solely on the natural environment can be limiting. The AKU considers the importance of physical, human, social and cultural environments as part of its ongoing commitment sustainability. This broader definition should and can be used by other institutions of higher learning in Muslim countries, as they too recognise the Qur'anic ideals and values outlined in the section on environmental stewardship. However, the scope of the AKU's approach is not limited to the Muslim world; as the AKDN is non-denominational and operates in countries which are not predominantly Muslim, this broader approach to sustainability can be applied in non-Muslim contexts. The notion of sustainable development within the values of an institution becomes an ethical notion which all universities can incorporate within their own guiding principles. The specific example of how universities engage with and expand the definition of sustainable development highlights their ethical responsibilities to the societies within which they exist.

References

Aga Khan University (n.d.). *Aga Khan University*. Retrieved January 15, 2012, from http://www.aku.edu/Pages/home.aspx?saveCookie=no.

Cortese, A.D. (1999), Education for sustainability: The need for a new human perspective. *Second Nature*. Retrieved January 12, 2012, from http://www.second nature.org.

Hasnain, N. (2006), Religion-Environment Interface: Islamic Environmentalism. In Chaudhury, S.K., *Culture, ecology, and sustainable development.* Mittal, New Delhi.

His Highness the Aga Khan (2008). *Where hope takes root democracy and pluralism in an interdependent world.* Douglas & McIntyre, Vancouver, BC.

Holy Qur'an. Trans. Yusuf Ali, A. (1987). Elmhurst, NY: Tahrike Tarsile.

Institute for Educational Development (n.d.). *Aga Khan University*. Retrieved January 15, 2012, from http://www.aku.edu/collegesschoolsan.

Interfaith Center for Sustainable Development (n.d.). *The Interfaith Center for Sustainable Development*. Retrieved March 23, 2012, from http://www.interfaith sustain.com.

Kassam, T. (2003), The Aga Khan Development Network: An Ethic of Sustainable Development and Social Conscience. In Foltz, R., Denny, F. and Baharuddin, A. (Eds.) (2003), *Islam and Ecology*. Harvard University Press, Cambridge, MA.

Khalid, F.M. and O'Brien, J. (1992), *Islam and Ecology*. Cassell, New York.

Matin, I. (2010), *Green Deen what Islam teaches about protecting the planet*. Berrett-Koehler Publishers, San Francisco, CA.

Marinova, D., Hossain, A. and Hossain-Rhaman, P. (2008), Islam on Sustainability. *The Journal of Environmental, Cultural, Economic and Social Sustainability* 4 (5), pp. 123-130.

Nasr, S.H. (1976), Man and nature: the spiritual crisis of modern man. Unwin Paperbacks, London.

Rippin, A. (1990), *Approaches to the history of the interpretation of the Qur'ān*. Clarendon Press, Oxford.

Robinson, J.B. and Dale, A. (1996). *Achieving sustainable development*. UBC Press, Vancouver BC.

UN Habitat (2003), *Global Report on Human Settlement: The Challenge of Human Settlement*. Earthscan, London.

World Commission on Environment and Development (1987), *Our common future*. Oxford University Press, Oxford.

Chapter 27

Incorporating Sustainability in Curricula while Meeting Entrenched Traditional Learning Outcomes

Matthew Rockall

Abstract

Many post-secondary educators are intent on incorporating sustainability into their curricula but are impeded by the many challenges, including breaking into discipline-based curricular silos, adding to established learning outcomes, consulting with the many stakeholders, and lack of available time or resources.This paper examines the process by which sustainability is being incorporated into post-secondary communications course curricula at the British Columbia Institute of Technology (BCIT). It explores the Institute's three-part sustainability plan, including operational, curricular, and community strategies. At the curricular level the Pacific Spirit Project (PSP), a series of professional development workshops, supports development of 'sustainability' projects by bringing key stakeholders together to collaborate. At BCIT, learner curricula are focused heavily on specialized training for job readiness and other internal and external requirements. Meeting these priorities can pose significant challenges to adding new learning outcomes or content. The paper goes on to describe how these pedagogical challenges were navigated as sustainability was incorporated into an Introductory Business Communications Course for a Marketing Management Program in the BCIT School of Business. It then closes with some recommendations for committed educators in other post-secondary institutions who wish to create ESD for their own courses as well as making suggestions in support of a consensus process aimed at more widespread and lasting curricular change.

Introduction: Sustainable Curricula at BCIT

This paper examines the process by which education for sustainable development (ESD) is being incorporated into post-secondary communications course curricula at the British Columbia Institute of Technology (BCIT). This is a process that was assisted by the Pacific Spirit Project (PSP), an annual professional development workshop at BCIT designed to foster relationships and support faculty development of curricula for sustainability.

Faculty who take part in the PSP are encouraged to develop an ESD project that could be implemented into their teaching practice. The project described here is one of those. Online forums with topics built around sustainability and citizenship were set up in COMM1100, an introductory business course. The course is delivered annually to a cohort in Marketing Management, a two-year diploma program in BCIT's School of Business. The discussion forums were first implemented in fall 2009 and have run every year since.

In particular, the paper explores some of the barriers and preconditions for curriculum development in the multi-stakeholder environment at BCIT, one of the largest polytechnic post-secondary institutions in Canada. It looks at the institutional context and at the nature of course evolution in the BCIT Communications Department, which delivers service courses to other departments and programs. Next, the actual sustainability discussion forums are examined. These discussion forums and topics result in learner activities that clearly intersect with existing program learning outcomes even though sustainability itself is not a component of those outcomes. Some opportunities for the further integration of ESD at BCIT are also addressed.

The Institutional Context

With two years to go, the United Nations' Decade of Education for Sustainable Development is nearly finished. With the mandate to "integrate the principles, values, and practices of sustainable development into all aspects of education and learning" (UNESCO 2005), this initiative has built on the objectives set out during the Rio Conference on sustainable development. Chapter 36 of Agenda 21 (UN 1992) calls for the integration of environmental and developmental concepts into all educational programs. Specifically, the objectives of the DESD, as outlined in UNESCO's International Implementation Scheme for the Decade, are to

- facilitate network linkages, exchange and interaction among stakeholders in ESD;
- foster increased quality of teaching and learning in ESD;
- help countries make progress towards and attain the Millenium Development Goals through ESD efforts;
- provide countries with new opportunities to incorporate ESD into education reform efforts (UNESCO ¶3).

Though progress has been made, this is still the looming challenge for post-secondary education as the pace of anthropogenic environmental decay accelerates (Turner 2008; Capra 1996; 1999). Orr has argued that, on the whole, post-secondary courses and programs develop and reinforce unsustainable paths of progress. In 1992, when describing the economic decline of Youngstown, Ohio, he referred to the perils of business schools and economics departments where learning "did not include the value of good communities or the human costs of a narrow destructive economic rationality that valued efficiency and economic abstractions above people and community" (Orr 1991: 57). In a later article he warned of corporate training supplanting colleges and universities as "young people discover alternatives that are far cheaper, faster and better adapted to economic realities. The rub is that corporations will not educate liberally. Instead, they will offer something much more akin to high-tech job training" (Orr 1992: 10).

On the BCIT website (BCIT 2012), the logo taglines of "BCIT It's your career. Get it right." and "BCIT Works" echo this shift signaled by Orr two decades ago. BCIT's philosophy has been to offer 'job readiness' to both students and employers; practical training is the primary consideration in curriculum and program development and a commitment to environmental health has only recently been included in formal planning.

This is consistent with many other post-secondary institutions around the world, where proponents of ESD continue to face barriers to meaningful and lasting curriculum change (Pearson et al. 2005; Payne 2006; Leal Filho and Wright 2002). The holistic, multi-disciplinary nature of ESD is part of the reason, as it challenges established discipline-based curricular silos. This lack of engagement is particularly true in business schools, where there has been much debate about the nature of ES (Leal Filho 1999; 2010), in some cases little progress in acknowledging the relevance of the concepts (Springett and Kearns 2001), and a conflict with profitability goals (Wong et al. 2010).

Despite much discussion and some progress as to the nature of sustainable development, the degree of implementation of ESD in higher education has generally been underwhelming, and Leal Filho advocates for applied sustainability, "an action-oriented and project-based approach, which uses principles of sustainable development and applies them to real contexts and real situations" (Leal Filho 2011: 15). It is useful to examine some of the good work being done at BCIT on this front, the areas of potential improvement, and one small strategy for inclusion of applied ESD as an add-on to an existing program. Ann Evans (2005) cautions that "implementing lasting change in schools is like turning around an ocean liner" (p. 251) adding that ongoing vigilance is required to maintain the new path of progress. Such a metaphor is well suited to BCIT, with its size and complex diversity of programs, but small course changes accumulate.

So, how do we shift the direction of such a large and complex social system? The Canadian National Round Table on the Environment and the Economy (NRTEE) offers an insight into support of complex consensus processes (Cormick et al. 1996), observing that achieving sustainability "is about dealing with people and their diverse cultures, interests, visions, priorities and needs" (p. 3). They identify several principles for building consensus, among them that such processes must include key stakeholders. In a large multi-stakeholder social system such as a post-secondary institution, these key stakeholders are perhaps the "areas of responsibility" that Donella Meadows suggested must be located when encountering systems and looking for leverage points (2001: 199).

Faculty intent on developing and incorporating ESD will face challenges and find opportunities that vary over time and for each school course, program, and department. However, by engaging the cooperation of key stakeholders one can access opportunities to incorporate sustainability into individual courses. The following explores the challenges and opportunities specific to the structure of BCIT and the role of the Communications Department within that structure, as well as describing some of the larger issues in play at BCIT in 2009 when the discussion forums project described here was originally implemented. It is hoped that this may inform the work of educators with similar goals in similar circumstances.

Teaching communications at BCIT

BCIT is the largest polytechnic college in the province of British Columbia, Canada, with 17,422 full-time learners and 45,781 part-time learners enrolled in 2009/2010 (BCIT, 2011). BCIT delivers 345 programs in business, computing, health sciences, applied science, trades, and engineering, including such diverse examples as carpentry, marketing management, food science, radio and television arts, and many more.

Naturally, graduates from these programs have widely varying communications training needs. At BCIT, the Communications Department is responsible for developing and delivering that training. As of May 2010, according to an internal departmental estimate, 24 full-time faculty members taught 66 courses to 3970 day-school learners in 35 different programs, or 'technologies' as they are commonly referred to at BCIT. Typically, day-school learner cohorts are divided into 'sets', groups of 30 or fewer learners who have 'labs' together to complete practice activities to build skills and apply theory covered in lectures.

This complex set of offerings presents challenges to communications faculty striving for course consistency and relevance. Adding to the challenge, some courses, such as the common *Comm1100 – Introduction to Business Communications* discussed here, are taught across multiple programs with similar communications training needs but very different learner cohorts. Also, in some programs there are too many learner sets for one instructor to teach all the labs. In these cases, team teaching is required, with a single 'course leader' delivering lectures and developing content for several instructors to use in the labs. All these elements complicate implementation of new content.

In all cases, it is the responsibility of faculty members to negotiate with all the various stakeholders in this equation; to liaise with program heads in the technology they are teaching into; to produce courses that enable learners to meet the established learning outcomes; to produce lab plans that they or their fellow lab instructors can effectively deliver; and, of course, to meet the needs and expectations of different learners within and between cohorts.

Adding content or changing learning outcomes

At BCIT, course or program revision and renewal is generally instructor driven but it can be triggered by other drivers, such as advisory committees of industry representatives from the technology areas, program heads seeking changes, student course and instructor evaluations that provide new insights, or through professional development initiatives or workshops designed to inspire. Such course reviews can result in additions or adjustments to course learning outcomes, but this can bring up thorny questions. If two or more instructors are teaching the same course in different programs, with the same course code and the same established learning outcomes, do the outcomes have to evolve in lockstep? If so, how many stakeholders must be involved in the process? How much variance from the learning outcomes list is allowable? If such variation is significant, what does that do for pre-requisites, external accreditation, and/or facilitating students being able to transfer into other programs?

More commonly, instructors engage in course evolution and review that improves or adds existing course content, such as new activities, more current readings, etc., without impacting established learning outcomes. Such course content change must still be crafted with sensitivity to the stakeholders, but these are less numerous. Arguably, most important among these stakeholders are the technology program heads with overall responsibility for the programs for which the courses are delivered. It is these people with whom instructors must maintain good working relationships. In fostering such relationships, professional development events such as the PSP offer particular value.

Budgetary restraint and a five-year plan

2008 was a year of general economic upheaval and BCIT was not spared the effects, incurring a significant budget deficit. This financial state was faced by a relatively new management team with a new college President and a new Vice-President of Education, among others. At the same time, these administrators were developing a five-year strategic plan aimed at crafting a vision and direction for the Institute. Both the budget shortfall and the five-year plan had ramifications for any instructors intent on incorporating ESD.

In response to the budget issues, the Institute made a number of difficult decisions to rein in costs, including lay-offs for some long-serving staff members and freezing funding for program development. During a Communications Department end-of-term workshop in June, 2009, when the VP Education was asked about support for incorporating sustainability into curricula, he indicated that management was in favour of such efforts but that there was no funding specifically earmarked for incorporating sustainability into curricula at the time. Furthermore, he felt that rather than a top-down program, such developments would be best served by a grassroots, 'bottom-up' model, where committed instructors would bring these issues into their courses and programs (P. Dangerfield, personal communication, June 3, 2009). This verbal assurance of support in principle was supported in writing in the Institute's five-year strategic plan for 2009-2014.

BCIT's vision statement, "Integral to the economic, social, and environmental prosperity of British Columbia" (BCIT 2009: 6) leads the strategic plan. With the explicit mention of "environmental prosperity" in this statement, a foundational policy brick is placed for incorporating sustainable development into curricula at the Institute. Further reading reveals the following specifics. From the description of the planning context in the introduction, under Environmental Sustainability:

> "The new BC Climate Action Program and other global initiatives focused on creating a sustainable environment will require us to incorporate the principles of environmental sustainability in our programming whenever appropriate. Graduates will be expected to understand and apply sound environment principles in their chosen careers. The Institute will be expected to model these principles in an exemplary fashion." (2009: 14)

Strategic Initiative 1, Objective 4 (BCIT 2009: 18), states that "[BCIT will] Produce graduates who excel in their roles as practitioners, leaders, and citizens; who act in a professional, social, environmental and ethical manner." Taken together, these elements provide the policy foundation for faculty intent on incorporating sustainability in curricula, but the flexible caveat "whenever appropriate" should be noted. It should also be recognized that there are six different "Schools" at BCIT and that they have varying degrees of commitment to sustainability, with the School of Construction and Environment (SoCE) leading the charge: establishing a framework for sustainability; initiating the PSP, described below in detail; and designating a full-time staff-member mandated to develop strategies to promote sustainability. With significant successes in the SoCE, Dr Jennie Moore was designated BCIT's Director of Sustainable Development and Environmental Stewardship.

Dr Moore and others have championed the development of a three-pronged approach to sustainability at BCIT: curriculum development in the six Schools; operational and infrastructural initiatives to reduce the Institute's ecological footprint; and a community-building sustainability committee with sub-committees in each school to channel ideas and develop local strategies and initiatives. These three levels are illustrated in Fig. 1, below.

At the curricular level, this work consists of new courses and programs built around sustainability or integration of ESD into existing courses, as with the discussions project described below. As other researchers have found, adapting these discipline-based curricular silos is a particular challenge requiring multi-stakeholder consultation, finding time in packed class schedules, and additional workload for instructors. More lasting and widespread adoption of these grassroots activities would benefit from additional support of top-down policy specifically aimed at integration of ESD into learning outcomes over a specific timeframe to build on efforts like the PSP. The consensus process developed by NRTEE (Cormick et al. 1996) may offer insights that can support the success of such policy. In the "islands of excellence" model described by for Canadian secondary schools in Manitoba, "committed teachers initiate isolated offerings and extracurricular activities to bring environmental education to their schools" (Metz et al. 2010: 5). This is similar to what is happening in many Schools at BCIT. Though valuable as a source of ESD strategies, it is not the wholesale change advocated by many (Capra 1999; Orr 1992; Louv 2005; Thomashow 1995; Stone and Barlow 2005).

Despite the challenges, the three-pronged approach described above offers great potential for the Institute over the coming years, particularly at committee level, as committees act as a communications conduit for sharing ideas for applied ESD course projects and strategies between the "islands of excellence", thus building capacity for adaptation to top-down policy decisions.

Fig. 1: Three levels of sustainability integration at BCIT

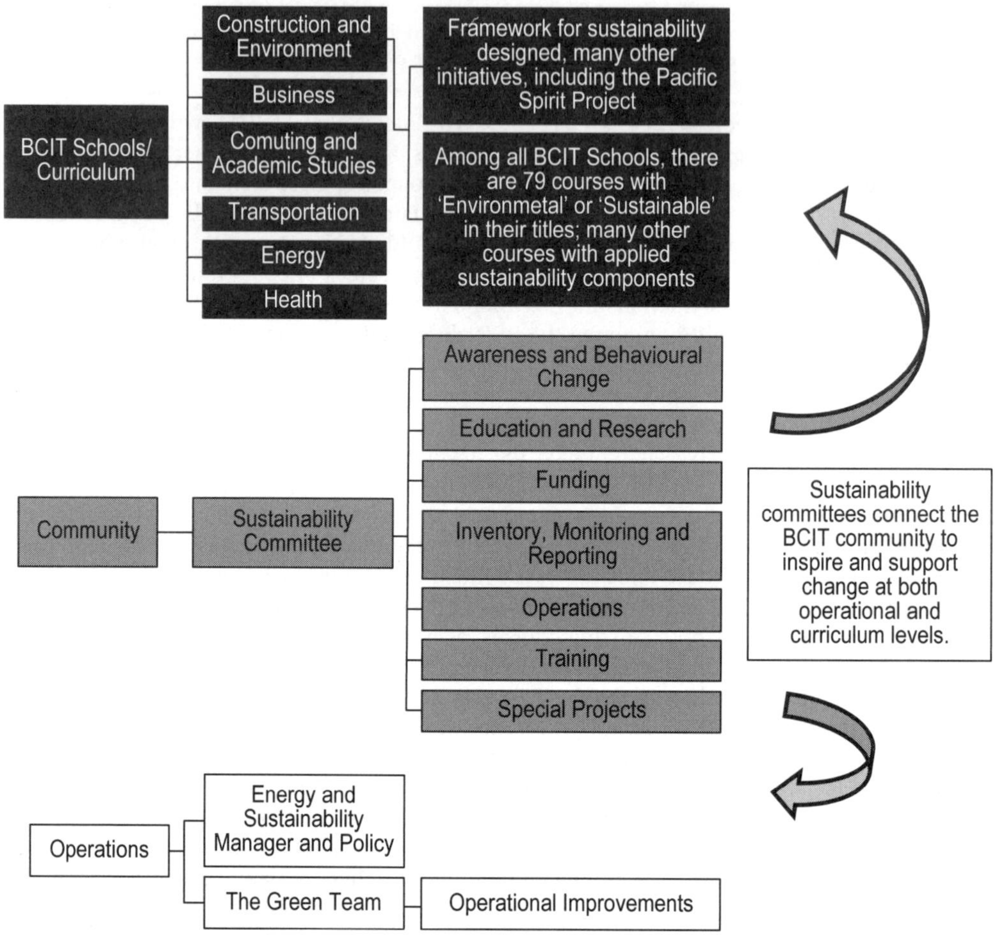

Pacific Spirit Project

The PSP was modeled on North Arizona University's Ponderosa Project (1998; Chase and Rowland 2004) and introduced at BCIT to bring together and support faculty interested in incorporating sustainability into their curricula. In this annual event, participants begin with an informal dinner to get to know each other before a two-day workshop that facilitates development of sustainability initiatives. In June 2009, key faculty participants included this author and the program head responsible for the Marketing Management Program. Over two days, workshop participants developed their own sustainability project ideas in a collaborative atmosphere where small teams worked to define underlying goals, share ideas, and make suggestions. Engaging in early project brainstorming with key stakeholders was particularly important. As a result of this early collaboration, they greeted introduction of online discussion forums into the COMM1100 course with enthusiasm, understanding and a certain shared pride.

Typically, these are applied projects, with an emphasis on job training. For example, for one project sustainable lumber was sourced and bought for the carpentry shop. In addition to building awareness of the materials and the desirability of their use, this allowed training around the sources and costs. The discussion forums described below also have an applied focus aimed at building capacity and understanding around green marketing.

PSP Project for Marketing Management

The forums were developed using BCIT's Desire to Learn (D2L), an e-learning tool that offers an online space for students to engage in asynchronous text-based discussion forums. Instructors pose questions for discussion by student groups, as determined by set-up options that include discussion group (DG) size and duration of visibility or access. It was assumed in the planning of this project, due to the heavy workload of the learners, that smaller discussion groups of four to six students would be enough participants per group to allow a lively discussion without overburdening learners. They would be able to read the comments more easily than a larger group with more contributions.

Four discussion questions are posted every term, one every three weeks, with each of them staying open for participant comments until the next topic becomes visible. All the discussions remain visible for the duration of the term. 10% of the total course mark is allocated to evaluating learner participation in these discussions. For top marks, learners:

- participate actively and regularly and post to the DGs as appropriate or required;
- critically think about the topic at hand;
- draw from the assigned readings;
- reflect on the postings of other learners and possibly even integrate ideas from more than one learner;
- suggest appropriate websites or other resources to move the discussion to a further level;
- do not place a burden on the rest of the group by including unnecessary information
- keep individual postings brief;
- encourage others in group to participate;
- respect other points of view;
- raise interesting and appropriate questions as conversation starters or activators.

Having developed the structure for these discussions on D2L, the next and most critical step was developing discussion topics to build learner understanding and awareness around sustainability and citizenship.

What is sustainable content at BCIT?

The Institute's strategic plan calls for graduates to be "… practitioners, leaders and citizens; who act in a professional, social, environmental and ethical manner." The sustainability framework set out by the BCIT School of Construction and Environment also offers guidance for instructors in other BCIT Schools. This framework consists of six themes that "reflect the interrelatedness of ecological, social, and economic interests needed to meet the challenge of sustainability" (BCIT 2012, ¶4). These themes are to "protect and strengthen assets, balance use and renewal of resources, account for costs and benefits, reduce waste and eliminate toxics, ensure safety and access to services and support opportunities for improvement and enjoyment" (BCIT 2012, ¶5).

Discussions in Marketing Management

Students in the Marketing Management program are in their first term of a BCIT School of Business diploma, a two-year (four-terms) program that can ladder into further education, such as BCIT's Bachelor of Business Administration (BBA), or lead straight to the job market. In their second year, these students specialize in one of five options, Marketing Communications, Real Estate, Professional Sales, Tourism, or Entrepreneurship. These are future sales people, realtors, event planners, hoteliers and entrepreneurs. There is great variation within the groups, but they are typically a younger cohort than some at BCIT, with many learners coming straight from their secondary schools with little or no work experience.

For this project, existing learning outcomes could be connected with the nature of the task for learners. Specifically, in the introductory business communications class for marketing, it was argued that participation in these discussion forums supported achievement of the following course outcomes:

For COMM1100 for Marketing:

- organize information effectively in routine written business messages;
- use an appropriate and effective style and tone in written business messages;
- use language, grammar and punctuation correctly, simply and effectively in a business-like manner, as required for a business workplace audience.

Beyond satisfying course-specific learning outcomes through the nature of the activity, the Institute's strategic objective of graduates who "excel in their roles as practitioners, leaders, and citizens" is supported through fostering greater understanding of sustainable development, consumerism, corporate social responsibility and other discussion topics that relate to their professional interests. These topics also needed to be relevant to the learner cohorts in the context of their programs.

The topics are as follows:

1. Etiquette for voicemail and texting in business
2. Academic honesty and intellectual property
3. Marketing and consumerism
4. 'Green-washing'.

All these topics relate to aspects of business communication for marketing but Topic 3 is where the discussions specifically concern environmental issues. Learners are asked to watch a controversial online video, The Story of Stuff (Leonard 2007). This informal animated process description makes some provocative claims about the extraction, manufacturing, distribution and disposal of consumer goods, describing a system in which high consumption levels are exceeding global carrying capacity. The topic challenges learners to critically comment on the video, particularly to respond to the contention that marketers, as the cause of high consumption levels, are responsible for environmental decay. This contention is then explored in the lecture that follows the closing of the online discussion topic.

Table 1: Online Discussion Forum Excerpt 1

Topic 3	**Student A**	**Student B response**
What do you say to those who blame marketers for unsustainable consumption levels and resultant environmental decay?	*"Marketing definitely plays a huge role in keeping consumerism alive. Marketing is a very powerful tool that can convince people that they need this and they should want that. I think we can't totally blame marketing for all of this though. At the end of the day its still up to the people to decide what to do with their lives and what decisions they should make. I think as consumers we should really think about our buying decisions and the impact in can make not only to the environment but to other people as well. As marketers we should be responsible enough to communicate only real messages and to promote products that we believe can be better for society."*	*"But Marketing is the driving force that has created this consumer society. Marketers are the people working for the big corporation to make people believe they need to spend their money. We live in a society where more is never enough. Maybe we as future marketers can change the way people consume."*

Students A and B (2009)

In the end, as an industry that specializes in influencing consumer behaviour, marketers have the potential to be valuable leverage points for change, shifting people to more sustainable buying decisions. The goal is to empower the learners with the sense that their actions as marketers *will* have an impact on consumerism, but that it can be positive. Topic 4 builds on this with an exploration of the topic of 'green-washing,' that is, the false marketing of something as eco-friendly. Learners read an article identifying seven 'sins' of green-washing and are asked to discuss it, identifying the 'worst' sin.

Engagement in the discussion forums has shown that some of the students have very little awareness of environmental issues, organic products, fair trade goods, or corporate social responsibility. Other students have much greater awareness. They may have traveled or worked extensively, or already have a post-secondary credential. It is this contrast of learners that makes the exchange of ideas on the discussion forums so interesting and valuable as a learning tool. Consider the following excerpts from the 2010 discussions of Topics 3 and 4.

These are complex, ethical, and empowered conclusions that build on each other, rich material to bring back to the cohort as a whole particularly as it comes from them in the first place. Such mentoring is illustrated even more clearly in Table 2, below, with an exchange generated by the fourth topic. For future marketers, there is great relevance in examining and discussing the ethics of advertising green products.

Table 2: Online Discussion Forum Excerpt 2

Topic 4	Student C	Student D	Student E
What is the 'worst' sin of green-washing?	*"I find it alarming that many products on the market that are known to release toxins. This is prevalent in cleaning supplies, cosmetics, packaging, and even toys. Companies are saving money by using cheaper materials to create products, rather than more expensive, all-natural materials. As a result, our exposure to toxic chemicals has increased; leading to more cases of cancer, asthma, allergies, autism, and other chronic diseases."*	*"I agree, it is very alarming. I read an article before on how most baby bottles have this chemical called BPA thats found in the plastic that they use to make them. When they would heat the bottles to sterilize them, the chemical would leak out and mix with the milk. Its horrible because babies who are most vulnerable to sickness are exposed to all these harmful chemicals from such a young age. Marketers should really be responsible about how they promote products instead of just slapping on a label that reads "organic" and "eco friendly" to just about anything they can think of these days when really they aren't."*	*"Canada recently declared BPA (Bisphenol A) toxic. BPA is found in plastics, aluminum cans, receipts, and in 91% of Canadians. I've also been reading articles on this non-profit organization, Healthy Child Healthy World. They are dedicated to protecting children from harmful chemicals. Check out their website! http://healthychild.org"*

Students A, B, and C (2009)

In these three excerpts we see an idea build from curiosity and some awareness in the first entry to a much more nuanced understanding by the third entry. Adding to that, Student E provided an Internet link for more information. Research has not been done on these student groups to gauge the impact of these activities, but the Marketing Management program accepts 256 learners every year. They engage on these topics in 50 small groups of four, five or six, and many of the exchanges show similar development of ideas. These excerpts show learners sharing different insights based in different life experiences as they grapple with issues of great complexity and gravity; together they are building a deeper understanding.

Conclusions

There is a significant role for post-secondary education in working toward sustainable development, but implementing and maintaining change in such large and complex social systems presents great challenges. This is proving particularly true when overcoming the silos of discipline-based curricula to incorporate ESD with its essentially multi-disciplinary nature. Further, with so many stakeholders involved in the evolution of these disciplines, consensus processes for incorporating sustainability or making any other changes tend to be prolonged and complex.

The BCIT Communications Department, serving other departments as it does, including the BCIT School of Business, illustrates this complexity. In addition to considering internal departmental priorities, curriculum change processes have to consider the priorities of other departments and programs, students, and external accreditation bodies that may differ from program to program. The lack of specific funding and/or policy to support such change processes can present additional barriers, as can heavy instructor workloads and student schedules that are already tightly packed with curricular priorities. Although there have been important and laudable high-level policies related to sustainability developed for the Institute for the current planning period, 2009-2014, incorporation of ESD into all courses and programs has not been required by policy. Rather, it is encouraged by management for program reviews. However, it is generally only developed at a grassroots level by instructors who consider it important and relevant enough to dedicate some of their limited curriculum development time to the task. Further, such content is usually designed to fit into entrenched outcomes.

Despite these challenges, BCIT's three-part strategy for sustainability of curriculum, operations and community is very important. At the curricular level, committed instructors are doing good work at BCIT and they are getting some support from the PSP workshops. These workshops, with their focus on applied ESD, have had successes: they bring stakeholders together, support inter- and multi-disciplinary collaborations, and foment creativity. As operational initiatives are implemented, they raise the profile of sustainability and provide opportunities for student engagement and awareness building. Finally, at community level, the sustainability committees provide a critical conduit for communication.

Recommendations for educators in other post-secondary institutions

The discussion forums project described here has been valuable for learners and rewarding to implement. However, if we are to educate our way to sustainable paths of progress that meet the needs of today without further jeopardizing the ability of future generations to meet their needs and aspirations, a more wholesale revamping of our post-secondary institutions may be warranted (Capra 2002; Orr 1994). Progress made by committed instructors working off the side of their desk to bring ESD into their courses as incidental content can be lost when they leave, change courses, or face negative reviews from other stakeholders. For more widespread and lasting curricular change, the BCIT executive may need to implement course and program review strategies and processes to require that ESD be included as a learning outcome element for all learners.

Considering the complexity of the stakeholder systems involved in outcome evolution, such an objective would require a consensus process that considers and involves internal and external stakeholders, from students and faculty members to employers and accreditation bodies. The sustainability committees and sub-committees may offer important channels for conversations between stakeholder groups. Further, widespread ESD incorporation would be supported by a continued commitment to applied sustainability. The practical benefits of applied ESD described by Leal Filho (2011) and encouraged by the PSP closely align with the "BCIT Works" philosophy of the Institute and should appeal to the pragmatism of its various stakeholders.

In the meantime, work should continue on developing smaller-scale, shorter-term strategies for incorporation of ESD, such as the project described herein, both for their immediate impact on learners and for the curricular content generated. At management level, resources could be allocated to support initiatives like the PSP, which affords the time and space to develop applied projects, as well as building relationships between like-minded faculty members and program stakeholders. At the individual faculty level, the project described here shows that those committed to ESD can overcome obstacles to development of sustainability content. However, such success may be supported by

- connecting content development with high-level policy where possible;
- bringing in key faculty stakeholders as early as possible, preferably when brainstorming, and communicate regularly with them on an ongoing basis;
- working within the constraints of existing learning outcomes; and
- embracing an applied focus to support relevance of the content for the students and their programs.

Challenges to incorporating ESD curricula into post-secondary curricula abound. The human element can be particularly challenging, as it requires dealing with complex systems of stakeholders and stakeholder groups with varying motivations, priorities, and requirements. However, following the recommendations above can help garner and maintain ongoing support for implementing ESD in such a multi-stakeholder working environment.

References

BCIT (2009), British Columbia Institute of Technology, *Five-year Strategic Plan 2009-2014*. BCIT, Burnaby.

– (2011), *Fact Sheet*. Retrieved 01 20, 2012, from BCIT: http://www.bcit.ca/about/facts. shtml.

– (2012, January 20), *BCIT Home Page*. Retrieved 01 20, 2012, from BCIT: http://www. bcit.ca.

– (2012, 01 22), *Sustainability Framework*. Retrieved 01 22, 2012, from BCIT.ca: http://www.bcit.ca/construction/sustainability.

Capra, F. (1996), *The Web of Life*. Anchor Books, New York.

– (1999, March 20), Ecoliteracy: The Challenge for Education in the Next Century. *Liverpool Schumacher Lectures*. Center for Ecoliteracy, Berkeley, California, USA.

– (2002), *The Hidden Connections, Intergrating the Biological, Cognitive, and Social Dimensions of Life Into a Science of Sustainability*. Doubleday, New York.

Chase, W. and Rowland, P. (2004), The Ponderosa project: Infusing sustainability into the curriculum. In Bartlett, P. and Chase, G., *Sustainability on Campus, Stories and Strategies for Change,* pp. 91-105. Massachusetts Institute of Technology Press, Cambridge.

Cormick, G., Dale, N., Emond, P., Glenn, S.S. and Stuart, B. (1996), *Building consensus for a sustainable future: putting principles into practice*. National Round Table on the Environment and the Economy, Ottawa.

Evans, A. (2005), Changing Schools: A Systems View. In Stone, M. and Barlow, Z., *Ecological Literacy: Educating our children for a sustainable world,* pp. 250-257. Sierra Club Books, San Francisco.

Leal Filho, W. (Ed.) (1999), *Sustainability and University Life*. Peter Lang, Frankfurt.

– (2010) (Ed.), *Sustainability at Universities, Opportunities, Challenges, and Trends*. Peter Lang Scientific Publishers, Frankfurt.

– (2011), Applied Sustainable Development. In Leal Filho, W. (Ed.), *World Trends in Education for Sustainable Development,* pp. 11-29. Peter Lang, Frankfurt.

Leal Filho, W. and Wright, T. (2002), Barriers on the path to sustainability: European and Canadian Perspectives in higher education. *International Journal of Sustainable Development and World Ecology 9* (2), p. 179.

Leonard, A. (2007), *Story of Stuff Project*. Retrieved 01 22, 2012, from Story of Stuff: http://www. storyofstuff.org.

Louv, R. (2005), *Last Child in the Woods*. Workman Publishing, New York.

Meadows, D. (2001), *Dancing With Systems*. Retrieved December 12, 2008, from Sustainability Institute: http://www.sustainer.org/pubs/Dancing.html.

Metz, D.M., McMillan, B., Maxwell, M. and Tetrault, A. (2010, February), Securing the place of educating for sustainable development within existing curriculum frameworks: A reflective analysis. *Canadian Journal of Environmental Education* 15, pp. 150-159.

Northern Arizona University (1998, 09 18), *The Ponderosa Project at Northern Arizona University*. Retrieved 1 22, 2012, from NAU.edu: http://www2.nau.edu/~ponder-p.

Orr, D. (1991, Winter), What is education for? *The Learning Revolution* (27), p. 52.

– (1992, October), Environmental Literacy: Education as if the Earth Mattered. *Twelfth Annual E.F. Shumacher Lectures*. E.F. Schumacher Society, Great Barrington, Massachusetts, USA.

– (1994), *Earth in Mind.* Island Press, Washington.

Payne, P.G. (2006), Environmental education and curriculum theory. *Journal of Environmental Education* 37 (2), pp. 25-35.

Pearson, S., Honeywood, S. and O'Toole, M. (2005), Not Yet Learning for Sustainablility: The Challenge for Environmental Education in a University. *International Research in Geographical and Environmental Education* 14 (3), pp. 173-186.

Springett, D. and Kearins, K. (2001), Gaining Legitimacy? Sustainable Development in Business Curricula. *Sustainable Development* 9 (4), pp. 213-221.

Stone, M. and Barlow, Z. (2005), *Ecological Literacy, Educating Our Children for a Sustainable World*. Sierra Club Books, San Francisco.

Students A and B (2009, Oct 26), *Discussion Topic 3 – Marketing and consumerism, Group 1 [Msg 2 and 3]*. Retrieved from Messages posted to https://learn.bcit.ca/ d2l/lms/discussions/messageLists.

Students C, D, and E. (2009, Nov 9), *Discussion Topic 4 – "Greenwashing", Group 2 [Msg 1, 2, 3]*. Retrieved from Message posted to https://learn.bcit.ca/d2l/lms/discussions/ messageLists.

Thomashow, M. (1995), *Ecological Identity, Beocming a Reflective Environmentalist.* The MIT Press, Cambridge, Massachusetts.

Turner, G. (2008, June), *A Comparison of the Limits to Growth with Thirty Years of Reality.* Retrieved December 12, 2008, from The Commonwealth Scientific and Industrial Research Organisation: http://www.csiro.au/files/files/plje.pdf.

UN (1992), The UN Conference on Environment and Development: a Guide to Agenda 21. UN Publications Service, Geneva.

UNESCO (2005), *United Nations Decade of Education for Sustainable Development (2005-2014): International Implementation Scheme*. Section for Education for Sustainable Development. UNESCO, Paris.

– (n.d.) *ESD: Frequently Asked Questions*. Retrieved February 12, 2012, from UNESCO Bangkok: http://www.unescobkk.org/education/esd-unit/faq/#irfaq_2_ d9c97.

Wong, A., Fu, L. and Elankumaran, S.S. (2010), Business students' perception of corporate social responsibility: the United States, China, and India. *Corporate Social Responsibility and Environmental Management* 17 (5), pp. 299-310.

Chapter 28

Teaching Sustainable Management to US Undergraduates: Learning to be Change Leaders

Diana C. Cooper

Abstract

Teaching about sustainability in undergraduate management programs is under-researched, but presents an opportunity to educate aspiring business-people to lead the charge of sustainable development in their future careers. Prompted by undergraduate exit surveys that found a lack of connection between sustainability, management, and corporate responsibility at the University of Minnesota, the author/lecturer offered a constructivist approach to teaching sustainable development in the standard MGMT1001 curriculum. The author leveraged the student "book club" to provoke critical thinking among students outside of class, inspiring them to examine their role in leading future organizations sustainably. The purpose of this qualitative study is to interpret the writing assignments of first-year undergraduate management majors at the University of Minnesota to inform instructors about student perceptions of sustainability in management and analyze student learning from the concepts presented in a required introductory management course. The findings suggest that a constructivist approach to undergraduate learning motivates students to develop knowledge, skills and abilities to be change agents, and to form values-driven goals in leading sustainable development in future organizations.

Introduction

> A paradigm shift is underway – one that necessitates the transformation of the conduct of commercial enterprise and the content of business curricula. The new paradigm incorporates a sustainability mandate, refuting clearly the old thinking of limitless resources, unbounded growth, and technologically derived solutions.
>
> (Marshall et al. 2010)

> I personally believe that for all the talk of metrics and guidelines and benchmarks and indices, that the heart as opposed to the head may well be the best place to find the real drivers of change. Inspiring people to look inside themselves – for the passions and the ideas and for the mission and the beliefs that we can all make a difference – is really what matters most.
>
> (Hollender 2004)

There is a sustainability crisis in management education.[1] The rate at which organizations consume natural resources and damage the environment is unsustainable (Marcus and Clark 2010; Rands 2009; Stead and Stead 2010). This situation calls for a change in the way managers and leaders make decisions (Hind et al. 2009; Mitroff 2004; Navarro 2008). The transformation needs to start with teaching undergraduate business students about sustainability, inspiring them to break away from conventional business models to find new approaches for managing organizations sustainably in the long run (Kearins and Springett 2003; Shrivastava 2010). The

1 The definition of sustainability in the context of this study is found in the Brundtland Report (1987): the systematic, long-term use of natural resources so that these are available for future generations; including ways by which environmental conservation can be systematically pursued, internationally, paying due attention to economic, social and political considerations.

purpose of this qualitative study is to examine how US undergraduates learned about sustainability in their introductory management course at the University of Minnesota.

The scope and magnitude of the environmental degradation suffered over the past century is well documented, from climate change (IPCC 2007) to decline of biodiversity and deforestation (UNFAO 2010), trauma to marine environment, such as overfishing, (Harvey 2011), and air and water pollution (Krugman 2011). The current environmental debacle "largely caused or facilitated by public and private sector organizations – creates a special need for businesses and business schools to address environmental concepts and issues … the only actions that are acceptable for organizations are [to be] simultaneously economically profitable (or at least break-even), environmentally sound, and socially just" (Rands 2009). Daly argued (1977) that it is unsustainable for organizations to persist in operating in competition for a shrinking pool of resources, advocating profit at all costs, without continual decline in quality of life and negative environmental outcomes. Thus, it is crucial for business leaders and management educators to reform current business models and inspire students to pursue innovative, sustainable, culturally appropriate (McKeown 2002; Wagner 2009) solutions in managing organizations in concert with societal needs and environmental conservation.

Although the concerns surrounding sustainability have been mounting over the past 40 years (Daly 1977), and management scholars have been prolific on the subject (e.g. Hollender 2004; Kearins and Springett 2003; Leal Filho 2000; Marcus and Clark 2010; Porter and Kramer 2006; Rands 1993a; 1993b; Shrivastava 2010) actual delivery of sustainability in management education programs has emerged slowly and is still evolving (Marcus and Clark 2010; Stead and Stead 2010). Changes in the curriculum are largely ad hoc, relying on the passion and initiative of individual faculty members to introduce sustainability into the curricula of business schools, often course by course (Down 2006; Naeem and Peach 2010; Stubbs and Schapper 2011; Tilbury et al. 2004). Individual scholars experience lack of institutional support and feelings of "isolation and resistance to change at their home campuses" in the process of advancing the sustainability business education agenda (Stead and Stead 2010). Most faculty are focused on their silos of research and not well versed in sustainability, while administrators are hesitant to invest the time to overhaul the curriculum by integrating sustainability cases and content across silos (MacVaugh and Norton 2011). The Aspen Institute's "Beyond Grey Pinstripes" Business and Society program reports that only 150 business schools worldwide devote attention to sustainability in their curricula; the "type and amount of this attention varies significantly from one school to another" (Marcus and Clark 2010).

Meanwhile, in organizations, according to Porter and Kramer (2006), managers are navigating uncharted territory and need to develop sustainable solutions specific to each organizational context in a complex world. A new business model needs to be developed to promote a "shared value" of organizations and society and "find ways not only to foster economic and social development but to change the way companies and society think about each other. NGOs, governments, and companies must stop thinking in terms of 'corporate social responsibility' and start thinking in terms of 'corporate social integration'" (Porter and Kramer 2006: 92). In the same vein, Bradbury (2003) proposes business leaders understand the organizational embeddedness in ecologically and socially complex systems. Seeing the "big picture" will steer managers away from the tension of "choosing sides" between profit and sustainability, adopting innovative strategies that gain a competitive advantage and at the same time benefit society and/or the environment (Elkington 1997).

The Principles for Responsible Management Education (PRIME) were established in 2007 to offer a framework and call for management schools to embed corporate responsibility and sustainability into the mainstream of business-related education (Naeem and Peach 2010; Rasche and Escudero 2009). According to a study of European and Latin American institutions, "the orientation towards sustainability … is considered to be, at many universities, an important element of quality, an indicator of their degree of commitment to providing their graduates with advanced and up-to-date training" (Leal Filho and Santos 2005 in Geli de Ciurana and Leal Filho 2006: 82).

Rands (2009) suggests developing young business leaders with the knowledge, skills and attitudes to implement sustainability into mainstream business operations by integrating knowledge on SD throughout the business and management curriculum, with opportunities for students to apply their learning to organizations and develop new skills and attitudes that are transformative for themselves, their future employers and stakeholders. Students need to go one step further than systems thinking to expand their understanding of the complexities of business strategy integral to healthy societies, learning the interdependencies and "interconnections between business disciplines and [with] other disciplines" (Marshall et al. 2010: 479). Kearins and Springett (2003) suggest the development of "reflexivity" and self-awareness. That is, future business leaders need the ability to step back and identify potential outcomes of personal and organizational actions in order to identify more environmentally sustainable scenarios.

In teaching sustainable business, the "business case" for sustainability is only a starting point for students to learn the shared value of sustainable strategies for the environment, society and organizations. Sterling (2001) advocates "a change of educational culture which both develops and embodies the theory and practice of sustainability in a way which is critically aware" to the point of a paradigm shift in the way we think about business. Marshall et al. (2010) argue: "To teach our students…in the pursuit of a more sustainability-inspired future we, as educators, will need to accept the anomalies in the old system and find a new set of principles – one that will serve as guideposts for a journey toward a newly inspired sustainability-based business education" (484). Shrivastava (2010) encourages students to be passionate about pursuing sustainable management, which means *faculty need to be passionate* in inspiring students to learn about sustainability, too. Finally, education on sustainability should be locally relevant and culturally appropriate, such that each country can reach its own interpretation of curricula that aim to transform leaders and organizations (McKeown 2002; Wagner 2009).

With the new paradigm for business and management education in mind, the current study explores student learning after the curricular revision of an introductory management course for undergraduate business majors. Much of the research on education on sustainable development (ESD) has focused on graduate programs (Benn and Dunphy 2009; Collins and Kearins 2010; Stubbs and Schapper 2011). This article contributes to the literature as a qualitative analysis of how undergraduates learn about sustainability through a constructivist teaching approach.

The Research Questions

How do US undergraduate management students learn about sustainability in MGMT1001 at the University of Minnesota, Carlson School? And what ideas do they have about applying sustainable practices in their career decisions, goals, and/or daily life after their first course in the management curriculum? This question is posed for three sections of MGMT1001, enrolled with 135 students.

The Method

The Carlson School of Management at the University of Minnesota Twin Cities offers one Master's level course focused on sustainability. The standard introductory course for undergraduate management majors at the University, *MGMT1001: Contemporary Management*, includes a two-week section on corporate social responsibility (CSR) and ethics. Feedback from management majors in their graduation exit surveys revealed that the students lack understanding of the connection between management, corporate responsibility and ethics in the curriculum. This situation provided incentive for the author to emphasize the concept of sustainability through revising lectures and assignments for undergraduates in MGMT1001, which is the first management course requirement for business majors.

The author engaged students in a combination of teaching and learning techniques to promote critical thinking about sustainability in relation to a range of management theories and models covered in their MGMT1001 course. Experiential and social learning theories suggest that group discussion, case studies, project work, and building on past experience enhance the learning process (Bandura 1986; Brew 2003; Kolb 1984). As a lecturer at the university, the author employed these strategies to teach students about the urgency of sustainable development and inspire them to lead the charge in responsibly managing the environmental and social impact of organizations globally.

This study hinges on one unique aspect of the MGMT1001 course at the University of Minnesota: a student "book club," which involved students working in groups of three to five to discuss and write journals about one of ten books selected by the professor. This assignment was a major point of differentiation for the author's classes compared to those of other instructors elsewhere in the first-year management curriculum. Historically, professors offered one or two books related to sustainability on the book club list, whereas, in fall of 2011, the author offered five, all with sustainability as the central theme. In teams, the students were required to discuss management topics from the course and textbook in relation to the concepts introduced in their assigned book club book (the students only read one book for their book club). This meant 52 of the 135 students across three sections opted to discuss a book about sustainability on a weekly basis, and write journal entries about their reactions, ideas and goals from their reading and group discussion in the context of management. The list of the selected five "book club" books that were focused on sustainability is presented in Table 1. The five books selected for the book club project focusing on sustainability describe challenges of environmental degradation, animal abuse, changes to ecosystems, energy supply and resource depletion. The books did not explicitly make the connection with management practices that would contribute to or could alleviate these problems. The lecturer did not address these issues specifically in class, in order to allow the students to reach their own conclusions and make up their own minds about future solutions.

Table 1: MGMGT 1001: Contemporary Management

Book Club List Focused on Sustainability
Teams of three to five students selected one book for "outside of class" discussion, journals, a final group presentation, and final paper. Forty percent (52) of the students enrolled in the author's three sections of MGMT1001 chose books with the sustainability theme from the list below. Total students across three sections N = 135.
Capra, F. (2003). *The Hidden Connections: Integrating The Biological, Cognitive, and Social Dimensions of Life into a Science Of Sustainability*, New York, NY: Doubleday.
Hawken, P., Lovins, A. and Lovins, L.H. (1999). *Natural Capitalism: Creating the Next Industrial Revolution*, New York, NY: Little Brown & Company.
Suzuki, D. and Dressel, H. (2002). *Good News for a Change: How everyday people are helping the planet*, Vancouver, BC, CAN: Greystone Books.
Pollan, M. (2006). *The Omnivore's Dilemma: A natural history of four meals*, New York, NY: Penguin Press.
Homer-Dixon, T.F. (2006). *The Upside of Down: Catastrophe, Creativity and the Renewal of Civilization*, Washington D.C.: Island Press.

The significance of this study is found in the pedagogical approach as well as the analysis of student learning. The concept of sustainability was presented in such a way as to make students think about it and discuss it outside of class in small "book club" groups. The instructor did not explicitly guide students through the process of learning the societal, ethical or environmental impact of current management theory, models of organizations, or specific measures of CSR. Thus, the contribution of this study to the literature in higher education is twofold: 1) it introduces a constructivist approach to integrating sustainability into an introductory management course, easily deliverable by faculty lacking expertise on the subject and requiring minimal time for course redesign, while placing the bulk of the "work of learning" on the students; and 2) the study employs a qualitative methodology to interpret the learning of business majors related to the concept of sustainability, analyzing the knowledge gained and goals formulated about sustainability during students' early college years. There are few qualitative studies on education for sustainable development (ESD) in undergraduate business curricula.

In the field of education, phenomenography is a methodology which was developed to pursue three lines of inquiry: general aspects of the student learning phenomenon, individual students' understanding of the subject matter learned, and the conceptions people have of the world (Marton 1981). Each of these research purposes is germane to the current study of sustainability, management education, and the student learning experience. The author analyzed student writing assignments, sorting and resorting data in narratives in an iterative process, to interpret categories of description (Akerlind 2005) that represent student knowledge and learning experiences about sustainability in their first undergraduate management course.

The Context for the Study

In the fall of 2011, the author integrated sustainability into the undergraduate curriculum at the University of Minnesota, Carlson School of Management. Lecturers may modify up to one tenth of the standard *MGMT1001: Contemporary Management* course content, based on the instructor's area of research and academic interests. Historically, the MGMT1001 curriculum presented the topic of corporate and social responsibility (CSR) in the final two weeks of the 15-week course. Thus, three or four lectures typically presented ethical issues in management in assigned readings, possibly using examples of sustainable organizations, such as in Patagonia. The author of this study, revised lectures throughout the course to include weekly, short video clips shown in class to showcase companies which achieved profitability in concert with a commitment to sustainability in their corporate mission. These videos presented some basic measures of SD while the lecture focused on current management models, structures and theories used by these organizations, including the business case for sustainability.

All students were required to write weekly journal entries – one after reading their book club book independently, and one entry after each meeting with their group. The students were instructed not to summarize the content of the book, but to reflect on what they had learned by reading and discussing the book, and how this knowledge might affect their future research and careers. The instructor (or a teaching assistant) graded these journals at three intervals during the 15-week course. The instructor and TAs provided feedback on the journal entries, encouraging students to relate their book club readings to contemporary management topics introduced in the course. In order to integrate the connection to sustainability into student learning without explicit mention of the term, the instructor urged them to *think critically, again and again, to dig deeper to integrate and synthesize new meaning from the knowledge they had learned while considering current social, political and environmental themes presented in the book.*

The students wrote individual papers, in addition to making a 20-minute group presentation at the end of the term to complete the book club project. These assignments represented 40% of their course grade. The final individual paper tasked students to think critically about themes from the book they agreed with, or not, and what "new ideas," personal views, and goals they formulated in book club discussions and independent analysis. It was this assignment, the culmination of the book club project, which the author analyzed and interpreted in the context of the research question: How did students learn about sustainability?

The Analysis

Student narratives showed critical thinking in their book club discussions and journals as they tried to reconcile the perceived clash of goals between their understanding of the profit motive and environmental conservation and social responsibility. Many students wrote about their frustration in trying to find any connection between the course material and their non-business book, initially maintaining that management theory and practice involved stakeholders, motivations, and goals in conflict with sustainability. Almost every student experienced this epiphany:

> [My book club] convinced me that the movement towards sustainability is relevant to management, and businesses can realize major benefits by embracing this change …
>
> I was surprised to learn how some companies truly neglect their surrounding environment and only focus on generating profits (MGMT1001 student, fall 2011).

In their final papers, students explained how they learned the importance of integrating management concepts into organizations, economic and political and systems in order to be effective in sustainable development. All but two students of the 52 who read books on sustainability expressed a change in their perception of the role of managers and their desire to advocate sustainability in their lives as students and in their career plans. In the learning process, many students indicated that they conducted *outside research* – not required for their course – to examine the sustainability of companies they might consider as employers after graduation.

The following categories of learning emerged from the analysis of writing assignments of book club members whose book dealt with sustainability:

- The Priority of Ethics and CSR: A Values-Driven Personal Investment Plan
- The Need for a Change in Mindset: Motivation to Teach and Model the Transformation
- Leadership Vision: Innovative Goals for Sustainable Development.

The themes of ethical decisions and the value of sustainable practices were almost nonexistent in the papers written by the 82 students in MGMT1001 *who did not read books focused on sustainability*, even though these students were exposed to the concepts through cases and lectures.

Ethics and CSR: A Values-Driven Personal Investment Plan

Students expressed attitudes they had developed about the importance of values shared with coworkers and the integrity of their employer. Based on their learning, their priority was to invest their time as employees, and their earnings, in a sustainable organization. They planned to research organizations in order to exercise their rights and influence as stakeholders to invest only in socially and environmentally responsible companies. The following is a collection of examples of this attitude, where students were keen to "look beyond the numbers" to know if the values of a company were consistent with their own:

> It is not uncommon that stakeholders are interested mostly in simply watching their profits grow, and with that in mind I would rather be in a position where the stakeholders I work for have the same beliefs and aspirations for the business as I do.

> It is important that we understand what our local companies are doing and either support them for doing a good job or send a clear message … that we demand better.
>
> Finding a sustainable company is not as easy as analyzing a company's balance sheet and income statement. To successfully execute my plan I need to look past the numbers, and find the true mission of the companies. I plan to pursue this goal by researching an array of companies, ranging across different sectors, and finding a sustainable stock portfolio with the number one goal of supporting the environment. This would be done through analyzing company mission statements and values.

The ethical compass of these students is clear in their narratives. They are already using their knowledge to question management models and structures in their undergraduate curriculum that challenge their position on the importance of sustainability of organizations and society.

A Need for a Change in Mindset: Motivation to Teach and Model the Transformation

Learning about sustainability transformed students' behavior and views of organizations, as they started to think about opportunities to integrate sustainability into their life and career. They saw the need to change the status quo "mindset" of business. They were inspired to be evangelists, influencing peers at work and the community in the future as well as being role models for that change in their individual daily lives.

> It's clear a change toward becoming eco-friendly needs to be made … I will be able to help clean up the environment, as well spread the message of becoming eco-friendly and supporting sustainable policies to fellow students. I have begun to recycle on a regular basis and am more conscious of my energy consumption. Also, I walk, ride a bike, or use public transportation instead of driving and wasting gas. This transition will be a difficult one because it not only involves a change in actions, but also **a change in mindset**. One has to be constantly aware of how actions are affecting others, as well as the environment. These small changes seem insignificant when viewed individually, but they add up and that is often very hard to comprehend … With our current values we will leave behind a completely depleted energy supply, and an extreme gap between the rich and poor. **But our world is on the verge of an extreme change.**
>
> I feel organizations can do better. I practice the most sustainable and green lifestyle that I can as a poor college student and plan to continue with it the rest of my life; I would like to continue that in my professional life … I would like to make it my goal to educate the people I work with in the future about these benefits.

In their narratives, the students see themselves as educators and mentors for transforming management to be in line with sustainable development. This position shows their knowledge of systems thinking and learning beyond the study of silos of business functions. These students were inspired to think critically, making connections between related social and environmental issues that were not specifically brought up in lectures on global management issues.

Leadership Vision: Innovative Goals for Sustainable Development

Many students explained how they were developing strategic plans to integrate business and management concepts with sustainability in their careers. They could see the big picture of systems affected by management decisions and recognized the need to set goals, act upon them, and become leaders in the change process themselves. Many of them communicated a clear vision for the future:

> I want to create a business that is as environmentally responsible and efficient as it can be. I would want it to operate on a system that eventually can be fully vertically integrated. In production for New Belgium Brewery, they are able to reuse wastewater and create methane gas for energy use in the process. I feel that a creative solution like this one would be something to shoot for when starting my own business. My energy and environmental conservation efforts will have to be unique and tailored to my business because each type of industry is different.
>
> I realized that if I want to be successful I need to make sure that I am passionate about what I am working on … I have yet to set specific, obtainable goals … I learned that if you want to see drastic re-

> sults, you need to be able to make drastic changes and not doubt yourself … I have learned that when pursuing my business career I may need to take risks in order to succeed.

The students integrated concepts from operations management, employee motivation and strategic planning into their vision for sustainability. Their goals show their inspiration to lead a new approach to managing organizations in the future.

The Findings

> I learned that the success of a business entails much more than a measurement of profits and revenues, and can include a large variety of factors. Success can mean that the business is producing the least carbon emissions, or that the company is heavily involved in its surrounding community. Success has different meanings depending on the eyes of the beholder, but to me, success means producing profits while participating in sustainable practices and being socially responsible (MGMT 1001 student, fall 2011).

This study suggests the revision of an undergraduate management course at University of Minnesota provided a catalyst for transformative student learning about sustainability. They developed leadership skills, the incentive for independent research, critical thinking, and passion to create a sustainable community. This learning inspired students to clarify their priorities and values. They held up current environmental and social problems comparing sustainability concepts with many management models and theories from class, which led them to conclude the current business "mindset" was unsustainable. Their papers explained in detail this analysis and their innovative plans to be change agents in sustainable management of organizations. According to Ardichvili and Kuchinke (2002), leadership can focus on achieving a purpose, goal or vision, as in transformational leadership, or can stress the benefits of imparting a task to others, as in transactional leadership. Students in MGMT1001 book clubs which focused on sustainability described their goals both as transformative strategies for their employers and specific daily activities in pursuit of sustainability. They saw the "big picture" and a vision for the future, which is not always the outcome from ESD (MacVaugh and Norton 2011: 83).

The difference between the pedagogical approach used by the author and others presented in the sustainability literature for business and management (MacVaugh and Norton 2011; Naeem and Peach 2010; Stubbs and Schapper 2011), is that the students in this study were *not directed to apply their learning specifically to sustainable outcomes* in society, such as measuring their carbon footprint or assessing their personal values, skills and motivation to pursue or support sustainability (Stubbs and Schapper 2011). Students were urged to think about their book club reading and relate it to management topics. The instructor did not mention sustainability, or the pillars of sustainable development, in the assignment for the final papers analyzed for this study. However, the students were instructed that, "Saying there was 'no relationship' between the book club reading and material covered in management class was not an option." She *allowed the students to muddle through* discussions and journals, trying to make sense of how sustainability and business could co-exist or produce a mutually beneficial outcome, and how this might affect their goals and concept of "success" as future managers. The onus was on students to integrate their learning about sustainability with principles of management.

Conclusion

> [The book club project] has altered the way I think about business, and has prompted me to establish a set of goals and actions that I intend to pursue so that I can make a positive impact on the environment (MGMT1001 student, fall 2011)

The study of the education for sustainable development course suggests the constructivist teaching approach encouraged critical and systems thinking in student learning about sustainability, and inspired them to be change agents. Students developed the knowledge, skills and attitudes preparing them to transform management practice and act as change agents in the future, committed to sustainable development in their organizations. The students formed tangible, specific goals and action plans to integrate sustainability into their lives. They were passionate about educating peers and future employers to engage in strategies for sustainable development. Future research could include investigation of the longitudinal effect of this educational process on students in their lives and careers throughout their university experience and after graduation. A survey of the elective courses they enroll in would show their commitment to learning about sustainability. Also, post-graduation interviews with students and graduates would ascertain the extent to which these students continue to integrate sustainable practices into their daily lives, employment, and selection of future employers and their role in organizational transformation.

References

Äkerlind, G.S. (2005), Variation and commonality in phenomenographic research methods. *Higher Education Research and Development* 24 (4), pp. 321-334.

Ardichvili, A. and Kuchinke. P. (2002), Leadership styles and cultural values among managers and subordinates: a comparative study of four countries of the former Soviet Union, Germany, and the US. *Human Resource Development International* 5 (1), pp. 99-117.

Aspen Institute (2009), *Aspen's Global 100: Beyond Grey Pinstripes 2009-2010.* Aspen Institute Center for Business Education, New York, NY.

Bandura, A. (1986), *Social foundations of thought and action: A social cognitive theory.* Prentice Hall, Englewood Cliffs, NJ.

Benn, S. and Dunphy, D. (2009), Action research as an approach to integrating sustainability into MBA programs. *Journal of Management Education* 33 (3), pp. 276-295.

Bradbury, H. (2003), Sustaining inner and outer worlds: A wholesystems approach to developing sustainable business practices in management. *Journal of Management Education* 27, pp. 172-187.

Brew, A. (2003), Teaching and research: new relationships and their implications for inquiry-based teaching and learning in higher education. *Higher Education Research & Development* 22 (1), pp. 3-18.

Brundtland (1987), *Report of the World Commission on Environment and Development: Our Common Future.* Available at: www.un-documents.net/wced-ocf.htm (accessed 7 January 2012).

Collins, E.M. and Kearin, K. (2010), Delivering on sustainability's global and local orientation. *Academy of Management Learning and Education* 9 (3), pp. 499-506.

Daly, H.E. (1977), *Steady-state economics.* W.H. Freeman and Company, San Francisco, CA.

Down, L. (2006), Addressing the challenges of mainstreaming education for sustainable development in higher education. *International Journal of Sustainability in Higher Education* 7 (4), pp. 390-9.

Elkington, J. (1997), *Cannibals with forks: The triple bottom line of 21st century business.* New Society, Gabriola Island, British Columbia, Canada.

Geli de Ciurana, A.M. and Leal Filho, W. (2006), Education for sustainability in university studies: Experiences from a project involving European and Latin American universities. *International Journal of Sustainability in Higher Education* 7 (1), pp. 81-93.

Harvey, F. (2011), Fishing industry attacks EU plan to end discards. *The Guardian.* Available at: http://www.guardian.co.uk/environment/2011/may/03/fishing-industry-eu-discards-plan (accessed 7 March 2012).

Hind, P., Wilson, A. and Lenssen, G. (2009), Developing leaders for sustainable business. *Corporate Governance* 9 (1), pp. 7-20.

Hollender, J. (2004), What Matters Most: Corporate Values and Social Responsibility. *California Management Review* 46 (4), pp. 111-119.

Intergovernmental Panel on Climate Change (IPCC) (2007), *Climate change 2007: Impacts, adaptation and vulnerability*. Cambridge University Press, Cambridge, UK.

Kearins, K. and Springett, D. (2003), Educating for sustainability: developing critical skills. *Journal of Management Education* 27 (2), pp. 188-204.

Kolb, D. (1984), *Experiential Learning*. Prentice-Hall, Englewood Cliffs, NJ.

Krugman, P. (Dec. 2011), Springtime for toxics. *New York Times*. Available at: http://topics.nytimes.com/topics/reference/timestopics/subjects/w/water_pollution/index.html? (accessed 15 January 2012).

Leal Filho, W. (2000), Dealing with misconceptions on the concept of sustainability. *International Journal of Sustainability in Higher Education* 1 (1), pp. 9-19.

MacVaugh, J. and Norton, M. (2011), Introducing sustainability into business contexts using active learning. *International Journal of Sustainability in Higher Education* 13 (1), pp. 72-87.

Marcus, A. and Clark, T. (2010), In Search of Sustainability in Management Education. *Academy of Management Learning and Education* 9 (3), pp. 377-383.

Marshall, S., Vaiman, V., Napier, N., Taylor, S., Haslberger, A. and Andersen, T. (2010), The end of a 'period': Sustainability and the questioning attitude. *Academy of Management Learning and Education* 9 (3), pp. 477-487.

Marton, F. (1981), Phenomenography: Describing conceptions of the world around us. *Instructional Science* 10, pp. 177-200.

McKeown, R. (2002), *Education for a Sustainable Development Toolkit*. Available at: www.esdtoolkit.org (accessed 15 January 2012).

Mitroff, I.I. (2004), An open letter to the deans and the faculties of American business schools. *Journal of Business Ethics* 54, pp. 185-189.

Naeem, M. and Peach, N. (2010), Promotion of sustainability in postgraduate education in the Asia Pacific region. *International Journal of Sustainability in Higher Education* 12 (3), pp. 280-290.

Navarro, P. (2008), The MBA curricula of Top-ranked business schools: A study in failure? *Academy of Management Learning and Education* 7 (1), pp. 108-123.

Porter, M.E. and Kramer, M.R. (2006), Strategy and society: The link between competitive advantage and corporate social responsibility. *Harvard Business Review*, December 2006, pp. 77-92.

Rands, G.P. (1993a), Preparing students to work for sustainability: Teaching as if the earth's future mattered. *Journal of Teaching in International Business* 5, pp. 19-46.

– (1993b), What's the meaning of greening? Goals & considerations in environmentalizing management curricula. In Pasquero, J. and Collins, D. (Eds.), *Proceedings of the 1993 annual meeting of the International Association for Business & Society*, pp. 538-542. San Diego, CA.

– (2009), A Principle-Attribute matrix for environmentally sustainable management education and its application: The case for change-oriented service-learning projects. *Journal of Management Education* 33 (3), pp. 296-322.

Rasche, A. and Escudero, M. (2009), Leading Change: The Role of the Principles for Responsible Management Education. *Journal for Business, Economics & Ethics* 10 (2), pp. 244-250.

Stead, J.G. and Stead, W.E. (2010), Sustainability comes to management education and research: A story of coevolution. *Academy of Management Learning and Education* 9 (3), pp. 488-498.

Sterling, S. (2001), *Sustainable Education Re-visioning Learning and Change*. Green Books, Totnes, Devon, UK.

Shrivastava, P. (2010), Pedagogy of Passion for Sustainability. *Academy of Management Learning and Education* 9 (3), pp. 433-455.

Stubbs, W. and Schapper, J. (2011), Two approaches to curriculum development for educating for sustainability and CSR. *International Journal of Sustainability in Higher Education* 12 (3), pp. 259-268.

UNFAO (2010), *Deforestation in decline but rate remains alarming*. Available at: http://www.un.org/apps/news/story.asp?NewsID=34195 (accessed 7 January 2012).

Wagner, M. (2009), National Culture, Regulation and Country Interaction Effects on the Association of Environmental Management Systems with Environmentally Beneficial Innovation. *Business Strategy and the Environment* 18, pp. 122-136.

II.
Implementing Sustainable Development in University Activities

Chapter 29

Implementing Sustainability at the Campus – Towards a better Understanding of Participation Processes within Sustainability Initiatives

Antje Disterheft, Sandra Sofia Ferreira da Silva Caeiro, Ulisses Manuel de Miranda Azeiteiro and Walter Leal Filho

Abstract

Participation is seen as a fundamental pre-requisite for the achievement of sustainable development. Applied to the university framework, participation refers to student and faculty involvement, giving the institutional community the opportunity to shape an institutional transformation process toward a more sustainable campus.

This ongoing research project intends to analyse how universities involve students and faculty in their efforts towards campus sustainability, and how these efforts are assessed. It aims to contribute towards a better understanding of the complexity inherent to sustainable development and the participation processes used in higher education for promoting sustainability practices and for fostering citizenship and democratic values.

This paper presents the project's methodological approach, based on an intensive literature review about participation and sustainability assessment tools, with a focus on tools applied in the higher education sector. Eleven of these tools, of which some use indicators, were selected, systemised and verified against the extent to which the participation of the campus community is captured and evaluated.

The results are used as a starting point for further discussion and research that shall lead to the development of an assessment tool for participatory approaches.

Introduction

Sustainable development (SD) and the question of how to overcome global challenges such as climate change, social inequality, loss of biodiversity, overpopulation, and lack of resources – to name but a few – have been discussed at the highest international political level for about four decades. Within this debate, the education sector has been attributed a key role in promoting SD (UNEP 1972; UNESCO 1998).

Due to their high societal impact, universities have been challenged to take a leadership role in disseminating sustainability principles. Universities are seen as multipliers with ethical obligations to systemically integrate SD in their institutions and to provide best-practice examples (Cortese 2003; Čiegis and Gineitiene 2006; Lozano 2006a; Alshuwaikait and Abubakar 2008; Leal Filho 2009) A growing number of higher education institutions have adopted declarations about Campus Sustainability (e.g. the Talloires Declaration, the Halifax Declaration and the recently updated Copernicus Charter). The current UN Decade of Education for Sustainable Development (2005-2014) has originated a vast number of projects related to SD in the tertiary education sector (UNESCO 2010). A research team associated to the UNESCO Chair in Higher Education for Sustainable Development (Leuphana University of Lüneburg) recently published a proposal for an indicator set evaluating education for sustainable development for the geographical regions Austria, Germany and Switzerland (Di Giulio et al. 2011).

Agenda 21 stresses the importance of public participation as a "fundamental pre-requisite for the achievement of sustainable development"(UNCED 1992), as does the OECD's governance strategy "Citizens as Partners" (OECD 2001) and the Aarhus Convention (UNECE 2001). Furthermore, Agenda 21 gave the impulse to develop sustainability indicators (UNCED 1992, Ch. 40) in order to strengthen the implementation of SD to be able to evaluate progress and to have a solid basis for decision-making.

Universities have started to recognise the use of assessment and reporting tools, as these tools constitute a helpful guideline for SD implementation. They make policy and charter statements more operational by identifying best-practice examples and striving for continuous improvement (Shriberg 2002). Furthermore, they enable more effective communication about the complexity of sustainability. Several assessment tools, of which some are indicator-based, have been developed for universities to assess their sustainability performance, and carried out on their strengths and weaknesses (Shriberg 2002; Cole 2003; Chambers 2009; Laroche 2009; Fonseca et al. 2011; Madeira et al. 2011). The dimension of participation in these assessment tools, if included at all, is approached in different ways, and there is a paucity of studies dealing with integrated approaches to SD involving faculty and students.

To help fill this gap, this ongoing research project focuses on campus sustainability and its assessment tools, and on participation processes within sustainability initiatives in particular, with the final objective of developing a measurement tool for participation. The overall aim is thereby to contribute towards a better understanding of the complexity inherent to SD and the means of participation processes in higher education for promoting sustainability practices and for fostering citizenship and democratic values.

To achieve these objectives, at the initial stage of the research project environmental management systems (EMS) were analysed as one group of assessment tools that have been adopted in many campuses around the world. This stage of the study examined whether the implementation followed a top-down or a participatory approach, and which activities were carried out in relation to EMS on campus (Disterheft et al. 2012). Case studies showed that EMS can be used beyond operations ends and give opportunities for research and teaching embracing a participatory dimension (ibid.). The results of the study pointed to the necessity of deepening the research into the participatory dimension of campus sustainability and investigating further assessment tools. This paper constitutes a continuation of the previous research, and focuses on the participatory dimensions within sustainability initiatives on campus and the related assessment procedures. Therefore, current practices were examined, which led to the selection of eleven sustainability tools that have been used in the university context. Some emphasis was given to indicator-based tools, since these allow the presentation of condensed information in a more comprehensive and traceable way. These tools were analysed with respect to the applied measurement approach (Dalal-Clayton and Bass 2002) and the extent to which the participatory dimension of the campus community is captured. A preliminary evaluation is presented in order to gain a better overview and understanding of the current situation, and allows some conclusions to be drawn for the ongoing research.

Participatory approaches are in general considered to be positive, as they can increase acceptance, achieve consensus, enhance the understanding of SD and may result in a higher level of awareness that in turn may contribute to an overall improvement of institutional sustainability performance (Bass et al. 1995). At the same time, participation has become a catch-all term with a multifaceted use and different understandings depending on the context. Some reflections on the term *participation* are given in the first section and linked later to the field of assessment, since the underlying understanding of participation influences the choices for indicator variables, and forms the basis for reflecting on the participatory dimension to existing sustainability tools.

Defining participation – a catch-all term

Promoting SD is closely linked to areas such as public participation and citizens' involvement. *Participation* and *empowerment* are two terms associated with the development of key competencies for SD. The first term refers to a continuous learning perspective, as pointed out by Howell et al. (1987): "Individuals must be provided with numerous opportunities throughout their lives to acquire the information and skills necessary to enact the citizen role". The second describes a multidimensional process of learning to think critically and to effect change in the personal life and in the community (Florin and Wandersman 1990). Particularly the latter aspect calls on citizens to be personally involved in decision-making processes (ibid).

The important commitments at the highest political level, as expressed in Agenda 21, the Aarhus Convention and the OECD strategy, have strengthened participatory approaches, but have also led to an inflationary use of the term *participation*: it has become a catch-all term and, similarly to the term *sustainable development*, it appears that the same word is understood in different ways; a universal definition does not exist. Therefore, in order to be able to conduct an analysis of the participatory dimension within sustainability assessment tools it became necessary firstly to understand what participation means or can mean. This reflection helps to define criteria for assessment ('what to measure') and helps find these criteria in existing assessment tools. From this reflections it is possible to move on to the question of 'how to measure', which provides an overview of future steps in this ongoing research project. In order to help clarify the different connotations around the term *participation*, the authors provide a brief resume of the theoretical context and the main streams of the current academic discussion.

Theoretical context

The recently concluded two-and-a-half-year project "Pathways through Participation", carried out in the United Kingdom by the National Council of Voluntary Organizations (NCVO), in cooperation with the Institute of Volunteering Research (IVR) and Involve[1], provides a useful summary of the huge amount of literature related to participation (Brodie et al. 2009) and gives insights into its complex dimensions from theoretical and practical perspectives. Below, some of the most important aspects connected to the research topic have been identified.

Since participation is linked to the understanding of democracy and the relationship between citizens and state, democratic theories have served as an analytical tool to further develop the research in this field. The two most important strands are the theories of *representative democracy* and *participative democracy*. Both theories see "individual participation as essential to democratic governance and in creating legitimate institutions", even though the relationship between civil society and state is perceived differently in each strand (Keohane 2002; Brodie et al. 2009). Based on these theories, and influenced by the preoccupation about the 'democratic deficit' that many Western societies are confronted with (Smith 2005), new forms of participation methods and techniques have emerged, such as participatory budgeting, citizens' juries and partnership governance (Brodie et al. 2009). In particular, participatory democracy, with its demand for "involving the majority of people in decisions that affect their lives" (ibid.), is seen as an imperative way of revitalising the concept of democracy, to keep communities agile and public institutions accountable (Potter et al. 1994; Roberts 2004).

1 Involve is a charity-funded organization that carries out research in the field of public participation: http://www.involve.org.uk/about/.

Agenda 21 aligns with this view and requests integrating participation on all societal levels (UNCED 1992; Ch. 1, 28, 36 ff). The positive implications of this participatory approach seem to be evident and are not questioned in the literature, but some authors criticize the fact that the principles of participatory democracy are not translated into practice, and fail to shift existing power relationships (Brodie et al. 2009).

Theories about civil society, social capital, social networks and movements can complement the understanding of representative and participatory democracy because they reflect on the power relationships between individuals, groups and wider society. Since a detailed discussion of these theories would exceed the scope of this paper, the authors have selected only single aspects of the broad discussion that are considered important for the understanding of participation: (i) the provision of space for voices of different stakeholders to associate is a critical component of democracy (Dahl 1989); (ii) joining and taking part in local organizations helps to foster trust in others and to develop a sense of values (Putnam 1995); (iii) the presence or absence of public engagement impacts on the quality of governance, democratic institutions and public life (Stoker 2004). Furthermore, these theories depict questions about social and socioeconomic inequality. Recent studies show, for example, a relationship between social status (class) and likelihood to engage (Brodie et al. 2009). Social movement theories shift the emphasis from organizational to social networks, where individuals are no longer *members*, but *participants*, who "have a sense of being involved in a collective endeavour" (ibid.). These movements are where personal involvement, individual investment, new cultural modes, relationships and world views are experienced and shaped, and can be seen as a predictor for individual participation (della Porta and Diani 2006).

In this context, several authors offer reflections on the question of power, characterizing power by its "public, hidden and insidious face" (Lukes 1974, 2005, in Brodie et al. 2009), and on how these forms of power relate to the space for participation and the different levels (local to global) of power (Gaventa and Cornwall 2006). Understanding these dimensions of spaces, the levels and forms of power as "separate yet interrelated dimensions" permits us to link them analytically together and to identify "obstacles and different entry points towards changing power balances in new forms of governance" (Gaventa and Cornwall 2006; Brodie et al. 2009). in this way, some light may be shed on the question of why some people are routinely and perpetually excluded from some form of participation.

Levels, forms, typologies and scope of participation

Societal levels

Participation has different connotations, depending on the societal level, and can be looked at from different perspectives. Due to the complexity of participation, it is helpful to have a clear picture of the societal level one is referring to when speaking about participation, since each level deals with specific questions and problems.

Participation as called for in Agenda 21 refers to the macro, meso and micro levels of society, and references to the importance of participation can be found throughout the entire document. With regard to the educational sector, at the macro-level it is, for example, required that participation be incorporated into the international and national framework of educational policymaking; at the meso-level institutions are challenged to embed the participatory dimension in their organizational structure and governance model; and at the micro-level it refers to the concrete learning settings and spaces for participation provided in institutions and their communities (Fig. 1).

Fig. 1: Societal levels of participation (adapted from UNCED 1992)

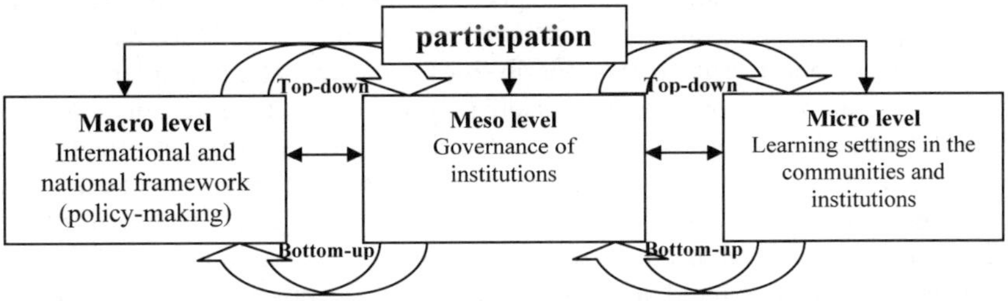

Since participation is not a static concept, all levels are interconnected and influence each other, either in a top-down or a bottom-up process.

Forms and typologies

Another helpful distinction is to categorize participation by *public*, *social* or *individual* participation (Brodie et al. 2009), though these boundaries often overlap.

A widely accepted perception of *public participation* is "the practice of consulting and involving members of the public in the agenda-setting, decision-making and policy-forming activities of organizations or institutions responsible for policy development" (Rowe and Frewer 2004). It is also often referred as political or civil participation or participatory governance. *Social participation* can be understood as collective activities in which individuals are involved on a regular basis. It is also referred to as 'civil' or 'community' participation (Brodie et al. 2009). *Individual participation* "covers the choices and actions individuals make as part of their life and that are statements of the society they want to live in (ibid.; Ginsborg 2005).

Some features and characteristics are common to any type of participation (Brodie *et al.*, 2011): (i) it is voluntary and chosen freely; (ii) it involves action; (iii) it can be collective or connected: even when the action is individual, a sense of common purpose exists and the act itself has a collective impact or ambition; (iv) it is purposeful: all participants are concerned about doing something that is worthwhile on their own terms, and every participatory act has, and is intended to have, consequences.

Another important aspect to consider when analysing participation is underlying interests, as White (1996) warns, because "if participation is to mean more than a façade of good intentions, it is vital to distinguish more clearly what these interests are". In her study she distinguishes between *nominal*, *instrumental*, *representative* and *transformative* forms of participation (Table 1). Even though her framework is based on experiences from the development policy field, it can be translated to other contexts as well, including the higher education sector, which is reflected in the examples given below.

Table 1: Typologies of participation and underlying interests toward participation from a top-down and bottom-up perspective (adapted from White1996)

Form / Type	**Top-down (governmental / institutional perspective)**	**Bottom-up (participants' perspective)**	**Function (What is the participation for?)**
Nominal	Legitimation	Inclusion	Display
Instrumental	Efficiency	Cost	Means
Representative	Sustainability	Leverage	Voice
Transformative	Empowerment	Empowerment	Means / End

A **nominal form of participation** can seek, for example, *legitimation* for continuous funding of a project or programme (institutional perspective); the participants may see advantages of being part of a project or programme (*inclusion*) because they benefit for example from personal recognition, or because it creates possibilities for personal future plans (e.g. financial loan, scholarships, etc.): the participation itself may merely fulfil a function of *display*. An **instrumental form of participation** may be based on the idea of cost-effectiveness (*efficiency*) from the institutional perspective, for example participation as a necessary component for providing/establishing services or facilities. From the participants' perspective this form of participation can be perceived as a *cost* (e.g. of time), and "its function is *a means* to achieve cost-effectiveness on the one hand, and a local facility on the other" (White 1996). **Representative participation** aims to 'give a voice' to the people involved, and by doing so the executing party (government or institution) can develop better structures for the long-term perspectives of a programme or project (*sustainability*), avoiding errors and misconceptions. For the participants, this form of participation allows *leverage* for a better recognition of their interests and needs. In a **transformative form of participation**, *empowerment* is at the central focus of both the institutional and the participants' perspective. Institutions might seek empowerment for several reasons, for example because of a general wish to improve performance or because of 'solidarity motivations' (e.g. with disadvantaged or disfavoured groups). Participants might perceive the positive impacts of empowerment when seeing their interests taken into account. Participation becomes a means to empowerment and an end in itself (ibid.). White stresses that empowerment will challenge existing power relations ("[governments and institutions] may find it rather uncomfortable when empowerment actually occurs", ibid.) and that any participation process is dynamic, as it is continuously influenced by a mix of interests.

Scope

A subject for further analysis within the participation discussion is the scope and depth of participatory processes. A classification still relevant today was made by Arnstein (1969) who developed a "ladder of participation", moving from *non-participation* to *citizen control* by differentiating between scopes of participation. Based on her work, the International Association of Public Participation (IAP2) presents a spectrum in which public participation is divided into five levels (no participation to high participation): the level of participation and the public impact increase when activities or methods are directed towards involvement and empowerment (Fig. 2).

Fig. 2: Scope of participation related to the spectrum of participation (adapted from International Association for Public Participation, 2007)

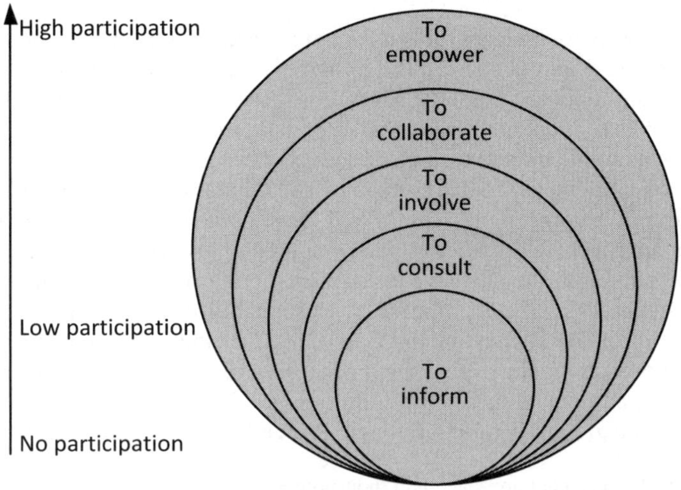

This spectrum can overlap with the previously presented forms and typologies of participation, but it links in a very explicit way the *intention of actions* within participatory processes to the *outcomes*, and offers a useful classification to analyse the scope of participation.

Sustainability assessment tools within the university context

The literature about sustainability assessment tools, and about indicators in particular, is vast, which only shows the importance and necessity of this research field. As Meadows (1998) remarked in 1998, when those indicators were emerging: "[…] the process of finding, implementing and improving sustainable development indicators will not be done right at first. Nevertheless, it is urgent to begin."

Sustainability indicators are the measurable part of a system (*variables*) (Dalal-Clayton and Bass, 2002) and allow information to be simplified, clarified and summarized, making the complexity of dynamic systems more transparent and understandable. They are therefore very useful for assisting decision-making (United Nations 2007; Singh et al. 2009). Furthermore, they help to visualize phenomena, to highlight trends and to provide early warning to prevent economic, social and environmental setbacks (ibid.). Sustainability indicators differ from traditional indicators, which are usually one-dimensional; SD indicators should go beyond for example growth indicators, and report "about efficiency, sufficiency, equity and quality of life" (Meadows 1998). They aim to capture the four dimensions of sustainable development, namely the economic, social, environmental and institutional dimensions, in order to help the reflection of the overall concept of SD (United Nations 2007; Singh et al. 2009). A large variety of indicator lists exist, aggregated in the form of *indices*, which differ in the particular selection of 'representative' indicators of the four dimensions of SD and the related sustainability concerns (Bartelmus 2008).

At international level, and following the requirement of Chapter 40 in Agenda 21, SD indicators were initiated by the UN Commission for Sustainable Development (CSD) after the Earth Summit in Rio de Janeiro (1992). They were developed during a five-year period (1994-2001) – since when they have been applied, tested and revised – and became known as the *CSD indicators* (United Nations 2007). Other well-known international indices are, for example, the *Index of sustainable and economic welfare* or the *Human Development Index*. A useful overview of the main international SD indicators and indices is given by Sing *et al.* (2009).

Not all sustainability assessment tools use or are based on indicators. Approaches to measurement can be divided into (1) accounts, (2) narrative assessments, and (3) indicator-based; however, assessment tools can combine several of these approaches (Dalal-Clayton and Bass 2002; Lozano 2006b). *Accounts* means that raw data are constructed and converted into a common unit, for example monetary, area or energy, as used in the Ecological Footprint. *Narrative assessments* combine text, maps, graphics and tabular data, sometimes using indicators though these are not a cornerstone. The World Development Report can be considered one example of this approach. The *indicator-based* approach also includes texts, maps, graphics and tabular data, similar to the narrative assessments, but groups them around indicators. The Well-Being Assessment or the Dashboard of Sustainability are examples for an indicator-based approach (ibid.). Dalal-Clayton and Bass (2002) attribute different strengths and weaknesses to each approach by classifying them according to their potential for (a) transparency, (b) consistency, (c) participation, and (d) usefulness for decision-making (Table 2). These criteria are based on the ground-breaking Bellagio Principles (Hardi and Zdan 1997).

Table 2: Types of Measurement approach and their potential (adapted from Dalal-Clayton and Bass 2002)

Measurement approach	*Accounts*	*Narrative*	*Indicator-based*
Potential for Transparency	Low	Medium	High
Potential for Consistency	High	Low	high
Potential for Participation	Low	High	medium
Usefulness for decision-making	Medium	Medium	high

Participation in this context refers to the potential scope of engagement of non-experts. Dalal-Clayton and Bass (2002) specify: "[...] the more technical the method, the less scope of participation".

For the higher education sector, Orr (2000, in Shriberg 2002) proposes that an ideal campus SD assessment tool should address the following questions: (1) What is the consumption of material goods on a per capita basis? (2) What are the university's policies regarding operational management (waste, recycling, purchase, energy and building)? (3) Does the curriculum strengthen the development of ecological literacy? (4) Does the outreach of a university financially support the creation of sustainable regional economies? (5) What do the graduates do in the world? This list, though it includes a broad range of topics and an intergenerational outlook (students' activities in the future) of extreme importance, excludes several aspects of the social dimension of SD and focuses more on the institutional impacts on the environment and economy. In contrast, Lozano (2006b) bases his criteria for an ideal assessment tool on the different parts of a university system (Cortese, 2003), and argues that sustainability indicators should cover systemically (i) education, (ii) research, (iii) campus operations, (iv) community outreach, and (v) assessment and reporting. Shriberg (2002) considers it essential "to identify issues with broad effects and influences", as well as to move beyond eco-efficiency, to measure processes and motivations and to include a large range of stakeholders. These different approaches demonstrate the complex and difficult task of defining *what to measure* when assessing campus sustainability.

Over the past 20 years, several authors have dedicated studies to measuring sustainability in higher education institutions, also using in some cases indicator-based tools (Roorda 2001; Lozano 2006b; Rode and Michelsen 2008; Chambers 2009; Laroche 2009; Brinkhurst et al.

2011; Lozano 2011). Some assessment tools were developed exclusively for universities to try to give an answer not only to the question of *what* to measure but also to the question of *how* to measure. These tools include for example the *Audit Instrument for Sustainability in Higher Education (AISHE)* (Roorda 2001); the *CSAF – Campus Sustainability Assessment Framework* (Cole 2003; Sierra Youth Coalition 2012); the *Graphical Assessment of Sustainability in Universities (GASU)* (Lozano 2006b); *STARS – Sustainability Tracking Assessment & Rating System* (Association for the Advancement of Sustainability in Higher Education (AASHE 2012); *STAUNCH – Auditing University Curricula in Higher Education* (Lozano 2010); and the *Sustainability Report Card* (Sustainable Endowments Institute 2011). Other tools, such as the *Ecological Footprint*, *GRI – Global Reporting Initiative Guidelines*, international environmental standards like *ISO 14001*, *EMAS* or the social responsibility standard *ISO 26000*, have been adapted to the higher education context and are currently being successfully implemented at many universities (Disterheft et al. 2012).

These tools have been assessed in terms of their strengths and weaknesses (Shriberg 2002; Cole 2003; Laroche 2009), and some were evaluated in case studies on specific campuses (see Glover et al. 2011, for *STAUNCH*; Beringer 2006, for *CSAF*; Flint 2001, and Venetoulis 2001, for the Ecological Footprint; and Disterheft et al. 2012, for case study examples of Environmental Management Systems at European Campuses). Furthermore, a still-small but increasing number of higher education institutions use Sustainability Reports, of which some follow the Global Reporting Initiative Guidelines (GRI) (Lozano 2011; Disterheft et al. 2012).

Methodological approach

For the first time, the dimensions of participation and their assessment have been analysed within sustainability initiatives in higher education institutions and compared with existing assessment tools.

This research is based on an exhaustive literature review about participation and sustainability assessment tools. Starting from Shriberg's (2002) and Cole's (2003) reviews about sustainability assessment procedures within higher education institutions, the list was updated to apply to the current situation, including some international standards (two ISO standards and EMAS). Then, eleven assessment tools that have been used in higher education institutions were selected, based on their complexity, timeliness and accessibility. These tools were systemized following Dalal-Clayton and Bass's (2002) categorization of measuring approaches (Table 4). Based on the literature review about participation (Chapter 2) the authors formulated preliminary criteria for the assessment of the participatory dimensions and analysed how participation is reported in the selected campus sustainability assessment tools. The criteria for this analysis were: (i) Participation possibilities are assessed (yes/no); (ii) Participation possibilities are differentiated by subgroups (students, faculty, staff, external community); (iii) The assessment of participation possibilities is either quantitative or process-oriented (or combined); (iv) Differentiation between participation forms is made (yes/no), and if affirmative, (v) Which differentiation between forms of participation are made? Finally: (vi) Participation processes themselves are assessed (yes/no).

Results

A large number of universities have opted for different tools to assess sustainability on campus. In the USA and Canada, many campuses use CSAF, STARS or the Sustainability Report Card, whereas in Europe an increasing number of universities publish sustainability reports following the GRI guidelines. Table 3 shows a very brief characterization of the tools that were selected for the present analysis.

Table 3: Characteristics of assessment tools applied in higher education institutions

Assessment tool	Characteristics
AISHE – Auditing Instrument for Sustainability in Higher Education	An instrument developed for the managerial board/ administrative experts as well as for education experts (faculty) and students, based on a model for quality management and using the Plan-Do-Check-Act-Cycle to assess up to which level sustainability principles are incorporated into the curriculum (education) and institution (operations) (Roorda 2001)
CSAF – Campus Sustainability Assessment Framework	An academically developed standardized audit tool for the Canadian university landscape. It uses 169 indicators in total to report on the "eco-system" (air, water, land, energy, material) and on the "people-system" (community, governance, knowledge, health & well-being, economy & well-being) (Cole 2003; Beringer 2006)
GASU – Graphical Assessment of Sustainability in Universities tool	Adds the dimension of education and research to the GRI- Global Reporting Initiative Guidelines. Consists of charts where the user can grade a list of indicators referring to the different dimensions of sustainability. The tool allows benchmarking over time and comparison with other institutions (Lozano 2006b)
STARS – Sustain-ability Tracking, Assessment & Rating System	Developed by the Association of the Advancement of Sustainability in Higher Education (AASHE), this tool uses indicators, based on the environmental, economical and social dimension of SD and divides these into four categories related to campus activities, such as Education & Research, Operations, Planning, Administration & Engagement, Innovation (Association for the Advancement of Sustainability in Higher Education (AASHE) 2012). Applied mainly in US and Canadian universities, but there has started recently as well an international pilot project.
STAUNCH – Sus-tainability tool for Auditing Univer-sities Curricula in Higher Education	This tool audits universities' curricula holistically by applying a two-tiered balance of SD. Based on four main aspects this tool calculates numerically the balances and strength of the curricula, providing a snapshot of how SD is addressed by the institution (Lozano 2010)
Sustainability Report Card	A survey based instrument, sending surveys to administrators and students' leaders to collect data that are translated to 52 indicators about campus operations, dining services, endowment investment practices and student activities. It is carried out by the non-profit Sustainable Endowments Institute and universities sign up to participate in the annual report (Sustainable Endowments Institute 2011)
Ecological Footprint	Measures how much land and water area a human individual or population requires to produce the resources it consumes and to absorb its carbon dioxide emissions. Developed in 1990 by Wackernagel and Rees and meanwhile adopted scientifically with differences in its applications (Global Footprint Network 2012)
EMAS – Eco-Management and Audit Scheme	Standardized management tool developed by the European Commission for companies and other organizations to evaluate, report and improve their environmental performance. It requires clear and quantified goals as well as a verified environmental declaration to obtain a final certification (European Commission 2010). As an environmental management systems it derives from quality management systems and follows the Plan-Do-Check-Act-cycle
ISO 14001 (International standardisation Organisation)	Most well-known and internationally recognized environmental standard; can be implemented with or without a final certification. Derives as well from quality management systems and is based on the Plan-Do-Check-Act-Cycle (ISO 2011b)

Assessment tool	Characteristics
ISO 26000 (International standardisation Organisation)	An assistant tool to incorporate social responsibility and to go beyond legal compliance with regard to sustainability issues. It offers guidance, but is not standardized nor does it offer certification (ISO 2011a)
GRI – Global Reporting Initiative Guidelines	Provide guidelines to companies and organization to report about their sustainability performance. They are structured in five sections (vision and strategy, organization's profile, governance structure, GRI content index, performance indicators). They have been developed by a non-profit organisation and aim to promote a long-term stakeholders' dialogue (Global Reporting Initiative 2012)

Six of the selected assessment tools were specifically developed for the higher education context; five originate from models for corporations and organizations, but have been used in universities as well (Table 3). The majority of the assessment tools are indicator-based; only two (AISHE and ISO 26000) follow a narrative assessment approach. With regard to the subsystems relevant for higher education institutions, namely the economic, educational, environmental, institutional and social dimensions, only three tools, CSAF, GASU and STARS, report on all subsystems. When attributing the respective potential for transparency, consistency, participation and usefulness for decision-making, the authors closely followed Dalal-Clayton and Bass's (2002) scheme (Table 4), but these classifications should be interpreted as an indicative reference and may vary from situation to situation.

Table 4: Measurement approaches of sustainability indicators used in Higher Education Institutions

		Scope of subsystems					**Measuring approach**				
Assessment tool	**Context**	**environmental**	**economical**	**social**	**institutional**	**educational**	**Acc/Narr/Ind**	**potential for transparency**	**potential for consistency**	**potential for participation**	**usefullness for decision-making**
AISHE	HEI	–	–	x	x	x	Narr	medium	low	high	medium
CSAF	HEI	x	x	x	x	x	Ind	high	high	medium	high
GASU	HEI	x	x	x	x	x	Ind	high	high	low	medium
STARS	HEI	x	x	x	x	x	Acc/Ind	high	high	medium	high
STAUNCH	HEI	–	–	–	x	x	Ind	high	high	medium	high
Sustainability Report Card	HEI	x	x	x	x	–	Narr/Ind	high	medium	medium	high
Ecological Footprint	ORG	x	–	x	–	–	Acc/Ind	medium	high	low	medium
EMAS	ORG	x	x	x	x	–	Acc/Ind	medium	high	high	high
ISO 14001	ORG	x	–	–	x	–	Acc/Ind	medium	high	low	high
ISO 26000	ORG	x	–	x	x	–	Narr	medium	low	high	medium
GRI	ORG	x	x	x	x	–	Acc/Ind	high	high	medium	high

HEI = Higher Education Institutions; ORG = Organisations; Acc = Accounts; Ind = Indicator-based; Narr = Narrative assessment

Table 5: Dimensions of participation within sustainability assessment tools used in Higher Education Institutions

Assessment tool	Dimensions of participation									
	participation assessed (x=yes; –=no)	**includes students**	**includes faculty members**	**includes staff**	**Includes external community**	**quantitative oriented**	**process-oriented**	**Differentiation between forms of participation**	**how?**	**assessment of participatory processes**
AISHE	x	x	x	x	x	–	x	x	Interactive learning methods for the academic community (not students-focused)	x
CSAF	x	x	x	n.d.	x	x	–	x	Volunteerism, voter turnout, community engagement within policy-making	–
GASU	x	x	x.	x.	x	x	–	x	Report about capacity building, course "Educate the educators in SD", research related to SD, partnerships on local level	–
STARS	x	x	x	n.d.	x	x	–	x	Co- curricular education, volunteerism and community service, partnerships on local level	–
STAUNCH	n/a	n/a	n/a	n/a	n/	n/a	n/a	n/a	n/a	n/a
Sustainability Report Card	x	x	x	x	n.d.	x	–	x.	Employee outreach opportunities, different forms of students involvement	–
Ecological Footprint	–	n/a	n/a	n/a	n/a	x	–	n/a	n/a	n/a
EMAS	x	n/a	n/a	n/a	n.d.	x	x	x	Differentiation between top-down and bottom-up governance, stakeholders engagement in diverse forms	–
ISO 14001	–	n/a	n/a	n/a	–	x	x	n/a	n/a	n/a
ISO 26000	x.	n/a	n/a	n/a	x	x	x	x.	Stakeholders' engagement	n/a
GRI	x	n/a	n/a	n/a	x	x	x	x	Differentiation between top-down and bottom-up governance, stakeholders engagement in diverse forms	n.d.

n/a= not applicable; n.d.= not defined

As explained before, Dalal-Clayton and Bass's *potential for participation* refers to the involvement of the public (non-experts) within the measurement process. In this paper, *participation possibilities* refers to the *space* (in a non-physical sense) given by the institution to its community in order that participation can *take place*. Table 5 shows how the dimensions of participation according to the methodological approach used in this paper are captured in the selected tools. The subgroups of the academic community, namely students, faculty, staff and external community, are distinguished. Often, the term faculty can include teaching staff and administrative staff at the same time; cases without a clear differentiation were classified as "not defined". Only AISHE and GASU distinguish clearly between all four subgroups and report on different forms of participation. The Sustainability Report Card differentiates between students, staff and faculty, and different forms of participation, but does not include the external community. STARS and CSAF do not explicitly define the subgroup "staff", but also differentiate between forms of participation such as volunteerism, community service, voter turnout and partnerships at local level (with businesses, NGOs, etc.); however, there is a focus on students' involvement. Among the tools designed for companies, EMAS and GRI include reports on different forms of participation that are based on a stakeholder dialogue. ISO 26000 constitutes a particular case, since it is not a management system with concrete requirements, but is to be understood more as a 'guideline'. The participatory dimension is, strictly speaking, not assessed, but was included in this evaluation because of its high potential for participation within the stakeholder dialogue as well as its increasing popularity (Pojasek 2011).

With the exception of AISHE, none of the tools considers the assessment of the participatory processes themselves.

Discussion: How can campus sustainability assessment tools contribute to a better understanding of participation?

It is largely agreed that participatory processes are indispensable for promoting sustainable development, as requested in Agenda 21 and again underlined in the UN Decade Education for Sustainable Development (2005-2014). Participation is addressed in some way in most of the assessment tools analysed above, but with different focuses, for example on community engagement, volunteerism, stakeholder dialogues or voter turnout. It remains unclear to what extent the internal community as a whole, as well as the external community, is considered; also how effective the different participation options and processes are and what their impact is with regard to institutional, academic, professional and personal life.

The multifaceted use of the term, blurred boundaries between individual, social and public participation, and unclear differentiation between participation at macro, meso or micro-level can turn the assessment of participation into a very challenging task. But since assessment tools, and in particular indicator-based approaches, allow complex and dynamic processes to be made more comprehensible, they can be a supporting tool for making the participatory dimensions within the complexity of sustainable development more transparent and tangible. Assessment tools are linked to values, because, according to Meadows (1998), "we measure what we care about and we care about what we measure". Considering the participatory dimension in a more integral way would demonstrate its significance to the university's community and could lead at the same time to an improvement of participation processes.

In their recent study, Brinkhurst et al. (2011) point out that faculty and staff members are important leaders in achieving lasting progress towards campus sustainability, but that their support is often overlooked and not sufficiently recognized. An assessment tool that looks at the participation of the entire community could help to reduce this imbalance. Dahl (2012) defends the inclusion of ethics and values in assessment tools, because "building awareness of values is an important part of the process of change towards sustainability". He identifies the lack of an indicator "to evaluate individual action or commitment" (ibid) as a gap in current sustainability indicator sets, because "sustainability (or the lack of it) depends on the individual actions of over 6 billion human beings, the choices they can and do make, the lifestyles they adopt, and their decisions on family size, consumption patterns, etc., recognizing that poverty greatly limits choice" (ibid). An assessment tool for participation, in the specific context of a higher education institution, might help visualize the impacts of individual, social and public participation. This would not only improve the institutional performance, but also contribute to increasing sustainable practices among the internal and external university communities' members, foster citizenship and democratic values, and to build sustainable development both for the present and future generations.

Conclusion

Participation, considered to be essential within efforts to create sustainable universities, has become a buzzword with different meanings to different users. In the present analysis of eleven assessment tools applied in universities, the authors were able to verify that participation is approached in distinct ways and, though assessed in most tools, that the perception of the dimensions of participation is limited. Only two tools, namely AISHE and GASU, differentiate between the subgroups of the internal academic community (students, faculty and staff) as well as the external community. All tools that have been developed for the university context distinguish between different forms of participation, such as volunteerism, community engagement and voter turnout, but put a focus on students' involvement. The participatory processes themselves are not assessed by the tools, with the exception of AISHE, and therefore it is very difficult to evaluate the effectiveness of any of the forms of participation.

The results, though still preliminary, show a research gap in terms of the need for a broader consideration of the dimensions of participation. This paper can therefore form a starting point for further discussion and reflection on how to develop a measurement tool for participatory processes within campus sustainability initiatives.

Acknowledgement

The main author wishes to thank Elona Hoover for her valuable comments on this work.

References

Alshuwaikait, H.M. and Abubakar, I. (2008). An integrated approach to achieving campus sustainability: assessment of the current campus environmental management practices. *Journal of Cleaner Production* 16, pp. 1777-1785.

Association for the Advancement of Sustainability in Higher Education (AASHE) (2012). "STARS". Available at https://stars.aashe.org/ (accessed 10 February 2012).

Bartelmus, P. (Ed.) (2008), *Encyclopedia of Earth*. Environmental Information Coalition, National Council for Science and the Environment, Washington, DC.

Bass, S., Dalal-Clayton, B. and Pretty, J. (1995), *Participation in strategies for sustainable development*. Environmental Planning Issues (7 (May 1995). International Institute for Environment and Development, London.

Beringer, A. (2006). Campus Sustainability audit research in Atlantic Canada: pioneering the campus sustainability framework. *International Journal for Sustainability in Higher Education* 7 (4), pp. 437-455.

Brinkhurst, M., Rose, P., Maurice, G. and Ackerman, J.D. (2011). Achieving campus sustainability: top-down, bottom-up or neither? *International Journal of Sustainability in Higher Education* 12 (4), pp. 338-354.

Brodie, E., Cowling, E., Nissen, N., Paine, A.E., Jochum, V. and Warburton, D. (2009), *Understanding participation: A literature review*. Available at http://pathwaysthroughparticipation.org.uk/wp-content/uploads/2009/09/Pathways-literature-review-final-version.pdf (accessed 18 January 2012).

Brodie, E., Hughes, T., Jochum, V., Miller, S., Ockenden, N. and Warburton, D. (2011), *Pathways through participation: What creates and sustains active citizenship?* NCVO, Institute for Volunteering Research, Involve, London.

Chambers, D. (2009), Assessing & Planning for Environmental Sustainability – A Framework for Institutions of Higher Education. In Leal Filho, W. (Ed.), *Sustainability at Universities – Opportunities, Challenges and Trends,* pp. 287-297. Peter Lang, Frankfurt.

Čiegis, R. and Gineitiene, D. (2006), The Role of Universities in Promoting Sustainability. *Universitetų vaidmuo skatinant darnųjį vystymąsi* 48 (3), pp. 56-62.

Cole, L. (2003), *Assessing sustainability on Canadian University campuses: development of a campus sustainability assessment framework.* Royal Roads University, Victoria.

Cortese, A.D. (2003), *The critical role of Higher Education in Creating a Sustainable Future. Planning for Higher Education.* Available at http://www.aashe.org/documents/resources/pdf/Cortese_PHE.pdf (accessed 2 November 2011).

Dahl, A.L. (2012). Achievements and gaps in indicators for sustainability. *Ecological Indicators* 17 (0), pp. 14-19 (doi: 10.1016/j.ecolind.2011.04.032).

Dahl, R.A. (1989). *Democracy and its Critics*. Yale University Press, New Haven.

Dalal-Clayton, B. and Bass, S. (2002). *Sustainable Development Strategies. A Resource Book.* Earthscan, Paris, New York.

della Porta, D. and Diani, M. (2006). *Social Movements: An Introduction*. Blackwell Publishing, Malden and Oxford.

Di Giulio, A., Ruesch Schweizer, C., Adomssent, M., Blaser, M., Bormann, I., Burandt, S., Fischbach, R., Kaufmann-Hoyoz, R., Krikser, T., Kuenzli David, C., Michelsen, G., Rammel, C. and Streissler, A. (2011), *Bildung auf dem Weg zur Nachhaltigkeit*. Interfakultaere Koordinationsstelle für Allgemeine Oekologie, Bern.

Disterheft, A., Caeiro, S.S., Ramos, M. and Azeiteiro, U.M. (2012), *Environmental Management Systems (EMS) implementation processes and practices in European Higher Education Institutions – top-down versus participatory approaches*. Journal of Cleaner Production (in press) (doi: 10.1016/j.jclepro.2012.02.034).

European Commission (2010), *EMAS*. Available at http://ec.europa.eu/environment/emas/index_en.htm (accessed 15 February 2012).

Flint, K. (2001), Institutional ecologial footprint analysis – A case study of the University of Newcastle, Australia. *International Journal for Sustainability in Higher Education,* 2 (1), pp. 48-62.

Florin, P. and Wandersman, A. (1990), An introduction to citizen participation, voluntary organizations, and community development: Insights for empowerment through research. *American Journal of Community Psychology* 18 (1), pp. 41-54.

Fonseca, A., Macdonald, A., Dandy, E. and Valenti, P. (2011), The state of sustainability reporting at Canadian universities. *International Journal for Sustainability in Higher Education* 12 (1), pp. 22-40.

Gaventa, J. and Cornwall, A. (2006), Challenging the Boundaries of the Possible: Participation, Knowledge and Power. *IDS Bulletin* 37 (6), pp. 122-128.

Ginsborg, P. (2005), *The Politics of Everyday Life. Making Choices, changing Lives*. Yale University Press, New York.

Global Footprint Network (2012), Global Footprint Network – Advancing the science of Sustainability. Available at http://www.footprintnetwork.org (accessed 15 February 2012).

Global Reporting Initiative (2012), *What is GRI*. Available at https://www.global reporting.org/ (accessed 15 February 2012).

Glover, A., Peters, C. and Haslett, S.K. (2011), Education for sustainable development and global citizenship – An evaluation of the validity of the STAUNCH auditing tool. *International Journal of Sustainability in Higher Education* 12 (2), pp. 125-144.

Hardi, P. and Zdan, T. (1997), *Assessing sustainable development – Principles in Practice*. International Institute for Sustainable Development, Winnipeg.

Howell, R.E., Olsen, M.E. and Olsen, D. (1987), *Designing a citizen involvement program: A guidebook for involving citizens in the resolution of environmental issues*. Western Rural Development Center, International Association for Public Participation (2007), Spectrum of public participation. Available at http://www.iap2.org/associations/4748/ files/IAP2%20Spectrum_vertical.pdf (accessed 27 January 2011).

ISO (2011a), *ISO 2600:2010*. Available at http://www.iso.org/iso/iso_catalogue/catalogue_tc/ catalogue_detail.htm?csnumber=42546 (accessed 10 February 2012).

– (2011b), *ISO 14000 essentials*. Available at http://www.iso.org/iso/iso_catalogue/management_ and_leadership_standards/environmental_management/iso_14000_essentials.htm (accessed 2 January 2011).

Keohane, R. (2002), Governance in a Partially Globalised World. In Held, D. and McGrew, A. (Eds.), *Governing Globalization: Power, Authority and Global Governance*. Polity Press, Cambridge.

Laroche, D.-C. (2009), *Tracking progress: Development and use of sustainability indicators in campus planning and management*. Arizona State University, Phoenix.

Leal Filho, W. (Ed.) (2009), *Sustainability at Universities – Opportunities, Challenges and Trends* (Vol. 31). Peter Lang, Frankfurt.

Lozano, R. (2006a), Incorporation and institutionalization of SD into universities: breaking through barriers to change. *Journal of Cleaner Production* 14 (9-11), pp. 787-796.

– (2006b), A tool for a Graphical Assessment of Sustainability in Universities (GASU). *Journal of Cleaner Production* 14 (9-11), pp. 963-972.

– (2010), Diffusion of sustainable development in universities curricula: an empirical example from Cardiff University. *Journal of Cleaner Production* 18 (10), pp. 637-644.

– (2011), The state of sustainability reporting in universities. *International Journal for Sustainability in Higher Education* 12 (1), pp. 67-78.

Madeira, A.C., Carravilla, M.A., Oliveira, J.F. and Costa, C.A.V. (2011), A methodology for sustainability evaluation and reporting in higher education institutions. *Higher Education Policy* 24 (4), pp. 459-479.

Meadows, D.H. (1998), *Indicators and Information Systems for Sustainable Development*. Donella Meadows Institute (former Sustainability Institute), Vermont.

OECD (2001), Citizens as partners – OECD Handbook on information, consultation and public participation in policy-making. In OECD (Eds.), *Governance*. Available from http://www.ezd.si/ fileadmin/doc/4_AKTIVNO_DRZAVLJANSTVO/Viri/Citizens_as_ partners_hanbook_oecd.pdf.

Pojasek, R.B. (2011), ISO 26000 guidance on social responsibility. *Environmental Quality Management* 20 (3), pp. 85-93.

Potter, D., Goldblat, D., Kiloh, M. and Lewis, P. (Eds.), (1994), *Explaining Democratization*. Polity Press, Cambridge.

Putnam, R. (1995), Bowling Alone: America's Declining Social Capital: An interview with Robert Putnam. *Journal of Democracy* 6 (1), pp. 65-78.

Roberts, N. (2004), Public Deliberation in an Age of Direct Citizen Participation. *The American Review of Public Administration* 34 (4), pp. 315-353.

Rode, H. and Michelsen, G. (2008), Levels of indicator development for education for sustainable development. *Environmental Education Research* 14 (1), pp. 19-33 (doi: 10.1080/13504620 701843327).

Roorda, N. (2001), *Auditing Instrument for Sustainability in Higher Education*. In Roorda, N. (Eds.). Available from http://www.eauc.org.uk/audit_instrument_for_sustainability_ in_higher_educ.

Rowe, G. and Frewer, L.J. (2004), Evaluating Public-Participation Exercises: A Research Agenda. *Science, Technology, & Human Values* 29 (4), pp. 512-557.

Shriberg, M. (2002), Institutional assessment tools for sustainability in higher education – Strengths, weaknesses, and implications for practice and theory. *International Journal of Sustainability in Higher Education* 3 (2), pp. 254-270.

Sierra Youth Coalition (2012), *Sierra Youth Coalition*. Available at http://www.syc-cjs.org (accessed 10 February 2012).

Singh, R.K., Murty, H.R., Gupta, S.K. and Dikshit, A.K. (2009), An overview of sustainability assessment methodologies. *Ecological Indicators* 15 (1), pp. 281-299 (doi: 10.1016/j.ecolind. 2011.01.007).

Smith, G. (2005), *Beyond the Ballot: 57 Democratic Innovations from Around the World*. A report of the Power Inquiry. The Power Inquiry, London.

Stoker, G. (2004), New Localism, Progressive Politics and Democracy. *The Political Quarterly* 75, pp. 117-129.

Sustainable Endowments Institute (2011), *Sustainability Report Card*. Available at http://www.greenreportcard.org/ (accessed 18 June 2011).

UNCED (1992), *Agenda 21*. United Nations, Rio de Janeiro.

UNECE (2001), *Aarhus Convention – Convention on Access to Information, Public Participation in Decision-making and Access to Justice in Environmental Matters*. Available at http://www.unece.org/env/pp/ (accessed 10 February 2011).

UNEP (1972), *Declaration of the United Nations Conference on the Human Environment*. Paper presented at the United Nations Conference on the Human Environment, Stockholm.

UNESCO (1998), *World Declaration on higher education for the twenty-first century: Vision and Action*, Paris.

– (2010), *UN Decade of Education for Sustainable Development, Education*. Available at http://www.unesco.org/en/education-for-sustainable-development/decade-of-esd/ (accessed 20 February 2011).

United Nations (2007), *Indicators for Sustainable Development: Guidelines and Methodologies*. New York.

Venetoulis, J. (2001), Assessing the ecological impact of a university: The ecological footprint for the University of Redlands. *International Journal for Sustainability in Higher Education* 2 (2), pp. 180-197.

White, S.C. (1996), Depoliticising development: the uses and abuses of participation. *Development in practice* 6 (1), pp. 6-15.

Chapter 30

A National Programme to Support Education for Sustainable Development in the United Kingdom: University Experiences of the HEA Green Academy Programme

Simon Kemp, Stephen Scoffham, Peter Rands, Alastair Robertson, Zoe Robinson, Sarah Speight, Antonius Raghubansie and Heather Luna

Abstract

During a challenging period for higher education in the United Kingdom (UK), the Higher Education Academy (HEA) has launched a timely new programme, 'Green Academy: curricula for tomorrow'. This programme builds on the success of the HEA Change Academy, and aims to help universities embed sustainability across the institution, with an initial focus on the curriculum. Eight institutions representative of the different types of university that form the UK HE sector were selected to participate in the first cycle of the programme.

This paper analyses the experiences of a representative sample of five of the eight university project teams as they sought to embed sustainability across their institutions. The common themes that emerged from the individual universities and the programme are the importance of positive challenges, senior management, critical friends, and multi-university partnerships guided by a central coordinating body. Initial findings suggest the Green Academy approach provides a change model which, through the use of critical friends and the development of a collegial network driving friendly competition between partner institutions, can lead the way towards whole institutional change for sustainability.

Introduction

The Higher Education (HE) landscape in the United Kingdom (UK) is in the midst of profound changes that will have wide-ranging implications for students and universities for years, possibly even generations, to come. The removal of the majority of public funding and the move to a market-focused HE sector driven by student choice has created considerable turbulence in the UK academic community. However, whilst considerable challenges undeniably exist, this level of change also generates opportunities for improvement in key areas across the sector. With over 80% of first-year UK HE students viewing sustainability skills as important to future employers, and the majority believing their university should prepare them for employment (Bone and Agombar 2011), it is considered that Education for Sustainable Development (ESD) is one area that might benefit from the move to increased student accountability.

Against this backdrop, the UK Higher Education Academy (HEA) developed a new institutional change programme in partnership with the National Union of Students (NUS) and the Environmental Association for Universities and Colleges (EAUC) to help universities embed ESD across their institutions, with an initial focus on the curriculum. UK universities were invited to apply to be pilot institutions in the programme. The eight successful applicants were the University of Bristol, Canterbury Christ Church University, Keele University, the University of Nottingham, the University of Southampton, Swansea University, the University of Wales Trinity Saint David, and the University of Worcester. Each project team had to comprise academics, senior management (Pro-Vice Chancellors), estate managers, and students to ensure an appropriately holistic approach to ESD.

Sustainability in universities

Attempting to successfully embed sustainability in universities is a process fraught with a multitude of difficulties. Possibly the most fundamental problem is that sustainability involves philosophical, theoretical and ethical dimensions, and these involve patterns of thought which can sit rather uneasily alongside some aspects of current academic thinking. This can be illustrated at surface level by the contested nature of the definition of sustainability. Although the Brundtland definition, 'development which meets the needs of the present without compromising the ability of future generations to meet their own needs' (WCED 1987), is broadly accepted as the universal definition, the meaning and application of this concept have been and continue to be the subject of much conjecture and debate (Eichler 1999; Jacobs 1999; Leal Filho 2011, amongst many others).

The discursive exploratory nature of academia makes the framing of a 'wicked problem' (Rittel and Webber 1973) such as sustainability even more problematic when attempting to reach a resolution on a workable definition that will enable the application of the concept in a university setting. Leal Filho (2011) outlines the range of international agreements and conventions signed by over 600 universities globally, along with the unquantified number of other institutions developing their own programmes, many with varying success. This mixed progress is likely to be partly due to confusion over the definition within the context of universities (Leal Filho 2000), along with the many recognised barriers to implementing sustainable development in higher education (Holmberg and Samuelsson (Eds.) 2006).

Sustainability in universities has frequently been tackled through a piecemeal approach. Many institutions have focused on the practical approach of environmental management in estates. This focus has distinct benefits when realigned so campus operations are combined with academia, whether focused on teaching or research, as a form of sustainability institutional transformation (Beringer and Adomβent 2008; Zhang et al. 2011). There are many further initiatives focused on the curriculum and pedagogical change, with a considerable range of literature under the names of Education for Sustainable Development (ESD) or Education for Sustainability (EfS) published as journal articles, academic reports, and policy documents (Corcoran and Wals 2004; Sterling 2004; Tilbury 2004; Dawe et al. 2005; Martin et al. 2006; HEFCE 2009).

Whilst there might be a unifying philosophical stance underpinning ESD, there is a lack of uniformity to the approaches taken in applying ESD within universities. Different institutional constructs have different requirements. There is a growing recognition of the value of interdisciplinarity when teaching ESD in HE (Parker 2010) with some practitioners advocating a fundamental reconstruction of pedagogical approaches and curricula to meet the challenge of sustainability (Sterling 2004). Many universities have implemented, or are in the process of developing, stand-alone sustainability modules as part of their curriculum. Although it might be argued that this is detached from interdisciplinarity and embedding ESD across curricula, these single modules can play an important role in introducing sustainability to large numbers of students (Hegarty et al. 2011).

To assist in the embedding of sustainability in the curriculum, Wiek et al. (2011) have advocated a framework for key competencies in sustainability. Whilst the development of a uniform approach to sustainability has been stated to be problematic, the discussion of key competencies forms part of the wider process of sharing knowledge and approaches to sustainability, a tactic that is of value to the higher education community. It is through this premise of universities tackling the 'wicked problem' of sustainability in a collegial manner that the HEA Green Academy programme sought to address the challenge of sustainability in universities, but through the initial focused approach of ESD in the curriculum to make the scope of the project manageable.

Methodology: the Green Academy programme

The conceptualisation and development of the methodology for the Green Academy programme was by the HEA and involved a scoping exercise through a desk-based study of global ESD-related Educational Change Programmes, interviews with key ESD experts, and institutional visits. The outcomes of the scoping study were used to design a draft programme for Green Academy which was tested through further expert interviews and a focus group. This research and proof of concept resulted in a well-informed, tested structure and programme content for Green Academy.

The institutional change programme was centrally coordinated by the HEA, but it was recognised that the majority of the work was to take place within the participant institutions over a varying number of years. The HEA-led element of the programme was a set of meetings (Table 1), as well as supporting email and telephone communications.

Critical to the potential success of Green Academy was the support for the participant teams provided by a set of 'critical friends' (Costa and Kallick 1993) from other UK universities renowned for leading practice in ESD. These academic practitioners, with successful records of institutional sustainability change, cast an independent critical eye over the ideas of the project teams, along with providing encouragement through meetings. As each critical friend had led different forms of institutional change, it was recognised that each of the Green Academy participants would have their own unique set of aims and objectives. It was also evident from the outset that many of the participants were not intending to limit the sustainability work to ESD in the curriculum.

This methodology of a collaborative multi-university approach led by a central coordinating body, and supported by independent leaders in ESD pedagogy and practice is innovative within the UK HE sector. This methodological approach is valuable as it encourages shared expert knowledge and solutions to the challenge of embedding sustainability in universities. Colleagues from across the sector can learn much of great value through partnership approaches that overcome the traditional competitive barriers between institutions. Through combining experience and expertise, new approaches to institutional sustainable development may be found.

Table 1: HEA Green Academy leadership and support meetings

HEA Leadership and Support	Aims
Initial team leaders meeting (One day February 2011)	• Explore team leadership and dynamics • Introduce team leaders to models of organisational change • Identify support for the residential meeting • Facilitate a collective understanding of the evidence available to inform changes in policies and practice that enhance the student learning experience • Introduce the institutions to the group of 'critical friends'

HEA Leadership and Support	Aims
Programme Teams residential meeting (Two and a half days March 2011)	• Identify implications for each institution • Introduce a range of tools and approaches for further project development • Share experiences with, and learn from, other teams participating in the programme • Produce action plans to develop a change initiative
Final team leaders' meeting (One day May 2011)	• Reflect on progress, share learning, and outline future plans • Consider issues around impact and evaluation • Review lessons learned and generate recommendations for the wider HE sector
Institution evaluation visits (One day per HEI, February 2012)	• External (non-HEA) ESD expert evaluators to visit each HEI • Meet all team participants, other staff and students • Assess progress towards institutional change management plans • Provide critical and supportive feedback to each participant

It is not realistic within one article to satisfactorily investigate eight HEI ESD change management programmes. This article will evaluate the Green Academy programme as a potential change agent for ESD through five case study examples (Table 2), which have been selected to represent the different types of university in the UK and within this project.

Table 2: Summary comparison of the sample Green Academy universities

University	Primary focus and university grouping	Student numbers 2010 (HESA 2012)	University status	The Complete University Guide UK Ranking 2012
Canterbury Christ Church University (CCCU)	Teaching focused Research active – Million + group	17980	2005	100
University of Worcester	Teaching focused Research active	9545	2005	106
Keele University	Interdisciplinary teaching and research focused	10470	1962	44
The University of Nottingham	Research intensive, teaching focused - Russell Group	34120	1948	17
University of Southampton	Research intensive, teaching focused - Russell Group	23735	1952	14

Canterbury Christ Church University

At Canterbury Christ Church University (CCCU), participation in the Green Academy programme coincided with the development and adoption of a new Strategic Plan. This commits the institution to a whole-university approach in order that all students and staff understand the need for sustainable development and address the challenges to our common future. In implementing this strategy there was a concern that a focus on the scale and complexity of environmental problems would be too negative and alienate significant proportions of the University community. A positive focus on the future seemed to offer a better prospect for engagement, regardless of discipline.

Participation in the Green Academy programme allowed the project team to reflect on hard-earned practical lessons. It also guided the team to the academic literature, which stresses the importance of organisation change, multi-disciplinary thinking and holistic learning. There was a particular attraction to the notion of 'wicked problems' (Rittel and Webber 1973) as these can be approached with integrity from many perspectives suggesting that they can also help to integrate different subjects. This resonated strongly at CCCU.

The University is now implementing a 'Futures Initiative' that uses the Green Academy model to bring together disparate groups in 'curriculum development centres' to work with 'wicked problems' that are pertinent to their own academic discipline. This is seen as enhancing student experience and learning opportunities, without compromising the core content of the curriculum. Emphasis is also being given to a whole-curricula approach, with a focus on informal and co-curricular learning.

Strategic leadership has been followed by funding to pump-prime the 'Futures Initiative'. The first development centres attracted 12 groups from departments across the University.

- The History team are using community archaeology and working with museums to give students the opportunity to understand how history can enhance social cohesion and a positive 'sense of place' that can help lift a community out of a cycle of negative perceptions and deprivation.
- A Business Studies module centres on the behaviours of individuals and groups, and the interactions between them in an organisational context.
- The Faculty of Education has invited a group of students to plan a module on sustainable schools for the future in which they will do most of the teaching. This is the first time that students within the Faculty have taken the lead in devising a validated course.

The experience of CCCU to date is that focusing on the future has generated considerable interest from a wide range of colleagues. By giving sustainability education a distinctive angle support has been galvanised. The challenge will be to maintain this momentum and see that the changes that are starting to happen are embedded and effective in the longer term.

University of Worcester

Participation in the Green Academy programme has led to a number of positive outcomes at the University of Worcester, building on a strong existing base of sustainability achievements. The first notable advantage from participation has been involvement in sector-wide, cross-institutional knowledge exchange with the critical friends and other participants. The critical friends have proven to be an excellent resource, and continue to be so. For example, the Plymouth Model (Campus, Curriculum, Community, Culture) (Jones et al. 2010) has been useful in guiding the institutional holistic approach to sustainability. Shared knowledge and experience from other participants has been a key element in the sustainability work, and is likely to continue to be so in the future.

The requirement for student representation in the project team has led to their enhanced involvement in curriculum design. Worcester had two students on its team of five, including the President of the Students' Union, and they made significant contributions in devising strategies to engage students in sustainability. Various student societies and STaRs (reps) have also been involved in a diverse range of sustainability-related projects. At recent student recruitment events, potential students were asked about sustainability issues and curriculum models for their future study.

Participation in the residential meeting allowed the team the space and time to think and act collaboratively. This enabled the team to design an entire curriculum and modules away from the daily routine of work life. The team were able to identify colleagues already championing work in their institutes to locate benchmarks and encourage support. Upon their return, the team received peer support, pedagogic tools, and resources to build on the work achieved at the residential meeting.

The most notable outcomes of the project have been the development of new academic programmes, and a recognisable Green Academy brand to raise the profile of sustainability amongst staff and increase awareness at top management. The knowledge and advice gained from the project has guided the discussion beyond environment and facilities to the curriculum. This was perhaps the most important contribution of the Green Academy programme to the University of Worcester. The Undergraduate Review will result in a major change for all programmes of study with sustainability as one of the cross-institution offers to students. It is hoped this will impact on the entire institution for many years to come.

Keele University

Prior to participation in Green Academy Keele University had in place a number of sustainability initiatives, including significant funding from the Higher Education Funding Council for England for the renovation of a dilapidated farm building on the University campus to serve as a 'Hub' for sustainability activity, and a new position of Pro-Vice Chancellor for Environment and Sustainability. There were pockets of ESD activity, including dedicated sustainability undergraduate and postgraduate programmes, and a sustainability-related module available to all first-year undergraduate students, but there was little coherent strategy for ESD and no clear integration with other sustainability activities.

The first major outcome of Keele's involvement in the Green Academy initiative was that 'Sustainability' was made one of three key themes (alongside Internationalisation and Employability) within a major institution-wide curriculum review project. Inclusion of 'Sustainability' into this project has immediately increased the exposure of all, and in particular academic, staff to the role of Sustainability in HE, even if initially they are unsure of its implications and relevance to their own disciplines or roles.

Keele's Green Academy team outlined a 'multi-strand' vision to inserting ESD into the University's activities, ranging from embedding sustainability within all programmes, inclusion of opportunities in the co-curriculum, options within postgraduate taught and research student training, and inclusion in Continuing Professional Development. Staff development has formed a major part of activities, including workshops for the Programme Directors of all the institutions' programmes, along with new training for academic staff. Alongside these ESD developments there have been other major University sustainability developments, including the launch of the University's new Strategic Plan, which places strong emphasis on the environment and sustainability in the University's Mission Statement, Vision and Strategic Aims. The institution has also unveiled a new logo, which includes a green stripe to represent the University's commitment to the environment and sustainability (along with its physically green campus), and the appointment of its new Chancellor, the environmentalist and writer Jonathon Porritt. This provides evidence of strong top-level support, but also increased internal pressure to deliver on these commitments.

Progress on the multi-strand strategy developed through Green Academy sessions is subject to resourcing. The case for this was helped by reference to the resources put in place in other Green Academy institutions. Attainment of the vision developed during Green Academy is a long way off, but that outlined vision provides a blueprint to return to in order to check progress and direct initiatives.

Overall, as a result of involvement in Green Academy, education has become a much stronger part of the sustainability agenda at Keele, and potentially, through emphasis on the whole student experience, even a unifying theme.

The University of Nottingham

For the University of Nottingham, the Green Academy programme presented an opportunity to progress an ESD agenda from supportive statements in various strategy documents (e.g. Environment, Teaching and Learning, Knowledge Transfer) to practical action. Green Academy provided crucial networking opportunities with colleagues at different stages of their own institutional journeys (the 'critical friends' being the most advanced) and dedicated time for a team to forge a shared vision that would take forward the embedding of ESD in the curriculum but also in the behaviours of staff and students.

The initial aim was to create an engagement strategy using a variety of methods to prioritise ESD within the institution. This strategy included provision of teaching resources (primarily for undergraduates), and active learning opportunities including employer-led and community projects. It included the development of a research agenda around the pedagogy of ESD and the foregrounding of ESD as a global higher education priority area. Crucially, it also involved 'piggy-backing' on other initiatives where quick gains could be made.

There were two central props to the strategy. First, the establishment of ESD as a 'Grand Challenge' for the University. Grand Challenge status signalled senior management buy-in, central administrative support, a mechanism for consulting with academic schools, and access to internal funding (existing Grand Challenges existed for 'Assessment in the Digital Age' and 'Internationalisation of the Curriculum').

Secondly, the Nottingham team's shared vision was based upon the University of Plymouth's '4C's' model (Jones et al. 2010). This model allowed the team to sell ESD as shared understanding rather than agreed definition – it meant that existing activity could be labelled as ESD and that academic staff could feel part of the project rather than its target. The '4C's model also moved the sustainablity discussion away from a primarily environmental perspective to a broader one, perhaps effectively summed up as a '5th C' – *contribution* towards a longer-term goal.

ESD was successfully harnessed to the Nottingham Advantage Award, an extra-curricular programme to support students in their career and personal development. Sustainability literacy and ESD became strong themes, supported by external funding to develop new open educational resources. A Lord Dearing Fellowship was awarded to produce a video project exploring different understandings of sustainability and ESD across faculties, disciplines, and stakeholders (an idea 'borrowed' from a Green Academy 'critical friend').

One year on from the Green Academy, Nottingham sees sustainability literacy as an important graduate capability and is debating different models for supporting it within the student experience – embedded, integrated, bolt-on? It sees the work as being about content, pedagogy and process. It knows it is on a journey, and that the journey is without end!

University of Southampton

The University of Southampton has a long, successful and diverse engagement with sustainability, yet this was historically fragmented, with mixed collaboration between different academic groupings. This likely resulted in many unquantified missed opportunities and duplication of effort, and did not make best use of the teaching and research expertise available across the University. Involvement in the Green Academy programme was intended to bring together the disparate activities to provide a coherent strategy for sustainability practice across the University. The long-term vision to appear from the Green Academy process is for the University of Southampton to become a recognised leader in the HE sector with sustainability in the CORE (Curriculum Operations Research and Experience) of the institution, reflecting the University Strategy.

The first difficulty encountered in the Green Academy project was reaching consensus on how sustainability related to the institution. Participation in the residential event allowed the team to work with other participants and the 'critical friends' to establish the institution's position. The event also gave time to work on the development of the University's sustainability plan, focusing on objectives, key performance indicators and completion dates in the four broad areas of embedding sustainability in the curriculum, as well as leading behaviour change to reduce operational impacts, enhancing cross-disciplinary sustainability research, and improving the student sustainability experience.

Progress at the University focused on engaging highest-level management support. Involvement in the Green Academy programme eased this process as it was possible to highlight the progressive work of comparable competitor institutions. This proved to be a beneficial tool whilst constructing executive-level board reports and business case documents for circulation to the wider university community. The support of the Environment and Sustainability Advisory group and key individuals led to endorsement of the strategy at the highest-level. This led to funding of a dedicated post, beginning in March 2012 to assist with embedding sustainability in the CORE of the University, with a particular focus on behaviour change.

Clearly this is only the beginning of the long and possibly endless path towards sustainability. However, being part of a collegial programme of other universities has provided vision and given the project considerable leverage at significant junctures by being able to highlight the work of partner yet competitor institutions.

Conclusion

This paper has demonstrated that the Green Academy programme has led to notable progress in the pursuit of sustainability in the case study institutions. A number of common themes have emerged. There has been recognition of the problematic nature of gaining consensus on a focus and definition for sustainability in the context of the institution. This reflects the wider difficulties of agreeing upon the meaning of sustainability, and remains indicative of its weakness (Leal Filho 2000). Approaches have focused on the positive nature of these 'challenges' rather than the traditional negativity associated with sustainability 'problems'. The need for strategic engagement with senior management has been demonstrated to be critical in securing resources for initiatives. This engagement has been shown to be strengthened when evidence can be provided of student engagement, their importance in acting as change agents (Myers and Beringer 2010), and the successes of competitor institutions. The role of the critical friends' has been proven to be important in helping all participants realise and link their existing successes, as well as assisting in their individual visions for sustainability in their institutions.

The most successful outcome of this programme has been the collegial approach adopted by the HEA, critical friends, and the institution teams. Participants have developed their own informal ESD change management network where they exchange ideas, request help, provide assistance, offer encouragement, and share resources that have helped in their institutions. Team leaders have also delivered joint workshops to delegates in the 4th International Conference on ESD and Global Citizenship 2011, and the 2012 EAUC national conference, extending the reach of this change management programme across the UK HE sector. This approach will be continued as the individual programmes mature, hopefully resulting in further positive sustainability stories.

Whilst this project has developed into an active network of universities for the purpose of pursuing sustainability in HE, the distinctiveness of individual institutions and the diverse nature of the higher education community should be recognised when considering adopting identical practices and strategies. No two universities are the same. Each is distinguished from others by factors such as size, geography, community relationships, infrastructure, financial resources, student community, strategic priorities, curriculum structure and content, research output, and the expertise of the academic community. All these factors have a potentially complex bearing on the approach to sustainability within a university. Although the unique nature of each university is part of the richness of the higher education community, this does make the construction of a standardised framework for achieving sustainability in universities an almost impossible task.

The main lessons learnt from this project can be summarised: as the necessity to focus on positive challenges rather than negative problems; the criticality of engaging senior management for resource support; the value of critical friends; and the progression that can be achieved through multi-university partnerships driving friendly competition. The combination of these separate elements into a successful interweaved programme for sustainability in universities has only been made possible through the neutral guidance of a central coordinating body.

Perhaps the greatest challenge for this programme lies in its future prospects as a model for change across the wider UK HE sector. The HEA intend to repeat the Green Academy programme in 2013 with the recruitment of new institutions. Only by analysing the outcomes of the second and future cycles of the Green Academy programme will it be possible to accurately determine whether this model does produce an operational change model that moves away from a piecemeal towards a collegial network approach, focused on knowledge and practice sharing along with friendly competition to drive not only ESD but whole institution change for sustainability.

References

Beringer, A. and Adomβent, M. (2008), Sustainable university research and development: inspecting sustainability in higher education research. *Environmental Education Research* 14 (6), pp. 607-623.

Bone, E. and Agombar, J. (2011), *First-year attitudes towards, and skills in sustainable development.* Higher Education Academy. York. UK.

Corcoran, P.B. and Wals, A.E.J. (2004), *Higher education and the challenge of sustainability: problematics, promise, and practice.* Kluwer Academic Publishers, Dordrecht.

Costa, A.L. and Kallick, B. (1993), Through the Lens of a Critical Friend. *Educational Leadership* 51 (2), pp. 49-51.

Dawe, G., Jucker, R. and Martin, S. (2005), *Sustainable Development in Higher Education: Current Practice and Future Developments.* A report for the Higher Education Academy. Higher Education Academy. York, UK.

Eichler, M. (1999), Sustainability from a feminist sociological perspective: a framework for disciplinary reorientation. In Becker, E. and Jahn, T. (Eds.), *Sustainability and the social sciences: a cross-disciplinary approach to integrating environmental considerations into theoretical reorientation*, pp. 182-206, Zed Books, New York.

Hegarty, K., Thomas, I., Kriewaldt, C., Holdsworth, S. and Bekessy, S. (2011), Insights into the value of a 'stand-alone' course for sustainability education. *Environmental Education Research* 17 (4), pp. 451-469.

HEFCE (Higher Education Funding Council for England) (2009), *Sustainable development in higher education, Statement of Policy February 2009/03*. HEFCE, Bristol.

HESA (Higher Education Statistics Agency) (2012), *Headline statistics – Institutions 2009/10*. www.hesa.ac.uk (accessed 26 February 2012).

Holmberg, J. and Samuelsson, B.E. (2006), *Drivers and Barriers for Implementing Sustainable Development in Higher Education*, UNESCO Education for Sustainable Development in Action Technical Paper No. 3. UNESCO.

Jacobs, M. (1999), Sustainable development as a contested concept. In Dobson, A. (Ed.), *Fairness and futurity: essays on environmental sustainability and social justice*, pp. 21-45, Oxford University Press, Oxford.

Jones, P., Selby, D. and Sterling, S. (2010), *Sustainability Education: Perspectives and Practice Across Higher Education*, Earthscan.

Leal Filho, W. (2000), Dealing with misconceptions on the concept of sustainability. *International Journal of Sustainability in Higher Education* 1 (1), pp. 9-19.

– (2011), About the Role of Universities and Their Contribution to Sustainable Development. *Higher Education Policy* 24, pp. 427-438.

Martin, S., Dawe, G. and Jucker, R. (2006), Embedding Education for Sustainable Development in Higher Education in the UK. In Holmberg, J. and Samuelsson, B.E. (Eds.), *Drivers and Barriers for Implementing Sustainable Development in Higher Education*. UNESCO Education for Sustainable Development in Action Technical Paper No. 3, UNESCO.

Myers, O.E. and Beringer, A. (2010), Sustainability in Higher Education: Psychological Research for Effective Pedagogy. *Canadian Journal of Higher Education* 40 (2), pp. 51-77.

Parker, J. (2010), Competencies for interdisciplinarity in higher education. *International Journal of Sustainability in Higher Education* 11 (4), pp. 325-338.

Rittel, H. and Webber, M. (1973), Dilemmas in a General Theory of Planning. *Policy Sciences* 4, pp. 155-169.

Sterling, S. (2004), Higher education, sustainability, and the role of systemic learning. In Corcoran, P.B. and Wals, A.E.J. (Eds.), *Higher Education and the Challenge of Sustainability: Problematics, Promise and Practice*, pp. 49-70, Dordrecht: Kluwer Academic Publishers.

Tilbury, D. (2004), Environmental Education for Sustainability: A force for change in Higher Education. In Corcoran, P.B. and Wals, A.E.J. (Eds.), *Higher Education and the Challenge of Sustainability: Problematics, Promise and Practice*, pp. 97-112, Dordrecht: Kluwer Academic Publishers.

Wiek, A., Withycombe, L. and Redman, C.L. (2011), Key competencies in sustainability: a reference framework for academic program development. *Sustainability Science* 6, pp. 203-218.

World Commission on Environment and Development (WCED) (1987), *Our Common Future*. Oxford University Press.

Zhang, N., Williams, I.D., Kemp, S. and Smith, N.F. (2011), Greening academia: Developing sustainable waste management at higher education institutions. *Waste Management* 31 (7), pp. 1606-1616.

Chapter 31

Greening the Campus: The Economic Advantages of Research and Dialogue

Eric Pallant, Kelly Boulton and David McInally

Abstract

The goal of our research was to assess the impact that student-faculty research teams in Allegheny College's department of Environmental Science had in initiating a broad move toward campus sustainable development. The following paper adds to this literature by explaining how Allegheny College changed its culture to one wherein nearly all campus constituents – students, faculty, administrators, trustees, and maintenance staff – came to consider sustainability within the purview of their regular activities, despite a comparatively small endowment and limited investment in personnel dedicated to promoting sustainability. The authors used a qualitative approach, semi-structured interviews, and a review of Allegheny's efforts – both failed and successful – to identify significant strategies and partners. When bottom-up efforts and top-down leadership converged, a rapid shift towards an integrated sustainability culture took place. Student research, written reports, development of partnerships, dialogue, and the integration of sustainability principles into existing projects and budgets created a comprehensive climate for sustainability. This case study identifies both effective and ineffective practices that have transformative value for Allegheny's continuing sustainability efforts as well as lessons for similar higher education institutions.

Introduction

Higher education institutions (HEI) are important actors (along with government, business, the military, etc.) in moving societies toward more sustainable practices (Orr 1992; 1994). Colleges and universities are in the business of teaching critical thinking about sustainability and their primary clientele, students, are simultaneously transient and destined for post-graduate positions of leadership (Carlson 2006; Emanuel and Adams 2011). Campuses can create a continuous flow of graduates interested in and capable of promoting sustainable development (Bartlett and Chase 2004; Keniry 1995; Leal Filho 2002; van Weenen 2000 Ward, 1999).

To achieve success – i.e. to become a campus that does more than just teach sustainability, but actually takes theory and puts it into comprehensive practice – requires considerable effort and focus (Krizek et al. 2012; Miller et al. 2011; Moore 2005; Shi 2008; Shriberg 2002). Barriers to creating sustainable HEIs include perceived cost, lack of knowledge of potential savings (Dahle and Neumayer 2001), aging infrastructure (Roberts and Murphy 2011), up-front costs (Pelletier 2008), uncertain economic conditions (Carlson 2008), the lack of commitment from top management (Lozano et al. 2011; Williamson 2012), existing literature on HEI sustainability that tends to focus on self-aggrandizing case-studies (Corcoran et al. 2004), cumbersome bureaucratic barriers (Sharp 2009), and a narrow understanding of the university's environmental impacts (Christensen et al. 2009; Velazquez et al. 2005).

It has been suggested that overcoming institutional barriers to sustainable development requires investment in social capital (Evangelinos and Jones 2009; Steiner 2008), top-down support from administrators and trustees (Lozano et al. 2011; Williamson 2012), a grassroots or bottom-up approach (Brinkhurst et al. 2011), the active involvement of staff (Sibbel 2009), and a comprehensive approach that insists on all of the above (Pittman 2004; Richardson and Lynes 2007; Sharp 2002).

Hands-on research by teams of students and faculty directed at real world issues of sustainable development has proven to be an effective educational tool (Brundiers and Wiek 2011; Shepherd 2008; Stauffacher et al. 2006; Steiner and Posch 2006). The goal of our research was to assess the impact that student-faculty research teams in Allegheny College's department of Environmental Science had in initiating a broad movement toward campus sustainable development and recognition as a national leader in environmental responsibility (see Table 1 for list of commitments).

Table 1: Abbreviated list of sustainability actions on Allegheny College's campus

Sustainability Development at Allegheny College
• Goal of climate neutrality by 2020 included in the strategic plan • Committed to the US Department of Energy's Better Buildings Challenge • 100% of electricity purchased with wind renewable energy credits • Campus grounds maintained with organic practices • On-campus compost operation uses food wastes to fertilize lawns, flower beds and sports fields • All new construction required to achieve LEED Silver or better • Purchasing policies require ENERGY STAR appliances and EPEAT certified computers

A first try: A failure

In the early 1990s, Allegheny College began planning for a major expansion of its athletic facilities, including the addition of a performance arena, a fieldhouse large enough for four, full-size indoor recreational courts, an elevated indoor track, and more than a dozen office spaces. The ES department advocated ground source heat pumps (GSHP) to heat and cool the facility, but being inexperienced in the process of on-campus advocacy, prepared no reports. In the absence of demonstrative research and reports submitted to the administration early in the planning process, a conventional HVAC system was chosen by default, since the purchase, costs, operation and performance were within the expertise and expectations of the Physical Plant staff. The resulting structure's high ceilings and large, open space has proven to be extravagantly expensive to heat and cool.

Had the ES department submitted reports and case studies of success at comparable facilities, rather than simply verbally advocating GSHP, this story might have ended differently. The same is true of a struggling recycling program begun in the 1989s, which was barely functioning by the mid 1990s, and of campus construction projects, where the College failed to select green architects and builders for several major building initiatives. These three cases – a meager recycling program, the failure to invest in GSHP, and missed opportunities in new construction – illustrated the challenges Allegheny faced in supporting lasting sustainability programs, and prompted the ES department to reach the following conclusions with respect to the obstacles it would need to overcome.

First, research is a prerequisite. The ES department began making successful recommendations for environmental sustainability only after sound research first demonstrated that alternative technologies and materials were available, acceptable, and high performing.

Second, cost matters. The most successful and swiftly adopted suggestions emphasized efficiency and cost savings.

Third, some personnel were impediments by virtue of their position of authority. On more than one occasion sustainability advocates were frustrated by personnel who either did not understand or did not accept the notion that there was an advantage to green methodologies over conventional methodologies. Over time the ES emphasis on efficiency and cost savings, and to a lesser extent a national and international discussion on sustainability, led to the hiring of new personnel and continuing education for sustainability of existing personnel.

Methods

Adopting an emancipatory usage of the case study model, the authors sought to consider how Allegheny's process and learning could illuminate issues, ideas and techniques that may be useful at other institutions of higher education (Corcoran et al. 2004). The process and results of this case study have informed further institutional change at Allegheny College. The authors pursued a qualitative approach, using semi-structured interviews with a past President, VPs of Finance and Planning, Physical Plant staff, and ES faculty. The authors also observed and reviewed Allegheny's historic efforts and student research, giving equal value to a consideration of failures and successes as sources of lessons learned. All three of the authors have been critically involved in the progression of sustainability at Allegheny for over a decade, and therefore witnesses to the transformation from a college that supported environmental sustainability in principle into a college whose actions embody those principles.

Campus Constituencies

Physical Plant

One year after high energy bills for the newly renovated fieldhouse began arriving, Allegheny was offered the opportunity to apply for a grant to support the purchase of an in-vessel composter. The ES department dedicated a semester-long seminar for junior majors to the project (Pallant 2001). Their report was delivered to the administration just prior to the arrival of the composter and suggested ways to make the most cost-effective use of the machine (Pallant et al. 2002). When the grant was funded and the composter arrived, the college's expenses increased because of the addition of an employee required to collect compostable materials on a daily basis from the campus dining halls and to operate the composter. Nevertheless, these expenses were offset entirely by savings to the campus. First, the college saved money by avoiding waste hauling fees to the landfill. Second, the college no longer needed to purchase topsoil and fertilizers. Additionally, and significantly, the Physical Plant director recognized substantial positive publicity for the actions undertaken by the college (Spencer 2008; Sullivan 2010).

The next major construction project on campus (opened in 2006) was the addition of a 106-bed dormitory called North Village I. The ES department employed a junior seminar of 14 students to find ways to make the building greener (Pallant et al. 2004). Working with the architects and Physical Plant, the students suggested the building be LEED certified and again recommended the installation of ground source heat pumps (GSHP). This time, a year-long senior thesis analyzed the payback time expected for GSHP: about ten years (Davis 2004). GSHP was installed and the economic savings exceeded student projections. Other student recommendations, such as porous paving parking lots (as opposed to expensive, conventional retention systems for storm water), the use of green cleaning products to enhance indoor air quality, and extensive skylighting also proved to be both financial and societal successes.

Finance and Planning

As the director of Physical Plant came to learn from experience that student and faculty reports and recommendations to increase sustainability were beneficial to his budget and ease of operation, he passed along analyses of his successes to his bosses in the administration. While demonstrative reports from the ES department had helped convince the Physical Plant, financial analyses of successful implementation of certain measures helped garner new support in Finance and Planning. These successes then prompted Finance and Planning to seek additional sustainability opportunities.

When the long-serving Director of Physical Plant retired, his superiors in Finance and Planning demonstrated their serious support of sustainability by advertising for a replacement who could promote sustainability across campus as part of his job. Had the VPs of Finance and Planning not been included in many sustainability conversations, and had they not reviewed many analyses of how efficiency measures improved budget scenarios, any progress on developing relationships and implementing sustainable practices could have been lost in the retire/rehire process. Since they had been educated and convinced of the budgetary and moral values of sustainability, they were able to ensure support for sustainability increased with a new hiring rather than fading away with a retirement. The new Director of Physical Plant not only sits in on junior seminars in ES as a participant, but is now one of the leading advocates on campus for environmental sustainability.

Initially, Physical Plant and Finance and Planning support for sustainability centered primarily around financial considerations. GSHP units were used in two large residence hall constructions and one major renovation for offices. Despite the upfront cost, GSHP units were included in the scope of work because of the anticipated and then proven financial payback due to significant reductions in natural gas consumption for heating. As the campus' overall consumption of electricity and natural gas per square foot declined, operational and administrative support for sustainability increased. However, since sustainability is about much more than the finances, there was still room for the conversation and institutional support to evolve.

Admissions

With student interest in campus sustainability on the rise, the Admissions Office expressed interest in highlighting various sustainability features when they were slated to move to an old building being renovated for their use. Physical Plant was already committed to installing GSHP because of excellent returns on investment following installation in North Village I. Recommendations prepared by senior theses in ES were shared with the Admissions team, Physical Plant and Finance and Planning. One senior thesis analyzed the cost-effectiveness of bioswales to capture rainwater and snowmelt (Silling 2009). These were built during the renovation and wetlands of native plants now filter runoff from the (porous) parking lot and roof. A second thesis recommended that furniture inside the building be locally sourced from FSC-certified (Forest Stewardship Council) timber (Conti 2009). That thesis also suggested that solar tubes be installed on the second floor of the building. The FSC furniture was purchased; the solar tubes were not (but they have been added to subsequent projects and become so desirable that other departments are asking for and receiving them).

As 64% of high school seniors now state that they would value knowing about campus sustainability commitments as they select a college or university (The Princeton Review 2011), the Admissions Office was pleased to highlight readily identifiable sustainability features in the first building prospective students visit, including native plants all around, porous paving, waterless urinals, recycled carpet (and recycled building), student-designed bioswales, and FSC-certified furniture. When students later proposed preparing flyers, campus signage, green tours, and virtual green tours, Admissions directors were quick to offer support (Elick 2011).

The President

In 2005, Second Nature, a non-profit NGO, began plans to gather a small cadre of college and university presidents willing to commit their institutions to carbon neutrality to become inaugural members of the "American College and University Presidents' Climate Commitment" (ACUPCC). This agreement binds participating institutions to a carbon budget analysis, a climate reduction plan, and a goal of climate neutrality. Anyone who has given serious thought to making such an agreement understands how large an undertaking this entails (Swearingen White 2009).

The logic of the ACUPCC is to "accelerate progress towards climate neutrality and sustainability by empowering the higher education sector to educate students, create solutions, and provide leadership-by-example for the rest of society" (ACUPCC.org). President Richard Cook, then Allegheny College's president, was approached and asked to be a member of the founding circle, as well as a charter signatory.

The authors asked now-retired President Cook why he agreed to put Allegheny, a college with fewer financial resources than other founding signers, in a position that could be quite costly. He explained the ES faculty had always pushed the college to stay on the cutting edge of environmental sustainability, and most importantly, their recommendations were always accompanied by sound research. Reports prepared by their students and in their classes, he said, were reliable and persuasive. History had taught the college administration that sound research, could be used in support of morally important goals.

In short order, the ES faculty, other top administrators, and the Board of Trustees expressed their support for a Presidential signature. The Chair of the Board supported the initiative because the Board had previously approved college-wide Environmental Guiding Principles. In short, by 2005 Allegheny College was in a position where there was a confluence of a national (and international) movement toward greater campus sustainability (Leal Filho 2002), broad internal support for campus greening as a consequence of years of discussion, and powerful leadership from the President and his immediate staff. President Cook summarized his decision by noting: "There is a risk in making this decision. It is going to be very difficult to meet the commitment, but it is something we must do. There is also a risk in not moving on this issue and to fail to engage our students in this mission would have been a big mistake."

Building on the momentum of the national dialogue on issues of sustainability in higher education, the ES department spent two years lobbying President Cook to establish a college-wide committee of students, faculty, staff, and administrators to oversee issues of campus sustainability. ES students prepared reports detailing similar committees at other institutions of higher education working in conjunction with a campus sustainability coordinator. Their research confirmed that the combination of coordinator plus committee was definitely the most successful model for promoting campus sustainability. President Cook refused, saying the college budget was insufficient to hire a coordinator; human resources were insufficient to support such a committee.

In its place he appointed a Task Force for Environmental Responsibility (TER) with an 18-month lifespan, chaired by an ES professor. The TER again recommended a campus committee and a sustainability coordinator. It got the coordinator position via an agreement between the Director of Physical Plant and the Chief Financial Officer, who together created the position by rearranging existing jobs in anticipation of pending retirements. In other words, the college recognized the potential value of joining the national progression toward hiring sustainability coordinators, but did it in a way that did not cost additional money.

Food Services

Allegheny College contracts with an independent service to provide dining services in two dining halls as well as a catering service for campus functions. For many years, Allegheny's provider was reluctant to make more than minimal efforts to supply local, organic, or fair-trade products, since changes not anticipated in the original contract might increase their costs. When this contract was up for renewal in 2008 a thorough review of students' desires led the administration to build sustainability values, along with freshness, directly into the contract. The new contract ensures a level of sustainability, but also establishes a strong foundation for further progress as a result of campus and student demand (see Table 2).

Table 2: Sustainable dining practices at Allegheny College

Food Services
• 30% of food sourced from regional growers and suppliers • Milk is rBGH free and eggs are cage free • Seafood sourced according to Monterey Bay Aquarium guidelines • DeHart Local Foods Dinner sponsored annually • Campus-grown herbs used in dining halls • Breads/rolls purchased from local bakery • Baked goods purchased from local farm/home kitchen • Free filtered water and discounts for using reusable mugs/bottles provided to all students • All pre- and post-consumer wastes are composted • All used fryer oil donated to City for conversion to biodiesel • Coffee is Fair Trade

Housekeeping

At the recommendation of the ES seminar researching the LEED process, Allegheny implemented a green cleaning program to protect the indoor air quality in North Village I (Pallant et al. 2004). For several years, North Village I was cleaned in an environmentally sensitive way while the remainder of the campus still used conventional cleaning chemicals and processes. Collaborating with St. Moritz, the college's contracted housekeeping company, the sustainability coordinator developed a comprehensive green cleaning program for every building on campus that would continue to demonstrate improvement over time as better products reached the market and employee training improved (see Table 3).

Table 3: Sustainable housekeeping practices at Allegheny College

Housekeeping
• Converted to greener cleaning chemicals for both the health of the environment and the individuals who clean as well as work, sleep and study in these spaces • Developed a training program to help support transition to green cleaning program • Purchased concentrated cleaners to limit amount of packaging and shipping weight • Converted to100% recycled-content toilet paper and paper towels • Employs triclosan-free (an antimicrobial) liquid hand soaps • Uses trash can liners containing recycled plastic • Diverts about 30% of campus waste through collecting recyclable material • Conducts regular audits to ensure that all spaces have adequate, well-labeled recycling receptacles

A complete greening of our housekeeping proved to be more complex than expected. While St. Moritz buys their own cleaning products and materials, Allegheny College's Physical Plant purchases all the toilet paper, paper towels, can liners, liquid hand soap, and trash and recycling receptacles mostly from a local business – DeSantis Janitor Supply. To switch to recycled-content paper products and triclosan-free hand soaps required dialogue between Physical Plant and DeSantis to explore other options and budget impacts. With St. Moritz, Physical Plant, and DeSantis involved in the decisions affecting our housekeeping program, it has been essential to dig into the details of how things work together and use the expertise and influence of each to leverage more sustainable practices.

Students

In recent years, the work of student groups such as Students for Environmental Action, Edible Allegheny Campus, and Allegheny Student Government have done much to promote sustainability initiatives on campus. In the absence of the sustainability committee recommended by the Taskforce for Environmental Responsibility, the sustainability coordinator developed an informal Sustainability Coordinating Committee, a monthly opportunity for these student groups, individual students, and campus stakeholders (such as Physical Plant, dining services, Finance and Planning, ES department) to meet to discuss current work, opportunities for collaboration, and how to combine their efforts to effect greater sustainable change.

Students are most effective when they combine serious dialogue and advocacy with fun methods of educating and engaging their peers. This "Fun Theory" (Volkswagen 2009) approach has proven to be extremely successful in motivating students and gathering a larger audience than simply the usual suspects. Students turned the annual Energy Challenge into a month-long series of fun events, including a bonfire, acoustic concert, and night games like glow bocce and hide-and-seek, all with the intent of educating and promoting responsible energy behaviors. Students also explore waste minimization with an annual Trashion Show. Decked out in fashions crafted from what otherwise might have been considered waste, students walk the runway to an increasingly large crowd while sharing facts and tips on how to minimize daily wastes. Other student groups promote local foods, sustainable living, and consumerism with bike-powered blenders, gardening workshops, and clothes swaps, making sustainability fun.

Conclusions

This research project clarified the practices that were most – and least – successful in transforming Allegheny from a school of good intentions to an institution of effective action. Several lessons emerged that are replicable for other institutions of higher education.

First, sustainability initiatives should be grounded in *faculty-student research*. Many students and professors advocate environmental responsibility, but moving the entire institution to action requires the sort of intellectual rigor that should be the hallmark of higher education in any case. Institutions can leverage the investments that already exist in curricula, faculty expertise, and laboratory resources by placing these real-world problems in the context of academic teaching, learning, and research.

Second, a successful program requires *education and reinforcement* for campus decision-makers. Many good suggestions are rejected not for lack of merit, but because of ignorance on the part of those in authority. Because it is unlikely that busy people will delve deeply into climate studies and environmental best practices on their own volition, it is vital to create early success stories – beginning with modest ones if necessary – and reinforce the decision-makers even if they were initially reluctant participants. When a facilities director is recognized by peers at a regional conference for reducing energy use, or a budget officer is thanked by trustees for managing utility expenditures, he or she is more likely to welcome new sustainability initiatives.

Third, a combination of *bottom-up* and *top-down* approaches is essential to building and sustaining momentum. This is where the power of research – labeled "bottom up" for the purposes of this study – and dialogue intersect to create lasting change. Even when backed by convincing evidence, excessively strident voices can lead to intransigence among those in authority, and any chief financial officer worth his or her salt can find a way to block new initiatives by miring them in the complexity of financial systems that are generally not well understood by the campus community. Genuine dialogue is required in order to approach problems from a variety of perspectives, where each participant can bring his or her expertise to particular dimensions of a problem – financial, physical, social, political, moral, and educational. No matter how laudatory a new idea may be, it is unlikely that any one individual or department possesses the expertise in the variety of fields needed to achieve comprehensive and lasting solutions.

Fourth, institutions must address *structural questions* in order to create lasting change. These questions include: Is environmental responsibility part of the institution's mission? Are the curricula, staffing priorities, and organizational charts aligned with environmental goals? Is sustainability included in the strategic plan? Do institutional marketing and communications reflect the organization's commitment? What will an institution do when short-term, fast-payback projects have been completed? Answering these questions is essential to move from a loose partnership of like-minded individuals to a sustainable organization, where cross-departmental coalitions of people are clearly charged with responsibility for environmental sustainability, and where research and dialogue are valued components of the campus culture.

After nearly twenty years of intermittent progress toward sustainability, Allegheny College experienced a tipping point. As the bottom-up and top-down messages converged, a fairly rapid shift took place. Beginning in the late 2000s, sustainability projects began to appear all over campus with great frequency. Interestingly, for the first time, most of these projects were initiated without provocation or even awareness on the part of ES faculty. This momentum led to the kind of structural changes that demonstrate fundamental transformation, including establishing the sustainability coordinator position, purchasing 100% wind power, renovating a facility into a high-profile home for the ES department, and incorporating an ambitious climate-neutrality goal into the strategic plan.

Sustainability is a long-term, multi-dimensional goal. No matter what resources are available in the short term, all institutions can consider these structural questions and reorganize their systems to align with their missions, and to emphasize the role of research and the opportunities for dialogue. Over time these structures will influence job descriptions, hiring decisions, and budget priorities, potentially leading to the tipping point where environmental sustainability is fundamental to the campus culture.

References

Bartlett, P.F. and Chase, G.W. (Eds.) (2004), *Sustainability on campus: Stories and strategies for change.* MIT Press, Cambridge, MA.

Brinkhurst, M., Rose, P., Maurice, G. and Ackerman, J.D. (2011), Achieving campus sustainability: top-down, bottom-up, or neither? *International Journal of Sustainability in Higher Education* 12 (4), pp. 338-354.

Brundiers, K. and Wiek, A. (2011), Educating students in real-world sustainability research: vision and implementation. *Innovation in Higher Education* 36, pp. 107-124.

Carlson, S. (2006), The sustainable university. *The Chronicle of Higher Education*, October 20.

Carlson, S. (2008), News Analysis: Financial Woes May Shelve Sustainability. *The Chronicle of Higher Education,* October 9.

Conti, A. (2009), *Incorporating green materials into Allegheny's Admissions House. Unpublished senior thesis, Allegheny College.* Retrieved from http://webpub.allegheny.edu/dept/envisci/ESInfo/comps/2009abstracts.html#conti.

Corcoran, P.B., Walker, K.E. and Wals, A.E.J. (2004), Case studies, make-your-case studies and case stories: a critique of case-study methodology in sustainability in higher education. *Environmental Education Research* 10 (1), pp. 7-21.

Christensen, P., Thrane, M., Herreborg Jorgensen, T. and Lehmann, M. (2009), Sustainable development: assessing the gap between preaching and practice at Aalborg University. *International Journal of Sustainability in Higher Education* 10 (1), pp. 4-20.

Davis, T. (2004), *A Feasibility Study on the Use of Geothermal Ground Source Heat Pumps for Allegheny College's North Quad.* Unpublished senior thesis, Allegheny College. Retrieved from http://webpub.allegheny.edu/dept/envisci/ESInfo/comps/2004 abstracts.html#davis.

Dahle, M. and Neumayer, E. (2001), Overcoming barriers to campus greening: a survey among higher educational institutions in London, U.K.. *International Journal of Sustainability in Higher Education* 2 (2), pp. 139-160.

Elick, A. (2011), *Allegheny College Sustainability Tour. Unpublished senior thesis.* Allegheny College. Retrieved from http://alleghenysustain.weebly.com/index.html.

Emanuel, R. and Adams, J.N. (2011), College students' perceptions of campus sustainability. *International Journal of Sustainability in Higher Education* 12 (1), pp. 79-92.

Evangelinos, K.I. and Jones, N. (2009), An analysis of social capital and environmental management of higher education institutions. *International Journal of Sustainability in Higher Education* 10 (4), pp. 334-342.

Keniry, J. (1995), *Ecodemia: Campus environmental stewardship at the turn of the 21st Century.* National Wildlife Federation, Washington, D.C.

Krizek, K.J., Newport, D., White, J. and Townsend, A.R. (2012), Higher education's sustainability imperative: how to practically respond? *International Journal of Sustainability in Higher Education* 13 (1), pp. 19-33.

Leal Filho, W. (Ed.) (2002), *Teaching Sustainability at Universities: towards curriculum greening.* Peter Lang, Frankfurt am Main.

Lozano, R., Lukman, R., Lozano, S., Huisingh, D. and Lambrechts, W. (2011), Declarations for sustainability in higher education: becoming better leaders, through addressing the university system. *Journal of Cleaner Production* 30, pp. 1-10.

Miller, T.R., Munoz-Erickson, T. and Redman, C.L. (2011), Transforming knowledge for sustainability: towards adaptive academic institutions. *International Journal of Sustainability in Higher Education* 12 (2), pp. 177-192.

Moore, J. (2005), Seven recommendations for creating sustainability education at the university level: a guide for change agents. *International Journal of Sustainability in Higher Education* 6 (4), pp. 326-339.

Orr, D.W. (1992), *Ecological Literacy*, State University of New York Press, Albany, NY.

– (1994), *Earth in Mind.* Island Press, Washington, DC.

Pallant, E. (2001), *Compost, Compost, We love you!* Unpublished syllabus. Allegheny College. Retrieved from http://webpub.allegheny.edu/dept/envisci/epallant/coursehome/composting/ES589_CompostSyllabus_01.htm.

Pallant, E. and nine student authors (2002), *Compost, Compost, We Love You! The final report.* Unpublished report. Allegheny College. Retrieved from http://webpub.alle gheny.edu/dept/envisci/epallant/coursehome/composting/CompostingFinalReport.htm.

Pallant, E. and fourteen student authors (2004), *Allegheny College North Village Project.* Unpublished report. Allegheny College. Retrieved from http://webpub.allegheny.edu/dept/envisci/epallant/coursehome/JrSem%20Green%20Buildings/AC_North_Village_Project.pdf.

Pelletier, S. (2008), Sustainability: What is the Trustees Stake? *Trusteeship* 16 (5), pp. 9-14.

Pittman, J. (2004), Living sustainably through higher education: a whole systems design approach to organizational change. *Higher Education and the Challenge of Sustainability*, pp. 199-212.

The Princeton Review (2011), *The Princeton Review's Guide to 311 Green Colleges 2011 Edition*. Retrieved from http://www.princetonreview.com/uploadedfiles/sitemap/home_ page/green_guide/princetonreview_greenguide_2011.pdf.

Richardson, G.R.A. and Lynes, J.K. (2007), Institutional motivations and barriers to the construction of green buildings on campus: a case study of the University of Waterloo, Ontario. *International Journal of Sustainability in Higher Education* 8 (3), pp. 339-354.

Roberts, B. and Murphy, J. (2011), The Benefits of Retrofits. *Business Officer* 44 (9), pp. 16-22.

Sharp, L. (2002), Green campuses: the road from little victories to systemic transformation. *International Journal of Sustainability in Higher Education* 3 (2), pp. 128-145.

Sharp, L. (2009), Higher education: the quest for the sustainable campus. *Sustainability: Science, Practice & Policy* 5 (1), pp. 1-8.

Shepherd, K. (2008), Higher education for sustainability: seeking affective learning outcomes. *International Journal of Sustainability in Higher Education* 9 (1), pp. 87-98.

Shi, D. (2008), Sustainability 101. *Trusteeship* 16 (2), pp. 21-25.

Shriberg, M. (2002), Institutional assessment tools for sustainability in higher education: strengths, weaknesses, and implications for practice and theory. *International Journal of Sustainability in Higher Education* 3 (3), pp. 254-270.

Sibbel, A. (2009), Pathways towards sustainability through higher education. *International Journal of Sustainability in Higher Education* 10 (1), pp. 68-82.

Silling, A. (2009), Allegheny College's Admissions Building Rooftop Runoff Rain Garden Proposal. Unpublished senior thesis, Allegheny College. Retrieved from http://webpub.allegheny.edu/dept/envisci/ESInfo/comps/2009abstracts.html#silling.

Spencer, R. (2008), Composting helps anchor campus's climate commitment. *BioCycle* 49 (5), p. 28. Retrieved from http://www.jgpress.com/archives/_free/001644.html.

Stauffacher, M., Walter, A.I., Lang, D.J., Wiek, A. and Scholz, R.W. (2006), Learning to research environmental problems from a functional socio-cultural constructivism perspective: the transdisciplinary case-study approach. *International Journal of Sustainability in Higher Education* 7 (3), pp. 252-275.

Steiner, G. (2008), Supporting sustainable innovation through stakeholder management: a systems view. *International Journal of Innovation in Learning* 5 (6), pp. 595-616.

Steiner, G. and Posch, A. (2006), Higher education for sustainability by means of transdisciplinary case studies: an innovative approach for solving complex, real-world problems. *Journal of Cleaner Production* 14, pp. 877-890.

Sullivan, D. (2010), Colleges scrape the plate, close the loop. *BioCycle* 51 (7), pp.44. Retrieved from http://www.jgpress.com/archivers/_free/002116.html.

Swearingen White, S. (2009), Early participation in the American College and University Presidents' Climate Commitment. *International Journal of Sustainability in Higher Education* 10 (3), pp. 215-227.

Van Weenen, H. (2000), Towards a vision of a sustainable university. *International Journal of Sustainability in Higher Education* 1 (1), pp. 20-34.

Velazquez, L., Munguyia, N. and Sanchez, M. (2005), Deterring sustainability in higher education institutions: an appraisal of the factors which influence sustainability in higher education institutions. *International Journal of Sustainability in Higher Education* 6 (4), pp. 383-391.

Volkswagen (2009), Home Page. *Thefuntheory.com*. Retrieved from www.thefuntheory.com

Ward, H. (Ed.) (1999), *Acting Locally: Concepts and models for service-learning in Environmental Studies*. American Association for Higher Education, Washington, D.C.

Williamson, S.R. (2012), A systems approach to reducing institutional GHG emissions. *International Journal of Sustainability in Higher Education* 13 (1), pp. 46-59.

Chapter 32

"Don't Think Twice – This is the Opportunity of a Lifetime!": Students' [Passionate] Engagement with Processes of Greening the Campus

Beatriz Acevedo, Romas Malevicius, Sarah Johnson and Charlotte Bonner

Abstract

This paper presents the experience of students at Anglia Ruskin University working as eco-auditors of the university environmental strategy in the context of a module on Environmental Management. Since 2010 17 students have trained as eco-auditors for the Green Impact initiative, a pro-environmental programme implemented in 65 universities institutions in the United Kingdom and coordinated by the National Union of Students. This case study represents an improvement in efforts to integrate the university environmental strategy with the development of innovative strategies in the curriculum concerning management studies. The authors argue that engaging students with sustainability in a way that stimulates their passions and empowers their actions requires changes in the means through which traditional lectures are delivered. This paper discusses some potential alternatives in within that aim: firstly, it assesses the relevance of action-learning opportunities and student engagement with processes of greening the campuses; secondly, it discusses how the use of art-based techniques can support the development practical skills for eco-auditing and self-enquiry; thirdly, it evaluates the usefulness of pedagogical action research in furthering education for sustainable development. Finally, the authors discuss the difficulties and potentialities of replicating these types of experiences in other higher education institutions and private organisations.

Introduction

Since 2001, British institutions of higher education have been adopting ideas regarding education for sustainable development (ESD), following the global trend of universities across the world (Leal Filho 2009; Lozano 2006; Shriberg 2002; Wright 2002; Velazquez et al. 2006), and driven by the increasing regulation of climate change in the United Kingdom (HEFCE 2010; Stern 2006). While universities across the country are establishing environmental management and systems in order to green the campuses, very few initiatives have linked these programmes with changes in the curriculum. The experience of a number of universities have demonstrated that students can contribute to and benefit from these processes of greening the campus, through audits and other participatory schemes (Sharp 2002; Bardati 2006; Beringer 2006; Ferreira et al. 2006; Pike et al. 2003; Barth et al. 2007). This seemingly obvious option had not previously been tested in the United Kingdom, and although universities are increasingly developing environmental management systems, the link between the administrative and the academic realms has not been fully developed. In business schools the consideration of ESD often involves "minor amendments to a typical curriculum pattern of providing a grounding in core knowledge, across a wide variety of domains, through relatively traditional teaching methods" (Hopkinson and James 2010: 366). In contrast, there are few attempts at developing "pedagogy of passion for sustainability" aiming at integrating "physical and emotional or spiritual learning with traditional cognitive learning about sustainable management" (Shrivastava 2010: 443). Moreover, students in higher education in the United Kingdom had previously disregarded sustainability as part of their careers and personal lives (Stir 2006; Jucker 2002; Kagawa 2007). Recent research has, though, demonstrated that this attitude is shifting towards the increasing realisation by students that skills in sustainable development are significant for future employability (Bone and Agombar 2011).

This paper argues that the lack of passionate commitment in the students' approach to sustainability can be explained partially due to the absence of a practical and engaging approach in the teaching strategy. In particular, this paper concerns the development of a pedagogical action research documenting the transformation of the module on Environmental Management and Sustainability as part of the International Business and Strategy degree, and it focuses on the promotion of opportunities for students to acquire practical knowledge of topics related to environmental management through their work as eco-auditors of the environmental strategy at Anglia Ruskin University.

The growing literature on education for sustainable development offers a generous starting point for the development of this research. The question for this research was: "How to design and implement a module on environmental management in which the participants (i.e. the students and lecturer) have the opportunity to engage passionately with sustainable practices." For this article, the authors explored three areas concerning ESD: firstly, recent publications concerning the path developed by universities on sustainability (Gough and Scott 2007; Ferrer-Balas et al. 2008) and the experiences of "greening the campus" (Allen 1999; Koester et al. 2006; Fisher 2003; Sharp 2002), focusing on some articles addressing the potential of students' engagement with this type of initiative (Stir 2006; Ferreira et al. 2006; Beringer 2006; Pike et al. 2003); secondly, the pedagogical and educational research linking teaching and research (Healy 2005; Burke and Rau 2010; Markides 2007; Tormey et al. 2008), particularly articles referring to changes in the curriculum in relation to sustainability (Juarez-Najera et al. 2006; Hopkinson et al. 2008; Holmberg and Samuelsson 2006; Leal Filho 2009; Roberts and Roberts 2007;); and thirdly, articles focused on sustainability education in management studies (Stubbs and Cocklin 2008; Wankel and Stoner 2009).

Sustainability at Anglia Ruskin University

Since 2003 Anglia Ruskin University has implemented programmes of energy consumption, recycling, responsible procurement, and travelling arrangements, hence achieving important accreditations for their progress (Anglia Ruskin University 2011). The department of Environmental Management, which is in charge of these programmes, has determined that besides the infrastructural aspect of the environmental investment, they need to create ways of changing behaviours among the staff and students. Campaigns such as 'Switch-Off' and 'Green Impact', initiated by the National Union of Students in universities across the country, began to be implemented in Anglia Ruskin campuses and residence halls; these were external initiatives with almost non-existent links with curricula and academic practices. At the same time, students studying the module mentioned have been encouraged to gain practical insights into environmental management by focusing on household changes while discussing global issues on climate change and sustainable development. The encounter of the three worldviews of administrative strategy, student-led initiatives and practical inquiry provided the opportunity of developing the experience that forms the basis of this case study.

The context for this collaboration is the Green Impact initiative, a pro-environmental behaviour change strategy coordinated by the National Union of Students and Student Force in 35 universities across the country. It was proposed that a group of students taking the module could become involved as eco-auditors, receiving extra training and documenting their practice as part of their assignment. Altogether, 18 students trained as eco-auditors, collaborating with 14 teams of 200 staff members in the two campuses of Cambridge and Chelmsford during the two years of this experience (2010-2011). The experience has been praised as "good practice" by the National Union of Students, since it is the first time that the work of students has been supported by an academic context (National Union of Students 2010). In addition, the module has simultaneously evolved in order to develop the abilities and knowledge necessary in supporting eco-auditing processes in a wider sense: analysing processes of organisational change concerning environmental management systems, and promoting discussions on wider topics of sustainability in organisations and businesses.

This paper argues that the university environmental strategy represents an ideal context in which students can gain first-hand experience on how environmental management is implemented in practice. The original contribution of this paper is the practical experience of articulating three main areas in ESD: firstly, the integration of students' practice in the context of the university environmental efforts, as environmental action learning. Secondly, the authors observed that the use of art-based techniques in teaching environmental management can effectively enhance the dimensions of the educational experience for sustainability: the personal dimension through processes of self-enquiry, and the practical aspects linked to eco-auditing supported by visual research methodologies. The practical aspects of the experience promote a constant process of self-inquiry. Moreover, the use of art-based techniques is aimed at balancing the power aspects and the trend to use 'structured definitions' in favour of more intuitive and personal views on environmental issues. Thirdly, this paper argues that pedagogical action research is a suitable methodology for implementing changes in both the teaching and the learning experiences of students, while addressing research issues on education for sustainable development.

Following these ideas, this paper discusses the main pillars of our experience, illustrated as follows:

Fig. 1: Three pillars in the education for sustainable development

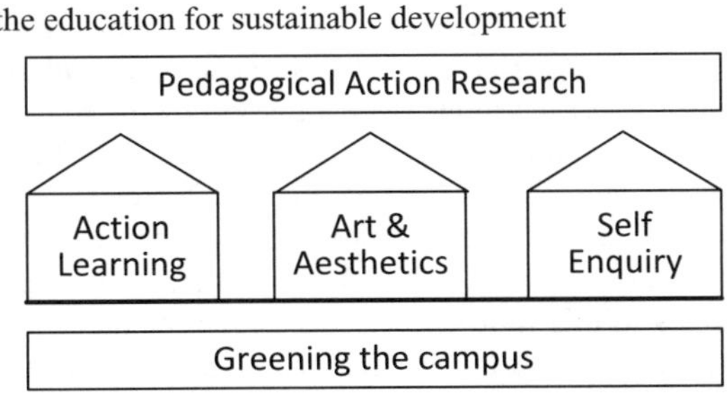

Methodology

The investigation follows a pedagogical action research. It focuses on investigating systematically the teaching and learning practices produced by the interaction of student engagement in processes of greening the campus; this occurs alongside the development of pedagogical strategies in education for sustainable development. The research originally aimed at improving teaching practices while contributing to the theoretical knowledge in order to benefit student learning (Norton 2009: 59). Action research has been defined as a form of self-reflective enquiry undertaken by participants in social situations to improve the rationality and justice of their own practices (Carr and Kemmis 1986: 186). Reason and Bradbury (2001) have emphasised the aim of action research in developing practical knowledge in the "pursuit of worthwhile human purposes" through participation. Further, Bradbury (2001: 312) acknowledges the value of action research in "building capacity for, and in the study of, efforts of sustainable development."

In order to find out what students in business and management thought about education on sustainability and the role of universities in this regard, in 2009 the authors formulated an open-ended question; it was distributed to 200 third-year students of the Business School and had a response rate of 40%. The answers collected and analysed in terms of their content evidenced two main issues: firstly, a demand to know what the university is doing about sustainability, perceived as a mainly environmental issue; secondly, the necessity of linking the concepts seen in class with practices and real-life examples.

The research was developed in two phases. It began with an exploratory approach based on three group interviews with students implemented at the end of each academic year (2009, 2010, and 2011); it concerned the evaluation of the module for that year and encouraged recommendations from the students. The interviews were transcribed and the content was codified into three categories: (a) the priority given by the student to environmental issues and sustainability for business and management; (b) the level of knowledge about environmental issues and sustainability provided in the curriculum of business and management, and (c) the potential for professional development in areas regarding environmental issues and sustainability. From these evaluations, the authors realised that students had not received previous information on environmental issues and sustainability, and that they were surprised to see how relevant these topics were for management and businesses. Few students manifested that they were interested in furthering studies in this area. A final category was to ask students to nominate their 'favourite' session, and how it contributed to their enthusiasm on sustainability. The authors chose to use 'favourite' instead of 'most suitable', 'best', or 'most relevant' in order to emphasise the intuitive and pleasurable aspects of the research. It was evident that exercises aimed at encouraging lateral thinking, such as those based on art and creativity, were favoured by the students as suitable ways to develop engagement with sustainability.

Following these findings, the second level of this research focused on the experience of the students working as eco-auditors in 2010 and 2011. Two group interviews with the students participating in the Green Impact were recorded and transcribed, and the results were codified into the following categories: (a) learning; (b) evaluation of the training; (c) relationship with the staff audited. In addition, 18 semi-structured interviews with staff members participating in the Green Impact addressed, among other issues, the staff perception of the students' professionalism and performance as eco-auditors. As before, the interviews were recorded and transcribed. In addition to primary research, secondary sources, such as an analysis of corporate documents, relevant articles in this field and an assessment of the results of the National Student Survey of the last three years for this particular module, have also been incorporated into this investigation.

The limitations of this study can be established as follows: first, the research covers only two years of implementation and the number of students participating is relatively low. The sample of the interviews was selected based on the students' voluntary participation and the authors did not include a 'control group'. The reason behind this decision was the exploratory character of this research. However, for future research it is important to establish a control group, determine a base line, and offer the opportunity of comparison (Entwistle and Tait 1990). On the other hand, the selection of the students who participated in the experience may present problems as regards the transparency of communication, since there are issues of power and authority that influence students' responses (Shevlin et al. 2000). In order to address this, the authors ensured the anonymous character of the interviews, and a friendly atmosphere was created by emphasising some of the pitfalls and 'gaffes' of the module as a starting point to encourage a more critical response to the questions.

Overall, the framework provided by the pedagogical action research allowed a constant dialogue between practice and reflective enquiry among the participants. The authors consider that this approach can be replicable for other settings; indeed, this approach allows the establishing of a relationship between teaching practice and research. However, it is important to consider some criticisms pointing out the subjective dimension to this methodology and the difficulties in generalising certain findings (Tormey et al. 2008). Such a subjective approach is acknowledged here. The combination of different sources, actors and discourses aims at presenting a rich picture of worldviews and perceptions.

Action Learning: a *"glocal"* approach

A considerable proportion of articles on ESD stress the potential of action or experiential learning opportunities. Kolb and Kolb (2005: 194) define experiential learning as "the process whereby knowledge is created through the transformation of experience. Knowledge results from the combination of grasping and transforming experience." Several experiences in North American universities demonstrate the relevance and benefits of this approach: Pike et al. (2003) assess the impact of problem-based activities on the learning process in the case of recycling in the students' halls, and Bardati (2006) discusses the problems in linking the processes of auditing the campus with academic dynamics. These suggest that linking the local actions of the university may result in a forum for the discussion of global issues. In this sense, Collins and Kearins (2010) argue that the local context is a mirror of global issues, thus a "glocal" approach would improve the students' understanding of sustainability. In our view, "glocality" may be interpreted as focusing on the university as an organisation undergoing processes of change towards environmental practices that may reflect many of the challenges, barriers and potentialities of the global progress on sustainable development. The authors are aware that linking sustainability to environmentalism is a limited approach; nevertheless, it is considered that environmental issues can become a stepping-stone in the process of developing any sustainable organisation.

Students' engagement with environmental management processes in universities brings fresh knowledge and enthusiasm to the implementation of these initiatives amongst staff members and other stakeholders (Ferreira et al. 2006; Beringer 2006; Bardati 2006). During the two years of this initiative, information from the staff members participating in Green Impact concerning the work of the students has been collected. For some, the students should go beyond the 'eco-auditing' by becoming facilitators and offering recommendations. For others, the experience has contributed to enhancing the sense of community:

> I thought that it was really *smart* that the students were doing the auditing because it is just right, they are the most important people for us. (Semi-structured interview with Green Impact team leader 2010)

For the students this is an opportunity to go beyond the classroom and learn a practical skill with potential implications for their future employment:

> I actually *learned something practical instead of just reading books* and it is also good for my CV. It's like a professional job. (Video-recorded interview with Aaron Yang, China 2010)

Furthermore, the opportunity of understanding complex topics such as environmental management in practice requires changes in the students' perception of classes and lecturers. As one of the students enthusiastically put it:

> What I would say [to students] is don't doubt your abilities, you can make a difference in every part. Doing this course and being auditors opens your eyes a lot more. I would say just go for it, join. Learn

> as much as you can and you would be surprised what you can achieve. Because it is an exciting, exciting journey. Don't think twice: this is the opportunity of a lifetime! (Video-recorded interview with Lucky Hussein, United Kingdom 2010)

The National Student Surveys of the last three years show that the level of satisfaction with the module has remained constant. However, the level of engagement of students in responding to the survey has increased in the last two evaluations (National Students Survey 2011). Most importantly, an increasing number of students are involved in the ongoing development of the initiative; for instance, former participants are assuming the role of coordinators for the coming group of students; the module is linked with student-based groups such as the Global Responsibility Leadership initiative and the Sustainability Group. Finally, the Facebook page – administered by the students – remains a hub for the exchange of ideas and new proposals for the ongoing development of the module.

In supporting the action-learning opportunities, the authors also implemented a number of activities related to non-traditional approaches in the contexts of management and accountancy, as well as arts and aesthetics. Some of these are outlined in the following section.

Aesthetic, art-based techniques and the environment

The literature on ESD emphasises the necessity of developing analytical skills, holistic thinking and innovation, and art may provide a vehicle towards developing a sustainability consciousness (Bathurst and Edwards 2009). Further, Jacoby and Ji (2010) argue that artists and the arts have an important role to play in environmental leadership; for example, in processes of urban planning. In our experience, the use of art-based techniques and aesthetic approaches has proved a suitable avenue for the development of skills in three main realms: the illustration of essence, projective techniques, and skills transfer (Taylor and Ladkin 2009).

The module began by including a number of films and videos to explain particular issues and to promote debate (Tejeda 2008; Champoux 2006). Not all individuals everywhere, especially international students, have full access to such material. Films such as *The Age of the Stupid* (2009), *An Inconvenient Truth* (2006), and *The End of the Line* (2009), based on the book by Charles Clover, were presented in order to promote the discussion on climate change and resource depletion. As acknowledged by Clemens and Hamakawa (2010), the use of films in the classroom implies considerable effort from the lecturer; it is vital to prepare extra information in advance, such as definitions, political issues, further readings, and selected web-pages for additional consultation. The application of games such as the 'Climate Challenge' (developed by the University of Oxford and the BBC), which challenge players to take decisions regarding climate change, sustainability and policymaking with limited resources and complex public opinion responses, were also implemented. This type of simulation promotes systemic and holistic thinking in an engaging and interactive way (Dieleman and Huseingh 2006).

During the tutorial sessions, certain exercises on drawing as a projective technique were proposed: participants were asked to draw their idea of what constitutes sustainability. The drawings were accompanied by a general discussion on how these ideas were enacted in the respective student's daily practice. The sessions focused on the students' development of the drawings and what they meant in practice; in that way, drawing became a way of producing data for the research (Warren and Vince 2012). The group discussed the fact that there is a 'gap' between the 'discourse' on sustainability and the personal practice. The exercise actually questions the traditional means of evaluation, by which the cycle of formalism and superficiality tends to be reproduced; in this case, the traditional questions produced articulated responses that did not always show the real engagement of the participants with the values and implications of complex subjects such as sustainability and environmentalism. By tapping into alternative methodologies such as drawing and visualisation, such divergences are easily evidenced.

Finally, the process of auditing was supported by the use of Eco-mapping© (Engel 2002) and photography. The participants were encouraged to draw a map of the departments audited, identifying environmentally significant aspects such as radiators, lightbulbs, appliances, computers, etc. The tool was complemented with the use of pictures as a form of visual evidence and a tool for diagnoses.

> The use of pictures made it easier for me to describe the whole process of how recycling was handled [...] So pictures give a more visual process and make it easier to see where to improve. It is also a very effective way to prove conditions before and solutions after the improvement. (Student-written evaluation, Christian Salling, Denmark 2011)

The use of photographs refers not only to the strictly visual; participants were encouraged to take photographs of environmentally significant aspects such as 'noise', 'heating', and 'pollution'. As expressed by one of the students:

> Here I decided to take a picture of the window to show that the office is near a factory, thus the smell and the noise are environmental problems. I would not have thought about this without the [photography] workshop. (Focus group evaluation, Souvenir Oyawale, Nigeria 2011)

In this experience, the adoption of visual research methodologies involving pictures both as a tool for diagnosis and as evidence has proved useful in the learning process, since it allows the students to gain an actual picture of the premises while identifying environmentally significant aspects. It creates double-loop learning processes in which the students collect information *in situ*, and later analyse visual material in order to offer concrete recommendations. The use of aesthetic and art-based techniques represents a key tool in ESD, facilitating an exploration of sensory and intuitive approaches to the concept of sustainability and similar complex concepts (Adler 2006). Apart from drawing and photography, the authors promoted additional sensory activities in the context of visits to the Botanic Garden. Intended initially as an example of 'organisation and nature', the actual impact of smelling, touching, hearing, and feeling the lush garden became the summit of the event.

As regards synthesising, it is not easy to pinpoint which of these techniques has the greatest impact in terms of learning and development. The synergy between the different elements, the attitude of the participants, the inter-organisational collaboration, and the diverse methodologies is what counts. Still, it is important to design these activities with a clear pedagogical approach informed by the advances in social research and scholarship in organisational aesthetics (Strati 1999), visuality (Pink 2007; Rose 2007; Warren 2002), sensuality (Pink 2010) and emotional engagement. In this way, the students appreciate that these are not just 'random experiments'; there is indeed a solid basis for the implementation of the activities.

Conclusions

This research has demonstrated that in order to encourage 'passion' and commitment among the students, it is important to transform and subvert current approaches to teaching and learning (Shrivastava 2010; Rands and Stairk 2009). The authors' original contribution consists in focusing on three main pillars: action learning, art-based methodologies, and processes of self-enquiry, in the context of 'greening the campus'. The paper has evidenced the potential of aesthetics and art-based practices in promoting lateral thinking, creativity and self-inquiry. A growing field in visual methodologies and sensual/sensorial research represents a rich avenue in relation to ESD.

Notwithstanding, initiatives such as that described here face a number of challenges and may be hampered by limitations. For example, major efforts must be made in maintaining the interest and the motivation of all involved (Dawe et al. 2004; Ramus 2001). Besides, the administration and coordination of this experience require extra efforts from the participants, as well as a dedicated budget. In a climate of austerity, these resources may become scarce, thus hampering the efforts. Another important aspect of this experience is the personal attitude of the participants and their desire to work in a collaborative fashion. Issues regarding trust and social capital need to be considered when attempting to promote inter-organisational collaborative action research (Konstantinos and Jones 2009). The authors acknowledge that their personal attitude toward this collaboration was a key factor, and that the convergence of factors involved in this initiative may not be replicable in other settings (Shevlin et al. 2000).

The idea has considerable potential. The authors are currently examining the possibility of expanding the work of the students as eco-auditors in three companies related to the car racing industry. If companies such as these, not traditionally associated with 'green practice', are enthusiastically taking part in this pilot, it means that there is hope for the future. In addition, universities need to start addressing the gap in skills created by the growing green market and the transformations of different industries from 'brown' practices toward environmental practices.

This paper has, to summarise, described the main benefits of integrating the processes of greening the campus and curriculum development. This notwithstanding, it is clear that one single course has little, if any, impact on the overall structure of the department. In the authors' view, it is not sufficient to create courses and modules *on* sustainable development; rather, it is crucial to promote means for education *as* sustainable development, involving changes in the current mechanisms and structures of traditional lectures and power dynamics among staff members, academics, administrators, students, communities, and stakeholders.

As a conclusion, the authors believe that the British context of higher education is ripe for experimentation and innovation, and the necessity of ensuring competitiveness to graduates implies the enhancing of employment options through placements, training and work experience. Indeed, the authors urge those who are experimenting and testing ideas, sometimes through adventurous or unorthodox approaches, to maintain the systematic recording and analysis of the different aspects of this type of experience using methodologies such as action research, and echoing the quoted student: "don't doubt your abilities, you can make a difference in every part". The present authors hope that this paper encourages readers to experiment in the creation of alternatives in the aim of a sustainable future.

Acknowledgements

The authors would like to thank Souvenir Oyawale, Christian Salling, Lucky Hussein, and Aaron Yang, who kindly agreed to be identified as the students mentioned in the article. The experience would not have been possible without their enthusiasm and commitment. We also would like to thank Alison Craig for her ideas regarding education for sustainable development and education as sustainable development.

References

Adler, N. (2006), The Arts and Leadership: Now that we can do anything, what will we do? *Academy of Management Learning and Education* 5 (4), pp. 466-499.

Allen, A. (1999), Institutional change and leadership in greening the campus. In Leal Filho, W. (Ed.), *Sustainability and University Life*, pp. 105-28. Peter Lang, Frankfurt am Main.

An Inconvenient Truth (2006), Film, directed by Davis Guggenheim, USA: Paramount Classics.

Anglia Ruskin University (2011), *Environmental Management at Anglia Ruskin University*. Available at: http://www.anglia.ac.uk/ruskin/en/home/microsites/environmental _management.html (accessed 1 November 2011).

Bardati, D. (2006), The integrative role of the campus environmental audit: experiences at Bishop's University, Canada. *International Journal of Sustainability in Higher Education* 7 (1), pp. 57-68.

Barth, M., Godemann, J., Rieckmann, M. and Stoltenberg, U. (2007), Developing key competencies for sustainable development in higher education. *International Journal of Sustainability in Higher Education* 8 (4), pp. 416-30.

Bathurst, R. and Edwards M. (2009), Developing a Sustainability Consciousness Through Engagement with Art. In Wankel, C. and Stoner, J. (Eds.), *Management Education for Global Sustainability*, pp. 115-138, Information Age Publishing, Charlotte, North Carolina.

Beringer, A. (2006), Campus sustainability audit research in Atlantic Canada: Pioneering the campus sustainability assessment framework. *International Journal of Sustainability in Higher Education* 7 (4), pp. 437-455.

Bone, E. and Agomber, J. (2011), *First year attitudes toward, and skills in, sustainable development*. Report by University of Bath, National Union of Students, Student Force and the Higher Education Academy. Higher Education Academy Publications, York.

Bradbury, H. (2001), Learning with The Natural Step: Action Research to Promote conversations for sustainable development. In Reason, P. and Bradbury, H. (Eds.), *Handbook of Action Research*, pp. 307-313. Sage, London.

Burke, L. and Rau, B. (2010), The Research-Teaching Gap in Management. *Academy of Management Learning and Education* 9 (1), pp. 132-143.

Carr, W. and Kemmis, S. (1986), *Becoming Critical: Knowing Through Action Research*. Falmer Press, Lewes.

Champoux, J.D. (2006), At the cinema: Aspiring to a higher ethical Standard. *Academy of Management Learning and Education* 5 (3), pp. 386-390.

Clemens, B. and Hamakawa, C. (2010), Classroom as cinema: Using film to teach sustainability. *Academy of Management Learning and Education* 9 (3), pp. 561-63.

Collins, E. and Kearins, K. (2010), Delivering on Sustainability's Global and Local Orientation. *Academy of Management Learning and Education* 9 (3), pp. 499-06.

Dawe, G., Vetter, A. and Martin, S. (2004), An overview of ecological foot printing and other tools and their application to the development of sustainability process. *International Journal of Sustainability in Higher Education* 5 (4), pp. 340-71.

Dieleman, H. and Huisingh, D. (2006), Games by which to learn and teach about sustainable development: exploring the relevance of games and experiential learning for sustainability. *Journal of Cleaner Production* 14 (9-11), pp. 837-47.

Engel, H. (2002), *EMAS Toolkit of SMEs*. Available at: http://ecomapping.com/en/links/disclaimer.html (accessed 1 November 2011).

Entwistle, N. and Tait, H. (1990), Approaches to learning, evaluations of teaching and preferences for contrasting academic environments. *Higher Education* 19 (2), pp. 169-94.

Ferreira, A.J.D., Lopes, M.A.R. and Morais, J.P.F. (2006), Environmental management and audit schemes implementation as an educational tool for sustainability. *Journal of Cleaner Production* 14 (9-11), pp. 973-82.

Ferrer-Balas, D., Adachi, J., Banas, S., Davidson, C.I., Hoshikoshi, A., Mishra, A., Motodoa, Y., Onga, M. and Ostwald, M. (2008), An international comparative analysis of sustainability transformation across seven universities. *International Journal of Sustainability in Higher Education* 9 (3), pp. 295-316.

Fisher, R. (2003), Applying ISO 14001 as a business tool for campus sustainability. A case study from New Zealand, In *International Journal of Sustainability in Higher Education* Vol. 4 (2), pp. 138-151.

Gough, S. and Scott, W. (2007), *Higher education and sustainable development: paradox and possibility*. Routledge, New York.

Healy, M. (2005), Linking research and teaching to benefit student learning. *Journal of Geography in Higher Education* 29 (2), pp. 183-201.

HEFCE (2010), *Carbon Reduction Target and Strategy for Higher Education in England*, Higher Education Funding Council for England, Bristol.

Holmberg, J. and Samuelsson, B.E. (Eds.) (2006), *Drivers and Barriers for Implementing Sustainable Development in Higher Education, Education for Sustainable Development in Action*, Technical Paper No. 3, UNESCO, Paris.

Hopkinson, P. and James, P. (2010), Practical pedagogy for embedding ESD in science, technology engineering and mathematic curricula. *International Journal of Sustainability in Higher Education* 11 (4), pp. 365-379.

Hopkinson, P., Hughes, P. and Layer, G. (2008), Sustainable graduates – linking formal, informal and campus curricula to embed ESD in the student learning experience. *Environmental Education Research* 14, pp. 435-454.

Jacoby, J. and Ji, X. (2010), Artists as Transformative Leaders for Sustainability. In Redekop, B, and Olson, S. (Eds.) *Leadership for Sustainable Development*, pp. 133-144, Routledge, London.

Juarez-Najera, M., Dieleman, H. and Turpin-Marion, S. (2006), Sustainability in Mexican Higher Education: towards a new academic and professional culture. *Journal of Cleaner Production* 14 (9-11), pp. 973-982.

Jucker, R. (2002), 'Sustainability? Never heard of it!' – Some basics we shouldn't ignore when engaging in education for sustainability. *International Journal of Sustainability in Higher Education* 3 (1), pp. 8-18.

Kagawa, F. (2007), Dissonance in students' perceptions of sustainable development and sustainability. *International Journal of Sustainability in Higher Education* 8 (3), pp. 317-338.

Koester, R. Eflin, J. and Vann, J. (2006), Greening the Campus: a whole systems approach. *Journal of Cleaner Production* 14 (9-11), pp. 963-72

Kolb, A.Y. and Kolb, D.A. (2005), Learning Styles and Learning Spaces: Enhancing Experiential Learning in Higher Education. *Academy of Management Learning and Education* 4 (2), pp. 193-212.

Konstantinos, E. and Jones, N. 2009, An analysis of social capital and environmental management in higher education institutions. *International Journal of Sustainability in Higher Education* 10 (4), pp. 334-342.

Leal Filho, W. (Ed.) (2009), *Sustainability at Universities – Opportunities, Challenges and Threats*. Peter Lang, Frankfurt.

Lozano, R. (2006), Incorporation and institutionalization of Sustainable Development into universities: breaking through barriers to change. *Journal of Cleaner Production* 14 (9-11), pp. 787-796.

Markides, C. (2007), In search of ambidextrous professors. *Academy of Management Journal* 50 (4), pp. 762-68.

National Student Survey (2011), *The National Student Survey*. Available at: http://www.thestudentsurvey.com/institutions.html (accessed 1 November 2011).

National Union of Students (2010), *Green Impact – Evaluative Meeting with Stakeholders*. Power Point presentation. Coordinated by Charlotte Bonner, National Union of Students, London, 2 September, 2010.

Norton, L. (2009), *Action Research in Teaching and Learning: A practical guide to conducting pedagogical research in universities*. Routledge, London.

Pike, L., Shannon, T., Lwrimore, K.J.A, Taylor, M. and Lamoreaux, G. (2003), Science education and sustainability initiatives. A campus recycling case study shows the importance of opportunity. *International Journal of Sustainability in Higher Education* 3 (3), pp. 218-29.

Pink, S. (2007), *Doing Visual Ethnography*. Sage, London.

– (2010), *Doing Sensory Ethnography*. Sage, London.

Ramus, C. (2001), Organizational Support for Employees: Encouraging creative ideas for environmental sustainability. *California Management Review* 43 (3), pp. 85-105.

Rands, G. and Starik, M. (2009), The short and glorious history of sustainability in North American management education. In Wankel, C. and Stoner, J. (Eds.), *Management education for global sustainability*, pp. 19-50. Information Age Publishing, Charlotte, North Carolina.

Reason, P. and Bradbury, H. (Eds.) (2001), *Handbook of Action Research*. Sage, London.

Roberts, C. and Roberts, J. (2007), *Greener by Degrees*. University of Gloucester, Gloucester.

Rose, G. (2007), *Visual methodologies: an introduction to the interpretation of visual materials*. Sage, London.

Sharp, L. (2002), Green Campuses: the road from little victories to systematic transformation. *International Journal of Sustainability in Higher Education* 3 (2), pp. 128-45.

Shevlin, M., Banyard, P., Davies, MNO, and Griffiths, M. (2000), The validity of student evaluation of teaching in higher education: love me, love my lectures? *Assessment and Evaluation in Higher Education* 25 (4), pp. 397-405.

Shriberg, M. (2002), Institutional assessment tools for sustainability in higher education. *International Journal of Sustainability in Higher Education* 3 (3), pp. 254-70.

Shrivastava, P. (2010), Pedagogy of Passion for Sustainability. *Academy of Management Learning and Education* 9 (3), pp. 443-55.

Stern, N. (2006), *Report on the Economics of Climate Change, HM Treasury*. Office of Climate Change. Available at: http://webarchive.nationalarchives.gov.uk/+/http://www.hm-treasury.gov.uk/stern review_index.htm (accessed 10 November 2011).

Stir, J. (2006), Restructuring teacher education for sustainability: student involvement through a 'strengths model'. *Journal of Cleaner Production* 14 (9-11), pp. 830-36.

Strati, A. (1999), *Organization and Aesthetics*. Sage, London.

Stubbs, W. and Cocklin, C. (2008), Teaching sustainability to business students: Shifting mindsets. *International Journal of Sustainability in Higher Education* 9 (3), pp. 206-221.

Taylor, S. and Ladkin, D. (2009), Understanding Arts-Based Methods in Managerial Development. *Academy of Management Learning and Education* 8 (1), pp. 55-69.

Tejeda, M.J. (2008), A resource review for diversity film media. *Academy of Management Learning and Education* 7 (3), pp. 434-39.

The Age of the Stupid (2009), Film, directed by Franny Armstrong, UK: Spanner Films / Passion Pictures.

The End of the Line (2009), Film, directed by Rupert Murray, UK: Dartmouth Films.

Tormey, R., Liddy, M., Maquire, H. and McCloat, A. (2008), Working in the action/ research nexus for education for sustainable development: two case studies from Ireland. *International Journal of Sustainability in Higher Education* 9 (4), pp. 428-40.

Velazquez, L., Munguia, N., Platt, A. and Taddei, J. (2006), Sustainable University: what can be the matter? *Journal of Cleaner Production* 14 (9-11), pp. 810-19.

Wankel, C. and Stoner, J. (Eds.) (2009) *Management Education for Global Sustainability*. Information Age Publishing, Charlotte, North Carolina.

Warren, S. (2002), Show me how it feels to work here: Using photography to research organizational aesthetics' ephemera. *Theory and politics in organization* 2 (3), pp. 224-45.

Warren, S. and Vince, R. (2012), Participatory visual methods. In Cassell, C. and Gillian, S. (Eds.), *The Practice of Qualitative Organizational Research: Core Methods and Current Challenges*, pp. 362-85. Sage, London.

Wright, T.A.A. (2002), Review of definitions and frameworks for environmental sustainability in higher education. *International Journal for sustainability in Higher Education* 3 (3), pp. 203-20.

Chapter 33

Change-Agents in Sustainability Governance: Institutional Transformation at Three Institutions of Higher Education

Valentin Tappeser and Arian Meyer

Abstract

How do you get complex institutions like universities, with their special responsibility as drivers of education and research, to operate and educate on a sustainable basis? This paper investigates the role of students in the transformation of institutions of higher education in general, as well as focusing in particular on their role in the establishment of sustainability governance structures. To this end, three case studies are investigated: Maastricht University, where a student initiative successfully transformed into a student-run university department managing the institution's sustainability portfolio; the University of California Santa Cruz, where a broad coalition of students and staff members support the university's sustainability movement; and Leuphana University Lüneburg, where sustainability governance is pushed by the faculty in collaboration with higher management. Ethnographic research performed before and one year after the establishment of Maastricht University Green Office as a university department, as well as qualitative interviews with key stakeholders and drivers of the sustainability movement at UCSC (2010) and Leuphana University Lüneburg (2011), highlight the importance of students in instigating processes of transformation at institutions of higher education. As institutional entrepreneurs, they have the potential to drive sustainability governance and effectively cope with the problem of embedded agency, if sufficiently empowered.

Introduction

The increasingly apparent deficiencies of our global societies and economies, in terms of the socio-economic and environmental consequences they produce, have made the need for sustainable development a widely acknowledged fact in government, businesses, and societal organizations. Yet, despite this realization, so far only little has been done to really operationalize this concern into systemic reform and institutional practice (Holt and Barkemeyer 2012; McCool and Stankey 2004). This is also the case for universities, which, as institutions of higher education, have a special responsibility in this regard, for they are not only role models for other public institutions but also creators of knowledge and educators of future decision-makers (Michelsen et al. 2008; UNESCO 1997; Wright 2002).

Aiming to address an underrepresentation of the role of students in current literature on sustainability in higher education (Arbuthnott 2009; Burki and Tappeser 2010; Stubbs and Cocklin 2008), this paper seeks to answer the following question: How can student participation influence institutional structures and thereby sustainability governance at universities? Drawing on ethnographic research at Maastricht University (UM), the University of California Santa Cruz, and Leuphana University Lüneburg, it is being argued that students can play a pivotal role as potential change-agents and institutional entrepreneurs who, if sufficiently empowered, can provide an effective solution to the problem of embedded agency in a university context. Here the notion of tempered radicalism (Meyerson and Tompkins 2007) is being invoked as a useful concept to understand this process of institutional change.

The remainder of the paper is organized as follows: The first section deals with the theoretical approach. Here, the authors pay specific attention to varying levels of deliberation and hierarchy in governance, as well as underlying conceptions of sustainability on the one hand (van Zeijl-Rozema et al. 2008) and, on the other, agency-structure considerations in an institutional context (Meyerson and Tompkins 2007). Secondly, the methodology will be explained. Qualitative interviews with key stakeholders, tracing the development of sustainability governance structure at three different universities, were combined with literature research in the fields of sustainability studies, governance, and institutional theory. The third chapter examines the cases of the Universities of Maastricht, Santa Cruz, and Lüneburg. Fourthly, the subsequent discussion will compare the cases and elucidate them by means of the theoretical frameworks introduced in the second section. Finally, the paper concludes with the statement that deliberate student-driven approaches towards sustainability governance at universities seem to be an innovative and functioning concept that has already produced remarkable results. Whether it is a sustainable concept in itself, however, remains to be seen while providing interesting avenues for further research.

Between sustainability governance and change agents – a theoretical approach

Sustainability governance

Van Zeijl-Rozema, Cörvers, Kemp and Martens (2008), developed a radically simplified model of different types of governance for sustainable development (SD) to make sense of the concept. They identified two major perspectives on sustainable development and two principal modes of governance as sensible distinctions categorizing a complex spectrum of approaches. This resulted in the 2 x 2 matrix of four idealized types of governance for SD as depicted in Fig. 1.

On the horizontal axis, a distinction is made between the ecological sustainability perspective and the well-being perspective of sustainable development. The former, characterized by an understanding of SD as defined by scientifically determinable ecological boundaries (e.g. Hueting and Reijnders 2004), allows the setting of clear objectified goals that are easily measurable while neglecting complexities inherent in the dovetailing of environmental, societal and economic processes. The latter, in contrast, sees goals of SD as societally rather than physically determined, allowing for a multiplicity of goals to exist alongside each other, based on societal preferences subject to normativity and local interpretation (e.g. Brand and Karvonen 2007; Gibson et al. 2001; Mccool and Stankey 2004).

Fig. 1: Conceptual framework of governance for sustainable development (van Zeijl-Rozema et al. 2008)

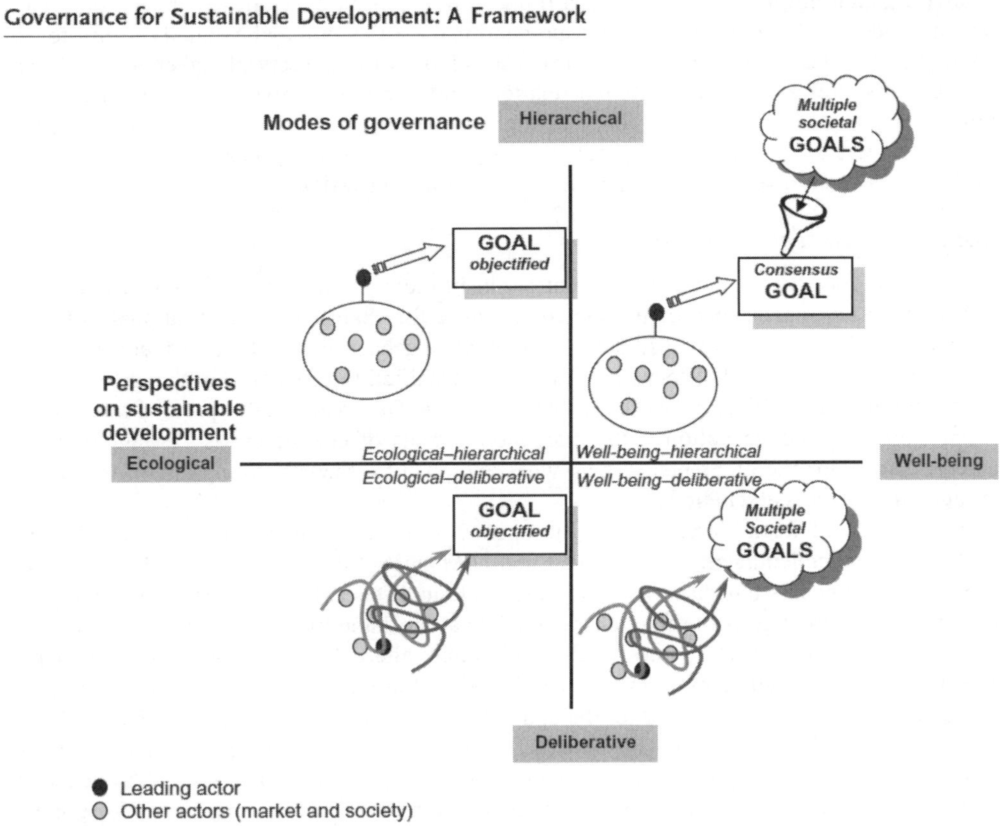

The vertical axis has hierarchical and deliberate modes of governance as its idealized types or endpoints, which is based on a review of similar types of governance distinguishing between hierarchical and self-governance (Kooiman 2003), hierarchical and non-hierarchical coordination (Börzel et al. 2005), hierarchy and heterarchy (Smismans 2008), etc. The results of this cross-section are the four idealized types of governance for SD, as depicted above: the ecological-hierarchical type, the ecological-deliberative type, the well-being-hierarchical type and the well-being deliberative type. While in the ecological perspective goal setting remains straightforward, deliberative governance allows for more diversity and experimentation in face of uncertainty with regards to the means to achieve these goals than a more hierarchical approach. In either case, however, a unidimensional viewpoint and focus on technological fixes potentially stands in the way of systemic reform. The well-being perspective on SD makes goal setting more difficult. Its socially constructed, interpretative, and normative character leads to a multiplicity of goals that, in the case of a hierarchical mode of governance may be boiled down to one consensus goal for the sake of implementation, while more deliberate forms allow multiple goals to exist alongside one another. Yet, while the former approach may lead to societal discontent and some level of inflexibility in face of uncertainty, the latter may hinder implementation for a lack of direction and decisiveness (van Zeijl-Rozema et al. 2008).

This typification and framework for analysis is used to provide a sense of direction in the discussion of sustainability governance structures in the three universities that are the objects of investigation in this study. While this framework has been developed with territorial entities in mind, such as regions, states, or supranational entities, it may be equally useful for institutions of higher education. As complex institutions with a variety of stakeholders, decentralized structures and various external relations and liabilities, university governance structures bear many resemblances to those of higher-level entities on regional or national level. However, there are also many elements of corporate governance and characteristics unique to the higher education sector that require some further consideration.

Institutional entrepreneurship

Governance for sustainable development in institutions of higher education requires the commitment and persistence of individuals pushing for change in an institutionalized environment. With processes of institutionalization having been covered extensively in institutional theory in the past, the focus has increasingly shifted to processes of change as well as deinstitutionalization (Oliver 1992). As Meyerson and Tompkins (2007) pointed out, this led to an increasing preoccupation with endogenous agents of change or, as DiMaggio (1988) calls it "institutional entrepreneurs" who work and push for the transformation of existing institutional structures and beliefs.

In the field of institutional theory, the agency structure problem manifests itself in the problem of embedded agency (e.g. Powell and DiMaggio, 1991; Seo and Creed 2002). Institutional entrepreneurs thus have to square the circle in a sense, for they need to be embedded enough to have the power to effect changes while at the same time remaining critical enough to have an interest in change and be able to imagine alternatives, the capacity for which is fundamentally constrained by the very idea of an institution. It has been suggested that competing institutional logics can create the space for innovation and agency for change. Being embedded in multiple institutional contexts, some individuals are exposed to the contradictions between an institutional logic and their own interests or identities shaped by schemata alien to that logic, which allows them to envision alternatives and become institutional entrepreneurs. Meyerson and Tompkins (2007) suggest the notion of tempered radicals as individuals who are affiliated with interest or identity based communities that are at odds with the dominant institutional logics of their organization, thus being predisposed to effectuate and push for change. Well-versed in the institutional structures and beliefs of their organization, they are in a position to legitimize the questioning and transformation of institutional norms and practices if they manage to maintain their dual identities. By means of targeted framing, they can express goals in ways that enhance their legitimacy in the existing institutional logic and influence meaning-making.

Establishing sustainability governance structures at universities requires very similar processes of institutional change, and the notion of tempered radicals fits very well with the institutional entrepreneurs who instigated these changes at the institutions under investigation. Yet, the complex nature of sustainable development, with its multiple definitions and changing goals, complicates the picture and, as our case studies indicate, produces a more dynamic pattern of institutional change.

Methodology

The current study is based on qualitative research conducted between May 2010 and December 2011 at three different universities. The aforementioned theoretical frameworks derived from extensive literature research in the fields of sustainability studies, governance, entrepreneurship, organizational studies, and sustainability in higher education serve as useful tools to understand the generated data. While the sustainability governance model (van Zeijl-Rozema et al. 2008) allows for a comparison of the cases along the axes of hierarchy and deliberation on the one, and differing definitions of sustainable development on the other hand, the discussion of institutional entrepreneurship and tempered radicalism (Meyerson and Tompkins 2007) enables a discussion of agency-structure considerations in an institutional context. In combining in-depth research of three rather successful, yet, rather different institutions in terms of their sustainable development approaches with insights from institutional theory and sustainability governance, the authors hope to advance the theoretical depth of the debate on sustainability in higher education as well as to instigate a discussion on the role of students as potentially crucial agents of change in the transformation of universities.

A total of 18 in-depth interviews with a variety of stakeholders including university management, faculty, staff and students central to the establishment of sustainability governance structures at their respective institutions were conducted by the authors and serve as a basis for the study's claims and propositions. All interviews were semi-structured and between 30 and 60 minutes long. Interview guides with open-ended questions covered the development of the respective change initiatives, the change-agents and their impact on the institution as well as the two in relation to each other.

However, the research partially presented in this paper also faces some considerable limitations. While the theoretical framework used provides significant insights, it can only partially account for the richness of the data. Furthermore, the choice of three young, small to medium sized universities cannot be seen as representative for the entire university landscape. The same holds for the sampling of interviewees. Using a mixture of snowball and purposive sampling and an overall limited sample size a certain bias may arise. Lastly, the active involvement of the authors in one of the cases, while possibly enabling a more in-depth understanding of the case itself and the subject matter in general, may also have lent itself to some bias in the overall course of the research.

Case studies

Maastricht University Green Office

Students started the Green Office as an initiative in 2010. It developed into a student-led university department that manages the university's sustainability portfolio. Staffed by seven student employees and the environmental advisor of the university, it initiates and coordinates sustainability projects at UM, including the areas of education, research, operations, and the broader university community with the help of volunteers as well as a broad network of stakeholders within and outside the institution. In the first year of its existence (2010-2011), the Green Office conducted the first comprehensive baseline assessment on sustainability at UM (The Climate Action Report: Maastricht University 2011) and implemented a major Green ICT project in collaboration with ICT-Services (Ctrl-Alt-Delete Emissions Project: Maastricht University Green Office 2011) besides providing support to student initiatives and hosting a variety of events. Currently, the Green Office is busy with the development of the sustainability policy 2012-2014 as well as the monitoring and the development of implementation plans for UM.

The establishment of the Green Office as a student-driven bottom-up initiative, and its endorsement by top-level management as a means to connect grassroots innovators and institutional efforts to sustainable development "so that the ideas and enthusiasm of grassroots initiatives would also have an impact on the overall institutional behavior of the university" (A. Postema, Vice-President Maastricht University, personal communication, November 28, 2011), was an attempt to really operationalize a more deliberative and pluralistic approach to SD within the institutional structures. The operationalization of this deliberative approach within a hierarchical structure seems to be working well so far, as students, administrators and faculty alike see the Green Office as a success story (U. Scharf, A. Zeijl-Rozema, A. Postema, H. Hospers, F. Spira, personal communication 2011).

While an increasing professionalization and a pragmatic, result-driven approach of the team has been mentioned as a crucial success-factor by all interviewees, Green Office members, administration, and faculty alike see the participatory nature of the GO to be threatened by this logic. Caught in between the institution and the dynamic sprawl experienced in their student lives, the Green Office student employees have the potential to act as tempered radicals and institutional entrepreneurs (Meyerson and Tompkins 2007). Given a positive further development of the Green Office; as an institutionalized change-agent it would be unique, or, as Harm Hospers, Dean Faculty of Humanities and Social Sciences, claims, "There is no other student-led, university funded organization that has goals about how to run this place and connects both students and other people working and the institution [...] I don't know of any other example" (personal communication, October 19, 2011).

University of California Santa Cruz

Through a major campaigning effort, the California Student Sustainability Coalition (CSSC) triggered a system-wide policy on sustainable practices that was passed by the regents and subsequently the President of the University of California System, which besides UC Santa Cruz (UCSC) comprises nine other campuses throughout the state. This policy laid the ground for increasing administrative engagement on campus level and, through a joint effort of students and staff, the establishment of the Counsel for Sustainability and Stewardship in 2006 (A. Winslade, personal communication, November 2010). Today, the sustainability office has four full-time staff members, ten student employees and more than thirty interns who coordinate a broad range of sustainability activities guided by the Campus Sustainability Plan (UCSC 2011), which is a collaboratively developed strategy document for the years 2011-2013.

While formal governance structures are still very much hierarchical and intransparent, especially at the level of the regents, efforts to increase sustainable development below this level are very much driven by deliberative modes of governance. It therefore seems that the deliberative nature of the process driving sustainability at UCSC is very much an achievement of the students themselves that, rather than being specifically endorsed by the administration, was achieved despite the relative inertia of hierarchical administrative bodies.

In either case, the University of California Santa Cruz is a role model of campus sustainability that, having won numerous awards in this domain (C. McDaniels, personal communication, November 2010.), successfully implemented an approach to sustainability governance between its ecological and well-being deliberate modes, as specified by van Zeijl-Rozema et al. (2008). The pivotal role of students in driving this process, who through dedicated training and support are apparently able to be effective change-agents and institutional entrepreneurs, is clearly visible in the case of UCSC. By alluding to a longstanding commitment to the environment and successfully reaching out to staff and administrators, they have been able to transform at least parts of their institutional culture to commit to the imperatives of sustainable development.

Leuphana University Lüneburg

The process of developing sustainability governance structures at the Leuphana University Lüneburg was strongly driven by faculty. With two major research projects on what sustainability means in a university context ("Agenda 21 and the University" as well as "Sustainable University"), they managed to integrate organizational concerns into the core of their work as researchers, thereby creating a space for active engagement in terms of time and resources, the effects of which would outlast the individual projects. By integrating sustainability into the domain of teaching and research, relatively little additional administrative structures were needed to operationalize sustainability, also on a campus operations level. Through their institutional legitimacy as high-ranking professors and members of the academic senate, the main change-agents were able to use existing hierarchical governance structures for systemic change towards sustainability while opening up some means of deliberation, mainly within educational structures. Two administrative working groups, one on the environment and the other one on health, drawn from student, staff and faculty bodies as well as the environmental coordinator specifically hired to address environmental and sustainability issues, is what exists in terms of dedicated, formal structures specifically related to sustainable development on an administrative level. In charge of the environmental and health management systems as well as the sustainability report, they ensure progress monitoring and implementation processes. Yet, the majority of impetus for change comes from teaching and research, initially through the two research projects, but then also through project seminars in which students develop and implement their own ideas. Yet, while students participate extensively through educational projects as well as by setting up own initiatives implementing e.g. solar panels on some university buildings or organic food options at university canteens (M. Adomßent, personal communication, November 30, 2011), their involvement in university management and thus in the larger strategic or visionary questions remains limited.

While there seems to be consensus that the students are an important motor for change, identifying problems and "pegging away" ("Die lassen auch nicht locker", M. Adomßent, personal communication, November 30, 2011) at trying to solve them, their attempts to constructively collaborate ("konstruktiv mitarbeiten", J. Frey, personal communication, November 30, 2011) with university management in governance questions are barely successful or appreciated, or as Frey describes it: "participation yes, but on the very lowest level" (ibid.). This observation is further supported in the book "Sustainable University" (Michelsen et al. 2008), presenting the outcomes of the second major research project on the issue mentioned above. Comparing student participation in Lüneburg with that of North American Universities, Michelsen et al. see more of an intrinsic motivation of the students in comparison with their American counterparts where such engagement is actively supported, thus leading to a situation where the potential of students in that regard is not fully put to use.

Discussion of the case studies

In light of sustainability governance

The importance of student participation emerged as an overarching theme in the case studies. While in Maastricht as well as in Santa Cruz, students were the pivotal change-agents that started the whole process of establishing sustainability governance structures, in Lüneburg they were seen as "a very, very crucial factor from the beginning" ("Ein ganz, ganz wichtiger Faktor von Anfang an": M. Adomßent, personal communication, November 30, 2011), acting as a driver of all kinds of initiatives but also as a catalist for a broader movement among students, staff and administrators. Surprisingly, even though the importance and the need for more students as co-producers (McCulloch 2009) or active participants in university governance is widely discussed in the literature of the field (e.g. Bergan 2003; Zuo and Ratsoy 1999), and of stakeholder involvement in sustainable development, a well-known imperative (Coenen et al. 1998), the role of students in sustainability in higher education literature is often marginalized. Rather than as an active participant in the shaping of policies, the student is often theorized as an object that needs to be changed in order to become sustainable (e.g. Arbuthnott 2009; Stubbs and Cocklin 2008). The data from the three cases in this study clearly tells us otherwise. In fact, students may be key in the process of making university structures more open to organizational learning as an imperative of sustainable development (Michelsen et al. 2008).

As Institutional Entrepreneurs

The success of the respective change-agents in Maastricht, Santa Cruz and Lüneburg significantly relied and continue to rely on the various entry points for instigating change that could be leveraged within the existing institutional logistics. By understanding sustainability challenges through the institutional culture the change-agents are at least partially embedded in, as well as through strategically framing these issues in terms compatible with this logic, they leveraged support and drew on existing resources to accomplish their goals (Benford and Snow 2000). As a result of this, the emerging cultures of sustainability at the different campuses are clearly defined by the preexisting cultures, norms and values of their local context. A focus on learning, innovation, critical thinking, and especially student-to-student teaching embodied in Maastricht's university-wide problem based learning approach enabled the envisioning as well as the adoption by university management of a concept like the Green Office. Long standing traditions in environmental stewardship and student activism at UCSC allowed for the development of its vibrant, environmentally focused sustainability movement, and Lüneburg, with its history in teacher education and cultural studies, made the self-reflective research effort that lead to its impressive transformation possible.

The empirical evidence and the discussion above clearly suggest that every case study can be allocated to one of the fields of the 2 x 2 matrix describing governance of sustainable development. The case studies share one common theme, namely the involvement of students and their consequent influence on institutional structures, as well as on sustainability governance at the respective universities. Maastricht University Green Office can be allocated to the lower right corner, featuring a deliberative approach and focusing on well-being. UCSC's characteristics suggest an allocation to the lower left corner due to its focus on ecological issues while being mostly deliberative in its sustainability governance structures. The Leuphana University, Lüneburg, is to be located more towards the upper right corner as the approach was driven by faculty, exhibiting a more hierarchical approach with a focus on well-being.

Due to the fact that students (or students and staff) had an influence on the alteration of the given institutional structure in the three cases, an element of institutional entrepreneurship becomes evident in all cases. If one combines both theoretical frameworks, the Maastricht University Green Office, albeit representing the youngest and therefore least developed approach to sustainability governance in the current study, comes closest to the theoretically most ideal situation. The UMGO exhibits both features of being student-driven (deliberative), and a focus on well-being (addressing various societal relevant issues) while being embedded into the university structure. Through involvement in various institutional contexts, the students in the Green Office, as tempered radicals of sorts, can be effective institutional entrepreneurs, able to challenge the institutional logics of the university.

However, it remains to be seen whether the position and effectiveness of the Green Office as an institutionalized change-agent can be sustained over revolving generations of students over time.

Conclusion

In conclusion, the examination of the development of sustainability governance structures at the universities of Maastricht, Santa Cruz and Lüneburg strongly suggests that students can, and should, play a pivotal role in this process. Engaged faculties which ware able to direct their research onto the question of sustainability in the university itself, as a transdisciplinary endeavor using their own institutions as a place for experimentation, as in the case of Lüneburg, probably represent the most effective type of change-agent. However, this can only ever be a minority approach of a few small institutions with a strong sustainability focus. While a lot can be learned from Lüneburg's successes, it apparently fails to give students a sense of empowerment in (sustainability) policy questions, leading to a perception of green washing from the side of the university board, which underlines the importance of participation in higher education sustainability governance as an underrepresented issue in the current literature.

The cases of Santa Cruz and Maastricht demonstrate the power of students in this matter. In Santa Cruz, students have built strong and independent organizational structures in order to be able to influence and to work with administrative structures. In Maastricht, students have been able to establish themselves at the very core of sustainability policy making. While the former system already proved its functionality, at least regarding environmental concerns. The latter, however, will have to stand the test of time to prove its effectiveness in the long run. If it does, Maastricht University Green Office could truly be a model for tempered radicalism institutionalized as a way to overcome the problem of embedded agency in the university context, providing a new level of participation and empowerment.

The further development of these cases and especially Maastricht University Green Office, provides a highly interesting avenue for further research, especially regarding the question of embedded agency and institutional entrepreneurship, as does the linkage between pre-existing institutional logics and emerging sustainability cultures at universities observed in the present cases. Furthermore, with all three universities exhibiting a tension between hierarchical and deliberate modes of governance at different stages of the continuum, a more detailed analysis and further mapping of university sustainability governance approaches within the framework proposed by van Zeijl-Rozema et al. (2008), could provide significant further insight and means of comparison.

References

Arbuthnott, K.D. (2009), Education for sustainable development beyond attitude change. *International Journal of Sustainability in Higher Education* 10 (2), pp. 152-163.

Benford, R.D. and Snow, D.A. (2000), Framing processes and social movements: An overview and assessment. *Annual review of sociology* 26, pp. 611-639.

Bergan, S. (2003), *Student participation in higher education governance, Council of Europe*. Available at: http://www.coe.int/t/dg4/highereducation/Governance/SB_student_participation_FR.pdf (accessed 10 December 2011).

Börzel, T.A., Guttenbrunner, S. and Seper, S. (2005), *Conceptualizing new modes of governance in EU enlargement*, NEWGOV project, Deliverable Ref.12, D1.

Brand, R. and Karvonen, A. (2007), The ecosystem of expertise: complementary knowledges for sustainable development. *Sustainability: Science Practice and Policy* 3 (1), pp. 21-31.

Burki, M.-F. and Tappeser, V. (2010), *Making Sustainability Sustainable.* Unpublished Work. University College Maastricht.

Coenen, F., Huitema, D. and O'Toole, L. (1998), *Participation and the quality of environmental decision making.* Kluwer Academic Publishers, Dordrecht.

DiMaggio, P.J. (1988), Interest and agency in institutional theory. In Zucker, L.G. (Eds.), *Institutional patterns and organizations: culture and environment*, pp. 3-21. Ballinger Publications, Cambridge, Massachusetts.

Gibson, R.B., Angolin, J., Lawrence, D., Robinson, J.B., Tansey, J. and Watson, J. (2001), Specification of sustainability-based environmental assessment decision criteria and implications for determining "significance" in environmental assessment. *Origins,* September 2001, pp. 7-8.

Holt, D. and Barkemeyer, R. (2012), Media Coverage of sustainable development issues – attention cycles or punctuated equilibrium? *Sustainable Development* 20 (1), pp. 1-17.

Hueting, R. and Reijnders, L. (2004), Broad sustainability contra sustainability: the proper construction of sustainability indicators. *Ecological Economics* 50 (3-4), pp. 249-260.

Kooiman, J. (2003), *Governing as governance.* Sage Publications Ltd., London.

Maastricht University (2011), *Climate Action Report.* Available at http://greenoffice maastricht.nl/wp-content/uploads/2011/11/Climate-Action-Report_Final-Version.pdf (accessed 10 December 2011).

Maastricht University Green Office (2011), *Ctrl-Alt Delete Emissions Project.* Available at: http://greenofficemaastricht.nl/ctrlalt-delete-emissions/ (accessed 3 January 2012)

Mccool, S.F. and Stankey, G.H. (2004), Indicators of sustainability: challenges and opportunities at the interface of science and policy. *Environmental Management* 33 (3), pp. 294-305.

McCulloch, A. (2009), The student as co-producer: learning from public administration about the student-university relationship. *Studies in Higher Education* 34 (2), pp. 171-183.

Meyerson, D. and Tompkins, M. (2007), Tempered radicals as institutional change agents: The case of advancing gender equity at the University of Michigan. *Harvard Journal on Law and Gender* 30 (2), pp. 303-322.

Michelsen, G., Adomßent, M. and Godemann, J. (2008), *Sustainable University" Nachhaltige Entwicklung als Strategie und Ziel von Hochschulentwicklung.* Verlag für Akademische Schriften, Frankfurt am Main.

Oliver, C. (1992), The antecedents of deinstitutionalization. *Organization Studies* 13 (4), pp. 563-588.

Powell, W.W. and DiMaggio, P.J. (1991), *The new institutionalism in organizational analysis.* University of Chicago Press, Chicago.

Seo, M.G. and Creed, W.E.D. (2002), Institutional contradictions, praxis, and institutional change: A dialectical perspective. *Academy of Management Review* 27 (2), pp. 222-247.

Smismans, S. (2008), New modes of governance and the participatory myth. *West European Politics* 31 (5), pp. 874-895.

Stubbs, W. and Cocklin, C. (2008), Teaching sustainability to business students: shifting mindsets. *International Journal of Sustainability in Higher Education* 9 (3), pp. 206-221.

UCSC (2011), *Campus Sustainability Plan 1.5-2011-2013.* Available at: http://sustainability.ucsc.edu/sites/sustainability.ucsc.edu/files/UCSC.Sustainability.office.booklet.final_.pdf (accessed 2 January 2012).

UNESCO (1997, 8-12 December), *Educating for a Sustainable Future: A Transdisciplinary Vision of Concerted Action.* Paper presented at the Conference Environment and Society: Education and Public Awareness for Sustainability, Thessaloniki.

Wright, T. (2002), Definitions and frameworks for environmental sustainability in higher education. *International Journal of Sustainability in Higher Education* 3 (3), pp. 203-220.

van Zeijl-Rozema, A., Cörvers, R., Kemp, R. and Martens, P. (2008), Governance for sustainable development: a framework. *Sustainable Development* 16 (6), pp. 410-421.

Zuo, B. and Ratsoy, E.W. (1999), Student participation in university governance. *The Canadian Journal of Higher Education* 29 (1), pp. 1-26.

Chapter 34

Sustainable Development of Universities and the "University of Sustainable Development": Experience from Russia

Natalia Tihomirova, Siarhei Zenchanka and Siarhei Malchenka

Abstract

Sustainable development covers all areas of human activity. The most important activity, however, is educating people in sustainable development. This paper examines the concept of a transition in Russia from classical or applied types of universities to universities of sustainable development. The considered concept consists of two stages: the first stage involves the transformation of university management, which should be based on international management standards, and the second stage involves the introduction of educational programmes on sustainable development into the education process. The paper covers the issues of reforming Russian education, and presents the experiences of Moscow State University of Economics, Statistics and Informatics in developing the principles of sustainable development.

Introduction

In December 2002 the United Nations General Assembly adopted Resolution 57/254, which proposed a United Nations Decade of Education for Sustainable Development (DESD) spanning from 2005 to 2014 (UN 2002).

The basic values of education for sustainable development (ESD) are related to present and future generations, to the planet and its resources. ESD must teach us to adopt new behaviour and practices in order to secure our future.

The DESD envisages the following changes to the traditional educational scheme:

- Interdisciplinary and holistic learning rather than subject-based learning;
- Value-based learning;
- Critical thinking rather than memorizing;
- Multi-method approaches: word, art, drama, debate, etc.;
- Participatory decision-making;
- Locally rather than merely nationally relevant information.

The DESD is aimed at changing the approach to education in order to integrate the principles, values and practices of sustainable development.

The sustainable development of universities is usually considered either in terms of the implementation of courses on sustainable development into the education process (Leal Filho 2010) or in terms of the development of scientific research in this field. Simpson (2010) writes that "university researchers were the first to alert the world to global warming, and researchers worldwide are working to help society find social and technical solutions to environmental challenges".

Sometimes universities include a programme of environmental management in their code of practice: for example the University of Bristol certified its environmental management system in accordance with ISO 14001; the University of Applied Sciences Zittau/Goerlitz certified its environmental management system in accordance with Environmental Management and Audit scheme.

Leal Filho (2011) notes that "as with all educational strategies, the inclusion of sustainable development into university programmes can only succeed if schemes aimed at informing and mobilizing people are combined with relevant structural measures such as campus greening, a robust sustainability research programme or a set of concrete, practical demonstration projects".

The process of the "greening of education" was started more than 10 years ago and is closely connected with the greening of industry. Duffell (1998) described objectives and contents for environmental courses for engineers.

The Higher Education Funding Council for England pointed out that university campuses can be a model of how to be more sustainable and efficient, for example in reducing consumption of fossil fuels (Simpson 2010).

Nevertheless, some authors have pointed out that "only relatively limited steps have been implemented to achieve global learning for sustainable development" (Anderberg et al. 2009), and Wright (2010) suggests that "the majority of university presidents are well versed in the concept of sustainable development, but less familiar with the concept of a sustainable university".

Russia's accession to the Bologna Process in 2003 has set it the task of reforming the education system at all levels. In an attempt to solve the problem of modernizing Russian education, the national project "Education" (2005) has been adopted, which notes that the implementation of an innovative university programme should lead to a qualitative change in the level of education and scientific research, and in the effectiveness of their implementation. The objectives of reforming the education system are also considered in the Strategy of National Safety of Russia to 2020 (2009) and in the Strategy of Innovative Development of the Russian Federation to 2020 (2011). To ensure national security, all the efforts and resources of the Russian Federation will be focused on such priorities of sustainable development as science, technologies, education, health and culture, the ecology of living systems and environmental management, the maintenance of which will be achieved through the balanced use and development of advanced technologies and the appropriate reproduction of the potential of natural resources.

In accordance with the documents, the educational priorities will be the restructuring of higher education aimed at the development of research at universities, intensifying the cooperation of universities with the leading companies of the real economy and scientific organizations, the major expansion of the international integration of Russian institutions of higher education in educational programmes and in research and development, the strengthening of academic mobility, and the development of the network of educational and research programmes.

The necessity of reforming the system of higher education is also due to the low rating given to Russian universities.

Methodology of transforming

The methodology of transforming the classic or applied university into the "University of Sustainable Development" includes two stages. The first stage is the incorporation of international management standards into university practices. This stage helps universities to make the management process more comprehensible. The second stage is the implementing of new programmes on sustainable development in the education process. As it is impossible to share out these stages over time, changes must take place simultaneously.

The originality is in a general approach to university transformation combining the change of university management, the introduction of new programmes on Sustainability Science, and the wide implementation of information technologies that help to achieve sustainability in management as well as in the education process.

Implementing such an approach will help to answer three questions:

How to manage?
How to teach?
What to teach?

Sustainable development of universities

The sustainable development of a university involves, first of all, its stable financial positioning, the development of new educational programmes, the introduction of sustainable development courses in different curricula, the alignment of the university management structure to international management standards (including quality management), environmental management, occupational health and safety management, security and information safety, and social accountability standards. Implementation of these standards, along with modern information and communication technologies in university practices, allows the creation of campuses of sustainable development.

Quality management system. A university is a complex hierarchical structure. Implementation of the quality management system (QMS) in accordance with international standard ISO 9001:2008 allows the definition of processes and their interaction, i.e. creating a university process model (ISO 2008). The combination of quality management principles (ISO 2005a), plus a traditional approach to guaranteeing quality of education, engenders specialist training of a new and higher quality.

QMS, developed in accordance with ISO 9001, is a management subsystem which is based on a structured set of documents regulating the key aspects of an organization's activities in terms of consumers' demands, responsibility of administration, resource management, product realization and measurement, analysis and improvement (ISO 2008; 2009a).

The characteristic feature of the university-to-consumer relationship is that the same person passing through the entrant-student-graduate chain is simultaneously a consumer of services, a participant in the process and a product of the process. As a result of such transformations, the consumers' requirements and their satisfaction may not correspond to each other. This is primarily because of the rather long duration of the process and, secondly, due to the changes in environmental factors. This process requires greater flexibility in educational standards and management.

Environmental management system. There is a growth of understanding worldwide that it is necessary to support and improve environmental quality in order to protect people's health. More attention is being paid to the potential impact of the results of different organizations' and institutions' activities, products and services on the environment.

Educational institutions negatively impact on the environment too. First of all, different types of waste, emissions and discharges are created in the course of certain kinds of research and in the day-to-day running of study laboratories, computers and copy devices. They therefore need to develop their access to competent experts in the field of ecology and environmental management systems (EMS).

The key factor in spreading knowledge on ecology and environmental protection is the use of environmental management methods and tools by the university in accordance with the requirements of ISO 14001(ISO 2004).

Occupational health and safety management systems. It is necessary to consider two areas – the implementation of professional safety systems at universities, and professional safety training.

The former is very important in classical, technical and applied universities, which widely engage in lab work and research in natural and applied sciences. Such work is usually associated with professional risk. Computerization of the Humanities also involves the creation of a professional safety system which helps to mitigate the effects of electromagnetic radiation and other factors on students.

Systems of professional safety that are developed need to be in accordance with OHSAS 18001 standard (OHSAS 2007), which establishes the requirements for occupational health and safety management systems.

Corporate Social Responsibility is a prerequisite for sustainable development. The university is an important social institution in which educational services are implemented and a set of programmes is carried out for the development of social infrastructure.

There are several approaches to social responsibility – a system that meets the requirements of international standards of Social Accountability SA 8000 (SA 2008), or ISO 26000 (ISO 2010), or joins the UN programme on Global Social Responsibility (Global Compact) (UN 1999). In particular, Global Compact requires the fulfilment of ten principles concerning human rights, labour conditions, environmental protection and the fight against corruption.

Information security implies preservation of information confidentiality, availability and integrity (ISO 2009b).

In order to comprehensively protect the information resources in institutions and enterprises, information security systems are developed. Information security management systems (ISMS) are developed on the basis of international standard ISO 27001 (ISO 2005b).

The availability of an information security system is an important factor in the protection of intellectual property at a university.

Financial management. As has already been noted, the foundation for the stability of any organization is determined by its financial state. Financial management (FM) of an organization is based on International Financial Reporting Standards (IFRS 2011), taking into account a regional perspective.

Integrated Management System. The aforementioned international standards allow an integrated management system to be established (IMS) (Fig. 1).

Fig. 1: Components of an Integrated Management System

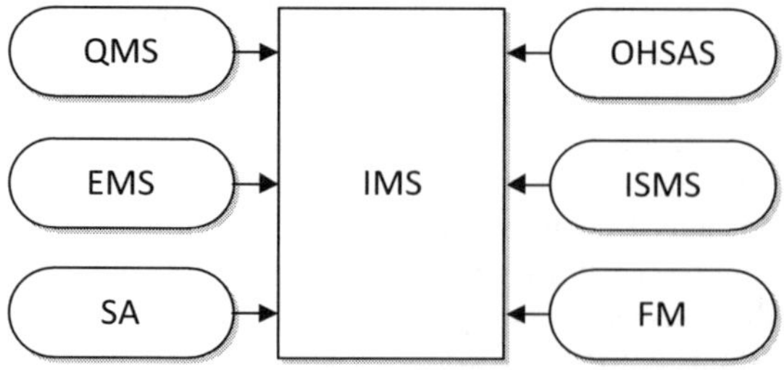

Information support of an integrated management system. The present level of development of the information society is characterized by the complexity of technical and social infrastructure. As a result, there has been a significant increase in the mobility of knowledge and a significant acceleration of the upgrade of various information resources (Appiah 2007). Under these conditions, data, information and knowledge become strategic resources together with the traditional ones.

The following modules based on the requirements of IMS are offered for the information system of the (Fig. 2):

Fig. 2 Modules of information system ISM of the University

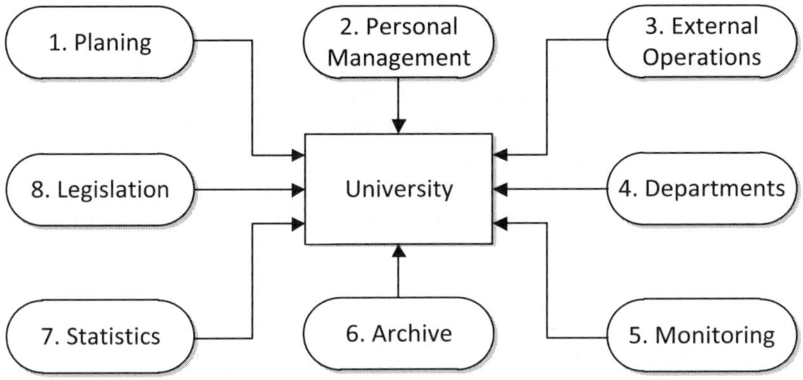

1. *Planning,* containing information on all aspects of the university's activities, an integrated policy, performance indicators and information on development programmes is used by the university administration;
2. *Personnel management,* containing information on the responsibilities of staff and professional development etc., is used in the department of personnel management;
3. *External operations,* containing information on the interaction with interested outside organizations, contractors, public authorities, mass media, NGOs, etc., as well as information on responses to external interactions, is used in the economic service;
4. *A Structural Unit (Department)* is used by the university administration in the areas of responsibility of the corresponding divisions and departments, operations and processes.
5. *Monitoring,* containing information on monitoring and measurement at the university, indicators of productivity, information on the status of equipment, and responses to current legislation;
6. *Archive,* containing the results of internal and external audits, plus information on corrective and preventive actions;
7. *Statistics,* used for integrated and statistical analysis with access to all databases of information systems;
8. *Legislation* is a database of laws and internal regulations used by other modules.

Analytical modules of IMS should provide effective decision-making for specific tasks:

- Forecasting of dynamic parameters for university development;
- Statistical analysis of the university performance indicators;
- Analysis of the student contingent in real time and academic performance;
- Dynamic analysis of the skills of teaching staff;
- Curricula development, optimal planning and consideration of workload in accordance with the curriculum.

Bologna Process. In 1988, the rectors of leading European universities signed the *Magna Charta Universitatum* (Magna Charta 1988). The Bologna Process started in 1999 with the signing of the Bologna Declaration (1999). The aim of the Bologna Process is to increase the competitiveness of European education and the creation of the European Higher Education Area.

At the conferences in Budapest and Vienna on 11 and 12 March 2010 the European Higher Education Area was officially launched (Declaration on the European Higher Education Area 2010).

A university system of smart management. The implementation of the integrated management system (IMS), the Bologna principles (BPP), the accounting requirements of national education standards (NES) and the widespread introduction of information technology (IT) will create a sustainable campus and the foundation for a university of sustainable development. This approach can be defined as "smart management" or "green management" (Fig.3). The creation of smart management at universities provides an answer to the first question, "How to manage?"

Fig. 3: Components of smart management

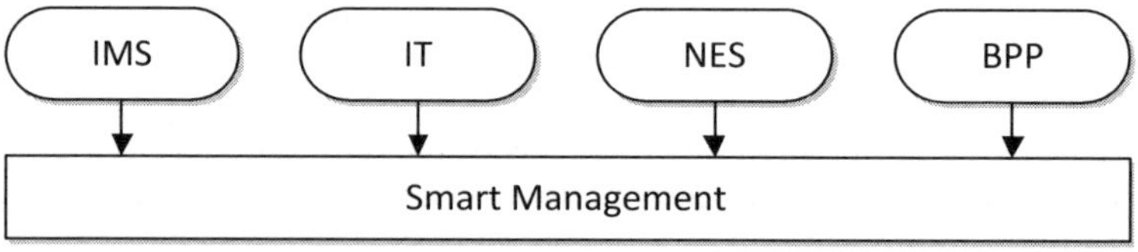

The university of sustainable development

Smart education. The introduction of smart education should be the next stage of transforming a university into a university of sustainable development. One of the areas of scientific knowledge as proposed in Agenda 21 is the creation of new capacities and capabilities for development (Agenda 1992).

The education and professional training of scientists and teachers should ensure a strong basis in natural systems, ecology and resource management. A training course should be created for specialists able to work in interdisciplinary programmes related to environmental topics and development, including programmes in applied social sciences. Thus, the introduction of the Bologna principles (BPP), the development of scientific schools, the training of teachers and experts in information technology (teacher training) and the widespread introduction of IT in the learning process all contribute to smart education (Fig. 4).

Fig. 4: Components of smart education

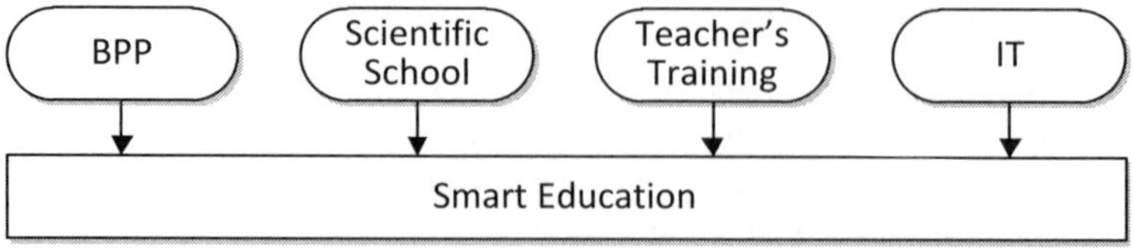

The concept for the implementation of smart education requires the creation of new training tools based on modern IT technologies and new knowledge, professionalism of teachers and staff, and common approaches to the competence of graduates. Transforming the education system of a university to correspond to smart education helps answer the second question, "How to teach?"

The basic paradigm in the creation of the "University of Sustainable Development" is the transition from specialization in education to universalization.

Whilst accepting that we need highly specialized professionals in mathematics, physics, chemistry, other natural sciences and the humanities, it is becoming increasingly clear that sustainable development requires the training of specialists able to perceive the world as a whole. Such education cannot be extensive but should be elite. The Bologna Process concept is a three-level training system – bachelor-master-doctor – which allows the training of professionals with narrow profiles as well as specialists possessing universal knowledge.

At the first level – Bachelor's degree – it is necessary to introduce the compulsory subject "Introduction to Sustainable Development" into the curriculum. This will help to prepare graduates for studying at the second level: Master's degree.

A Master's degree in Sustainability Science should aim to educate universal specialists with a knowledge of the full range of issues of sustainable development, i.e. economics, environmental and social sciences.

To realize the process of problem-based learning for Master's degrees in Sustainability Science, it is reasonable to enrol graduates of various undergraduate programmes, i.e. natural, economic and social science.

Thus, there will be three levels of training specialists at a university of sustainable development:

- Bachelor's degree with core training typical for classical universities;
- Master's degree as interdisciplinary training in sustainable development. This will be characterized by the enrolling of graduates of Bachelor's degrees in various disciplines – economics, ecology, natural science, mathematics, social studies – and their extended training in sustainable development. In particular, the second level of training – Master's classes – will be implemented at a university of sustainable development;
- PhD education will be focused on training for work in particular areas of sustainable development.

In conclusion, the "University of Sustainable Development" must be realized by implementing the principles of smart management (SM), smart education (SE) and Sustainability Science education (SSE) (Fig. 5). The implementation of these principles helps answer the third question, "What to teach?"

Fig. 5: Concept of the "University of Sustainable Development"

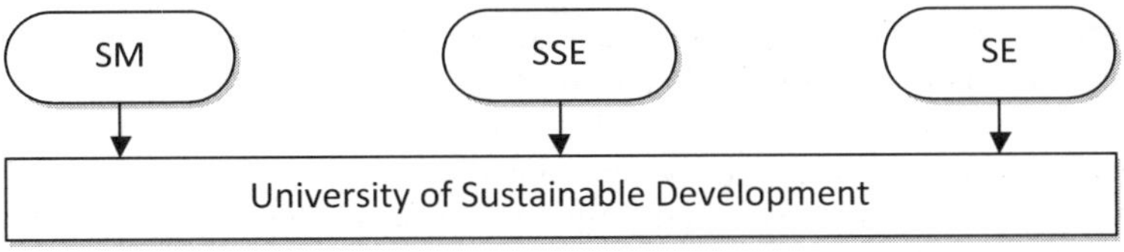

Implementing the principles of sustainable development in the Russian system of higher education

With the accession of Russia to the Bologna process, about 50% of higher education experts expected the process to yield large benefits, in particular an increase in the academic mobility of students and teachers and an expansion of academic relationships. In practice, in order to secure these benefits certain conditions had to be met: for example a system of easily recognized and comparable degrees had to be adopted, credit systems introduced, quality of education guaranteed, and curricula developed. The other 50% of experts thought that it was impossible to drop the existing education system, believing that a transition to European degrees would increase the brain drain from Russia.

In order to improve the quality of higher education in environmental aspects, the Decade of Education for Sustainable Development requires the support of the participating countries in achieving their Millennium development goals through promoting education in sustainable development and providing new features and tools for educational reform.

Smart management. Most universities have now introduced a quality control system. The representatives of students' and employers' associations are included in the process of state accreditation of educational or research institutions. Information on the implementation of management systems meeting international standards ISO 14001, OHSAS 18001 and SA 8000, or integrated management systems, is not available. Only the work on QMS is assessed by independent experts.

The ENQA Steering Committee decided to admit the Russian Accreditation Agency to the European Association of Quality Assurance. The Russian Accreditation Agency is a member of the International Network of Quality Assurance Agencies in Higher Education.

In moving towards the "green campus", universities should pay special attention to the introduction of integrated management systems.

Smart education. The integration of Russian higher education into the global system, while maintaining and developing its achievements and traditions, is one of the principles of public education policy, which is fixed by the Federal Law "On Higher and Postgraduate Professional Education" (Law 1996).

This law establishes the levels of higher professional education – Bachelor's degree, Master's degree or specialist training. Thus, Russia does indeed have a three-level system, since, after qualifying as a "specialist", a graduate may continue his or her education by taking a Master's programme. People already possessing a specialist diploma, who then receive a Master's degree, are seen as having a second higher education.

Currently, more than 80 higher education institutions have introduced credit units. This increases the autonomy of the university in improving the planning and organization of the educational process. It also strengthens the role of independent students' work, and optimizes the workload of teachers. Nevertheless, the state educational standards impose strict requirements on the content of training, which limits the university's ability to develop programmes of two diplomas with European universities.

State educational standards of the third generation make it possible to obtain an application for a higher education diploma similar to a European application.

The plan of action for implementing the Bologna Declaration provides for the formation of an institutional and individual grants system aimed at increasing academic mobility within Russia as well as abroad (Ministry of Education and Science 2008). The system of measures aimed at attracting European students and teachers to Russian universities is no less important than the creation of mechanisms and conditions for the mobility of Russian students and teachers. Intercollegiate exchange of experience and cooperation is a prerequisite for the further training of teaching and administrative staff, the quality of education.

Training in various areas of sustainable development is carried out mainly in universities: in geography, biology, agriculture, etc. These areas of training are identified in the Russian Classifier of Professions by Education (2007). The concept of sustainable development is included in the description of such postgraduate specialist subjects as economics, ecology, natural resources, and technogenic safety. For example, the Department of Geography and Geophysics at the University of St. Petersburg offers Master's programmes, as does the Department of Geography at Moscow State University and the Siberian Federal University.

But it should be noted that there is practically no training in the interdisciplinary field of knowledge.

With reference to climate change education, Feher et al. (2010) pointed out that "It will be a long road to resolve remaining problems, but the first steps have already been taken". This statement can be applied to Sustainability Science as a whole. Russian universities have a lot to do in order to convert themselves to universities of sustainable development.

Implementation of sustainable development principles in MESI

In 2004, MESI became the first Russian university to implement and certify a quality management system in accordance with ISO 9001, which is currently applied in its branches. An integrated management system based on the international standards is now being developed. For its implementation, an assessment of the environmental aspects of the university was carried out (Ahafonenka and Zenchanka 2010). An analysis of compatibility with the requirements of ISO 9001, ISO 14001 and OHSAS 18001 management systems was also performed.

The university adopted a code of corporate ethics providing social and employment rights for the staff and students.

MESI is a member of the Baltic University Programme focusing on sustainable development issues. Within the framework of this programme the academic mobility of students and teachers is implemented and students of MESI and some teachers participate in summer schools held at the Hamburg University of Applied Sciences. Academic mobility is also implemented according to the bilateral agreements with European universities – the Second University of Naples and the University of Calabria, Italy, the University of Auvergne Clermont-1, France, as well as universities within the Russian-Dutch programme. Students also participated in an online Teach-in-Day on sustainable development in 2010. The School of Sustainable Development conducts training programmes within the Baltic University Programme and offers a special "Environmental Management" course. This was introduced into the curricula of Bachelor degree courses in management.

MESI has wide experience in training specialists. The programmes "Management" and "Crisis Management" passed the international certification of quality education in 2011 and were recognized as the best project in the "Best Programme of Innovative Russia" competition.

In developing educational programmes and e-learning, MESI has cooperated with many European and international educational associations: the European Association of Distance Learning Universities (EADTU), the European Association of Universities (EUA), the International Council for Open and Distance Education (ICDE), the World Education Consortium (IMS), and the European Network of Distance Education and E-learning (EDEN), among many others.

To commence the e-learning process, MESI developed the concept of the "quick start", implying the use of cloud services that allow (Tikhomirov 2011):

- Improvements in the effectiveness of management activities;
- Avoidance of decisions related to the complexity in choice of technology in training;
- Significant reduction in the financial and time costs of performing research and development activities;
- Improvements in the skills of teaching staff;

- Getting the e-learning experience;
- Preparations within the organization for the full implementation of e-learning technologies.

In cooperation with the Hamburg University of Applied Sciences, the Master's programme in Sustainability Science was developed and presented for participation in TEMPUS.

The European Seminar on Sustainable Development (ESSD 2011) was held by MESI in cooperation with the Hamburg University of Applied Sciences and the Baltic University Programme. Some of the presentations are available to view at http://www.mfmesi.ru/essd/.

Conclusion

An attempt has been made to implement a general approach to the transformation of the classical and/or applied university into the "University of Sustainable Development".

Two stages of the transformation have been considered in this paper: the development of smart management at the university on the basis of international standards of management and IT, and the realization of educational programmes in Sustainability Science.

In moving towards sustainability, universities use different approaches – "green campus", QMS and EMS implementation, introduction of courses on Sustainable Development – but they never use them together.

A brief analysis showed that such an approach was not realized to the same extent in Russian universities as it was in world universities. Waheed et al. (2011) write that "economic development, social equity, and education in sustainability are the major drivers for achieving sustainability in HEI, while health and safety issues, energy requirements, institutional enhancement, and international research and development trends are the less significant driving forces".

The suggested concept of transformation based on smart management and smart education principles helps to make university management more transparent. The analysis of the implementation of the proposed concept in Russian universities shows the need for further work in this direction. It is necessary to agree with Anderberg et al. (2009) that "only relatively limited steps have been implemented to achieve global learning for sustainable development".

The experience of MESI shows its continuous development towards becoming a university of sustainable development, that the components of smart management and smart education are being implemented, that a number of training programmes on sustainable development are being carried out, and that research studies are being conducted in sustainability science.

References

Agenda (1992), *UN Programme of action from Rio*. Available at: http://www.un.org/esa/dsd/agenda21/index.shtml (accessed 10 January 2012).

Ahafonenka, V. and Zenchanka, S. (2010), Environmental Aspects of University Activity. *Annals of the "Constantin Brâncuşi" University of Târgu Jiu, Letters and Social Sciences Series* (3), pp. 49-64.

Anderberg, E., Nordén, B. and Hansson, B. (2009), Global learning for sustainable development in higher education: recent trends and a critique. *International Journal of Sustainability in Higher Education* 10 (4), pp. 368-378.

Appiah A. (2007), Building Post Roads of a New Millennium. *University of Pennsylvania Almanac* 53 (34). Available at: http://www.upenn.edu/almanac/volumes/v53/n34/c-bac-ka.html (accessed 10 January 2012).

Bologna Declaration (1999), *The Bologna Declaration of 19 June 1999*. Available at: http://ec.europa.eu/education/policies/educ/bologna/bologna.pdf (accessed 10 January 2012).

Declaration on the European Higher Education Area (2010), *Budapest-Vienna Declaration on the European Higher Education Area*. Available at: http://www.ond.vlaanderen.be/hogeronderwijs/bologna/2010_conference/documents/Budapest-Vienna_Declaration.pdf (accessed 10 January 2012).

Duffell, R. (1998), Sustainable development, Engineering Education and Activity. In Leal Filho,W. (Ed.), *Environmental engineering: international perspectives*, pp. 123-140. Peter Lang, Frankfurt am Main.

ESDD (2011), *Sustainable development: education, society, technologies, economy, ecology*, Minsk, 16-19 March 2011, Minsk: GIUST, p. 224 (in Russian).

Feher, A., Hauptvogl, M., Trangosova, P. and Svetlanska, L. (2010), How to educate for a Healthy Climate at University? An Intergenerational Cooperation (A Case Study from Slovakia). In Leal Filho, W. (Ed.), *University and Climate Change*, pp. 137-149. Springer, Heidelberg, Dordrecht, London, New York.

IFRS (2011), *Report of the Trustees' Strategy Review. IFRSs as the Global Standard: Setting a Strategy for the Foundation's Second Decade*. Available at: www.ifrs.org (accessed 10 January 2012).

ISO (2004), *ISO 14001:2004. Environmental management systems – Requirements with guidance for use*, Second edition. 2004-11-15.

– (2005a), *ISO 9000:2005 – Quality management systems – Fundamentals and vocabulary*. Third edition. 2005-09-15.

– (2005b), *ISO/IEC 27001:2005, Information security management systems – Requirements*.

– (2008), *ISO 9001:2008 – Quality management systems – Requirements*. Fourth edition. 2008-11-15.

– (2009a), *ISO 9004:2009 – Managing for the sustained success of an organization – A quality management approach*. Third edition. 2009-11-01.

– (2009b), *ISO 27000 – Information technology – Security techniques – Information security management systems* – Overview and vocabulary.

– (2010), *ISO 26000:2010 – Guidance on social responsibility*.

Leal Filho, W. (Ed.) (2010), *University and Climate Change. Introduction Climate Change to University Programmes*, p. 283. Springer, Heidelberg, Dordrecht, London, New York.

– (2011), Role of Universities and Their Contribution to Sustainable Development. *Higher Education Policy* 24, pp. 427-438.

Law (1996), Federal Law *On higher and postgraduate professional education (with changes and additions to 2011),* (in Russian). Available at: http://base.garant.ru/ 135916/1/#100 (accessed 10 January 2012).

Magna Charta (1988), *Magna Charta Universitatum*. Available at: http://www.magna-charta.org/library/userfiles/file/mc_english.pdf (accessed 10 January 2012).

Ministry of Education and Science (2008), *Bologna Process in Russian universities*, (in Russian). Available at: http://mon.gov.ru/obr/pri/4508/ (accessed 10 January 2012).

National project *"Education"* (2005) (in Russian). Available at: http://rost.ru/projects/education/ed6/docs.shtml (accessed 10 January 2012).

OHSAS (2007), *OHSAS 18001:2007 – Occupational health and safety management systems – Requirements*. Second edition. Available at: http://www.ohsas-18001-occu pational-health-and-safety.com/ (accessed 10 January 2012).

Russian Classifier of Professions by Education (2007) (in Russian). Available at: http://www.kadrovik.ru/docs/okso.htm (accessed 10 January 2012).

SA (2008), *Social Accountability 8000*. Available at: http://www.iqnet-ltd.com/userfiles/SA8000/2008StdEnglishFinal.pdf (accessed 10 January 2012).

Simpson, J. (2010), *A shared vision for sustainable development in higher education*. Available at: http://sd.defra.gov.uk/2010/09/a-shared-vision-for-sustainable-develop ment-in-higher-education/ (accessed 10 January 2012).

Strategy of National Safety of Russia (2009) (in Russian). Available at: http://www.scrf.gov.ru/docu ments/99.html (accessed 10 January 2012).

Strategy of Innovative Development of the Russian Federation to 2020 (2011) (in Russian). Available at: http://mon.gov.ru/files/materials/4432/11.12.08-2227r.pdf (accessed 10 January 2012).

Tikhomirov, V. (2011), World on the way to smart education. New possibilities for development. *Open Education,* Special Issue 4 (10) (in Russian).

UN (1999), *United Nation Global Compact*. Available at: http://www.unglobalcompact.org/About TheGC/index.html (accessed 10 January 2012).

– (2002), *UN Resolution 57/254. United Nations Decade of Education for Sustainable Development.* Available at: http://www.un-documents.net/a57r254.htm (accessed 10 January 2012).

Waheed, B., Khan, F.I. and Veitch, B. (2011), Developing a quantitative tool for sustainability assessment of HEIs. *International Journal of Sustainability in Higher Education* 12 (4), pp. 355-368.

Wright, T. (2010), University presidents' conceptualizations of sustainability in higher education. *International Journal of Sustainability in Higher Education* 11 (1), pp. 61-73.

Chapter 35

Integrating Sustainability throughout an Institution: Eden Hall Campus and the School of Sustainability and the Environment

David M. Hassenzahl, Molly Mehling and Esther Barazzone

Abstract

Chatham University has a long-standing commitment to women's leadership, the environment (a legacy of being the alma mater of Rachel Carson), and globalization. The 2008 gift of the Eden Hall Campus, a 157 hectare (388 acre) farm about 31 kilometers (19 miles) from Chatham's historic city campus in Pittsburgh, has provided an opportunity for Chatham to expand and deepen those long-standing commitments, especially with regard to the environment, as well as position the University as a regional and world thought leader in the area of sustainability. This paper describes the process that Chatham has undertaken in designing an energy positive, regenerative campus master plan; develop sustainable agriculture education, demonstration, research and outreach; and create a new School of Sustainability and the Environment. Topics include the philosophy that ties together facilities, curriculum and research across the campus locations, as well as the challenges and opportunities associated with developing new sustainable facilities and programs while simultaneously deepening commitment to sustainability of existing facilities and programs. Finally, it reports on the current state of the campus, faculty, and programs, and explores opportunities for the near and distant future.

Introduction

Chatham University has served as a Pittsburgh, PA, USA academic institution for over 140 years. Originally founded as a college for women, it is now a university with four academic units: Chatham College for Women, the College for Graduate Studies, the College for Continuing and Professional Education, and the most recent addition, the School for Sustainability and the Environment. Chatham's academic mission focuses on three principal areas: women's leadership, global and intercultural education, and sustainability and the environment.

The alma mater of environmental scientist and writer Rachel Carson, Chatham has a long-standing commitment to the environment and, more recently, sustainability. In 2008, in recognition of Chatham's leadership in environmental education and advocacy, Pittsburgh's Eden Hall Foundation gave Chatham the 157 hectares (388 acres) of farmland and forest now known as Chatham University's Eden Hall Campus. Chatham has committed to building there an energy positive campus that incorporates regenerative landscaping, on-site wastewater management, and sustainable living and growing. The development of the new campus, the first campus in the world to be designed from the ground up on sustainable principles, relates not just to buildings and grounds but to all aspects of campus life, as well as the relation of the Campus to the rest of the world.

Context: sustainability in Higher Education

Over the past few years, the need for sustainable practices and sustainability-trained professionals in a variety of fields has increased markedly (Muro et al. 2011; Vincent 2010), and student interest in these fields has increased (Vincent 2010). In the United States, the number of sustainability-related degree programs has risen from a few dozen in 2009 to over 100 in 2012 (Vincent 2012). At the same time, institutions of higher education are applying the values of sustainability across the board, including in facilities and purchasing, in research, and in curricula (see generally Dade 2010 and Jones et al. 2010). Campuses are actively lowering their carbon footprint, reducing their energy demands, and decreasing consumption through such activities as trayless dining and recycling programs. Organizations including the Association for the Advancement of Sustainability in Higher Education (with 855 member institutions [AASHE 2012]), the International Campus Sustainability Network, and the Society for College and University Planning, as well as journals such as the International Journal of Sustainability in Higher Education and Sustainability, the Journal of Record, demonstrate the extent to which sustainability is recognized and pursued by institutions for higher education.

Cortese (2010) situates sustainability as "the Mission of Higher Education" (Cortese 2010). However, while sustainability programs have grown rapidly over the past several years, some scholars question the pace and scope of adoption. Elder (2008), for example, argues that sustainability curricular changes have been slow to unfold and are usually not comprehensive. Others complain that sustainability inappropriately manifests itself as add-ons to disciplinary curricula (Brundiers and Wiek 2011). Lozano (2011) finds that institutes of higher education as a class have done far less sustainability assessment than has taken place in corporations. Nonetheless, many interdisciplinary environmental education programs are adding sustainability as a core curricular principle (Vincent 2010). In order to effectively improve the process of education for sustainability in higher education there are several reoccurring recommendations that surface and can be lumped under the movement to transform higher education into a transdisciplinary experience (see Sipos et al. 2008).

Sustainability at Chatham

Chatham University's efforts seek to respond to these concerns about adopting sustainability across campus activities. Activities and decisions will represent an experiment in campus emersion in sustainability; here we document major efforts to date, which we intend as a record against which to compare future accomplishments and challenges.

Commitment to sustainability at Chatham flows both from top to bottom and bottom to top, and includes all aspects of campus life, including academics, student and faculty life, as well as buildings and facilities. There are numerous examples of this, including: integration of principles of sustainability into new student orientation and academic courses across the curriculum; converting campus vans to biodiesel; eliminating plastic bottles on campus and installing water stations in their place; and hiring a Campus Sustainability Director.

Chatham University's leadership has been supportive of and engaged in the process. For example, in 2008 Chatham University's Board of Trustees made sustainability one of the University's three Mission Initiatives (joining women's leadership, the environment, and global and intercultural education) and has been closely involved in the planning for Chatham's Eden Hall Campus. President Esther Barazzone was an initial signatory of the American College and University President's Climate Commitment (ACUPCC) and an early member of its Steering Committee, and Chatham's reporting for ACUPCC demonstrates that it is making successful and consistent progress towards its climate goals. Other University-wide commitments include active membership in the International Sustainable Campus Network and the Association for the Advancement of Sustainability in Higher Education. The latter held its annual 2011 conference planning meeting at the Eden Hall campus.

Sustainability is also fostered by individuals and programs across the Chatham community. For example, the Director of the Counseling and Psychology program maintains a blog on sustainability and mental health; a member of the music faculty has been recognized on several occasions for his efforts to improve bicycle transportation options in the Pittsburgh area (resulting in Chatham becoming the first university in Pennsylvania to be named a Bike Friendly University by the League of American Bicyclists), and the Creative Writing program – among Chatham's hallmark academic institutions – has established expertise in environmental writing.

Examples of sustainability-oriented activities at Chatham's main campus include a major lighting retrofit at the Athletics & Fitness Center, the installment of solar thermal panels at Woodland Hall (Rachel Carson's former dorm), and the decision in 2002 to discontinue chemical pesticides and fertilizers on campus (in honor of the 40th anniversary of Carson's *Silent Spring*), and the designation of the campus as an arboretum in 1998. Such work has been acknowledged by *Princeton Review*, which in 2010 gave Chatham a green rating of 98, the highest score awarded to any college or university in Pittsburgh.

Sustainability activities have not been restricted to the University's historic city campus. Chatham also owns an industrial building 1.3 km east of the main campus; this building houses several of the university's graduate programs. Lighting in that building has been switched out to substantially reduce demand. Furthermore, to accommodate transportation between the two locations and between other nearby education institutions, Chatham purchased shuttles that run on biodiesel purchased from a local supermarket chain.

The Eden Hall campus plan

Upon receipt of the Eden Hall property, members of the campus community (including the board, senior administrators, faculty, staff, students and alumni) undertook a year-long learning and imagining process to determine how the property could best further Chatham's educational mission. Initial activities included a several learning expeditions: one by senior administrators (the President, Assistant Vice President for Planning, Vice President for Academic Affairs, and Vice President for Finance) to visit schools with farms (Berea, Warren Wilson) or nature preserves (Catawba) – followed later by a tour of many of the same institutions by a half dozen trustees; another, in the midst of a driving winter snowstorm, by trustees and administrators to visit Freshwater Institute and the US Fish & Wildlife Service's National Conservation Training Center; and a third by senior administrators to the University of Guelph (Canada). The process also included reviewing academics, farming, and facilities at many other programs around the world, and visits to a number of college and university campuses in the United States (e.g. Duke's Nicholas School of the Environment, Yale's School of Forestry and Environmental Studies). The leadership group also consulted with campus and community stakeholders. As a result of this process, the trustees decided (1) to establish a School of Sustainability and the Environment (SSE), and (2) to make Eden Hall Campus a comprehensive residential academic campus that would demonstrate principles of sustainability – and eventually house SSE – as well as becoming the first campus in the world to be built sustainably from the ground up.

Fig. 1: Eden Hall Campus master plan

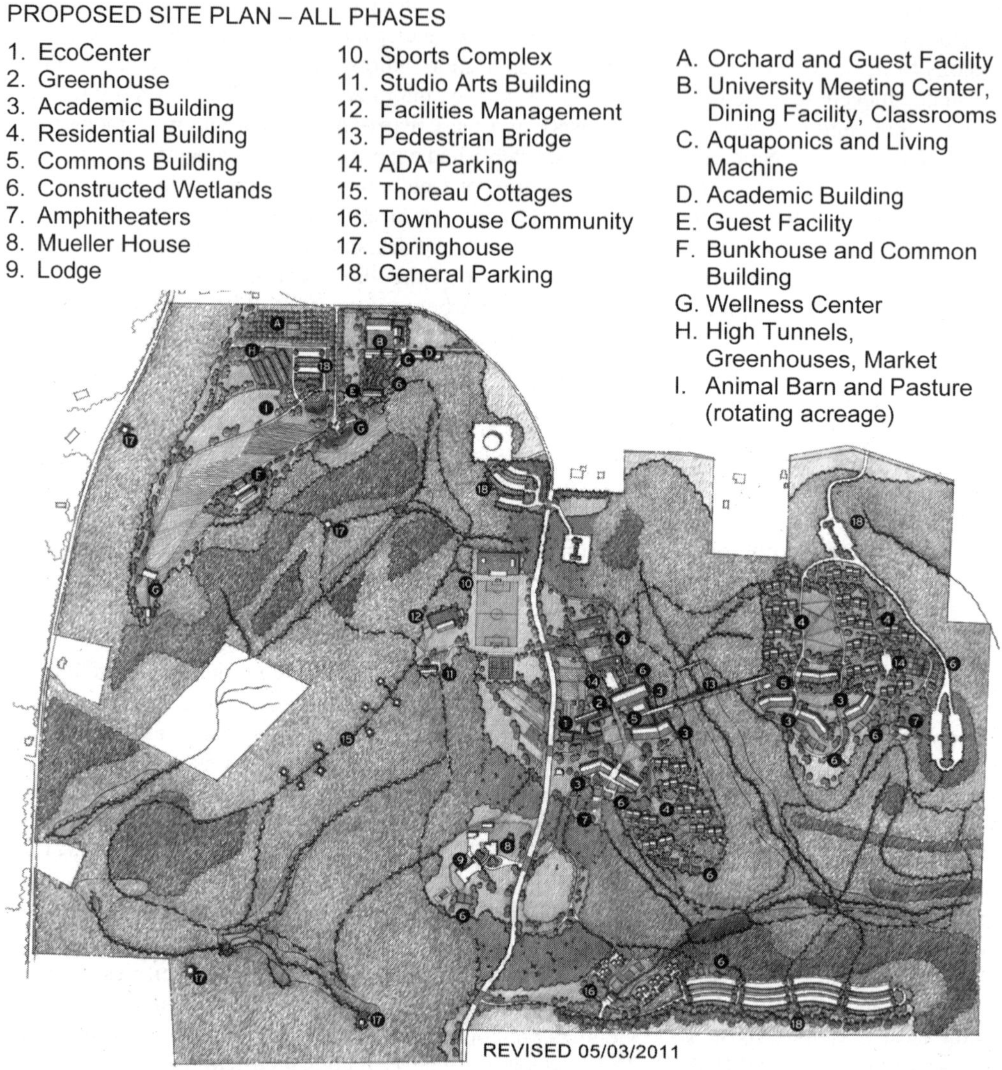

This master plan, approved by Richland Township (Pennsylvania, US), represents a climate positive campus with 1,200 residential students, on-site ecological wastewater treatment, experimental building, grounds, and living.

Two grants, one from PNC Bank Foundation and the other from the Richard King Mellon Foundation, provided funding to (a) produce a master plan for the new campus and (b) run a search for a Dean for the new school. BNIM (Kansas City, MO), a leader in sustainable architecture and design (and winner of the 2010 top firm award from the Institute of American Architects) and Andropogon (Philadelphia, PA), a leader in environmentally sensitive landscape architecture, were hired to create the Master Plan. BNIM had primary responsibility for buildings and infrastructure, while Andropogon took the lead on the landscaping and grounds. The master planning process took place over a year and a half, with numerous charettes involving the extended Chatham community (students, staff, faculty and administration), BNIM, Andropogon, and numerous guests. The master plan was approved by Richland Township in early 2011 (see Fig. 1).

The master plan represents a residential campus of 1,200 students, with a twenty year build-out plan. Salient features include regenerative landscaping (much of the current land has been used for non-organic agriculture and silviculture); climate positive construction; on-site, ecosystem-based wastewater treatment, with nutrient cycling through constructed wetlands, and mosaic garden, and orchards; demonstration agriculture; an outreach center; Thoreau Cottages for visiting scholars and artists; and opportunities for wellness.

Subsequent to master plan approval, a second competition was held for design of the initial building phase. Based on their extensive experience with sustainable campus planning and exciting design concepts, Mithūn (Seattle, WA), another leader in sustainable building and design, was selected to oversee the first phase of site development. Mithūn has partnered with engineers, Pittsburgh-area architects and vendors, and others to work with Chatham on an initial set of buildings that include:

- An ecocenter that will serve as the campus welcome and outreach center, along with some office and classroom space
- A large barn redeveloped as a performance/conference venue
- A dairy barn re-envisioned as a coffee shop
- A field lab/wastewater treatment/aquaculture facility
- A 90-person residential facility.

In addition, first phase campus development will include

- A 250 seat outdoor amphitheater
- A district energy loop
- Organic agriculture activities
- Extensive infrastructure development.

Consistent with the campus mission and master plan, the initial construction is designed to generate more energy than it uses, through a combination of solar photovoltaics, solar thermal, and ground-effect geothermal energy (see Fig. 2). The facility will be tied to the electrical grid, with the expectation of net sales back to the local electrical system. Furthermore, all wastewater will be processed on site using natural processes, with nutrient cycle back to a mosaic garden and orchard. Aesthetics of purpose and place are also an essential component of campus design, which reflects the relationship of the old, traditional farm to the new campus/farm. Funding for several components has been secured, and Chatham continues to fundraise with the goal of completing the first round of construction by fall semester 2013.

Fig. 2: Energy diagram for initial Eden Hall construction

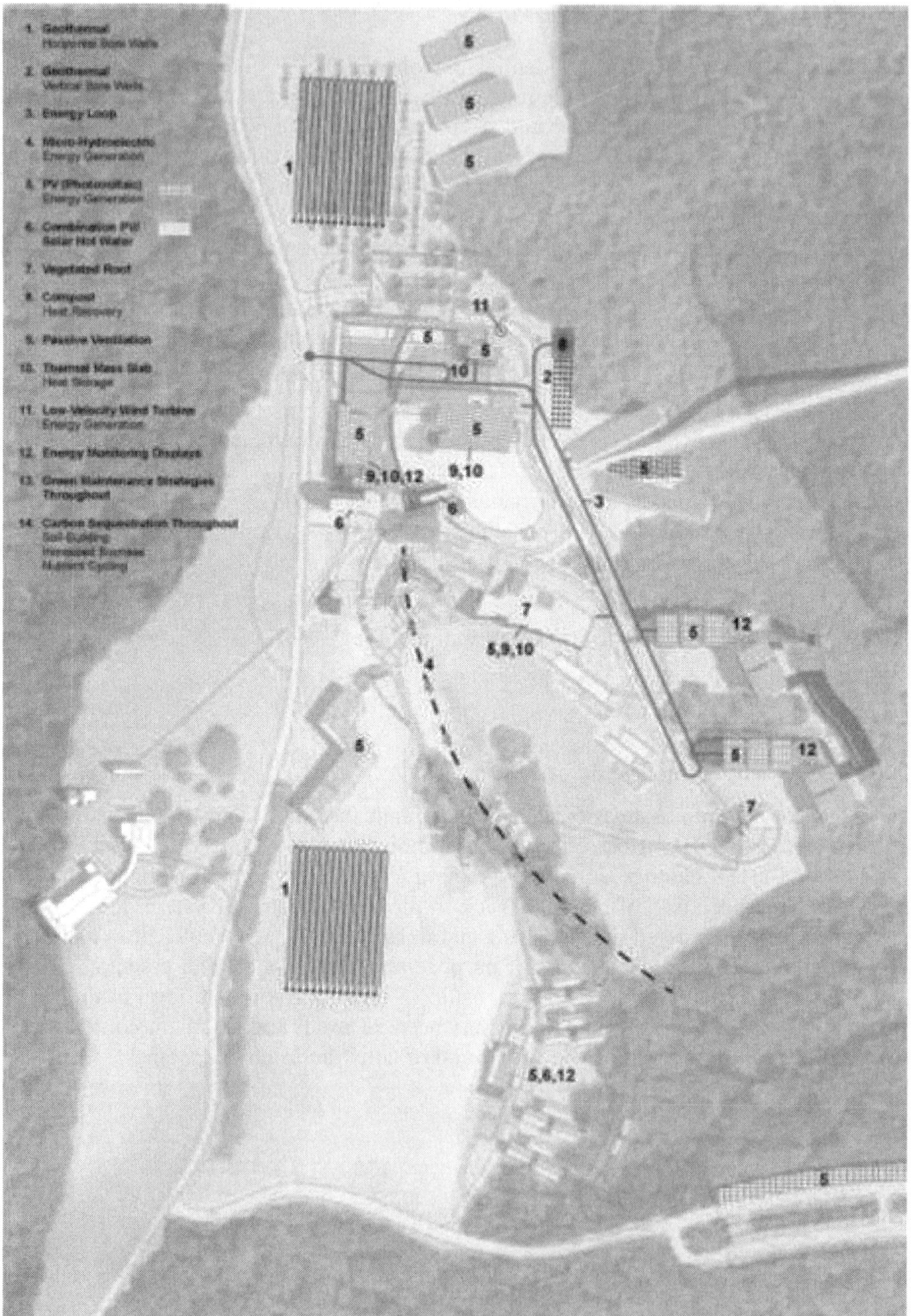

The first phase of construction at the Eden Hall Campus will include a number of energy innovations, including solar electric, solar thermal, ground-effect geothermal, wind, and microhydropower. In combination, these should allow the campus to generate more energy than it uses over the course of the typical year.

The School of Sustainability and the Environment

The School of Sustainability and the Environment is founded on the principle that sustainability requires that we simultaneously attend to economic, social, and environmental wellbeing. In this context, sustainability is best understood as a process, not an endpoint. Assessing sustainability should focus more on whether changes in conditions are expected to enhance or damage social conditions, economic development, and the biophysical environment. Sustainable options or choices can only be effective if they are preferable – that is, they are only likely to be adopted if they are understood to enhance quality of life.

Understanding sustainability requires a systems perspective. Changes in society, economic conditions, or the biophysical environment can be expected to lead to second, third, and greater order effects that are often difficult or impossible to predict. Diverse sources of knowledge and wisdom are necessary to understand systems. Consequently, members of the SSE community must be able to work across disciplinary boundaries to identify and develop appropriate knowledge and approaches to knowledge.

Finally, knowledge developed by the SSE community should be experienced, practiced, and shared. Knowledge not shared beyond the community is of no value. While the Eden Hall campus is a valuable resource for Chatham and the SSE, it should both attract and reflect the world, providing an understanding of sustainability that is useful on an individual and global level.

a) Academic Programs

The SSE began recruiting for its first program – a Master of Arts in Food Studies – in late 2009, and admitted a first cohort of 24 in fall 2010. This program, led by sociologist and food expert Alice Julier, is designed to promote student learning about food access, systems, security, and sovereignty. Areas of focus include marketing, communication, sustainable agriculture, and food systems.

A second academic program began in January 2012: a Certificate in Environmental Management. This certificate is intended as a credential for sustainability professionals, and includes training in both the fundamentals of sustainability and business/management. It is offered entirely online. Students completing the certificate will be able, through some additional course work and a limited residency program at Eden Hall, to complete an Executive Masters in Sustainability Leadership.

The third academic program is a Master of Sustainability; the inaugural cohort of students will join in fall 2012. This program begins with a year of intensive courses on the fundamentals of sustainability, followed by a summer experience that could include research, a professional placement, internship, or other activity. The second year is devoted to specialization, including programs offered jointly with existing academic units at Chatham. Specialization will include communication, business, education, psychology, and food studies as well as areas designed by students.

Eventually, the SSE expects to provide programming from the undergraduate to doctoral levels, as well as life-long learning opportunities.

b) Faculty

The search for a Dean led to the hire of David M. Hassenzahl, a scholar in the field of risk and uncertainty and former Chair of the Department of Environmental Studies at the University of Nevada, Las Vegas. In his interviews and conversations with Chatham, Dr Hassenzahl presented a vision of sustainability that resonated with the Board of Trustees, administrators, and faculty. Dean Hassenzahl split time between Chatham and the National Council for Science and the Environment in Washington DC between August and December 2010, and has been full time at Chatham since then.

Four new faculty members were added in the middle of 2011. These faculty members were selected for their academic excellence, commitment to transdisciplinary and community-based research, and ability to teach a diverse set of students. Dr Michael Finewood, a geographer with expertise in human and economic geography; Dr Molly Mehling, an ecologist with interest in environmental communication and health; Dr Crystal Fortwangler, an environmental anthropologist with interests in social justice and land use; and Allen Matthews (MS), a fifth-generation farmer with extensive experience in sustainable agriculture. Currently, the SSE is searching for two additional faculty members, one in energy and water resources management, and one in food systems. These are expected to join the Chatham community in fall 2012.

Conclusion

Chatham University has a long-standing commitment to sustainability, anteceding the widespread recognition of that term in public discourse. Its current sustainability efforts span its academic programs, campus life, operations, and infrastructure. The 2008 gift of the Eden Hall Campus inspired a reprioritization of sustainability, and as the campus develops, it will serve as an example of sustainable living and learning. Chatham is currently reaching out to potential partners – academic, non-Governmental, public sector and private sector – across the globe; its participation in the International Conference on Sustainability in Higher Education and this paper represent a key opportunity to that end.

Recently, Krizek et al. (2012) note the challenges that the University of Colorado at Boulder has faced in trying to establish sustainability across campus functions. They identify that campus sustainability "may evolve through a series of phases labeled: grassroots; executive acceptance of the business case for sustainability; the visionary campus leader; and fully self-actualized and integrated campus community. The Chatham University approach is designed to recognize those phases, and seeking to allow sustainability to thrive through them all simultaneously. As campus, curricular, and research progress, we will continue to report progress in this context, hoping to learn and report on lessons for ourselves and other institutions of higher education.

Acknowledgements

We thank Sean Coleman for his thoughtful editing and contributions, Walter Leal for his encouragement of this project and more broadly for his leadership in sustainability in higher education, and two anonymous reviewers for thoughtful suggestions that improved this chapter.

References

AASHE (2012), http://www.aashe.org/membership/member-directory. Accessed February 8, 2012.

Brundiers, K. and Wiek, A. (2011), Educating students in real-world sustainability research: Vision and implementation. *Innovative Higher Education* 36, pp. 107-124.

Cortese, A. and Hattan, A.S. (2010), Research and Solutions: Education for Sustainability as the Mission of Higher Education. *Sustainability: The Journal of Record* 3 (1), pp. 48-52.

Dade A. (2010), *The impact of individual decision making on campus sustainability initiatives.* Doctoral Dissertation, University of Nevada. Las Vegas.

Elder, J. (2008), Think Systemically, Act Cooperatively: the Key to Reaching a Tipping Point for the Sustainability Movement in Higher Education. *Sustainability: The Journal of Record* 1 (5), pp. 319-328.

Jones, P., Selby, D. and Sterling, S. (Eds.) (2010), *Sustainability Education: Perspectives and Practice Across Higher Education*. Earthscan, Washington DC.

Krizek, K., Newport, D., White, J. and Townsend, A. (2012), Higher education's sustainability imperative: how to practically respond? *International Journal of Sustainability in Higher Education* 13 (1), pp. 19-33.

Lozano, R. (2011), The state of sustainability reporting in universities. *International Journal of Sustainability in Higher Education* 12 (1), pp. 67-78.

Muro, M., Rothwell, J. and Saha, D. (2011), *Sizing the Clean Economy: A National and Regional Green Jobs Assessment.* Brookings Institute Press, Washington DC.

Sipos, Y., Battisti, B. and Grimm K. (2008), Achieving transformative sustainability learning: engaging head, hands and heart. *International Journal of Sustainability in Higher Education* 9 (1), pp. 68-86.

Vincent, S. (2010), *Interdisciplinary environmental education on the nation's campuses: Elements of field identity and curriculum design. A research study conducted by The Council of Environmental Deans and Directors of the National Council for Science and the Environment.* Washington DC (39 pp).

– (2012), *Report to the Council of Environmental Deans and Directors 2012 winter meeting.* Washington, DC. January 21, 2012.

Chapter 36

Transformation to a Sustainable-Led University: Insights from the South-East Asia Region

Mostafa Nejati, Amirul Shah Md Shahbudin and Azlan Amran

Abstract

Following the announcement that the years 2005-2014 will be the United Nations Decade of Education for Sustainable Development, there has been growing interest among universities to integrate sustainability values within their respective organizations. However, there is a dearth of research works to investigate the implementation results, as well as the changes which have happened at the universities to implement sustainability. Hence, this paper is an attempt to tackle this gap by studying one of the universities at the forefront of sustainability in the South-East Asia region (namely, Universiti Sains Malaysia) and providing practical insights on the process of implementing sustainability agenda there. It investigates what changes have happened in the selected case schools of the university in order to practice and disseminate sustainability vision in the different aspects of teaching, research, community engagement, and institutional arrangements/operations. By applying a qualitative research design and adopting the theory of New Institutional Sociology (NIS), the reactions of organizational members to the proposed changes were explored and the institutional factors that influence universities to follow a sustainability vision were examined. The research findings are of interest to the decision-makers and administrators of universities and institutes of higher learning, as they provide invaluable insights from the South-East Asia region. The research findings can be a suitable guide for other institutions of higher learning throughout the world to adjust and manage their practices.

Introduction

The cliché that universities are merely to provide quality education to the learners no longer applies in this century. Universities are now looked upon more holistically and rather as institutes that could nurture minds and talents, empowering them to resolve the issues faced by society while instilling human values in an academic environment. Hence, unlike the traditional perspective whereby universities were known as centres for providing quality education, teaching, and research, they now have far bigger responsibilities and play a special role in society (Wright 2004). They should take action, play their role in enhancing society at large (Lukman and Glavic 2007), and should become pioneers in acting toward and promoting sustainability, engaging with the community, and avoiding global environmental degradation – they have the potential and expertise to be leaders in all fields of research, teaching and learning, sustainability and community engagement (Moore 2005; Stephens and Graham 2010), and can facilitate the creation of an equitable and sustainable future (Cortese 2003).

Institutionalisation of sustainability at the universities and educational institutes allows them to reach new experiences and previously unexplored opportunities, as well as allowing them to prepare for unborn challenges (Adomssent et al. 2007). Moreover, through such comprehensive educational changes at all educational levels (from primary to higher learning and life-long education), a modification of social habits towards a more sustainable lifestyle could be expected (Geli de Ciurana and Leal Filho 2006).

Nowadays, and in response to the growing demands by various stakeholders, universities worldwide are changing their mission, vision, and educational practices to cope with the growing concerns about social and environmental issues (Md Shahbudin et al. 2011). They are implementing sustainability practices within the scope of a sustainable university, namely teaching, research, outreach programmes, and in-campus operations, as well as evaluation and reporting (Lozano 2010; Velazquez et al. 2006). This commitment of universities towards the sustainability agenda was further reinforced by the announcement that the years 2005-2014 will form the United Nations Decade of Education for Sustainable Development (DESD) to incorporate the principles, values, and practices of sustainable development into research, teaching and learning areas (Sanusi and Khelghat-Doost 2008; UNESCO 2005). The objective of the education for sustainable development has been to challenge people to undertake new behaviours and practices, respect the rights of the fellow generations, and to utilise the core competencies of universities to further support such a move.

Now, as we get closer to the end of the Decade of Education for Sustainable Development, it is imperative to reflect on how universities worldwide have performed with regards to promotion and building public understanding, as well as implementation of sustainable development. However this has remained an under-researched area that needs extra attention to explore the process of transition to a sustainable university, as well as implementing and reporting sustainability within the higher education setting (Lozano 2011; Velazquez et al. 2005). As such, the current study investigates one of South-East Asia's foremost universities in relation to sustainability (namely, Universiti Sains Malaysia) and specifically looks at the process of institutionalisation and implementation of a sustainable development focus at its selected schools to explore what changes have occurred at the university to support its sustainability agenda. This allows an in-depth insight on the current implementation status of sustainability in a benchmark higher education institute in the region. Furthermore, the application of the theory of New Institutional Sociology (NIS) to contemplate the institutional factors pertaining to sustainability makes this research and its findings stand out. The findings are therefore expected to not only contribute to the body of knowledge, but also assist the decision-makers and administrators in the institutes of higher learning in adjusting their policies and strategies based on the status quo.

Methodology

The subject of the current study concerns further exploring the university's efforts to integrate sustainability focus in the organisational setting. Due to the exploratory and explanatory nature of the research topic and in order to get in-depth understanding of the subject matter, this study applied an interpretive case study as its main research strategy. As such, qualitative methods including in-depth literature review, document review, and semi-structured interviews, along with researcher's observations, were used to collect necessary data for the research. The significance of qualitative research lies in the fact that it can contribute to the advancing of the field through reaching unique and deep findings (Gephart 2004) and understanding the *hows* and *whys* of phenomena (Davy and Valecillos 2009).

The conducted interviews were of the semi-structured types and were conducted with organisational members from the selected schools as well as some of the selected university's key decision-makers from the university management team and its centres of excellence. The organisational members were of significant importance to this study because, as Blackburn (2007) argues, academic organisations are only as good as the individuals inside the organisation. In fact organisational members are the ones who should modify their attitudes and become more responsible so as to contribute effectively to their university goal of implementing sustainability; hence enabling it to become a sustainable organisation (Lozano 2006).

The review of documents including schools' annual reports and other publications, reports, memos, list of curriculum and courses syllabuses, progress reports, and other related documents were also used to collect and generate further data. In addition, researcher observations played an important role in better understanding and analysing the field and collecting necessary data to answer the research questions.

Triangulation of data was done through applying at least two different methods to collect the necessary data (namely, semi-structured interviews, researcher observations, and document review), hence ensuring the reliability and validity of collected data.

A total of 42 individuals were interviewed, mostly face-to-face with the exception of a few individuals who were interviewed via email. All interviews were transcribed and analysed using the thematic network by Attride-Stirling (2001) to organise the collected inquiries and ideas, and draw meanings out of them. After following several steps for the purpose of data reduction, summarising, coding, and conceptualising the data, the arising themes and their connections were identified. Document review results and literature review also complemented the analysis process.

Case study of USM

Known as one of the most active universities in terms of contribution to sustainability in the South-East Asia region, Universiti Sains Malaysia (USM) has been one of the forefront universities in pursuing sustainability on campus and implementing the principles of Education for Sustainable Developments. USM has gained the internationally acclaimed status of ASEAN's only Regional Centre of Expertise on Education for Sustainable Development accorded by the United Nations University in 2005, which happened along with the introduction of UN Decade of Education for Sustainable Development. Moreover, USM's setting up of the Centre for Global Sustainability Studies (CGSS) has been another important step in realising the university's mission to mainstream sustainability into the fabric of its activities. The centre was officially launched in December 2009 to contribute to the field of sustainable development through education for sustainable development, monitoring and assessment, research, capacity-building and other means to resolve the sustainability problems faced locally and globally.

USM has also started to transform its actions to be more in-line with sustainable development concept, including revamping most of its activities related to nurturing and learning, research and innovation, services, students and alumni, society, and the management of the university as a whole, in pursuit of sustainability-led growth as its destination (USM-APEX 2008). Overall, there have been various changes occurring in line with the university's sustainability vision, which will be discussed below.

Findings and Discussion

All actors can influence institutional factors and the new social structures to some extent; thus, their support for the proposed institutional changes can assist the organisation in achieving its targets. As mentioned previously, the current research was conducted to explore what changes have taken place in the selected school in order to support the university's sustainability vision. These changes have been investigated in different aspects, namely teaching, research, community engagement, and institutional arrangements/operations:

Teaching

In order to revise the teaching materials to accommodate necessary changes and adapt to the sustainability requirements of the university, certain changes have taken place at the schools' level. Namely, the schools have tried to embed the concepts and issues relevant to sustainable development within the existing courses and enriching them with *sustainability flavour*.

Given the fact that to formally change course contents or introduce new courses a proposal has to go to the university senate and be approved there, which is a rather time-consuming process, the schools and lecturers prefer to enhance their existing courses and embed the sustainability concepts and related materials into the current curricula. As echoed by one of the school deans:

> [...] So if we train them now, say our graduate, undergraduate and postgraduates, we train them that this is the concept of sustainability that we want to introduce is concerning the environment, concerning about the jobs, concerning about your future, concerning about health, everything. So and also about the profession; you know, sustainable profession. [...] So that's why, in the school we are now embedding all this curriculum on sustainability concepts into the teaching, curriculum and also part of the research, so that the students will benefit, and the students will carry the message to the rest of the world.

Apart from that, the university itself is in the process of introducing a new course on "sustainable development" to be taught to all undergraduate students as a compulsory course. Such a course would create a basic understanding and awareness of sustainability-related subjects amongst students.

Some of the schools have also come up with degree programmes to cover the inadequacy of specialists and a skilful workforce in certain fields in the country; though this could indirectly apply to the majority of existing degree programmes at university level which are ultimately producing graduates who can apply their knowledge and expertise in society. Furthermore, some schools have embarked on organising educational workshops that directly or indirectly cover sustainability topics.

Research

In terms of conducting research work related to sustainability, all the interviewed faculty members confirmed they are taking into account the agenda of sustainability when conducting research. This has been done through doing research that directly or indirectly supports sustainable development and improves the quality of life for the many people at the bottom of society. Moreover, some faculty members have gone one step further by encouraging their students to conduct sustainability-related research. As highlighted by one of the faculty members:

> [...] I am active in terms of research in sustainable construction. So I have done a couple of research. It's basically more towards increasing the awareness of the developers about sustainable construction and the importance of it. How they want to achieve it. So I am helping interns of research as well. And then for the students, we do encourage them to bring this sustainable campus within their research. Most of my research students [do research] related to green or sustainable [concepts].

Among other plans of the schools to further enhance sustainability research and support the university's efforts to operationalise sustainability is setting up centres of excellence that strategise and guide the research on sustainability in the context of the school's programmes.

Community engagement

By having a specific role at each school as the Deputy Dean of the Industry & Community Network, the schools at USM are better geared towards addressing the needs of society and active industries in their respective communities through using the expertise of university members. This is done through applying the university's core competency in teaching, research, problem-solving and solutions in an attempt to increase the well-being of society.

There have been different community engagement projects which have been completed or are ongoing in the schools examined. Moreover, some of the faculty members are also active in raising the awareness about the importance of a sustainable-led university by informing their peers from different universities about the achievements of USM in regard to its vision of *transforming higher education for a sustainable tomorrow*:

> [...] Sometimes, I consider myself as a small ambassador when I go out from USM. When I went to KL and met other universities. I will try to promote USM saying that USM is like this or that; we make sure to try to conserve energy; we abolish this white coffin so we cannot use polystyrene. So I've been telling others what USM has done. So in a way I did play my role.

Besides, at the university level, the Division of Industry and Community Network is committed to expand the engagement with the community and industry to benefit the well-being of the "bottom billions" and enhance social cohesion. This division promotes effective and sustainable relationship collaboration between USM and industry and community.

Institutional arrangements/operations

The sustainability vision of the university has affected the schools' operations, structure, and systems to some extent. The most noticeable change at the schools has been the minimisation of paper usage. In fact, all the schools investigated have made an effort to minimise use of paper by adopting a paper-less strategy or at least trying to minimise use of paper. This has happened through use of alternative communication means, especially electronic media to disseminate information among staff.

As for recycling, this is done voluntarily by interested faculty members, since there is no-one in charge of it, and no mechanism exists to monitor and promote recycling. It is worth mentioning that some schools used to have recycling projects, but after some time these could not be sustained and were stopped. Therefore, recycling is now done on a voluntary basis, as there is no mechanism to monitor and promote recycling.

Overall the current practices and programmers at the schools in support of the university's sustainability vision and mission have been identified and summarised in Table 1. It is necessary that these programmes are further enhanced and run on a continuous basis.

Table 1: Schools' programs in support of sustainable development

- Empowering the university graduates with the necessary skills required in the society
- Financial supports to poor students
- Sustainability-related/enriched courses
- Community Projects
- Organizing sustainability-related events
- Introducing new sustainability-related degree programs
- Promoting university overseas to attract students from low-income countries to join the university
- Assisting SMEs
- Tackling country's brain drain
- Setting up centres of excellence to align the research works on sustainability in the context of the school's programs
- Environmental sustainability programs (e.g. minimizing the use of papers, recycling, etc.)

Theoretical underpinnings

There is growing pressure on organisations to be committed to their social responsibilities and be concerned about sustainable development. These pressures come in different institutional forms, including rules and regulations put forth by the government or other entities, standards set by authorities and accreditation bodies, decisions and activities of peer organisations, social norms and the demands of stakeholders in the society, or even cultural and logical structures and beliefs of society. These institutional factors may affect the way an organisation shapes itself in response to the growing global demands that we act in a more socially responsible way.

Nowadays, and specially following the recognition of the decade from 2005 to 2014 as the Decade of Education for Sustainable Development announced by the United Nations, universities are under growing coercive pressure by international rules and regulations that push them to undertake their sustainable development responsibilities. The growing demands from various stakeholders, NGOs, and governments, all pressurise universities to incorporate community-related programmes into their organisations and contribute to the well-being of people in society.

Of course, universities are gaining a deeper understanding of their role and responsibility towards society and (either willingly or forced by the pressures mentioned above) seek social credibility by following benchmarks and other leading universities in acting more responsibly toward society. This is recognised by the mimetic pressure which pushes the university to shape itself and its processes to resemble other benchmark universities and symbol organisations.

The theory of New Institutional Sociology (NIS) has been selected as the main theoretical approach to explicate the findings in the current study and to investigate the changes in the norms, rules, activities, programs and routines within the university context, following the introduction of sustainability focus. Through the lens of NIS theory, the roles of institutional pressures in pushing the organisation to adapt its values, structures, and sub-systems according to sustainable development principles have been analysed. Moreover, the role of different actors in the organisational context in accepting and supporting, or rejecting the new proposed institutional changes toward sustainability could be investigated (Scott 2008).

Apart from various pressures on organisations to act sustainably, the sustainability focus can be considered as an institutional factor in itself under the new institutional theory, because there is growing attention on the importance of sustainable development as a social demand and requirement to be taken into consideration by organisations which can shape organisational structures, values and strategies, as well as individuals' beliefs and mind-sets. In that sense, the organisation members play a big role in supporting this institutional factor by strengthening the sustainability focus of the university and helping to achieve it throughout the organisation.

In the context of our selected case, the institutional factors pertaining to sustainability can be categorised into *rules & regulations*, *standards*, and *norms* which are discussed further below:

Rules and regulations

These include the rules and regulations put forth by the government and other authorities to monitor and control the universities' performances and their contributions to society. Universities are expected to get involved with their respective communities and elevate the socio-economic status of people in society. In fact, this is in line with the Malaysian Deputy Higher Education Minister in our face to face interview with him, when he points out the role of universities in pioneering sustainability thinking and solutions:

> As far as the ministry is concerned, this (sustainable development) is something we like to promote. In fact we want to encourage professors and students to be more involved in this area, simply because it is from the universities that I believe we can offer the government new approaches for sustainable development. We are still grasping with ideas on green technology. We are still grasping with ideas on conservation. And we can only do this if we have continuity not only in doing things but also in sustaining things.

Standards

The standards set by authorities and accreditation bodies, as well as decisions and activities by peer organisations are the main institutional factors in this group. In the context of our selected case schools, there is a requirement for accreditation by external accreditation bodies which necessitates the compliance of the school and its educational programme with certain requirements, including sustainability agenda. As one of the faculty members echoes:

> [...] We all have professional affiliations and bodies at each school that accredit our programmes. So whatever we do has to comply with the requirements of the professional bodies.

Moreover, universities are pressurised by social obligations to adhere to sustainability issues through various voluntary certificates and accreditation programs, certifications, and memberships (e.g. Talloires Declaration, Halifax declarations, Swansea declaration, COPERNICUS Charter, Kyoto Declaration, ISO 14001:2004).

Norms, cultural and logical structures

Other forms of external institutional factors affecting universities are the common norms and demands imposed by the various stakeholders in society. Overall, the expectations of the public from the university have been raised to a whole new level. In fact, there is an overwhelming pressure on academic organisations to respond to the social demands (Sanusi and Khelghat-Doost 2008). In Malaysia, the public understanding of sustainability and environmental issues is being elevated and the public demands from the universities are on the rise.

Conclusions

Sustainable development is relevant to institutes of higher learning from different perspectives. Universities have the potential to accelerate initiatives for promoting and implementing sustainability not only on campus, but also throughout society. Hence, it is of utmost importance for them to align their organisational goals with their social responsibilities. To do so, a holistic consideration of university's functions and institutional structures is essential. Moreover, the support from individuals in the organisation is crucial in order to successfully apply new changes in the university environment. As Easterby-Smith (1987) argues, changes in the higher education at different organisational levels are related to each other, and the implementation of changes at lower levels of the system depends greatly on changes taking place in the wider institutional structures and systems.

Current research has looked into the sustainability vision of Universiti Sains Malaysia (USM), as one of the foremost universities for sustainability in the South-East Asia region, in an attempt to investigate the changes that have happened at the school levels in order to implement the agenda of sustainability in the university. In addition, the role of institutional factors in pushing the university towards a sustainability-led organisation has been investigated.

The study results showed that all selected schools have been in the process of implementing sustainability in all aspects of teaching, research, community engagement and operations. However, the organisation has not reached a complete "institutionalisation" stage. There are still some flaws and deficiencies that need to be addressed gradually, before transformation to a sustainable university can be achieved smoothly.

The research findings are of interest to the decision-makers and administrators of universities and institutes of further education, as they provide invaluable insights of a pioneer university. Thus, the findings can be a suitable guide for various institutions of further education throughout the world to adjust and manage their practices, and use the sustainability strategies of USM as their benchmark. Future studies may examine how the mind-sets of the organisational members have been shaped following the introduction and implementation of a sustainability agenda in the university. Future research may also investigate the factors which facilitate, as well as those which act as barriers to, sustainability in the university.

Acknowledgement

The authors would like to acknowledge Universiti Sains Malaysia (USM) for their support in the publication of this paper through the USM Fellowship and for supporting the presentation of this research work through the USM Institute of Postgraduate Studies (IPS) Graduate Fund.

References

Adomssent, M., Godemann, J. and Michelsen, G. (2007), Transferability of approaches to sustainable development at universities as a challenge. *International Journal of Sustainability in Higher Education* 8 (4), pp. 385-402.

Attride-Stirling, J. (2001), Thematic networks: an analytic tool for qualitative research. *Qualitative Research* 1 (3), pp. 385-405.

Blackburn, W.R. (2007), *The Sustainability Handbook: The Complete Management Guide to Achieving Social, Economic and Environmental Responsibility*. Environmental Law Institute, Washington, DC.

Cortese, A.D. (2003), The critical role of higher education in creating a sustainable future. *Planning for higher education* 31 (3), pp. 15-22.

Davy, D. and Valecillos, C. (2009), *Summary of a literature review of qualitative research in technical communication from 2003 to 2007*. IPCC, 2009 IEEE International Professional Communication Conference, pp. 1-7.

Easterby-Smith, M. (1987), Change and innovation in higher education: a role for corporate strategy? *Higher Education* 16 (1), pp. 37-52.

Geli de Ciurana, A.M. and Leal Filho, W. (2006), Education for sustainability in university studies: Experiences from a project involving European and Latin American universities. *International Journal of Sustainability in Higher Education* 7 (1), pp. 81-93.

Gephart, R.P. (2004), Qualitative Research and the Academy of Management Journal. *Academy of Management Journal* 47 (4), pp. 454-462.

Lozano, R. (2006), Incorporation and institutionalization of SD into universities: breaking through barriers to change. *Journal of Cleaner Production* 14 (9-11), pp. 787-796.

– (2010), Diffusion of sustainable development in universities' curricula: an empirical example from Cardiff University. *Journal of Cleaner Production* 18 (7), pp. 637-644.

– (2011), The state of sustainability reporting in universities. *International Journal of Sustainability in Higher Education* 12 (1), pp. 67-78.

Lukman, R. and Glavic, P. (2007), What are the key elements of a sustainable university? *Clean Technologies and Environmental Policy* 9 (2), pp. 103-114.

Md Shahbudin, A., Nejati, M. and Amran, A. (2011), Sustainability-based knowledge management performance evaluation system (SKMPES): Linking the higher learning institutes with the bottom billions. *African Journal of Business Management* 5 (22), pp. 9530-9540.

Moore, J. (2005), Seven recommendations for creating sustainability education at the university level: A guide for change agents. *International Journal of Sustainability in Higher Education* 6 (4), pp. 326-339.

Sanusi, Z.A. and Khelghat-Doost, H. (2008), Regional Centre of Expertise as transformational platform for sustainability: A case study of Universiti Sains Malaysia, Penang. *International Journal of Sustainability in Higher Education* 9 (4), pp. 487-497.

Scott, W.R. (2008), *Institutions and Organizations: Ideas and interests*, 3rd ed. Sage Publications, Thousand Oaks, CA.

Stephens, J.C. and Graham, A.C. (2010), Toward an empirical research agenda for sustainability in higher education: exploring the transition management framework. *Journal of Cleaner Production* 18 (7), pp. 611-618.

UN Decade of Education for Sustainable Development. Retrieved December 30, 2011 from http://www.unesco.org/en/esd/decade-of-esd/

UNESCO (2005), *UN Decade of Education for Sustainable Development 2005-2014*. Division for the Promotion of Quality Education, France. Available at: http://unesdoc.unesco.org/images/0014/001416/141629e.pdf (accessed 4 January 2012).

USM-APEX (2008), *Universiti Sains Malaysia Transforming Higher Education for a Sustainable Tomorrow*. Universiti Sains Malaysia (USM). Available at: http://www.usm.my/ver4en/apex/get_doc/USM@APEX%20University(intro).pdf (accessed 12 May 2009).

Velazquez, L., Munguia, N., Platt, A. and Taddei, J. (2006), Sustainable university: what can be the matter?" *Journal of Cleaner Production* 14 (9-11), pp. 810-819.

Velazquez, L., Munguia, N. and Sanchez, M. (2005), Deterring sustainability in higher education institutions: An appraisal of the factors which influence sustainability in higher education institutions. *International Journal of Sustainability in Higher Education* 6 (4), pp. 383-391.

Wright, T. (2004), The evolution of sustainability declarations in higher education. In Corcoran, P.B. and Wals, A.E.J. (Eds.), *Higher Education and the Challenge of Sustainability: Problematics, Promise and Practice*, pp. 7-19. Kluwer Academic Publishers, Netherlands.

Chapter 37

Exploring the Usefulness of Problem-Based Learning in Education for Sustainable Development: A Case Study

Lukas Paul Fesenfeld, Charlotte Filiz Merten, Christoph Michael Schulze, Felix Tilmann Vahle and Dennis Redeker

Abstract

This paper considers to what extent problem-based learning (PBL) is a useful learning method for Education for Sustainable Development (ESD). According to literature available on this subject, effective ESD requires the development of certain skills. PBL is said to foster specific skills which can be linked to the aforementioned skills. Therefore, PBL seems to be a useful teaching environment for ESD. In order to illustrate this link, an introductory course on Sustainable Development at Maastricht University served as a case study. Quantitative and qualitative research was conducted using a questionnaire and interviews with staff and students. The findings suggest that in general PBL is a useful method to teach skills for Sustainable Development (SD). However, not all skills were equally strengthened through PBL. While the most indicative results were obtained with regards to critical thinking, anticipatory thinking and analyzing complex systems, it could not be shown that PBL had any beneficial effect on students' ability to use media or their frustration tolerance. Further research on a comparative level is indispensable to shed more light on the relationship between PBL and ESD.

Introduction

> "Our biggest challenge in this new century is to take an idea that seems abstract – sustainable development – and turn it into a reality for all the world's people."
>
> Kofi Annan, Former Secretary General of the United Nations

The declaration of the UN Decade of Education for Sustainable Development highlights the importance of learning and teaching in the process of creating a sustainable future. Research suggests that Sustainable Development (SD) requires people to command a set of core skills such as critical and anticipatory thinking (De Haan and Harenberg 1999; Rieckmann 2011; Wiek, Withycombe and Redman 2011). Education for Sustainable Development (ESD) seeks to convey these skills. For this purpose, new, innovative learning methods should be taken more strongly into account (Tilbury 2011). An example of such an innovative method could be problem-based learning (PBL), which is used as the principal learning method by a growing number of universities around the world, including Maastricht University (UM).

The purpose of this case study is to scrutinize to what extent PBL as a learning method is useful for Education for Sustainable Development, as no substantial research has been conducted in this direction until now. In order to explore this question, both quantitative and qualitative research has been carried out at UM. Analysis surrounding to what extent staff members and students perceive PBL to strengthen skills for SD was conducted. As a case study, this research wants to shed light on the efficacy of using PBL for ESD.

The paper is structured as follows: in the first section, a list of core skills for SD will be presented, which is based on an extensive literature review. Thereafter, PBL, as employed at

UM, is defined. After illustrating the link between the two concepts, the research methods and results are presented. Finally, the paper discusses both qualitative and quantitative findings as well as the limitations of the study and avenues for further research.

Defining ESD & PBL

Education for Sustainable Development

For the purpose of this study an extensive literature review on skills[1] for SD was conducted. In order to explore skills for SD, studies from De Haan and Harenberg (1999), Rieckmann (2011), and Wiek, Withycombe and Redman (2011) were reviewed. Concerning specific skills, the research of further authors was taken into account.[2] The literature review has shown that there is a large overlap of skills for SD found by the various researchers. As Rieckmann's Delphi study (2011) shows, for experts from both Europe and Latin America the most important key skills for SD are analyzing complex systems, anticipatory thinking and critical thinking. The experts found that these key skills are specifically important for SD due to the complexity, uncertainty, risks and high velocity of (global) societal change (ibid.). For the purpose of the present study the following questions from the questionnaire contain the 18 core skills for SD as derived from the literature review, above all from Rieckmann (2011) (see Table 1).

Table 1: The questionnaire, based on the 18 competences for SD

1. In this class, PBL has enhanced my ability to analyze complex systems across different domains (such as society, environment, economy).
2. In this class, PBL has enhanced my ability to analyze complex systems across different scales (local to global).
3. In this class, PBL has helped me to better understand complex problem situations.
4. In this class, PBL has helped me to think more critically about issues of Sustainable Development.
5. In this class, PBL has increased my anticipatory thinking (the ability to think future- oriented under conditions of uncertainty).
6. In this class, PBL has helped me to better recognize and analyze problems of unsustainable development.
7. In this class, PBL has been a suitable method to convey factual knowledge about Sustainable Development.
8. In this class, PBL has helped me to recognize my own role in the global community.
9. In this class, PBL has increased my concern for justice.
10. In this class, PBL has helped me to better cooperate in groups.
11. In this class, PBL has helped me to better communicate with other students of the course.
12. In this class, PBL has increased my ability to empathize with others and my ability to adopt other perspectives.
13. In this class, PBL has helped me to work together with people from different academic and personal backgrounds.

---

1 For the purpose of this research, competence is defined as "a functionally linked complex of knowledge, skills, and attitudes that enable successful task performance and problem solving" (cf. Baartman et al. 2007; De Haan 2006; Rieckmann 2011; Spady 1994).

2 "Critical thinking": Loorbach and Rotmans 2006; McArthur and Sachs 2009; Orr 2002; Rowe 2007; Willard et al. 2010 "Understanding complex systems competence": Clayton and Radcliffe 1996; Crofton 2000; Gibson 2006; Kates et al. 2001; Porter and Cordoba 2009; Sterling 1996.
"Anticipatory thinking competence": Bammer 2005; Loorbach 2007; Sarewitz and Kriebel 2010.
"Normative competence": Rockstrom et al. 2009; Swart et al. 2004.
"Interpersonal competence": Byrne 2000; Crofton 2000; Kelly 2006; Kevany 2007; McKeown and Hopkins 2003; Ospina 2000; Sipos et al. 2008; Sterling and Thomas 2006; Van Dam-Mieras et al. 2008.

14. In this class, PBL has helped me to connect relevant knowledge from different disciplines in respect to Sustainable Development.
15. In this class, PBL has helped me to improve my ability to deal with information.
16. In this class, PBL has helped me to make better use of media.
17. In this class, PBL has improved my evaluation skills.
18. In this class, PBL has increased my frustration tolerance.

Problem-based learning

PBL is an inquiry-based learning method (Loyens and Rikers 2011) that has been developed in medical education as a response to students' insufficient abilities to apply knowledge acquired during their theoretical education to real-world conditions (Albanese and Mitchell 1993; Barrows 1996). PBL facilitates the connection of theoretical knowledge to real-world situations because it is driven by complex problems, with which students are confronted in small tutorial groups. As the tutor merely functions as a facilitator, PBL is a student-centered and self-directed pedagogy. Lastly, PBL is a self-reflective process because students monitor their own learning process and are responsible for adjusting their learning strategies (Gijbels et al. 2005).

Research findings on advantages of PBL for students' education are not entirely conclusive and depend, among other factors, on the quality of the problem definition, the tutor, and the composition of the group. The main conclusive findings are that PBL increases problem-solving skills (Dochy et al. 2003; Gallagher 1992; Kuhn 1998; Lohman and Finkelstein 1999; Woods 1996), and enhances self-directed learning (Blumberg and Michael 1992; Schmidt and van der Molen 2001; Schmidt et al. 2006; Van den Hurk et al. 1999; Woods 1996). Moreover, PBL enhances the performance in higher order thinking, i.e. critical and analytical thinking as well as metacognitive skills (Hmelo 1998; Kaufman and Mann 1996; Schlundt et al. 1999; Sheperd 1998). Regarding the effect of PBL on content retention as compared to other learning methods, the results are contested. While Gijbels et al. (2005) maintain that PBL enhances content retention, others find this effect as being limited to long-term content retention (Dochy et al. 2003; Eisensteadt et al. 1990; Mårtenson et al. 1985). Further studies show that traditional learning methods yield equal or even better results for content retention (Dochy et al. 2003; Eisensteadt et al. 1990; Gallagher and Stepien 1995; Norman and Schmidt 1992).

Given the inconclusive research findings and the lack of a universal definition of PBL, this paper uses UM as an example for the practical application of PBL on a large scale. At UM, tutorial groups usually meet twice a week for two hours, which can be complemented by lectures. Groups consist of a tutor and 10-14 students, among them a note taker and a student guiding the discussion in terms of content and process. In the PBL literature by UM, the different steps involved in this process are termed Seven-Jump (Schmidt 1983) or Seven-Steps (van Til and van der Heijden 2009) approach (see Fig. 1).

Fig. 1: The PBL cycle

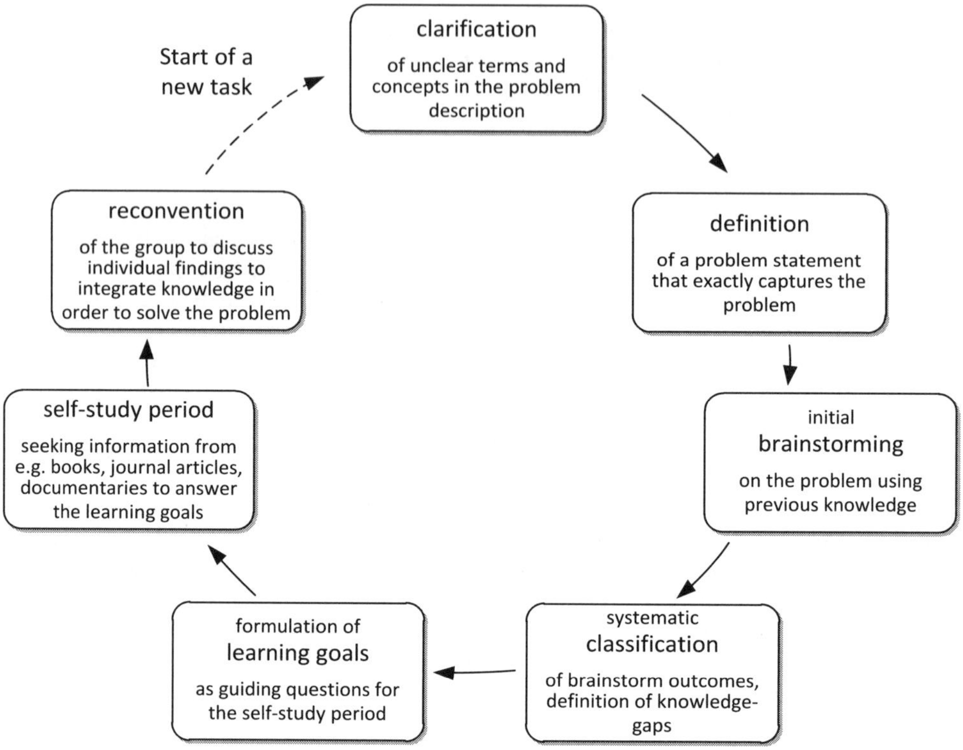

Linking ESD and PBL

The purpose of this study is to examine to what extent a PBL environment fosters skills that are associated with ESD. Based on the literature review of the core skills for SD and the definition of PBL, it is reasonable to assume that PBL is a useful framework for ESD. This overlap of the purported strengths of PBL and the required skills for SD can be illustrated by three examples: Firstly, problems of SD require analyzing complex systems (Rieckmann, 2011). PBL confronts students with ill-structured, authentic and complex problems. By analyzing these problems, students discuss underlying patterns and mechanisms. This should generally benefit their ability to think in a more systemic and complex manner. Secondly, SD necessitates critical thinking. PBL tutorial groups are centered around student-led discussions on topics that often lack one clear solution. In consequence, students critically engage with academic literature and question each other's arguments and points of view. Thirdly, SD needs interdisciplinary thinking. PBL encourages discussions between students with different personal and academic backgrounds. The more heterogeneous the group, the more interdisciplinary thinking is fostered. Additionally, the ill-structured nature of problems in PBL contributes to the students' capacity for interdisciplinary thinking, such as required by most real-world problems. In principle this overlap between the purported strengths of PBL and the required skills for SD holds for all 18 skills. For instance , similar connections between PBL and ESD can also be drawn concerning anticipatory thinking and interpersonal skills. Having initially established several theoretical overlaps between ESD and PBL, the study now proceeds to empirically explore the link of PBL to all eighteen core skills for SD.

Method

Participants

The participants of this study were undergraduate students enrolled in the course "Introduction to Sustainable Development" at University College Maastricht, the liberal arts and sciences bachelor program of UM. The response rate was n = 75 out of 90 students enrolled in the course. 61% of the participants were female, 28% male and 11% did not indicate their gender. The majority were Dutch (21%) or German (37%). The remaining participants represented another 13 nationalities from all over the world. 47% of the participants were aged 18, 19 or 20 respectively, 20% were aged 21 and 18% 22 or 23. The rest were older or did not indicate their age. A further seven students participated in semi-structured focus group interviews. A total of five academic staff members were interviewed, four of which teach and research in the field of sustainability at UM, and one who is researcher on PBL at the UM's Education Research Department.

Table 2: Demographics of the participants

Age		Gender		Nationality	
18-20	35 (47%)	female	46 (61%)	Dutch (%)	16 (21%)
21-23	29 (38%)	male	21 (28%)	German (%)	28 (37%)
>23	8 (11%)	no answer	8 (11%)	other	31 (42%)
no answer	3 (4%)				
Participants total	75				

Measures

This study employs a multi-method approach: while quantitative data depicts general trends and comparisons between sub-groups, qualitative data allows for a better understanding of specific cases. A questionnaire was designed, asking to what extent PBL enhanced the 18 skills for SD during the eight weeks of the course. Answers could be indicated on a 5-point Likert Scale (from 1 = "strongly disagree" to 5 = "strongly agree"). Also included were partially open-ended yes/no questions regarding changes in personal behavior due to taking the course, the application of PBL during the course and regarding demographics. The order of the items concerning skills was mixed to avoid a response set bias. For the qualitative section, two semi-structured focus group interviews were conducted with three and four students respectively. Furthermore, interviews were conducted with academic staff to gain complementary insights from the teacher's perspective. Since staff members have experience from outside of UM they supplemented an outside perspective on PBL. All interviews were guided by the questionnaire and recorded. No separate analysis for the interviews will be presented in this paper, but data will be used to enrich and clarify results from the quantitative data in the discussion section of this paper.

Procedure

Questionnaires were distributed during the second to last meeting of all eight tutorial groups (see Table 1). Following a brief oral introduction on the purpose of the research by the researchers, all students present completed the questionnaire. Fifteen out of ninety students were not present in the tutorial. Students were not compensated for the completion of the questionnaires. On this occasion, all interested students were recruited for a voluntary participation in a focus-group meeting. Two researchers were present during each of the two focus group meetings, which lasted around 45 minutes. Students were compensated with five euros each. For the staff members' interviews, the interviewees also participated on a voluntary basis.

Results

The overall response rate of participants to the questionnaire was 83%. One-tailed one sample t-tests were conducted. For every test H_0 was set at the mean value "3" as this implies that PBL does not have any significant impact at the specific competence for SD in this class. The alternative hypotheses were $H_a>3$. All tests were conducted with a significance level of .05. Moreover, 95% confidence intervals for the true mean were calculated.

Results of the tests indicate that students perceived almost all skills to be positively affected by PBL at a statistically significant level. The findings with the highest means and most statistically relevant results concerned the following skills: the skills "complex systems" (see Table 1, "across different domains": $p < .0001$, CI [3.694, 4.012]; "across different scales": $p < .0001$, CI [3.575, 3.885]; "better understanding of complex problems": $p < .0001$, CI [3.295, 3.691]); "anticipatory thinking" (see Table 1, $p < .0001$, CI [3.603, 3.997]) and "critical thinking" (see Table 1, $p < .0001$, CI [3.808, 4.219]) appeared to be strongly influenced by PBL. In contrast, students did not perceive PBL as having enhanced their use of media to find information on Sustainable Development ($p = .2319$, CI [2.864, 3.296]) nor their frustration tolerance ($p = .9328$, CI [2.593, 3.056]).

Table 3

Competence		**Mean**	**t-statistic**	**95% CI**	**p**
Hyp. 1:	critical thinking	4,01	9.83	$3.808 < \mu < 4.219$	(<).0001*
Hyp. 2:	complex systems across different domains	3.85	10,69	$3.694 < \mu < 4.012$	(<).0001*
Hyp. 3:	work diff. academic & personal backgrounds	3,83	11,11	$3.678 < \mu < 3.975$	(<).0001*
Hyp. 4:	anticipatory thinking	3,80	8.11	$3.603 < \mu < 3.997$	(<).0001*
Hyp. 5:	recognize problems of unsustainable development	3,76	8,77	$3.587 < \mu < 3.933$	(<).0001*
Hyp. 6:	complex systems across different scales	3.73	9.39	$3.575 < \mu < 3.885$	(<).0001*
Hyp. 7:	role in global community	3,73	6,1	$3.489 < \mu < 3.964$	(<).0001*
Hyp. 8:	knowledge from diff. disciplines	3,66	6,71	$3.002 < \mu < 4.219$	(<).0001*
Hyp. 9:	empathize & adopt perspectives	3,63	7,07	$3.450 < \mu < 3.803$	(<).0001*
Hyp. 10:	factual knowledge**	3,56	5,2	$3.345 < \mu < 3.775$	(<).0001*
Hyp. 11:	communication with other students	3,53	5,19	$3.328 < \mu < 3.738$	(<).0001*
Hyp. 12:	understand complex problems	3.49	4.97	$3.295 < \mu < 3.691$	(<).0001*
Hyp. 13:	cooperation in groups	3,43	4,16	$3.222 < \mu < 3.631$	(<).0001*
Hyp. 14:	concern for justice	3,27	2,21	$3.027 < \mu < 3.514$	.015*
Hyp. 15:	ability to deal with information	3,21	2,01	$3.002 < \mu < 3.425$	.0241*
Hyp. 16:	evaluation skills	3,20	2,15	$3.015 < \mu < 3.385$	.0174*
Hyp. 17:	use of media to find information	3,08	0,74	$2.864 < \mu < 3.296$	.2319
Hyp. 18:	frustration tolerance	2,82	-1,51	$2.593 < \mu < 3.056$	.9328

*Note: * reject H_0 at α=.05; **statistically significant difference between German and Dutch students*

A majority indicated that they changed their personal behavior after having taken the class (yes: 68%, no: 32%). Overall, the greater part indicated that PBL was a good teaching method for this class (yes: 81%, no: 12%, N/A: 7%) and that PBL was applied as defined above (yes: 77%; no: 19%; N/A: 4%). The majority of students indicated that the tutor facilitated the learning process (yes: 72%, no: 23%, N/A: 5%).

A possible impact of the factors "nationality" (difference between German and Dutch students) and "gender" was tested via a two-variable t-test. For one hypothesis, statistically significant differences between Dutch and German students were found (alpha = 0.1): German students expressed that PBL helped to increase their factual knowledge on sustainable development whereas Dutch students did not (*p =.0045; t-statistic = -3.08)*. No statistically significant differences between genders were found.

Discussion

The preceding results show that students perceive PBL especially supportive in regards to skills such as critical thinking, anticipatory thinking and analyzing complex systems. In the following discussion the qualitative research results underline these findings. To start with, students and staff stated that critical thinking was strengthened through the exchange of opinions and knowledge in a PBL group situation. The constructive criticism of others was seen as beneficial to supporting higher order thinking. One staff member remarked critically that "[...] PBL has the potential to increase [critical thinking...] but it can do even more" (P. van den Bossche, personal communication, December 15, 2011). Furthermore, students found that the general advantages of PBL on higher order thinking were associated with improvements of their anticipatory thinking, as PBL trained them to be more aware of the way they think and act. Regarding their competence to analyze complex systems, students in the focus groups mentioned the importance of discussing issues of SD in a PBL group setting. Especially the different academic backgrounds of students were seen to be beneficial to analyze complex systems. However, this might be owed to University College Maastricht as a liberal arts program rather than to PBL alone.

Students mentioned that they profited from learning in a PBL environment together with other students from different disciplines. However, some commented that they had missed a stronger academic interdisciplinary setting in their tutorials through which more varied discussions could have taken place. This could be amended by having courses organized by an interdisciplinary committee of course coordinators, as has been the case at UM in the past.

Generally, students perceived PBL to be a useful method for conveying factual knowledge about SD. Marc Dijk (personal communication, December 12, 2011) stated that in comparison to more traditional forms of instruction in higher education, PBL requires students to read continuously, which he considered beneficial for students' long-term activation of knowledge. However, students and staff stressed the risk of talking about complex concepts of SD, such as climate change, without a deeper understanding of the subject. In order to accumulate factual knowledge, one course coordinator highlighted the importance of well-informed and motivated students (A. Offermans, personal communication, December 14, 2011). She proposed supplementing tutorials with interesting lectures and other audiovisual materials to raise students' interest.

The statistical results support the claim that PBL is beneficial for interpersonal skills. Results were positive for cooperation in groups and collaboration with people of other backgrounds as well as for general communication skills. Regarding the impact of PBL on students' ability to empathize and change perspectives, positive results were found for the complete sample. Generally, qualitative data show that both students and staff perceived PBL as highly beneficial for students' interpersonal skills, based on the interviews; however, it cannot be taken for granted that intercultural communication and group learning occur automatically as soon as a tutorial group encompasses diverse backgrounds. More generally, it appears that a well-functioning group process cannot be taken for granted. "PBL asks for a lot of interpersonal behavior, but the problem is that this is often not part of [the] discussion" (P. van den Bossche, personal communication, December 15, 2011). Therefore, more attention should be paid to group processes while training students and staff for PBL, since conscious efforts are required to create an atmosphere where everybody feels comfortable enough to participate.

Interesting results were found in regard to evaluation skills. Although the Seven-Steps approach to PBL as used at UM does not incorporate evaluation, the statistical analysis indicated that students perceived PBL as beneficial for their evaluation skills. It thus seems that PBL is beneficial for students' evaluation skills in another way. Notably though, both staff and students strongly suggested that there is still a lot of potential that could be realized by conducting an evaluation of the group process at the end of every PBL cycle.

Another result of the interviews is that the tutor's role as facilitator should be stressed more during the training of the tutor. Students repeatedly stated that the level of intervention of their tutor in the discussions was either too high or too low. Both had adverse impacts on the learning outcomes. These insights coincide with concerns raised in the literature (e.g. Dolmans et al. 2005). Overall, it could perhaps be concluded from this that the benefits of PBL depend a lot on its actual implementation and specific context.

The quantitative analysis confirmed the claim that PBL helps students to deal with information. The interviews support these findings. It was stated that connections between pieces of information as well as their relation to the larger context are easier to grasp in a group. More generally, students perceive the PBL process as facilitative to structuring large amounts of information. One student stated for instance that PBL helps to understand all the connections and digest the information. In contrast, the statistical analysis did not confirm the claim that PBL trains students in the use of media to find information. Taking into account the course manual as well as the interviews illuminates this link. The literature for all tasks is indicated on a reading list, disposing of the need for independent search for information during the self-study period. Consequently it is not possible to make any statement about the potential of PBL in this regard based on this study.

The assumption that PBL elevates students' frustration tolerance could not be confirmed. Statements made during the interviews might explain the findings in the quantitative analysis: students indicated that at times they perceived the topic of sustainability as very frustrating. The chief reason was that merely problems but hardly any solutions were discussed, leaving students overwhelmed and disillusioned. Academic staff equally recognized this challenge, pointing to the need to include successful examples of SD in courses.

Limitations

There are some limitations to this study that need to be mentioned. First of all, the data is based on a self-reported questionnaire. Therefore, there is no objective measurement of how useful PBL is in teaching SD. Ideally, the study's set up would also include a control group with students studying SD in a non-PBL setting. This is especially relevant as most students in this study have no experience with other learning methods in tertiary education.

Secondly, it must be acknowledged that University College Maastricht is a selective liberal arts program with an open curriculum. In this context, staff members repeatedly highlighted that most students are display above average levels of motivation. Furthermore, due to the open curriculum, students with different academic backgrounds can be enrolled in the same course. Therefore, interdisciplinary aspects are likely to be stronger than in other programs.

Further research

It has to be highlighted that this study only had the limited objective of showing a possible benefit of PBL as a learning method regarding skills for SD in the frame of a case study. Hence, this research does not claim to show that PBL is a useful method for ESD in general. In order to further investigate the potential PBL has for ESD, comparative approaches are of vital importance. Comparative studies should, for instance, research differences between PBL and non-PBL courses on SD and the impact of a liberal arts system on students' skills for SD. Students' academic and cultural backgrounds are likely to have a considerable impact on the effect of PBL for ESD. More research is needed as well into the effects differently composed literature lists may have on students' skills. It would be potentially fruitful to investigate the difference of the benefit of PBL for SD classes at Master's and Bachelor level. Lastly, attention should be directed to the impact of tutor behavior, assessment methods and problem design on PBL in the context of ESD.

Concluding remarks

This paper has presented a case study on the usefulness of problem-based learning to convey skills for sustainable development. For that purpose, quantitative and qualitative research was conducted on an introductory course on sustainable development at Maastricht University. Some of the findings suggest that Problem-based learning is a useful method to teach skills for sustainable development. However, results have shown that not all skills were equally strengthened through problem-based learning. While the most indicative results were obtained with regards to critical thinking, anticipatory thinking and analyzing complex systems, it could not be shown that problem-based learning had any beneficial effect on students' ability to use media for finding information and their frustration tolerance. Further research, especially on a comparative level, is indispensable to shed more light on the issue raised in this first case study.

Acknowledgements

The authors would like to thank Prof. Rene Kemp (International Centre for Integrated assessment and Sustainable development [ICIS], UNU-Merit) who supervised the research process within the scope of University College Maastricht's undergraduate research project PEERS, organized by Oscar van den Wijngaard. Many thanks to Wilfred van Dellen (University College Maastricht) who was enthusiastic about helping in the process as well as to the interviewees for their perspective on the subject: Piet van der Bossche (Department of Educational Research and Development of the Maastricht University School of Business), Carijn Beumer (ICIS), Alvarro Moreno (ICIS), Marc Dijk (ICIS), Astrid Offermans (ICIS). The final version of the paper was greatly improved by the comments and recommendations of Christopher Newman and Heiner Salomon. Finally, our special thanks go to Prof. Harm Hospers who has sacrificed much time and provided us with very insightful criticism.

References

Albanese, M.A. and Mitchell, S. (1993), Problem-based learning: a review of literature on its outcomes and implementation issues. *Academic Medicine* 68, pp. 52-81.

Baartman, L.K.J., Bastiaens, T.J., Kirschner, P.A. and Van der Vleuten, C.P.M. (2007), Evaluation assessment quality in competence-based education: a qualitative comparison of two frameworks. *Education Research and Review* 2, pp. 114-129.

Bammer, G. (2005), Integration and implementation sciences: building a new specialization. *Ecology and Society* 10 (2), Art. 6.

Barrows, H.S. (1996), Problem-based learning in medicine and beyond: a brief overview. In Wilkerson, L. and Gijselaers, W.H. (Eds.), *Bringing Problem-Based Learning to Higher Education: Theory and Practice*, pp. 3-12. Jossey-Bass, San Francisco, CA.

Blumberg, P. and Michael, J.A. (1992), Development of self-directed learning behaviors in a partially teacher-directed problem-based learning curriculum. *Teaching and Learning in Medicine* 4 (1), pp. 3-8.

Byrne, J. (2000), From policy to practice: creating education for a sustainable future. In Wheeler, K.A. and Bijur, A.P. (Eds.), *Education for a Sustainable Future: a Paradigm of Hope for the 21st Century*, pp. 35-72. Kluwer/Plenum, New York, NY.

Clayton, A.M.H. and Radcliffe, N.J. (1996), *Sustainability: a Systems Approach*. Westview, Boulder, CO.

Crofton, F. (2000), Educating for sustainability: opportunities in undergraduate engineering. *Journal of Clean Production* 8 (5), pp. 397-405.

De Haan, G. (2006), The BLK '21' programme in Germany: a 'Gestaltungskompetenz'-based model for education for sustainable development, *Environmental Education and Research* 12 (1), pp. 19-32.

De Haan, G.D. and Harenberg, D. (1999), *Bildung für eine nachhaltige Entwicklung: Gutachten zum Programm von der 72. Bund-Länder-Kommission für Bildungsplanung und Forschungsförderung* (BLK).

Dochy, F., Segers, M., van den Bossche, P. and Gijbels, D. (2003), Effects of problem-based learning: a meta-analysis. *Learning and Instruction* 13 (5), pp. 533-568.

Dolmans, D.H.J.M., De Grave, W., Wolfhagen, I.H.A.P. and van der Vleuten, C.P.M. (2005), Problem-based learning: future challenges for educational practice and research. *Medical Education* 39 (7), pp. 732-741.

Eisensteadt, R.S., Barry, W.E. and Glanz, K. (1990), Problem-based learning: cognitive retention and cohort traits of randomly selected participants and decliners. In Anderson, B. (Ed.), *Research in Medical Education 1990: Proceedings of the Twenty-Ninth Annual Conference*, pp. 11-12. Association of American Medical Colleges, Washington, DC.

Gallagher, S.A. and Stepien, W.J. (1996), Content acquisition in problem-based learning: depth versus breadth in American studies. *Journal for the Education of the Gifted* 19 (3), pp. 257-275.

Gallagher, S.A., Stepien, W.J. and Rosenthal, H. (1992), The effects of problem-based learning on problem solving. *Gifted Child Quarterly* 36 (4), pp. 195-200.

Gibson, R. (2006), Sustainability assessment: basic components of a practical approach. *Impact Assess Project Appraisal* 24 (3), pp. 170-182.

Gijbels, D., Dochy, F., van den Bossche, P. and Segers, M. (2005), Effects of problem-based learning: a meta-analysis from the angle of assessment. *Review of Education Research* 75 (1), pp. 27-61.

Hmelo, C.E. (1998), Problem-based learning: effects on the early acquisition of cognitive skill in medicine. *Journal of Learning Sciences* 7 (2), pp. 173-208.

Kates, R.W., Clark, W.C., Corell, R., Hall, J.M. and Jaeger, C. (2001), Sustainability science. *Science* 292 (5517), pp. 641-642.

Kaufman, D.M. and Mann, K.V. (1996), Students' perceptions about their courses in problem-based learning and conventional curricula. *Academic Medicine* 71 (1), pp. S52-S54.

Kelly, P. (2006), Letter from the oasis: helping engineering students to become sustainability professionals. *Futures* 38 (6), pp. 696-707.

Kevany, K. (2007), Building the requisite capacity for stewardship and sustainable development. *International Journal of Sustainability in Higher Education* 8 (2), pp. 107-122.

Kuhn, G.J. (1998), *Designing problem-based instruction to foster expertise in emergency medicine residents*. PhD, Wayne State University, Detroit, MI.

Lohman, M.C. and Finkelstein, M. (1999), *Segmenting information in PBL cases to foster the development of problem-solving skill, self-directedness, and technical knowledge*. Unpublished manuscript. Florida State University, Tallahassee, FL.

Loorbach, D. (2007), *Transition Management: New Mode of Governance for Sustainable Development*. International Books, Utrecht.

Loorbach, D. and Rotmans, J. (2006), Managing transitions for sustainable development. In Olshoorn, X. and Wieczorek, A.J. (Eds.), *Understanding Industrial Transformation – Views from Different Disciplines*, pp. 187-206. Springer, Dordrecht.

Loyens, S.M.M. and Rikers, R.M.J.P. (2011), Instruction based on inquiry. In Mayer, R.E. and Alexander, P.A. (Eds.), *Handbook of Research on Learning and Instruction*, pp. 361-381. Routledge, New York, NY.

Mårtenson, D., Eriksson, H. and Ingelman-Sundberg, M. (1985), Medical chemistry: evaluation of active and problem-oriented teaching methods. *Medical Education* 19 (1), pp. 34-42.

McArthur, J.W. and Sachs, J. (2009), Needed: a new generation of problem solvers. *Chronicle of Higher Education* 55 (40), pp. A64-A66.

McKeown, R. and Hopkins, C. (2003), EE = ESD: defusing the worry. *Environmental Education and Research* 9 (1), pp. 117-128.

Norman, G.R. and Schmidt, H.G. (1992), The psychological basis of problem-based learning: a review of the evidence. *Academic Medicine* 67 (9), pp. 557-565.

Orr, D. (2002), Four challenges of sustainability. *Conservation Biology* 16 (6), pp. 1457-1460.

Ospina, G. (2000), Education for sustainable development: a local and international challenge. *Prospects* 30 (1), pp. 31-40.

Porter, T. and Cordoba, J. (2009), Three views of systems theories and their implications for sustainability education. *Journal of Management Education* 33 (3), pp. 323-347.

Rieckmann, M. (2011), Future-oriented higher education: Which key competencies should be fostered through university teaching and learning? *Futures* 44 (2), pp. 127-135.

Rockström, J., Steffen, W., Noone, K., Persson, A., Chapin III, F.S., Lambin, E.F., Lenton, T.M., Scheffer, M., Folke, C., Schellnhuber, H.J., Nykvist, B., de Wit, C.A., Hughes, T., van der Leeuw, S., Rodhe, H., Sörlin, S., Snyder, P.K. and Constanza, R. (2009), A safe operating space for humanity. *Nature* 461, pp. 472-475.

Rowe, D. (2007), Education for a sustainable future. *Science* 317 (5836), pp. 323-324.

Sarewitz, D. and Kriebel, D. (2010), *The Sustainable Solutions Agenda: Consortium for Science, Policy and Outcomes*. Arizona State University and Lowell Center for Sustainable Production, University of Massachusetts, Lowell, MA.

Schlundt, D.G., Flannery, M.E., Davis, D.L., Kinzer, C.K. and Pichert, J.W. (1999), Evaluation of a multicomponent, behaviorally oriented, problem-based 'summer school' program for adolescents with diabetes. *Behavior Modification* 23 (1), pp. 79-105.

Schmidt, H.G. (1983), Problem-based learning: rationale and description. *Medical Education* 17 (1), pp. 11-16.

Schmidt, H.G. and van der Molen, H.T. (2001), Self-reported competency ratings of graduates of a problem-based medical curriculum. *Academic Medicine* 76 (5), pp. 466-468.

Schmidt, H.G., Vermeulen, L. and van der Molen, H.T. (2006), Long-term effects of problem-based learning: a comparison of competencies acquired by graduates of a problem-based and a conventional medical school. *Medical Education* 40 (6), pp. 562-567.

Shepherd, N.G. (1998), *The Probe Method: A problem-based learning model's effect on critical thinking skills of fourth and fifth grade social studies students*. PhD, North Carolina State University, Raleigh, NC.

Sipos Y., Battisti, B. and Grimm, K. (2008), Achieving transformative sustainability learning: engaging heads, hands and heart. *International Journal of Sustainability in Higher Education* 9 (1), pp. 68-86.

Spady, W.G. (1994), *Outcome-Based Education: Critical Issues and Answers*. American Association of School Administrators, Arlington.

Sterling, S. (1996), Education in change. In Huckle, J. and Sterling, S. (Eds.), *Education for Sustainability*, pp. 18-39. Earthscan Publications, London.

Sterling, S. and Thomas, I. (2006), Education for sustainability: the role of capabilities in guiding university curricula. *International Journal of Innovation and Sustainable Development* 1 (4), pp. 349-370.

Swart, R.J., Raskin P. and Robinson, J. (2004), The problem of the future: sustainability science and scenario analysis. *Global Environmental Change* 14 (2), pp. 137-146.

Tilbury, D. (2011), *Education for Sustainable Development: An Expert Review of Processes and Learning*. UNESCO, Paris.

Van Dam-Mieras, R., Lansu, A., Rieckmann, M. and Michelsen, G. (2008), Development of an interdisciplinary, intercultural master's program in sustainability: learning from the richness of diversity. *Innovation for Higher Education* 32 (4), pp. 251-264.

Van den Hurk, M.M., Wolfhagen, I.H.A.P., Dolmans, D.H.J.M. and van der Vleuten, C.P.M. (1999), The impact of student-generated learning issues on individual study time and academic achievement. *Medical Education* 33 (11), pp. 808-814.

Van Til, T. and Van der Heijden, F. (2009), *PBL Study Skills: An Overview*. Datawyse – Maastricht University Press, Maastricht.

Wiek, A., Withycombe, L. and Redman, C.L. (2011), Moving forward on competencies in sustainability. *Environment-Science Policy Sustainable Development* 53 (2), pp. 3-13.

Willard, M., Wiedmeyer, C., Flint, R.W., Weedon, J.S., Woodward, R., Feldmand, I. and Edwards, M. (2010), *The Sustainability Professional: 2010 Competency Survey Report*. International Society of Sustainability Professionals.

Woods, D.R. (1996), Problem-based learning for large classes in chemical engineering. In Wilkerson, L. and Gijselaers, H. (Eds.), *Bringing Problem-Based Learning to Higher Education: Theory and Practice*, pp. 91-99. Jossey-Bass, San Francisco, CA.

Chapter 38

Effective Structures for Sustainability Programs in Higher Education

Lauren C. Edmonds and John P. Swaddle

Abstract

This paper presents qualitative and quantitative approaches to campus sustainability research, through which we identify factors most strongly associated with the success of sustainability programs at select institutions of higher education. Our comparative case study of eight colleges and universities identifies variables we then correlate to outcomes measured by a modification of the Sustainable Endowments Institute's Green Report Card score. Data analysis comprised of a principle components analysis to identify correlated groupings of variables and a subsequent regression analysis using the principle components. The variables most strongly correlated with outcomes were: relative funding, reliability of funding, bottom-up initiation, physical space, and program size. These variables had positive correlations, leading to the conclusion that campuses aiming for successful sustainability programs should make significant investments in their programs while fostering grassroots campus support.

Our work contributes a quantitative analysis of the factors leading to success of sustainability programs and provides a model for both implementation and also verification with a larger sample of institutions. We focused on identifying variables and a measurement of success to create a model; this model now needs to be applied to a larger sample size to test its reliability. This research was motivated by the need to produce practical recommendations for sustainability programs in higher education. The clear identification of variables makes it simple to translate this research into practice. Already, The College of William and Mary has received a series of recommendations from the application of this research and has begun implementing a selection of those.

Introduction

In the 1990s, institutions of higher education experienced a shift and began to acknowledge their role as participants in environmental issues, rather than as observers. Before this time, universities frequently studied the environmental impacts of others, but they rarely turned this analysis onto themselves (Creighton and Cortese 1992; Eagan 1992). It was in 1992 that David Eagan noted "campus environmentalism is starting to alter perceptions about what higher education is for, what is worth learning, and what graduates ought to understand and do about environmental concerns" (Eagan 1992). The way universities operated, the way they were constructed, the way they managed their own environmental responsibilities, among a myriad of other actions, imparted lessons to students about their relationship with the environment (Orr 1991; Rohwedder 2004; Cohen 2007).

Colleges and universities have an enormous immediate environmental impact simply from their institutional size and population. Reducing this high environmental impact is often the primary concern of campus sustainability programs. Beyond this immediate impact, institutions of higher education influence the way their graduates live the rest of their lives. As one campus sustainability leader told us about Clemson University's sustainability goals, "It's probably more important to change the culture of the students in the long run, of all the students who graduate, than it is for Clemson University to be carbon neutral because our students go out and change the world" (Sill 2011). The influence of colleges and universities extends far beyond their campuses through the values they teach their graduates. Sustainability is a perfect example of how students can learn outside of the classroom. Perhaps the idea has been best articulated by David Orr: "Students are taught in various and subtle ways beyond the content of courses" (Orr 1991). The concept echoes through much of the published sustainability research (Pittman 2004; Rohwedder 2004; Bacow and Moomaw 2007; Thompson and Creighton 2007). The way in which the institution conducts its business and teaches its students instills them with the lessons to become the leaders of tomorrow.

For these reasons, the environmental sustainability of colleges and universities has become a major focus of higher education research and operations. Campus sustainability has been gaining traction as the prevailing trend in higher education. If it is to become more than a trend, however, colleges and universities need to successfully institutionalize their sustainability efforts.

Our research focuses on the structure of campus sustainability programs and how the different structures are associated with the different sustainability outcomes. The ultimate goal is to identify the institutional structures most strongly associated with positive campus sustainability outcomes. We use the term "program" inclusively, as to encompass sustainability initiatives of every shape and size. Some schools have chosen to establish an office to coordinate and lead their sustainability efforts. Other schools dedicate one staff position to the sustainability projects on campus. On some campuses, sustainability is still an afterthought, a responsibility added on to existing positions. These programs vary widely in structure and also in outcomes.

This paper focuses on the research design of a comparative case study and the results of that study. We begin by explaining the case selection and data collection. We discuss the variable identification and measurement and the outcome measurement in depth. Finally, we present the results of the quantitative analyses and discuss the implications of the patterns in the data.

Methodology

Our research employs an innovative quantitative method to the study of sustainability programs at institutions of higher education. In the data analysis, we identified twenty variables across sustainability programs and created a system for scoring these variables on either numerical or ordinal scales. We also developed a system for scoring outcomes of sustainability programs, adapted from the Sustainable Endowments Institute's Green Report Card. With numerical values for the variables and outcomes, we were able to conduct a statistical analysis of the data and develop an empirical model. Correlations with statistical significance were possible, leading to tangible recommendations. The predominant research on sustainability programs has been qualitative, relying on descriptive case studies, and our research contributes an example of a quantitative study. Our work advances the discussion on sustainability in higher education by demonstrating the potential for quantitative methods and empirical models.

Research Approach

We adopted a comparative case study approach to collect information about campus sustainability programs (Tight 2003; Walker et al. 2004). Case studies are frequently used to thoroughly study a campus' institutions for any issue area; "Externally, the evaluator may compare institutions in an effort to identify practices that work and those that do not. This work is particularly valuable for those attempting reform in their own institutions" (Walker et al. 2004). The comparative analysis of the studies is integral to the research in order to provide perspective on the information collected for the separate institutions.

Case Selection

For this research we chose eleven campuses for study, though only eight were ultimately analyzed. The cases were chosen with the specific goal of providing useful policy recommendations to The College of William and Mary, a public institution located in southeastern Virginia. Appropriate comparisons are integral to identifying structures and policies that occur in one institution, but that could be effectively emulated elsewhere (Tight 2003). To offer cases relevant to The College, we began with the entire list of peer institutions identified by the College Provost's Office and the list of public colleges and universities in Virginia. Thirty-seven schools appeared on this original list. We then narrowed this list by several criteria to ensure that any successful structures identified would be transferable to The College.

Only schools in rural or suburban areas were considered. Variation in settings had the potential to affect a campus' options for sustainability efforts in ways outside the focus of this study so we restricted the focus to settings similar to William and Mary's location in Williamsburg, Virginia, a city with a population of under 12,000 residents.

The student population and endowment are important factors in the operation of the institution. The student population is indicative of the college or university's overall size. Larger institutions typically have greater resources available because of their scale. An institution's endowment is representative of its financial situation.

The final element in case selection was ensuring variation in structures and outcomes. Evaluating schools with only one form of program would fail to highlight the differences possible in the structure of sustainability institutions. If all of the schools chosen had successful programs, it would be impossible to correlate any structure with poorly ranked outcomes.

Through this process, ten colleges and universities, in addition to William and Mary, were selected for the comparative analysis.

1. Brandeis University
2. Clemson University
3. Rutgers University
4. University of California – Irvine
5. University of California – Santa Barbara
6. University of Georgia
7. University of New Hampshire
8. Wake Forest University
9. The College of William and Mary
10. Virginia Military Institute
11. Virginia Polytechnic Institute and University.

As research progressed, the University of California campuses in Irvine and Santa Barbara were excluded from the study because they did not receive scores for the Sustainable Endowments Institute Green Report Card, which was ultimately used to measure the success of programs. The University of Georgia was unavailable for the data collection interview and was also excluded from the analysis.

Data collection: personal interviews and further research

Because campus sustainability research is relatively new, little raw data is available for comparative analyses. Information released publicly is either synthesized in independent reviews of institutional practices or selectively released by the institutions themselves. In early 2011, there was no sector-wide system for reporting sustainability data at the level necessary for a comparative analysis. Most campuses have yet to incorporate sustainability measures into the data they collect through their institutional research offices (Litten and Terkla 2007). These circumstances make personal interviews the most practical method to gather specific sustainability data from colleges and universities. Personal interviews are also considered a highly appropriate method of gathering data related to campus sustainability for comparisons among institutions of higher education (Tight 2003).

The sustainability program for each college and university selected was first contacted via email. The email was sent to the primary contact associated with sustainability on the institution's website. In some cases, the interview solicitation was sent to a program-specific address, others provided personal staff or faculty addresses as sustainability contacts. After we made contact with a representative, arrangements were made for a telephone interview.

The telephone interview, conducted by LCE, consisted of fifteen questions regarding sustainability at the institution. The conversation was conducted using a web-based conference call system which enabled us to record interviews for transcription afterward.

The University of New Hampshire was the only institution to deny our request for an interview. However, the University of New Hampshire released a book detailing its program in 2009: *The Sustainable Learning Community*. Instead of conducting a personal interview, we answered the same questions using their book and website for the information. Throughout this report, we use the term interview but include this research as well.

After personal interviews were conducted to collect specific information on institutions' sustainability practices, further information was obtained from their official publications and websites and from data publicly released by the Sustainable Endowments Institute.

Data analysis

After conducting the interviews, we identified twenty variables related to campuses' sustainability programs. Some variables had numeric values, such as the school's endowment; others were assigned ordinal values on defined scales. We measured outcomes using a modification of the Sustainable Endowments Institute's (SEI) Green Report Card score.

Variable identification and measurement

The basic variables for comparing the institutions were its *endowment*, its distinction as either a *public or private* school, and its *enrollment*. We used the endowment data provided by the National Association of College and University Business Officers (NACUBO) – Commonfund Study of Endowments for 2010, with the exception of Virginia Military Institute, which came from the Sustainable Endowments Institute. The Sustainable Endowments Institute's listing of a school as either public or private was also used. The enrollment data was taken from the *Princeton Review* information on colleges and universities.

Unless otherwise noted, all subsequent variables were scored on scales of zero to three using information gathered through the interviews and further research.

The source of a program's initiation is an indicator of its campus-wide support. Some campuses had a primarily grassroots support to begin a sustainability program. Other programs were instigated by members of the upper administration. Recognizing there can be a call for sustainability from all levels of the campus, we used two distinct variables identified as *bottom-up initiation* and *top-down initiation*.

We assigned scores for institutions' current *staff involvement*, *faculty involvement*, and *student involvement* in the sustainability program.

Beyond an institution's endowment, other financial variables were considered, such as funding across the colleges and universities. The variable of *relative funding* compares the funding amount provided to sustainability efforts on the campus relative to the other institutions. The *funding reliability*, measures the consistency of funding available. The existence of a *student fee* may provide a program with additional funding, and its presence was scored as a binary character.

Providing essential information to answer our research questions, the structure of a college or university's sustainability program was measured in three variables: *physical space*, *scope*, and *program size*. A dedicated physical space for sustainability on campus has a dual impact through the efficiency of sustainability efforts and the cultural norms suggested (Fugazzotto 2009). The scope of a program has the potential to have an extensive impact on its success level. On the ordinal scale, zero indicated the school did not have a sustainability program, one showed that a school had a facilities-based program, two that it has an academic scope, and three that the scope was within the senior administration. Program size is a relative measure of the positions (full-time equivalents) enveloped in sustainability programs across campuses.

Age of the program was another variable considered to influence its success.

Campus policies regarding sustainability were identified by an institution's public commitments or organization memberships. These variables were scored as a zero if the policy or action did not exist on the campus or a one if it did. The existence of a *climate action plan* or *sustainability action plan* were scored in this binary manner. These plans are institution-specific and encompass campus policies on sustainability. The campus' status as an *AASHE member*, *STARS participant*, or *PCC signatory* was a variable of an institution's commitment to sustainability and involvement in the larger movement. These programs also all offer resources to campuses intended to improve their sustainability programs.

Outcome measurement

To measure outcomes, we chose to use the information produced by the Sustainable Endowments Institute (SEI). The Institute is a non-profit organization that releases an annual Green Report Card in which they score colleges and universities on a variety of sustainability measures. Each institution receives an overall score represented with a letter grade, but SEI also releases scores for the components of schools' sustainability programs. SEI's emphasis on transparency in their data and scorings (unless the information is withheld by the college or university, SEI publishes all survey data collected) makes their data exceptionally useful for campus sustainability research.

Because the presence of an office and sustainability staff members affects an institution's Green Report Card score, we needed to modify this measure. Using the component scores provided by SEI, we recalculated a sustainability score using their scoring for the non-structural categories only. We also excluded the scores related to the institution's broader investment procedures.

SEI only provides letter grades for the component scores, but we converted these on a 4.0-scale for the reassessment. Using the scores provided for: climate change and energy, food and recycling, green building, transportation, and student involvement, we created a modified SEI score that was an unweighted average. This gave us a measurement of sustainability outcomes over time.

Table 1 is an example of the modified SEI score for The College of William and Mary.

Table 1: Recalculated SEI green score for the College of William and Mary

Year	Climate Change & Energy	Food & Recycling	Green Building	Student Involvement	Transportation	Modified Score	New Grade
2008	0.0	2.0	2.0	n/a	0.0	1.0	D
2009	2.0	3.0	3.0	3.0	2.0	2.6	C+
2010	2.0	3.0	3.0	4.0	3.0	3.0	B
2011	3.0	4.0	3.0	4.0	3.0	3.4	B+

Results

A comparison of the recalculated SEI green scores for each campus is shown in Fig. 1.

Fig. 1: A comparison of the recalculated SEI green scores, 2007 through 2011

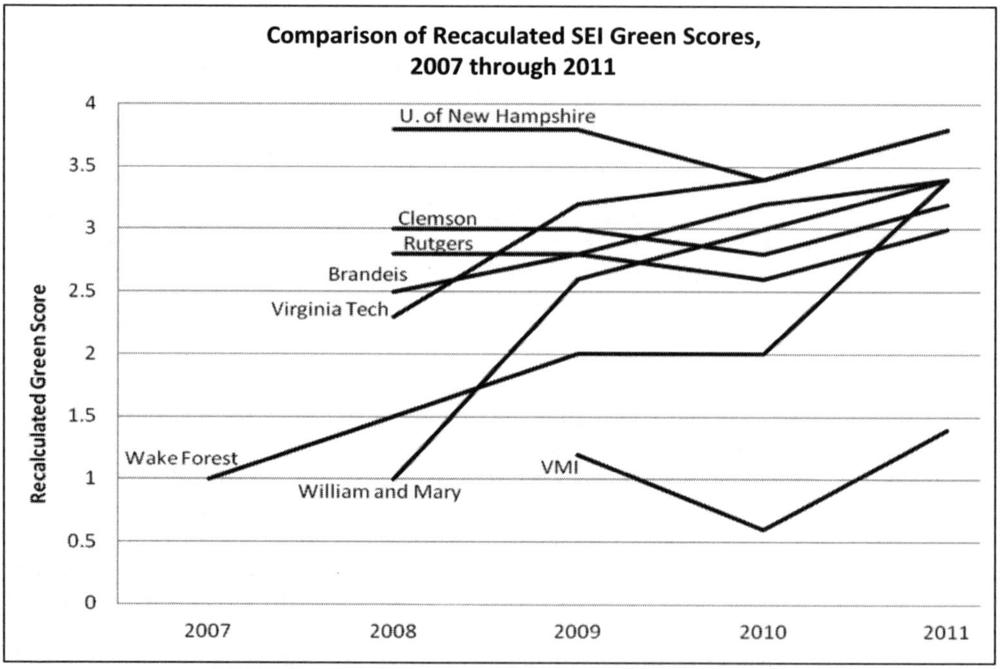

The highest scoring schools were the University of New Hampshire and Virginia Tech. Each received a score of 3.8 on a 4.0 scale for 2011. In the second tier, Brandeis, William and Mary, and Wake Forest all scored 3.4 out of 4.0.

The university with the second-to-lowest score was Rutgers University, which scored 3.0. Virginia Military Institute was the school with the lowest score. In 2011, Virginia Military Institute received a score of 1.4.

The institutions with the greatest increases over time were William and Mary, which went from a score of 1.0 in 2008 to a score of 3.4 in 2011, and Wake Forest which went from a score of 1.0 in 2007 to a score of 3.4 in 2011.

Quantitative analysis

A principle components analysis (PCA) was performed to extract correlated groupings of variables. The loading factors for each original variable in each principle component are shown in the Table 2. Values close to +1 or -1 indicate a strong association (positive or negative, respectively) with that component score.

The PCA produced three components that each had an eigenvalue of greater than 1, and were used in further analyses. The variables most strongly associated with variation in each of these first three components are identified in the table.

We used a multiple regression analysis to examine how each of the principle components independently explained variation in the schools' modified 2011 SEI Green Scores, and the rate of change in this score over time.

Table 2: Component matrix

	Principle component (Percent of original variation explained)					
Original variable	**1 (35.0%)**	**2 (19.9%)**	**3 (14.0%)**	**4 (12.7%)**	**5 (7.94%)**	**6 (7.38%)**
2010 Endowment	0.394	-0.620	-0.266	0.450	0.424	0.071
Public/Private	0.484	-0.532	0.272	0.505	0.211	-0.327
Enrollment	-0.101	0.402	-0.810**	-0.140	0.189	-0.028
Involvement, Staff	0.564	-0.528	-0.156	-0.572	-0.165	-0.112
Involvement, Faculty	0.294	0.826**	0.085	-0.154	-0.126	0.403
Involvement, Students	0.830**	-0.182	0.459	0.213	-0.049	0.080
Funding, Relative Amount	0.939**	0.122	0.278	0.019	0.108	-0.113
Funding, Reliability	0.971**	0.090	-0.076	-0.083	-0.082	0.147
Student Fee	0.270	-0.326	-0.001	0.438	-0.754	0.058
Top-down Initiation	-0.565	0.219	0.713**	-0.075	0.338	0.044
Bottom-up Initiation	0.755**	0.086	-0.570**	0.099	-0.261	0.100
Physical Space	0.765**	-0.257	0.229	-0.438	0.272	0.151
Scope	0.018	0.069	0.286	0.475	0.244	0.785
Program Size	0.806**	0.305	-0.233	-0.296	0.296	0.007
Age	0.500	0.550	0.362	-0.516	-0.141	0.104
AASHE	0.555	0.284	-0.452	0.555	0.070	0.250
STARS	0.843**	0.121	0.108	0.049	0.298	-0.385
PCC	0.192	0.594	0.452	0.424	-0.262	-0.218
Climate Action Plan	0.268	0.793**	0.072	0.298	-0.083	-0.439
Sustainability Action Plan	-0.335	0.654**	-0.303	0.306	0.322	-0.178

The first component consisted of seven variables: student involvement, relative funding, reliability of funding, a bottom-up initiation of the sustainability program, physical space, the program size, and participation in STARS. This first component explained 51% of the variation in the outcome; as the component increases, the outcomes measured by the modified 2011 SEI Green Score increase as well (Wald Chi-square = 4534.0, $r^2 = 0.512$, $P < 0.0001$). However, under closer inspection, some of these variables appear less meaningful than others. Student involvement was measured by the original SEI score for later years, so student involvement is simply an endogenous variable and not a significant influence on the actual outcomes. A school's participation in STARS is likely more a factor of the program's size and funding and may appear in the first principle component due to that correlation rather than its impact on outcomes. STARS is meant to be a reporting system for institutions, and while four of the case study schools are participating, only one has reported so far. It is safe to assume that a school's participation in STARS better indicates their funding levels and their program size, because of the ability to pay the membership fee and the availability of staff to gather and submit the data. Therefore, we interpret the first principle component as being related to variables concerning relative funding, reliability of funding, a bottom-up initiation of the sustainability program, physical space, and the program size.

The second component was composed of the involvement of faculty and the existence of planning documents, specifically a climate action plan or a sustainability action plan. 17% of the variation in outcomes could be explained by this second component. The relationship shows that as faculty involvement increases and the institution commits to more planning documents, the modified SEI green score for 2011 increases (Wald Chi-square = 1518.6, $r^2 = 0.171$, $P < 0.001$).

Enrollment and the variables measuring top-down and bottom-up initiation comprised the third component. This third component explains 12% of the variance in the outcomes observed. Higher modified SEI green scores are associated with larger enrollments and less emphasis on top-down initiatives (Wald Chi-square = 1095.4, $r^2 = 0.124$, $P < 0.001$).

Discussion

For the institutions included in our analyses, the most effective institutional structure for campus sustainability would exhibit five characteristics: high levels of funding, reliable funding, a large program size, a dedicated physical space, and bottom-up initiation. This requires that a college or university make a significant investment if its sustainability program is to thrive but also that it builds from grassroots initiation and includes participation from students and faculty.

In the interviews, perhaps not surprisingly, funding was often cited as the limiting factor for sustainability programs. The economic downturn has made the creation or expansion of programs like sustainability particularly challenging. Little money is available for high levels of funding for sustainability programs, and in an uncertain economic environment it is difficult for schools to commit to reliable funding at any level. This is ultimately one of the most significant variables in a program's outcome. In some ways, times of financial constraint are the best opportunity to begin a program; sustainability is largely about increasing efficiency and doing more with fewer resources (Simmons and Moody 2010). Making a commitment to sustainability funding can be the most important step an institution takes to improve the sustainability program as well as the finances of the entire campus.

The number of personnel permanently associated with a program is a major determinant of the capacity of a sustainability program. Programs with a large staff will have a greater ability to address complex sustainability challenges and to implement wide-ranging policies. The potential for division of labor within the staff can also make a program more efficient. With only a small staff or a volunteer committee, the same few people will be left overseeing programs, implementing specific initiatives, and administering any budget they have, as well as numerous other tasks.

A dedicated space for the sustainability program gives a physical center for the campus' efforts and also serves as a tangible display of the school's commitment to sustainability. Schools with a physical office space or public area dedicated to sustainability are able to use that location to host programming that supports their sustainability goals. In institutions where the sustainability program does not have a specific, permanent location or where the space is limited to an individual office, the broader campus can have difficulties engaging the program.

Bottom-up initiation is a strong factor in a program's success and shows the campus-wide support for the program. The grassroots effort for creating or expanding a sustainability program must be met with a positive response from the administration in order to create an official sustainability program, but the real significance rests on the grassroots efforts. In only two schools did the administration begin a sustainability program without inspiration from the campus community, and neither program was among the highest scoring. Though the correlation was weaker, top-down initiation did have a significant relationship with lower scoring sustainability programs. This emphasizes the strong need for broad support and investment in the program rather than policies imposed from the upper administration.

Planning documents had an interesting effect on outcomes; they correlated with more effective sustainability programs overall, but they also were somewhat linked to slower improvements over time (Wald Chi-square = 3.81, $r^2 = 0.204$, $P = 0.051$). This may indicate that sustainability plans or carbon action plans exist in a mutually exclusive relationship with actual projects and programming, at least during the initial phases of a program. Multiple schools interviewed articulated this decision, saying they had spent the early years of the program crafting plans without taking much action.

Some characteristics of programs had no significant impacts on the outcome. The defined scope of a program did not appear to influence the outcomes of the program. There are exceptional cases where the scope of the program did impact the program's outcomes. Brandeis University did experience limitations due to its scope, however, and the University of New Hampshire had found it necessary to revise the scope of its program in order to have the most successful outcomes.

A program's age was unrelated to its modified sustainability score; relatively new sustainability programs apparently have no trouble catching up to their more established peers. Perhaps this is not surprising, given the relative youth of the majority of sustainability programs that we studied.

In the end, the most successful programs are found on campuses where both the institution and the grassroots have made a strong commitment to sustainability, investing money, time, and labor in the program. This institutional support is crucial for enacting effective sustainability initiatives.

Conclusion

Institutions of higher education can use the model we developed to determine the focus of efforts to improve their campus sustainability programs. Success does require serious investments in the sustainability program, and in a world of limited resources, our work can help colleges and universities ensure they are acting as effectively as possible in accomplishing their goal of sustainability.

In the spring of 2011, we presented The College of William and Mary with a series of recommendations derived from the model we created. From the five characteristics correlated to the most successful sustainability programs, we suggested that the administration: increase funding for the sustainability program, improve the reliability of sustainability funding, expand the size of the program, establish a dedicated space for sustainability on campus, and continue supporting grassroots involvement in sustainability efforts. With each recommendation we provided details for the application of our research in the context of The College of William and Mary. As of 2012, multiple of these recommendations have been adopted, and others are under consideration. This experience highlights the value of our research and sets an example for other campuses.

Our conclusions have strong potential to form the foundation for future research. Now that the variables characterizing campus sustainability programs have been identified among the institutions studied here, future research can evaluate this model on a broader scale. Our case studies provided an opportunity to analyze programs in detail, but we suggest that a wider survey of institutions of higher education would be the best next step in determining the most effective campus sustainability structures.

The variables identified could also be studied in greater detail. Funding, in particular, is a broad characteristic and institutions of higher education could benefit from better understanding the relationship between finances and sustainability programs. Our distinction of funding amounts and funding reliability is a start, but future research could consider what specific items receive funding through sustainability programs, or how programs are able to fund projects outside of their regular budget.

References

Bacow, L. and Moomaw, W. (2007), The University Case for Sustainability. In Litten, L.H. and Terkla, D.G. (Ed.), *Advancing Sustainability in Higher Education,* pp. 37-40. Jossey-Bass, San Francisco.

Cohen, B. (2007), Developing Educational Indicators That Will Guide Students and Institutions Toward a Sustainable Future. In Litten, L.H. and Terkla, D.G. (Eds.), *Advancing Sustainability in Higher Education,* pp. 83-94. Jossey-Bass, San Francisco.

Creighton, S.H. and Cortese, A.D. (1992), Environmental Literacy and Action at Tufts University. In Eagan, D.J. and Orr, D.W. (Eds.), *The Campus and Environmental Responsibility,* pp. 19-30. Jossey-Bass, San Francisco.

Eagan, D.J. (1992), Campus Environmental Stewardship. In Eagan, D.J. and Orr, D.W. (Eds.), *The Campus and Environmental Responsibility,* pp. 65-76. Jossey-Bass, San Francisco.

Fugazzotto, S.J. (2009), Mission Statements, Physical Space, and Strategy in Higher Education. *Innovative Higher Education* 34 (5), pp. 285-298.

Litten, L.H. and Terkla, D.G. (2007), Editor's Notes. In Litten, L.H. and Terkla, D.G. (Eds.), *Advancing Sustainability in Higher Education,* pp. 1-5. Jossey-Bass, San Francisco.

NACUBO-Commonfund Institute (2011), *US and Canadian Institutions Listed by Fiscal Year 2010 Endowment Market Value and Percentage Change* in Endowment Market Value from FY 2009 to FY 2010.* Available: http://www.nacubo.org/Documents/research/2010NCSE_Public_Tables_Endowment_Market_Values_Final.pdf (last accessed 15 February 2011).

Orr, D. (1991), What Is Education For? *In Context* 27, pp. 52-56.

Pittman, J. (2004), Living Sustainably through Higher Education: A Whole Systems Design Approach to Organizational Change. In Corcoran, P.B. and Wals, A.E.J. (Eds.), *Higher Education and the Challenge of Sustainability*, pp. 199-212. Kluwer Academic Publishers, Boston.

Rohwedder, R. (2004), The Pedagogy of Place: The Environmental Technology Center at Sonoma State University. In Corcoran, P.B. and Wals, A.E.J. (Eds.), *Higher Education and the Challenge of Sustainability*, pp. 293-304. Kluwer Academic Publishers, Boston.

Sill, B. (2011), *Campus Sustainability Interview.* Interviewed by Lauren Edmonds [phone interview]. Blacksburg, Virginia, Feb 22, 2011.

Simmons, C. and Moody, J. (2010), Tales from Year One: Launching a Sustainability Office in Lean Times. In: Association for the Advancement of Sustainability in Higher Education, *AASHE 2010*. Denver, Colorado, 2010.

Thompson, G. and Creighton, S.H. (2007), Metrics and Measurement for Environmental Sustainability. In Litten, L.H. and Terkla, D.G. (Ed.), *Advancing Sustainability in Higher Education,* pp. 41-54. Jossey-Bass, San Francisco.

Tight, M. (2003), *Researching Higher Education*. Berkshire, England: Society for Research into Higher Education and Open University.

Walker, K.E., Wals, A.E.J. and Corcoran, P.B. (2004), The Practice of Sustainability in Higher Education: An Introduction. In Corcoran, P.B. and Wals, A.E.J. (Eds.), *Higher Education and the Challenge of Sustainability*, pp. 229-234. Kluwer Academic Publishers, Boston.

Chapter 39

Enhancing the Sustainability Element in Chemical and Biological Engineering Problem-Based Laboratories by Partnering with Community Groups

Royann Jean Petrell and Susan Anne Baldwin

Abstract

In a senior-level problem-based learning laboratory (PBL), small groups are given a problem and six weeks to carry out a research plan for solving it. A plan was devised to isolate PBL problems related to community sustainability for which solutions would benefit both the local community and student learning. Through the process, two long-term community service learning (CSL) PBLs were created. Sixteen out of 20 groups chose them over other PBLs, but due to limited resources, they were carried out by five groups. A lecture course was, however, modified to support one of the CSL-PBLs so that all of the senior-level students could learn how applying theory would enhance an ecosystem surrounding a local community. The new CSL-PBLs required more planning by faculty and students, and longer project timelines. Students developed a new teaching method which permitted project continuity and a decreased need for teaching resources in the future.

Introduction: introducing sustainability via community service learning

Incorporating sustainability concepts into the University of British Columbia engineering educational programs is essential because it furthers the overall campus plan, is an essential element in an accredited program, and knowledge and application of sustainability is expected in professional practice. One goal of the University of British Columbia's (UBC) strategic plan is to "foster social sustainability through teaching, research, and community engagement that promote vibrant human interaction and community cohesion". This goal is accomplished with present economic realities in mind: "Prudent with financial resources and mindful of its mandate to society, UBC supports those initiatives that will ensure the long-term resilience of the University and its ability to serve for generations to come." According to the Canadian Engineering Qualifications Board (2006), education in sustainability is crucial: "firstly, for engineers so that they will say "no" when "no" needs to be said; secondly, to be participants of bodies constituted to formulate environmental laws and their enforcement; and thirdly, for the public so that they see professional engineers as true stewards who have viable, knowledge-based solutions." As engineers in training, post graduates would be tasked and responsible for exercising professional judgment related to "cooperating with colleagues, clients, employers, decision-makers and the public in the pursuit of sustainability" (APEGBC, 1993). The above guidelines suggest that the Department of Chemical and Biological Engineering be tasked at UBC to incorporate sustainability into the classroom in such a way as to give students a meaningful and low cost experience with a community partner.

The engineering context of sustainability involves the design and management of sustainable technology, research into environmental and social impacts, as well as limitations, living within those limitations outlined, and management of resources from cradle to cradle (Boyle

2004). Pearce (2006) makes a compelling case for community service learning (CSL) as a direct way to learn engineering sustainability contexts. In CSL, students employ their expertise and discipline-specific theoretical knowledge to help further the goals of a community, which most of the time comes in the form of a non-governmental organization (NGO). In turn, students gain access to a rich set of community-based experiences in which to develop essential professional competencies such as teamwork, communication skills, creative problem-solving and critical thinking. To be a significant project, the CSL project should represent at least 15-20% of a course grade. CSL has been carried out in higher education for over twenty years in the United States in many disciplines. Engineering departments and Canadian Universities have more recently become active in CSL as it was shown to enhance learning in engineering. In a comparison between CSL and non-CSL engineering students, CSL students had more accurate task analysis, clearer strategic planning, information handling, and client interviewing skills (Lemons et al. 2010).

Engineering departments with their focus on serving industry and consultants find the sourcing of suitable CSL projects a difficult task. In fact, finding suitable projects is often cited as being the most challenging aspect of any CSL endeavor (Al-Khafaji and Morse 2006; Duffy et al. 2008; Lima and Oakes 2006; Ward and Zlotkowski 1999). In addition, suitable projects are difficult to find due to the long-term nature of the solution process in engineering. Before a solution can be formulated and carried out, the CSL project could finish due to course end or community group disintegration. To address these CSL project issues at the Massachusetts Institute of Technology, Jue (2011), presents advice on how to pick a solid community group, plan for continuity, and identify short and long term goals. Another constraint for project identification in engineering relates to the requirement that it fall within a particular discipline. The engineering code of ethics demands this. In summary, CSL has the potential to teach concepts of sustainability to chemical and biological engineering students, but a process for identifying suitable communities and projects, project timelines and deliverables needs to be developed and tested before CSL can become a permanent fixture of the curriculum.

This paper is used to describe how long-term community partnerships and projects were developed for inclusion into an existing senior-level chemical and biological engineering (CHBE) laboratory course to increase awareness of social, economic and environmental sustainability and implementation challenges.

The teaching environment and resources

The laboratory course, which runs from September until April, had the potential to provide a meaningful CSL experience without the need for the creation of a new course. The laboratory course consists of two types of laboratories, namely problem-based learning and regular. The regular laboratories are between three and five hours long and are used to illustrate key concepts taught in theoretical courses. The problem based learning laboratories (PBL), which represent over 60% of the course grade were targeted for CSL as they would permit a significant CSL experience. In PBL, groups of four or five are given a problem and six weeks to plan and carry out research for solving it. Typical of PBL projects elsewhere, teaching resource requirements are high: one teaching assistant or instructor assigned to every PBL group (Gómez-Ruiz et al. 2009). There are four deliverables, namely an oral proposal, written proposal, progress memo and a final report. Roles assigned by the group rotate three times, and unlike most PBL projects where roles relate to a team discussion leader, recorder, reporter and accuracy coach (Gómez-Ruiz et al. 2009) in the CHBE laboratory course they reflect actual employment roles (project engineer, process engineer, design engineer, environment and safety engineer, and quality assurance engineer). In all the deliverables the sections pertaining to the different roles are graded separately to assess individual team performance. PBLs have been in existence in this department since 2005 when student and teaching manuals were developed. A 2007 survey indicated that students' problem solving and critically thinking skills improved for both female and male students after a PBL laboratory (t-test, $n = 37$, $p = 0.05$). PBL has also proven to be very effective in teaching engineering design (Dym et al. 2005).

Equipment is available for most groups, but there have been groups assigned to design and test new equipment. Funds are greatly limited for the implementation of new designs, and to apply for funds, a group must submit a detailed plan and budget. The budget is US$200/group for labs with existing equipment.

Methodology

The following tasks were undertaken to develop long-term community partnerships and PBLs:

- A list of environmental groups was developed.
- A web-based search was completed in order to discover if any of the groups were involved in projects related to topics in sustainability of interest to chemical and biological engineering (hazardous waste, green chemistry, alternative energy, etc.).
- Acting as a volunteer for an environmentally focused group was found to be the most effective way to gain trust and an appreciation of the local issues. Central Vancouver Island was chosen as many well-organized environmental groups are based there. Moreover, the community is very concerned with sustainability as the former major employment bases (forestry, fishing and mining) are down 53% from 1995, and fewer children are being born (BC Stats 2010). R. Petrell, who resides there at weekends, found volunteering activities were announced in the local paper, and after volunteering once, these announcements were sent to her via email.
- Meetings were arranged. The local branch of the Canadian Sierra Club (the Comox Valley Sierra Club) volunteered to coordinate meetings between UBC faculty and the various local environmental groups. Meetings took place in local coffee shops and/or private homes.
- Problems and projects were discussed in one or two hour long meetings. Usually a member of a community group volunteered to record the minutes. Before a project was selected as a PBL the following criteria had to be met.
    - The project had to have the potential to be long-term. This is important for the sustainability of a particular laboratory exercise.
    - The project had to be able to produce results in the short and long-term.
    - The project had to require the expertise of the students.
    - The project had to involve existing equipment or be of low cost.
    - It was desirable that students could visit in person with their community group, and the community group would help with the travel costs by providing transportation, meals and lodging. This help is important because Vancouver Island is connected to the mainland where the University is located via a ferry. The ferry costs are just within budget for most students.
- Preference testing was undertaken to examine student acceptability of projects at the beginning of the academic year.
- Implementation
- Continual assessment and remedial actions.

The meetings that took place from March to December 2011 resulted in many project ideas (Table 1). Due to the sensitive nature of some of the issues, the names of the community groups are not shown. Some of the ideas did not generate suitable PBL problems due to expertise or because they could not be resolved via experimentation in the laboratory course. These unsuitable projects were passed onto professors running other project based courses at UBC. To ensure that a project would fall within the expertise area of chemical and biological engineering, literature reviews were conducted. Refereed papers were located which indicated where within the general subject of the community project idea chemical and biological engineers published (these would be used to develop the problem statement, new laboratories and as reading material supplied to students). Once key reference papers were found, the basic equipment was sourced. Equipment proposals were developed and submitted, but due to budgetary constraints, no funds were available for major capital outlays. Some professors donated equipment from their research laboratories, while other pieces derived from storage areas. Sites for the new experiments had to be procured, and this involved many administrative units (e.g. physical plant, safety, head, Office of the Dean).

Table 1: Project ideas stemming from various community groups concerned about sustainability

Community group	**Project ideas**
One	Validation of CO_2 sequestration via eelgrass beds
One	Rehabilitation of an old coastal sewage treatment lagoon to support fisheries
Two	Concerns about leachate from old coal mining waste piles on fisheries
Three	Concerns about a proposed new coal mining operation on drinking water quality and fisheries
Four	Remediation of old wood waste landfills for the development of new industries and protection of fisheries
Five	Remediation of a pile of by-products from an old coal cleaning operation for real estate development
Six	Concerns about a new gas service or petrol station located next to sensitive aquatic habitat

Once a basic problem statement, references and resources were procured, students were polled to determine the acceptability of the new PBLs. The projects were briefly described to the students, as well the need for prerequisite courses. Student groups were asked to rank their preferences for a PBL problem. Further meetings took place with the community groups. In these meetings, the laboratory projects had to be explained, and the scope made relevant to the community goals. At these meetings, the level and type of support a community group could give to the students was defined.

Results

Campus support for CSL is high; equipment and sites were eventually procured for two new CSL PBLs. One would run three times, and the other, twice during the 2011-12 academic year to give five student groups a significant CSL experience (1/4 of the total). Sixteen out of 20 student groups selected community service learning (CSL) PBLs over the others (CSL were given first place rankings). With the addition of sustainability and community elements, the new PBLs meet or exceed the following graduate attributes associated with an accredited engineering program in Canada: problem analysis, investigation, design, use of engineering tools, individual and team work, communication skills, impact of engineering on society and the environment (the concepts of sustainable design and development, and environmental stewardship), lifelong learning, and economics and project management (Canadian Engineering Accreditation Board 2008).

The sponsoring community group in one of the PBLs is the Comox Valley Project Watershed Society, and the project relates to the Society's blue carbon pilot and eelgrass restoration project. Blue carbon is an expression relating to how much CO_2 is stored in coastal aquatic environments in sediments due to the actions of plants such as eelgrasses and sedges (Laffoley and Grimsditch 2009). A blue carbon offset is a credit for greenhouse gas reductions achieved through uptake by plants and sediments. The society hopes to obtain Provincial carbon offset funds from establishing new eel grass beds which they will use to plant more eelgrass. Eelgrass beds are important habitat for birds, fish, shell fish and aquatic animals, but unfortunately eelgrass beds have greatly diminished in the Puntledge River Estuary of the Comox Valley, and elsewhere around the world due to habitat deterioration, over supply of nutrients and other factors related to human disturbances (de Boer 2007). This project will not only reduce atmospheric CO_2 levels, but help restore eelgrass beds. PBL groups will work towards the development of a method to verify the CO_2 sequestration rate by eelgrass beds. This is important because to receive carbon offset funds the carbon capturing must be verified by a third party. The following are emails from two groups sent in support of this PBL:

> "Lab H – Kinetics of CO2 sequestration by coastal eel grass is a community service learning lab that revolves around a real-life problem…. We believe that our team has the most qualifications to conduct this lab based on the following:
>
> 1. Motivation and drive for sustainability;
> 2. Skill sets and ability to design the experiment;
> 3. Industrial experience and team working skills.
>
> Our team is highly motivated by the aim of this experiment – to improve the carbon dioxide sequestration in a real community. Sustainability is the key to engineering design. Our team has worked together for many projects, and sustainability has always been a focus no matter what project we worked on. The goal of this lab aligns with our team's objective and passion. With this focus, our team would be entirely driven to solve the problem through our busy course loads in fourth year.
>
> Our team also has a great combination of skill sets to develop and execute this problem-based lab. Our team consists of three biology option students and one process option student. The biology students have completed CHBE 381 (bioprocess engineering) with outstanding grades and the process student has excelled in CHBE 455 (Reaction engineering). All of us have completed CHBE 356 process control with distinction …."

The second email message follows:

> "For our PBL, our preferences for the labs are as follows: *1. Kinetics of CO_2 sequestration by coastal eel grass. We were hoping to be able to pursue this PBL, as after discussion, the four of us decided that this would be the most ideal for our group in terms of interest as well as skill set. Regarding the requirements, two of our members are in the Biology option and have both passed CHBE 381 (Bioprocess Engineering), a course which I am also taking in the next semester as a technical elective. We are also all taking CHBE 455 (Reaction Engineering) this semester and have all passed CHBE 356 (Process control) from the previous year. Regarding our interests, Y. has already completed a BSc. in Biology and has expressed deep interest in tackling this specific PBL. Additionally, as students in the environmental option, A. and I were hoping to apply the theory of carbon capture and sequestration studies and methods throughout the course of this specific lab. Overall, our group is eager to participate in Lab Q and would be highly motivated to complete it if we were given the opportunity."

Chemical engineers carry out research on eelgrass CO_2 sequestration modeling and other industrial uses of eelgrass. This year the students are experimenting with numerical modeling (Suzuki et al. 2003) and setting up experimental eelgrass beds complete with tidal flows. The roles of eelgrass, and bacteria in the sediment and water column in CO_2 sequestration must be examined under different lighting and temperature conditions. The experimental eelgrass system is vital, because in nature, eelgrass beds are exposed for inspection only briefly during the lowest tides, which generally occur every other month in the Comox Valley. The experimental system would permit the development and testing of cost effective carbon monitoring methods based on a good understanding of the sequestration model. In addition, the students will apply process control theory as they design the artificial eelgrass beds. Flows, tank shape and size, salinity, pH, nutrients, aeration and lighting must be considered in the design of the artificial beds. In essence, students in the eelgrass project will use most of the knowledge obtained from course work, in addition to knowledge obtained from reading and digesting literature, setting up experiments, and analyzing the results toward the development of a new monitoring system.

The other CSL PBL relates to understanding the chemistry and reactions taking place in abandoned mining waste dumps. Students carry out bench and pilot scale extraction experiments to estimate total contaminate levels and contamination periods given local climate data, and develop neutralization methods. As in the eelgrass project, the sediments contain bacteria so knowledge of bioprocesses is important. Laboratory safety is also very important, as the extracts are toxic. Controls must be in place to collect and safely dispose of the waste products. The reclamation of old mining waste dumps brings up sensitive issues relating to environmental regulations, financial and economical arrangements, and ownership; students undertaking this PBL learn about these sensitivities and how to work around them. The hope is the information derived from the experiments can be used to develop effective remediation plans for brownfields. A brownfield is defined as "an abandoned, vacant, derelict or underutilized commercial or industrial property where past actions have resulted in actual or perceived contamination and where there is an active potential for redevelopment" (ANON 2012). It has been estimated that there are 4,000 to 6,000 brownfields, including urban, rural and waterfront locations across British Columbia. This PBL is long-term in that there are so many communities in need the help of the students to resolve the contamination issues. The following is an email message from one of the teams in support of this PBL:

> "We chose the extraction topic as our number one topic because of its ties to the Vancouver Island community and the learning opportunities it encompassed. We share their concerns regarding the toxicity of the effluents produced by the coal mine waste piles and the environmental impacts it may have on the area. Also, we will likely get to learn about procedures such as the toxic characteristics leaching or the mine water leaching procedures. By completing this PBL, in addition to just gaining valuable real life design experiences from the planning to the scale up stages, we would also be able help out a community in need."

Implementation challenges and emergent teaching methods

The eelgrass PBL officially started in October of 2011. However, the project actually started in September, 2011 when a senior-level biological science student started a year-long directed study under Dr Petrell. He was instrumental in conducting a lengthy literature review, and providing the necessary biological support for all of the Chemical and Biological Engineering students. His participation decreased the learning curve time for the groups. Interdisciplinary teams have been suggested as ideal in CSL projects (Schmitz et al. 2010). The addition of just one biological science student to the team is a testimony to that ideal. In the future, students in oceanography, microbiology or biological sciences will be sought out, and on this large campus, finding the students would require an advertising scheme. The UBC Community Learning Initiative (UBC-CLI) is a central unit on campus which seeks to engage students, faculty, staff, and community in building their capacity to collaborate toward the resolution of complex community-based challenges. The UBC-CLI works closely with faculty and staff across the UBC Vancouver Campus to integrate Community-Based Experiential Learning (CBEL), including Community Service Learning and Community-Based Research, into academic courses and to ensure that meaningful CBEL opportunities are available outside the context of coursework. This extensive cross-campus network can be leveraged to locate students for inclusion on interdisciplinary teams.

Despite the addition of the biological student, the learning curve was still considered steep. Dr Baldwin offered to modify her senior-level required reaction engineering course (CHBE 455), so that the basic theory behind the sequestration model and community service component could be learned. All 88 students examined a CO_2 eelgrass kinetic model to isolate the optimal conditions for the process. The leader of the eelgrass project from the Comox Valley Watershed group (Dr Paul Horgen) presented a one hour lecture, and a two hour tutorial was setup to illustrate the theory. Support for the visit and tutorial development came in part from the UBC-CLI and the Applied Science Dean's Office. Students in the reaction engineering course were asked to reflect on the community service activity in such a way as to "gain further understanding of the course content, a broader appreciation of the discipline and an enhance sense of civic responsibility" (Lima and Oakes 2006).

> "In completing this exercise, I learned that real life problems can be modeled using engineering equations, given valid assumptions. As an engineer, one must be able to apply the engineering concepts learned in class to a variety of real life application[s] to solve problems. As this tutorial is one example of solving a community problem using engineering knowledge, the same sort of idea can be applied to other situations in society. One can use his or her engineering background to [improve] ... society, whether it is through the community, the environment, or the people."

> "This tutorial exemplifies how concepts learned in the course can be applied to real world applications. The complex system encountered in the Courtenay River Estuary project can be represented by a simplified plug flow reactor model. The tutorial enables students to practice their critical thinking skills. The problem [statement had] to be studied in detail and useful information [was] extracted from the article. For instance, students [had] to read the paragraphs and determine which chemicals and biological species are to be selected to form overall stoichiometric equations. Also, it is not a straightforward exercise to input the relevant information into the PFR design equation in order to reach to the final answer which [was] the amount of carbon dioxide taken up within 100 years. Apart from that, students [were] required to work in groups of three to formulate the solution to the problems. Often in real world situations, team work is essential for engineering groups to solve complicated design problems."

> "This exercise demonstrated how the material studied in CHBE 455 can apply to real world systems even with non-conventional systems. In this example, an estuary is defined as the reactor while eelgrass, algae, and CO2 are the primary reactants. The final product of interest is the carbon absorbed into the estuary. The residence time is also non-conventional at 100 years. This shows that chemical engineering students can use their education for a broad range of fields; from working in a processing plant of a large company, to working environmental projects with a small community."

In general, the CSL exercise required the CHBE 455 students to become more engaged with understanding the problem statement in order to apply the methods that they had learned in the course. In this way they acquired a better appreciation for how engineering design can be used in a broad range of applications including those that are complex to understand and describe. Whereas most engineering learning activities are focused on industrial settings, in the CSL activity students found out that their skills can be useful for small communities and society as a whole.

The first eelgrass group had available to them four long wooden troughs, and miscellaneous items for setting up aquatic systems. The group in their design process decided to divide the artificial eelgrass system into component parts. Each part was designed and tested. Early on it was discovered that the troughs had to be modified, and pumps, valves and control systems purchased. The team produced a detailed plan complete with design matrices, sample calculations and theoretical considerations. The plan costing $5,000 (Can) was again funded collaboratively by the UBC-CLI and the Applied Science Dean's Office. Parts were ordered, but would not arrive in time to be assembled before the end of the experimental period.

Fig. 1: Eelgrass team 1 from left: Anastasia Gumulia, Wendy Cheng, David Eng, Heather Kempthorne, and Cherie Tan

The first eelgrass group (see photo) came up the idea to mentor the next group, firstly, because they felt that the learning curve due to the nature of the project would be too great, and secondly, because they wished to be involved in building the design they had created. The group was then given the option to replace a regular laboratory with an additional PBL progress memo if they agreed to mentor the next group for two weeks. The first group provided mentorship in two novel ways namely, one-on-one theoretical sessions and one-on-one action sessions. In the one-on-one theoretical sessions, students assigned to identical roles were paired up. Students from the first group taught the second group for approximately one hour. Then the students in the second group were given a ten question quiz on project goal and objectives, theoretical concerns, and issues that the first group had (Table 2). The first group then used the quiz results to refocus their mentoring efforts in another hour long session. In the one-on-one action sessions, students again paired up for two weeks of laboratory work devoted to constructing and understanding the design plan. The second group felt that the mentorship that emerged from this CSL project was crucial for project continuity. The third group of students (beginning in March of 2012) will receive the same type of mentorship from the second group. The mentorship scheme created by the first eelgrass group not only ensures project continuity (issues with CSL projects in general), but lowers the cost of PBLs (teaching assistant hours can be reduced or applied to develop another CSL project).

Besides the project timelines related to the experimental periods, there were timelines associated to community interactions. The community group had agreed to collect the sediments and eelgrass for the students during the low tide periods (which during the project time period occur in the evenings). The students had to communicate in a timely matter their requirements and changes in plans. The eelgrass was collected by both the students and their community at 11.00 p.m., January 21, 2012, cleaned by the students, and placed into aerated nursery tanks in a cool environment to await final transplantation into the artificial eelgrass bed.

Table 2: Quiz given to the second eelgrass group after receiving mentorship from the first group (short answer)

1.	The overall goal of the project is
2.	The objective(s) of the research are
3.	The rationale for the experimental design is
4.	The flow diagram for the experimental setup is below
5.	Describe the fundamental theory behind the research, and how the experiments relate to it.
6.	What are the major safety hazards, and how will they be addressed.
7.	What are outstanding issues related to finishing and testing the experimental setup.
8.	What are the outstanding issues related to testing and proving the CO2 verification monitoring system?
9.	What is the design engineer expected to do and deliver in this project?
10.	What is the quality control engineer expected to do and deliver in this project?

Future challenges will be how to carry out the project in the summer when the University is not in session, and how to secure equipment for measuring carbon in tissue, sediment and water next fall.

Lessons learned from the eel grass CSL PBL were applied to the second CSL project that started in January 2012. The first group set up the equipment and secured the samples from a brownfield. They plan to mentor the next group the same way the first eel grass mentored the second eel grass group, for the same reasons and under the same terms. Problems concerning confidentiality arose with their community group. In the future, confidentially agreements will be set up well in advance of the laboratory exercises in the hope that trust can be built between University and community groups.

Conclusions

A process was developed and found effective for isolating community groups and CSL related to sustainability in engineering. Volunteering for local environmental groups was an effective way to identify long-term a community service learning projects having strong sustainability elements. Building mutual trust was an important element in forming long-term relationships. From the process, two long-term community service learning problem-based laboratories were created that provided a significant learning experience in sustainability for chemical and biological engineers. In one project, students are helping a community begin a new industry surrounding eelgrass and CO_2 sequestration. In the other project, students are helping a community remediate a brownfield thereby addressing water quality concerns affecting fisheries and providing new job opportunities related to the development. A new teaching method emerged during the implementation of the eelgrass project. Students became the teachers of the next group, and by doing so, project continuity was assured and the need for teaching resources decreased. Modifying the content in a key lecture course so that it reflects a new laboratory project is crucial for decreasing learning curves and providing CSL for all students. Future prospects depend on funding mechanisms for new laboratory equipment and carrying out the projects outside of the course period. The vital connections made with the Comox Valley Sierra Club will be used in the future to develop new CSL projects in sustainability with the Chemical and Biological Engineering Department.

Acknowledgments

We deeply appreciate the support from CSL Coordinator Alaya Boisvert, Chemical and Biological Engineering (Head, workshop, safety officer and laboratory technician), Lorraine Chan from the UBC Public Affairs Office, and our two community leaders, Mike Bell and Paul Horgen.

References

Al-Khafaji, K. and Morse, M.C. (2006), Learning Sustainable Design through Service. *International Journal for Service Learning in Engineering* 1 (1), pp. 1-10.

ANON (2012), *A Community Resource Guide For Brownfields Redevelopment: Case Studies*, Government of British Columbia. www.brownfieldrenewal.gov.bc.ca.

APEGBC (Association of Professional Engineers and Geoscientists of British Columbia) (1993), *Guidelines for Sustainability*. Association of Professional Engineers and Geoscientists of British Columbia, Vancouver, Canada.

BCStats (2010), *A Guide To the B.C. Economy and Labour Market 2010*. B.C. Ministry of Advanced Education and Labour Market Development.

Boyle, C. (2004), Considerations On Educating Engineers In Sustainability. *International Journal of Sustainability in Higher Education* 5 (2), pp. 147-155.

Canadian Engineering Accreditation Board (2008), *Accreditation Criteria and Procedures 2008*, Canadian Engineering Accreditation Board, 1100-180 Elgin Street, Ottawa.

Canadian Engineering Qualifications Board (2006), *National Guideline on Environment and Sustainability*. Canadian Council of Professional Engineers. Ottawa, Canada.

De Boer, W.F. (2007), Seagrass-Sediment Interactions, Positive Feedbacks and Critical Thresholds For Occurrence: a Review. *Hydrobiology* 591, pp. 5-24.

Duffy, J., Barrington, L., Moeller, W., Barry, C., Kazmer, D., West, C. and Crespo, V. (2008), Service-Learning Projects in Core Undergraduate Engineering Courses. *International Journal for Service Learning in Engineering* 3 (2), pp. 18-41.

Dym, C.L., Agogino, A.M., Eris, O., Frey, D.D., Leifer, L.J. (2005), Engineering Design Thinking, Teaching, and Learning. *Journal of Engineering Education*, January, pp. 103-120.

Gómez-Ruiz, S., Pérez-Quintanilla, D. and Sierra, I. (2009), Problem-Based Learning: An Approach to Chemical Engineering Education within the EHEA. *Technology, Education and Development,* InTech. pp. 178-190 (<a href= “http://www.intechopen.com/ books/technology-education-and-development” title=”Technology Education and Development”>Technology Education and Development</a>).

Jue, D. (2011), Improving the Long-Term Sustainability of Service-Learning Projects: Six Lessons Learned from Early MIT IDEAS Competition Winners. *International Journal for Service Learning in Engineering* 6 (2), pp. 19-29.

Laffoley, D. and Grimsditch, G. (Eds.) (2009), *The Management of Natural Coastal Carbon Sinks*. International Union for Conservation of Nature, Rue Mauverney 28, 1196 Gland, Switzerland.

Lemons, G., Carberry, A., Swan, C. and Jarvin, L. (2010), The Effects of Service-Based Learning on Meta-Cognitive Strategies During an Engineering Design Task. *International Journal for Service Learning in Engineering* 6 (2), pp. 1-18.

Lima, M. and Oakes, W.C. (2006), *Service-Learning: Engineering In Your Community*. Great Lakes Press, Inc. Okemos, USA.

Pearce, J.M. (2006), Service Learning in Engineering and Science for Sustainable Development. *International Journal for Service Learning in Engineering* 1 (1), pp. 1-4.

Schmitz, C.L., Stinson, C.H. and James, C.D. (2010), Community and Environmental Sustainability: Collaboration and Interdisciplinary Education. *Critical Social Work* 11 (3), pp. 83-100.

Suzuki, Y., Fujii, M., Casareto, B.E., Furuta, A. and Ishikawa, Y. (2003), CO_2 Sequestration and Fate of Organic Matters Within Seagrass (*Zostera marina*) Ecosystem. *The Society of Chemical Engineers, Japan* 36, pp. 417-427.

Ward, H. and Zlotkowski, E. (Eds.) (1999), *Acting Locally: Concepts and Models for Service-Learning in Environmental Studies*. Stylus, Sterling, USA.

Chapter 40

An University-Based Assessment of the Levels of Sustainability of Maritime Transport Systems in the Baltic Sea Region: Experiences from Warsaw University

Ljubov Kohutnicka

Abstract

With sustainable development becoming more and more popular in different fields, it is important to find ways to implement the concept and put it into practice. The faculty of geography at the University of Warsaw has begun to undertake research that aims to assess the level of sustainability of the maritime transport systems in the Baltic Sea Region with the help of the sustainability indicators, the statistical data for the region, and geographical information systems to give the assessment a degree of visualization. In the Baltic Sea Region the maritime transport systems are the connection between the countries in the region and are irreplaceable for economic and social development, whilst influencing the use of fossil fuels and the environmental pollution.

Since not much research regarding sustainable development in maritime transport systems has been carried out, this innovative paper might interest those working with transport systems and looking for ways to make them more sustainable and part of university research programs.

Introduction

Nowadays, transport systems are an important part of everyday life. Even if one might not pay attention to them, it is important to understand that the daily functions that transport systems fulfill directly influence the people's well-being. With natural resources being situated in very different parts of the world, transportation systems serve as a link between different countries and regions, thereby providing what each region is lacking –, influencing GDP due to trade, providing social services by the means of transporting inhabitants and influencing the environment through the use of fossil fuels.

Maritime transport systems allow easier access to natural resources of further located continents, which all together makes the world more inter-connected. Maritime transport systems, which include not only the means of transportation, but also the land-based facilities and the networks of transportation, have become increasingly more important in the economy and the social mechanisms of any country that has access to bodies of water. With the international (in many cases) nature of the maritime transport systems, it is at times very difficult to analyze them. Not only do they cover different areas with different geographical conditions, they also cover countries with different legislation, traditions and customs. But for that, analysis of different data can lead to problems, as in case of maritime transport systems, quite a lot of different actors have to be taken into consideration. Not only the locations and directions of transport routes have to be analyzed, but also the geographical conditions of the areas, size of the means of transportation, the goods transported (or passengers), the locations of harbors and their influence on the social functions and the environment.

The Baltic Sea Region has unique environment due to its natural conditions – it has brackish waters with almost no inflow of the salinity from the North Sea. While around 1950's the Baltic Sea was considered environmentally healthy, by now the situation has changed dramatically. The nitrogen and phosphorus input into the Baltic Sea has caused a change in the environment and has made the environment more unstable. Maritime transport in the Baltic Sea Region is highly dense and is the main source of connections between countries and faster way of transporting goods from country to country, as well as import goods from outside the region and export the local produce. Being the source of a percentage of GDP, maritime transport, although being rather environmentally friendly (compared to air and road transport) is the source of pollution for the environment and, in case of accidents, the damage to the environment can be extreme.

The University of Warsaw in Poland, as part of the Baltic University Programme (a network of universities coordinated by Uppsala University in Sweden focusing on sustainable development, environmental protection and democracy in the Baltic Sea Region), conducts courses on environmental protection and sustainable development, spreading awareness regarding the vulnerable state of the region and the need for the protection of the environment. Although maritime transport systems and their components are taught at the universities, as well as the environmental protection of the marine environments and sustainable development, the idea of the combining the research in the fields of maritime transport systems and sustainable development is novel. As an attempt to assess the level of sustainability of maritime transport systems the undertaken research is aimed to take a look at the maritime transport systems in the Baltic Sea Region both from the perspective of the sustainability of the process and find the indicators for the proper assessment, as well as use the help of Geographical Information Systems as a means of presentation of levels of sustainability of the countries in the region and the region itself.

Sustainable development as a balanced development of social, environmental and economic aspects

As was adopted in IUCN Programme 2005-2008 (IUCN 2004), sustainability is seen as an interaction of three interlocking circles, or pillars (Fig. 1).

As can be seen, sustainable development appears at the balanced intersection of environmental, social and economic development. If the development of all three dimensions is not equal, sustainable development would also be shifted correspondingly. However, if sustainable development is considered as a balance, then in order for this balance to exist, it needs to be on the exact intersection of development of all spheres and cannot be shifted in any direction.

There are many links between the environmental, social and economic aspects. Environmental sphere provides global support for the economic system and is the source of resource input, while economic system has its impacts on biodiversity and uses environment as a waste sink (Hanley at al. 2001). The demand for natural resources has the direct influence on the economic development, which in turn is influencing the state the environment is left in. Economic development also has direct impact on the social conditions, however, economy is not just about resources – it is a social system established to produce, exchange, distribute and consume those resources, a system, established for interaction between humans. More than that, environment is also a source of amenity (Hanley et al. 2001) – people get aesthetic pleasure, which can hardly be put into monetary value, but which is invaluable to human happiness.

Fig. 1: The three pillars of sustainable development. From left to right: the theory, the present state of things and the change needed to better balance the mode

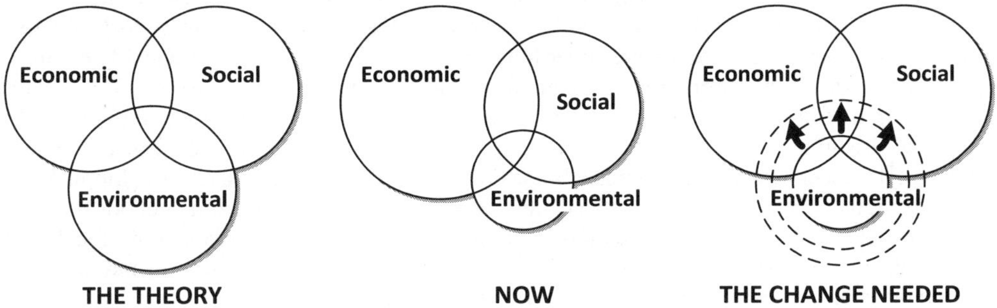

Transport systems

"The ideal transport mode would be instantaneous, free, have an unlimited capacity and always be available. It would render space obsolete. This is obviously not the case. Space is a constraint for the construction of transport networks. Transportation appears to be an economic activity different from the others. It trades space with time and thus money" (translated from Merlin 1992, as quoted in Rodrigue et al. 2006).

Moving people and freight has always been one of the most crucial factors of maintaining a solid and efficient economic system through the history of development of societies and regions. Transport and economic development are very closely linked with socio-economic changes around the world. Countries in the nowadays inter-connected world can hardly survive without international or regional trade, as no place on the planet has all the resources to satisfy its needs, thus local, regional and international transport systems are fundamental to economic growth and development of the region.

The main conditions, constraining the possibilities of moving people and goods are physical distance, topography, hydrology, climate and natural hazards (Rodrigue et al. 2006). With trade and mobility threatened by the environment, it has always been man's goal to overcome those obstacles which are created consequently. Paving roads, bridging rivers, cutting pathways over mountains – all of these represent an effort to overcome distance in an attempt to create the possibility of uninterrupted road transportation. As for maritime transport, the earliest developments came through development of canal locks to improve the waterway transportation by overcoming the adverse natural gradients of inland waterways. Further improvements made the increase of speed, range and capacity possible in the case of ocean transport. This effectively demonstrates that environmental conditions can complicate, postpone or prevent the activities carried out by the actors of transport industry.

In addition to that which is mentioned previously, transport systems influence the economic development, provide a social service for population, as well as influence the environment with the need for energy, which for now can mostly be derived from the fossil fuels. Therefore, the paradigm of sustainable development is well placed to be integrated within transport systems, as the social, economic and environmental spheres can be influenced to enable the system to reach the level of sustainable development. With this in mind, the approach should be geared towards balanced development; the social sphere should be supported with the provision of access to services for the inhabitants, as well as the growth of economic development and the reduction of the negative impact of the transport systems on the environment.

Maritime transportation

Because of the physical properties of water, and the fact that approximately 72% of the Earth is covered by water, most of which is navigable, maritime transportation is the most cost-effective mode to move large freight quantities over long distances, as well as being relatively environmentally friendly at the same time. Around the world, depending on the area, maritime routes are composed of oceans, coasts, seas, lakes, rivers and channels. The construction of channels, locks and dredging are attempts to facilitate maritime circulation by reducing discontinuity, which might occur when land separates bodies of water. The only drawbacks of shipping and maritime industry in general are the speed and delays, which are not uncommon in ports around the planet. Also, emissions from ship fuels are in part responsible for the amount of greenhouse gases in the atmosphere, and in the case of large catastrophes the size of disasters and the influence on the environment has the potential to be devastating.

Transport demand is a key issue that must be prioritized because it influences several environmental issues. Transport demand originates outside the transport sector itself and should be addressed there. In the field of transport policy, more research should be conducted.

However, if one tries to assess the transport systems from the perspective of the sustainable development paradigm, how can one do it? One of the possibilities is to assess them through the indicators of sustainable development. The term "indicators" might not be the best one for the sustainable development, as the term itself has many definitions, however, the OECD definition might be good for the purpose of the research conducted. Sustainable development indicator: "A statistical measure that gives an indication on the sustainability of social, environmental and economic development" (OECD 2002). With the help of this definition one can see that the indicator is more of a statistical measure, generally, a number, with the help of which a certain area can be assessed. While the term "indicator" can be viewed like that, there are no universal indicators for sustainable development. As much as the definition of sustainable development is broad, there are equally many indicators for the assessment of sustainable development. As the author wants to assess the maritime transport system, more specific indicators should be chosen.

Geographical area of the research

When speaking about the Baltic Sea Region from a geographical point of view, it is usually considered that it covers the sea itself, but also the countries that surround it – Germany, Denmark, Sweden, Finland, Russia, Estonia, Latvia, Lithuania and Poland. However, when taking into consideration the basin of the sea itself, even more countries need to be factored in – because many rivers run into the Baltic Sea, meaning that the countries the rivers run through should also be considered in addition to those with coastlines bordering the Baltic Sea. All the same, for the purpose of this research, only First Level NUTS (Eurostat 2011) of the countries surrounding the area were chosen. Although the goods transported are distributed around areas larger than this, as the research is focused on the maritime transport systems and harbors, the indicators and the data for them were chosen for the First Level NUTS of the countries surrounding the Baltic Sea.

Although the construction of roads presents many topographical obstacles on the land, there are also similar issues to be considered in the sea. As much as there are borders on land, countries that have access to bodies of water, and that share those bodies of water with other countries, have their own "territorial waters" and then, further out lie the "international waters", which is where problems can arise. Territorial waters, or a territorial sea, as defined by the 1982 United Nations Convention on the Law of the Sea (UN 1982) is a belt of coastal waters extending at most 12 nautical miles (22 km; 14 miles) from the baseline (usually the mean low-water mark) of a coastal state. The territorial waters are considered to be the sovereign territory of the state and even though foreign civilian and military ships are allowed an "innocent passage". According to the UN Convention on the Law of the Sea (UN 1982), "Passage is innocent so long as it is not prejudicial to the peace, good order or security of the coastal State. Such passage shall take place in conformity with this Convention and with other rules of international law" (UN 1982).

International waters can be described as bodies of water outside the realms of territorial waters meaning that ships are free to navigate them. However, as the coastlines are never straight or easy to establish, the shape of the coastline directly influences the international waters and directly influences the possible routes that the ship can take. Another major influence on the navigable routes of the sea is natural resources and their protection. Fishing areas are not navigable, as well as protected nature areas, which pose an additional restriction on maritime routes. Considering all of the issues mentioned above, it is easy to see that maritime routes present almost as many restrictions as road and rail transportation on land.

Indicators of sustainable development for the maritime transport systems in the Baltic Sea Region

As there is no universal list of indicators of sustainable development, the choice must come down to specific situations and the problem at hand. In the case of the research conducted, few issues were decisive ones; vulnerable environmental areas in the region, the demand for maritime transport as a crucial part of GDP, social conditions for work aboard ships, etc.

As the field of maritime transport systems has not been researched before on the subject of sustainable development, when choosing the indicators of sustainable development, the following issues were taken into consideration:

- Overall indicators of sustainable development;
- Sustainability indicators for the transport systems;
- Sustainability indicators for the Baltic Sea Region (based "An Agenda 21 for the Baltic Sea Region – Baltic 21") (Baltic 21, 1998);
- Strategic goals and recommendations for the EU's maritime transport policy until 2018 (DG ET, 2009).

The research of the literature on the subject showed the following indicators of sustainable development in the economic sphere to be relevant for the maritime transport systems in the Baltic Sea Region:

- Total freight transport in tons and kilometers,
- Tax revenues from transport,
- Total investment in maintenance costs,
- Fuel prices in real terms,
- GDP per capita (Moldan and Hak 2004);
- Share of import and exports,
- Volume of freight transport relative to GDP,
- Port handling costs,
- Shipping costs.

The social indicators of the sustainable development of transport systems include the following indicators:

- Safety (EC 2005);
- Ratio of sector employment rate to overall employment rate;
- Medical cost attributed to transportation;
- Serious accidents at work (EC 2005);
- Number of cases with serious health effects;
- Life expectancy in the region;
- Population growth rate;
- Population with access to safe drinking water;
- Proportion of population living in households affected by noise and pollution;
- Total employment growth (EC 2005);
- Work with high level of job strain/stress (EC 2005);
- Percentage of population living in coastal areas (UN 2007).

The environmental indicators of the sustainable development include:

- Energy consumption by transport and GDP at constant price (energy consumption of transport) (EC 2005);
- Greenhouse gas emissions (including CO2, NOx, CVOC, Particulates),
- Noise pollution (Litman 2005);
- Resource efficiency (Litman 2005);
- Maritime oil spills,
- Protection of fisheries (Litman 2005);
- Toxic materials (Litman 2005);
- Hazardous waste (Litman 2005);
- Infrastructure influence on ecosystems and habitats ("fragmentation") and proximity of transport infrastructure to designated sites (EEA 2002);
- Energy efficiency for passenger and freight transport per pass-km, and per ton km and by mode;
- Use of renewable energy versus total energy use;
- Water quality;
- Dredging and impact to aquatic resources;
- Introduction of non-native species;
- Number of threatened species (mammals, birds, higher plants);
- Load of nitrogen and phosphorus to the Baltic Sea;
- Baltic Sea water transparency;
- Land areas where acidifying load is above critical levels;
- Population served by waste water facilities versus total population;
- km exploited/natural coastline;
- Bathing water quality;
- Contaminants in fish;
- Damage to designated and protected areas;
- Demand and supply of public water;
- Designated and protected areas;
- Discharges from nuclear installations and nuclear power generation;
- Estuarial water quality;
- Native species at risk;
- River quality – chemical and biological;
- Special waste;

- Illegal discharges of oil by ships at sea;
- Accidental discharges of oil by ships at sea;
- Hazardous materials transported by mode;
- Average age of fleet;
- Land take by transport infrastructures (EEA 2002).

Use of Geographic Information Systems to research maritime transport systems

When talking about the use of Geographic Information Systems in maritime transport systems, it should be mentioned that a lot of work has been done already in separate harbors to improve security and standard of general port operations. In certain harbors (depending on the size and the need), different port operations are reflected in the GIS of the harbors, including the locations of the goods and vessels, tracking of berth occupancy and the flow of goods in and out of harbor. But for that, with the help of GIS it is possible to increase the security of the harbor and improve control of all of conducted in the harbor. As GIS are able to stream all of the different operations into one system, it is possible to assess situations with much more accuracy and draw more efficient conclusions, which, in turn, improves the decision-making process.

However, the harbors are only a small part of the maritime transport system and while its contribution and importance are invaluable, systems also include the routes, the ships and analysis of the whole system.

The digitalized representation of the routes in the area can give an overview of the density of the traffic. The use of statistical data of the sizes of ships, as well as the cargo they carry and the reflection of the data in the system can give an accurate interpretation of the flows of the region. The reflection of the routes in the system connected with the environmental data can give the overview of the danger the transport system poses to the environment and to natural resources.

But for that, the analysis of harbor locations presents the possibility of analyzing the transportation system as a whole, which would allow, in turn, data on the bottle-necks of the system to be collected and conclusions to be drawn on the positive improvement of the system.

The main tools exposed and used in GIS in research of transport systems are generally derived from graph theory and spatial interactions models forming the part of cartographical methods of research. The geo-visualizations of such processes involve flow mapping taking into account the directions, volume and quality of movements.

As GIS can be used for graphical presentation of data, the countries in the Baltic Sea Region can be compared based on indicators and statistical data. Sustainable development of the region depends on the equal development of all of the countries and components of the system, so with the help of indicators and data the region's level of sustainability can be assessed.

Conclusion

Problems in assessing the level of sustainability of any sphere are quite common, as methods are not always available for the correct assessment. When trying to assess complex maritime transport systems, it was important to find the appropriate indicators for the region. Also, research shows that the use of Geographical Information Systems can be useful when presenting and comparing results for the region, as to achieve the common sustainable development, the same level of sustainability should be achieved in the countries which form the rest of the region. Although the region's level of sustainability can be assessed in the ways described above, more work can be done in the sphere: as the connections inside the systems are complicated, as well as the connection between the social, environmental and economic spheres, more links can be found between the three and an attempt at finding a single co-created indicator can be made. As maritime transport systems are an important part of life in the Baltic Sea Region and protection of the environment is the priority of all of the countries involved, the research conducted could be used both as an example of a university-based project into sustainable development, but also introduced into curricula to give deeper meaning to the maritime systems and sustainable development.

References

Baltic 21 (1998), *An Agenda 21 for the Baltic Sea Region – Baltic 21.* Adopted at the 7th Ministerial Session of the Council of the Baltic Sea States, Nyborg, June 22-23, 1998. Available at: http://www.cbss.org/Environment/baltic-21.

DG ET (2009), *Communication from the Commission to the Council, the European Parliament, the European Economic and Social Committee and the Committee of the Regions.* Strategic goals and recommendations for the EU's maritime transport policy until 2018. Available at: http://ec.europa.eu/transport/citizen/doc/2009_maritime_policy_space_en.pdf.

EEA (2002), *Are We Moving In The Right Direction?: Indicators On Transport And Environment Integration In The EU, European Environment Agency.* Available at: http://reports.eea.eu.int/environmental_issue_report_2002_24.

Eurostat (2011), *The NUTS Classification.* Available at: http://epp.eurostat.ec.europa.eu/portal/page/portal/nuts_nomenclature/introduction.

EC (2005), *Measuring progress towards a more sustainable Europe: sustainable development indicators for the European Union.* Data 1990-2005. Available at: http://epp.eurostat.ec.europa.eu/cache/ITY_OFFPUB/KS-68-05-551/EN/KS-68-05-551 -EN.PDF.

Hanley, N., Shogren, J.F., White, B. (2001), *Introduction to Environmental Economics*, pp. 3-4. Oxford University Press.

IUCN (2004), *IUCN Programme 2005-2008. Many Voices, One Earth.* Available at: http://cmsdata.iucn.org/downloads/programme_english.pdf.

Litman, T. (2005), *Well Measured. Developing Indicators for Comprehensive and Sustainable Transport Planning.* Victoria Transport Policy Institute.

Moldan, B., Hak, T. (2004), *Composite Indicators of Environmental Sustainability. OECD World Forum on Key Indicators.* Available at: www.oecd.org/dataoecd/43/48/33829383.doc.

OECD (2002), *Strategies to Reduce Greenhouse Gas Emissions from Road Transport Analytical Methods.* OECD Publications Service.

Rodrigue, J.-P., Comtois, C. and Slack, B. (2006), *The Geography of Transport Systems*, pp. 1-11. Routledge.

UN (1982), *United Nations Convention on the Law of the Sea.* Available at: http://www.un.org/Depts/los/convention_agreements/texts/unclos/closindx.htm.

– (2007), *Indicators of Sustainable Development: Guidelines and Methodologies.* Available at: http://www.un.org/esa/sustdev/publications/indisd-mg2001.pdf.

Chapter 41

The Campus as a Teaching Tool: A Case Study of Macalester College's EcoHouse

Suzanne Savanick Hansen and Christopher W. Wells

Abstract

The Macalester College EcoHouse opened as a sustainable living-and-learning housing option for four students in 2007. In this paper, the authors analyze both the effectiveness and challenges of using the EcoHouse as a teaching tool. Students worked with an environmental studies faculty member to develop criteria for the renovation, oversaw the actual renovations, and authored and implemented a civic engagement grant designed to educate students and community members about green building and sustainable living. Since the initial renovations, students have held sustainability internships and conducted independent study projects while living in the house. Challenges include easily accessing data, developing realistic expectations for volunteers, and maintaining institutional knowledge in the face of annual turnover among house residents.

Introduction

Active planning for Project EcoHouse at Macalester College, located in St. Paul, Minnesota, began in fall 2006, and the first residents moved in in fall 2007. Macalester College is a small liberal arts college situated in urban St. Paul, Minnesota, United States. The surrounding neighborhood is made up of older homes, built between 1910 and 1960. The region is renowned for its cold winters, which require significant energy use for residential heating. The modest, college-owned house, which was built in the 1950s immediately adjacent to the campus boundary, underwent significant "green" renovations in the summer of 2007. Renovations included installing a domestic solar hot water system, a recyclable steel roof, blown-in wall insulation, and new high-efficiency appliances with ENERGY STAR certification.

Macalester College has a long history of student, faculty, and staff interest in sustainability issues (Table 1). The environmental studies program started in 1973 and in 2004 it became an official department of the college. The college Facilities Services Department has focused on energy efficiency since the 1960's and 1970's. The college also has a special emphasis on internationalism, multiculturalism, and service to society. Despite the long-standing international focus, however, nearly all of its sustainability efforts prior to 2007 were entrepreneurial efforts launched by interested students, staff, or faculty, and ran parallel to international sustainability efforts such as Agenda 21 (United Nations 1993), the Earth Charter (Earth Charter Initiative 2000) and the UN Decade for Sustainability Education (UNESCO 2011), rather than engaging them directly. The lone exception came in 2000 with the signing of the Talloires Declaration, the first official statement made by university administrators advocating a commitment to environmental sustainability in higher education (ULSF 2011). Macalester College is one of only two higher education institutions in Minnesota to sign that commitment (ULSF 2011).

A major shift in sustainability efforts at Macalester College occurred in 2007, when the college president signed the American College and University Presidents' Climate Commitment, committing the college to achieving climate neutrality (ACUPCC 2012). As a result, the Sustainability Office was established with the hiring of the first Sustainability Manager in 2008. In 2009, the college adopted a comprehensive Sustainability Plan with specific goals, including climate neutrality by 2025 and zero waste by 2020, and outlined how the college intends to meet all these sustainability goals (Macalester College 2009). The Sustainability Plan came about through a participatory strategic planning process as well as an environmental studies senior seminar class project. The class developed recommendations for climate neutrality; the strategic planning project entailed input from over 400 people and built the format for the Sustainability Plan (Hansen et al. 2011).

Macalester College's environmental studies faculty actively pursues using the campus as a sustainability teaching tool. Student papers for environmental studies classes on campus sustainability issues dating back to 2004 are currently archived on the department's website (Macalester College Environmental Studies 2010). Students in environmental studies senior seminar classes developed the college's first greenhouse gas emissions report, recommended actions for the college's climate action plan, developed zero waste recommendations, and developed communications materials for the college's Katherine Ordway Natural History Area, as well as developed grant proposals for the EcoHouse. The first senior seminar project that developed the initial greenhouse gas emissions report is detailed in Wells et al. (2009) and Savanick (2008). A fall 2011 environmental studies class used campus sustainability as a theme for a course titled "Cities, Sustainability and the Campus."

Table 1: Key Sustainability Milestones and Timeline

1967-2000

Education:
- Establish Katharine Ordway Natural History Study Area (1967)
- Create Environmental Studies Program (1973)
- Create Civic Engagement Center (1988)

Operations:
- Implement campus-wide recycling (1970s)
- Undertake numerous campus energy efficiency projects (1960s and 1970s)

2000-2007

Policy:
- Sign Talloires Declaration (2000)
- Sign American College and University Presidents' Climate Commitment (2007)
- Master Plan includes sustainability (2005)
- Join Worker's Rights Consortium (2001)

Education:
- Establish Environmental Studies Department (2004)
- Open EcoHouse Student Residence (2007)
- Establish Lealtad-Suzuki Center (2002)

Operations:
- Install wind turbine on campus (2003)
- Plant native prairie on campus (2004)
- Create new position for a sustainability student worker in Facilities Services & Purchasing (2005)
- Install green roofs on two campus buildings (2006)
- Create Clean Energy Revolving Fund (2006)
- Create bike share program: MacBike (2006)

2008-2009
Staff:
- Hire sustainability manager
- Establish Sustainability Student Worker Network

Policy:
- Launch sustainability strategic planning process
- Subsidized bus pass program
- Develop and implement green cleaning policy
- Create designated suppliers program

Education:
- Environmental Studies class audits campus greenhouse gas emissions and recommends actions for Climate Action Plan.

Operations:
- Markim Hall designed and built to LEED Platinum standard
- "MacFreeSwap" starts to reduce waste

2009-2010
Policy
- Develop Sustainability Plan
- Education
- Environmental Studies Senior Seminar students focus on Zero Waste projects

Operations
- Develop Move Out waste reduction program
- Food waste to pig food program started
- Energy recommissioning report for all main campus buildings

2010-2011
Policy
- First sustainability update report
- Sustainability course definition adopted
- Bottled water free policy developed

Education
- Campus life includes sustainability in educational outcome goals

Operations
- Sustainable Landscaping Master Plan Developed

Source: Macalester College (2011a)

The EcoHouse project was originally conceived as a "living-and-learning laboratory" for students interested in environmental sustainability and interested in incorporating sustainability into their daily life. This approach is not unique, as 103 residence halls or houses are listed as "green buildings" in the Association for the Advancement of Sustainability in Higher Education's list of green buildings in the United States (AASHE 2012). In addition, "buildings as a teaching tool" has become a commonly used phrase since David Orr originally wrote "Architecture of Pedagogy" in 1993 (Orr 1993). A well-developed literature now exists on using buildings as teaching tools (Orr 1997; Wright 2003; Gelfand and Freed 2010; LPA 2009; Ford 2007).

In addition, the most common green building rating system, the United States Green Building Council's Leadership in Energy and Environmental Design (LEED) green building rating system for schools includes points for using the building as a teaching tool under "LEED Schools IDc3: The School as a Teaching Tool" (LEED User 2011). Articles about specific green buildings often describe how they have included touchscreens, building dashboards and other educational plans for their buildings. For example, Oberlin College's Adams Lewis Science Center was built expressly to be a teaching tool and a section of the building's website includes research and education by using the building (Lucid Design Group 2012; Orr 2006). The University of Maine at Farmington has also explicitly connected their LEED building with K-12 curriculum (Eason 2012).

In addition, resources are available to support using the campus as a sustainability teaching tool.

The United States Green Building Council has developed a series of educational resources for K-12 and higher education in their Center for Green Schools (Meehan 2010; Emmanuelle et al. 2010; USGBC 2011). The Science Education Resource Center also includes a website module for faculty, "Campus-Based Learning" that details how to use a campus as a geoscience teaching tool (Savanick 2007).

The need for education for sustainability is well documented (United Nations 1993; Orr 2004; UNESCO 2011; Tilbury 2004), as is the case for Higher Education's special role in education for sustainability (AASHE 2010; Cortese 1999; Cortese and McDonough 2001) and descriptions of how to incorporate sustainability in higher education (Rusinko 2010; Comm and Mathaisel 2003; Creighton 1998). Currently the buildings and grounds of colleges and universities are widely used as teaching tools, either by default or by design (Orr 1997). For example, in traditional classrooms students learn that energy is not important, nor is the locality where the education building is located. Some recent buildings, however, are designed and constructed with sustainability education in mind. However, only a small amount of analysis of the effectiveness of using campus building as teaching tools is available (Savanick et al. 2008; Owens and Halfacre-Hitchcock 2006).

This paper analyzes the effectiveness and challenges of using the EcoHouse as a teaching tool during its first four years with student residents. The paper firstly describes the vision and process for the EcoHouse renovation. Next the authors describe the educational opportunities arising from the project. Finally, the authors analyze the successes and challenges of the project and offer advice for other schools starting similar housing options.

The vision

The initial planning team for the EcoHouse saw it as a unique opportunity to give interested students the ability to explore practical green-living lifestyles, to test the effectiveness of new green technologies, and to work with community partners to develop better resources on the subjects of affordable "green" home renovations and sustainable lifestyles. In addition, the team hoped that the house would function as a "live-in" laboratory for testing the energy savings and cost-effectiveness of various renovations and green technologies, with the aid of a sophisticated monitoring system that would collect empirical data on the performance of various house systems. Team members also hoped that the house could function as an information clearing house that would foster broader discussions about sustainability by connecting interested homeowners with relevant community organizations and efficiency resources. In its original vision, EcoHouse residents would host various events, skill-shares, and workshops on these subjects, and would develop a comprehensive web site to help build connections between campus and neighborhood resources.

Planning

The EcoHouse began as a 1950s ranch home that had been used as faculty rental property. In fall 2006, Professor Chris Wells (Environmental Studies) initiated a conversation among the Department of Environmental Studies, the Dean of Students, and Facilities Services about the possibility of developing a new special interest housing option for students, the EcoHouse. After the administration approved the idea and identified a suitable college-owned property, Facilities Services allocated US$50,000 for renovations over the summer of 2007. During the spring 2007 semester, a volunteer EcoHouse Advisory Committee, comprised of students and Prof. Wells, developed criteria for the renovation and researched various sustainable features to include during renovations. During that same semester, members of the environmental studies senior seminar class, taught by Prof. Wells, wrote four different grant proposals for the EcoHouse as their class project. The civic-engagement grant was successful and provided US$5,000 for outreach programs. During the summer of 2007, another student, employed as a summer worker by Facilities Services, worked with the contractors to implement the renovations.

EcoHouse renovations

All renovations focused on making affordable changes that would dramatically improve the energy efficiency of the house, while also investing in a few showcase items. Renovation costs are detailed in table two.

Table 2: EcoHouse renovation expenses

Project	Units	Price ($USD)
Bathroom Aerator	1	4
Bathroom/Kitchen – Labor	1	820
Concrete Work	1	6,257
Countertops	1	4,950
Dishwasher	1	599
Energy Audit Analysis	1	175
Gerber Dual Flush	1	350
Glass Block Windows	8	215
Insulation	1	1,985
Kill-A-Watt	1	30
Landscaping	1	1,000.00
LEED EB manual	1	207.5
Metal Roof	1	14,216.00
Painting	1	2,150.00
Sears Fridge	1	647.00
Showerhead	1	25.00
Small Stove	1	490.00
Solar Hot Water	1	11,000.00
Solar Trench and Backfill	1	2,071.65
Solar Skylights	1	480.00
Wall Removal	1	46.30
Worm Composter	1	87.00
Worms	1	42.25
TOTAL		$49,602.70 USD

Key operating principles for the renovation were:

- Focusing on cost-effective, energy-saving improvements;
- Leaving working parts of the house in place;
- Using recycled and reused material wherever possible, and using sustainable, recyclable materials when recycled or reused materials were unavailable or impractical.

The final renovations included a new steel roof with a 70-year life guarantee, as well as new gutters, attic ventilation, and aluminum-wrapped fascia and soffits. Two solar-thermal panels were installed on the south-facing roof of the garage, providing for the house's domestic hot water needs. Blown-in cellulose insulation in all exterior walls dramatically increased the efficiency of the building's envelope. All incandescent lighting was replaced with compact fluorescent bulbs. In addition, the interior was completely repainted with low-volatile-organic-compound paint.

In the kitchen, renovations included installing a new ENERGY STAR-rated refrigerator, dishwasher, stove, and vent hood, and a new sink made from recycled aluminum. Paperstone counters, locally made from recycled paper and cashew resin, replaced the old counters. Existing cabinets remained in place, with one bank of cabinets being removed to make room for a dishwasher, a new recycling center, and extra counter space. The removed cabinets were converted to a new kitchen island. In a subsequent renovation in 2009, a small broom closet in the kitchen was converted into a fully functional pantry, and in late 2011 a new student-designed backsplash, made from recovered materials, was installed.

In the bathroom, renovations included installing a new low-flow aerator in the faucet and a new low-flow showerhead, along with a new dual-flush toilet (1.1 gallons per flush down, 1.6 up). Paperstone counters, a large reused bathroom mirror acquired from the Reuse Center (a local used buildings supply store), a solar tube light fixture in the ceiling, and an energy-efficient vent fan completed the bathroom renovations.

The basement includes an ENERGY STAR-rated front-loading washing machine, equipment for the solar-thermal domestic hot water system, an ENERGY STAR-rated chest freezer and, as of fall 2011, an ENERGY STAR-rated dehumidifier. In addition, the basement houses a clothesline for use during dry months and a vermicomposting bin for composting food scraps.

Staff and Faculty Support

In 2007-2008, a formal EcoHouse Advisory Committee – comprised of a faculty member, the Sustainability Manager, the staff responsible for community relations, the Director of Facilities Services, the Dean of Students, and the residents of the house – met regularly to discuss EcoHouse-related activities and to oversee the implementation of the small civic-engagement grant. In 2009, the original board disbanded and an environmental studies faculty member assumed the role of the EcoHouse advisor.

Living in the House

Since fall 2007, the EcoHouse has had four student residents, who are selected by the Environmental Studies Department based on a competitive application process. Students are required to plan two to three house events every semester, have communal meals several times a week, write a reflection essay for addition to the house journal, and host occasional tours. In addition to these basic requirements, each group of residents has worked out its own goals, standards, and house rules in collaboration with a faculty advisor. In 2007-2008, the group focused on grant-supported civic engagement activities and began to think about the relationships between food and community-building among house residents. In 2008-2009, residents took on a range of individual projects and house-related internships, and placed a large emphasis on planning and cooking communal meals. In 2009-2010, interpersonal conflicts among residents created a trying atmosphere for everyone involved in the house. In 2010-2011, after implementing a new group-based application procedure, in which students could apply in groups rather than as individuals, residents again focused on community-building within the house through planning and cooking meals, and one student took on a house-related internship while also developing a sustainable landscaping plan for the house. In 2011-2012, residents have returned to the initial focus on conducting skill-shares, events, and tours, along with the ongoing focus on community-building among housemates, especially through communal meals. One student resident is currently developing a zero-waste related independent study project.

Educational Opportunities

The EcoHouse has been used in the formal curriculum as the subject for a class project, as an internship site, and as the focus of independent study projects. It has also provided a number of informal educational opportunities for house residents, student workers, staff, faculty, and interested community members.

Education in the curriculum

The 2007 Environmental Studies Senior Seminar course, taught by Prof. Wells, generated four student-authored grant proposals seeking funding for various components of Project EcoHouse, including the successful civic-engagement grant to the Xcel Foundation, mentioned above. Two of the other three proposals generated research and arguments that ultimately resulted in major changes in the house. One grant, for example, led to the decision to meet the house's domestic hot-water needs by using the renovations budget to purchase solar-thermal panels – the type of solar technology with the shortest "pay-back" period. Another grant, for a cutting-edge energy-monitoring system, was initially unsuccessful but later became the basis for a revised proposal that secured the donation and installation of an energy-monitoring system in the house. The course's final proposal, which sought funding for a "geo-thermal" heat-pump system to heat and cool the house, led to the discovery that the bedrock under campus makes it a less-than-ideal site to drill, in an economically efficient manner, the multiple wells necessary to install a geothermal system.

Four EcoHouse residents developed successful internships using the house as a project and one student used the EcoHouse as an independent study project (Macalester College 2011b). The 2008 internships outcomes included a guide to eating sustainably on a budget, "Fostering Sustainability in the Kitchen: A Mac-Groveland Guide to Conscious Communal Eating," by Zoe Hastings. During the same semester, a second student developed sustainability educational materials for middle school and elementary school age children for sustainability tours of the campus, "School Curriculum," by Annie Pasbrig. A third student developed the successful proposal for the donation of an energy monitoring system, "Energy Monitoring System," by David Schmoeller. A fourth student, Meg O'Halloran, interned with the District 7 neighborhood association, where she did research on how to live a sustainable lifestyle on a tight budget. One purpose of this research was to inform a plan to purchase foreclosed houses, renovate them sustainably, and then make these new "Eco Houses" available to low income families.

Since 2011, two independent study projects used the EcoHouse as a study area. Ben Francis used EcoHouse data for an independent study project in statistics, "Energy Efficiency:

Energy Use of the EcoHouse". He analyzed three years of energy data for the EcoHouse and the similarly-sized, but not renovated, Japanese House, and found that the EcoHouse uses just 57% of the natural gas that the Japanese House uses. He also found a strange spike in electricity use in July, when the house remained unoccupied, which prompted an investigation. It turned out that a maintenance worker had placed an old, inefficient dehumidifier in the basement to handle a summertime moisture problem. However, the dehumidifier was never turned off after the summer. The authors also discovered that the venting for the clothes dryer was loose, causing it to vent humid air from the dryer directly into the basement, which meant that the old dehumidifier was sent into overdrive. Because of the research project, the authors were able to identify the problem, fix the venting, and solve the moisture problem with a new ENERGY STAR-rated dehumidifier for the house, eliminating the summertime spike in electrical use and significantly lowering the house's electricity consumption during the winter months. A second independent study project is currently underway. EcoHouse resident Camille Erikson is analyzing all of the waste generated in the house and developing strategies for producing zero waste.

Informal education

The EcoHouse has also offered informal educational opportunity for interested students, house residents, student workers, staff, faculty, and interested community members. In the planning stage, the Student Advisory Committee developed criteria for renovating the house and researched potential house features and modifications. Four groups of students have also lived in the house over the past four years, organizing a variety of EcoHouse-sponsored events and offering numerous open houses and tours. Student residents learn from organizing events and tours and by using the sustainability features of the house.

In addition, three student workers were paid to work on projects related to the house. The first student worker spent the summer of 2007 as the project manager overseeing the house's initial renovations. This practical experience persuaded him to embark on a career path in green building. A second student worker was hired in 2008 for civic-engagement and outreach activities. He developed text for a website, organized outreach materials, and organized events. A third student (who was also a house resident) developed outreach materials and a draft landscaping plan as part of her summer student employment in the Sustainability Office.

Tours, open houses, the house's website, and various outreach activities have all helped make the house a community educational opportunity. The house has been part of the regular tour designed for alumni during reunion weekend, and has regularly been a site on the Twin Cities Solar Tour, which is open to the entire community. In its first year, the EcoHouse project generated significant press coverage. The Minneapolis Star-Tribune newspaper, The Villager newspaper, Channel 9 television news, Minnesota Public Radio, and various Macalester publications published stories about the project. In addition, house residents and Prof. Wells attended several community fairs, seminars, and meetings to present information about the EcoHouse. In 2012, Sustainability Manager, Suzanne Savanick Hansen, used the EcoHouse as an example in a staff training event, "Saving Energy (and Money) at Home and at Mac."

EcoHouse successes

The first success was the ability to designate the house as a special living option and to renovate it with sustainable features. The budget came from Facilities Services, but students and faculty staff made all decisions about how to renovate the house. In addition, the successful external grant generated funding for outreach events in the house's first year, which helped establish its place on campus (and in the community) from the outset. The media attention the project received early in the project shows the interest among neighborhood groups and residents for information on sustainability in housing.

The EcoHouse has also provided ongoing educational benefits. Four EcoHouse residents have developed successful internships using the house (and their experiences in it) as the basis for projects, and two students developed independent study projects with data from the house. The EcoHouse has also provided informal educational benefits in the form of events, skill-share activities, tours, and open houses. With the exception of one cohort of residents with interpersonal conflicts unrelated to the house itself, the students who have lived in the house have described their experiences there in overwhelmingly positive terms.

Each of the house's clear successes has had at least one of three common characteristics: a concrete link to the curriculum, a defined budget, and clear oversight (Table 3). Students earning course credit for their work – whether in the form of a regular class like the senior seminar, or an alternative form of credit such as an internship or independent study – have done particularly good work in generating ideas, written work, and other documentation of sufficient quality and quantity to earn course credit. Likewise, good work with careful documentation has also flowed from projects associated with defined budgets and clear accounting responsibilities, as with the renovation of the house itself and with the work associated with the grant for civic engagement work. Funding, in other words, has generated accountability as well as the ability to spend money on particular projects. In addition, when a formal EcoHouse Advisory Board met on a regular basis (once a week or bi-weekly), projects stayed on track and in line with the long-term vision for the house.

Table 3: EcoHouse success characteristics

EcoHouse Success	Concrete Link to the Curriculum	Defined Budget	Clear Oversight
Renovation	Yes	Yes	Yes
Formal curriculum focused on the house (class project, internships, independent studies)	Yes	No	Yes
Informal educational benefits (events, skill-share activities, tours, open houses)	No	Sometimes	Sometimes

Residents have also reported significant learning experiences from self-consciously focusing on and discussing questions about sustainable living as a group. Students report that developing common standards for a range of decisions that are not typically part of life in the dorms, such as collecting food scraps to put in the worm composting bins, deciding on the temperatures to program into the house's thermostat, imposing time limits on showers, deciding whether to line-dry or machine dry clothes, and so forth, have all been interesting and instructive. Nearly all residents report, however, that by far the most important set of issues they have had to deal with and think about have revolved around food. Setting budgets, planning menus, making decisions about processed foods versus whole foods (and organic versus conventional produce) have sparked serious conversations, additional reading, the development of new skills (ranging from basic knife skills to elaborate freezing and canning routines), and a clear sense of community and common cause.

Rather than focusing on the house or its "green" features, in other words, most of the residents have looked at living in the EcoHouse as an opportunity to think more deeply about what "sustainable living" means, and the implications that it has for their everyday activities, and relationships, within the house.

EcoHouse Challenges

For all its successes, the EcoHouse remains an under-utilized opportunity. The house regularly hosts a variety of small-scale events, but the use of the house as a broader community resource has been much more limited than the house's founders originally envisioned. Most events are student-oriented, but students are not renovating houses and, perhaps unsurprisingly, seem to care more about lifestyle issues than the nuts and bolts of home renovations. Despite initial media exposure, connecting staff and community members who are interested in green renovations and opportunities for improved home energy efficiency with the experiences and resources of the EcoHouse has been much less effective, thus far, than the authors had hoped. Challenges also include landscaping and data accessibility, as well as time, communications, interpersonal conflict, and student turnover.

The house's landscaping is currently not well maintained. Two students developed extensive sustainable landscaping plans, but those plans have been only partially implemented in an ad hoc, volunteer-based manner. The first plan lacked substantial funding, and despite several successful volunteer "planting parties" to get a variety of native plants in the ground, no one followed through to implement the rest of the plan after the original organizer graduated. In addition, since there was no designated gardener charged with maintaining the plants, they received only sporadic attention from students whose interests waxed and waned over the course of the years. The second student's plan was not implemented, primarily due to concerns about assigning responsibility for maintaining the landscaping and what training would be needed. As a result, the EcoHouse landscaping is not currently well maintained and resolving questions about its ongoing maintenance remains a top priority.

Data from the house is difficult to access, despite a monitoring system that was installed in 2010 that tracks water, natural gas, and electricity use. Students in the house do not routinely access the energy data, so they are unaware of the impact of their lifestyle choices, such as thermostat temperature. The original goal of analyzing the cost effectiveness of renovations has not been met, except through one independent study project. Without easily accessible and organized data, faculty members are unlikely to include data from the house in their courses, so this lack of analysis is unlikely to change without better data availability.

Another significant challenge is developing realistic expectations about how many events, activities, and analysis a group of volunteers can organize on a regular basis. Without a significant academic component or staff dedicated to the project, volunteers are limited in how much time they can devote to projects. Civic-engagement work on the scale originally envisioned requires significant demands on the volunteer time of the students and faculty involved. The original press coverage of the project generated a barrage of questions, invitations to give presentations, and requests for detailed information, which volunteer students and their busy faculty sponsor found challenging to handle promptly. Even this article would not have been written, if it were not part of the Sustainability Manager's job description to analyze the effectiveness of sustainability projects on campus. In addition, the ability to analyze the effectiveness of the house's renovations requires more time and expertise than typical student residents of the house possess.

Maintaining institutional knowledge and communication during the inevitable turnover in students has also proved a challenge. The original summer student who oversaw the renovations misplaced much of the original Student Advisory Committee's recommendations and research, which meant reinventing some of their research and missing out entirely on the rest of it. In the four years that the house has had student residents, the binder containing the reflections of all the previous residents, which residents decided to keep in hard copy in order to preserve it better and give it a concrete link to the house itself, has been lost. In addition, once the students learn enough about the house to give an informed tour, they are often done with the academic year and another group is ready to move into the house. Current house residents are beginning to address this problem by heavily annotating the house and its many features with a variety of small signs and captions.

Finally, the house requires students to live in a communal setting and also organize and participate in outreach events for the house. For most residents, however, living in the house is often their first experience living away from a dorm or home. Many of the students have never had to negotiate and organize communal meals, shopping, cleaning, and other household chores prior to living in the house. Holding training sessions for students before they move in has proven much more difficult than anticipated, since students often do not arrive on campus and move into the house more than a day or two before the semester begins.

Analysis

As the authors noted above, each of the house's clear successes has had at least one of three common characteristics: a concrete link to the curriculum, a defined budget, and clear oversight (Table 3). To effectively use a building like the EcoHouse as a teaching tool, the authors recommend developing a link to the curriculum by requiring an independent study or internship as part of living in the house, if possible. Students who have not designed internships or projects around the house report that it is difficult to find substantial time to devote to house activities in the face of other demands; students who have designed internships or projects around the house, on the other hand, report that it allowed them to devote significant time, thought, and attention to the experience in a broader context.

A second characteristic of the clear successes of the EcoHouse was clear oversight. To foster oversight, the authors recommend:

- *Offering staff support:* Find a way to include the project as part of a staff position or part of a faculty research project. Sustainability offices, where available, can collect data for use in classes, analyze energy and water data, and develop training materials. If a faculty member can use the house as a research site, then more substantial analysis is likely to occur.
- *Continually training new students.* Develop a standard packet of information for each group of new students with event checklists, college and neighborhood media outlet information, and standard language for events and suggested formats. Develop permanent signage with basic information about the house for new house residents and visitors.

The successes of living-and-learning laboratories like this one hinge on developing realistic expectations of volunteers. To assist with developing realistic time commitments, the authors recommend:

- *Organizing events early in the semester.* Time-strapped students typically have more capacity early in the semester, if the events are organized on a volunteer basis.
- *Using a larger house.* A larger pool of students would be able to share the workload for events more evenly, and would justify the regular attention of student life professionals. If a small house is all that is available, the authors strongly recommend having the students apply as a group to reduce the potential for serious interpersonal difficulties. When the EcoHouse moved to group applications, interpersonal conflicts eased.
- *Connecting with EcoHouses at other schools.* Several EcoHouse-type residences have been established in the United States (at least 22 at the time of this publication), but currently there is no established network for residents in these houses to connect with one another. A regional or national campus sustainability organization would be well suited to connect EcoHouses in this way and share sustainable living successes or engage in healthy energy or waste competitions.

Finally, students at the EcoHouse reported significant learning and interest in practical sustainable lifestyle choices. Students are not typically involved in the renovation of houses, but they are buying food and making decisions on thermostat temperature. The authors recommend:

- *Proactively developing communal living skills.* Assisting students with communal living skills could help gain a more cohesive group and blunt. A required orientation where house chores such as cooking and cleaning are detailed and discussed would help with this transition. At this same time, a thoughtful discussion of sustainable food issues, energy use and waste generation should be part of this early discussion.
- *Connecting with community partners.* Students are not involved in the renovation of houses, but neighbors and college staff are. A strategic partnership with a community organization that specializes in sustainable home renovation could assist with the development and maintenance of an information clearinghouse about the house, but not require continually changing students to become experts in home renovation.

Conclusion

Project EcoHouse at Macalester College has been successful in that students and faculty staff successfully renovated a college house that has provided ongoing educational opportunities, both formal and informal. Over four years, the house has provided academic projects for one class, four internships and two independent study projects. Student residents have developed outreach programs and have developed deliberate ways of living sustainably. Tours, open houses, a comprehensive website, paid student employment opportunities, staff training, and community outreach also provided informal education. Students have reported that they have been able to learn practical sustainable living skills. The clear successes of the EcoHouse had the following characteristics: concrete links to the curriculum, a defined budget, and clear oversight.

Yet the project remains under-utilized. The original goal of analyzing the cost effectiveness of the renovations has not been met and data from the house is not easily accessible. Implementing sustainable landscaping at the house is a challenge, despite two thoughtful student designs. Challenges also remain in developing appropriate expectations for how much time volunteers can devote to organizing events, in developing consistent lines of communication for all people involved in the project, and in the lack of prior communal living experience among the house residents.

Based on Macalester's experiences with the EcoHouse, other institutions developing similar projects should consider a curricular component, develop clear oversight by offering staff support and continually train new students. To assist in developing realistic time commitments, other institutions should organize events early in the semester, use a larger house, and connect with ecohouses at other schools. Finally, students are not typically renovating houses, but they are buying food and making decisions on thermostat temperature. The authors recommend proactively developing sustainable living skills and connecting with community partners. Community partners that focus home renovation can assist with the development and maintenance of a home renovation information clearinghouse.

Education is crucial for meeting the challenges of the transition to a sustainable future. Higher Education institutions are the main way that future leaders will learn the tools necessary to meet upcoming global sustainability challenges. If Macalester College can improve the way its buildings are used as positive sustainability teaching tools, then the students will take what they learn with them as they meet future global sustainability challenges. If the experience of the EcoHouse project at Macalester College can be used to better design buildings and offer better sustainability education both inside and outside the curriculum.

References

AASHE (2010), *Sustainability Curriculum in Higher Education: A Call to Action*. Available at: http://www.aashe.org/files/A_Call_to_Action_final%282%29.pdf (accessed 30 January 2012).

– (2012), *Green Residence Halls*. Available at: http://www.aashe.org/resources/green-residence-halls (accessed 6 February 2012).

ACUPCC (2012), *American College and University Presidents' Climate Commitment*. Available at: http://www.acupcc.org/ (accessed 10 February 2012).

Comm, C.L. and Mathaisel, D.F.X. (2003), Less is more: a framework for a sustainable university. *International Journal of Sustainability in Higher Education* 4 (4), pp. 314-323.

Cortese, A. (1999), *Education for Sustainability: The University as a Model of Sustainability*. Available at: http://secondnature.org/pdf/snwritings/articles/univmodel. pdf (accessed 30 January 2012).

Cortese, A. and McDonough, W. (2001), Education for Sustainability: Accelerating the Transition to Sustainability Through Higher Education. *Environmental Grantmakers Association News & Updates*, pp. 11-14.

Creighton, S. (1998), *Greening the Ivory Tower: Improving the Environmental Track Record of Universities, Colleges, and Other Institutions*. The MIT Press, Cambridge.

Earth Charter Initiative (2000), *The Earth Charter*. Available at: http://www.earthcharterinaction. org/content/pages/Read-the-Charter.html (accessed 3 February 2012).

Eason, G. (2012), *LEED Certified Buildings as Teaching Tools: University of Maine Farmington's 'Sustainable Architecture that Teaches' Curriculum*. Available at: http://www.climateneutralcampus. com/landing.php?whitepaper=leed-certified-buildings-as-teaching-tools-university-of-maine-farmingtons-sustainable-architecture-that-teaches-curriculum (accessed 30 January 2012).

Emmanuelle, H., Rebecca, O. and Leo Pierre, R. (2010), *Roadmap to a Green Campus*. Available at: http://centerforgreenschools.org/campus-roadmap.aspx (accessed 18 April 2012).

Ford, A. (2007), *Designing the Sustainable School*. Images Publishing Dist Ac, Victoria.

Gelfand, L. and Freed, E.C. (2010), *Sustainable School Architecture: Design for Elementary and Secondary Schools*. Wiley, 1sted, Hoboken.

Hansen, S.S., Bucki, J. and Lee, J. (2011), Engaging the Campus Community through Participatory Sustainability Strategic Planning. *Sustainability: The Journal of Record* 4 (2), pp. 75-79.

LEED User (2011), *LEED Schools 2009 IDc3 The School as a Teaching Tool | LEEDuser – LEED Certification Toolkit and Forum*. Available at: http://www.leeduser.com/ credit/Schools-2009/IDc3 (accessed 6 February 2012).

LPA, E. at. (2009), *Green School Primer: Lessons in Sustainability*. Images Publishing Dist Ac.

Lucid Design Group (2012), *Adam Joseph Lewis Center*. Available at: http://oberlin. edu/ajlc/ (accessed 31 January 2012).

Macalester College (2009), *Macalester College Sustainability Plan*. Available at: http://digital commons.macalester.edu/cgi/viewcontent.cgi?article=1004&context=susdocs (accessed 6 February 2012).

– (2011a), *Macalester College Sustainability Plan, 1st Revision*. Available at: http://www.macalester. edu/sustainability/sustainabilityplan/MacalesterSustainabilityPlanDec2011.pdf (accessed 6 February 2012).

– (2011b), *EcoHouse – Student Projects*. Available at: http://www.macalester.edu/ecohouse/student projects/projectshome.htm (10 February 2012).

Macalester College Environmental Studies (2010), *Student Projects*. Available at: http://www. macalester.edu/environmentalstudies/students/projects/projects.htm (accessed 6 February 2012).

Meehan, P. (2010), *Hands-On LEED: Guiding College Student Engagement*. Available at: http://www.centerforgreenschools.org/docs/Hands%20on%20LEED_web_FINAL.pdf (accessed 18 April 2012).

Orr, D. (1993), Architecture as Pedagogy. *Conservation Biology* 7 (2), pp. 226-228.

– (1997), Architecture as Pedagogy II. *Conservation Biology* 11 (3), pp. 597-600.

– (2004), *Earth in Mind: On Education, Environment and the Human Prospect*. Island Press, Washington, DC.

– (2006), *Design on the Edge: The Making of a High-Performance Building*. MIT Press, Cambridge.

Owens, K.A. and Halfacre-Hitchcock, A. (2006), As green as we think? The case of the College of Charleston green building initiative. *International Journal of Sustainability in Higher Education* 7 (2), pp. 114-128.

Rusinko, C.A. (2010), Integrating sustainability in higher education: a generic matrix. *International Journal of Sustainability in Higher Education* 11 (3), pp. 250-259.

Savanick, S. (2007), *Campus-Based Learning*. Available at: http://serc.carleton.edu/introgeo/campusbased/index.html (accessed 26 January 2012).

Savanick, S. (2008), *Campus Greenhouse Gas Emissions Inventory*. Available at: http://serc.carleton.edu/introgeo/campusbased/examples/greenhouseinv.html (accessed 6 February 2012).

Savanick, S., Strong, R. and Manning, C. (2008), Explicitly linking pedagogy and facilities to campus sustainability: lessons from Carleton College and the University of Minnesota. *Environmental Education Research* 14 (6), pp. 667-679.

Tilbury, D. (2004), Environmental Education for Sustainability: A Force for Change in Higher Education. In Corcoran, P.B. and Wals, A.E.J. (Eds.), *Higher Education and the Challenge of Sustainability*, pp. 97-112. Dordrecht, Kluwer Academic Publishers.

ULSF (2011), *Talloires Declaration Institutional Signatory List*. Available at: http://www.ulsf.org/programs_talloires_signatories.html#US (accessed 31 January 2012).

UNESCO (2011), *Education for Sustainable Development | Education | United Nations Educational, Scientific and Cultural Organization*. Available at: http://www.unesco.org/new/en/education/themes/leading-the-international-agenda/education-for-sustainable-development/ (accessed 30 January 2012).

United Nations (1993), *Agenda 21: Earth Summit – The United Nations Programme of Action from Rio*. Available at: http://www.un.org/esa/dsd/agenda21/ (accessed 3 February 2012).

USGBC (2011), *Guide to Transforming Your Campus, Community & Career*. Available at: http://centerforgreenschools.org/docs/USGBC_StudentsGuide.pdf (accessed 7 February 2012).

Wells, C.W., Savanick, S. and Manning, C. (2009), Using a class to conduct a carbon inventory: A case study with practical results at Macalester College. *International Journal of Sustainability in Higher Education* 10 (3), pp. 228-238.

Wright, J. (2003), Introducing sustainability into the architecture curriculum in the United States. *International Journal of Sustainability in Higher Education* 4 (2), pp. 100-105.

Chapter 42

Teacher Preparation Programs: Integrating Sustainability Education Components Using a Social Justice Framework

Louise Gonsalvez

Abstract

The purpose of this paper is to conceptualize a social justice framework for institutionalizing sustainability education in teacher preparation programs; the conceptualization will include searching for a place for instituting such a framework (e.g. a required course), the significance of such a pedagogical framework, and recommendations for further research on the placement of sustainability education in pre-service teacher programs. Many current teacher preparation programs focus on best practices to improve reading, writing and mathematical test results and citizenship education which focuses primarily on history, geography, government and civic virtues. Current sustainability education in schools often fails to address the policy ecology of sustainability issues (e.g. political tensions, confrontations, and competing discourses) as the subject is often distributed to specific disciplines (e.g. biology, geography, and civics) that focus on technical aspects of sustainability (e.g. examining the impact of relocating animals to ecosystems). The paper will address some of the gaps in sustainability education and conceptualize a social justice framework for teaching sustainability (e.g. cultural, social, environmental and economic) within pre-service teacher training.

Introduction

> "Sustainability is a concept, a goal, and a strategy. The concept speaks to the reconciliation of social justice, ecological integrity and the well being of all living systems on the planet. The goal is to create an ecologically and socially just world within the means of nature without compromising future generations. Sustainability also refers to the process or strategy of moving towards a sustainable future." (Moore 2005b: 78)

According to Hegarty et al. (2011), "there has been an urgent call for universities to engage with sustainability education" (p. 451), subsequently there is a large body of research on how universities are embedding sustainability concepts, knowledge and principles into university curricula, policies, and operations (Barth et al. 2007; Brundiers and Wieks 2009; Ciurana and Leal Filho 2006; Hegarty 2008; Huckle 2004a, 2007; Leal Filho 1999; Leal Filho, Dzemydiene and Sakalauskas 2006; Rusinko 2010; Sipos, Battisti and Grimm 2008). Unfortunately, there is a gap in campus initiatives and sustainability education. Although initiatives are often impressive (e.g. greening campuses, embedded curricula pedagogies, and sustainability committees), many campus programs are not addressing "the broader concept of sustainable development" (Venkataraman 2009: 8). Newer and more campus sustainability initiatives are needed to holistically address the complexities of sustainability scenarios and to advance sustainability education both in universities and K-12 schools.

There is a growing field of "'education for sustainability' (EFS), which gives primacy to best practice in learning and teaching as the point of entry to sustainability skills and knowledge" (Hegarty et al. 2011: 452). Educational facilities are expected to play a strong role in the mission to create a more sustainable world, and expectations continue to rise. Venkataraman (2009) states, "The question now facing the educational community is how can ESD [Education for Sustainable Development] be translated into practice so that it can be effective in transforming society to a more sustainable future?" (Venkataraman 2009: 8). The task can seem be daunting, as illustrated in the following scenario. According to Stafford et al. (2010):

> "Only about 5 percent of the 3 gigatons of materials that we use annually to fuel the US economy are renewable, and we release about 7 gigatons of carbon annually into the atmosphere, which causes increases in carbon dioxide that contribute to climate change. This is what we refer to as the gigaton problem. To solve these gigaton problems, we will need to develop solutions that have gigaton effects. Accordingly, we need to examine the social, economic, and political drivers that underlie these problems, and we need to develop environmental technologies that can be brought to scale to solve our environmental challenges" (p. 44).

Determining how universities can best implement sustainability education is an important and significant research area. "How can universities realize their powerful role in dealing with issues of over-consumption, environmental degradation and social injustice?" (Moore 2005c: 326). This paper seeks to conceptualize a course where broad-based sustainability issues can be the focus of inquiry and a lexicon for developing K-12 education for sustainability (EFS). However it recommends only one dynamic to improve campus sustainability initiatives. Post-secondary pre-service education programs are in a unique situation, as they can bring students from different disciplines (e.g. biology, business, and social sciences) together to explore the underpinnings, complexities (e.g. Tainter 2006), and layers of the broader sustainability ecology. Students can also learn effective strategies to teach K-12 students about the interlocking aspects of sustainability and to lead their schools in sustainability initiatives.

How could a faculty of education capstone course for future teachers address the broader and more political elements of sustainability issues that often require critical reflection, analysis, and dialogue, and what framework is recommended to address this gap in sustainability education in a more holistic fashion? This is the research question being addressed in this paper. To critically, democratically, and justly address the more political aspects of sustainability issues critically, democratically, and justly, a social justice framework (e.g. Freire 1970, 1998; Gewirtz 2001; Kose and Shields 2010; McLaren 2005; Schamber and Mahoney 2008) is recommended. Students need to learn how to systematically identify, critically analyze, and confront the many barriers (e.g. resistance to change, hegemony, communication gaps) that impede democratic, sustainable decision-making (e.g. British Columbia Ministry of Education 2008). Cadres of new student teachers can be prepared to carry forward sustainability education skills into K-12 classrooms, schools, and communities. This inquiry explores the potential of a capstone sustainability course for undergraduate education students, the gap in education for sustainability (EFS) such a course would address, and the use of a social justice approach (e.g. British Columbia Ministry of Education) to frame the learning experience.

Towards an education for sustainability

> "I am excited to be part of a new generation of educators that has access to a wide range of teaching and learning models. I am concerned that transformative learning and sustainability education will become buzzwords and that academics will not recognize the support necessary for personal changes of this magnitude to take place [embracing a more sustainable way of thinking, being, and doing]. If we're truly interested in social transformation toward a more sustainable future, then we need to consider the entire system of university education." (Moore 2005b: 76)

According to UNESCO (2004) environmental education "is a well-established discipline which focuses on humankind's relationship with the natural environment and on ways to conserve and preserve it and properly steward its resources," but education for sustainable development is far more encompassing as it is set "in the broader context of socio-cultural factors and the socio-political issues of equity, poverty, democracy and quality of life" (p.16), and subsequently it includes education for interrelated topics such as poverty alleviation, peace and security, gender equality, human rights, governance, corporate social responsibility and cultural diversity. UNESCO's decision to subsume environmental education into education for sustainable development was not arbitrary, but rather, it was in response to the sustainable development movement of the 1970s and 1980s that noted alarmingly unsustainable patterns of consumption occurring that were resulting not only in environmental degradation, but substantial socio-cultural and economic consequences (e.g. unequal distribution of resources, health problems related to environmental toxins, and migrations due to so-called 'development') (UNESCO, 2005a; 2005b). "The term, sustainable development, was popularized in Our Common Future [also referred to as the Brundtland report], a report published by the World Commission on Environment and Development in 1987," and in 1992, "the principles of sustainable development" were established at the United Nations Conference on Environment and Development, in Rio de Janeiro, Brazil" (Drexhage and Murphy 2010: 2). In 2000, the Earth Charter was launched to inspire a global shift towards an ethic of sustainability.

After the 1992 meetings in Rio de Janeiro some insightful scholars recognized a gap in sustainability education and noted a need for students to critically and reflectively engage in the controversial dialogue spaces where sustainability solutions are reached. These writers (e.g. Cochran-Smith 2006; Pritchard and Sanderson 2002; Robbotom and Hart 1993) recognized that the 'political' discourse landscapes that were traditionally avoided (e.g. challenging the status quo) enabled uncritical, unjust, and unsustainable dialogues to be perpetuated, thus diminishing the learning and practice of democratic purpose, principles and the promotion of democratic life. They recognized the need to embed democratic, ethical, and moral discourse into education for sustainability.

Robbotom and Hart (1993) challenged hegemonic educational structures and encouraged a paradigm shift in education whereby social, political and philosophical questioning was an integral aspect of environmental education and the newly forming sustainability education. Robottom and Hart (1993) wrote, "What is a more appropriate form of environmental education research? … [I]t is one which includes consideration of both human consciousness and political action and thus can answer moral and social questions about educational programs which the dominant form cannot" (p. 51). They recognized that sustainability issues require critical socio-economic and cultural analysis, as well as socio-political reconstructions that may be met with considerable resistance.

Pritchard and Sanderson (2002) later noted that sustainability issues involve political dimensions at multiple levels and that "[p]olitics is without question an important consideration in the quest for sustainability and enlightened environmental management. The social world after all, is organized politically from the globe to the village" (p. 147). Shortly after, Wals and Corcoran (2006) recognized that making democratic sustainability decisions often occur within an emotional and oppositional deliberation space that can't be circumvented, and stated,

> "Deep democracy offers a way of thinking about difference, as opposed to consensus. Democracy, from this perspective, depends on difference, dissonance, conflict, and antagonism, so that deliberation is radically indeterminate (Goodman and Saltman 2002). The conflicts that emerge in the exploration of sustainable development, for instance, the inevitable tensions among the three P's (People, Planet, Profit) or the three E's (Efficiency, Environment, Equity), are prerequisites for rather than barriers to higher learning." (p. 103)

Robottom and Hart (1993), Pritchard and Sanderson (2002), and Corcoran and Wals (2004) recognized the complexities, layers, tensions, underpinnings, hegemonic architectures, policies, and processes that undermine sustainability and the need to engage in the difficult dialogue that shapes successful sustainability initiatives, policies, and projects.

Education faculties have the capacity to capture sustainability issues in their broader context and address sustainability education holistically. By using a social justice framework, they can provide a safe, secure forum for students to discuss the complexities of the sustainability ecology (e.g. globalization, global warming, and human rights). According to Sensoy and Diangelo (2009):

> "[A] social justice framework is one that actively address[es] the dynamics of oppression, privilege, and isms, [and recognizes] that society is the product of historically rooted, institutionally sanctioned stratification along socially constructed group lines that include race, class, gender, sexual orientation, and ability [among others]. Working for social justice in education means guiding students [and often being guided by students] in critical self-reflection of their socialization into this matrix of unequal relationships and its implications, analysis of the mechanisms of oppression, and the ability to challenge these hierarchies." (p. 350)

Current sustainability education in schools often fails to address the more contested aspects of sustainability issues (e.g. political tensions, confrontations, and competing discourses) (e.g. Corcoran and Walls 2004; Pritchard and Sanderson 2002; Robbotom and Hart 1993), but a requisite sustainability education course would prepare teachers to not only embed sustainability education into all aspects of their curriculum, pedagogy, and institutional arrangements (e.g. policies, practices, and programs), but also build competencies to advance sustainability initiatives in society. The course would incorporate a "scenario-based approach" (Raskin 2010: 2645) and other sustainability-related approaches such as dialoguing (e.g. Mizobe 2011), collaborating (e.g. Zuniga et al. 2007), and the conceptualizing of futures (e.g. Mannermaa 1995; Robertson 2005).

Integrating sustainability education on university campuses

> "What we teach, what we don't teach and how we teach are all considered when creating sustainability education. Sustainability education is a process of creating a space for inquiry, dialogue, reflection and action about the concept and goals of sustainable development." (Moore 2005b: 78)

According to Rusinko (2010), "there seems to be a relative shortage of broader, more general frameworks for integrating sustainability into curricula in higher education despite "a growing number of studies on how to integrate sustainability in higher education (e.g. Benn and Dunphy 2009; Cusick 2009; Fien 2002; Hegarty et al. 2011; Minuet et al. 2011; Lozano 2006; Moore 2005; Rathzel and Uzzell 2009; Roome 2005; Venkataraman 2009)" (p. 250-251). Early sustainability initiatives rejected slapstick stand-alone courses and promoted embedding sustainability; the two were initially presented as rival opposites. As Hegarty et al. (2011) state, "[O]ne of the key discussions within the EFS project relates to the value of so-called 'bolt on' or 'stand alone' courses/ subjects ... taught by staff with expertise in, and likely a strong ethical commitment to, sustainability" (p. 452). Sterling (2004) suggests that,

> "Sustainability is not just another issue to be added to an overcrowded curriculum, but a gateway to a different view of curriculum of pedagogy, of organizational change, of policy and particularly of ethos. At the same time, the effect of patterns of unsustainability on our current and future prospects is so pressing that the response of higher education should not be predicated only on the 'integration of sustainability' into higher education because this invites a limited, adaptive, response." (p. 57)

A diverse array of EFS approaches could be advantageous (e.g. broad-based to specialized; stand alone to embedded; socio-cultural to technical). A stand alone course for future educators could provide students with the opportunity to address complex dialogue spaces (e.g. conflicting stakeholder interests, competing world views, systemic hierarchies of oppression) that are crucial to democratically resolving sustainability issues. Such a dialogue space (Schoem et al. 2001; Schoem 2003) could be very beneficial (e.g. Chasin et al. 1996; Ellis and

Weekes 2008) to sustainability education and to teaching in general. As Cochran-Smith (2006) notes "to improve teacher education, we will have to debate values, ideas, ideals, and ...we will have to have models of analysis that account for all of these rather than models that eschew politics as barriers to good policy and practice" (p. 194).

A holistic and sophisticated examination of sustainability at the postsecondary level will require critical reflective thinking, dialogue, and re-thinking. The long quote below requires extensive critical analysis, reflection, and critique as does a holistic study of sustainability, sustainable issues, and sustainable education. O'Riordan (2009) summates the Network of European Environment and Sustainable Development Advisory Council's interpretation of sustainable development:

> "Sustainable development seeks to ensure that humankind cares for the needs and interests of all people, among different nations and between generations, in such a way that all are treated fairly and with justice. Such a robust society will act in such a way as to maintain and enhance the life support functions of the planet, and will establish an economy designed to foster livelihoods that create both prosperity and a fundamental sense of personal and collective wellbeing. This trajectory encompasses not only all citizens alive today, but all generations to come. Their wellbeing should be intrinsically "our" wellbeing. Sustainable development requires transformation and innovation in the public and private sectors, creativity in society, the capacity to anticipate future effects, to act with precaution and prevention, and to make responsible decisions affecting the vitality of the future." (p. 40)

An education course that brings together students, who will be future K-12 teachers, policy-makers, and leaders, can begin to address a major issue in sustainability education – how to address the complexities and tensions involved in multi-stakeholder democratic sustainability decision making through discourse and dialogue. In essence, through education, it would holistically address the broader, more encompassing aspects of sustainability.

This paper advocates a stand-alone sustainability education course for future K-12 teachers who have completed a prior degree or a significant portion of their defined major, and such a course is recommended as part of other sustainability initiatives that the university may be undertaking. Such a course seeks to address a gap in sustainability education – the preparation of K-12 teachers to address broad scale sustainability issues that can often be political, conflict-ridden, and power-based. A social justice framework that seeks to identify, analyze, and confront social justice issues (BCMOE) using emancipatory theories (e.g. critical, queer, feminist, decolonizing, and race) to confront and overcome injustices (e.g. racism, sexism, heterosexism, speciesism, resource exploitation, and environmental degradation) is recommended for such a course. Faculties may choose to integrate social justice, citizenship and sustainability studies into one course to increase their odds of initiating some form of sustainability course.

Conceptualizing a way forward: an education for sustainability course for teachers

A social justice framework is well suited to sustainability issues, sustainability education, and the broader, socio-political aspects, of sustainability-related decision-making. Quin (2009) states:

> "The work of social justice educators – explicitly aims to be anti-oppressive through *seeking to empower educators and learners* to *act* in anti-oppressive ways *for* social justice. The 'aim' is not a fixed state, space or point that can be 'arrived at'. The pursuit of social justice will always of necessity be a consistent process of anti-oppressive ways of being against oppressive reconstructions in progressive, 'moving about' (Trinh 1988) contexts. Social justice education in this discourse is therefore about *being and becoming* a social-justi*cer* – not just someone trying to behave in a 'just' or 'fair' manner ... The particular challenge we face is to be able to facilitate the growth and development of social justice educators from among a broad spectrum of educators who may be hampered by their own internalised subordination and/or internalised domination in one or more of their/our social identities – identities with enormously varied social and educational capital and discursive norms."

A critical and transformative social justice framework for teaching a generalized sustainability course is recommended for a broad-based post-secondary sustainability course for teachers that places an emphasis on critical democratic dialogue.

A social justice framework is suggested for a pre-teacher's EFS course, but the scope of a social justice framework would include all aspects of cultural, environmental, economic, and social sustainability. The selection of a 'social' justice framework (e.g. Gewirtz 2001; Kose and Shields 2010; Sensoy and DiAnglelo 2009), as opposed to an economic or environmental framework, was selected so that the course could emphasize critical thinking, democratic principles, and critical discourse (Freire 1970, 1998; Giroux 1988; 2005; Kincheloe 2008; McLaren 2005) – all are very suitable for both K-12 and postsecondary educational structures and would be helpful when contending with broad-scale sustainability issues that require substantial discourse analysis and dialogue.

The following scenario will provide a basis to explain how a social justice framework could be helpful when preparing such a course. Hawken (2007) states,

> "In a peer-reviewed paper published in the International Journal of Occupational Environmental Health in 2004, the authors report significantly elevated levels of childhood leukemia between the ages of birth and four years in the oil-producing areas of the Amazon basin of Ecuador. (Benzene is a well-known cause of leukemia.) ... The juggernaut of growing corporate power cannot last forever. In Bolivia, an Aymara Indian, President Juan Evo Morales, a coca leaf farmer without a high school diploma, made good on his campaign promises and took charge of all gas fields in May 2006, declaring that any country had the sovereign right to control its natural resources. Oil and gas companies had six months either to leave the country or to sign new contracts more favourable to the Bolivian people." (p. 112)

The above scenario demonstrates the intersectionality of social, cultural, environmental, political and economic sustainability and how, for example, multinational corporations play a significant role in sustainability. But, it also demonstrates how democratic processes, dialogue, and policy-making altered socio-cultural, environmental, economic and political landscapes (e.g. Bowers 2001); in this case unfavourably for the locals in Ecuador and favourably and more sustainably for the people in Bolivia. An educational forum to discuss the complexities of multinational resource extraction in developing worlds and/or on Indigenous lands, and its differing impacts on various stakeholders would be an ideal case study for upstart teachers, to explore sustainability using a social justice framework.

The focus of this paper is not to define what social justice education framework should be used, but it should be noted that such a framework would not exclude eco-justice criteria (Bowers 2001) or spiritual and/or philosophical elements (Dantley and Tillman 2006). Bowers (2001) states,

> "Any definition of social justice that does not take account of how human demands on the natural environment are affecting the lives of future generations is fundamentally flawed. Indeed it seems incomprehensible to write about social justice for women, minorities and the economic underclass without considering ways in which the Earths' ecosystems are being rapidly degraded." (p. 3)

A social justice framework is selected for the proposed course, so as to promote critical discourse, democracy and decision-making (e.g. Thomas 2009; Wals and Jickling 2002) in the sustainability program for new teachers; such a framework recognizes and includes, if not prioritizes, the environmental aspects of a holistic approach to sustainability education.

Researchers believe that sustainability education should challenge students to explore their own assumptions, world views, and their relationship to the environment (e.g. Bowers 1993; Huckle and Sterling 1996; Jickling 1994; Moore 2005b; Stubbs and Cochlin 2008). Sustainability education also requires recognition and concern for future generations.

> "[H]uman society now consumes natural resources faster than they can be replenished, and this is compromising the ability of future generations to meet their needs. Current and future generations are inheriting a world that is biologically impoverished, has fewer resources, and suffers from more pollution than ever before. Sustainability challenges present day humans to consider the well-being of future generations, to view their needs as worthy of our moral concern." (Warner 2009, n.p.)

An inquiry into (e.g. Jickling 2004; Singer 2002; Warner 2009; Weakland 2009), moral imperatives (Giroux 2005), and democratic decision making (e.g. Scott 2011) should be incorporated into an introductory course on the broader aspects of sustainability; such sustainability competencies (e.g. Wiek et al. 2011) are especially important while conducting a holistic sustainability analysis on emergent issues, for example. Such competencies rely heavily on self reflection, education, and paradigm shifts in thinking and behaving. In regards to sustainability, "Many have concluded that what we need is a major shift in our core values and dominant culture – in effect, the evolution of a new consciousness" (Leiserwitz and Fernandez 2007, n.p.). Senge et al. (2005) state, "When it is all said and done, the only change that will make a difference is the transformation of the human heart" (p. 6). Both social justice and sustainability education rely on pedagogies that include a "heart, hands, mind" pedagogy (e.g. heart: critical self and cultural reflection; hands: sustainable practices; mind: social sciences, engineering, and bio-sciences) (Sipos et al. 2008).

A social justice framework takes advantage of critical theory and explores ways to pragmatically advance (e.g. through advocacy, social mobilization, and solidarity) a rights-based solution that most often takes into consideration socio-cultural, economic, environmental, species, and democratic rights. Critical theorists (e.g. Habermas 1991; Harvey 2001, McLaren and Kincheloe 2007) "question the social implications of the very assumptions" on which systems are based, engage in a "political inquiry" that questions "dominant ideologies;" and they aim to "refine or reform a system" (Rey 2011, n.p.). They seek to expose systemic oppression, inequities, and hegemonic policies that ultimately discriminate against and disadvantage those who are in a position of less power (e.g. marginalized populations) and to offer alternative ways of looking at issues and resolving issues (e.g. Huckle 2004b). The Australian Commonwealth Government National Action Plan on Education for Sustainability listed the following principles to advance EFS: "transformation and change, education for all lifelong learning, systems thinking, envisioning a better future, critical thinking and reflection, participation, and partnerships for change" (Australian Government Department of the Environment, Water, Heritage and the Arts, 2009: 451). There are a number of sustainability indicators that a course could use to assess its merit and student performance (e.g. Barth et al. 2011; Sollart 2005).

Sustainability education involves teaching students to challenge and transform existing structures, processes, and policies (e.g. Barnett et al. 2001; Moore 2005a; Raskin 2008; Rathzel and Uzzell 2009) as they may be perpetuating unsustainable, inequitable, and discriminatory practices (e.g. Sachs 2008). Hill (2011) states, "Evolving criticality implies approaches to research which are no longer satisfied with the status quo and seek to uncover new ways to identify dominant power relations and challenge these in transformative ways (p. 91). Through reflexive, progressive, and political understandings, an "evolving criticality" can inform a student's role in a sustainability-related project (Hill 2011: 91). McLaren and Kincheloe (2007) state that one who learns to use ongoing criticality becomes "an awkward detective always interested in uncovering social structures, discourses, ideologies, and epistemologies that prop up both the status quo and a variety of forms of privilege" (p. 306). Hill (2011) expands up this:

> "The role of an 'awkward detective' can be further elucidated by Creswell (2002) who suggests critical researchers: are politically minded people who seek to change society, recognise that research is value laden and therefore celebrate their biases, challenge the status quo, make connections to broader structures of social power, and seek to engage research collaborators in dialogue." (p. 91)

Thus, sustainability education at post-secondary schools would even require students to critique the sustainability of the educational courses, pedagogies, and institutional arrangements on their own campuses. Even educational institutions "wear blinders that limit the incorporation of new information and perspectives in problem processing" (May et al. 2011: 7) and to the realities of the current global situation, thus they need to continually explore how to ontologically, epistemologically and paradigmatically make sustainability part of their institutional way of being.

It is recommended that the following four elements exist in any social justice education framework that is selected: a) students are introduced to the various social justice/sustainability terms, concepts, and histories; b) students learn how to identify social injustices and unsustainable practices; c) students learn to critically analyze social injustices/sustainability practices and dialogue about such issues; and d) students learn how to confront social injustices and unsustainable decision-making and practices through dialogue and democratic avenues (British Columbia Ministry of Education 2008). This framework parallels the British Columbia Ministry of Education (2008) Social Justice 12 course and is well suited to teaching sustainability concepts, theories, and issues. This framework would help address the political dimensions (e.g. Barry 2008; Smith and Sterling 2010; Wals 2010) of sustainability and address a major gap in sustainability education (e.g. interdisciplinary dialogue on broad-based sustainability issues). It encourages critical reflection, analysis, and the development of discourse strategies (e.g. letter writing, campaigning, and social media) to counter unjust and/or unsustainable decision-making and practices.

Conclusion

> "What we teach, what we don't teach and how we teach are all considered when creating sustainability education. Sustainability education is a process of creating a space for inquiry, dialogue, reflection and action about the concept and goals of sustainable development." (Moore 2005b: 78)

According to Ferreria et al. (2006), "Teachers hold the key to change in schools. This has been recognised by international agencies such as UNESCO who have identified the professional development of teachers in Learning for Sustainability as 'the priority of priorities'" (p. 10). Compartmentalized studies of cultural, economic, technical, environmental and social elements of sustainability are important (e.g. compartmentalized cognitive constructs), but these elements often collide as various stakeholders (e.g. actors) compete to ensure their interests are protected, and in these competitions power often trumps all interests (especially Mother Earth's and the many diverse species who inhabit her), the marginalized become further marginalized, and the disenfranchised and dispossessed are cast aside. A sustainability education course designed for future K-12 educators, which uses a social justice framework, focuses on dialogue, and addresses a gap in sustainability education (e.g. addressing broad-based sustainability issues and/or scenarios), is recommended.

References

Australian Government Department of the Environment, Water, Heritage and the Arts (2009), *Living sustainability: The Australian Government's national action plan for education for sustainability.* Sydney: Australian Commonwealth. Retrieved from http://www.environment.gov.au/education/publications/pubs/national-action-plan.pdf.

Barnett, K., McCormick, J. and Conners, R. (2001), Transformational leadership in schools – panacea, placebo or problem? *Journal of Educational Administration* 39 (1), pp. 24-46.

Barry, J. (2008), Towards a green republicanism: Constitutionalism, political economy, and the green state. *The Political Economy of the Good Society* 17 (2), pp. 1-11.

Barth, M., Adombent, M., Albrecht, P., Burandt, S., Franz-Balsen, A., Godemann, J. and Rieckmann, M. (2011), Towards a Sustainable University: scenarios for sustainable university development. *International Journal of Innovation and Sustainable Development* 5 (4), pp. 313-332 (doi: 10.1504/ IJISD.2011.043322).

Barth, M., Godemann, J, Rieckmann, M. and Stoletenberg, U. (2007), Developing key competencies for sustainable development in higher education. *International Journal for Higher Education* 8 (4), pp. 416-430.

Bowers, C.A. (2001), *Educating for eco-justice and community*. University of Georgia Press, Athens, Georgia.

British Columbia Ministry of Education (BCMOE) (2008), *Social Justice 12: Integrated resource package*. Victoria: BCMOE. Retrieved from http://www.bced.gov.bc.ca/irp/pdfs/social_studies/ 2008socialjustice12.pdf.

Brundiers, K. and Wiek, A. (2009), Real-world learning opportunities in sustainability: from classroom into the real world. *International Journal of Sustainability in Higher Education* 11 (4), pp. 308-324 (doi: 10.1108/14676371011077540).

Chasin, R., Herzig, M., Roth. S., Chasin, L., Becker, C. and Stains, R.R. Jr. (1996), From diatribe to dialogue on divisive public issues: Approaches drawn from family therapy. *Mediation Quarterly* 13 (1), pp. 1-19.

Ciurana, A. and Leal Filho, W. (2006), Education for sustainability in university studies: Experiences from a project involving European and Latin American universities. *International Journal of Sustainability in Higher Education* 7 (1), pp. 81-93.

Cochran-Smith, M. (2006), Teacher education and the need for public intellectuals. *The New Educator* 2 (3), pp. 181-206 (doi: 10.1080/15476880600820136).

Corcoran, P.B. and Wals, A.W. (2004), *Higher education and the challenge of sustainability: Problematics, promise and practice*. Kluwer Academic Publishers, Norwell, Massuchetts.

Dantley, M. and Tillman, L. (2006), Social justice and moral transformative leadership. In Marshall, C. and Oliva, M. (Eds.), *Leadership for Social Justice: Making revolutions in Education,* pp. 16-30. Pearson, New York.

Drexhage, J. and Murphy, D. (2010), *Sustainable development: From Brundtland to Rio 2012* (Prepared for consideration by the High Level Panel on Global Sustainability). United Nations, New York.

Ellis, E. and Weekes, T. (2008), Making sustainability 'real': Using group-enquiry to promote education for sustainable development. *Environmental Education Research* 14 (4), pp. 482-500.

Ferreira, J., Ryan, L. and Tilbury, D. (2006), *Whole school approaches to sustainability: A review of models for professional development in pre-service teacher education*. Australian Research Institute in Education for Sustainability, Sydney, Australia.

Fien, J. and Wilson, D. Promoting sustainable development in TVET: The Bonn Declaration. *Prospects* 3 (3), pp. 273-288.

Freire, P. (1970), *Pedagogy of the oppressed.* Continuum, New York.

– (1998), *Pedagogy of freedom: ethics, democracy, and civic courage*. Rowman and Littlefield, Oxford.

Gewirtz, S. (2001), Rethinking social justice: A conceptual analysis. In Demaine, J. (Ed.), *Sociology of Education Today,* pp. 49-64. Palgrave, New York.

Giroux, H. (1988), *Schooling and the struggle for public life. Critical pedagogy in the modern age.* University of Minnesota Press, Minneapolis.

– (2005), *Translating the future and the promise of democracy. Address to Convocation Memorial University*. Retrieved from http://www.henryagiroux.com/awards/convoca tion_address.htm.

Habermas, J. (1991), *The structural transformation of the public sphere: An inquiry into the category of bourgeois society*. MIT Press, Boston.

Harvey, D. (2001), Time space compression and the postmodern condition. In Waters, M. (Ed.), *Modernity: Critical concepts* 4, pp. 77-98. Routledge, London, Ontario.

Hawken, P. (2007), *Blessed unrest*. Penguin Books, New York.

Hegarty, K. (2008), Sustaining Learning, Learning to Sustain: educating to envisage sustainable futures. In Orr, D., Danaher, P., Danaher, G. and Harrevald, R. (Eds.), *5th International Lifelong Learning Conference*. Central Queensland University Press Yeppoon, Qld.

Hegarty, K., Thomas, I., Kriewaldt, C., Holdsworth, S. and Bekessy, S. (2011), Insights into the value of a 'stand alone' course for sustainability education. *Environmental Education Research* 17 (4), pp. 451-469 (doi: 10.1080/13504622.2010.547931).

Hill, A. (2011), *Re-envisioning the status quo: Developing sustainable approaches to outdoor education in Aotearoa New Zealand.* (Dissertation). Dunedin, New Zealand: University of Otago. Retrieved from http://otago.ourarchive.ac.nz/bitstream/handle/10523/1954/HillAllenM2011PhD.pdf?sequence=1.

Huckle, J. (2004a), Citizenship education for sustainable development in initial teacher training. *Citizen and Teacher Education (citizED),* November, 1-23. Retrieved from: http://www.citized.info/pdf/induction/john_huckle.pdf.

– (2004b), Critical realism: a philosophical framework for higher education for sustainability. In Corcoran, P. and Wals, A. (Eds.), *Higher Education and the Challenge of Sustainability,* pp. 33-46. Kluwer Academic Publishers, Norwell, Massuchetts.

– (2007), Policies and Practices for Sustainable Development: a briefing paper for trainee teachers. *Citizen and Teacher Education (citiZed)*, December, 1-11. Retrieved from http://john.huckle.org.uk/download/2878/Policies%20and%20Practices%20for%20Sus tainable%20Development.pdf.

Huckle, J. and Sterling, S. (1996), *Education for Sustainability.* Earthscan, Oxford.

Jickling, B. (2004), Making Ethics an Everyday Activity: How Can We Reduce the Barriers? *Canadian Journal of Environmental Education* 9, pp. 11-30.

– (1994), Studying sustainable development: Problems and possibilities. *Canadian Journal of Education* 19 (3), pp. 231-240.

Kincheloe, J. (2007), Critical pedagogy in the twenty-first century: Evolution for survival. In McLaren, P. and Kincheloe, J.L. (Eds.), *Critical pedagogy: Where are we now?* pp. 9-42. Peter Lang, New York.

– (2008), *Knowledge and Critical Pedagogy: An Introduction.* Springer Publishing, London.

Kose, B.W. and Shields, C. (2010), Ecological and social justice: A leadership framework for sustainability. *Journal of Environmental, Cultural, Economic & Social Sustainability* 6 (3), pp. 275-288.

Leal Filho, W. (1999), *Sustainability and university life.* Peter Lang Scientific Publications, New York.

Leal Filho, W., Dzemydiene, D. and Sakalauskas, L. (Eds.) (2006), *Citizens and Governance for Sustainable Development.* Technica Press, Vilnius.

Leiserwitz, A.A. and Fernandez, L.O. (2007), Toward a new consciousness: Values to sustain human and natural communities. *Environment: Science and Policy for Sustainable Development.* Sept/October. Retrieved from http://www.environmentmagazine.org/Archives/Back%20Issues/September-October%202008/Leiserowitz-Fer nandez-full.html.

Mannermaa, M. (1995), Alternative futures perspectives on sustainability, coherence and chaos. *Journal of Contingencies and Crisis Management* 3 (1), pp. 27-34.

May, P., Jochim, A.E. and Pump, B. (2011), *Political limits to the processing of policy problems.* Centre for American Politics and Public Policy University of Washington, Washington.

McLaren, P. (2005), Critical pedagogy, political agency and the pragmatics of justice: The case of Lyotard. *Educational Theory* 44 (3), pp. 319-340 (doi: 10.1111/j.1741-5446. 1994.00319.x).

McLaren, P. and Kincheloe, L. (Eds.) (2007), *Critical pedagogy: Where are we now?* Peter Lang, New York.

Mizobe, M.N. (2011), *Intergroup dialogue: One tool for social justice education and action.* Simon Fraser University: Centre for Education, Law & Society. Retrieved from http://cels.sfu.ca/current Issues/intergroup_dialogue_one_tool_for_social_justice_education_action.html?KeepThis=true&TB_iframe=true&height=460&width=720.

Moore, J. (2005a), Barriers and pathways to creating sustainability education programs: policy, rhetoric and reality. *Environmental Education Research* 11 (5), pp. 537-555.

– (2005b), Is higher education ready for transformative learning?: A question explored in the study of sustainability. *Journal of Transformative Education* 3 (1), pp. 76-91.

– (2005c), Seven recommendations for creating sustainability education at the university level: A guide for change agents. *International Journal of Sustainability in Higher Education* 6 (4), pp. 326-339.

O'Riordan, T. (2009), On the politics of sustainability a long way ahead: Sustaining Europe for a long way ahead: Making long-term sustainable development policies work. *Environment* 51 (2), pp. 40-43.

Pritchard, L. and Sanderson, S.E. (2002), The dynamics of political discourse in seeking sustainability. In Gunderson, L.H. and Holling, C.S. (Eds.), *Panarchy: Understanding transformations in human and natural systems*. Island Press, Washington.

Quin, J. (2009), Growing social justice educators: A pedagogical framework for social justice education. *Intercultural Education* 30 (2), pp. 109-125 (doi: 10.1080/14675 980902922192).

Raskin, P. (2008), World lines: A framework for exploring global pathways. *Ecological Economics: The Transdisciplinary Journal of the International Society for Ecological Economics* 65, pp. 461-470 (doi: 10.1016/j.ecolecon.2008.01.021).

– (2010), The century ahead: Searching for sustainability. *Sustainability* 2, pp. 2625-2651 (doi: 10.3390/su2082626).

Rathzel, N. and Uzzell, D. (2009), Transformative environmental education: a collective rehearsal for reality. *Environmental Education Research* 15 (3), pp. 263-277).

Rey, P.J. (2011, September 20). Critical Theory: Useful distinction or unconscious smugness. *The Society Pages*. Retrieved from: http://www.popmatters.com/pm/moving-citations/article/151889-critical-theory-use ful-distinction-or-unconscious-smugness.

Robottom, I. and Hart, P. (1993), *Research in environmental education: Engaging the debate*. Deakin University Press, Geelong, Australia.

Robertson, Susan L. (2005), Re-imagining and rescripting the future of education: global knowledge economy discourses and the challenge to education systems. *Comparative Education* 41 (2), pp. 151-170.

Rusinko, C.A. (2010), Integrating sustainability in higher education: A generic matrix. *International Journal of Sustainability in Higher Education* 11 (3), pp. 250-259.

Sachs, J. (2008), *Common wealth: Economics for a crowded planet*. Penguin Books, Toronto, Canada.

Schamber, J. and Mahoney, S. (2008), The development of political awareness and social justice citizenship through community-based learning in a first-year general education seminar. *The Journal of General Education* 57 (2), pp. 76-99.

Schoem, D. (2003), Intergroup dialogue for a just and diverse democracy. *Sociological Inquiry* 7, pp. 212-227.

Schoem, D., Hurtado, S., Sevig, T., Chesler, M. and Sumida, S. (2001), Intergroup dialogue: Democracy at work in theory and practice. In Schoem, D. and Hurtado, S. (Eds.), *Intergroup Dialogue: Deliberative democracy in school, college, community, and workplace,* pp. 1-21. University of Michigan Press, Ann Arbor.

Scott, W. (2011), Sustainable schools and the exercising of responsible citizenship – a review essay. *Environmental Education Research* 17 (3), pp. 409-423 (doi: 10.1080/ 13504622.2010.535724).

Senge, P.M., Scharmer, and Jaworski, J. and Flowers, B.S. (2005), *Presence: An exploration of profound change in people, organizations, and society*. Currency Doubleday, New York.

Sensoy, O. and DiAnglelo, R. (2009), Developing social justice literacy: An open letter to colleagues. *Phi Deltan Kappan* 90 (5), pp. 345-350.

Singer, P. (2002, October 11). Navigating the ethics of globalization. *The Chronicles of Higher Education* 49 (7), pp. B7-B10.

Sipos, Y., Battisti, B. and Grimm, K. (2008), Achieving transformative sustainability learning: Engaging head, hands and heart. *International Journal of Sustainability in Higher Education* 9 (1), pp. 68-86.

Smith, A. and Sterling, A. (2010), The politics of social-ecological resilience and sustainable socio-technical transitions. *Ecology and Society* 15 (1): p. 11. Retrieved from http://www.ecology andsociety.org/vol15/iss1/art11.

Sollart, K. (2005), *Framework on indicators for education for sustainable development: Some conceptual thoughts*. Netherlands Envirionmental Assessment Agency (MNP). Retrieved from www.unece.org/env/.../Framework%20onESD%20 indic%20NL.doc.

Stafford, S.G., Bartels, D.M., Begay-Cambell, S., Bubier, J.L., Crittenden, J.C., Cutter, S.L., Travis, J. (2010), Now is the time for action: Transitions and tipping points in complex environmental systems. *Environment* 52 (1), pp. 38-45.

Sterling, S. (2004), Higher education, sustainability, and the role of systemic learning. In Corcoran, B.P. and Wals, A. *Higher education and the challenge of sustainability: Problematics, promise, and practice,* pp. 49-70 (doi: 10.1007/0-306-48515-X_5).

Stubbs, W. and Cocklin, C. (2008), An ecological modernist interpretation of sustainability: the case of Interface Inc. *Business Strategy & the Environment,* 17, 512-523.

Tainter, J. (2006), Social complexity and sustainability. *Ecological Complexity* 3 (2), pp. 91-103 (doi: 10.1016/j.ecocom.2005.07.004).

Thomas, I. (2009), Critical thinking, transformative learning, sustainable education, and problem-based learning in universities. *Journal of Transformative Education* 7 (3), pp. 245-264.

UNESCO (2004), *United Nations decade of education for sustainable development 2005-2014: Draft international implementation scheme*. UNESCO, Paris.

– (2005a), *Guidelines and recommendations for reorienting teacher education to address sustainability: UNESCO Education for sustainable development in action, Technical paper* No. 2. UNESCO Section for Education for Sustainable Development: Division for the Promotion of Quality Education, Paris.

– (2005b), United Nations decade of education for sustainable development 2005-2014*: Draft International Implementation Scheme*. UNESCO, Paris. Retrieved from http://unesdoc.unesco.org/ images/0013/001399/139937e.pdf.

United Nations (1992), Results of the World Conference on Environment and Development: Agenda 21. United Nations Conference on Environment and Sustainability (UNCED), Rio de Janeiro, United Nations, New York.

Venkataraman, B. (2009), Education for sustainable development. *Environment* 51 (2), pp. 8-10.

Wals, A.E.J. (2010), Between knowing what is right and knowing that is it wrong to tell others what is right: on relativism, uncertainty and democracy in environmental and sustainability education. *Environmental Education Research* 16 (1), pp. 143-151.

Wals, A.E.J.. and Corcoran, B. (2006), Sustainability as an outcome of transformative learning. In Holmberg, J. and Samuelsson, B.E. (Eds.), *Drivers and barriers for implementing sustainable development in higher eduation: Technical Paper*, No. 3, pp. 103-110. UNESCO, Paris.

Wals, A.E.J. and Jickling, B. (2002), "Sustainability" in higher education: from doublethink and newspeak to critical thinking and meaningful learning. *International Journal of Sustainability in Higher Education* 3 (3), pp. 221-232.

Warner, K.D. (2009), *A short course in environmental ethics: The ethical dimensions of sustainability (part 4*). Markkula Centre for Applied Ethics, Santa Clara. Retrieved from http://www.scu.edu/ ethics/practicing/focusareas/environmental_ethics/lesson4. html.

Weakland, J.P. (2009), The Earth Charter in higher education for sustainability. *Journal of Education for Sustainable Development* 3 (2), pp. 151-158.

Wiek, A., Withycome, H., Redman, C. and Mills, S.B. (2011), Moving Forward: On competence in sustainability research and problem solving. *Environment* 53 (2), pp. 3-12.

Zuniga, X., Nagda, B.A., Chesler, M. and Cytron-Walker, A. (Eds.) (2007), Intergroup dialogue in higher education: Meaningful learning about social justice. *ASHE Higher Education Report* 32 (4), pp. 1-128.

Chapter 43

Developing a First-Time Sustainability Report for a Higher Education Institution

Natalia Moudrak and Amelia Clarke

Abstract

While municipal and corporate sectors benefit from step-by-step guidance on the process of developing a sustainability report from scratch, such guidance is lacking for higher education institutions. In this chapter, select municipal and corporate sustainability reporting processes and existing sustainable campus literature are examined alongside empirical evidence from the University of Waterloo (Waterloo, Canada) to arrive at the seven-step process for developing a first-time sustainable development report for a higher education institution. The steps are: 1) review Higher Education (HE) sector sustainability reports; 2) review HE sector sustainability assessment tools; 3) develop guiding principles; 4) engage the stakeholders; 5) collect sustainability performance information; 6) verify report accuracy; and 7) publish the report and welcome feedback. This chapter is particularly relevant for higher education institutions that are considering preparing a sustainability report for the first time.

Introduction: sustainable development and higher education institutions

Higher education (HE) institutions play a significant role in advancing the sustainable development movement. Often compared to small cities or corporations due to their size, HE institutions have a responsibility to integrate sustainability into their operations in order to reduce their environmental footprint (Stafford 2011). Moreover, as HE institutions are considered to be the incubators of tomorrow's leaders and decision-makers, they are urged to deliver learning and research opportunities to advance knowledge in the area of sustainable development (Velazquez et al. 2006). In recognition of their unique role, some higher education institutions have signed sustainability-related declarations that commit them to various actions to drive the movement forward (Wright 2002). Similarly, some of these institutions have undertaken sustainability performance assessments, created sustainability coordinator positions, and assigned sustainability committees to develop their sustainability plans and manage their sustainability initiatives (Bardati 2006; Herremans and Allwright 2000). However, the uptake of sustainability reporting within the higher education sector has been slow (Fonseca et al. 2010; Lozano 2011).

Sustainability reports are documents that present relevant performance information. Whereas sustainability reporting has experienced significant growth in the past decade within corporations, it is has not become a common practice among the majority of higher education institutions (Fonseca et al. 2010; Global Reporting Initiative 2011). This is unfortunate, as a number of benefits from sustainability reporting within the higher education sector exist, including better stakeholder communication, and improved environmental management (Bardati 2006; Walton et al. 1997).

One of the barriers to sustainability reporting among higher education institutions may be the lack of sector-specific, step-by-step guidance for developing sustainability reports in the higher education sector (Lozano 2011). In contrast, academic and practitioner literatures provide such guidance for the corporate and municipal sectors (e.g. Commonwealth of Australia, 2000; Maclaren, 1996; Global Reporting Initiative 2011; Mitchell et al. 2008). Campus sustainability literature covers the role of sustainability coordinators (e.g. Herremans and Allwright 2000), conducting assessments (e.g. Beringer 2006), and involving students (e.g. Helferty and Clarke 2009), which is all part of the process of preparing a sustainability report, but the extant literature does not detail the complete picture. Drawing from both the practitioner and academic literature, as well as from the empirical evidence of developing the first sustainable development report at the University of Waterloo (Waterloo, Ontario, Canada), this chapter provides a seven-step process for developing a first-time sustainability report for a higher education institution. The specific research question is:

RQ: What is the process of developing a first-time sustainability report for a higher education institution?

The chapter begins with a literature review of existing campus sustainability literature related to the process of developing a report, followed by an introduction to the corporate and municipal literature on the topic. Next the methodology section introduces the University of Waterloo case study. The combined results and discussion section detail the seven steps in relation to both the literature and the case study findings. The chapter ends with conclusion and future research section.

Sustainable development reporting in the Higher Education sector

HE institutions that are considered leaders in sustainable development usually hire full-time personnel to co-ordinate sustainability efforts and report on the progress to senior governing bodies (Herremans and Allwright 2000). These personnel (sustainability coordinators) typically conduct the sustainability assessment (Beringer 2006; Viebahn 2002), which involves students during the data collection stages (Bardati 2006; Ferreira et al. 2006; Helferty and Clarke 2009). The ISO 14001 standard and the GRI Sustainability Reporting Guidelines can be used to guide the assessment process (Clarke and Kouri 2009; Lozano 2011). However, one of the challenges with using these tools is the lack of sector-specific guidance for the higher education sector (Clarke and Kouri 2009; Lozano 2011; Taddei-Bringas et al. 2008). Thus, campus sustainability coordinators often draw from a variety of sustainability assessment tools designed specifically for higher education institutions, including in North America, the Campus Sustainability Assessment Framework (CSAF), and the Sustainability Tracking, Assessment, and Rating System (STARS) (Cole 2003; Shriberg 2002; AASHE 2012). These campus sustainability assessment frameworks can be used to identify most pertinent data needed in order to communicate sustainability performance findings to the institutional decision-makers (Bardati 2006; Beringer 2006). To this end, the establishment of sustainability principles, a selection of few yet effective indicators, and clear communication of sustainability assessment results through an organized reporting structure are beneficial (Troschinetz et al. 2007). However, aside from the sustainability assessment frameworks, practical guidance for the development of sustainability reports within the higher education sector is limited.

While Bardati's three-stage process of environmental assessment outlines possible steps prior to the report's creation, the focus of her discussion remains on the environmental audit course structure (Bardati 2006). Viebahn's environmental management model for universities provides broader insight into the set-up of environmental management systems, with only a brief discussion on reporting (Viebahn 2002). A recent student paper entitled *Developing an Annual Sustainability Report for WPI* shares the authors' experience in writing the first-ever sustainability report for Worcester Polytechnic Institute (Alden et al. 2010). One may deduct that the process the authors employed consisted of the following: a review of GRI and STARS sustainability reporting frameworks; an examination of nine higher education sector sustainability reports' contents; interviews with staff and faculty members; and online publication of the report (Alden et al. 2010). Good insight can be obtained from this report, yet there is still no explicit step-by-step guidance for a sustainability report development process proposed.

Sustainable development reporting in the corporate and municipal sectors

In the corporate sector, investor pressures to disclose non-financial performance is becoming a significant driver for sustainability reporting in the corporate sector. Material risks to investors from inadequate environmental performance disclosure were identified as an issue and disclosure is now mandatory in some jurisdictions (Ioannou and Serafeim 2011). To track sustainability performance, the corporate sector relies mostly on the International Standards Organization (ISO) environmental management standards and on the Global Reporting Initiative (GRI) Sustainability Reporting Guidelines. The ISO 14001 standard is implemented by more than 200,000 organizations in 155 countries (International Standards Organization 2011). The GRI Sustainability Reporting Guidelines now constitute the most used sustainability reporting framework in the world (Global Reporting Initiative 2011). Stakeholder engagement is the forefront focus of the corporate sector sustainability assessment and reporting process (Adams and Frost 2008). The following process exemplifies the steps involved in corporate sustainable development reporting:

1. Determine which stakeholders should be involved.
2. Empower stakeholders to select sustainability performance indicators and to suggest a course of action to improve organization's sustainability performance.
3. Ensure commitment from organization's leaders from the get-go.
4. Determine the extent of the sustainability performance evaluation.
5. Include social and equity issues in the analysis.
6. Focus on local sustainability issues. Select indicators that lead to practical actions, using generic sustainability indicator sets, such as the GRI Sustainability Guidelines, only as a guide.
7. Emphasize a problem-based focus of sustainability reporting, with the stakeholders making suggestions on appropriate actions to enhance sustainability.
8. Create links between the sustainability reporting process and the organization's everyday activities.
9. Collaborate with other organizations in achieving sustainability goals.
   (Adapted from Mitchell et al. 2008: 73)

In the public sector, influenced by their publically visible nature and by the increasingly popular corporate models for sustainability performance measurement and accountability, local governments have begun to monitor and evaluate their policies based on economic, social and environmental trends (Seasons 2003). Popular sources for guidance on municipal sustainability reporting include the Global Reporting Initiative Public Agency Sector Supplement and International Council for Local Environmental Initiatives (ICLEI) publications (CPA Australia 2007). However, the uptake of the GRI Sustainability Reporting Guidelines has also been slow in the municipal sector due to the lack of direction on methods of reporting community-based data (Leeson et al. 2006). Furthermore, due to the differences in local planning context and a multitude of interpretations of the term 'sustainable development', there is no consensus on the optimal sustainability assessment approach or measurement tools (Tanguay et al. 2009). Nevertheless, taking Canada as an example, a broad range of community-based 'state of the environment', 'healthy city', 'quality of life', and 'sustainability' reports have been undertaken (Pembina Institute 2002). Maclaren provides step-by-step guidance for the development of an urban sustainability report:

1. Defining the urban sustainability goals for which indicators are needed
2. Scoping
3. Choosing an appropriate indicator framework
4. Defining indicator selection criteria
5. Identifying a set of potential indicators
6. Evaluating the indicators and selecting a final set
7. Collecting data and analyzing the indicator results
8. Preparing and presenting the urban sustainability report
9. Assessing indicator performance
   (Adapted from Maclaren 1996: 198-203)

Methodology: Sustainable development reporting – University of Waterloo experience

This research uses a case study approach. The University of Waterloo served as an information-rich case to investigate the intricacies of developing the first ever sustainability report in a higher education institution without an official sustainability plan, policy, goals or a sustainability coordinator.

Since its creation in 1957, the University of Waterloo has become one of Canada's leading comprehensive universities. It is home to 30,000 students and the largest post-secondary co-operative education program in the world. The University of Waterloo was named Canada's most innovative university in the Maclean's annual university rankings for 19 years in a row since 1991; and the Faculty of Environment is the oldest faculty of its kind nationwide (University of Waterloo 2011). In terms of sustainability education, the environment and business undergraduate program has been ranked number one in Canada by Corporate Knights magazine (Waterloo Environment 2010).

In 2009, the University of Waterloo signed the Council of Ontario Universities (COU) sustainability pledge, *Ontario Universities: Committed to a Greener World* (University of Waterloo Daily Bulletin 2009). The pledge committed the university "to assist in finding solutions to the challenges of environmental sustainability; to share knowledge about sustainability and climate change; and to incorporate, wherever possible, principles of sustainability into our own operations" (COU 2009: 1). The University of Waterloo's Faculty of Environment Dean's Advisory Council thus initiated the creation of the *University of Waterloo Sustainable Development Report 2010* to aid the university in its fulfillment of the pledge. The University hired Natalia Moudrak (a Master's student) to do the task as part of her thesis. For the final report, see: http://uwaterloo.ca/accountability/documents/july152011finaluniversityofwaterloos dr2010.pdf.

Results and discussion: the seven step process

The process of developing the *University of Waterloo Sustainable Development Report 2010* consisted of the seven steps outlined below:

1. Review higher education sector sustainability reports.
2. Review higher education sector sustainability assessment tools.
3. Develop a draft 'sustainable development' definition and associated guiding principles.
4. Engage the stakeholders.

5. Collect sustainability performance information.
6. Verify report accuracy.
7. Publish the report and welcome feedback.

These steps are deemed essential for any higher education institution undertaking sustainable development reporting for the first time.

Step 1: Review higher education sector sustainability reports

Stakeholder identification and engagement are usually the first steps in municipal and corporate sustainability report development (e.g. Mitchell et al. 2008). Alternatively, the seven-step process for first-time sustainability report development in a HE institution, suggests to first review higher education sector sustainability reports and sustainability assessment tools; engaging stakeholders does not occur until step 4. This difference is due to the first-time nature of the report being developed. In particular, the assumption is that the person in charge of sustainable development report creation needs to familiarize themselves with higher education sector sustainability reports and sustainability assessment tools in order to adequately lead the report development process. Indeed, Step 1 and Step 2 could be omitted if the person in charge of sustainability report development is already familiar with the optimal report structure and the commonly-suggested performance indicators. Nevertheless, one may regard undertaking Step 1 and Step 2 as good due diligence practice. For example, the authors of the Worcester Polytechnic Institute sustainability report have also reviewed higher education sector sustainability reports and sustainability assessment tools as their first steps (Alden et al. 2010). The authors found that the reports were generally 20 to 50 pages long, profiled approximately 20 sustainability performance indicators, and made extensive use of pictures and graphs. An introduction, letter from a sustainability representative or president and recommendations were other key components (Alden et al. 2010).

At the University of Waterloo, in order to understand what constitutes a superior sustainability report structure, a random sample of 17 sustainability reports from colleges and universities as well as a large number of corporate and municipal reports were reviewed. These included seven sustainability reports by Canada's largest universities (Fonseca et al. 2010). The following components were identified as integral to a well-designed sustainability report:

1. Title page
2. Message from the president
3. Table of contents
4. Introduction to the report, including its purpose, reporting period and scope
5. List of guiding principles and key performance areas being reported on
6. Executive summary of findings
7. Sustainability performance analysis
8. Conclusion and recommendations
9. Appendix
10. Contact information of a person/office to address report inquiries.

Step 2: Review of CSAF, STARS and GRI sustainability assessment tools

There is a diversity of sustainable development assessment tools available, with varying emphasis placed on the environmental, social and economic aspects of sustainability performance measurement and on the level of reporting detail (Shriberg 2002). Of these tools, CSAF, STARS and GRI have gained popularity (Alden et al. 2010; Fonseca et al. 2011; Lozano 2011). Selecting sustainability performance indicators from existing professionally-designed assessment tools ensures scientific validation of the process (Cloquell-Ballester et al. 2006).

At the University of Waterloo, to determine the most common sustainability performance measurements for the higher education sector, lists of core indicators from CSAF, STARS and the GRI 3.0 Sustainability Reporting Guidelines were analyzed. Since there is no GRI sector supplement for higher education institutions, Lozano's modification to account for the "academe" dimension of sustainability performance was used (Lozano 2011). Table 1 details this analysis.

Table 1: Comparison of CSAF, STARS and GRI core indicators

Environment

CSAF - Core		STARS - Core		GRI - Core	
Energy					
E-1	Renewable Energy: Buildings	OP Credit 7	Building Energy Consumption	EN3	Direct energy consumption by primary energy source.
E-8	Reduction in Energy Consumption	OP Credit 8	Renewable Energy	EN4	Indirect energy consumption by primary source.
Water					
W-1	Potable Water Consumed	OP Credit 22	Water Consumption	EN8	Total water withdrawal by source.
W-7	Efficiency of Fixtures	OP Credit 23	Stormwater Management	EN21	Total water discharge by quality and destination.
W-9	Wastewater Produced				
Land/Grounds/Biodiversity					
L-1	Managed Greenspace	OP Credit 9	Integrated Pest Management	EN11	Location and size of land owned, leased, managed in, or adjacent to, protected areas and areas of high biodiversity value outside protected areas.
L-3	Pesticides	OP Credit 1	Building Operations and Maintenance	EN12	Description of significant impacts of activities, products, and services on biodiversity in protected areas and areas of high biodiversity value outside protected areas.
M-1	LEED Certified Base Buildings	OP Credit 2	Building Design and Construction (eg. LEED)		
Waste/Materials					
M-3	Paper Consumption	OP Credit 6	Food Purchasing (eg. composting)	EN1	Materials used by weight or volume.
M-4	Recycled Content of Paper	OP Credit 10	Computer Purchasing	EN2	Percentage of materials used that are recycled input materials.
M-7	Local Food Production	OP Credit 12	Office Paper Purchasing	EN22	Total weight of waste by type and disposal method.
M-9	Solid Waste and Recyclables Produced	OP Credit 13	Vendor Code of Conduct	EN23	Total number and volume of significant spills.
M-11	Recyclables Being Landfilled	OP Credit 17	Waste Reduction	EN26	Initiatives to mitigate environmental impacts of products and services, and extent of impact mitigation.
		OP Credit 18	Waste Diversion	EN27	Percentage of products sold and their packaging materials that are reclaimed by category.
		OP Credit 19	Construction and Demolition Waste Diversion		
		OP Credit 20	Electronic Waste Recycling Program		
		OP Credit 21	Hazardous Waste Management		
Air Quality					
A-7	Chemical Free Cleaning	OP Credit 11	Cleaning Product Purchasing	EN16	Total direct and indirect greenhouse gas emissions by weight.
E-4	Greenhouse Gas Emissions: Buildings	OP Credit 3	Indoor Air Quality	EN17	Other relevant indirect greenhouse gas emissions by weight.
E-5	Greenhouse Gas Emissions: Commuting Transport	OP Credit 4	Greenhouse Gas Emissions Inventory	EN19	Emissions of ozone-depleting substances by weight.
C-25	Affordability of Public Transit	OP Credit 5	Greenhouse Gas Emissions Reduction	EN20	NOx, SOx, and other significant air emissions by type and weight.
		OP Credit 14	Campus Fleet		
		OP Credit 15	Student Commute Modal Split		
		OP Credit 16	Employee Commute Modal Split		

Society

CSAF - Core		STARS - Core		GRI - Core	
Health					
HW-1	Recreation Space			LA7	Rates of injury, occupational diseases, lost days, and absenteeism, and number of work-related fatalities by region.
HW-3	Diet Types			LA8	Education, training, counseling, prevention, and risk-control programs in place to assist workforce members, their families, or community members regarding serious diseases.
HW-5	Organic, Non-GMO, Fair Trade Food				
HW-9	Physical Health Care Practitioners				
HW-12	Mental Health Care Practitioners				
HW-17	Accessible Greenspace				
Equity and Diversity					
C-7	Faculty With Disabilities	PAE Credit 6	Diversity and Equity Coordination	HR4	Total number of incidents of discrimination and actions taken.
C-8	Staff With Disabilities	PAE Credit 7	Measuring Campus Diversity Culture	LA1	Total workforce by employment type, employment contract, and region.
C-9	Students With Disabilities	PAE Credit 8	Support Programs for Under-Represented Groups	LA2	Total number and rate of employee turnover by age group, gender, and region.
C-10	Faculty of Ethnic Minorities	PAE Credit 9	Support Programs for Future Faculty	LA4	Percentage of employees covered by collective bargaining agreements.
C-11	Staff of Ethnic Minorities	PAE Credit 10	Affordability and Access Programs	LA5	Minimum notice period(s) regarding significant operational changes, including whether it is specified in collective agreements.
C-12	Student of Ethnic Minorities			LA13	Composition of governance bodies and breakdown of employees per category according to gender, age group, minority group membership, and other indicators of diversity.
C-13	Faculty Gender			LA14	Ratio of basic salary of men to women by employee category.
C-14	Staff Gender				
C-15	Student Gender				
C-16	Equity of Indigenous Peoples: Faculty				
C-17	Equity of Indigenous Peoples: Staff				
C-18	Equity of Indigenous Peoples: Students				
Employee Training					
K-1	New Faculty Orientation	PAE Credit 13	Staff Professional Development in Sustainability	LA10	Average hours of training per year per employee by employee category.
K-4	Faculty Sustainability Training	PAE Credit 14	Sustainability in New Employee Orientation		
		PAE Credit 15	Employee Sustainability Educators Program		
		ER Credit 1	Student Sustainability Educators Program		
		ER Credit 2	Student Sustainability Outreach Campaign		
		ER Credit 3	Sustainability in New Student Orientation		
		ER Credit 4	Sustainability Materials and Publications		
Community Outreach					
		PAE Credit 19	Community Sustainability Partnerships	SO1	Nature, scope, and effectiveness of any programs and practices that assess and manage the impacts of operations on communities, including entering, operating, and exiting.
		PAE Credit 20	Inter-Campus Collaboration on Sustainability		
		PAE Credit 21	Sustainability in Continuing Education		
		PAE Credit 22	Community Service Participation		
		PAE Credit 23	Community Service Hours		

Economy

CSAF - Core		STARS - Core		GRI - Core	
EW-2	Student Debt Load	PAE Credit 16	Committee Socially Responsible Investment	EC1	Economic value generated and distributed, including revenues, operating costs, employee compensation, donations and other community investments, retained earnings, and payments to capital providers and governments.
EW-7	Wage Gap	PAE Credit 17	Shareholder Advocacy	EC2	Financial implications and other risks and opportunities for the organization's activities due to climate change.
EW-17	Ethically and Environmentally Sound Investments	PAE Credit 18	Positive Sustainability Investments	EC3	Coverage of the organization's defined benefit plan obligations.
EW-15	Locally Purchased Goods and Services	PAE Credit 11	Sustainable Compensation	EC4	Significant financial assistance received from government.
				EC6	Policy, practices, and proportion of spending on locally-based suppliers at significant locations of operation.
				EC7	Procedures for local hiring and proportion of senior management hired from the local community at significant locations of operation.
				EC8	Development and impact of infrastructure investments and services provided primarily for public benefit through commercial, in-kind, or pro bono engagement.

Academe

CSAF - Core		STARS - Core		GRI - Lozano	
Sustainability Courses					
K-17	Courses With Applied Learning	ER Credit 5	Sustainability Course Identification	CU1	Number and percentage (in respect to the total) of courses related to sustainability concepts
		ER Credit 6	Sustainability-Focused Courses	CU2	Number of students enrolled in sustainability-related courses
		ER Credit 7	Sustainability-Related Courses	CU3	Number of courses with some content on SD themes
		ER Credit 8	Sustainability Courses by Department	CU6	List with course titles and SD theme contained
		ER Credit 9	Sustainability Learning Outcomes	CU4	Specific course to 'Educate the Educators' in SD
		ER Credit 10	Undergraduate Program in Sustainability	CU7	Course structure, goals and duration
		ER Credit 11	Graduate Program in Sustainability	CU5	Management procedures to monitor incorporation of SD themes into Curricula
		ER Credit 12	Sustainability Immersive Experience	CU11	Number and percent of departments and colleges including sustainability courses and curricula
		ER Credit 13	Sustainability Literacy Assessment		
		ER Credit 14	Incentives for Developing Sustainability Courses		
Sustainability Research					
K-11	Research Collaboration - For Profit	ER Credit 15	Sustainability Research Identification	RE1	Research in the area of sustainability
		ER Credit 16	Faculty Involved in Sustainability Research	RE6	List issues addressed: Renewable energies, ecological economics, urban planning, etc
		ER Credit 17	Departments Involved in Sustainability Research	RE2	Percentage of graduate students doing research in sustainability
		ER Credit 18	Sustainability Research Incentives	RE7	List of knowledge field involved.
		ER Credit 19	Interdisciplinary Research in Tenure and Promotion	RE3	Percentage of faculty doing research in sustainability issues
				RE8	List of faculty members and Departments or Centres to which they belong
				RE4	Institutional support and management procedures for multidisciplinary and interdisciplinary research in sustainability
				RE5	Number of research projects that are multidisciplinary and interdisciplinary in the area of sustainability
				RE11	Total revenues from grants and contracts specifying sustainability-related research
				RE12	Published research with focus on sustainability-related issues
				RE13	Number and function of centres on campus providing sustainability-related research or services

Table 2 below identifies key sustainability indicators for the four main dimensions of sustainability performance at an institution of higher education – environment, society, economy and academe. In deriving this table, recurring themes from the CSAF, STARS and GRI were identified and where such themes were common to at least two of the three assessment tools, the item was added to the table. There were two exceptions, the 'Employee Injury and Severity Rates' and 'Economic Value Generated' indicators that were only listed under the GRI 3.0 Sustainability Reporting Guidelines but are included in Table 1. These two indicators were deemed as two important omissions from the other two sustainability assessment tools because of the central role that they occupy within corporate and municipal sustainability reports. The table's last column illustrates which indicators were addressed in the *University of Waterloo Sustainable Development Report 2010*.

Table 2: Key sustainability indicators from GRI, CSAF and STARS

	CSAF	STARS	GRI	Addressed?
1. Environment				
Energy Consumption	■	■	■	Yes
Water Consumption	■	■	■	Yes
Wastewater Discharge	■		■	Yes
Biodiversity/Habitat/Green Spaces	■		■	Yes
Pesticide Use	■	■		Yes
LEED Certified Buildings	■	■		Yes
Waste and Recyclables Produced	■	■	■	Yes
Office Paper Consumption/Other Purchasing (such as Food)	■	■		Yes
Greenhouse Gas Emissions	■	■	■	Yes
2. Society				
Employee Injury and Severity Rates			■	Yes
Gender, Age, Persons with Disabilities and Indigenous Groups	■	■	■	Yes
Student, Staff and Faculty Sustainability Training	■	■		N/A
Community Outreach Programs		■	■	Yes
3. Economy				
Economic Value Generated			■	Yes
Socially Responsible Investment	■	■		Yes
Purchasing Considerations	■	■	■	Yes
4. Academe				
Sustainability-Related Courses	■	■	■	No
Sustainability-Focused Research	■	■	■	Yes

In addition to these key sustainability indicators, the *University of Waterloo Sustainable Development Report 2010* assessed student health issues by tracking the number of student visits to the on-campus clinic, included information on fundraising efforts, research grants, pension and endowment fund management, and profiled student engagement performance. While the table above indicates the importance of tracking sustainability-related courses, at the time the *University of Waterloo Sustainable Development Report 2010* was prepared, there were no resources to create such a database. It is also important to note that all three sustainability assessment tools analyzed stress the importance of reporting on sustainability commitments, including an institution's sustainability policies and plans. These items were not included in the table above because they do not directly relate to sustainability indicator selection, but to recognizing sustainability within the organizational structure of an institution.

Step 3: Develop a draft 'sustainable development' definition and associated guiding principles

The way sustainability is defined and which guiding principles are selected determines what parameters of sustainable development are being measured and reported on. Differences in priorities and values at all levels of decision-making explain why most sustainability tools favour one of the standard dimensions of sustainability more strongly – economic, social, or environmental (Wilson et al. 2007). Prior to any reporting effort, stakeholder consensus must be reached on the 'sustainability' definition reflective of their core values and priorities (Cloquell-Ballester et al. 2006). There are merits of providing a draft document with possible sustainability goals and indicators to the stakeholders to drive the discussion (Valentin and Spangenberg 2000).

As the University of Waterloo does not have an overarching sustainability policy, it was important to develop a definition of 'sustainable development' reflective of the University of Waterloo stakeholder values and parameters of sustainability most pertinent to its operations. The creation of guiding principles was driven primarily by the findings from the GRI, CSAF and STARS indicator comparison. The wording of the guiding principles suggested which key sustainable development performance areas would be addressed in the report, thus aiding the selection process for sustainability indicators. On campuses with an existing policy, this step would entail using this document to help with indicator selection instead.

Step 4: Engage the stakeholders

Stakeholder consensus-building on the definition and principles of sustainable development helps to identify an agreed vision of sustainability. To this end, the merits of bringing stakeholders and subject-matter experts together in a workshop discussion were outlined by Donnelly et al. (2007):

a) better way to communicate and explain detailed ideas and opinions,
b) good forum for interactive discussion and allows trains of thought to be continuous rather than stopping and starting as with other forms of communication such as the internet,
c) having all stakeholders in the one place allowing people to openly discuss issues with individuals over coffee or at breaks.

Donnelly et al. 2007: 167

The inclusion of subject-matter experts in particular (Reed et al. 2006) and high-level decision-makers at this point is important for the report recommendations to be actionable later on (Valentin and Spangenberg 2000).

At the University of Waterloo, the aim of this step was to draw on stakeholder values and on the sustainability subject matter experts' practical knowledge of the field when finalizing the sustainable development definition and associated guiding principles. To this end, the Sustainable Development Advisory Committee was formed and a workshop-format was chosen to allow for dynamic discussion and a real-time consensus building on the University of Waterloo's sustainable development definition and guiding principles. At the end of the workshop, the participants further agreed to limit the number of guiding principles to four and recommended to focus data collection efforts on key sustainability performance indicators.

The size of the Sustainable Development Advisory Committee had to remain small to ensure quality discussion. Thus, a multi-disciplinary team of 11 people comprised of subject matter experts from private, public and NGO sectors, university professors, undergraduate and graduate student representatives were called for the consultation. While the merits of including the University of Waterloo senior administration in the workshop discussion were understood – to ensure the linkage of performance indicators to concrete organizational objectives (Donnelly et al. 2007; Wilson et al. 2007) – it was deemed inappropriate to do so in this case. The workshop was held whilst the project was still in its infancy, during which there was no formal support for sustainability reporting initiative on behalf of the university. Therefore, it was important to first ensure the feasibility of fulfilling the task, prior to approaching the senior administration. To this end, preliminary information to be collected was determined based on the workshop discussion and then support for the report from the university staff responsible for tracking this information was established. The senior administration was approached after the first report draft was prepared.

While the municipal and corporate processes (e.g. Maclaren 1996; Mitchell et al. 2008) do not explicitly call for a draft document to be distributed to stakeholders prior to any discussions, the rationale for doing so was advocated for in Valentin and Spangenberg's (2000) work on community sustainability indicators. A more productive discussion among the stakeholders can be achieved if they are provided with a common starting point for discussion. This step and the subsequent steps suggested in the municipal and corporate sectors are identical to the proposed seven-step process: engaging stakeholders in the selection/development of sustainability goals and indicators, collecting data, analyzing results, and publishing the report (Maclaren 1996; Mitchell et al. 2008).

Step 5: collect sustainability performance information

Employing purposeful sampling and conducting face-to-face interviews during the data collection stage allows for attainment of in-depth information (Alden et al. 2010; Patton 2002). Snowballing, or asking for references to obtain further information, is a useful technique (Alden et al. 2010; Patton 2002). During this stage it is important to gather both positive and negative stories to ensure transparent reporting and to maintain concise information, avoiding unreadable reports (Alden et al. 2010). In fact, condensation of data must take place from detailed and scientifically-oriented information to a short and user-friendly format to entice public interest in the report (Shields et al. 2002).

At the University of Waterloo, where possible, sustainability performance information for a five-year period was collected to determine trends. Purposeful sampling was employed to conduct in-depth face-to-face interviews with the university staff. Moreover, a snowball sampling strategy was used during the interview process to identify additional interviewees. In addition, publically available documents published by the university were used. In terms of interviews, the general interview guide approach was used; this involves outlining a set of issues to be explored and using that outline as a checklist during the interview (Patton 2002). The set of questions for the interview was shared with the interviewee prior to the interview to allow the interviewee time to prepare.

Step 6: verify report accuracy

Using member-checking, peer debriefing, and an external auditor can help ensure data accuracy (Patton 2002). Every section of the *University of Waterloo Sustainable Development Report 2010* was verified for content accuracy by the staff and faculty members who provided relevant information. Additionally, the entire report was reviewed by the workshop participants to provide any high-level recommendations. The report was then edited by a technical writer to ensure proper grammar and punctuation and by the *University of Waterloo Communications and Public Affairs* personnel.

Step 7: publish the report and welcome feedback

Since minimizing the environmental footprint is one of the common sustainability goals (Stafford 2011), it is best to publish sustainability reports online and have only a limited number of copies available in print (Alden et al. 2010). Once the report is available for viewing, it is important to establish a feedback mechanism to accommodate comments, queries and recommendations. This will improve reporting efforts and help meet stakeholder needs more adequately in the future.

The *University of Waterloo Sustainable Development Report 2010* was written to create awareness about sustainability performance on the main campus among the university's key stakeholders; these include current and prospective students, staff, faculty, administration, alumni and community members. The report was made available online at the *Public Accountability* http://uwaterloo.ca/accountability and *Sustainability* http://www.sustainability.uwaterloo.ca web pages on August 3, 2011. A news article announcing the release of the report was published on August 8, 2011: http://www.bulletin.uwaterloo.ca/ 2011/aug/08mo.html. To minimize its potential environmental footprint, no copies of the report were printed. Contact details were provided at the end of the report for people who wished to give feedback on its content.

Conclusion and future research

Sustainability reporting in the higher education sector is perhaps limited as a result of the lack of step-by-step sector-specific guidance. By drawing on the literature aimed at the corporate and municipal sectors, and by entwining the campus sustainability literature about assessments, this chapter addresses this gap by presenting the seven-step process for developing a first-time sustainable development report for a higher education institution. The new 2012 initiative by the United Nations asks that universities report regularly as part of the Commitment to Sustainable Practices of Higher Education Institutions for Rio+20 (UNCSD 2012), so could use this chapter's insights to help guide the universities who are new to reporting.

Aside from the specific process differences between the proposed seven-step process and the reviewed literature from the municipal and corporate sectors, there are other considerations that arise from campus sustainability literature. The latter often suggests employing a group of students to carry out data collection (Bardati 2006; Helferty and Clarke 2009). Emphasis on having a sustainability coordinator to guide the effort is also noted (Alden et al. 2010; Herremans and Allwright 2000). While there are educational merits of engaging students in the sustainability report creation, there might be no system in place, or willingness to organize such effort. There also might be administrative pushback to allocate resources to establish a sustainability coordinator position at the institution. This study demonstrates that sustainability reporting task can be carried out by one graduate student. However, it is important to note that the student must have sufficient guidance, which in this case was achieved by the researcher having two Master's supervisors and Sustainable Development Advisory Committee. Equally important is to establish support for sustainability reporting among key information gate-keeper staff in charge of the sustainability performance data. The university's previous commitment to the *Ontario Universities: Committed to a Greener World* pledge helped facilitate this process.

The *University of Waterloo Sustainable Development Report 2010* provides an example of possible report content and format when following the seven-step process. The four important practical lessons learned from following this process are:

1. It is possible for one person to carry out sustainability reporting on a scale of a large institution in under twelve months.
2. The key determining factor to the successful completion of the *University of Waterloo Sustainable Development Report 2010* was the early support of the staff who managed to acquire pertinent information. This is because the majority of information needed for the report was not publically available.
3. Due to the unofficial nature of the initiative, the University of Waterloo senior administrators were only approached once the first draft of the report was prepared. While this approach worked well at the University of Waterloo, due to the close-knit relationships between senior administrators and the rest of the staff which ensured that the information collected was relevant to the decision-makers, this approach might not work for other institutions.
4. The most significant driver of support for the report was the short and targeted list of most pertinent indicators for which sustainability performance information was collected. This finding arose from the discussions with the University of Waterloo staff about reasons that prevented sustainability reporting at the university in the past and reasons why this time the report was welcomed. Some sustainability assessment frameworks, with their hundreds of indicators, are hard to justify.

In terms of future research, drawing from the University of Waterloo experience, a set of about 30 indicators would be the optimal strategy for a continual reporting effort on behalf of the institution. The new effort on behalf of the STARS, the Princeton Review, Sierra magazine and Sustainable Endowments Institute to reduce survey fatigue and collect one set of information has helped standardize indicators in North America (AASHE 2012). However, more research is needed to find a short list of indicators that are applicable on a global scale.

Another area where further research would be valuable would be to gauge whether mandatory sustainability reporting across HE institutions is a good idea and what topics would be included. The move to require disclosure of sustainability details for investors in the corporate sector could be replicated by governments requiring HE institutions to complete sustainability assessments and reporting. In Canada, some aspects of sustainability reporting (in its social dimension) are already mandatory nationwide, as is the case with employee fatalities, and injury frequency and severity rates. The province of British Columbia also requires public institutions to report on greenhouse gas emissions (Glor-Bell and Clarke 2011). Other critical aspects of sustainability performance could be of value for mandatory reporting.

References

AASHE (2012), *Reducing survey fatigue: A collaboration with AASHE, Sustainable Endowments Institute, the Princeton Review, and Sierra Magazine*. Retrieved March 5, 2012, from: https://stars.aashe.org/pages/news-events/stars-webinars/2012-stars-webinars.html.

Adams, C.A. and Frost, G.R. (2008), Integrating sustainability reporting into management practices. *Accounting Forum* 32, pp. 288-302.

Alden, N., Gomez, J. and Shang, S. (2010), *Developing an annual sustainability report for WPI*. An interactive qualifying project for the Worcester Polytechnic Institute, Massachusetts.

Bardati, D.R. (2006), The integrative role of the campus environmental audit: Experiences at Bishop's University, Canada. *International Journal of Sustainability in Higher Education* 7 (1), pp. 57-68.

Beringer, A. (2006), Campus sustainability audit research in Atlantic Canada: Pioneering the campus sustainability assessment framework. *International Journal of Sustainability in Higher Education* 7 (4), pp. 437-455.

Clarke, A. and Kouri, R. (2009), Choosing an appropriate university or college environmental management system. *Journal of Cleaner Production* 17 (11), pp. 971-984.

Cloquell-Ballester, V.-A., Monterde-Diaz, R. and Santamarina-Siurana M.-C. (2006), Indicators validation for the improvement of environmental and social impact quantitative assessment. *Environmental Impact Assessment Review* 26 (1), pp. 79-105.

Cole, L. (2003), *Assessing sustainability of Canadian university campuses: Development of a campus austainability assessment framework.* Unpublished Master's Thesis. Royal Roads University, Victoria, Canada.

Commonwealth of Australia (2000), *A Framework for public environmental reporting: An Australian approach*, Environment Australia. Canberra. Retrieved June 5, 2011 from: http://www.envirore porting.com/others/Australian%20PER%20Guidelines.pdf.

COU (2009), *Ontario Universities: Committed to a Greener World.* Retrieved November 20, 2010, from: http://www.cou.on.ca/issues-resources/key-issues/more/pdfs/commit ted-to-a-greener-world---a-pledge-from-execu.aspx.

CPA Australia (2007), *Sustainability reporting in local government: Systemic change or greenwash?* CPA Australia Ltd. Melbourne. Retrieved June 5, 2011 from: http://www.cpaaustralia.com.au/ cps/rde/xbcr/cpa-site/sustainability-reporting-in-local-government. pdf.

Donnelly, A., Jones, M., O'Mahony, T. and Byrne, G. (2007), Selecting environmental indicator for use in strategic environmental assessment. *Environmental Impact Assessment* 27 (2), pp. 161-175.

Ferreira, A.J.D., Lopes, M.A.R. and Morais, J.P.F. (2006), Environmental management and audit schemes implementation as an educational tool for sustainability. *Journal of Cleaner Production* 14 (9-11), pp. 973-982.

Fonseca, A., Macdonald, A., Dandy, E. and Valenti, P. (2010), The state of sustainability reporting at Canadian universities. *International Journal of Sustainability in Higher Education* 12 (1), pp. 22-40.

Global Reporting Initiative (2011), *Reporting framework overview.* Retrieved April 29, 2011, from: http://www.globalreporting.org/ReportingFramework/ReportingFrame workOverview.

Glor-Bell, J. and Clarke, A. (2011), Community engagement in university of British Columbia's climate action plan. In Leal Filho (Ed.), *World Trends on Education for Sustainable Development,* pp. 39-60. Peter Lang Scientific Publishers, Frankfurt.

Helferty, A. and Clarke, A. (2009), Student-led campus climate change initiatives in Canada. *International Journal of Sustainability in Higher Education* 10 (3), pp. 287-300.

Herremans, I. and Allwright, D.E. (2000), Environmental management systems at North American universities: What drives good performance? *International Journal of Sustainability in Higher Education* 1 (2), pp. 168-181.

International Standards Organization (2011), *International Standards Organization.* Retrieved April 30, 2011, from About Us: http://www.iso.org/iso/about.htm.

Ioannou, I. and Serafeim, G. (2011), *The consequences of mandatory corporate sustainability reporting*. Harvard Business School. Retrieved June 3, 2011 from: http://www.hbs.edu/research/ pdf/11-100.pdf.

Leeson, R., Ivers, J. and Dickinson, D. (2006), Sustainability reporting by the public sector: Momentum changes in the practice, uptake and form. Special GRI-G3 Launch Conference Edition, pp. 56-59.

Lozano, R. (2011), The state of sustainability reporting in universities. *International Journal of Sustainability in Higher Education* 12 (1), pp. 67-78.

Maclaren, V.W. (1996), Urban sustainability reporting. *Journal of the American Planning Association* 62 (2), pp. 184-203.

Mitchell, M., Curtis, A. and Davidson, P. (2008), Evaluating the process of triple bottom line reporting: Increasing the potential for change. *Local Environment* 13 (2), pp. 67-80.

Patton, M.Q. (2002), *Qualitative research and evaluation methods* (3rd edition). Sage Publications, Inc., Thousand Oaks.

Pembina Institute (2002), *A conceptual framework for monitoring municipal and community sustainability in Canada*. Retrieved June 4, 2011, from: http://www.anielski.com/Documents/Community%20Sustainability.pdf.

Reed, M.S., Fraser, E.D.G and Dougill, A.J. (2006), An adaptive learning process for developing and applying sustainability indicators with local communities. *Ecological Economics* 59 (4), pp. 406-418.

Seasons, M.L. (2003), Monitoring and evaluation in municipal planning: Considering the realities. *Journal of the American Planning Association* 69 (4), pp. 430-440.

Shriberg, M. (2002), Institutional assessment tools for sustainability in higher education: Strengths, weaknesses, and implications for practice and theory. *International Journal of Sustainability in Higher Education Policy* 3 (3), pp. 153-167.

Shields, D.J., Solar, S.V. and Martin, W.E. (2002), The role of values and objectives in communicating indicators of sustainability. *Ecological Indicators* 2 (1-2), pp. 49-60.

Stafford, S.L. (2011), How green is your campus? An analysis of the factors that drive universities to embrace sustainability. *Contemporary Economic Policy* 29 (3), pp. 337-356. Retried June 23, 2011 from: http://web.wm.edu/economics/wp/cwm_wp77 rev1.pdf.

Taddei-Bringas, J.L., Esquer-Peralta, J. and Platt-Carrillo, A. (2008), ISO 14001 and sustainability at universities: A Mexican case study. *Management of Environmental Quality: An International Journal* 19 (5), pp. 613-626.

Tanguay, G., Rajaonson, J., Lefebvre, J.F. and Lanoie, P. (2009), *Measuring sustainability of cities: A survey-based analysis of the use of local indicators*. CIRANO. Scientific Publications, No. 2009s-02. Retrieved June 4, 2011 from: http://webcache. googleusercontent.com/search?q=cache:http://www.cityfutures2009.com/PDF/29_Tanguay_George_A.pdf.

Troschinetz, A.M., Mihelcic, J.R. and Bradof, K.L. (2007), Developing sustainability indicators for a university campus. *International Journal of Engineering Education* 23 (2), pp. 231-241.

UNCSD (2012), *Higher education sustainability initiative for Rio+20*. Retrieved March 5, 2012 from: http://www.uncsd2012.org/rio20/index.php?page=view&nr=341&type=12 &menu=35.

University of Waterloo (2011), *About Waterloo*. Retrieved June 17, 2011 from: http://uwaterloo.ca/aboutuw.

University of Waterloo Daily Bulletin (2009), *Green: UW and other universities make pledge*. Retrieved June 17, 2011 from: http://www.bulletin.uwaterloo.ca/2009/nov/ 30mo.html.

Velazquez, L. Munguia, N., Platt, A. and Taddei, J. (2006), Sustainable university: What can be the matter? *Journal of Cleaner Production* 14 (9-11), pp. 810-819.

Valentin, A. and Spangenberg, J.H. (2000), A guide to community sustainability indicators. *Environmental Impact Assessment Review* 20, pp. 381-392.

Viebahn, P. (2002), An environmental management model for universities: From environmental guidelines to staff involvement. *Journal of Cleaner Production* 10 (1), pp. 3-12.

Walton, J., Alabaster, T., Richardson, S. and Harrison, R. (1997), Environmental reporting for global higher education institutions using the World Wide Web. *The Environmentalist* 17, pp. 197-208.

Waterloo Environment (2010), *Waterloo's SEED tops Corporate Knights sustainability ranking for third year*. Retrieved November 20, 2010 from: http://www.environment. uwaterloo.ca/seed/documents/SEEDNewsReleaseJuly2010.pdf.

Wilson, J.T., Tyedmers, P. and Pelot, R. (2007), Contrasting and comparing sustainable development indicator metrics. *Ecological Indicators* 7 (2), pp. 299-314.

Wright, T.S. (2002), Definitions and frameworks for environmental sustainability in higher education, *International Journal of Sustainability in Higher Education* (3), pp. 203-220.

Chapter 44

Promoting Transparency of Sustainability Factors through Responsibility Reporting: A Case Study

Arja Sinkko

Abstract

The Bologna Declaration and the UN Decade for Education for Sustainable Development (DESD) have provided the guidelines for the promotion and implementation of ESD in Finnish Higher Education. The perspectives of responsibility and sustainability are understood as a significant part of quality assurance in education. Among the key factors in managing the academic community towards responsibility are shared values and trust of each other. There is also a need to facilitate an inspiring and innovative working and learning environment and atmosphere to ensure motivation and commitment to the joint set vision, strategies, goals and objectives as well as pedagogic and daily operations. The stakeholder's needs and expectations form an essential factor in the enhancement of the regional influence. The responsibility indicators and self-evaluation tools, as well as criteria, were created in a study in 2008-2009 to assist evaluating the responsibility within quality assurance in Higher Education Institutes (HEIs). Development, research and innovation activities within the field of responsibility and sustainability were included, too. The profile of Kymenlaakso University of Applied Sciences (KyUAS) is based on expertise of socially, ecologically and technologically sustainable and safe development and innovation processes. KyUAS describes its sustainability performance by responsibility reporting. Annually published report is based on the Sustainability Reporting Guidelines (GRI-G3) and it is used for internal and external communication. The report aims to openly and transparently describe the social responsibility undertaking and the results of the work. Responsibility reporting is becoming more and more common in Finnish industry and commerce. However, among Finnish Universities responsibility reporting is unusual.

The Bologna Process Guidelines

The Bologna Process started in 1999 and its influence on reforming Finnish and European higher education has been very strong ever since. In the period 1999-2010, all the efforts of the Bologna Process members were targeted at creating a coherent and attractive European Higher Education Area (EHEA), which became reality with the Budapest-Vienna Declaration of March, 2010. EHEA consists of comparable educational system as well as enhancement of student and staff mobility. The lifelong learning strategy and quality assurance in higher education including internal and external evaluation and dissemination of the results are examples of the tools which are created in order to enable international comparison of education. Higher Education is also a public responsibility.

"The Bologna Process 2020 – The European Higher Education Area in the new decade", the fifth follow-up conference in 2009 in Belgium established the priorities for the coming decade to face the consequences of a global financial and economic crisis:

> "In order to bring about sustainable economic recovery and development, a dynamic and flexible European higher education will strive for innovation on the basis of the integration between education and research at all levels." "Striving for excellence in all aspects of higher education, we address the challenges of the new era. This requires a constant focus on quality." "All students and staff of higher education institutions should be equipped to respond to the changing demands of the fast evolving society." [The Bologna Process 2020[1]].

Finnish Higher Education

In Finland the higher education sector is polarised into academic universities and universities of applied sciences with different tasks and profiles. The demand for supporting lifelong learning, achieving regional and societal influence and quality assurance of education and daily activities is, however, common to the entire higher education sector in Finland [FINLEX[2]]. The reform of higher education legislation in Finland has been executed and it came into operation on 1 January 2010. The perspectives of responsibility and sustainability are recognized as a significant part of quality assurance in education (ENQA 2009). The progress of the Bologna Process will cause pressure to take sustainable development into account as a natural part of international perspective in Higher Education, too. In Finland many actions have already been taken to support sustainable development through education and to strengthen education for sustainable development. Sustainable development has been integrated into the curriculum of Finnish preliminary and secondary education since 2004. It is voluntary to promote responsibility and sustainable development, although, little by little, the ongoing degree and curriculum reform will integrate sustainability into Higher Education, too (Kalliomäki 2007).

How to indicate responsibility and sustainability in Higher Education Institutes (HEIs)?

According to the Finnish higher education institution internationalisation strategy, international co-operation is the best way to improve quality. Increasing education and research co-operation based on national and international networking is the way of attaining the goals of the UN Decade of Education for Sustainable Development 2005-2014 (DESD). Promoting global responsibility indicates ethically sustainable activities at HEIs and support for students' competence to function in a global environment. (The Ministry of Education 2009).

The study "Education for Global Responsibility" ("Kasvaminen maailmanlaajuiseen vastuuseen" in Finnish) was executed by The Ministry of Education from 2007 to 2009. The aim of the study was to increase the quality and influence of global education in Finnish society. In connection with that, a sub-study "Global responsibility and sustainable development – basic database" ("Globaalivastuu ja kestävän kehityksen tietopohja" in Finnish[3]) was carried out between 2008 and 2009. As results of the sub-study, the responsibility indicators and self-evaluation tools, as well as criteria, were created to assist evaluating the responsibility within quality assurance in higher education institutes (Fig. 1). The subproject concentrates on develop-

1 The Bologna Process 2020. The European Higher Education Area in the new decade. http://www.ehea.info/Uploads/Declarations/Leuven_Louvain-la-Neuve_Communiqu%C3%A9_April_2009.pdf

2 FINLEX Legislation Data Bank in Finland. http://www.finlex.fi/en.

3 The final report can be read on the Internet (in Finnish): http://www.minedu.fi/OPM/Julkaisut/2009/Globaalivastuu_ja_kestava_kehitys_koulutuksessa._Kehittamisen_ja_seurannan_tietopohja.html?lang=fi&extra_locale=fi.

ing indicators and evaluation tools based on the Auditing Guidelines of 2008-2011 set by the Finnish Higher Education Evaluation Council (FINHEEC). This was an essential baseline for the project to ensure easier acceptance of responsibility and sustainable development being a natural part of quality assurance in HEIs. The indicators with lists of questions on the subject are useful when a smaller-scale or wider internal audit is implemented (Virtanen 2009).

Fig. 1: Responsibility indicators – general structure

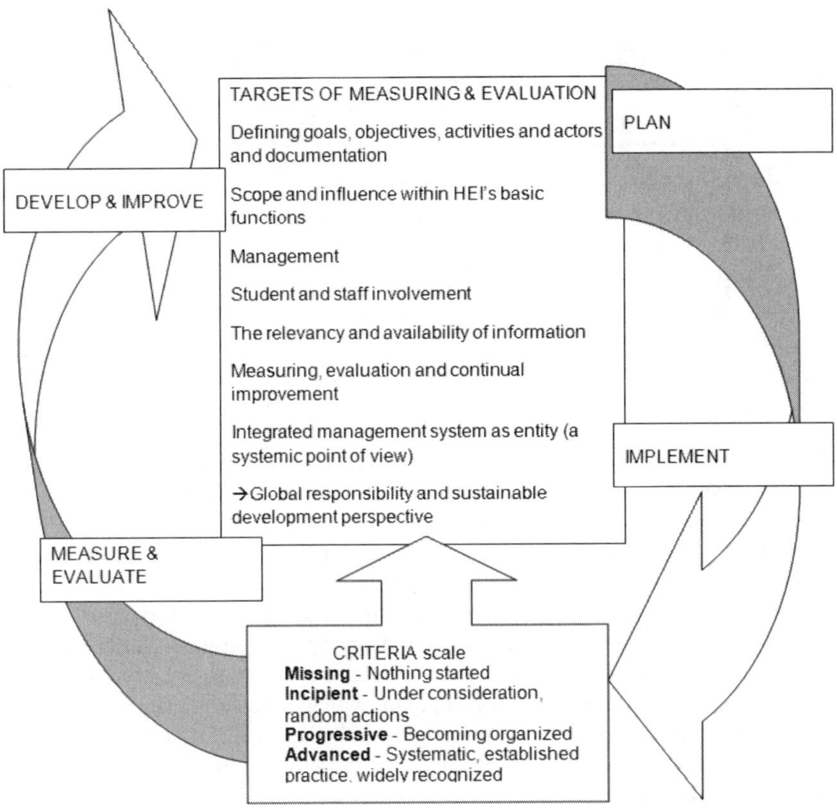

Virtanen (2009)

On the sub-project, the procedure of goal-setting follows the well-known quality assurance model, so called Deming's PDCA cycle:

- **P**lan: establish the objectives and processes necessary to deliver results in accordance with the institute's policy.
- **D**o: implement the process.
- **C**heck: monitor and measure processes against policy, objectives, targets, legal requirements. Report the results.
- **A**ct: take actions to continually improve performance of the management system.

Among the key factors in managing the academic community towards responsibility are shared values and trust of each other. There is also a need to create an inspiring and innovative working and learning environment, as well as creating an atmosphere to ensure motivation and commitment to the joint set vision, strategies, goals, objectives and operations (Fig. 2).

Fig. 2: Activities for sustainable development in HEIs

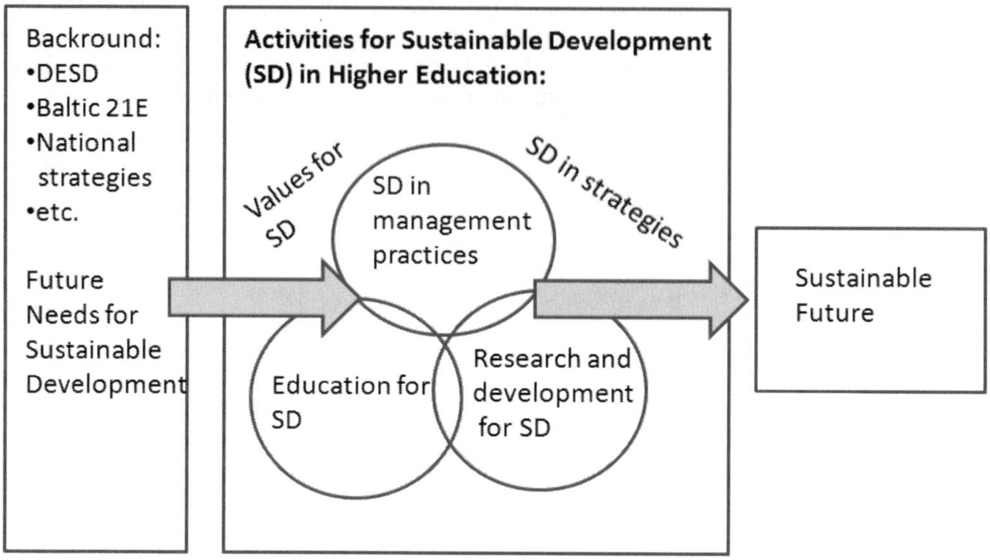

Virtanen et al. (2007)

The Higher Education Institute's specific goals need to focus on teaching methods and contents as well as research, development and innovations (R&D&I). The indicators show with measurements, how well or badly the goals are achieved. According to the responsibility indicators, the aspects of evaluation are, for example, management systems in use, shared values, student thesis and work training as well as the number of academic publications and R&D&I activities within the field of responsibility and sustainability. The indicators also measure daily actions. The consumption of electricity and energy per student, staff, sustainable procurements and carbon footprint of the HEI are good and easily measured factors.

There are many various sustainable development management and evaluation systems in use both nationally and internationally. For example, UNECE[4], SEED[5], ENSI[6] and DHO[7] have established well-known systems. In Finnish public administration the CAF-model[8] is used, and for NGOs, primary and secondary schools there are systems created especially to fulfil their needs. In the educational sector ISO-standards and other systems (ISO 9001, EFQM, ISO 14001, EMAS[9], Green Office/WWF, GRI Global Reporting Initiative) are also in use.

Regularly running self-estimation and internal audit processes are prerequisites for continuous improvement and development at higher education institutes. The indicators, created in this sub-study, can help HEIs to perceive responsibility and find targets for development.

---

4 United Nations Economic Commission for Europe UNECE. http://www.unece.org/env/esd/SC.EGI.htm.
5 School Development through Environmental Education, SEED.
6 Environmental and School Initiatives, ENSI.
7 Dutch Foundation for Sustainable Higher Education, DHO (AISHE Assessment Instrument for Education for Sustainable Development).
8 Common Assessment Framework, CAF.
9 Eco-Management and Audit Scheme, EMAS.

Case study: Responsibility reporting at Kymenlaakso University of Applied Sciences

Kymenlaakso University of Applied Sciences (KyUAS) was established in 1996. At the same time with the recently established institute of higher education the systemic development of quality management and assurance were started. The Malcolm Baldrige and Finnish Quality Prize Models were used and the very first external audit was accomplished in 2000 by the Finnish Higher Education Evaluation Council (FINHEEC). The criteria consisted of structural scope and establishment in practice as well as wide commitment and participation of the entire academic community. To improve the quality and overall performance, the EFQM[10] model with the BSC[11] management system have been taken as guiding systems in KyUAS. The latest certified external audit (in 2005) was valid for six years. Upcoming external quality audit will take place in March, 2012.

Kymenlaakso University of Applied Sciences started to include sustainable development in all functions (education, R&D, daily activities) in 2003 when the environmental management system (EMS) was launched. The EMS was based on ISO 14001 standard and the EMAS Regulation. Initially, the EMS mainly consisted of the environmental impacts of KyUAS. During the past decade, the advantages of integrating quality management and sustainable development in education have become clear. However, the integrated management system itself does not ensure change in the organisation. A change in the organisational culture is also needed. The change to see things differently starts with strategies, vision and goals (Sinkko et al. 2009).

As stated in the Responsibility Report 2010 the vision for 2015 is: "Kymenlaakso University of Applied Sciences is an achiever of success: a first-rate, modern, international higher education institution, where we succeed together."

At KyUAS, the managerial commitment is demonstrated in the updated strategy, too. It defines that the profile of expertise is based on socially, ecologically, and technologically sustainable and safe development according to the goals and objectives of KyUAS. Sustainability and responsibility are also one of the main components in the integrated management system (EFQM, BSC, ISO 14001 + GRI). The GRI guidelines offer a generally accepted framework and a tool for reporting HEI's economic, environmental, and social performance. The transition towards responsible organisational culture and governance proceeds step-by-step. The process is continuing (GRI 2009).

The challenge is how to turn the strategy into reality. The PDCA quality assurance protocol simplifies the task: a) *Plan: Say what you do*; b) *Do: Do what you say*; c) *Check: Prove it*; d) *Act: Improve it*. The Sustainable Development Policy was established to explain to the general public what KyUAS intends to do. According to the policy, the governance has been modified towards sustainability, as well as pedagogical and daily action processes. Sustainability is built into the degree programmes (curriculum development). Sophisticated information and communication technologies are used both in teaching and administration (e-learning facilities, e-library, thin client workstations, video meetings, electronic communication and documentation, etc.). The daily actions include, for example, waste management (recycling); efforts to measure and reduce energy, water, and electricity consumption; and the use of "green energy" (wind energy, district heating). Excellent quality together with responsibility and sustainability performance are the corner stones of good economic balance. Cost-efficiency and productivity reflects to the number of degrees and structural and process development. The important part of a responsible

10 EFQM, European Foundation for Quality Management.

11 BSC, Balanced ScoreCard.

organisation is the willingness and ability to take care of the physical and mental health and well-being of the students and staff. Regional and social influence is a central criterion for KyUAS's performance, too. All the actions need to be measured to certify the progress and to set new targets for continuous improvement. To profit from the achievements, they need to be announced and disseminated.

At KyUAS, the sustainable development processes are integrated into the electronic quality assurance system (QualityMoodle). According to the internal audits and evaluations completed in 2010, the strengths of QualityMoodle lie in systemic documentation and the structure of the quality assurance system. The information is transparent, available and up-dated. The biggest challenges can be found from more efficient enforcement: the system is not familiar to all members of the academic community. The actions to utilise all of the collected information in daily actions are definitely of high priority.

The quality assurance system conforms to the EFQM model, and the BSC management system produce information that is basically in line with the GRI guidelines. The guidelines offer a generally accepted framework for reporting HEI's economic, environmental, and social performance. Responsibility reporting based on the GRI Guidelines discloses outcomes and results in the context of the HEI's commitments, strategy and management approach. Reporting can be seen as a supporting tool for continuous improvement in performance. It provides balanced and understandable information for internal and external stakeholders.

Conclusion

The demand for quality assurance in Higher Education is based on Finnish legislation. For Higher Education, one main characteristic for quality is sustainability as an integrated part of the curriculum and management systems. The quality in education could be defined as satisfaction of the students, stakeholders and customers as well. Kymenlaakso University of Applied Sciences wants to improve its overall performance by ensuring that the strategy is functional, target-oriented and measurable. Capability in demonstrating and communicating the efficiency is important, too. During the past decade, the advantages of integrating quality management and sustainable development in the education system have become clear. Sustainable development has been integrated into the curriculum of Finnish primary and secondary education since 2004. In tertiary education, promoting responsibility and sustainable development is done on a voluntary basis, but the ongoing degree and curriculum reform will integrate sustainability into Higher Education as well. Moreover, in the Finnish higher education internationalisation strategy for 2009-2015 aims to develop a strong, internationally appealing, high-standard higher education and research community in Finland. Support for multicultural society and promotion of global responsibility are highlighted in the strategy, too.

The systems mentioned in this study do not ensure change at the university. A change in the organisational culture is also needed. It means that the entire academic community needs to be committed to the shared goals.

Mission statement (excerpt): "Kymenlaakso University of Applied Sciences focuses on promoting positive changes in the industrial structure of the Kymenlaakso province and the development of innovative, responsible industries in the region, as well as reinforcement of Kymenlaakso's logistic position and management of changes in the population structure."

As the only independent HEI in the Kymenlaakso region, the KyUAS has an important role in systematic regional development by improving the overall level of education and professional skills. Promoting sustainable development in its activities, the university is supporting regional sustainable development in Kymenlaakso, too. In the region, KyUAS is engaged in productive co-operation with companies, organisations and other players (regional, national and international stakeholders), as well as boosting Kymenlaakso's appeal up. KyUAS for its part promotes structural change as well as cultural, social and operational harmonisation in the region.

The sustainability achievements at KyUAS are announced by internal and external communication. An annually published Responsibility Report is structured as closely as possible according to the guidelines of the GRI. The Report consists of reviews, strategic course and profile, and above all, the results and influence of activities relating to responsibility (technologically, socially, ecologically and economically sustainable development). The report accumulates all KyUAS activities and makes the operational performance transparent. It is a tool used to demonstrate and communicate the efficiency of KyUAS's activities and highlights that their attitude towards taking responsibility is standard practice. This documentation based on actions taken is a response to the stakeholders' expectations and aspirations.

In accordance with the principles of sustainable development, KyUAS publishes only an online version of the Annual Responsibility Report. The Report can be read on the internet (in English) on the website: www.kyamk.fi/responsibilityreport.

References

European Association for Quality Assurance in Higher Education, ENQA (2009), *Standards and Guidelines for Quality Assurance in the European Higher Education Area*. Available at http://www.enqa.eu/pubs.lasso.

GRI, The Global Reporting Initiative (2009), *Sustainability Reporting Guidelines (G3)*. Available at https://www.globalreporting.org/resourcelibrary/G3.1-Guidelines-Incl-Technical-Protocol.pdf.

Kaivomäki, A. (2007), Higher Education for Sustainable Development – International and National Guidelines. In Kaivola, T. and Rohweder, L. (Eds.), *Towards Sustainable Development in Higher Education – Reflections*. Publications of the Ministry of Education 2007:6, pp. 14-18. Helsinki. http://www.minedu.fi/export/sites/default/OPM/ Julkaisut/2007/liitteet/opm06.pdf?lang=en.

Kymenlaakso University of Applied Sciences (2011), *Annual responsibility report 2010*. Available at http://www.kyamk.fi/responsibility.

The Ministry of Education (2009), *Strategy for the Internationalisation of Higher Education Institutions in Finland 2009-2015*. Publications of the Ministry of Education 2009:23, Helsinki. Available at http://www.minedu.fi/OPM/Julkaisut/2009/Korkeakoulujen_kansainvalistymis strategia_2009_2015.html?lang=fi&extra_locale=en (accessed 18 October 2011).

Sinkko, A., Suntioinen, S. and Tapola, H. (2009), Vastuullisuus ja laadunhallinta korkeakouluissa. In Virtanen, A. and Kaivola, T. (Eds.), *Globaalivastuu ja kestävä kehitys koulutuksessa. Kehittämisen ja seurannan tietopohja*. Publications of the Ministry of Education 2009:56, pp. 77-84. Helsinki.

Virtanen, A. (2009), Korkeakoulujen vastuullisuuden tavoitteet sekä seurannan, arvioinnin ja kehittämisen menetelmät ja indikaattorit. In Virtanen, A. and Kaivola, T. (Eds.), *Globaalivastuu ja kestävä kehitys koulutuksessa. Kehittämisen ja seurannan tietopohja*. Publications of the Ministry of Education 2009:56, pp. 155-180. Helsinki.

Virtanen, A. and Rohweder, L. (2007), Teaching and Learning for Sustainable Development. In Rohweder, L. and Virtanen, A. (Eds.), *Learning for the Sustainable Future. Innovative Solutions From the Baltic Sea Region,* pp. 60-65. The Baltic University Press, Uppsala. Available at http://www.balticuniv.uu.se/index.php/recent-publications (accessed 23 November 2011).

Chapter 45

The Ethical Dimension of Sustainability in Higher Education: Applying the Principles of the Earth Charter in Malaysia and Beyond

Brandon P. Hollingshead, Peter Blaze Corcoran, Adnan A. Hezri, Kanayathu Chacko Koshy, Joseph P. Weakland and Zakri Abdul Hamid

Abstract

This chapter addresses environmental education and education for sustainable development in Malaysia, with a particular focus on Malaysian institutions of higher education. It begins with a review of the global context of sustainability in higher education. It analyses the education policy context in Malaysia. The chapter considers the ethical dimension of sustainability and how the Earth Charter aligns with Universiti Sains Malaysia's transformational vision (Campbell 2010; Dzulkifli et al. 2010). The chapter concludes with policy recommendations for Universiti Sains Malaysia, for education policy in Malaysia, and for education in the context of Rio+20 and beyond. The global review of higher education and the Malaysian educational policy analysis indicate a timely and necessary need to attend to the ethical dimension of sustainability (Dzulkifli 2010). The Earth Charter is used as an aspirational and practical guide to ethical considerations. It is a declaration of fundamental ethical principles for building a just, sustainable, and peaceful global society in the 21st century (Bosselmann and Engle 2010; Mackey 2008). The Earth Charter can be used to deepen USM's already substantial efforts in its transformation to sustainability.

Review of the global context of sustainability in higher education

The preamble of the Earth Charter (2000) declares that, "We stand at a critical moment in Earth's history, a time when humanity must choose its future. As the world becomes increasingly interdependent and fragile, the future at once holds great peril and promise" (Paragraph 1). The "peril" that the Earth Charter describes resides in large part within the struggle of social and economic globalization, with significant implications for education for sustainable development (Clugston 2010; Weakland and Corcoran 2009). Launched in the year 2000, the Earth Charter characterizes the challenges associated with globalization as follows:

> "The dominant patterns of production and consumption are causing environmental devastation, the depletion of resources, and a massive extinction of species. Communities are being undermined. The benefits of development are not shared equitably and the gap between rich and poor is widening. Injustice, poverty, ignorance, and violent conflict are widespread and the cause of great suffering. An unprecedented rise in human population has overburdened ecological and social systems. The foundations of global security are threatened. These trends are perilous – but not inevitable." (Preamble, paragraph three)

The Earth Charter is often described as a concise statement of the necessary ethical principles to achieve sustainable development that is, according to the well-known definition provided by the Bruntland Commission, "development that meets the needs of the present without compromising the ability of future generations to meet their own needs" (WCED 1987: 8).

The Earth Charter's sixteen main principles and accompanying supporting principles are divided into four parts that form the body of the Charter and comprise fundamental ethical guidelines for a sustainable way of life. The Earth Charter Initiative involved the most open and participatory consultation process ever conducted in drafting an international document. Thousands of individuals and hundreds of organizations from all regions of the world, different cultures, and diverse sectors of society participated. The Earth Charter was shaped by scientific experts, government and civil society leaders, students, representatives from indigenous groups, and grassroots communities. It is a fundamental expression of the hopes and aspirations of the emerging global society.

The drafting of the Earth Charter was part of the unfinished business of the 1992 United Nations Conference on Environment and Development in Rio de Janeiro (the Earth Summit). In 1994, Maurice Strong, Secretary General of the Earth Summit and Chairman of the Earth Council, and Mikhail Gorbachev, the President of Green Cross International, launched a new Earth Charter Initiative in The Hague. It was undertaken with the encouragement of former Prime Minister Ruud Lubbers and Her Royal Highness Queen Beatrix of The Netherlands, and with financial support from the Dutch government.

The United Nations Educational, Scientific, and Cultural Organization (UNESCO) manages the United Nations Decade of Education for Sustainable Development (DESD 2005-14) and has adopted a resolution that recognizes the Earth Charter as ethical principles for sustainable development and as a valuable educational resource. UNESCO's 32nd General Conference resolved to "recognize the Earth Charter as an important ethical framework for sustainable development, and acknowledge its ethical principles, its objectives and its contents, as an expression that coincides with UNESCO's vision" (UNESCO 2003). DESD has outlined the role of the university in the transition to sustainability. The implementation scheme for the Decade states:

> Universities must function as places of research and learning for sustainable development ... Higher education should also provide leadership by practicing what they teach through sustainable purchasing, investments and facilities that are integrated with teaching and learning ... Higher education should emphasize experiential, inquiry-based, problem-solving, interdisciplinary systems approaches and critical thinking. Curricula need to be developed, including content, materials and tools such as case studies and identification of best practices. (UNESCO 2004: 22-23)

Most recently, the Bonn Declaration referenced the Earth Charter when it states that education for sustainable development "is based on values of justice, equity, tolerance, sufficiency and responsibility. It promotes gender equality, social cohesion and poverty reduction and emphasizes care, integrity and honesty, as articulated in the Earth Charter" (UNESCO 2009, Principle 8).

With an extensive rhetorical commitment to sustainability, where do universities today stand in integrating the concept into their core functions? If we are to meet the sustainability challenges of our time, colleges and universities will need to teach and practice a deeper form of sustainability than just ecoefficiency and "greening" a small set of courses. The Earth Charter points to what this deeper form of sustainability might be by illuminating the ethical and spiritual as well as the environmental, social, and economic dimensions of the sustainable development challenge. The promise of the Earth Charter is that it provides a vision of sustainability built on a broad global consultation. It can assist us in articulating a new framework for economic and social policies and practices that is not primarily for short-term economic gain, but rather for the full-flourishing of all life. Imagining viable solutions to the social and ecological challenges of our time will involve looking at subjects as diverse as "development, justice, peace and conflict, human rights and dignity, and the intrinsic value of other species and whole ecosystems" (Wals and Jickling 2002, paragraph two).

The Earth Charter and education for sustainable development

Since its inception, the Earth Charter and Earth Charter Initiative prioritized education as a key concern. We believe the Earth Charter can be a critically important text for learning in several areas: resolving the multiple definitions of sustainable development, putting forth ethics for sustainability while navigating the challenges of values education, and guiding ethical and sustainable actions for the university community (Weakland and Corcoran 2009: 153). If the leaders of the major institutions and disciplines do not make sustainability a central academic and organizational focus, it will be impossible to create a just, equitable and sustainable future. Higher education can realize sustainability through policy-making, capacity building, technology transfer, science and research on its campuses. Highereducation institutions must also accept their moral responsibility to assist in the creation of democratic, equitable and ecologically sound societies.

In 2001, not long after the launch of the Earth Charter, the Earth Charter Initiative conducted an online global forum on Earth Charter education. The synthesis report of the consultation (Mackey 2002) highlighted two overall educational objectives of the Earth Charter, and also a set of educational principles that provided the basis of a philosophy to guide the development of Earth Charter educational materials. The first objective was to use the Earth Charter to promote education for sustainable living through rasing levels of consciousness, in the application of values and principles, and as a call for action. The second was to promote a global dialogue on ethics for sustainability.

Beginning in 2005, the Earth Charter Initiative (ECI) underwent an organizational shift to leverage a decentralized, broad-based, voluntary, civil society effort to promote and implement the values and principles of the Earth Charter. Part of this reorganization was the establishment of an Earth Charter Center for Education for Sustainable Development at University for Peace in Costa Rica to focus on Earth Charter education. The Center works to meet the education goals of Earth Charter International, which are, "To promote the use of the Earth Charter as an ethical guide" and "To encourage and support the educational use of the Earth Charter in schools and universities." Additionally, the Center works toward the strategic objectives of ECI, "to create and disseminate high quality communications and educational materials to different target groups" and "to position the Earth Charter… so that its ethical framework can be used as a guide in efforts to address urgent challenges" (Earth Charter Handbook 2008: 9-10).

Most recently, an Earth Charter Scholarship Program, under the leadership of Richard M. Clugston, has begun to address the need for high quality scholarship on Earth Charter education and education for sustainable development in higher education (Clugston 2010). To this end, a group of international Earth Charter scholars convened on Sanibel Island, Florida in 2009 to review high-quality Earth Charter educational resources, organize them into an annotated bibliography, and determine what additional resources should be developed. They developed a "Guide for Using the Earth Charter in Education." That report lists ways the Earth Charter can support a variety of educational goals. These include goals such as:

1. Raising awareness and understanding of critical global problems.
2. Promoting dialogue on the values and principles for a sustainable way of life.
3. Promoting individuals' ethical development.
4. Inspiring a spirit of collaboration, cooperation, and action.
5. Encouraging a bio-sensitive outlook.
6. Applying values and principles.
7. Facilitating an understanding of the relationship between the Earth Charter, public policy, and international law.
8. Assisting educational institutions and systems in reorienting their teaching and operations toward sustainable ways of living.

The Earth Charter has been put to use in a variety of ways in higher education institutions. More than 370 universities and institutions of higher education, representing 33 nations from all regions of the world, have endorsed the Earth Charter (Earth Charter Initiative, "Endorse" 2011). Formal endorsement of the Earth Charter signifies a commitment to the spirit and aims of the document and is an indication that universities intend to utilize the Earth Charter in ways that are appropriate given their situation.

Analysis of the education policy context in Malaysia

Education has been a key area of Malaysian policy since the country gained independence. It has also been used as an important strategy in the development process of the nation (Abdul Rahman and Mahani 2007). Among other objectives, the education system is expected to provide accessible and equal educational opportunity for all Malaysians, promote national unity in a pluralistic society, and to fulfill the manpower demands of the economy. On average, the Malaysian government spends 4.5% of its GDP on education. This is a significant percentage compared to emerging economic powers such as India, Korea, and China, where government expenditure on education amounts to 3.2, 4.2, and 2.3% of GDPrespectively.

That the higher education system in developing countries faces many challenges different to the developed world is well documented. Poorly rewarded and under-qualified staff, under-developed curriculum, insufficient autonomy and underfunded operations are among the key problems facing developing countries. Malaysia is not immune to these challenges, and it responded with a decentralization exercise through 'corporatization' of public universities. The other notable challenge facing the higher education system in Malaysia is the occasional disconnection between national and international education objectives. Driven mainly by economic factors, the internationalization of higher education in many countries is characterized by mobility of students, scholars, and researchers, joint research between different countries, and other academic cooperation. Nonetheless, the tension between globalization and national interests presents an opportunity for positive changes to be implemented. As global economic forces require graduates with new values and skills to become active in the new economy, there is an opportunity for the infusion of new educational values in Malaysia's higher education system. As ethical principles are legitimized by supporters across the globe, the Earth Charter may resonate with the principles of Malaysia's education system, particularly with the oft-cited National Education Philosophy (*Falsafah Pendidikan Kebangsaan*). For example,

> "Education in Malaysia is a continuous effort toward enhancing individual potentials in a comprehensive and integrated manner to create intellectually, spiritually, emotionally, and physically balanced and harmonious human-beings based on the belief and submission to God. This effort aims at producing Malaysian citizens who are knowledgeable, courteous, responsible and able to achieve not only self wellbeing but also capable of contributing constructively to a prosperous and harmonious family, society and nation." (Wan Zahid 1993: 175)

Universiti Sains Malaysia's transformational agenda in sustainability

Since its inception in 1969, Universiti Sains Malaysia has undergone different waves of change. The first among them was the consolidation of teaching and training in a "school" based interdisciplinary approach during the early two decades. This was followed by a second wave of change that focused on achieving excellence in research, which resulted in the university being recognized by the Ministry of Higher Education as a Research University (RU) in 2007.

USM stands at the top of Research Universities in Malaysia. As the second oldest government university, USM has been at the forefront in supporting major national development initiatives aimed at improving the well being of its citizens while preserving ecological and social harmony. This commitment took on a more consolidated expression at the turn of the century with a number of focused initiatives, four of which may now be seen as representing university's third wave of "transforming higher education for a sustainable tomorrow."

These include the Kampus Sejahtera (Campus Well-Being) Programme, USM as a Regional Centre of Expertise for education for sustainable development, the University in a Garden scenario, and Malaysia's Accelerated Programme for Excellence (APEX) status. Among these, the APEX award – the sole award under this scheme to any university by the Ministry of Higher Education in Malaysia – is specifically focused on "transforming higher education for a sustainable tomorrow."

The Universiti Sains Malaysia Scenario Planning in 2005-2006 was organised to create the most advanced and proactive futuristic scenario for USM, looking ahead to almost 20 years. After almost 15 months of university-wide consultations and soul-searching "The University in a Garden" scenario – which is now the University's slogan,[1] was adopted. This model is in keeping with the development model followed by the university since the start of the millennium and is designed to depict the close affinity between the role and function of the University as an institution of higher learning and nature as part of the global ecological setting. The flora, fauna, aquatic elements, and other natural creations are dynamically linked in the exploration of knowledge into the nature of existence. The concept is an invitation to value, preserve and nurture the campus ambiance as part of the efforts to create and sustain an intellectually conducive setting in order to kindle the spirit and practice of symbiotic co-existence. It is about touching the hearts and minds of each campus citizen and appreciating the natural surroundings as a source of inspiration as bequeathed to us by the Creator.

The University in a Garden scenario takes into account a harmonious blend of various vistas in the search for answers to further illuminate who we are, how we attain insights, and how we should fashion our future survival in a world where, on the one hand we have extreme wealth, and on the other almost a billion people are disempowered and are desperately strivingto provide even the most basic necessities in life for themselves and their families. USM's Garden scenario, thus, has a major focus on United Nation's Millennium Development Goals.

As part of the APEX award, USM has adopted a new vision to *Transforming Higher Education for a Sustainable Tomorrow*. This is part of a larger agenda to redefine "excellence" in line with the challenges of the future under Malaysia's Accelerated Programme for Excellence (APEX) of which USM is currently the sole recipient. In translating the vision into action, a new mission now reads: "USM is a pioneering, transdisciplinary research intensive university that empowers future talents and enables the bottom billions to transform their socio-economic well-being." This is essentially outlining the university's efforts in advocating the process of sustainable development as a new paradigm for a more lasting future (Dzulkifli et al. 2010). The following guiding principles currently lead USM in its endeavours to become a sustainability-led university of world-class standing:

- To enhance a rigorous intellectual culture with a worldview that creates a sense of "oneness" of knowledge through a transdisciplinary approach;
- To serve societal needs without compromising their socio-cultural integrity and sensitivities based on appropriate and relevant knowledge;

1 See Universiti Sains Malaysia's publication Constructing Future Higher Education Scenarios – Insights from Universiti Sains Malaysia, 2007.

- To demonstrate excellence in leadership premised on strong principles, scholarly pursuits and wisdom of thoughts (Universiti Sains Malaysia, "Guiding Principles").

The Earth Charter and Universiti Sains Malaysia's sustainability transformation

The thrust of the Earth Charter has much in common with Universiti Sains Malaysia's sustainability transformation agenda. It also has much in common with Malaysia's socially-inclusive and environmentally sustainable economic growth policies and strategies.

To be explicit, USM's sustainability focus may be summarized in its current vision and mission and the values and thrusts associated with them. The vision is "Transforming Higher Education for a Sustainable Tomorrow." The mission is "USM is a pioneering, transdisciplinary research intensive university that empowers future talents and enables the bottom billions to transform their socio-economic well-being." The values are equity, availability, accessibility, affordability, and appropriateness. The thrusts are the future, uniqueness, sustainability, humanity, universality, change, and sacrifice.

USM's roadmap for sustainability mainstreaming across its entire range of operations factors the major sustainability challenges that span across the three pillars of sustainability – economy, environment, and society – into the university's mission activities in the three pillar areas of education – teaching, research, and community engagement. There are many possible direct uses of the Earth Charter at Universiti Sains Malaysia. They are based on the premise that USM's work is already consistent with the Earth Charter. Earth Charter principles can be set in the USM context to define sustainability across its many initiatives and academic programs.

The Earth Charter can explicitly bring the ethical dimension of sustainability concerns to light. Vice Chancellor Dzulkifi Abdul Razak has written, "what we are talking about here does not merely involve ethics …. rather it reaches beyond into the fundamental role of the university to nurture and shape the character of human beings to transform them into ethical citizens of the world" (Dzulkifli 2010: 4). The Earth Charter provide core values that are life-affirming, promote human dignity, advance social, economic, and environmental justice, respect cultural and ecological diversity and integrity. They are a perfect place to begin to create ethical citizens. Then they can be contextualized within the Malaysian setting, the Islamic setting, and the USM mission in order to expand current ethical horizons.

The Earth Charter can provide a solid conceptual foundation for USM's commitment to sustainability. The document is based on 40 years of international treaties, covenants, laws, and people's declarations. The unique contribution of the Earth Charter is to provide an integration of the values and principles contained in these. Thus, it is the ideal framework for making clear the ethical basis for development.

Earth Charter ethics are consistent with the Islamic worldview in which we, in His surroundings, are servants of Allah, according to his guidance. The Muslim holistic view of creation and the universe are embodied in the Preamble's view of the "vast evolving universe." Many Muslims were involved in the writing of the Earth Charter. They are also much involved in the ongoing implementation of it, such as with the 2010 *Dead Sea Declaration*.[2] It also seems that several of the ten principles of the Islam Hadhari, or civilizational Islam, are developed in the Earth Charter.

---

2 The Dead Sea Declaration was agreed to and signed by representatives from ten Arab nations at a meeting in the Hashemite Kingdom of Jordan in 2010. They are Bahrain, Egypt, Iran, Lebanon, Jordan, Palestine, Saudi Arabia, Syria, Tunisia, and United Arab Emirates.

Universiti Sains Malaysia is driven by powerful metaphors that act as inspiration for action programs – the "blue ocean", the "university in a garden", the "Koranic garden". The Earth Charter vision can be an extraordinarily powerful driver, as well. The Earth Charter could provide the ethical vision for the action program for humanising the University. We need fundamentally different visions that are inspirational. We need ecological metaphors – for our challenging time – that are aspirational. The Earth Charter is such a source.

The Earth Charter touches upon all areas of human endeavor and, as such, provides many entry points for academics across the full range of studies at the University. In our opinion, this is especially helpful in including the humanities and the social sciences. It is also helpful for students to document their learning on sustainability in many fields.

The Earth Charter can be used as a set of ethical indicators. USM could use the Earth Charter Assess process to benchmark best practices on campuses. It would be an important exercise to analyze what USM is doing in reference to the Earth Charter principles. This principle-by-principle analysis will reveal many areas of possibility.

The Earth Charter movement, at all levels, from grassroots to government, is a global movement of civil and political views. USM's global leadership role could be enhanced by providing academic leadership in this community. For example, the nation of Malaysia and the Universiti Sains Malaysia could play major roles at Rio+20 in 2012 in advancing the moral and ethical framework.

In addition to these direct uses, there are two conceptual uses of the Earth Charter to be considered. First, USM has adopted a cluster approach to addressing sustainability issues, striving to strike a balance among science and technology, social sciences, and humanities – especially in research and innovation. Humanities such as art, literature, culture, language, history, and philosophy are playing an increasingly important role in promoting sustainability. The emotive power of the humanities can cut across disciplines, sectors, and borders, to bring meaning and purpose to daily life. In other organizations, including higher education institutions, the Earth Charter has been used in a similar way across traditional boundaries. These models of practice could be a great boost to USM teamwork and the multi- inter- and trans-disciplinary approaches needed to succeed in sustainability project implementation and sustainability capacity building at all levels.

The model of collaboration within the Earth Charter movement can also be instructive. Given the fact that the Earth Charter Initiative lacks expensive central institutional mechanism for coordination, large scale funding, and yearly reviews to ensure implementation success, its limited resource model could provide a model for USM. The operationalization of the USM sustainability agenda is expected to work the same way, with very minimal central control but with a good measure of moral and ethical appeal to core values and ethics decentralized to participants.

Universiti Sains Malaysia, while already being a global leader in its level of commitment and practice in sustainability, could benefit from applying the principles of the Earth Charter. This application can be both direct and conceptual. Our belief is that through the use of the Earth Charter, USM could enhance its current projects. We also believe that new areas of substantive sustainability activity could be engaged across the comprehensive and integrated content of the Earth Charter's Parts and Principles.

Conclusion: Policy recommendations for applying the principles of the Earth Charter in Malaysia and beyond

The intersection of Earth Charter ethics, Universiti Sains Malaysia's transformational vision in sustainability, the policy context of Malaysia, and ethical dimension of sustainability in higher education is ripe for analysis and implementation. We conclude this chapter by proposing policy recommendations that can be taken up at Universiti Sains Malaysia, in Malaysia, and beyond. These sets of recommendations were vetted in two stakeholder meetings, the first at the Malaysian Industry Group for High Technology (MIGHT) in Cyberjaya on March 4, 2011, and the second at Centre for Global Sustainability Studies (CGSS@USM) on March 11, 2011. The recommendations are based upon the conceptual linkage between education for sustainability and the catalytic contribution of the Earth Charter.

Across the scholarly literature, the Earth Charter is seen as a transformative tool for examining, expanding, and deepening educational practice in environmental education and education for sustainable development. We believe the Earth Charter is a valuable tool for transformative education and that university educators can find value in applying such a comprehensive set of ethical principles. The possible application of these principles is delineated in "Part A: Recommendations for Universiti Sains Malaysia to apply the principles of the Earth Charter." Further recommendations speak to the application of Earth Charter principles to larger educational policy beyond USM. These recommendations are delineated in "Part B: Recommendations for education policy in Malaysia applying the principles of the Earth Charter." We also recommend how Universiti Sains Malaysia can take a leadership role, guided by the ethical vision of the Earth Charter, at the Rio+20 conference in 2012. These recommendations are delineated in "Part C. Recommendations for Universiti Sains Malaysia in the context of Rio+20 and beyond."

Part A: Recommendations for Universiti Sains Malaysia to apply the principles of the Earth Charter

As an integrated ethical framework for sustainability, the Earth Charter is a useful tool for education, discussion, and reflection concerning the critical dimensions of university life. The Earth Charter reaffirms USM's past and ongoing initiatives on education for sustainable development. The following actions may enhance USM's genuine commitment to the transformation towards a sustainable future:

1. Assess the consistency of the Earth Charter with the sustainability mission and vision of USM. Universiti Sains Malaysia could endorse the Earth Charter, following the commendable step taken by over 200 universities across the world.
2. Consider the principles of Earth Charter as areas USM could benchmark against in the development of inter-disciplinary modules and in designing extra-curricula activities related to sustainability. In addition, an objective study of the Earth Charter may help set requirements for specific subjects and learning areas in Universiti Sains Malaysia.
3. Create a campaign to popularize the Earth Charter widely among university students. Strategies for promoting more fully embodied sustainability values may be explored through internships and volunteer activities in the sustainability enterprise. Activities may be organized around one or more specific principles or supporting principles of the Earth Charter, emphasizing the "learning by doing" approach.
4. Appraise the performance of physical operations and infrastructure in Universiti Sains Malaysia using the Earth Charter's holistic principles as a directional tool to measure progress toward the concept of sustainable campus and greening of university operations.

5. Explore how and where sustainability could be included as a focus of holistic research and scholarship using the Earth Charter as a benchmark. As an ethical framework promoting shared global values, the Earth Charter may support initiatives such as SAINS@USM (Science and Art Innovation Space) in blurring the divide between natural sciences and the humanities.
6. Enhance the University's existing service to communities by holistically promoting the ecological metaphor and other principles such as care for the community of life, social or economic justice, and in peace building that inform the Earth Charter. By inviting a diverse collection of people to come together and discuss and reflect on the Earth Charter, Universiti Sains Malaysia may help the country in building a more accessible, equitable and inclusive process of governance, to forge an alliance between the academe and civil society through social discourse.
7. Adapt from the principles of Earth Charter relevant parameters for faculty and staff assessment such as in the development of Key Performance Indicators (KPI) and Key Intangible Performance (KIP) for academic appraisal. The integrated principles of the Earth Charter could also be used to assess the readiness of students and staff of Universiti Sains Malaysia to adopt sustainability-oriented attitudes and behavior.

Part B: Recommendations for education policy in Malaysia applying the principles of the Earth Charter

Education for Sustainable Development (ESD) as an organizing principle in higher education is still a relatively new policy focus. Therefore, sustainability provides Malaysia's education system an opportunity to confront its core values regarding the environment and sustainability. As a holistic, ethical framework for sustainable development, the Earth Charter may enrich Malaysia's National Education Philosophy (*Falsafah Pendidikan Kebangsaan*) through reflections on the practices, entrenched pedagogies, and the way the education system uses resources. The following actions may be particularly relevant for education authorities in Malaysia.

1. Assess the synergy between the ethical fount of the Earth Charter and the *National Education Philosophy* and *the National Higher Education Action Plan*, with emphasis on the interconnected challenges to inculcate deep sustainability as part of ethical values. Where possible, Malaysia could use the Earth Charter as a definition of sustainability in order to contribute to sustainable development to the best of its ability.
2. Rationalize the implementation of sustainability component of the New Economic Model and subsequent government transformation plans by referring to the Earth Charter, as a higher order means to strengthen the legitimacy of these documents.
3. Organize an in-service educators training seminar to inform about the Earth Charter text, its principles and philosophy and introduce the methodological possibilities of integrating the Earth Charter in teaching different subjects at universities and schools.
4. Identify best practices on sustainable campus efforts and mainstream these in other universities in Malaysia by considering the guiding principles of the Earth Charter.
5. Empower an inter-varsity sustainability network through SURE.Net to embrace the Earth Charter as a basis to consolidate common grounds among Malaysian universities in areas of ecological integrity, care for the community of life, social or economic justice, and in peace building.
6. Develop a home-grown university appraisal system for Malaysia around the principles of the Earth Charter, focusing on the development of human person (*insan).*

Part C: Recommendations for Universiti Sains Malaysia in the context of Rio+20 and beyond

The United Nations Conference on Sustainable Development will focus on two themes. The first is a green economy in the context of sustainable development and poverty eradication. Higher education can make a difference to the transition towards a green economy and the creation of green jobs. The second is the institutional framework for sustainable development. Higher education can contribute to the conceptual development of this framework. Yet there is almost no mention of higher education in the Rio+20 preparatory processes and the accompanying documents. The following actions are therefore necessary to bring higher education onto the agenda of Rio+20.

1. As one of the global leaders in the area of education for sustainable development, and guided by the ethical vision of the Earth Charter, USM should take a leadership role in articulating the importance of higher education in the transition to a green economy. USM should collaborate with the following institutions in doing so.
   a) International Association of Universities (IAU) provides an international centre of co-operation among 800 member universities and institutions of higher education. IAU developed and adopted the Kyoto Declaration on Sustainable Development in 1993.
   b) The United Nations Educational, Scientific and Cultural Organization (UNESCO) and the Decade for Education on Sustainable Development (DESD) Programme. UNESCO is also the task manager for the implementation of Chapter 36, "Education, Public Awareness and Training", of Agenda 21.
   c) COPERNICUS-Campus, formerly a Programme of the Association of European Universities, is responsible for the University Charter for Sustainable Development, signed to date by over 290 university heads in 36 European countries.
2. USM should work with the Association of University Leaders for a Sustainable Future (ULSF) to update and revitalize the Talloires Declaration principles for discussion and adoption at Rio+20.
3. USM should strategize its leadership role by collaborating with key players and participating in key events in the pre- and post-Rio+20 processes. Fig. 1 illustrates the plausible milestones for Universiti Sains Malaysia to position itself as an active player in the Rio+20 process.

We believe the implementation of the actions and activity proposals in this policy discussion paper will result in stronger sustainability at Universiti Sains Malaysia, an improved policy milieu for sustainability in Malaysia, and beyond. We hope that readers of this chapter will assist in implementing the recommendations.

Fig. 1: The road to Rio+20 and beyond: centre for global sustainability studies milestones

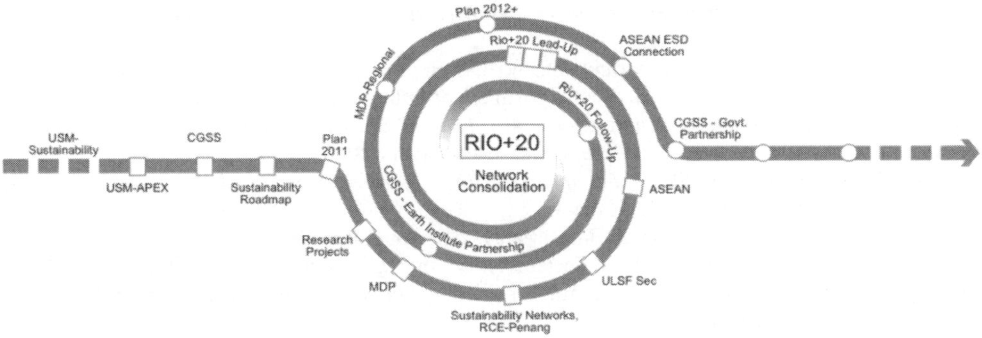

Note: ASEAN stands for Association of South East Asian Nations. CGSS stands for Centre for Global Sustainability Studies. MDP stands for Masters in Development Practice. USLF stands for University Leaders for a Sustainable Future. USM-APEX stands for University Sains Malaysia Accelerated Programme for Excellence

References

Abdul Rahman, H.I. and Mahani, M. (2007), Sejarah perkembangan dan pembangunan pendidikan tinggi di Malaysia sehingga 2007. In Zailan, M. (Ed.), *50 Tahun Pendidikan Tinggi di Malaysia 1957-2007,* pp. 1-37. Penerbit Universiti Sains Malaysia, Penang.

Bosselmann, K. and Engle, J.R. (Eds.) (2010), *The Earth Charter: A framework for global governance.* Royal Tropical Institute (KIT) Publishers, Amsterdam.

Campbell, J. (2010), *Understanding reform and the Universiti Sains Malaysia agenda: Discussion and critique.* APEX Intelectual Discourse Series. Universiti Sains Malaysia Press, Pulau Pinang, Malaysia.

Clugston, R.M. (2010), Earth Charter education for sustainable development. *Journal of Education for Sustainable Development* 4 (2), pp. 157-166 (doi: 10.1177/09734082 10004 00202).

Dzulkifli, A.R. (2010), *In search of a world-class university of tomorrow: The importance of the APEX initiative.* Oxford Fajar, Shah Alam, Selangor, Malaysia.

Dzulkifli, A.R, Zakri, A.H, Zainal, A.S, Koshy, K. (2010), *Transforming Higher Education for a Sustainable Tomorrow: A Case of Learning by Doing at Universiti Sains Malaysia.* Book on DESD, Tomorrow Today: Learning to build a sustainable future. UNESCO-Tudor Rose: United Kingdom.

Earth Charter International (2000), *The Earth Charter.* Earth Charter International Secretariat, San Jose, Costa Rica. Available online at http://www.earthcharter.org.

– (2008). *The Earth Charter Initiative handbook.* Earth Charter International Secretariat, San Jose, Costa Rica. Available online at http://www.earthcharter.org.

Earth Charter Initiative (2011), *Endorse the Earth Charter.* Retrieved from http://www.earthcharterin action.org/content/pages/Endorse.html.

Mackey, B.G. (2002), Synthesis: Summary of the Earth Charter education advisory committee inaugural online forum. *International Journal of Curriculum and Instruction* 4 (1), pp. 81-96.

– (2008). The Earth Charter, ethics, and global governance. In Westra, L., Bosselmann, K. and Westra, R. (Eds.), *Reconciling human existence with ecological integrity,* pp. 61-72). Earthscan, London.

UNESCO (United Nations Educational, Scientific, and Cultural Organization) (2003), *Records of the General Conference.* 32nd Session. 29 September to 17 October 2003, Paris. Available online at http://unesdoc.unesco.org/images/0013/001331/1331/13317e. pdf (accessed 16 June 2009).

– (2004), *United Nations Decade of Education for Sustainable Development (2005-2014): Draft International Implementation Scheme*. UNESCO, Paris.

– (2009), *Bonn Declaration of the World Conference on Education for Sustainable Development*. UNESCO, Paris.

United Nations Conference on Environment and Development (1992), *Agenda 21: Programme of action for sustainable development.* UN Department of Public Information, New York.

United Nations Development Programme (2010), *40-year Trends Analysis Shows Poor Countries Making Faster Development Gains*. Press release 4 February 2010.

Universiti Sains Malaysia (2004). *The university in a garden: Special edition*. Universiti Sains Malaysia Press, Pulau Pinang, Malaysia.

– (n.d.), *Guiding Principles*. Universiti Sains Malaysia, Penang, Penerbit.

Wals, A.E.J. and Jickling, B. (2002), 'Sustainability' in higher education: From doublethink and newspeak to critical thinking and meaningful learning. *International Journal of Sustainability in Higher Education* 3 (3), pp. 221-232.

Wan Zahid, M.N. (1993), *Wawasan Pendidikan: Agenda Pengisian*. Nurin Enterprise, Kuala Lumpur.

Weakland, J.P. and Corcoran, P.B. (2009), The Earth Charter in higher education for sustainability. *Journal of Education for Sustainable Development* 3 (2), pp. 151-158 (doi: 10.1177/097340820900300210).

WCED (World Commission on Environment and Development) (1987), *Our Common Future*. Oxford University Press, New York.

Zainal, A.S. et al. (2010), *Sustainable Universities: Principles and Practices. Fundamental Research Grant Scheme (FRGS) Report*. Universiti Sains Malaysia.

Chapter 46

Evolving and Implementing Green Campus Strategies in Indian Higher Educational Institutions for Sustainable Development

Gopalsamy Poyyamoli, Arun Prasath Ramaswamy, Muthu Nandhivarman and Golda Arpudhalin Edwin

Abstract

In Western countries, many colleges and university campuses are embracing sustainability goals in an effort to improve their environmental responsibility, as well as maintaining their facilities in a more sustainable, environmentally friendly way. Areas targeted for improvement include campus energy/water use, landscapes, biodiversity, food service and waste streams. However, third world countries have lagged behind in these efforts. Though India has made a beginning in this direction recently, the achievements so far have not been either significant or desirable, since the institutions involved have mostly focused on just one or two issues – mostly covering water, energy and solid wastes – whilst largely failing to incorporate much needed inter and/or trans-disciplinary approaches. Almost any administrative or operational function on campus influences the use of resources/energy and the generation of waste to some degree. Hence, there is an urgent need to critically review the lessons learnt from the best practice case studies of other countries to derive valuable lessons for India. This paper aims to achieve this goal by evolving an action framework for green campus in the Indian context. This paradigm shift would help us to move closer to the globally/nationally accepted/ implemented principles/ programmes on sustainability as higher educational institutions can be used as living laboratories as well as for community outreach.

Introduction

Sustainable development is one of the biggest challenges of the 21st century. Of late, several universities have begun the debate about the content of this concept and the ways in which to integrate it into their university policy, organisation and activities (Weenen 2000; Sharp 2002). Since universities possess access to the most up-to-date knowledge of both environmental problems and technical solutions, they have the responsibility to lead society towards environmentally sustainable policies and practices (Uhl and Anderson 2001). Owens and Halfacre-Hitchcock (2006) emphasize that universities may be seen as "microcosms" of society, and therefore their experiences may inform efforts for change at the societal level. In this context, sustainability efforts are defined broadly to include changes in campus operations, financial and administrative planning and/or policy, and/or academic curricula and research that facilitate positive environmental changes.

Universities can nowadays be regarded as 'small cities' due to their large size, population, and various complex activities, which have some serious direct and indirect impacts on the local environment. The environmental pollution and degradation caused by universities in the form of energy and material consumption via activities and operations in teaching and research, provision of support services and in residential areas could be considerably reduced by an effective choice of organizational and technical measures (Alshuwaikhat and Abubakar

2008). However, the traditional practices and regulations of addressing environmental issues in a reactive (Brix et al. 2006) project and ad hoc manner in the campuses have become highly inefficient and cannot guarantee sustainability.

In spite of a number of sustainable development (SD) initiatives and an increasing number of universities becoming engaged with SD, most higher education institutions (HEIs) continue to be traditional, and rely upon Newtonian and Cartesian reductionist and mechanistic paradigms (Lozano et al. 2010). Many universities in the developing world especially, still focus only on direct environmental aspects like paper use and waste handling, even though the main tasks of the universities, namely education, research and co-operation with the surrounding society, that is the indirect aspects, are likely to have a considerable environmental impact (Wall et al. 2010).

The principle of sustainability was endorsed by Agenda 21 (United Nations 1992) and was globally confirmed again in the World Summit in Johannesburg (United Nations 2002). As a result of the conference, the years from 2005 to 2014 will be the United Nations Decade of Education for Sustainable Development (DESD) (http://iefworld.org/desd.htm). DESD has to be strategically viewed in relation to other international initiatives that are already in place, in particular the Millennium Development Goal (MDG), Clean Development Mechanism (CDM), Global Environmental Facility (GEF) and International Panel on Climate Change (IPCC) in order to facilitate campus sustainability on a wider scale.

The purpose of this paper is to provide a roadmap by providing the back ground and a broad conceptual framework for planning/implementing green campus strategies for sustainability in higher education institutions (HEIs) of India, based on best practice case studies elsewhere (Comm and Mathaisel 2005; Chambers 2009; Wall et al. 2010; UC Berkeley 2011).

Background

There are eleven of the most widely accepted SD initiatives in higher education, which relate to the university system: the Talloires Declaration, the Halifax Declaration, the Swansea Declaration, the Kyoto Declaration the Global Higher Education for Sustainability Partnership (GHESP), the Copernicus Charter, the Lüneburg Declaration, the Declaration of Barcelona, the Graz Declaration, the Turin Declaration, and the Abuja Declaration. They are all summarized and their relevance for ESD is discussed by Lozano et al. (2010). Of those mentioned, it is the Talloires Declaration which requires special attention.

As of February 10, 2012, over 400 university leaders in over 50 countries that include ten HEIs from India (nine of them from North India) have signed the Talloires Declaration that was composed in 1990 at an international conference in Talloires, France (http://www.ulsf.org/programs_talloires_td.html). This is the first official statement made by university presidents, chancellors, and rectors of a commitment for a ten-point action plan to incorporate sustainability and environmental literacy in teaching, research, operations, assessment, reporting and outreach at colleges and universities. In India, there is a wide-spread growing desire to address sustainability issues on college/university campuses; however, many lack the effort necessary to effect real and significant change. Part of the reason for their lack in effort is that many institutions do not understand what is at stake.

Though sustainability has become the mantra for development for virtually all sectors, HEIs are different. While HEIs have stakeholders, not shareholders, as in the case of corporate sectors, they are otherwise not immune to the many external drivers behind the corporate shift to sustainability (Bardaglio and Putman 2009; Lubin and Esty 2010). Yet implementing green/sustainability practices across all walks of the campus often proves to be considerably more challenging than implementing such practices in a corporate environment, because campuses have additional pressures (Walton and Galea 2005; Ferrer-Balas et al. 2008; Bardaglio and Putman 2009). These include:

- HEIs are typically tasked with the trinity of education, research and service; these three and sometimes competing orientations may dilute a focused orientation and create competing priorities.
- Different university constituents demand different services – students, faculty, staff and alumni – each of whom have varying, and sometimes competing, priorities in terms of sustainability. For instance, recycling efforts offer some experiential learning for students, but may contribute little towards advancing formal curricula. Likewise, cross-cutting sustainability-oriented courses are on the rise, but may be insufficient to advance campus renewable energy installations. It is often difficult to uncover cross-constituency synergies between these constituents.
- New domestic competition from for-profit enterprises increasingly commodifies educational products and cuts into market share. Campuses may be less likely to move forward on sustainability if programs and revenues are shrinking.
- With recession economies the "new normal," campus professionals are asked to cut costs, increase employee productivity but not payroll, and recruit and cultivate a new generation of administrative leaders to supply quality leadership succession.
- The typical structure of universities – including power concentrated at several levels and a philosophy of protecting tradition and academic freedom – hinders the sweeping change that is often required for the paradigm shift. By contrast, corporations can often streamline and focus on core products or services with comparative ease. University structures account for many of their good attributes, but their inherent structure may impede the wide-ranging changes that are often needed.

Therefore, there is an urgent need for a professional and systematic approach to revamp the curricula/syllabi across the disciplines, institutionalize the sustainability principles as well as to reduce the consumption of resources and negative impacts of the various campus operations. Unfortunately, this approach is generally lacking in most universities, especially in the developing world, and therefore, achieving sustainability is not reported to be easy (Elizabete et al. 2005).

Revamping the curriculum for greening the campus

HEIs are typically tasked with the trinity of education, research and service; these three and sometimes competing orientations may dilute a focused orientation and create competing priorities. Therefore, a complete revamping/reorientation of the curriculum are the essential vehicles for the implementation of ESD in higher education. To rearrange current curricula in higher education, educators will be required to address critical issues and to promote integrative and analytical thinking with problem-solving approaches (Ceulemans and De Prins 2010; Lambrechts and Ceulemans 2011).

The breadth and depth of environmental issues/problems/concerns have prompted greater international interest in the need for education for sustainable development (ESD) at all levels. The inherent multidisciplinary nature of sustainability provides unique challenges and opportunities for the development of institutional targets. HEIs are better placed to promote ESD in the campus and their neighborhoods due to their fundamental characteristics such as (Niu et al. 2010):

- Higher education tends to be more innovative and spontaneous and can therefore promote the emerging ESD agenda most effectively and train others to carry out its aims, particularly the future generations of teachers in primary and middle schools.
- Higher education has the expertise to offer far more scientific and technological support in energy saving and emissions reduction and to contribute to the development of indicators and assessment systems and the implementation of strategies for new technologies.
- Higher education educates students in all professions, trains future government leaders and thus influences future policy development. Its laboratories and demonstration projects also serve as examples of good practice for the public as well as for students on campus.

Several workers have suggested that higher education should and can act as a change agent for sustainability (Stephens et al. 2008; Corcoran and Wals 2004; Shriberg 2002b). To that end, campus sustainability programs have been the subject of recent case studies, conferences, and dissertations (Clugston and Calder 1999; Dahle and Neumayer 2001; Carpenter and Meehan 2002; Shriberg 2002b; Wright 2004; Price 2005; Hunting and Tilbury 2006; Nicolaides 2006; University of Calgary 2007; University of Toronto 2007; Ferrer-Balas et al. 2008; Savanick et al. 2008a, 2008b; SustainAbility 2008; Rauch and Newman 2009; The Princeton Review 2010; US Green Building Council 2010; Kevin et al. 2012). These studies have proposed and assessed current models (Shriberg 2002a, 2002b; Williamson 2012), identified barriers (Shriberg 2002b, 2003; Velazquez et al. 2005, 2006; Lozano et al. 2010), examined specific strategies of implementation such as environmental management systems (Noeke 2000; Carpenter and Meehan 2002; Spellerberg et al. 2004; Sammalisto and Arvidsson,2005; Nicolaides 2006; Jain and Pant 2010) and the role of collaboration (Moore et al. 2005). Price (2005) asserts that universities need to move from the perception that environmental management systems are marginally important and a "soft control issue", and separate from campus sustainability operations to a more integrated system within university sustainability.

The interdisciplinary approach to involvement of students in education and practice in the area of sustainability is well documented (Bhasin et al. 2003; Cortese 2005; Domask 2007; Sammalisto and Lindhqvist 2008; Stubbs and Cocklin 2008; Walker and Seymour 2008; GVSU 2009; Chhokar 2010; Jones et al. 2010). Students have been found to be formidable ambassadors for furthering sustainability both on and off campus through experiential learning. This engagement has allowed students to see the linkages between other areas of studies beyond the sciences, such as social science, community development, and political science, thus facilitating an interdisciplinary/trans-disciplinary outlook (http://www.khanya-aicdd.org/, http://www.gvsu.edu/sustainability/grantapp.htm, http://wscsd.org/). Students have also been able to learn valuable skills that have led to personal job creation and entrepreneurship.

In spite of these emerging trends, HEIs have not adopted sustainability strategies at the same rate, creating an increasing disparity between market needs and academic deliverables in the area of sustainability, including adopting sustainability strategies, producing intellectual capital, and knowledge workers (Wall et al. 2010).

The barriers

The earlier workers have identified the following barriers to work towards campus sustainability (Richardson and Lynes, 2007; Brinkhurst et al., 2011; Krizek et al., 2012; Lozano et al., 2010) that are very much relevant for Indian universities too:

- Lack of awareness on the multiple dimensions of SD due to the legacy of ad hoc campus development programs.

- Over-crowded curricula with little if any room for additional load and academic conservationism/traditions that tie universities to old mechanistic mental models.
- Lack of opportunities/encouragement of credit-linked green campus activities/action research.
- The fundamental incompatibility of sustainability's systems-thinking; and/or an interdisciplinary approach at odds with the reductionist, turf-heavy culture that earmarks the academe's traditional paradigm; insecurity and threat to academic credibility of teachers.
- Lack of motivation among some faculty members/researchers as they think that sustainability is not "hardcore" science. Not all faculty members identify their academic fields as related to sustainability, and even if they do, rarely does this extend to investing in implementation or activist efforts at their own university; if a project is to command a serious investment of time and energy in academia, it must either align with academic interests or at least be something respected by other academic professionals.
- Limitations posed by the heavy workloads faculty members' "capacity to address new problems – such as campus sustainability vs. commitments/priorities to requirements of their positions.
- Poorly-coordinated silos with Cartesian thinking.
- Resistance from/apathy of unsupportive supervisors/administrative heads or project partners who may not be aware of the potential advantages, lack of time and/or authority, and feeling disempowered as agents of change within a complex institution.
- Constituencies/activities leftover from the phase one acting as road blocks.
- Time intensive bureaucratic and consultative phases of advocating desirable sustainable changes within an institution due to bureaucracy.
- Lack of communication and collaboration between faculty and operations staff.
- Cycles of resource availability (both financial and human), shifting individual and organizational priorities, and fluctuations in the success of the initiatives themselves frequently challenge university sustainability initiatives.
- Scarcity of time/funds to participate/ lead the initiatives.
- Lack of support/encouragement for initiatives; skeptical supervisors can block changes or deflate entrepreneurial efforts; if initial attempts are unsuccessful or suppressed, it can lead to feelings of disempowerment within the institution. Moreover, once deterred by bureaucratic complexity or fear of criticism from more powerful or influential individuals or groups, faculty and staff may feel compartmentalized, powerless to effect change, and choose to not identify strongly with the university community as a whole.
- Poor/no adoption of environmental management systems.

However, such barriers have to be perceived as guidelines so that we can try to avoid or reduce these bottle necks by analyzing the root causes for them, which can help in facilitating effective solutions.

The way forward

Increasingly, tertiary education institutions are taking steps to address their environmental impact and particularly their contribution to climate change, including the implementation of formal or informal environmental management systems (Clarke and Kouri 2009; Koester et al. 2006; Jain and Pant 2010).

Clarke and Kouri (2009) highlight many different frameworks of Environmental Management Systems (EMSs): ISO14001; Higher Education 21 (UK); the EMS Self-Assessment

Checklist (USA); the Auditing Instrument for Sustainability in Higher Education (Netherlands); the Osnabrück University model (Germany) and the Sustainability University Model (Mexico). They argue that colleges and universities require EMS frameworks that are specific to their sector yet consider individual institution drivers. Fisher (2003) highlights EMS as a way to bring the business world into the classroom, which may provide usefulness for environmental education. Spellerberg et al. (2004) argue that an environmental policy and implementation plan can be just as appropriate and effective as an EMS and illustrates this through a case study of Lincoln University in New Zealand.

Notable among the global initiatives on ESD were reviewed by several workers: Stefanovic (2008) on the University of Toronto and the establishment of a RCE; Keen and Baldwin's (2004) research on collaboratively designed research between institutions of higher education and the community; the Grand Valley State University (GVSU) Sustainable Community Development Initiative (SCDI); Student Sustainability Partnership and Proactive Implementation Model for Sustainability (PIMS) (GVSU Sustainability Recognitions, 2010; GVSU STARS in sustainability efforts, 2011); RecycleMania (a national competition) (RecycleMania is Here! April 10, 2010).

Sustainable procurement from local sources (including fair trade sources) that mandates focus on sustainable practices in farming, sustainable waste management and recycling, grey water management, rain gardens, sustainable landscaping practices, development and monitoring of city and community sustainability indicators reports as well as assisting with developing sustainable neighborhoods and communities are some significant strategies that are adopted for campus sustainability (Brown 2009).

It is important that institutes of higher education leadership support sustainability by setting the tone and helping to create awareness, interest, and involvement (Velazquez et al. 2005; Stafford 2010). Increasing numbers of campuses across North America are taking steps to integrate sustainability into their operations and philosophies. For example, the Association for the Advancement of Sustainability in Higher Education (AASHE) reported 783 institutions as active members in North America, including 52 in Canada, in the summer of 2010 (AASHE 2010). However, sustainability is not just another issue to be added to an overcrowded curriculum, but a gateway to a different view of curriculum, of pedagogy, of organizational change, of policy and particularly of ethos (Sterling 2009).

Both administrators and students are powerful agents of change (top-down and bottom-up respectively) who can be effective at building enthusiasm and leading institutional change. However, the faculty and staff (the middle-out) are not simply adversarial or passive audiences to "engage", as they are often portrayed (UBC 2007). In many cases, the potential of faculty and staff as the makers of lasting change is far greater than that of administrative or student leaders.

Within universities, visionary faculty and staff have the opportunity to be social intrapreneurs (an extension of the "social entrepreneur" who creates positive change through entrepreneurial solutions). Like their counterparts in business, they have "less celebrated, but perhaps more reliable" and "political savvy, tact, teamwork and patience" that makes them effective at identifying appropriate and innovative opportunities for change. Case studies from institutions across North America allude to the important work of faculty and staff in areas ranging from facilities design and maintenance, program and curriculum design, food services, purchasing and administrative policies, student engagement and encouragement, fundraising, and networking campus groups with other sustainability action organizations (Savanick et al. 2008a; Brinkhurst et al. 2011). The majority of these strategies are not prohibitively complex or expensive, and their potential impact on campus sustainability can be significant.

The Ministry of New and Renewable Energy of the government of India has recently initiated several proactive programmes for wider adoption of renewable energy technologies/devices such as the National Action Plan on Climate Change (GOI 2008), the Jawaharlal Nehru National Solar Mission (JNNSM) (2009), the National Mission for Enhanced Energy Efficiency, solar/green campus/cities, the Special Area Demonstration Project Programme, the Energy Park scheme, the National Policy on Biofuels, policy support for grid interactive renewable power, and Green Special Economic Zones (SEZ). In addition, there are funding programmes offered by the Ministry of Environment and Forests of the government of India, such as the Environmental Research Programme (ERP), the Ecosystems Research Scheme (ERS), and the National Natural Resources Management System (NNRMS). They can be effectively used by HEIs in India to meet the essential seed funding for designing/implementing green campus strategies.

Top-down and bottom-up campus sustainability efforts can both have strengths, but the greatest potential for long-term change comes from active intrapreneurship by faculty and staff. When reviews of sustainability initiatives are framed only as top-down or bottom-up, or focus primarily on student-led or administratively championed projects, they obscure the important roles played by faculty and staff members. In order to encourage further innovation by the institutional "middle," the efforts of these leaders should be explicitly acknowledged and celebrated. Sharing the successes of faculty and staff intrapreneurs will do much to overcome barriers created by faculty and staff's own reluctance to realize their potential as leaders of change. Experiences of universities around the world, some of which have been highlighted here, demonstrate that faculty and staff are the most important and yet unrecognized champions of campus sustainability. Supporting their efforts is an essential part of achieving lasting progress towards sustainable universities (Brinkhurst et al. 2011).

Existing buildings account for the vast majority of the building stock on campuses. As such, retrofitting and renovating existing buildings offer the greatest opportunity for institutions to cut energy and water consumption, save money and reach their sustainability goals as most of the HEI buildings in India have not been built using green building principles. For the new/proposed buildings in Indian HEIs, green building norms should become mandatory that can be regulated by the town country planning departments of respective states.

Based on the best practice case studies and the distilled lessons learnt throughout the world, the following strategies are suggested to work towards campus sustainability in India:

- Add environmental sustainability to the existing mission and vision statement of the University;
- Incorporate creativity in curriculum design, campus-wide coordination, promotion/encouragement of sustainability in curricula/research, and community outreach;
- Address sustainability problems on campus through course credit;
- Encourage faculties by funding agencies and administrators "to perceive opportunities for alignment of their teaching and research efforts through practical experimentation and application within the campus system" such as greenhouse gas inventories, sustainability assessments, or green design projects;
- Empower and reward "sustainability champions" in the campus – can be students, faculty or staff/administrator –; provide proper academic recognition of the importance of multidisciplinary and transdisciplinary teaching, research and community outreach;
- Use pilot projects on bio-mimicry/sustainability to build capacity;
- Realize maximum energy/resource savings through technological innovation;

- Provide official permission and encouragement and the development and institutionalization of faculty/staff involvement in decision making as key strategies for empowering and motivating trained and skilled faculty/staff intrapreneurs;
- Use participatory goal-setting tools, such as online surveys, charrettes and campus-wide forums; solicit stakeholders to develop a set of environmental goals and related objectives that address the organizational, financial, operational, physical and cultural needs of the campus;
- Facilitate flexible and adaptive sustainability initiatives;
- Set up a task force with representatives from teaching, research, financing, facilities planning, utilities, construction management, housing, grounds and waste, purchasing and transportation;
- Promote green purchase, operations, and maintenance. Promote conservation efficiency/sustainability through effective management and incorporate them into the day-to-day activities in the university experiences;
- Develop conservation/sustainability awareness through educational activities, dissemination of training materials and media promotions to facilitate applied behavioral change; foster trans-disciplinarity; make SD an integral part of the institutional framework; create on-campus life experiences; 'Educate-the-Educators';
- Mobilize financial sources/incentives through various strategies such as: Sustainability Hero nominations and awards given to faculty and staff at George Mason University (George Mason University 2009); for faculty at the University of Indiana (Indiana University 2009); Climate Action Fund, the Green Initiative Fund, certified green departments, Sustainability Week and Earth Week activities at the University of California (UC Berkeley 2011). Green Power Partnership (GPP), a voluntary environmental program (VEP) initiated by the US Environmental Protection Agency (EPA), energy revolving funds, Energy Savings Performance Contracting (ESPC) for buildings, renewable energy-based power purchase agreements, sustainability fees, and mandatory student fees for renewable energy and energy efficiency adopted in several US colleges and universities; American College and University Presidents Climate Commitment (ACUPCC 2010);
- Network with all the relevant local/regional/national/international stakeholders for capitalizing on the collective wisdom as well as for outreach;
- Identify and amalgamate the campus development elements in to the emerging national/international development priorities/principles/programmes for sustainability so as to ensure a continuous flow of funds;
- Promote an integrated approach to campus planning, design, operations and maintenance
- Build a comprehensive framework by which to achieve the goal, placing faculty/administrative personnel in place to take action before the actual goal is announced;
- Articulate, define, recognize and reconcile inherent risks;
- Institutionalize the participation, cooperation, understanding and commitment of the campus community to ensure responsibility, accountability and long-term sustainability;
- Adopt widely used and credible certification and rating systems such as the Sustainability Tracking, Assessment and Rating System (STARS®) – a transparent, self-reporting framework for colleges and universities to gauge relative progress toward sustainability developed by AASHE (Wigmore and Ruiz, 2010); US Green Building Council – Leadership in Energy and Environmental Design (LEED) Green Building Certification System; Indian Green Building Council – LEED India; Green Rating for Integrated Habitat Assessment (GRIHA) of India:

- Periodic assessment/monitoring and reporting using key sustainability indicators to be disseminated through a dedicated sustainability website;
- Build sustainability research networks that encourage collaboration and dialogue across universities.

In adopting such green strategies, sustainability becomes the "Golden Thread" that permeates throughout the university system. Finally, in the "race to the top" to green campuses, the enhanced image as a certified green/sustainable campus provides competitive advantage over peer institutions, and attracts prospective students and faculty, and positive media attention. It is even more relevant as India will soon have to face the entry/competition of foreign universities (Shri Kapil Sibal, Minister of Human Resource Development 2010; Sudhakar 2011).

Conclusion

The suggested strategies in this review are neither prohibitively complex nor expensive, but their potential impact on campus sustainability can be highly significant. Whereas institutional transformation cannot be imposed solely from the top-down, it will not thrive without the support, vision, and involvement throughout the top, bottom, and middle. We need to kick start a series of intensive long term strategies for facilitating commitment, demonstration, and goal-setting on campus sustainability. In order for HEIs to embrace/integrate the principles of campus sustainability into their day to day affairs more effectively, we need a dedicated office of sustainability with a dean/coordinator, with separate staff to investigate the multiple dimensions of campus sustainability. This office should allow for the inclusion of student, staff, and faculty input, as these individuals on campus are most able to see where improvements can be made and where weakness can be remedied. Such an office will be central in providing a detailed action plan with set goals and a specific timeline of when those goals should be reached. Furthermore, this office would be the main driver in implementing and keeping notes upon the policies/planning in place. Fortunately, sustainability is an important funding objective for Indian/foreign donors and hence, the faculty may be encouraged to approach appropriate funding agencies through this office. In fact, the anticipated cost savings through campus sustainability strategies will be more than sufficient to pay the salary for the office of sustainability staff. Their hands can be effectively strengthened by student volunteers and sponsored internships from philanthropic industries through corporate social responsibility.

HEIs in India have the capacity and responsibility to play a unique role in addressing the unprecedented challenge of climate change and sustainability. They are also on the forefront of shaping life-styles and world-views of future generations, as well as influencing economic development and civil societies. The recently emerging green campus initiatives, as effective bottom-up approaches, could create test-beds and incubators for innovative solutions in various fields. Their collective efforts are of tremendous educational and demonstrative value, especially when coupled with research education and outreach activities. The methodology and strategies for sustainable green campus practices can be applied to other built environment settings, such as communities and cities, thus facilitating university campuses as living laboratories for working towards sustainability.

References

ACUPCC (2010), *ACUPCC Reporting System – Tangible Action Statistics.* Available at: http://acupcc.aashe.org/stats/tangible-actions/ (accessed 20 November 2010).

Alshuwaikhat H.M. and Abubakar, I. (2008), An integrated approach to achieving campus sustainability: assessment of the current campus environmental management practices. *Journal of Cleaner Production* 16, pp. 1777-1785.

AASHE (2010), *AASHE Institutional and System Office Members.* Available at: www.aashe.org/membership/members/institutional_members (accessed 5 July 2010).

Bardaglio, P. and Putman, A. (2009), *Boldly Sustainable: Hope and Opportunity for Higher Education in the Age of Climate Change.* National Association of College and University Business Officers, Washington DC.

Bhasin, B.S., Bjarnadottir, T., Das, V.N., Dock, M.M., Pullins, E.E., Rosales, J.R., Savanick, S., Stricherz, D.M. and Weller, L.A. (2003), Passport to Earth Summit 2002. A case study in exploring sustainable development at the University of Minnesota. *International Journal of Sustainability in Higher Education* 4 (3), pp. 239-249.

Brinkhurst, M., Rose, P., Gillian, M. and Ackerman, J.D. (2011), Achieving campus sustainability: top-down, bottom-up, or neither? *International Journal of Sustainability in Higher Education* 12 (4), pp. 338-354.

Brix, A., Brydon, T., Davidian, E., Dinse, K. and Vidyarthi, S. (2006), *Toward sustainable campus communities: evaluating alternative development scenarios.* MSc. project, University of Michigan, USA.

Brown, M.J. (2009), Campus Triple Bottom Line: College Sustainability and Grand Valley State University. *McNair Scholars Journal* 13 (1), Article 4. Available at: http://scholarworks.gvsu.edu/mcnair/vol13/iss1/4.

Carpenter, D. and Meehan, B. (2002), Mainstreaming environmental management: Case studeis from Australasian universities. *International Journal of Sustainability in Higher Education* 3 (1), pp. 19-37.

Ceulemans, K. and De Prins, M. (2010), Teacher's manual and method for SD integration in curricula. *Journal of Cleaner Production* 18 (7), pp. 645-651.

Chambers, D. (2009), Assessing and planning for environmental sustainability: a framework for institutions of higher education. In Leal Filho, W. (Ed.), *Sustainability at Universities: Opportunities, Challenges and Trends.* Peter Lang, Frankfurt.

Chhokar, K.B. (2010), Higher education and curriculum innovation for sustainable development in India. *International Journal of Sustainability in Higher Education* 11 (2), pp. 141-152.

Clarke, A. and Kouri, R. (2009), Choosing an appropriate university or college environmental management system. *Journal of Cleaner Production* 17 (11), pp. 971-84.

Clugston, R.M. and Calder, W. (1999), Critical dimensions of sustainability in higher education. In Leal Filho, W. (Ed.), *Sustainability and University Life*, pp. 31-46. Peter Lang, Frankfurt.

Comm, C.L. and Mathaisel D.F.X. (2005), An exploratory study of best lean sustainability practices in higher education. *Qual. Assurance Edu.* 13, pp. 227-240.

Corcoran, P.B. and Wals, A.E.J. (Eds.) (2004), *Higher Education and the Challenge of Sustainability: Problematics, Promise, and Practice.* CERC Studies in Comparative Education, Springer, Berlin.

Cortese, A.D. (2005), Integrating sustainability in the learning community. *Facilities manager* 21 (1), pp. 28-35.

Dahle, M. and Neumayer, E. (2001), Overcoming barriers to campus greening: A survey among higher educational institutions in London, UK. *International Journal of Sustainability in Higher Education* 2 (2), pp. 139-160.

Domask, J.J. (2007), Achieving goals in higher education: An experiential approach to sustainability studies. *International Journal of Sustainability in Higher Education* 8 (1), pp. 53-68.

Elizabete, M., Seifferta, B. and Lochb, C. (2005), Systemic thinking in environmental management: support for sustainable development. *Journal of Cleaner Production* 13 (12), pp. 1197-1202.

Ferrer-Balas, D., Adachi, J., Banas, S., Davidson, C.I., Hoshikoshi, A., Mishra, A., Motodoa, Y., Onga, M. and Ostwald, M. (2008), An international comparative analysis of sustainability transformation across seven universities. *International Journal of Sustainability in Higher Education* 9 (3), pp. 295-316.

Fisher, R.M. (2003), Applying ISO 14001 as a business tool for campus sustainability: a case study from New Zealand. *International Journal of Sustainability in Higher Education* 4 (2), pp. 138-150.

George Mason University (2009), *Office of sustainability.* Faculty and Staff Awards. Available at: http://sustainability.gmu.edu/for_faculty/sustainabilityaward.html (accessed 31 August 2009).

GOI, Prime Ministers Council on Climate Change (2008), *National Action Plan on Climate Change 2008*, pp. 50 (accessed at pmindia.nic.in/Climate%20Change.doc on 25th Feb 2012).

GVSU (2009), *Grand Valley State University Making it Happen, Applied Sustainability at GVSU.* Retrieved April 11, 2010 http://www.gvsu.edu/cms3/assets/1ACDDEF0-A15A-67B1-F268BE06 B2416593/2655-App_Sus_Broch_2[1].pdf.

GVSU STARS in sustainability efforts (2011) at http://www.mlive.com/business/west-michigan/index.ssf/2011/03/grand_valley_state_university_3.html (accessed 21 January 2012).

Hunting, S.A. and Tilbury D. (2006), *Shifting towards sustainability: Six insights into successful organisational change for sustainability.* Australian Research Institute in Education for Sustainability (ARIES) for the Australian Government Department of the Environment and Heritage. ARIES, Sydney.

Indiana University (2009), *New sustainability teaching awards recipients announced.* Available at: http://newsinfo.iu.edu/news/page/normal/9799.html (accessed 31 August 2009).

Jain, S. and Pant, P. (2010), Environmental management systems for educational institutions. *International Journal of Sustainability in Higher Education* 11 (3), pp.236-49.

Jones, P., Selby, D. and Sterling, S. (Ed.) (2010), *Sustainability Education-Perspectives and Practice across Higher Education*, pp. 384. Earthscan, UK.

Keen, C. and Baldwin, E. (2004), Students promoting economic development and environmental sustainability: An analysis of the impact of involvement in a community-based research and service-learning program. *International Journal of Sustainability in Higher Education* 5 (4), pp. 384-394.

Koester, R., Eflin, J. and Vann, J. (2006), Greening of the campus: a whole-systems approach. *Journal of Cleaner Production* 14 (9-11), pp. 769-79.

Krizek, K.J., Newport, D., White, J. and Townsend, A.R. (2012), Higher education's sustainability imperative: how to practically respond? *International Journal of Sustainability in Higher Education* 13 (1), pp. 19-33.

JNNSM (2010), *Jawaharlal Nehru National Solar Mission (JNNSM) (2009).* pib.nic.in/archieve/others/2010/jul/jnnsm.pdf (accessed 27 February 2012).

Lambrechts, W. and Ceulemans, K. (2011), *Sustainability in Higher Education: Experiences using the Auditing Instrument for Sustainability in Higher Education (AISHE).* Brussels, HUB Research Paper 2011/11 at https://lirias.hubrussel.be/bit stream/123456789/4822/1/11HRP11.pdf (accessed 6 March 2012).

Lozano, R., Lukman, R., Lozano, F.J., Huisingh, D. and Zilahy, G. (2010), *Jumping sustainability meme: SD transfer from society to universities. Proceedings of the 6th Conference 'Environmental Management for Sustainable Universities' (EMSU)*, Delft, The Netherlands, 25-29 October 2010.

Lubin, D.A. and Esty, D.C. (2010), The big idea – the Sustainability Imperative. *The Harvard Business Review*, pp. 10. www.hbr.org.

Moore, J., Pagani, F., Quayle, M., Robinson, J., Sawada, B., Spiegelman, G. and Van Wynsberghe, R. (2005), Recreating the university from within: collaborative reflections on the University of British Columbia's engagement with sustainability. *International Journal of Sustainability in Higher Education* 6 (1), pp. 65-80.

Nicolaides, A. (2006), The implementation of environmental management towards sustainable universities and education for sustainable development as an ethical imperative. *International Journal of Sustainability in Higher Education* 7 (4), pp.414-424.

Niu, D., Dahe Jiang, D. and Li, F. (2010), Higher education for sustainable development in China. *International Journal of Sustainability in Higher Education* 11 (2), pp. 153-162.

Noeke, J. (2000), Environmental management systems for universities: a case study. *International Journal of Sustainability in Higher Education* 1 (3), pp. 239.

Owens, K.A. and Halfacre-Hitchcock, A. (2006), As green as we think? The case of the College of Charleston green building initiative. *International Journal of Sustainability in Higher Education* 7 (2), pp. 114-128.

Price, T.J. (2005), Preaching what we practice: experiences from implementing ISO 14001 at the University og Glamorgan. *International Journal of Sustainability in Higher Education* 6 (2), pp. 161-178.

Rauch, J.N. and Newman, J. (2009), Institutionalizing a greenhouse gas emission reduction target at Yale. *International Journal of Sustainability in Higher Education* 10 (4), pp. 390-400.

RecycleMania is Here! Retrieved April 10, 2010, from http://www.gvsu.edu/facilitiesservices/index.cfm?id=7A35702E-D089-7913-B495E8C24D8EB962.

Richardson, G.R.A. and Lynes, J.K. (2007), Institutional motivations and barriers to the construction of green buildings on campus: a case study of the University of Waterloo, Ontario. *International Journal of Sustainability in Higher Education* 8 (3), pp. 339-354.

Sammalisto, K. and Arvidsson, K. (2005), Environmental management in Swedish higher education: Directives, driving forces, hindrances, environmental aspects and environmental co-ordinators in Swedish universities. *International Journal of Sustainability in Higher Education* 6 (1), pp. 18-35.

Sammalisto K. and T. Lindhqvist, I. (2008), Integration of Sustainability in Higher Education: A Study with International Perspectives, *Innovative Higher Education* 32 (4), pp. 221-233.

Savanick, S., Strong, R. and Manning, C. (2008a), Explicitly linking pedagogy and facilities to campus sustainability: lessons from Carleton College and the University of Minnesota. *Environmental Education Research* 14 (6), pp. 667-679.

– (2008b), Using a class to conduct a carbon inventory: a case study with practical results at Macalester College. *International Journal of Sustainability in Higher Education* 10 (3), pp. 228-238.

Sharp, L. (2002), Green campuses: the road from little victories to systemic transformation. *International Journal of Sustainability in Higher Education* 3 (2), pp. 128-45.

Shriberg, M. (2002a), Institutional assessment tools for sustainability in higher education: strengths, weaknesses, and implications for practice and theory. *Higher Education Policy* 15, pp. 153-67.

– (2002b), Sustainability in US higher education: organizational factors influencing campus environmental performance and leadership. Unpublished dissertation, University of Michigan.

– (2003), Is the 'maize-and-blue turning' green? Sustainability at the University of Michigan. *International Journal of Sustainability in Higher Education* 4 (3), pp. 263-76.

Shri Kapil Sibal, Minister of Human Resource Development (2010), *The foreign educational institutions (regulation of Entry and operations) bill, 2010 – A Bill*, Bill (57) of 2010, GMGIP MRND-1854LS-21.04.2010.

Spellerberg, I.F., Buchan, G.D. and Englefield R. (2004), Need a university adopt a formal environmental management system?: Progress without an EMS at a small university. *International Journal of Sustainability in Higher Education* 5 (2), pp. 125-132.

Stafford, S.L. (2010), *How Green Is Your Campus? An Analysis of the Factors that Drive Universities to Embrace Sustainability.* Contemporary Economic Policy. Published Online: October 18, 2010 (doi: 10.1111/j.1465-7287.2010.00231.x).

Stefanovic, I.L. (2008), Educational Alliance for a Sustainable Toronto: The University of Toronto and the City's UNU Regional Centre of Expertise. *International Journal of Sustainability in Higher Education. Special issue on Regional Centres of Expertise on Education for Sustainable Development* 9 (4), pp. 416-427.

Stephens, J.C., Hernandez, M.E., Román, M., Graham, A.C. and Scholz, R.W. (2008), Higher education as a change agent for sustainability in different cultures and contexts. *International Journal of Sustainability in Higher Education* 9 (3), pp. 317-338.

Sterling, S. (2009), Towards ecological intelligence. In Stibbe (Ed.), *The Handbook of Sustainability Literacy – Skills for a changing world.* Green Books, Dartington.

Stubbs, W. and Cocklin, C. (2008), Teaching Sustainability to business students: shifting mindsets. *International Journal of Sustainability in Higher Education* 9 (3), pp. 206-221.

Sudhakar, J. (2011), Impact of Foreign Universities on Indian Higher Education. *Journal of Commerce and Management Thought* 2 (1), pp. 119-127

SustainAbility (2008), *The Social Intrapreneur: A Field Guide for Corporate Changemakers, Skoll Program Report*. Available at: www.sustainability.com (accessed 15 July 2009).

The Princeton Review (2010), *The Princeton Review's Guide to 286 Green Colleges.* Available at: www.princetonreview.com/green-guide.aspx (accessed 10 November 2010).

UBC (2007), *The UBC Sustainability Report 2006-2007*. Available at: www.sustain.ubc.ca (accessed 31 August 2009).

Uhl, C. and Anderson, A. (2001), Green destiny: universities leading the way to a sustainable future. *BioScience* 51, pp. 36-42.

United Nations (1992), *Agenda 21, the Rio Declaration on Environment and Development.* Available at: www.un.org/esa/sustdev/documents/agenda21/index.htm (accessed 1 June 2003).

– (2002), *Johannesburg Summit*. Available at: www.johannesburgsummit.org/and www.un.org/esa/sustdev/index.html (accessed 1 June 2003).

University of Calgary (2007), *Campus sustainability assessment: curriculum and research*. Available at: www.ucalgary.ca/sustainability/ (accessed 31 August 2009).

University of Toronto (2007), *Annual Sustainability Report 2007*. Available at: http://sustainability.utoronto.ca/Assets/Sustainability+Digital+Assets/Research+and+Publications/Annual+Reports/annual2007.pdf (5 July 2010).

UC, Berkeley (2011), *Campus Sustainability Report, Snapshots of Sustainability,* pp. 42. UC, Berkeley.

US Green Building Council (2010), *Roadmap to a Green Campus*, pp. 112. US Green Building Council, Washington, DC.

Velazquez, L., Munguia, N. and Sanchez, M. (2005), Deterring sustainability in higher education institutions: an appraisal of the factors which influence sustainability in higher education institutions. *International Journal of Sustainability in Higher Education* 6 (4), pp. 383-391.

Velazquez L., Munguia, N., Platt A. and Taddei, J. (2006), Sustainable University: what can be the matter? *Journal of Cleaner Production* 14, pp. 810-819.

Walker, J.B. and Seymour, M.W. (2008), Utilizing the design charrette for teaching sustainability. *International Journal of Sustainability in Higher Education* 9 (2), pp. 157-169.

Wall, A., Birou, L. and Wall, J.L. (2010), *Sustainability Strategies in Universities: Developing a Conceptual Framework.* 2010 Oxford Business and Economics Conference Program, pp. 26. St. Hugh's College, Oxford University, Oxford, UK.

Walton, S.V. and Galea, C.E. (2005), Some considerations for applying business sustainability practices to campus environmental challenges. *International Journal of Sustainability in Higher Education* 6 (2), pp. 147-160.

Williamson, S.R. (2012), A systems approach to reducing institutional GHG emissions *International Journal of Sustainability in Higher Education* 13 (1), pp. 46-59.

Weenen, H.V. (2000), Towards a vision of a sustainable university. *International Journal of Sustainability in Higher Education* 1 (1), pp. 20-34.

Wigmore, A. and Ruiz, M. (2010), Sustainability assessment in higher education institutions. The stars system. *Ramon Llull J Appl Ethics* 1, pp. 25-42

Wright, T.S. (2004), Consulting stakeholders in the development of an Environmental Policy Implementation Plan: a Delphi Study at Dalhousie University. *Environmental Education Research* 10 (2), pp. 179-194.

http://iefworld.org/desd.html (accessed on 6th March 2012).

http://www.gvsu.edu/sustainability/grantapp.htm (accessed on 6th March 2012).

http://www.khanya-aicdd.org/ (accessed on 6th March 2012).

http://wscsd.org/ (accessed on 6th March 2012).

http://www.ulsf.org/programs_talloires_td.html (accessed on 6th March 2012).

III. Implementing Sustainable Development in Research and in Outreach Programmes

Chapter 47

Creating and Assessing a Campus Culture of Sustainability: The University of Michigan Experience

Robert W. Marans and Michael Shriberg

Abstract

With the focus on operational campus greening, climate commitments and environmental education, the sustainability in higher education movement has paid limited attention to engendering deep cultural change among university stakeholders. In fact, systematic efforts towards and evaluations of cultural change programs for faculty, staff and students are almost completely absent. To help fill this void, this paper presents an overview of the University of Michigan's (U-M) experience in creating and monitoring a culture of sustainability, with a model and methods that are broadly applicable. The paper begins by describing two key "cultural change" initiatives at U-M: the Planet Blue Ambassadors (a peer-to-peer "EcoReps" program) and "Sustainability & the Campus" (a course that links students and operational staff). It then describes a groundbreaking new program – the Sustainability Cultural Indicators Project (SCIP). The program aims at monitoring U-M's progress in moving towards a culture of sustainability by collecting indicators on understanding (literacy), commitment (motivation/attitudes), and behaviors of students, faculty and staff. The program design – which was developed through a unique collaborative "Integrated Assessment" – includes following a cohort of undergraduate students through their U-M experience as well as annual surveys of graduate students, staff and faculty. The indicators can serve as a model for campuses striving to complete the self-evaluation for AASHE (Association for the Advancement of Sustainability in Higher Education) STARS (Sustainability Tracking, Assessment & Rating System) or other campuses interesting in monitoring progress toward advancing a culture of sustainability.

Introduction

The scientific and technical challenges of sustainability on which universities and colleges, as well as practitioners, have focused the bulk of their efforts represent only part of the necessary intellectual and social transition to a sustainable society. Institutions of higher education play a pivotal role in addressing the more difficult yet powerful part of the sustainability transition. That role is in creating and maintaining a "culture of sustainability" among members of the university community. A culture of sustainability has been defined as "a culture in which individuals are aware of major environmental (and social/economic) challenges, are behaving in sustainable ways, and are committed to a sustainable lifestyle for both the present and future" (Marans et al. 2010). To achieve this ideal state within institutions of higher education, Sharp (2002) calls for a rethinking of organizational action and actors that questions the prevailing assumptions of organizational rationality that stays within the confines of the current systems. Similarly, Senge (2000) stresses the importance of cultivating a "learning organization," rather than a "knowing organization" since change at higher education institutions is a "complex learning and unlearning process for all concerned" (Scott 2004). Therefore, nothing less than a paradigm shift in organizational thinking is needed for colleges and universities to promote cultural transformation.

This organizational transformation is needed in all sectors of society. Yet institutions of higher education can and should be at the forefront with the collective mission of fostering sustainability through our actions and through cultivating future sustainability leaders. To date, however, most campus sustainability efforts stop either at "greening" or at the level of institutional commitments to eco-efficiency, climate (and waste) mitigation and increasing environmental education. Though calls for institutional and cultural transformation are multiplying at a rapid rate, rarely do institutions address the deeper cultural change necessary to transform into sustainable organizations which empower citizens with a sustainability perspective; instead, focus is often on implementing many individual projects, isolated initiatives, or broad commitments (Sharp 2002; 2009). This is partly attributable to the lack of guidance for institutions attempting to follow this more uncertain and uncomfortable path.

In this paper, we hope to fill this void by offering a case study documenting what is happening at the University of Michigan (U-M). Following an overview of the challenge faced by U-M and past achievements, we review two recent cultural change initiatives in the context of their transformative potential. Next, we describe a program aimed at monitoring university progress in moving toward a culture of sustainability. Finally, we discuss the potential for U-M's approach to assessing the culture of sustainability to be replicated in other organizations and institutions of higher education.

A Challenge to Cultural Change and Past Achievements – The U-M Experience

One of the most challenging aspects of cultural change on campuses is size. U-M's Ann Arbor campus is analogous to a diverse and expansive small city. The university has over 80,000 students, faculty, and staff who vary in age, background, and experience. They learn, conduct research, study, and work in more than 450 buildings occupying over 31 million square feet spread across more than 3,000 acres. Although the size and diversity of the population and scale of the campus need to be addressed in creating change in the culture of sustainability on campus, it has not impeded the university's earlier efforts to address environmental concerns.

U-M began to embrace environmental protection more than 40 years ago. For example, the university held the first mass Earth Day celebration in 1970, the same year U-M initiated a campus-wide recycling program. In the early 1960s the university's power plant converted from coal to natural gas cogeneration. These early initiatives were sparked in part by student activism combined with leading environmental education programs. U-M presently offers 10 undergraduate degrees and 11 Master's degrees on some aspect of sustainability with nearly 400 sustainability-oriented courses available to students. Moreover, U-M's School of Natural Resources & Environment is one of the oldest educational and research outlets for sustainability in the US, is internationally well respected, and serves as the hub for the nearly 400 faculty members involved in sustainability-related research.

Currently, U-M continues to advance its operational sustainability efforts. For example, the university maintains the largest fleet of alternative fuels vehicles amongst US universities; its hybrid busses are amongst those offering free service to over nine million riders per year. U-M's energy efficiency efforts are headlined by a "Planet Blue" program that decreased overall energy usage by 12% annually in 71 buildings, with an energy cost avoidance of nearly US$3.7 million each year.[1] The university has also adopted a LEED Silver building standard for major new non-clinical building construction and renovation projects.

1 Plant Blue was launched in 2008 as a way of engaging building occupants along with operational personnel in reducing building energy costs and increasing recycling. In 2011, Planet Blue became part of

In fall 2011, President Mary Sue Coleman launched a Presidential Initiative in order to further U-M's commitment to environmental protection to sustainability. At the core of the initiative is a series of goals described later in the paper (see also Table 1). In her public address, President Coleman stated: "Today we begin an important new chapter – one that will alter the face of our campus and, more important, the character of our teaching, research and impact as a global leader" and "Sustainability defines the University of Michigan … A great university such as ours does not blink when presented with difficult challenges."[2] An underlying priority that runs throughout the presidential initiative is the goal of creating a culture of sustainability among U-M's students, faculty and staff. Among the many campus activities aimed at establishing this goal are the Planet Blue Ambassadors program and an undergraduate course, 'Sustainability & the Campus'.

Planet Blue Ambassadors program

Perhaps the most direct efforts to create a culture of sustainability at institutions of higher education come from "Eco-Rep" programs (Erickson 2010). These programs typically rely on peer-to-peer methodologies and use social-based marketing to promote sustainable practices and lifestyles. U-M's version of this emerging model is called the "Planet Blue Ambassadors" program, which operates under the following mission: "Through hands-on learning and peer-to-peer engagement, the Planet Blue Ambassadors (PBA) program fosters environmentally and socially responsible thinking and action to advance a campus-wide culture of sustainability." The program was launched in 2011 as part of the presidential initiative. It includes 25 undergraduate students living in the residence halls as well as 15 staff volunteers from throughout the university.

The backbone of the PBA program is a seminar designed to provide the operational skills and intellectual framework for students and staff to be effective in cultural change initiatives. Through a series of workshops, the seminar provides PBAs with tools to create their own methods for inducing behavioral change for sustainability. Faculty and staff serve as guides for PBAs, helping to bring appropriate resources, models and contacts to students while playing a mentorship rather than a traditional teaching role. PBAs are encouraged to learn from each other, with a particular emphasis on the McKenzie-Mohr "Fostering Sustainability Behavior" model (2011). Students meet for four hours per month, while staff members meet for two hours each month, including a joint session with students. In this way, students and staff co-learn through their experiences and insights.

The activities that PBAs engage in include organizing sustainability programs, auditing living and work environments for sustainability, conducting focus groups in connection with the monitoring effort described later, planning social activities to build "community", engaging in informal and formal discussions, establishing and tracking of a personal sustainability goal, and preparing self-reflective papers on leadership development and sustainability transformations. Moreover, the students and staff are jointly proposing a sustainability certification program – one for students in residence halls and one for staff in office environments. While the full results are yet to be determined (though they are being tracked and monitored closely), the initial enthusiasm for the program and receptivity of other U-M stakeholders holds strong promise for beginning to fulfill the cultural transformation mission.

U-M's overall sustainability initiative. For a discussion of the Plant Blue program and its origins, see: http://opsteams.plantops.umich.edu/ and Marans and Edelstein (2010).

2 For the complete text of President Coleman's sustainability address, see: http://sustainability.umich.edu/news/going-green-staying-blue-sustainability-michigan.

"Sustainability & the Campus" course

"Sustainability & the Campus" aims to create a culture of sustainability on campus by linking faculty, students, and staff in a shared mission through a project-based course. Initiated in 2001, the course is currently offered every semester to approximately 40 undergraduate students. "Sustainability & the Campus" is project-focused, directly linked with operational units that submit project proposals to have groups of four to six students work with them. These projects are complemented by lectures and activities on organizational change and leadership from the instructor as well as guest lecturers from operational units. Using the campus as a living-learning laboratory, the students also engage in site visits and other means of "hands-on" learning.[3]

An extensive review of the course was recently conducted using focus groups, interviews, or surveys with current or former students and with project sponsors (Shriberg and Harris 2012). While most of the comments were directed at specific outcomes – such as the professional skills and relationships that the course engendered – or fond memories of hard work and project success, one striking finding dealt with the impact of the course on the culture of sustainability on campus. This outcome emerged organically from a combination of joint investment in the work (it was mostly voluntary from all parties) and an unusually close mixing of students and staff, with faculty providing guidance and expertise. Students and staff learned that they have complementary needs and skills which could be brought together effectively with the right structure and time commitment, helping to create a "culture of sustainability". Students, staff and faculty live in very different cultural worlds at U-M. Often termed "loose coupling of organizational systems", and particularly prevalent on large or decentralized campuses, the problem is that, for example, the inner workings of the philosophy department are only loosely connected to those of the power plant operators. This separation can lead to a dynamic of competition or, at least, lack of understanding that deters the cultural transformations necessary for deep organizational changes. "Sustainability & the Campus" helps break this barrier through close collaboration with students and staff, often exposing mutual needs and a shared sense of purpose.

Another unexpected but particularly strong finding was that the course serves to facilitate the integration or convergence of views among students, faculty and staff through class-based projects. Specifically, students benefit greatly from the "real-world" view of university staff, learning about complexity, responsibility, and the types of analysis and arguments that can lead to high-level support. The idealism of students often collides with the dim view of change shared by many of the staff. Staff members benefit from the students' perspective, enthusiasm and hard work, often using interest on the part of the university's "customers" to advance projects that might have been untouched by university officials. For example, one sponsor described the key benefit as "increasing the engagement" and "buy-in" of people on campus, further explaining that projects that are directed at students but conducted by staff require student input up front in order to be successful.

While the outcomes of student projects vary widely, both students and staff reported high levels of satisfaction with skills and confidence acquisition as well as a mutual understanding of each other's perspective. Course projects have launched collaborative initiatives such as campus gardens, a published booklet called "How To Be a Green Wolverine", and several efforts to improve recycling and composting around campus. While small compared to the scale of the prob-

3 For the most recent course syllabus and a listing of past project reports, see: http://www.graham.umich.edu/education/campus.php.

lem and campus, these collaborative initiatives through one course are key underpinnings to creating a campus culture of sustainability.

Fig. 1: Organizational framework for campus sustainability integrated assessment

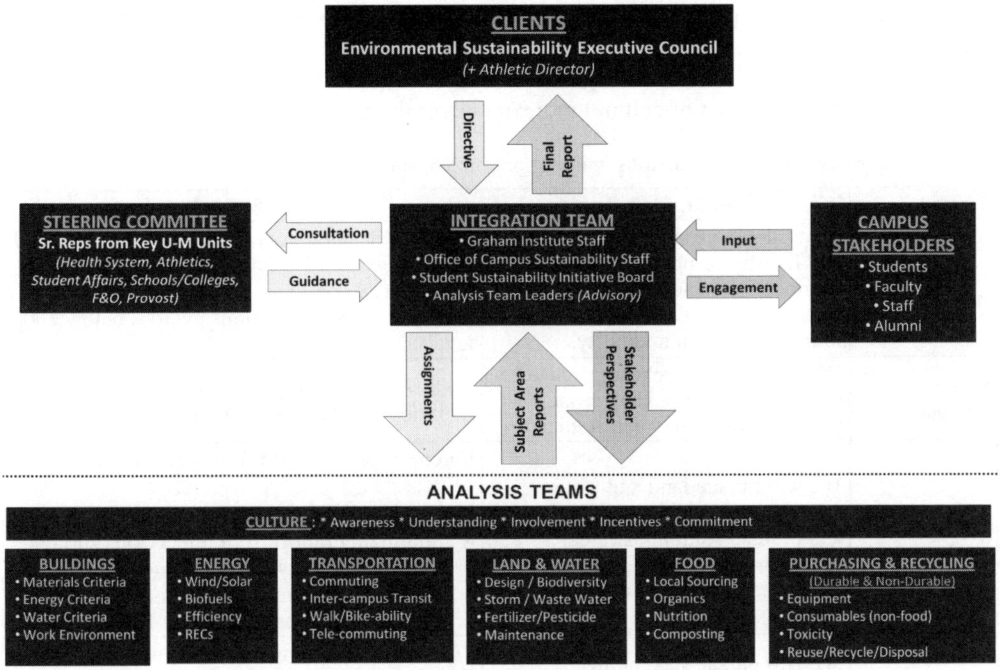

Monitoring progress in moving toward a culture of sustainability

The Planet Blue Ambassadors program was an outgrowth of a comprehensive campus-wide sustainability integrative assessment (IA) conducted on U-M operational campus sustainability from 2010-2011.[4] An IA is a stakeholder process to "collectively define problems, address diverse perspectives, use best-available information, and establish partnerships with the goal of analyzing options for making positive change." The organizational framework and operations of the IA shown in Fig. 1 demonstrate how the intense stakeholder process focused around key themes of buildings, transportation and travel, purchasing and recycling, food, energy, land and water, and culture. With U-M's Environmental Sustainability Executive Council as a client, and an iterative process with multiple levels of the organizations, key data and recommendations for each area were developed.

The integration of this data and these recommendations was based on the collective work of faculty-led teams of students covering each topic. Subsequently, the integration resulted in four broad themes – climate action, waste prevention, healthy environments, and community – with guiding principles and targeted goals established for each by 2025 (Table 1).[5] Cultural change initiatives stem from the principles outlined under community awareness. They indi-

4 For a discussion of U-M's integrative assessments, see: http://www.graham.umich.edu/ia/ and Lund et al., (2011).

5 The Integrated Assessment Report which incorporates the report of the Culture Team is found on: http://www.graham.umich.edu/pdf/CampusIA-FinalReport.pdf.

cate that U-M will "pursue evaluation strategies toward a campus-wide ethic of sustainability" as articulated in President Coleman's September 2011 sustainability speech announcing the sustainability goals. Specifically, she stated that "we will scientifically measure and report our progress and behavior as a community … ISR (Institute for Social Research) researchers will measure the sustainability attitudes and activities of students, faculty and staff, as well as identify where we can improve." The evaluation strategies involve a groundbreaking program for monitoring U-M's progress in moving toward a culture of sustainability. Progress would be determined by tracking a set of cultural indicators over time.

Table 1: Integrative assessment themes, guiding principles, and goals

Theme	Guiding Principle	2025 Goals
Climate Action	We will pursue energy efficiency and fiscally-responsible energy sourcing strategies to reduce greenhouse gas emissions toward long-term carbon neutrality.	Reduce greenhouse gas emissions *(scopes 1&2)* by 25% below 2006 levels. Decrease carbon intensity of passenger trips on U-M transportation options by 30% below 2006 levels.
Waste Prevention	We will pursue purchasing, reuse, recycling, and composting strategies toward long-term waste eradication.	Reduce waste tonnage diverted to disposal facilities by 40% below 2006 levels.
Healthy Environments	We will pursue land and water management, built environment, and product sourcing strategies toward improving the health of ecosystems and communities.	Purchase 20% of U-M food from sustainable sources. Protect Huron River water quality by: • minimizing runoff from impervious surfaces *(outperform uncontrolled surfaces by 30%)*, & • reducing the volume of land management chemicals used on campus by 40%
Community Awareness	We will pursue stakeholder engagement, education, and evaluation strategies toward a campus-wide ethic of sustainability.	*There is no stretch goal recommendation for this theme. However, the report recommends investments in multiple actions to educate our community, track behavior, and report progress over time.*

Conceptual framework

Concepts of culture of sustainability range from the world-view of sustainability among members of the university community to their involvement in sustainability practices on a daily basis. Measuring these concepts over time requires both direct and indirect contact with students, faculty, staff, and even alumni. The annual reporting of these measures or cultural indicators is intended to complement the campus-wide environmental indicators that have been reported by the University since 2004, which are very comprehensive in scope yet focused on traditional measures like water and energy use, transit ridership, procurement, etc.[6]

Direct contact with members of the university community will involve surveys (individual indicators) while indirect measures will be obtained through university records or field observations (generic or aggregate indicators). Examples of the latter include the proportion of campus bike racks that are occupied, proportion of university parking spaces occupied, number of students involved in sustainability organizations both on- and off-campus, etc. Surveys with members of the university community will generate cultural indicators that reflect levels

6 Among the key environmental indicators by the university are building energy use, greenhouse gas emissions, and waste disposal. For a complete listing of indicators and annual sustainability reports, see: http://www.ocs.umich.edu/reporting.shtml.

of understanding (literacy), commitment (motivation), and behaviors of students, faculty, staff and alumni.

Fig. 2: Cultural indicator operational framework

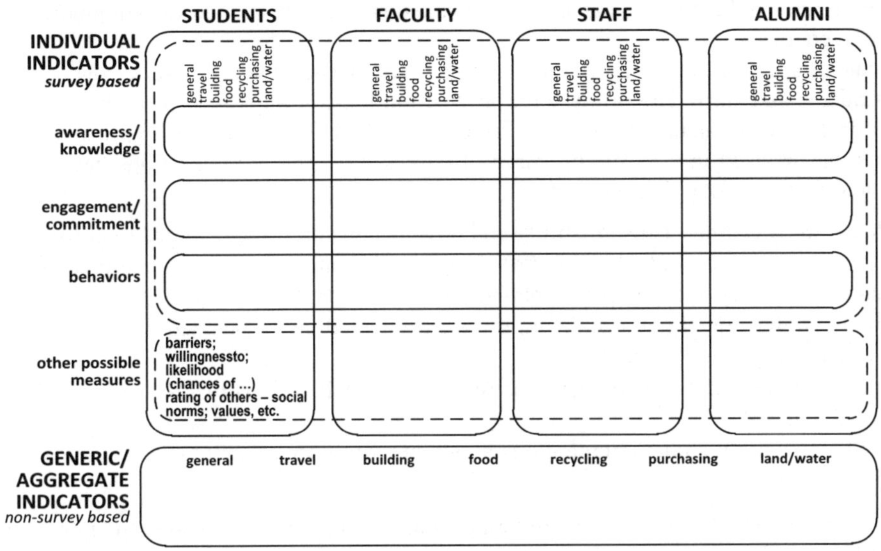

Fig. 2 presents an operational framework for the individual and aggregate cultural indicators covering buildings, travel, land and water, food, and purchasing and recycling The top set of four boxes represent the survey component used to measure individual indicators for each population group while the bottom box shows the aggregate indicators representing each topical area that reflect all members of the University community. Since the survey components offer opportunities to collect other kinds of information besides cultural indicator data, the figure shows other possible topics that could be measured as part of the questionnaires administered to students, faculty, staff, and alumni (i.e., barriers, willingness to pay, ratings of others, and so forth). The operational framework is intended to serve as a guide to the detailed planning needed to develop and measure the cultural indicators.

The process

The program for monitoring the culture of sustainability on campus including the operational framework for measuring and reporting indicators was submitted to U-M officials in summer 2011.[7] The overall program was approved including an authorization of funding (as a portion of the US$14 million in new funding that U-M committed to sustainability following the IA) to gather a set of baseline indicators through surveys in fall 2012. Funds have also been authorized to collect and report follow-up indicators in 2013 and 2014.

As an initial step to carry out this work, a steering committee was established consisting of representatives from the Graham Environmental Sustainability Institute, the Office of Campus Sustainability, faculty and students, and the Institute for Social Research, the organization that

7 The program submitted to U-M administrators did not include the development and collection of indicators for alumni. In the future, we plan to develop indicators covering alumni as these indicators will become increasingly important when current and future students graduate.

will conduct the surveys and analyze the data. We also created an external advisory panel with expertise in environmental literacy and campus sustainability to provide feedback to the steering committee. To date, the steering committee has prepared a detailed schedule for the administration of the surveys and the reporting of their findings including baseline data covering the cultural indicators. It has also reviewed sustainability-related questionnaires used at other universities.

Focus groups: A key activity in the planning phase of the project is a series of focus groups with students and staff. The primary purpose of the focus groups is to provide input into the design of the questionnaires to be administered in fall 2012. A secondary purpose is to gather insights that will aid in interpreting survey findings. Five research questions have been used to guide the focus group discussions:

1. What is the meaning of sustainability and how has this meaning changed over time?
2. To what extent are students and staff aware of the university's sustainability efforts and what roles do they see themselves playing in these efforts?
3. How is knowledge about sustainability communicated and processed through interactions between students and others?
4. How are student and staff perceptions reflected in their actions vis-à-vis sustainability on-campus and off?
5. To what extent do student perceptions about sustainability relate to career choices and life after college?

Approximately 10-12 focus groups covering undergraduate and graduate students, residence hall and private housing residents, student athletes, students involved in sustainability organizations, and staff from different parts of the university are planned over a two-month period. The focus groups are being facilitated by the Planet Blue Ambassadors and monitored by senior research staff. Findings from the focus group sessions along with our review of past sustainability-related questionnaires will be used in drafting questions to be asked of U-M students, faculty and staff.[8]

Questionnaire Design: As a guide for drafting questions and writing questionnaires, an analytical model was prepared showing relationships to be explored between different types of cultural indicators and contextual and antecedent conditions. The model shown in Fig. 3 has several parts. Firstly, it suggests that indicators could consist of possibly two or more questions. For example, an indicator capturing the travel behavior of students could consist of responses to questions about automobile ownership, mode of travel to campus, and intra-campus bus use. Secondly, the number of indicators reflecting knowledge ($\mathcal{K}$), commitment ($\mathcal{C}$), and behaviors ($\mathcal{B}$) and the number of variables reflecting contextual and antecedent conditions will vary. Thirdly, the model suggests that contextual variables such as type and place of residence, and distance from residence to campus may be related to selected cultural indicators. Similarly, antecedent conditions such as length of time on campus and academic background in high school and college could affect selected cultural indicators. Finally, the model suggests possible relationships between cultural indicators. The questionnaires will include items that capture contextual and antecedent conditions as well as those used in creating the different types of indicators.

8 Instead of conducting focus groups with faculty, the research team with be testing concepts and draft questions with selected faculty using a cognitive interview approach.

Fig. 3: Analytic model for investigating determinants of cultural indicators

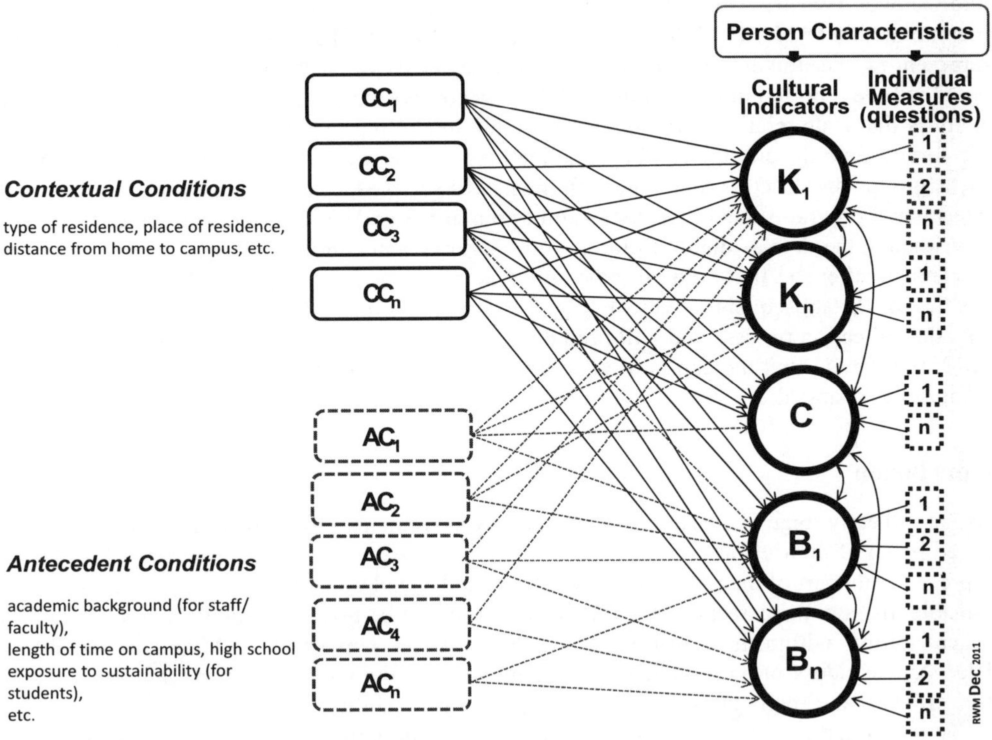

Questionnaire administration: Following extensive questionnaire pretesting including indicator development, plans are being made to launch web surveys soon after the beginning of the fall semester 2012. The plan calls for a selecting samples that would produce 4000 undergraduate students, 400 graduate students, and 1500 faculty and staff members. For the targeted 4000 students, more than 1000 email addresses will be selected from each list of incoming students (freshmen), sophomores, juniors and seniors. Because we are interested in measuring change in the indicators reflecting the culture of sustainability, we are using a panel design such that in fall 2013, approximately a quarter of the 2012 entering class will be contacted again and asked to complete the same questionnaire. These students will be re-contacted in 2014 and 2015, the latter being their senior year at the university. A similar sample design will be applied to the 2012 sophomore, juniors, and seniors with a third form each cohort re-contacted during their each of their remaining years at the university.

In addition to the panel design, a cross-sectional design will be employed; for subsequent years (2013 and 2014), a sample will be selected from each freshmen class to yield 330 respondents. Similarly, a new sample producing 330 respondents from the sophomore and junior classes will be selected. This combination of sample designs will enable us to track cultural indicators for individual students who move through their four years at the university to see the extent of change in their level of understanding of sustainability issues, their degree of commitment to dealing with environmental problems, and their actions. It will also enable us to track changes in cohorts of students as well as faculty and staff.

Next steps: The collection of indicator data in September 2012 is expected to continue until mid-October. By early December, a preliminary set of cultural indicators will be available and incorporated in U-M's 2012 Annual Environmental Report. The final set of cultural indicators representing the university's baseline measures against which future measures can be compared will be reported in the spring 2013. This report will include findings from in-depth data analysis that tests relationships suggested by the model presented in Fig. 3. Interpretation of the survey findings and their policy implications will be guided in part by the results of the focus groups discussed earlier. These findings will serve as an evaluation mechanism for our educational programs and other sustainability initiatives and can be used in altering them.

Although the data available from the 2012 surveys will continue to be scrutinized by U-M researchers after the first cultural indicators report is issued, we expect to begin planning for the first of the subsequent surveys that will take place in fall 2013. These surveys will replicate the collection process and will produce data that, when compared to the data collected in 2012, will be the first indication of whether or not the culture of sustainability at on the U-M campus has changed.

Conclusion

In this still early stage of the campus sustainability movement, universities and colleges have focused much of their efforts on the scientific and technical challenges of sustainability, which represent only part of the intellectual, structural, and social transition necessary to attain a sustainable institution. Increasing attention can and should be given to establishing a "culture of sustainability" within these institutional settings. At the University of Michigan, the Planet Blue Ambassadors program and the "Sustainability & the Campus" course are two leading two efforts that directly target cultural change for sustainability.

To measure the success of these and other specific initiatives, a new evaluation program is being implemented to monitor U-M efforts to create a culture of sustainability. This program involves the development of sets of "cultural indicators' that measure sustainability literacy, commitments to being a more environmentally and socially responsible members of society, and pro- sustainable behaviors among students, faculty, and staff. The indicator program will enable U-M to track its progress as it moves toward becoming more sustainable by changing its culture. At the same time, it will provide insights that can be used to strengthen existing campus sustainability efforts and launch new ones. We believe that this program could be replicated at other institutions of higher education and could guide them in measuring the effectiveness of their efforts to create a more sustainable university. In particular, the University of Michigan program and its indicators can serve as models for campuses striving to complete the self-evaluation for AASHE (Association for the Advancement of Sustainability in Higher Education) STARS (Sustainability Tracking, Assessment & Rating System) or other campuses interesting in monitoring progress toward advancing a culture of sustainability. Finally, measuring cultural indicators in a number of institutional settings offers opportunities for comparative research on topics related to the human dimensions of sustainability.

References

Erickson, C. (2010), *Peer to peer sustainability outreach programs: The interface of education and behavior change.* Working paper, University of Vermont, Burlington, 18 February 2012.

Lund, K., Dinse, K., Callewaert, J. and Scavia, D. (2011), Benefits of using integrated assessment to address sustainability challenges. *Journal of Environmental Studies & Science* 1 (4), pp. 289-295.

Marans, R.W., Levy, B., Bridges, B., Keeler, K., Avrahami, T., Bennett, J., Davidson, K. Goodman, L., Holdstein, B., Smith, R. (2010), *Campus sustainability integrated assessment: Culture team phase I report.* Available at: http://www.graham.umich.edu/pdf/culture-phase1.pdf (accessed 10 January 2012).

Marans, R.W. and Edelstein, J. (2010), Human Dimensions of Energy Conservation and Sustainability: A case Study of the University of Michigan's Energy Conservation Program. *International Journal of Sustainability in Higher Education* 11 (1), pp. 6-18.

McKenzie-Mohr, D. (2011), *Fostering Sustainability Behavior: An Introduction to Community-Based Social Marketing.* New Society Publishers, British Columbia.

Scott, G. (2004), *Change matters: Making a difference in higher education.* Available at: http://www.uws.edu.au/__data/assets/pdf_file/0007/6892/AUQF_04_Paper_Scott.pdf (accessed 10 January 2012).

Senge, P. (2000), *The Fifth Discipline: The Art and Science of the Learning Organization.* Crown Business, New York, NY.

Sharp, L. (2002), Green campuses: The road from little victories to systemic transformation. *International Journal of Sustainability in Higher Education* 3 (2), pp. 128-145.

– (2009), Higher education: the quest for a sustainable campus. *Sustainability: Science, Practice & Policy* 5 (1), pp. 1-8.

Shriberg, M. and Harris, K. (2012, in Press), Building sustainability change management & leadership skills in students: Lessons learned from "Sustainability & the Campus" at the University of Michigan. *Journal of Environmental Studies & Science.*

Chapter 48

Developing and Implementing Energy Sustainability Research in Business Education

Vlasios Voudouris and John Sedgwick

Abstract

The sustainability agenda is becoming an important element in business education. This is because there is a growing understanding of the need to develop strategic thinking that safeguards the long-term sustainability of business. To that end the Agent-based Computational Economics of the Global Energy System (ACEGES) project is introduced as a novel framework for developing a new generation of long-term decision scenarios. At a more general level, the project raises awareness of sustainable thinking in long-term business planning by means of controlled computational experiments. This paper describes the ACEGES tool, which has been introduced at the Business School of London Metropolitan University to educate students in the art and science of sustainable strategic thinking.

Introduction

The links between sustainability and business in a higher education context are very strong ones, but even though some works have been performed to date (e.g. Christensen et al. 2007; Rusinko 2010), there is a paucity of publications related to the inclusion of sustainability in the context of business education. One of the exceptions is a survey performed by Naeem and Neal (2012), administered to business schools in the Asia Pacific region. In addition to measuring the number of courses and programs integrating sustainability, the study solicited qualitative observations by respondents, to provide information and insight into the issues.

In a recent publication by the management consultants Arthur D Little, its authors make four assertions: 1) the political momentum towards a low carbon future is unstoppable; 2) that carbon is a real cost to business; 3) that investors increasingly understand the significance of a low carbon future and behave accordingly; and 4) that the time to act is now.[1] Conceiving sustainability as the ability of a business to keep growing over the long term, the authors are concerned with how firms maintain, or obtain, a competitive advantage given the changing environment, arguing that in all cases it will involve step changes in the strategic thinking of businesses and the business models they unfurl. The Arthur D Little approach is one based on a capitalist reality, in which firms compete by obtaining competitive advantage – it is a scenario of business-as-usual within the context of a changing environment in which companies survive through adaptation and innovation connected to the changing circumstances of time and place.

Other perspectives challenge the possibility that such a definition of 'sustainability' is sustainable, maintaining that the only way in which world average temperature rises can be restricted is to reduce consumption and production across the globe. The problem here is that capitalism as a system does not do 'going backwards' very well.

---

1 See 'Ensuring Survival: Business Models in a Low-Carbon World' www.adl.com/ensuringsurvival (accessed on 1 November 2009.

Energy sources and security are key to this debate – perhaps the single most critical issue facing our society in the 21st century. Sustainable energy supply is also one of the most important strategic issues facing business, particularly for those businesses involved in the production (e.g. EDF, Shell, BP) and high-level consumption of energy services (e.g. airline companies). Given that in many industrialized societies the transition to a low-carbon global economy has begun, a radical change that promises to transform the way we live and produce goods can be expected.

The provision of secure, affordable and abundant energy sources presents a set of physical, regulatory, and market risks, as well as potential opportunities. The implication is that tomorrow's leaders need to incorporate the challenge of sustainable energy supply into the core of their thinking today.

The challenge of sustainable energy supply thus creates an urgent need for educational institutions, such business schools, to prepare tomorrow's managerial workforce to effectively manage this 'transition' and develop new business models, such as ones based upon 'collaborative advantage' rather than 'competitive advantage'. Therefore, tomorrow's managerial workforce will need new conceptual frameworks and analytical skills in order for them to play more effective roles as leaders in government, business and civil society within a low-carbon economy.

Although few institutions of higher education currently possess all of the resources and experience necessary to educate a diverse range of professionals in various sectors with appropriate skills and knowledge, the Centre for International Business and Sustainability (CIBS) at the Business School (LMBS) of London Metropolitan University has established a multidisciplinary team of academics and practitioners to delivery the MSc in Sustainable Business Energy and Finance and the CIBS Doctoral Training Programme. These two initiatives were designed to educate tomorrow's workforce giving them new analytical and research tools, notably through the Agent-based Computational Economics of the Global Energy System (ACEGES) project, while at the same time developing project-based partnerships with businesses in this field, and facilitating student access to internships and online learning.[2]

This paper concentrates on a novel approach to energy modeling as a key dimension in scenario planning. Section 2 presents the ACEGES project (Voudouris 2011; Voudouris et al. 2011) which runs as a component of the energy and investment research dimension of CIBS. Section 3 discusses how the outputs of the ACEGES project has been integrated in two modules designed around a new generation of long-term energy scenarios to test the future resilience of business. Section 4 concludes the paper.

The ACEGES Project

The overall aim of the ACEGES proposal is to develop, test and disseminate a computational laboratory for the systematic experimental study of the global energy system through the mechanism of energy scenarios. In particular, the intention is to show how Agent-based Computational Economics (ACE) models (Tesfatsion 2006) can be applied to help leaders in government and business understand better the challenging outlook for energy by means of controlled computational experiments.

It is evident today that the global energy system is under acute strain, involving developmental, trust and industrial dilemmas. A key challenge is to investigate how the macroscopic tensions in the global energy system might play out, especially when strains within the system cannot be resolved quickly because of the lengthy timescales involved in bringing on new ca-

2 See http://www.londonmet.ac.uk/lmbs/research/cibs/courses.cfm.

pacity – think, for example, of the economic, political and security-type problems which might affect the oil capacities of Canada, Venezuela and Nigeria respectively, and how these might impact upon the global energy system, and beyond.

The research objectives of the ACEGES project are:

- To develop a computational laboratory (ACEGES) in a substantive social science research area, namely the global energy system;
- To provide critical assessment through the establishment of a comprehensive database, which draws upon existing energy-related records from various national and international sources;
- To raise awareness of, and give access to, modern numerical and computational methods in social science in general, and energy economics in particular;
- To provide expertise, technical infrastructure and training for agent-based simulations of the global energy system targeted at policy makers, advisors and strategists in government, and business and civil society.

The usefulness of the proposed project is two-fold. Firstly, it addresses a perceived need for accurate data on energy systems, which are not readily available at the moment. Secondly, it provides the leaders in government, business and civil society with the means to assess several alternative sets of policy measures before any operational innovations are introduced. In particular, users of the computational laboratory will be able to adjust the parameters of energy scenarios (e.g. temporal rate of oil and gas extraction, primary-energy demand) and ascertain their emergent effects on the global energy system using a bottom-up approach. This is to say that simulations of changes in the energy system generate surprising and unanticipated outcomes because of multiplier effects that are initially difficult to identify and even harder to contain. This approach emphasizes the idea that the energy system is better modelled not as a black-box associated with 'the invisible hand' idea of markets, but as a sophisticated 'computational' device whose performance is determined by its 'programming', the 'decoding' of which became the goal of research of the ACEGES project.

Fig. 1 shows the ACEGES toolkit. The ACEGES toolkit is used in one undergraduate and one postgraduate course at LMBS. The aim is to support the cognitive abilities of students in developing energy scenarios as a way of exploring challenges faced by the energy system. In other words, the ACEGES toolkit is a computational laboratory that enables students to conduct thought experiments with the help of computational software. Using the 'model tab', users of the ACEGES toolkit can develop a range number of simulated scenarios by adjusting any of the key uncertain variables (more important and uncertain driving forces), namely:

- Resource estimates – fossil fuels originally present before extraction.
- Demand growth – this can be estimated based upon population growth, economic growth as well as estimated upon the energy intensity of a country.
- Production growth – this is the average maximum production between years. This production growth encapsulates both above ground and below ground factors that can limit production.
- State of depletion at peak – this defines a tipping point in the production of fossil fuels with significant economic consequences if not managed properly.

Fig. 1: The ACEGES toolkit

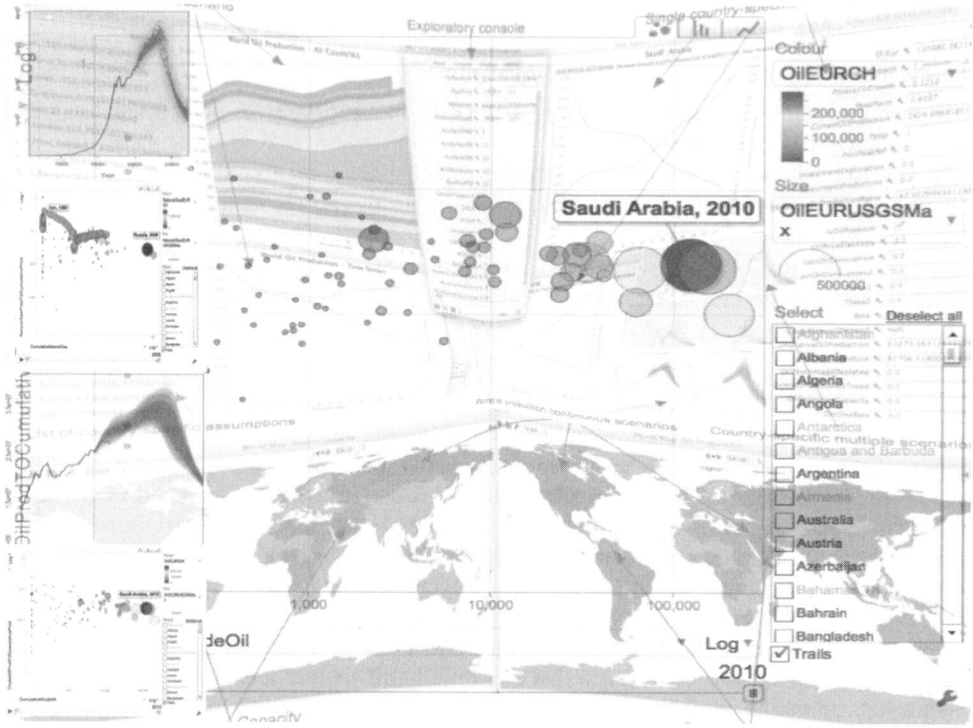

Integrating ACEGES in undergraduate and postgraduate education

In CIBS the vital role that business education plays in shaping and disseminating best practice for the future leaders of a low-carbon economy is fully understood. To enable tomorrow's workforce to think sustainably, it has adopted a 'project-based' approach. This means that students, as part of a group, address a real-world problem using real-world data.

Although CIBS recognizes that developing scenarios is primarily a non-mechanistic cognitive process, the main categories from which scenarios are derived were identified by Schwartz (1991) as:

- Societal Change
- Technological Innovation
- Economics
- Politics
- Environment.

The process of developing scenarios using the ACEGES tool relies upon interactive graphics and computer models and is structured around the following variables (Schwartz, 1991):

- *Key forces in the local environment:* Given the focal point of interest (defined by the assessment component of the module), students identify and list all the key factors that can influence the decision that needs to be taken. In the case of oil scenarios, these include resources originally present before any extraction, demand growth, production growth and peak/decline point.

- *Driving forces:* Based upon the configuration of key forces, students identify the driving forces that affect the factors derived from the five categories listed above. For example, to get an idea of the forward-looking production growth rate students need investigate the up-stream investment plans of major and national oil companies, particularly those in the MENA (Middle East and North Africa) region.
- *Ranking of most important and uncertain key forces:* Students then need establish a risk profile associated with those factors that impact on the focal point of interest. For example, to test the effectiveness of energy efficiency policies, demand growth can be fixed as a pre-determined force, while the other key forces can be regarded as either more or less risky building blocks of the scenarios – see for example Hallock *et. al.* (2004) for an example of building blocks for oil scenarios.
- *The scenario logic:* While the ranking of the key forces defines the scenario drivers or the axes along which the scenarios differ, the process of ranking grouping the key forces determines the logic of the scenario. In the case of a two-dimensional scenario, using the ACEGES tool can explore the full range of the scenario space and define the scenario logic based upon the location of the scenario in a two-dimensional matrix. The key advantage of the ACEGES-based scenarios is that higher dimensional scenarios can be readily explored to help policy makers frame their way of thinking. Caution is needed not to develop scenarios for every conceivable uncertainty as the whole process of developing scenarios as a policy tool can dissipate. Scenarios should be developed in ways that make a difference to the policy makers.

The points outlined above provide a structured way to develop scenarios during the *exploration stage* of the scenario development process. During the *confirmation stage*, students focus on highlighted areas of interest and inspect patterns identified in exploration in order to assess rigorously the implications of the scenarios.

During the *synthesis stage*, the team step back from detail in order to provide a 'big picture' and select the leading indicators and signposts. This is very important as these leading indicators and signposts will enable decision makers to i) monitor the scenarios in an ongoing way and ii) know as soon as possible which of the selected scenarios is closest to the course of history. During the presentation stage, well-designed scenarios with clear and convincing messages are developed.

By exposing its students to the art and science of ACEGES-based scenarios, CIBS develops a new way of thinking about long-term strategic planning and help them think the unthinkable. According to Wack (1985) the key aim of the scenarios to challenge managers' personal microcosms. By also exposing students to the scenario development process, CIBS also challenges their own personal microcosms and help them to incorporate 'sustainability' as a core concept of their decision framework.

Conclusion

Many of the pitfalls in developing scenarios for long-term strategic planning are linked to specific failures to identify a focal point of interest. The main challenge is to build scenarios that enable executives to rehearse the future in a way that transforms the identified factors and key uncertainties into new perceptions – a new microcosm.

The need to develop scenarios as a tool for long-term strategic thinking is a straightforward idea, although one that is often complex to operationalize. However, we claim that this complexity can be managed rigorously through the use of computational laboratories.

This chapter has discussed how the main output of the ACEGES project has been built into the syllabuses of two courses running in the Business School of London Metropolitan University. The aim was to develop the analytical skills of students and get them thinking in a long-term way about the business world they would soon be working in.

Given that the fossil hydrocarbons of coal, conventional oil and natural gas, which have fuelled much of human progress and economic growth since the industrial revolution, are depleting rapidly we set out to help our students understand the concept of 'sustainability' by looking into the concepts of 'peak oil' and 'peak gas' through the lens of agent-based computational economics.

The authors believe that, for decades, the more advanced industrial societies have consumed virtually uninterrupted energy supplies and the services related to this. The prospect now looms of this assumption no longer holding, owing to of the finite nature of these fossil hydrocarbons and the increased demand for flexible energy services such as electricity and transport fuels. This is likely to promote energy and investment as a key theme of the 21st century, particularly during periods of uncertain supply and volatile prices. Therefore, CIBS is preparing its students for the transition to a low-carbon economy by developing their skills and intelligence as business planners in a world in which collaborative advantage is likely to become increasingly important. Perhaps a more balanced approach to globalization is also required: one that focuses on a more sustainable path.

Scenario planning forces participants to identify key building blocks and the dynamic relationship between them. As a consequence, it is an approach that should lead to improved business decision-making and strategic thinking, focusing as it does on the sustainability of future plans. Scenarios can, therefore, be a powerful tool for shifting awareness of leaders and managers, and the basis for management development training programmes.

References

Christensen, L.J., Peirce, E., Hartman, L.P., Hoffman, W.M. and Carrier, J. (2007), Ethics, CSR, and Sustainability Education in the Financial Times Top 50 Global Business Schools: Baseline Data and Future Research Directions. *Journal of Business Ethics* 73 (4), pp. 347-368.

Hallock, J., Tharakan, P., Hall, C., Jefferson, M. and Wei, W. (2004), Forecasting the limits to the availability and diversity of global conventional oil supply. *Energy* 29, pp. 1673-1696.

Naeem, M. and Neal, M. (2012), Sustainability in business education in the Asia Pacific region: a snapshot of the situation. *International Journal of Sustainability in Higher Education* 13 (1), pp. 60-71.

Rusinko, C. (2010), Integrating Sustainability in Management and Business Education: A Matrix Approach. *The Academy of Management Learning and Education (AMLE)* 9 (3), pp. 507-519.

Schwartz, P. (1991), *The art of the long view: Planning for the future in an uncertain world*. Doubleday, New York.

Tesfatsion, L. (2006), Agent-based computational economics: a constructive approach to economic theory. In *Handbook of Computational Economics* (Vol. 2): Agent-Based Computational Economics. Elsevier, Amsterdam.

Voudouris, V. (2011), Towards a conceptual synthesis of dynamic and geospatial models: fusing the agent-based and Object – Field models. *Environment and Planning B: Planning and Design* 38 (1), pp. 95-114.

Voudouris, V., Stasinopoulos, D., Rigby, R. and Di Mai, C., (2011), The ACEGES laboratory for Energy Policy: Exploring the production of crude oil. *Energy Policy* 39 (9), pp. 5480-5489.

Wack, P. (1985), Scenarios: Uncharted Waters Ahead. *Harvard Business Review* September-October.

Chapter 49

The Impact of University Policies for Sustainable Development on Students: a Comparison of Students from two Institutions, One in the UK and One in Portugal

Chris Shiel and Arminda do Paço

Abstract

If we are to achieve sustainable development and secure targets for carbon reduction then universities have a role to play in educating students to address the issue of sustainable development (SD) and influencing behaviour change. Some universities have already developed approaches to address SD within the curriculum, on campus and in the community; others have done far less, or very little. The assumption might be that students from a university which is formally addressing SD, should exhibit different attitudes and behaviours to students from a university where SD is not a concern. This paper will compare the environmental consciousness level and "green" behaviour of students from a university in Portugal which has no formal sustainable policies, nor a structure supporting the development of global citizenship, with a university in the UK where a formal policy has been implemented which embraces a holistic approach to SD and global citizenship. Analysis of survey data from the two student populations reveals significant differences in environmental attitudes, with students from the UK university displaying greater concern about the environment, greater belief that their actions can influence change and more concern with conservation and buying locally sourced food. The Portuguese students are more concerned about future generations. The conclusion suggests that formal policies do have an impact; however, the cultural context must also be considered.

Introduction

The role of universities in securing a more sustainable future and contributing to targets for carbon reduction is undoubted; they have a critical role to play as agents of change (McMillan and Dyball 2009; Marans and Edelstein 2010; Cortese 2003), particularly where they adopt an integrative approach to SD which embraces curriculum, campus, community and research. However while it may seem perfectly obvious (at least to advocates of SD) that higher education bears '*a profound moral responsibility to increase the awareness, knowledge, skills and values needed to create a sustainable future*' (Cortese 2003: 17), evidence of holistic approaches to sustainability is patchy. Few universities have successfully embedded education for sustainable development across the entire curriculum; too many universities continue to advance '*the kind of thinking, teaching and research that leads to unsustainability*' (Wals and Blewitt 2010: 70). There are however changes afoot: the place of sustainability is '*slowly shifting from one of campus greening and curriculum integration to one of innovation and systemic change in the whole university system*'. However as Sterling and Scott (2008) comment, environmental management (led by environmental managers) has made far greater progress in the UK because of legislation and financial incentives; curriculum change continues to be a much tougher challenge. A similar situation is evident in Australia (Jones et al. 2010) and North America; progress in Europe is seen as '*variable but rarely spectacular*' (Wals and

Blewitt 2010: 60). In Canada, there is similar variability with some universities opting into campus climate initiatives which include renewable energy production, reducing greenhouse gas emissions, behaviour changes, and institutional involvement (Helferty and Clarke 2009).

It is hardly surprising that most writers on SD continue to bemoan a lack of progress, while emphasising the transformation needed within higher education, and the potential of education, to lead to the type of transformational learning which might address sustainable development. The underlying assumption (although rarely explicit) is the power of education for sustainable development (ESD) to develop a generation who will tackle un-sustainability and who will presumably be equipped to work towards a more just and sustainable future. Such graduates will naturally need to be quite different to those who have been through the education system previously. As Helferty and Clarke (2009) suggest, this target group (students) is really important specifically in regard to the environmental facet of SD since the youth of today already feel some affinity towards the environmental movement and consider environmental problems as an important issue.

Despite a substantial literature on sustainability, there is a lack of empirical work which focuses on the difference it (ESD) makes to students. A number of questions merit consideration, for example, if a university adopts a strategic approach to sustainability which embraces the curriculum, what is the real impact on students? Will they have different attitudes and behaviours, for example, to students who have studied at a university which has not taken up the sustainability challenge? It is easy to espouse ideology (in terms of what higher education should be doing in relation to SD) but where is the evidence to suggest that enactment of this ideology has such a powerful influence on learning outcomes? Such questions suggest a need for both comparative and longitudinal studies which seek to understand the impact of ESD on students; such studies will no doubt emerge in the fullness of time as more universities seek to develop sustainability literate graduates. However, as a tentative start, this paper offers an exploratory study which seeks to compare students from two institutional contexts: University A in England and University B in Portugal. University A has undertaken considerable work to progress sustainable development in a holistic way; University B has no formal approach and has achieved far less. As the problem is complex and SD is a multidimensional concept, in this specific research the focus of the analysis will be largely on environmental issues.

The institutional contexts of Universities A and B will be presented briefly, followed by the details of a questionnaire that was undertaken to explore whether the English students differed in their responses, to those from Portugal. Discussion will then attempt to draw conclusions and suggest areas for further research.

Educational and institutional contexts

Within the UK, the government identifies education as a major vehicle for securing sustainable development (DEFRA 2005; DfES 2006). The Higher Education Funding Council for England (HEFCE) with a clear vision and strategy for higher education (HEFCE 2005; 2008) has been instrumental in urging universities to engage with sustainability. The Funding Council emphasises the importance of education for sustainable development but has, however, steered away from making it compulsory. As a result, substantial progress has been made by some institutions but progress across the sector is 'disparate'. It continues to be the case that, '*the higher education sector is one of the hardest sectors in which to institutionalize sustainability*' (Junket and de Ciurana 2008: 764). While very few English universities have adopted a holistic approach to sustainability, most have made strides forward in environmental management and carbon reduction strategies to meet legislation and financial incentives. The emergence of 'League Tables' and benchmarking exercises which assess 'green credentials' has also triggered change.

University A is an example of an English institution which has attempted to adopt a holistic and strategic approach to the agenda which embraces campus, curriculum and community (Shiel 2007). In its ambition to develop global perspectives across the university, substantial work has been undertaken to ensure that the curriculum seeks to cultivate global citizens who understand the need for sustainable development. Strategy is in place and a set of curriculum guidelines were developed in 2005, and then later revised in 2008. These guidelines require all Course Teams to address sustainable development at Course Design and in Course Review. In essence it is suggested that the curriculum will, among other things,

- enable students to understand the links between their own lives and those of people throughout the world;
- increase understanding of economic, social and political forces which shape life;
- develop skills, attitudes and values to enable people working together to bring about change for the 'common good' and to take control of their own lives;
- provide the learner with the knowledge and skills to work towards a more just and sustainable world where power and resources are more equitably shared.

Student surveys reveal that substantial progress has been made at University A over five years; students have a better understanding of SD and global citizenship. External indicators of success also confirm that the institution has also made great progress in terms of its campus and environmental management. The University is one of only six universities in the UK to have achieved 'gold (standard), or above' in the Eco-Campus Award. It has also achieved other national awards such as 'Green Gowns,' and is fairly high (although not at the top) of the 'Green League Table'. The University has been a 'Fairtrade' University for a number of years, engages with partners in the region to progress sustainable development, and is committed to social responsibility.

In Portugal discussion around the role of Universities in relation to sustainable development has been almost non-existent, and the few events which have been organised have been limited to an environmental perspective. This lack of engagement is illustrated by a situation where before 2005, just one institution (the University Nova of Lisboa), had signed up to the Talloires Declaration. Since then, although some Portuguese universities have been taking forward sustainability initiatives (e.g. University of Algarve, Aveiro, Porto, Nova of Lisboa, Técnica of Lisboa), there is a gap in terms of co-ordination and communication at a national level, which could have detrimental consequences (Couto et al. 2005). In this sense, the creation of an organisation or body to coordinate issues in relation to sustainable development within higher education is crucial, but not evident in Portugal.

An explanation of why there is less higher education activity in relation to sustainability in Portugal is because Portugal has been 'behind the game'. The OCDE Report "Good Practices in the National Sustainable Development Strategies of OECD Countries" highlights that of the 30 OCDE countries, 23 of them had prepared formal plans in the field of national sustainable development strategies; some (Australia, United Kingdom, France, Japan, Finland, Luxemburg, Holland, Sweden and Switzerland) formulated strategies very early and had already revised those strategies. Other countries (including Portugal) had prepared their strategies more recently (OCDE 2006).

The themes of the Portuguese National Sustainable Development Strategy are detailed in a set of documents approved by the Government. One of the four principles of the strategy '*is to progress towards a society of solidarity and knowledge, including through interventions to strengthen the citizen components of education and greater access to information and participation in decision-making*' (OCDE 2006).

In the education sector, the adoption of a National Strategy for Development Education presents a great challenge for the country. The main intention of this plan is to strengthen the inter-institutional cooperation mechanisms between educational agents; develop tools to promote global citizenship by means of learning processes, and raise consciousness of development related aspects in Portuguese society. Although the idea of this national strategy is to promote development education at all levels of education, learning and training, the reality is that its implementation in higher education is still very incipient. Thus, the involvement of higher education in the area of education for citizenship and development education remains to be enforced. In the pre-school, basic and secondary levels, investment in education for citizenship has been progressed. However, there is a long way to go to overcome obstacles, which include the frequent non consideration of development education in the context of education for citizenship, especially in the training of professionals, the lack of pedagogical materials to support the learning, the financial constraints, and the difficulties that teachers have in working as an interdisciplinary team (IPAD 2009).

In order to gain a better picture of the situation in Portugal, Schmidt et al. (2011) examined the results of a questionnaire sent to 15,000 public and private schools (kindergarten and first cycle of basic education; second and third cycles of basic education; and third cycle and secondary education) involved in projects associated with both environmental education and education for SD. The authors concluded that environmental education and education for SD are internally focused, rarely involve the community, and projects which involve the whole scholarly community, are difficult to find.

The projects tend to emphasise ecological issues rather than broader engagement: "This means there is too much environmental education and not enough education for sustainable development in the schools of Portugal" (Schmidt et al. 2011: 174).

The projects also give preference to younger students instead of post-adolescent or pre-adult students. The biggest problem Schmidt et al. (2011) identified was the short term nature of the projects and lack of continuity. Given the national context, when students arrive at university, it is unlikely that they will have had the same exposure to sustainable development as UK students. It is within this context that University B operates.

University B has been in existence since 1986. One of the most interesting physical features of this university is that its estate comprises old buildings with historical, cultural and architectural value, which have been repaired and conserved. At the same time that these historical landmarks have been re-constructed, they have been revitalized into teaching and investigation spaces. In this way the institution has been promoting both sustainable construction and building conservation. However the University has no formal policy or strategy in place for sustainable development. It has made some effort related to energy, water saving and recycling but not as part of an overarching strategic approach. In relation to the curriculum, there is no drive to incorporate sustainable development into formal education. There are a few postgraduate courses that partially address sustainability but beyond that, education for sustainable development is not being considered across disciplines.

Research methodology

Given the different approaches to sustainable development at Universities A and B the assumption might be that students from A (a university which is formally addressing sustainable development) should exhibit different attitudes and behaviours than the students from B (a university where sustainable development is much less of a concern). The question which this paper seeks to address is thus:

- Do the students from a university which has a formal approach to sustainable development respond differently to those from an institution which has not adopted this approach, when asked questions related to sustainability?

In seeking to address this question which has been largely unexplored, data was used from a survey/questionnaire, which had already been designed and implemented for a wider study. As part of the wider study there was already a data set, which allowed English students to be compared to Portuguese. The original questionnaire sought to test a model to consider the relationship between green attitudes and values and behaviour, in relation to green consumer behaviour. The data collected as part of the earlier wider study yielded the opportunity to test whether the existence, or non-existence, of SD policies in a university, do influence the environmental attitudes and behaviours of students as part of the wider survey study a questionnaire was designed, piloted and then adapted, to enable data to be collected from various countries. The survey took the form of a self-administered questionnaire which was made available on line. As University A has a considerable number of international students, when the survey was released, only UK students were invited to respond.

The questionnaire was designed to include several scales to enable information to be gathered about values (Man Nature Orientation and Loyola Generativity Scale), attitudes (New Environmental Paradigm and Perceived Consumer Effectiveness) and behaviours (ENVIROCON and Ecologically Conscious Consumer Behaviour). Additionally, two questions were included about fair trade and locally sourced goods, all measured on a seven point scale. Finally some questions to gather demographic information (age, gender, nationality, course and year of frequency) were included.

The scales make reference to five dimensions or constructs:

- Man Nature Orientation (MNO): states that people should behave according to the way of nature and respect the world where they live in (Chan 2001);
- Generativity (GE): the concept is related to the belief that an individual regards the future as important, and as such there is an obligation to secure it for future generations (Urien and Kilbourne 2011);
- Environmental Concern (EC): includes concerns related to the limits to growth, pollution, steady-state economy and recourses conservation (Dunlap and Van Liere 1978);
- Perceived Consumer Effectiveness (PCE): is based on the idea that peoples' responses to environmental appeals are linked to the belief that they can positively influence and contribute to solve environmental problems (Ellen et al. 1991);
- Conservative Behaviour (CB): is related to conservation activity – dispositional actions, recycling, preservation of resources, etc. (Pickett et al. 1995);
- Buying Behaviour (BB): covers topics such as purchasing green products, the attention given to packaging, energy-efficient equipment, polluting or recycled products (Straughan and Roberts 1999).

Data was collected over a six-week period in 2010. The final sample includes 612 respondents: 301 university students from Portugal and 312 from the UK.

After collection, the data was statistically analysed and interpreted using the statistical software SPSS version 20.0.0. Descriptive analysis, t-tests and discriminant analysis were used.

Results presentation and discussion

The sample from the UK has a median age of 25 years (mode 18; standard deviation 8,981; minimum 18; maximum 56); 28% are male; many of the respondents are studying health sciences (21.8%), miscellaneous (17.3%) and business and economics (15.4%); 42.0% are in their first year at the university, 26.9% in their second and 16.7% in their third, whilst 87.5% are principally at the undergraduate level.

In turn, the Portuguese sample has a median age of 22 years (mode 20; standard deviation 4,731; minimum 17; maximum 50); 50% are male; mostly study business and economics (29.6%), sports (16.9%) and engineering (16.6%); 28.9% are in their first year at the university, 38.2% in their second and 30.2% are in their third year, whilst 85.7% are principally at the undergraduate level.

Table 1 shows the descriptive statistics for both samples regarding the variables, related to environmental issues, included in the study.

Table 1: Descriptive analysis

Variables	Nationality	N	Mean	Std. Deviation
Man nature orientation (MNO)	Portuguese	301	4.8600	1.07877
	English	312	5.7064	0.99320
Generativity (GE)	Portuguese	301	4.8403	0.93213
	English	312	4.6391	0.97711
Environmental concern (EC)	Portuguese	301	4.7763	0.87140
	English	312	5.4874	0.87661
Perceived consumer effectiveness (PEC)	Portuguese	301	4.2113	1.03403
	English	312	5.4127	1.13825
Conservative behaviour (CB)	Portuguese	301	4.4233	0.96591
	English	312	5.3114	0.93891
Buying behaviour (BB)	Portuguese	301	4.1300	1.13177
	English	312	4.6167	1.35145
Buying fair-trade products (BFP)	Portuguese	301	4.53	1.515
	English	312	4.59	1.703
Buying locally sourced food (BLS)	Portuguese	301	4.43	1.645
	English	312	4.73	1.621

In relation to the table above we notice similar means in some factors (e.g. GE, BB, BFP, BLS) but some accentuated discrepancies in MNO, PEC and CB, with all these having higher mean values in the English sample. But in relation to data dispersion, higher dispersion values can be observed in PEC, BB, BFP and BLS. In general the standard deviation statistic value is higher in the UK sample, which indicates more dispersion of the data in this case.

The next step was to test for statistically significant differences between the two samples. Table 2 presents the results of that test for the variables included in the study.

Table 2: Independent samples test

	Levene's Test for Equality of Variances		t-test for Equality of Means			
Variables	**F**	**Sig.**	**t**	**df**	**Sig. (2-tailed)**	**Mean Difference**
Man nature orientation (MNO)	4.355	0.037	-10.111	611	0.000	-0.84642
Generativity (GE)	1.941	0.164	2.606	611	0.009	0.20116
Environmental concern (EC)	0.479	0.489	-10.071	611	0.000	-0.71116
Perceived consumer effectiveness (PCE)	2.148	0.143	-13.664	611	0.000	-1.20141
Conservative behaviour (CB)	0.204	0.652	-11.543	611	0.000	-0.88806
Buying behaviour (BB)	9.569	0.002	-4.825	611	0.000	-0.48669
Buying fair-trade products (BFP)	4.577	0.033	-0.472	611	0.637	-0.062
Buying locally sourced food (BLS)	0.333	0.564	-2.290	611	0.022	-0.302

Considering a confidence level of 95%, MNO, BB and BFP show differences in the variances of the two groups of students. All constructs, except BFP which is not significant, show a different mean between UK and Portuguese students, that is, there are significant differences in the students regarding GE, EC, PCE CB and BLS (note that MNO, BB and BFP do not verify the condition of the homogeneity in the variances and by this they do not need to be taken into account in the t-test).

To complement the analysis, in order to discover which variables were more significant to differentiate/distinguish the two groups of students, a discriminant analysis was carried out (Table 3). By observing the results of F statistics all the variables, when considered individually, are significant for differentiating between the groups, except BFP ($p = 0.637$). The Wilks' lambda test statistic suggests that the variable PCE is the one that provides the greatest difference between the means of the two groups of students, since it presents the lowest score. After this and in descending order of their discriminatory power come the variables CB, MNO, EC, BB, GE and BLS.

Table 3: Tests of equality of group means

Variables	**Wilks' Lambda**	**F**	**df1**	**df2**	**Sig.**
Man nature orientation (MNO)	0.857	102.241	1	611	0.000
Generativity (GE)	0.989	6.793	1	611	0.009
Environmental concern (EC)	0.858	101.419	1	611	0.000
Perceived consumer effectiveness (PCE)	0.766	186.691	1	611	0.000
Conservative behaviour (CB)	0.821	133.239	1	611	0.000
Buying behaviour (BB)	0.963	23.283	1	611	0.000
Buying fair-trade products (BFP)	1.000	0.223	1	611	0.637
Buying locally sourced food (BLS)	0.991	5.244	1	611	0.022

The discriminant analysis that was undertaken made it possible to find one discriminant function (Table 4). The differences between the groups may be analysed on the basis of the loadings of this function.

Table 4: Canonical discriminant function

Functions	Eigenvalue	Canonical correlation	Wilks' lambda	Chi-square	df	Sig.
1	0.605	0.614	0.623	287.224	8	0.000

As can be seen in the Table 4, referring to the canonical discriminant function, estimated on the mean of the groups (centroid), by squaring the canonical correlation coefficient, the percentage of variance explained by the function is 38%. The statistical significance of the function is represented by the value of the Wilks' lambda test statistic, which, when transformed into a Chi-square, has a significance level of 0.000. This shows that the function is significant for discriminating between the two groups of students.

In view of the statistical significance observed between the groups, it is useful to examine the individual contribution of the variables to the discriminant function, which can be observed in Table 5.

Thus, the highest correlations can be observed in the first four constructs (PCE, CB, MNO and EC). In contrast, BLS and BFT present the lowest scores in terms of correlation with the discriminant function.

In synthesis, the results show that the students from University A and the students from University B do exhibit significant differences in their responses to the questionnaire in the following dimensions: generativity, environmental concern, perceived consumer effectiveness, conservative behaviour, and buying locally sourced food. The means founded are superior in all dimensions for the English sample, except in the case of generativity where a higher value is presented by the Portuguese students. A possible explanation could be that within some sectors of Portuguese society, there has been and continues to be, a traditional concern with the heritage that will be left to their sons. However the results were surprising, as the English students should have been exposed to the concept of 'future generations' as an aspect of SD. Again cultural explanations might explain the difference and the impact of a societal trend (in the UK) to 'have it all now' rather than defer gratification.

Table 5: Structure matrix

Variables	Function
Perceived consumer effectiveness (PCE)	0.711
Conservative behaviour (CB)	0.600
Man nature orientation (MNO)	0.526
Environmental concern (EC)	0.524
Buying Behaviour (BB)	0.251
Generativity (GE)	-0.136
Buying locally sourced food (BLS)	0.119
Buying fair-trade products (BFT)	0.025

Note: largest absolute correlation between each variable and the discriminant function

Perceived consumer effectiveness is the variable that presents the greatest difference between the English and the Portuguese students, that is, the former feel more strongly that their actions will have an impact on the environment, and that they will make a difference. It seems that while they may be less concerned about future generations, they do believe that their actions will contribute to protecting the environment.

The overall higher means (generally) and specifically in relation to consumer decisions might be taken as the fruit of the effort made by University A in terms of its SD policy. The inference might be that the message has been passed on to students. However while the results are interesting in the substantial differences they reveal, it is not possible to suggest that raising the profile of SD through a strategic and holistic approach is the sole contributory factor.

A further anomaly which suggests a need for wider explanations related to the concept of fair-trade; on this dimension, both groups of students exhibit similar responses. This was a curious result because fair trade has not been extensively promoted in Portugal. Indeed, it is the norm that finding fair-trade products outside of the big centres is very difficult. It has however been extensively promoted in the UK, and particularly at University A. A possible explanation might be that some of the Portuguese respondents did not really know the meaning of 'fair trade', or confused it with another concept.

Conclusions

This paper has sought to consider whether students from a university which has made considerable efforts to embed sustainable development through a formal approach, respond differently when asked questions which relate to sustainability issues, to students from a university that has done far less. Data has been used from a wider study to make this comparison. This naturally places some limitations on the results, as the wider study was designed to test a particular model, rather than to explicitly look for difference in this way. It is also necessary to underline that this study focused only on a single aspect of SD (the environmental one) and that a further limitation is that although the samples are of a fairly equal size they are not an exact match, in terms of age, gender, course studied. Such limitations suggest a need for caution, yet the results do, suggest that differences are apparent and some are significant.

The study highlights that there are significant differences between the two student populations on five dimensions. In summary, the students from University A, where there is a formal approach to sustainability, demonstrate greater concern with environmental issues (environmental concern); are more likely to believe that their actions can positively influence and solve environmental problems (perceived consumer effectiveness); are more likely to engage in behaviours related with the conservation of resources (conservative behaviour); and to buy locally sourced food. The students from Portugal however place greater importance on their obligation to consider the needs of future generations (generativity).

The results could be taken as evidence that embedding sustainable development within a university does make a difference. Unfortunately it is not possible to draw such a conclusion: the data may simply be evidence that differences occur due to nationality and cultural factors, rather than strategy and intervention. This research has not explored the extent to which students' attitudes and behaviour are influenced by the wider social/cultural/political context and media influences within their nation state.

Further studies are necessary to show that education for sustainable development makes a difference. In the future research might seek to explore differences before and after institutional interventions (within a single institution); between institutions (within a single country) and between institutions (within different countries). It will be important in these studies to have the ability to control variables which may influence the data. Longitudinal studies might also seek to show whether students' responses vary as they progress through their studies but also beyond the university into employment.

It is quite easy for an institution to show that it has reduced its utility bills, increased recycling and reduced its carbon footprint. It is much more difficult to show that an education experience which embraces sustainable development impacts on students to such an extent that they will be able to secure a sustainable future. Collecting the proof presents a number of challenges, but the evidence might increase engagement across the sector.

Acknowledgements

NECE – R&D Centre funded by the Multiannual Funding Programme of R&D Centres of FCT (Portuguese Foundation for Science and Technology), Ministry of Science, Technology and Higher Education.

References

Chan, R. (2001), Determinants of Chinese consumers' green purchase behavior. *Psychology & Marketing* 18 (4), pp. 389-413.

Cortese, A.D. (2003), The critical role of higher education in creating a sustainable future. *Planning for Higher Education*, March-May, pp. 15-22.

Couto, A.P., Alves, M.C., Matos, A.F. and Carvalho, P.G. (2005), Universidade na transição para a sustentabilidade: tendências, estratégias e práticas. In Bryan, N., Gonçalves, L. and Sanchez, O. (Orgs), *Los Desafios de la Gestión Universitária hacia el Desarrollo Sostenible,* pp. 25-48. UNA/UNICAMP, Costa Rica.

DEFRA (2005), *Securing the Future: The UK Government Sustainable Development Strategy. The Stationery Office, London.* Available at www.defra.gov.uk/publications/ files/pb10589-securing-the-future-050307.pdf(accessed 4 October 2011).

DfES (2006), *Sustainable Schools for Pupils, Communities and the Environment: Government Response to the Consultation on the Sustainable Schools Strategy.* Available at https://www.education.gov.uk/publications/eOrderingDownload/Government%20Response%2004294-2006.pdf (accessed 7 October 2011).

Dunlap, R.E. and Van Liere, K.D. (1978), The New Environmental Paradigm. *Journal of Environmental Education* 9 (4), pp. 10-19.

Ellen, P., Wiener, J. and Cobb-Walgren, C. (1991), The role of perceived consumer effectiveness in motivating environmentally conscious behaviours. *Journal of Public Policy and Marketing* 10 (2), pp. 102-117.

HEFCE (2005), *Sustainable Development in Higher Education: Strategic Statement and Action Plan.* HEFCE, Bristol.

– (2008), *Sustainable Development in Higher Education: Consultation on 2008.* Update to Strategic Statement and Action Plan. HEFCE, Bristol.

Helferty, A. and Clarke, A. (2009), Student-led campus climate change initiatives in Canada. *International Journal of Sustainability in Higher* Education 10 (3), pp. 287-300.

IPAD (2009), *National strategy for Development Education (2010-2015).* http://www. ipad.mne.gov.pt/SociedadeCivil/educacaodesenvolvimento/EstrategiaNacionalENED/Documents/National%20Strategy%20for%20Development%20Education.pdf (accessed 1 December 2011).

Jones, P., Selby, D. and Sterling S. (Eds.) (2010), *Sustainability: Perspectives and Practice Across Higher Education.* Earthscan, London.

Junyent, M. and de Ciurana, A.M.G. (2008), Education for sustainability in university studies: a model for reorienting the curriculum. *British Educational Research Journal* 34 (6), pp. 763-782.

Marans, R.W. and Edelstein, J.Y. (2010), The human dimension of energy conservation and sustainability. A case study of the University of Michigan's energy conservation program International. *Journal of Sustainability in Higher Education* 11 (1), pp. 6-18.

McMillan J. and Dyball, R. (2009), Developing a whole of university approach to educating for sustainability: linking curriculum, research and sustainable campus operations. *Journal of Education for Sustainable Development* 3 (1), pp. 54-64.

OCDE (2006), *Good Practices in the National Sustainable Development Strategies of OECD Countries*, Sustainable Development Studies. http://www.oecd.org/dataoecd/ 58/42/36655769.pdf (accessed 1 December 2011).

Pickett, G.M., Kangun, N. and Grove, S.J. (1995), An examination of the conservative consumer: implications for public formation policy in promoting conservation behaviour. In Polonsky, M.J. and Mintu-Wimsatt, A.T. (Eds.), *Environmental Marketing: Strategies, Practice, Theory and Research*, pp. 77-99. The Haworth Press, New York.

Schmidt, L., Nave, J.G., O'Riordan, T. and Guerra, J. (2011), Trends and dilemmas facing environmental education in Portugal: from Environmental problem assessment to citizenship involvement, *Journal of Environmental Policy & Planning* 13(2), pp. 159-177.

Shiel, C. (2007), Developing and embedding global perspectives across the university. In Marshall, S. (Ed.), *Strategic Leadership of Change in Higher Education: what's new*?, pp. 158-173. Routledge, London and New York.

Sterling, S. and Scott, W. (2008), Higher education and ESD in England: a critical commentary on recent initiatives. *Environmental Education Research* 14 (4), pp. 386-398.

Straughan, R. and Roberts, J. (1999), Environmental segmentation alternatives: a look at green consumer behaviour in the new millennium. *Journal of Consumer Marketing* 16 (6), pp. 558-575.

Urien, B. and Kilbourne, W. (2011), Generativity and self-enhancement values in eco-friendly behavioural intentions and environmentally responsible consumption behaviour. *Psychology & Marketing* 28 (1), pp. 69-90.

Wals, A.E.J. and Blewitt, J. (2010), Third-wave Sustainability in Higher Education: Some (Inter)-national Trends and Developments. In Jones, P., Selby, D. and Sterling, S. (Eds.), *Sustainability: Perspectives and Practice Across Higher Education*, pp. 55-74. Earthscan, London.

Chapter 50

Manifestos for Sustainable Development: Student Visioning of Change through a Geography Course

Michael D. Lee

Abstract

In recognition of the fact that 2005-2014 was declared to be the International Decade of Education for Sustainable Development by the United Nations, since 2007 students at the California State University East Bay have been able to satisfy part of their undergraduate general education requirement by passing a geography course on sustainable development. An important element of this course is that students must participate in a visioning process to articulate priorities for sustainability across five interconnected and interdependent geographic and societal scales from the individual, through the corporation, community, nation, and up to a global level. Through exposure to a variety of key readings on sustainable development, individual research on the meaning and practice of sustainability within a given discipline or sector of society, and a careful analysis of their ecological footprints, students rapidly develop an understanding of humanity's principle adversity challenges and their potential solutions. The course and this key learning outcome, the ability to identify and articulate concrete steps to achieve sustainability, provides a useful example of how geographers can and should take a leadership role in education for sustainable development and in achieving the successful transition to a sustainable future. It illustrates the role stand-alone courses and the faculty who develop them can play in any discipline to spur education for sustainability across the broader curriculum.

Introduction: geography and sustainable development

At the 2002 Rio+10 Conference on Sustainable Development in Johannesburg, South Africa, 2005-2014 was declared to be the International Decade of Education for Sustainable Development (IDESD). The primary goal for the IDESD was laid out in the United Nations General Assembly Resolution 59/237, which encourages governments to consider the inclusion of measures to implement the decade in their respective education systems and to promote public awareness of and wider participation in this decade (United Nations 2005). The IDESD seeks to integrate the principles, values, and practices of sustainable development into all aspects of education and learning. According to UNESCO, education for sustainable development "calls for a re-orientation of educational approaches – curriculum and content, pedagogy and examinations – to ensure that change in public attitudes and educational approaches keep pace with the evolving challenges of sustainable development" (UNESCO 2005a). The IDESD was designed to encourage changes in behavior that will create a more sustainable future in terms of environmental integrity, economic viability, and a just society for present and future generations (UNESCO 2005b).

McManus (2004) stated that geography is an ideal discipline for the advancement and promulgation of the concept of sustainable development due to the long historical involvement of geographers in investigating environmental processes and nature-society relations. However, he also suggested that there is a sense that geographers have missed the opportunity to be at the forefront of research and teaching on sustainable development. This assertion was recently restated by Liu (2011) in his paper "Where in the World of Sustainability Education is US Geography?" In it Liu paraphrases Manning's 1990 observations that … geographers somehow missed the opportunity to be the central unifying discipline for the environmental movement in the 1960s … and that … sustainable development presents a major challenge to geography to demonstrate its capacity and relevance to what may be the most important public issue of our time (Manning 1990: 299, 290). Liu points out that in a review of sustainability-related course inventories provided by the Association of Advancement of Sustainability in Higher Education (AASHE) for 34 institutions, geography courses had a very limited presence making up only 4.4% of the sustainability courses listed and with less than half the institutions having at least one geography course listed in their inventory (Liu 2011). A review of the more than 60 voluntary associations formed within the Association of American Geographers shows that there is, as yet, no specialty or affinity group in sustainability or sustainable development (AAG 2012). This is surprising given that following a year of discussions and deliberation by delegates to the International Geographical Union's Commission on Geographical Education (IGUGCE), the Lucerne Declaration on Geographical Education for Sustainable Development was signed in July 2007. It promotes the integration of sustainable development into the teaching of geography at all levels and in all regions of the world based on the belief that knowledge, skills, attitudes and values learned in geography inspire decisions and actions that support sustainability and the goals of education for sustainable development (ESD) (Reinfried 2009).

The challenge identified by Manning (1990) is not, of course, geography's alone. In their 2011 paper on sustainability education, Hegarty et al. (2011) make a case for all disciplines to embrace sustainability across the curriculum, stating that "deep integration of EfS (education for sustainability) is crucial as we seek to build sustainable futures. But part of the necessary change process must be that sustainability scholars find ways to introduce EfS principles into their courses and perhaps into elements of those of colleagues, as we seek to build the change that will lead to integration". This certainly seems to be the position taken by the Lucerne Declaration for geography programs. Reinfried (2009) indicates that "through the Lucerne Declaration, geography educators can turn ESD into a reality in geography classrooms around the world. Designing new curricula, in which the criteria, principles and practices highlighted in the declaration are taken into account, is a first step toward this goal. The general frame of such curricula should focus on students' learning outcomes instead of syllabus-oriented teaching; be individualized and personalized; emphasize higher-order skills such as problem solving, communicating, critical thinking and creativity." It is vital, therefore, that geographers who have developed such curricular innovations promulgate these courses to the profession to stimulate a wider uptake and focus on education for sustainability both within the discipline itself, and as part of a growing trend to develop sustainability-focused degrees, minors and certificates in which geography can be a core focus, as detailed by Liu (2011). This paper, therefore, introduces one such innovative course that might provide inspiration for other educators; GEOG 4330 Sustainable Development, which has been offered at the California State University East Bay (CSUEB) since 1999. This paper describes its genesis, its main characteristics, and how its objective of engaging students in visioning sustainability across interconnected scales fits well with the prevailing wisdom concerning the required learning outcomes for today's graduates if we are to pursue and achieve a sustainable future (Haigh 2010; Wiek et al. 2011; Rieckmann 2012). It concludes with suggestions as to steps needed by institutions like CSUEB to use stand-alone courses within their curriculum as a basis for promoting education for sustainability across the institution and its curriculum in an effort to achieve the goals of this movement and the IDESD.

Sustainability and higher education

A growing body of literature is developing with respect to the role of universities in achieving sustainable development (for example: van Weenen 2000; Martin and Jucker 2005; Lozano 2006; Leal Filho 2010; Wiek et al. 2011; Rieckmann 2012). Within this literature exists an expanding core of publications concerned with the pedagogic aspects of sustainability (Barth et al. 2007; Cotton et al. 2009 ; Bacon et al. 2010. Ceulemans and De Prins 2010; Clark and Button 2011) and the role pedagogy plays in higher education's efforts to help create a society in which individuals are aware of the impact of their actions on others and on the planet, and that they take responsibility for those actions (Hegarty et al. 2011). Within this body of literature, papers are appearing dedicated to explaining the nature of courses developed on the topic of sustainability and sustainable development both to inspire others to adopt similar approaches and to share effective pedagogic practices within the wider academic community (for example: Hegarty et al. 2011). This is necessary because, as pointed out by Leal Filho (2010) "there is still a deficiency, in the literature, of empirical works which have tried to ascertain, first hand, what is hoped, expected or otherwise among those in charge of sustainability policies, from attempts to promote sustainability in the framework of institutions of higher education". As stated by Cortese more than a decade ago, it is difficult to imagine the individual and societal changes needed to ensure a sustainable future could occur without the assistance of institutions of higher education (Cortese, 1999) and by extension, therefore, without a systematic effort by educators of all disciplines to address education for sustainable development across the curriculum. There is thus an urgent need to share effective pedagogic practices amongst motivated faculty and expose the wider academic community to the possibilities for effective incorporation of sustainability into the curriculum. Echoing this sentiment, Leal Filho (2010) stated that "proportionally little emphasis seems to be given to sustainability by large universities. Indeed, many universities perform activities in the area of environmental protection and regard this as their contribution to sustainability, neglecting the fact that campus operations are indeed important contributions, but by far not the only ones". He details six key areas in which universities must integrate sustainability, a key one being "the implementation of principles of sustainable development as an intrinsic part of university teaching programs" (Leal Filho 2010). Haigh (2010) explains how sustainable development is missing from many degree programs because of the cross-disciplinary nature of the topic; few real-world problems of sustainability rest comfortably within single disciplines. Liu (2011) points out how for geography, for example, human geographers and physical geographers seem to have both assumed that sustainability falls mostly into the camp of the other and thus neither have fully embraced it. This silo problem has been circumvented in many higher education institutions by adding short, usually introductory level stand-alone courses at the early stages of the undergraduate curriculum, offering learners only a small package of ideas on the needs for sustainable development. The problem is that these packages are often seen as something apart from the learner's main education, a discrete package of knowledge that may not be seen as relevant to their main course of study (Haigh 2010). In Haigh's opinion, there remains only one solution: the full-scale "curriculum greening" in which the whole academic enterprise is suffused with education for sustainable development topics (Haigh 2010).

CSUEB, as a mid-sized university of approximately 14,000 students, currently exemplifies Leal Filho's observation and, like the majority of United States educational institutions, is yet to achieve Haigh's objective. The university has excelled in recent years in efforts to implement sustainable resource management policies in campus operations and possesses one of the largest solar photovoltaic energy systems of any campus in the United States. It also houses a state-of-the-art hydrogen fuel cell through a partnership with the regional electrical utility. This provides electricity to the regional grid while a retrofitted co-generation system uses the fuel cell's waste heat to heat the adjacent campus swimming pools, resulting in considerable annual energy savings. CSUEB has also made great strides in water conservation, waste reduction and recycling, energy efficiency, and purchase of zero-emission vehicles as part of a 2009, 20-year facilities master plan to create a sustainable campus. However, in the area of pedagogy, sustainability currently occupies a much lower profile, as is the case in many other institutions. A meta search of the 2010-2011 and 2011-12 academic catalogs indicated that out of over 2000 undergraduate courses and over 1000 postgraduate courses listed, only ten individual courses use the words sustainable or sustainability in their title or their catalog description (although clearly, there are many courses who's subject material support and promote environmental, social and economic topics that are critical to sustainability). The program with the most courses (three) directly focused on sustainability is geography, one of the academic disciplines most clearly aligned with the sustainable development field as indicated by the Lucerne Declaration (Reinfried 2009). These courses are GEOG 3000 Sustainable Resource Management, GEOG 6820 Seminar in Sustainable Cities, and the aforementioned GEOG 4330 Sustainable Development, the focus of this paper. Each of these courses is taught by the author.

Advancing the Lucerne Declaration at CSUEB: GEOG 4330 Sustainable Development

Pre-dating the Lucerne Declaration and UNDESD, the senior-level (fourth year) sustainability-focused course, GEOG 4330 Sustainable Development, has been taught in the geography program at the California State University East Bay (CSUEB) since the late 1990s. Coincidentally, to the discussions being held by the IGUCGE to adopt the Lucerne Declaration and in light of the UNDESD, the faculty member responsible for this course, in responding to feedback in course evaluations from students, determined that sustainability should not just be the domain of geography majors but, if the goals of the UNDESD and Lucerne Declaration are to be achieved, should be opened up to a wider audience. It was thus proposed in 2006 that GEOG 4330 be granted general education (GE) credit so as to attract students from a wider variety of disciplines who might take it as an elective, regardless of their degree path. In order to graduate from CSUEB, students must complete 72 credit units out of their total requirement of 180 by taking courses outside their major subject from across the curriculum, the objective being to assure that graduates have made "measurable progress toward becoming *truly educated persons* for a diverse society" (CSUEB 2010). One of the key purposes of the CSUEB GE program is to provide the means whereby graduates:

> "acquire appreciable knowledge about their own bodies and minds, about how human society has developed and how it now functions, about the physical world in which they live, about the other forms of life with which they share that world, and about the cultural endeavors and legacies of their civilizations" (CSUEB 2010).

As the author sought this wider exposure of CSUEB students to sustainable development, the University Committee on Instruction and Curriculum (CIC) that grants GE status to courses was provided with background information on the UNDESD as part of the justification for assigning GE credit. The CIC agreed that there was a good fit between CSUEB's GE goals and those of the IDESD and thus GEOG 4330 was approved by the Academic Senate for GE credit in the 2007/08 academic calendar and thus became eligible as an elective for any major seeking to fulfill upper-division Social Science GE requirements.

For non-geography students who take it, GEOG 4330 fits what Hegarty et al. (2011) refer to as the "bolt-on" or "stand-alone" course model, taught by a faculty member with a strong ethical commitment to sustainability. Unlike Hegarty et al.'s course Sustainability, Society and Environment at RMIT University in Australia, which is an introductory course, GEOG 4330 at CSUEB is a senior undergraduate class taken near the end of a student's studies. However, like the RMIT course, it is available as an elective that could be taken by any undergraduate at CSUEB, regardless of their major, thanks to its designation as a general education course for upper division social science. For most non-geography students who take it for GE credit, it will be the only course they get that focuses explicitly on sustainability.

GEOG 4330 meets twice per week over ten weeks for two-hour seminar sessions (with asynchronous online discussions in between times) and is a survey of the processes by which sustainable levels of development might be achieved across the human and natural geographical dimensions of time and space. It provides an assessment of the personal, corporate, local, national and global dimensions of sustainability with special emphasis on policy questions and the identification of options and practical alternatives for the future at each scale. As recommended for geography courses that meet the goals of the Lucerne Declaration (Reinfried 2009), it takes a systematic approach to addressing the various interconnected scales at which decisions must be taken and policies must be made to achieve a level of societal development at which the welfare of today's generation can be met without compromising the welfare of future generations. Building off a body of literature that has expanded exponentially since the Brundtland Report (WCED 1987), the course taps into the views of key thinkers about sustainability and synthesizes important issues from the individual to the global scale. The course presents a triple-bottom line perspective (advanced by Elkington 1997; Hawken et al. 2000; McDonough and Braungart 2002a and others) on the world, one in which the joint concerns of economic viability, environmental integrity, and social equity are effectively balanced in the decision-making process through which our human systems are designed and implemented. As such, it provides a valuable case study of how this complex topic can indeed be addressed in a comprehensive, yet condensed manner to provide students that are new to the topic with a chance to acquire vision and insight into society's many challenges and opportunities.

Students engage in guided readings (for example, Berry 2001; Brown 2009; Cairns 2004; Curtis 2003; Hart 1997; Hawken 1997; Johnston et al. 2007; Kates et al. 2005; Mazurkewich 2004; McDonough et al. 2003; McDonough and Braungart 2002b; McGranahan and Satterthwaite 2003; Moran et al. 2008; Myers 1997. Redefining Progress 2009; Pillarisetti and van den Bergh 2007; Rockström et al. 2009; WCED 1987), participate in lectures, construct, evaluate and model changes to their ecological footprints, and conduct individual research designed to relate the topic of sustainability to their own degree programs and/or projected career interests (e.g. sustainable food systems, sustainable design, sustainable tourism, etc.). A key component in the course relates to intensive class and online discussion on the subject of sustainable development at the five scales, the goal being to reflect on and fashion a collective set of fundamental actions to be taken if society is to advance to sustainability. The most recent model for this evolving class has the students divided into groups of 4-5 students, assigned group names such as groundbreakers, innovators, pioneers, trailblazers and trendsetters. At the end of each lecture sequence on the five sustainability scales, each student must vision and articulate three actions they believe are needed to advance sustainable development at that scale and present them as manifesto prescriptions to the group in an online threaded discussion. In articulating the prescription, they are instructed to be clear what action is being proposed, which actor is responsible for its implementation, and why it leads us toward future sustainability (note that based on student feedback, in the next iteration of this class students will also be required to specify an appropriate indicator/metric by which progress toward sustainability can be gauged). At the subsequent classroom session, the students break into their groups to discuss the suggested prescriptions and to vote on their top three. The class then reconvenes and each group presents their priorities, which are discussed and, as appropriate, modified and reworded to best capture and articulate the action from the broader class perspective. During the discussion the opportunity is taken to draw connections between prescriptions offered at a smaller scale and/or think ahead to identify obstacles at larger scales that might make implementation difficult or ineffective. The result is a powerful manifesto of between 12 and 20 prescriptions for each scale and 60-100 for the set of five. The students are presented with the full manifesto at the end of the course.

The learning outcome; a manifesto for sustainability

The following section provides examples from the set of prescriptions offered by a single group of students who took the class in order to qualitatively illustrate their broad range and level of insight.

Manifesto for a sustainable individual

The following prompt was provided to solicit student prescriptions on individual-scale sustainability: While we do not and probably cannot know the exact parameters that must be met for individual lifestyles to collectively result in a sustainable world system, we offer the following as some critical characteristics that a sustainable lifestyle must logically include and which must be adopted and replicated, first and foremost, by all Americans and others living a materially-intensive, industrial-style life across Europe, Japan, Australia and the other OECD regions and nations. This list, it must be noted, must also be applied to the elite, affluent minorities of the developing world as well as to the growing ranks of middle-class in those nations that are swelling with new members due to the growing prosperity of their emerging economies, especially China and India. Three of the manifesto prescriptions provided and endorsed by the class in response to this prompt included:

Practice voluntary simplicity: Every individual should seek to practice a form of voluntary simplicity – minimizing their ecological footprints from the use of materials, energy and so forth by practicing frugality (using only what is necessary) and efficiency (when using them, using only the minimum necessary to satisfy the particular objective).

Seek services not products: Every individual should seek out services rather than products in their satisfaction of needs and wants. Thus they should seek to become customers, rather than consumers, allowing our needs to be met by corporations seeking to satisfy those needs with the highest degree of reuse, thus placing value on cradle-to-cradle efficiency, longevity, repairability and so forth, and not on disposability and rapid obsolescence.

Shorten food chains: Every individual should eat as low down the food chain as possible, securing most or all of their nutrition from producer organisms rather than consumer organisms. The foods consumed in this low energy input diet should come, as much as possible, from local sources, including home gardens, with minimal processing and grown in season.

Manifesto for a sustainable business

The following prompt was provided to solicit student prescriptions on corporate-scale sustainability: While we do not and probably cannot know the exact best management practices and standards that the business sector must adopt to help consumers pursue more sustainable lifestyles and for communities to function with a sufficiently small ecological footprint to stay within the earth's carrying capacity of renewable resource production and waste assimilation, we propose the following actions that could help shift corporations involved in retail, manufacturing, service provision and so forth in a direction toward sustainability. Three of the manifesto prescriptions provided and endorsed by the class in response to this prompt included:

Reduce disposability: Businesses should minimize disposability in all aspects of their operations – internal or external. For example, internally they should eliminate single-use items in cafeterias or for special events (or, at a minimum, replace these with biodegradable items made from renewable materials that are designed to be composted – they would also have to have an effective green waste separation program for these items). And externally, for example, they should eliminate the manufacturing of single-use or substitute reusable or compostable ones for non-compostable items.

Practice social responsibility: Businesses should be socially responsible, facilitating employees to make positive contributions to society by being active in their own communities and in the community in which the business is located or in which its products or services are employed. Resources (time, matching funds, material support, etc.) should be made available to employees to encourage volunteerism and other beneficial individual behaviors whether independently or as part of corporate-directed collective activities.

Incentivize workers: Employees should be offered direct incentives to shift to greater personal sustainability especially in relation to commuting, i.e. reducing the energy cost of getting workers to their workplace which is part of the ecological backpack of the products and services produced. Workers should be assisted financially or materially so that they can take public transport, live closer, car-pool, drive more fuel-efficient vehicles, and so forth.

Manifesto for a sustainable community

The following prompt was provided to solicit student prescriptions on community-scale sustainability: A community can be defined as a geographic area in which a collection of individuals operates in informal and formal organizational structures. Communities occupy a pivotal point in the continuum of scales over which the sustainability paradigm must be shifted. The informal community is the place and ways in which individuals exert personal decisions and behavior in a bottom-up manner that will be very significant, due to the collective effects, in achieving sustainability. The formal community is where many top-down decisions take place that stimulate or organize different types of individual decisions that help create unsustainable or sustainable forms of societal organization and actions. They are also the scales at which decisions are focused and implemented, having been made at higher levels of decision-making, for example at a regional or national level. If the informal and the formal dimensions of community, the bottom-up and top-down do not match, then sustainability is unlikely to be achieved. The following actions could help move a community, a population of individuals acting informally (in neighborhoods, for example) and formally (within a city governed by locally elected officials and administered by public servants), toward sustainability. Three of the manifesto prescriptions provided and endorsed by the class in response to this prompt included:

Facilitate non-vehicular transportation: Communities must maximize the opportunities within their community boundaries for non-vehicular modes of transport to reduce vehicle miles driven while at the same time raising exercise levels. Communities should install bike paths, pedestrian walkways and other physical infrastructure modifications while encouraging a culture that respects (e.g. sharing the road by drivers) and promotes non-vehicular uses.

Make public spaces functional and productive: Communities should make community public spaces more functional in ways that enhance their environmental, economic and social benefits. Underutilized space should be put to optimal use, as applicable, as community gardens, community forests (carbon sinks/carbon-neutral fuel sources), urban runoff filters, and the like. For example, areas of public parks could be used as public gardens run by volunteers and/or parks staff directing community service individuals (high school students, misdemeanor offenders, and so forth) with harvests used to supply schools, community hospitals, shelters or other food outlets, to be shared with volunteers proportional to their labor inputs, and/or as sources of revenue, to be sold to the community at a community-based farmer's market.

Favor imbyism over nimbyism: The informal community must favor "imbyism" (in my back yard) over "nimbyism" (not in my back yard), the informal community being the private citizens that band together to influence the formal community, i.e. decision-makers, to exclude undesirable land-uses or other such factors from the geographic area the community occupies. Community members must be encouraged to confront the broader consequences of excluding controversial land or resource uses rather than accepting them, and taking those same relatively strong powers to resist (and therefore foist said problem onto a less-able community assuming that the object of protest will indeed find a community home somewhere) and using them to force those responsible to eliminate the causes of their concerns.

Manifesto for a sustainable nation

The following prompt was provided to solicit student prescriptions on national-scale sustainability: *There are over 200 entities in the world that we recognize as nations, ranging from the population giants of China and India, both with over one billion citizens, to the minnows, the tiny island nations or city-states like the Marshall Islands and Andorra that might number only a few tens of thousands. Collectively, they provide the framework with which the seven billion people that currently occupy the planet are geographically organized and how they are governed. In organizing and governing these peoples, nations have a key top-down role to play in the shift to sustainable development in almost every conceivable area. Collectively, the nations of the world, with their respective ecological footprints and rates of population growth, are critical determinants of global ecological sustainability and will thus determine whether humanity stays within or exceeds the carrying capacity of this planet. Nations, through their ability to legislate, to tax, and to spend, have a key role to play in aligning the behavior of their regions, communities, corporations and individual citizens with the broader goals of sustainable development. It must be recognized that all nations are not of equal importance in this; individual nations have very different roles and abilities to determine the gross outcomes of the multi-level sustainable development process. Unilateral actions by nations can fundamentally alter the course we take and in myriad ways, from the catastrophic, for example by triggering a nuclear war, to the chronic incremental, longer-term processes such as the failure to switch individuals and corporations away from fossil fuels and towards energy conservation and renewable energies. Similarly, due to their small size or relatively lowly economic status, other nations are virtually powerless to influence global outcomes yet suffer disproportionately from the changes that have and will occur in global economic, social and environmental conditions, for example the small island nations that produce little of the world's greenhouse gases but on whom the effects of climate change and sea level rise will be disastrous. They could become fully sustainable, but it wouldn't make much difference if their bigger fellow-nations do not follow suit. In light of this understanding, we suggest the following actions that could help move every nation, but especially our own, the United States of America, and the oth-*

er wealthy and powerful economies toward sustainability. Three of the manifesto prescriptions provided and endorsed by the class in response to this prompt included:

Apply domestic laws internationally: Nations should ensure that their citizens and corporations conduct themselves and their business operations internationally in compliance with their own domestic laws and requirements or with the laws and requirements of their host nations, whichever is more rigorous/sustainable. Similarly, domestic laws on products made for internal consumption – related to safety, pollution, energy efficiency, etc. – would apply to all products made for export to nations without such laws, i.e. it would not be permissible to produce goods injurious to sustainability goals/standards established for the nation of origin for use in other nations without such goals/standards.

Best practice leadership: National governments should take the lead in implementing best practices in all areas related to sustainability acting as a role model, trendsetter and catalyst for sustainable development. Based on the fact that some 20-40% of national GDP is government spending and with a high percentage of the service sector in public ownership, the installation of best practices in energy use and conservation, green/sustainable products, and so forth can provide a massive stimulus to bring best technologies to the forefront, encourage innovation, and thus pull the national economy over obstacles that might inhibit the profit-driven private sector of the economy. Governments can more easily set goals for themselves such as 100% green energy, 100% recycling, and a 90% reduction in energy use and implement the steps required to achieve them.

Tax for sustainability: Nations should levy taxes on the basis of sustainability considerations, taxing harmful goods or services and using the revenues to stimulate conservation and efficiency and the consumption of beneficial but otherwise more expensive substitutes. Taxation should be used to approach the full-cost pricing of goods and services that reflect and internalize their environmental and social costs and benefits. Taxation should be geared to consumption, not production, and rather than tax labor, tax burdens should be shifted to consumption but in a manner that is progressive and reflects the reality by which different forms of consumption contribute to the advancement of human welfare and support sustainability.

Manifesto for a sustainable global system

The following prompt was provided to solicit student prescriptions on global-scale sustainability: *While society does not yet know the exact parameters associated with global sustainability, for example exactly how much carbon dioxide could safely be emitted before catastrophic anthropogenic climate change is unavoidable, or exactly how many people the earth could support without decline (which is dependent on many factors not least future technological advancements), etc., it nevertheless is possible, using the concepts at our disposal, especially the notions of strong and weak sustainability, the precautionary principle, the natural step, planetary boundaries and just plain old common sense, to prescribe key requirements/conditions for global sustainability. The following represent a list of key prescriptions that seem critical for indefinite global health and sustainability. Note these are described in a manner that attempts to capture the full global scale, recognizing that they also apply to and will partly or wholly be achieved by specific actions and changes at smaller scales and by specific entities. Some may require new global institutions and mechanisms to be created.* Three of the manifesto prescriptions provided and endorsed by the class in response to this prompt included:

Global governance: In order to achieve sustainable development we should establish true, effective global governance with the force of law. If global sustainability is to be achieved, it seems unavoidable that national sovereignty be subordinated to a greater degree to a global governance authority. A body that can make policies and draft laws that apply to all nations and peoples and which have appropriate funding and enforcement mechanisms is required. Global conventions such as CITES, the Montreal Protocol, or the Basel Convention, previously arrived at through painful and protracted, voluntary, ad-hoc mechanisms would be formalized and systematized. Thus we would have a binding, universal Law of the Sea, Law on Global Greenhouse Gases, etc., that would be formulated to address global problems at a holistic, global scale under which all piecemeal, national-level actions could and should be organized.

Enhance development assistance: In order to enhance global equity, richer nations should provide development assistance to poorer nations based on their ability to pay and this should be dispersed to nations on their relative needs (on some objective criteria such as the UNDP Human Development Index rankings). Thus some form of means test should be used at both ends – ability to pay and relative need.

Financing linked to sustainability: Any form of global financing through a new global governance system or the current organizations such as the IMF, World Bank and so forth must come with sustainability strings attached – a triple bottom line approach must be used in implementing such financing which should meet sustainable development criteria in social equity and environmental integrity areas as well as in terms of economic outcomes. Agreeing to sustainability actions should be a condition of receiving loans and evidence of their implementation and adherence to sustainability principles must be a condition of renewing such financing or financing renegotiation.

Scaling up to full-blown education for sustainability

Haigh points out in calling for a "curriculum greening" the persistent belief that our present educational structures are less appropriately geared to meeting the needs of the future than reinforcing the destructive characteristics of the current, unsustainable age (Haigh 2010). This echoes Sterling (2001) who stated that for nearly 30 years, education has been identified as the key to addressing environment and development issues and latterly to achieving a more sustainable society but that most education daily reinforces unsustainable values and practices in society through competition and consumption. Scott (2002) suggested four key responsibilities of educators with respect to the pedagogy of sustainability: to demonstrate to our students why sustainability is important to them, to help them gain a diversity of perspective on what sustainability means, to provide learning opportunities through appropriate curricular choices, and to help them understand what they have to do keep their own and others' options open over time in a changing world.

According to Sterling (2001), realization of a sustainable education paradigm requires vision, image, design and action at all levels and from all who are concerned with achieving healthy societies and ecologically sustainable lifestyles. Rieckmann (2012) writes that given the "development state of the world society", education should enable individuals to reflect on their own actions by taking into account their current and future social and environmental effects – from a global perspective – and to intervene productively in shaping them in a sustainable manner. As pointed out by Rieckmann (2012), one important component of Education for Sustainable Development is envisioning a better future. Citing the work of Adomssent et al. (2007), Rieckmann states that the purpose of this visioning is to establish a link between long term goals and immediate actions, and motivate people to action by harnessing their deep aspirations (Rieckmann 2012). The learning outcomes from such an effort are that students should acquire the values, knowledge, and skills for sustainable living and participation in society. Other observers too stress the importance of visioning the future as a core competency for graduates prepared to address sustainability challenges. Hicks (2012) defines *Future Studies* as teaching and research so as to "discover or invent, examine, evaluate and propose possible, probable and preferable futures (Hicks 2012: 4)," and suggests that geography instruction embrace this field to add to its traditional responsibilities of describing and explaining the way the world currently is. He believes that *Future Studies* are well aligned with the broader objectives of education for sustainability and lists nine key concepts associated with futures education. In these he includes a component he calls "Visions for the Future" asking "What needs to be left behind and what taken forward? In particular, what visions of a better future are needed to motivate active and responsible citizenship in the present?" and concludes that students therefore need to develop their skills of visioning and use of the creative imagination (Hicks 2012: 8). Rieckmann (2012) examined the views of 70 experts in education for sustainable development from "northern" nations (the UK and Germany) and "southern" nations (Chile, Ecuador and Mexico) concerning core objectives and core student competencies to be achieved through education for sustainable development. He offers that universities have to create particular learning settings in which students can improve their competencies for understanding complexity and long-term effects of present-day actions as well as for questioning common assumptions. For both groups of experts, the most important objective to achieve was creating and changing student values, attitudes and awareness and the most important competency to promote was the systemic thinking and handling of complexity which is made all the more important by "the complexity, uncertainty, risks and the high velocity of societal (global) change" (Rieckmann 2012: 133). On a similar theme, Wiek et al. (2011) state that sustainability competence is composed of five components: systems-thinking competence, anticipatory competence, normative competence, strategic competence, and interpersonal competence. Anticipatory competence is perhaps the one most aligned with the visioning process that forms the backbone of GEOG 4330 described in this paper. Wiek et al. (2011) state that anticipatory competence involves fostering the capacity to think systematically about the future and future generations and requires, among other things, an understanding of the different types of futures, i.e., possible futures (based on notions of plausibility), probable futures (those determined "likely" to occur), and desirable futures (value-laden; based on sustainability principles).

The question, then, is how best are the full set of sustainability competencies to be achieved and what role can courses like GEOG 4330 and geography as a discipline play in the process? McManus (2004), identified three different ways for geographers to incorporate sustainable development into their teaching; to address it within an existing environmental course or within a course focused on development or management, to develop a new course wholly focused on sustainable development, or to develop a new program of study that focuses on sustainable development, with the latter often requiring a merger with another program such as environmental studies. At CSUEB, the Geography and Environmental Studies Department adopts all three. According to Hegarty et al. (2011), "stand-alone" courses, such as RMIT's Sustainability, Society and Environment and CSUEB's GEOG 4330 add a wide range of value to the sustainability goals of universities. Such courses foster crucial, transferable skill sets such as critical and integrated thinking as they seek to locate knowledge within disciplinary spheres and situate their learning objectives in complex, real-world contexts (Hegarty et al. 2011). For example, the vision statements crafted through student discussion in GEOG 4330 illustrate how empowered students can become to critically assess sustainability challenges and craft proposed solutions. Similarly, Hegarty et al. (2011) indicate that at RMIT, their introductory course offers students a vision of empowerment in the hope that this shapes their sense of possibility about the future of the world and their efficacy in it. They noted, however, that in confronting their values and preconceptions concerning sustainability problems, discomfort was evident amongst participating students. In an earlier work, Sterling had observed that "too much environmental knowledge (particularly relating to the various global crises) can be disempowering, without a deeper and broader learning process taking place" (Sterling 2001: 19). He suggests that full engagement of students involves their critiquing much current thinking and practice, but also visioning and designing a credible and practical alternative. Experience in teaching GEOG 4330 certainly aligns with his observation; the visioning process of creating their own manifesto appears to have empowered students to embrace the notion of change and challenge the status quo not just at a personal level, but also at the higher levels of societal organization that they may one day be in a position to influence through their career choices.

Conclusion: stand-alone courses are a foundation for curriculum change

Can a stand-alone course really make a substantive contribution to education for sustainability and advance sustainability learning outcomes? A qualitative evaluation of the outcomes from GEOG 4330 shows that each new offering of the course produces a different set of class manifestos demonstrating both the ability of a set of students largely new to the subject to quickly grasp and prioritize a sophisticated set of prescriptions for action, and the heterogeneity of student opinion on the subject of sustainable development and what it takes to achieve it across our society. In part, this heterogeneity reflects the considerable diversity of the student body at CSUEB which is ethnically diverse, has a growing proportion of international students, has a majority of female students, and has a wide range of ages from teenage high-school graduates to returning mid-career adults seeking retraining. Everyone's experiences and values are different and inform their perceptions of what sustainability involves and requires. This is to be expected and of course tracks to the fact that no universal definition of sustainability exists across academia or in society. That said, with respect to the visioning process from the individual scale through to the global, students appear to find it relatively easy to articulate their ideas concerning sustainability for the micro- and meso-scales of individual lifestyles and corporate practices for sustainability, levels that they are most intimately engaged with. They find it harder to do so as they progress to the global scale and thus exhibit some discomfort and frustration with the challenge. However, that notwithstanding, it is remarkable the degree to which they can still develop interlocking and complimentary actions up through increasing levels of scale. Responses from students suggest that scaling the visioning up from the local to the global strengthens their ability to articulate each new level of prescriptions even though they have little direct experience and only limited knowledge of the systems that operate at the larger scales. By working systematically through the issue of sustainable development by looking at how adversity trends and the forces of unsustainability present themselves at increasingly complex and aggregated scales of human organization, the pieces of the puzzle and the alignment of actions can fall readily into place as recurring obstacles and opportunities become recognizable. Though sustainable development is now an enormous field, as demonstrated by the more than 40 million English language web pages devoted in some way to it by late 2011, the experience of teaching GEOG 4330 suggests that with an appropriate pedagogic approach, it can nevertheless be readily grasped and understood as a societal challenge in a single "stand-alone" course.

The flaw is, of course, that GEOG 4330 and the other handful of courses at CSUEB that focus directly and centrally on the subject of sustainability are currently taken by only a small minority of the student body who elect to follow this as yet non-mainstream avenue of study. The AASHE STARS (Sustainability Tracking, Assessment and Rating System) recognizes institutions that have formal education programs and courses that address sustainability because they help equip their students to lead society to a sustainable future. Out of the ten rating categories related to curriculum, the lion's share of points are determined by the abundance of courses meeting criteria that qualifies them as sustainability focused and sustainability related (AASHE 2012). Maximum points go to those institutions with at least 10% of their courses judged to be sustainability focused and an additional 30% or more judged to be sustainability related. A multi-disciplinary campus sustainability committee must have formally established criteria for what qualifies as each. All courses must be judged one or the other, or neither, but not both. Currently, the majority of higher educational institutions in the United States lack such formal criteria and have much fewer sustainability-focused courses. That said, a benefit of a "stand-alone" sustainability-focused course like GEOG 4330 is that although few in number, as sustainability becomes an institution's objective, they can become springboards for broader inclusion of sustainability across the curriculum as their presence in the classroom and the actions of the faculty who teach them and students who take them permeates campus culture through a variety of mechanisms. As pointed out by Hegarty et al. (2011), "stand-alone" courses can become a point of entry for sustainability studies within an institution. Developments at CSUEB illustrate this quality. In 2011, efforts began to more fully embrace education for sustainability as an explicit part of the university mission beginning with the convocation address of the new President who pledged sustainability as one of the priorities of his tenure. Supported by the new Provost, campus-wide efforts to develop explicit institutional learning outcomes (ILO) for all graduates will be completed by the end of the 2011-12 academic year, the final draft set including a proposed ILO that all graduates be able to "act responsibly and sustainably at local, regional and global scales". Having joined ASHE in the fall trimester of 2011, the Provost's office followed this up in the winter trimester of 2012 with the inauguration of CSUEB's first faculty learning community (FLC) on the subject of "Sustainability & Pedagogy" to begin promoting the development of new sustainability focused courses like GEOG 4330 in other disciplines. The use of FLCs to promote education for sustainability has been adopted at other institutions such as Emory (Eisen and Bartlett 2006) and Temple (Barnett et al. 2010) universities. The CSUEB FLC seeks to achieve what Ceulemans and De Prins (2010) suggest is the most promising approach for college graduates to integrate sustainable development and its principles in their future professional lives, that of having sustainability intertwined as a concept in regular disciplinary courses all across the different majors offered in the curriculum (Lozano 2010). The new courses developed will place sustainability more centrally and visibly into degree programs at CSUEB compared to its current ad-hoc and piecemeal presence largely invisible within the curriculum. An expanded list of sustainability-focused courses makes feasible the emergence of an interdisciplinary certificate, minor or major program in sustainability studies for interested students as the profile of sustainability is raised on a campus. This is an increasing trend at American universities as identified by Liu (2011) and will be accelerated by such efforts.

It is imperative that such efforts succeed and more discipline-based courses like GEOG 4330 be created at higher education institutions. As pointed out by Hegarty et al. (2011), our communities' futures depend on the active endorsement of sustainability by all professional fields and disciplines. Without this, the goals of the UNDESD will likely not be realized. With respect to sustainability and the discipline of geography, according to Liu (2011), sustainability and geography programs benefit each other, and many geographers are well positioned to participate in sustainability degree programs by creating a sustainability concentration in geography or specifying geography as a core area in a sustainability program. As most American universities still do not have sustainability degree programs, Liu believes that geographers in particular should play a key role in developing the programs to increase geography's participation and leadership (Liu 2011). This key role for geography has been demonstrated at CSUEB where the Geography and Environmental Studies Department have leveraged GEOG 4330 to create a suite of courses that are either sustainability focused or clearly sustainability related (roughly 40% of its courses could be reasonably judged to be sustainability focused or sustainability related based on their content), permitting the development of a certificate in Sustainable Resource Management. This certificate can be added to a student's degree, not just by geography majors, but also by any major across campus completing a five-course set of elective classes. It provides the nucleus, therefore, for a step-up to an option, minor and eventually a major in sustainability. Other disciplines, not least geography, should be encouraged by this example, assured that just a few dedicated courses focused on sustainability can become an effective beachhead in a broader institutional effort to transform the curriculum and thus meet the aspirations of the UNDESD sooner rather than later, one university and college at a time.

References

AAG (2012), *AAG specialty and affinity groups*. Association of American Geographers. Available at: http://www.aag.org/cs/membership/specialty_groups (accessed 26 March 2012).

AASHE (2012), *STARS Sustainability Tracking, Assessment and Rating System Version 1.2 Technical Manual*, February 2012. Association for the Advancement of Sustainability in Higher Education. Available at: http://www.aashe.org/files/documents/STARS/stars_1.2_technical_manual.pdf (accessed 26 March 2012).

Adomssent, M., Godemann, J. and Michelsen, G. (2007), Transferability of approaches to sustainable development at universities as a challenge. *International Journal of Sustainability in Higher Education* 8 (4), pp. 385-402.

Bacon, C.M., Mulvaney, D., Ball, T.B., Du Puis, E.M., Gliessman, S.R., Lipschutz, R.D. and Shakouri, A. (2010), The creation of an integrated sustainability curriculum and student praxis projects. *International Journal of Sustainability in Higher Education* 12 (2), pp. 193-208.

Barnett, P., Fiore, S. and Wingert-Playdon, K. (2010), Interdisciplinary goals, disciplinary faculty: bridging the gap with faculty learning communities. *Academic Leader* 26 (10), pp. 3-5.

Barth, M., Godemann, J., Rieckmann, M. and Stoltenberg, U. (2007), Developing key competencies for sustainable development in higher education. *International Journal of Sustainability in Higher Education* 8 (4), pp. 416-430.

Berry, W. (2001), The idea of a local economy. *Orion* 20 (1), pp. 28-37.

Brown, L. (2009), *Plan B 4.0 Mobilizing to Save Civilization*. W.W Norton & Co, New York, NY.

Cairns, J. Jr. (2004), *Choosing Model Nations to Set Examples for Achieving Sustainability*. Available at: http://www.mnforsustain.org/model_nations_selecting_cri teria_cairns_0404.htm (accessed 26 March 2012).

Ceulemans, K. and De Prins, M. (2010), Teacher's manual and method for SD integration in curricula. *Journal of Cleaner Production* 18, pp. 645-651.

Clark, B. and Button, C. (2011), Sustainability transdisciplinary education model: interface of arts, science, and community (STEM). *International Journal of Sustainability in Higher Education* 12 (1), pp. 41-54.

Cortese, A.D. (1999), *Education for sustainability – the university as a model for sustainability, Second Nature, Boston*. Available at: http://secondnature.org/pdf/ snwritings/articles/univmodel.pdf (accessed 26 March 2012).

Cotton, D., Bailey, I., Warren, M. and Bissell, S. (2009), Revolutions and second-best solutions: education for sustainable development in higher education. *Studies in Higher Education* 34 (7), pp. 719-733.

CSUEB (2010), *What are the general education-breadth requirements?* California State University East Bay. Available at: http://www20.csueastbay.edu/ecat/files/docs/previous-catalogs/20102011.pdf (accessed 26 March 2012).

Curtis, F. (2003), Eco-localism and sustainability. *Ecological Economics* 46, pp. 83-102.
Eisen, A. and Bartlett, P. (2006), The Piedmont Project: fostering faculty development towards sustainability. *Journal of Environmental Education* 38 (1), pp. 25-36.
Elkington, J. (1997), *Cannibals With Forks: the Triple Bottom Line of 21st Century Business.* New Society Publishers, Gabriola Island, BC; Stony Creek, CT.
Haigh, M. (2010), Greening the university curriculum: Appraising an international movement. In Chalkley, B., Haigh, M. and Higgitt, D. (Eds.), *Education for Sustainable Development*, pp. 25-42. Routledge, London.
Hart, S.L. (1997), Beyond Greening: Strategies for a Sustainable World. *Harvard Business Review* Jan-Feb, pp. 66-76.
Hawken, P. (1997), Natural Capitalism. *Mother Jones* 22 (2), pp. 40-52.
Hawken, P., Lovins, A. and Lovins, L.H. (2000), *Natural capitalism: creating the next industrial revolution.* Little, Brown and Co., Boston, MA.
Hegarty, K., Thomas, I., Kriewaldt, C., Holdsworth, S. and Bekessy, S. (2011), Insights into the value of a 'stand-alone' course for sustainability education. *Environmental Education Research* 17 (4), pp. 451-469.
Hicks, D. (2012), The future only arrives when things look dangerous: reflections on futures education in the UK. *Futures* 44, pp. 4-13.
Johnston, P., Everard, M., Santillo, D. and Robèrt, K.-H. (2007), Reclaiming the definition of sustainability. *Environmental Science and Pollution Research* 14 (1), pp. 60-66.
Kates, R.W., Parris, T.M. and Leiserowitz, A.A. (2005), What is sustainable development? Goals, indicators, values, and practice. *Environment: Science and Policy for Sustainable Development* 47 (3), pp. 8-21.
Leal Filho, W. (2010), Teaching sustainable development at university level: current trends and future needs. *Journal of Baltic Science Education* 9 (4), pp. 273-284.
Liu, L. (2011), Where in the world of sustainability education is US geography? *Journal of Geography in Higher Education* 35 (2), pp. 245-263.
Lozano, R. (2006), Incorporation and institutionalization of SD into universities – breaking through barriers to change. *Journal of Cleaner Production* 14, pp. 787-796.
– (2010), Diffusion of sustainable development in universities' curricula: an empirical example from Cardiff University. *Journal of Cleaner Production* 18, pp. 637-644.
Manning, E. (1990), Presidential address: sustainable development, the challenge. *The Canadian Geographer* 34 (4), pp. 290-302.
Martin, S. and Jucker, R. (2005), Educating earth-literate leaders. *Journal of Geography in Higher Education* 29 (1), pp. 19-29.
Mazurkewich, K. (2004), A wealth of happiness: Tashi Wangyal. *Asian Wall Street Journal*, 8 October 2004, pp. 1.
McDonough, W. and Braungart, M. (2002a), *Cradle to Cradle: Remaking the Way We Make Things.* North Point Press, New York, NY.
– (2002b), Design for the triple top line: new tools for sustainable commerce. *Corporate Environmental Strategy* 9 (3), pp. 251-258.
McDonough, W., Braungart, M., Anastas, P.T. and Zimmerman, J.B. (2003), Design through the 12 principles of green engineering. *Environmental Science and Technology* 37 (5), pp. 94A-101A.
McGranahan, G. and Satterthwaite, D. (2003), Urban centers: an assessment of sustainability. *Annual Review of Environment and Resources* 28, pp. 243-74.
McManus, P. (2004), Geography. In Blewitt, J. and Cullingford, C. (Eds.), *The sustainability curriculum: the challenge for higher education*, pp. 218-231. Earthscan Publications, London.
Moran, D.D., Wackernagel, M., Kitzes, J.A., Goldfinger, S.H. and Boutaud, A. (2008), Measuring sustainable development – nation by nation. *Ecological Economics* 64 (3), pp. 470-474.
Myers, N. (1997), Consumption in relation to population, environment and development. *The Environmentalist* 17, pp. 33-44.

Pillarisetti, J.R. and van den Bergh, J.C.J.M. (2007), *Sustainable nations: what do aggregate indicators tell us?* Tinbergen Institute Discussion Paper TI 2008-012/3, Amsterdam. Available at: www.tinbergen.nl/discussionpapers/08012.pdf (accessed 26 March 2012).

Redefining Progress (2009), *Ecological footprint quiz.* Available at: http://www.myfoot print.org/en/ (accessed 26 March 2012).

Reinfried, S. (2009), Education for sustainable development and the Lucerne Declaration. *International Research in Geographical and Environmental Education* 18 (4), pp. 229-232.

Rieckmann, M. (2012), Future-oriented higher education: which key competencies should be fostered through university teaching and learning? *Futures* 44, pp. 127-135.

Rockström, J., Steffen, W., Noone, K., Persson, Å, Chapin, III F.S., Lambin, E.F., Lenton, T.M., Scheffer, M., Folke, C., Schellnhuber, H.J., Nykvist, B., de Wit, C.A., Hughes, T., van der Leeuw, S., Rodhe, H., Sörlin, S., Snyder, P.K., Costanza, R., Svedin, U., Falkenmark, M., Karlberg, L., Corell, R.W., Fabry, V.J., Hansen, J., Walker, B., Liverman, D., Richardson, K., Crutzen, P. and Foley, J.A. (2009), A safe operating place for humanity. *Nature* 461 (24), pp. 472-475.

Scott, W. (2002), Education and sustainable development: challenges, responses and frames of mind. *The Trumpeter* 18 (1), pp. 1-6. Available at: http://trumpeter. athabascau.ca/index.php/trumpet/article/view/123/135 (accessed 26 March 2012).

Sterling, S. (2001), *Sustainable Education: re-visioning learning and change*, Schumacher Briefing (6, Green Books) The Schumacher Society, Totnes.

UNESCO (2005a), *Guidelines and recommendations for reorienting teacher education to address sustainability*, Technical Paper No. 2. United Nations Educational, Scientific and Cultural Organization Education for Sustainable Development in Action, Paris.

– (2005b), United Nations Decade of Education for Sustainable Development (2005-2014): international implementation scheme. ED/DESD/2005/PI/01, United Nations Educational, Scientific and Cultural Organization Education Sector, Paris.

United Nations (1993), *Agenda 21: the Earth Summit strategy to save our planet.* Daniel Sitarz (Ed.), EarthPress, Boulder, CO.

– (2005), 59/237. *United Nations Decade of Education for Sustainable Development.* United Nations General Assembly, New York NY 24 February 2005. Available at: http://daccess-dds-ny.un.org/doc/UNDOC/GEN/N04/490/48/PDF/N0449048.pdf?Open Element (accessed 26 March 2012).

van Weenen, H. (2000), Towards a vision of a sustainable university. *International Journal of Sustainability in Higher Education* 1 (1), pp. 20-34.

WCED (1987), *Our common future.* World Commission on Environment and Development. Oxford University Press, New York, NY.

Wiek, A., Withycombe, L., Redman, C. and Banas Mills, S. (2011), Moving forward on competence in sustainability research and problem solving. *Environment* 53 (2), pp. 3-12.

Chapter 51

Green Energy in Higher Educational Institutions for Sustainable Development: Potentials and Challenges in India

Arun Prasath Ramaswamy, Gopalsamy Poyyamoli, Muthu Nandhivarman and Golda Arpudhalin Edwin

Abstract

The extensive and ever expanding use of fossil fuel for energy has triggered several global issues such as pollution, climate change and energy crisis. The green energy sources (solar, wind, biomass, hydro, geothermal, ocean, etc.) are considered to have a higher potential to solve these major global issues. India can play an important role in the development and utilization of green energy, blessed with *a strategic geographical location, the world's largest young trained/trainable manpower and abundant green energy resources*. India experiences solar radiation of 4-7 kWh/m^2 per day, which means its potential to harvest solar energy is very high; its tropical location means biomass can be used to generate bio-energy at a higher rate; there are hilly areas for hydro energy, seasonal factors advantageous for wind energy, long coastlines for ocean energy, etc. However, there are several constraints on the development of green energy in the country. For example there is a lack of awareness, skilled personnel, research and development, proper linkage between public and private sectors, and access to appropriate technology and appropriate policies. Costs are also higher and financing capabilities are inadequate; there are tariff disparities in green energy projects, issues in uploading green power to grids, and subsidies still given to conventional energy sources. Though considerable work has already been done on the application of green energy in rural and urban India, green energy application in higher educational institutions is scarce. We strongly believe that higher educational institutions in India have to play an important role to promote green energy, such that they will be instrumental in facilitating action research, co-curricular activities and community out-reach on green energy technologies.

Introduction

India is a highly secular democratic country located in South Asia. The country lies to the north of the equator between 8°4' and 37°6' N latitude and 68°7' and 97°25' E longitude (India Year Book 2007). It is surrounded by sea on three sides with the majestic Himalayan mountain range in the north, and shares its borders with Pakistan, Bangladesh, Myanmar, China, Bhutan, and Nepal. The seventh largest country and the second most populated country in the world with 1.22 billion people, India is the second fastest growing economy in the world after China, with GDP growth at the average rate of 8% for 2002-2011 and expected to remain at this level in the 12th five-year plan, 2012-2017 (Planning Commission 2011). As GDP grows, India's demand for energy is also growing at the same pace. Currently, the gap between demand and supply of energy resource is about 12-15%. About 40% of the population, particularly the rural population has little or no commercial energy access. Furthermore, others who have access to energy often have to cope with poor and erratic energy supply. The per capita consumption of electricity in India during 2011 was 779 kWh (CEA 2011), much lower than the world average, which stands at 2,782 kWh. India's 2020 targets envisage per capita consumption of electricity of 2460 kWh, still well below the world average (Gupta 2002). If this target is to be achieved, power generation capacity has to be increased to 2000 GW, compared to 187.6 GW as on 31 January 2012. To sustain its economic growth and provide decent energy supply, India has to increase its energy production to reduce the energy gap between its demand and supply. This can be achieved by harnessing its abundant naturally available green energy such as solar, biomass, wind, hydro, waste-to-energy, ocean, geothermal, etc.

India has to target its higher educational institutions for development and implementation of green energy technologies to fully exploit its real green energy potential. The country has one of the largest higher education systems in the world next to China and US English, as the primary language for teaching and research in the country, has played a significant role in building a large, high quality Higher Education System (HES) in the 21st century knowledge race. The HES of India offers the facility of education and training in almost all aspects of human creative and intellectual endeavors. The framework of the educational institutions consists of universities established by an Act of Parliament (Central Universities) or of a State Legislature (State Universities); Deemed Universities (universities with authority to award their own degrees through central government notification); Institutes of National Importance (prestigious institutions awarded said status by Parliament); institutions established by a state legislative act; and colleges affiliated to universities (both government-aided and unaided). According to the annual report (2009-2010) of the Ministry of Human Resource Development (MHRD 2011) on higher educational institutions, there are 504 universities and university-level institutions, of which 243 are state universities, 53 are state private universities, 40 are central universities, 130 are deemed universities, 33 are institutions of national importance established under Acts of Parliament, and five are institutions established under various state legislations. In addition, there are around 25,951 colleges, including 2,565 ladies' colleges. At the beginning of the academic year 2009-2010, the total number of students enrolled in the formal university and college system was reported at 13.64 million. In the technical education system there are 65 centrally funded institutions, 15 Indian Institutes of Technology (IITs), 20 National Institutes of Technology (NITs), seven Indian Institutes of Management (IIMs), five Indian Institutes of Science Education & Research (IISERs), four Indian Institutes of Information Technology (IIITs), four National Institutes of Technical Teachers Training & Research (NITTTRs), one Indian Institute of Science (IISc), and nine other technical institutions in the country (2009-2010) providing various undergraduate degrees, postgraduate degrees, and research programs. In addition, the technical institutions under the All India Council for Technical Education (AICTE) include postgraduate and undergraduate degrees and diplomas across the whole spectrum of technical education. The government estimates that the share of enrolments of private, unaided higher education institutions will be around 51%. The present GER (Gross Enrollment Ratio) in higher education in India is around 13.8% (world average 23.2%) and the Indian government wants to increase the GER to 21% by 2017, and 30% GER by 2020. Higher education in the country has experienced significant growth over the last two decades. Universities and colleges in India have grown at a CAGR (Compound Annual Growth Rate) of over five and 6% respectively since independence (MHRD 2011). Considering it has the second largest population in the world, with 50% of its population below 35 years of age, the country has the potential to become a hub of trained manpower provided the young population is given the right education and training, particularly in the field of green/renewable energy. To this end, the Higher Educational Institutions (HEIs) should be in a position to provide all students with knowledge in green energy technologies, sustainable development, and general environmental issues.

The development of green energy depends upon the availability of adequate trained manpower and appropriate resources. The skill requirements for green energy technologies are unique in nature and vary widely across different green energy sectors. The Indian green energy sector suffers mainly due to non-availability of trained manpower across different green energy sectors. The AICTE database shows that only 52 out of 1,346 engineering colleges offer courses on Energy Management at postgraduate level in which green energy is one of the major electives. India is slowly addressing the problem by providing green energy degree courses at postgraduate level in universities/colleges. As of now, the level of green energy education is not adequate to meet demand. Because green energy education has not been accorded sufficient importance, Indian higher education institutions are yet to incorporate green energy technologies in a bigger way. There are some universities/colleges, particularly private universities/colleges, that have implemented green energy demonstration projects such as solar street lighting, solar hot water for hostels, solar thermal cooking, electrical mobility, mobile chargers, etc., which have been publicized in the media. In order to develop and promote green energy in a bigger way the HEIs in India should adopt strong policies. However, although India's education policy is embedded with several principles of sustainable development, implementation faces the same challenges due to a lack of inter-disciplinary competence among staff and students, and traditional methods of assessment in higher education (Kiran, 2010). This is because Indian HEIs lag behind in the adoption of environment management systems (EMS) compared to the developed world. There are, however, some institutions, such as IITs (Kanpur, Madras, and Delhi), IIMs (Calcutta and Ahmedabad), and universities (Madras, Jabavpur, Hyderabad, Pune and Jammu) which have managed to establish sustainable education programs/courses to promote sustainable development in India. TERI University in Delhi is a model green campus in terms of building and energy conservation, including promoting renewable energy on campus, water conservation and waste-water recycling, etc., whilst also including sustainability concepts in most of its courses offered (Kiran 2010). It is pertinent to recall that the Rashtrapati Bhawan (office-cum-residence of the President of India) has asked to adopt its "Roshni" model for sustainable management of large campuses of Raj Bhawans and institutes of higher education, especially central universities, IITs and IIMs. The government's commitment to the promotion of energy efficiency in buildings through the Green Rating for Integrated Habitat Assessment (GRIHA) has attracted some HEIs in India to adopt and promote green energy through Green Buildings (GB). Some of the projects currently registered with GRIHA are from the Centre for Environmental Science and Engineering in the Indian Institute of Technology (IIT) Kanpur, five Indian Institutes of Science Education and Research campuses, the National Institute of Fashion Technology in New Delhi, and IIT Bhubaneswar.

Universities and colleges across developed nations have accomplished notable developments in sustainable education on campuses. Brown (2009), in his review article on higher education sustainability, explored campus sustainability programs that have been the subject of recent case studies (reduction of energy, water and waste; awareness of energy efficiency and renewable energy, etc.). The main targets of higher education sustainability focused on greenhouse gas (GHG) reduction by reducing energy use and specifically targeting green electricity purchase. The procurement of Renewable Energy Credits (RECs) and building retrofits are two main popular methods of reduction of GHGs at colleges and universities (ACUPCC 2010). Several US universities and colleges are members and are participating in the Green Power Partnership (GPP), a voluntary environmental program initiated to promote use of green energy power among US organizations by the US Environmental Protection Agency (EPA) in 2001. Among US universities, the University of Pennsylvania ranks first, with a capacity of annual green energy power usage of 201,841,600 KW, about 48% of the university's total power use. The University prides itself on the Green Power Leadership Award 2002 and Partner of the Year Award 2003 and 2008 (Soma 2011). Although India has a large number of engineering and technology institutions both in the public and private sectors, they concentrate on basic research rather than the application of science and technology in sustainable development. Only very few initiatives are in progress in locations scattered all over India where green energy technologies are being adopted on a sufficiently large scale to have a significant impact on the environmental and economic sustainability of campuses. In spite of the wider adoption of green energy technologies in rural and urban India, institutions of higher education have not adopted green energy strategies at the same rate, creating an increasing disparity between market needs and academic deliverables in the areas of green energy technologies and the production of intellectual capital and knowledge workers in green energy. The purpose of this paper is to critically analyze the strengths, weaknesses, and opportunities with regard to adopting green energy technologies in HEIs of India, and the role that HEIs across the country have to play to develop green energy technologies for sustainable development. The potentials, challenges and the key success factors for implementation will be discussed.

Energy, climate change, and sustainability

Energy is vital for humans. Without it human life on earth could not go on. All over the world, the demand for energy has increased in an alarming way due to population growth coupled with the developments of the human race. In spite of the increasing popularity of green energy sources/technologies, we still rely heavily on fossil fuels for our development. Currently, the US is the world's largest consumer of energy, followed by China, Russia, and India (IEO, 2011). The use of fossil fuel for energy and its related application around the world is above 80%, which is severely damaging to the environment because of the release of greenhouse gases (e.g. CO_2, NO_x, SO_x, CH_4, CFCs, halons, aerosols, etc.) into the atmosphere. In 2011, the concentration of CO_2 was about 389 ppmv in the atmosphere, compared to 280 ppmv in the mid-1800s, and it is increasing year by year. Furthermore, a significant increase has occurred in the levels of methane (CH_4) and nitrous oxide (N_2O) in the atmosphere (IEA, 2011). In its Fourth Assessment Report (IPCC, 2007), the IPCC concluded: "Most of the observed increase in global average temperatures since the mid-20th century is very likely due to the observed increase in anthropogenic greenhouse-gas concentrations". Nearly 80% of anthropogenic greenhouse gases (GHGs) are due to the use of fossil fuel for energy production and utilization. Among the greenhouse gases emitted into the atmosphere, CO_2 alone contributes 94%. Industrialized nations showed higher emission levels of GHGs compared to developing nations. In the near future, energy service to households and productive activities in developing countries will expand in a very rapid manner, in an attempt to improve the living standard of millions. It is predicted that the fast-developing countries such as China, India and South American countries could consume a larger energy share of the use of fossil fuels and emit more than 50% of GHGs by 2030. IPCC projects an average global temperature rise of around 5-6°C in the next hundred years. If this proves correct, catastrophic damages such as sea level rise, drought, floods, cyclones, hurricanes, storms, typhoons, etc., could occur in most parts of the world. The developed nations, developing nations and under-developed nations are doing their bit to reduce emission levels of GHGs by opting for green energy. As an energy-starved nation, India is seriously looking for greener renewable energy to sustain its economic growth for a sustainable future. Green energy resources can help the country to provide secure energy, increase its economic growth, reduce adverse environmental impacts and carbon footprint, balance regional development, promote sustainable development, and help it become a leader in green energy technologies. India encourages and promotes the development of green energy sources to accelerate the shift towards sustainable consumption and production for its sustainable future. Green energy sources and technologies offer a chance to protect the atmosphere by reducing GHG emission. In addition, switching over to green energy will reduce the rate of depletion of natural resources and, in turn, support sustainable development across the country.

Overview of global green energy

Currently 80% of the world's energy requirement is largely met by fossil fuels. Green energy is a promising prospect for the future, as it can substitute fast-depleting fossil fuel-based energy, and therefore provide a permanent solution to the energy crisis and environmental degradation. The total cumulative green power installed capacity (excluding hydro power) across the globe was around 312 GW in 2010 (REN21 2011). Global green energy production has seen its use grow sharply in recent years. It has been estimated that renewable energy capacity grew at rates of 10-60% for several green technologies during the last one-and-a-half decades, including solar (grid and off-grid), solar thermal, wind, geothermal, bio-fuel, and small hydropower. Given the phenomenal growth of green energy worldwide, jobs in green energy have rapidly increased during recent years, and the sector will witness even higher growth in future. It is estimated that there were more than 3.5 million people employed globally in the green energy sector in 2010. The biomass sector, which is manpower intensive, is expected to grow faster and would be the major contributor of the total power generation (about 59%), followed by solar PV (31%) and wind (10%) by 2030. The global investment in green energy for the year 2010 stood at USD 211 billion, up 32% from USD 160 billion the previous year. India ranks fourth with regard to installed power generation capacity in the field of green energy sources (wind, small hydro power, bio-energy, waste-to-energy, solar, etc.). India is considered as one of the most ideal investment destinations in the world for green power. The country's current status of green energy sources is as follows.

Potential and current status of green energy in India

The strategic geographical location of India, by which it is gifted with abundant green energy resources for solar, wind, biomass, waste-to-energy, and small hydro energy, means that the country can overcome its energy deficit by harnessing these resources. It could, in fact, help in particular to cover the energy requirements in rural areas, where nearly 400 million live without electricity in over 90,000 villages. India recognized the importance of green energy in the early 1970s. India was actually the first country in the world to establish a separate department to promote green energy, the Department of Non-Conventional Energy Sources, in 1982, which was upgraded in 1992 to become the Ministry of Non-Conventional Energy Sources (MNES) and subsequently renamed as the Ministry of New and Renewable Energy (MNRE). The recent report (Akshay Urjoa 2011) from MNRE showed the potential of green energy from wind, small hydro and bio-energy to amount to 87 GW excluding solar energy; the report estimates about 20-30 MW7km^2 for solar power. The accurate potential of India's solar energy remains largely elusive because of the non-availability of credentials for its solar radiation database. Interestingly, the World Bank South Asia Energy Unit report estimated the green energy potential for India to be 150 GW. There are independent reports which claim that reassessment is needed to estimate the green energy potential of India by keeping pace with improving technologies in the field. This new research indicates that India can generate more than 1000 GW, fives time more power than current production/consumption. Currently, 21.1 GW power is produced through green energy, which accounts for nearly 11.3% of the total power (187.6 GW) generated in India (CEA 2011). Fig. 1 shows the share of renewable energy from the total installed capacity connected through the grid as of December 2011.

Fig. 1: Total electricity installed capacity: 187.6 GW (data source: CEA 2011)

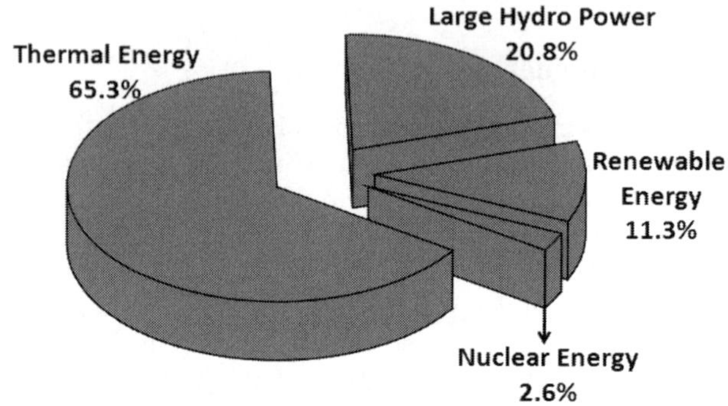

The green energy sector has shown the highest growth rate, registering an annual growth rate of 18% in the Eleventh Plan Period (2007-2012), compared to just 5% witnessed in the non-green sector. By knowing the potential of green energy, major thermal energy companies like the National Thermal Power Corporation Limited, Tata Power Company, and Reliance Power, have begun investing in a big way in green energy business. This clearly indicates that India is determined to become one of the world's leading green energy producers. Fig. 2 shows the percentage of installed power capacity by various green energy sectors in India, with grid-interactive as of August 2011 according to MNRE (MNRE 2011a).

Fig. 2: Grid-interactive green power in India (data source: MNRE 2011)

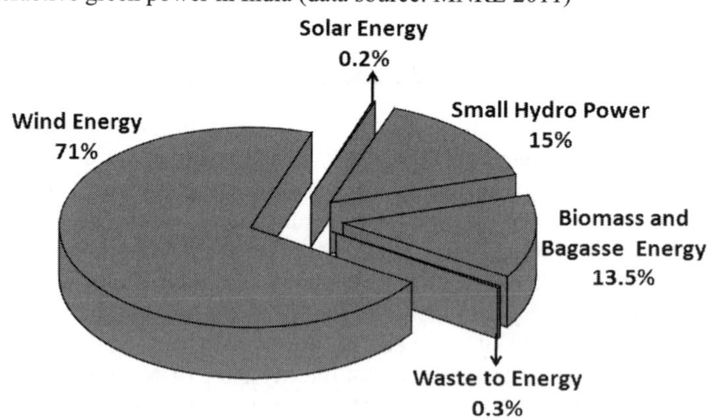

Solar energy potential

India has one of the world's largest solar energy programs, which includes research & development, demonstration and utilization, testing and standardization, and industrial and promotional activities. India possesses abundant solar energy resources as it receives an average incident solar radiation of 4-7 kWh/m^2/day, with 250-300 clear sunny days in a year. The country receives solar energy equivalent to 5,000 trillion kWh/year. Solar thermal and photovoltaics technologies have been used since the 1960s to harness solar energy. Though the solar energy potential is huge, only 46.16 MW is produced via grid-connected and about 72.5 MW is produced by off-grid, as was recorded in August 2011 (MNRE 2011a). The National Action Plan on Climate Change, which was announced in June 2008 with eight missions for climate protection, includes a solar mission as its first mission, which represents a major step for the utilization of solar energy for power generation and other purposes. The solar mission, Jawaharlal Nehru National Solar Mission (JNNSM), was launched to boost solar energy in India with a target of 20 GW of grid-connected solar power (based on both solar photovoltaic [SPV] technologies and solar thermal power generating systems), 2 GW of off-grid capacity including 20 million solar lighting systems and 20 million m² of solar thermal collector area by 2022 (MNRE 2011b). The mission will be implemented in three phases. The first phase will take three years (up to March 2013), the second phase will last until March 2017 and the third phase will continue until March 2022. The first phase of the project allocation was done to set up 1.1 GW of grid-connected solar panels, 100 MW on rooftops, 200 MW capacity equivalent off-grid solar applications and seven million m² of solar thermal collector area. The state, Rajasthan, Gujarat, Andhra Pradesh, Tamil Nadu and Karnataka are major players in developing and implementing solar energy projects through the national solar mission and their own state policy. In 2010, employment figures in solar PV on-grid was estimated to be around 4,000. Considering the JNNSM targets, it is estimated that the solar PV on-grid sector will employ 39,000 people by the year 2017 and 152,000 by the year 2022. The estimated employment for solar PV off-grid was about 72,000 in 2010, and the sector should employ about 140,000 people by 2017 and 225,000 by 2022 at the growth rate of 10% (CII-MNRE 2010).

Wind energy potential

India ranks as the fifth largest country for wind power capacity in the world, and the industry is much more developed than other green energy technologies. Interest in wind energy in the country began to develop from the 1990s, and has progressed steadily in the last few years. The total installed capacity is 14,989 MW as of August 2011, which accounts for 71% of the country's total installed green energy (refer Fig. 2). The major wind power capacity is in the states of Tamil Nadu, Gujarat, Maharashtra, Karnataka and Rajasthan. Tamil Nadu tops the list in wind energy production with a capacity of 5,904 MW. The nation's wind energy resources have been mapped by the Centre for Wind Energy Technology in Chennai, Tamil Nadu. The Centre for Wind Energy Technology provides technical support to the MNRE in the implementation of its wind energy programs across India. The estimated potential of wind energy by MNRE is around 49,130 MW (Akshay Urjoa 2011). However, the actual potential will be much higher due to technology development, and offshore potential is not taken into consideration. The World Institute for Sustainable Energy (WISE) estimates the potential could be as high as 100 GW. The long coastline (about 7,500 km) of India could promote the development of offshore wind energy in this area, an option which has not yet been explored until now. It is estimated that there are more than 40,000 people employed in the wind sector in 2010 and it is expected to have about 75,000 jobs by 2020 at the medium growth rate of 10% (CII-MNRE 2010).

Bio-energy potential

Biomass energy is another of India's very important green energy mixes, which has predominantly been used in rural households. Biomass resources are abundant in India, and it has been estimated that about there is the potential for around 51,000 MW-17,000 MW from agro-residue (straw, stalk and husk can be the main contributors) and around 34,000 MW from energy plantation in waste land (MNRE 2011c). The MNRE expects the launch of the National Bioenergy Mission (NBEM), which will target a cumulative biomass-based generation of around 10,000 MW by the end of 2022. Nearly 70% of the population of India lives in villages and their main source of income is agriculture. Biomass, as agricultural residues produced in these villages, has been estimated at 600-700 million tons per year. The biomass can be harnessed for production of commercial energy, including fuel, power and heat. However, it has been found that more than 50% of the biomass residue is not harvested due to various reasons. At the level of less than 50% harvesting efficiency, it is possible to generate 7,000 MW power based on using various conversion technologies (combustion, gasification, pyrolysis, bio-methanation, etc.). Technology choices for the development of KW range from gasification-based systems to both small (1-2 MW) and large-scale (10-30 MW) combustion-based systems. With an increased level of efficiency in these technologies, the potential would further increase. According to the Planning Commission 14, "India has approximately 50 million hectares of degraded wasteland that lie outside the areas demarcated as national forests, and another 34 million hectares of protected forest area, of which tree cover is severely degraded. A massive program of energy plantation consisting of fast-growing tree crops such as bamboo, casuarinas, and eucalyptus, etc., can serve as the raw material for a national network of small, decentralized biomass power plants" (Gupta 2002).

Biofuels are another important resource for India to supply its energy needs, particularly for the vast rural population without grid connectivity and the transport sector. Biofuels are derived from green bio-mass resources and, therefore, provide a strategic advantage to promote sustainable development. The biofuel sector has been driven by the MNRE's national policy on biofuels (NPBF 2009). The goal of the policy is to ensure that a minimum level of biofuels become readily available in the market to meet the demand at any given time. An indicative target of 20% blending of biofuels, both for bio-diesel and bio-ethanol, by 2017, has been proposed (NPBF 2009) by the MNRE along with other departments such as the Ministry of Environment & Forests, the Ministry of Petroleum & Natural Gas, the Ministry of Rural Development and the Ministry of Science & Technology, in order to deal with different aspects of biofuel development and promotion in the country. These departments promote biofuel development from non-edible oilseed-bearing plants to produce second generation biofuels including the conversion of ligno-cellulosic materials to ethanol – such as crop residues, forest wastes and algae, biomass-to-liquid (BTL) fuels, bio-refineries, etc. The country will launch the National Biofuel Coordination Committee, headed by the Prime Minister of India, to promote bio-fuels in India. In 2010, the employment in the bio-energy sector (bio-mass on-grid and off-grid, biogas) is estimated to be around 142,000, and is predicted to grow at 10% in the coming years. It is also estimated that the bioenergy sector will employ 561,000 by the year 2020 (CII-MNRE 2010).

Waste-to-energy potential

India's waste generation capacity is increasing enormously due to its population growth, urbanization and industrialization, coupled with rising standards of living. As industrial, urban and biomass wastes contain organic matter, these wastes can be used to generate power and solve the problem of huge waste management. The Ministry of Urban Development recently estimated that about 64 million tons of solid waste and about 40,000 million liters of liquid municipal waste is generated per day. MNRE aims to produce 1,500 MW of power from current municipal solid waste in the country and is focused to achieve 5,000 MW of power from waste in future with economic developments (Arijit 2011) using various waste-to-conversion technologies.

Small hydro power potential

India is blessed with an immense amount of hydro-electric potential and currently 20.5% of the total electricity power of India comes via large hydro power plants. The total hydroelectric power potential of the country is estimated to be about 150,000 MW, equivalent to 84,000 MW at 60% load factor. The potential of Small Hydro Power projects (SHP) is estimated to be about 15,000 MW. In India, the large hydro power projects are under the control of the Ministry of Power, while the small hydro power developments are under the control of the Ministry of New and Renewable Energy. As of August 2011, 3,154 MW is produced through small hydro power, which is around 15% of the country's green energy (MNRE 2011a). The large potential of SHP is largely untapped because of the location of potential sites in remote and mostly under-developed hilly regions of northern states. The employment status in SHP stood at 12,500 in 2010 and it is estimated that this figure will be closer to30,000 by 2020 at a 10% growth scenario (CII-MNRE 2010).

Challenges and drivers of green energy sector for India

In India, green energy can contribute a significant proportion of the nation's energy. Though the country has a greater potential in harnessing green energy resources, there are several challenges that need to be overcome: (1) high capital costs, (2) lack of primary solar radiation/wind speed/biomass data, (3) lack of clarity on technologies (as technologies are still evolving), (4) inadequate infrastructure, (5) lack of technical support after sale service network, (6) lack of technological advancement, (7) incentive challenges, (8) difficulty in raising finance, (9) optimal pricing of power generated from the green energy sources- affordability issues, (10) technical feasibility to generate biogas from human excreta versus social acceptability, (11) quality and consistency issue of green power arising from the intermittent nature of electricity from wind and small hydro power, (12) costs of technology development and production need to be reduced significantly from current levels, (13) lack of coherent long-term policies/policy instruments, (14) inadequate R&D sector, etc.

One of the important measures identified in NAPCC involves increasing the share of green energy in total electricity consumption in the country. There are several opportunities for users of green energy technologies/devices in India. The following green energy drivers will encourage people to adopt green energy technologies/devices: (1) decline in the cost/kWh for green-based power generation, (2) recent advances in material science and green energy technologies, (3) the government (both central as well as state) is encouraging investments in solar energy by providing various incentives and duty concessions, (4) capital subsidy in the case of semi-conductor-based units, (5) low import duty for several raw materials and components required for solar projects, thus bringing down the cost, (6) 100% Export Oriented Units are permissible, and it also permitted to import raw materials and components duty free, thus bringing down the cost, (7) soft loans to users, intermediaries and manufacturers, (8) Renewable Energy Certificate (REC) Mechanism, (9) Clean Development Mechanism (CDM) projects, etc. In addition to the government's green energy targets, initiatives are underway across India that include the development and implementation of energy-efficiency improvements for appliances, buildings, power generation, and industry; a revised national policy to upgrade and "smarten" much of India's power grid; and emerging leadership in states like Gujarat, Rajasthan, Maharashtra, and Tamil Nadu to adopt and implement progressive green energy policies. Civil society and the private sector are increasingly active in implementing green and smart energy projects across many parts of the country.

Green energy education and its dissemination in India: future role

India's India's green energy installation is expected to grow three-to-four fold in the next ten years, which in turn will create a number of job opportunities in this sector. However, a major identified bottleneck is the availability of trained manpower for planning, installation, and maintenance of green energy devices. To remedy this situation, selected institutions can organize training and capacity building programs in the form of training courses and workshops, as well as offering professional diploma courses. The training and capacity building programs should aim to provide a better understanding of the principles and operations of green energy technologies. A similar facility will also play an important role as a resource for teaching, applied research to develop the technology, and outreach to educate community through community colleges on green energy. To generate a large pool of skilled manpower in green energy, it is been recommended that AICTE and MNRE should encourage more engineering colleges to offer green energy courses. In order to extend green energy education across several HEIs in India, mandatory undergraduate/postgraduate green energy courses should be in place to draw attention to the urgent need of green renewable energy technologies for sustainable development. In addition, the government should make green energy jobs as attractive as jobs in other sectors. The real concern is that the actual number of students opting for green energy jobs is much lower than the sector requires: this is due to (1) unattractive salary compared to IT sector, (2) wide difference in salary package within the green energy sectors, (3) very few green energy students moving into the industry through campus selection, due to the fact that reputed renewable energy companies are non-existent or minimal, (4) career prospects in the green energy sector are not attractive, (5) subsidies given to green energy industry really kills development and jobs, (6) awareness of green energy prospects among the students is low, (7) lack of skilled and knowledgeable trainers, (8) lack of proper course syllabi and laboratories to address the requirement, (9) non-availability of adequate scholarships, (10) inadequate R&D laboratories, and (11) inadequate co-ordination between different disciplines in this multi-disciplinary sector. In order to attract students, there is an urgent need to address the above concerns to increase the number of students on green energy courses in various universities/colleges/institutions. In addition, the due-recognition and encouragement should be provided by the administration of the concerned institutions to promote green energy courses.

The CII-MNRE report (2010) recommends various strategies, such as MNRE scholarships for postgraduate students opting for green energy courses; funding reputed central universities/institutions to develop/procure necessary infrastructural facilities such as lab equipment, training prototype models, etc., for imparting green energy education; increasing campus recruitment in HEIs; promoting industry with HEI linkages; interaction and collaboration with world-class international universities; setting up a cell to organize green energy awareness programs including experts from the industry and institutions; and facilitating the creation of green jobs in sector-specific skill gaps in various green energy technologies. There are exciting possibilities for integrating the concepts of green energy technologies into various disciplines, such as engineering and technology, physics, chemistry, biology, ecology, as well as in social sciences. However, very little has been done towards this goal, and in order to overcome this drawback major revision of the curricula and syllabi by including leading experts from academia and industry in the field of green energy technology is needed.

In India, HEIs ideally should have an Office of Sustainability (OOS) or separate Cell for Sustainable Development (CSD) to ensure the long-term future of campus activities and practices in sustainability by facilitating cross-disciplinary dialogues, collaborations, co-operation, and closer involvement of all the key stakeholder groups; administration, faculty, students, alumni, and civil society. The same can be facilitated by opting for green energy awareness programs in line for administration staff, academic staff, student, alumni and the local society through seminars, conferences, and courses. The OOS/CSD can network with the ongoing regional/national programs such as solar campus, solar city, JNNSM, Jawaharlal Nehru National Urban Renewal Mission (JNNURM), CDM, etc. Therefore, universities and higher education institutions can serve as living laboratories for resident student communities as well as for demonstration of appropriate Green Energy Technologies (GET) in their nearby areas and neighborhoods. Such a horizontal transfer of technology is very much desirable and feasible in India, taking into consideration the presence of major institutions of higher learning in strategic locations with higher concentrations of emerging institutions. The HEIs in India should be in a position to generate a "Green Fund" to implement green energy demonstration projects in line with environmental demonstration projects, similar to those of HEIs in developed nations. This "Green Fund" should be utilized to develop ambitious green energy projects, and the education institutions should strive to achieve targets related to a certain percentage of green energy utilization to reduce the use of fossil fuel energy.

The country's HEIs need to take bold steps to implement green energy in their campuses; their contribution can help to achieve the government's mission in implementation of renewable energy projects to ensure sustainable growth of the nation. We believe that strong policies are needed to implement green energy initiatives in HEIs to help HEIs to move forward with the nation's mission of 50% renewable energy/green energy by 2050. To attain this dream, the HEIs need to come up with clearly defined energy policies by initiating energy auditing, followed by implementation of green energy in the campus. The policies should encourage, (1) the utilization of shadow-free rooftop areas for solar photovoltaics (available from hostels/departments/centers/buildings), (2) solar thermal heaters for hot water generation and steam for cooking/cleaning purposes, (3) hybrid green energy systems (solar-wind-bio, etc.), (4) the use of kitchen waste and biomass produced in the campus can be utilized for biogas production/power generation, (5) use solar street lights, (6) use electrical mobility, (7) encourage the use of energy efficient systems to reduce the use of energy, (8) encourage green buildings with higher GRIHA rating, (9) establishment of a green energy park can be recommended for each institution to promote green energy, (10) sell extra green energy to the public during vacation period, etc.

Pondicherry University initiative in green energy projects

Pondicherry University, a central university established in 1985 by the government of India, is located 14 km north of Puducherry city. The university has emerged as India's fastest growing and is a collegiate university with a jurisdiction spread over the Union Territories of Pondicherry, Lakshadweep and Andaman and Nicobar Islands. The main campus of the university has an area of 780 acres spread on either side of the East Coast Road, with the large chunk on the landward side of the road. The teaching faculty consists of more than 400 active and erudite members in different schools, departments and centers. Currently, there are around 4,500 students pursuing their postgraduate and doctoral programs on the campus. In addition, several thousand students are enrolled at the university through the Community College, the Directorate of Distance Education and 85 affiliated colleges contribute to the diversity of the university's learning experience in frontier areas of Arts and Sciences and engineering courses. Moreover, the Semester in India Programme (SIP) at the university attracts several foreign students. It was the first university in India to provide students with a choice of elective courses in addition to their major subject area under the so-called Choice Based Credit System (CBCS). Extension activities are also an integral part of the university's activities.

In 2010, Pondicherry University took a lead in creating a Centre for Green Energy Technology (CGET) to conduct teaching and research in the fields of environmentally benign "Green Energy Technologies". This step was taken in order to develop the much needed manpower as well as provide research and development in renewable energies for sustainable development. Currently, the center conducts postgraduate-level (M.Tech, 26 students per year) and research-level programs in green energy technologies. Dr. RAP from CGET and Dr. GP from the Department of Ecology-Environmental Sciences of Pondicherry University are collaborating on various issues related to green campus projects (energy, grey water management, and solid waste management). Pondicherry University has already achieved many campus sustainability milestones and has implemented many of the common best green practices, such as (1) installation of solar water heating systems in ladies and gents hostels, (2) installation of approx. 160 solar street lights, (3) electrical mobility, (4) fitting of all street lights with energy efficient lighting systems, and (5) new upcoming buildings with GRIHA ratings. Moreover, we have initiated detailed energy /environmental auditing, planning for a green energy park and solar campus and institutional biogas plants using kitchen wastes from the hostel canteen, etc. Recently our university has been working on preparing a major proposal for using its shadow-free rooftops for solar energy harvesting through rooftop grid-interactive inverters. It is estimated that about 78,469 square meters of shadow-free rooftops are available on the campus to use for solar power harvesting. The project aims to achieve a target of approx. 30% green energy in the next five years (2013-17) through rooftop grid-interactive inverters on selected rooftops on the campus. The university has seventeen hostels on the campus: eleven for men, five for women students/research scholars, and one for foreign students with an intake of some 2,650 students. It has been estimated that roughly 2,200 kg of kitchen organic waste is generated per day from the gents and ladies canteens. Recently, we have initiated a proposal to use the kitchen organic waste generated to produce biogas through two separate 100 m^3 biogas anaerobic digester plants. These initiatives demonstrate our university's commitment towards (1) reduction in carbon emissions, (2) reduction in the use of fossil fuels, (3) sustainably utilizing biomass including kitchen waste from mess for energy generation, (4) implementation of green energy projects to ensure sustainable development of the campus, (5) demonstration of the use of green energy in the campus, (6) exposing faculty and students to green energy technologies, and (7) evolving the model for sustainable development of the campus.

Conclusion

India's abundant natural green energy potential and its large young population of trainable manpower means the country is potentially able to harness the green energy resources to meet its ever-growing energy needs for sustainable development. To exploit its full potential in green energy, the country has to address the challenges faced in the light of lop-sided policies. India's vision to achieve 50% of its energy through greener sources by 2050 can be realized if the challenges are met with confidence. To overcome the challenging issues, the country's higher education institutions have to be utilized to develop and promote green energy technologies such that they will be instrumental in facilitating action research, co-curricular activities and community out-reach programs on green energy technologies. The main challenge faced by the green energy sector is the non-availability of trained manpower; to fill this gap the Indian government should encourage several universities and engineering colleges to offer green energy courses at both undergraduate and postgraduate level, whilst also encouraging students/faculty through generous support. Furthermore, green energy courses should have various disciplines of science, engineering, and technology integrated into their respective curricula. This can be done through major revision of the curricula and syllabi by including leading experts in the Board of Studies from academia and industry in the field of green energy technology. This will help institutions of higher education to adopt green energy strategies at a faster rate, creating an effective balance between market needs and academic deliverables in the area of green energy technologies, producing intellectual capital, and knowledge workers in green energy. The creation of a "Green Fund" and separate cells of sustainable development with sustainability-aware leaders in every educational institution will promote green energy technologies through facilitating cross-disciplinary dialogue, collaborations, cooperation, and closer involvement of all the key stakeholder groups – administration, faculty, students, alumni, and civil society.

References

ACUPCC (2010), *Reporting system. American College and University Presidents' Climate Commitment*. Available at: http://acupcc.aashe.org.

Akshay Urjoa (2011), *Renewable Energy. Ministry of New and Renewable Energy*. Government of India, Vol. 5, Issue 2.

Arijit, S. (2011), Renewable Watch 2 (1), pp. 50-51.

Brown, M.J. (2009), Campus Triple Bottom Line: College Sustainability and Grand Valley State University. *McNair Scholars, Journal* 13 (1), Article 4.

CEA (2011), Ministry of Power, Government of India, Central Electricity Authority, *Load Generation Balance Report for 2011-2012.*

CII-MNRE (2010), *Human resource development strategies for Indian renewable energy sector*. Ministry of New and Renewable Energy, Government of India.

Gupta, S.P. (2002), *India Visions 2020*. 14th Planning Commission, Government of India.

IEA (2011), *CO_2 emission from fuel combustion*. IEA Statistics 2011 Highlights.

IEO (2011), *International Energy Outlook 2011*. http://205.254.135.7/forecasts/ieo/

India Year Book (2007), Compiled and edited by Research, Reference and Training Division, Ministry of Information & Broadcasting, Government of India (ISBN 81-230-1423-6).

IPCC (2007), *Climate Change 2007. Fourth Assessment Report*. Available at www.ipcc.ch.

Kiran, B.C. (2010), Higher education and curriculum innovation for sustainable development in India. *International Journal of Sustainability in Higher Education* 11 (2), pp. 141-152.

MHRD (2011), Ministry of Human Resource Development, Government of India, *Annual Report for 2009-2010.*

MNRE (2011a), Ministry of New and Renewable Energy, Government of India. http://www.mnre.gov.in.

– (2011b), Ministry of New and Renewable Energy, Government of India, *Annual Report 2009-2010.*

– (2011c), Ministry of New and Renewable Energy, Government of India, *Draft Recommendations of Sub-Group on 'Bioenergy Mission' for 12th Five Year Plan (2012-17).*

NPBF (2009), Ministry of New and Renewable Energy, Government of India, *National Policy on Biofuels.*

Planning Commission (2011), *An approach to the 12th five year plan draft, "Faster, Sustainable and more inclusive growth"*. Government of India.

REN21 (2011), *Renewables 2011. Global status report.*

Soma, G. (2011), Participation in the Green Power Partnership – An analysis of higher education institutions as partners in the program. *International Journal of Sustainability in Higher Education* 12 (4), pp. 306-321.

Chapter 52

A Sustainable Strategy for Gauging Rural Vulnerability and Promoting Sustainable Rural and Urban Lifestyles

Norizan Md. Nor, Salfarina Abd Gapor, Govindran Jegatesen and Hamoon Khelghat-Doost

Abstract

An important aspect of sustainability pertaining to a university is to ensure that the university's knowledge and expertise is readily accessible to its surrounding community, thus strengthening its ties with the community it is located in. Embracing this concept, Universiti Sains Malaysia (USM) via its sustainability institute, the Centre for Global Sustainability Studies (CGSS), has initiated several projects for implementing sustainability practices in its outreach and community engagement programs. The primary objectives of these projects are to improve rural development pathways through enhanced livelihood skills and community empowerment initiatives together with better training, up-skilling and retooling approaches for urban development – especially aimed at low income groups and the *bottom billion*. It was for this reason that the two areas of Kuala Nerang and Balik Pulau (in the northern region of Malaysia) were chosen, due to their respective problems of annual floods and rural poverty.

The Kuala Nerang project aims to identify the complications and issues of flood catastrophes which frequently occur in the area. Subsequently, this project will examine flood-related issues in Kuala Nerang by utilizing the vulnerability and adaptation method to identify groups that are most vulnerable to flood disasters. The second project, i.e. the Balik Pulau research project comprises the three research clusters of Socio-Economic, Socio-Cultural and Community Biodiversity. These clusters are aimed at revitalizing the "home garden" concept for household use among Balik Pulau's villages, promoting the "differences" and roles of each ethnic group in shaping the community's overall identity and culture. This paper aims to explore the innovative methodologies, participatory approaches and scientific research conducted in both Kuala Nerang and Balik Pulau toward reducing flood vulnerability and increasing adaptability as well as teaching villagers to tap into ecosystem goods and services toward strengthening social wellbeing and building a sustainable economy.

Introduction

Universities have been traditionally defined as centers for gaining knowledge, teaching and research. Generally (through their teaching activities), universities offer specialized training for different sectors of society, as well as the education essential for personality development. Additionally, university education boosts the theoretical knowledge among different divisions of society as well as offering practical solutions to deal with societal dilemmas. Conventionally, the traditional framework of a university consisted of a close circle of students and lecturers – usually referred to as the "ivory tower" (Pierce and Cameron 2009). As an elite component of society, this circle had been able to channel the collective will of the community in order to suit a particular agenda for a considerable period of time in history. In recent centuries, however, we have witnessed the once closely knit relationship among universities, knowledge and ethics increasingly fading away.

However, in light of the needs of the new millennium, universities all around the world are beginning to realize their rapidly changing roles in a globalizing world. The full benefits of a university can only be made manifest when both the university and the society it is located in are organically linked together. In other words, a university's activities must be flexible enough to factor in the needs of its society – given society's rapidly changing needs and trends.

Recognizing the role which education can play in the development of societies that are more equitable and sustainable, the United Nations launched the Decade of Education for Sustainable Development in 2005. As mentioned in Section 6 of the 2009 Bonn Declaration, "Education for sustainable development is setting a new direction for education and learning for all. It promotes quality education, and is inclusive of all people. It is based on values, principles and practices necessary to respond effectively to current and future challenges."

"With large pools of disciplinary experts, high quality research facilities, state-of-the-art infrastructures and a cohort of students with varied academic interests, universities have considerable comparative advantages in promoting sustainability within the communities they serve." (Hassan et al. 2000).

Universities are rarely discussed from the perspective of the "tragedy of the commons." (Brown 2000). This however is the very plight affecting universities and societies (business communities, public communities, etc.) due to the effects of globalization. Although universities have an excellent record in serving the needs of both the university and the outside community, they are too often trapped in reconciling conflicts of interest which ultimately condemn universities as dumping grounds, scapegoats or testing platforms for different problems – ranging from the environmental to the social as well as the economic. Fortunately, certain better-informed approaches have already been introduced to prevent this situation from reoccurring and change it for the better. Various new reformations are taking place at different universities which ultimately produce new concepts such as quality assurance, policy revisions, university corporatization, etc.

Although some of these approaches have been partially successful in their respective efforts, what remains unchanged in most of these approaches is the element of un-sustainability. This is most often seen at;

- The implementation process which creates weak synergies and coherencies, ending in an inability to recognize the diversity that exists among universities,
- The incomprehensiveness of the solution – resulting in more *ad-hoc* and reactive approaches.

An important aspect of sustainability pertaining to a university is to ensure that the university's knowledge and expertise is readily-accessible to its surrounding community, thus strengthening its ties with the community it is located in. As is clearly outlined in the Ninth Malaysia Plan (2005), it is vital "to improve the standard and sustainability of quality of life" in Malaysian society. Embracing this concept, Universiti Sains Malaysia (USM) via its sustainability institute, the Centre for Global Sustainability Studies (CGSS), has initiated several projects for implementing sustainability practices in its outreach and community engagement programs. This paper will provide a detailed insight of the projects which have been conducted by CGSS@USM in terms of community engagement – both rural and urban. The primary objectives of these projects are to improve rural development pathways through enhanced livelihood skills and community empowerment initiatives together with better training, upskilling and retooling approaches for urban development – especially aimed at low income groups and the bottom billion.

According to Chambers and Conway (1992), "green practices such as recycling and food/garden waste composting at schools, residential areas and work should be primary goals toward attaining and maintaining sustainability; unfortunately, such goals are often absent among urban inhabitants." Consequently, these issues are challenges to be resolved if sustainable lifestyles are to be achieved among the communities surrounding USM. This project aims to achieve sustainability integration by encouraging and integrating sustainable lifestyles among communities together with USM and relevant stakeholders via a campaign on sustainable lifestyles as the main agenda of the project. Programs such as the University in A Garden, Kampus Sejahtera, White Coffin (ban on polystyrene food packaging) and Say No to Plastic were initiated by the campus community and were pivotal in transforming USM into a more livable and sustainable campus. As it complements USM's Accelerated Programme for Excellence (APEX) mission i.e. 'Transforming Higher Education for A Sustainable Tomorrow,' this project will certainly impact surrounding communities by promoting and inculcating sustainable lifestyles.

This issue will be further discussed with regard to the core missions of higher education in providing services to the community as was defined by UNESCO. According to UNESCO (1998), the four goals of EfSD in higher education, i.e. basic quality education, reorienting education for EfSD, creating and raising awareness on EfSD, and training for EfSD, have already provided a holistic framework of action to replace the reactive and *ad hoc* problem-solving approaches that are currently used by most universities around the world.

CGSS@USM and community engagement

University academics, researchers and students alike can all play important roles in spreading knowledge and expertise beyond university walls. They are able to impart to the community the skills needed to solve their problems while simultaneously exposing them to new ideas and innovations. Additionally, such efforts will assist the university in strengthening its ties with the communities surrounding it.

Incidentally, an important step in bridging communities and universities is the concept of indigenous knowledge. This concept forms one of the primary methods by which CGSS@USM tries to approach its community engagement framework. According to UNEP, indigenous knowledge (IK) can be broadly defined as the body of knowledge that an indigenous (local) community accumulates over generations of living in a particular environment. This definition encompasses all forms of knowledge – technologies, know-how skills, practices and beliefs – that enable the community to achieve stable livelihoods in their environment. A number of terms are used interchangeably to refer to the concept of IK, including traditional knowledge (TK), indigenous technical knowledge (ITK), local knowledge (LK) and indigenous knowledge system (IKS).

As a field, indigenous knowledge and practices have in the past few decades seen a steady decline in popularity in a world that is fixated on laboratory statistics, chemical engineering and intensive modernization. Stereotypes exist regarding indigenous knowledge and practices as a field of pseudo-science that is concerned with shamanism, witch doctors and faith healings as opposed to modern medication – and where culture is concerned, as a 'backward' collage of rituals and practices that conflict with modernization efforts.

However, when indigenous knowledge is successfully integrated into the fabric of modern society, it is able to contribute significantly to the richness of a society, giving it history and a sense of identity in a world whose lines are increasingly blurred by rapid globalization and ongoing rigorous endeavors to separate indigenous knowledge from secular mainstream knowledge. In addition to this, "the incorporation of indigenous knowledge into our daily life might even edify and further strengthen modern practices and aspects of society to our advantage. In such an event, universities can act as an ideal platform for researchers to combine Indigenous knowledge and modern sciences for the greater benefit of humanity." (Jalil 2010).

In addition to the above matter, the issue of how to enable universities to better contribute to innovation processes has become an important issue with regard to the international agenda. Transferring the results of university research to industries may take several forms and can thus be achieved in different ways, e.g. patenting, licensing, spin-off firms, etc. Although the transfer of results from academic research into the industry is widely accepted as a crucial contributing factor for industrial growth and competitiveness, this does not mean that it is an effortless or simple linear process of direct knowledge transfer from academia to the industry.

The following sections will describe CGSS@USM's efforts in creating sustainable linkages between its pool of expertise and the needs of the society; specifically, the three projects of Enhancing Sustainable Living within USM and Its Neighboring Communities, Flood Disaster Risk Management in Kuala Nerang and Showcasing Balik Pulau as a Sustainable Village will be discussed in the following sections.

Enhancing sustainable living within USM and its neighboring communities

In light of global sustainability issues, USM has envisioned 'sustainability' as a core development theme of the university. In becoming a world-renowned university for sustainability, USM facilitates sustainability attainment through (i) innovations and technology, and (ii) changes in lifestyles. Subsequently, CGSS has launched a project that engages its surrounding communities via the dissemination of USM's sustainability lifestyle practices – focusing on the areas of recycling and composting involving awareness campaigns at stakeholders' venues. It was initiated by conducting situational analyses with stakeholders, after which specific sustainability awareness/practices campaigns were then devised, encompassing the setting-up and monitoring of composting units and recycling bins at six schools (SMK Bukit Jambul, SMK Bukit Gambir, SMK Dato' Haji Mohd Noor Ahmad, SK Minden Height, SK Sungai Gelugor and SK Bukit Gambir), commercial entities (Giant Supermarket), residential areas (Taman Brown, Sungai Gelugor and Taman Tun Saadon) and an SME industry (Cincaria Sdn.Bhd), all located around USM's vicinity.

This project was initiated to provide a platform to assess the level of sustainability integration across USM and its neighboring communities via community engagement practices. Participation from various community groups in the study area will also facilitate better understanding in finding ways to promote sustainability through awareness campaigns, knowledge transfer and community engagement on various levels.

This research project aims to integrate and extend the university's efforts to achieve two major goals: (a) to be a world-renowned university for sustainability and (b) to be a sustainability-led university. Realizing that the university has a key role to play in searching for ways to enhance sustainable living, this project attempts to extend what was promoted in USM's campus to neighboring communities, as these communities are a good showcase for mini-urban lifestyle activities given the diverse mini-urban land uses among them.

The main objectives of the project are:

a) to promote and diffuse USM's sustainable lifestyles to surrounding neighborhoods;
b) to engage participants from various community levels into the sustainable agenda;
c) to create an enabling environment toward realizing a sustainable community;
d) to explore the challenges facing comprehensive sustainability implementation.

This project is divided into three main phases:

Phase 1 – Situational analysis activities to produce digital maps, collate recommendations/ ideas from stakeholders and prepare sustainable lifestyle kits for 'USM's Sustainability Neighborhood Area.'

Phase 2 – Data collection and community engagement would increase awareness and education activities via the project launch, followed by roadshows, campaigns & seminars/ workshops. Monitoring and assessment will be conducted during this phase.

Phase 3 – Analyses and assessments will provide findings on project implementation and suggest improvements for the project. Dissemination of research findings to various stakeholders via newsletters, progress and project reports will follow.

Research method

A participatory research framework based on the pilot project approach was used in this study with an emphasis placed on action research strategies. It was designed by involving major stakeholders (experts, urban municipalities, schools, households, industries and leaders) in order to develop community skills for recycling and composting. Action research emphasizes collaboration between researchers and practitioners (Avision et al. 1999) which aims to address the practical concerns of people in an immediately problematic situation. In other words, it focuses on addressing problems through collaborations (Avision et al. 2010).

(i) Participants: The participants of this research involve twenty identified stakeholders from USM's neighboring community area selected from education institutions, governmental/ NGOSs, residential areas, commercial, and industrial premises. These twenty stakeholders are situated within an eight kilometer radius from CGSS and were selected via a situational analysis exercise based on the authors' research objectives.

(ii) Focus Group Discussion (FGD) and Logical Framework Analysis (LFA); Two qualitative methods were used in this project: Focus Group Discussion (FGD) and Logical Framework Analysis (LFA). These qualitative methods were used to identify problems and setbacks on issues facing recycling and composting activities. The employment of both methods is crucial in understanding issues, challenges and the necessary solutions.

(iii) Target Groups' Survey Distribution; A quantitative survey was conducted to assess the community perception and level of understanding towards sustainability elements. A pilot study involving 85 participants was also conducted during the community outreach Program held in Bukit Gedung with 437 questionnaires randomly distributed to stakeholders and later analyzed using the SPSS software. Interviews were also conducted during the survey. An interview with the Director of MPPP contributed information/data regarding Penang's recycling programs while the Director of PPSPPA identified composting methods used by the organization. Data on school recycling initiatives was also collected in a similar manner.

In terms of the fieldwork, the allocation of recycling bins and composting stations were identified prior to the project's launch. The location of recycling bins and composting stations can be found in Tables 4 and 5. Initially, campaigns were conducted to identify and assess community perception and awareness regarding Penang's recycling and composting programs. The first awareness campaign was held in the residential area of Bukit Gedung and a demonstration using THM was also conducted by CGSS together with a pilot study. A campaign was also held at Giant Hypermarket involving consumers as respondents to disseminate recycling information. Similar awareness campaigns on recycling and composting were held at schools.

Findings and discussions

1. Sustainable Lifestyle Awareness Campaign

As a part of this project, surveys aimed at identifying the level of urban sustainable lifestyle awareness were distributed during the awareness campaign to 437 respondents and the findings analyzed using SPSS Version 19. The age groups of the respondents, with respondents aged below 20 at 46.7% within the age range of 12-20 years, indicates that most of the respondents were school students. 20.6% of the respondents were aged between 20-29 years, 15.6% of the respondents were between 30-39 years, 11.9% of the respondents were between 40-49 years and finally, 5.3% of the respondents were aged between 50-59 years.

Fig. 1 depicts the sustainability awareness of respondents with regard to various education levels. Respondents were varied and included those at the university level to respondents who were classified as non-educated. The data shows us that while most respondents are aware and understand the concept of sustainable development, there remained a small percentage of respondents who were uninformed of the concept of sustainable development.

Fig. 1: Awareness on sustainable development

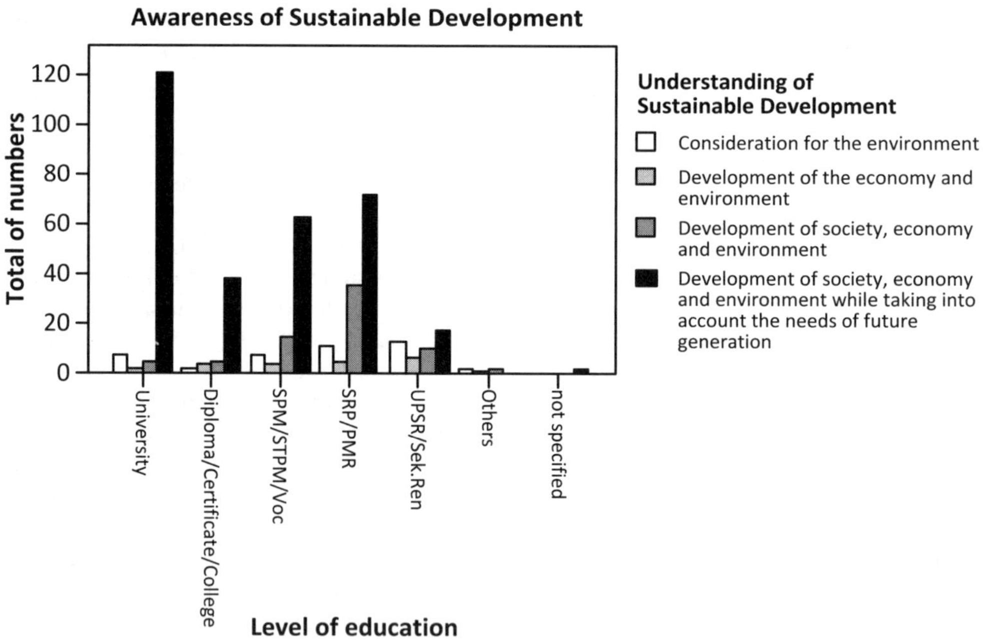

Awareness campaigns were conducted at selected schools and community outreach programs, with recycling facilities holding the highest criteria percentage in our study area. 296 out of the 437 respondents (67.7%) agreed that recycling facilities were available in their residential areas, while 310 from the 437 respondents (70.9%) agreed that planting of shade-providing trees is sustainable. On the other hand, composting initiatives were still rather low as only 157 per 437 respondents (35.9%) had experience of composting, with the remaining respondents lacking all knowledge of composting. 320 out of 437 respondents (73.2%) agree that they live in a community imbued with sustainability elements, i.e. (1) an active neighborhood and (2) an enthusiastic community.

The percentage describing the different mediums through which consumers obtain their information on sustainable development shows that 30.7% (134) of the respondents gained such information through newspapers, magazines and educational books – largely because the respondents were mostly university educated. The Internet constituted 28.1% (123) of the respondents, followed by the 'Others' 20.8% (91) category which consists of other kinds of media.

Ultimately, the elements of sustainable lifestyle practices have elicited various outcomes. 182 respondents (41.6%) now observe water-saving measures, such as ensuring faucets are completely turned off and utilizing pails when washing cars. In waste management practices, 173 respondents (39.6%) now separate recyclables such as papers, plastic bottles, glass and aluminum cans while utilizing recyclable products such as reusable bags when grocery shopping. Unfortunately, the transportation element is still rather low with 155 respondents (35.5%) opting to utilize their own vehicles instead of public transportation systems.

2. *Recycling*

With regard to the recycling initiative, SMK Bukit Gambir generated the highest income (RM 38.95), followed by SMK Bukit Jambul (RM 21.05) and SK Sungai Gelugor (RM 20.30). The most recycled items were mixed and white papers at 56 kg and 55 kg respectively. Plastic was at 18 kg, newspapers at 13 kg, metals at 10 kg, boxes at 6 kg and aluminum cans at 0.5 kg. Results indicate that recycling rates were still low due to a lack of recycling bins and awareness programs.

Table 1: Respondents' feedback on sustainable lifestyle practices

	Never practiced		Rarely practiced		Practiced sometimes		Often practiced		Most often practiced	
Elements	**Frequency**	**%**	**Frequency**	**%**	**Frequency**	**%**	**Frequency**	**%**	**Frequency**	**%**
a) Energy saving	17	3.9	52	11.9	118	27	170	38.9	80	18.3
b) Waste management	17	3.9	40	9.2	90	20.6	173	39.6	117	26.8
c) Water saving	17	3.9	41	9.4	80	18.3	182	41.6	117	26.8
d) Transportation	21	4.8	53	12.1	127	29.1	155	35.5	81	18.5
e) Consumption	19	4.3	35	8	123	28.1	161	36.8	99	22.7

Table 2: Composting activity in Universiti Sains Malaysia and its neighboring communities

	Material (kilogram)				Estimated end product (composted manure)
Method/Location	**Soil**	**Chicken feed**	**Food/garden waste**	**Cow dung**	
			Universiti Sains Malaysia		
a) Takakura method	0.5kg	0.25kg	0.5kg	–	≤ 1kg
b) Heap composting	–	–	128kg	10kg	≤ 32kg
c) Cage composting	–	–	120kg	10kg	≤ 30kg
Total			**748kg**		**≤ 63kg**

Method/Location	Material (kilogram)				Estimated end product (composted manure)
	Soil	Chicken feed	Food/garden waste	Cow dung	
	Schools				
a) SMK Bukit Jambul (cage composting)	–	–	12kg	2kg	≤ 3kg
b) SMK Bukit Gambir (cage composting)	–	–	14kg	2kg	≤ 3.5kg
c) SMK Datuk HMNA (heap composting)	–	–	17kg	2kg	≤ 4.25kg
d) SK Minden Height (cage composting)	–	–	18kg	2kg	≤ 4.5kg
e) SK Bukit Gambir (heap composting)	–	–	14kg	2kg	≤ 3.5kg
Total			**75kg**		**≤ 18.75kg**

3. Composting

As Agamuthu et al., (2009) argue, "food waste is non-recyclable and is usually sent to landfills, but when managed efficiently via composting, the amount of food waste generated can be reduced." Various composting techniques were implemented at USM and selected schools utilizing Takakura Home Method (THM), Heap Composting and Cage Composting. THM was introduced by Solid Waste and Sanitary Management Board of Penang (PPSPPA) while heap and cage composting were both introduced by a visiting professor – Prof. Sultan Ismail from The New College, India. Different composting methods are used for different types of waste, e.g. cage and heap composting methods are suitable for garden waste but THM is more suited to kitchen waste. THM is challenging on a larger scale as it requires many ingredients and procedures, while cage and heap composting are both practical in a school setting and uses cow feces as the main source of microorganisms to degrade solid waste. The five schools chosen as case studies are SMK Bukit Jambul, SMK Datuk Haji Muhammad Nor Ahmad, SMK Bukit Gambir, SK Minden Height and SK Bukit Gambir. Cage composting was applied to all schools except SK Bukit Gambir. The total weight of solid waste used at Universiti Sains Malaysia was 748 kg in contrast to those used at schools (75 kg) due to differences in space area and waste composition. In all composting methods, the estimated weight of the product is lower than that of the initial waste used due to the degradation process of solid waste.

4. Conclusion

Recycling and composting initiatives are crucial steps in promoting sustainability in urban environments/communities – consequently, public participation is crucial in sustaining such projects. 3R contributes greatly in reducing the amount of waste produced daily and while 3R implementation is possible, a change in societal attitude and lifestyles is indisputably necessary. Reduction in daily resource consumption entails a lower consumption of energy and natural resources for a more sustainable future; notably, a growth of 22.0% in recycling and composting is required if Malaysia is to reach its 2020 recycling goal. The project discussed in this paper is still in the midst of implementation and it is hoped that it will attain its desired goals with long-term monitoring and continuous improvement within a pre-determined time frame.

Flood Disaster Risk Management in Kuala Nerang

The Kuala Nerang project aims to identify groups which are vulnerable to floods and the current adaptation strategies practiced by these groups. This project also aims to analyze the factors that contribute to the increased vulnerability of certain communities with regard to the flood problem and to implement programs that seek to reduce the discomfort and danger of floods while improving the adaptation strategies of communities located in flood-prone areas. Activities include a preliminary workshop to identify groups and areas vulnerable to floods, a Logical Framework Assessment (LFA) workshop to identify potential and strategic solutions for flood mitigation and increase adaptation, a pilot study conducted in Changlun to test the questionnaire method as the main tool in identifying vulnerable areas and their adaptation strategies, as well as surveys which were conducted in 11 *mukim* (divisions) in the district of Padang Terap covering 684 respondents and a focus group discussion with 32 village leaders.

The study is comprised of two types of research, namely a survey and a social action research program. The survey utilized primary data which was collected using the 'vulnerability and adaptation' methods. Vulnerability and adaptation (V&A) frameworks provide a structure for examining the potential impacts of natural disasters such as floods and adaptation by combining both quantitative and qualitative approaches. For the purpose of this project, the quantitative approach was carried out via a survey while the qualitative approach involved in-depth interviews and focus group discussions. Quantitative data was obtained by utilizing a structured questionnaire to enable profiling, causes of vulnerability, and the villagers' vulnerability and adaptation strategies. Qualitative approaches, on the other hand, involved the use of a semi-structured guide to obtain in-depth and richer data relating to the problems of vulnerability with regard to flooding and food security issues. Part of the goal of the action research component is to develop a close rapport with flood-affected communities through visitations and dialogues.

Sampling

The quantitative method utilized the state of Kedah as a case study for the purpose of purposive sampling with the study area then narrowed down to 11 *mukim* (districts) with particular focus placed on 65 villages. The total number of sampling was derived from all 65 villages, to achieve a final total of 684 respondents.

The qualitative method involved three focal group discussions (FGDs), one in each village with a focus on flood security and adaptation strategies. 32 respondents were then chosen for the FGD which consisted of community leaders, farmers and representatives who were selected according to gender and socio-economic status background. The dialogues obtained from the action research formed an integrated component in FGD.

Analyses and presentations

PASW was utilized to analyze quantitative data with several tests run including frequency tests (to identify patterns, trends, etc.), T-tests (to rank the vulnerability indicators) and cross tabulation and correlation (to establish the associations between demographic factors and the villagers vulnerability and adaptations strategies). Qualitative data, on the other hand, was analyzed using content analysis methods, cognitive mapping and a comparative matrix.

The study established that vulnerable areas varied according to the source of the flood (different flooded rivers affect different areas); therefore, the areas badly affected in 2005 were not necessarily affected in 2010. However, two areas were badly affected in both 2005 and 2010 – Padang Temak and Kurong Hitam. In the 2010 floods, there were 1,436 households affected (equivalent to 5,615 victims) with two casualties, mainly due to negligence. Other implications of floods include property, crop and infrastructural damage and a decrease in land value. Some of the adaptation strategies include building storage facilities at a higher location – especially electrical appliances, by building high-placed shelves or scaffolding, increasing the level of reciprocity especially with regard to helping the elderly, infirm, the disabled, women and children. Simultaneously, local authorities had also made arrangements with local shopkeepers to ensure that locals can obtain food supplies from local groceries before help is delivered to the area (mainly for trapped, inaccessible locales).

More action-oriented activities include Education for Sustainable Development (ESD) initiatives through awareness campaigns on danger and adaptation strategies in six schools within the district as well as provision of flood assistance programs such as providing relief items – including boats with engines, jet wash, tents, kitchen equipment and electricity generators. Capacity building initiatives include counseling for trauma victims, training for handling and maintaining boats, accredited flood rescue training and the setting-up of a formal committee consisting of village leaders, local authorities and CGSS staff. USM was responsible for operating, managing and maintaining relief items and a Students' Sustainable Club was developed to organize and manage volunteers as well as other flood mitigation and activist related programs.

Showcasing Balik Pulau as a Sustainable Village

This project consists of the following detailed activities which will ultimately culminate in the declaration of Balik Pulau as a model for a sustainable village:

Community and biodiversity project

The main activity under this project is the '12 herbal plants for 1 household' program, with the objective of revitalizing the 'home-garden' concept in the 13 villages of Balik Pulau. Various activities were organized in order to implement the program successfully, such as surveys, socialization processes, herb planting, the continuous monitoring of planted plants and visits to herbal factories and related agencies – such as the Department of Agriculture. So far, 22 types of herbal plants were chosen by the participants, such as ginger, torch ginger, stevia, lemon grass, lime, etc. The program has thus far covered two villages involving 122 households, whereby 53 households were from Kampung Permatang Pasir and 69 from Kampung Sungai Pinang. At this point in the project, Kampung Permatang Pasir has been monitored and all 17 households have undergone initiatives to propagate herbal plants; the program will continue to be carried out in 11 other villages. Participants are also exposed to high-end marketing by introducing them to herbal companies and agencies involved in the processing of herbal products. Education for Sustainable Development (ESD) was used as a mode through which to create awareness via a campaign that focused on teachers as well as children aged between four and six years of age at four kindergartens and one preschool, involving a total of 134 children and 6 teachers. In line with the project's objectives, seven types of herbal plants were planted around the kindergartens' compounds and relevant activities were chosen to stir interest among the children, such as games and songs related to herbal plants and their usage, puzzles, 'poison boxes,' identification of different types of herbs, etc. A mascot, '*Mat Herba,*' was also used in order to attract and maintain the interest of the children.

Fig. 2: Community and biodiversity project

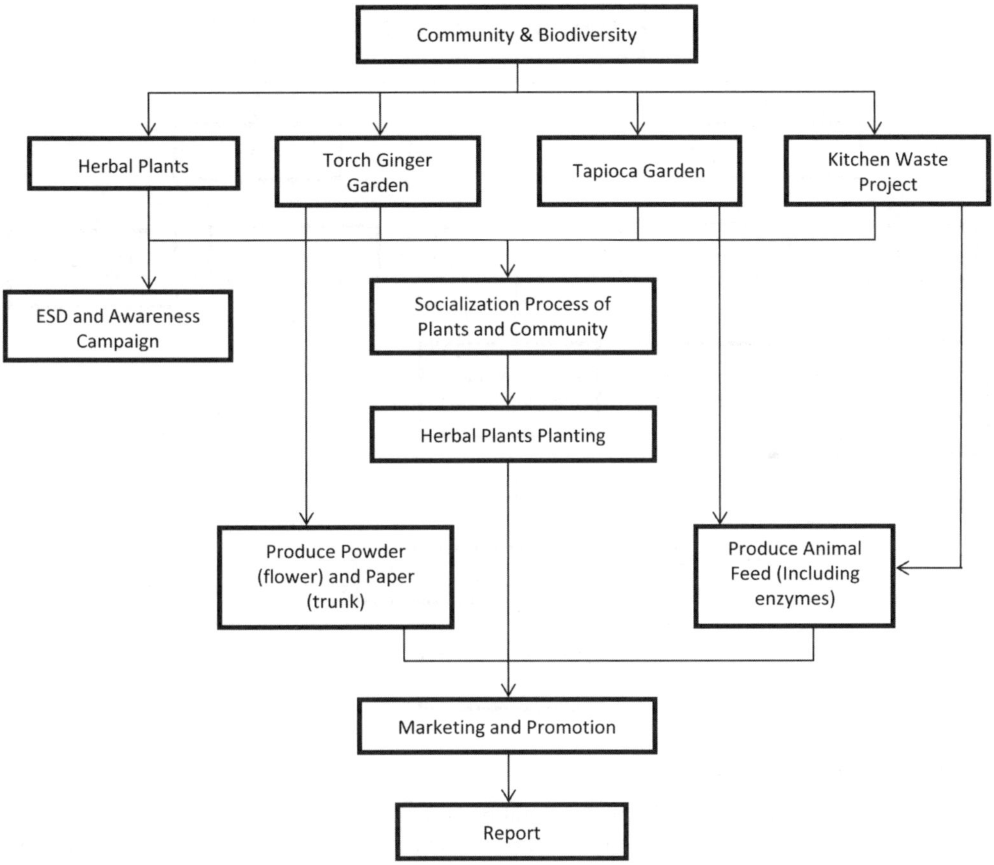

Economic empowerment project

The focus of this project is on the development of downstream products utilizing bananas and the banana tree. This project takes into consideration the huge banana plantations owned by Balik Pulau farmers. Consequently, the main purpose of the project is to provide an alternative platform for banana producers to sell their harvest. This project is worth a special mention as it involves cooperation between the private sector, CGSS in collaboration with the School of Social Sciences, USM and the villagers.

CGSS acts as a liaison between the market (private sector) and the villagers, while assisting in initiating R&D and providing capacity-building efforts and facilities in the form of machineries and buildings for the villagers to utilize in the production of banana flour and paper, as well as identifying potential high-end markets as part of its marketing strategies. So far, 30 participants were trained to produce banana flour, and ten participants for banana paper. The building provided by USM at Kampung Perlis is well-equipped with machinery that meets the standards of factory specifications including slicers, oven dryers, disk mills, separators/filters and sealers.

Fig. 3: Economic empowerment project

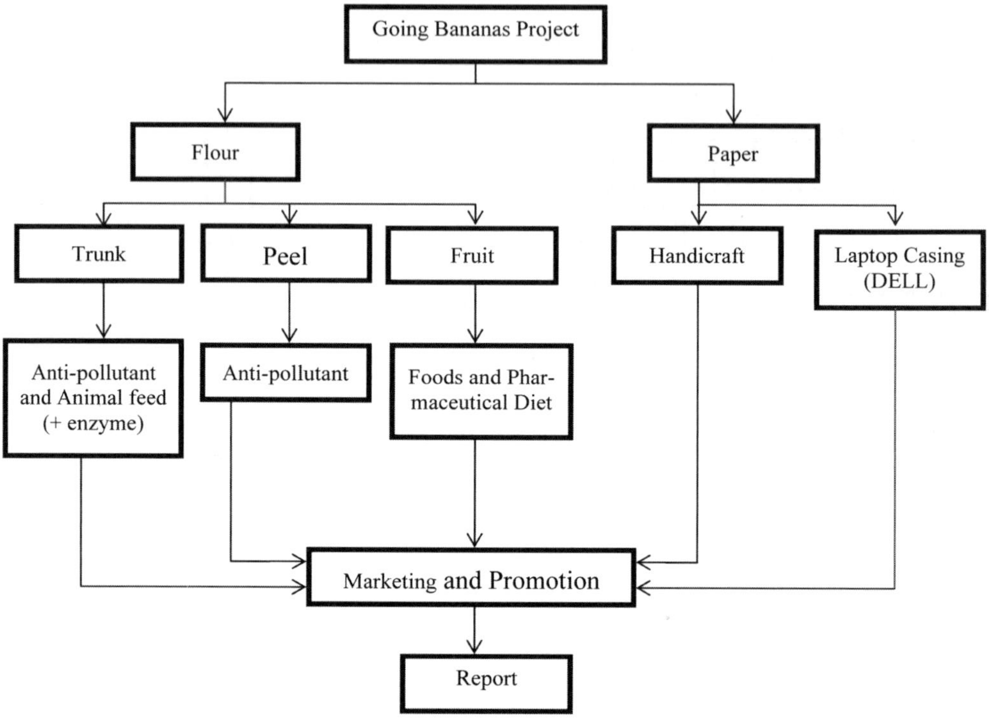

Enhancing socio-cultural sustainability

This project is concerned with the documentation of endemic cultures and their revitalization (wherever possible) to enable the preservation of such cultures. Activities within this project entail the documentation of historical, socio-cultural activities and small-scale business activities (SMEs products). The oral history documentation component was carried out and completed in 13 villages and resulted in the identification of 40 SMEs and the development of two homestay programs. The Homestay Program is a government initiative that enables local urban tourists and foreigners alike to experience village living by staying with a rural village family. The host family will then benefit from the supplementary income that this initiative provides. The activities involved under the Community Biodiversity project include the previously mentioned homestay programs, Women Empowerment Workshops, Cultural Diversity Programs and the transformation of Balik Pulau into a Cultural Heritage Village.

The 'Generating Cultural and Sustainable Tourism' program was developed from documenting and identifying the culturally diverse heritage of Balik Pulau. As cultural tourism focuses on an area's culture, including its lifestyle, heritage, identity and activities, Balik Pulau holds a great amount of potential to succeed in such an initiative because of its diverse nature. Furthermore, these touristic activities were developed around local communities and operated by these communities. Collaboration will be undertaken with stakeholders (homestay operators, USM, SME producers) to develop a range of information resources that will connect visitors to the local cultural scene which has been developed as cultural tourism packages.

Fig. 4: Enhancing socio-cultural sustainability

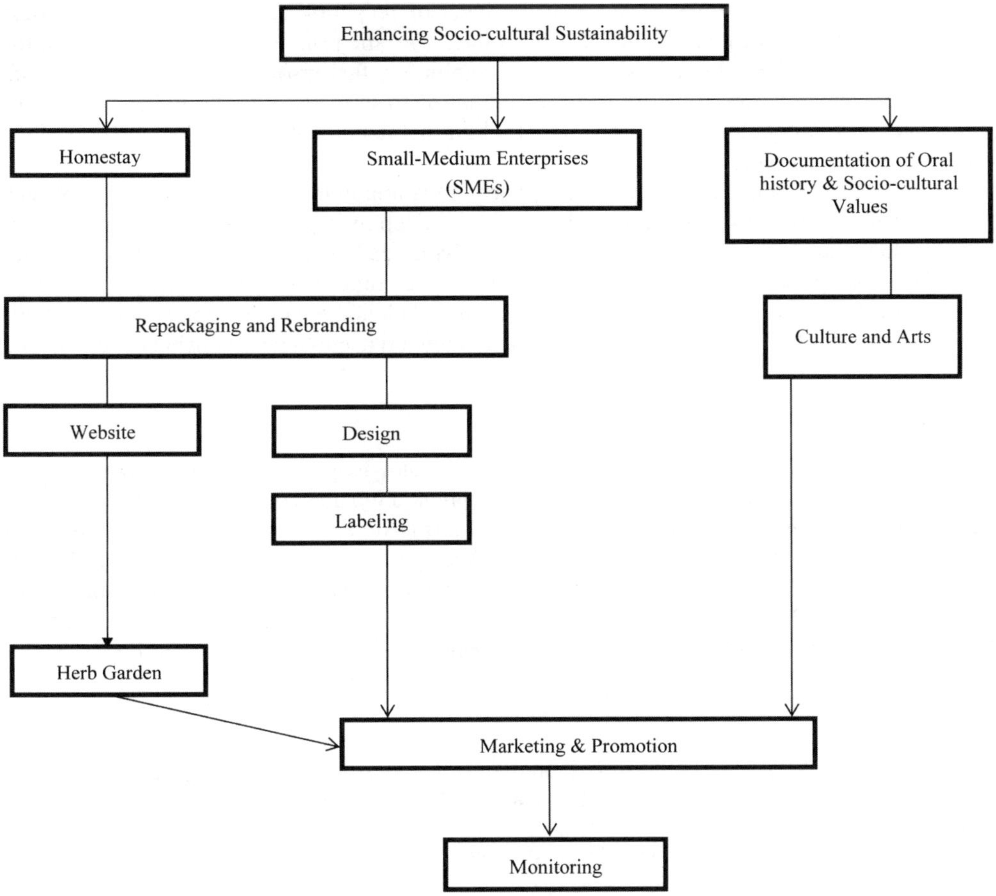

Conclusion

The case studies above illustrate USM's attempts to make a difference at various levels, starting from the campus, to areas neighboring the campus, as well as both rural and urban communities located away from the campus. The promotion of sustainability encapsulates three dimensions; socio-economic, socio-cultural and environmental. Continuous efforts in community engagement require a strong understanding of public participatory approaches and the concept of sustainable livelihoods.

To recap briefly, the Kuala Nerang Project involves the efforts of the USM team in developing a database that identifies the vulnerability of rural villagers with regard to flood-security, and to understand the dynamics in the transformational process from flood-vulnerability to flood security for each group. For the action research project, areas that were prone to natural disasters (particularly flooding) were studied. This project was successful in educating and instilling awareness among communities who face these issues. Similarly, although dealing with cultural and economic issues, the Balik Pulau project aims to assist rural populations in achieving an increased quality of life and a better future for their children.

The project 'Enhancing Sustainable Living within USM and its Neighboring Communities' on the other hand seeks to increase the level of recycling and composting awareness among communities located around USM's vicinity with the primary argument being that the generation of refuse should be lessened with a community that grows increasingly savvy with regard to its appreciation of nature and the amount of trash produced. The indicators for this project are currently being worked on and a tool kit that documents the implementation of the project will be released.

It should also be added here that the component most important to ensure the success of such projects is the continuous support from the university's top management in promoting sustainable development. Top-level management must share the passion and commitment to see through and support such projects and initiatives. Strong commitments from staff and students, as well as extensive cooperation between grassroots stakeholders, private sectors and the public sector will undoubtedly facilitate the transformation toward achieving sustainable development.

Acknowledgement

The authors of this paper would like to thank and acknowledge the following individuals for providing invaluable input and support toward the preparation and completion of this paper: Dr. Suzyrman Sibly, Dr. Asyirah Abd. Rahim, Fera Fizani Ahmad Fizri, Muhammad Abd Malik Daud, Syarilla Mohamad Suhaili, Siti Mariam Abdul Kadir, Nur Amalina Tersiden, Ridzlie Ridzuan, Muhamad Azahar Abas, Mohd Sukor Abd Karim, Muhd Zulhafiz Said, Mohd Nazri Samian, Sharifah Ismail, Hafizah Rosli, Daniel Ugih Anak Echoh, Zuziirmayanti Shahari, Liyana Romli, Nor Hafizan Habib Sultan, Kumarashwaran Vadevelu and Ruzaini Ijon.

References

Agamuthu, P., Hamid, F.S. and Khidzir, K. (2009), Evolution of Solid Waste Management in Malaysia: Impacts and Implications of the Solid Waste Bill 2007. *Journal of Matter Cycles Waste Management* 11, pp. 96-103.

Avision, D., Lau, F.P., Neilsen A. and Myers, M. (1999), Action Research Projects. *Communication of ACM* 42 (1), pp. 94-97.

Avision, D., Baskerville, R. and Myers, M. (2010), Controlling Action Research Projects. *Information Technology and People* 14 (1), pp. 28-45.

Bonn Declaration (2009), *UNESCO world Conference on Education for Sustainable Development*. Available at: www.esd-world-conference2009.org/ (accessed 18 December 2011).

Brown, J.R. (2000), Privatizing the University – the New Tragedy of the Commons. *Science* 290 (5497), pp. 1701-1702.

Chambers, R. and Conway, G.R. (1992), *Sustainable Rural Livelihoods: Practical Concepts for the 21st Century*. IDS Discussion Paper, No. 296, Brighton.

Jalil, Md. A. (2010), Sustainable Development in Malaysia: A Case Study on Household Waste Management. *Journal of Sustainable Development* 3 (3), pp. 91-102.

Hassan, M.N., Rakmi, A.R., Theng, L.C., Zulina, Z. and Muhammad, A. (2000). Waste Recycling in Malaysia: Problems and Prospects. *Journal of Waste Management & Research* 18, pp. 320-328.

Ninth Malaysian Plan (2005), The Malaysian Government, Putra Jaya, Malaysia. Available at: www.parlimen.gov.my/news/eng-ucapan_rmk9.pdf (accessed 12 December 2011).

Pearce, J. and Cameron, S. (2009), *The Ivory Tower and Beyond: Bradford University at the Heart of its Communities.* The Bradford University's REAP Approach to Measuring and Evaluating Community Engagement.

UNESCO (1998). *United Nations Agenda 21 Official Website.* Retrieved April 2011. Available at: http://www.undocuments.net/a21-36.htm (accessed December 2011).

Chapter 53

New Approaches in Graduate Education for Sustainable Development: Modalities of Field Training in the Master's in Development Practice Programs

David Nugent, Lucia Rodríguez and Jairo García

Abstract

In this paper the authors report on the experiences of the Master's in Development Practice (MDP) Program, an innovative new graduate degree in sustainable development currently offered at 23 universities worldwide. Created under the auspices of the John D. and Catherine T. MacArthur Foundation, and based on the Final Report of the International Commission on Education for Sustainable Development Practice (ICESDP 2008), the MDP seeks to train a new generation of development practitioners who will have the interdisciplinary skills and understandings necessary to confront the complex challenges of implementing sustainable solutions. Toward that end, students in the MDP program receive classroom training in four core areas: the natural sciences, social sciences, health sciences, and management. They also receive field training that focuses on the integration of these core areas in real-world field settings. In the present work the authors describe the strategies that have evolved at the 23 MDP universities to provide students with integrated field training. The authors selected a qualitative, exploratory, and cross-sectional method and used surveys, interviews, and examined documentary evidence produced by the schools that describe their field activities. The authors used Moustakas's (1994) analysis of significant statements for the analysis of data. The information collected showed that the institutions in the MDP network currently employ three modalities of field training: an embedded practitioner, a subcontractor, and an individual entrepreneur. The authors identified the characteristics of the different modalities, devoting special attention to the strengths and weaknesses of each, and described how the schools have sought to address the weaknesses of the particular field model they have employed. The authors conclude that it is a specific constellation of institutional resources, organizational networks, faculty expertise, and student interest that determines which model of field training is best suited for any particular school. The results of this study may provide relevant information to the field of experiential learning, specifically in describing how universities are integrating field training into their curricula. The results may also contribute to the field of higher education for sustainable development practice. Finally, the study may be useful to program directors interested in designing or implementing interdisciplinary programs in sustainable development practice that focus on field training.

Introduction

An increasing number of universities and research centers around the world are creating interdisciplinary programs in sustainable development. Many students are drawn to these programs because they are afforded the opportunity to work in addressing complex challenges of sustainable development such as environmental degradation, disease, hunger, and poverty. Among the most innovative aspects of these degree programs is their field training. Aspiring professionals in sustainable development find the prospect of "learning by doing" to be especially attractive.

According to the theory of *Social Constructivism*, knowledge is "constructed when individuals engage socially in talk and activity about shared problems or tasks. Making meaning is thus a dialogic process involving persons-in-conversation, and learning is seen as the process by which individuals are introduced to a culture by more skilled members" (Driver et al. 1994: 7). This approach locates the acquisition of knowledge within a social context in which learning is shared culturally in "ways of understanding and talking about the world and reality" (Merriam et al. 2007: 292).

Situated Learning (e.g. Lave and Wenger 1991) extends social constructivism by emphasizing that learning occurs through actions within a *community of practice*. According to this theory, training by abstraction is of little use as action (i.e., practice) is necessary (Anderson et al. 1996). Thus, advocates of situated learning reject, or at least downplay, the importance of learning as an individualistic transformation, rather, they view learning as a practical experience within a special social context, which is a community of practice (Fry et al. 2003).

Importance of Practical Experience in Higher Education: Experiential Learning

Experiential learning or 'learning by doing' is the process of learning through experience gained through life, education, and work (Fry et al. 2003). Experiential learning is based on the notion that "understanding is not a fixed or unchangeable element or thought but is formed and re-formed through 'experience'" (p. 14).

Kolb (1984) describes experiential learning as a cycle. This learning cycle comprises four stages: *concrete experience*, *reflective observation*, *abstract conceptualization*, and *active experimentation*. During the concrete experience stage, learners are fully involved in the new experience. This stage is followed by a period of reflection about their experience. In the reflection stage, learners are strongly influenced by feedback from peers and professors. This feedback allows learners to "form and re-form, process their ideas, take ownership of them and integrate their new ideas into sound, logical theories" (Fry et al. 2003: 14). During the abstract conceptualization stage, learners take these new ideas and make decisions to solve concrete problems (i.e., active participation). Following this active participation, learners will begin the learning cycle again. All four stages of the learning cycle are necessary to achieve effective learning.

Beaty (2003) describes two basic techniques of making experiential learning an integral part of course design. The first technique is by providing opportunities for experience inside the university through pre-planned practical work in a controlled environment using the following methods: (1) Laboratory experiments, (2) simulations, (3) case studies (including Problem Based Learning), (4) micro-teaching, and (5) projects.

The second technique of integrating experiential learning in the curriculum is by providing students with the opportunity to learn from experience in a 'naturalistic environment' (i.e., outside the university) using three methods: (1) field trips, (2) placements, and (3) work-based learning projects.

Beyond offering a 'naturalistic setting', field trips allow learners to meet many secondary learning objectives such as interacting with practitioners or local people, collecting information not provided in class, and practicing field research and cooperation skills; however, to be effective, experiential learning within a naturalistic environment "involves an intentional cycle of action and reflection" (p. 146).

Placements usually give students a temporary job within an organization. Employees of this organization have the function of supervising the students. Universities need to be sure that organizations agree to help students gain appropriate work experience and receive adequate mentorship. In work-based learning projects, students are clients of an organization: they are not employees. Students work with a particular project either in a design capacity or as a consultant. The success of the outcome using this method relies on the roles and responsibilities of the organization, the students, and the faculty.

In the section that follows, the authors discuss the ways in which the global network of MDP programs has attempted to draw upon the insights of the experiential learning and situated learning theories to provide students with field training. The MDP program relies predominantly on field placements within a community or practice (i.e., situated learning); however, not all field placements are created equally. The discussion focuses on different kinds of field placements, and the ways in which each kind structures experiential learning in a distinct manner.

The Global Network of Master's in Development Practice

In 2007, The Earth Institute, Columbia University, brought together a group of 20 distinguished development experts from around the world to form the International Commission on Education for Sustainable Development Practice (ICESDP). The charge of the commission was two-fold: To evaluate the strengths and weaknesses of existing graduate programs in development practice and, based on its findings, to design a new program of study that would train practitioners to confront the challenges of implementing sustainable solutions to complex development problems.

After more than a year of research and analysis, the commission arrived at two conclusions (ICESDP 2008): Firstly, although many existing programs in development practice were of very high quality, the training they offered focused on a specific discipline. The commission concluded that practitioners trained in programs with a disciplinary emphasis did not speak a common language, making it difficult to work collaboratively to achieve shared goals.

A second conclusion of the commission was that few existing programs in development practice included in-depth field training that transcends specific disciplines. Rather than provide students with direct exposure to the complexities of fieldwork, most programs paid relatively little attention to the challenges of implementing integrated solutions to on-the-ground development problems. As a result, many practitioners began their professional lives with only a limited understanding of the difficulties involved in working effectively in real-world settings.

Based on these findings, the ICESDP designed a new graduate degree in sustainable development, a program that the commission referred to as the Master's in Development Practice (MDP). To overcome the predominantly disciplinary nature of existing graduate programs, the ICESDP proposed an interdisciplinary curriculum designed to expose students to key development-related insights and understandings across four core areas: natural sciences, social sciences, health sciences, and management. To provide more adequate preparation in the applied-interdisciplinary dimensions of development practice, the commission recommended that all students receive approximately three months of intensive field training. During their field experience, students would: (a) learn the practical skills of on-the-ground development work, (b) be exposed to the real-world challenges and complexities of implementing development projects, and (c) learn field skills that would allow them to integrate knowledge across the core areas of the MDP program.

The Final Report of the ICESDP (2008) also suggested creating a global network of MDP programs, so that its innovative approach to development training could be disseminated as quickly as possible. By December 2011, 23 universities on six continents had launched MDP programs (see Table 1).

The MDP Field-Training Program

One central finding of the ICESDP was that few graduate programs in development provide students with field training that is sufficiently in-depth, broad, and interdisciplinary so that it can transcend the boundaries of specific disciplines. As a result, relatively few practitioners begin their careers with the kind of field training necessary to implement integrated solutions to real-world development problems in a manner that is sustainable. The MDP field-training program is designed to address this problem.

Table 1: MDP institutions

MDP Institutions
1. BRAC Development Institute, BRAC University (Dhaka, Bangladesh)
2. CATIE (Turrialba, Costa Rica)
3. Columbia University (New York, New York)
4. Emory University (Atlanta, Georgia)
5. Institute of Political Sciences (Sciences Po) (Paris, France)
6. James Cook University (Cairns and Townsville, Australia)
7. TERI University (New Delhi, India)
8. Trinity College Dublin and University College Dublin (Dublin, Ireland)
9. Tsinghua University (Beijing, China)
10. Universidad de los Andes (Bogotá, Colombia)
11. Universidade Federal Rural do Rio de Janeiro, UFRRJ (Rio de Janeiro, Brazil)
12. University of Arizona, Tucson, Arizona
13. University of Botswana (Gaborone, Botswana)
14. University of California, Berkeley (Berkeley, California)
15. University of California, Davis (Davis, California)
16. University of Cheikh Anta Diop, UCAD (Dakar, Senegal)
17. University of Denver (Denver, Colorado)
18. University of Florida (Gainesville, Florida)
19. University of Ibadan (Ibadan, Nigeria)
20. University of Minnesota (Minneapolis, Minnesota)
21. University of Peradeniya (Peradeniya, Sri Lanka)
22. University of Waterloo (Waterloo, Ontario, Canada)
23. University of Winnipeg (Winnipeg, Canada)

The MDP field training program focuses on the cultivation of interdisciplinary, cross-sector skills, and understandings that cannot be learned in the classroom. Students learn these skills by working closely with experienced practitioners, community members, and faculty mentors in the context of ongoing development projects.

The research

The authors were interested in the following two research questions:

1. What strategies have MDP institutions developed in an effort to provide students with field training that can integrate across the four MDP core areas?
2. What are the strengths and weaknesses of these strategies and what are MDP institutions doing to overcome the weaknesses?

The authors selected a qualitative, exploratory, and cross-sectional method, and used surveys, interviews, and examined documentary evidence produced by the schools that describe their field activities. The population selected consisted of 16 MDP universities that have provided students with integrated field training. It was not possible to validate informant responses by means of field verification. Consequently, the authors have assumed that informants responded accurately to all questions. The authors employed Moustakas's (1994) analysis of significant statements for the analysis of data.

Results

Based on this analysis, the authors identified three distinct field-training approaches (i.e., models): 1. the embedded practitioner's model, 2. the subcontractor model, and 3. the individual entrepreneur model. In the pages that follow the authors present a description and analysis of these models.

1. Embedded Practitioner (EP) Model.

Under this model, universities send small teams of students to a large number of dispersed field sites to work with partner organizations involved in development projects (e.g. NGOs and government ministry offices). The large number of sites involved makes it difficult for MDP faculty to be directly on hand to supervise the field training of their students. Instead, the schools that use this model come to rely on the personnel of their partners to oversee the field training of MDP students.

The fact that MDP students join existing development projects means that the EP model has both strengths and weaknesses in terms of its ability to provide students with field training in sustainable development practice. A major strength of this model is that it offers students first-hand experience in how development projects are carried out in the field. As the students are headed for careers in sustainable development practice, this exposure to the realities of project implementation must be considered the strongest characteristics of the EP model. By the time the students have finished their field experience, development is no longer a hypothetical exercise that they have read about and critiqued. Rather, it is an activity that they have seen and experienced first-hand, one that they know from the perspective of insiders.

The EP model comes with certain disadvantages, mainly pedagogical. They concern the potential difficulty of exposing students to forms of development practice that are cross-sectoral and succeed in integrating across the four core areas of the MDP program. Because the MDP students join ongoing development projects implemented by organizations that have their own specific agendas, the students have little control over the content or goals of these projects. Nor do the students have input into the management of the projects. In most cases, the students have little to say about their own day-to-day activities. Rather than structure their own activities, the students must adapt themselves to the conditions they encounter in the projects that have agreed to take them on.

Because the MDP program is seeking to train students in a new kind of development practice, it is almost inevitable that there will be a gulf the goals of the partner organization and the goals of the MDP program. As a result, one major set of challenges for schools that employ the EP model is how to overcome this gulf.

An important dilemma is how to be sure that the students learn the skills of cross-sectoral integration even as they are exposed to development projects that do not employ integration, or that do so to a lesser degree than desired by the MDP.

MDP schools that employ the EP model have been quick to adapt to these limitations. They have done so by adjusting their curriculum during the period immediately before and after field training. One popular approach is that during the semester prior to fieldwork, schools research as much as possible about each of the field projects in which their students will be involved, paying special attention to the degree to which the projects are cross-sectoral. To the extent that the projects do involve multiple sectors, students can learn directly from the ways that the projects integrate across sectors. With projects that are not cross-sectoral, students can analyze the factors that prevent the projects from working across sectors more effectively. They can also draw upon their experience with nonintegrated or partially integrated programs to devise ways of redesigning their projects so as to make them more integrated. During the semester after returning from the field, students can present the results of their findings concerning strengths and weaknesses to their professors and their peers.

A second set of disadvantages associated with the EP model is logistical. As noted above, in this approach to field training students learn the skills of the practitioner by joining ongoing field projects. Difficulties may arise from the fact that the organization personnel that agree to take on MDP students already have full-time jobs, in that they are responsible for the implementation of a major development project. Considering this situation, it is very important that steps be taken to ensure that the MDP students do not become a burden to project personnel and that project personnel have a strong vested interest in supervising the activities of the MDP students. These are especially important considerations because project personnel have a great deal to do in implementing the development project for which they are responsible. Supervision of graduate students is unlikely to be part of their original set of work responsibilities.

To avoid the possibility that students become a burden to project personnel, and to ensure that the latter have a strong vested interest in overseeing the activities of the former, it is important that several precautions be taken. First, it is crucial for the MDP school and the partner organization to identify in advance a specific set of work activities that the partner needs to have done in the course of one of its ongoing development projects. In other words, it is important to work out specifically how the MDP students will advance the work of the project in question. Second, it is important that these agreements about student work activities be made with the specific person(s) who is responsible for implementing the project, on the ground. Finally, it is essential to ensure that the individual(s) who is responsible for overseeing the work of the MDP students has the incentive, the time, and the experience needed to supervise the MDP students. To ensure that this is the case, the MDP schools must have discussions with the supervisor, and with his or her superiors in the partner organization, prior to the beginning of the field-training period.

Table 2: Embedded practitioner model

Embedded Practitioner Model	
Characteristics	• Small teams of students are sent to a large number of dispersed field sites throughout the world or a region to work with partner organizations (NGOs of various kinds, government ministry offices). • MDP faculty do not supervise directly the field training of their students. • Day to day activities of the students are overseen by personnel of partner organization • MDP faculty usually play a larger role during the period immediately before and after field training
Field projects	• MDP faculty work with partner organizations to identify field sites suitable for students. • Students join already existing development projects

Embedded Practitioner Model	
Supervision	• The personnel of the partner organizations oversee the field training of MDP students
Strengths	• Students receive hands-on experience in how development projects are actually carried out in the field
Weaknesses	• There may be some disconnect between the goals of the partner organization and the goals of the MDP program. • Students may not learn the skills of an interdisciplinary or cross-sectoral approach to sustainable development. • Partner organization personnel have full-time jobs, and may not have the time needed to provide adequate guidance and advice to the MDP.

2. The Subcontractor (SC) Model

This approach to field training differs from the EP model in a number of important ways – so much so that it may be thought of as the inversion of the EP approach. The key differences between the two models revolve around the number of field sites involved, the role of MDP faculty in supervising field training, and the degree to which MDP students are compelled to adapt to the regimen of an existing development project.

Schools that employ the SC model send their students in relatively large groups to a small number of field sites. Because field training is focused on relatively few field sites, it is possible for MDP faculty to go with their students, and to supervise their training directly for the duration of the summer field-training period.

"Subcontracting" is an apt term to characterize this approach because of the role of the MDP faculty in structuring and implementing the field training process. In this model, MDP students do not find it necessary to adapt themselves to the routines of an existing development project. Instead, an MDP faculty supervisor arranges with a partner organization to do a specific project for the organization, or in tandem with it.

The SC model differs from the EP approach in that the MDP faculty leaders have considerable autonomy to determine the daily activities of their student field trainees. This greater degree of autonomy in turn means that it is possible for the MDP faculty to craft field training experiences according to the needs and prerogatives of the MDP program. Ultimately, the MDP team must produce a set of deliverables for the partner organization with which the team has subcontracted. However, it is the responsibility of the MDP team itself to decide how it will produce these deliverables.

The strengths of the SC model revolve around the ability of MDP faculty leaders to determine the everyday activities of field training, and in the process to teach the skills of integration directly to their students. In this sense, the advantages of the SC model are the reverse of those of the EP approach. So too, however, are the disadvantages. The most important weakness of the SC model is that it does not provide students with direct, firsthand exposure to the everyday operation of development organizations.

Schools that employ the SC model are aware of this limitation, and have taken steps to ameliorate it. Some schools are doing so by building into the summer field training a series of interactions with the personnel of the partner organization, to provide the students with a sense for how these organizations operate. Another option that some schools are pursuing is to involve some of their students in SC activities and others in EP activities. In the semester after field training, each group of students compares their experiences, and each group can learn from the experiences of the others.

Table 3: The subcontractor model

The Subcontractor Model	
Characteristics	• MDP faculty members usually accompany students on their field training experience • MDP faculty makes arraignment with partner organizations to carry out a specific project by the MDP students. • Large teams of students are sent to a small number of field sites.
Field Projects	• MDP faculty work with partner organizations to identify field sites where students may do their field training • Students join development projects that are customized to their needs by the MDP faculty or the partner organization • MDP faculty has a greater degree of autonomy determining their students' daily activities • MDP faculty, in discussion with the partner organization, works on developing the set of deliverables which are agreed upon and produced by the MDP students
Supervision	• MDP faculty accompanies the students on the field experiences and supervises the students
Strengths	• MDP faculty supervisor determines the everyday activities of the field training experience ensuring that the MDP student receive an integrative approach to the challenges of sustainable development
Weaknesses	• Model does not provide students with direct, firsthand exposure to the everyday operation of development organizations, instead experience is one developed specifically for the MDP students

3. The Individual Entrepreneur (IE) Model

This approach to field training differs in important ways from each of the other two. The differences revolve around the number of field sites involved, the role of MDP faculty in supervising field training, and the ways in which MDP students are compelled to adapt to the regimen of an existing development project.

The most salient characteristic of the IE model is that it is individual students who are expected to assume the lead role in arranging for and structuring their own field training. In this model, MDP faculty do not arrange to have small groups of students placed with a variety of partner organizations (as in the EP model). Nor do MDP faculty lead student groups, after arranging with partner organizations to carry out a specific piece of work on behalf of the partner (as in the SC model). Rather, in the case of the IE approach, there may be as many field sites as there are MDP students. Furthermore, it is the responsibility of each individual student to arrange for his or her own summer field training. Students do so by contacting an organization of their choice, which is working on a development problem of interest to the student, and by convincing the organization in question that the student can make a productive contribution to the work of that organization. Having done so, the students arrange the terms of their work on their own. They negotiate the challenges of the workplace on their own. They also work out what deliverables they will be responsible for on their own.

As is the case with EP, MDP faculty are not directly on hand to supervise these students during their field training. The role of the MDP program is to review and approve the project that the student proposes to carry out prior to fieldwork, and to help the student assess and evaluate the experience after fieldwork.

IE's greatest strength is that it provides students with the closest experience to what they will encounter as development professionals. In other words, the IE approach does the best in providing students with what might be thought of as "on the job training." It is true that the EP approach also provides students with a sense of what it is like to work with a development organization; however, the success of the model depends upon MDP faculty intervening with partner organizations, and with project directors, prior to (and in many cases during) fieldwork. The MDP faculty do so to ensure that the key personnel in their partner organizations can supervise MDP students effectively. None of this is true of the IE approach.

As is the case with the other two approaches to field training, the IE model's strengths are also its weaknesses. The fact that students join an ongoing project being carried out by a development organization means that they will often find it necessary to adapt themselves to the work of that organization. The fact that there are no MDP faculty on hand to supervise their activities means that, just as is the case with the embedded practitioner model, students may have little control over the content of the project they join. Nor are the students guaranteed any control over the project's goals, or over its organization and management. Nor can students be sure that they will have much direct input into their own everyday activities.

To the extent that students find themselves unable to control these aspects of their field training, they may not be exposed to field practices that emphasize integration across the four core areas of the MDP. As the authors have seen, the MDP program is seeking to train students in precisely this area. As a result, there may be a lack of fit between the goals of the partner organization and the goals of the MDP program. As in the case of schools that adopt the EP approach, one major set of challenges for schools that employ the IE model is how to overcome the gulf between the goals of the MDP program and those of its partner organizations.

MDP schools that employ the IE model have been quick to adapt to this problem. They have done so by adjusting their curriculum during the period immediately before and after field training. One popular approach is that during the semester prior to fieldwork, schools research as much as possible about each of the individual field projects in which their students will be involved, paying special attention to the degree to which the projects are cross-sectoral. To the extent that the projects do involve multiple sectors, students can learn directly from the ways that the projects integrate across sectors. To the extent that the projects are not multisectoral, students can analyze the factors that prevent the projects from working across sectors more effectively. Students can also draw upon their experience with non-integrated or partially integrated projects to devise ways of redesigning the projects so as to make them more integrated. During the semester after they return from the field, students can present the results of their findings concerning strengths and weaknesses to their professors and their peers.

Table 4: The individual entrepreneur model

The Individual Entrepreneur Model	
Characteristics	• Individual students are expected to assume the lead role in arranging for and structuring their own field training.
Field Projects	• There may be as many field sites as there are MDP students • Students contact partner organizations of their own • Students negotiate the terms, the deliverables and challenges of the workplace.
Supervision	• Students mainly supervise themselves once in the field • MDP faculty reviews and approves the students project and helps the student to assess and evaluate the experience after fieldwork.
Strengths	• This model comes closest to providing students with "on the job training."
Weaknesses	• No MDP faculty supervision; students may have little control over the content of the project they join; and little direct input into their own everyday activities • There may be a lack of fit between the goals of the partner organization they work on and the goals of the MDP program.

4. The Hybrid Model

Some MDP institutions provide students with the possibility of choosing from among the above-mentioned field-training models. The strength of this "hybrid" approach is that students can select the model that best suits them. However, the ability to offer students a range of options depends on the ability of the institution to train and educate faculty and students about the advantages and disadvantages of each model. Indeed, with each additional model that an institution offers, the task of providing a viable field-training program becomes progressively more complicated.

Conclusion

The information that the authors have collected for this study shows that MDP directors have developed several strategies to provide their students with field training that integrates across the four competencies of the program. The authors refer to these as the embedded practitioner, subcontractor, and individual entrepreneur models. The authors have sought to identify the characteristics of each model, paying special attention to its strengths and weaknesses. The authors have also sought to draw attention to the particular constellation of institutional resources, organizational networks, and faculty expertise that leads the various schools in the MDP program to adopt one of these three models as the core of its program.

What is the relevance of these findings for sustainable development programs more generally, and more specifically for those grounded in an interdisciplinary approach? As universities around the globe incorporate sustainable development into their curricula, among the central problems they face is how to provide students with the practical skills of the development professional. Because many programs in sustainable development have a professional orientation, it is especially important for students to complete their studies with a firm grasp of the skill sets of the development practitioner.

Whereas some of these skills may be taught in the classroom, it is essential that students have the opportunity to practice them in the field. It is for this reason that many sustainable development programs are incorporating field training into their curricula. The three models of field training outlined above are intended to identify important variations in the way that experiential learning may be structured. It is the authors' hope that this analysis will be of use to other schools as they go about the task of designing field training, and of preparing the next generation of development professionals.

Programs of sustainable development practice can be more effective by incorporating the following principles:

Take an Interdisciplinary Approach. The challenges of sustainable development are complex. Individual areas of knowledge such as agriculture, economics, or health are not sufficient on their own to solve hunger, disease, or poverty. Knowledge from the various disciplinary areas and experience in the field is necessary for a broad and holistic understanding of the challenges of sustainable development.

Centralize the Field Experience. The field program is what integrates the MDP curriculum and acts as its core. It thus cannot be relegated to a sight-seeing experience that is separate from the broader curriculum; instead field programs in sustainable development should be integrated directly into the class-based curriculum. Field training should be rigorous, in-depth, and interdisciplinary.

Support Faculty Members. Faculty members need to be provided with the support necessary to allow them to integrate the field experience into their curriculum on sustainable development and vice versa. For example, a specialist in economics, global health, or education, may find it difficult to make the connections to other discipline areas. Carefully crafted weekly seminars, guest speaker series, and other activities can assist with this challenge.

Choose Partners wisely. Partners for embedded, subcontractor or individual entrepreneur models should understand the intended goals and outcomes of the field program and allow for flexibility, so that students may experience as much as possible of a holistic approach to sustainable development.

Create Sustainable Partnerships. Sustainable development requires long-term partnerships among different stakeholders such as governments, NGOs, communities, the private sector, and others. Because the challenges of hunger, disease, and poverty will not be eradicated overnight, the partners should also be committed to deep and meaningful partnerships.

Provide Support for the Field Program. The field program is what integrates the MDP program. Therefore, it is important to ensure that all students participate in an MDP field experience. Similar programs in sustainable development should allocate human and financial resources specifically to the field experience.

Deepen the Field Experience. The MDP field training programs are at least three months in duration. In addition, field activities are embedded in a broader pedagogical structure that follows Kolb's (1984) experiential learning cycle, which includes preparation before, and reflection after, the field training. Only in this way will students be able to make the most of their time in the field as an element of experiential learning.

References

Anderson, J., Reder, L. and Simon, H. (1996), Situated Learning and Education. *Educational Researcher Journal* 25 (4), pp. 5-11.

Beaty, L. (2003), Supporting Learning from Experience. In Fry, H., Ketteridge, S. and Marshall S. (Ed.), *A Handbook for Teaching & Learning in Higher Education* (2nd ed.), pp. 134-147. Taylor & Francis e-Library.

Driver, R., Asoko, H., Leach, J., Scott, P. and Mortimer, E. (1994), Constructing Scientific Knowledge in the Classroom. *Educational Researcher Journal* 23 (7), pp. 5-12.

Fry, H., Ketteridge, S. and Marshall, S. (2003), Understanding Student Learning. In Fry, H., Ketteridge, S. and Marshall S. (Ed.), *A Handbook for Teaching & Learning in Higher Education* (2nd ed.), pp. 9-25. Taylor & Francis e-Library.

ICESDP (2008), *International Commission on Education for Sustainable Development Practice: Final Report.* The Earth Insitute, Columbia University and The John D. and Catherine T. MacArthur Foundation, New York, NY.

Kolb, D. (1984), *Experiential Learning.* Prentice-Hall, Upper Saddle River, NJ.

Lave, J. and Wenger, E. (1991), *Situated learning Legitimate peripheral participation.* Cambridge University Press, Cambridge, UK.

Merriam, S., Caffarella, R. and Baumgartner, L. (2007), *Learning in adulthood: A comprehensive guide* (3rd ed.). Jossey-Bass, San Francisco, CA.

Moustakas, C. (1994), *Phenomenological research methods.* SAGE, Thousand Oaks, CA.

Chapter 54

Research of Sustainable Development for a Better Quality City[1]

Ružica Bogdanović, Ranka Gajić and Svetlana Batarilo

Abstract

When it comes to the concept of sustainable development in city planning and managing in Serbia, there is a problem concerning a primarily declarative approach with lack of incorporation into urban plans.

The elective course "Sustainable Development and Environment Protection" was launched in 2009 as part of Master's course at the Faculty of Transport, University of Belgrade. Its general goals are formed as a direct consequence of the urban sustainability concept's position in Serbia: 1. Informing students about sustainable development; 2. Raising the students' awareness about the necessity of permanent education, the entire society's development and environment protection and 3. Professional training for practical activity in urban planning. This represents a pioneering endeavor which is why the work on the course is designed in a traditional manner – theoretical and practical parts along with active discussions and analysis of comparative examples. The focus is on the development of skills for argument-based assessment of the concept. Along with a detailed description of the curriculum, this paper will feature students' works with presentations of key themes each of them deals with, and their conclusions on what they believe they have learned and became capable of thanks to the course.

Introduction: Urban sustainability – implementation through urban planning

As Prof. Leal Filho stated, one of the misconceptions regarding sustainability that results in insufficient studying of this subject at universities is represented by the opinion that *"the sustainability concept is too broad"*, but *"...one can apply the principles of sustainable development to different parts of the university life, contextualizing it"* (Leal Filho 2011). The focus on urban sustainability problems, especially concerning implementation of sustainable development concepts in domestic practice of urban planning and their further realization, represents a contextualization of this widely established issue for the necessities of Serbia and the concrete profile of transport engineers.

Urban sustainability assumes a wide variety of themes with numerous publications, and at our course it is narrowed primarily to the implementation through urban planning. "Livable Cities: The Benefits of Urban Environmental Planning", a book created jointly by the Cities Alliance, UNEP and ICLEI, is one of the works dealing with issues of "*city-environment interactions and how to integrate the Environment in urban planning strategies*" (UNEP et al. 2007). Strategic Environmental Assessment (SEA) is stated among planning instruments for integration of the environment in urban planning as *"a process that ensures that the environmental effects caused by policies, plans and programs are identified and assessed"* (UNEP et al. 2007). In addition to this, our course focuses on the practical elaboration of the Strategic

1 Course: "Sustainable Development and Environment Protection" – Master's, Faculty of Transport and Traffic, Belgrade, Serbia.

Environmental Assessment (SEA) or Strategic Environmental Impact Assessment (SEIA), as stated in Serbia's legislation (*Law on Strategic Environmental Impact Assessment* 2004, 2010).

There are numerous publications/works about the methodology, structure and essence of possibilities provided by SEIA. However, our course deals particularly with those that emphasize European and/or international perspective (Schmidt et al. 2005; Therivel and Wood 2005; Jones 2005; Therivel 2010; Sadler et al. 2010). Academic papers and publications on energy and sustainability issues are also in the scope of the course (e.g. from the web-site of the World Nuclear Association).

Our team is very much active in dealing with issues of insufficient education of our experts on sustainable development as well as raising public awareness about the state of the environment in Serbia and possibilities of public influence in order to improve it. We have worked on important research projects in this field: *"Integrated Approach for Solving Traffic Related Problems from a Sustainable Aspect"* (Serbian Town Planners Association and ProAktiva, Macedonia 2004/2005); *"Improving the Practices of Public Participation within Spatial and Urban Planning in Serbia concerning Environmental Matters",* (Serbian Town Planners Association 2006); *"Promotion of a Sustainable Urban Lifestyle"* (Serbian Town Planners Association 2007/2008).

Preconditions – social framework (Position of sustainable development within the practice of urban planning in Serbia)

Situation overview

Serbian city planning and managing is still facing the same problem, even after almost 30 years since the concept of sustainable development was defined. Namely, the approach to dealing with this issue remains mainly a declarative one: the idea itself is present in the *National Sustainable Development Strategy of Republic of Serbia*, which was adopted in 2008. (The Government of the Republic of Serbia 2008); the Law on Planning and Construction cites principles of sustainable development (*Law on planning and construction* 2009, 2011), but it is not sufficiently incorporated in planning documentation/shortage of methodologies and concrete actions.

At the same time, the general public is insufficiently informed on the principles of sustainable development, possibilities of energy saving and the fact that citizens' behavioral patterns in urban systems are essential causes of harmful influences and that a change of those patterns (habits) represents a key step towards the realization of "healthy cities".

This is illustrated by the results of a survey carried out by the Serbian Town Planers' Association in 2004 for the project "Integrated Approach for Solving Traffic-Related Problems from a Sustainable Aspect", which showed that there is a pronounced awareness about the fact that automobiles belong to the biggest air polluters in the central city area – 95% responded affirmatively, but at the same time 50% claimed to use public transportation only because they do not own an automobile. Approximately 80% stated that they only rarely ride a bicycle, while 51% said that they would not ride a bike more often even if there were more marked bicycle lanes in Belgrade (Serbian Town Planners Association and ProAktiva, Macedonia 2004/2005). All of this indicates, *inter alia*, irresponsible behavior – denying the fact that the problem we notice (polluted city air caused by automobiles) is caused by ourselves.

Imprudent and inefficient spending of different energy sources also represent consequences of Serbian citizens' rooted bad habits when it comes to the environment, which is best illustrated by the spending of energy for heating purposes: around 250 kw/h is spent per m^2 a year, whereas, for example, Austria spends 120 kw/h whilst developing projects to decrease spending to 70 kw/h (*Politika* [daily newspaper], 2006).

The situation is even more complex bearing in mind that certain professionals involved in drafting documents and the process of decision-making and implementation in the field of city planning are undereducated about the possibilities provided by the concept of sustainable development. In expert circles, one can hear the opinion that sustainable development is something that we have been "implementing forever", which alludes to currently present environmental regulations in planning documents and participation of the Institute for Health Protection representatives in the plan drafting, without taking into account the complex nature of a concept which demands change of behavior with the aim to achieve a balance of economic, ecological, morphologic and social factors, while relying on the present tradition – cultural roots. All of this calls for additional permanent education and systematic changes of goals, sub-goals and activities in our plans.

Key problems

In the processes of drafting, adopting and implementing urban plans, but also on a wider scale, when it comes to environmental issues in Serbia (particularly in Belgrade – with a population of 1,700,000 the largest urban agglomeration in the country), the problems regarding sustainable development can be divided in two basic groups. These are: unavailability of relevant information (which would provide a sufficient number of those interested in education), and consequently insufficient knowledge/education.

In the first group, the following problems appear:

- lack of information of the broader public about the importance of sustainable urban development;
- inadequate participation of the public in the process of drafting, adopting and implementing of urban plans;
- insufficient availability of information about the influence of activities, as planned by spatial and urban plans, about the environment (within all significant aspects of sustainable development of the spatial entities encompassed by the plan: economical, ecological, social and cultural heritage).

When it comes to the level of knowledge, the following problems appear:

- insufficient education of experts about the idea of sustainable urban development (on a theoretical plan of defining/understanding the term and the essence of the idea, as well as in relation to practical activities used to achieve sustainable urban development);
- insufficient education of the broader public on ways of achieving urban development/necessity of altering behavioral patterns in everyday life.

Therefore, it is essential to implement permanent education of professionals in the field of city planning and functioning in Serbia, as well as to establish a network of mutual cooperation. It is expected that further education of inhabitants/citizens will only have a concrete effect after the influence on professionals. It is expected that the best results will be achieved through education of students of relevant profiles. That is to say, future young experts who will begin their practical work with the required knowledge but free of prejudice (which exists with the majority of older professionals) need to be trained and developed.

Strategic Environmental Impact Assessment (SEIA) as a tool for supporting the implementation of the idea of sustainable urban development

The implementation of declaratory adopted standpoints and strategic orientations for sustainable development of the Republic of Serbia within urban planning is aimed at realization also by drafting the "Strategic Environmental Impact Assessment" (SEIA) (*Law on strategic environmental impact assessment* 2004, 2010), determining that "strategic environmental impact assessments" were part of urban planning documents, formulates the methodology of their drafting, manner of public monitoring and adopting by authorized bodies.

One of the problems is also seen in public participation in the course of drafting and adopting this document, i.e. the manner in which the citizens are informed about planning proposals for the spaces they inhabit. Namely, in the process of drafting and adopting the plan, the *Law on Planning and Construction* (*Law on planning and construction* 2009, 2011) requires participation by the public through a "public debate", clearly defined by the procedure. At public debates, citizens and users have insight only in the final planning document, as well as the report on SEIA for the same location, while the procedure for insight to the course of the plan's drafting (and possible reactions to solutions harmful to the environment) is not formulated by the law.

This was the subject of our team's work on the research project "*Improving Public Participation in Spatial and Urban Planning in Serbia Relating to Environmental Matters*" (Serbian Town Planners Association 2006), with the aim to identify SEIA's contribution in drafting plans and programs, as well as the level of participation of all actors in the decision-making process.

During the work on this project, we have noticed shortcomings when it comes to level of information on the state of environment in Serbia. One such weakness is that Serbia has a relatively weak "network" of experts. Serbia also suffers from a lack of connection and understanding between relevant institutions in this respect. Another potential problem is the lack of citizens' level of information. Citizens of Belgrade are to a large extent uninformed about the means of communication with authorities on environmental problems. As part of the survey carried out during the research on the state of environment and knowledge about control mechanisms, when asked whether they know who to address when noticing environmental violation, 75% said "no", 15% said "yes", while 10% stated that they were not interested in the problem – which, along with the 75% of those who do not know, makes up a large number of Belgrade citizens who do not understand and/or do not know their rights in this field.

Partly due to the results of this project, Serbia ratified the "*Convention on Access to Information, Public Participation in Decision-Making and Access to Justice in Environmental Matters*" (UNECE 1998) in 2009. However, problems in the field of urban planning remain present to date. We anticipate that this project will be a continuous one, which is closely related to the fact that the problem of public participation and accessibility of information in spatial and urban planning concerning the environment is still present in Belgrade's planning practice.

During the last few years, the methodology used for drafting and adopting SEIA was subject to expert debates (particularly given the fact that the period of almost eight years in which the drafting of this document has become a legal obligation was long enough to notice certain omissions). Some of the more significant changes to the original law are: 1. the condition that SEIA is now carried out for every planning document (in the first version it was left to the authorized bodies to determine whether it was necessary or not) and 2. the condition that SEIA is to be realized synchronized with the plan itself. These conditions provide that every urban plan in Serbia is accompanied by considering the influence of planned activities on the environment, as well as active participation in changing (adjusting) these activities (when necessary), i.e. influencing the abandoning of certain harmful activities.

The next step is to provide citizens with a better opportunity to become educated about opportunities given by the Arhus Convention on the one hand, and the existing regulations at the municipal/community level, on the other. Such informational measures will hopefully encourage people to demand access to study SEIA while taking part in "public debates" on particular plans and to actively participate in the decision-making process. It will also create a planning process that is more open to the public, especially regarding environmental matters, allowing the citizens to freely access information that concerns the quality of their lives and to actively participate in the decision-making processes, while developing plans for their cities.

It is imperative to conduct education of future urban planning professionals in Serbia with the aim of understanding the necessity of including the public in all planning stages, commitment of multidisciplinary cooperation and affirmation of quality of the plans carried out in that manner, by realizing a healthy, sustainable environment of responsible persons.

As a long-term output, the planning authorities will develop sustainable urban plans which are simultaneously socially, environmentally and economically responsible. An informed citizenry will begin to change its perception of environmental matters and will be capable to hold a firm position in defending its rights when it comes to sustainability.

Course – Sustainable Development and Environmental Protection (Faculty of Traffic and Transport, University of Belgrade, Serbia)

Resources

In domestic scientific circles, there is undisputed orientation toward considering/researching concepts of sustainable development, especially within the issue of urban sustainability. Therefore, besides widely available foreign literature, students also have access to literature on theoretical analysis of this subject in their mother tongue (Janić 1996; Bajić-Brković 1999; Bogdanović 1999, 2000, 2003).

As stated in the introductory part of this paper, members of our team are experienced experts, both in theory and in practice of dealing with sustainability issues in urban planning processes in our country, and "*can adequately pass on the knowledge from this field, which is also of great importance when it comes to education on sustainable development*" (Leal Filho 2011).

Specificities of the Course – methodology

An elective Course – "Sustainable Development and Environment Protection" was launched as part of studies for a Master's at the Faculty of Transport and Traffic, University of Belgrade (in the academic year 2009/10). In reference to the *Earthscan* webinar "*Sustainability Education and Training*" (Webinar/Earthscan 2011), during the education on sustainable development it is important to achieve: 1. awareness (on the necessity of changing previous habits which caused negative influence on the urban environment), 2. motivation (taking part in changes), 3. empowerment (building a basis and certainty that the changes are positive), 4. knowledge (essential understanding along with knowledge), 5. skillful means (personal skills for using what one has learned; competence) and 6. practice (taking part in practice). Thus, the general goals of the course are formed as a direct consequence of the position that the concept of urban sustainability has in Serbia.

The basic goal is raising awareness about the need for changes along with motivation and empowerment of the candidates to implement them on their own and, along with understanding and knowledge of the subject, advocate sustainable solutions in their future professional practice. Achieving competence (skills for using what one has learned) and empowering for direct practical involvement, during a one-semester course (15 weeks) in concrete social circumstances (where not even all professionals fully understand the use of sustainability concepts), is not a realistic goal. However, the course does provide cooperation with relevant teams (Institute of Urbanism, CEP/Bureau for Urban Planning Development). The candidates also write their term papers by analyzing existing urban plans for the territory of Belgrade, therefore gaining the necessary confidence for their future work.

This represents a pioneering endeavor; it is necessary for the candidates to master the basics, which is why the methodology of the course is designed in a traditional manner – theoretical and practical parts, along with active discussions and comparative analyses of examples.

Within the theoretical part, through a series of lectures by relevant experts (along with active discussions), the candidates acquire a broad image of the meaning, aims and principles of sustainable development and become familiar with significant international and domestic documents, charters and agendas established for the concept. In addition to this, the candidates give one of the lectures on their own (on selected international documents related to sustainable development) and discuss the possibilities of implementation in Serbia. A part of the program is carried out through visits to institutions which realize plans and study domestic planning practice, as a real context of future professional activity.

As part of their practical work and training in understanding the process of planning while using principles of sustainable development, the candidates produce an individual paper as the result of analysis of urban planning and environmental legislature in Serbia. The final paper represents a critical overview of the Strategic Environmental Impact Assessment on a specific urban plan on the territory of Belgrade (each candidate produces an analysis of a selected plan). The focus in this part is on the development of capability to understand causes and effects within the issues of urban planning and finally, through the analysis of a previously adopted plan, ensuring that each student is capable of argument-based assessment of the concept and whether, and to which extent, the activities proposed by the plan are suited for achieving sustainable urban development.

The method of the course also includes a specific approach developed by the authors in order to meet students' capabilities for producing simulation of SEIA, since "*...impact prediction, evaluation and description in Strategic Environmental Assessment cannot take place without some kind of framework and methodology: without 'tools'*" (Therivel and Wood 2005). Among the commonly used SEIA tools stated by Therivel and Wood we choose simplified versions of impact matrices and causal effect diagrams as appropriate for students' capabilities (Therivel and Wood 2005).

An innovation in the broad methodology of the course is represented by its integrative scope, as it combines knowledge obtained by students on the course focused on urbanization during previous years of studies (the subject of urbanism is obligatory at the basic studies at the Faculty of Traffic and Transport in Belgrade) with a holistic approach to influences on urban space through urban planning processes.

It is of great importance that the subject relates to traffic engineers who, working during the course, realize connectivity and integrated approach, since otherwise, through their education, they tend not to observe traffic in connection with other city functions. The education of traffic engineers is necessary, since there are usually other profiles of researchers in the team for SEIA. That is why this methodology – despite its rather traditional form – is very much needed when considering the education of engineers involved in urban planning.

As it represents one of the first steps in the education about sustainable development at Serbian faculties, this methodology has great importance as a catalyst for other faculties to adopt this subject as part of their educational processes.

Curriculum in detail

A. Goal of the Course

- Informing students about the concept of sustainable development;
- Raising the students' awareness about the necessity for permanent education, developing of the entire society and environment protection; and
- Implementing professional training for students in practical activities in urban planning.

B. Outcome of the course

Upon completing the course, the student will be capable to:

- Analyze concrete city locations by using multi-criteria matrixes of problems and potential;
- Determine/create activities which achieve sustainable urban development of the area;
- Argue and defend established principles for achieving sustainable urban development;
- Define and use principles of sustainable development, particularly in the field of sustainable transport, as an associate in the team for creating urban plans – regarding the field of transport.

Table 1: Structure of exercises and topics covered

	Phase of work	**What is expected from the phase**	**Topics covered**	**Final presentation**	**Class**
1.	Introduction to the Course and selection of topic that will be processed through the Term paper/1	• Comprehension of the working process during exercises • Selection of the topic for Term paper/1 • Introduction to method for making Term paper/1	• Sustainable Transport • Urban activities /accessibility • Park and Recreation • Urban Morphology • Cultural and historical ambience within sustainable development	/list of topics is given in advance/	1
2.	Preparation of Term paper/1	• Understanding the theoretical base of sustainable development and practices	• Notion/Definitions • Examples from practice • Suggestion for the implementation /for Belgrade's practice	As required / optional: text, sketches …	1
3.	Preparation of Term paper/1 and gathering the data necessary for further work	• Understanding the theoretical base of sustainable development and practices • Understanding relevant legislation /from the environmental field	• Law on Planning and Construction • Law on the Environment • National Sustainable Development Strategy for Serbia • Strategic Environmental Impact Assessment (SEIA)	As required / optional: text, sketches, maps…	2
4.	Analysis of the urban plan/part I (every student has a different Plan)	• Understanding the potential of the given site/location in accordance with the idea of sustainable development	• Establishing Indicators of Sustainable Development for the plan's location • Analyzing problems and potentials of the site through sustainable development indicators	notes tables, text ... / **DELIVERY** completion of Term paper/1	2
5.	Analysis of the urban plan/part II	• Understanding the elements/basis of the SEIA • SEIA proposal for the given plan	• Elements/basis of Strategic Environmental Impact Assessment (SEIA)	notes tables, synopsis...	2

Table 1: Structure of exercises and topics covered – *continued*

	Phase of work	What is expected from the phase	Topics covered	Final presentation	Class
6.	SYNTHESIS /part I: Correction of the urban plan's proposals after the SEIA suggestions	• Understanding the relation URBAN PLAN: SEIA	• The legislative sphere / elaboration • Impacts indicators matrix	tables, matrix notes	3
7.	SYNTHESIS /part II: DEBATE on observed facts; designing the ideas	• Creation of a proposal /recommendations for correcting the Plan in accordance to SEIA suggestions	• Activation of the potential of the investigated area with the goal of sustainable urban development.	tables, notes synopsis, sketches, maps…	1
8.	Presentation and finalization of the Term paper/2	• Developing the ability for quality assessment of urban plans in accordance with the concept of sustainable development	• Discussion substantiated with arguments on works of all the participants in group	tables, notes synopsis, sketches, maps… / **DELIVERY** completion of Term p./2	3

C. Content of the course

Theoretical part:

- Concept of sustainable development – meaning, goals, principles; basic theoretical principles and achievements in urban practice;
- International and national documents on sustainable development – UN, European Union, Republic of Serbia; Sustainable development on the local level – Local Agenda 21;
- Implementation of sustainable development principles in urban and spatial planning;
- Transportation and the environment: conflicts between city functioning and transportation;
- Indicators and criteria of sustainable development – regional, local, transportation; systems of managing environmental protection – ISO 14000 standards;
- Environment and spatial planning (regional and urban); strategic impact assessment of development policies, plans and programs on the environment.

Practical part (the main phases):

- Selection of a topic which will be processed through the first term paper (on a subject relating to sustainable development with examples from international practices);
- Analysis part of the selected urban plan through indicators of sustainable development and according to criteria of conflict points – identifying spatial limitations and possibilities;
- Discussion/debate on the facts that have been discovered;
- Synthesis part: correction of the urban plan's proposals by means of SEIA, activating potentials of the location with the goal of sustainable urban development;
- Presentation/public exhibition and discussion among candidates.

A detailed scheme of exercise structure, topics covered and type of presentation for every phase is presented in Table 1, "*Structure of Exercises and Topics Covered*".

D. Grading

The grading concept for evaluating the candidates' achievements is based on the standpoint that 70% of the grade relates to his/her activities during the semester, so therefore the activity during the course is scored with a maximum of 10%, quality of the initial term paper (homework) also with a maximum of 10%, final term paper with a maximum 40% and project presentation with a maximum of 10% out of a total of 100 points. The written exam taken by the candidates at the end of the semester is graded with a maximum of 30%, with the possibility for the candidate to be relieved of the written exam by performing above average activities during the semester.

Interaction / dialogue / work with students – *examples*

Problems and challenges

During the work, special focus was placed on determining indicators, which was an entirely new way of thinking and work for the students. The indicators established in the Strategy of Sustainable Development of the Republic of Serbia were helpful to them only as support in understanding and ways of thinking (a relatively small number of Strategic Environmental Impact Assessments has been carried out in Serbia and the students received one of the first as a representative example – as a methodological pattern for drafting their simulation of the Strategic Environmental Impact Assessment (SEIA), rather than a concrete set of data which would be useful; the given examples did not contain indicators in detailed fashion, because the *National Sustainable Development Strategy for Serbia* (The Government of the Republic of Serbia 2008), contains a system of indicators for the entire state level, while a system of indicators for more detailed planning levels in Serbia is not yet complete).

Students were given valid plans (for the territory of Belgrade) and used them to perform strategic impact assessment of predicted activities on the environment, including economic, social and morphologic elements of those influences. The plans they were given were on the level of an urban plan for a certain part of the city, therefore it was not expected that they could directly use the indicators from the Strategy of Sustainable Development of the Republic of Serbia. Rather than that, they were in a position to use the manner of thinking from the strategy in order to determine and profile the indicators which would monitor the influences of the plan's activities. The plans mostly involved reconstructions of residential zones with some specific features: proximity of a gas station, hospital, transportation node or city highway, need for school, proximity of a river or the presence of a loess deposit at the location.

In the first phase of the work, the students were determining the activities predicted by the plan, separating problem activities within four aspects: morphologic, sociological, economic and ecological. Following this, they separated those which are considered the most serious, i.e. most harmful to the environment. The key value criteria for spatial activities' influence on the environment are the level and significance of the influence. The level of the influence represents the illustration of damage or gain (the influence can be negative or positive) and the estimated influence on the environment (a very positive influence; activities with a small positive or negative influence; critical influence), while the significance of the influence represents the scale of a physical space that can be exposed to effects of change in the area (from local, city, regional, state, to global influence of activities). Influences, i.e. consequences, are graded from -3 to +3, according to level of changes (-3 marking the most negative changes). Therefore, the selected activity represented a basis for finding indicators by grading according to *criteria for level of the influence* and *criteria for grading the scale of the activities' influence;* influences with the highest number of negative consequences were identifies, which was followed with determining indicators that would measure and monitor the situation. Developing indicators on any of the mentioned levels (local, city and even regional) helps create a system of continuous environmental monitoring of the situation, according to all the aspects on a specific territory. According to the determined values of the indicators of sustainable development, bases are formed for defining possible directions and priorities of activities with the aim to annul negative influences.

At the end, students gave recommendations for ways of reacting to possible negative consequences to the environment and quality of life (air quality, possibility of relocation, employment, rise of real estate prices, etc.), i.e. concrete measures.

Upon completion of the course, students acquired new knowledge which means that, beside concrete analyses of location with the use of problem matrixes and potentials and creating activities for achieving sustainable urban development of the analyzed space, they were also capable to use principles of sustainable development as transport engineers and associates in a team for creating urban plans and to argue a defence in favour of the principles they establish, as well as using them for achieving sustainable development in a concrete area.

A particularly significant result of the course is raising the students' awareness about the necessity for permanent education on environmental protection and principles of sustainable development of the entire society, especially bearing in mind that the biggest problem the author faces is the fact that the students are focused mainly on ecological problems and that we spent the entire semester raising their awareness about the complexity of the problems, i.e. their integral nature. In other words, we deal with *"'awakening' and explaining that environmental problems should be viewed holistically, taking into account activities and conse-*

quences on the economic and social levels, rather than just ecological" (Leal Filho 2011). We have reached this conclusion through work with the students during the semester and it was eventually confirmed by the students themselves when asked what they had learned while working on this subject.

Why did they decide to take this course?

Students who decided to take part in the program "Sustainable Development and Environment Protection" were asked at the beginning of the course to explain their choice. With no concrete, detailed and clear knowledge about the program (they received a detailed program of the course, usually handed out at the beginning, upon the completed survey), they gave their replies based on which we divided them in several groups, according to the motive of their choice. However, what later turned out to be a common motive in answers of all 'groups' (including answers such as: "I don't know enough about this, but I want to learn since it is a current and important problem", or "I have heard much talk about it in the media, so I wanted to learn more", or "I know it has to do with laws", or "rift between what is written in laws and what is really being implemented", or "I feel responsibility for the environment as a future transport engineer," etc.), is their *orientation to think about ecology and environmental protection*, as reason for their participation in the course.

Students gave different answers but they mostly related to a way of thinking which implied a sense of responsibility for the environment in a world with an increasing number of ecological catastrophes as consequence of human activities, while the activities are mostly branded "professional". These students pointed out that, as future transport engineers, they should take on a greater responsibility when it comes to environment protection, given the fact that there are increasingly intensive global effects of transportation: global warming, ozone depletion, depleting non-renewable resources and "since more than 60% of transportation is taking place in cities, every measure to decrease it on the local level contributes significantly to the preservation of non-renewable resources and quality of the environment on all levels". They go in detail while describing negative influence of transportation, by stating influences: fuels, lead, vehicle emissions, noise and vibrations, exhaust gas, transportation accidents, fires, explosions, etc. They wish to consider negative influences of activities they, as experts in the future, will be able to influence, i.e. to learn in advance both positive and negative possible influences of predicted activities in order to act preventively; that is to say to use principles of sustainable development, particularly sustainable transportation, as future associates in a team for drafting urban plans, relating to transport...

> "Aware of the fact that sustainable development is one of the basic aspects of all forms of quality actions and managing, and that the tendency for development and a healthy environment are firmly connected, I think that prevention and decrease of pollution are key elements of sustainable development." (Seminar papers of students, 2011.)

Later, through their final paper, they tried to explain their motivation for choosing the course.

> "Assessing the environment has two main goals: to minimize or avoid negative influence on the environment before they occur and to include environmental factors in the decision-making process. The assessment of the environmental issues in a certain project needs to be done as soon as possible, during the stages of planning and project proposal, so that the analysis is useful for decision-makers and the measures of decreasing negative effects are included in the plans. Timely and efficient assessments of the environment have a result in making decisions based on information which support sustainable development. There are following gains from the analysis of environmental influences and measures to minimize negative influences in the early planning stages: a chance to include the public in the decision-making process, increased protection of human health, sustainable use of natural resources, decreased risks of ecological catastrophes and increased responsibility of the authorities."

"Responsibility for quality of the environment, which means clean air, sufficient amount of quality and hygienically adequate drinking water, preservation of agricultural land, ecosystem and biological diversity; Achieving rational spatial organization, development and protection by harmonizing its use with possibilities and limitations in managing natural resources and created values, i.e. optimal managing and use of natural resources; Combating further degradation of the environment by determining the current state, priorities of protection and conditions of sustainable spatial use; Taking adequate preventive measures while establishing control systems of all forms of pollution; Raising and strengthening the level of ecological awareness, as well as information and education on ecological problems; Including the public in the decision-making process when it comes to measures of environmental protection."

"Until the emergence of man and his active relationship with nature, the living world was ruled by mutual and harmonic connection and dependence in an ecological balance. The violation of ecological balance is the result of human activity, i.e. the fact that human beings "took over" nature and began creating products. Knowledge about ecological laws, processes and occurrences in nature enriches humans to view nature as a whole. While organizing it according to their own needs, human beings must not bring its processes to a halt, nor disturb the functional balance which exists between living creatures and their environment. This can be avoided only by good knowledge of ecological laws and provided that humans act according to them. Therefore, ecological awareness represents a pattern of life, which respects and harmonizes the circulation of matter, spending of energy and renewal of life with natural laws, at the same time encouraging the standpoint that nature should be deprived of only as much necessary for providing basic human needs. That is why ecological ethics represents the ecological relation of man towards the environment, relating to the ethical relationship between the human/technosphere and the natural/biosphere. A new relationship towards the environment, as well as a transformation of the modern work sphere spirit becomes imperative. The concept of sustainable development offers the possibility of harmonic development."
(Seminar papers of students, 2011.)

What they had learned taking this course?

The majority of students we recognized and characterized as the "ecological" group had similar answers to this question. The replies were mostly based on the acquired width in grasping the term, methodology and importance of sustainable development; the majority of students mentioned the integral nature of the procedure as an important new addition to their way of thinking about sustainable development – the integral nature of the sociological, economical and morphological aspects. These students point out that they have learned to think "in a new way", which implies all of the previously stated – responsibility, consciousness, thinking ahead, prevention, integration. It seems that the most important new addition to their way of thinking is the fact that they have learned that sustainable development, besides thinking about ecology, also implies other aspects: morphological, economic and social.

"One can reach the conclusion that the three characteristic principles of sustainable development are based on ecological sustainability, socio-cultural sustainability and economic sustainability. Ecological sustainability provides development compatible with sustaining ecological processes, biodiversity and natural resources. Socio-cultural sustainability provides increasing human monitoring of their own lives, while economic sustainability provides successful economic development. The overlapping of global, regional and national, i.e. local dimensions of sustainable development is one of the main features of sustainable development."

..."A new relationship towards the environment, as well as transformation of the modern work sphere spirit, becomes imperative. The concept of sustainable development offers the possibility of harmonic development."
(Seminar papers of students, 2011.)

Some other answers to this question:

> "Every day, we face talk about environmental protection, but only after taking this Course I have acquired a broader knowledge about the issue. I have understood that it is important for the planet to be preserved both for current as well as future generations in an acceptable manner. As well as the fact that the current generation's right to using resources and a healthy environment must not jeopardize the same right for generations to come."

> "The importance of this subject lies in the fact that sustainable development represents integral economic, technological, social and cultural development, harmonized with the necessity of protection and advancement of the environment, enabling current and future generations satisfaction of their needs and improvement of quality of life."

> "By working on this subject, I have acquired necessary knowledge about harmful influences on the environment. I have often been a witness of talk about sustainable development in the media, without being aware what it actually concerns. This work has helped me in many ways, in a sense of understanding the meaning of sustainable development. As a future transport engineer, I think that environmental protection is very important, as well as future planning of the space we inhabit."

> "Through this paper, as well as during the entire Course, I learned about the term of sustainable development, as well as about national laws concerning the issue of environment (primarily the Law on Strategic Environmental Impact Assessment)."

> "The knowledge gained during this course can be used in determining potential positive and/or negative influence of human activities on the environment. With the knowledge about possible results of certain planning activities, it possible to perform corrections of plans or to take measures which can increase or decrease potential positive or negative influences on the environment. This knowledge can be implemented on the example of building a new transport route or enlargement of a part of the existing transportation network."

> "I have noticed that in Serbia, integral managing of the environment is not satisfactory; adopted laws and other regulation are insufficiently used, or not used at all, while the economic instruments of environmental protection remain underdeveloped. The existing technological processes are not adequate to the demands of environmental protection and the education in this field is insufficiently organized and programmatically directed. Due to the lack of an integral information system concerning the environment, the level of information on this issue remains insufficient, slow and inefficient."
> (Seminar papers of students, 2011.)

Currently, 18 students (third generation since the course was established, in 2009) are taking the course. Since this is an elective course, it is indicative that the number of candidates increases (in the first generation there were five students, in the second one there were 12) which could indicate that in addition to the growing interest about the topic in Serbia, candidates have recommended the course to their younger colleagues because it offers interactive, inspiring work and because it deals with a current issue.

Conclusion

- Serbia's version of city planning and managing features a problem which is reflected in the primarily declarative approach to the concept of sustainable development; the idea exists, but the level of incorporating it into the planning documentation is low. As there is a shortage of methodologies and concrete actions, the general public is insufficiently informed on the principles of sustainable development, possibilities of energy saving and importance of appropriate behavior, which is illustrated in the paper with help of statistics, questionnaires, and other examples dealing with energy wasting, private car use, etc.
- When it comes to *education* on sustainable development, there is an insufficient level of information and insufficient education and knowledge on the subject, which induces the essential idea to implement permanent education of professionals (starting with students of

relevant profiles, where the best future results are expected) in the field of city planning and functioning in Serbia. That way, the further education of citizens can be influenced, as well.

- The "Sustainable Development and Environment Protection" *course work* is organized in a traditional manner – theoretical and practical parts, along with active discussions and comparative analyses of examples. Within the theoretical part, the candidates acquire a broad image of the meaning, aims and principles of sustainable development and within their practical work; they produce an individual paper as a result of analysis of urban planning and environmental legislature in Serbia. In this paper, the authors have presented detailed description of the curriculum, along with the students' works: key themes presentations with their conclusions on what they believe they had learned through education in this course.
- General *goals* of the course are formed as a direct consequence of the position that the concept of urban sustainability has in Serbian society; the focus is mainly on raising the students' awareness about the necessity for change and also on motivating, encouraging and empowering them to use and implement their knowledge, and therefore act professionally in the field of urban planning.
- The initial ideas with which we started the course coincide with students' *conclusions* on the question what is/should be learnt during the course. At the end of course, the student – as a future young expert who will begin practical work with the required knowledge – is capable of defining and using sustainable development principles, especially in sustainable transport, e.g. transport engineer – an associate in the team for the development of urban plans and to argue a case in defence of principles established and used for achieving sustainable development of specific areas. Also, since they were previously mostly ecologically orientated, during this course the students are taught to think "integrally" – they are 'awakened' about the fact that environmental problems should be viewed holistically, in accordance with sustainable development, taking into account the effects and consequences on the economic, social and morphological levels, not just environmental.

  At this point, the authors would like to emphasize that it is of great *importance* that the subject relates to transport engineers who, by taking this course, implement a certain connectivity and an integrated approach, because otherwise, through their education, they tend to observe transport regardless of other city functions. Since there are usually other profiles of researchers in the team for SEIA, the education for transport engineers is necessary.

  This work is significant and *useful*. First of all it is useful in Serbia, with the previously described specific circumstances, concerning a declarative approach and insufficient implementation. However, it is also important on a broader scale – for other societies with similar backgrounds, preconditions/social frameworks and a similar position of sustainable development within the practice of urban planning, in which even professionals do not/or rarely apply these principles, let alone the general public. Therefore, it is necessary to consequently change the way of thinking and acting through education of students, professionals and the public. Thus, education of future urban planning professionals is imperative, in order to understand the need for involving the public in all phases of planning, the need for multidisciplinary cooperation and confirmation of the quality of implemented plans.
- New answers, from the third generation of students will be received in March 2012. Regarding the future of the course, aim is to adjust the manner of work, methodologies, as well as topics with the current situation and changes in society, in terms of knowledge, opinions and needs.

References

Bajić-Brković, M. (Ed.) (1999), *City and Sustainability* (Grad i održivost), Faculty of Architecture, Belgrade, Serbia.

Bogdanović, R. (Ed.) (1999), *Strategies and Methods for Improving the Quality of Compact Cities (Strategije i metode za unapredjenje kvaliteta života u kompakt gradovima)*, Urban Planning Association & Town Planning Institute of Belgrade, Serbia.

Bogdanović, R. (Ed.) (2000), *Sustainable Development of Settlements in Serbia: Principles and Practice (Principi i praksa održivosti u razvoju naselja u Srbiji)*, Urban Planning Association, Belgrade, Serbia.

Bogdanović, R. (2003), *Toward a Sustainable City; Strategies and Methods for Improving the Quality of the Environment in Cities*, Faculty od Transport and Traffic, Belgrade, Serbia.

Janić, M. (1996), *Sustainable Development of Countries in Transition (Održivi razvoj zemalja u tranziciji)*, Juginus, Beograd, Srbija.

Jones, E.C. (2005), *Strategic environmental assessment and land use planning: an international evaluation*, Earthscan, Taylor & Francis Group, UK, USA.

Law on environmental impact assessment (Zakon o proceni uticaja na zivotnu sredinu) (2004, 2009), Službeni glasnik RS. No. 135/04 and 36/09.

Law on planning and construction (Zakon o planiranju i izgradnji) (2009, 2011), Službeni glasnik RS. No. 72/2009 and 24/2011.

Law on strategic environmental impact assessment (Zakon o strateskoj proceni uticaja na zivotnu sredinu*)* (2004, 2010), Službeni glasnik RS. No. 135/04 and 88/10 (in English at the: http://www.putevi-srbije.rs/strategijapdf/zstratzseng.pdf (accessed 18 January 2012).

Leal Filho, W. (2011), About the Role of Universities and Their Contribution to Sustainable Development. *Higher Education Policy* 24: 427-438. Available at: www.palgrave-journals.com/hep (accessed 9 November 2011).

Politika /daily newspaper/ (2006), No. 43, 12 December, p. 8.

Sadler, B., Aschemann, R., Dusik, J., Fischer, T., Partidario, M. and Verheem, R. (Eds.) (2010), *Handbook of Strategic Environmental Assessment*, Earthscan, Taylor & Francis Group, UK, USA.

Schmidt, M., Joao, E. and Albrecht, E. (Eds.) (2005), *Implementing Strategic Environmental Assessments. Series: Environmental Protection in the European Union*, Vol. 2. Springer-Verlag: Berlin.

Seminar papers of students (2011), from master Course of the Faculty of Traffic and Transport Engineering, University of Belgrade, Subject "*Sustainable Development and Environment Protection*", class of 2010/11, winter term; ending quotes are from the seminar papers of students (the order they are quoted): Natasa Markovic, Tijana Maric, Vladimir Pavlovic, Nebojsa Micic, Marko Šipovac and Goran Dakić; all remaining quotes are from the seminar paper of student Tanja Parezanovic.

Serbian Town Planners Association and ProAktiva, Macedonia (2004/2005) *Project: Integrated Approach for Solving Traffic Related Problems from Sustainable Aspect*, Supported by the Regional Environmental Centre for Central and Eastern Europe with financial support from the European Commission, Web site of the project: http://www.uus.org.rs/sh/projekti/saobracaj (accessed 18 January 2012).

Serbian Town Planners Association (2006), *Project: Improving the Practices of Public Participation Within the Spatial and Urban Planning Field in Serbia Relating to the Environmental Matters*, Supported by the Regional Environmental Centre for Central and Eastern Europe with financial support from the Ministry of Foreign Affairs of the Netherlands, Web site of the project: http://www.uus.org.rs/sh/projekti/aarhus (accessed 18 January 2012).

– (2007/2008), *Project: Promotion of Sustainable Urban Lifestyle*, Implemented by means of the program SECTOR Regional Centre for Environment, the program funded by Swedish International Development Cooperation Agency (SIDA), Web site of the project: http://www.uus.org.rs/sh/projekti/ee/ (accessed 18 January 2012).

The Government of the Republic of Serbia (2008), *National Sustainable Development Strategy for Serbia*, Belgrade, Serbia.

Therivel, R. and Wood, G. (2005), Tools for SEA. In Schmidt, M, Joao, E. and Albrecht, E. (Eds.), *Implementing Strategic Environmental Assessment, Series: Environmental Protection in the European Union*, Vol. 2, Springer-Verlag, Berlin, pp. 349-363.

Therivel, R. (2010), *Strategic Environmental Assessment in Action*, Earthscan, Taylor & Francis Group, UK, USA.

UNECE (1998) *Convention on access to information, public participation in decision-making and access to justice in environmental matters*, Aarhus, Denmark.

UNEP, Cities Alliance, and I.C.L.E.I. (2007*), Livable Cities: The Benefits of Urban Environmental Planning*, The Cities Alliance, Washington.

Webinar /Earthscan (2011), *Sustainability Education and Training.* 27 January (speakers: Desha, C., Prof. Sterling, S. and Prof. Murray, P.) On Earthscan web-site: http://sd.defra.gov.uk/2011/01/webinar-sustainability-education-and-training.

World Nuclear Association web site – issues:

Energy Balances and CO2 Implications.
http://world-nuclear.org/info/inf100.html (accessed 24 January 2012).

Renewable Energy and Electricity,
http://world-nuclear.org/info/inf10.html (accessed 24 January 2012).

Sustainable Energy,.
http://world-nuclear.org/info/inf09.html (accessed 24 January 2012).

Chapter 55

The Baltic University Programme: Twenty Years of Experience

Linas Kliucininkas, Lars Rydén and Paula Lindroos

Abstract

This paper provides insights of international cooperation between universities towards sustainable development in the Baltic Sea region. The authors present experiences gained during twenty years of the Baltic University Programme's (BUP) activities in the field of sustainable development studies. The introduction of the concept of sustainable development in many higher education institutions in the Baltic Sea region began in the last decade of the 20th century. The BUP was one of the first in the region to raise understanding and offer courses in sustainable development. The programme focuses on questions of sustainable development, environmental protection, and democracy in the Baltic Sea region. The aim is to support the key role that universities play in democratic, peaceful and sustainable development. This is achieved by developing university courses, and by participation in projects in cooperation with authorities, municipalities and other social partners. The courses consist of lectures, seminars, excursions, and case studies; all accompanied by a series of course materials. The individual parts of course materials can be studied as separate courses.

Introduction

Regional cooperation[*] between universities involved in education for sustainable development (ESD) has large advantages. International regional cooperation provides prerequisites for universities to learn that sustainability is part of their core business to encourage them to begin taking action to incorporate it in their missions (Hesselink et al. 2000). Students are provided with the broad concept of sustainable development and are acquainted with the best practices from the region; they relate their future professions not only with the development issues of their home country, but also with the long-term development of the region.

ESD in higher education is vital in those learning and clarifying processes which help to support future actions, learning and empowerment in societies. Authors from the Regional Centre of Expertise (RCE) (Skane, Axelsson et al. 2008) argue that initiation of the idea of sustainable development as a cognitive, competence or experiential learning will require development and restructuring of sustainable development knowledge itself as well as elaboration of concrete applications.

The role of dedicated individuals or key actors acting as forerunners and norm setters is crucial when starting regional cooperation between the universities. The implementation of joint activities would not be possible if the institutional involvement was not present in creating norm-supporting structures at the local universities. The norm support given by the leaders of the universities is necessary for local implementation of developments (Wickenberg 2006).

---

[*] In this paper international cooperation refers to the wider geographic entity.

Fig. 1: The Baltic University Programme (BUP) network

The Baltic University Programme (BUP) was established in 1991 as a regional university network. At present the network unites more than 200 universities and other higher education institutions in 14 countries which belong to the Baltic Sea drainage basin (see Fig. 1). The network is coordinated by the Baltic University Programme Secretariat, as part of the Centre for Sustainable Development at Uppsala University, Sweden. The programme plans, produces and coordinates undergraduate and graduate courses using a modern educational setting, and runs joint research projects. The chosen subjects are of common concern for the entire region, and international cooperation is of key importance. At present almost 9,500 students participate in the programme each year (BUP 2012).

The network widely should be Information and Communication Technologies (ICT) for communication between course groups. During the first years, satellite TV was used to reach all partners, and to create interactivity. Today, video, audio and computer conferencing connects study groups in the network. Case studies, such as text as and videos, illustrate the courses and give regional perspective. English is the common language, and the courses are problem-oriented and thematic. Several courses are available in distance mode, and others have considerable web support.

The programme regularly offers its students conferences, summer courses and sailing cruises to promote contacts, international understanding and democracy development. For the close to 500 active academic teachers and researchers the programme offers teachers conferences and multiple other occasions to meet.

The programme cooperates with municipalities and national authorities in the region in applied projects. This model for cooperation benefits both professionals at municipalities and authorities, as well as universities, and shows the direction for future implementation of sustainability strategies in the region.

Considering that implementation of the principles of sustainable development is impossible without active public support and participation, the BUP also has a strong commitment towards public education and promotion of environment friendly lifestyle.

Studies for sustainable development: undergraduate level

In January 1991 the BUP invited university teachers from eastern and western Baltic Sea region countries to exchange views on further development of the region. It should be kept in mind that event took place after the fall of the Berlin Wall, after the victory of Solidarity movement in Poland and after the declarations of independence of the Baltic States. It seems that the only agreeable topic for further cooperation was environmental degradation of the Baltic Sea. Thus, BUP in cooperation with the partners at the universities in the region, took the initiative to develop the course *The Baltic Sea Environment*. The course was started in 1991 and dealt with the environmental situation of the Baltic Sea and its drainage area, conveyed an overview of topics important for understanding and improving the situation. It consisted of ten two-hour satellite TV sessions, now available on video, and for each session a 35-page booklet. The last topic in the series was 'The Prospect of a Sustainable Society', which could be considered as a pioneering undertaking of further developments within the Baltic University Programme.

In 1997 the course *A Sustainable Baltic Region* (7.5 ECTS credits (European Credit Transfer and Accumulation System)) was introduced to the university teachers; the conference bringing together more than 100 university teachers from the Baltic region was held at Kaunas University of Technology, Lithuania. The course deals with sustainable use and management of natural resources, the long-term protection of our environment and the sustainable organizations of human societies from the level of households to those of municipalities, cities and countries. The course has been organized as ten "packages". There is an introductory part explaining the fundamental notions of sustainability and its historical background. Then follow four themes of more technical and natural science character. The second part consists of four themes of social science character and the course is finished by a topic of implementation and Agenda 21. Important issues covered are energy, material flows, economy and ethics, as well as industry, agriculture, transport, community development and Agenda 21 practices.

A great number of academic contributors provided examples of good sustainability practices from all Baltic Sea region countries. Till now the course is one of the most popular amongst those developed by the BUP (Sörlin 1997; Salay 1997; Karlsson 1997; Ebbersten and Bodin 1997; Strahl 1997; Tengström and Thynell 1997; Andersson and Berg 1997; Zylicz 1997; Rydén 1997).

The course material consists of ten 50 pages booklets, the TV series Mission Possible with over 80 individual reports and a database useful for student work. A computer conferencing system, First Class, was used for direct interaction between the students in different countries and locations throughout the region. The TV series of ten 45 minute programmes called Mission Possible was produced by a consortium of 17 public TV companies in ten countries in the region. The TV production consists mostly of reports on cases of good practice. It shows how new models for sustainable management is developed in cities, at farms, in factories etc. in the countries in the region.

The TV series was broadcast to the universities over satellite during spring 1997. Whole or parts of the productions have also been broadcast to the general public in Finland, Latvia, Lithuania, and Poland, as well as on a number of cable TV networks.

In 2011 there was an initiative to update the course materials and make it available on the BUP Internet portal. In spring term 2012 the course materials will be tested by a small group of universities, and after receiving feedback, the course materials will be offered to the whole BUP network. In its nature the updated course will be very interdisciplinary and will cover areas from humanities and social sciences to natural sciences and technology. The basic course on sustainable development will consist of twelve thematic packages (historical background and concepts, energy and climate change, resources and limits to growth, urbanization, sustainable production and consumption, Life and land, food and fibre, mobility, human welfare and sustainable life style, political incentives, economy, changes and management, education) with each one subdivided into several parts. Each part will have an introductory text followed by links to material for further studies. The material will include book chapters, films, power point presentations, etc. The material will be divided under three headings: basic, medium and advanced. Dependent on the context of course within the study programme, the teachers responsible will be free to decide exactly what to use from this wealth of information in the particular course.

Studies for sustainable development: postgraduate level

The introduction of environmental management approaches in industry and other organizations are of great importance for developing a healthy environment. Environmental management is seen a general approach encompassing a long series of issues such as management tools, economic and legal policy instruments, industry related tools such as cleaner production and waste minimization, but also more traditional tools as Environmental Impact Assessment, as well as eco-design, Life Cycle Assessment, certification and green labelling.

However, well-functioning possibilities for learning and training environmental management is lacking at many universities and in particular in eastern European countries. The BUP has developed *Environmental Management* course materials, which consist of four parts: Policy Instruments for Environmental Management (7.5 ECTS), Cleaner Production and Technologies (7.5 ECTS), Product Design and Life Cycle Assessment (7.5 ECTS) and Environmental Management Systems and Certification (7.5 ECTS) (Klemmensen et al., 2007; Nilson et al., 2007; Zbicinski et al., 2007; Weiß and Bentlage, 1997).

This set of four courses address management in all kinds of organizations, but they are of particular relevance for the industrial sector. They are well adapted when developing skills for professionals. University-industry contacts are used to allow visits on a regular basis to specified industries as part of the studies, and in general stimulate university-industry cooperation.

In the industrial sector, technologies for removing pollutants from effluents and flue gases are today enforced by increasing fines and taxes for pollution, and solid waste deposition on landfills, but in the long run this approach has limitations. A simple fact is that too much waste is produced. Technical approaches, which address this fundamental fact, include eco-design and cleaner production. The management tools include the systems of ISO 14001 and EMAS introduced to an ever larger extent.

The Sustainable Water Management course focuses on issues of sustainable use of water and water resources in the Baltic Sea region. The course materials could be studied either as three separate courses or as one larger course. The full course, Sustainable Water Management (15 ECTS), consists of three parts: The Baltic Waterscape (4.5 ECTS), Water Use and Management (4.5 ECTS) and River Basin Management (6 ECTS) (Lundin 2000a; 2000b; 2000c).

The courses intend to impart professional knowledge on the state of water resources in the Baltic Sea region, their present use and management, and the challenge of creating sustainable water use practices in the different countries in the Baltic Sea catchment area. With the use of video, audio and data conference systems the participating students engage in common seminars.

The courses are organized as an international cooperative program among universities in the Baltic Sea region. Since 2000 the courses are offered as campus courses by the participating universities. However, since 2003, there is also as an internet based distance learning course at Uppsala University. The latter course is open to all students anywhere in the world.

Students should preferably have a background in e.g. hydrology, geology, physical geography, agronomy, soil science, forestry or environmental engineering; other study areas e.g. biology, law with specialization on water issues, etc., may also be relevant. The water management courses are best placed in the final part of the studies. Students are recommended to have studied previous Baltic University Programme courses, in particular The Baltic Sea Environment.

The Sustainable Community Development and Urban Planning course (15 ECTS) is intended for students with previous courses in relevant fields, such as architecture, landscape architecture, geography, or environmental science. The course focuses on sustainable community development, and urban planning with special emphasis on the conditions in the Baltic Sea region. The course covers both social science, architectural and cultural aspects of community development and intends to convey practical knowledge. The course consists of three parts: The City (4.5 ECTS), Building sustainable community (4.5 ECTS), Sustainable urban management (4.5 ECTS) (Granvik and Rydén 2002; Rydén 2003a; 2003b; 2003c). The first part describes the urbanisation of the Baltic Sea region and the basic functions of a city. City administrative cultures are described and discussed; the focus is given on the concepts of sustainability and development. The second part provides understanding on how to build sustainable cities from the building blocks to the neighbourhood. The focus is on sustainable architecture, social space, and principles of urban planning. The rebuilding of cities is treated with reference to former industrial, residential, and military areas. The third part describes the totality of urban management with reference to land, buildings, and inhabitants. The sustainability indicators in the environmental, economic, and social fields are included. Economic, cultural, and social development of a city is treated with several examples. International aspects including the Habitat principles and Agenda 21 conclude the course.

The Baltic University Urban Forum (BUUF 2000) is a platform for cooperation between cities/towns and universities in the Baltic Sea region to promote sustainable development.

During the period 1999-2002 a total of 35 case studies were conducted in the cities, concerned with urban planning, democracy and environmental management. These were published in a series of reports and TV cassettes. In its present phase the BUUF project aims to develop strategies for sustainable development for cities and towns. 20 cities and towns, and 15 universities in nine countries have taken part in the project. Through the Union of Baltic Cities project results were available for an additional 100 cities in the region.

University teacher support

The BUP has provided materials to facilitate teaching of the courses. The materials include teachers' guides, ECTS documents, films, MS PowerPoint presentations and other.

The courses provide a holistic view to the perception of sustainable development and require an interdisciplinary nature. Co-operation within the university but also with other institutions in society, e.g. local governments, trade and industry and associations is needed. High priority is given to connecting theoretical research with practical reality in creating a sustainable society. Another aspect is to learn from each other, to spread "best practices", such as from projects that have been successful. This way of thinking and working is in line with the decisions from the United Nations conference on Environmental and Development (UNCED) in Rio de Janeiro 1992. Discussions about sustainable development are new and not very well known in our societies. The innovative research and technology developments should be "translated", understood and presented both to in academic circles and to society. Universities need to discuss how to deal with this concept. Universities and other institutions of higher education, which have expertise in all fields of research, have the capacity to develop an understanding of the concept of sustainable development and apply it to specialties. In the field of sustainable development not only understanding and theoretical knowledge is needed, but also practical skills are essential.

Although there is a common framework for the courses, each university may develop individual modifications. The course groups have flexibility to plan a course that suits students best. Academic teachers from several disciplines and expert teams from local and national departments are usually involved in education, to handle the interdisciplinary material. The guidance gives ideas on how to teach the course, which methods to use, what to discuss, whilst also providing a list of student tasks, reference literature, etc. Suggestions for teaching are given for each week of study.

Education for Change (EduC) is a BUP project aiming at enhancing the competence of both school and university teachers on how to implement the concept of sustainable development in their education. The method developed as an effective means of informal education is well known as Study Circle (EduC 2008).

Conclusions

The Baltic University Programme could serve as a model for regional co-operation between universities and societies in promotion putting the principles of sustainable development into practice.

- Regional cooperation between universities involved in education for sustainable development (ESD) has large advantages. These include exchange of experiences, sharing development and use of educational material, cooperation in curricula development, and mobility of students and teachers.

- At the same time there are obstacles in getting regional cooperation to function fully, such as the language barriers, different traditions in education both with regard to content and how to study, and different structures such as study periods (semesters) and allocation of teaching hours in the time schedule.
- Cooperation between universities offering education in sustainable development in the Baltic Sea region is confronted with the differences between eastern and western traditions and practices. The Baltic University Programme has been addressing two of these differences. The first is the attitude towards reactive and proactive approaches to the issues of sustainable development; the second the different practices with regard to close cooperation with social partners.
- Increase of the role of research and more effective use of research and development (R&D) results in the education process could be seen as a driving force to foster culture of understanding and implementing principles of sustainable development.
- Implementation of the principles of sustainable development is impossible without active public support and participation. The BUP has a strong commitment towards public education, cooperation with social partners and promotion of environmentally friendly lifestyle.

A regional approach to sustainable development promotes not only increased international collaboration across a variety of disciplines and sectors but also provides prerequisites for regional shift.

References

Andersson, H. and Berg, P.G. (1997), *Community development – sustainable cities and habitation.* Uppsala University, Uppsala.

Axelsson, H., Sonesson, K. and Wickenberg, P. (2008), Why and how do universities work for sustainability in higher education (HE)? *International Journal of Sustainability in Higher Education* 9 (4), pp. 469-478.

Baltic University Programme (BUP) (2012). Available at: www.balticuniv.uu.se (accessed 24 February 2012).

Baltic University Urban Forum (BUUF) (2000). Available at: www.balticuniv.uu.se/index.php/teaching-materials (accessed 24 February 2012).

Ebbersten, S. and Bodin, B. (1997), *Food and fibres – sustainable agriculture, forestry and fishery.* Uppsala University, Uppsala.

Education for Change (EduC) – *Teaching Materials* (2008). Available at www.baltic univ.uu.se/index.php/component/content/article/124-education-for-change-teaching-materials/336-the-educ-teaching-materials (accessed 24 February 2012).

Granvik, M. and Rydén, L. (Eds.) (2002), *Basic Patterns of Sustainability.* Baltic University Press, Uppsala.

Hesselink, F., van Kempen, P.P. and Walls, A. (Eds.) (2000), *ES Debate: International debate on education for sustainable development.* IUCN Publication Services Unit, Cambridge.

Klemmensen, B., Pedersen, S., Dirckinck-Holmfeld, K.R., Marklund, A. and Rydén, L. (2007), *Environmental Policy – Legal and Economic Instruments.* Baltic University Press, Uppsala.

Karlsson, S. (1997), *Man and materials flows – towards sustainable materials management.* Uppsala University, Uppsala.

Lundin, L.C. (Ed.) (2000a), *The Waterscape.* Uppsala University, Uppsala.

– (2000b), *Water Use and Management.* Uppsala University, Uppsala.

– (2000c), *River Basin Management.* Uppsala University, Uppsala.

Nilson, L., Persson, P.O., Rydén, L., Darozhka, S. and Zaliauskiene, A. (2007), *Cleaner Production Technologies and Tools for Resource Efficient Production.* Baltic University Press, Uppsala.

Rydén, L. (1997), *The foundations of sustainable development – ethics, law, culture and the physical boundaries*. Uppsala University, Uppsala.

– (2003a) (Ed.), *Building and Re-building Sustainable Communities*. Baltic University Press, Uppsala.

– (2003b), *Public Participation and Democracy*. Baltic University Press, Uppsala.

– (2003c), *Urban Environmental Management*. Baltic University Press, Uppsala.

Salay, J. (1997), *Energy – from fossil fuels to sustainable energy resources*. Uppsala University, Uppsala.

Sörlin, S. (1997), *The road towards sustainability – a historical perspective*. Uppsala University, Uppsala.

Strahl, J. (1997), *Sustainable industrial production – waste minimization, cleaner technology and industrial ecology*. Uppsala University, Uppsala.

Tengström, E. and Thynell, M. (1997), *Towards sustainable mobility – transporting people and goods in the Baltic Region*. Uppsala University, Uppsala.

Weiß, P. and Bentlage, J. (2007), *Environmental Management Systems and Certification*. Baltic University Press, Uppsala.

Wickenberg, P. (2006), Norm supporting actors and structures at the very local level of implementation of sustainable development in higher education. In Holmberg, J. and Samuelsson, B.E. (Eds.), *Drivers and barriers for implementing sustainable development in higher education* (111-120), pp. 111-119, UNESCO, Paris, France.

Zbicinski, I., Stavenuiter, J., Kozlowska, B. and van de Coevering, H. (2007), *Product Design and Life Cycle Assessment*. Baltic University Press, Uppsala.

Zylicz, T. (1997), *Ecological economics – markets, prices and budgets in a sustainable society*. Uppsala University, Uppsala.

Chapter 56

Working with Stakeholders for a Successful Research on Sustainable Development

Alline Storni, Valentina Castellani and Serenella Sala

Abstract

Sustainability science explores the interactions between human activities on Earth's life support systems. Unless we understand these interactions, we will not be able to design a path towards sustainable development. The education system in relation to sustainability is based, partly at least, on the following three aspects: knowledge, training and orientation. The ultimate goal is thus not the mere transfer of knowledge but the development of empowerment and capacity building.

In this context, the research activities focused on bridging the gap between theory and practice, from problem understanding towards solution developing with a methodology covering three areas – content, involvement and efficacy – in order to innovate the current approach to sustainability education and contribute to the debate on sustainable development in higher education.

The paper aims at presenting the key drivers of education evolution, from environmental to sustainability education. The criteria for sustainability education are defined, and the role of the university as a key stakeholder is explored through the analysis of case studies of multi-stakeholders' education activities developed by the Research Group on Sustainable Development (GRISS) of the Department of Environmental Science at the University of Milano Bicocca in Italy.

Introduction – Sustainability science to support sustainability education

Sustainability science focuses on the dynamic interactions between nature and society (Clark and Dickson 2003). As a new discipline, it points the way to a sustainable global society by facing challenges that existing disciplines have not yet fully addressed. It requires the evolution of two paradigms: a scientific one, towards greater integration of disciplines for problem solving, and a social one, towards a new culture on natural-humankind interaction (Sala and Castellani 2010). In this context, it has been widely recognized that collaboration and partnerships with and across different stakeholder groups are critical conditions for sustainability science (SSc) and its real-world impacts (Blackstock et al. 2007; Spangenberg 2011; Talwar et al. 2011, Wiek et al. 2012). In fact, applying SSc means developing solution-oriented research activities aimed at problem solving. Education for sustainability plays a key role in the evolution of the current cultural context, based on a thorough technical-scientific knowledge, embedded in a systemic and holistic environment that allows people to grasp the cause-effect relationship between elements that compose it, and between environment, economy, society, culture, traditions and history.

This has to be reflected in the evolution of sustainability: from mere environmental education (compartment-oriented, based on observation/knowledge of environment) towards ecological (function of ecosystems-oriented, based on observation but also on practical activities) and sustainability education (SE) (highly interdisciplinary, behaviour-oriented, multidimensional and multi-stakeholders). The growing interest in "competence" and "competency" in sustainability education circles is based partly on the presumed lack of relevance of current educational provision and the need to produce "change agents" within society to address key challenges of sustainability science (Kates et al. 2001; Kates 2011).

The paper aims at presenting the key drivers of education evolution, from environmental towards sustainability education, focusing on the role of universities and research. Criteria for SE are defined, and the role of universities are explored as a key stakeholder, through the analysis of case studies of multi-stakeholders' education activities developed by the Research Group on Sustainable Development (GRISS) of the Department of Environmental Science from the University of Milano Bicocca in Italy.

Environmental, ecological and sustainability education from Rio 92 to Rio+20

From environmental to ecological education

The potential of environmental education (Env-E) as an instrument of cultural change for a sustainable future has already been known for some time, but in order that this is widely applied, we should expand the horizons of education, introducing a series of multidisciplinary and global perspectives, which can affect thinking, knowledge, but also behaviours and social relations (Falchetti and Caravita 2005). Specifically, the emphasis should be on the difference between Env-E and ecological education (Ec-E). Although often used interchangeably, they are not actually. Ec-E exceeds the mere description of environmental compartments and the transfer of knowledge. It provides the integration of some basic ecological principles, both in content and methods, that should drive the development of new paradigms provided below, based on the so-called "eco-literacy". With eco-literacy, the belief is that the current ecological crisis is a crisis in part related to educational systems and contents of modern societies (Stone and Barlow 2005) and it teaches students to identify connections between apparently distinct problems, to imagine ""patterns" instead of disjointed issues, to live together in learning communities in interrelation with the whole web of life.

The basic concepts of Ec-E refer to the patterns and processes by which nature sustains life: networks, nested systems, cycles, flows, processes of development and evolution, and dynamic balance (Centre of Ecoliteracy 2009; Capra 1996).

These basic ecological principles have to be interconnected also with social and economic dimensions (multidimensional) and to be integrated in education/communication to different stakeholders (multi-stakeholders/civil society).

From ecological education to sustainability education and a new concept of "citizenship"

The idea of sustainability covers at least four dimensions (IUCN/UNEP/WWF 1991): environmental issues are covered as well as economic, social and institutional ones. This vision requires, therefore, involving the social and economic system in the transition towards sustainability. The transition from education to Env-E and sustainability is embodied is this involvement. To increase the overall ecological awareness means to stem "the careless and rapacious attitude toward life" (Lorenz 1974) that characterizes the archetypal compulsive consumer citizen. If we imagined the current economic crisis as a limiting factor of the development, and we took for example the dynamics of ecological systems to understand what will happen to the "population", natural selection would teach us that the strongest will be the ones who developed major abilities to adapt with new resources, in order to be able to continue to meet their needs. If this occurs in nature by instinct, in a society of men it is also the result of knowledge and of new knowledge.

It may seem very idealistic, but "knowledge" has to become the keystone, and training, education and research in the science of ecology and sustainability the commanding heights of the evolution of our era, the Anthropocene.

Since the primary characteristic of the biosphere is to sustain life, a sustainable human community should be such as to use its cultural, social and technological potential to honour, support and cooperate with nature in its role of supporting life (Centre of Ecoliteracy 2009).

That's why ecological principles that ecosystems have developed to support the web of life are the basis for developing a new concept of citizenship: if to know the "Constitution" is one of the elements to be a conscious inhabitant of a nation, to understand the ecological principles is one of the elements to be a conscious inhabitant of the Earth and to share transcultural common values.

From sustainability education to action: how to promote sustainable lifestyles

The priorities for the sustainability policy agenda have evolved since Rio 92 both in regard to the main themes and to the strategies to be adopted. As far as education is concerned, what is more evident during this period is a shift concerning responsibility for action: during the 90s, sustainability policies were more concerned about the environmental impacts from the production side (e.g. air and water pollution, environmental hazards from chemicals and so on), while since the Johannesburg meeting in 2002 more attention was paid to consumption and to the impacts coming from the use stage in the life cycle of products, that can be influenced also by the consumers' choices and lifestyles (Clark 2007; Scott 2009). Following this call, a framework of programmes on sustainable consumption and production (Marrakech Process and UNEP 2010) and UN Decade of Education for Sustainable Development (DESD 2005-2014) was launched. This builds on the increasing awareness of the role of formal learning processes in providing knowledge, awareness and competencies aimed at enabling individuals and social groups to be the key actors of change towards more sustainable consumption behaviour.

Now, as we are in the Rio+20 era, there is an emerging demand of concrete actions (Schrader and Thögersen 2011) and stronger implementation of the existing policies and international resolutions and for concrete actions to meet the targets set in the previous year (AA.VV. 2011).

UNEP (2010) states that "Education for Sustainable Consumption (ESC) is essential to empower individuals and social groups with appropriate information on the impacts of their daily choices as consumers, as well as for workable solutions and alternatives". It implies that there is the need to translate the results of scientific research into useful and understandable guidelines for sustainable lifestyles grounded on scientifically based information. Therefore SE cannot be performed or developed only by researchers and professionals in the education sector, but needs to be developed jointly also with scientific experts that can provide information about global changes and related sustainability solutions. Stronger cooperation between environmental scientists and other players (experts from several disciplines and decision makers and representatives of different part of society) is considered essential (Agenda 21, Chapter 35, Tamburini 2005) and this requires also adequate intermediaries. There is a need for professionals who have relationships with key stakeholders and are able to explain information to them so that it easily translates into action.

The environmental scientist cannot limit his work to the definition of the problem (e.g. climate change), which in turn leaves the decision as to how to solve this issue entirely to others. The solution has to be evidence-based, following an environmental and socio-economic perspective, promoting a mutual and continuous exchange both among experts from several disciplines and among experts and stakeholders (Castellani 2010; Reid et al. 2010, Turnhout et al. 2007).

The education system in relation to sustainability should be based, therefore, at least on three aspects:

- knowledge (ecological principles on the one hand, and environmental impacts of products and processes on the other hand, including socio-economic dimension)
- training (capacity building to work for problems, understanding the complexity and development of creativity for the development of a wide variety of solutions)
- orientation of actions (based on the stimulation of cognitive activities that the subjects found to build their knowledge and to be aware of their behaviours).

University and society: experience of multi-stakeholders relationship

There has been a growing awareness in national and international policies of the need to integrate sustainability into both business and educational arenas. To reach this objective a multi-stakeholder network is essential. Actors from educational institutes, business, civil society and public authorities come together in order to find a common approach to an issue that affects them all and that is too complex to be addressed effectively without collaboration (Roloff 2008).

In this context, GRISS focused its research activities on bridging the gap between theory and practice, from problem understanding towards solution. The methodology developed by GRISS regards three areas, innovating the current approach to SE and contributing to the debate on sustainable development in higher education: content, involvement and efficacy.

On the *content* side, mainstreaming sustainability and life cycle thinking to enhance the completeness of scope in formal and informal education, in line with the SE principles discussed above;

On the *involvement* side, tailoring the content to target stakeholders, boosting bidirectional relationship between research and society. To follow this aim, during the last years GRISS established multi-stakeholder bidirectional relationships in its activities, within formal and informal education. In fact, universities can set up feasible applied researches taking into account the complexity of society and the interactions that may affect the feasibility of proposed solutions and stakeholders can take decisions which are evidence-based and that are the result of scientific assessment.

Fig. 1: Multistakeholder bidirectional relationship in the activities of GRISS.

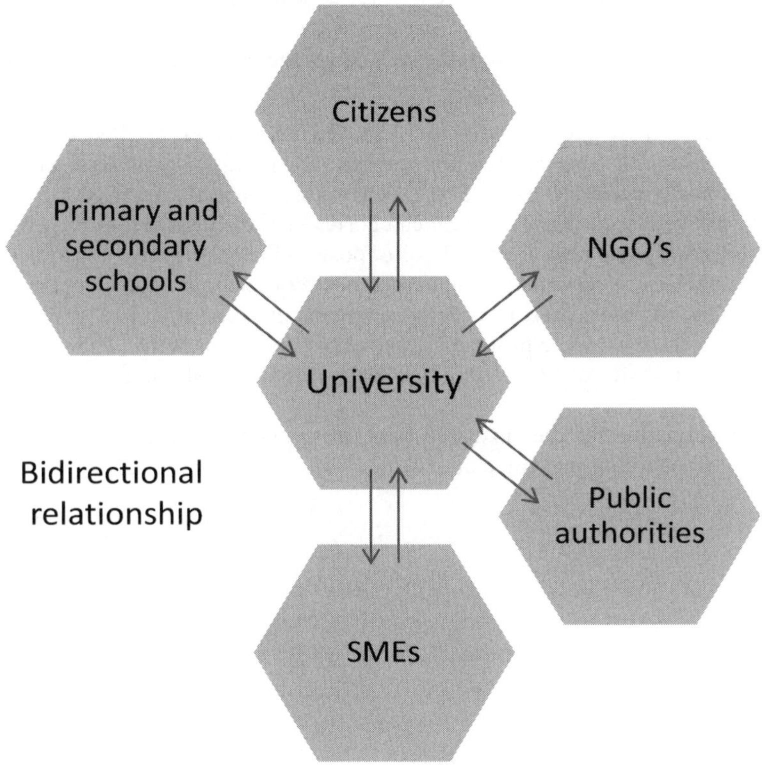

Undergraduate and postgraduate students, as well as researchers, are within the "university" hexagon. The relationship is bidirectional because the results of the activities were used as a basis for further applied research both on specific sustainability issues and on education.

On the *efficacy* side, monitoring results against established criteria of evaluation. Relying on the three aspects outlined in the section "From sustainability education to action: how to promote sustainable lifestyles" and on the recommendations for education for sustainable consumption (UNEP 2010) about what aim SE should have today and how it should be structured and targeted. With this in mind, the authors propose the following criteria against which to evaluate SE initiatives:

A: Inter-disciplinarity, trans-disciplinarity and ability to transfer the key messages of the different disciplines (e.g. carrying capacity and finiteness of resources, life cycle thinking, equity, etc. Enhancing cooperation between professionals from diverse disciplines in order to develop integrated approaches to education for sustainable consumption (rec 5, UNEP 2010).

B: Collective responsibility of all the stakeholders, being producers or users, researchers or students, since the entire civil society has to evolve and move towards sustainability and trans-cultural values.

C: Creation of a solid interface between experts and education targets, also involving appropriate intermediaries for each kind of target; strengthening connections between researchers, lecturers, teacher trainers and socio-economic actors and stakeholders (rec. 4, UNEP 2010).

D: Ability to go beyond theory and to provide guidelines to translate it into action, adding also practical examples during education activities, in which stakeholders are directly involved as actors (rec. 10, UNEP 2010).

E: Choice of the most suitable communication language and means to reach and to involve the targets of education.

According to these three methodological areas, GRISS coupled applied research activities with scientific dissemination tailored and targeted for multi-stakeholders. The multi-stakeholders initiatives consist of a series of educational projects focused on sustainability science and life cycle thinking. The main categories of stakeholders involved were: students ranging from primary and secondary school to postgraduates, citizens-consumers, enterprises (especially SMEs), NGOs and public authorities. For each of these categories, ad hoc activities have been experimented in order to increase awareness and to provide a basis for decision making on sustainable production and consumption patterns. The ultimate goal was thus not the mere transfer of knowledge but the development of empowerment and capacity building.

Activities are briefly described in Table 1 and are evaluated against the strategic criteria for education for sustainability presented above.

Table 1: Overview of multi-stakeholders activities carried out by GRISS

Target stakeholder	Activity	Primary objectives	Outcomes	Other Stakeholders involved	Criteria				
					A	B	C	D	E
Postgraduate students	Master Gesal. Master on sustainable development at the local scale (planning and management)	Developing interdisciplinary competence and skills and creating an expert's network able to connect the academic with the public and the private sectors.	3 edition of a postgraduate master on sustainable development, taught by environmental, economic, social experts. Inter and trans disciplinary experience. Over 80 post graduate students.	SME, enterprises, NGOs and Public Administration	H	M	H	M	H
Postgraduate students	Master Macsis. Master on scientific communication for sustainability	Developing communication skills with existing and emergent media towards mainstreaming and dissemination of sustainability.	The Master is at its second edition, promoting cultural exchange of students.	NGOs , Public Administration, Museums, journalist associations	H	M	H	M	H
Graduate students	Green Days	Yearly appointment with interdisciplinary seminars across the different Faculties within the university. Practical and demonstrative activities. conference on specific topics involving different disciplines within the campus. To inform university students about SD issues and practices. To integrate the standard educational role of the university with more action-oriented educational methods. To promote the greening of behaviours and daily practice inside the university campus . To create a network of research groups on SD coming from several disciplines and to promote the integration of their research work.	Three workshops realized about sustainability themes (Sustainability science and sustainability practice; Sustainable Production and Consumption; Sustainable use of water resources) targeted at students with highly differentiated background. Results were two books published on interdisciplinary research activities.	Launched by the Department of the Environmental Science but in partnership with economy, sociology, science of education. Non academic stakeholders for presenting case studies. Univeristy's Mobility manager office, Bicocca Campus management offices.	H	M	L	M	M
Undergraduate and graduate students	UNESCO DESS seminar	Yearly interdisciplinary seminar focused on the topic of the UNESCO-DESS initiative (mobility, waste, water etc).	Three edition realized, mainly integrating different research fields.	NGOs, Public Authorities	H	M	M	M	M

Table 1: Overview of multi-stakeholders activities carried out by GRISS – *continued*

Target stakeholder	Activity	Primary objectives	Outcomes	Other Stakeholders involved	Criteria				
					A	B	C	D	E
Undergraduate students	Laboratory on Sustainable tourism	Supporting the students of Sociology Faculty in understanding interdisciplinary issues related to tourism sustainability (environment, socio-economic aspects, cultural heritage).	Practical exercises for developing indicators for decision support and planning in tourism sectors.	Department of Environmental Science and Sociology	M	L	M	M	M
Primary (5-10 year-old) and secondary schools students (11-17 year-old)	Sustainability education activities with schools	Introducing students to sustainability issues, interacting with them through theorical and practical activities, fostering life cycle thinking towards sustainable consumption behaviors.	Several initiatives in primary and secondary schools involving teachers for promoting better integration in schools' programs, and distributing questionnaires to families.	Eco-innovative firms providing information and materials about innovative technologies; families/citizens, teachers	M	L	M	M	M
Secondary schools	Gjusti project: Green Jobs – University, Schools, Community, Enterprises	Supporting stúdents of high school in better understanding potentiality related to green economy, green jobs and eco-innovation through an holistic perspective. To promote the idea that every job can be green, or greener, if planned following SD principles and that every kind of school should include SD in its teaching program.	3 high schools involved; 12 green jobs projects developed by the students. One-day event in which the students presented their projects to firms.	Departments of Sociology and of Economy of the Bicocca University, Local councils of the Municipality of Milan, Green firms.	M	H	M	M	M
Secondary school	"Energia all'Opera" School contest about sustainable energy use	Practical activity in which each school should prepare an artwork about energy themes, to improve consciousness about energy saving and energy efficiency among children.	Several schools took part in the activities, creating their own models for energy production.	Consumers associations, schools.	L	M	M	H	M
Primary and secondary schools and citizen	Exhibit "Il ciclo del riciclo" on material and waste cycle	To promote among government, businesses and citizens the logic of the life cycle of products and materials. Producers of green products made of recycled materials directly involved for exhibited objects.	20 editions realized, with more than 20000 visitors (see www.ciclodelriciclo.com).	Chamber of Commerce, local Municipalities, SMEs	H	H	M	M	M

Table 1: Overview of multi-stakeholders activities carried out by GRISS – *continued*

Target stakeholder	Activity	Primary objectives	Outcomes	Other Stakeholders involved	Criteria				
					A	B	C	D	E
Consumers and NGOs for consumer protection	several educating project targeted to consumers/family Involving NGOs for the protection of consumers	Informing citizens about sustainable consumption practices, giving tools and concepts for better understanding their consumption choices increasing their responsibility and ecolabels knowledge.	More than 400 citizens involved in the public meetings. Questionnaires of satisfaction and capacity building to evaluate the progress made through the seminars/activities.	Schools, SMEs, Local Authorities,	H	H	M	M	M
SMEs	Sectorial handbook for furniture sectors	Pilot project on SMEs producing furniture as a basis for sector specific guidance document. Handbook distributed during events, exhibits and fairs.	Production of prototypes following guidance on sustainable production pattern for the specific sector.	Consortium of SMEs, trade association, local retailers	H	H	M	H	M
SMEs	Training courses to SMEs about eco-innovation and eco-design	Disseminating eco-innovation and eco-design concepts among Italian SMEs and to inform entrepreneurs about: main benefits achievable through eco-innovation; competitive advantage reached by frontrunners in eco-innovation; tools and methods for eco-design and eco-innovation in SMEs.	About 300 participants to informative events at local offices of the chamber of commerce+ about 80 participants to the courses. (2 editions – 12 hours for each edition). Book about eco-innovation strategies, tools and good-practices, collected best practices of SME's and trade associations (Sala and Castellani, 2011).	Local Chamber of commerce, SMEs associations, other research institutes	H	H	M	M	M
Activities with public authorities	Development of public participated projects, mainstreaming sustainability in local planning	Supporting projects such as Local Agenda 21, European Chart of Sustainable Tourism in protected areas, Energy Sustainability Planning (as for the recent EU initiative – Covenant of Mayors – EU 2008).	Mainstream of sustainability concept beyond the mere involvement of the environmental policy department of the Public Authorities involved.	PA, local citizen association, SMEs	H	M	M	M	M

A, B, C, D, E refers to criteria reported in section above, H (High), M (Medium), L (Low) refers to the degree of fulfillment of the criteria already experienced within each specific activity evaluated

Monitoring and evaluating educational interventions is clearly not a new phenomenon, but in the context of multi-stakeholder partnerships for education it is relatively new (Marriott and Goyder 2009).

The purpose of establishing criteria and evaluating them is to facilitate communication among partners about how well a multi-stakeholder partnership for education initiative is performing. It is designed to help partners identify areas that need attention, as well as areas of success.

Conclusions and ways forward

Barriers on sustainability mainstreaming and implementation are often referred to stakeholders and effective strategies should be created in order to avoid "one size fits all" solutions (Williams and Dair 2007). To overcome these barriers, several issues have to be addressed: contextualization and dialogue with key stakeholders (Luskin and Del Matto 2007); wide sharing of knowledge in order to increase awareness and evidence-based decision making (Holmberg and Samuelson 2006); strengthening interaction between contents and practical activities and human/emotional involvement; measuring the results of education, building a set of criteria related to sustainability.

Case studies have been shown to reflect both a range of different types of activities (with different emphases across the recommendations proposed by Agenda 21, Chapter 6) and the importance of involvement with multi-stakeholders in educational activities. The authors do not expect to find a single model for success but rather a range of approaches, based on the local context where the activity has been developed.

The experiences showed that some issues still need to be explored and optimized in order to fulfil the basic criteria for more robust sustainable education, towards concrete actions:

- Identifying specific career pathways for the postgraduate students, following highly interdisciplinary masters, establishing permanent activities with stakeholders for mainstreaming sustainability instead of spot activities.
- Engaging public authorities on a greater scale as partners or as direct actors of sustainability initiatives.
- Involving specific population targets (families, and elderly). So far, they were, mainly involved through the schools (being parents/family of students at which education activities were directed) but not as direct target of actions.
- Defining the most suitable means of involvement of enterprises. Very good involvement in the exhibits and the training course but difficulties in widespread diffusion of knowledge. SD issues need to be integrated in professional training.

It is important that university researchers are aware of their responsibility to educate for sustainability and of its limitations. University team should meet with the multi-stakeholders and decide what exactly different sets of conditions are needed for success and which sets are appropriate under which circumstances.

To date some of the objective stated in Agenda 21 (Chapter 36) in 1992 has been partly achieved (e.g. diffusion of information and consciousness about SD issues), while others still require some attention in the future policies for SE (e.g. SD issue integration in professional training, integration of SD in the curricula of universities).

References

AA. VV. (2011), Natural Resources Forum. *Section Viewpoints* 35, pp. 334-342.

Blackstock, K.L., Kelly, G.J. and Horsey, B.L. (2007), Developing and applying a framework to evaluate participatory research on sustainability. *Ecological Economics* 60, pp. 726-742.

Capra, F. (1996), *The web of life.* Doubleday – Anchor Book, New York.

Castellani, V. (2010), *Development of methodologies and indicators to assess sustainability in tourism and agriculture*. PhD Thesis, Department of Environmental Science, University of Milano-Bicocca. Retrieved from http://boa.unimib.it/handle/10281/10332.

Centre of Ecoliteracy (2009), *Smart by nature*. Available at: http://www.ecoliteracy.org/ (accessed 20 January 2012).

Clark, G. (2007), Evolution of the global sustainable consumption and production policy and the United nations Environment Programme's (UNEP) supporting activities. *Journal of Cleaner Production* 15, pp. 492-498.

Clark, W.C. and Dickson, N.M. (2003), Sustainability Science: The emerging research program. *Proceeding of the National Academy of Science* 100 (14), pp. 859-860.

Falchetti, E. and Caravita, S. (2005), *Per una ecologia dell'educazione ambientale.* Scholè Fut. Ed., Torino.

Holmberg, J. and Samuelson, B.E. (Eds.) (2006), Drivers and Barries for Implementing Sustainable Development in higher education. *Technical Paper* No. 3. Unesco Education Sector.

IUCN/UNEP/WWF (1991), *Caring for the Earth. A strategy for sustainable living.* IUCN, UNEP and WWF, Gland, Switzerland and Earthscan, London.

Kates, R.W. (2011), What kind of science is sustainability science? *Proceeding of the National Academy of Science* 108 (49), pp. 19449-19450.

Kates, R.W. et al. (2001), Sustainability Science. *Science* 292 (5517), pp. 641-642.

Lorenz K. (1974), *Civilized Man's eight deadly sins*. Harcourt Brace, New York.

Luskin, J. and Del Matto, T. (2007), Introduction to the special issue on sustainable production and consumption: making the connection. *Journal of Cleaner Production* 15 (6), pp. 489-491.

Marriott, N. and Goyder, H. (2009), *Manual for monitoring and evaluating Education Partnerships.* International Institute for Educational Planning/UNESCO, Paris.

Reid, W.V., Chen, D., Goldfarb, L., Hackmann, H., Lee, Y.T., Mokhele, K., Ostrom, E., Raivio, K., Rockstrom, J., Schellnhuber, H.J. and Whyte, A. (2010), Earth System Science for Global Sustainability: Grand Challenges. *Science* 330 (6006), pp. 916-917.

Roloff, J. (2008), Learning from multistakeholder networks: Issue-Focussed stakeholder management. *Journal of Business Ethics* 82, pp. 233-250.

Sala, S. and Castellani, V. (2010), *Significato e prospettiva della sostenibilità.* Tangram Edizioni Scientifiche, Trento.

– (2011), *Atlante dell'eco-innovazione.* Franco Angeli editions, Milano.

Scott, K. (2009), *A Literature Review on Sustainable Lifestyles and Recommendations for Further Research*. Stockholm Environment Institute, Project Report.

Schrader, U. and Thøgersen, J. (2011), Putting sustainable consumption into practice. *Journal of Consumer Policy* 34, pp. 3-8.

Spangenberg, J.H. (2011), Sustainability science: a review, an analysis and some empirical lessons. *Environmental Conservation* 38, pp. 275-287.

Stone, M.K. and Barlow, Z. (2005), *Ecological Literacy: education our children for a sustainable world*. Sierra Club Books, San Francisco.

Tamburini, P. (2005), Da Rio 1992 a Johannesburg 2002. In Beccastrini, S. and Cipparone, M. (Eds.), *Tutto è connesso. Voci, idee, esperienze per l'educazione, l'ambiente, la sostenibilità*, pp. 65-74. Arpa Sicilia, Sicilia.

Talwar S., Wiek, A. and Robinson, J. (2011), User engagement in sustainability research. *Science and Policy* 38, pp. 379-390.

Turnhout, E., Hisschemoller, M. and Eijsackers, H. (2007), Ecological indicators: Between the two fires of science and policy. *Ecological Indicators* 7 (2), pp. 215-228.

UNEP (2010), *Here and Now! Education for Sustainable Consumption. Recommendations and Guidelines*. Available at: http://www.unep.fr/scp/marrakech/taskforces/pdf/H&NMay 2010. pdf (accessed 17 January 2012).

Wiek, A., Farioli, F., Fukushi, K. and Yarime, M. (2012), Sustainability science: bridging the gap between science and society. *Sustainability Science* 7 (1), pp. 1-4.

Williams, K. and Dair, C. (2007), What is stopping sustainable building in England? Barriers experienced by stakeholders in delivering Sustainable Developments. *Sustainable Development* 15 (3), pp. 135-147.

Chapter 57

Connecting Universities and Stakeholders through Integrated Climate Services at Local and Regional Level: The EU-FP7 ECLISE Project Experience

Mihaela Sima, Dan Bălteanu and Ines Grigorescu

Abstract

Climate change is expected to have a significant influence on a variety of societal sectors and expose people to a wide range of natural and technological hazards. Thus, any climate change mitigation and adaptation strategy should be based on increasing the involvement and awareness of different stakeholders in order to better cope with the associated impacts and have a better response to a climate-triggered event. In this respect, climate services have proved to be an important tool in dealing with the diversity of stakeholders' needs in providing accessible and usable climate information adapted to vulnerable sectors of society. For this purpose, universities have key capabilities to develop strong user-adapted partnerships, as well as to offer quality interdisciplinary research. This paper aims to highlight the new developments of the "climate services" concept and emphasize the role of universities and research institutes as key actors in providing climate services to stakeholders, as well as to outline the experience of an EU-funded project – FP7 ECLISE – in providing climate information services adapted to local users' needs.

Introduction

Comprehensive scientific assessments have clearly demonstrated that the Earth's climate system is changing in response to direct/indirect consequences of human actions. This has led the broad scientific community to the acceptance of climate change as a multi-faceted environmental issue that requires global regulations (IPCC 2007) and coordinated responses able to mitigate or adapt to current and anticipated consequences (NRC 2010). Therefore, climate change is one of the key issues of the sustainability debate.

According to regional projections, the European (IPCC 2007) climate will change substantially, mainly in South-Eastern Europe, with major environmental and economic consequences, especially in areas which currently significantly rely on their climate. Precipitation will very likely be subject to major changes in different regions of the world and temperature and evaporation will rise, which will lead to increased droughts and decreased water availability, with an impact on a variety of economic sectors (UNISDR 2008).

Under the given circumstances, in order to meet the complex challenges of climate variability and change, there is a need for active involvement of users at all levels (from households, communities, countries and regions, to international forums), as well as for access to the best climate information through climate services (Sivakumar et al. 2011).

This paper aims to highlight the new and emerging developments of the "climate services" concept and emphasize the role universities play as key actors in providing climate information to stakeholders, as well as outline the experience of an EU-funded project – FP7 ECLISE – in providing climate knowledge tailored to local users' needs.

Towards a global framework for climate services

The issue of providing climate services in developing countries has been addressed through the activities of the Regional Climate Outlook Forum (RCOF 2008; Semazzi 2011) which had subsequently contributed to the development of a Global Framework for Climate Services (GFCS). As a result, during the World Climate Conference-3 (WCC-3), hosted by the World Meteorological Organisation (WMO) in 2009 the international scientific community stressed the importance of providing *climate services* seen as "climate information prepared and delivered to meet users' needs" and which is to be harmonised globally through a Global Framework for Climate Services (WMO 2011). Climate services should not only communicate relevant information, but also 'translate' it to non-experts in a way that it can be acted upon. Therefore, climate services involve the provision of climate information (both past and future climate data) enabling early warning systems to help the global community better cope with impacts of climate variability and change (Graham et al. 2011), generally using downscaling to provide information on regional and local scales (Martinez 2011).

Subsequently, another main upshot of the WCC-3 was the establishment of the *Global Framework for Climate Services (GFCS)*, which was able to "strengthen production, availability, delivery and application of science-based climate prediction and services" (WCC-3 Declaration 2009; Martinez 2011). Under the given circumstances, the scientists stress the importance of an accurately measured global set of climate observations, which can be shared without restrictions through providing and developing adaptation and mitigation measures.

Therefore, the WMO initiative is considered an important contribution to adaptation in the framework of UN climate negotiations lined up with the *European Commission White Paper on Adaptation – Adapting to climate change: Towards a European framework for action,* which aims to strengthen the EU's resilience in mitigation and adaptation to climate change (European Commission 2009).

Climate services should meet the expanding needs of society through a variety of measures and procedures and should be able to link data providers (universities, research centres, etc.) with regional and local end users through: developing a new language between scientists and stakeholders, enabling more accessible and comprehensible climate data and proposing understandable tools (graphs, maps, tailored meetings) (Ruti 2011). For that reason an effective climatic service has to be integrated into an end-to-end system that serves the target users with valuable and easy to use information, thus going beyond the simple production of information (Srinivasan et al. 2011).

Currently, the development of climate services for sustainable development in the context of changing climate is one of the main concerns of the WMO's Commission for Climatology (CCl). Its key objectives are to coordinate and promote better practices of climate data in terms of observation, gathering, management, analysis and monitoring of variability and change, developing prediction services and ultimately providing climate information to vulnerable sectors such as agriculture, water resources and health through best capacity building tools (Sivakumar et al. 2011).

Bridging the gap between universities and society by integrated climate services

Over the last few years, due to the increasing occurrence of extreme weather events, bridging the gap between science and policy has become extremely important, especially when discussing the regions most vulnerable to climate change and vulnerability (e.g. the Mediterranean), considered among the most prominent and vulnerable "hot spots" affecting sustainable development (Giorgi 2006; UNEP-Plan Bleu 2011). This situation could not be surmounted through effective exchange of information between universities/research institutes/data providers on one hand and stakeholder communities on the other.

Given that climate change adaptation and mitigation measures concern all political, administrative, economic sectors and levels, thus addressing the interest of different groups; it requires the integrated and joint involvement of different public and private actors (decision-makers, stakeholders, etc.) (Hilpert et al. 2007).

When referring to "stakeholders" acting in the climatic domain, one must take into consideration individuals or larger groups having a two-way information transfer with a certain system (regions, industries, or communities) vulnerable to the potential impacts of climate change (Gardner et al. 2009). On the other hand, "decision-makers" should integrate climate change information into long-term planning and policy in order to provide reliable climate services in terms of impact, vulnerability and adaptation information. Thus, they might become hands-on actors being able to provide legislative regulatory instruments to manage, limit or prevent undesirable activities, support cooperation and linkage between different institutions, thus determining changes in human behaviour through information, education and participation (Hilpert et al. 2007).

Therefore, in relation to the major climate-vulnerable systems (cities, agro-ecosystems, forests, waters, etc.) a set of potentially relevant end-users (both stakeholders and decision-makers) can be put forward in terms of local and national authorities, government departments, agribusiness, forest services, emergency inspectorates; non-government organizations etc.

The main challenge for stakeholders is to understand climate change impacts and to develop and implement policies able to provide and guarantee an optimal level of adaptation through climate-driven management and strategies by means of providing adequate climate information (climate services) applicable to a broad range of vulnerable sectors and environments (WMO 2011).

Academics provide the improvement of data observation, time-range and expertise climate predictions and quality control climate data (Sivakumar et al. 2011) through applicable research that can be effectively communicated to the wider community of stakeholders and the general public whose activities are sensitive to climate conditions (Hanson et al. 2006).

The role of universities in developing user-centred information

Although there are connections between academics and users in the climate change-exposed communities, seeking the best means to integrate their cooperation more effectively becomes necessary (Yohe et al. 2007) in order to link scientific tools with stakeholders needs on climate variability and change adaptation for sustainable development.

The relationship between academics and users was formerly tackled through several local case-studies conducted in the Baltic Sea Region or in the Mediterranean, which focused on assessing the regional political decision maker's perception on climate-induced changes, the communication and use of regional climate knowledge and the implementation of possible measures for adapting to climate change (BSR INTERREG III B ASTRA project, 2005-2008; Eisenack et al. 2007; Reckermann et al. 2011; CLIM-RUN project, 2011-2013). Besides the regional level, stakeholder integration in analysing the climate change impacts, vulnerability and adaptation was tackled in the CLIMSAVE project (2010-2013) focusing on European level, which includes procedures for selecting the appropriate stakeholders as well as different tools for communicating with them (through workshops or an Integrated Assessment Platform) (Gramberger et al. 2011).

These studies revealed significant information on stakeholders' preferences for the type of information sources and providers, their experience in using climate data in regional decisions making, perceptions of the necessity of adaptation measures, etc. (Martinez and Bray 2011). Therefore, the stakeholder participation in a climate assessment is required to be integrative and proactive throughout the entire research in order to guarantee the supply of useful and valuable results to end-users through stakeholder/end-user interactions (meetings, workshops, etc.) (Hanson et al. 2006; O'Connor et al. 2000).

Fig. 1: Functional elements and relationships of a National Climate Service

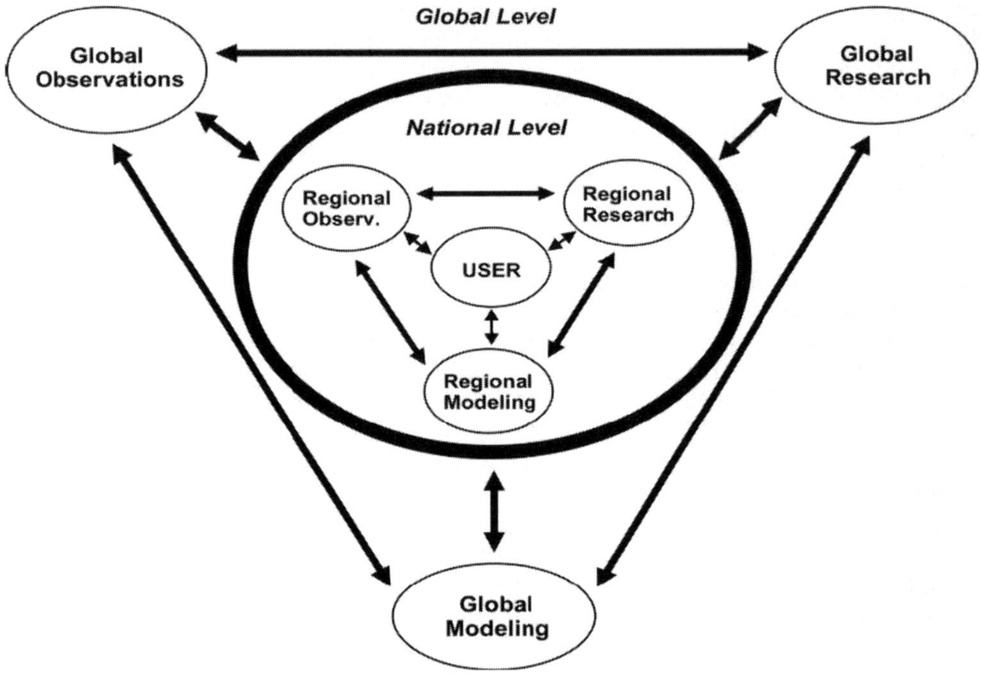

Miles et al. (2006)

The crosscutting issue of "bridging gaps" or "gapping" among data providers and users is continuously tackled by both academics and users, who try to establish connections through meetings, protocols, and even scientific projects. Therefore, over the last period, various collaborations between – amongst others – research centres, national hydro-meteorological services, and national and local decision-makers occurred in order to assist the most vulnerable countries in establishing end-to-end climate information application systems for the reduction of climate change risks. However, a great number of weaknesses were identified in terms of: matching temporal and spatial scales of forecasts to users' needs, poor mechanisms to encourage users in specifying the requisite type of information, a lack of experience in communicating and processing probabilistic scientific information, etc. (Srinivasan et al. 2011).

Climate information needs a sustainable instrument acting as a national climate service (NCS) to connect and translate climate data to end users and support building regional and national capacity to develop predictions and adapting strategies to climate variability and change (Miles et al. 2006) (Fig. 1).

The NOAA Regionally Integrated Sciences and Assessments (RISA) Program developed in the USA has shown the successful development of regional climate services as well as the interaction with different stakeholders in providing climate information data. The RISA promoted a partnership between universities in supplying climate services, universities being mentioned as an ideal environment for interdisciplinary research and for integrating data into practice. In this context, RISA indicates several key capabilities of the universities in managing climate services. These include expertise in climate science, traditional regional stakeholder partnerships, ability to work with multiple stakeholders and partners, flexibility to adapt to the stakeholders needs; rapid development of climate models and technology as well as of a partnership with private sector (Anderson et al. 2009).

Academic institutions are essential in dealing with climate services as they provide skilled academics (researchers, professional masters etc.) and fundamental and applied research on different related topics such as: climatology, hydrology and their impacts on social and ecological systems (Rood et al. 2009), mathematics, statistics, modelling and GIS, remote sensing, etc. All of these would provide up-to-date knowledge, methods and proficiency to be translated into comprehensive and easy-to-use climate information for different user categories and levels. Academic institutions also have the capacity to develop and run training programs and courses on different climate change-related topics, thus reaching the end-user community through finding the best ways to identify sustainable means to improve the interface between scientific research outcomes and society (Rood et al. 2009).

Therefore, academia would improve climate information and services through providing knowledge on climate drivers and variability at various time scales, producing future climate projections (based on probabilistic forecasts), transposing regional climate models to local conditions (Buizer 2011).

Providing scientific information to serve both society and decision-makers' needs is also one of the priorities of the decade of research to address five "Grand Challenges in Global Sustainability Research", which was recently launched by the ICSU (International Council for Science). These challenges refer to forecasting, observing, confining, responding and innovating with the scope to identify and mobilize the research to be carried out in order to address them at global, regional, national and local level (ICSU 2010).

The EU FP7 – ECLISE project experience in providing climate services to local users

Currently, at European level, observational and data services are mainly connected to the Group on Earth Observations (GEO)/ Global Earth Observation System of Systems (GEOSS) and the Global Monitoring for Environment and Security (GMES) programmes context. As a consequence, the key actions related to climate modelling and prediction are gathered under *Activity 6.1 Climate Change, pollution and risks of the FP7 Environment Programme* also designed to foster global collaboration in climate research, to which, the FP7 cooperation area allocated EUR 1.8 billion out of the total amount of EUR 32.4 billion for the 2007-2013 time span (European Commission 2010; Tilche 2011).

In line with the EU research priorities, as major outcome, the socio-economic dimension of climate change is expected to provide significant decision support measures for key economic sectors. Under the given circumstances, the European Commission considers the establishment of a European Climate Service of strategic importance and it is projected to strengthen this initiative in the forthcoming FP7 calls of the Environment Programme. Furthermore, it is expected that the 2012 call will include a major bottom-up initiative on *Seasonal-to-decadal climate predictions, considered as a European contribution to climate services.* The central aim of this initiative will lead to increasing society's resilience to climate change impacts and seasonal extremes with special focus on climate sensitive economic sectors (Tilche 2011).

ECLISE project will rely on previous climate knowledge provided by several relevant EU and international projects (e.g. ENSEMBLES, CIRCE, ACQWA, MICE). Furthermore it is projected to cooperate and interact with other on-going projects (e.g. CMIP5, COMBINE, CORDEX, CLIM-RUN funded in framework of the same call) in order to make available innovative concepts for the development of a pan- European Climate Service.

The FP7 project "*Enabling climate information services for Europe (ECLISE)*" (2011-2014) aims to support future European climate services for selected regions in Europe in close cooperation with local stakeholders and to develop the concept for the European-wide climate services. The project is a cooperation network between 12 partners (research institutes and universities) from eight European countries, coordinated by the Royal Netherlands Meteorological Institute. The other project's partners are the Swedish Meteorological and Hydrological Institute, the Climate Service Center (Germany), the Institute for Coastal Research (Germany), the Wageningen University and Research Center (Netherlands), the Institute of Atmosferic Sciences and Climate (Italy), the Uni Research (Norway), the Norwegian Meteorological Institute, the National Institute of Hydrology and Water Management (Romania), the Institute of Geography, Romanian Academy, the Technical University of Crete (Greece) and Newcastle University (UK).

Within ECLISE project, researchers, in close cooperation with users, develop local climate services to support climate-related policies by providing climate services for several vulnerable regions in Europe, set up at on the following key domains: coastal defence, cities, water resources and energy production. In this respect, the main objectives of ECLISE are (www.elise-project.eu):

- To provide a European perspective to the functionality of climate services and to develop and integrate national services into a broader European approach;
- To take a first step towards the realisation of a European Climate Service, addressing climate information needs of the EU and its member states;
- To capitalize on previous research projects and bring together European climate research expertise and the needs of public and private organizations engarding future climate information;
- To further develop local climate services in four areas: coasts, cities, water and energy, and to make concrete demonstrations of the utility of these services in support of local climate adaptation policies; also to provide an outline and proof of concept for future European-wide climate services.

Both providers and users of climate services in the ECLISE project belong to the key domains which include local and regional case studies, as well as several public or private organisations. The climate models used for the ECLISE project originate from different regional and global models obtained in previous EU-funded projects (PRUDENCE, CECILIA, CLAVIER, ENSEMBLES, WATCH, SCENES and COMBINE). Therefore, model runs at high spatial resolution (10 km across Europe and up to 1-4 km for selected user cases) will be available for selected case studies in Europe.

The Institute of Geography, Romanian Academy is involved in the ECLISE project with 3 climate impact studies within the "cities" and "water" work-packages, referring to: the assessment of past and future climate variability and climate change trends in the Baia Mare Urban System, north-western Romania as well as socio-economic and health impacts associated to climate change; the impact of climate change on agriculture and future land-use patterns in the south-eastern part of the Lower Danube Plain (Bărăgan Plain) and the assessment of floods and landslides related hazards and risks in the Vrancea Seismic Region (Fig. 2).

Fig. 2: The location of the ECLISE project's case studies in Romania

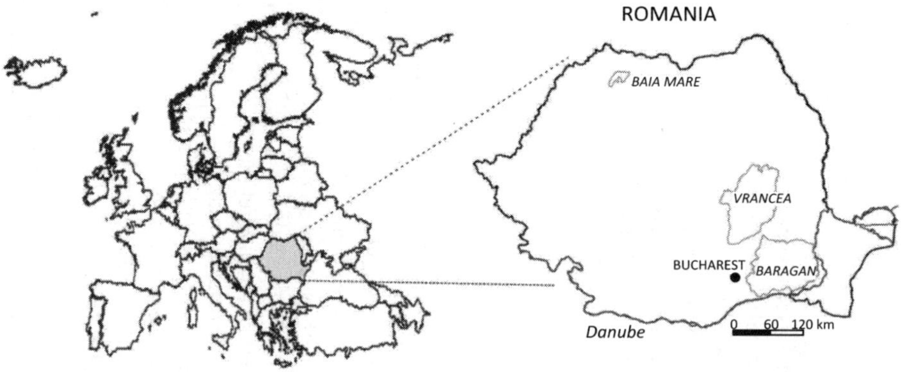

The project will provide adequate practice-oriented climate information on a local scale through comprehensive and easy-to-use climate products for the selected case-studies. Therefore, it will offer tailored climate data to each area's environmental idiosyncrasies on one hand, and users' needs on the other.

The climate information (observation, analysis, modelling and prediction) will be put forward through tables, charts, graphs and maps for the key parameters (temperature, rainfall, wind, extreme climate events, etc.), prediction models, risk and vulnerability maps for climate-sensitive regions etc.

Table 1: A user-interaction summary for the selected case-studies in Romania

Case-study	**Users' profile**				**Data providers**
	Potential users…	**Type of climate-driven exposure**	**What kind of climate information do they need and for which decision?**	**Is the user aware of the uncertainty of future climate projections?** **Do they have the capacity to receive, interpret, and translate climate data into sectoral impacts?**	
Baia Mare Urban System	Baia Mare City Council North-West Development Region Public Health Directorate	Heat and cold waves, floods, fog and acid deposition	Impact studies report, graphs, hazard maps for short, medium and long-term spatial planning, for determining the biophysical vulnerability (e.g. heat waves, wind intensity and direction etc.) and further delineating hazard prone areas	During a workshop, users have been explained about the uncertainty of climate models, particularly at local and regional scales. Most of the users do not have the knowledge and expertise to interpret the climate data, thus they required the impact reports with model data to be translated in an accessible language.	National Meteorological Administration Other climate information suppliers (e.g. Universities, Research Institutes) National Agricultural Research and Development, Fundulea, Romania
Baragan Plain	Local Agency of Payments and Intervention for Agriculture (APIA) National Agricultural Research and Development, Fundulea, Romania South Development Region	Drought, aridity, heat waves, strong winds (blizzard), hail storms, tornadoes	Impact studies, future agro-climate scenarios, maps for better agricultural land management, crop productivity, irrigation necessity		

Table 1: A user-interaction summary for the selected case-studies in Romania – *continued*

Case-study	Users' profile				Data providers
	Potential users…	**Type of climate-driven exposure**	**What kind of climate information do they need and for which decision?**	**Is the user aware of the uncertainty of future climate projections?** **Do they have the capacity to receive, interpret, and translate climate data into sectoral impacts?**	
Vrancea Region	Vrancea County Council	Heavy rains triggering floods, landslides	Detailed maps, impact studies for the management of extreme events, the improvement of intervention procedures in case of extreme events (floods, landslides, mudflows)		
	Civil Protection				
	South-East Development Region				
	Local municipalities				

The climate change studies performed by the National Administration of Meteorology (NAM) have revealed an increasing air temperature trend in Romania, of approx. 0.5°C for the timeframe between 1901-2005 (at an average of 0.05°C/decade). This trend has been more obvious in the north of the country, the Black Sea coast and locally in the west of the country (by about 0.7°C). In terms of precipitation, the same approaches point to a heterogeneous trend, showing a slight long-term decrease; however, at some weather stations, increases are to be expected. Seasonally, during winter and spring the precipitation amounts tend to fall, slowly increasing in summer and autumn. During the period 1961-2007, some changes in the regime of extreme pluviometric events were registered, mainly related to longer intervals of rain-free days in the south and significantly more frequent rainy days in autumn, especially in the northern half of the country, and also in its western and south-eastern regions (Busuioc et al. 2010).

The IPCC climate scenarios for the future (2090-2100) estimate an increase of the annual mean temperature by 1.8-4°C (IPCC 2007). The NAM climate change projections in Romania were made for two future periods (2021-2050 and 2071-2100) against the standard interval (1961-1990), based on global and regional climate models under IPCC A1B emission scenario. The results indicate a temperature rise of around 1.4°C for 2021-2050 timeframe, significantly higher, however, is the prediction of 3.1°C for 2071-2100. In terms of precipitation, the climate signals are somehow weaker for the first projected period as compared to the second one, when changes become more obvious, given that winter and summer quantities as well as spring and autumn ones are considerably lower (Busuioc et al. 2010).

Developing integrated vulnerability-oriented case studies

As a user-oriented project, one of the main tasks of the ECLISE project is to identify the most suitable end-user for each proposed case-study, which could benefit from the results and use the provided climate information in their further studies and decisions. Therefore, a strong interaction with the potential users was integrated into the workshops, questionnaires and other communication tools in order to understand the users' profiles, their needs and expectations in terms of climate data and select the most appropriate ways to supply it according to the decision for which the climate information is required. An important issue was devoted to explaining the users the potential uncertainty of the existing climate models and the way future climate data (as raw data, graphs, tables, maps, reports) are to be required in order to meet their needs. The selected users for the Romanian case studies required an interpretation of the climate model data, thus climate impact studies will be performed by the partner institute and delivered through accessible deliverables (reports, graphs and maps). A summary of users' profiles and needs for the three selected case-studies is presented in Table 1.

The decision for which the climate information is needed refers to short and medium-term intervals (from years to decades), thus considering as adequate future projections for climate models two reference years (2030 and 2060). Periodical meetings are planned, permitting the users to give and receive feedback, understand the research approach and share the ideas with the researchers, as well as for the partner institute to adapt and tailor the research methodology and outcomes to users' needs and to establish further research priorities.

One of the case-studies in the project refers to Baia Mare Urban System, located in the north-western part of Romania with a population of around 200,000 inhabitants, a typical medium-size Romanian town. For Baia Mare, one of the main ECLISE project's objectives is to analyse the potential connections between future climate and society, in order to offer the decision-makers reliable scientific services to be considered in their decision as well as in the climate adaptation strategy of the town. Romania has prepared the National Strategy on Climate Change for the period 2012-2020, with an "Adaptation" component, that includes the needs for larger municipalities (over 200,000 inhabitants) to design their own adaptation strategies which should include the specific adaptation measures of the vulnerable societal sectors. Thus, the ECLISE project case-study on towns will offer valuable inputs for the Baia Mare Climate Adaptation Strategy.

In this respect, several indicators on climate, socio-economic and health aspects will be aggregated to determine the potential threats of extreme climate events (heat waves or cold, snaps heavy rainfall, etc.) to human health. This is particularly relevant when considering also the industrial background of the town Baia Mare, one of the air-pollution "hot-spots" in Romania due to traditional metal mining activities. The selected sustainability indicators will be used to meet the study's objectives, on the one hand, and to be representative for the stakeholders on the other. The indicators are meant to offer the stakeholders good documentation on how certain environmental drivers affect human health, allowing them to find solutions and develop long-term strategies (McCool and Stankey 2004).

The human health-centric approach of the project will be investigated through analysing the effects of climate change on health in relation to the physical and socio-economic environment. The disease categories considered relevant to linking climate and environment are as follows: respiratory (asthma, bronchitis) and cardio-vascular ones. Geo-statistical and GIS methods will be used to aggregate the indicator categories (socio-economic, epidemiological and environmental), the most relevant ones being summarized in Table 2.

Table 2: Relevant sustainability indicators for the relationship climate-health-environmental quality

Epidemiological indicators	**Socio-demographic indicators**	**Environmental quality indicators**	**Climatic indicators**
Incidence of cardiovascular diseases Incidence of respiratory diseases etc.	– population income – number of cars	Air quality (temporal series)	temperature
	– number of medical units – monitorised patients – healthcare access	– greenhouse gas emissions (CO2, SO2, NOx, NH3, NMVOC, ozone)	Relative humidity
	– living space per person; – green space per person	– suspended particles (PM10, PM2,5)	Heat and cold waves
	– population by age and sexes – employed population	– heavy metals (Pb, Cd, Hg etc.)	Thermal comfort indices

The study-area is exposed to a variety of extreme hydro-climatic events which trigger climatic hazards, characteristic to both cold and warm periods of the year, among which *snow layer, winter heat waves producing floods, heavy rainfalls, fog and acid depositions and hail storms* ranks first.

This leads to the need to communicate useful climate information on relevant meteorological phenomena to local end-users through translating climate data ranges into comprehensive products and services: descriptive materials (reports, technical guides etc.), statistical analysis (tables, graphs, charts), spatial data (digital maps, modelling/simulations, web-based networks) etc. Therefore, considering that one of the leading climatic hazards of the study area is related to precipitation-related phenomena, relevant data processing of the rainfall parameters is required (e.g. the mean monthly/annual precipitation amounts and the frequency of average monthly precipitation amounts on class intervals) using different graphical/visual representations such as spreadsheets, tables, charts and maps (Fig. 3).

Fig. 3: Climate data and services provided to local end users – e.g. communicating precipitation data through different visual tools: spreadsheet (A) tables (B, D), bar chart (C), cell chart (E) and map (F)

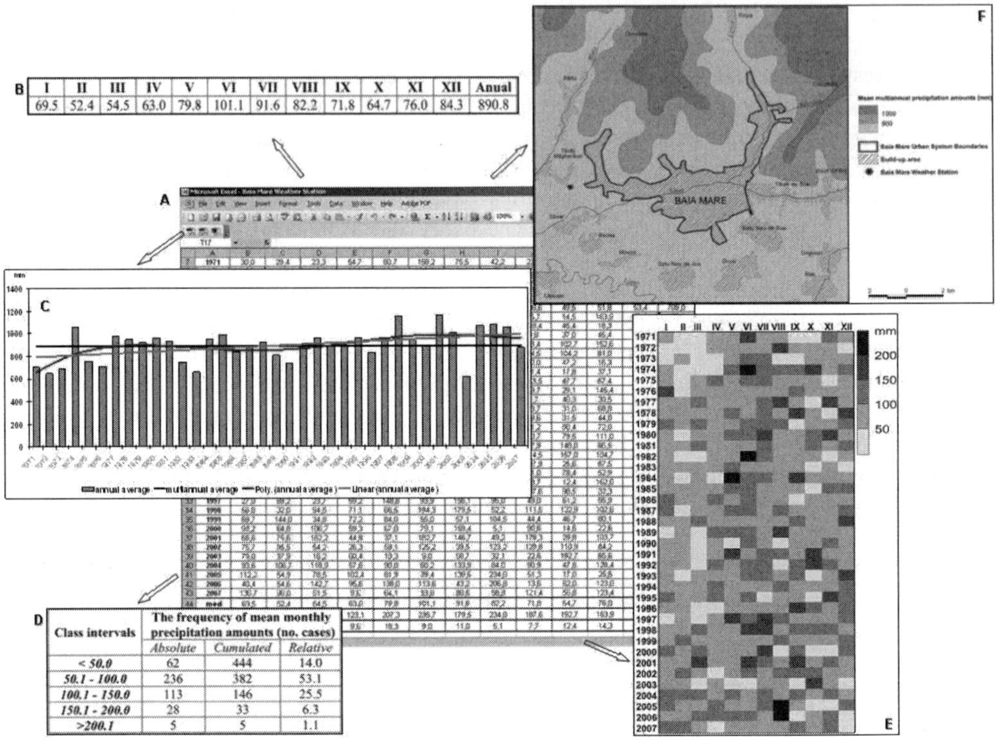

Baia Mare Weather Station (1971-2007) (ECLISE project Report 2011)

Conclusions

There is a strong demand for future climate information at regional and local level to support different decisions and effective actions as a response to climate change. As recognized by scientists, there is a real challenge to offer good resolution climate data, due to the large uncertainty of climate models usually performed globally or regionally and downscaled to local scale as well as the doubts whether the new climate services can deliver actionable and reliable information to users (Kerr 2011). However, there is a consensus that, in a changing world, where climate is an important component, decision-makers should rely on scientific climate information for their decisions and for designing the most suitable mitigation and adaptation measures. For this purpose, universities and research institutes are recognised as important actors in providing climate data, in promoting the partnership with the stakeholders and in offering quality interdisciplinary research. There is a need for adapting the universities' curricula to the current society's demands by introducing dedicated courses on climate change, climate services and communication with stakeholders.

Considering that climate services rely on processed and tailored climate data which is able to meet the user's requirements, universities must play an essential role in providing climate education and research, through preparing skilled experts and owning quality logistics (laboratories, hardware, software, etc.) to undertake climate-related studies. Consequently, academic institutions will contribute to climate services with education, research assessment, problem solving as well as direct access to local-to-regional decision-makers (Rood et al. 2011). Therefore, it is highly recommended to include explicit climate change-related academic activities able to translate climate data into accessible information in the national climate services for each country.

In less developed countries (e.g. Romania), several gaps in designing climate services and in offering interdisciplinary impact studies adapted to users' needs were identified: low access to historical climate data usually owned by the national meteorological services, stakeholders lack of awareness/concern/expertise in considering climate information for their decisions, mismatching the scale of climate forecasts to users' needs, poor experience in involving the stakeholders in the scientific projects as potential beneficiaries, lack of devoted university courses on climate change and climate services related topics.

A further step within ECLISE project would be to elaborate the downscaled good resolution models suitable for local scales (5-10 km), offer the users information on what the future climate in their area may look like and identify the potential impacts in the sectors addressed by the Romanian case studies (landslides/floods, agriculture and cities). Simply providing raw information on climate change to users is not enough to support their adaptation decisions, therefore a cross-disciplinary and multi-stakeholder approach is recommended. The final research outcome will be then adapted and improved based on users' feedback, due to the fact the stakeholders hold a major responsibility in implementing sustainability programs for the benefit of their local community.

Acknowledgments

This chapter was elaborated in the framework of FP7 ECLISE project "Enabling Climate Information Services for Europe", no. 265240 (2011-2014), coordinated by the KNMI (Royal Netherlands Meteorological Institute).

References

Anderson, C., Cayan, D., Dettinger, M., Dow, K., Hartmann, H., Jones, J., Miles, E., Mote, P., Overpeck, J., Shafer, M., Udall, B. and White D. (2009), *Climate Services: the RISA Experience*. Available at: http://hazards-climate-environment.org/yahoo_site_admin/assets/docs/New_Pub_RISANationalClimateServiceMay09_finalV3b.15412644.pdf.

BSR INTERREG III B ASTRA project (2005-2008), *Developing Policies & Adaptation Strategies to Climate Change in the Baltic Sea Region*, http://www.astra-project.org/.

Buizer, J. (2011), *Climate Services and University Research*. Sixth Southwest Hydrometeorology Symposium, Arizona State University, 27 September 2011. Available at: http://www.wrh.noaa.gov/psr/pns/symposium/presentations/buizer.pdf.

Busuioc, A., Caian, M., Cheval, S., Bojariu, R., Boroneanţ, C., Baciu, M. and Dumitrescu, A. (2010), *Variability and climate Change in Romania,* 226 p. Pro Universitaria, Bucuresti (in Rom.).

CLIM-RUN project (2011-2013), *Climate local information in the Mediterranean region responding to user needs, FP7 project*. www.climrun.eu.

ECLISE project Report (2011), *The elaboration of database necessary for the project's models, the methodology and relevant indices and indicators,* 64 p. Internal report, Romanian Academy, Institute of Geography.

Eisenack, K., Tekken, V. and Kropp, J.P. (2007), Stakeholder Perceptions of Climate Change in the Baltic Sea Region. *Coastline Reports* 8, pp. 245-255.

European Commission (2009), *White Paper on Adaptation – Adapting to climate change: Towards a European framework for action*, Brussels, pp. 1-17. Available at: http://eur-lex.europa.eu/LexUriServ/LexUriServ.do?uri=COM:2009:0147:FIN:EN:PDF.

– (2010), *European Research Framework Programme, Research on Climate Change*. Prepared for the Third World Climate Conference (WCC-3) and the UNFCCC Conference of the Parties (COP-15), Directorate-General for Research Environment. Available at: http://ec.europa.eu/research/environment/pdf/cop-15.pdf.

Gardner, J., Dowd, A.-M., Mason, C. and Ashworth, P. (2009), *A framework for stakeholder engagement on climate adaptation*. CSIRO Climate Adaptation Flagship Working paper No. 3. Available at: http://www.csiro.au/resources/CAF-working-papers.html.

Giorgi, F. (2006), Climate change hot-spots. *Geophys. Res. Lett.* 33, L08707 (doi: 10.1029/2006 GL025734.

Graham, R.J., Yun, W.-T., Kim, J., Kumar, A., Jones, D., Bettio, L., Gagnon, N., Kolli, R.K. and Smith, D. (2011), Long-range forecasting and the Global Framework for Climate Services. *Clim. Res.* 47, pp. 47-55.

Gramberger, M., Kok, K., Eraly, E. and Stuch, B. (2011), *Report on the first CLIMSAVE European stakeholder workshop*. CLIMSAVE project, 45 p. Available at: http://www.climsave.eu/climsave/doc/Report_on_the_first_European_workshop.pdf.

Hanson, C.E., Palutikof, J.P., Dlugolecki, A. and Giannakopoulos, C. (2006), Bridging the gap between science and the stakeholder: the case of climate change research. *Clim. Res.* 31, pp. 121-133.

Hilpert, K., Mannke, F. and Schmidt-Thomé, P. (Eds.) (2007), *Towards Climate Change Adaptation in the Baltic Sea Region, Developing Policies and Adaptation Strategies to Climate Change in the Baltic Sea Region*. Geological Survey of Finland, Espoo.

ICSU (2010), *Earth System Science for Global Sustainability: The Grand Challenges,* 24 p. International Council for Science, Paris.

IPCC (2007), *Climate change 2007: The physical science basis contribution of working group to the 4th assessment report of the Intergovernmental Panel on Climate Change*, 996 p. Cambridge University Press, Cambridge, UK and NY.

Kerr, R.A. (2011), Time to adapt to a warming world, but where's the Science? *Science* 334, pp. 1052-1053.

Martínez, R.G. (2011), Building sustainable regional climate information systems. *Climate Research* 47, pp. 41-45.

Martinez, G. and Bray, D. (2011), *A survey of the perceptions of regional political decision makers concerning climate change and adaptation in the German Baltic Sea region.* International BALTEX Secretariat Publications, Publication No. 50, Geesthacht.

McCool, S.F., Stankey, G.H. (2004), Indicators of sustainability: challenges and opportunities at the interface of science and policy. *Environ Manage* 33 (3), pp. 294-305.

Miles, E.L., Snover, A.K., Whitely Binder, L.C., Sarachik, E.S., Mote, P.W. and Mantua, N. (2006), An approach to designing a national climate service. *PNAS* 103 (52), pp. 19616-19623.

NRC (2010), *Advancing the Science of Climate Change*, 528 p. National Research Council, National Academies Press, Washington, DC.

O'Connor, R.E., Anderson, P.J., Fisher, A. and Bord, R.J. (2000), Stakeholder involvement in climate assessment: bridging the gap between scientific research and the public. *Clim. Res.* 14, pp. 255-260.

RCOF (Regional Climate Outlook Forum) (2008), *RCOF review 2008*. Concept note. Available at: http://www.wmo.int/pages/prog/wcp/wcasp/documents/RCOFReview2008 ConceptNote.pdf.

Reckermann, M., Langner, J., Omstedt, A., von Storch, H., Keevallik, S., Schneider, B., Arheimer, B., Meier, M. and Hunicke, B. (2011), BALTEX – an interdisciplinary research network for the Baltic Sea region. *Environ. Res. Lett.* 6, 11 pp.

Rood, R.B., Scavia, D., Burton, A., Michalak, A., Nadelhoffer, K., Lemos, M.C. (2009), *White Paper: Federal Climate Services and Academic Institutions*. 11 p. . Available at: http://climateknowledge.org/openclimate/doclink/20090515_NOAA_RFI_Climate_Services_final_v1.pdf.

Semazzi, F.H.M. (2011), Framework for climate services in developing Countries. *Clim. Res.* 47, pp. 145-150.

Sivakumar, M., Bessemoulin, P., Peterson, T.C. and Asrar, G. (2011), Changing climate and demands for sustainable development. *Clim. Res.* 47, pp. 3-4.

Srinivasan, G., Rafisura Kareff May and Subbiah, A.R. (2011), Climate information requirements for community-level risk management and adaptation. *Clim. Res.* 47, pp. 5-12.

Tilche, A. (2011), *Climate Services, a European perspective.* WCRP Joint Scientific Committee (JSC-32) UK Met Office Exeter, 4-8 April 2011.

UNISDR (2008), *Strengthening the Hydrometeorological Services in South Eastern Europe.* South Eastern Europe Disaster Risk Mitigation and Adaptation Programme, 85 p. Available at: http://www.unisdr.org/files/7650_StrengtheningHydrometeorological SEE1.pdf.

UNEP (2011), *Mediterranean Strategy for Sustainable Development* follow-up: main indicators 2011 update. Plan Bleu, 31 p. Available at: http://www.planbleu.org/publications/Indicateurs_SMDD_2011_EN.pdf.

WCC-3 (World Climate Conference-3) (2009), *High Level Declaration.* Geneva, 31 August – 4 September 2009. Available at: http://www.wmo.int/wcc3/documents/WCC3_declaration_en.pdf.

WMO (2009), *World Climate Conference-3 (WCC-3) Conference Statement.* Geneva, 31 August – 4 September 2009. Available at: http://www.wmo.int/wcc3/documents/ WCC-3_Statement_04-09-095PM.pdf.

– (2011), *Climate knowledge for action: A Global Framework for Climate Services – Empowering the most vulnerable.* The Report of the High-level Taskforce for the Global Framework for Climate Services, Report WMO, No. 1065, Geneva, Switzerland.

Yohe, G.W., Lasco, R.D., Ahmad, Q.K., Arnell, N.W., Cohen, S.J., Hope, C., Janetos, A.C. and Perez, R.T. (2007), Perspectives on climate change and sustainability. Climate Change 2007: Impacts, Adaptation and Vulnerability. In Parry, M.L., Canziani, O.F., Palutikof, J.P., van der Linden, P.J. and Hanson, C.E. (Eds.), *Contribution of Working Group II to the Fourth Assessment Report of the Intergovernmental Panel on Climate Change*, pp. 811-841. Cambridge University Press, Cambridge, UK. http://www.climateservices. gov.

Chapter 58

Water Management and Reuse Strategies at Pondicherry University: Sustainable Alternatives

Golda Arpudhalin Edwin, Gopalsamy Poyyamoli, Arun Prasath Ramaswamy and Muthu Nandhivarman

Abstract

Pondicherry University is one of the fastest growing central universities in India. The university has a sprawling area of 315 hectares spread on either side of the scenic East coast road, with a large proportion on the landward side of the road. The main objective of this study is to critically analyze and propose innovative and long-term sustainable water management strategies for the campus. The major sustainable practices discussed in this study are gray water management using eco-technologies, roof-top gardening, and the harvesting of rainwater. It is concluded that gray water reclamation and reuse, along with rainwater harvesting could not only reduce the extraction of groundwater resources, but also protect the coastal urban areas from saline water intrusion. Moreover, implementing these appropriate technologies at campus level not only offers a science-based practical tool, but also supports the transition to a sustainable ecological and economic model of mutually reinforcing social cohesion, the creation of wealth and environmental protection.

Introduction

The purpose of this study is to determine the strategies for sustainable water management to maximize water security at a campus level. Pondicherry University (PU) campus, located in the Union territory of Puducherry, India, has been selected for this study. The university is currently fully dependent on the groundwater resources for all its water needs and any wastewater generated is sent directly to the municipal sewage (Nandhivarman et al. 2011a). This causes several undesirable problems such as an unsustainable rate of groundwater extraction, contamination and saltwater intrusion, etc. (Lenin et al. 2008). In order to reduce the pressure on the groundwater and increase campus sustainability, this study focuses on three major aspects of water management such as grey water treatment (GWT), rooftop gardening (RTG) and rain water harvesting (RWH). As part of this study, we have estimated the current amount of groundwater withdrawal, potential for amount of water generated through RWH, amount of water reclaimed through gray water treatment and the potential reduction in withdrawal of groundwater with the help of global water tool template provided by the World Business Council for Sustainable Development and the Global Reporting Initiative (WBCSD 2007). For freshwater, the water footprint concept developed by Hoekstra and Hung (Hoekstra and Hung 2002) and Hoekstra and Chapagain (Hoekstra and Chapagain 2008) has been applied. It has been estimated that by implementing these strategies in the residential areas of the campus alone, about 46.25% of reduction in the groundwater extraction can be achieved.

Background of Pondicherry University

Pondicherry University is a central university located in the Union Territory of Puducherry, a prominent coastal town in India. The university was founded in 1985 by the Government of India. Since then, it has emerged as one of the India's fastest growing universities. It is an affiliating university with its jurisdiction spread over the Union Territories of Puducherry, Lakshadweep, Andaman and Nicobar Islands. Fig. 1 shows the layout map of PU. Since its inception, the PU campus has grown from a few academic buildings to a multifaceted community showing a dramatic increase in campus community and infrastructure, particularly over the last four years (2007-2011). The University has two directorates, twelve schools, 14 departments, 14 centers and an Institute for Community Development. It has 35,000 students in its 87 affiliated colleges in addition to 5,500 students enrolled on the main campus and another 27,000 students on a distance learning program (PU Statistics 2011).

Fig. 1: Layout Map of Pondicherry University

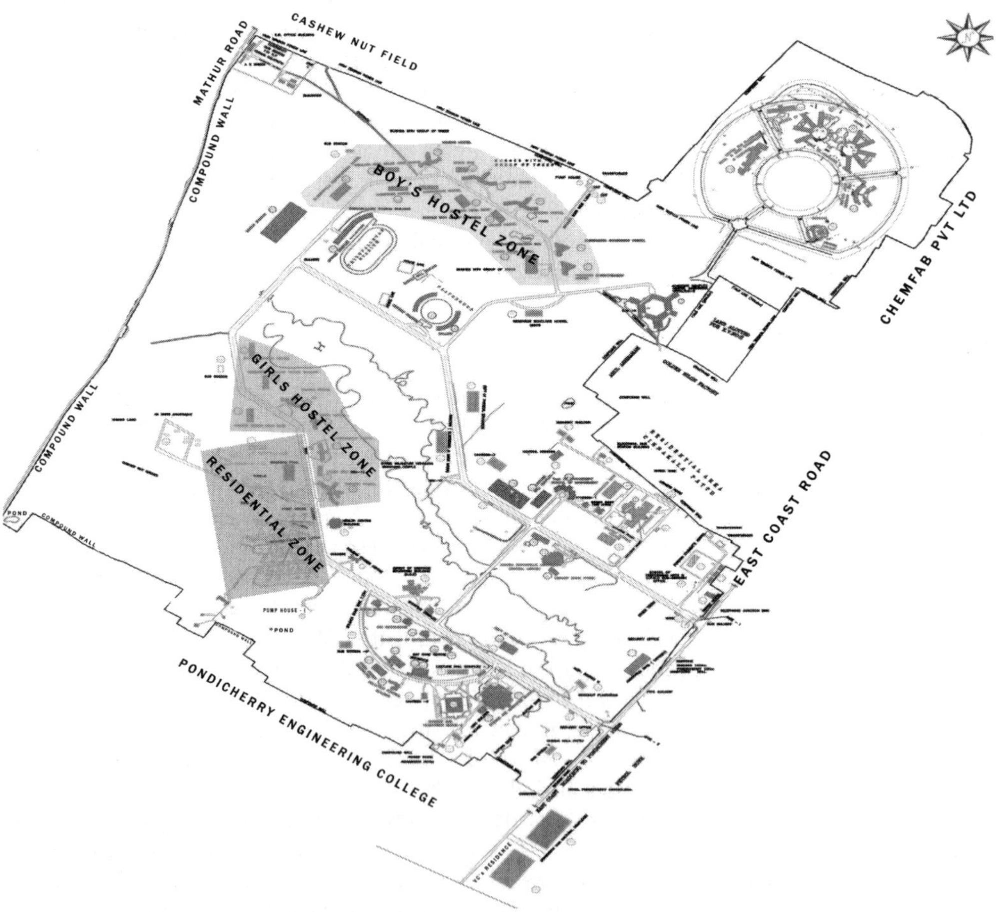

The water consumption trend observed at the campus quarters, hostels, schools, departments, health center, public utility centers, administrative block, horticulture department, cafeteria and civil construction sites have become increasingly unsustainable over the years leading to irreversible impacts on the groundwater aquifers. The impact varies from reduced freshwater percolation, increased infiltration of pesticide and pathogens, declined levels of groundwater and increased salt water intrusion from the ocean, as well as the pollution of freshwater bodies caused by drainage problems and monsoons (Nandhivarman et al. 2011b).

Role of educational institutions in addressing environmental challenges

For a projected world population of 9.2 billion by the year 2050 (MDG 2007), there is reason for profound concern in several regions and countries with limited water resources. In view of this, the government of Puducherry's vision for 2020 clearly describes the water management problems of the coastal regions, including groundwater depletion, salt water intrusion, industrial pollution, sewage pollution, etc., with utmost priority, and has called for the attention of educational and research institutions to supplement their effort on minimizing state water-related issues by regulating groundwater consumption, conducting periodic water quality monitoring, implementing groundwater conservation measures and improving the quality of drinking water (Pondicherry Vision-2020, 2002).

Methodology

This study is supposed to provide a strategic plan for sustainable water management for phase I of the implementation of the much broader Green Campus Initiative undertaken at the campus, whilst phase II will be for the schools, departments, health center, public utility centers, administrative block, horticulture department, cafeteria and civil construction sites. At present, only the residential zones of the university (Fig. 1), which includes the hostels and quarters, are considered in this study. Fig. 2 shows a framework for the assessment of groundwater consumption (GWC) and the potential for rainwater harvesting (RWH), gray-water treatment (GWT) and rooftop gardening (RTG) for selected zones of the PU Campus. The residential buildings and the student's cafeteria accounted for much of the water consumption and generation of waste water on campus, whilst the hostels and quarters contributed more when compared to the service sectors (Nandhivarman et al. 2011b). For the initial phase of implementation of sustainable strategies, the residential buildings inside the campus, including the hostels and quarters, are considered for an in-depth study.

Groundwater consumption: present scenario and challenges

According to the World Resources Institute, the term "water stress" areas will have less than 1,700 cubic meters per capita and "water scarcity" areas will have below 1,000 cubic meters per capita of renewable fresh water available annually(Carmen et al. 2000). Annual per capita water availability in the Puducherry region is roughly 200 cubic meters per person which shows that the Puducherry is in a "water stress" condition (Puducherry Hydrology Project II, 2006). Groundwater is the main source of water for drinking, irrigation, and industrial needs in the entire Puducherry region. Alluvium sandstone, Cuddalore sandstone, vanur Ramanathapuram sandstone are the main aquifer systems, whilst Ottai granular sandstone zones and Kadaperikuppam sandstone aquifers also contribute to some extent (Vasanthakumar 1999). Since alluvium aquifer covers about 90% of the Puducherry region, water level remains shallow, ranging from 12-14 m below ground level. Fig. 3 indicates the saltwater intrusion using real time data.

Fig. 2: Assessment of water footprint for selected blocks of the Pondicherry University campus

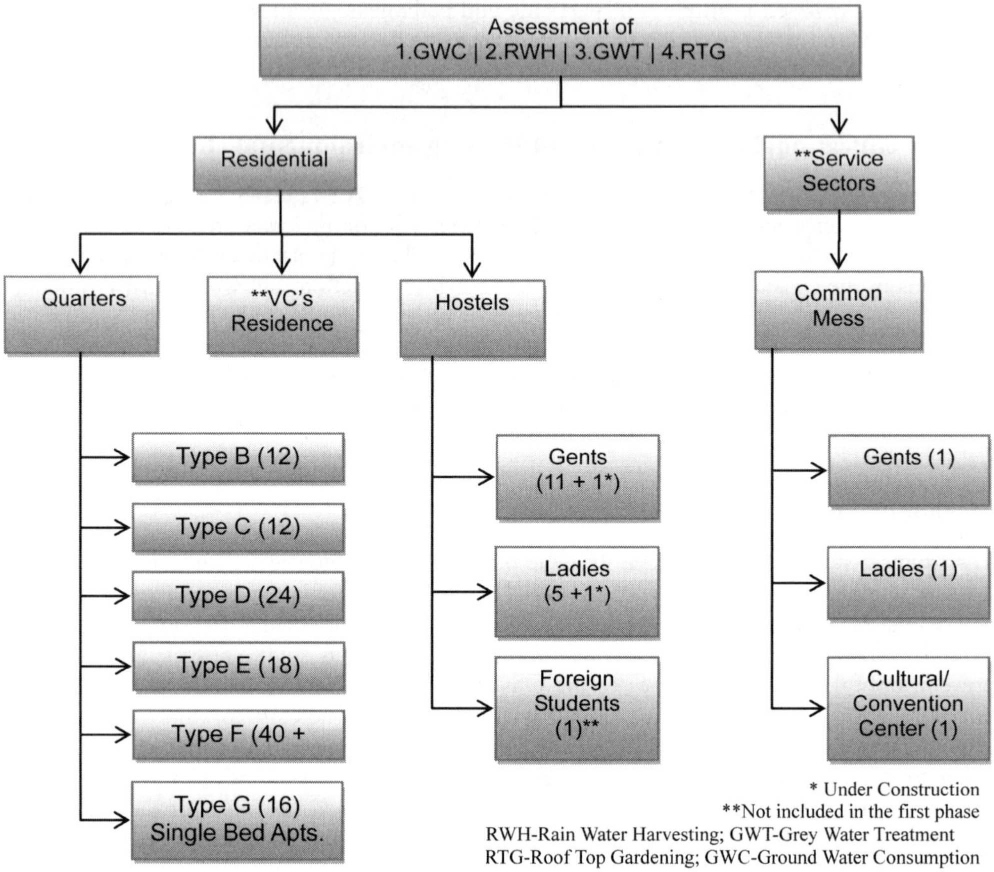

Modified from water footprint of Pondicherry University, Nandhivarman et al. (2011b)

As per the groundwater recharge estimation conducted during 1995-96 by the Puducherry central groundwater board and the state groundwater unit of the agriculture department, the utilizable groundwater recharge resources at 85% of the gross potential is assessed at 155 MCM. However, the present groundwater draft to meet the needs of all the sectors is in the order of 169 MCM (Water Year 2007). Therefore, it is possible that the maximum limits of the utilization of groundwater would have exceeded the availability of resources by this time. The decline of water is in the order of 15-30 m in the west and 7 m in the east of Puducherry. This aggravates the saltwater intrusion, impacting the establishments close to the seashore as reported by Lenin et al. (2008).

Fig. 3: Saltwater intrusion in the Puducherry region

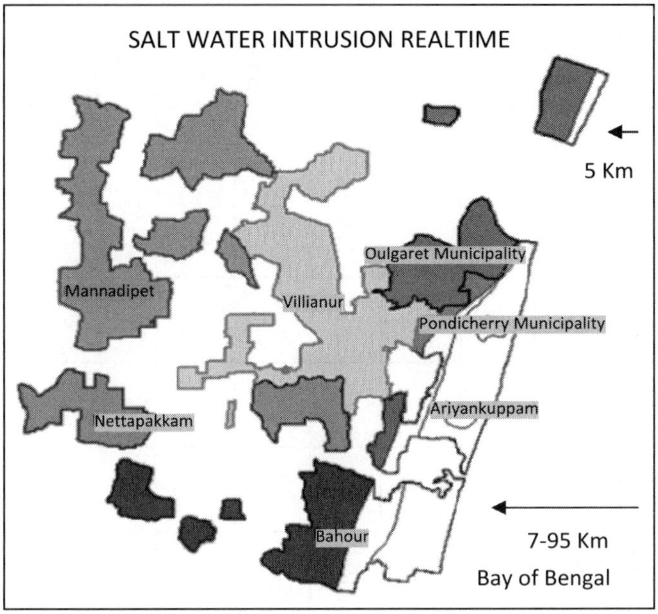

Source: Puducherry Hydrology Project II (2006)

Table 1 shows the location of the bore wells and their services for the university campus. The entire campus is supplied by two overhead tanks, with a capacity of 200,000 liters, which run continuously during peak hours. As reported by the public works department and the Civil Engineering department of the university, tank number one is located opposite the health center and supplies the southern side of the campus and campus territories such as the Vice Chancellor's residence, the university guest house and the cultural cum convention center. Tank number two supplies the northern side of the campus including the administrative block (Nandhivarman et al. 2011b). Based on the growth rate of PU in the last few years, the pressure on groundwater resources more than tripled, triggering the increase in campus groundwater consumption with diversified degree of pollution. Since the university is completely dependent on the groundwater and is also situated along the seashore, the impact is very high and therefore campus life cannot be sustained in the long term unless adequate measures are taken to address the challenges.

Rainwater harvesting potential

The annual rainfall in the Puducherry region is approximately 1,250 mm in about 50 rainy days, received result of both the southwest and northeast monsoons. Throughout Indian history, there have been several kingdoms where special emphasis was given to the management of rainwater and harvesting systems (Deep et al. 2003). In Puducherry and the surrounding states, the people in the past have created *Ooranies*, a segregated catchment area where rainwater in stored separately for drinking, bathing and other domestic purposes. They also formed percolation tanks or ponds, for the purpose of recharging irrigation or domestic wells.

Table 1: Details of main bore wells and their water supply for the campus

S#	Type	Location	Depth (m)	Tube (m)	HP	Status	Supply	Speed	6am-2pm (Hrs)	2pm-10pm (Hrs)	10pm-6am (Hrs)	Tank Capacity (l)	Service To
1	Bore well	Opposite to Health center	43	34	20	Operational	Main	High	*8	*8	**2-3	2,00,000***	
2	Bore well	Quarters pump house	43	34	10	Operational	Main	High					Southern side of the campus
3	Bore well	Sathyanarayana	43	34	10	Operational	Emergency	Medium					
4	Bore well	Substation	43	34	15	Operational	Main	High	**4.5	**4	**2.5	2,00,000	
5	Bore well	Substation	43	34	7.5	Operational	Emergency	Medium					Northern side of the campus
6	Bore well	Pasic bore	43	34	20	Operational	Emergency	High					
7	Bore well	Kamban hostel	43	34	3	Operational	Emergency	Medium				****	Gardening

* *Never over flows*
** *overflows after specified hour*
*** *1,00,000 lts underground sump is also available*
**** *Every building is having separate bore well for gardening/ horticulture activities apart from the above mentioned high speed bore wells.*

"Rainwater harvesting" is a widely used term covering all those techniques whereby rainwater is collected and used in areas of close proximity to where it was collected. In general the term has been applied when rainwater has caused the soil to percolate and recharge the aquifer, or runs off into small reservoirs for micro-irrigation purposes (Thomas and Martinson 2007). The rooftop rainwater harvesting system is more focused on domestic consumption, with a very few examples of centralized catchment areas and storage in cisterns.

As part of this study, the rainwater harvesting potential of PU is estimated to be around 19,822 m^3 by utilizing only the rooftops of the residential building of the campus. It was determined that such a system could potentially save PU money, decrease the demand on groundwater, and create water reserves for times of emergency. It may therefore be concluded that it would be significant enough to not only reduce the dependency on groundwater, but also to make a difference to the overall water consumption on the campus.

Potential of rooftop gardening in campus

Roofs with a vegetated surface and substrate otherwise called "green roofs", provide ecosystem services in urban areas which includes the regulation of building heat (Karen and Bas 2003), the improvement of fire resistance (Köhler 2003), the improvement of storm-water management (Carter and Jackson 2007), a reduction in energy consumption (Christian and Petrie 1996), an increase in sound insulation (Dunnett and Kingsbury 2004) and an increase in the urban wildlife habitat. Other potential benefits include green-space amenity, air-quality improvement, and reduction of the urban heat-island effect (Getter and Rowe 2006). Implementing such infrastructure at PU would take advantage of the unused roof space and could provide many of the same benefits. Analysis of the published data across the world suggests that an extensive green-roof system can significantly lower the temperature and save on energy consumption through a reduction in air conditioning (Christian and Petrie 1996; Del Barrio 1998; Karen and Bas 2003).

Grey water-treatment potential: reclamation and reuse

As the campus continues to grow and expand, there is an increase in the need for fresh water supply thereby resulting in the further extraction of groundwater resources. Seeking measures to reduce the consumption of freshwater and to recycle and reuse the wastewater will benefit the university. So far there is a lack of community engagement in water conservation and the goal is to raise awareness of the environmental impact, financial cost and advantages of recycling and reuse of wastewater for irrigation and other building usage. Wastewater from the bathrooms and kitchen (gray water) can be treated and recycled with relatively little effort making it fit to use for specific reuse purposes such as flushing, gardening, etc.

There are several eco-technologies for gray water reclamation and reuse, of which a constructed wetland system is more suitable and economical for developing countries. Constructed wetland is one of the series of engineered and managed natural systems that is receiving worldwide attention for wastewater treatment and recycling as reported by Bastian and Hammer (1993). The use of the constructed wetlands systems (CWs) in urban areas on both a small and large scale is now being recognized across the world due to their good treatment performances and low construction and operating costs (Kadlec et al. 2000; Sonavane et al. 2008). CWs for gray water treatment designed for campuses or small communities can reclaim up to 70%of their wastewater, and use them for various purposes including flushing toilets, for cleaning purposes, landscaping, etc. (Friedler 2008). By recycling and reusing the gray water, the potable water needs can be reduced greatly, thereby relieving the demand on groundwater. Treated water from CWs can also be used for restricted or unrestricted agricultural irrigation of crops depending on its effluent standard, and it can also serve as a tool for recharging groundwater (Rousseau et al. 2008). The reuse of gray water primarily reduces the freshwater requirement, reduces the generation of sewage and helps close the water and nutrient cycle to a greater degree at the source itself.

The retrofitting of existing buildings can also prove to be a simple ,method in achieving great results in water saving. Installing dual flush toilets, low-flow faucets in shower head fixtures, automatic control for urinals and flush tanks or water-free urinals will definitely reduce the water demand.

PU campus planning and development should consider the gray water reclamation and reuse through CWs or other cost-effective eco-technologies such as slow-sand filtration and soil bio technology. The university should also create a policy that mandates water-sensitive urban design in all new developments. Therefore, in the long term an integrated water management approach will take into consideration the water usage, production, recycling and reuse.

Results and discussion

Data has been collected from the residential buildings of PU campus and an estimate has been prepared for a current groundwater withdrawal rate, the potential for rainwater harvesting, potential for gray-water reclamation and reuse and potential reduction in the withdrawal of groundwater.

Fig. 4: Potential saving in groundwater withdrawal for boys hostels at Pondicherry University

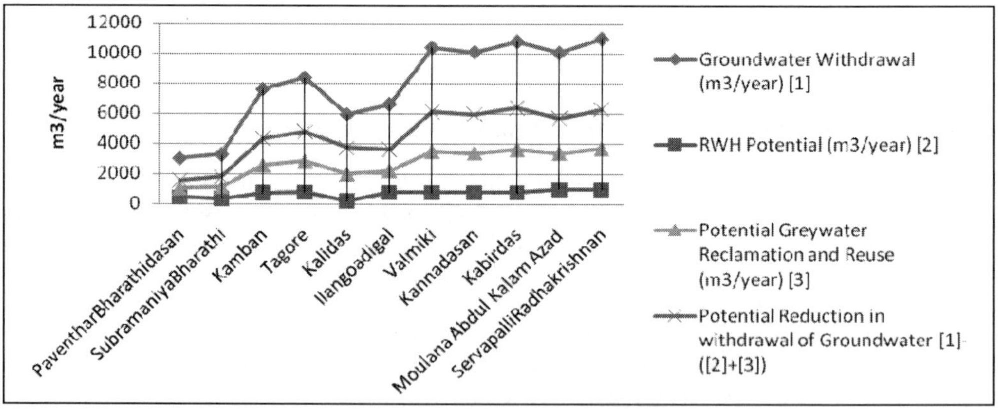

The estimates are prepared separately for gents hostels (Fig. 4), ladies hostels (Fig. 5) and staff quarters (Fig. 6). Table 2 illustrates the overall potential saving in groundwater withdrawal that can be achieved at all hostels and quarters at PU.

The current rate of groundwater extraction to satisfy the needs of the residential activity is assessed to be 200 million liters of water per year with the help of the Global Water Tool. Out of this, 160 million liters constitutes wastewater, gray water constitutes approximately 112 million liters which is discharged into the Bay of Bengal, passing through the adjacent Oulgaret Municipality limit. Each day more than 550,000 liters of water is extracted from the aquifer for the residential activities alone, which constitutes the main blue water footprint of the university, excluding the service sectors and other agricultural activities. Out of this, around 80% contributes to the gray water footprint of the campus.

Fig. 5: Potential saving in groundwater withdrawal for ladies hostels at Pondicherry University

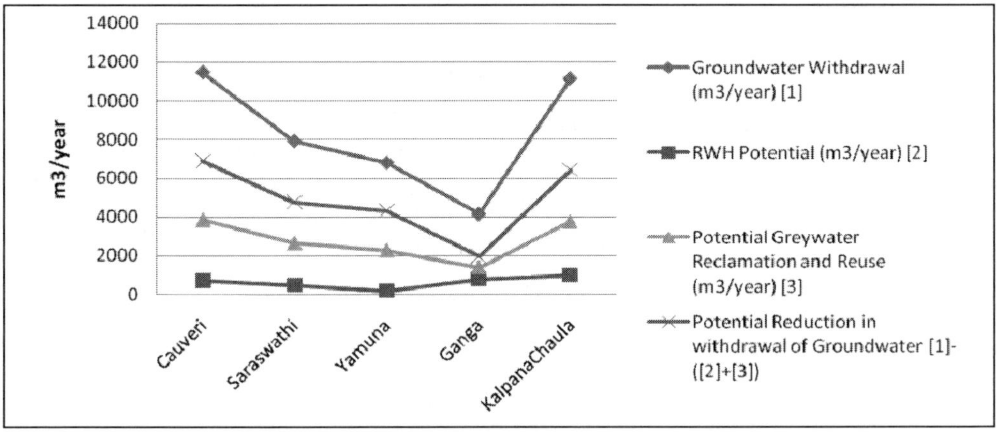

Fig. 6: Potential savings in groundwater withdrawal for residential quarters at Pondicherry University

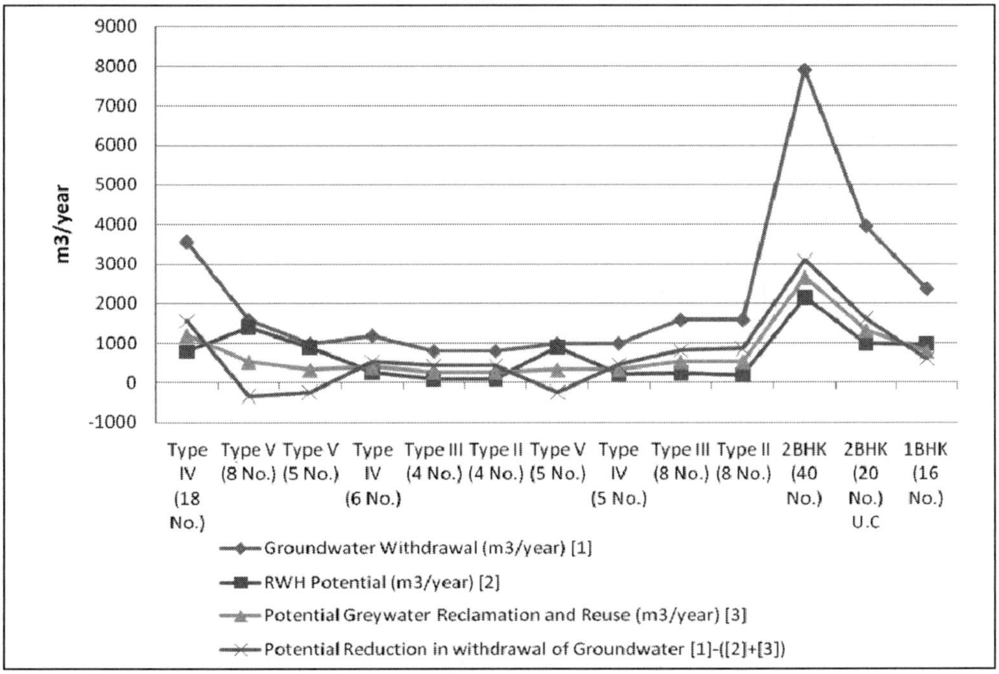

The water consumption for an average Indian, as officially reported is 135 l pcd (Govt. of India 1997, 2002). At present there are no water savings in the form of wastewater management and rainwater harvesting in the university. The total wastewater generated as a result of residential activities alone constitutes about 156,744,000 liters per year. Almost 80% of the water supplied for domestic use comes out as wastewater. Out of this, around 52,665,000 liters of gray water per year (Nandhivarman et al. 2011b), with an expected reclamation level of 60% (Kujawa-Roeleveld and Zeeman 2006) is realizable within the campus.

Table 2: Potential savings in groundwater withdrawal for hostels and quarters at Pondicherry University

Building	**Groundwater Withdrawal (m^3/year) [1]**	**RWH Potential (m^3/year) [2]**	**Potential Greywater Reclamation and Reuse (m^3/year) [3]**	**Potential Reduction in withdrawal of Groundwater [1]-([2]+[3])**	**Potential % Reduction in withdrawal of Groundwater**
Gents Hostel	87069	7390	29255	50424	42.09%
Ladies Hostel	41490	3185	13940	24364	41.28%
Staff Quarters	28185	9247	9470	9468	66%
Total	156744	19822	52665	84257	46.25%

Water produced within the PU campus is estimated at 19,822 m^3 per year through rainwater harvesting, and 52,665 m^3 per year through gray water reclamation and reuse. Gray water reclamation and reuse along with rainwater harvesting could not only reduce the extraction of groundwater resources, but also protect the coastal urban areas from saline water intrusion. Data suggests that there could be a potential reduction of 46.25% or a saving of 72,487 m^3 per year (Fig. 7) as a result of the withdrawal of groundwater at the campus, which would be a significant achievement contributing towards environmental sustainability. The systems and infrastructure should be maintained carefully and the administration should ensure that the produced water be tested for quality and only be used for purposes recommended by the established standards.

Fig. 7: Anticipated reduction in groundwater withdrawal rate

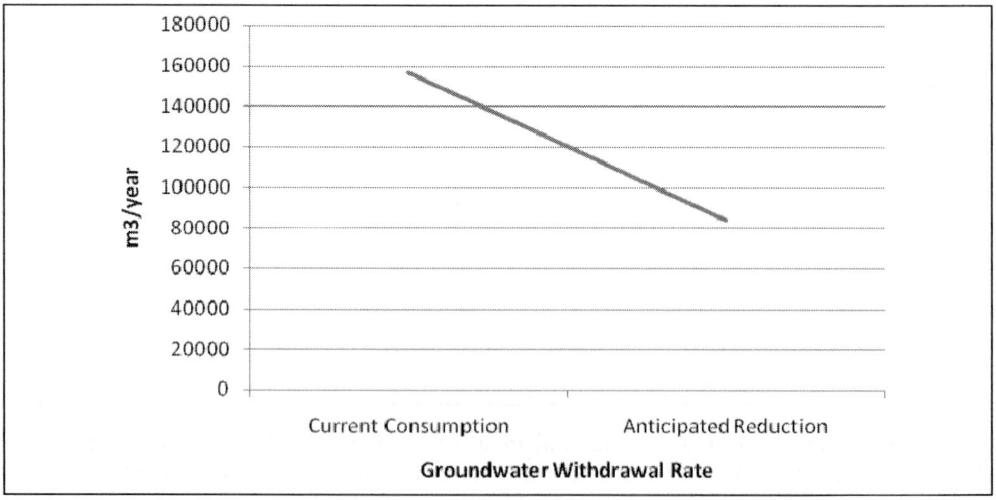

Therefore the concept of environmental sustainability is based on formulating development plans that incorporate an integrated approach to the management of the urban water cycle. The challenge ahead is to implement the strategies through a collaborative approach that excites and engages the campus communities into delivering improved environmental sustainability for the present and future generations.

Conclusion

Universities around the world should take the lead role in implementing sustainable strategies for a smaller water footprint and water security. Costal urban cities like Puducherry can reduce the salt water intrusion by effective and sustainable management of its groundwater resources along with other water conserving strategies like gray water reclamation and reuse, rainwater harvesting, water conserving fittings, etc. Our study has concluded that by adopting water management strategies, the overall groundwater withdrawal can be reduced and a considerable amount of water could be saved leading to an environmentally sustainable campus.

References

Bastian, R.K. and Hammer, D.A. (1993), The Use of Constructed Wetlands for Wastewater Treatment and Recycling. In Moshiri, G.A. (Ed.), *Constructed Wetlands for Water Quality Improvement*, pp. 59-68. CRC Press, Boca Raton, FL.

Carmen, R., Jake, B., Norbert, H., Richard, P. and Ken, K. (2000), *Pilot analysis of global ecosystems: Freshwater Systems*, pp. 1-78. World Resources Institute.

Carter, T. and Jackson, C.R. (2007), Vegetated roofs for stormwater management at multiple spatial scales. *Landscape Urban Plan* 80, pp. 84-94.

Christian, J.E. and Petrie, T.W. (1996), Sustainable Roofs with Real Energy Savings. Proceedings of the Sustainable Low-Slope Roofing Workshop. Ed. Desjarlais, A., pp. 99. Oak Ridge National Laboratory, Oak Ridge, Tennessee.

Deep, N.P., Anil, K.G. and David, M.A. (2003), Rainwater harvesting as an adaptation to climate change. *Current science* 85 (1), pp.

Del Barrio, E.P. (1998), Analysis of the green roofs cooling potential in buildings. *Energy and Buildings* 27, pp. 179-193.

Dunnett, N.P. and Kingsbury, N. (2004), *Planting Green Roofs and Living Walls*. Timber Press, Portland, OR.

Friedler, E. (2008), The water saving potential and the socio-economic feasibility of greywater reuse within the urban sector – Israel as a case study. *International Journal of Environmental Studies* 65 (1), pp. 57-69.

Getter, K.L. and Rowe, D.B. (2006), The role of green roofs in sustainable development. *HortScience* 41 pp. 1276-1286.

Govt. of India (1997), *Ninth Five-Year Plan 1997-2002* – Vol. II. Planning Commission. New Delhi.

– (2002), *Tenth Five-Year Plan 2002-2007* – Vol. II. Planning Commission. New Delhi.

Hoekstra, A.Y., Chapagain, A.K. (2008), *Globalization of water: Sharing the planet's freshwater resources*. Blackwell Publishing, London.

Hoekstra, A.Y. and Hung, P.Q. (2002), *Virtual water trade: a quantification of virtual water flows between nations in relation to international crop trade*. Value of Water Research Report Series, No. 11. UNESCO-IHE, Delft, The Netherlands.

Kadlec, R.H., Knight, R.L., Vymazal, J., Brix, H., Cooper, P. and Haberl, R. (2000), *Constructed wetlands for pollution control. Process, performance, design and operation*. IWA Scientific and Technical Report, No. 8.

Karen, L. and Bas, B. (2003), *Thermal performance of green roofs through field evaluation, Proceedings for the First North American Green Roof Infrastructure Conference,* pp. 1-10. Chicago.

Köhler, M. (2003), *Plant survival research and biodiversity: Lessons from Europe*. Paper presented at the First Annual Greening Rooftops for Sustainable Communities Conference. Awards and Trade Sho, pp. 20-30.

Kujawa-Roeleveld, K. and Zeeman, G. (2006), Anaerobic treatment in decentralized and source separation based Sanitation concept. *Rev. in Env. Sci. and Bio-Tech.* 5, pp. 115-139.

Lenin, K.S.V., Dinesh, G., Ravikumar, G. and Govindarajalu, D. (2008), Vulnerability assessment of seawater intrusion and effect of artificial recharge in Pondicherry coastal region using GIS. *Indian Journal of Science and Technology* 1 (7), pp. 1-7.

MDG (2007), *The Millennium Development Goals Report.* Available at www.un.org/ millenniumgoals/ pdf/mdg2007.pdf (accessed 20 December 2011).

Nandhivarman, M., Poyyamoli, G. and Edwin, G.A. (2011a), *Environmental Sustainability – a conceptual frame work for Pondicherry University, Puducherry, India – a case study.* Paper presented at the 11th International Congress of Asian Planning Schools Association, 19 September 2011, Tokyo, Japan.

– (2011b), *Evolving a water policy framework 2011 for Green Campus Initiative – a case study of Pondicherry University, India.* Paper presented at the 8th International conference on urban watershed management, 7 September 2011, Beijing, China.

– (2011c), *Water foot print as a tool for sustainable management of ground water resources in Pondicherry University, India.* Paper presented at the 8th International conference on urban watershed management, 7 September 2011, Beijing, China.

Pondicherry Vision-2020 (2002), *Water resources management and urban environment.* Available at http://unpan1.un.org/intradoc/groups/public/documents/apcity/unpan013202.pdf (accessed 12 December 2011).

PU Statistics (2011), *Vital Statistics of Pondicherry University.* Available at http://www.pondiuni.edu.in/content/vital-statistics (accessed 20 December 2011).

Puducherry Hydrology Project II (2006), Available at www.pdywaterinfo.org/ (accessed 5 November 2011).

Rousseau, D.P.L., Lesage, E., Story, A., Vanrolleghem, P.A. and De Pauw, N. (2008), Constructed wetlands for water reclamation. *Desalination* 218, pp. 181-189.

Sonavane, P.G., Munavalli, G.R., Ranade, S.V. (2008), Nutrient removal by root zone treatment systems: a review. *Journal of Environmental science and engineering* 50 (3), pp. 241-248.

Thomas, T.H. and Martinson, D.B. (2007), *Roofwater Harvesting: A Handbook for Practitioners.* (Technical Paper Series No. 49). IRC International Water and Sanitation Centre, Delft, The Netherlands.

Vasanthakumar, R. (1999), *Perspectives plan of water resources in Pondicherry.* Public Works Department, Government of Pondicherry, pp. 1-8.

Water Year (2007), *Report on the Ground water brochure of Puducherry region. Central Ground Water Board.* Available online at http://cgwb.gov.in/District_Profile/Puduchery/Puducherry.pdf (accessed 5 November 2011).

WBCSD (2007), *Global Water Tool.* World Business Council for Sustainable Development, Conches-Geneva, Switzerland. Available at www.wbcsd.org. (accessed 5 November 2011).

Chapter 59

Integrated Organic Kitchen Waste Management for Campus Sustainability: A Case Study of Pondicherry University, India

Muthu Nandhivarman, Golda Arpudhalin Edwin, Arun Prasath Ramaswamy and Gopalsamy Poyyamoli

Abstract

Universities performing leadership roles in higher education have an additional responsibility – environmental sustainability. Pondicherry University has become India's fastest growing university, with a growth rate of 300% over the last four years. As a result of the rapid expansion and growth, proper planning is of utmost importance to address the challenges of sustainable development. The recently initiated strategic green action plan for campus sustainability is in tune with the principles of sustainable development, and affirmed and confirmed the commitment to the full implementation of Agenda 21 along with the other millennium development goals. This program focuses on the campus commitment towards sustainability encompassing social equity, environmental management, and economic prosperity. As environmental futurists, we believe that sustainable campuses can be evolved only when sustainable win-win strategies are adopted at a grass-roots level to foster their contribution to environmental, social and economic problems. This paper is oriented towards research and planning for reducing fuel consumption through biogas production from organic waste generated within the campus. It has been estimated that around 43 kg per day of biogas can be produced from the kitchen wastes of one common boys' hostel mess alone. Through this action, there will be a reduction of 28.25% of liquefied petroleum gas and a reduction of about 131.22 kg per month of CO_2 emitted through the petroleum gas delivery vehicle. The same model, when replicated in the 87 affiliated colleges of the university, is expected to generate huge environmental and economic benefits.

Introduction

Comprehensive organic kitchen waste management is one of the greatest challenges for achieving campus sustainability (Danielle et al. 2010). The aim of this study is to determine the amount of organic waste generated, its composition and the current management practices within the key campus operational areas, and to develop an action framework in order to provide recommendations to the university administration on strategies for waste minimization, higher rates of energy and nutrient recovery, thereby improving the overall sustainability of the campus waste management program. This framework was developed based on the research studies done globally in various universities and higher educational institutions.

Ecologically sustainable campus development is the environmental component of sustainable development. Precautionary Principle 15 of the Rio declaration states that "where there are threats of serious or irreversible environmental damage, lack of full scientific certainty should not be used as a reason for postponing measures to prevent environmental degradation". This plays a major role in facilitating campus sustainability in developing countries such as India. Supported by the principle of intergenerational equity, the present generation should ensure that the health, diversity and productivity of the environment is maintained or enhanced for the benefit of the future generations. The conservation of biological diversity with ecological integrity and improved valuation, pricing and incentive mechanisms for the up-gradation of environmental factors should also be included in the valuation of assets, liabilities and services. Since 2002 the international gathering of the Earth Summit on Sustainable Development held in Johannesburg, confirmed the commitment to the "full implementation" of Agenda 21, along with the other Millennium Development Goals to be achieved by universities and higher education institutions. To facilitate this within the Indian scenario, a conceptual framework for Pondicherry University's action plan towards environmental sustainability has been developed focusing on four main strategies within the campus (Nandhivarman et al. 2011a).

Fig. 1: Map showing the four geographically disconnected regions of Puducherry

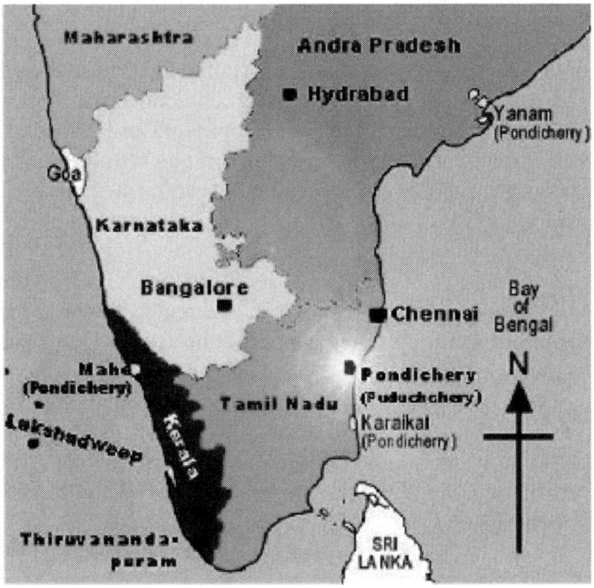

Methodology

Research site Selection

Puducherry is one of the union territories of India, situated on the Coromandal coast, 160 km south of Chennai, (11° 56' N; 79° 53'E) comprising of four geographically disconnected regions (Fig. 1). Pondicherry University is a central university, founded in the year 1985 by the government of India. Since then, it has emerged as India's fastest growing central university. As an affiliating university its jurisdiction is spread over the union territories of Puducherry, Lakshadweep, Andaman and Nicobar Islands. Fig. 2 shows the location map of the university, on the northern border of the state.

Fig. 2: Location map showing the Pondicherry University

Study design

The activities of universities promoting environmental sustainability is highlighted in a number of significant declarations, such as the Talloires Declaration (1990), the Halifax Declaration (1991), the Swansea Declaration (1993), the Kyoto Declaration (1993), the Copernicus Charter (1993) and Students for a Sustainable Future (1995), IISD (2002) (IISD Report, 2011). A number of environmental sustainability and related development policies have been promulgated by universities in many parts of the world (Mason et al. 2003; Keniry 1995). A successful comprehensive waste management program often starts at a grass-roots level (Keniry 1995). Higher education institutions (HEI) generate environmental impacts through direct and indirect activities (Loerincik et al. 2003) and as our university is the one and only central institute of our union territory, it has more responsibilities towards research and sustainability, in order to cater to the needs of the present and future generations in a sustainable and holistic manner. Conducting an organic kitchen waste audit is the first critical and crucial step in successful waste management and for overall campus sustainability. Amudham mega mess (boy's hostel cafeteria), which is encircled by the boy's hostels and sports complex ring road within the university campus, is selected as the principle area for the implementation and scaling up for campus sustainability, due to the fact that it is the largest cafeteria and is strategically located.

Approach

Campus students' strength

The university has two directorates, twelve schools, 14 departments and 14 centers and an Institute of Community Development. It has 35,000 students in its 87 affiliated colleges in addition to 5,500 students enrolled on the main campus and another 27,000 students enrolled in long-distance learning programs (PU Statistics 2011). The strength of the ladies' hostel has increased by 220% and the men's hostel by 198% over three years since 2008, obviously with a steep increase in the waste generation as well. Fig. 3 indicates the overall growth in student admission and the hostel strength over the last three academic years. The amount of kitchen organic waste generated was multiplied based on the campus growth, which posed challenges for the proper disposal of waste. As the waste is currently dumped at a site within the campus, this has a high potential to create nuisance and ill effects such as vomiting, nausea, diarrhea, etc.

Fig. 3: Students numbers

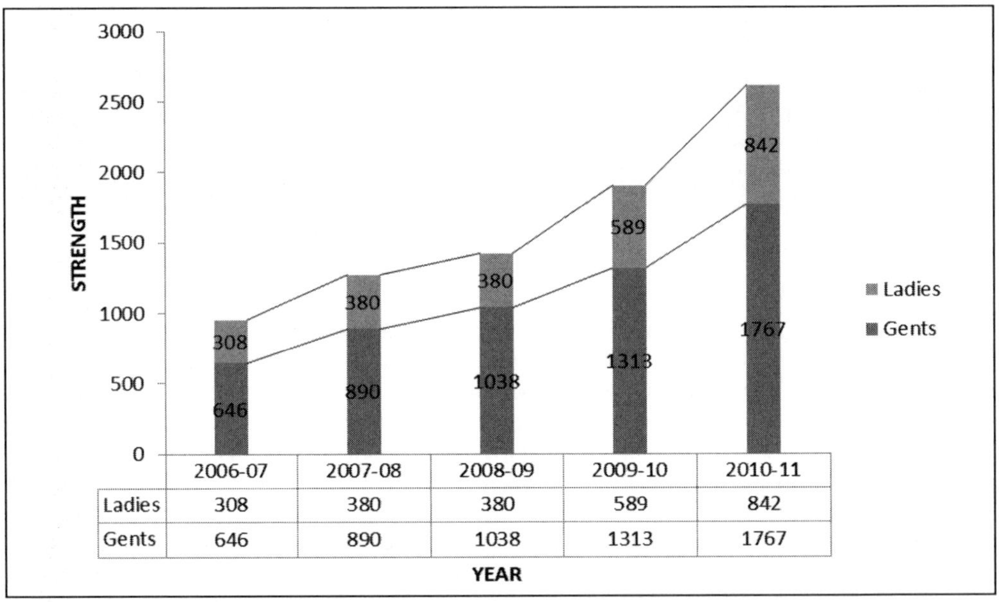

	2006-07	2007-08	2008-09	2009-10	2010-11
Ladies	308	380	380	589	842
Gents	646	890	1038	1313	1767

Source: Registrar's Secretariat, Pondicherry University (2011), pers. Comm., 20 Dec

Apart from this, the generated gray water, along with the monsoon runoff, carries the organic wastes downstream, thereby expanding the levels of pollution to the neighboring lands, as shown in Fig. 4, 5, 6 and 7.

Campus environmental impacts assessment

Current management practices

Waste, especially organic waste, when dumped in open places causes severe environmental pollution to soil, groundwater and surface waters. Food waste is currently collected and dumped behind the Amudham mega mess or transported to a land fill site outside of the campus by four wheelers (Nandhivarman et al. 2011a). Dumped kitchen food waste in turn generates putrid smells, contaminates surface water and seeps into groundwater aquifers.

Waste stream analysis

Effective solid waste management (SWM) requires a complete understanding of the composition of a waste stream as well as the activities that determine its origin and generation in the first place. Generated preparatory food residues like vegetable peelings, leafy vegetables in the kitchen are collected in gunny bags (primary collection), whilst in the dining room, uncooked food residues and uneaten food waste is collected and dumped in an open place behind the dock area of the Amudham cafeteria, which at present acts as a breeding ground for various insects, pathogenic micro-organisms and a gathering site for mammals such as dogs, cats, cows, buffalos and birds who feed on this waste. The remaining waste from this point is transported to the dumping sites outside the university by four wheelers. For this transportation the caterer is paying 150 rupees per day.

Soil analysis by an X-ray fluorescent spectrometer

To assess the degree of point source contamination and the environmental impact of current management practices, two sites were selected, namely the contaminated site A, behind the Amudham mega mess, and normal site B, which is 1000 m away from the principle site. The contaminated site A is the place where kitchen gray water and left-over residues pass through, and normal site B is where there are no anthropogenic activities. The soil samples were collected from two locations at a depth of 15 cm, shade dried, and ground into fine powder using a ceramic mortar and pestle. After sieving through a fine mesh, the samples were analyzed for major and trace elements with the help of X-ray fluorescent spectrometer XRF at the central instrumentation facility lab, Pondicherry University. Eleven major elements and 24 trace elements were analyzed and the results are presented in Fig. 4 for the elements Al, Fe, Si, TI (%); Fig. 5 for the elements Ca, K, Mg, Mn, Na, P, S (%); Fig. 6 for the elements Zr, V, Cr, Ba, Ce, La (ppm); and Fig. 7 for the elements Sc, Co, Ni, Cu, Zn, Ga, As, Rb, Sr, Y, Nb, Mo, Sn, Sb, Cs, Pb, Th, U (ppm), respectively.

The degree of contamination from organic kitchen waste in site A was very high when compared with the isolated area site B. Through the implementation of the integrated organic kitchen waste management strategies, such point-source pollution is eliminated at the start, and in due course the lost nature of the native soil will also get restored through bio fertilizer. Aside from this, based on the XRF values it is guaranteed that the heavy metals leaching into the ground water aquifer also will be conserved for future generations.

Fig. 4: XRF Analysis for the elements Al, Fe, Si, TI (%)

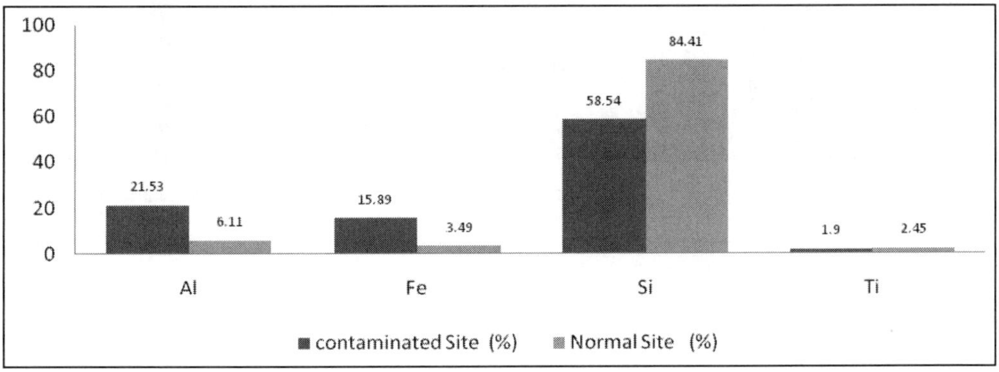

Fig. 5: XRF Analysis for the elements Ca, K, Mg, Mn, Na, P, S (%)

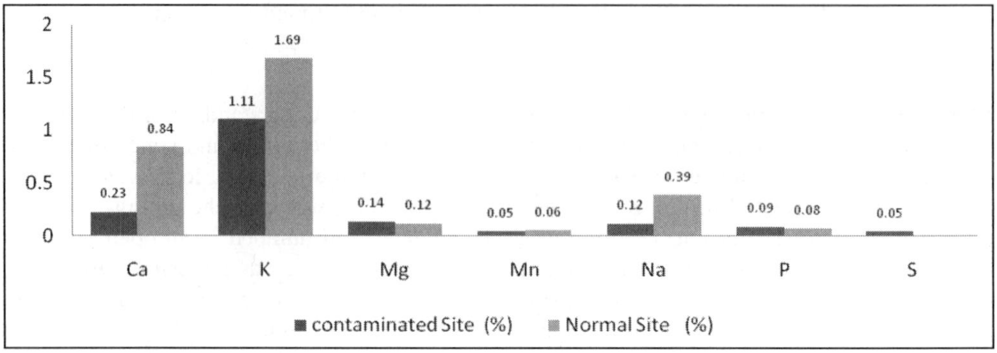

Fig. 6: XRF Analysis for the elements Zr, V, Cr, Ba, Ce, La (ppm)

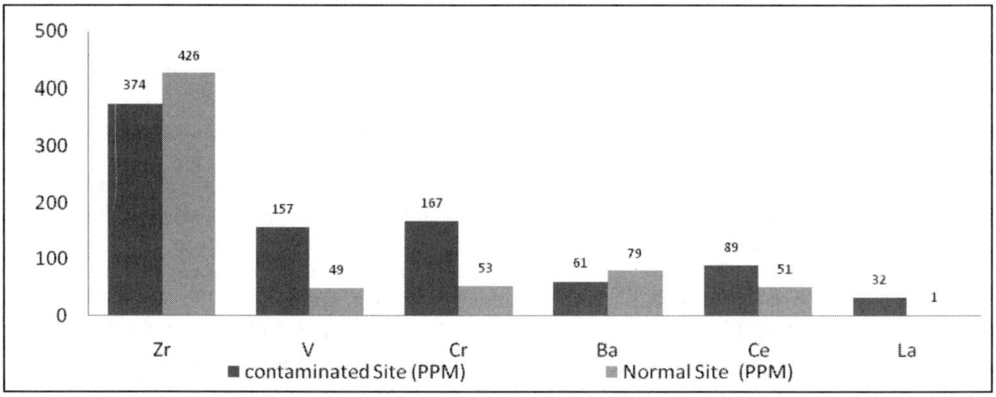

Fig. 7: XRF Analysis for the elements Sc, Co, Ni, Cu, Zn, Ga, As, Rb, Sr, Y, Nb, Mo, Sn, Sb, Cs, Pb, Th, U (ppm)

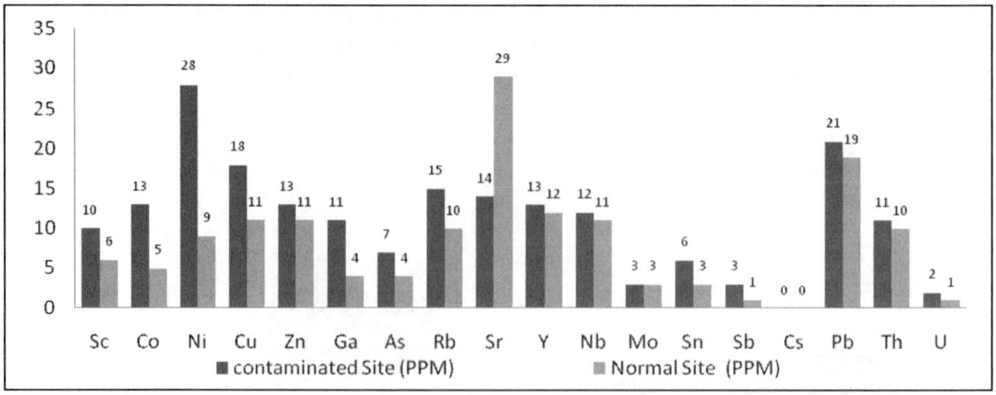

Soil analysis by chromatogram technique

The work of Ehenfried Pfeiffer has developed the simple and useful circular paper chromatography for the analysis of compost and agricultural products (Lederer and Lederer 1957). The same system has been modified and developed by the MCRC group, as an "alternative analytical technology for soil nutrient analysis through image processing chromatograms". In order to visualize the available soil nutrients the samples from the two sites A and B were analyzed. After the completion of the procedure, the chromatogram was analyzed for zonations, color, distance, shape, pattern and the number of spikes developed in the reacting area for interpretation.

Fig. 8: Contaminated Site A, rich in organic matter

Fig. 8 shows well-developed radiating spikes with three different colors in the inner zone, and in case of middle zones, distinguished spikes with sharp edges were found. This indicates that the given sample is rich in organic matter and the soil is thus highly fertile.

Fig. 9: Normal site B, poor in organic matter

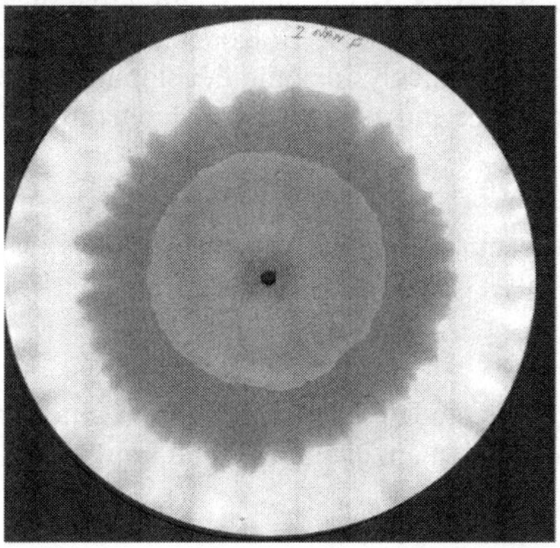

Fig. 9 shows under-developed or no radiating spikes, with two different colors in the inner zone. Clear spikes were not found in middle zone. This indicates that given sample has low organic matter.

Waste characterization

Examining waste generation source is very important, as the characteristics and composition of solid waste vary according to its source (Tchobanoglous et al. 1996). SWM programs based on origin and generation of the source are far more successful than mimicked programs that have been implemented elsewhere (Armijo de Vega et al. 2008). Conducting a waste characterization study is a critical first step in successful waste management planning and advancing for the overall sustainability of the educational institution.

Waste quantification

Based on the kitchen waste audit conducted over a period of 28 days, with 20 full working days and eight weekends, it was estimated that during the academic year 2010-2011, the Amudham mega mess alone generated an average of 0.7 tons of organic waste per day, and without proper waste management strategies it is either dumped or buried (Nandhivarman et al. 2011b).

Evolving best management practices

To manage these scenarios, a detailed action plan was developed for segregating inorganic waste for recycling and safe disposal, followed by the recycling of organic waste for the generation of biogas for kitchens was proposed in the first phase. By the year 1962, the first biogas plant, KVIC floating dome model, popularly known as Indian model was commissioned in India. In 1979, the first fixed dome model (Janata) was commissioned. By 1982, the Indian government launched a countrywide "National Biogas Program" through the ministry of non-conventional energy sources. There is an estimated potential for 12 million domestic biogas plants in India. By the end of 2005 only 3.8 million existed, about 31.66% of the total estimate. According to the census report of India, 2001, 76.32% plants are in rural India, whilst 23.68% are in urban India (Biogas Promotion 2006). Biological treatment has already been demonstrated as one of the most advantageous methods for maximizing recycling and recovering its nutrient components. Anaerobic digestion (AD) of an assorted organic fraction of solid wastes, especially food wastes, is the utmost attractive alternative and the most cost-effective technology (Rao and Singh 2004; Bolzonella et al. 2005; Bouallagui et al. 2005). The biogas programs developed quickly in some developing countries, only because of substantial support from governments and aid agencies (Marchaim 1992; Martinot et al. 2002).

Data recording and analysis

To facilitate the action plan, campus community participation is essential, while periodical data collection, recording and analysis are also mandatory.

Limitations of the study

Summer holidays, fluctuation in student strength, different ratios of organic solids, water, uncooked waste, oils and fats are the main challenges and limitations.

Feasibility study report

In our university we have eleven men's hostels, five ladies' hostels and hostel for foreign students, with three common dining hall facilities respectively. Inadequate management not only leads to a polluting of surface and groundwater through leachate, but also further promotes the breeding of flies, mosquitoes, rats and other disease-bearing vectors, as well as emitting unpleasant odors and methane, which is a greenhouse gas contributing to global warming (Nandhivarman et al. 2011a). Among the cost-effective treatment methods, anaerobic digestion of biodegradable organic waste is the most popular, due to the high energy recovery and low environmental impact (Mata-Alvarez 2000). The benefits of this anaerobic digestion include: dilution of potential toxic compounds, improved balance of nutrients, synergistic effects of microorganisms, increased load of biodegradable organic matter and a better biogas yield (Poggi-Varaldo and Oleszkiewicz 1992). Food waste also has the potential to be converted into methane by anaerobic digestion (AD), and some studies on this have been carried out worldwide (Lai et al. 2009). The thermophilic AD process could be applied to the kitchen waste, increasing the methane recovery rate of biogas yield (Lai et al. 2006).

Methane has more than 20 times the global warming potential than carbon dioxide and the concentration of it in the atmosphere is increasing by 1-2% per year with about 3-19% of anthropogenic landfills (Kumar et al. 2004). The total biogas yield will be increased from 50% to 90% during the operation process (Lissens et al. 2004) and is also a highly recommend option for new and renewable energy resources (Taleghani and Kia 2005). Apart from this, bio fertilizer production from the biogas slurry makes it distinct (Murphy et al. 2004). Anaerobic digestion (AD) is a suitable and easily available method for the treatment of wastewater and organic waste (Jantsch and Mattiasson 2004).

Two biogas plants are proposed to be installed: one to generate 5 kw of power for 20 hours per day with a saving of INR 54,825 per month (this calculation is based on the subsidized power for our region when compared with the neighboring states), and the other to produce biogas substituting 67.5 cylinders per month with a potential reduction in petroleum gas of 25-30% with a saving of INR 101,250 per month (Nandhivarman et al. 2011b).

Reduction of fossil fuel consumption

100 m^3 of plant generates	= 43 kg of biogas per day
Generated 43 kg of gas	= 2.26 No. of cylinder (commercial)
	= 2.96 No. of cylinder (residential)
No. of cylinders used per day	= 8
No. of cylinders generated	= 2.26 per day
Percentage of reduction in fossil fuel consumption	= **28.25%**

Cost benefit analysis for bio gas generation

100 m^3 plant costs	= INR 1,500,000
100 m^3 plant saves	= 25-30% of fossil fuel
100 m^3 plant saves	= 67.5 cylinders per month
100 m^3 plant saves	= INR 101,250 per month
Money back period	= **14.81 months**

Cost benefit analysis for bio energy generation

100 m^3 plant generate	= 5kw power for 20 hrs
Cost of 5kw power	= INR 1825/day @Rs 365/1kw
Cost benefit for 1 month	= INR **54,825** (inc. of ser + min charge)
Total unit cost	= INR 2,000,000
Money back period	= **36.47 months**

Calculated carbon emission reduction

Carbon footprint of the delivery vehicle	
Nature of fuel used	= Diesel
Total kms run by the vehicle	= 270 kms per month
Total diesel consumed	= 18 lts@15 kms/lit
1 liter of diesel emits	= 2.7 kg of CO_2
18 liters of diesel emits	= 131.22 kg of CO_2
vol. of CO_2 reduction per month	= **131.22 kg**

Vermicompost

Composting is gaining attention for treating organic food waste from various agricultural products in different systems (Faucette et al. 2001; Das et al. 2003). The slurry from the digester can be composted using indigenous earth worms and the vermicompost can be used in the kitchen garden, minimizing the application of synthetic fertilizers and leading towards habitat restoration and conservation of biodiversity. Furthermore, considerable income could be generated from the slurry in the form of organic manure and vegetables.

Outreach

As part of this green campus initiative (GCI) for campus sustainability and also in order to set an example for the state and country, the university's GCI is extended to the neighboring schools, colleges and other campus as an outreach activity for our study and action research. This includes Pondicherry Engineering College, Pondicherry Law College, Jawahar Navodhaya Vidhyalaya (residential school), Kendriya Vidhyalaya (central school) and Central Prison (Puducherry). Incidentally, we have also obtained letters of willingness and cooperation from the above partners and work is in progress with Pondicherry Engineering College.

Discussion and recommendations

Under the Environment Protection Act (EPA 1986), the central government has given wide power to take any measures as it deemed necessary for protecting and improving the quality of the environment either by initiating the concepts for sustainability in the residential, campus and regional level, and to comprehend the measures adopted for sustainability in the international gathering. Keeping this as the main agenda, Pondicherry University as a liberal educational and research institution of the Puducherry region, is committed to the environmental stewardship in the sense of citizenship and moral duty, with a desire to teach and lead by setting an example in all its activities for sustainability and outreach. The university has recently started to set a high priority for the campus environment by integrating environmental awareness, personal responsibility, respect and care for the environment, sustainable living, intergenerational responsibility and health and safety into every aspect of life and these are among the fundamental values that guide the planning, decision making, and other procedures as an offshoot of the green campus initiative, on which we have been working for the past two years. Operational research with integrated strategies provide recommendations to the university administration on waste minimization, higher rates of recycling and composting and improving the overall sustainability of the campus waste management program.

Fig. 10: Schematic diagram of integrated organic kitchen waste management

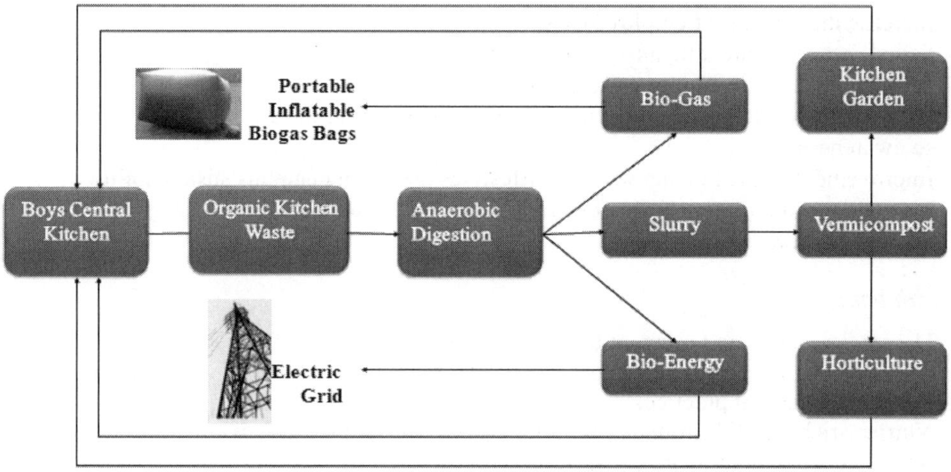

Benefits associated with biogas generation

Strengths, weaknesses, opportunities, and threats to the campus, with reference to both the internal and external factors that influence the stated objectives, are critically evaluated. The SWOT analysis is further used to develop the scenario planning for the integration and advancement of sustainable development concepts (Nandhivarman et al. 2011c). Based on the SWOT analysis is the framework for an integrated organic kitchen waste management focusing on the economic, social and health benefits of the campus. The framework includes the collection of operational resource use statistics, followed by monitoring and investigating waste treatment and management activities, gray water management, potentials for saving electricity and fossil fuel consumption, biodiversity conservation through horticultural activities and kitchen gardens. In the long term, the lost habitat will be restored along with the upgrading of top soil, water-aquifer protection and conservation. The following are the anticipated economic, social, and health benefits:

Economic benefits

1. Segregating organic solid waste for aerobic digestion process would minimize the municipal load to about 50-60%.
2. Methane production would supplement the petroleum gas consumption to about 25-30%.
3. Converting excess biogas so that it generated energy would reduce the dependency on conventional energy.
4. Treatment of organic solid waste with a long-term follow-up plan would usually protect and conserve soil and water.
5. Biogas slurry management along with green litters would increase the availability of easily available nutrient-rich bio fertilizer.
6. Bio fertilizer would decrease the campus dependency on synthetic fertilizer.
7. Generation of income through compost and energy.
8. Improved soil and agricultural productivity as a result of long-term treatment of soil structure and fertility.
9. Long-term benefits on biodiversity conservation, habitat restoration, campus food security.

Social benefits

1. Increase the number of employed staff.
2. Female empowerment thanks to motivating local self-help groups.
3. Waste recycling ethic among the kitchen workers.
4. Earn-while-you-learn schemes for students, which would increase not only interest but also awareness.
5. Improvement in the kitchen worker's lifestyles promoting campus sustainability.
6. Live, day-to-day practical demonstrations for sustainable kitchen waste recycling by the kitchen workers.

Health benefits

1. Decrease of smell and pungent odors.
2. The feeding grounds of scavenger mammals, rodents, and birds are eliminated during the waste management practices.
3. Nutrient-rich landfill grounds Fig. 8 for pathogenic and hazardous microbes get converted into bio diverse organic gardens.
4. 60-70% of the microbial load will be minimized during anaerobic digestion.
5. The pathogenic salmonella sp. found in the kitchen waste will be eliminated by the ninth day of digestion (Kunte et al. 2000).
6. Associated diseases like malaria, cholera, typhoid will be prevented.
7. E. coli contamination, vomiting and headaches, which are common during the monsoon season, are prevented.
8. Cross contamination of soil during monsoons is prevented.
9. Degraded beneficial soil flora and fauna will be restored in due course.
10. Directly stops water contamination and indirectly prevents leachate entering the ground water aquifer.
11. Contributes directly to the campus and global environmental health.

Policy formation

A "University's Sustainable Policy & Planning Committee (USPPC)" is proposed for facilitating campus sustainability, built on a governing body which includes environmental sustainability coordinators, an Information Resources Department (IRD), an Institutional Review Board (IRB), a Safety-In-The-Workplace Committee (SWC), research supervisors and coordinators, research scholars, regional, national and international collaborators and volunteers. This strategic planning is to be seen as a guide to assist in the development of a comprehensive environmental sustainability process that suits the campus. The USPPC serves as the central executive body for all the campus governance, and along with the Green campus Initiators (GCI), advises the vice chancellor and registrar with regard to strategic environmental planning, taking comprehensive steps to address the significant issues which affects the campus environment. The evolved and implemented strategic campus action plan, fostering organic waste management along with environmental education and outreach, will be of significant contribution to the United Nations Decade of Education for Sustainable Development, 2005-2014.

Conclusion

Anaerobic digestion is an attractive, cost-effective, eco-friendly route to minimize landfill dumping that generates a high-quality renewable fuel, and reduces GHG emissions. Co-digestion strategies are widely applied in order to enhance the methane production in agricultural biogas plants. For instance, in Denmark (Raven and Gregersen 2007) and Germany (Weiland 2010) the digestion of manure and organic waste is a well-established technological practice. We at our university have developed a novel "five-in-one" innovative model that combines renewable energy and organic horticultural activities. It consists of a biogas digester for biogas to be stored in portable inflatable bags, a bottling plant, biogas to bio-energy unit, slurry composting and a vermicomposting unit and a greenhouse for horticultural activities, therefore called "five-in-one". It is established based on the principles of the environment, economy and systematic engineering, fostering campus environment sustainability.

References

Armijo de Vega, C., Ojeda Benitez, S. and Ramirez Barreto, M.E. (2008), Solid waste characterization and recycling potential for a university campus. *Waste management* 28, Supplement I, pp. S21-S26.

Biogas Promotion (2006), *Biogas Promotion – India*. Available online at http://www.hedon.info/docs/20060406_Biogas_promotion_India.pdf(accessed 20 December 2011).

Bolzonella, D., Pavan, P., Mace, S. and Cecchi, F. (2005), *Dry Anaerobic digestion of differently sorted organic municipal solid waste: a full scale experience*. 4th International Symposium of Anaerobic digestion of Solid Waste (Copenhagen-Denmark), Vol. 1, pp. 85-92.

Bouallagui, H., Touhami, Y., BenCheikh, R. and Hamdi, M. (2005), Bioreactor performance in anaerobic digestion of fruit and vegetable wastes. *Proc. Biochem*. 40, pp. 989-995.

Danielle, P.S., Arthur, L.F. and Annie, L.B. (2010), Reducing solid waste in higher education: The first step towards 'greening' a university campus. *Resources, Conservation and Recycling* 54 (11), pp. 1007-1016.

Das, K.C., Tollner, E.W. and Eiteman, M.A. (2003), Comparison of synthetic and natural bulking agents in food waste composting. *Compost Science & Utilization* 11 (1), pp. 27-35.

EPA (1986), *The Environment (Protection) Act*. Published in the Gazette of India. Available at http://envfor.nic.in/legis/env/env1.html (accessed 5 November 2011).

Faucette, B., Das, K.C. and Risse, M. (2001), University tests in-vessel composting of food residuals. *Biocycle* 42 (1), pp. 68-70.

IISD Report (2011), *Declarations for Sustainable Development: the Response of Universities*. Available at www.iisd.org/educate/declare.htm (accessed 7 December 2011).

Jantsch, T.G. and Matttiason, B. (2004), An automated spectrophoyometric system for monitoring buffer capacity in anaerobic digestion processes. *Water Research* 38, pp. 3645-3650.

Keniry, J. (1995), *Ecodemia: Campus Environmental Stewardship at the Turn of the 21st Century – Lessons in Smart Management from Administrators. Staff and Students*. National Wildlife Federation, Washington, DC.

Kumar, S., Gaikwad, S.A., Shekdar, A.K., Kshirsagar, P.K. and Singh, R.N. (2004), Estimation method for national methane emission from solid waste landfills. *Atmospheric Environment* 38, pp. 3481-3487.

Kunte, D.P., Yeole, T.Y. and Ranade, D.R. (2000), Inactivation of Vibrio cholera during anaerobic digestion of human night soil. *Bioresource Technology* 75, pp. 149-151.

Lai, C.M., Chen, S.W., Chen, K.H., Lee, C.C., Liu, K.I. and Wei, C.B. (2006), The composting of household food waste in Teipai city. *Biomass Energy Soc China* 25, pp. 53-64.

Lai, C.M., Ke, G.R., Chung, M.Y. (2009), Potentials of food wastes for power generation and energy conservation in Taiwan. *Renewable Energy* 34, pp. 1913-1915.

Lederer, E. and Lederer, M. (1957), *Chromatography* (2nd ed.). Elsevier Publishing Co, Amsterdam

Lissens, G., Thomsen, A.B., Baere, L., Verstraete, W. and Ahring, B. (2004), Thermal wet oxidation improves anaerobic biodegradability of raw and digested biowaste. *Environmental Science and Technology* 38, pp. 3418-3424.

Loerincik, Y., Suh, S. and Jolliet, O. (2003), *Evaluation of the environmental impact of the Internet Infrastructure within a University*. Proceedings of the 11th LCA case study symposium, pp. 81-84.

Marchaim, U. (1992), Biogas Processes for Sustainable Development. *Bull. FAO Agric. Services, Rome* 95, pp. 165-193.

Martinot, E., Chaurey, A., Lew, D., Moreira, J.R. and Wamukonya, N. (2002), Renewable energy markets in developing countries. *Annual Review of Energy and the Environment* 27, pp. 309 -348.

Mason, I.G., Brooking, A.K., Oberender, A., Harford, J.M., Horsley, P.G. (2003), Implementation of a zero waste program at a university campus. *Resources, Conservation and Recycling* 38 (4), pp. 257-269.

Mata-Alvarez, J., Macé, S., Llabrés, P., (2000), Anaerobic digestion of solid wastes. An overview of research achievements and perspectives. *Bio resource Technology* 74, pp. 3-16.

Murphy, J.D., McKeogh, E. and Kiely, G. (2004), Technical/economic/environmental analysis of biogas utilisation. *Applied energy* 77, pp. 407-427.

Nandhivarman, M., Poyyamoli, G. and Edwin, G.A. (2011a), *Environmental Sustainability – a conceptual frame work for Pondicherry University, Puducherry, India – a case study*. Paper presented at the 11th International Congress of Asian Planning Schools Association, 19 September 2011, Tokyo, Japan.

– (2011b), *Evolving a water policy framework 2011 for Green Campus Initiative – a case study of Pondicherry University, India*. Paper presented at the 8th International conference on urban watershed management, 7 September 2011, Beijing, China.

– (2011c), *Water foot print as a tool for sustainable management of ground water resources in Pondicherry University, India*. Paper presented at the 8th International conference on urban watershed management, 7 September 2011, Beijing, China.

PU Statistics, (2011), *Vital Statistics of Pondicherry University*. Available online at http://www.pondiuni.edu.in/content/vital-statistics (accessed 20 December 2011).

Poggi-Varaldo, H.M. and Oleszkiewicz, J.A. (1992), Anaerobic co-composting of municipal solid waste and waste sludge at high total solid level. *Environmental Technology* 13, pp. 409-421.

Rao, M.S., Singh, S.P. (2004), Bio energy conversion studies of organic fraction of MSW: kinetic studies and gas yield-organic loading relationships for process optimization. *Bioresource Technology* 95, pp. 173-185.

Raven, R.P.J.M. and Gregersen, K.H. (2007), Biogas plants in Denmark: successes and setbacks. *Renewable and Sustainable Energy Reviews* 11, pp. 116-132.

Taleghani, G. and Kia, A.S. (2005), Technical-economical analysis of the Saveh biogas power plant. *Renewable Energy* 30, pp. 441-446.

Talloires Declaration (1990), *Report and Declaration of the Presidents' Conference*. Talloires, France. Available at www.ulsf.org/programs_talloires_report.html (accessed 7 December 2011).

Tchobanoglous, G., Theisen, H. and Vigil, S. (1996), *Integrated Solid Waste Management*. McGraw-Hill.

UNDP Report (1997), *Energy After Rio: Prospects and Challenges*. Available at www.hubrural.org/pdf/pnud_energy_after_rio.pdf/// (accessed 7 December 2011).

Weiland, P. (2010), Biogas production: current state and perspectives. *Applied Microbiology and Biotechnology* 85 (4), pp. 849-860.

Chapter 60

Studying Sustainability through Research with the Floating University Program

Lev Karlin, Mikhail Shilin, Tatjana Eremina, Alexandra Ershova and Alexey Suzyumov

Abstract

The role of the innovative UNESCO program "Floating University", and specifically its component in the Baltic Sea, the Baltic Floating University (BFU), is described within the system of professionals' training in the areas of study, research, management and sustainable development of coastal regions. The "Floating University" is a long-term multidisciplinary educational program started by UNESCO in 1991. BFU was initiated by the Russian State Hydrometeorological University and launched in 1993 (with the support of HELCOM), becoming well known internationally for high-quality training as well as research focused on supporting the needs of qualified marine specialists and oriented towards the sustainable development of coastal regions. Over the years it has been successfully implementing the "training-through-research" principle, combining the advantages of the formal training of students with experiences from advanced research in environmental and marine sciences, under the supervision of leading scientists. Between 1993 and 2009, a total of over 270 students from 28 countries, mostly from the Baltic region, participated in the BFU cruises. This paper presents a set of innovative ideas that represent the core of the project, and shows the effectiveness of their practical realization in research cruises. Lessons learned through the BFU project are also presented.

Introduction

The Intergovernmental Oceanographic Commission of UNESCO (IOC) was founded in 1960 with the recognition that "the oceans, covering some 70% of the Earth's surface, exert a profound influence on mankind and upon all forms of life on Earth. In order to properly interpret the full value of the oceans to mankind, they must be studied from many different points of view" (UNESCO 1960). Since the oceans are central to both global environmental change and sustainable development, it has become of increased interest to all countries to develop ocean research and develop a truly global and mutually supportive activity. Training, education and mutual assistance have to be an integral and important part of any major program in marine science.

Shipboard training has always been on the IOC agenda (Suzyumov 2010). However, throughout the years the approach has changed from occasional, non-systematic training provided mostly in the use of modern oceanographic equipment and new research technologies, to long-term capacity development and multi-disciplinary educational programs.

The "Floating University" is an example of such a program started by UNESCO in 1991 and becoming an integral part of the IOC program in 1996. This was a time of notable changes in the global political climate as a result of Mikhail Gorbachev's "perestroika" that led to the speedy development of new ideas and approaches in science and education. New political realities required urgent training of a whole new generation of Russian scientists, stakeholders and managers with "new thinking" in all areas of science, including marine and environmental sciences.

To foster an international community of young scientists who would be well-trained and open to the outside world, come from various societies and cultures but appreciate these differences and gain from them, speak openly to each other, be capable of handling advanced equipment manufactured in various countries of the world, manage high-quality scientific data, collected and interpreted under the guidance of leading professionals from universities and academic institutions – all of this was the basic idea of the "Floating University" program and its "training-through-research" (TTR) approach. In addition to IOC and UNESCO, many countries have provided financial, material and intellectual support to the program, including Belgium, Estonia, Finland, France, Germany, Greece, Israel, Italy, Monaco, Morocco, the Netherlands, Portugal, Russia, Spain, Turkey, the Ukraine and the United Kingdom. The program has several branches: in the Atlantic Ocean it has been focused on the Geosphere-Lithosphere-Hydrosphere coupling processes at continental margins; in the Baltic Sea region on environmental matters; in the Caspian Sea region on the awareness and protection of the marine environment, and in the Indian-Pacific region on the training of national specialists from island states.

The "Baltic Floating University" Project (BFU) was initiated by the Russian State Hydrometeorological University (RSHU, St. Petersburg) and started in 1993 (Suzyumov 1995; Sychev 1995). Soon the Helsinki-based Baltic Marine Environment Protection Commission (HELCOM) gave its co-sponsorship to BFU. The RSHU partners who collaborated with the BFU events were the Hydrographic Service of the Russian Navy, the World Wide Fund for Nature (WWF) and the Gulf of Finland Environment Society "Suomen Lahti" (SULA) (Komaritsin and Sytchev 1996).

During the 25 years of its successful operation, BFU has become a powerful and effective tool in the work and collaboration of universities in the Baltic countries (Estonia, Finland, Germany, Latvia, Lithuania, Poland, Sweden) as well as when it comes to the development of a sustainable future within the Baltic Region (Eremina et al. 2005; Karlin and Suzyumov 2005; Karlin et al. 2008).

Methodological aspects: main principles and innovations

Since 1993, the BFU has provided training to a young generation of researchers collecting and analyzing hydro-meteorological, oceanological and environmental data. The BFU combines the advantages of the formal training of students with experiences from advanced research in environmental and marine sciences under the supervision of leading scientists from European universities, thus implementing the TTR approach. At the core of the program are annual marine expeditions and post-cruise scientific conferences.

A recognized BFU innovation is a two-ship operation. During its cruises, the BFU uses two training platforms in parallel: a big oceanographic vessel – such as, for example, the RV "Sibiryakov", belonging to Hydrographic Service of the Russian Navy (2,700 tons) – for studies in the open Baltic Sea (Fig. 1), and a sailing catamaran "Centaurus" (11 m and 4.2 tons) for investigations in shallow coastal waters (Fig. 2). In addition, there is also the possibility for underwater research using SCUBA divers (especially useful for investigations around small islands).

Fig. 1: RV "Sibiryakov" (Hydrographic Service of the Russian Navy) with the BFU expedition onboard

Fig. 2: Sailing catamaran "Centaurus" (www.centaurus.ru)

The key ideas of the BFU program are:

- the "training-through-research" principle (TTR);
- multi-disciplinary approach and team work in small research groups;
- combining shipboard training with studies of terrestrial coastal ecosystems under anthropogenic pressure;
- effective transfer of skills and experiences;
- developing intercultural contacts;
- orientation on the practical use of research results for the support of sustainable development of coastal zones (since 1999);
- cyclic training.

The usual annual TTR cycle of the BFU program includes:

- preparation of an international cruise by the executive committee and definition of scientific tasks;
- a BFU cruise that combines an extensive research program; shipboard data collection, processing and analysis; on-the-job training; daily lectures and seminars;
- site visits, excursions, study tours;
- after the cruise – exchange of the participants for data processing and analysis;
- a post-cruise conference to present and discuss the results and to co-ordinate with other regional studies;
- preparation of scientific reports and publications in the special "BFU Research Bulletin".

Between 1993 and 2009, over 270 undergraduate and postgraduate students from 28 countries of Europe, Asia, Africa and America participated in the BFU project. These included all Baltic countries, countries from other parts of Europe as well as Brazil, China, Columbia, Mexico, Mozambique, Namibia and Venezuela.

The annual BFU cruises are devoted to the investigation of hydrophysical, hydrochemical and hydrobiological processes of the Baltic Sea and their variability in the Baltic Proper and coastal parts of the sea, geoecological monitoring in the Gulf of Finland and on the Kaliningrad shelf. All students work in different laboratories (geology, chemistry, hydrology, biology, meteorology and data processing) (Fig. 3-5), attend lectures and seminars (Fig. 6, 7), make presentations of their own scientific projects and contribute to the discussion of preliminary results of the expedition.

Fig. 3. Sampling of environmental information

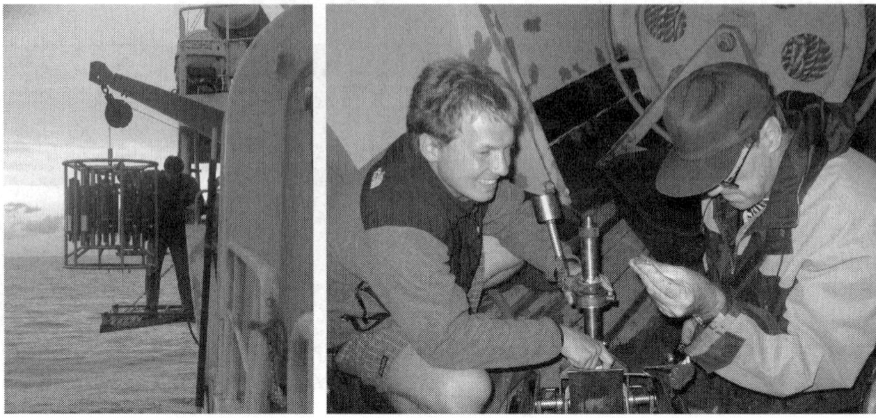

Fig. 4: Sampling material for hydrochemical research

Fig. 5: Analysis of plankton samples

Fig. 6: Onboard seminar by professor Alexey Nekrasov (RSHU)

Fig. 7: Onboard seminar by professor Lars Rydén (Uppsala University, Sweden)

Students can choose where to work themselves, and the choice depends on their career interests. The main idea, however, is to let everyone work and at the same time see all kinds of work done onboard. The official language of the BFU is English, but foreign students also attend Russian language classes. Social activities play an important role during the expeditions.

A reasonable amount of practical work, a rather intensive course of lectures, presentations and scientific discussions, combined with good rest and a unique mixture of cultures creates an unforgettable atmosphere onboard the vessel and supports the specific spirit of the BFU. The final point of the BFU cruises is the city of St. Petersburg. Many of the BFU students come to St. Petersburg – a beautiful and historical city and a home city of the BFU project – for the first time. The Russian students guide them through the city and St. Petersburg's world-class museums. As a result of these activities and personal contacts the BFU project contributes in a substantive way to the "Promoting intercultural dialogue" program of UNESCO.

1998 became an outstanding year for the BFU project as when celebrating the International Year of the Ocean the BFU-98 cruise went as far as the North Atlantic and visited the World Ocean Exhibition (Lisbon, Portugal). In many respects this was a unique cruise as trainees on board were (mostly) schoolchildren from Finland, France, Georgia, Germany, Greece, Lithuania, Netherlands, Russia, Turkey and the United Kingdom – the subject of the cruise was to teach them about the fragility of the Ocean's ecosystem and the need to protect it from uncontrolled human activities and irresponsible attitude to the living resources. The BFU experience of realizing the TTR principle as part of the training of young professionals in marine sciences was summarized and presented at the exhibition, and received the IOC and UNESCO honorary certificates.

It was shown that the Baltic region currently faces numerous challenges concerned with the changing marine and coastal environment. These include the eutrophication of water, hypoxia, chemical pollution and the destruction of natural biotopes as a result of anthropogenic activities like hydrotechnical construction, dredging, etc. To overcome the complex problems of the marine and coastal environment, the scientific basis for the decision making should be created by qualified experts. Training and education in marine environmental sciences becomes a top strategic priority for the sustainable development of coastal regions.

On the basis of the collected experience, the BFU have developed a partnership and co-operation with the new UNESCO Unit for Coastal Zones and Small Islands (CSI UNESCO) dating back to 1999. UNESCO priorities for action have currently taken a new direction, focusing on sustainable development in coastal regions. Coastal management was recognized as one of the key challenges facing maritime nations. Some of the technical and environmental issues to be considered in the process of management and decision making included:

- loss of biodiversity and decrease of ecosystem productivity in coastal areas;
- environmental pollution and degradation;
- depletion of renewable and non-renewable natural resources;
- natural disaster reduction, including coastal instability problems.

In addition, a multitude of social, economic and cultural factors have to be considered in order to achieve lasting solutions. As a tool for sustainable decision making, the Integrated Coastal Zone Management was proposed.

As a result of the co-operation with the CSI UNESCO, the BFU program was reoriented, with a revision of activities and priorities. The development of socio-economic and cultural components of the BFU program was started, involving both natural and social sciences (human geography, anthropology, sociology, economics, law etc.) for developing the cross-sectoral, integrated approaches to problems of coastal regions. Since then, the main scientific goal for the BFU research program is to understand the principles and to form the scientific base for the sustainable development of the coastal zone of the Baltic Sea.

The 2009 global crisis became an obstacle for training at sea. The number of research cruises significantly diminished. Under the IOC umbrella only a small-scale BFU cruise in the coastal Baltic was carried out in the summer of 2009. However, it is notable that even despite the severely limited funding and to the limited opportunity for the TTR cruises at present, the interest of students in this program is constantly growing; students keep applying to participate in the BFU cruises.

BFU scientific component

Contributing to the HELCOM efforts in monitoring the environmental situation in the Baltic Region, BFU carries out standard complex observations at the so-called HELCOM monitoring stations. Using oceanographic data from the Baltic Sea, BFU contributes to the Baltic Ocean Observing System (BOOS, part of the Global Ocean Observing System, a program of the Intergovernmental Oceanographic Commission – IOC of UNESCO), as well as to the BED – Baltic Environment Database (Stockholm University, Sweden). Through the project, RSHU initially contributes to a number of national programs, the Federal Target Program "The World Ocean". The project works were financed by the Ministry of Industry and Science of the Russian Federation, Ministry of Education of the Russian Federation and the State company "Sevmorgeo".

The main research subjects during the BFU activities were:

1. Study of eutrophication processes and "water blooms" in relation to nutrient dynamics and composition of phytoplankton. Revealing relationships between chlorophyll "a" concentrations and chemical variables (Eremina and Lange 2003).
2. Effect of alien species invasions to the Baltic Sea (*Cercopagis pengoi*) (Panov et al. 1996).
3. Monitoring of benthic communities in the Gulf of Finland and dynamics of distribution of the "lifeless zone" (Eremina and Karlin 2006; Eremina and Lange 2007).
4. Mapping of radioactive traces of the Chernobyl catastrophe in the Luga – Kopor'e area (Kankaanpaa 1995).
5. Seabed and benthos sampling and photography of small islands in the Russian sector of the Gulf of Finland. Results were used as an information basis in the process of organizing the new marine protected area "Ingermanlandsky" in 2010.
6. Study of pollution of bottom sediments in the Gulf of Finland (Rybalko and Fedorova 2007).
7. Research into the influence of dumped chemical munitions on the state of marine environment in deepwater areas of the central and southern Baltic and in the Bornholm deep (areas of World War II chemical munitions dumping) (Eremina et al. 2002).
8. Study of the effect of the North Sea water inflow to the Baltic Sea on the ecosystem (Eremina and Lange 2007).
9. Perspectives of sustainable tourism in the coastal zone areas and small islands off the Gulf of Finland (Grigor'eva, 2005).
10. Assessment of the ecological situation in the areas where the construction of the anthropogenic macro-objects in the coastal zone of the Baltic Sea takes place (Ust-Luga harbor, Primorsk harbor, Flood Barrier, Nordstream pipeline, Batareynaya Bay) (Shilin et al. 1996).

As a direct result of the long-term BFU activity, the database on the state of the Baltic Sea ecosystem was created including hydrological, hydrochemical, hydrobiological data and information about socio-economic activities. The principles of environmental safety and sustainable co-operation between basic stakeholders in the coastal zone are revealed in this database. The number of pilot projects in the areas of new harbors and hydro-technical objects have been also been started, oriented towards improving relations between stakeholders.

In addition to on-board facilities, this project benefits from the utilization of the UNESCO computer-based learning modules on the application of coastal and marine image data from satellites and other sensors. In 2006 an agreement was signed between UNESCO and RSHU for the establishment at RSHU of the UNESCO/IOC Chair in Remote Sensing and Modeling in Oceanography.

BFU program publications

The BFU Research Bulletin (published annually by RSHU with support from the IOC) is a forum for researchers and students participating in the Baltic Floating University to present the results of their research carried out during the BFU cruises and to share their experiences and views. This annual publication contains papers based on BFU data, presentations by students, and also articles with students' impressions of BFU cruises. Students from different countries and universities are encouraged to work together (many published articles are written by two or three students). For many students this is their first experience of writing research papers.

A pilot issue of the BFU Research Bulletin was produced in May 1995, presenting the results of the first two years of operation. The second issue, published in 1996, presented an overview of the BFU activities in 1995. It was followed by an IOC publication "Baltic Floating University: Training Through Research in the Baltic, Barents and White Seas – 1997" (UNESCO 1998). The third BFU Research Bulletin was issued in 1999 and was devoted to the International Year of the Ocean – 1998. From 2002 onwards, the BFU Bulletin has become an annual publication. In 2007 the ninth issue was published (co-sponsored by IOC), presenting the 2006 activities. Results of other cruises have been published on the BFU website, hosted by RSHU (http://www.rshu.ru/eng/bfu/).

Lessons learned from the BFU project

The "Baltic Floating University" and its "training-through-research" principle has become an international trademark of training and high-quality research in order to support the sustainable development of coastal regions. Moreover, the program has promoted regional and interregional cooperation between universities that teach marine science, integrated coastal zone management and concepts of sustainable development. Close collaboration has been established with a number of universities in Western Europe who regularly send their students on the BFU cruises (Aveiro, Portugal/Cadiz, Spain/Plymouth, UK/Riga, Latvia and Klaipeda, Lithuania), as well as with the Baltic University Programme in Uppsala, Sweden.

The following guidelines can be identified to make the long-year program successful:

- Exciting and challenging problems of current interest that need to be identified for each cruise, as this will attract the interest of students, leading researchers and funding agencies;
- Programs need to have a clear and concise structure that ensures professional training at sea by experts in the discipline;
- The student mix must be suitably diverse to accommodate many of the countries in the region;
- Cruises should be constructed to ensure a strong interdisciplinary approach;
- Time should be allocated to give students a fair idea of the many different aspects of sustainability;

- The onboard training is essential but not the only component: data processing and analysis, presentation and discussion of results at national and international forums and elaboration of recommendations for local stakeholders, administrators and decision-makers should all be realized;
- The collaboration and interaction with other international programs is important, such as with the Baltic University program initiated by the Uppsala University (Sweden) and the emphasis on the distance learning (Shilin 2007).

Conclusions

Several final remarks can be outlined to conclude the activities in frames of the BFU project:

- The long-term success of the BFU project is due to a number of innovations, such as the use of the TTR principle, parallel work of a big research vessel and a sailing catamaran in coastal waters, orientation on the coastal zone and using the principles of Integrated Coastal Zone management (ICZM);
- The BFU project became an effective tool for training experts in marine sciences; several BSc and MSc theses were successfully defended by students from RSHU, as well as from the UK and Portugal universities, based on the BFU cruise data;
- Such projects at BFU allow the integrating of knowledge, skills, and approaches, and promote the diversity of cultures, approaches and at last all forms of diversity;
- At the same time such projects are extremely dependent on crisis events in the society as they require consistent funding (provision of TTR cruises, equipment and personnel);
- Data obtained during the BFU cruises can be considered of interest not only for experts in marine sciences (many scientific papers have been published throughout the program), but these data represent valuable pieces of information for numerous local stakeholders promoting the sustainable optimization of their activities;
- In general, such program as the BFU project can be considered a valuable attribute of modern sustainable society.

The authors believe that the BFU, through its activities, contributes in a substantive way to the "Promoting Interculturral Dialogue" program of UNESCO and to the sustainable development of coastal zones and regions. In its own way the project enhances the development of the universal culture of peace and tolerance – fundamental objectives of the United Nations.

References

Grigor'eva, V. (2005), Some aspects on perspectives of sustainable tourism in the coastal zone areas and small islands of the Gulf of Finland. *BFU Res. Bull.* 7, pp. 20-23.

Eremina, T., Karlin, L. and Fedorova, N. (2005), *Results of the Ecological Investigations of the Baltic sea within the Framework of the Baltic Floating University Project*. Proc. Int. Environmental Forum "Baltic Sea Day", p. 130-132. St. Petersburg.

Eremina, T. and Karlin, L. (2006), *Hydrochemical conditions of the Baltic sea deep layers in 2002-2005 and their impact on eutrophivation process*. Proc. Int. Environmental Forum "Baltic Sea Day", pp. 191-192. St. Petersburg.

Eremina, T., Korneev, O., Rybalko, A., Fedorova, N. and Frumin G. (2002), Preliminary results of dumped chemical munitions research in Bornholm Deep. *BFU Res. Bull.* 4-5, p. 54-56.

Eremina, T. and Lange, E. (2003), Estimation of changes in the ecosystem state from observations in the Gulf of Finland". *BFU Res. Bull.* 6, pp. 15-19.

Eremina, T.R. and Lange, E.K. (2007), The effect of inflow from the Baltic sea on the ecosystem of the Eastern part of the Gulf of Finland. *BFU Res. Bull.* 9, pp. 29-33.

Kankaanpaa, H. (1995), Radioactive Cs-137 in sediment cores from the Luzhskaya-Koporskaya area, eastern Gulf of Finland. *BFU Res. Bull.* 1, pp. 7-10.

Karlin, L.N., Eremina, T.R. and Ershova, A.A. (2008), *Baltic Floating University Programme – a way to international integration of marine education*. Proc. Int. Environmental Forum "Baltic Sea Day", pp. 248-249. St. Petersburg.

– (2010), *IOC UNESCO "Baltic Floating University" Programme – towards international integration of marine education*. 50 Years of Education and Awareness Raising for Shaping the Future of the Oceans and Coasts. Proc. int. conf., pp. 189-190 St. Petersburg.

Karlin, L.N. and Suzyumov, A.Ye. (2005), Baltic Floating niversity: a dozen of years at service of marine science, education and training. *BFU Res. Bull.* 8, pp. 4 -5.

Komaritsin, A. and Sytchev, V. (1996), The Hydrographic Service of the Russian Federation Navy and International Programmes. *BFU Res. Bull.* 2, pp. 88-90.

Panov, V., Krylov, P. and Telesh, I. (1996), The Caspian predatory cladoceran *Cercopagis pengoi* invades the Gulf of Finland. *BFU Res. Bull.* 2, pp. 80-82.

Rybalko, A.E. and Fedorova, N.K. (2007), State of the geoenvironment of the Baltic Sea: geological hazards and exogenic processes under the increased anthropogenic stress. *BFU Res. Bull.* 9, pp. 46-48.

Shilin, M.B. (2007), Inter-Programme Cooperation. *BFU Res. Bull.* 9, pp. 62-70.

Shilin, M.B., Pogrebov, V.B., Nitishinsky, M.A. and Galjushina, N.P. (1996), Shoreline monitoring sites for measuring anthropogenic pressure on the eastern Gulf of Finland environment. *BFU Res. Bull.* 2, pp. 68-71.

Suzyumov, A.E. (1995), The Floating University facility and Training-through-Research. *BFU Res. Bull.* 1, p. 2.

– (2010), *Capacity development through training at sea: history and lessons learned from the last 20 years*. 50 Years of Education and Awareness Raising for Shaping the Future of the Oceans and Coasts. Proc. int. conf., pp. 287-290. St. Petersburg.

Sychev, V.I. (1995), The Baltic Floating University and the Global Faculty Network. *BFU Res. Bull.* 1, pp. 4-6.

UNESCO (1998), *The Baltic Floating University: Training Through Research in the Baltic, Barents and White Seas – 1997*. IOC Technical Series, No. 53.

UNESCO Document NS/167 (1960), *The Intergovernmental Conference on Oceanographic Research*. Copenhagen, 11-16 July, p. 6.

Chapter 61

Change from Below: Student Initiatives for Universities in Sustainable Development

Moritz A. Drupp, Alejandro Esguerra, Lena Keul, David Löw Beer, Simon Meisch and Felix Roosen-Runge

Abstract

To advance the discussion on how the transformation of universities may lead to sustainable development may be achieved, the authors examine an often-overlooked actor category: student initiatives. By drawing on a case study of the initiative Greening the University e.V. (University of Tübingen, Germany) the authors conceptualize the role of student initiatives for institutional change through the notion of "communicative arena". The authors argue that student initiatives are crucial for the transformation process toward a university in sustainable development for three reasons: they facilitate change by acting as (1) an institutional innovator, (2) a "boundary agent", and (3) creator of social and institutional learning spaces. Further, the authors discuss factors that limit a student initiative's potential to unfold its transformative power. The article concludes with recommendations for policy-makers, university leaders and students on how to foster this "change from below".

Introduction

Universities play an important role in bringing about change for sustainable development (SD). They contribute expertise required to understand unsustainable processes, develop alternative sustainable strategies and also educate future decision-makers. Educational politics is well aware of universities' special responsibility for SD. Since the Talloires Declaration in 1990, many international documents suggested strategies to include SD issues into university policies (see e.g. Leal Filho 2011). Universities are not only places of research and learning but also places of living and working, thus leaving "ecological footprints". A sustainable university not only realizes an efficient use of resources and overall ecological viability, but also enables more sustainable work and lifestyles.

In this paper the authors are concerned with the role of student initiatives (SIs) for universities in SD. The authors argue that SIs form an important but overlooked actor category in discussions on sustainable universities. While admitting that students are to be included in designing SD policies, key documents on SD at universities keep silent who "the students" are in organizational terms. In most cases students are referred to only as addressees of education for SD (ESD), but not as participants in the formation of new SD knowledge, as organizers of their own educational success or as initiators of new SD curricula and university-wide institutional change. Among those key documents, only the 2009 Bonn declaration outlined "the commitment, solidarity and potential of youth and their organizations and networks in enhancing ESD", thereby highlighting student organizations.

Within the scientific literature, SIs' potential as change agents is also not fully acknowledged. Furthermore, the scientific literature has not fully acknowledged the potential of SIs as actors in the transformation towards sustainable universities (for notable exceptions see Chapter 2). The central question this paper addresses is: *what role do student initiatives play in the institutional transformation process toward a university in sustainable development*? To theoretically capture the ways in which SIs can foster institutional change, the authors draw on approaches that take discourse, communication, and knowledge seriously.

Over the last decade, discursive approaches have proved useful to the study of environmental politics and institutional change (Hajer and Versteeg 2005). Among others, scholars elaborated on the notion of discourse coalitions (Hajer 1995), interpretive communities (Johnstone 2005) or epistemic communities (Haas 1992). Vivien Schmidt assembles the variety of concepts under the term "discursive institutionalism" (Schmidt 2012). The authors follow this school of thought with a special interest for the epistemic and communicative practices within a community.

In particular, the authors probe the analytical power of the term *communicative arenas* to examine how a particular SI has changed its university toward becoming a university in SD. That is to say, *communicative arenas* function as an analytical lens to study the transformative process.

Briefly defined, *communicative arenas* are spaces of informal or formal exchange in which a multitude of social actors may meet. Often centered around a specific topic, *communicative arenas* may be organized as spaces with the immediate presence of social actors or also with more mediated forms of communication such as exchanging emails, producing papers, or organizing campaigns (Knoblauch 2001). The authors claim that the ability of SIs to create, maintain and skillfully control communicative arenas is important for institutional change from below, a change toward a university in SD.

In what follows, the authors shed light on three characteristics of communicative arenas by providing empirical evidence from the work of the SI *Greening the University e.V.* at the University of Tübingen (Germany). Firstly, for SIs, invoking communicative arenas is a means to work as an institutional innovator. Secondly, by transcending the boundaries of the university, SIs may engage in "boundary work", including social actors outside the classical disciplines into the communicative arena (Star and Griesemer 1989). Thirdly, communicative arenas function as learning spaces for all actors of a university.

Having elaborated on the potential of SIs for institutional change, the authors examine conditions governing SIs' potential to enduring and successful engagement. The authors focus on possible strategies for SIs as well as enabling conditions to be implemented by university leaders and politicians. The authors conclude with a brief summary of the potential role of "change from below" initiatives.

Toward a performative understanding of sustainable development: Communicative arenas and student initiatives

The role of student initiatives does not feature prominently within the discussion on sustainability at universities. However, it has been recognized that SIs need to be included in a transformation process as important agents (Lonzano 2006; Schneidewind 2009). Further studies emphasize the potential role of SIs as initiators of sustainability processes (Newport et al. 2003) or even as central agents for a full SD process (Krizek et al. 2012). Sharp (2002) puts an in-depth focus on SIs and argues that they are particularly capable to start a SD process because they do not follow the usual rationalities in a university. In addition Sharp (2002) as well as Helferty and Clarke (2009) provide specific advice for SIs in SD processes as well as for universities that wish to support such SIs. This paper adds a detailed case study to this literature and provides a theory-driven discussion of the role of SIs in the institutional transformation process toward a university in sustainable development. By drawing on theoretic approaches that take discourse, communication, and knowledge seriously, this paper adds new depth and insight to the discussion as it can both theoretically explain and empirically demonstrate channels through which SIs can foster institutional change toward sustainable universities.

Student initiatives as institutional innovators

Analyzing SIs' potential to enact innovations via the creation of communicative arenas, this section describes the evolution of the *StudierendenInitiative Greening the University* (SIGU).[1]

To understand the emergence and early successes of SIGU it is helpful to recall three distinct discursive events in Germany and its educative system five years ago. Firstly, after the fourth IPCC Report in 2007 and the scientific evidence of the anthropogenic climate change, public actors such as universities found themselves increasingly pressured to address questions of SD.[2] This process was, secondly, paired with the promotion of ESD as a guiding principle for educational institutions during the UN-Decade for ESD (2005-2014). Thirdly, the German federal government initiated a reform process called "Excellence Initiative"[3], in which German universities were encouraged to apply for federal grants with an encompassing individual profile.

In this context, a group of students with diverse academic backgrounds founded SIGU at the University of Tübingen (Germany). In an early position paper, the initiative developed a *storyline*, linking the climate change discourse to the university's individual responsibility as an educational institution (Esguerra 2006).[4] Two central themes for *structural change* were put forward by the initiative: firstly, the reduction of the university's consumption of natural resources and production of waste; for this theme the *metaphor* of "ecological footprint" was appropriated (Hajer 2009; Wackernagel and Rees 1996). Secondly, the initiative called for an integrated understanding and promotion of SD within teaching and research.

Having learned about the importance of intense cooperation from the study of sustainability processes at other universities, SIGU created a network with members from the university administration, the student union and academic staff. In addition, the initial members worked toward expanding the group, and institutionalized the initiative as a non-profit association.

Earlier efforts from (student) groups with similar interests were mainly centered around single issues, such as the use of recycling paper. Since these failed to structurally work toward the reduction of the university's "ecological footprint", SIGU decided at an early stage to target the university's directors and relevant local and regional politicians directly. The efforts to seed the sustainability process culminated in the organization of a two-day symposium. The momentum for change was caused by the production of "usable knowledge" for the transformation process (Haas 2004),[5] and a successful discursive of "self entrapment" (Risse-Kappen et al. 1999): a festive and public evening talk followed by a panel discussion with prominent disputants manifested the spin and raised the bar for the subsequent discussions within topical keynote lectures

1 For a more detailed account of the SIGU-case study, see Roosen-Runge et al. (2012).

2 To avoid misunderstandings, the authors do not claim that scientific consensual knowledge alone brought the climate change discourse to the forefront. In fact, other mediators, such as Al Gore, were necessary to make this claim authoritative.

3 The "Excellence Initiative" aims at improving research conditions, promoting research between disciplines and scientific competition as well as international cooperation and thereby profiling Germany as science location (http://www.wissenschaftsrat.de/1/fields-of-activity/excellence-initiative/).

4 Also, the "storyline" drew explicitly on the University's concept for the federal "Excellence Initiative". For an elaboration on the concepts of storyline see Hajer (2009).

5 The outcomes of the symposium were documented in a book (SIGU 2009) and have become a useful reference tool for SIGU, not least as a signal of commitment and seriousness toward e.g. the university's directors and funding bodies.

and workshops. This event finally set the stage for intended decisions that would have been impossible without the imprinted public interest. Shortly afterward, the university directors implemented the Eco-Management and Audit Scheme (EMAS). For SIGU, this success helped to build confidence and acceptance of SIGU as a serious and trustworthy actor. SIGU consolidated this acceptance by participation at all stages of the EMAS validation process.[6]

The symposium also laid the foundations for changing the university's research and teaching program toward SD. In this respect SIGU followed two strategies:

On the one hand, SIGU arranged numerous ESD courses. In cooperation with the "Studium Professionale" (interdepartmental course program for key skills) SIGU created the "Studium Oecologicum". This program encompasses seminars concerned with questions of SD, covering theoretical foundations of SD as well as the practice of SD and its connection to scientific and socio-political discourses. Examples range from "understanding predictions on climate change", "sustainable city development" and "intercultural aspects of SD" to "Gender@Nature". The program has become well known and is, with roughly twenty courses offered each semester, an established and expanding part in the interdisciplinary teaching program. Furthermore, using innovative teaching methods and including young professionals and scientists as lecturers, the program also aims for performative and methodological conformity with ESD.

To further change teaching and research towards SD, SIGU organized a public lecture series on "Science for Sustainable Development" with various well-known scientists of different academic disciplines.[7] This series and a related book with topical essays (SIGU 2012), address (i) the role of SD in the science-system and (ii) the relationship of single disciplines with SD – i.e. how SD influences research and how the specific discipline may contribute to realizing SD.

On the other hand, SIGU worked for the institutional inclusion of ESD. Therefore, SIGU rallied for the establishment of a university's advisory council for SD. The council consists of academic staff, members of the university administration and students and is chaired by one of the university directors. Established in late 2010, this council is commissioned to draft a comprehensive SD strategy for the university. The council provides recommendations to the university's directorate on SD-related issues, and promotes and coordinates teaching and research for SD. In 2011, e.g. for example, a sustainability price for Bachelor's and Master's theses was awarded.

The intensive cooperation with the student unions has also been important for anchoring the initiative in existing institutional structures. In particular, SIGU ensured a seat for a member of the student union in the university's working group for environmental management and five seats for students in the advisory council on SD. This representation of all actors of the university is essential for a growing legitimation of the sustainability process as a whole.

A general theme of these examples is the successful establishment of several communicative arenas driven by the engagement of SIGU. Within these arenas, also partly represented by new institutional structures, SIGU was able to put into practice many innovative ideas, ranging from new teaching and learning concepts to more sustainable institutional procedures.

Student initiatives as "boundary agents"

The authors have already touched upon the ways in which SIs play an important role in communicating and enacting a sustainability program. Along these lines the authors now reflect

6 Among other things, SIGU was coauthor of the university's environmental policy of the university and contributed ambitious goals for the EMAS process.

7 The notion of "science" encompasses in German natural science, social science, and the humanities.

about SIs' capacity to assemble diverse transdisciplinary actor groups.[8] In general, students and SIs are not bound to hierarchies and overall fixed schedules. This flexibility allows SIs to translate, negotiate, mediate and simplify in order to work with students, university administration, academic staff and other local actors outside the academic boundaries. Scholars in science studies have termed such communicative efforts "boundary work" (Star and Griesemer 1989). Let us consider three examples from the case study.

As one of SIGU's subgroups, the project Bunte Wiese ("colorful lawn") has dedicated itself to foster biodiversity in areas that have traditionally been shaped by monocultures such as public parks or green verges.[9] Since these public spaces are deployed and governed by a variety of actors, transformation requires a transdisciplinary approach. Students and administrative university staff work with experts from the city and regional parliament to develop and implement a blue print for public areas to increase biodiversity.

On the local and regional level, SIGU co-organizes an annual "sustainability day", a platform for various local initiatives, the university and local stakeholders to present themselves and to explain how their own work relates to SD.[10] A sustainability lecture by a prominent speaker ensures both the visibility of the "sustainability day" within the region, and the presence of high-ranking university officials. These 'dramaturgical considerations' proved important since the politics of science in a university are as much staged as they are in the classical public domain (Hilgartner 2000) and Sustainability Science as an emerging crosscutting approach depends particularly on careful staging since it is not established as a legitimate discipline, yet (Clark 2007).

Considering the federal level, SIGU is a co-founder of the "Netzwerk studentischer Nachhaltigkeitsinitiativen". This network of SIs working toward SD in Germany has published, e.g. for example, a position paper with a detailed catalog of suggestions and demands for students, scientists, presidents of universities, and the federal government to create enabling conditions for the pathway toward a university landscape in SD.[11] Additional efforts of SIGU were talks at hearings, for example in a regional Laender parliament, and several topical conferences to share experiences and usable knowledge with other actors.

All three cases exemplify how SIGU in particular and SIs in general engage on the local, the regional, and the federal level in creating distinct communicative arenas. Depending on the specific context, their role changes: they may appear as activists, work toward the local enactment of policies, or get diverse actors together and thereby act as boundary agents.

Student initiatives as learning spaces

The notion of learning is the final category to be examined as a distinct feature of communicative arenas created by SIs. While SIGU was still in emergence, the first meetings were productive communicative arenas focused on central questions: how have others defined a university in sustainable development? How does the initiative envision such a university? What might be an appropriate strategy? Who might fund projects the initiative would like to realize? In addressing these questions collectively, students learn about different approaches toward SD, and engage in discussions with people from various disciplinary backgrounds.

---

8 In the understanding of the authors, transdisciplinarity includes actors who hold non-academic forms of knowledge, whereas interdisciplinary research stays within the boundaries of classical research institutions (Hadorn et al. 2008; Esguerra and Roosen-Runge 2012).

9 For more information see http://www.greening-the-university.de/index.php/bunte-wiese.

10 For a broad coverage of the sustainability day including videos see http://www.uni-tuebingen.de/einrichtungen/internationales-zentrum-fuer-ethik-in-den-wissenschaften/forschung/natur-und-nachhaltigkeit/nachhaltigkeitstag-der-hochschulregion-tuebingen-hohenheim.html.

11 The position paper is accessible both in German and English at: www.nachhaltige-hochschulen.de.

To be politically effective, SIs also need to gain insight about the organization of the university administration, about hierarchies and opportunities for intervention. SIs must develop adequate methods for participation in different contexts of communicative arenas, and learn to be sensitive to the attitudes and acting modes of other actors (Barth 2007: 57). If successful, the participation in such an initiative creates a deep sense of self-empowerment of learners who realize that they can shape their own, sustainable future (Rieckmann and Stoltenberg 2011: 125).[12]

While the above notion of learning is mostly concerned with a self-reflexive gaining of knowledge, the authors also want to stress a second form of learning: SIs promote communicative arenas in which participation of several actors enables a reflective and discursive redefinition of aims and meanings of SD. To consider two examples, the symposium organized by SIGU offered seven intense workshops in which various actors discussed core themes of a university in SD with a focus on Tübingen but also on more general aspects (SIGU 2009). Secondly, in a recent transdisciplinary workshop in the "Studium Oecologicum", a former SIGU member, an environmental economist, and the university's sustainability manager brought together staff and students of universities from the region. The aim of the workshop was, first, to enable participants to develop solutions to reduce environmental impacts of universities. Secondly, visions of a university in SD were collectively developed, including recommendations for the next sustainability program of the University of Tübingen, and drawing on experiences of university staff and fresh ideas from the student body.

Both cases – the symposium and the workshop – exemplify at least two further notions of learning. The understanding of SD is not given *a priori* but evolves with people's conceptions and perceptions. Hence, self-empowerment and awareness for participation is generally important to enable *social learning processes*. In this sense, SIs open learning spaces for the entire university, in which methods, as well as aims and meanings of SD, enter a participatory discourse. On this basis, SIs can be regarded as an end in itself based on a discursive understanding of SD (Martello and Jasanoff 2004). In addition, both cases show how these communicative arenas foster *institutional learning* (Haas and Haas 1995). A university in SD is a learning institution facing a twofold challenge: the content of SD has to be incorporated into the university's curriculum and the organizational procedures, which is unavoidably interwoven with the establishment of suitable means to achieve the goal of a participatory and discursive understanding of SD. In other words, open communicative arenas as promoted by SIs performatively construct learning institutions in SD.

The role of student initiatives in context

Based on the theory-driven discussion of successful SI's achievements, the preceding chapter elaborated on the importance of SIs for the transformation of universities via creating *communicative arenas*. This section discusses how SIs' potential to catalyze institutional change can be supported and put on a lasting basis.

As it is SIGU's main idea to initiate a process that leads to an SD transformation of the University of Tübingen, two preliminary considerations have to be kept in mind. Firstly, the process itself is intended to result in enduring formal and informal institutions allowing for participation of all university members in order to reach a university in SD. Secondly, although SIs can achieve much, they are not panceas. It is neither the basic principle of an SD transformation process that it should be maintained by SIs alone instead of a broad supportive movement from all university groups, nor is it feasible for an SI.

12 In German, this ability is called Gestaltungskompetenz (de Haan and Harenberg 1999) often translated as 'shaping competence' (e.g. de Haan 2006).

Limits

There are a range of challenges and obstacles that limit an SIs' ability to initiate and uphold a structural transformation process themselves. Foremost, SIs have to come into being in the first place. Like-minded students have to meet, formulate objectives of their shared commitment and agree on a strategy to foster institutional change.

Already at this stage of internal agenda setting, the two major challenges of student engagement become apparent: firstly, activists desire not only to seed projects, but also to experience an outcome within their period of engagement. Secondly, several projects require knowledge and expertise on general concepts and specific contexts before they can be understood and are able to contribute to the addressed issue. Consequently, activist short-term projects might be easier to agree on than long-term strategical engagement – at least at first glance.

Even after a SI's formation, a crucial challenge lies in the constantly changing membership structure due to short curricula, studying abroad and exam periods. If the initiative's agenda and strategies cannot be communicated to new members, the latter might feel constricted in their activities by projects started by previous members. Work-intensive projects with tight schedules need the attention and coordination of more members in order to bridge vacation and exam periods. Finally, participation in more than one – possibly long-term – project demands a rather professionalized working structure with both internal management and project tasks that are time-consuming.

The challenges referred to above not only render the internal discourse complicated but student commitment is voluntary, unpaid and subject to severe time and expertise constraints, while other professionals in administration and the public sector are established and educated in their specific fields-of-activity. Students lack experience in administration and law, and have to learn most of the necessary knowledge from scratch to allow for a constructive participation in decision-making processes. Furthermore, members of SIs change frequently, while contact persons in university and the public sector complete longer tenures, which complicates building up reliable communication, trust and reputation.

Strategies

As SIGU's example shows, it is possible to overcome many limitations. In order to engage in long-term processes, it is important to develop a clear vision of a university in SD, concrete short- and long-term aims and the strategic local knowledge to implement them.

Starting with the initiative's internal process, a regular, reflexive evaluation and redefinition of the aims is necessary to acknowledge success, prevent dead-ends of engagement and allow all members to agree on the initiative's agenda. In order to put engagement on a fruitful basis, internal management and project tasks should be fairly distributed and kept open as much as possible for everybody. Retreats may serve as forums for assessment, feedback, strategic planning and team building. Documentation of objectives, strategic planning of programs and results as well as open access to information within the group are crucial as they reduce the challenge of a changing membership and facilitate participation.

For long-term strategic projects, one very basic rule of thumb may be of help: do not act without a success-representing intermediate-term exit strategy. From the beginning, possible forms of future self-sustaining institutional settings have to be brought into discussion and projected. In particular, the gradual transformation of existing structures within collaborations has proved successful to implement the SIs' aims into enduring structures. Naturally, working with established structures, which stand in clear contrast to SIs' vision of a university in SD, is undoubtedly accompanied with compromises and may lead to opposition of fellow students who aim for more. For the viability of the SI, however, such transformative catalysis frees power to tackle further projects.

Practically speaking, focus events with prominent, convincing speakers help to get a project started and sometimes even financed. Having finished one starting project successfully increases the regard as a trustworthy and constructive SI, which in turn facilitates the interaction with other stakeholders. When known to partners, further doors open for thematic and strategic partners in new projects.

Enabling conditions

University officials and politicians can support SIs' potential to contribute innovative work. Apart from the very obvious precondition that they have to take seriously what students are able and willing to contribute to a university in SD, they can open experimental spaces that allow for the development of best practice examples and the nucleus of new institutions. Furthermore, financial support and organizational backing could be effective, e.g. in the form of service centers and start-up financing grants for SIs. University presidents can also help via their own commitment to SD and by encouraging university members to engage in SD processes. Moreover, working in SIs for SD advances crucial ESD competencies such as participation, cooperation or motivation and enables perceived self-efficacy, self-organization and self-learning (de Haan and Meisch 2012). Given tight schedules, it remains challenging for students to actually find the time for extra curricula activities. The authors thus strongly recommend less obligatory courses so that students have space to pursue innovative paths; in addition, universities must think about fostering and crediting this learning as part of the regular course program in order to do both: facilitate student engagement and promote a more comprehensive learning.

Conclusion

To advance the discussion on how the transformation of universities toward sustainable development may be achieved, this paper has focused on an often-overlooked actor category in the discourse on universities in SD: student initiatives.

Based on the example of the SI *Greening the University e.V.* (University of Tübingen, Germany), and by drawing on theories of knowledge, discourse and communication, this paper has shown how SIs can catalyze institutional change toward a university in SD. The basic mechanism of this bottom-up initiated change is conceptualized by the creation of communicative arenas, i.e. spaces of informal or formal exchange centered around a specific topic in which a multitude of social actors may meet.

More specifically, the authors argue that SIs are able to play a vital role for a university in SD for three reasons:

1. SIs invoke communicative arenas that address innovative ideas and give birth to *envisioned projects* and new concepts. Coupled with a long-term strategy, these efforts induce changes of the university's *institutional implementation* of SD.
2. Not bound to hierarchies, SIs play the important role of '*boundary agents*' in communicating and enacting a SD program to the "outside world", i.e. actors in politics, society and economy.
3. Finally, SIs not only provide *individual learning* spaces for the members, in which competencies enabling participation are acquired, but more generally, SIs bring forward both *social and institutional learning* in the entire university allowing for a participatory and discursive institutional implementation of SD.

Despite these promises, it is neither desirable nor feasible for a SI to initiate and uphold a SD transformation on its own. SIs and their communicative environment have to be clear, reflective and open about the scope and the available means of a SI. Accepting the own limitations, fresh, innovative and passionate student engagement is able to flourish instead of being suppressed in hierarchical and inflexible structures due to lacking expertise and the restricted time horizon of student engagement. The process of university-wide institutional change toward SD, must therefore be a participative process involving a broad supportive movement.

In summary, this paper has shown that SIs provide a unique potential for universities in their quest to realize a participatory understanding and implementation of SD. It is essential that strategies and actions by politicians and university leaders recognize and support SIs to foster innovative change from below that is indispensible for a transformation towards universities in SD. Students should feel encouraged to use their innate abilities to make changes toward sustainable universities take place.

References

Barth, M. (2007), *Gestaltungskompetenz durch Neue Medien? Die Rolle des Lernens mit Neuen Medien in der Bildung für eine nachhaltige Entwicklung*. Berliner Wissenschafts-Verlag, Berlin.

Clark, W.C. (2007), Sustainability science: a room of its own. *Proceedings of the National Academy of Sciences of the United States of America* 104 (6), pp. 1737-1738.

de Haan, G. (2006), The BLK '21' programme in Germany: A 'Gestaltungskompetenz' based model for education for sustainable development. *Environmental Education Research* 12 (1), pp. 19-32.

de Haan, G. and Harenberg, D. (1999), *Bildung für eine nachhaltige Entwicklung. Gutachten zum Program,* Materialien zur Bildungsplanung und zur Forschungsförderung, Vol. 72. Bonn: BLK. www.blk-bonn.de.

de Haan, G. and Meisch, S. (2012), Bildung für nachhaltige Entwicklung. Bildung für eine zukunftsfähige Volluniversität. In StudierendenInitiative Greening the University e.V. (Eds.), *Wissenschaft für nachhaltige Entwicklung! Multiperspektivische Beiträge zu einer verantwortungsbewussten Wissenschaft*, pp. 47-70. Metropolis-Verlag, Marburg.

Esguerra, A. (2006), *Mit "High Tech und Hegel" in eine nachhaltige Zukunft. Eine Nachhaltigkeitsstrategie für die Universität Tübingen*. http://fu-berlin.academia.edu/ AlejandroEsguerra/Papers.

Esguerra, A. and Roosen-Runge, F. (2012), Nachhaltigkeitswissenschaft – die undisziplinierte Disziplin. Transdisziplinäre Herausforderungen nachhaltiger Entwicklung. In StudierendenInitiative Greening the University e.V. (Eds.), *Wissenschaft für nachhaltige Entwicklung! Multiperspektivische Beiträge zu einer verantwortungsbewussten Wissenschaft*, pp. 135-155. Metropolis, Marburg.

Haas, P.M. (1992), Introduction: Epistemic Communities and International Policy Coordination. *International Organization* 46 (1), pp. 1-35.

– (2004), When does power listen to truth? A constructivist approach to the policy process. *Journal of European Public Policy* 11 (4), pp. 569-592.

Haas, P.M. and Haas, E.B. (1995), Learning to learn: improving international governance. *Global governance* 13, pp. 255-284.

Hadorn, G.H., Biber-Klemm, S., Grossenbacher-Mansuy, W., Hoffmann-Riem, H., Joye, D., Pohl, C., Wiesmann, U. and Zemp, E. (2008), The emergence of transdisciplinarity as a form of research. In Hadorn, G.H., Hoffmann-Riem, H., Biber-Klemm, S., Grossenbacher-Mansuy, W., Joye, D., Pohl, C., Wiesmann, U. and Zemp, E. (Eds.): *Handbook of transdisciplinary research*, pp. 19-39. Springer-Verlag, London.

Hajer, M.A. (1995), *The politics of environmental discourse: ecological modernization and the policy process*. Oxford University Press, New York.

Hajer, M.A. (2009), *Authoritative Governance: Policy-making in the Age of Mediatization*. Oxford University Press, New York.

Hajer, M.A. and Versteeg, W. (2005), A decade of discourse analysis of environmental politics: achievements, challenges, perspectives. *Journal of Environmental Policy & Planning* 7 (3), pp. 175-184.

Helferty, A. and Clarke, A. (2009), Student-led campus climate change initiatives in Canada. *International Journal of Sustainability in Higher Education* 10 (3), pp. 287-300.

Hilgartner, S. (2000), *Science on stage: expert advice as public drama*. Stanford University Press, Stanford.

Johnstone, I. (2005), The power of interpretive communities. In Barnett, M. and Duvall, R. (Eds.): *Power in global governance*, pp. 185-248. Cambridge University Press, Cambridge, UK.

Knoblauch, H. (2001), Communication, Context and Culture. A communicative constructivist approach to intercultural communication. In Di Luzio, A., Günthner, S. and Orletti, F. (Eds.): *Culture in communication: analyses of intercultural situations*, pp. 3-33. J. Benjamins, Amsterdam.

Krizek, K.J., Newport, D., White, J. and Townsend, A.R. (2012) Higher education's sustainability imperative: how to practically respond? *International Journal of Sustainability in Higher Education* 13 (1), pp. 19-33.

Leal Filho, W. (2011), About the Role of Universities and Their Contribution to Sustainable Development. *Higher Education Policy* 24, pp. 427-438.

Lozano, R. (2006), Incorporation and institutionalization of SD into universities: breaking through barriers to change. *Journal of Cleaner Production* 14 (9-11), pp. 787-796.

Martello, M.L. and Jasanoff, S. (2004), Introduction Globalization and Environmental Governance. In Jasanoff, S. and Martello, M.L. (Eds.), *Earthly Politics. Local and Global in environmental Governance,* pp. 1-29. MIT Press, Cambridge, USA.

Newport, D., Chesnes, T. and Lindner, A. (2003), The "environmental sustainability" problem: Ensuring that sustainability stands on three legs. *International Journal of Sustainability in Higher Education* 4 (4), pp. 357-363.

Rieckmann, M. and Stoltenberg, U. (2011), Partizipation als zentrales Element von Bildung für eine nachhaltige Entwicklung. In Heinrichs, H., Kuhn, K. and Newig, J. (Eds.): *Nachhaltige Gesellschaft. Welche Rolle für Partizipation und Kooperation?* pp. 117-131. VS Verlag für Sozialwissenschaften, Wiesbaden.

Risse-Kappen, T., Ropp, S.C. and Sikkink, K. (1999), *The power of human rights: international norms and domestic change*. Cambridge University Press, Cambridge, UK.

Roosen-Runge, F., Drupp, M.A., Keul, L. and Esguerra, A. (2012), Bottom-Up Engagement für Nachhaltige Entwicklung – Erfahrungen, Chancen und Grenzen. In StudierendenInitiative Greening the University e.V. (Eds.), *Wissenschaft für nachhaltige Entwicklung! Multiperspektivische Beiträge zu einer verantwortungsbewussten Wissenschaft*, pp. 91-118. Metropolis, Marburg.

Schmidt, V. (2012), Discursive Institutionalism: Scope, Dynamics, and philosophical Underderpinnings. In Fischer, F. and Gottweis, H. (Eds.): *The argumentative turn revisited: public policy as communicative practice.* Duke University Press, Durham.

Schneidewind, U. (2009), *Nachhaltige Wissenschaft: Plädoyer für einen Klimawandel im deutschen Wissenschafts- und Hochschulsystem*. Metropolis, Marburg.

Sharp, L. (2002), Green campuses: the road from little victories to systemic transformation. *International Journal of Sustainability in Higher Education* 3 (2), pp. 128-45.

SIGU (StudierendenInitiative Greening the University e.V.) (Eds.) (2009), *Greening the University – Perspektiven für eine nachhaltige Hochschule.* oekom, Munich.

– (2012), *Wissenschaft für Nachhaltige Entwicklung! Multiperspektivische Beiträge zu einer verantwortungsbewussten Wissenschaft*. Metropolis, Marburg.

Star, S.L. and Griesemer, J.R. (1989), Institutional Ecology, 'Translations' and Boundary Objects: Amateurs and Professionals in Berkeley's Museum of Vertebrate Zoology, 1907-39. *Social Studies of Science* 19 (3), pp. 387-420.

Wackernagel, M. and Rees, W.E. (1996), *Our Ecological Footprint: Reducing Human Impact on the Earth.* New Society Publishers, Gabriola Island, British Columbia, Canada.

Chapter 62

Community Leadership within Universities: A Guide to Sub-institutional Sustainability Planning

Melissa Goodall and Allison Potteiger

Abstract

Universities are both well-positioned to offer analytically rigorous approaches to global environmental concerns, and to provide students the opportunity to participate in working models of sustainability. Higher education institutions around the globe are developing sustainability strategies that include ambitious, yet achievable goals. There is frequently a divide, however, between plans on the institutional scale and the various intellectual and administrative communities that comprise a university. In 2009, the Yale Office of Sustainability launched a process to draft a sustainability strategic plan for the university. At the same time, the Office established a project to develop sustainability plans at sub-institutional levels. The project began with a one-year pilot in three professional schools of Yale's 13, then transitioned into planning for the Yale Medical Campus. The planning processes and materials were also used to develop a "roadmap" that would allow any unit on campus to develop and implement their own sustainability strategy. The Roadmap is currently in the pilot phase, with expected completion in June 2012. This paper describes the theoretical foundations of the project, and chronicles the pilot initiative. It then offers quantitative and qualitative insights on the efficacy of various elements of the first set of plans. This is followed by an overview of the Yale Sustainability Planning Roadmap. The article concludes with insights on the replicability of this initiative and possible next steps.

University Sustainability Planning

In the 40 years since the Stockholm Conference on the Human Environment, universities have steadily increased their commitment to sustainability (Calder and Clugston 2003). Higher education institutions around the world have been signing international declarations such as the Talloires Declaration and the Halifax Declaration, which has resulted in new or updated operational policies and practices. As it is becoming more common for universities to have sustainability initiatives and personnel, universities have started to develop sustainability plans with measurable goals.

This transition from operational change to forward-looking planning is vital for two reasons: firstly, these plans clearly define the depth and breadth of the university's commitment, and secondly, they help to frame the requisite policies and initiatives for implementation (Wright 2003). According to Hitchcock and Willard, a sustainability plan determines "… what is important to you (as an institution) and builds a long-term plan to get there … it is usually an iterative process where you revisit your goals and priorities, check on progress and revise the plan year after year" (Hitchcock and Willard 2008).

The typical format of a sustainability plan begins with a letter from the president, introduction and layout of the plan, followed by a description of a set number of priorities. The letter from the president stresses the importance of preparing for the future of the institution and its commitment to the environment. The priorities typically reflect broad categories related to the university's operations, and include goals, metrics, and opportunities to effect change. Common priorities include transportation, procurement, waste management, buildings and construction, education and research, food and dining, energy, water, and land management. A three- to five-year year timeline is common, and consistent with the recommendations of groups such as the International Institute for Sustainable Development and the World Business Council for Sustainable Development (International Institute for Sustainable Development et al. 1992).

Arguably the most important component to a sustainability plan is buy-in and support from the community. The front cover of the *Yale Sustainability Strategic Plan 2010-2013* concludes, "Success depends on the active contribution of every member of the Yale community" (Yale Office of Sustainability 2010). The first page is a letter of endorsement from President Levin. The implication of these two elements is that buy-in from the top and participation of staff, students, and faculty are both essential to the realization and implementation of its goals. Pennsylvania State University successfully institutionalized its plan by incorporating sustainability into the school's mission statement. In 2000, while Penn State Biology Professor Christopher Uhl and two students were drafting an "ecological" mission statement, Penn State President Bryce Jordan also showed his support of the plan by mentioning sustainability in his public speeches. When the time came he quickly approved of the new mission (Barlett and Chase 2004).

Community engagement with sustainability should involve all levels of a university. Top university administrators are responsible for communicating the institution's mission and goals within the organization to ensure collaboration and contribution at every level. According to the *Business Strategy for Sustainable Development: Leadership and Accountability for the 90's*, "The board of directors, the chief executive officer and other senior executives must be actively involved in the [organizational change] process. They need to lead by example, and to set the tone for the rest of the organization" (International Institute for Sustainable Development et al. 1992).

The drive for institutional sustainability in a university setting frequently comes from the energy and enthusiasm of the students, and can often be connected to lessons learned in the classroom, as seen in Penn State's efforts while developing the Mueller Report: Going beyond Sustainability Indicators to Sustainability Action (Penn State Green Destiny Council 2001). It is critical to both cultivate this spirit and empower the students, as they are tomorrow's workforce and idea-generators (Johnston and OECD 2007).

Faculty members have also been instrumental in advancing campus sustainability and planning, as has particularly been the case at the University of British Columbia (University of British Columbia, n.d.). Professors and researchers may generally be best suited to offer intellectual leadership on sustainability, rather than to assist with the execution of a strategy: mainly because faculty must prioritize teaching and research and are not likely to have the time or attention for implementing sustainability initiatives. Faculty members have an important role to play in enhancing and advancing the culture of sustainability, however, through applied research initiatives such as Campus as a Living Lab (Cortese 2010).

Non-faculty staff members are perhaps the most critical audience for a sustainability strategy. It is they who are empowered to affect enduring change by collecting data over time and implementing programs. Staff members play an essential role in maintaining everyday life at colleges and universities – they manage resources, order supplies, set policies, and manage systems. While the student population is larger, that group is, by nature, transient. Staff members tend toward longevity and can thus have a long-lasting impact.

Most sustainability plans use the triple bottom line as a foundation, meaning a balance between environmental protection, social well-being and economic development (Stenzel 2010). Plans in developed countries – such as those of the University of Hong Kong (The University of Hong Kong 2004), Griffith University (Griffith University 2002), and the University of Gotenburg (University of Gotenburg 2010) – tend to spell out tactics for reducing energy use, reducing waste, and otherwise streamlining systems. As a result there is a strong financial rationale for many items likely to be contained in such a plan. Other items – such as sourcing sustainable food and products – may come at a premium and therefore may also pose budgetary challenges. This may be overcome with tactics such as refining systems to use fewer or different products as well as coaching staff members to look for efficiencies that lead to additional cost savings, such as the University of British Columbia's sustainability coordinator program (University of British Columbia, n.d.).

The social aspect of sustainability plans is often the least emphasized. This element is critical, however, as fostering community and recognizing the different cultures that exist within a university, to meet the needs of the different cultures, creates authentic engagement and behavior change (Bergquist and Pawlak 2008). In addition, sustainability plans can provide an opportunity to foster community service within the institution and the surrounding community. Part of the mission of the Portland's State University District Framework Plan, for example, is "a commitment to sustainable neighborhood development" by focusing on producing strategies for future development in Portland (Portland State University 2010).

Yale Sustainability Strategic Plan 2010-2013

There are many ways to approach the development of a sustainability plan. Yale University focused on many of the aspects mentioned above. In May 2009, Yale University's President Richard Levin asked key leaders within the Yale community to form a Sustainability Task Force, which was "charged with developing a set of sustainability goals that would be both visionary and achievable over the next three years" (Yale Office of Sustainability 2010). The co-chairs of the Task Force were the Vice President and Secretary of Yale University and the Director of the Office of Sustainability. The Task Force included 16 representatives from various units throughout the university, each of whom was well-positioned to influence and effect change. A group of seven student Sustainability task force Fellows offered support to the members of the Task Force. The Task Force recommended a set of 43 goals divided into four priority areas: campus systems, earth systems, administrative systems, and education and engagement.

At the time of this writing, the Yale Office of Sustainability has had 18 months to measure the effectiveness of the *Yale Sustainability Strategic Plan 2010-2013*. In many ways, the Yale Sustainability Strategic Plan has energized the Yale community: President Levin's commitment and the plan's emphasis on measurable achievements offered an essential foundation for action. The results have included several shifts in staffing, multiple large-scale initiatives related to the built environment, and an increasingly assertive set of outreach campaigns and educational programming. In turn, elements such as these have all contributed to a robust culture of sustainability.

Challenges

While the Yale Sustainability Strategic Plan has been relatively effective thus far, there have been some challenges and barriers to implementation, including financial constraints; poor or inconsistent data; and disconnects between the institutional-scale planning and the day-to-day functioning of the various units, departments, and schools. For example, a custodian's everyday duties are drastically different from a finance manager's obligations. This can make it easy for a member of the community to ignore the plan because they feel, "it doesn't apply to them." Similarly, the culture of the art department is likely entirely distinct from that of the economics department.

In pursuing the "active contribution of every member of the Yale community," it is vital to recognize intellectual and functional differences between departments or units and tailor those differences into the approach and language used when engaging particular groups (Yale Office of Sustainability 2010). Goals must be translated into scales and concepts that hold meaning to individuals and figure into the disciplines and functions of various campus units.

Professional School Sustainability Planning at Yale: a pilot exercise

Yale's 13 Professional Schools are self-supported entities that are housed within the University, but function with a high degree of financial autonomy. These schools vary dramatically in their culture, discipline, and values. In 2009, the Office of Sustainability received grant funding to launch an initiative aimed at developing a set of sub-institutional sustainability strategies. As a pilot effort, the Office team invited the deans of the Yale School of Forestry & Environmental Studies (FES), the Yale School of Management (SOM), and the Yale Divinity School (YDS) to participate in a one-year drafting process in parallel with the Yale University's overarching sustainability strategic plan drafting process. Each dean agreed, and each took a very different tactic in the execution of the program.

The School of Forestry & Environmental Studies already had an environmental stewardship committee that was chaired by a faculty member and comprised of staff and faculty. For the purpose of developing the plan, the committee took on two student assistants who collected all of the metrics, wrote the report, and presented suggested priorities to the committee and the dean.

The Dean of the Divinity School established a committee and assigned staff and faculty members to that committee. Student volunteers were also invited to participate. The effort was coordinated by a student who was paid by the Office of Sustainability. The committee started by identifying priority areas and then divided into subcommittees by priority. Each subcommittee drafted a set of goals and projects. The paid student then collected the sections and compiled them into a plan.

The School of Management team was comprised of five students who completed the planning exercise as a joint independent study. They established priorities and divided the topic areas between team members. Throughout the course of the year, they collected metrics and identified opportunities for change. The committee then offered the dean a plan of action.

Students played an integral role in the development of all three plans. Without their participation, the plans would not exist or be implemented. Numerous schools utilize student assistance to progress and promote plans, similarly to the University of British Columbia's use of student participation in workshops to develop alternative ideas for sustainability education (Moore 2005).

Table 1 offers highlights of the strengths and weaknesses of each process and plan.

Table 1: Strengths and Weaknesses of Pilot Plans

	Strengths	**Weaknesses**
School of Forestry & Environmental Studies	The activity revitalized the environmental stewardship committee The dean was inspired to demonstrate leadership on the execution of the Yale Sustainability Strategic Plan	The final plan was idealistic and included too many "priorities," rendering its execution overwhelming

	Strengths	Weaknesses
Divinity School	The activity called attention to the breadth of sustainability initiatives already under way	Those who were "assigned" to the committee were not authentically committed The final plan did not reflect all buildings associated with the School
School of Management	Formatted in a clear, logical way that made it nicely accessible as a model to other groups on campus	Staff members were not directly involved with the drafting process The priority areas were well-described, but the goals and tactics vague

In each case, the most significant outcome was the direct engagement of the staff – particularly the business officers – in enhancing the culture of sustainability in their unit. The next year, each of the professional schools created two or three student positions to implement activities in support of the plans. In addition, the plans and their subsequent activities have fostered communication and knowledge-transfer between the staff and students of the three schools. For example, the School of Forestry & Environmental Studies launched a program to "rent" reusable to-go containers to students. This initiative was successful and had a noticeable impact on the waste stream, so the School of Management followed suit. The School of Management students are also partnering with the Divinity School students on a project that will highlight the sustainability practices of the most-used caterers in their parts of campus.

This pilot program was innovative and significant in two major ways. Firstly, before embarking on this project, the student researchers researched this concept of sub-institutional planning, and found no other examples of a project that had institutionalized sustainability planning that was scaled to fit the physical and academic cultures of departments. Secondly, the three schools that participated in the pilot nicely reflected the triple bottom line of sustainability, as is illustrated in Fig. 1 below:

Fig. 1: The three pilot schools in the context of the priorities of sustainability

Next Steps

The development and execution of these plans inspired two separate sustainability planning initiatives. First, a commitment to develop a sustainability strategy for the Yale Medical Campus; second, the pilot of a program that is intended to offer any group on campus all of the tools and information needed to develop a sustainability strategy without the direct involvement of the Office of Sustainability.

Yale Medical Campus

In terms of square footage, people, and dollars, the Yale medical campus is roughly half of the Yale campus. Separated from the "central" campus by a busy throughway, the medical campus includes the School of Public Health, School of Medicine, School of Nursing, and is vitally affiliated with the Yale-New Haven Hospital. The School of Medicine is not only entirely distinct in its culture, but its operations also include laboratory and clinical space – neither of which had been addressed in the pilot phase. The School of Medicine Sustainability Steering Committee is chaired by the Director of the School of Medicine Institutional Planning & Communication department and the Deputy Dean of School of Medicine Finance & Administration Office along with the Assistant Director of the Office of Sustainability, the Office of Sustainability Project Manager, the Director of Planning at the Medical Campus, and the Facilities Director of Sustainability Initiatives.

The Steering Committee oversees and guides the development of the subcommittees and the overall Sustainability Strategic Plan. This group determined the priorities of the plan: energy, procurement, and waste. The Steering Committee set the agenda and the timeline, and made decisions on subcommittee inputs. With input from the group, the Steering Committee established four subcommittees: Animal Resources and Building Operations, Administration, Clinical (Ambulatory Patient Care), and Laboratories. Each subcommittee is responsible for developing and evaluating recommendations and opportunities based on the three priorities. To this end, each group is to identify initiatives that can be measured and to suggest goals as well as communication tactics that are likely to resonate with members of their communities. Fig. 2 offers an overview of the subcommittees. Because many of the facets of the Yale Medical School are closely linked with the Yale School of Public Health and the Yale-New Haven Hospital, the subcommittees are also partnering with key professionals from those areas.

Fig. 2: Yale Medical School subcommittee and priority area organization

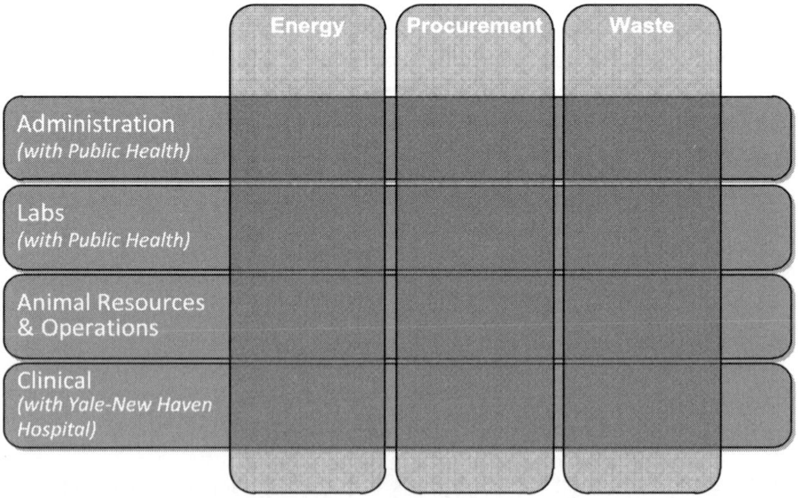

This planning process is started with the academic year and is slated to result in a Yale Medical Campus Sustainability Strategic Plan by the end of the academic year. It will contain goals in support of the Yale Sustainability Strategic Plan 2010-2013 as well as recommendations for tactics and communications initiatives that reflect the unique culture of the medical campus. As with the professional school pilot program, this planning process has fostered dialogue both within the Yale School of Medicine and with other entities in that part of campus. Because the School of Medicine is intimately linked with Yale-New Haven Hospital, it has also sparked a partnership with the Hospital's sustainability champions or leaders. When finished, this plan will be distinct from those of the other schools in three ways: it will include laboratory and clinical space; due to its size, it will include a number of sub-school plans; it will have language that connects to an external partner.

Sustainability Roadmap

While the development of sub-institutional plans is rewarding and stimulating, it is also time-consuming for the staff of the Office of Sustainability. Interest from a broad range of units on campus inspired an effort to foster that enthusiasm without over-committing the time of the Office of Sustainability. Based on the experiences from the pilot phase, the Office of Sustainability compiled a comprehensive "Roadmap to Sustainability" that enabled interested parties to create their own sustainability plans. The Roadmap to Sustainability includes practical templates, tools, and resources applicable to all departments seeking to develop their own sustainability plans. Table 2 offers an overview of the elements included in the *Roadmap to Sustainability*:

Table 2: Tools included in the *Roadmap to Sustainability*

Tool	Purpose
Sustainability Strategic Plan	"Fill in the blanks" plan for any unit
Timeline Guide	Timeline to keep unit on track of developing sustainability plan
Annual Reporting Template	Example report of accomplishments and projects to be sent to dean or head of unit
Sustainability & You Exercise	Tool to relate sustainability to specific fields
Presentation on Roadmap to Sustainability	Presentation to share with groups interested in piloting the roadmap
Sustainability Committee First Meeting Notes	Example itinerary and discussion for initial meeting
Sustainability Survey	Survey for members of the community to track their behaviors relating to sustainability
Sustainability Audit Tool	Audit to record current energy, transportation, procurement, spatial relations, waste, water, and lab initiatives
Waste Audit Tool	Table to chart waste and recycling within a unit
Metrics Gathering Tool	Template to assist in gathering metrics within a unit
Department Scale Equivalencies	Chart that translates quantitative metrics to relatable concrete terms

There are currently 11 groups from a diverse set of units across campus participating in the Roadmap pilot program, including the library system, two museums, a laboratory, and several academic and administrative divisions. The pilot began in September 2011, with the aim of establishing a draft plan for each unit by April 2012. The Roadmap will then be revised based on input from the participants, and the full program will be made public in September 2012.

The pilot participants meet monthly to share their successes and struggles, and discuss next steps. The meetings are organized by an Office of Sustainability student research assistant who answers questions, maintains momentum, and provides support for the leaders. As with the schools in the pilot program, the exercise of developing these strategies has fostered community and engagement, which bodes well for the execution of the plans. An interesting and unanticipated benefit to the process has been that several participants who had previously felt unnoticed were empowered by the support from their unit directors. This speaks again to the nature of the overarching Yale Sustainability Strategic Plan – it sent a clear message to unit heads that sustainability is a priority for the institution, which in-turn cut a path for those wishing to effect change at the community scale.

Lessons Learned

These activities have uncovered a vital set of qualities for successful sustainability planning:

- The active participation of the business manager is important: these are the staff members who understand and the financial realities of sustainability initiatives. They have the capability to assess and advance the economic viability of a plan and its particular initiatives. They also tend to be well-connected with the division leader and other key members of the community.
- The identification of sustainability "champions" or leaders within units – whether or not their jobs relate directly to sustainability. Authentic enthusiasm will always be more effective than anonymous policy.
- The support and engagement from all levels of authority. Sometimes the staff members with the lowest rank are those who have the highest impact, and therefore the best opportunity for improvement.
- The communication must be based on the priorities of the audience, not on the priorities of the planners. For example, rather than suggesting that community members, such as custodians, should care about green cleaning for the sake of the environment, consider pointing out that "traditional" chemicals are harmful to humans – particularly children.
- The "why" of sustainability should be clearly stated. Despite the increasing prevalence of sustainability as a concept, it is a term that is often met with confusion ranging the idea that it is synonymous with recycling or environmentalism to its association with higher costs for things like organics.

Finally, the costs associated with implementation can be a barrier in two ways: first, those who are predisposed to prioritizing sustainability may be troubled by the tradeoffs that budget restrictions will require; second, those not well-versed in sustainability may assume that sustainability measures will automatically generate expenses. As such, it is crucial to include clear financial rationale, return on investment reports for example, in order to attain buy-in from the top. Without buy-in, it can be very difficult to obtain support and engagement from the university community.

Replicability

It is important to note that these initiatives were developed and executed in a private institution in a developed country. While many of the tools and tactics could be replicable in public universities, those institutions may be restrained by their reliance on government systems and policies. Institutions in developing countries may be able to draw on the overarching framework of this initiative, but it is unlikely that the content would be directly transferrable: where the Yale plans speak to reducing wasted energy, food, and supplies, developing country universities are more likely concerned about promoting access to those same resources. The next phase of this project may seek to test the feasibility of applying this model in a public university within the United States. Following that, an adjusted model may be developed for application in a developing country.

Conclusion

There are more than 20,300 universities in the world (Pérez and Fernández 2012). Each of these institutions has the capacity to advance sustainable development through teaching and research. Each also has the opportunity to create a culture of sustainability for its students, thereby setting the expectation of sustainability in tomorrow's leaders. By linking sustainability into the distinct academic cultures of a campus, this initiative strives to inspire authentic and enduring engagement.

Today, it is increasingly de rigueur for universities to have a sustainability program – whether this programming comes in the form of student coalitions or official staffing and departments (Leal Filho 2009). Yale University has had a robust sustainability program since 2005. When President Levin committed to the *Yale Sustainability Strategic Plan 2010-2013*, it inspired action and commitment throughout the university, but it also highlighted the diverse ways that the university is being used by the various academic and administrative units. The professional school plans are designed in a way that takes into account how the students, staff, and faculty of those communities function as well as their intellectual interests. Similarly, the Roadmap is structured in a way that those developing plans for their units can "personalize" the priorities, goals, and programs to suit their populations.

Universities have important roles to play in advancing analytically rigorous sustainability solutions and training tomorrow's leaders. It is critical that lead administrators continue to commit to goals and develop strategies. The successful execution of those strategies will be reliant on the active contribution of all parts of a campus. It is critical to recognize that many universities are actually collections of a diverse set of communities, and that expectations, understanding, and values will vary by discipline and administrative function. In offering a structure to tailor sustainability to the unit-level, this initiative offers a creative and compelling means to connect local action to global environmental issues.

References

Barlett, P.F. and Chase, G.W. (2004), *Sustainability on Campus: Stories and Strategies for Change.* Massachusetts Institute of Technology Press, Cambridge, MA.

Bergquist, W.H. and Pawlak, K. (2008), *Engaging the Six cultures of the Academy: revised and expanded edition of The four cultures of the academy.* John Wiley and Sons, Inc., San Francisco, CA.

Calder, W. and Clugston, R.M. (2003), International Efforts to Promote Higher Education for Sustainable Development. *Planning for Higher Education* 31 (3), pp. 30-44.

Cortese, A. (2010), The Campus as Living Laboratory". Available at: http://www.fastcompany.com/1567840/the-campus-as-living-laboratory (accessed 2 January 2012).

Griffith University (2002), *Griffith University Sustainability Policy.* Available at: http://www62.gu.edu.au/policylibrary.nsf/xmainsearch/a128037eefd3dc144a2579240082f6b1?opendocument (accessed 2 January 2012).

Hitchcock, D. and Willard, M. (2008), *The Step-by-Step Guide to Sustainability Planning: How to Create and Implement Sustainability Plans in Any Business or Organization.* Earthscan, London, England.

International Institute for Sustainable Development, Deloitte & Touche, and the World Business Council for Sustainable Development (1992), *Business Strategy for Sustainable Development: Leadership and Accountability for the 90s.* International Institute for Sustainable Development, Winnipeg.

Johnston, A. and Organisation for Economic Co-operation and Development (OECD) (2007), *Higher Education For Sustainable Development: Final Report of International Action Research Project.* Available at: http://www.oecd.org/dataoecd/55/4/455 75516.pdf (accessed 2 January 2012).

Leal Filho, W. (2009), *Sustainability at Universities – Opportunities, Challenges and Trends.* Peter Lang International Academic Publishers 31, pp. 340.

Moore, J. (2005), Seven Recommendations for Creating Sustainability Education at the University Level: A Guide for Change Agents. *International Journal of Sustainability in Higher Education* 6 (4), pp. 326-339.

Penn State Green Destiny Council (2001), *The Mueller Report: Moving beyond Sustainability Indicators to Sustainability Action at Penn State.* Available at: www.p2pays. org/ref/17/16968.pdf (accessed 2 January 2012).

Pérez, J.A.P.V. and Fernández, M. (2012), *Ranking Web of World Universities 2011.* Available at: http://www.webometrics.info/ (accessed 2 January 2012).

Portland State University (2010), *Portland State University District Framework Plan.* Available at: http://www.fap.pdx.edu/planning/public_cppc_1/PSU_FrameworkPlan_June. pdf (accessed 2 January 2012).

Stenzel, P.L. (2010), Sustainability, the Triple Bottom line, and the Global Reporting Initiative. *Global Edge Business Review*, Ann Arbor, MI. Vol. 4 N. 6, pp. 1-2.

The University of British Columbia (n.d.), *Sustainability Coordinators Toolkit.* Available at: http://sustain.ubc.ca/campus-sustainability/getting-involved/sustainability-coordinators/sctoolkit (accessed 2 January 2012).

The University of British Columbia (n.d.), *Teaching and Learning.* Available at: http://sustain.ubc. ca/teaching-learning (accessed 2 January 2012).

The University of Hong Kong (2004), *University of Hong Kong's Sustainability Goals and Policy.* Available at: http://www.sustainability.hku.hk/files/sustainable-hku/report/sustainabilityreport 2004.pdf (accessed 2 January 2012).

University of Gotenburg (2010), *University of Gotenburg Action plan for Environment and Sustainable Development.* Available at: http://www.mls.adm.gu.se/digitalAssets/1331/1331963_ action-plan-for-environment-and-sustainable-development-2011-2015-1-.pdf (accessed 2 January 2012).

Wright, T.S.A. (2003), Definitions and Frameworks for Environmental Sustainability in Higher Education. *Higher Education Policy* 15 (2), pp. 105-120.

Yale Office of Sustainability (2010), *Yale University Sustainability Strategic Plan 2010-2013.* Available at: http://sustainability.yale.edu/sites/default/files/StragtegicPlan/StrategicPlanUpdate June2011.pdf (accessed 2 January 2012).

Chapter 63

Bridging Disciplinary and Professional Divides to Improve International Development Research at Universities

Emily Anderson, Kieran M. Findlater, Olivia E. Freeman, Jordan Levine, Cynthia Morinville, Megan Peloso, Lucy Rodina, Gerald Singh, Dawit Tesfamichael, Leila Harris and Hisham Zerriffi

Abstract

University research on sustainable international development (ID) faces two persistent challenges in its effective execution and application. First, ID research is most often topical rather than disciplinary. Researchers working on thematically and geographically related problems are often isolated from each other by disciplinary divides. Second, because researchers and practitioners are often disconnected, ID research is not effectively informed by, or applied to, development practice. The International Development Research Network (IDRN) at the University of British Columbia (UBC) in Vancouver, Canada provides a case study of one approach to addressing these challenges. As a graduate student-driven networking initiative, IDRN is well-positioned to facilitate communication and collaboration among researchers across disciplines and between researchers and ID practitioners. IDRN's early experiences have contributed to fostering interdisciplinary and inter-professional communication and collaboration, and to developing key competencies in sustainability research in its members. These experiences reveal a number of lessons, and suggest that, while persistent challenges remain, graduate student-led networking initiatives like IDRN have considerable potential to contribute to sustainability education through by helping to address the identified challenges in ID research and practice. Understanding more precisely the nature and extent of this potential at UBC and in different institutional contexts will require further evaluation, and will continue to emerge as IDRN and other approaches to addressing these challenges develop and evolve.

Introduction

International development (ID) research and practice rely heavily on solution-oriented approaches to study and address development problems, necessitating interdisciplinary and inter-professional collaboration. Unfortunately, as argued below, persistent institutional barriers and disciplinary traditions continue to inhibit interdisciplinary and transdisciplinary research, as well as bilateral cooperation between ID researchers and practitioners. Appropriate, university-based solutions may provide valuable opportunities to bridge these interdisciplinary and inter-professional divides, building on the potential of the university to anchor wide-reaching social networks. This case study draws on the experience of the International Development Research Network (IDRN) at the University of British Columbia (UBC), Canada, to exploit the potential for flexible, graduate student-led initiatives to bridge these persistent divides through a facilitative structure that supports the exchange of ideas and the establishment of cooperative partnerships.

Section 1 argues that both sustainability and ID solutions are inherently collaborative in nature, while Section 2 explores in more detail the divides that persist between ID researchers (particularly across disciplines), and between researchers and practitioners. Section 3 introduces IDRN, the unique institutional factors that led to its creation, its approach in connecting researchers with other scholars and practitioners, lessons from two years of IDRN activities, and a discussion of ongoing challenges and persistent questions. Section 4 offers conclusions about the strengths of such an initiative and its potential for expansion.

The Collaborative Nature of Sustainability and International Development

Discussions of ID inevitably lead to fundamental questions of definition: what is ID, what is ID research, and how does it relate to sustainable development (SD)? People from many different professions and disciplines engage with ID, impeding the establishment of a common definition, while the concept of "sustainable development" further complicates matters. First articulated in a report published by the Brundtland Commission in 1987, sustainable development is rooted in the idea that development should meet the needs of the present without compromising the ability of future generations to meet their own needs (WCED 1987). Of course, SD applies globally, encompassing both industrialized ("developed") and non-industrialized ("developing") countries, while ID is generally concerned only with those countries that continue to lag in terms of both economic and non-economic welfare measures. However, the concept of sustainable development has certainly had major implications for ID theory, research, and practice. In essence, sustainability in ID demands that both social and environmental aspects be considered in evaluating the success of any development intervention, along with other traditional indicators such as economic growth (Leal Filho 2011). The design and planning of ID projects increasingly includes a consideration of biophysical science, as well as social and ethical concepts such as "intergenerational equity".

While there are clear and strong relationships between ID and SD, they do not completely correspond. For the purposes of this discussion, International Development (ID) refers to research and practice designed to increase human welfare, focusing on those countries generally defined as developing or least developed (or as defined by the World Bank, countries that are not high income). Sustainable Development (SD) refers to the broad spectrum of research and practice motivated by and contributing to sustainable development as defined by the Bruntland Commission. In essence, they are overlapping fields – SD research and practice have clear implications for ID, and vice versa. Much of the literature cited below was written in the context of SD rather than ID, but to the degree that they share key characteristics – including their cross-disciplinary nature and the need to bridge academic and practice in terms of solutions – the SD literature cited applies to the ID issues discussed.

Since ID, and thus related studies at universities, includes economic, social, and physical challenges, there is a critical and well-recognized need for enhanced collaboration among academic disciplines, as well as for communication and cooperation between the academic researchers who study development problems and the practitioners who implement development programs (Anderberg et al. 2009; Bovill 2009; Harriss 2002; Hulme and Toye 2006; Kanpur 2002; Mikkelsen 2005; Sumner and Tribe 2008a, 2008b). Hulme and Toye (2006) argue that ID problems are complex and multi-faceted, and that the understanding of these issues therefore requires cooperation between specialists from different disciplines, each contributing their own expertise and addressing blind spots that inevitably arise from a single, disciplinary perspective (e.g. Liu 2011). As an example, consider that clean water is not uniquely a problem of physical distribution, but also necessarily includes social, economic, and institutional challenges that impinge on water quality, access, and use. As such, it is increasingly common to encounter hydrologists, engineers, geographers, and anthropologists working together on water issues.

While collaboration between researchers is often a key facet of ID work, there is also a need for close engagement between scholars and practitioners, and for interaction with communities. Implicit in these connections is the need to revisit conventional notions of expertise by facilitating exchange and dialogue with traditional or lay knowledges (Berkes 2004; Brosius 2006; Harris 2009; Turner 1997).

Strengths of the University in International Development

While many researchers have recognized that universities have a unique and important role to play in ID knowledge-generation, it is also clear that the disciplinary structures common to universities can inhibit cross-disciplinary work – due, for instance, to a scarcity of academic incentives for collaboration, the knowledge fragmentation common to 'silo-based' institutions (e.g. Sherren 2008), and the difficulty of reconciling the distinctive conceptual logics and languages that experts employ in different disciplines (Hulme and Toye 2006).

Many researchers have stressed the particularly valuable role that universities can play in addressing key societal needs, such as sustainability and ID. For instance, the Agenda 21 proceedings of the United Nations Rio Summit in 1992 emphasize "the role of education in the process of moving towards SD" (Ceulemans et al. 2011: 621). Similarly, Beringer and AdomBent (2008) emphasize the mandate of universities, individually and collectively "to position themselves in, and contribute to, the sustainability/sustainable development discourse, agenda, and practice" (p. 608). This imperative comes at a time when universities are struggling with ways to ensure that knowledge production is meaningful for broad societal goals, including social justice, sustainability, and economic development.

Although the literature on university engagement in ID remains heavily focused on curriculum development (Beringer and AdomBent 2008; e.g. Fenner et al. 2005; Jennings and Lund 2001; Kevany 2007; Kitamura and Hoshii 2010; Mochizuki and Fadeeva 2010; Perdan et al. 2000; Reid and Petrocz 2006; Taleghani et al. 2011), there is a clear sense that universities can and should focus on ID challenges by providing a fertile environment where academics and practitioners struggling with similar complex and interdisciplinary challenges can connect and build upon one another's work and experience.

As stated by Tiffin and Kunc (2011), "What universities can deliver extremely well, perhaps better than any other agency, is the networking, awareness and social capital building necessary to catalyze clusters" (p. 65). Universities have considerable potential to act as hubs for social networks that can promote knowledge flows through graduates and university-industry linkages (Ostergaard 2008). In addition to having the necessary social imperative and considerable social capital, universities can provide institutional legitimacy, along with logistical and financial support. Universities also provide a physical space in which people, knowledge, and resources are concentrated, which aid community building. This spatial, financial, and logistical facilitation, the academic independence given to faculties and departments, and the migration of graduates into government, private, and non-profit sectors provide an ideal intellectual and physical space for network creation. Furthermore, the forging of social relationships within universities can be an important factor in the development of intellectual capital (Nahapiet and Ghoshal 1998).

There is also a recognition that many valuable learning opportunities occur outside the lecture hall (as emphasized by concepts such as experiential learning, or participatory and informal learning (e.g. Livingstone 2001)). With respect to sustainability, Foster (2002) argues that there is a clear need for 'learning societies' that incorporate life-long learning into education and training, involving a "framework of new institutional relationships between higher education and the wider society in respect to teaching and learning …" (p. 36). Beyond content, enquiry and critical thinking enable students and researchers to transcend disciplinary boundaries and to develop new foci of study. For instance, with respect to other key theories of learning, several authors describe competencies of particular importance to sustainability and development research: (1) systems-thinking competence, (2) anticipatory competence, (3) normative competence, (4) strategic competence, and (5) interpersonal competence (Wiek et al. 2011; also see Raunch and Steiner 2006, or Barth et al. 2007 for similar lists).

Even with this clear convergence of mandate and potential, universities continue to fall short, at least in the domain of ID – this despite the fact that funding agencies across North America increasingly require that universities demonstrate their relevance in precisely these ways (e.g. the Social Sciences and Humanities Research Council in Canada, and the US National Science Foundation).

Discipinary and Professional Divides

Disciplinary Divides in Research

As argued above, effective ID research often requires a collaborative, interdisciplinary approach. However, disciplinary divides in academia can act as barriers between scholars working on similar questions, creating inefficiencies: "the syndrome of knowledge fragmentation" (Carley and Christie 1993). As fields of knowledge become more specialized, or as success is defined in uniquely disciplinary terms, tensions and challenges emerge with respect to interdisciplinary training and research, which may limit their effectiveness. Additionally, while ID challenges are often practical, instrumental, or problem-based, such research is not always considered "cutting-edge" with respect to theory or methodology.

However, ID's practical foundations also provide clear incentives for researchers to transcend conceptual and theoretical differences. Given the urgency of many ID challenges, broader institutional or knowledge barriers may be overcome by bringing together researchers in interdisciplinary teams to focus on more narrowly defined tasks or problems, rather than on questions of conceptual framework or worldviews. In an attempt to bridge disciplinary divides, many universities have established academic institutes focused on ID. Nevertheless, while these efforts have certainly provided some value (Zilahy and Huisingh 2009), the authors' experiences strongly suggest that it is not uncommon to find researchers working independently in different departments, even when they are engaged in research on similar or related issues. Many researchers studying specific aspects of global problems – whether through social science, technology, economics or physical sciences – also tend not to identify their own work as "development" work, and therefore may not be incentivized to bridge disciplinary divides. In addition, many researchers also continue to assume that practitioners and policy makers will make use of their research without further effort in disseminating the results, beyond publishing in the peer-reviewed literature.

Professional Divides Between Research and Practice

Limited interactions between researchers and ID practitioners can constrain the effective implementation of ID. These limits stem not only from disciplinary or conceptual distance, but also from the physical distance between many project implementation and research loci. Links between research and practice are particularly important in ID, since it is a problem-driven field (Kates et al. 2001). A stronger, more collaborative relationship is beneficial to both researchers and practitioners. For instance, it is important that research on the efficacy and unintended impacts of conventional and novel ID projects informs practice. Conversely, it is crucial that practitioners inform research priorities and design by communicating concerns arising from experience with the implementation of ID projects (Clark and Dickson 2003).

Several obstacles inhibit cross-fertilization between research and practice, and their relationship remains weak (van Kerkhoff and Lebel 2006). Writing primarily for academic, peer-reviewed journals, researchers may report their findings using language that is not always easily understood by practitioners (Rosenberg 2007). Conversely, the influence of ID practitioners and participating communities in establishing research priorities is limited by many factors, including the "publish or perish" mantra of academia (Baneyx 2008)[1], the need to seek research funding from granting organizations with pre-determined priorities, and fragmentation due to disciplinary specialization. This effect is magnified when research programs are conceived at institutions far from study sites. Insufficient communication increases the difference between local priorities and the challenges articulated during research planning, potentially increasing the mismatch at each subsequent level of dissociation. Properly conducted local consultation has helped to mitigate this problem (Tomás 2009); however, the success of many ID initiatives remains hindered by such divides (an example is provided in the Text Box).

TEXT BOX: Research, practice and priorities in a Red Sea fishing community

Semi-structured field interviews were conducted with 412 fishers in the Red Sea to assess the sustainability of fishery resources and to understand how fishing communities perceive initiatives for the sustainable development of local fisheries. The interviews revealed a long decline in the abundances of all of the main fishery stocks harvested by the communities, with sharks being worst hit (Tesfamichael, unpublished data). However, to date, no development project has considered shark stocks, either by attempting to understand their decline or its impact on the community, revealing a gap between research and practice. When the fishers were asked their opinion about the sustainable development initiatives currently being undertaken, they expressed their discontent, and described a lack of connection between ongoing projects and their daily realities and priorities. The communities have shown little interest in being involved in such initiatives, which may stem from the fact that they were not properly consulted, or do not believe that the initiatives address the most urgent issues that they face.

The results of the interviews also closely reflect results from a quantitative ecological model of the same area, as well as a multidisciplinary evaluation of the sustainability of the Red Sea fisheries (Tesfamichael and Pitcher 2006). This suggests that well-planned interviews can sometimes provide insights similar to independently run sampling and analysis that require substantially more resources, highlighting the importance of interdisciplinary approaches for effective and accurate research. Simultaneously, interviews involve local communities in the research process and provide an understanding of their perceptions: key factors that increase the likelihood of success in sustainable resource management initiatives (Fraser et al. 2006).

At the University of British Columbia

UBC is a microcosm of many of the challenges facing interdisciplinary research and practice in ID. Graduate students and faculty conducting ID research are distributed throughout the university in different faculties and departments. Like many universities, UBC has historically lacked an established platform to foster inter-departmental, and inter-professional communication and collaboration on ID issues. UBC also lacks a clearly defined "development studies" program to provide an intellectual home for interdisciplinary, development-related research.[2] Instead, development research is pursued and independently defined by myriad research groups across departments within the university. Some graduate students conducting

1 The implications of the "publish or perish" mantra are widely debated (Baumann 2002).

2 UBC Continuing Studies offers a Certificate of International Development; however, this program is geared towards professionals and practitioners, not ID researchers.

ID research have noted that they feel isolated, and that it can take years to become aware of and connect with other professors or students working on similar issues. Although departments across the university host numerous ID-related seminars, panel discussions, and conferences (e.g. in public health, political science, economics, mining engineering), they are often not specifically identified as such.

Additionally, while many professors maintain personal contact with development practitioners, there is no formalized mechanism to link ID research to practice, or to foster communication between researchers and practitioners. Although the BC Council for International Cooperation (BCCIC) connects Vancouver's ID community of practice, it tends not to attract and retain researchers to the community, suggesting an alarming divide between academics and practitioners based in the region. Regular events, including those hosted by the BCCIC (e.g. Development Drinks or the Fair Trade Show), explicitly create networking opportunities for government and non-government professionals, but academic researchers rarely attend; at best, universities are most strongly represented by active undergraduate student groups (e.g. Africa Canada Awareness Coalition, Engineers Without Borders). Graduate students note that it can be time-consuming and difficult to identify and connect with local ID organizations and activities relevant to their work.

The International Development Research Network: Part of the Solution?

There is a demonstrated need for a mechanism to encourage connections between ID-related people and organizations, both at UBC and in British Columbia. Perceiving these disciplinary and professional divides as a hindrance to ID research, a small group of graduate students and faculty, most affiliated with UBC's Liu Institute for Global Issues, founded the International Development Research Network (IDRN) in mid-2010. The following results are drawn from the personal experiences of the IDRN steering committee's members and faculty mentors, post-event feedback solicited from attendees, and an annual survey of member needs.

The Network's Roots

IDRN's founders were able to formulate a response to this challenge due to specific institutional conditions at UBC. The Liu Institute was especially central to the IDRN's emergence, being both a physical building and an intellectual space that fosters research and teaching on global and emerging issues. Although it has historically functioned primarily as a hub for faculty and post-doctorate researchers, under its latest board of executives the Liu Institute has launched the interdisciplinary 'Liu Scholars' program that aims to engage PhD students across diverse departments, working on at least one of the Institute's core themes: sustainability, social justice, and security. A hallmark of this program is that Liu Scholars are encouraged to present their ideas to the faculty-led Liu Executive Board for funding, particularly for the creation of new networks to foster a sense of intellectual community and to enrich the quality and quantity of student research. This institutional context allowed IDRN's founders to imagine, propose, and create what emerged as IDRN.

IDRN is currently led by a steering committee of ten graduate students with two faculty mentors. IDRN's growing formal membership exceeds 250 people, including students and faculty at UBC and other institutions, as well as others beyond academia. Although IDRN is entirely graduate student-led, faculty and staff at the Liu Institute provide both the logistical and regular financial support necessary to maintain momentum.

IDRN's Approach

IDRN's mission is to improve both ID research and practice by facilitating conversations and collaboration across disciplinary and professional divides – strengthening the linkages between disparate research efforts, and between research and practice.[3]

To this end, IDRN hosts events that provide opportunities for members to meet, discuss topical issues, and expand their awareness and knowledge of ID challenges, research, and practice. IDRN's approach builds on Sumner and Tribe's (2008a) three stages in fostering interdisciplinary and inter-professional collaboration:

Stage 1: *Awareness and understanding:* Disciplinary researchers and practitioners learn to communicate effectively, and understanding is built across disciplines, and between researchers and practitioners;

Stage 2: *Conceptualization:* Problem-based activities allow for the identification of relevant disciplines and information, as well as appropriate methodologies and approaches for tackling specific challenges;

Stage 3: *Collaboration:* Interdisciplinary teams are formed, including both researchers and practitioners, and leading to collaborative work.

The steering committee's primary focus to date has been to experiment with different organizational strategies and activities to achieve IDRN's mission. Such activities include panel discussions, individual speakers, a thematic speaker series, field debriefs, film showings, other social events, a graduate-student conference, and an upcoming ID team challenge.[4] Though heavily weighted towards Stage 1, IDRN's activities variously contribute to all three stages.

Stage 1: Awareness and Understanding

IDRN provides intellectual spaces and platforms to promote knowledge exchange, networking, and interdisciplinary understanding. IDRN thereby encourages the active exchange of knowledge between people otherwise unlikely to cross paths, as well as providing centralized information about ID-focused people and organizations, to improve networking and understanding. By creating networking opportunities and venues to share research, IDRN increases awareness and understanding of development problems and approaches, and helps to define a common language to communicate across diverse disciplinary domains. IDRN hosts a range of talks and workshops, including an ongoing speaker series examining current progress towards the United Nations' Millennium Development Goals; an interdisciplinary panel discussion reconciling the disparate disciplinary approaches to ID; and debriefs reporting on ID fieldwork.

The steering committee also perceives IDRN's website as crucial to networking efforts. The website's functionality is currently being upgraded to facilitate connections between researchers and practitioners, and to enrich and improve the relevance of its information resources. In the near future, IDRN will host an online platform through which members can search for tags of interest, including topics and regions, to locate and contact other members doing similar work.

Though local community-building at UBC and in Vancouver have been crucial to IDRN's early progress, the steering committee sees an international network as integral to the organization's aims, particularly given that IDRN's members are working in diverse regions across the globe. IDRN's long-term vision is not only to bridge disciplinary and professional divides in

---

3 IDRN's Mission and Vision are detailed in Appendix 1.

4 IDRN's various activities are described in Appendix 2.

British Columbia, but to extend the organization's networking capacity internationally and to become a hub for the creation of similar student-run groups on other campuses in the Global South and North. Towards this goal, during its 2011 graduate student conference, IDRN attracted students and faculty from 43 programs/departments, across 11 academic institutions (IDRN 2011). Additionally, steering committee members have instigated relationships with potential partner institutions while travelling internationally for fieldwork.

Stage 2: Conceptualization

IDRN's contribution to the identification of relevant information and methods for specific ID challenges has primarily been informal. Members have noted that it was refreshing to discuss their research at networking events with people who were approaching similar problems from different disciplinary or professional perspectives. The graduate student conference's group activities addressed this more explicitly by engaging small interdisciplinary groups of graduate students, faculty and practitioners in discussions that focused on student research problems and thematic ID challenges.

Stage 3: Collaboration

IDRN's online platform, currently under development, will provide a virtual space to further mitigate geographic constraints, and facilitate knowledge-sharing and collaboration. The steering committee aims to host a virtual space where like-minded researchers and practitioners from across the globe can maintain an online presence, communicate with one another, share insights, and solicit feedback from Northern and Southern colleagues alike. The effort will reach beyond research to non-governmental organizations (NGOs) and practitioners who are interested in collaborating with researchers and informing research design. This platform will connect individuals and institutes in the Global South and North to facilitate interdisciplinary and inter-professional collaborations, research and professional exchanges, and other mutually beneficial relationships.

The organization is also currently hosting an interdisciplinary challenge, drawing together teams of graduate students, researchers and practitioners to pitch creative and effective solutions to widely recognized ID challenges.

Competencies in Sustainability Research

With respect to Wiek et al. (2011) framework of competencies in sustainability research, IDRN contributes to each in varying degree. Most notably, IDRN's activities foster interpersonal competence in both steering committee members, as well as the general membership[5] – through panel facilitation (leadership, communication); field debriefs (communication, understanding across disciplines); and the 'Development Challenge' (problem-solving, leadership, and collaboration). All of these activities relate to collaboration, networking, the development of alliances, and team-building (cf. Svanstrom et al. 2008), as well as linked ideas of interpersonal competence – the capacity to understand, compare, and critically evaluate different positions, perspectives, and preferences ('epistemological pluralism' (Wiek et al. 2011)).

---

5 Interpersonal competence is defined as the ability to motivate, enable, and facilitate collaborative and participatory sustainability research and problem solving, involving advanced skills in communicating, deliberating and negotiating, collaborating, leadership, pluralistic and trans-cultural thinking, and empathy (Wiek et al. 2011).

Early Lessons

Having experimented with diverse networking strategies over the past two years, and having solicited feedback from IDRN's membership, this section shares the organization's experience in improving the effectiveness of different types of tools and increasing member engagement.

Different activities attract different communities. Panel discussions and speakers drew primarily on academic researchers, bringing together students and faculty, with a much smaller number of practitioners and other visitors. Field debriefs generally attracted researchers with narrower interests, often those using similar research approaches or working in the same region as the speaker. Social events tended to attract more students than faculty or practitioners, while film showings drew a broader audience, including visitors from the general public. IDRN's preliminary survey results suggest that the organization's members are most interested in attending conferences, while networking and speaker events are also appealing. IDRN members express considerably less interest in film showings and field debriefs, although consistent attendance (typically 15-30 people) indicates that there is clearly an audience for these events, particularly non-members in the case of film showings. Events with broad appeal to non-members may therefore increase IDRN's impact by raising awareness and understanding of ID issues.

Structured interaction improves networking outcomes. Conversations began more easily when events included an interactive element, and when such interaction was at least partially structured. Many of the events with less inherent audience participation, such as speakers and film showings, showed more potential for networking when they included an active component, such as a post-event audience-oriented discussion. Additionally, inherently interactive events benefited from explicit structure. For example, social events sparked more effective conversations when they included a structured activity designed to facilitate networking (e.g. ice breakers).

Personal invitations bring together graduate students, researchers, and practitioners. The graduate student conference was most successful at bringing together and facilitating discussion between graduate students, academic researchers, and practitioners. It was preceded by months of recruitment that benefitted from personal contact and relationship building initiated by steering committee members with individual researchers and practitioners. More practitioners engaged in the conference's "portfolio review" sessions than in any other IDRN activity in the first year. These sessions allowed for interdisciplinary and inter-professional discussions and feedback regarding students' research-in-progress, and were very well received by students, faculty and practitioners alike.

High quality conversation is more likely when topics are sufficiently focused. During the conference's portfolio review and workshop sessions, groups that were organized within narrower topics more easily found common ground and had more rewarding discussions. Other factors leading to improved session outcomes and productive interdisciplinary conversations included participant buy-in, as well as the perceived facilitation skills of the session leaders. As with other IDRN activities, some structure was deemed essential to foster good conversations and networking, even in small group interactions.

Face-to-face opportunities promote new relationships. When in other countries for conferences and fieldwork, steering committee members have endeavoured to meet with like-minded faculty, students and practitioners at development-oriented institutions to assess interest in becoming affiliated with IDRN. All researchers and NGO workers with whom face-to-face contact was established expressed interest in affiliating themselves with IDRN in some capacity, in contrast with initial contact by email. Particular themes of interest among researchers included the fostering of relationships that could allow for researcher exchanges, both South-to-North and North-to-South, as well as the development of mechanisms for mutual in-kind support to enable such exchanges. Among NGO workers, there was interest in creating a greater awareness of their organizations' aims using IDRN's website, ideally culminating in having more researchers pay heed to and actively study their ongoing efforts and achievements. These initial efforts have created opportunities for more formal relationships with institutes and organizations in Mexico, Iran, India, Sri Lanka, and Malaysia. IDRN plans to enrich and expand these initial efforts as the organization matures, enhancing the website's functionality and initiating a global research-oriented newsletter.

An appropriate online platform is crucial for effective ID networking. In the long term, as IDRN expands beyond the local ID community, the website will be of particular critical importance in connecting geographically separated individuals and organizations. IDRN's website should serve as a hub where people can find information about other development-focused researchers, practitioners, initiatives and opportunities, as well as a collaborative space where like-minded people can work together to tackle development challenges. The steering committee has struggled to find the best platform to accomplish these goals and meet the needs of IDRN members. The organization has been challenged to balance functionality with technical sustainability; the website design must meet the properly assessed knowledge and networking requirements of IDRN members, while accounting for the time constraints of the steering committee and budgetary constraints of the organization. Lessons from social media and successful online communities will likely be important in successfully expanding the IDRN website as a networking tool.

Ongoing Challenges

Continuity of Membership and Leadership

Since IDRN is run by graduate students, the organization's longevity requires the continual recruitment of new, dedicated members. Graduate students' academic work is already time-consuming, and the nature of ID research and fieldwork ensures that there are always several steering committee members overseas, increasing the workload of the remaining members. The organization's reliance on a steering committee of dedicated graduate students has proven extremely challenging in the maintenance of organizational cohesion, knowledge and dedicated labor. Following IDRN's first-year activities, the steering committee was overwhelmed with opportunities for collaboration on both nascent and established projects. Additionally, the Liu Institute showed great confidence in the organization, providing additional flexibility and the assurance of funding opportunities for novel events.

Unfortunately, the expansion of IDRN's activities, including the website and the exploration of international partnerships has been severely limited by the strong and inconsistent time constraints of its steering committee members. Despite the support of temporary, part-time conference and website assistants, this is currently the most serious limiting factor in realizing IDRN's long-term international vision – the most dedicated members have the necessary passion and vision, but one or more such individuals would likely need to devote themselves full-time to the endeavor. As it stands, the committed efforts of the current steering committee members have provided rich networking opportunities for the UBC community, but the future expansion of the organization's reach will likely depend on novel approaches to the management of volunteers drawn from the broader membership and to ensuring the continuity of organizational and operational knowledge and vision.

Persistent Questions

As IDRN moves forward, two other substantial questions remain. Firstly, any discussion of the contributions of such an initiative is inherently dependent on the evaluation of its successes and failures. But how should the organization evaluate success? And which criteria should matter most? When can the organization consider its objectives achieved, particularly given the nature of networking as an abstract and ongoing process? Secondly, while networking appears to be a valuable strategy to bridge interdisciplinary and inter-professional divides, will this alone be sufficient to foster sustainable ID research and practice? What else might be necessary?

Discussion

More experimentation and the formal evaluation of IDRN's efforts will be necessary to ascertain how to adequately address these divides and to further define the potential and proper role of university-based networks like IDRN. While it is evident that universities have an important role in engaging and promoting sustainable ID, outreach and collaboration beyond academia are also necessary to address many ID challenges. While research drawing on methods and insights from diverse disciplines is undoubtedly important, collaborative, interdisciplinary research that simultaneously involves practitioners and affected communities throughout the research process, from design to dissemination, perhaps holds the most significant potential for ID research at universities and its effective application to SD practices.

Academic institutions have made efforts to connect researchers across disciplines, and to connect researchers with practitioners (e.g. sustainability "consortia", Motloch et al. 2007). UBC, for instance, has undertaken two major shifts in six years in an effort to establish an institutional "home" for interdisciplinary research and graduate programs, particularly those focused on solution-oriented research (e.g. the Liu Institute for Global Issues; the Institute for Resources, Environment and Sustainability; and the Fisheries Centre). Unfortunately, the creation of mechanisms to support these activities – many of which are promoted in the university's own strategic vision – is challenged by disciplinary norms, academic incentive structures and other institutional barriers common to large organizations (e.g. conflicts over resources, etc.).

IDRN makes a unique contribution in addressing the identified disciplinary and professional divides by providing graduate students with a leading role. Graduate students are in a unique position to engage academics and professionals across disciplines and professions, since they are still in the process of forming their own conceptual frameworks, which necessarily involves questioning the boundaries between disciplines and professions. As a graduate student initiative, IDRN is both energetic and dynamic, with most of the organization's steering committee members not only actively engaged in research, but also focused on, and often also engaged in, ID practice. IDRN, as a graduate student organization, is therefore well positioned to address the challenges of interdisciplinary communication and academic-professional collaboration.

Perhaps it is no coincidence that IDRN's most active members are largely drawn from explicitly interdisciplinary graduate programs, including the Institute for Resources, Environment and Sustainability (IRES), and that IDRN was born of the explicitly interdisciplinary Liu Scholar program. Further, with the deep involvement of graduate student researchers, IDRN is well-positioned to make a unique contribution to the education of future ID researchers by providing these students with opportunities for more inclusive and more frequent interdisciplinary and inter-professional conversations and collaborations.

Conclusions

Persistent disciplinary and professional divides compromise the effective participation of researchers in furthering ID goals. It is imperative that the different facets of ID research and practice be integrated in a practical way, to capitalize on the core competencies and comparative advantages of the research community (e.g. knowledge generation, education, potential for social networking, logistics, physical and financial capital). IDRN represents one model, from within an academic setting, by which to bridge these divides.

University-based and primarily driven by graduate students, IDRN is well situated to address the collaboration and integration of different ID-related disciplines. Interdisciplinary ID research creates outputs (knowledge, publications, reports, and collaborative networks) that address ID problems in more comprehensive and innovative ways, complementing the work of scholars in specific disciplines. ID is an applied field by its very nature, as opposed to most traditional university research disciplines (in both natural and social sciences). While some topics may be addressed within traditional disciplines (e.g. design and testing of new technologies for water or energy), others simply transcend disciplines (e.g. understanding how and whether a new technology might meet the water and energy needs of a community, or the institutional mechanisms necessary for its diffusion). Even where a problem may be addressed adequately within a discipline, a network like IDRN can help other researchers locate those disciplinary researchers who may be able to provide expertise or context to help them to effectively approach a more interdisciplinary problem.

IDRN is also well positioned to address the professional divide between researchers and practitioners. Interdisciplinary research outcomes are inherently more applicable to ID problems faced by practitioners. Bilateral knowledge sharing and collaboration where research informs practice and vice versa strengthen both research and practice, making each more relevant to the other and improving ID outcomes. Practitioners can also facilitate the participation of local communities in the design and implementation of research programs; these communities harbor a treasure of knowledge simply because of their proximity to issues (e.g. natural resources) and their direct interest in the outcomes (e.g. improved livelihoods). Foregoing such collaborative opportunities compromises the success of ID research and project implementation. Interdisciplinary approaches may also ease communication with stakeholders by encouraging the use of shared language, while collaboration can imbue them with a sense of ownership, improving cooperation and uptake.

As the need for integrated ID knowledge and implementation has become increasingly prominent, research institutions and funding organizations have responded in an effort to maintain their relevance and increase the utility of research outputs, although the pace of their progress is open to criticism. Universities are struggling to identify how best to foster and support interdisciplinary research, and how best to bridge the gap between researchers and practitioners in interdisciplinary and inter-professional problem spaces.

IDRN is an example of an adaptive initiative that has the capacity to bridge different disciplinary, professional, cultural, and institutional frameworks within a university setting, as well as bridge existing divides while more institutionalized structures for addressing identified challenges in ID research and practice are created. The organization has the strength and flexibility to complement and enhance the efforts of the university at large, while initiating partnerships beyond the institution and beyond academia and contributing to the development of ID researchers. IDRN has great potential for learning and growth, yet faces major challenges related to the time constraints of its volunteer steering committee, and to the continuity of both leadership and membership. Though evaluation is still preliminary and on-going, IDRN's early experiences suggests that graduate student-led networking initiatives have considerable potential to contribute to sustainability education by helping to address identified challenges in ID research and practice. Understanding more precisely the nature and extent of this potential at UBC and in different institutional contexts will require further evaluation, and will continue to emerge as IDRN and other approaches to addressing these challenges develop and evolve.

References

Anderberg, E., Norden, B. and Hansson, B. (2009), Global learning for sustainable development in higher education. *International Journal of Sustainability in Higher Education* 10 (4), pp. 368-378.

Baneyx, A. (2008). 'Publish or Perish' as citation metrics used to analyze scientific output in the humanities: International case studies in economics, geography, social sciences, philosophy, and history. *Archivum Immunologiae et Therapiae Experimentalis* 56 (6), pp. 363-371.

Barth, M., Godemann, J., Rieckmann, M. and Stoltenberg, U. (2007), Developing Key Competencies for Sustainable Development in Higher Education. *International Journal of Sustainability in Higher Education* 8 (4), pp. 416-430.

Baumann, H. (2002), Publish and Perish? *Journal of Industrial Ecology* 6 (3-4), pp. 13-26.

Beringer, A. and AdomBent, M. (2008), Sustainable university research and development: inspecting sustainability in higher education research. *Environmental Education Research* 14 (6), pp. 607-623.

Berkes, F. (2004), Rethinking community-based conservation. *Conservation Biology* 18 (3), pp. 621-630.

Bovill, C. (2009), Rhetoric or reality? Cross-sector policy implementation at the UK government Department for International Development. *Policy & Politics* 37 (2), pp. 179-199 (doi: 10.1332/030557309X397919).

Brosius, P. (2006), What counts as local knowledge in global environmental assessments and conventions? In Reid, W., Berkes, F., Capistrano, D. and Wilbanks. T. (Eds.), *Bridging Scales and Knowledge Systems: Concepts and Applications in Ecosystem Assessment*, pp. 129-144. Island Press, Washington, DC.

Carley, M. and Christie, I. (1993), *Managing Sustainable Development.* University of Minnesota Press, Minneapolis, MS.

Ceulemans, K., De Prins, M., Cappuyns, V. and De Coninck, W. (2011), Integration of sustainable development in higher education's curricula of applied economics: Large-scale assessments, integration strategies and barriers. *Journal of Management & Organization* 17 (5), pp. 1-21.

Clark, W.C. and Dickson, N.M. (2003), Sustainability science: the emerging research program. *Proceedings of the National Academy of Science of the United States of America* 100 (14), pp. 8059-8061.

Fenner, R.A., Ainger, C.M., Cruickshank, H.J. and Guthrie, P.M. (2005), Embedding sustainable development at Cambridge University Engineering Department. *International Journal of Sustainability in Higher Education* 6 (3), pp. 229-241.

Foster, J. (2002), Sustainability, Higher Education and the Learning Society. *Environmental Education Research* 8 (1), pp. 35-41.

Fraser, E.D.G., Dougill, A.J., Mabee, W.E., Reed, M. and McAlpine, P. (2006), Bottom up and top down: Analysis of participatory processes for sustainability indicator identification as a pathway to community empowerment and sustainable environmental management. *Journal of Environmental Management* 78 (2), pp. 114-127.

Harris, L. (2009), Contested sustainabilities: assessing narratives of environmental change in southeastern Turkey. *Local Environment* 14 (8), pp. 699-720.

Harriss, J. (2002), The Case for Cross-Disciplinary Approaches in International Development. *World Development* 30 (3), pp. 487-496 (doi: 10.1016/S0305-750X(01)00115-2).

Hulme, D. and Toye, J. (2006), The case for cross-disciplinary social science research on poverty, inequality and well-being. *Journal of Development Studies* 42 (7), pp. 1085-1107 (doi: 10.1080/00220380600884050).

IDRN (2011), *IDRN Conference Report 2011. International Development Research Network*. Liu Institute for Global Issues, Vancouver, BC.

Jennings, P. and Lund, C. (2001), Renewable energy education for sustainable development. *Renewable Energy* 22 (1-3), pp. 113-118.

Kanpur, R. (2002), Economics, social science and development. *World Development* 30 (3), pp. 477-486 (doi: 10.1016/S0305-750X(01)00117-6).

Kates, R.W., Clark, W.C., Corell, R.C., Hall, J.M., Jaeger, C.C., Lowe, I., McCarthy, J.J., Schellnhuber, H.J., Bolin, B., Dickson, N.M., Faucheux, S., Gallopin, G.C., Grübler, A., Huntley, B., Jäger, J., Jodha, N.S., Kasperson, R.E., Mabogunji, A., Matson, P., Mooney, H., Moore III, B., O'Riordan, T. and Svedin, U. (2001), Sustainability Science. *Science* 292 (5517), pp. 641-642 (doi: 10.1126/science.1059386).

Kevany, K.D. (2007), Building the requisite capacity for stewardship and sustainable development. *International Journal of Sustainability in Higher Education* 8 (2), pp. 107-122.

Kitamura, Y. and Hoshii, N. (2010), Education for sustainable development at Universities in Japan. *International Journal of Sustainability in Higher Education* 11 (3), pp. 202-216.

Leal Filho, W. (2011), About the Role of Universities and Their Contribution to Sustainable Development. *Higher Education Policy* 24, pp. 427-438.

Liu, L. (2011), Where in the World of Sustainability Education is US Geography? *Journal of Geography in Higher Education* 35 (2), pp. 245-263.

Livingstone, D. (2001), Adults' informal learning: definitions, findings, gaps and future research. NALL Working Paper 21, New Approaches to Lifelong Learning (NALL), Ontario Institute for Studies in Education, University of Toronto, Toronto, ON.

Mikkelsen, B. (2005), *Methods for Development Work and Research: A New Guide for Practitioners* (2nd ed.). Sage Publications India Pvt Ltd, New Delhi, India.

Mochizuki, Y. and Fadeeva, Z. (2010), Competences for sustainable development and sustainability: Significance and challenges for ESD. *International Journal of Sustainability in Higher Education* 11 (4), pp. 391-403.

Motloch, J., Pacheco, P. and Vann, J. (2007), Sustainability for the Americas: building the American network of sustainability consortia. *International Journal of Sustainability in Higher Education* 8 (2), pp. 183-197.

Nahapiet, J. and Ghoshal, S. (1998), Social Capital, Intellectual Capital and the Organizational Advantage. *The Academy of Management Review* 23 (2), pp. 242-266.

Ostergaard, C.R. (2008), Knowledge flows through social networks in a cluster: Comparing university and industry links. *Structural Change and Economic Dynamics* 20 (3), pp. 196-210.

Perdan, S., Azapagic, A. and Clift, R. (2000), Teaching sustainable development to engineering students. *International Journal of Sustainability in Higher Education* 1 (3), pp. 267-279.

Raunch, F. and Steiner, R. (2006) School development through education for sustainable development in Austria. *Environmental Education Research* 12 (1), pp. 115-127.

Reid, A. and Petrocz, P. (2006), Understanding lecturers' understanding of sustainability. *Higher Education* 51, pp. 105-123.

Rosenberg, A.A. (2007), Fishing for certainty – an essay on science and politics. *Nature* 449 (7165), pp. 989-989.

Sherren, K. (2008), A history of the future of higher education for sustainable development. *Environmental Education Research* 14 (3), pp. 238-256.

Sumner, A. and Tribe, M. (2008a), What could Development Studies be? *Development in Practice* 18 (6), pp. 755-766 (doi: 10.1080/09614520802386603).

– (2008b), Development studies and cross-disciplinarity: Research at the social science-physical science interface. *Journal of International Development* 20 (6), pp. 751-767 (doi:10.1002/jid.1494).

Svanstrom, M., Lozano-Garcia, F.J. and Rowe, D. (2008), Learning outcomes for sustainable development in higher education. *International Journal of Sustainability in Higher Education* 9 (3), pp. 339-351.

Taleghani, M., Ansari, H.R. and Jennings, P. (2011), Sustainability in architectural education: A comparison of Iran and Australia. *Renewable Energy* 36, pp. 2021-2025.
Tesfamichael, D. and Pitcher, T.J. (2006), Multidisciplinary evaluation of the sustainability of Red Sea fisheries using Rapfish. *Fisheries Research* 78 (2-3), pp. 227-235.
Tiffin, S. and Kunc, M. (2011), Measuring the roles universities play in regional innovation systems: a comparative study between Chilean and Canadian natural-resource based regions. *Science and Public Policy* 38 (1), pp. 55-66.
Tomás, B.R. (2009), Development of regional sustainability indicators and the role of academia in this process: the Portuguese practice. *Journal of Cleaner Production* 17 (12), pp. 1101-1115.
Turner, N. (1997), Traditional Ecological Knowledge. In: *The Rainforests of Home: Profile of a North American Bioregion*, pp. 275-298. . Island Press, Washington, DC.
van Kerkhoff, L. and Lebel, L. (2006), Linking knowledge and action for sustainable development. *Annual Review of Environment and Resources* 31, pp. 445-477 (doi: 10.1146/annurev.energy.31.102405.170850).
WCED (1987), *Our Common Future,* World Commission on Environment and Development. Oxford University Press, New York, NY.
Wiek, A., Withycombe, L. and Redman, C.L. (2011), Key competencies in sustainability: a reference framework for academic program development. *Sustainability Science* 6 (2), pp. 203-218.
Zilahy, G. and Huisingh, D. (2009), The roles of academia in Regional Sustainability Initiatives. *Journal of Cleaner Production* 17 (12), pp. 1057-1066.

Appendix 1: IDRN's Mission and Vision

IDRN's Vision

The IDRN aims to be a network of development researchers who advance insightful, integrated research that effectively influences decision-making.

IDRN's Mission

- Facilitate collaboration among development researchers across social, geographic and disciplinary divides.
- Ensure well-informed decision-making on issues related to international development.
- Ensure a prominent voice for graduate students on issues of international development, both within the research community, as well as in decision-making, public discussions and institutions.

More information about IDRN can be found at www.idrn-ubc.org.

Appendix 2: IDRN Activities

Panel Discussions

Panel discussions bring together several experts from different fields – researchers and/or practitioners – to discuss or debate a topic or question in front of an audience. IDRN's inaugural event was a panel discussion involving researchers from four disciplines discussing the meaning of development in the research context and the potential to bridge interdisciplinary divides, with respect to perspective and methodology. During such a discussion panelists each present their position. They are then given an opportunity to respond to each other's remarks, and the discussion is subsequently opened to the audience.

Speakers

IDRN hosts experts from diverse fields, including both researchers and practitioners, speaking on myriad topics. The organization is additionally collaborating with the Liu Institute and UBC's Green College to host a speaker series critiquing progress on the United Nations' Millennium Development Goals (MDG).

Round Table Discussions

In conjunction with the MDG speaker series, described above, IDRN has organized a series of Round Table Discussions, which provide a smaller group of graduate students the opportunity to meet with the speaker in a more interactive forum. Students come prepared with questions and discussion topics. An IDRN facilitator leads an informal discussion between the speaker and the students.

Field Debriefs

Field debriefs involve a presentation by a researcher who has recently returned from international fieldwork. The presentation involves an exploration of the challenges encountered in conducting ID fieldwork, along with background information and the preliminary findings of the research project. IDRN hosts regular field debriefs, inviting presentations by both graduate students and faculty.

Film Showings

RN regular shows films related to international development challenges. Where possible, a ssion is facilitated after the film. If possible, someone involved the production of the film director) is invited to attend and take questions after the showing.

ts

informal opportunities for people to gather and initiate conversations. To vents have been hosted on campus and have taken many forms, with food ommon denominator. For instance, IDRN experimented with "Speed cipants have two minutes to introduce themselves to a colleague and before switching partners.

Conference

In May 2011, IDRN hosted a conference for graduate students involved in international development research. The conference was unconventional, focused on connecting people interested in similar ID research themes rather than on the presention of research papers. There were four main activities:

Keynote Speakers: There were two keynote speakers – experienced international development practitioners from two well-known organizations.

Professional Development Seminars: Students attended seminars where speakers presented on issues specific to ID research, such as field methods, working in conflict zones, and applying for research grants.

Portfolio Review Sessions: Students were divided into small groups based on their topical research interests. Each group had at least one experienced researcher and one experienced practitioner with some expertise on the group topic. Each student gave a short presentation of their research – background, progress, and key challenges – and then received feedback from the experienced researchers and practitioners and from the other students in the group.

Workshops: Students were divided into topical research interest groups. An IDRN facilitator led a discussion about key issues and challenges in their field, which were then presented to all participants in a final plenary session.

Development Challenge

In May 2012, IDRN will host a development challenge, in which participants form teams of interdisciplinary researchers and practitioners, including both students and experienced academics. Teams are to propose a creative, novel, effective solution to a specific ID challenge of their choice, and compete for funds to implement their solution.

Chapter 64

Putting Concepts of Sustainable Development into Practice at Higher Education Institutions: Experiences from the SKINT Project

Walter Leal Filho

Abstract

This paper discusses the need to promote less theoretical and more practical approaches in influencing future trends of education for sustainable development in higher education institutions. It also outlines the project SKINT as one example of the contribution universities may provide towards putting the principles of sustainable development into practice. In particular, it focuses on the sustainable management of water resources, a matter of central importance to both industrialising and developing nations.

Introduction

The quest for implementing sustainable development at university programmes is a challenge which entails many barriers, which have been extensively described elsewhere (e.g. Leal Filho 2011a, Leal Filho 2011b). As we head towards the end of the UN Decade on Education for Sustainable Development (UNDESD), it becomes evident that we have not yet reached a situation where satisfaction with the progress achieved may be expressed (Leal Filho, Goncalves, Pereira, Azeiteiro 2012). Indeed, we still see the same problems related to the implementation of sustainability at higher education institutions, which were observed at the beginning of the UNDESD, in particular:

- an acute lack of specific financial support to sustainability programmes;
- a lack of coordination about what is happening internationally;
- the lack of any system which may document, measure and disseminate significant developments, and catalyse more action.

This situation should not however be seen as completely negative. There have been many successful stories which show how rewarding sustainability initiatives in a higher education context can be, and how much can be achieved (Leal Filho and Manolas 2011). However, apart from the traditional problems posed by lack of institutional support or lack of funding to university-based sustainability programmes, a common trend seen is that a purely theoretical focus on sustainability often fails to convince decision-makers and university administrations about its real value. This is a problem which needs to be addressed and which this paper tries to tackle.

In order to succeed, sustainability **concepts** need to be put into **practice** and to engage and motivate the relevant stakeholders (Leal Filho and Schwarz 2008). In addition, if there is a real intention to pursue change and mobilise change agents as outlined by Moore (2005), then there is also a perceived need, reiterated by Brundiers, Wiek and Redman (2010), to take advantage of use real-world opportunities, both in a classroom context and elsewhere.

In their paper discussing the whole-person approach to educating about sustainability, Podger, Mustakova-Possardt and Reid (2010) did note the need to appeal to each person and the need to ensure that values and perceptions are emphasised. Only in this way can we catalyse a commitment which will subsequently be the basis for action, either in regard to teaching and learning in a higher education context, or indeed elsewhere.

In order to sustain the impacts of changes, both on a campus (Mitchell 2011) or elsewhere, there should be a clear perception on the extent to which changes will improve the state of play with regard to sustainability in some key areas such as:

a) campus culture,
b) university policies,
c) institutional priorities,
d) public perceptions.

Furthermore, there is a perceived need to carefully consider the extent to which such changes in higher education will be long-lasting and will not be diluted by the usual pressures on finances and resources. Achieving sustainability, be it via reform or transformation (Rees 1995), requires a new thinking and a new focus, which transcends traditional approaches and makes use of new, practical ones.

One such approach is the use of real projects, meaning projects which are based on sustainability principles and strategies on the one hand, but which also have a practical nature on the other. Projects where one can recognise the value of sustainable development, applied to a context which people deem important and with goals people see as worth pursuing. In the subsequent parts of this chapter, the experiences from one such project are described in detail, with a view to illustrating the usefulness of this approach.

Promoting sustainable water management: the SKINT project

Sustainable water management is one of the top priorities in Europe today. Legislation such as the European Water Framework Directive (WFD) and Floods Directive (FD) have clear goals about improving water quality and regulating water quantity in the European Union (EU). When coupled with climate change and urbanization, these create increasing pressures on water management, especially in urban areas. The North Sea Region, which involves Belgium, Germany, Denmark, Netherlands, Norway, Sweden and the UK, has a long tradition of solving problems related to ground and surface water in urban areas by means of optimising innovative technical solutions. However, the recent shift towards a more integrated approach in urban development, in which many different disciplines have to work together, demands a new and more sustainable approach to water management.

At present, it is often the case that innovative solutions to the problems related to sustainable water management are available on the one hand, but on the other hand, implementation is hindered by barriers to multidisciplinary work. There is therefore a need to foster awareness of the role of different disciplines and to build up capacity within the disciplines in order to make the required contributions. This will enable more rapid and sustainable responses to problems associated with economic regeneration, thus encouraging the inward investment that is needed in some regions.

Consistent with this, the project "North Sea Skills Integration and New Technologies" (SKINT), funded by the Interreg IVB North Sea Programme, aims to facilitate the implementation of sustainable urban land and water management, by improving the integration of water management in spatial planning processes. The goals of the Lisbon and Gothenburg agendas will be achieved more easily in a situation where water is seen as a beneficiary of regeneration and development projects, instead of a burden that has to be managed at the end of the pipe. Therefore, it will be more important than ever to improve the skills of water managers and to provide evidence to convince them of the need for sustainable water management in multidisciplinary projects. SKINT's website can be seen in Fig. 1.

Fig. 1: Home Page of the Project

Within SKINT's transnational partnership, every beneficiary has a specific expertise and experience in capacity building to share with others. Since the WFD and FD are import for all European member states, it is believed that the outputs of SKINT will be most effective as part of a transnational approach. The SKINT beneficiaries all come from the countries within the NSR which are extremely vulnerable to the water-related effects of climate change due to their location near the sea. SKINT deals with both groundwater and surface water and integrates both water quality and quantity. The main focus is to build up the capacity of the different professionals within multidisciplinary processes.

Heuckenlock: sustainable approaches at a local level in Hamburg

The regional projects that all project partners undertake are the cornerstones of the SKINT project. Regional projects have been undertaken across the partnership. In the case of Hamburg, the project "Towards integrated approaches to the preservation of surface waters – examples from the Nature Protection Area 'Heuckenlock'" was performed. This project is in line with SKINT in two main ways. Firstly, it elaborates on a strategy for the integration of environmental considerations in the urban development process of this sensitive site, with due consideration to the need to ensure the long-term quality of surface waters. Secondly, it tests and documents experiences aimed at showing the role of education and public awareness in the sustainable management of urban waters, and demonstrates technical approaches and methods to ensure the impacts of tidal flows and climate change are duly considered. The position of Hamburg in Germany is outlined in Fig. 2.

Fig. 2: Position of Hamburg in Germany

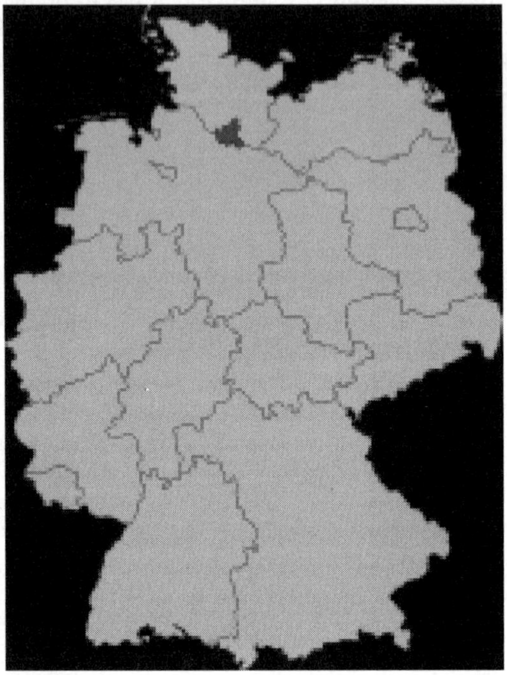

The pilot project has been executed by the HAW Hamburg in cooperation with various partners such as the Hamburg Ministry of Environment and the Elbe Tidal Centre Bunthaus.

The project area is the largest river island in Europe, the Elbe-Island Wilhelmsburg (Elbinsel Wilhelmsburg). The focus has thus been on the river Elbe and the "Elbe-Auen" in particular.

The Elbe divides the city of Hamburg into two branches, the *Norderelbe* and the *Süderelbe*; but these branches meet again opposite Altona, just west of the old city, to form the *Unterelbe*, which flows into the North Sea some 65 miles downstream from Hamburg. Two other rivers flow into the Elbe at Hamburg – the Alster from the north and the Bille from the east. Due to the fact that it is the place where the northern and southern branches of the River Elbe meet, the project site in Wilhelmsburg is especially sensitive and demonstrates the need for the types of action envisaged as part of the pilot project, "Towards integrated approaches to the preservation of surface waters – examples from the Nature Protection Area "Heuckenlock".

The protection and maintenance of the quality of Hamburg's surface waters is a matter of priority to Hamburg. In the district of Wilhelmsburg, one of the poorest in Hamburg and one with the highest number of immigrants, the need to integrate environmental protection and water management with urban development plans is particularly acute. As a result of various initiatives such as dredging and drying of some areas to make room for projects, less than 25% of the natural flooding surfaces remain. Fig. 3 and Fig. 4 offer some impressions of the site.

Fig. 3: Biergarten at the Ernst-August Canal

The current problems seen in the site are especially acute, since it hosts the Nature Protection Area "Heuckenlock", which is under pressure from current development plans. The government of the City of Hamburg made a decision to prioritise the regeneration of the Elbe-Auen, a decision which is in line with the Gotheburg agenda of promoting sustainable economic development. In addition, because it does relate to a set of innovative approaches which take into account economics, social elements and nature protection, the pilot project also has to be in line with the aims set at the Lisbon Declaration, in particular by focusing on nature conservation as one tool, via which the economic competitiveness of the site may be kept up to date and new jobs may be created.

There have been various projects in the city dealing with the impacts of the tidal Elbe, but only a few have dealt with the need for an integrated and sustainable development concept where the regeneration of the Elbe Auen is coupled with the protection of the nearby Nature Protection Area "Heuckenlock". This project has not only analysed the inherent conflicts related to the conservation of surface waters at the site, but also considered the extent to which the guidelines from the Water Framework Directive are being followed, whilst also suggesting ways to deal with the pressing need to protect its unique biodiversity.

Fig. 4: The Elbe Bridge over Wilhelmsburg

The key question addressed in this pilot project is how to make sure that environmental considerations are included in the urban development process of a sensitive site, with due consideration to the need to ensure the long-term quality of surface waters.

This pilot project is in line with the transnational aims of the project SKINT since it involves:

a) an analysis of stakeholders' views and attitudes related to the conservation of the Elbe Auen in the current urban development and planning process;
b) a concrete contribution to the water-portal tool by means of the documentation and dissemination of the experiences from the pilot project, which may be replicable elsewhere and can be fed into the Watertoets instrument;
c) the training of managers and other stakeholders;
d) a contribution to the review of projects in the SKINT network and partnership;
e) the education of water managers, urban planners and other stakeholders by running training and information sessions in cooperation with the Hamburg Ministry of Environment and the "Elbe-Tideauenzentrum Bunthaus" where training sessions and exhibits have been held.

The pilot project has been executed in cooperation with the IBA, the Hamburg Ministry of Environment and the "*Elbe-Tideauenzentrum Bunthaus*". Data from the national environment agency (*Umweltbundesamt*) has supported the training sessions. The role of the project partnership is outlined in Table 1.

Table 1: Project partnership in Hamburg

Organisation	Role	Project tasks
HAW Hamburg	Public University with teaching and research duties	Official project partner; overall project coordination and management in Germany
International Bauausstellung 2013 (IBA)	Local stakeholder responsible for the development of Wilhelmsburg	Links with the development of Wilhelmsburg and input in stakeholder engagement
Hamburg Ministry of Environment	Local environment authority responsible for the management of the Elbe	Input into the work packages, implementation of the strategy
"Elbe-Tideauenzentrum Bunthaus"	Local environment NGO	Information and training sessions
Umweltbundesamt	National Environment Agency	Technical inputs to the training sessions

The pilot project has led to solutions which are ecologically sound and economically viable. Since the project site has a more-or-less even distribution of males and females but a strong migrant community, it has paid a special emphasis to equal opportunities and has ensured its outputs and its activities are not only scientific sound, but also socially just and ethically acceptable.

Lessons Learned: sustainability at the local level

As part of this unique sustainability project, HAW Hamburg has had the opportunity to put into practice many of the principles of sustainable development the literature tackles, but which are seldom put into practice. These can be divided into gaining knowledge and experience, further transnational exchange and the integration of social issues. These will now be reviewed in turn.

In terms of **gaining knowledge and experience**, the pilot project has managed to:

i. cross-fertilise information and ideas on the means to promote the sustainable management of urban waters in line with the principles of the Water Framework Directive. In particular, the training sessions on water law were very popular and catalysed much interest;
ii. introduce, discuss and debate technical approaches and methods used in the partner cities to cope with the challenges regarding the protection of surface waters;
iii. disseminate the gathered transnational information and knowledge among local stakeholders, in a way which informed but also engaged them.

As far as further transnational exchange is concerned, the results were:

i. it fostered the exchange of ideas and experiences, which led to a better and mutual understanding of the methods used to foster the integration of environmental considerations in spatial planning processes seen in different countries;
ii. it discussed and benchmarked approaches seen in the North Sea region related to public participation and stakeholders engagement in dialogues related to water conservation;
iii. it provided information and training to a transnational audience based on topics of common interest and relevance.

In terms of the integration of social issues, the pilot project has managed to link areas such as employment, innovation and social policy, which are all central to the pilot project in Hamburg, and to the SKINT project as a whole. The project's focus anchored it in support of the Gothenburg Agenda for more sustainable patterns of development. Last but not least, a further result was the encouragement of more transnational and cross-sectoral thinking in urban development processes.

Conclusions

The impact of the SKINT project as a whole and the pilot project in Hamburg in particular, can be divided into local, regional, national and international. At the local level, inputs from the local SKINT project and its outputs such as local strategy on the inclusion of water issues in the urban development process will be used by the city administration as a "Blueprint" for future initiatives. In addition, the training provided will allow local managers to use integrative approaches in a more systematic way. Results from SKINT will be used in preparation for the International Construction Exhibition to be held in Wilhelmsburg in 2013.

At a regional level, results from SKINT and from the local case study will be used by local authorities in the Hamburg Metropolitan region as guidance and as a tool to be used in their own initiatives.

At a national level, data and experiences from Hamburg will be disseminated across Germany and used to encourage the use of integrative approaches to link water management and urban development plans.

At an international level, the experiences from the Hamburg case study may assist city governments across the North Sea region in their efforts towards implementing integrative approaches. Finally, the training provided will reflect in a better degree of preparedness of cities to implement the "public participation" component of the Water Framework Directive, thus ensuring the impacts of this local sustainability project will be felt for years to come.

References

Brundiers, K., Wiek, A. and Redman, C. (2010), Real-world learning opportunities in sustainability: from classroom into the real world. *International Journal of Sustainability in Higher Education* 11 (4), pp. 308-324.

Leal Filho, W. (Ed.) (2011a), *World Trends in Education for Sustainable Development.* Peter Lang Scientific Publishers, Frankfurt.

– (2011b), An Overview of ESD in European Countries: What is the Role of National Governments? In *Global Environmental Research* 14 (2), pp. 119-124.

Leal Filho, W., Goncalves, F., Pereira, R. and Azeiteiro, U. (Eds.) (2012), *Contributions to the UN Decade of Education for Sustainable Development.* Peter Lang Scientific Publishers, Frankfurt.

Leal Filho, W. and Manolas, E. (2011), Making Sustainable Development in Higher Education a Reality: examples from leading institutions. In *Higher Education in the World* (4). GUNI Series on the Social Commitment of Universities. Palgrave, London (ISBN 9780230535558).

Leal Filho, W. and Schwarz, J. (2008), Engaging stakeholders in a sustainability context. *International Journal of Sustainability in Higher Education* 9 (4), pp. 498-508.

Mitchell, R.C. (2011), Sustaining change on a Canadian campus. *International Journal of Sustainability in Higher Education* 12 (1), pp. 7-21.

Moore, J. (2005), Seven recommendations for creating sustainability education at the university level: a guide for change agents. *International Journal of Sustainability in Higher Education* 6 (4), pp. 326-339.

Podger, D.M., Mustakova-Possardt, E. and Reid, A. (2010), A whole-person approach to educating for sustainability. *International Journal of Sustainability in Higher Education* 11 (4), pp. 339-352.

Rees, W.E. (1995), Achieving sustainability: reform or transformation? *Journal of Planning Literature* 9 (4), pp. 343-361.

IV.
Latin America

Chapter 65

Fostering Sustainable Development at the Post-Graduate Level: Experiences from LATEC's Master's Programme in Systems Management

Osvaldo Luiz Gonçalves Quelhas, Marcelo Jasmim Meiriño, Sérgio Luiz Braga França, Valdir de Jesus Lameira and Leonardo Bezerra Pimentel

Abstract

In the midst of a process of change that is being experienced in all areas of human activity in this new century, there is a demand for new directions, concepts, values and paradigms to be observed in social, economic and environmental fields. Upon the emergence of patterns of sustainable economic growth, the LATEC/UFF identifies and tries to suggest how to answer the following problem: How to transform not-green businesses (for-profit, non-profit and public) into green projects? And how to characterize the modus operandi of a green business in different socio-economic environments? Since 1991, LATEC has developed courses and research projects to address the assumptions, issues and research projects related to sustainable development. Among other contributions, the LATEC allows the Brazilian government to recognize, promote and support the development of green production chains; to subsidize trade associations and research institutions in order to prioritize projects, actions and resources related to sustainable development; and to structure a database of opportunities for entrepreneurs interested in integrating green supply chains. A more detailed explanation of the project, as well as the description of several other important goals achieved by the project, can be obtained by reading the full article.

Introduction: Sustainable Actions of the University

The theme of corporate sustainability has generated growing debates in both the academic and business environments, gaining praise in the marketplace and within academia. Sustainability, in practice, can be seen as the art of conducting business in an interdependent world (Savitz and Weber 2006). However, this interdependence is consistent with interests that often diverge, and the method utilized to understand its importance is sometimes contradictory. No-one disputes the fact, however, that it takes a joint effort by all of the legitimate actors of society to understand and apply its precepts.

As a result of this discussion, the authors hope to understand how organizations, through their agents, may respond and contribute to this new human epic, which is to enable growing financial gain that is more harmoniously balanced with the social interests of economic growth. Another aspect to be balanced is to increase understanding of the subtle limits of the availability of natural resources, without further contributing to the existing environmental degradation. Corroborating this scenario, Tachizawa (2004) claims that "companies are living systems, whose understanding is not possible only through the economical perspective".

Finally, in addition to this balance, efforts must be coordinated in order to ensure that corporate decisions about economic, social and environmental approaches are set within a non-subordinate dynamism among the economic, social and environmental issues.

The main challenge here is to not just to require companies to adopt a new behavior towards the problems that lie ahead, but to contribute to a new society with broader commitments than those of today. To require companies to act sustainably is also a global demand. This global demand affects the constitution of new ethical, civic and moral commitments. Lacking timing and cooperation, it will be difficult to succeed and to be fair when demanding change on decisions and production within organizations.

However, one should not expect that companies would fully adhere to such desired sustainable behavior. Our understanding is that companies can play a key role in the planetary sustainability. The big problem is: how must companies change? What is the safest path? Where do these large changes begin?

This study is current and relevant in the sense that it describes the procedures in which LATEC/UFF (Technology, Business Management, and Environment Laboratory) has contributed for the making of a sustainable society.

In this context, the study identifies the following problem that is the genesis of LATEC's actions: How to fully transform the businesses (profit, non-profit, and public) from non-sustainable to sustainable? And, how to characterize the *modus operandi* of a sustainable organization in distinct socioeconomic environments?

The goal of this paper is to identify guidelines utilized in case studies within the areas of research, teaching and extension on organizational sustainability to undergraduate and post-graduate students.

The research method which has been utilized is based upon the review of the existing literature and on study cases in Brazilian Universities.

This article presents an account of the experience of education on sustainability in LATEC, which is a laboratory associated with the Fluminense Federal University and highlights the motivations for developing research, education and extension projects.

Sustainable Organizations and International Guidelines

Sustainable organizations: Organizations where one can see depolluting equipment, green products, and products derived from recycling as well as services like recycling and environmental consulting (Faria and Silva 1998). Such definition evolves and is dependent on the alignment of organizations and governments with sustainable practices, which have gradually grown in importance as the issue of survival in increasingly globalized markets. Companies and governments are aware of their survival and continued development. The opportunities and threats of the external environment, the development of the skills and abilities of their leaders, employees and suppliers are crucially important for achieving their goals and for experiencing longevity.

The challenge is posed: organizations and governments seek integration with research centers to get answers and enable planetary life and the continuity of human civilization's wellbeing. This quest is realized through the development of tools to foster dialogue with society and the improvement of business leaders in order to develop social and environmental sensitivity. Moreover, it is necessary to improve the multimedia communication systems in order to provide connectivity among businesses, society, government, non-governmental organizations and research centers. Technology, new management models and sustainable solutions are sought to harmonize the various interests of a complex and diverse world.

Companies and governments have been closely watched by citizens that are growing better informed and more aware of the hazard derived from the pollution, waste and gas emissions that cause global warming and have significant social impacts.

Business activities in the field of social and environmental responsibility are valued in the Stock Exchange by national and international funding agencies (BOVESPA 2012). Production, growth and development, supported by socioeconomic responsibility, guarantees the sustainability and longevity of organizations, in addition to the planet's wellbeing. This is a growing perception in the business world.

In this scenario, there are several questions lacking responses:

What is a sustainable organization? How can organizations contribute to minimizing social and environmental impacts on the planet?

What can be done to improve the efficiency of social and environmental responsibility education or the development of citizen and global awareness in society?

What needs to be changed? What paths, steps, organizational initiatives must be made in order to adapt to the changing times and economic transformations that lie ahead?

These, among many others, are crucial issues for sustaining life on the planet and the continuity of civilization. The link between universities, companies and governments seems to be the most appropriate way to build models applicable to the sustainable development of society as a whole. Finding more sustainable means of action is the proposed focus of this paper.

By observing the timeline, the development of various tools, indicators, guidelines, methodologies and concepts that address social and environmental issues, it is clear how urgent the discussions revolving around this project's proposed theme are. Action strategies generated in the most diverse spheres (business, government, etc.), and applicable in various fields (country, company, etc.) can already be seen.

The alignment between this paper's proposals and the most diverse international guidelines is latent. A more in-depth look at some of these guidelines allows us to highlight some aspects, as described below.

Global Pact

The global pact represents an agreement for businesses proposed by the UN, in July 2000, with the intention of reconciling market forces with the ideals of human rights and taking into account the social and environmental impacts caused by globalization (UN 2010a). It is worth mentioning the third principle, "Principles of Environmental Protection", which references:

- Supporting a preventative approach to environmental challenges
- Promoting environmental responsibility
- Encouraging technologies that do not harm the environment.

Millennium Development Goals

These were proposed by the UN in September 2000 and represent a set of goals for the development and eradication of poverty in the world, which must be adopted by the member states (UN 2010b). The principles identified below are directly related to this research project:

- Guarantee environmental sustainability.
- Establish a global partnership for development.

UN's Global Agenda 21

During the ECO-92 Conference in Rio de Janeiro, a plan of action was created internationally to be applied globally, nationally and locally in areas in which human action impacts the environment (UN 1992). As a guide for a new pattern of development in the 21st Century, the ideas translate into "Think globally, act locally". Among the issues addressed by Agenda 21, the following are worth mentioning:

- Partnership and awareness;
- Commitment to solutions;
- Social, environmental and economic issues.

Earth Charter

The Earth Charter Commission (1994), composed of 23 individuals, proposed the document after an extensive consultation into the global society's understanding of social and environmental problems. The following are among the principles directly related to the project proposed herein:

- Respect the Earth and life in all its diversity;
- Protect and restore the integrity of Earth's ecological systems, with special concern for biological diversity and the natural processes that sustain life;
- Prevent harm to the environment as the best method of environmental protection and, when knowledge is limited, adopt a precautionary approach;
- Adopt standards of production, consumption and reproduction that protect Earth's regenerative capacity, human rights and community wellbeing;
- Advance the study of ecological sustainability and promote the open exchange and extensive implementation of acquired knowledge.

OECD Guidelines (Organization for Economic Cooperation and Development)

The OECD Guidelines describe voluntary standards and recommendations for business conduct in a wide range of social and environmental issues (Torres et al. 2007). They are intended to ensure that business activities are in harmony with governmental policies, contributing to the sustainable development made by multinational companies. Among the topics covered by the guidelines, are:

- Environment
- Science and Technology
- Consumer's Interest.

Equator Principles

In June of 2010, principles were proposed by the ten largest banks in International Project Financing (EP, 2010). These are minimum criteria for granting credit, which ensures that financed projects are developed in a socially and environmentally responsible manner. Companies must incorporate the following into their projects, among other things:

- Environmental risk management, protection of biodiversity and adoption of mechanisms to prevent and control pollution;
- Evaluation of socioeconomic impacts;
- Efficiency of production, distribution and consumption of energy and water resources and the use of renewable energy.

Kyoto Protocol

Global agreement made by nations, in 1998, to reduce the emission of gases that cause global warming (UN, 1998). The issues covered include:

- Promote the use of renewable energy sources;
- Limit methane emissions in waste management and energy systems;
- Protect forests and other carbon sinks.

Dow Jones Sustainability Indexes (DJSI)

This New York stock exchange index was created in 1999, and monitors the financial performance of leading companies in the field of sustainable development (DJSI 2011). The DJSI was the first global reference model to evaluate the financial performance of global leaders in sustainability. The indexes are intended to evaluate the companies for investment purposes, taking into consideration their performance with sustainability, and allowing for a type of investment prioritization of companies, which differ based on their ability to take advantage of the opportunities and properly manage the risks associated with economic, environmental and social factors. Selecting the companies that make up the index is done through an extensive questionnaire that covers the following dimensions: social, economic and environmental (Quelhas et al. 2011). The following items of analysis are worth noting:

- Environmental policy – Eco-efficiency
- Environmental reports – Environmental Management System.

CSI (Corporate Sustainability Index of BOVESPA – São Paulo Stock Exchange)

CSI aims to create an investment environment compatible with the demands of contemporary sustainable development of society, as well as encouraging ethical corporate responsibility (BOVESPA 2012). The process of selecting companies implements an extensive questionnaire supported by a cluster analysis. The following are among the criteria and indicators considered in the CSI:

- Economic-financial, social and environmental dimensions;
- Environmental dimension;
- Corporate governance dimension.

Case Study

The Technology, Business Management and Environment Laboratory – LATEC

Comprised of faculty and researchers from diverse areas of expertise, it has sought to foster teaching, extension and research actions for organizational sustainability.

It is institutionally and physically connected to The School of Engineering and to Production and Civil Engineering departments.

Issue to be tackled in this case study

The case study proposes to answer: what are the guidelines for research, teaching and extension in organizational sustainability to postgraduate and undergraduate students?

Description of Guidelines for Research, Teaching and Extension on Organizational Sustainability

The guidelines to be followed during the construction of knowledge in the realm of organizational sustainability at LATEC are applied throughout extension and teaching according to the characteristics of each problem. This is a process that aims at gaining recognition for a sustainable organization. In addition to this main guideline, there are other specific objectives pursued by LATEC, such as:

1. Conduct an extensive, thorough study of the state-of-the-art characterization of sustainable organizations;
2. Characterize the concept of sustainability and evaluate its evolution with regard to the international new demands of society;
3. Evaluate and compare the various proposed methodologies, national and international, in order to assess a company's level of sustainability;
4. Develop a scientific research strategy that can characterize the methodology and test it in different socioeconomic environments;
5. Build a strategy for constructing an international network of scientific cooperation to support the research;
6. Train researchers at a masters and doctoral level within the international network of scientific cooperation;
7. Publish final graduation papers, scientific articles, dissertations, theses, technical reports and books aimed at publicizing the partial and final results of the research.

The research hypotheses to be developed are defined based on the needs of the market and the academy. The issues raised are intended to guide the definition of scope, the mapping of skills, and the development of research projects that unfold into products and processes, as can be seen in Fig. 1:

Fig. 1: Process model of knowledge development for sustainability at LATEC

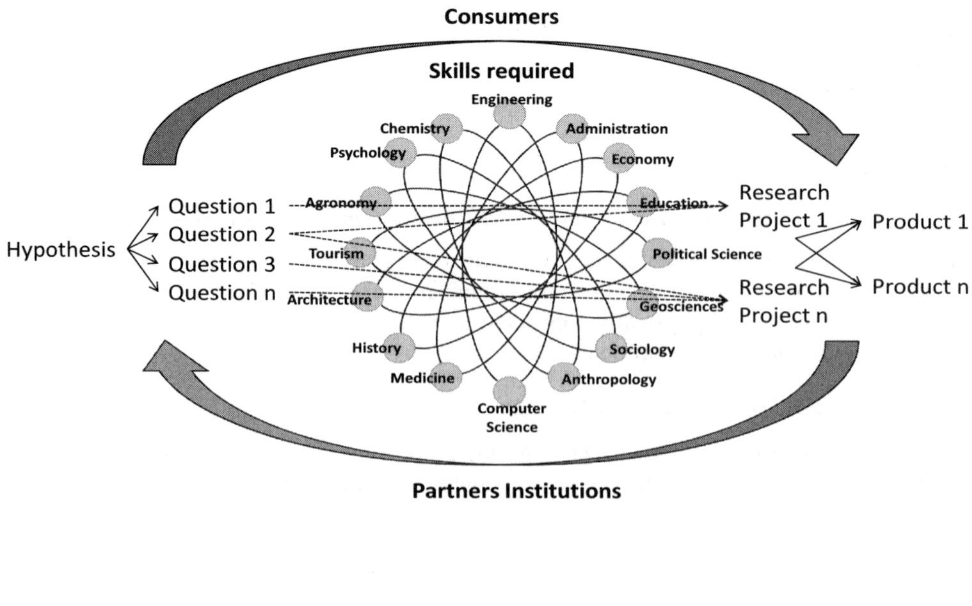

With regard to the hypotheses and their associated projects, the authors can highlight the following:

Hypothesis 1:

H1: Societies with a more advanced level of socioeconomic development are more likely to have sustainable organizations.

Questions

Q1: What is the most suitable methodology for measuring a society's socioeconomic level?
Q2: What are the criteria for evaluating a more advanced society?
Q3: What are the definitions/concepts of Sustainable Organizations?
Q4: How does one measure a society's willingness to invest in sustainable organizations?

Research Projects

RP1: Methodology for analyzing a society's socioeconomic level;
RP2: Methodology for identifying and classifying Sustainable Organizations;
RP3: Methodology for analyzing a society's willingness to invest in Sustainable Organizations.

Hypothesis 2:

H2: The social structure, the tolerance between groups, the standard of high human morals and values guarantee the success of green businesses.

Questions

Q1: What are the conditions for implementing sustainable organizations?
Q2: How to relate the conditions with the implementation of sustainable organizations?
Q3: What is the relation between the conditions and the successful implementation sustainable organizations?

Research Projects

RP1: Multi-criteria analysis for selecting the conditions for implementing sustainable organizations;
RP2: Methodologies for implementing sustainable organizations.

Hypothesis 3:

H3: Green businesses are vectors of socioeconomic development and can rescue and/or remove societies from backward practices.

Questions

Q1: What are the main vectors of socioeconomic development at present?
Q2: Do the vectors of socioeconomic development vary according to society's characteristics?
Q3: How can sustainable organizations guide society towards socioeconomic development?

Research Project

RP1: Methodology for implementing Green Businesses focused on socioeconomic development;
RP2: Analysis of Green Businesses for socioeconomic development.

Hypothesis 4:

H4: Sustainable Organizations have better financial, social and environmental results than sustainable organizations in diverse competitive environments.

Questions

Q1: How best to analyze the results of sustainable organizations, taking into account the financial, social and environmental dimensions?

Q2: What are the advantages and disadvantages in implementing sustainable organizations?

Q3: What is the competitive edge of sustainable organizations?

Research Projects

RP1: Comparative studies between the results of sustainable organizations and those of non-sustainable organizations;

RP2: Methodology for analyzing the results of Green Businesses.

Discussion of Results

Since 1991, courses, extension courses and research projects have been developed to address the hypotheses, questions and research projects related to this theme. The team of researchers, faculty and alumni is made up of faculty and professionals working in the market, having had both national and international experience. The education, research and extension, as shown in Fig. 2, are represented by three fronts that operate in an integrated manner: Relationship, Extension and Research.

Fig. 2: Systemic view of LATEC's actions in the dimensions of relationship, research and training

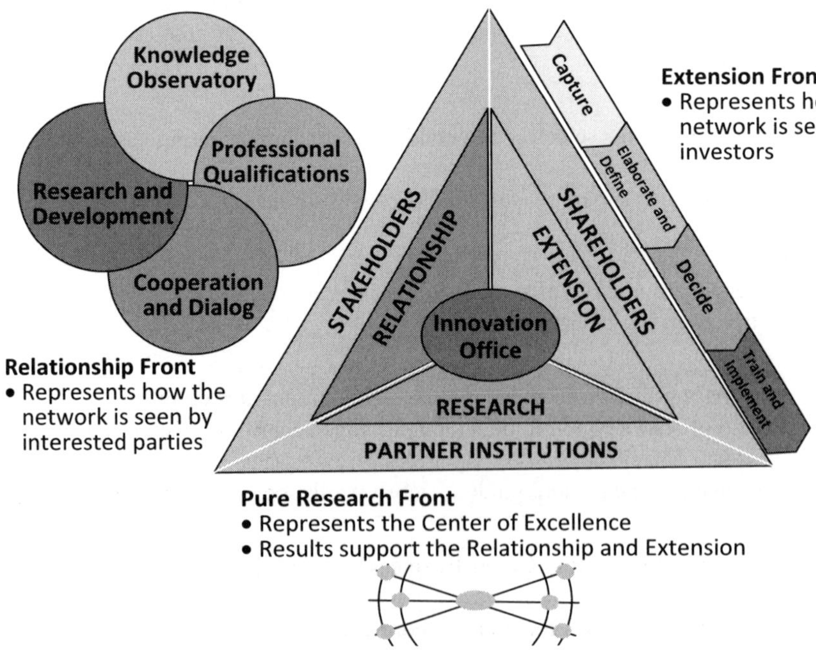

The Office of Integration is responsible for the coordination of all network initiatives, taking into account the relationship with the interested parties and the application of knowledge gained through research.

Relationship Front

Responsible for the relationship with interested parties, consisting of 4 cells:

Fig. 3: Dimension of relationship in LATEC

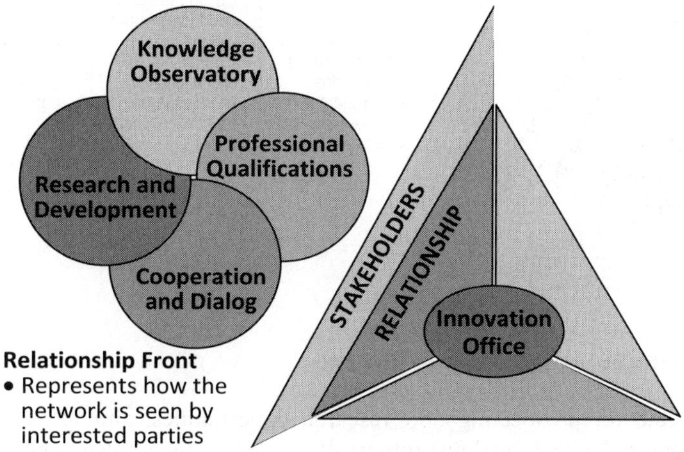

Knowledge Observatory: state-of-the-art practices, through periodicals, articles and course work (theses, dissertations and monographs), which are represented in the network's area of operation.

Academy: coordinates short, medium, and long-term courses intended to meet the market needs and the professional qualification of researchers and professionals working in the strategic, tactical and operational levels of institutions and organizations.

Cooperation and Dialogue: responsible for business consultancy support, focused on research and extension opportunities.

Research and Development: seeks expertise in various areas of knowledge, identifying opportunities and developing scientific publication through initiatives and projects developed by LATEC.

Extension Front

Responsible for the implementation of opportunities developed in the institutions and organizations. It is the representation of how the network is seen by the investors and it is comprised of 4 stages:

Fig. 4: Dimension of extension in LATEC

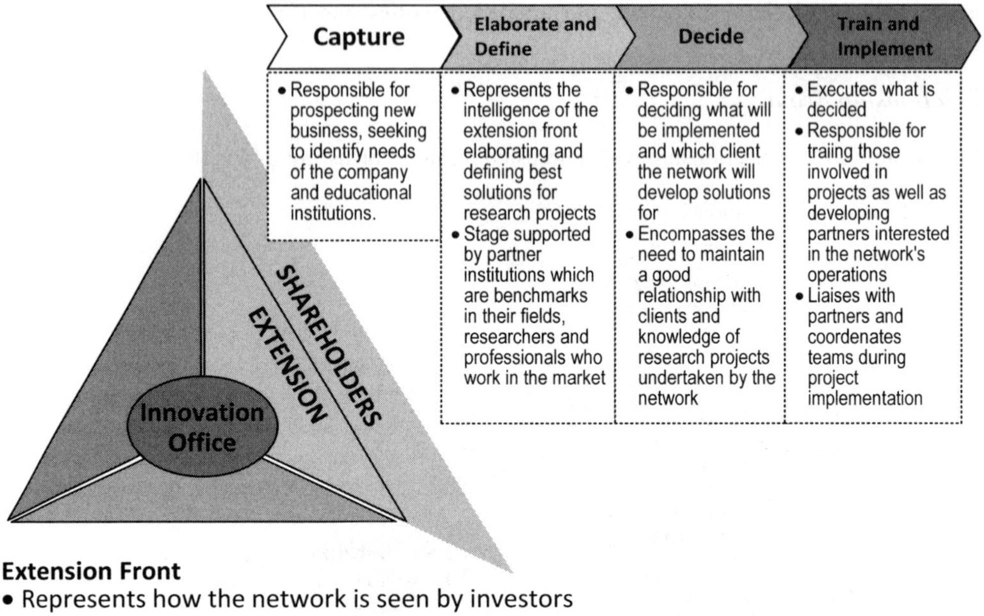

Collecting: responsible for prospecting new research opportunities, aiming to identify the needs of organizations and educational institutions. It constitutes the implementation of initiatives, along with other institutions and promotion agencies that are responsible for providing LATEC with visibility to entrepreneurs and society.

Developing and Defining: continuation of the previous step. It represents intelligence in the extension front, developing and defining the best solutions for the research projects regarding the needs identified in the market. This stage is supported by partner institutions that are benchmarks in areas of knowledge related to LATEC, researchers and professionals working in the market.

Deciding: Responsible for deciding what will be implemented, and to which organization LATEC will be developing its solutions. It covers the need to maintain a good relationship with clients and knowledge of the research projects developed by LATEC.

Training and Implementing: Execute that which has been decided. Responsible for training the people involved in the projects, research as well as developing training programs for stakeholders within LATEC's areas of operation. Articulates with partners and coordinates teams for project implementation.

Research Front

LATEC is responsible for the development of pure research, where the results support its relationship with stakeholders and extension fronts. Partner institutions collaborate with the activities carried out by LATEC (Fig. 5).

Fig. 5: The dimension of research in LATEC

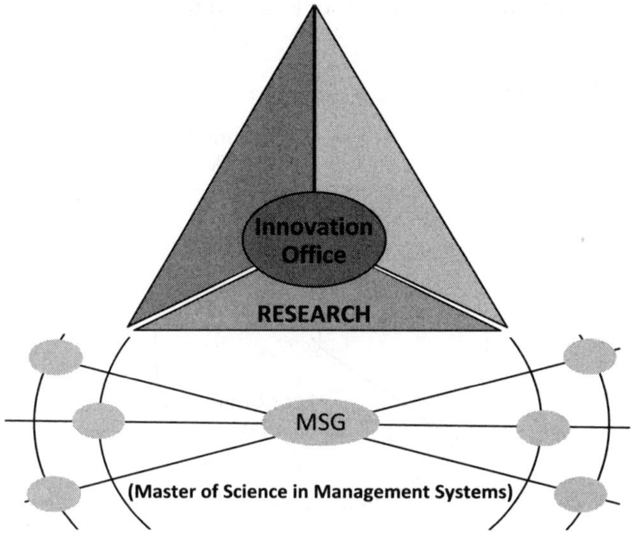

Academic research linked to TBL model dimensions

Over the last five years of the project, (2007-2011) with the creation of the Laboratory of Business and Environmental Management (LATEC/UFF), the Master's Program in Management Systems (MSG) has produced 350 dissertations, with 233 linked to aspects of the Triple Bottom Line (social, economic or environmental) (Brown et al. 2006). Fig. 6 shows how the studies were distributed among the aspects of the TBL from 2007 to 2011.

It appears that 2008 had the greatest number of dissertations defended (70) whose topics were linked to aspects of the TBL (Fig. 6). The figure also shows the variability of the students' preference in choosing these areas as a line of research.

Fig. 6: Number of dissertations aligned with aspects of the TBL

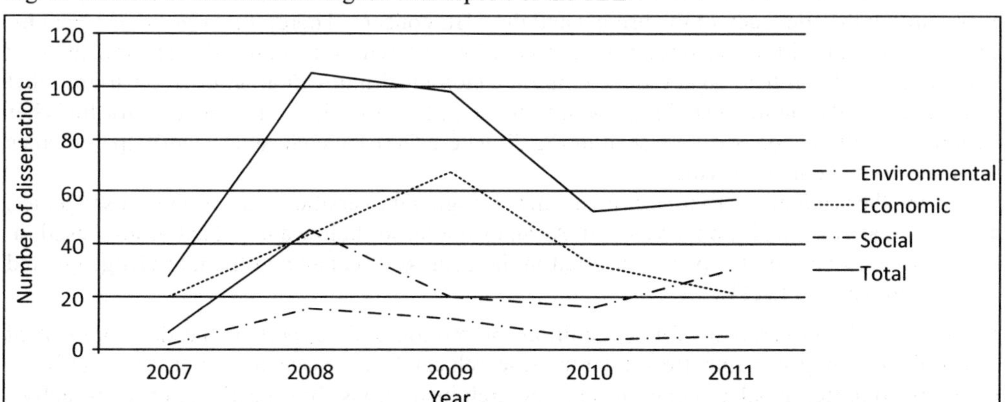

Source: LATEC (2012)

In Fig. 7, it is verified that in the last five years (2007-2011), there has been a preference for themes associated with economic issues (54%), followed by social issues (35%) and environmental issues (11%).

Fig. 7: Percentage of dissertations

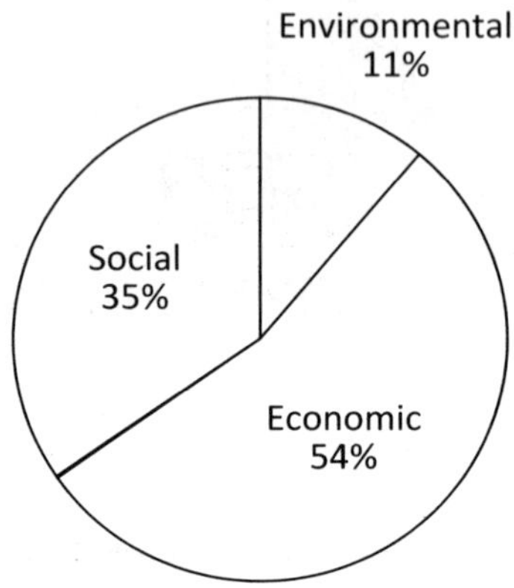

Source: LATEC 2012

The predominance of economic-results oriented projects, as one can see in the dissertations previously cited is a reflex of a market-oriented society where decisions are made mostly based on traditional aspects (quick payback, productivity and efficiency).

Conclusions

In the past few years, both nationally and internationally, there has been greater media coverage of the good examples of business strategy and management practices related to the sustainability of businesses. The organizations that incorporate the concept of the TBL to gain a competitive edge are well placed in the eyes of investors and consumers.

The current market demands sustainable projects, but there are still doubts among stakeholders regarding the adequacy of the project and the TBL concept. Therefore, large educational institutions must provide an academic environment for research, education and extension, in order to align concepts on the subject, developing technological innovation and demonstrating that it is economically viable to act based on this concept. LATEC/UFF sets a benchmark for other researchers will be led, as can be seen in Fig. 7, where 54% have chosen to work upon themes associated with economic issues.

Among the academic activities, which aim at bringing universities and corporations together, diverse initiatives can also be developed. As an example, see below a list of activities, based on the model presented in this paper, inserted in the context of enhancing the knowledge created and disseminated in LATEC:

- Grants to the government for recognition, promotion and support of the development of sustainable supply chains. Results of the research can be accessed at www.LATEC.uff.br.
- Grants to professional associations and research institutions for prioritizing projects, activities and resources.
- Structuring databases of opportunities for entrepreneurs interested in integrating sustainable supply chains.

- Understanding the major shortcomings of sustainable entrepreneurs, providing a basis for sectoral and multi-sectoral programs, and a network for developing skills and financing facilities, etc.
- Understand the production demands of the sustainable supply chain, based on, for example, the construction industry.
- Know the technology, materials and innovative services that differentiate the production process of sustainable supply chains.
- Recognize key indicators and methodologies respectively, in order to measure and design sustainable processes and projects.
- Structure a panel with the main areas of knowledge involved in sustainable processes and its configuration in product and service organizations.
- Organize literature and provide input to researchers and designers.
- Produce literature dedicated to the management and technology of production, services and product quality.
- Study the life cycle of products inherent to each production chain, methodologically structuring the manufacturing process in each organization and its interaction with other actors in the production chain.
- Study the implementation of GRI/G3 for evaluation/accountability to society for the characterization of the sustainable supply chain's results.
- Test the diverse methodologies (CSI, Dow Jones, Ethos Institute Indicators, etc.) in configuring management systems for sustainable supply chains.

This paper shows that there are several opportunities within education in order to build a more just society that is better aligned with the sustainable development concepts. Hence, university lab coordinators are challenged to identify what strategies and guidelines are most viable to academic activities so that it may arouse interest from students to seek sustainable solutions to problems faced by organizations and universities.

References

BOVESPA (2012), *Índice de Sustentabilidade Empresarial,* BM&FBOVESPA, São Paulo. Available at: http://www.bmfbovespa.com.br/indices/ResumoIndice.aspx?Indice=ISE& idioma=pt-br (accessed 23 January 2012).

Brown, D., Dillard, J. and Marshall, R.S. (2006), *Triple Bottom Line: A business metaphor for a social construction*, Document' s de Treball, núm. 06/2. Universitat Autonoma de Barcelona, Departament d' Economia de l' Empresa, Barcelon.

DJSI (2011), *Dow Jones Sustainability World Indexes Guide Book*, versão 11.6. Available at: http://www.sustainability-index.com/djsi_pdf/publications/Guidebooks/DJSI_World_Guidebook_11%206_final.pdf (accessed 24 January 2012).

Earth Charter Comission (1994). Available at: http://www.earthcharterinaction.org/invent/images/uploads/echarter_portuguese.pdf (accessed 25 January 2012).

EP (2010), *The Equator Principles association governance rules*. Available at: http://www.equatorprinciples.com/resources/ep_governance_rules_june_2010.pdf (accessed 24 January 2012).

Faria, H. and Silva, R. (1998), *Environmental Management Systems: Why invest?,* Production Engineering Post-Graduation Program. Escola Federal de Engenharia de Itajubá, Minas Gerais.

LATEC (2012), *LATEC's Dissertation Database* (Laboratório de Tecnologia, Gestão de Negócios e Meio Ambiente). Escola de Engenharia, Universidade Federal Fluminense, Niterói, RJ.

Quelhas, O.L.G., Meirino, M.J. and Neto, J.V. (2011), *Market Motivators for Environmental Performance in Organizations*, chapter 3 (1° Ed.). Campus Elsevier, Rio de Janeiro.

Savitz, W. and Weber, K. (2006), *The Triple Bottom Line: how today's best-run companies are achieving economic, social, and environmental success – and how you can too*. Jossey-Bass, San Francisco.

Tachizawa, T. (2004), *Gestão Ambiental e Responsabilidade Social Corporativa: estratégias de negócios focadas na realidade brasileira* (2a ed.). Atlas, São Paulo.

Torres, C., Bezerra, I. and Hernandes, T. (2007), *Responsabilidade Social de Empresas Multinacionais: diretrizes da OCDE*, Rio de Janeiro: CERIS/IBASE, 2004. 92 p. Available at: http://www.balancosocial.org.br/media/DiretrizesOCDE_Aplicacao_das.pdf.

UN (1992), *Earth Summit, Agenda 21, The United Nations Programme of Actions from Rio*. United Nations, Rio de Janeiro. Available at: http://www.un.org/esa/sustdev/documents/agenda21/english/Agenda21.pdf (accessed 24 January 2012).

– (1998), *Kyoto Protocol to the United Nations Framework Convention on Climate Change*. Site: http://unfccc.int/resource/docs/convkp/kpeng.pdf (accessed 24 January 2012).

– (2010a), *Environmental Stewardship Strategy – Overview and Resource for Corporate Leaders*. United Nations, Nova York. Available at: http://www.unglobalcompact.org/docs/issues_doc/Environment/Environmental_Stewardship_Strategy.pdf (accessed 24 January 2012).

– (2010b), *Millennium Development Goals*, United Nations. Available at: http://www.un.org/wcm/webdav/site/sport/shared/sport/pdfs/Backgrounders/Sport%20and%20the%20MDGs_FACT SHEET_February%202010.pdf (accessed 24 January 2012).

Chapter 66

Towards Sustainable Electricity Generation in Latin America: The Footprint of the REGSA Project

José Baltazar Salgueirinho Osório De Andrade Guerra, Youssef Ahmad Youssef and Veronika Schulte

Abstract

The international community is beginning to understand that measures to reduce greenhouse gas emissions and combat the negative environmental effects of global warming need to be seriously considered. Meanwhile, despite technological advancement and economic development worldwide, many communities, especially in the developing countries, still struggle to obtain very basic necessities such as clean water and electricity. A number of EU-sponsored mitigation initiatives have taken place with the aim of transferring knowledge to provide these communities with the necessary background and skills to allow them to develop. The REGSA project, sponsored by the EuropeAid agency, is one of these initiatives. The project aims to help small communities in three Latin American countries (Brazil, Bolivia and Chile) to generate electricity using renewable energy sources. This paper will show the results of a survey conducted in Latin America by the REGSA project, aiming to provide a descriptive analysis of the energy situation in the three Latin American countries.

Introduction: REGSA, the generation and use of renewable energy in South America (and its effects on the environment, security and poverty)

The need to find new forms of renewable energy has long been recognized worldwide. However, socioeconomic and climatic developments have turned that consensual purpose into a priority – it is crucial both in the short and the long term that financially sustainable power sources are developed that guarantee effective energy production in harmony with environmental considerations.

An increasing number of studies and scholars agree that the era of cheap oil and fossil fuel energy is coming to an end (Kunze and Busch 2011; IEA 2009; Scheer 1999; Energy Watch Group 2006; 2007; Heinberg 2007). There is therefore, a need for a shift to renewable energy sources for electricity production. Despite the failure of political effort and social and ecological responsibility to spark any major change in energy policy, a scarcity of resources and skyrocketing oil prices are likely to force many of the necessary changes. Kunze and Busch (2011) and Henseling (2008), however, observe that without assertive decisions, the transition will come at a high ecological cost. It should be underlined that these studies are concerned with the situation in Europe, where virtually every city relies on the supply of electricity. In Latin America, many communities are still living without electricity.

The current narrative on renewable energy in Europe emphasizes the potential to emancipate the countryside from its often-peripheral economic situation (Wuppertal Institute for Climate, Environment and Energy 2010). There are very many villages and rural cities in Latin America with no or limited access to electricity. The effects of this urban-rural dichotomy on both continents have been similar, with the young, skilled population choosing to move to more developed urban areas.

The causes of this phenomenon may be different in Europe and Latin America, but the consequences appear to some extent to be similar. Knowledge exchange between the continents would be extremely beneficial, especially for Latin America.

For this reason, REGSA research projects would be seen as a valuable contribution for establishing renewable electricity generation in Latin America. Renewable energy will give new economic opportunities, and may be considered a coping strategy to address the issue of economic development not only in areas in Latin America without electricity, but also in rural and remote areas in Germany. According to Busch (2010), renewable electricity production may contribute to local economies on both continents in three different ways: by creating jobs, by creating local tax income from energy sold, and by lowering energy costs. These three factors would allow communities to retain and create more jobs.

In Brazil, the Latin America's largest economy, for example, and according to Pereira, Reiche and Gouvello (2005), more than 12 million Brazilians in about 2.5 million households have no access to electricity. Although the country has recently made great strides economically, vast differences still remain between rural and urban populations, between states and between people of different income levels, with poor rural populations in the north and northeast of Brazil being the most affected by a lack of energy supply. In other Latin American countries the situation is similar.

The REGSA project aims to reproduce in Latin America some successful pilot initiatives from Europe, where some German regions began to invest in renewable energy production in the early 1990s. The outcome of these initiatives, as observed by Kunze and Busch (2011), was the often-correlative creation of regional value and positive economic development. Many local politicians in Germany are now drawing on the examples set by these pioneers. In this paper, the research limits its scope to the technical and political framework in Latin America, as one step towards the reproduction and adaptation of the European experience.

The REGSA project and context

The REGSA project (*Promoting Renewable Electricity Generation in Latin America*) was launched in partnership with representatives from four non-profit universities in Germany (Hamburg University of Applied Sciences), Bolivia (Universidad Catolica Boliviana/Instituto de Investigaciones Socio-Economicas), Brazil (UNISUL – Universidade do Sul de Santa Catarina) and Chile (Universidad de Chile).

The main objective of REGSA is to contribute to the implementation of renewable electricity generation in three Latin American countries, namely Bolivia, Brazil and Chile. Target groups and ultimate beneficiaries are defined as the power suppliers, local authorities, entrepreneurs and agriculturists (in the renewable energy segment), as well as the public in general, whose opinion is crucial for the success of these endeavors.

Research objective

The major contribution of this paper is to focus on understanding market structure, the social dimension and the impact of renewable electricity generation upon the surveyed block of countries as a whole, namely Argentina, Bolivia, Brazil, Chile, Colombia, Ecuador, Paraguay, Peru, Uruguay and Venezuela.

In these countries, the dynamic of the transition from non-renewable electricity generation to renewable electricity in macro-economic, political and sociological contexts has attracted little attention so far. Most research on renewable electricity generation in Latin America has been focused only on individual countries.

This paper aims to expand the academic discussion on renewable electricity generation in Latin America by presenting the results of a survey conducted in Latin American countries as part of the REGSA project – a multinational initiative between European and Latin American countries to enable a better understanding of this dynamic, and to begin to address the social and economic impacts of the development of renewable electricity generation in the pilot projects to be implemented in Brazil, Bolivia and Chile.

Research method

Since the authors are only aware of a handful of existing studies on renewable electricity generation in Latin America that take into consideration technical, political and social constraints, grounded theory was chosen instead of a quantitative approach (Glaser and Strauss 1998; Charmaz 2006; Flick et al. 2000). The research was carried out between early 2011 and mid December of the same year. A review of the literature including online material was a first stage. In order to consolidate and confirm information collected from secondary sources, phone interviews were conducted with relevant stakeholders including policymakers, members of local communities, business owners and informal opinion leaders. Snowball sampling led to market professionals and academics in the field of renewable energy being interviewed as well. It is important to emphasize that cooperation with local governments yielded important insights and contacts. From March 2011 to July 2011, internet research was conducted along with a review of the literature. The interviews were narrative and open to begin with; later on, an interview guide was used. 22 pre-structured telephone interviews were conducted with heads of electricity companies and policymakers. The researchers participated in ten meetings and public discussions on the topic, gathering data through participatory observation. The majority of material was documented through physical note-taking, since audio recording had been shown to make interviewees very cautious, to the extent of holding back crucial information. Some interviews were carried out by students at the Universidade do Sul de Santa Catarina, Universidad do Chile and the Universidad Catolica de Bolivia San Pablo during the spring and summer of 2011, though the authors did the majority.

Survey results and discussions

➢ Latin American energy markets

The survey results show that most of the countries investigated have similar energy market models. The different market designs and methods of electricity pricing are shown in Table 1.

Table 1: Market designs, methods of electricity pricing and existence of capacity payments in the surveyed countries

	Argentina	**Bolivia**	**Brazil**	**Chile**	**Colombia**	**Ecuador**	**Paraguay**	**Peru**	**Uruguay**	**Venezuela**
Market Model	Pool Financial BC	Pool Financial BC	Pool Financial BC	Pool Financial BC	Pool (PX) Financial BC	Pool Financial BC	Monopoly	Pool Financial BC	Pool Financial BC	Monopoly
Wholesale Market Operation Spot Price	Marginal Cost hourly	Marginal Cost hourly	Marginal Cost hourly	Marginal Cost hourly	Power Exchange Price hourly	Marginal Cost hourly		Marginal Cost hourly	Marginal Cost hourly	
Capacity Payment				Yes	Yes	Yes		Yes		

As can be seen in Table 1 the pool market model with bidding competitions is the most common one. In this model, capacity rights are acquired by electricity generators in competitive auctions held by public utilities. The wholesale price (the marginal cost) is set hourly by the generators, i.e. these markets are wholesale operated (Lennard 2003). This market model is used in Argentina, Bolivia, Brazil, Chile, Ecuador, Peru and Uruguay. Colombia, on the other hand, presents a different market model, where electricity is traded at the hourly power exchange price in place of the hourly marginal cost. Energy is traded at spot markets and the power exchange price is set hourly. Retailers function as intermediaries who buy and sell electricity at these spot markets. The customer can choose the retailer offering the cheapest energy (Posorski and Werner 2009).

Some countries have capacity payments, i.e. additional energy costs during peak periods that function as indicators of greater energy demand and therefore give an incentive for the creation of new capacity. Capacity payments exist in Chile, Colombia, Ecuador and Peru (Lennard 2003; Oren 2000).

Paraguay and Venezuela are the only two countries surveyed that still use the integrated model of a public monopoly.

Renewable energy support schemes in Latin America

Since the beginning of the 21st century almost all Latin American countries have come to understand that renewable energies are an inevitable part of their future energy supply (Batlle and Barroso 2011; Haselip et al. 2011). Numerous initiatives to promote renewable energy generation at both a national and local level have emerged. Moderate support for the use of renewable energies, such as fiscal and tax incentives, has been implemented in many countries. Some countries, however, have introduced more impelling measures, such as energy contract auctions, tenders, renewable portfolio standards and capacity payments. They have become an important, and in some countries, the major instrument for renewable energy market stimulation (Batlle and Barroso 2011; Haselip et al. 2011).

The following pages outline the main efforts towards the implementation of renewable energy in each country surveyed.

Argentina

In 2007, the Secretary of Energy enacted Law 26 190, which aims to increase the contribution of renewable energy sources to power generation by at least 8% by 2016. With a view of achieving that goal, the Ministry of Federal Planning, Public Investments and Services (Ministerio de Planificación Federal, Inversión Pública y Servicios) launched a program for the generation of RE (RENGEN) in 2010 (Posorski and Werner 2009; IEA 2010). The system operator CAMMESA (Compañía Administradora del Mercado Mayorista Eléctrico S.A.) and the public service corporation ENARSA play important roles here, as they hold energy contract auctions and guarantee energy purchase (Batlle and Barroso 2011).

Bolivia

In the National Development Plan (Plan Nacional de Desarollo) that was introduced in 2007, the Bolivian government committed itself to the provision of electricity for the whole country by 2025. Coordinated by the Vice-Ministry of Electricity and Alternative Energies, the government program "Electricity for Living with Dignity" (Electricidad para Vivir con Dignitad) was brought to life as part of efforts to reach that goal. This program also incorporates the promotion of renewable energy use. One example is the initiation of the rural electrification project IDTR (Infraestructura Descentralizada para la Transformación Rural), the biggest project within the program, which has a budget of 15 million dollars and exclusively promotes the implementation of solar home systems for rural electrification (Rutschman 2010).

Brazil

Within the framework of a program for wind and biomass energy, the PROINFA program (Programa de Incentivo a Fontes Alternativas de Energia Elétrica) in Brazil introduced a feed-in tariff in 2002 (Batlle and Barroso 2011). The Brazilian government had previously enacted Law 10 438, which was responsible for the start of PROINFA but was not specific to the fostering of renewable energies (IEA 2002). It was linked to the purchase of energy contracts with the state-owned company Eletrobrás, but was criticized for a lack of efficiency. Subsidized by the government, a second initiative to promote renewable energies was started by implementing a reduction in transmission and distribution tariffs beneficial to the consumer in 2007 (Batlle and Barroso 2011). Today the energy contract auctions are the main tool for renewable market stimulation and general market expansion in Brazil. Tax incentives and pre-investment appraisal subsidies have also been set up in Brazil.

Hydroelectric power generation accounts for the majority of Brazil's electricity supply and is treated preferentially in comparison to the other renewable energy technologies. Hence it must be considered separately.

Chile

Law 20 257 was passed in 2008 and aims at increasing the renewable energy share in energy generation to 5% by 2010 and 10% by 2024 (Wright and Adlerstein-Gonzalez 2009). Moreover, it defines penalties for electricity generation companies that do not fulfill these obligations by the prescribed time. Companies that fall short of the specified renewable shares in their energy production will be fined 28 USD/MWh, rising to 42 USD/MWh after a three-year violation period. The success of the renewable portfolio standard is, however, questionable, since not enough renewable energy projects are currently being planned. Electricity providers have called for alternative solutions, e.g. a feed-in tariff (Batlle and Barroso 2011).

Colombia

Colombia developed a national development plan for alternative energies (Plan de Desarrollo Nacional de las Energías Alternativas), which proposed strengthening capacities for RE, as early as 1995. The plan did not fulfill expectations however, as no action was taken over the following years (Projekt-Consult GmbH and Loy 2007). By issuing Law 697 in 2001 the Colombian government created a framework for the promotion of renewable energies that led to the development of the Programme of Rational and Efficient Use of Energy and Other Forms of Non-Conventional Energy, PROURE (Programa de Uso Racional y Eficiente de Energia y Fuentes no Convencionales). Since then the promotion of RE has made progress through tax incentives. The Ministry of Mines and Energy supervises the PROURE program (Projekt-Consult GmbH and Loy 2007). Resolution 180919, passed in 2010, follows an action plan from 2010 to 2015 for the further development of PROURE (Mojica 2011).

Ecuador

A feed-in tariff for photovoltaic electricity was introduced in Ecuador in 2000. The Ecuadorian government, however, never complied with its payment obligations. Minor efforts were made to promote PV technology, for example by setting up a few hundred isolated PV systems in 2003 and 2006. The Perva project for the electrification of rural homes in the Amazon basin is on-going, and aims to electrify 15,000 households using PV systems by 2013 (Batlle and Barroso 2011). With the regulation CONELEC 004/11 in 2011, the administrative institution CONELEC (Consejo Nacional de Electricidad) introduced a feed-in tariff for the next fifteen years for electricity generation from wind, sunlight, water, biomass and geothermal sources (Gipe 2011). Private investments are expected to be severely limited, however, because of the recent regulation passed by President Correa permitting the state to retain full control of the country's electricity system (Batlle and Barroso 2011).

Paraguay

In Paraguay, 99% of electricity generation capacity is hydroelectric. No incentives exist for the development of other renewable energies, and no initiatives are in progress (Batlle and Barroso 2011).

Peru

Law 28546, for the promotion and use of non-conventional energy resources in rural, isolated and frontier zones of the country, was enacted in 2005 (Peru 2005). In 2008, Decree 1002 was passed into law, which declared the development of renewable energies a matter of national interest and exigency. The decree mentions in particular the fostering of wind, solar, geothermal, hydraulic, and biomass energy generation. Hydroelectric energy only counts as renewable when produced by power plants of an installed capacity of up to 20 MW. The decree also obliges the Ministry of Energy and Mines to specify a target percentage every five years for the contribution of renewable energy in the national electricity supply. This ratio was set to 5% for the first five year period (SNMPE 2011). To meet this objective, Peru has now introduced energy contract auctions attempting to encourage renewable energy investments.

Uruguay

Decree 389/005, issued in 2005 by Uruguay's Ministry of Energy and Mines together with the national energy supplier UTE, represented a first effort to include renewable energy in the national energy mix. The decree allowed UTE to put out tenders for renewable energy generation projects; there were too many restrictions, however, and results were limited to only one small wind power project. Restrictions were reduced with Decree 77/006 in 2006, but inadequately (UNFCCC – CDM – Executive Board 2006). The policymaking process has since progressed along similar lines, with projects ever-increasing in size – recently, wind power projects of up to 150 MW have been launched (Batlle and Barroso 2011). A law promoting solar thermal energy generation (Law 18.585) was enacted in 2009. It is pledged that hotels, hospitals, public buildings and new buildings will generate 20% of the energy they need to heat water through solar thermal technology (Epp B 2010).

Venezuela

No incentives for the development of renewable energies exist and no initiatives of this kind are in progress in Venezuela. The so-called "Organic Law of the Electric System and Service" (Ley Organica del Sistema y Servicio Electrico) was passed in 2009. Article 21 states the intention to launch initiatives for the fostering of alternative energies in the future. However, no further action in this direction has yet been taken (Batlle and Barroso 2011; Venezuela 2010).

A summarized overview of the institutional and legal frameworks in the surveyed countries is given below (Tables 2, 3 and 4).

Table 2: Overview of the institutional frameworks in Argentina, Bolivia, Brazil, Chile and Colombia

	Argentina	**Bolivia**	**Brazil**	**Chile**	**Colombia**
Policies	Secretaría de Energía (SENER)	Vice-Ministry of Electricity and Alternative Energies (Ministry of Hydrocarbons and Energy)	Ministerio de Minas y Energía (MME)	Ministerio de Energía	Ministerio de Minas y Energía
Regulator	Ente Nacional Regulador de la Electricidad (ENRE)	Electricity Supervision and the Social Control Authority (AE)	ANEEL	Comisión Nacional de Energía (CNE)	Comisión de Regulación de Energía y Gas (CREG)
Supervisor	ENRE	Electricity Supervision and the Social Control Authority (AE)	ANEEL	Superintendencia de Electricidad y Combustibles (SEC)	La Superintendencia de Servicios Públicos Domiciliarios (SSPD)
System Operator	CAMMESA	National Committee for Charge Dispatch (CNDC)	Operador Nacional del sistema Eléctrico (ONS)	Centro de despacho Económico de carga (CDEC)	XM filial de ISA S.A
Planning	Ministerio de Planificación Federal, Inversión Pública y Servicios	Vice-Ministry of Energy Development (Ministry of Hydrocarbons and Energy)	Empresa de Investigación Energética (EPE)	Comisión Nacional de Energía (CNE)	Unidad de Planeación Minero-Energética (UPME)
Environmental Authority	Secretaría de Ambiente y Desarrollo Sustentable (SEMARNAT)	Ministry of the Environment and Water	Ministerio del Medio Ambiente	Ministerio del Medio Ambiente y CONAMA	Ministerio del Ambiente, Vivienda y Desarrollo Territorial
RE Promotion	Dirección Nacional de Promoción (DNPROM)			Centro de Energía Renovable (CER)	
Conflict Solving	ENRE			Panel de Expertos	Superintendencia de Servicios Públicos Residenciales

Table 3: Overview of the institutional frameworks in Ecuador, Paraguay, Peru, Uruguay, Venezuela

	Ecuador	**Paraguay**	**Perú**	**Uruguay**	**Venezuela**
Policies	Federal Ministry of Policies Ministerio de Electricidad y Energía Renovable (MEyER)	Viceministerio de Minas y energía	Dirección General de Electricidad	Ministerio de Industria, Energía y Minería	Ministerio del Poder Popular para la Energía Eléctrica (MPPEE)
Regulator	CONELEC	ANDE	Dirección General de Electricidad	Unidad Reguladora de Servicios de Energía y Agua (URSEA)	MPPEE
Supervisor	CONELEC	ANDE y Contraloría General De la República	Organismo Supervisor de la Inversión Privada en Energía - OSINERG		MPPEE
System Operator	CENACE	ANDE	Comité de Operación Económica del Sistema Interconectado Nacional (COES)	Administración del Mercado Eléctrico (ADME)	MPPEE
Planning	CONELEC	ANDE y Secretaría Técnica de Planificación	Ministerio de Energía y Minas	Dirección Nacional de Energía y Tecnología Nuclear (DNETN)	MPPEE
Environmental Authrity	Ministerio del Ambiente	Consejo Nacional del Ambiente (CONAM) y la Secretaría del Ambiente (SEAM)	Consejo Nacional del Ambiente	Dirección Nacional de Medio Ambiente (DINAMA)	MPPEE y Ministerio del Ambiente
RE Promotion	MEYER	ANDE	Fondo Nacional del Ambiente (FONAM)	DNETN	Dirección General de Energía Alternativa
Conflict Solving					

Table 4: Overview of the legal frameworks in the different countries

	Argentina	Bolivia	Brazil	Chile	Colombia	Ecuador	Paraguay	Peru	Uruguay	Venezuela
General Law	Law 24.065	Electricity Law 1604	Law 10 848	General Law of Electric Service (DCL4)	Laws 142 and 143	Law of the Electricity Sector Regime (LRSE, ROS 43)	Law 167/93	Law of Electric Concessions	National Electricity Law (Law N° 14.694)	Organic Law of Electric System and Service
Renewable Law	Law 26 190		Law 10438	Law 20.257	Law 697 and Resolution 180919	CONELEC – 004/11		Law N° 28546 and Decree 1002	Law 18.585 and Decree 77/06	

Renewable energy generation in Latin America

Although this research was conducted in 2011, the secondary data obtained in order to prepare this comparative analysis were collected by the local governments in Argentina, Colombia, Peru and Bolivia in 2009; and in Brazil, Ecuador, Uruguay, Chile and Germany in 2010.

In Fig. 1, the shares of the different renewable energy sources – water, wind, sunlight and biomass – in the energy generation of the surveyed countries is shown.

Fig. 1: Contribution of different energy sources in energy generation in the countries surveyed

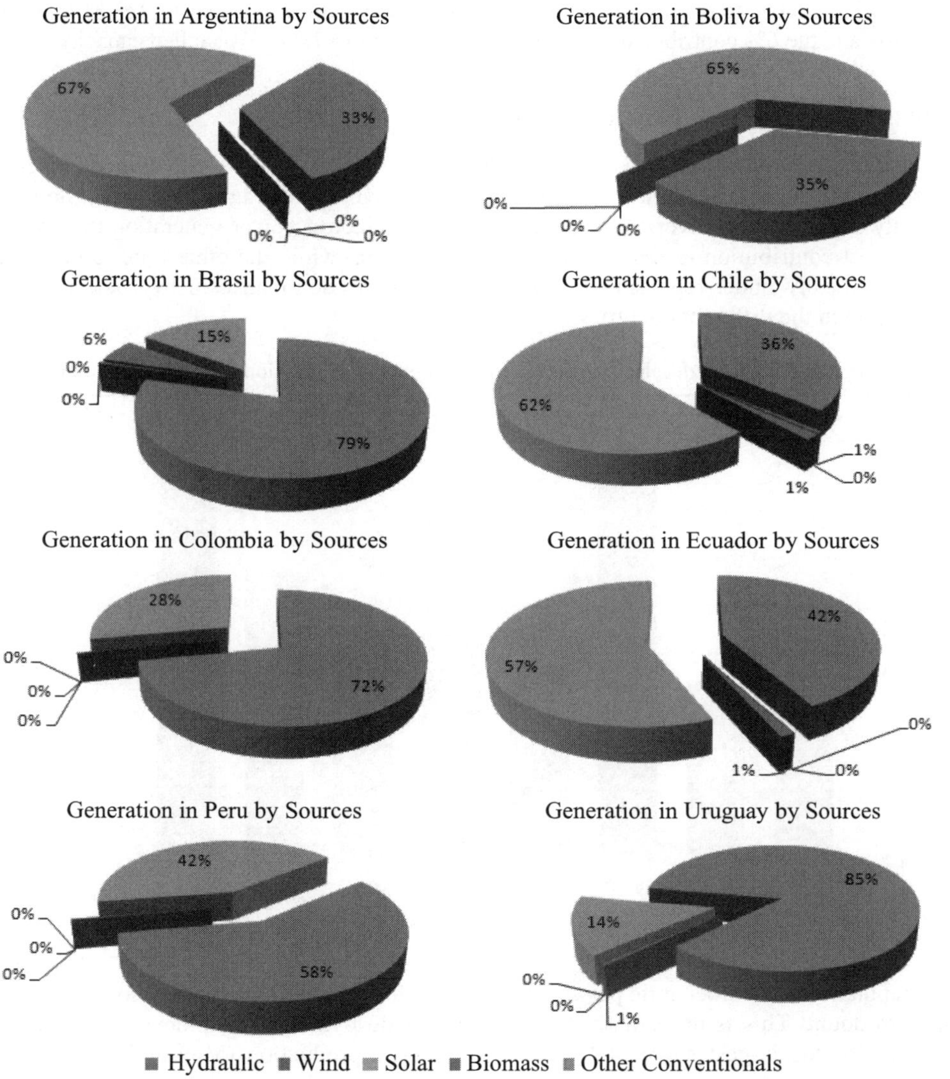

The clearest conclusion to be drawn from Fig. 1 is that hydroelectric power is the dominant renewable energy in Latin America. Argentina, Bolivia and Chile obtain about one third of their energy through hydroelectric power generation. Brazil and Colombia use it to cover approx.

three quarters of their energy demand and Uruguay its entire energy demand, while Ecuador and Peru are in between. All other renewable energy sources only make an insignificant contribution to countries' energy generation. The production of biofuels from biomass plays an important role in the transportation sector in Brazil, Chile and Ecuador, where it accounts for approx. 1% of national energy production. Wind power generation amounts to 0.5% of energy production in Brazil and Chile and 0.7% in Uruguay. Solar power production is virtually non-existent, accounting for less than 0.1% of energy generation in each of the surveyed countries. However, in a more comprehensive analysis, these numbers would have to be interpreted in relation to the total amount of energy that the individual countries produce. The 0.43% contribution of wind energy to energy production in Brazil, for instance, is equal to 927 MW installed capacity, and the 6% contributed by biomass is equivalent to 7.8 GW installed capacity.

The controversy of hydroelectric power

Compared to the situation in OECD countries, at first sight the contribution of renewable energy in the surveyed countries seems quite impressive, on average. This impression is potentially deceptive, however. As shown above, hydroelectric power generation makes by far the biggest contribution to renewable energy production, while the other renewable energies are dramatically under-represented. Fig. 3 illustrates the dominance of hydroelectric power production in the different countries.

Fig. 3: Share of RE and hydroelectric power in overall energy production in the different countries

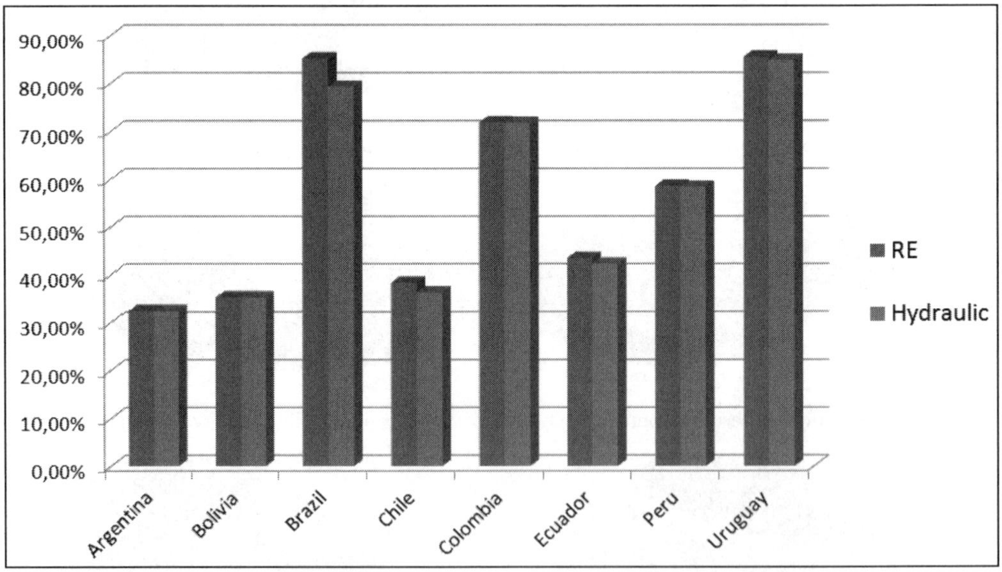

The reputation of hydroelectric power as a sustainable and renewable energy source is, in fact, open to doubt. This is particularly true for large-scale hydroelectric plants, which naturally account for the largest share of hydroelectric energy production (Meisen P. and Krumpel 2009).

There are two main arguments in this dispute. One is the fact that many Latin American countries have become quite reliant on hydroelectric power generation. They are therefore vulnerable to dry periods and the resulting decrease in water levels. The other argument refers to the severe environmental and social problems that have been experienced as a consequence of large-scale hydroelectric plants. The flooding of large areas of land has caused the disappearance of precious natural resources, as well as forcing indigenous people to relocate away from their traditional territory (UNFCCC – CDM – Executive Board 2006).

Although it is arguable to what extent these criticisms are valid, it is clear that some perceive large-scale hydroelectric as not the most adequate solution.

Conclusion

The REGSA partners' first step was to identify the technical and socioeconomic potential of Latin American countries for renewable energy production; secondly, they opened a dialogue between public and private entities in the sector to raise awareness of the potential of renewable electricity production. Sustainable renewable electricity production projects developed by pilot communities and regions will also be supported.

In order to achieve good and sustainable results through technology and knowledge transfer in the field of RE between the REGSA partners, it is first essential to investigate the current situation and explore existing structures. This research was conceived as part of the REGSA project Work Package 2 (WP2), which aims to direct the spotlight onto the political and institutional frameworks for RE development and the status quo analysis of RE generation in Latin America.

In the course of this research, as described above, specific roles were assigned to the various stakeholders as follows:

1. Energy suppliers, ministries and local authorities were essential sources of information for this research, participating in awareness and counseling actions, round tables and seminars, and were informed of the results accordingly. The research revealed the importance of the fact that many producers were not able to advance further in the use of renewable energy due to the lack of detailed information on the technical and economic potentials; nor were they necessarily able to afford the planning and implementation costs of projects. With regard to local authorities, the research points out that they may also be insufficiently informed about such potentials (which could even decrease poverty rates) and deprived of the political and financial means to support renewable energies.
2. In relation to the companies and agriculturists involved in this research, they delivered important contributions in raising awareness, and were also represented at the research meetings (round tables and seminars).

This paper represents a first step in a wider program of research in the context of the REGSA project, which will aim to deliver a foundation for advocating the increased inclusion of renewable energies in the extant electricity networks in the long term, also in rural decentralized sites mainly in Bolivia, Brazil and Chile. The results of this research will therefore lay the groundwork for future studies, particularly with regard to the identification and preparation of infrastructure projects once the main executors and decision-makers are given appropriate technological and economic information.

This paper has also presented evidence of similarities and differences between Latin American countries with regard to RE development, especially in the comparison of RE technology preferences where similarity was discovered. The fact that alternative RE sources play an insignificant role in energy production compared to hydroelectric, is certainly due partly to the level of complexity inherent in these RE technologies. It may well be suspected, however, that a lack of appropriate RE incentives is also a reason.

Furthermore, the analysis shows that there are initiatives underway for the development and promotion of RE in Latin America. However, despite the large potential of RE, large-scale implementation of RE generation has yet to be realized.

All in all, the research conducted in WP 2 shows that most of the countries surveyed perceive RE as an inevitable part of their future energy supply. It is therefore crucial that existing efforts are duly supported through appropriate knowledge and technology transfer.

References

Batlle, C. and Barroso, L.A. (2011), *Review of Support Schemes for Renewable Energy Sources in South America*. Survey of the Center for Energy and Environmental Policy Research, Cambridge (MA), USA.

Busch, H. (2010), *Local Success Stories: An analysis of the motivations of municipalities in Germany to support renewable energies*. Published Master thesis. Lund University, Lund.

Charmaz, K. (2006), *Constructing Grounded Theory: A Practical Guide Through Qualitative Analysis*. Sage Publications. Thousand Oaks (CA), USA.

Energy Watch Group (2007), Coal: *Resources and Future Production*. Retrieved from http://www.energywatchgroup.org/fileadmin/global/pdf/EWG_Report_Coal_10-07-2007 ms.pdf.

Energy Watch Group (2006), *Uranium Resources and Nuclear Energy*. Retrieved from http://www.energywatchgroup.org/fileadmin/global/pdf/EWG_Report_Uranium_3-12-2006ms.pdf.

Epp, B. (2010), *Solar Obligation in Uruguay*. Global Solar Thermal Energy Council. http://www.solarthermalworld.org/ (accessed 31 January 2012).

Flick, U., von Kardorff, E. and Steinke, I. (2000), *Qualitative Forschung: ein Handbuch*. Rowohlt-Taschenbuch-Verlag, Hamburg, Germany.

Gipe, P. (2011), *Ecuador Adopts Feed-in Tariffs*. Wind-Works, http://www.wind-works. org (accessed 31 January 2012).

Glaser, B. and Strauss, A. (1998), *Grounded Theory: Strategien qualitativer Forschung*. Verlag Hans Huber, Bern, Switzerland.

Haselip, J., Nygaard, I., Hansen, U. and Ackom, E. (Eds.) (2011), *Diffusion of renewable energy technologies: case studies of enabling frameworks in developing countries*. Technology Transfer Perspectives Series, UNEP Risø Centre, Denmark.

Heinberg, R. (2007), *Peak Everything: Waking Up to the Century of Declines*. New Society Publishers, Gabriola Island (BC), Canada.

Henseling, K. (2008), *Am Ende des fossilen Zeitalters: Alternativen zum Raubbau an den natürliche Lebensgrundlagen*. Oekom Verlag, München, Germany.

IEA (2002), *Programme of Incentives for Alternative Electricity Sources – Programa de Incentivo a Fontes Alternativas de Energia Elétrica – PROINFA*. IEA/IRENA Global Renewable Energy Policies and Measures database, http://iea.org (accessed 26 January 2012).

– (2009), *World Energy Outlook 2009*. Paris. Available at: http://www.worldenergyoutlook.org/media/weowebsite/2009/WEO2009.pdf.

– (2010), *Renewable Energy Generation Program (GENREN)*. IEA/IRENA Global Renewable Energy Policies and Measures database. http://iea.org (accessed 26 January 2012).

Kunze, C. and Busch, H. (2011), The Social Complexity of Renewable Energy Production in the Countryside. *Electronic Green Journal* 1 (31), Article 1, pp. 1-17.

Lennard, J. (2003), *Capacity payments may be the solution*. Utility Week, http://www.utilityweek.co.uk (accessed 25 January 2012).

Meisen, P. and Krumpel, S. (2009), *Renewable Energy Potential of Latin America*. Survey of Global Energy Network Institute, San Diego (CA), USA.

Mojica, K.Y.C. (2011), *Minister of Mines and Energy signed PROURE resolution*. Instituto de Planificacion y Promocion de Soluciones Energeticas para las Zonas, No. Interconectadas IPSE, http://www.ipse.gov.co (accessed 31 January 2012).

Oren, S.S. (2000), *Capacity Payments and Supply Adequacy in Competitive Electricity Markets*. VII Symposium of specialists in electric operational and expansion planning. Curitiba (PR), Brazil.

Pereira, O.S., Reiche, K. and Gouvello, C. de (2005), *Brazil, Background Study for a National Rural Electrification Strategy: Aiming for Universal Access*. Energy Sector Management Assistance Program (ESMAP), The International Bank for Reconstruction and Development/THE WORLD BANK, Washington (DC), USA.

Peru (2005), *Law of Promotion and Use of Non Conventional Energy Resources in Rural, Isolated and Frontier Zones of the Country*. The Congress of the Republic of Peru, Peru.

Posorski, R. and Werner, D. (2009), *Energy-policy Framework Conditions for Electricity Markets and Renewable Energies: 16 Country Analyses*. Energy-policy Framework Papers, Deutsche Gesellschaft für Technische Zusammenarbeit (GTZ), Section Energy and Transport, Eschborn, Germany.

Projekt-Consult GmbH and Loy, D. (2007), *Energy-policy Framework Conditions for Electricity Markets and Renewable Energies*. Deutsche Gesellschaft für Technische Zusammenarbeit (GTZ), Division Environment and Infrastructure, Eschborn, Germany.

Rutschman, I. (2010), Die unterschätzte Energiequelle. *Photon* (4), pp. 86-98.

Scheer, H. (1999), *Solare Weltwirtschaft: Strategie für die ökologische Moderne*. Verlag Antje Kunstmann, München, Germany.

Sociedad Nacional de Minería Petróleo y Energía – SNMPE (2011), *Peru Electricity Investment Handbook*. Sociedad Nacional de Minería Petróleo y Energía – SNMPE, Peru.

UNFCCC – CDM – Executive Board (2006), *Project Design Document*. UNFCCC – CDM – Executive Board.

Venezuela (2010), *Organic Law on the Electricity System and Service*. The national assembly of the Bolivarian Republic of Venezuela, Venezuela.

World Atlas. http://www.world-atlas.us/ (accessed 21 February 2012).

Wright, S. and Adlerstein-Gonzalez, S. (2009), *Sustainable Energy Development in Chilean Patagonia: A Comprehensive Critique of the August 2008 HidroAysén Environmental Impact Assessment*. Report of the Graham Scholars Program CEE 490: Sustainable Energy Development in South America, Ann Arbor (MI), USA.

Wuppertal Institute for Climate, Environment and Energy (2010), *Potentiell treibende Kräfte und potentielle für den Ausbau erneuerbarer Energien*. Wuppertal. Retrieved from http://www.wupper inst.org/uploads/tx_wiprojekt/driving-forces-final_de.pdf.

Chapter 67

Case Studies of University Contribution towards the Development of Renewable Energy in Chile

Luis Vargas, Guillermo Jimenez and Manuel Díaz

Abstract

Renewable energy is a key factor in promoting the sustainable development of South American countries. However, large-scale integration of this kind of technology into the power systems is still a pending task at regional level. With this in mind, the paper reviews selected initiatives that the university sector has proposed and developed in Chile. Specifically, the paper addresses the role of the University of Chile in disseminating renewable energies in three fields of application: education, demonstration projects, and regulatory proposals. The impact of international cooperation projects for the deployment of renewable energies in Bolivia, Brazil, and Chile is also discussed in the paper.

Introduction: Development of Renewable Energy and Higher Education Institutions

Higher education (HE) institutions play a significant role in promoting the deployment of renewable energies from different vantage points, such as new educational programs, research, and cooperation with other entities (i.e. NGOs, government agencies and private companies). In the case of education, it has been noted that in most engineering curricula the use of renewable energy technologies is not incorporated, and most engineers are not aware of the principles of sustainability. There is therefore a need to develop and implement new courses that prepare engineers, scientists, and energy planners to work with renewable resources for producing sustainable energy generation systems (Jennings 2009). Moreover, the renewable energy field is developing so fast that the short-training course area is likely to be crucial to the needs of industry over the next 20 years. Even recent graduates will need to update their knowledge on the job, and this will be done via continuing professional education courses (Jennings 2009).

Education in renewable energy has been addressed to motivate students to study energy topics related to renewable sources. Furthermore, in the field of renewable energies didactic competencies covering learning arrangements for educators, trainers, and lecturers in adult education have also been developed (Acikgoz 2011).

Other approaches have been based on the creation of centers oriented towards education for sustainable development (Itoh et al. 2008). These centers are characterized by an integrated approach where different target groups are involved (e.g. schools, NGOs and industry) and where a local university plays a catalyst role (Leal Filho and Schwartz 2008). This has promoted discussion about the roles and responsibilities of universities in the task of promoting sustainable resource energy development in society (Axelsson et al. 2008; Mochizuki and Fadeeva 2008).

A different course of action has been the promotion of active research on this topic in higher education, where the dual roles of the academic practitioner as a teacher and a researcher are brought together to integrate educational practice and research (Tormey et al. 2008). Novel demonstration projects have been promoted as campus initiatives (Krizek et al.

2012). These works have raised the level of awareness about environmental issues on campuses, and have increased the vested interest on the part of the students, as demonstrated by the dominance of environmental studies in the curriculum, student-run organizations dedicated to sustainability issues, and adoption of green fees (Ghosh 2011). The goal of these initiatives is to show policymakers ways of using current higher education partners as role models for promoting green policies, and as well as how to exploit the enormous potential impact colleges and universities can have in the development and deployment of renewable energy. However, there have been only a limited number of specific field projects, in which the university plays a leading role in the design, building and implementation of renewable energy projects focusing on satisfying the real needs of people and industry.

Given the previous background, this paper attempts to provide an overview of contributions in the promotion of renewable energy made by the School of Engineering, at the University of Chile, in three fields of application: education (Vargas 2010), demonstration projects (Palma-Behnke et al. 2010) and regulatory proposals. In the case of demonstration projects, this paper describes a successful project undertaken in an isolated community located in the Atacama Desert in the north of Chile, which shows a different scope of application when compared with other "on-campus" experiences. Finally, the impact of some international cooperation projects at the university level for the deployment of projects based-on renewable energies in the country is also discussed.

New Initiatives in Education on Renewable Energy

As it is discussed in the previous chapter, education in renewable energy has the capabily to produce two main outputs: a first one focused on a deeper knowledge about the technology itself and its applicability, and the second one that provides a more integral education by means of creating awareness about sustainability. In this section two examples of education initiatives related to renewable energies are presented: a minor in renewable energies and a Postgraduate Diploma in Non-Conventional Renewable Energies (NCRE).

Renewable Energy Minor

The aim of the minor in renewable energy is to increase the number of renewable energy topics in the current engineering degree programs at the Universidad de Chile (Vargas 2010). With this addition the students will have more competitive professional profiles, as they will be able to expand the vision and integration of renewable energy in all fields of engineering.

The minors are part of the undergraduate electives included in the curriculum of the School of Engineering, which have been in force since 2007. Students in the Civil Engineering department at the University of Chile can take any minor, as long as they comply with the course requirements.

Starting in the sixth semester there is a sequence of five courses dedicated to the minors. Two of them are in semester 7 and the other two in semester 8.

The minor in Renewable Energy was conceived to have a flexible organization; students may choose four courses from a given list, and the fifth at their own election (they may pick any course given in the entire faculty). So far, the list of courses is as follows:

- Electricity Generation with Renewable Energy
- Introduction to Meteorology and Oceanography
- Principles of Geothermal Energy
- Introduction and Applications of Solar Energy
- Renewable Energy with Biomass.

The minor channels initiatives on renewable energy of various programs at the University of Chile, such as the Domeyko Energy Program, an initiative whose main purpose is to encourage interdisciplinary study within the University of Chile, through the support, funding, and dissemination of scientific research projects related to the challenges of our country (http://www.uchile.cl/domeykoEnergia).

The implementation of the minor courses began in 2009. The enrollment of students since then has demonstrated great interest in the topic of RE, especially among the electrical engineers. There have been 605 students enrolled in RE courses, with an average of 55 students per semester.

Diploma in Non-Conventional Renewable Energies (NCRE)

This initiative began in 2011 in the Department of Electrical Engineering. The diploma is oriented towards managers, executives, engineers and lawyers. It addresses professionals who are interested in the technical, economic, and social aspects of NCRE.

The program is designed for industry, government agencies, and regulatory bodies that require professional training in topics related to NCRE, and whose needs are not met by one-week postgraduate courses, or by graduate programs which require a minimum stay of two semesters.

The program is structured into seven modules, each ranging in length from one to two weeks duration, in order to facilitate the attendance of practitioners. The entire program is divided into a schedule of 15 weeks duration, which includes 88 hours of instruction. The modules are structured as follows:

- Module 1: Chile's Energy Markets and Climate Change (8 hours)
- Module 2: Basics of Renewable Energy (12 hours)
- Module 3: Hydroelectric Generation (12 hours)
- Module 4: Wind Generation (12 hours)
- Module 5: Fuel Cells (12 hours)
- Module 6: Biomass (12 hours)
- Module 7: Smart Grids (16 hours)
- Module 8: Case Studies (4 hours).

The overall objective of the program is to deliver an updated overview of the key technologies of power generation based on non-conventional renewable energy resources (i.e. those other than power plants using fossil fuels and large hydro) with high potential for electricity power generation in the country.

The specific objectives of the degree are:

- To describe the major sources of non-conventional renewable energy applications to electric power generation.
- To develop an academic program geared to modern technologies of NCRE.
- To transmit the conceptual and practical skills required to identify the structure and use of non-conventional renewable energy resources in different regions of the country.
- To provide diagnostic tools, analysis, evaluation and project design-based technology NCRE.
- To promote understanding and analysis of the regulatory aspects in the various fields of natural resource use and how they affect the manner of implementation of the NCRE.
- To promote knowledge and use of study methods and assessment of technical and economic solutions in the area of power generation based on NCRE.

- To enable application of knowledge gained from actual case studies of projects based on NCRE, developing technical analysis of the particular problem, and economic evaluation of the proposed solution, in a context of efficient use of resources.

The first version in 2011 had 28 students enrolled. The professional profile (profession of origin) of them is as follows:

- 2 Lawyers
- 3 Economists
- 1 Geographer
- 5 Electrical Engineers
- 8 Industrial Engineers
- 2 Hydraulic Civil Engineers
- 3 Chemical Engineers
- 2 Mechanical Engineers
- 1 Molecular Biotechnology Engineer
- 1 Journalist.

This list shows the great diversity of professionals interested in the Diploma in NCRE.

Demonstration Projects

In this section two applied projects developed at the University of Chile in the field of Renewable Energies are presented. The first one is devoted to providing power supply solutions to isolated locations, while the second one focuses on developing micro hydro generators for rural communities. These two projects present a clear emphasis on real applications where not only the technological development and transfer plays an important role, but also shows the integration of communities as a key issue that finally guarantees the sustainability of these initiatives in the long term.

Smart Power Micro Grids to Isolated Locations (Huatacondo Case – ESUSCON)

The ESUSCON rural electrification project (Electrificación SUStentable CONdor acronym in Spanish, Condor Sustainable Electrification Project in English), is an initiative from the Energy Center, Faculty of Physical and Mathematical Sciences, Universidad de Chile, with a local mining company in the locality of Huatacondo, Tarapacá Region, Chile. It aims at developing an electricity supply solution for an isolated system under a scheme known as a mart power microgrid (Palma et al. 2010), where, by means of information technology, automation devices and power electronics, a group of different electric energy sources, storage devices, metering equipment, and network equipment is coordinated and optimized to provide the final user with a good-quality continuous supply of electricity.

The microgrid was conceived for a small, isolated village in the Atacama Desert, Chile, called Huatacondo (20° 55' 36.37" S 69° 3' 8.71" W). Its electric network was isolated from the interconnected system and was supplied only ten hours a day by a diesel generator. This microgrid takes advantage of the distributed renewable resources in the area, providing 24-hour electricity service. Because the village experienced problems with the water supply system, a management solution was also included in the microgrid. Additionally, a demand-side option to compensate for the generation fluctuations due to the renewable sources was also considered. The system is composed of two photovoltaic systems, a wind turbine, the existing diesel generator unit of the village (typical in isolated locations), an energy storage system

(ESS) composed of a lead-acid battery bank (LABB) connected to the grid through a bidirectional inverter, a water pump, and a DSM (loads). Fig. 1 summarizes the power schematic of this microgrid.

Fig. 1: Renewable-Based Microgrid: Power Schematic

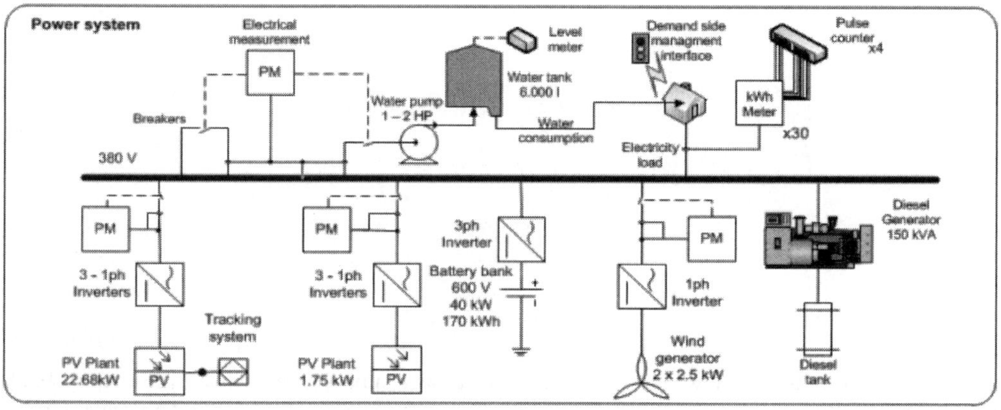

These elements are commanded by a central Energy Management System (EMS) that provides signals for optimizing their operation. The optimization considers load and resources forecasts in order to minimize the consumption of diesel and keep the power quality indicators close to optimal values. Additionally, it includes a demand side management system, which sends visual information about recommended daily load profiles to the customer, according to forecasted resource availability. Actual consumption data is recorded and sent back to the EMS through smart metering (Palma et. al. 2010 and 2011).

The main goals of the EMS are:

- To minimize the use of diesel fuel;
- To deliver active generation set points for the diesel generator, the ESS inverter, and the PV plant;
- To turn the water pump on and off in order to keep the elevated water tank level within predefined limits;
- To send signals to consumers promoting changes of use behavior.

The microgrid operation can be described based on operation results such as the one shown in Fig. 2.

Fig. 2: Smart Power Microgrid Operation

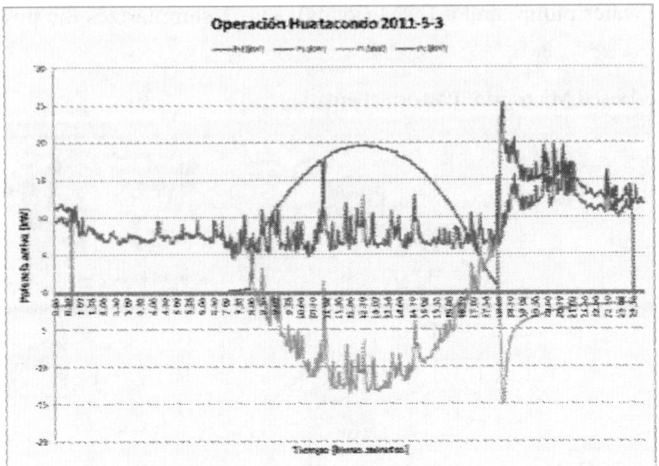

From the beginning of a new day (0:00 hrs.) the power supply comes from the battery bank (green line), following the load behavior (purple line). As is also observed, when the sun rises (approx. 7:00 hrs.) the PV power production (red line) begins to increase reaching values higher than the load demand. Given that PV production presents a surplus of energy, this surplus is used to recharge the battery bank again (green line goes on negative values). Basically, this characteristic continues while PV is higher than load. When the sun sets, power demand increases, so the diesel engine is turned on and produces energy, not only to cover the demand but also to finish the battery charge. It is also important to state that the diesel engine works in a more efficient way given that it provides power for peak periods and also a surplus for battery charging, which means that it operates at recommendable load levels.

Finally, there is also a joint community intervention program that runs in parallel with the technical solution. This community intervention process plays a key role when developing this kind of project, allowing the creation of trust between project developers and the community, working in a "co-construction" way (for instance, the community participates in the decision-making process) and providing enough tools, such as the Social SCADA concept (Palma et al. 2011), that allow the community itself to take care of the operation and maintenance of the microgrid, which finally ensures the sustainability of the project in the long term.

Smart Micro Hydro Plug & Play

Given the recent initiatives focused on the promotion of renewable energies and their integration into the electrical market, small hydro has experienced significant growth. Its continuous increase in the energy matrix is a consequence of different factors, including: technological maturity, competitiveness, and the great availability of resources exploitable from both technical and economical points of view. Additionally, it is important to consider that local experience gained by local generation companies and funding institutions has increased their willingness to support this kind of project. However, there are also minor hydraulic resources associated with rural electric consumption which are good enough to be exploited efficiently providing a sustainable energy solution.

Most of the available solutions for supplying isolated systems are often not capable of being synchronized with the power system unless an expensive conditioning is carried out, which is associated with both technological and economic barriers. Thus, when the power system reaches the micro-hydro location, these units can´t be operated easily and they end up being abandoned.

Chile and other countries of the region present high hydro-energy potential. In the case of Chile it is estimated that this solution would be suitable for about 3,000 potential locations. In addition, scale economies are not completely applicable to micro hydro power because some of the parts and pieces required come from large manufacturers who develop equipment for commercial or residential applications.

The limited supply and technological developments of low kW (1-100 kW) micro hydro solutions, together with the large potential, creates a clear interest in developing a micro hydro unit with the capability of operating in an automatic way, in both isolated and grid-connected conditions. Such development would be a good alternative to exploiting water resources located close to isolated places and to the power system as well. In Fig. 3 the size and design of the unit is shown.

Fig. 3: Micro hydro Unit: Schematic

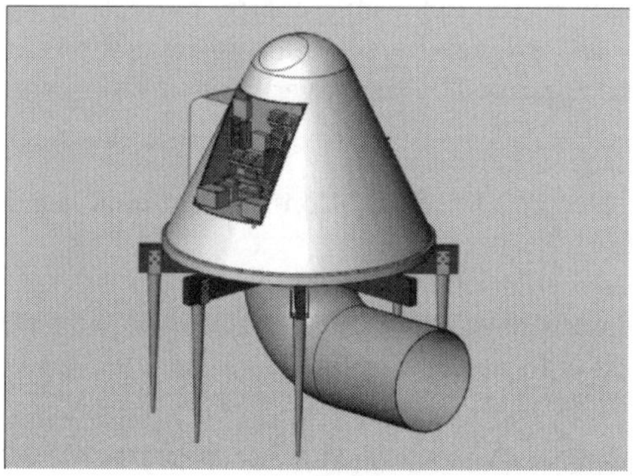

In this context, the development of the unit is based on improvements to be made in a standard micro hydro turbine. Improvements consist of expanding its functions and converting it into a smart unit with competitive costs, with easy installation and compatibility with the distributed generator integration schemes.

The capabilities described above are not seen in the actual market. A detailed description of them follows:

- Automation: A completely automatic unit makes possible its acquisition by users without technical backgrounds; when it is compared with other existing micro hydro units, this unit comes with the necessary equipment to operate in a 100% automatic way.
- Grid connection: Though it is capable of operating as an isolated generator, this micro hydro also has the capability of operating as a distributed generator in a distribution network. Actual energy prices and the estimated costs for development of this kind of unit make it possible that, for a potential of 10 KW and a reasonable water inflow, capital recovery would take about three to four years. The latter creates business opportunities for the concept of saving /selling of energy to people of other sectors who also have the resources.
- "Plug & Play" concept: Under this concept, the objective of this development is to achieve a product with easy and standardized installation, so as to be competitive with a standard combustion engine of about the same size. In this sense, the micro hydro unit has the advantage of not requiring fuel, and its maintenance is simple and carried out at low costs.

Regulatory Proposals – The University of Chile Experience

In the framework of cooperation initiatives, there are specific experiences of contributions from the university to other institutions. This is the case of the development of regulatory proposals for government institutions, where the knowledge of the general regulatory framework and renewable energy technologies endows the university as an adequate advisor when it is necessary to discuss the formal integration of these technologies into the power system and market. In fact, due to the continuous changes that renewable energy technologies have experienced during the last years, and the changes on the energy policy required to be able to incorporate this technological evolution, the university sector has proposed a number of changes on the regulatory framework to determine rules, definitions, procedures and standards that facilitate the integration of new technologies into the power system.

The University of Chile has had active participation in the design and formulation of standards and regulations associated with Distributed Generation[1] (DG) (Ministry of Economy, Government of Chile 2005; National Energy Commission, Government of Chile 2007). It is worth mentioning that many of the renewable energy projects in operation in Chile are integrated with the system as DG projects. A selection of regulations and standards proposed by the University of Chile and approved by the government are summarized as follows:

- Supreme Decree 244

Official name: Supreme Decree, No. 244 "Non-conventional generators and small generators", enacted Sept., 2005.

Summary: It establishes regulations applicable to companies that possess generators whose power surplus is lower than 9 MW (PMG, PMGD) and/or whose primary energy source is non-conventional (MGNC) with power surplus lower than 20 MW. These rules cover the following titles:

- Small Distributed Generators Connection Bylaws (PMGD)
  - i. Procedure and conditions for connection
  - ii. Maintenance of and intervention in facilities
  - iii. Determination of additional construction costs for connection
  - iv. Operation regime
  - v. Remuneration and payments
  - vi. Measurement and billing.
- Small generators Connection Bylaws (PMG)
  - i. Operation regime
  - ii. Remuneration and payments
  - iii. Measurement and billing
  - iv. Non-conventional generators (MGNC)
  - v. Classification according to energy source
  - vi. Payment exemption due to use of the main transmission systems
  - vii. Complaints and disagreements.

1 Distributed Generation (DG) can be defined as the integrated use of small generation units directly connected to a distribution system or inside the facilities of a customer (Ackermann et.al, 2001).

- Technical Standard on Connection and operation of PMGD in MV

Official name: RME, No. 24 “Connection and operation of small distributed generators in medium voltage facilities”.

Enacted: May, 2007.

Summary: It establishes the procedures, methodologies and other exigencies for the PMGD connection and operation in the medium voltage networks of distribution companies.

International Cooperation Projects

International cooperation projects have played a significant role in the incorporation of renewable energies in the curriculum as well as in the dissemination of information on clean technologies, and in the implementation of laboratories and demonstration projects. Two examples are presented in this section.

The first project is the Joint European-Latin American Universities Renewable Energy-JELARE project (*www.jelare-project.eu*), which is a co-operative project involving universities from Bolivia, Brazil, Chile, Germany, Guatemala, and Latvia, funded by the EU program, ALFA III. The aim of JELARE was to foster innovative labor, market-oriented educational, and research approaches in the field of renewable energies at Latin American and European institutes of higher education.

The JELARE project was key for implementing a number of initiatives with high impact in the promotion of renewable energies such as:

- The Renewable Energy Technology Demonstration Centre. This initiative was implemented at the Rural Academic Unit (UAC) of Batallas, Bolivia, and belongs to the Bolivian Catholic University.
- International Conference and Fair on Education, Research & Employment in the Renewable Energy Sector. This conference and fair was specifically aimed at covering areas such as funding, initiatives, and research projects in renewable energy, as well as work experience and technology transfer in the areas of renewable energy and climate change. It was organized by the Department of Electrical Engineering at the University of Chile in November, 2010.
- Publication of a book on renewable energy market needs. The book covers the results of the labor market survey undertaken in each JELARE country (Bolivia, Brazil, Chile, Germany, Guatemala, and Latvia). It also provides a transnational comparison of all the results, as well as a general overview of the renewable energy situation in each partner country.

The other project is Promoting Renewable Electricity Generation in South America-REGSA (*www.regsa-project.eu*), which is a cooperation scheme involving universities in Germany and South America, namely Bolivia, Brazil, and Chile.

The aim of REGSA is to work towards increasing the use of electrical energy generated by renewable energy sources in South America as a way of improving environmental conditions, and thereby enhance energy security, and alleviate poverty in the project areas. Additionally, it directly supports sustainable energy options in the partner countries. To achieve this goal, new electricity grids, which will integrate the use of renewable energy as a source of electrical energy, must be constructed. These grids are to be developed especially in rural areas. In addition, extensive baseline studies, as well as scenario building with regard to renewable electricity generation, must be carried out, including political, technological and socioeconomic aspects. Various activities to raise awareness and practical support will increase the planning and design of policy frameworks. Furthermore, infrastructure feasibility studies are planned to eventually lead to an increased use of clean energy technologies. Finally, REGSA will give direct support in the form of a feasibility study to 3 pilot communities by means of the pilot action: renewable electricity communities. This will facilitate regional dialogue and capacity building which can subsequently be used as a best-practice model for other regions in South America.

The project is funded by the "Thematic Programme for Environment and Sustainable Management of Natural Resources, including Energy", a program funded by the EU.

Conclusions and Future Research

Sustainability reporting in the higher education sector is limited perhaps as a result of the lack of step-by-step sector-specific guidance. This paper shows selected initiatives in the areas of education, demonstration projects, and regulatory proposals carried out by the University of Chile.

New trends in education and curriculum show the importance of providing complementary forms in specific areas related to new challenges to human development such as renewable energy. The minor on Renewable Energies, implemented in Chile is very popular with an average enrolment of over 100 students per semester. On the other hand, the Diploma in NCRE has attracted many professionals from different fields, which was one of the goals of the program.

Both the Minor and the Diploma in NCRE have proved to be excellent initiatives for disseminating knowledge related to renewable energy.

In the area of demonstration projects the paper presents two novel solutions for supplying isolated systems. A smart grid based on an efficient combination of renewable resources, and an economical and easy to operate plug-and-play micro hydro is presented. The main features of those systems and some learned-lessons are discussed and identified.

Outlines of regulatory bodies that were proposed by the university are also discussed in the context of the promotion of renewable energies.

Finally, international cooperation in the form of tangible projects, with high impact results, shows the importance of these types of initiatives in the promotion of renewable energies in different South American countries.

References

Acikgoz, G. (2011), Renewable energy education in Turkey. *Renewable Energy* 36 (2), pp. 608-611.

Ackermann, T., Andersson, G. and Söder, L. (2001), Distributed generation: A definition. *Electric Power System Research* 57 (3), pp. 195-204.

Axelsson, H., Sonesson, K. and Wickenberg, P. (2008), Why and how do universities work for sustainability in higher education (HE)? *International Journal of Sustainability in Higher Education* 9 (4), pp. 469-478.

Ghosh, S. (2011), Participation in the Green Power Partnership: An analysis of higher education institutions as partners in the program. *International Journal of Sustainability in Higher Education* 12 (4), pp. 306 -321.

Itoh, M., Suemoto, M., Matsuoka, K., Ito, A., Yui, K., Matsuda, T. and Ishikawa, M. (2008), Contribution of Kobe University to the Regional Centre of Expertise (RCE) on Education for Sustainable Development (ESD) Hyogo-Kobe. *International Journal of Sustainability in Higher Education* 9 (4), pp. 479– 486.

Jennings, P. (2009), New Directions in Renewable Energy Education. *Renewable Energy Journal* 34 (2), February, pp. 435-439.

Krizek, K.J., Newport, D., White, J. and Townsend, A.R. (2012), Higher education's sustainability imperative: how to practically respond? *International Journal of Sustainability in Higher Education* 13 (1), pp. 19-33.

Leal Filho, W. and Schwarz, J. (2008), Engaging stakeholders in a sustainability context: The Regional Centre of Expertise on Education for Sustainable Development in Hamburg and region. *International Journal of Sustainability in Higher Education* 9 (4), pp. 498-508.

Ministry of Economy, Government of Chile (2005), *Non-Conventional Generation Bylaws and Small Generation Systems. Supreme Decree 244, Chilean Electricity Law* (Regulatory framework in Spanish: Reglamento de Medios de Generación No Convencionales y Pequeños Medios de Generación).

Mochizuki, Y. and Fadeeva, Z. (2008), Regional Centres of Expertise on Education for Sustainable Development (RCEs): an overview. *International Journal of Sustainability in Higher Education* 9 (4), pp. 369-381.

National Energy Commission, Government of Chile (2007), *RME, No. 24 Connection and Operation of Small Generating Units in Medium Voltage Installations* (Regulatory framework in Spanish: "Conexión y Operación de Pequeños Medios de Generación Distribuidos en Instalaciones de Media Tensión").

Palma, R., Benavides, C., Aranda, E., Llanos, J. and Sáez, D. (2010), *Energy management system for a renewable based microgrid with a demand side management mechanism*. Proceedings of the IEEE Symposium on Computational Intelligence Applications in Smartgrid. IEEE-PES, Korea.

Palma-Behnke, R., Ortiz, D., Reyes, L., Jiménez-Estévez, G. and Garrido, N. (2011), *A social SCADA approach for a renewable based microgrid – the Huatacondo Project*. Proceedings of the IEEE Power and Energy General Meeting. IEEE-PES, Detroit, MI.

Tormey, R., Liddy, M., Maguire, H. and Mc Cloat, A. (2008), Working in the action/research nexus for education for sustainable development: Two case studies from Ireland. *International Journal of Sustainability in Higher Education* 9 (4), pp. 428-440.

Vargas, L.S. (2010), *A minor of Renewable Energies in Engineering Curriculum*. Proceedings of the International Conference and fair on Education, Research & Employment in the renewable energy sector – ICERE 2010, Santiago, Chile, Nov 11-12.

Chapter 68

The Institutionalization of Sustainability at the University of Sonora Mexico

Munguía Nora, Javier Esquer Peralta, Zavala Andrea and Velazquez Luis

Abstract

This paper presents the process for reaching a high degree of institutionalization of sustainability at the University of Sonora. It gathers and updates information previously published in a high impact journal and books on sustainability as proof of the institutionalization process at this institute of higher education. Although the principles of sustainability have permeated the institution, different attitudes toward sustainability practice remain present among the members of the university community. It has taken more than two decades to reach this degree of institutionalization, but sustainability is now part of the university's cultural values, resulting in positive attitudinal changes among students, professors, staff, and authorities. The end of the sustainability path is still a long way off; sustainability, however, is now the criterion that leads many initiatives on campus.

Introduction

Numerous researchers have studied the ways in which sustainability has been institutionalized in higher education institutions, and most commonly in universities (Leal Filho 2009). Institutionalization is frequently defined around values, norms, traditions, routines, interactions and relationships within organizations (Altintas et al. 2011). Although sustainability appears to be an important value in many organizational cultures (Esquer 2008), this is seen from different perspectives depending on the cultural values of each organization (Prugh and Assadourian 2003; Filho Leal 2000). As a result, its institutionalization differs at each university or campus, or even between academic departments; however, it is possible to find common sustainability practices (Velazquez 2006).

Sustainability offers the possibility of a much broader coalition for positive changes both within and among societies (Edwards 2005). Unfortunately, in most of the universities, faculty members conform with predominant tenets on campus and prefer incremental changes rather than radical changes, in order not to threaten autonomy (Aypa and Kalayce 2008). As a consequence, the institutionalization of sustainability in higher education institutions may take a long time.

As long as universities have the honest commitment to support sustainability initiatives, it is possible to call them sustainable learning organizations (Velazquez 2011) without regarding the time it takes for institutionalizing sustainability.

Kostova and Roth (2002) identified four patterns of attitudes toward the practice of any concept, which called: active, minimal, assent and ceremonial adoption. The ideal would be to find active persons who display the deepest levels of adoption, but on campus there are mixtures of all profiles and humans are just another of the factors that deter sustainability in higher education institutions (Velazquez 2005).

The purpose of this article is to track sustainability at the University of Sonora from its early stage to its high degree of institutionalization.

This is a public institution located in the State of Sonora in the north-west of Mexico. It is the biggest in the state and one of the most important universities at the region. Furthermore, thanks to its proactive sustainability efforts, it enjoys a global reputation.

Early institutionalization

Prior to the early 1990s, there were a few projects dealing with environmental issues, most of them with the quality of water. Nobody had in mind the actual dimensions of sustainability (Velazquez 1999) and discussions focused only on technical issues. Then by 1992, a group of professors in the Industrial Engineering Department created the Sustainable Development Group to promote sustainability on campus. During the following years, the main task consisted of disseminating the sustainability concept by organizing conferences, workshops, and seminars.

In the beginning, outcomes were not satisfactory at all, some professors opposed to accept sustainability as a part of the industrial engineering knowledge; others professors strongly criticized the organizations of events such as the annual celebration of the Earth´s day; and often, events were conducted with low attendance. However, more members of the university community were now aware of sustainability and this awareness prepared the campus for paving new paths towards sustainability.

Gaining momentum for institutionalization

A critical step towards the institutionalization of sustainability was taken in 1994, when a course in sustainable development was incorporated into the Industrial Engineering curricula (Velazquez 2002). This was the first mandatory course of sustainability into a Mexican engineering curriculum. By 1996, the course became a hand-on course by creating an educational model called "sustainable cell", in which the campus became a living laboratory for the implementation of sustainability projects. This has been the foundation for the incorporation of more formal courses related to sustainability not only into the Industrial Engineering Department, but also into many other departments in the institution.

As a natural consequence, the sustainability efforts evolved into research activities applying science to solve unsustainability problems in different areas such as cleaner production, pollution prevention, global climate, sustainable consumption, toxic use reduction, industrial ecology, design for environment, and hazardous waste.

By 1997, the authorities of the institution noted that the sustainability efforts on campus influenced groups of interest outside the university because the Sustainability Development group had become widely recognized as a sustainability icon in the region. At this stage some degree of institutionalization of sustainability in the University of Sonora was more than obvious.

Furthermore, a collaboration international agreement with the University of Massachusetts Lowell was signed in 1999 that initiated a collaborative network with other American and European universities that are sustainability leaders in their countries (Velazquez 2000).

By 2002, a good degree of institutionalization was reached that led to the systematic pursuit of the goal to become a more sustainable university day by day.

In order to reach this status, the strategy depicted in Fig. 1 was conceived with the goal of being a higher education institution, as a whole or as a part, that addresses, involves and promotes, on a regional or a global level, the minimization of environmental, economics, societal, and detrimental health effects in the use of their resources in order to fulfill its main functions of teaching, research, outreach and partnership, and stewardship, as a way of helping society make the transition to sustainable lifestyles.

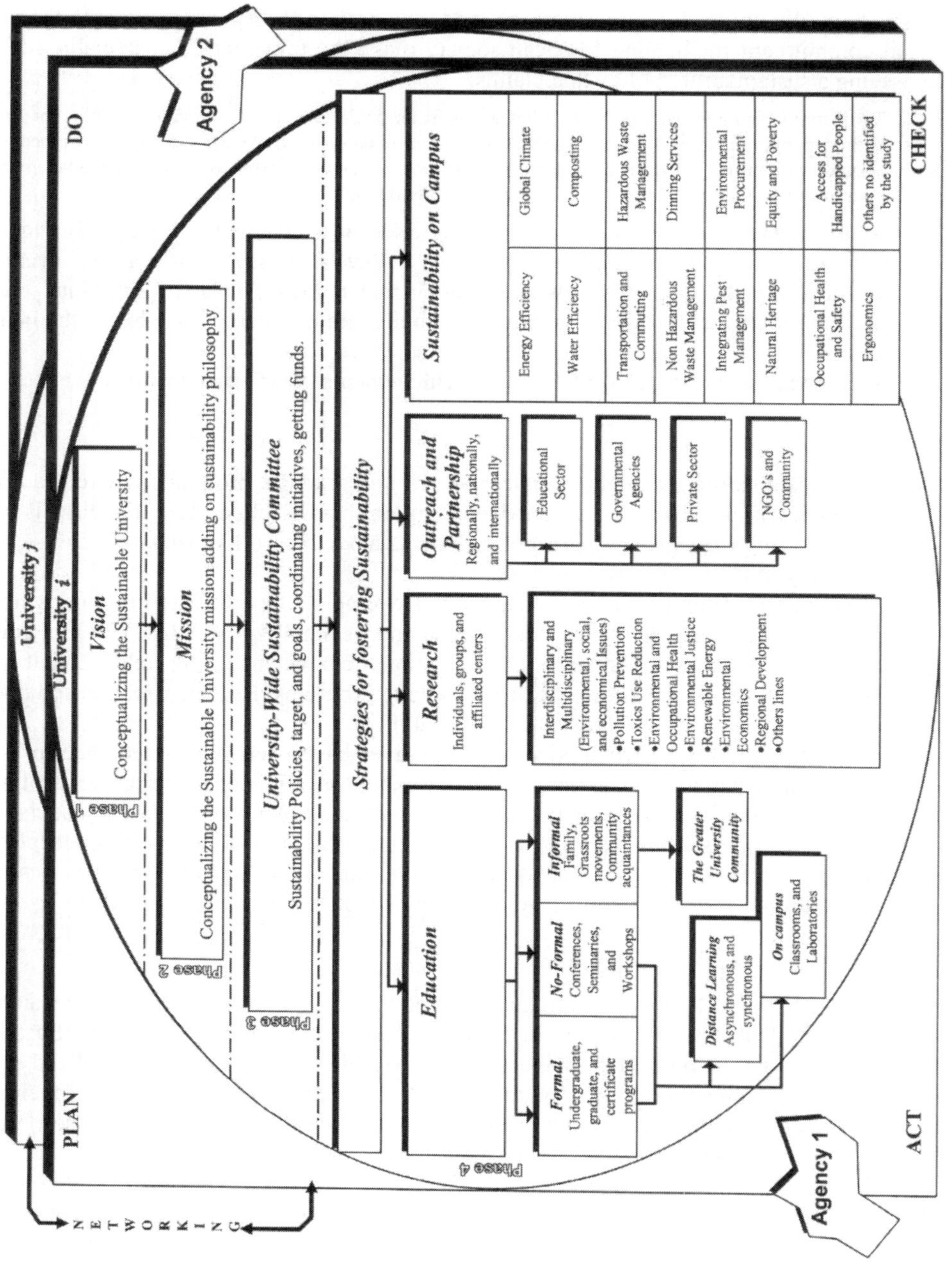

Fig. 1: Sustainable University Framework

The high degree institutionalization

With this degree of institutionalization; the strategy for becoming a sustainable university has continuously matured from a simple structure composed by individual and isolated projects, to a systematic and integrative Sustainability Management System (Velazquez 2009).

By 2004, the university authorities assumed their responsibility to demonstrate the institutional commitment for helping Mexican society make the transition to sustainable lifestyles by issuing a sustainability policy that claims:

> "The institution adopts a sustainability policy which fosters in the university community a culture for preventing, eliminating and/or reducing the negative environmental impacts generated when performed its substantive functions of teaching, research, and outreaching and partnership as well as its management operation tasks. Given as a result local and global initiatives into and outer of campus."

The establishment of a formal strategy and the issue of a sustainability policy brought new energy to all sustainability initiatives under this umbrella. Today, there are four avenues to tackling unsustainability patterns, with the hope that it brings positive attitudinal changes among the both university and the greater community and not merely to enhance the image of our university.

The following sections highlight the major achievements in each sustainability avenue:

Education

Currently, sustainability is taught at undergraduate and graduate programs. There is a diversity of contents for sustainability courses depending on the discipline where the class is taught. In the Engineering College, all engineering students have to enroll a mandatory course named "Sustainability in the Engineering Sciences".

At graduate level, it is offered a Sustainable Development Certificate Program which has been ranked by the Mexican Council on Science and Technology (CONACYT) as a graduate program of excellence with the status of international competence. This achievement is only held by less than 1% of the entire Mexican graduate programs. Recently, master and doctoral degrees programs have been launched in sustainability.

Most of the initial non-formal education initiatives have been kept on-going. Workshops, seminars, training sessions, meetings, and pilot projects are some of the initiatives still widely used to educate and increase knowledge among community and organizations leaders. However, actual outcomes differ from those at the beginning, sustainability is now accepted as a part of engineering by most, if not all, professors; and workshops and conferences attract many students and professors from several academic departments.

Research

The Sustainable Development Group that initiated the institutionalization of the sustainability concept in the University of Sonora in 1992, became an academic group called "Sustainable Engineering" with the purpose to prevent, eliminate and/or reduce environmental and occupational risks in organizations; mainly industrial organizations. However, because of their multidisciplinary approach to science, the researchers in this academic body work closely with researchers in other scientific fields.

A few years ago, the Sustainable Engineering Academic Body was ranked by the Mexican Public Education governmental Agency (SEP), via the Faculty Improvement Program (PROMEP), as a "Consolidated Academic Body" This is the maximum ranking granted to academic bodies by SEP.

Researchers usually perform applied science to solve practical problems in the region; however, basic research is also performed in order to expand knowledge of how better to reach sustainability. In actual fact, they maintain three research lines devoted to foster sustainability; the first is aimed at encouraging cleaner production practices in the industry, the second aims at the implementation of environmental management systems in organization and the third one promotes sustainability in higher education institutions. These research lines include a great variety of sustainability topics, such as cleaner production, pollution prevention, global climate, sustainable consumption, toxic use reduction, industrial ecology, design for environment, and hazardous waste, among others.

Outreach and partnership

During the previous years, this substantive function of the university has been strongly supported by the university´s rector. Therefore, the academic body has aimed its efforts at collaborating with communities, agencies, industry, non-governmental organizations among others groups to improve the quality of life of their members by given advising services, technical assistance, demonstration projects, and capacity building.

Physical Operations

In order to make the campus more environmentally friendly, the University of Sonora maintains its Sustainability Management System (SMS) based on the ISO 14001 framework. The SMS was certified in 2008 under the ISO 14001 standard, and after achieving three follow up audits, this system will be ready in 2012 for the re-certification process.

The goal of the SMS is the protection of natural resources and the prevention, reduction and/or elimination of environmental and occupational risks generated by the members of the university community when using resources in order to fulfill its substantive functions of teaching, research, outreach and partnership, and stewardship.

The EMS is an evidence of the institutionalization of sustainability on campus because it is maintained by students who monitor and maintain records daily to track environmental performance as described in the SMS manual. If these sustainability values and principles weren't permeated in the institution, this SMS would not be a reality.

International and National Alliances

The University of Sonora has broadened its reach through international collaborative agreements that have been signed with several of the most prestigious higher education institutions in North America, South America, and Europe.

The first one was signed in 1999 with the University of Massachusetts Lowell; then in 2006, collaborative agreements were signed with the German University of Applied Sciences (Hochschule Zittau/Görlitz) and with Erasmus University in Rotterdam, the Netherlands; those were followed by another international agreement with the Universidade Paulista in Sao Paulo Brazil. Additionally, joint initiatives have been carried out with the following universities: the Universitat Politecnica de Catalunya in Barcelona Spain, Linkoping University in Sweden, Opole University in Poland, the Universidad Nacional de General Sarmiento in Argentina, and Roskilde University in Denmark.

Collaborative networks with national universities such as the Autonomous University of Baja California, the Baja California Polytechnic University, The Chapingo Autonomous University, and others, have also been maintained

International and national alliances have generated research opportunities and facilitated student and professor exchanges.

Conclusions

The concept of sustainability started to be institutionalized in the University of Sonora in the early 1990´s, almost two decades after the concept itself was institutionalized. This fact is not by any means a declaration of the goal of being a sustainable university having been reached at all, however. This means that the principles of sustainability have permeated in the institution and they are widely understood as a conceptual framework that takes into account economic, social, and environment aspects.

As theorized early in this article; although there is a high degree of sustainability institutionalization on campus; it is possible to find different attitudes toward the practice of sustainability; yet active adoption profiles are still common.

Under this premise, most of the professors do not object any longer to sustainability, and students are taught not only about key methods and tools, but also about the skills to understand the externalities and side effects of their decisions and actions on society and environment well enough.

The four fronts to promote sustainability in the university have resulted in positive changes in attitudes among the both university and the greater community; in consequence, the conditions for the successful implementation of sustainability initiatives are better than twenty years ago.

The University of Sonora has not yet reached the end of the sustainability path; it is very likely that there is a still a long way to go; however, sustainability is now a very important criterion that leads the substantive functions and the administrative tasks of the institution.

References

Altintas, M.M., Vontris, D. Kaufmann, H.R. and Alon, I. (2011), Internationalization market forces and domestic sectoral institutionalization. *European Business Review* 23 (2), pp. 215-235.

Aypa, A. and Kalayci, S.S. (2008), Assessing institutionalization of educational reforms. *International Journal of Educational Development* 28, pp. 723-736.

Edwards, A. (2005), *The Sustainability Revolution: portrait of a paradigm shift.* New Society Publishers, Canada.

Esquer, J., Velazquez, L. and Munguia, N. (2008), Perceptions of Core Elements for Sustainability Management Systems (SMS). *Management Decisions* 46 (7), pp. 1027-1038.

Kostova, T. and Roth, K. (2002), Adoption of an organizational practice by subsidiaries of multinational corporations: Institutional and relational effects. *Academy of Management Journal* 45 (1), pp. 215-233.

Leal Filho, W. (2000), Dealing with misconceptions on the concept of sustainability. *International Journal of Sustainability in Higher Education* 1 (1), pp. 9-19.

– (Ed.) (2009), *Sustainability at Universities- Opportunities, Challenges and Trends.* Peter Lang, Frankfurt.

Prugh, T. and Assadourian, E. (2003), *What is sustainability, anyway?* World Watch Magazine, September/October.

Velazquez, L. (2002), A Mexican Model for Teaching Sustainability in Universities. In Leal Filho, W. (Ed.), *Environmental Education, Communication, and Sustainability.* Peter Lang. Frankfurt.

Velazquez, L., Munguia, N. and Platt, A. (2000), Fostering P2 Practices in northwest Mexico through inter-university collaboration. *Journal of Cleaner Production* 8, pp. 433-437.

Velazquez, L., Munguia, N., Platt, A. and Taddei, J. (2006), A sustainable University: what can be the matter? *Journal of Cleaner Production* 14, pp. 814-819.

Velazquez, L., Munguia, N. and Romo, M. (1999), Education for Sustainable Development: The engineer of the 21st century. *European Journal of Engineering Education* 24 (4), pp. 359-370.

Velazquez, L., Munguia, N. and Sanchez, M. (2005), Deterring Sustainability in Higher Education Institutions. *International Journal of Sustainability in Higher Education* 6 (4), pp. 383-391.

Velazquez, L., Munguia, N., Zavala, A. and Esquer, J. (2009), Sustainability Leadership by Implementing the ISO 14001Framewrok on a Latin-American Campus. In Leal Filho, W. (Ed.), *Sustainability at Universities – Opportunities, Challenges and Trends.* Peter Lang. Frankfurt.

Velazquez, L., Esquer, J. and Munguia, N. (2011), Sustainable Learning Organizations. *The Learning Organization* 18 (1), pp. 36-44.

Chapter 69

Practical Strategy for Sustainable Development in Education: The Green Office's Case of the Federal University of Technology of Paraná, Curitiba, Brazil

Eloy F. Casagrande Jr., Bruno V. Kobiski, João Gois and Libia Patricia Peralta Agudelo

Abstract

The Green Office was built last year at the Federal University of Technology of Paraná (UTFPR) in Curitiba, Brazil, in order to disseminate a plan for sustainable development within education. The main objective is to develop innovative green technology and to function as a research laboratory. It also aims to update educational strategies, in order to stimulate actions that result in a more sustainable campus. Great efforts were made to obtain financial support, which finally became viable when the university signed cooperation agreements with over 60 companies, who donated green materials and sustainable technologies. The Green Office was built using a wooden-framed structure instead of a conventional concrete construction, which made it possible to conduct a preliminary study of carbon footprint reduction during the entire construction process. Furthermore, energy efficiency was achieved by using renewable energy systems, thermal and acoustic insulation, natural lighting and LED illumination devices. Solar energy is supplied by 2100 Watt photovoltaic panels installed on a grid model, and 850 Watts in a stand-alone model. So far, the building has generated almost 1000 kWh since December 2011. The performance of thermodynamic solar panels in heating water is presented together with the benefits of green roofs and the water-saving gadgets installed in lavatories, as well as an efficient rainwater collection system and other inovations. The Green Office has also established cooperation agreements with independent researchers, who have developed an electric car and a hydrogen-fuel cell bicycle fed by the solar-powered electricity generated in the building. It is strongly believed that the Green Office's experience is proving to be more profitable – both for students and for researchers – than many of the theories dumped into classrooms.

Introduction

There is no doubt that education is one of the more important tools available to society, in so far as achieving sustainable development is concerned. However, universities are currently facing some important challenges to implement that. Traditional teaching and research methods are based on the fragmentation of knowledge amongst various departments. This creates obstacles in the way of establishing the needed interdisciplinarity, which is the basic instrument in addressing socio-environmental problems (Motloch and Casagrande Jr. 2010).

In 1972, the concept of "Environmental Education" was officially introduced during the UN Conference on the Environment held in Stockholm, Sweden. Since then society has become increasingly aware of the major environmental problems which occur at a global level. As a result, governments have understood that the only way to improve this situation is through education.

In 1977, definitions, objectives, principles and strategies to implement adequate actions of Environmental Education were set as a result of the Intergovernmental Conference in Tbilisi, Georgia – one of the main events at Environmental Education, organized by UNESCO and the United Nation's Environmental Programme – UNEP. It was then established that: "*The educational process should be directed at solving the concrete problems of the environment through interdisciplinary approaches and active and responsible participation of each individual within the community*".

According to Palmer (1998), throughout the 1970s, the focus of Environmental Education was based on the use of the natural environment for first-hand experiences, creation of centers for conservation education and urban studies of the built environment. In the 1980s nevertheless, it was possible to observe more concern about global education methods, introducing a wider view of global problems and also actions related to community problem-solving. In the 1990s, with the establishment of the Internet, the world was "empowered" to generate wider communication networks and, as a result of this, more participatory actions from nongovernmental organizations, volunteers, schools and the business sector.

However, during the 20th century, the scenario was a mix of great awareness, disillusion and a far-reaching search for new models to achieve more efficient Sustainable Education practices. Considering that by 2030, it is expected that almost five billion people will be living in urban areas and most of them will be poor (UNFPA 2007), the future of cities, particularly in developing countries, as well as the future of humanity, will depend very much on Environmental Education decisions made at the present.

Positive examples: Linking Green Theory and Practice at the University

One of society's institutions that can contribute to a more sustainable future development is academia. Most academics agree that a university campus has to be a positive and rich environment whilst, at the same time, promote and be a place where economic, social and environmental themes are discussed in a critical manner, making it possible to link "green theory and practice" in the university routine. However, most universities require a profound transformation in their structures to implement that effectively. In spite of this, some good initiatives can still be mentioned:

1. Greening the Campus

Since 1996, an annual interdisciplinary conference named "Greening the Campus" takes place in the United States, which allows people representing diverse segments of university communities to share information on environmental issues that range from the practical day-to-day management activities to the actual planning of a more "green" curricula development and the best use of campus resources. These strategies aim to achieve a common agenda for achieving environmental soundness through safe management of resources at universities. It is therefore expected that the campus community can become a "green model" for the local community by gathering and sharing information and by creating actions and policies that are tested and validated with success (BSU 2012).

2. The Association for the Advancement of Sustainability in Higher Education

The Association for the Advancement of Sustainability in Higher Education (AASHE) was founded in 2005 to help coordinate and strengthen campus sustainability efforts at regional and national levels, and to serve as the first North American professional association for those interested in advancing campus sustainability. Of the 1,088 members at the end of 2010, 839 were colleges and universities. Although AASHE continued working with the operations and administrative sectors of the campus, in 2010 it increased its emphasis on working with the academic side of the campuses to address education for sustainability in the curriculum (AASHE 2010).

In order to reach it, AASHE has created the "Sustainability Tracking, Assessment & Rating System (STARS) which is a transparent, self-reporting framework for colleges and universities to measure their sustainability performance. The system includes environmental, economic, and social indicators, which are divided into four categories related to campus activities: Education & Research, Operations, Planning, Administration & Engagement, and Innovation. All of the requirements for evaluating and scoring institutions are transparent and made publicly available. Points are earned in each category and lead to a STARS Rating.

3. Bucknell University Green Initiatives

In 1998, the power plant of Bucknell University, Pennsylvania, US, was converted from a conventional coal-burning facility to a co-generation power plant fueled by natural gas. The co-gen facility captures and reuses waste heat for an overall efficiency of 75-80%, which is exceptionally high compared to conventional power plants. The co-gen power plant has also led to a 40% reduction in the university's greenhouse gas emissions (Bucknell University 2012).

In 2000, Bucknell began purchasing wind power to supplement its own site power generation. In January 2008, wind power represented 100% of Bucknell's purchased energy, for a total of approximately 4 million kilowatt hours per year. In April 2008, the Campus Greening Council (CGC) was established as an official university committee at Bucknell. The CGC puts together students, faculty members and staff, and is responsible for policy recommendations to the Bucknell administration on matters relating to the environmental impact of the university.

4. Green Practices at Carnegie Mellon University

In 1990, Carnegie Mellon adopted a formal recycling policy and appointed campus Recycling Liaisons to implement the policy. The liaisons adopted the "Reduce, Reuse, Recycling" theme and promoted campus-wide efforts to reduce waste. This was the first step to a more sustainable campus. In 1999, as part of the Carnegie Mellon strategic initiative to enhance the scope and impact of our education and research programs related to the environment, the Environmental Practices Committee (EPC) was charged to develop a plan for an expanded environmental practices program on campus.

In 2001, Carnegie Mellon made a commitment to pursue Leadership in Energy and Environmental Design (LEED) Silver Certification for all new buildings and major renovations on campus were undertaken, which enabled it to become a member of the US Green Building Council.

Ten years after the creation of a campus-wide recycling program, the percentage of non-hazardous waste being recycled has increased from 5% to 13% by weight. In addition, a full-time position was created at the University for an Environmental Coordinator. In 2006, the students at Carnegie Mellon studied the weave of grass, dirt and bugs atop its new "living roof" at Hamerschlag Hall. In the class projects they studied how the building design can reduce storm water drainage and improve water quality (Carnegie Mellon University 2012).

5. The Solar Decathlon Challenge

Since 2002, the US Department of Energy Solar Decathlon has challenged collegiate teams to design, build, and operate solar-powered houses that are cost-effective, energy-efficient, and attractive. The winner of the competition is the team that best blends affordability, consumer appeal, and design excellence with optimal energy production and maximum efficiency (Solar Decathlon 2012).

Every two years, the Solar Decathlon is organized and aims to:

- Demonstrate to the public the opportunities presented by affordable homes that combine energy-efficient construction and appliances with renewable energy systems that are available today;
- Provide student participants with unique training to help fill jobs in our nation's clean-energy economy;
- Foster collaboration among students from different academic disciplines, including engineering and architecture students, who otherwise may not work together until they enter the workplace;

Promote an integrated, or "whole building design," approach to new construction, which differs from the traditional design/build process because the team considers the interactions of all building components and systems to create a more comfortable building, save energy, and reduce environmental impact.

The Green Office proposal at the Federal University of Technology of Paraná (UTFPR)

The UTFPR has been traditionally valued both nationally and internationally by its technological focus, and now it is proud to be the first in Brazil to propose a Green Office, which is responsible for the development and implementation of environmental policies at the Curitiba campus.

The Green Office's planning started in 2002, as an initiative of the TEMA Technology and Environmental Multidisciplinary Research Group that gathers teachers, students and technicians in a permanent program called TECSUS (Sustainable Technology), which aims to implement actions in accordance with Agenda 21.

As a result of this "greening process", the UTFPR recently signed an official document called "The Pact 21", which involves some of the higher education institutions of the State of Paraná and the Forum for Agenda 21 coordinated by the Environmental Secretariat of the State of Paraná (SEMA). Currently, the Green Office is taking further actions, namely:

- CAZA – Carbon Zero in the Academy: A program to promote initiatives towards green buildings in the university campus.
- REZTO – Zero Technological and Organic Waste: A program to separate all the waste produced in campus in order to send it on for adequate recycling.
- TRECO – Reuse of Electronic and Computer Waste: A program to take care of discarded electrical and electronic devices.
- BUYING GREEN – A manual that helps the administration to purchase every day materials observing the reduction of their environmental impacts.
- GREEN LABEL – To establish procedures for Green Certification for technologies, processes and materials targeting companies.

- GREEN CONSULTANCY – To organize a service which can be offered by UTFPR professionals and students to companies who want to be more sustainable.
- PRIZE "GREEN CITIZENSHIP UTFPR" – A prize given every year by the UTFPR to a person or a group that contributes to achieve sustainability.

Some other actions include:

1. The ECOConsultoria JUNIOR Enterprise

According to Breyman (2000), campus greening tangibly improves important physical and intellectual spaces. He writes: "Greens are able, though not sometimes without frustrating hurdles, to put their theories and values into action. Colleges and universities are both congenial and frustrating institutions for Greens. Tenured Green faculty are generally protected by academic freedom (untenured professors proceed at their own peril). Junior faculty needs mentors in powered positions. Green students are generally protected by the service they provide to their institutions. Their efforts are enhanced when embraced by student governments and college administrations".

Since the beginning of the Green Office construction, the direct participation of graduate and undergraduate students was encouraged. This initiative resulted in the formulation of the first interdisciplinary "Junior Enterprise" of the UTFPR, and named "Econsultoria". It generated a spontaneous teaching-learning process, where about twelve students, plus some teachers and technicians have interacted through all the Green Office's construction processes. Therefore, they are now capable of performing consultancy for people interested in green buildings, yet still under the close supervision of a senior professor or technician.

2. Hosting the RCE-Curitiba-Paraná devoted to Education for Sustainable Development (UN-DESD2005-2014) The RCE-Curitiba-Paraná

The Green Office is hosting the RCE-Curitiba-Paraná – "Regional Centre of Expertise on Education for Sustainable Development" linked to the United Nations University – UNU / IAS – Institute of Advanced Studies, which coordinates a network of existing formal, non-formal and informal education organizations, mobilized to deliver education for sustainable development (ESD) to local and regional communities. A network of RCEs worldwide constitutes the Global Learning Space for Sustainable Development (RCE-UNU 2012) and is working to achieve the goals of the UN Decade of Education for Sustainable Development (DESD 2005-2014), by translating its global objectives into the context of the local communities in which they operate (RCE-UNU 2012). There are currently 100 acknowledged RCEs worldwide and the RCE-Curitiba was the first to be approved and installed in Latin America in 2007.

A sustainable construction planned to disseminate Education for Sustainable Development: The Green Office headquarter

The Green Office building is in itself the result of a long search for innovation associated with educational transformation.

Initially, the basic proposal was to erect an innovative eco-efficient building. The Green Office of the UTFPR was designed with the objective of integrating planning and practices, such as a wide central room for multiple activities, where teachers, researchers, graduate and undergraduate students share not only the tables and chairs, but also knowledge and friendship. The coffee break room is also large, and has some comfortable furniture designed to facilitate interactions amongst people. The idea is to create a "think tank" environment, where dynamic professionals from different areas of expertise can plan and implement a new university.

The Green Office project aims not only to implement an environmental management program for the campus, but also to promote the integration of environmental education, new methodologies and practices capable of generating integration within the teachers and students at the campus, as well as with the community representatives, in order to develop effective sustainable projects and programs that can benefit the educational processes and the city itself. The Green Office headquarters were built with the following principles of sustainable construction:

- Bioclimatic guidelines applied in the architectural design;
- Wooden-framed System – double-walled structure in OSB (Oriented Strand Board) panels and treated pine;
- Thermal insulation acoustic blankets of recycled PET and recycled tire;
- Windows in wooden frames and double glazing;
- Green Roof – use of local vegetation in the two modules of the building;
- Natural lighting through skylights intelligent systems and LED lights;
- Use of thermodynamic solar heating system for water;
- Solar energy generation using photovoltaic panels to supply up to 80% of energy to be consumed (the building can also charge an electric car);
- Rainwater-collection system for toilets and cleaning;
- Use of raised floor made from recycled polypropylene covered by modular carpet of approximately 45% recycled content;
- Mezzanine floor of certified wood (FSC – Forest Stewardship Council):
- Outside deck made from "wood-plastic" (made up of sawdust and recycled plastic);
- Bamboo furniture for the coffee break area.

Why a Wood-Frame Building?

Brazil is a major producer of wood; however, most of its buildings are made of concrete, steel and brick. It is known that the traditional construction has a high environmental impact when considering its whole production chain. Most materials are made from nonrenewable natural resources that have a high carbon footprint, demand great use of energy and water, and produce excessive waste when installed.

According to the biologist José Geraldo Zenid (2012) from the Institute of Technological Research of São Paulo (IPT), there is still prejudice in Brazil against the more intensive use of wood in buildings' structures, as the result of an inherited tradition from the Portuguese colonizers. Compared to other structural materials, wood has environmental properties: low energy consumption during processing, high strength and endurance, provides thermal and electrical insulation and ease of handling and processing with machines or by hand.

Another factor that is increasingly important is wood properties for carbon sequestration that help in reducing global warming (Zenid 2009). Furthermore, wood has the possibility of sustained production in native and planted forests, where management techniques allow changes in the quality of the raw material according to the desired end use. Reforestation wood is renewable, energy efficient and has good potential for waste recycling, as well as contributing to the clean air (Barbosa and Ino 2001).

In southern Brazil, most wooden houses are built with simple boards with only 2.2 inches, mounted simply in vertical panels. Such building structures are considered inappropriate, which contributes in spreading prejudice against structural wood. Authors such as Giglio (2005) and Bogo (2003) reported the low thermal performance of these houses, in a study that compares wooden houses with simple panels with wooden-framed houses that use double panels in the city of Londrina, State of Paraná, showing that wooden-framed houses offer better thermal comfort compared to traditional brick walls (Giglio 2005). For Molina and Calil Junior (2010), the structure of wooden-framed houses allows fast installation, and allows total control of spending during the design phase; it is also comparable or even superior to masonry in terms of structural strength, thermal and acoustic comfort.

Solutions for Wood-Frame at the Green Office of UTFPR

The Green Office at UTFPR was designed to be a research laboratory, as well as a model for business offices within the principles of sustainable construction. The wood frame structure, using double OSB (Oriented Strand Board) panels, allows the thermal and acoustic insulation to be fixed. In this case, a 75 mm recycled sheet of *Polyethylene Terephthalate* (PET) film, as a well as a 5 mm sheet made from recycled tires that provides acoustic isolation. Each square meter of the PET film is obtained by recycling thirty PET bottles and each square meter of acoustic, three tires. By using these materials in the Green Office building it is calculated that about 7,000 bottles and 540 tires have not been dumped in the environment.

As internal coating, another layer of OSB panels was applied with a layer of pine finish (DecoWall) and cement slabs. For the outer covering, sheaths were used (Vinyl Siding) and OSB boards, which have an insulating resin on one side (SmartSide) (LP Brazil 2011).

Five days were taken to build the structure, with a crane operator and five people (Fig. 1), and the walls were executed in twelve days at the factory with virtually zero waste.

The thermal-acoustic insulation is directly linked to the concept of sustainable buildings when considering the energy savings and thermal efficiency – important factors to be considered when seeking a "green certification". The initial cost is slightly higher, but compensated by a reduction in financial costs of using and maintaining the building,

Less energy use, cost reduction, lower carbon emissions, higher productivity and greater satisfaction with the academic work environment are some of the positives results of a sustainable construction like the UTFPR's Green Office.

Fig. 1: Building system of Green Office of UTFPR

Source: Escritório verde (2011)

One of the studies/research to generate a Master's dissertation of the Postgraduate Program in Civil Engineering – PPGEC, supervised by Prof. Casagrande Jr., focuses the calculation of fixed carbon inventory throughout the system. According to Tavares (2006), brick, cement, clay products, lime, and steel require a high amount of energy to be ready for use in construction. This energy is associated with CO_2 emissions both directly and indirectly. Therefore, this study is of great importance to science, especially when discussing measures for greenhouse gases emission reductions and some ways to tackle climate change.

According to Kobiski and Casagrande Jr. (2011), who compared CO_2 emissions among different types of construction, the wood framing model emits up to 73% less carbon dioxide compared to the concrete construction model, considering the energy life cycle of the materials. Since the production of CO_2, in this case is indirectly connected to the energy demand for the production of materials, priority was given during the Green Office construction for materials which demanded low energy and produced low CO_2 emissions. The forestcarbon sequestration is linked to the process of CO_2 fixation by plants through a biological mechanism called photosynthesis. Through it, the CO_2 is absorbed by plants and then released into the atmosphere in the form of O_2 with the carbon atom incorporated in the timber. The Green Office of UTFPR has a great number of materials that have a large quantity of carbon incorporated into it, such as windows, the deck, walls, the structural model, and the stairs.

Double Glazed Wood Windows

The design of the Green Office facilitates the natural light entrance through a set of large windows, strategically placed, with doubled glass. Recent studies demonstrate the efficiency of these double glazed windows, considered expensive and thus not yet widespread in Brazil. Baltar (2006) shows the results of a study in hospitals where various types of glass are tested, considering air-conditioning spending, and concluding with a trial reduction of 7% over 12 months in energy costs for air conditioning (cooling and heating). The windows of the Green Office are made with Lyptus wood and double glazing, as shown in Fig. 2.

Fig. 2: Double glazed windows.

Source: Escritório Verde (2011)

The Use of Green Roofs

The international use of green roofs is already well known. According to Araújo (2007), countries like Germany, Austria and Norway have been widely using the concept of the green roof, due to the strong interest in reversing environmental degradation and the elimination of green areas. The Green Roof is a roof with a layer of soil with vegetation or substrate. The intensive models use plants with higher consumption of water, fertilizer and maintenance, whilst extensive models use plants that have a high resistance to rain and climatic variations, minimizing maintenance and structure (Laar et al. 2001).

The high surface temperature of the majority of roofs when exposed to sunlight is one of the problems of conventional construction. In Brazil, cracks and leaks are common in slabs over the years and the high temperature variations with summer, winter, sun and rain deteriorate the materials, particularly concrete. Onmura et al. (2001) showed that a slab exposed to sunlight at an environment temperature of 38°C during summer in Japan, has a difference of 30°C, except when covered by green roof. Vecchia (2005) reported that temperatures on the surface of roofs are much milder in the green roof, and recorded internally at a building with a green roof thermal amplitude average of 9.2°C, relative to the external of 21.4°C, and a thermal lag of about 4 hours, the time it takes to change internal when changes external temperature.

Thermal comfort with lower amplitude of variation of the internal temperature translates into savings. Mello et al. (2010) estimate a 40% reduction in energy consumption of air conditioning in the Faculty of Mechanical Engineering, UNICAMP, with the implementation of green roofs. Another important factor is the studies which show that green roofs improve sound-proofing, a substrate of 12 cm depth can reduce sound by 40 decibels (Oliveira and Ribas 1995).

Two green roofs were installed on the Green Office by students under the supervision of a professional, as shown in Fig. 3. One is in the coffee area that can be accessed by the mezzanine and another one is in the front of the building, the coordination area. This system used the modular system composed of residues of EVA (the polymer used in shoe soles) that comes with substrate for vegetation where grass was laid upon them (Ecotelhado, 2011). These modules are placed on the roof protected by a plastic blanket and a drain system. This will be the first building of UTFPR to employ green roof and therefore it will be studied and evaluated by Master's degree research.

Fig. 3: Students helping in the installation of EV's green roof UTFPR

Source: Escritório Verde (2011)

Solar power generation

The term "photovoltaic cell" is used by any device that converts light into electricity, such as those used in cameras. For the case of electricity production to consumption, the most common term is "photovoltaic cell"; the set of photovoltaic cells could be called module or photovoltaic panel board. As its energy source is sunlight, designated solar photovoltaic panels.

The solar panels can be connected by an inverter which converts the variable DC output of the solar panel into a utility frequency alternating current that can be fed into the commercial electrical grid or store energy in batteries that can be used when there is no sunlight, in this case additional equipment is required in the system. A charge controller is responsible for keeping batteries charged properly.

Fig. 4: Installation of Solar panels

Source: Escritório Verde (2011)

The Green Office is the first building with solar autonomy of the State of Paraná, with 2950 Watts installed (Fig. 4). 2100 Watts of panels are generating energy that is connected directly to the grid (grid tie model), and another 850 Watts are connected to a bank of batteries (standalone model). A small wind generator of 500 Watts was also installed for testing and observation.

Since the inauguration of the Green Office, on December 15, 2011, around 1000 kWh was generated by the grid tie system. Taking into consideration that we are in the summer season, it's believed that we are having maximum generation of the system.

Natural Lighting and LED lights

Energy efficiency offers a powerful and cost-effective tool for achieving a sustainable energy future. In the Green Office of UTFPR a combination of natural lighting and the use of LED lights (Fig. 5) can guarantee less consumption of electricity. According to Polidoro (2009), LED lights can last up to 50 times longer than traditional incandescent bulbs and can save up to 80%in energy.

LEDs (Light Emitting Diodes) are semiconductor electronic devices in the class of diodes (junction of P and N materials) that emit light when forward biased (positive on the P side and negative on the N side) and supplied with sufficient energy. Chemical and physical processes cause the energy to turn into heat and photons (light), and special design elements distinguish the common diode of the LED, where the energy is converted mainly into light (electroluminescence). Because LEDs use a direct current, converters (drivers) are mounted close to LEDs in luminaries.

Fig. 5: Pendant Lamps with LED lights

Source: Escritório Verde (2011)

Economic water heating system

The water heating of the Green Office UTFPR consists of a thermodynamic system using solar panels which does not need sunlight. Although requiring electric power source with a compressor, this system uses external energy to heat up, expending around a quarter or less of the energy required to heat with electric heaters, for example.

Solar panels made from aluminum are filled with an environmentally friendly refrigeration gas and are used to take heat from the air. The solar panels are mounted on the roof or walls of the property. The refrigeration gas passes through the solar panel and is heated by the outside air temperatures. The refrigerant, having been heated by the outside air temperature changes from a liquid to a gas vapor.

This gas vapor flows from the solar panels to a compressor. The compressor increases the pressure of the gas. The gas is now at a high temperature (100 to 120c). The gas is then sent to the heat exchanger where the water in the heating system is heated. The gas is then lowered in pressure and temperature and circulates back to the solar panel to restart the process.

The finished building

The Green Office was inaugurated December 15, 2011, and has been in full operation since March 2012 (Fig. 6). Its evaluation regarding energy efficiency, thermal and acoustic comfort, carbon footprint and water consumption are conducted by graduate students from different postgraduate programs of the UTFPR.

Fig. 6: Outside e inside view of the Green Office

Source: "Escritório Verde" (2011)

Future projects associated to the Green Office

I. Electric Car and Hydrogen Bike

With the electricity generated during the day by photovoltaic panels, the project also aims to feed an electric car and a bicycle powered by hydrogen. The idea is to unify projects like this to the proposal of the Green Office, creating environments for discussion and research within the field of sustainable mobility.

The Pompeo Eletric Car

The electric vehicle to be evaluated is the "Tricycle Pompeo" showed in Fig. 7, which uses rechargeable batteries. It's an electric car being developed in Paraná, the first genuinely Brazilian electric car and promises to be accessible to all, with low cost and simple technology. The "Pompeo" may have a range of up to 200 km, and will only need 1KWh to run 10 km at a cost of R$ 0.40 (~US$ 0.22), or R$ 0.04 / km (US$ 0.022 / km). The project has been developing since 2007, through partnerships with several companies in the automotive sector, and the tests with the prototype will be completed soon. Considering previous investments in its production, the estimated market price for a basic vehicle would be in the range of 30,000 "Reais" (US$ 17,000 US dollars). (Triciclo Pompeo, 2001)

Fig. 7: Illustration and prototype electric car POMPEO

Source: Triciclo Pompeo (2011)

Hydrogen Bike

The hydrogen bike is part of research started by a Master's student on the postgraduate program in Mechanical Engineering at the UTFPR. The hydrogen-powered bike uses this gas inside a cylinder made from innovative composite materials, including PET bottles and carbon fibers, lighter than conventional steel. The bottled hydrogen is used in a fuel cell that generates electricity that powers a motor. The idea is to develop a new business by generating the hydrogen at charging stations that collects rainwater and will be covered with photovoltaic solar panels (Fig. 8), so they have some autonomy. At the recharging station, the electricity generated will be used to electrolysis the water collected, generating hydrogen.

Fig. 8: Hydrogen bike

Source: Escritório Verde (2011)

Conclusion

Universities are one of the main institutions that can contribute directly to the sustainable growth of any country. Disseminating the knowledge accumulated, as well as stimulating new knowledge production, such high level educational institutions are able to play a key role in ensuring a nation's sustainable future. However, an urgent structural transformation is necessary to enable its evolution in parallel to communal sustainability, mainly through the preservation of important resources like energy and water. Unfortunately, many academic scholars think narrowly, produce narrow research and consequently teach too narrowly. This means students will graduate without a broad understanding of how the natural systems are dependent on human respect and care (AASHE 2011).

The creation of a Green Office within a university campus is supposed to open new doors for the multiplication of knowledge in an open system of interaction. Therefore, to implement education for sustainability on a large scale, represents a huge challenge and poses a great responsibility for academia. Finding new formulas to create environmental awareness and commitment in order to have an engaged society is not an easy task. It requires partnerships and strong networking, as well as efficient actions to improve training in environmental education and management, in parallel to the development of sustainable projects that can be replicated elsewhere in any region.

The Green Office at the UTFPR project follows these guidelines and intends to go on helping to raise awareness through sustainable-based projects. Since it is quite recent, its proposals and effectiveness are still being monitored and require validation of its results over the next five years. Only then will it be demonstrated if the environmental and socioeconomic gains of the applied technologies have been successful. Therefore, the construction is not only a place for sustainable assessment initiatives inside a university, but also a living laboratory which evolves towards the future with scientific research, innovative technology and human goodwill.

References

Araújo, S.R. (2007), *Telhados Verdes*. Monografia apresentada ao Curso de Engenharia Florestal, como requisito parcial para a obtenção do Título de Engenheiro Florestal, Instituto de Florestas da Universidade Federal Rural do Rio Janeiro. 22 p.

AASHE (2010), *Association for the Advancement of Sustainability in Higher Education*. Available at: http://www.aashe.org/ (accessed 22 April 2012).

– (2011), *The Sustainability Officer's Dilemma – Thoughts from AASHE 2011*. Available at: http://greeningthecampus.wordpress.com/ (accessed 22 April 2012).

Baltar, M.G. (2006), *Redução da Demanda de Energia Elétrica utilizando Parâmetros Construtivos visando ao Conforto Térmico*. Dissertação para obtenção do Título de Mestre em Engenharia Elétrica, 123 p. Pontifícia Universidade Católica do Rio Grande do Sul. Porto Alegre, RS, Brasil.

Barbosa, J.C. and Ino, A. (2001), Madeira, Material de Baixo Impacto Ambiental na Construção Civil – Análise do Ciclo de Vida. In *Anais do II Encontro Nacional e I Encontro Latino Americano sobre Edificações e Comunidades Sustentáveis (ENECS)*. Canela, RS, Brasil.

Bogo, A. (2003), Avaliação de Desempenho Térmico de Sistemas Construtivos de Paredes em Madeira em Habitações. in *Anais do VII Encontro Nacional sobre Conforto no Ambiente Construído*, Curitiba, Brasil.

Breyman, Steve (2000), *Greening the Campus*. Available at: http://www.greens.org/s-r/ 23/23-25.html (accessed 25 March 2012).

BSU – Ball State Universtiy (2012), *Greening the Campus – Conference Series Mission*. Available at: http://www.cms.bsu.edu/Academics/CentersandInstitutes/GOC.aspx (accessed 29 March 2012).

Bucknell University (2012), *Campus Greening Council (CGC)*. Available at: http://www.bucknell.edu/x49294.xml (accessed 10 April 2012).

Carnegie Mellon University (2012), *Carnegie Mellon Green Practices*. Available at: http://www.cmu.edu/greenpractices/news/Timeline%20of%20Accomplishments.html (accessed 10 April 2012).

Ecotelhado (2011), *Soluções em infraestruturas verdes*. Available at: http://www.ecotel hado.com.br. (accessed 25 March 2012).

Escritório Verde (2011), *Acompanhe o andamento da obra*. Available at: http://www.escritorioverdeonline.com.br (accessed in 15 March 2012).

Giglio, T.G.F. (2005), *Avaliação do desempenho térmico de painéis de vedação em madeira para o clima de Londrina – PR*. Dissertação de Mestrado, Programa de Pós-graduação em Engenharia de Edificações e Saneamento. Universidade Estadual de Londrina, Londrina, Brasil.

Kobiski, B.V. and Casagrande Jr., E.F. (2011), *Comparativo de Emissões de Casas de Alvenaria de Interesse Social e Casas Utilizando o Sistema de Construção Energitérmica Sustentável*. Universidade Tecnológica Federal do Paraná, Programa de Pós-Graduação em Engenharia Civil, Curitiba, Brasil.

Laar, M., Souza, C.G., Paiva, V.L.A., de Nisete, A., Tavares, S., Grimme, F.W., Gusmão, F., Kohler, M., Schmidt, M. (2001), Estudo de aplicação de plantas em telhados vivos extensivos em cidades de clima tropical. In *Anais do 6º Encontro Nacional de Conforto no Ambiente Construído-ENCAC*. São Paulo, Brasil.

LP Brasil (2011), *LP Building Products*. Available at: http://www.lpbrasil.com.br (accessed 15 December 2011).

Mello, G.B. de Pereira, Costa, M.D.P., Alberti, M.S., Freita Filho, R.D. de Gadelha (2010), Estudo da implantação de um telhado verde na Faculdade de Engenharia Mecânica. *Revista ciências do ambiente on-line* 6 (2) (dezembro). Faculdade de Engenharia Mecânica, UNICAMP, São Paulo, Brasil.

Molina, J.C. and Calil Jr., C. (2010), Sistema construtivo em wood frame para casas de madeira. *Seminário Ciências Exatas e Tecnológicas* 31 (2), pp. 143-156 (jul./dez), Londrina, Brasil.

Motloch, J. and Casagrande Jr., E.F. (2010), The US-Brazil Sustainability Consortium: local-global collaboration for a sustainable future. *International Journal of Environment and Sustainable Development* 9, pp. 228-240.

Oliveira, T.A. and Ribas, O.T. (1995), *Sistemas de Controle das Condições Ambientais de Conforto*. L Série Saúde & Tecnologia – Textos de Apoio à Programação Física dos Estabelecimentos Assistenciais de Saúde – Sistemas de Controle das Condições Ambientais de Conforto. Ed. Brasília, Brasília, 92 p. Available at: http://bvsms.saude.gov. br/bvs/publicacoes/sistemas_conforto.pdf. (accessed 7 April 2012).

Onmura, S., Matsumoto, M., Hokoi, S. (2001), Study on evaporative cooling effect of roof lawn gardens. *Energy and Buildings* 33 (7) (September), pp. 653-666.

Palmer, J.A. (1998), *Environmental Education in the 21st Century: theory, practice, progress and promise*. Routledge, London.

Polidoro, E. (2009), *Iluminação Eficiente com LEDs*. Palestra no 6º Congresso Brasileiro sobre eficiência energética, 21 e 22 de julho de 2009. Available at: http://www.metodoeventos.com.br/6eficienciaenergetica/palestras/22_07_workshop2/eduardo_polidoro.pdf. (accessed 12 April 2012).

RCE-UNU (2012), *Regional Center of Expertise – United Nations University*. Available at: http://www.ias.unu.edu/sub_page.aspx?catID=108&ddlID=183 (accessed 17 April 2012).

Solar Decathlon (2012), *US Department of Energy*. Available at: http://www.solardecathlon.gov/index.html (accessed 18 April 2012).

Triciclo Pompeo (2011), *Fotos*. Available at: http://www.triciclopompeo.com.br/ (accessed 17 March 2011).

Tavares, S.F. (2006), *Metodologia de análise do ciclo de vida energético de edificações residenciais brasileiras*. Tese de Doutorado, Universidade Federal de Santa Catarina, Programa de Pós-Graduação em Engenharia Civil, Florianópolis, 225 p.

UNFPA (2007), *The State of the World Population 2007*, The United Nations Population Fund. Available at: http://www.unfpa.org/swp/2007/english/introduction.html (accessed 30 March 2012)

Vecchia, F. (2005), Cobertura Verde Leve (CVL): Ensaio Experimental. In *Anais do VI Encontro Nacional de Conforto no Ambiente Construído (ENCAC) e IV Encontro Latino-Americano sobre Conforto no Ambiente Construído (ELACAC)*. Maceió, Brasil.

Zenid, G.J. (2009), *Madeira: uso sustentável na construção civil*, Instituto de Pesquisas Tecnológicas do Estado de São Paulo (IPT 3010) (2° Ed.). São Paulo, Brasil.

– (2012), *Madeira na construção civil*, Instituto de Pesquisas Tecnológicas do Estado de São Paulo – IPT – Divisão Produtos Florestais. 08 pp. (s/d). Available at: http://www.celsofoelkel.com.br/artigos/outros/Madeira%20na%20constru% E7%E3o%20civil.pdf (accessed 10 April 2012).

Chapter 70

Towards a Sustainable Campus at *Universidad de Chile*: The Key Role of Students

Francisco Bucchi, Martín Jacques-Coper and Pablo Sánchez

Abstract

Institutions of higher education are key entities for developing a new sustainable development project for humankind: a paradigm change that is sorely needed. These institutions have the capacity to create knowledge and form technical and professional human resources, thus deeply influencing society as a whole. Following the example of other educational institutions worldwide, Universidad de Chile signed the Talloires Declaration in 1990, formally undertaking a series of concrete actions in order to face major environmental challenges. However, for many years these goals have remained a simple declaration of intention, and have not been actively pursued. In this context, Oikos, a student group at the Faculty of Physical and Mathematical Sciences (FCFM) of Universidad de Chile, has been working since 2005 to make this initiative a reality. After gaining some experience developing projects at the FCFM Campus – for example in recycling, flora identification, proposing the creation of a minor in Renewable Energies (which was accepted and successfully implemented) and establishing a Solar Energy Workshop – this group has taken the first steps to demand that the faculty develop its own concept of campus sustainability, involving all the different dimensions that make up a university: teaching, extension, research and management of material resources. Following an agreement with the Faculty Council, a *Sustainable Campus Commission* (SCC) has been created, with the participation of different university bodies: professors, students and employees. In this paper the authors present the case study of FCFM and review the experience gained by Oikos since its foundation, as well as the maturation achieved during the planning and implementation phases of sustainable campus initiatives. The key role of students in implementation is also discussed, as well as the obstacles and opportunities faced in the particular idiosyncrasies of an Engineering and Science Faculty. Valuable lessons are drawn from this process, including regarding the development of an effective strategy for achieving the objective: to advance from small victories to systemic transformations. Finally, further steps towards the materialization of the sustainable campus are described.

Introduction

The role of educational institutions in disseminating the "sustainable development" concept and in creating awareness of the urgency of its implementation, are indisputable. In particular, Higher Education Institutions (HEI) are the key entities for elaborating a new sustainable development project for humankind: a paradigm change that is sorely need. These institutions have the capacity to create knowledge and form technical and professional human resources, and thus to deeply influence society as a whole (Sharp 2002; Conde Hernández et al. 2006). HEIs not only pursue knowledge, they also apply knowledge in order to solve the complex problems of society (Sharp 2002). The approach to sustainability of HEIs has been conceptualized in the Sustainable Campus, which addresses transformations in research, outreach, and teaching (ULSF 1990).

A systematic transformation leading to the materialization of a Sustainable Campus must be responsive to the existence of the three predominant university bodies – or *subcultures*, according to Sharp (2002) – which exist within universities: faculty, administrative staff and students. Evidence suggests that the greatest leverage in achieving institutional change occurs when all subcultures have a shared vision and a sense of organizational alignment (Sharp 2002). This seems logical, considering that the change process involves several levels of organization and a variety of actors. Nevertheless, the participation of these actors will not necessarily be synchronized in the pursuit of a Sustainable Campus, and their respective contributions will depend on their motivation, circumstances, and capacities (decision-making practices, time constraints, priorities, opportunities, threats and experiences). There are successful examples of Sustainable Campuses led by student organizations (bottom-up model), by authorities (top-down), and by academics and staff (middle-out) in different contexts, each having their own idiosyncrasies, but all sharing common obstacles and strengths (Owens and Halfacre-Hitchcock 2006; Brinkhurst et al. 2011).

Student-led campaigns can be catalysts for impressive change, by building support for new initiatives and applying pressure for change in university policies and operations, which are later implemented by the community (Brinkhurst et al. 2011). The scope of these initiatives varies widely, from specific targeted campaigns, such as recycling promotion, to calls for deep change in policies or systematic campus operations (Helferty and Clarke 2009).

Some common cultural elements shared by student-led campaigns are summarized by Sharp (2002): I) a shared sense that older generations and powerful institutions are failing to take action to address environmental and social issues that are of significant concern to the younger generations; II) a lack of timely access to information about the way the university works; III) little guidance on how students can most effectively influence organizational decisions; and IV), absence of effective means to ensure a continuum of organizational learning between multiple student generations. These elements could result in student campaigns being limited to focusing on short-term projects, and not advancing to long-term systemic transformation (Sharp 2002). When students assume a leadership role towards a Sustainable Campus, these cultural elements emerge as barriers. Similarly, Brinkhurst et al. (2011) considers that the main barriers in the bottom-up leadership experiences are: I) high turn-over rate and variable continuity, with dependence on individual personalities; II) lack of institutional knowledge and skills for navigating in the bureaucracy, and a short-term outlook, and III) ineffective design of the initiatives, due to narrow-focus projects and lack of systemic vision. However, successful examples exist, and lessons have been learned to overcome these barriers and to sustain student initiatives. Foremost among these lessons, Owens and Halfacre-Hitchcock (2006) and Helferty and Clarke (2009) highlight the following: I) developing a continuous learning curve for undertaking more effective campaigns; ii) analyzing stakeholder relationships on campus to define the best approach to accomplish the Sustainable Campus goals, and III), creating partnerships with more permanent members of the campus community or with members having specific expertise.

The objective of this paper is to review the experience gained during the planning and implementation of Sustainable Campus initiatives at Universidad de Chile, and to discuss the role of students in its implementation. The campus in question is Beauchef Campus, which houses the Faculty of Physical and Mathematical Sciences (FCFM), and which recently adopted the institutional goal of developing a Sustainable Campus, as the culmination of a process promoted since 2005 by Oikos, a student environmental organization. Valuable lessons are drawn from this process, including the development of an effective strategy for the maturation of the objective: to advance from small victories to systemic transformations.

Background

Following the example of other educational institutions worldwide, two Chilean universities – Universidad de Chile and Universidad de Santiago de Chile – subscribed the Talloires Declaration in 1990 (ULSF 1990), undertaking a series of concrete actions in order to face major environmental challenges, thus recognizing themselves as relevant actors in this regard. Until now however, the main aspects of the Talloires Declaration have not been implemented either at Universidad de Chile as a whole, or at FCFM.

The study case of Universidad de Chile

Universidad de Chile, founded in 1842, is the major and oldest institution of higher education in Chile. It is one of the 16 state-owned universities, and is characterized by its public and national vocation. It offers a broad range of undergraduate and graduate programs, covering all areas of knowledge. The relevance of Universidad de Chile also lies in its tradition of service to the country, as established by Andrés Bello (its founder and first rector) in the university's declaration of its mission. The student body numbers are approximately 27,000 undergraduates (2010) and ca. 8,600 graduate students. In addition, it is the country's leading research institution, concentrating 37% of the country's scientific publications (ISI) in 2010. Within Universidad de Chile, FCFM is one of the largest faculties, offering Engineering and Science careers, as well as Astronomy, Physics and Earth Sciences.

Nevertheless, one of the main shortcomings of Universidad de Chile is the lack of environmental policies. This situation has begun to change in recent years, as various initiatives have attempted to establish a Sustainable Campus at the university. The cases of three faculties stand out, namely those of Economy and Business (FEN), Architecture and Urbanism (FAU) and Physical and Mathematical Sciences (FCFM). They have all set out on the road towards sustainability by forming commissions whose aim it is to include Sustainability in each campus' development plan.

At FEN, a Sustainable Campus Commission (SCC) was created in 2011, made up of staff and students. Formally, it is hosted by the department in charge of corporate social responsibility. Its main achievements are related to teaching and waste management. On the other hand, at FAU, the academic staff are the main promoters of sustainability practices and policies, concentrating their efforts on including sustainability in the curriculum. Finally, at FCFM a SCC was created in 2010 as the result of an agreement by the Faculty Council to advise the Dean and the Council on the development of a Sustainable Campus. In fact, this was the result of a student initiative led by the Oikos group. Academic and administrative staff, as well as students, participate in this commission which has taken various concrete steps in relation to waste management, energy efficiency, and teaching.

The History of Sustainable Campus at FCFM

Oikos was born in 2005 as an organized group of students committed to implementing a recycling program at FCFM. This original motivation attracted the first members, who that year prepared a first proposal for regular competitive funding for students' projects at the faculty. The main objective was to "*create awareness in the student body about the importance of reducing the waste we produce, and about the related impact on our quality of life that can be achieved through this simple task ...*" The total sum requested was approximately USD 200, intended for purchasing ten waste bins for separating glass, aluminum and plastic waste. It is worth noting that, even at that early stage, the proposal stated that a crucial aspect was "*to ensure that this program will last and be recognized as a priority by future student associations and by the faculty administration*". In this way, the young group envisioned the necessary step of institutionalizing this important program. Oikos then installed "recycling points" in strategic locations within the campus and directly administered these waste bins during the initial stages: glass was collected and carried by students to external collecting points outside the FCFM buildings – which constituted a hazardous task. Plastic was directly donated to recycling companies that supported charity programs, while aluminum cans were sold – if there were any left, because they were dearly appreciated by cleaning personnel for selling to commercial recyclers. Furthermore, apart from an ambitious and creative information campaign, the group's commitment to the project included washing and maintaining the waste bins: every member of the group "adopted" one of the bins. The next aim of the project consisted of collecting and recycling paper. For this purpose, a large commercial company was contacted and small paper collectors were placed in classrooms, corridors and administrative offices. Paper was regularly collected by students and taken away by the commercial company. However, this was just a pilot plan, and organized in this manner it couldn't evolve beyond being a kind of anecdote. As stated in the first proposal, the project had to be institutionalized and administered by FCFM as part of an *integral waste management program*. Otherwise, the initiative was condemned to failure. For successful implementation, a serious analysis of the faculty's real situation had to be conducted. For instance, the informal methods of collecting paper and aluminum that were already employed by employees (in order to increase their incomes) had to be revised. As a result, in 2007 (i.e. two years later), a formal proposal for an integral recycling program was presented to the faculty by Oikos. This represented a great qualitative advance for the consolidation of the group and its impact on FCFM.

However, during the early years some restrictions were imposed by FCFM authorities, which limited the group's activities. The restrictions were related mostly to the group's representativeness (Oikos was just an organized group and not a part of a student association), and to the validation of its work. The first aspect was overcome by establishing a working relationship with the FCFM Student Union (CEI), for whom environmental issues were important. Thus, the initiatives were now organized by Oikos and supported by CEI. Concerning the validation of the group, Oikos learnt – after its initial naivety – how to formally develop projects related to Campus Sustainability and how to present them to FCFM authorities, always completing the necessary steps to successfully achieve its goals. In other words, the group presented every new initiative as a mature project, highlighting its relevance, goals, expected results and associated costs. Moreover, a detailed description of past activities and their impact were always presented, so that the achievements of the group were always taken into account. This procedure reflects the idiosyncrasy of a faculty devoted to Engineering and Sciences, such as FCFM.

The group gradually grew – mostly as a result of the impact of its activities, and of a number of welcome sessions organized every semester to attract interested students, plus its members reinforced the idea that recycling was just a small part of a major challenge: the *Sustainable Campus* (which actually began to be used as the group's slogan). The most relevant influences in this regard were reading the Taillores Declaration, and the revision of similar experiences worldwide, especially at Princeton University (2008), the University of California Santa Barbara (2008), Brown University (2012), and Tecnológico de Monterrey (2012) in Mexico. The discovery of specialized literature was highly inspiring: besides crucial papers by IJSHE, there were those of Sharp (2002), Conde Hernández et al. (2006), Sharon and Wright (2006) and Ortiz et al. (2008).

Oikos also "learned by doing", and gradually matured the objectives and the challenges it set for itself, for the short and long term. Among the short-term objectives were:

- To create an instance for students to exchange ideas related to environmental issues, and which would serve as a link to FCFM professors and researchers dedicated to related disciplines and concerns;
- To become an information channel for environment-related subjects inside the university community, facilitating access to the studies on ecology, the environment, etc., that are generated at our university;
- To develop our own concept of Campus Sustainability, which must include such aspects as an interdisciplinary approach, participation, and democracy, and to implement said concept.

The long-term tasks included the following:

- To become an environmental point of reference at university level, with the participation of students, academics, and employees;
- To develop critical and reliable points of view on environmental issues;
- To have Environmental Sustainability included in the study curricula, in order to ensure that this conceptual framework becomes familiar to students;
- To demand from FCFM and university authorities that they adopt strict standards concerning environmental aspects (e.g. management of energy and material resources, etc.).

As already mentioned, the group launched and developed other types of projects and activities related to environmental sustainability at FCFM, based on our understanding that, for a real Sustainable Campus to become a reality, major transformations had to take place in various dimensions: the material aspect (physical transformations of infrastructure, etc.); curricula; and research and extension (interaction between the university and society as a whole). The most successful projects are described in Table 1.

Table 1: Description of main student-oriented activities led by Oikos during its first years, aimed at raising environmental awareness

Project	**Description**	**Frecuency/Quantity**
Sustainable Library	With the aim of improving our theoretical basis, books were purchased to establish the Sustainable Library. Nowadays, it comprises 51 volumes, covering subjects such as Ecology, Sustainability, Waste Management, Renewable Energies, etc. Following the strategy of institutionalization of our projects, this collection was implemented as a part of the Central Library of FCFM.	7-8 books added each year. Books available permanently in the FCFM Central Library.
Parallel Courses	A strategy was used in order to review the different topics covered by titles in the Sustainable Library. It consisted of distributing the titles among the members of Oikos, so everyone could read one and prepare a presentation highlighting its main aspects. Those presentations were widely promoted among the FCFM Community. Later on, the parallel courses covered a wider diversity of subjects, not necessarily related to the available books at the Library.	3-4 parallel courses per year

Project	Description	Frecuency/Quantity
Vocational Talks	With the aim of giving the students of the first years an outlook on the different specializations available at FCFM, Oikos organized a Vocational Talk Row dedicated to the existing links between these specializations and environmental issues. Professors from different Departments were invited to describe how a particular professional should deal with these aspects.	3 Vocational Talks in total
Specialized Talks	Oikos organized presentations dedicated to discussing environmental subjects of national and global relevance. For this, specialists were invited to lecture to the Community.	2 Specialized Talks per year
Activism	Additionally to the described specialized talks, subgroups of Oikos were formed to divulge particular Chilean environmental conflicts among the FCFM Community and external society; such is the case of "Ahora es Cuando" ("Now is the time"), committed to discussing the impacts of the HidroAysén Dams project in the South of the country.	During 2009
Flora Identifica-tion	As part of a didactic initiative by the FCFM authorities, Oikos led a project to identify vegetation species in the Campus, which were described by means of small signs (a virtual tree map can be visited at http://www.publisiga.cl/produccion/uchile/mapa_arboreo/)	A total of 47 species were identified
Earth's Day	A major event is organized in April of each year, comprising talks by FCFM Professors and Researchers, and open-air activities such a mural painting showing the priority projects of the group and the planting of a tree	Annualy

Starting in the group's second year, members of Oikos aimed to establish alliances with academic partners interested in issues related to Campus Sustainability. As one of the early results of this "lobbying" effort, another crucial opportunity for Oikos presented itself in 2007: four Oikos members were invited by Prof. Roberto Román (Dept. of Mechanical Engineering) to participate in EVER07, a summer school on renewable energies organized by Chilean and German universities. There, the students learned how to build solar cookers and water heaters, and recognized the potential for divulging low-cost solar artifacts amongst the Chilean population. After a successful experience during the 2008 Voluntary Work Program organized by the University Student Union (FECh) in rural areas of Northern Chile, a Solar Workshop was established to continue this interesting initiative. The Solar Workshop has continued to develop the initiative as an independent organization (see Jacques-Coper et al. [2012], in this volume).

Also on this same subject of renewable energies, one of the most relevant achievements of the group consists of some changes introduced in the curricula. As part of a new curricular structure, minors began to be offered in 2009. This was an ideal opportunity to formalize a program devoted to renewable energies. Thus, after discussions with professors who were active in different aspects of this field, such as solar, wind, hydro, geothermal, and biomass energy, a detailed program was presented to FCFM. After the implementation of this minor in renewable energies, its success has been such that it has become the most popular among students (see Table 2 for statistics).

Following this approach of establishing strategic alliances, Oikos invited similar groups devoted to environmental issues at other Universidad de Chile faculties to come together to exchange experiences and to undertake common initiatives. Thus, EcoUchile was born. However, as Oikos had already found out, it was necessary to clearly establish the representativeness of this new organization. As a result, the Secretariat of Ecology and Environment (SECMA) was created as part of the University Student Union (FECh).

Table 2: Lectures comprised in the renewable energies minor proposed by Oikos jointly with academics. Note that, compared to the historic mean of 20 registered students per semester, the average enrolment in course GF3003 increased to approx. 350% due to this program

Lecture	Spring Term 2009	Autumn Term 2010	Spring Term 2010	Autumn Term 2011	Spring Term 2011	Accumulated Total
ME6000 Introduction and Applications of Solar Energy	–	40	–	–	40	80
BT4551 Renewable Biomass Energy	–	–	28	–	24	52
EL6000 Electricity Generation from Renewable Sources	50	64	25	24	–	163
GF3003 Introduction to Meteorology and Oceanography	81	86	55	92	45	359
GL4009 Principles of Geothermal Energy	–	–	40	–	–	40
Semester's total	**131**	**190**	**148**	**116**	**109**	

Sustainable Campus Commission (SCC)

The experience gained during the early years of designing and implementing projects aimed at transforming FCFM into a Sustainable Campus, gave Oikos sufficient maturity to recognize the need for a change of strategy, in order to have the projects successfully implemented as part of the programs undertaken by the institutional authorities of the faculty. Thus, a crucial step was to consolidate the alliance with academic partners.

As stated in the aims of Oikos, a democratic three-tiered body (i.e. including FCFM students, academic staff, and employees) was to be the logical configuration of the strategic organization that was needed to consolidate the road towards Campus Sustainability. A starting point for justifying the creation of the Sustainable Campus Commission (SCC) was the fact that Universidad de Chile had subscribed the Talloires Declaration and that, as a result, FCFM had a commitment to pursuing its objectives. Thus, in the mid-2010 Oikos and two allied professors drew up a proposal for faculty authorities to consider the creation of this commission, which was to have the form of an advisory committee presided by the dean. The proposal included the hiring of a professional as *Person Responsible for Campus Sustainability*, whose mandate was to consist of implementing the SCC resolutions. The crucial task of the SCC was to elaborate our own, original concept of Campus Sustainability. For this purpose, the first SCC activity was to review the international experience in this field. As a second task, an environmental assessment of FCFM was to be conducted, in order to evaluate how to proceed to achieve proper formal certification on environmental issues. Then, a schedule of short-, middle- and long-term tasks had to be developed in order to finally establish FCFM as a Sustainable Campus. Of fundamental importance is achieving the commitment of FCFM to incorporate these goals into the Faculty's Institutional Development Plan.

The commission was established in January 2011. It was first formed by students and academics, while employees would be included at the end of that year. However, the proposal for hiring a *Person Responsible for Campus Sustainability* was not accepted by the dean, who proposed that, instead, three students were to be paid to execute SCC projects during 2011. The approval of this proposal was a major achievement on the road to Sustainability at FCFM, and it reflected a process of maturation of the strategy of the Oikos group. Indeed, in order for the SCC to be established, it was necessary, apart from relying on the experience already acquired by Oikos, to achieve an in-depth understanding of the idiosyncrasy of FCFM. As such, and in a similar way to Oikos' earlier initiatives, the SCC had to be formulated as an engineering project, considering quantitative elements and clearly identifying the indicators that had to be improved. This aspect of the formulation of the project surely played a major role in the success of this initiative. On the other hand, the strategic alliance with academics contributed institutional knowledge that could not have been achieved in any other way, which supports the views of Brinkhurst et al. (2011).

Discussion

Maturing of objectives

Since 2005, most of the "green" initiatives created or proposed by students at FCFM have been carried out by Oikos. These initiatives have evolved from simple awareness-raising (recycling, green weeks, sustainable library, and conferences) during the initial stages, to include the Proposal for Institutional Sustainable Policies (e.g. Waste Management Program, SCC, courses on sustainability). This maturation of Oikos has been the result of long-term internal discussions guided by the concept of Sustainable Campus, which reached the conclusion that the true role of FCFM within a sustainable society could only be played through achieving systemic transformations. A key part of this maturation process was the study of the theoretical background of the Sustainable Campus and the analysis of successful experiences worldwide.

Student contributions

At FCFM, the main contributions made by students along the road to Sustainability have been (see scheme in Fig. 1):

I. The development of initiatives focused on students: these initiatives raise awareness and help divulge the concept of Sustainable Campus. Simultaneously, these initiatives validate Oikos in the eyes of authorities.
II. Proposals for changes in university policies, and applying pressure for their implementation: Students can go beyond criticizing existing conditions, since they can play a relevant role in finding solutions with a propositive-proactive attitude. Proposals are oriented to institutionalizing the initiatives, and demonstrating the need to make systemic transformations. On the other hand, the generation of a three-tiered instance of dialogue and decision-making (SCC) reflects the evolution towards a more democratic paradigm. Moreover, success is more likely and pressure on university community is exerted more effectively, through an instance such as SCC.
III. Uniting and coordinating efforts made by academics: In the past – and also at present –, some research and teaching was carried out in relation to sustainability, but not under the common theoretical framework of sustainability. The contribution of students consists of creating synergy and coordinating these efforts under the common objective of Sustainable Campus.

Fig. 1: Scheme of the various incidence levels of Oikos towards a Sustainable Campus at FCFM

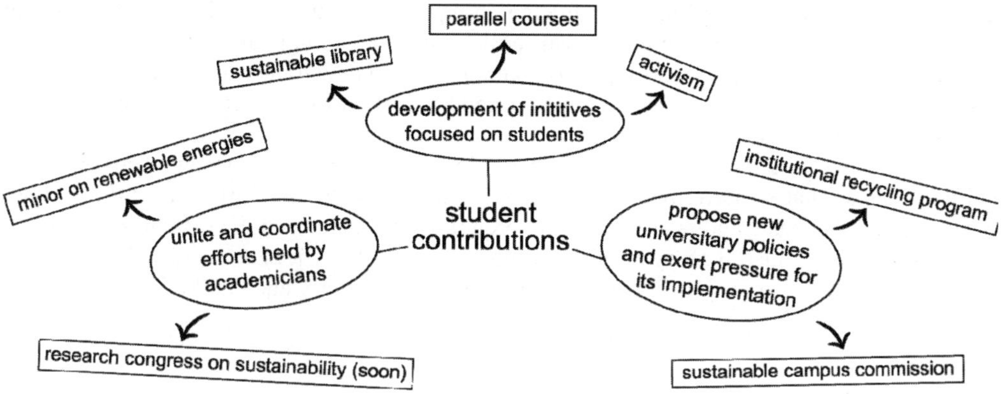

Strategy

The construction of a Strategy on the road toward a Sustainable Campus should consider the variety of subcultures, the development of alliances, and the local idiosyncrasies.

Each university and faculty has a particular idiosyncrasy, and a particular "niche" that it occupies in the ecology of higher education (Clugston and Calder 1999). In general, successful initiatives must fit in with this institutional identity. As an example of this, for our initiative to be supported by authorities at FCFM, it had to be formulated as an engineering project.

The materialization of the Sustainable Campus is a challenge for the whole community, and every single body and department of the university has to play its own, specific role. In fact, expectations can be either unique to each body, and/or common to the whole university. The same applies to the relevance of, and the valuation criteria for each Sustainable Campus initiative. In this sense, the main factors that determine each subculture's valuation criteria on initiatives, as evidenced in our experience, are: I. for the authorities: excellence, image-improvement, and economic benefits; II. for academics: scientific rigor and theoretical background of the initiatives; III. for employees and administrative staff: possible changes in their work conditions; IV. for students: improvement of university environmental issues and materialization of a common ideal. In a systemic transformation, such as the one needed to create the Sustainable Campus, better results are achieved if the different viewpoints are aligned. Therefore, the SCC is a space for dialogue and for sharing and discussing the implications of initiatives, and to reach consensus on a common strategy.

Keeping this in mind, strategic alliances have been formed with the faculty and Student Unions to achieve institutional-level implementation. In this sense, joint work with FCFM Student Union (CEI) led to institutionalization of the Recycling Program. Another example is the alliance created with academics to coordinate and develop the proposal for a minor in renewable energies.

Obstacles

Awareness-raising activities are one of the most common forms of student initiative, rather than the concrete changes in policy or planning needed to reform operational practices (Brinkhurst et al. 2011; Helferty and Clarke 2009). This reflects the barriers and asymmetries faced by students when they try to act as decision-makers. Barriers in the early years of Oikos were mostly related to its perceived representativeness and to the validation of its work. The alliance with the Student Union overcame the un-representativeness of Oikos. The validation process is a continuous task, mainly because the relevance, complexity, and implications of the sustainability proposals tend to increase, and to involve new topics and decision levels. In this sense, even though Oikos has obtained a good valuation by the authorities and students over six years, Oikos should be wary of resting on its laurels.

As Oikos-led initiatives have become institutionalized, the topics of new proposals have shifted to areas that so far remain unattended. Examples of this kind of topic shift, in other countries, are the waste audits and recycling programs in Canada, conducted ten years ago by students and now undertaken by administrative sustainability offices or mandated by the government, therefore changing the focus of student concerns (Helferty and Clarke 2009).

Conclusions

An effective process of transforming the FCFM Campus into a Sustainable Campus is being catalyzed by the student organization Oikos. As a result, Universidad de Chile (UCh) is now implementing the "sustainable development" concept and creating awareness of the urgent need for society to follow suit.

The initiatives undertaken by Oikos have evolved from simple awareness-raising activities (e.g. vocational talks, green weeks, etc.) during the initial stages, to concrete proposals for Sustainable Institutional Policies (e.g. Waste Management Program, etc.) aimed at systemic transformations in the Campus. In this maturation process, careful analysis of successful experiences worldwide and the study of theoretical aspects of the Sustainable Campus were key elements.

In synthesis, the main contributions made by students towards a Sustainable Campus have been: the development of initiatives focused on students; proposals for changes in university policies together with applying pressure for their implementation; and coordinating efforts made by academics. These contributions are oriented towards institutionalizing the initiatives and demonstrating the need to make systemic transformations in the Campus.

Main lessons learned about strategy

Some of the crucial lessons learned, and which should be taken into account when dealing with projects of a similar nature, are as follows:

- In initial stages, students should establish alliances with academic staff, Student Unions, and university employees to achieve collaboration, obtain feedback, and exert pressure on authorities.
- It is essential to develop an original concept of Sustainable Campus suited to the local reality. In this task, reviewing international experiences is a good starting point.
- In most of the campuses, various initiatives related to sustainability already exist. These should be identified and coordinated under the general concept of Sustainable Campus. By doing this, it will become evident that several short-term goals have already been accomplished.
- When dealing with authorities, submitted projects should be mature and take into account local idiosyncrasies. In the case of certain initiatives, it is useful to start by considering economic arguments or criteria, but the vision of systemic and holistic transformation should be always present.

- A shared strategy on the part of the university community is the only way to make systemic and holistic transformations. Different expectations, perspectives, and valuation criteria held by university bodies should be conciliated to build the common strategy.
- In the same sense, better results are achieved when different viewpoints are aligned. The Sustainable Campus Commission is a space for dialogue and for sharing and discussing the implications of initiatives, and therefore for reaching consensus on a common strategy.
- In order to make continuous progress, it is necessary to elaborate a realistic program of activities comprising short-, medium- and long-term achievable tasks which can be assessed, following pre-established valuation criteria.

Future perspectives for FCFM Sustainable Campus

Subsequent developments in FCFM Sustainable Campus should include progress in each major area:

- *Teaching*: I. reviewing existing curricula in order to diagnose specific needs or gaps regarding sustainability; II. creation of a "Sustainability in the Campus" course, oriented to second year students of the common program (to be given in the second semester of 2012).
- *Research*: I. similarly, reviewing existing research programs to diagnose specific needs or gaps regarding sustainability; II. organizing a Research Congress in Sustainability, to create synergy and improve communication among researchers at Universidad de Chile.
- *Campus infrastructure*: I. improving the existing recycling program; II. organizing 2012 competitive funding for sustainable initiatives at FCFM, following up on the succesful 2011 version.
- *Outreach*: I. ensuring permanent divulgation of the Sustainable Campus Project; II. establishing effective channels of communication with the community, in order to involve a greater number of people.

To achieve systemic transformations at FCFM, it is essential to draft an "Sustainable Campus Plan" for FCFM, with a schedule of short, medium and long-term tasks. It is of fundamental importance that FCFM commits itself to incorporating the Plan into the Faculty's Institutional Development Plan. Finally, a multidisciplinary vision of a Sustainable Society, developed by Higher Education Institutions in Chile, will only be achieved if this project is expanded to Universidad de Chile as a whole.

Acknowledgments

The various achievements recounted in this article have been possible thanks to the valuable work of an enormous group of people. This article is thus dedicated to all the members of Oikos through its seven years of existence, for their tireless dedication. The authors also want to thank Prof. Roberto Román, Prof. Rodrigo Palma B. and Prof. Luis Vargas, who have advised and supported various initiatives of Oikos; Prof. Judit Lisoni and Prof. Rodrigo Espinoza, for their motivation in the beginnings of the "FCFM Sustainable Campus Project"; and professors, students and staff who conform to the Sustainable Campus Commission. Many thanks to the Deanship and Vice Deanship of the FCFM for believing in this proposal. Last but not least, the authors are immensely grateful to Giancarlo Bucchi, who enthusiastically made possible the proper redaction of this article. F.B. would like to thank FCFM for the financial support to attend the WSSDU-2012 meeting.

References

Brinkhurst, M., Rose, P., Maurice, G. and Ackerman, J.D. (2011), Achieving campus sustainability: top-down, bottom-up, or neither? *International Journal of Sustainability in Higher Education* 12 (4), pp. 338-354.

Brown University (2012), *Brown is Green*. Available at: http://www.brown.edu/about/ brown-is-green/ (accessed 25 January 2012).

Clugston, R.M. and Calder, W. (1999), Critical dimensions of sustainability in higher education. In Leal Filho, W. (Ed.), *Sustainability and University Life*, pp. 31-46. Peter Lang, Frankfurt.

Conde Hernández, R., González Castillo, O. and Mendieta Márquez, E. (2006), Hacia una gestión sustentable del campus universitario. *Casa del Tiempo* 8 Ep. III, No. 93, pp. 15-25.

Helferty, A. and Clarke, A. (2009), Student-led campus climate change initiatives in Canada. *International Journal of Sustainability in Higher Education* 10 (3), pp. 287-300.

Jacques-Coper, M., Sánchez, P., Frodden, E., del Campo, F., Sánchez, J.I., Oyarzo, D., Macías, J., Bucchi, F. and Aguilera, I. (2012), Solar Power to the People: the Solar Energy Workshop at Universidad de Chile. In Leal Filho, W. (Ed.), *Sustainable Development at Universities: New Horizons*. Peter Lang, Frankfurt (this issue).

Ortiz, B., Rodriguez, I. and Guadarrama, S. (2008), *Gestión ambiental universitaria: hacia un campus sustentable. paper presented at Proceedings IV Seminario Internacional Universidad y Ambiente, Gestión Ambiental Institucional y Ordenamiento de los Campus Universitarios*, October 25-26, 2007, Bogotá, Colombia. Available at http://www.ariusa.org/memoriaseminarioIVudca.pdf (accessed 25 January 2012).

Owens, K.A. and Halfacre-Hitchcock, A. (2006), As green as we think? The case of the College of Charleston green building initiative. *International Journal of Sustainability in Higher Education* 7 (2), pp. 114-28.

Princeton University (2008), *A Sustainability Plan for Princeton*. Available at: http://www.princeton.edu/sustainability/plan/ (accessed 25 January 2012).

Sharon, T., Wright, A. (2006), Giving "teeth" to an environmental policy: a Delphi Study at Dalhousie University. *Journal of Cleaner Production* 14 (9-11), pp. 761-768.

Sharp, L. (2002), Green campuses: the road from little victories to systemic transformation. *International Journal of Sustainability in Higher Education* 3 (2), pp. 128-45.

Tecnológico de Monterrey (2012), *Programa Campus Sostenible*. Available at: http://campus-sostenible.mty.itesm.mx/ (accessed 25 January 2012).

ULSF (University Leaders for a Sustainable Future) (1990), *Talloires Declaration*. Available at: http://www.ulsf.org/pdf/TD.pdf (accessed 25 January 2012).

University Of California Santa Barbara (2008), *Campus Sustainability Plan*. Available at: http://sustainability.ucsb.edu/plan/ (accessed 25 January 2012).

Chapter 71

Sustainability as an Integration Proposal for Teaching: A Study of the Pedagogical Projects of the Management Courses at the Federal Institute of Education, Science and Technology of Rio Grande do Sul (IFRS)

Lisiane Celia Palma and Daniela Rodrigues da Silva

Abstract

The discussion about how to introduce concern about sustainability issues into students' education at different levels and in several areas of knowledge, is essential to the development of socio-environmental awareness. The present study intends to start a debate about how the issue of sustainability has been considered in the Federal Institutes of Education, Science and Technology (IFs) in Brazil, focusing on the management courses offered in the state of Rio Grande do Sul, and more specifically at the Federal Institute of Education, Science and Technology of Rio Grande do Sul (IFRS). Therefore, the study presents the main results and analyses from research on how sustainability has been treated in the pedagogical projects of the management courses at the IFRS. Research findings show that the inclusion of sustainability in the curricula of management-related courses is already happening at the IFRS, though not in all campuses. Nevertheless, it was observed that, of all courses analyzed, only a few mention sustainability in their pedagogical projects, generally in a superficial way. However, there are already some positive actions. In this sense, a project carried out at the IFRS Canoas with a view to fostering integration and making sustainability a core theme, is also presented.

Introduction

While management studies have been oversimplified by a pragmatic and deterministic orientation, the world's current state of affairs demands more complex solutions. For this reason, it is necessary to introduce new elements in business administration programs in order to prepare future managers to act in a more conscious and flexible way, given the complex and fast-changing character of the organizational environment, which is now dialogically guided by competitiveness and sustainability, creating a complex relationship between individuals, organizations and society.

From this perspective, educational institutions can contribute by treating the study of business administration in a broader sense, going beyond the economic sphere to encompass social, environmental and cultural issues as well. By doing so, future decision makers within organizations will be better prepared to broaden their views, thus expanding the organizational scope. As a consequence, organizations will aim at generating more than financial profit, that is, they will function more like sustainability agents.

The discussion about how to introduce, in different levels of the educational system, the concern with sustainability issues as a part of the formative process of business administration students is imperative. Although it is already possible to identify the insertion of topics related to this theme; in some business administration courses, the process is still incipient, limited, in most of the cases, to a single discipline (optional, at times) dealing with the subject or with related topics, like environmental management and corporate social responsibility. This is demonstrated in a study involving 40 Brazilian federal universities, which shows that only 13 business administration programs (around 33%) offer sustainability-related disciplines (Palma et al. 2011).

However, the issue seems to be increasingly more present in educational discussions. In this sense, it is important to mention the Federal Institutes of Education, Science and Technology (IFs). Created in 2008, their guidelines present some concern with the theme, such as "fostering an academic practice that contributes to the development of social, environmental and political awareness, forming professional-citizens" and "participating critically in projects that aim at all dimensions of local sustainable development" (IFRS 2011a).

Considering this, the present study's main objective is to identify how this issue has been addressed in the IFs, more specifically in the Federal Institute of Education, Science and Technology of Rio Grande do Sul (IFRS). To this end, a survey was carried out to verify if the theme of sustainability has been included in the Pedagogical Projects of the Courses (PPCs) related to the area of management offered by IFRS. The study also proposes an Interdisciplinary Project on sustainability for the Technical Course in Management Integrated to High School of the IFRS – Campus Canoas, in order to start an integration process with the pedagogical project of that course, which can be used as reference in other courses and teaching levels.

Sustainability and Education

The Industrial Revolution and, later, the beginning of the so-called "Ford society" and its mass production and consumption patterns, contributed to aggravating environmental problems, which became increasingly visible in society. As a consequence, all over the world, a series of events that helped raise awareness about the environmental problem took place (Souza 2000; Dias 2006; Barbieri 2007).

Among these was the Un's noteworthy creation of the World Commission on Environment and Development in 1983. In 1987, it published the report entitled "Our Common Future", also known as the "Brundtland Report", which officially recognized the concept of Sustainable Development – the commitment to meet present needs whilst granting future generations the right to satisfy their own necessities. This act emphasizes the importance of adopting a new global developmental pattern. The document was a reference guide for the debates at the United Nations Conference on Environment and Development, the ECO-92, which took place in Rio de Janeiro in 1992 and popularized the concept of sustainable development, thus linking inextricably environmental and developmental issues (Dias 2006).

Sustainability, however, means different things for different audiences. Some authors debate this question by presenting distinct definitions and interpretations on the theme (Gladwin et al. 1995; Leal Filho 2000; Hopwood et al. 2005), as well as the belief in different strategies to achieve sustainable development (Graaf et al. 1996; Veiga 2005).

Gladwin et al. (1995: 877) researched publications about this topic and were able to identify some key implicit components in most of the concepts: inclusion, connectivity, equality, prudence and safety. They claim that, in fact, development is not sustainable when human growth chooses to exclude, disconnect, promote inequality, reflect imprudence or increase insecurity.

Over the years, the pillars of sustainability (economic, social and environmental) have been complemented and enlarged. In this respect, Steurer et al. (2005) state that, at present, Sustainable Development is considered a process that also encompasses notions of transparency and participation, reflexivity, integration and equality between generations.

Leal Filho (2000: 16) presents some factors that influence attitudes towards sustainability (Table 1). It should be noted that many of these factors can be influenced by education.

Table 1: Some factors which influence attitudes towards sustainability

Knowledge	Information on the meaning of sustainability and its implications
Background	The nature of his/her training often influences an individual's degree of receptivity in relation to sustainability
Experience	Previous experience with environmental and social affairs facilitates understanding of the role of sustainability
Perception	The integrated view of environmental, political and economic elements enables a broader perception of sustainability
Values	Differing from the previous ones due to their high degree of complexity, an individual's values often determine whether his/her attitudes are favourable or otherwise
Context	Sustainability is not only related to ecological components *per se*, but also entails items such economics, politics and social matters. However, links with the latter are often ignored by universities.

Source: Leal Filho (2000)

In the ECO-92, education was considered one of the core forces to promote sustainable development in the 21st century. At present, the serious commitment to sustainability and education remains as a highly significant goal to be achieved. The increasing levels of poverty, insecurity and global inequality remind us that the targeted economic development is far from being achieved. Universities play an important role in this process, for they can no longer simply observe, criticize and evaluate what is going on outside the classroom or the campuses; they are global players that have become responsible for generating knowledge and wealth as well as fostering poverty and insecurity due to their increasing role as servants of the global economy. Higher education, therefore, helps shape the world we live in and also the ways through which we attempt to understand, reflect on and, perhaps, even change it (Blewitt and Cullingford 2004). It can be added that it is not only universities that play this role, but all educational instances bear the same responsibility, for they exert strong influence on every stage of the process of building individuals up.

According to Blewitt and Cullingford (2004), education towards sustainability should guarantee that the cognitive, affective and aesthetic dimensions of the learning process are not fragmented. Subjects whose critical thinking is stimulated during school years would engage with sustainability issues in a holistic way, examining their lived experiences and actions in a reflexive manner.

For this reason, the school environment is considered to be an adequate place to promote debates, build and share environment-related actions, given the urgency and importance of the issue. Reigota (1994: 25) states that "the school is a privileged place to develop environmental education, provided that there are plenty of opportunities for creativity".

Nowadays, one of the points of agreement about methodological procedures in environmental education is that they need to be thought in a more interdisciplinary fashion. The greatest challenge faced by both educators and researchers is to create a way to address environmental issues in all disciplines or, at least, in most of them, and also incorporate such notions in the daily pedagogical practice, regardless of field or level, be it in pre-school, elementary school, high school or university (Barcelos 2008).

As Jacobi (2003) puts it,

> The production of knowledge should necessarily comprise of the inter-relations between the natural and the social levels, including the analysis of the determinants of the process, the role of the various actors involved and the types of social organization that enhance the power of alternative actions of a new development, under a perspective that prioritizes a new developmental profile, with emphasis on socio-environmental sustainability (Jacobi 2003: 190).

Such discussion makes us consider that the great variety of aspects implied in this wide theme demands the engagement of many actors in the process, namely the government, citizens and organizations in general. Given the important operational role played by organizations in the concept of Sustainable Development, some ideas related to this aspect are briefly discussed in the next section.

Sustainability and Organizations

Organizations have been increasingly demanded to consider multiple factors when making decisions, forcing managers to take a broader view. In this sense, many organizational theorists state that contemporary problems will only be solved if social, economic, cultural and environmental aspects are considered interconnected, that is, as parts of a single system (Graaf et al. 1996; Starik and Rands 1995; Shrivastava 1995a; Gladwin et al. 1995). This academic debate and the empirical experience in organizations highlight how important it has become for managers, stakeholders and researchers to reach a consensus regarding the monitoring, planning and management of this system (Munda 2004; Kay et al. 1999; Mayumi and Giampetro 2006; Lehtonen 2004; Ravetz 2004; Shrivastava 1995b). When associating these topics to organizational decisions and strategies that have an impact on the ecosystem, the necessity emerges of shifting from a predominantly economic-oriented logic to another that considers the possibilities and consequences of decisions in systemic terms.

Therefore, managers today and in the future must be prepared to face this change and to make decisions in a complex and dynamic organizational environment. In this respect, Tenório (2006) declares that it is necessary to understand that because companies are inserted in complex environments, their activities influence various social agents, the community and society as a whole. Thus, it is no longer possible to think of a business whose purpose is to attend to the interest of shareholders only; instead, there is the need to incorporate social objectives as a way to integrate companies and society. The author adds that the industrial society sought, basically, to gain financial success; the post-industrial society, on the other hand, aims to: increase the quality of life, the emphasis on the human being, the respect for the environment, the multiple-objective business and the emphasis on social actions on the part of companies as well as individuals. In this regard, Toffler (1995) states that corporations must not focus on a single objective (the economic one); they should work towards becoming multiple-objective organizations, adapting to social and ecological environments.

In order to contribute to a more sustainable development, organizations need to have the logic of sustainability embedded in their actions. Therefore, it can be said that it is of paramount importance for managers to have an in-depth knowledge about this subject and take into consideration the many factors related to it when making decisions. They must understand that a complex world demands complex solutions, marked by a holistic, integrated view. That is why this theme must be dealt with during the preparation of such managers. Education has a key role to play on this process, for it can spread the concept, instruct and motivate actions towards sustainability, thus contributing to forming critical, active and change-oriented subjects.

Method

The present study intends to start a debate about how the issue of sustainability has been considered in the IFs, focusing on the management courses offered in the state of Rio Grande do Sul, more specifically at the IFRS. This is a significant and innovative project; the IFs aim to approach education in a more comprehensive way, one that seeks to prepare individuals to be not only technicians, but also critical citizens.

Thus, it is important to develop an exploratory study with a descriptive scope. According to Gil (1999), the main objective in exploratory research is to develop, clarify and modify concepts and ideas so as to formulate more precise questions to be investigated in further studies. They offer a more general view about a specific fact. For Roesch (2005), descriptive studies gather information about a particular population or phenomenon. They seek information necessary to act or predict.

In order to reach the proposed objective, research including all IFRS campuses was carried out to identify how the theme of sustainability is being developed in the Pedagogical Projects of the Courses (PPCs) related to management and which sustainability-related disciplines are offered. The basis for defining the courses were the National Catalogue of Technical Courses and the National Catalogue of University Courses in Technology of the Ministry of Education, which is responsible for offering technical and undergraduate courses in the field of technology, following the guidelines of national curricula. This is in accordance with the dynamics in the productive sector and with society expectations, grouping courses according to their scientific and technological characteristics. The courses listed in the technological area of Business and Management, which "encompasses technologies associated to the tools, techniques and strategies applied in the search for quality, productivity and competitiveness in organizations" (MEC 2011) were considered in the present study.

Firstly, by means of a search through the websites of the different IFRS campuses, all the existing courses in the area of Business and Management in each campus were identified. Subsequently, the respective PPCs of these courses were accessed and analyzed. Following data collection, which happened in October and November 2011, an analysis of both PPCs and the curricula of all courses found was performed, in order to identify whether or not the issue of sustainability – as well as any related topic – was present.

The main results are shown in the next section, followed by a brief introduction to IFs and IFRS (researched in the present study). Furthermore, an integration proposal for the Technical Course in Administration Integrated to High School is presented, with sustainability as its central theme. The proposal was designed by professionals in the area of basic and professional education who teach in the course and helped elaborate its PPC. The proposal will be implemented in the beginning of 2012 for second-year students.

The Federal Institutes of Education, Science and Technology (IFs)

In December 2008, Law 11.892 instituted the Federal Network for Professional, Scientific and Technological Education and created the Federal Institutes of Education, Science and Technology (IFs), which are

> higher, basic and professional educational institutions, pluri-curricula and multi-campuses, specialized in offering professional and technological education in different teaching modalities, based on the combination of both technical and technological knowledge and their pedagogical practice (Brasil 2008).

The sixth article of this law presents some characteristics and purposes of the IFs; some of the most significant ones are:

> I – offer professional and technological education at all levels and modalities, training and qualifying citizens to act professionally in various sectors of the economy, with emphasis on local, regional and national socioeconomic development;
> II – develop professional and technological education as an educational and investigative process that is able to generate and adapt technical and technological solutions to social demands and regional peculiarities; (…)
> IV – guide their training offer so as to foster the consolidation and strengthening of local productive, social and cultural arrangements, identified by the mapping of potential socio-economic and cultural development within the scope of the Federal Institutes; (…)
> IX – promote the production, development and transfer of social technologies, notably those aimed at preserving the environment (Brasil 2008).

The IFs offer the following teaching modalities (Brasil 2008):

- Professional technical education at medium-level, especially in the form of integrated courses, for elementary school graduates and for the public attending youth and adult education programs;
- Basic and continuing courses for workers;
- Undergraduate programs in the area of technology;
- Undergraduate programs, as well as special programs for pedagogical training;
- Bachelor programs and Engineering;
- Improvement and specialization latu sensu graduation programs;
- Master's and Doctoral strictu sensu programs.

It is important to note that, according to Edict 5.154/2004, the articulation between medium-level professional technical education and high school level education will be (Brasil 2004):

- Integrated, offered only to those students who have already finished elementary school; the program will be planned so as to provide students with a technical qualification at medium level, in the same teaching institution; each student will have only one registration;
- Concomitant, offered only to those students who have already finished elementary school or are still attending high school; the complementary character between medium-level professional technical instruction and high school presupposes distinct registrations to each program;
- Subsequent, offered only to those students who have already finished high school.

The creation of the IFs allows the implementation of the integrated courses because, at every term, at least 50% of their vacancies must necessarily be offered to medium-level professional technical courses, prioritizing integrated courses.

Law 11.892 instituted the creation of 38 IFs in Brazil. The present study focuses on the Federal Institute of Rio Grande do Sul (IFRS), which has twelve campuses, situated in the cities of Bento Gonçalves, Canoas, Caxias do Sul, Erechim, Farroupilha, Feliz, Ibirubá, Osório, Porto Alegre, Restinga, Rio Grande and Sertão.

Main Research Findings at the IFRS

Out of the twelve IFRS campuses, ten offer at least one course related to the technological field of Management and Business (except for Rio Grande and Sertão). On the whole, 17 courses were found. Table 1 presents the courses and modalities per campus, also showing the disciplines that deal with the issue of sustainability or related topics (such as environmental management and social responsibility, for example), the hour load of each discipline, the total hour load of the course and the amount of time each discipline represents in relation to the total time of the course.

It can be seen that, out of the 17 courses found, six (35%) do not offer any specific discipline related to the theme emphasized in the present study. The other eleven courses (65%) offer at least one discipline dealing directly with this subject, and the majority (nine out of eleven) offer, throughout the course, only one discipline related to the topic (according to their pedagogical program). On the whole, 14 disciplines that deal with the issue of sustainability or directly related topics are offered, with an average load of 48 hours each.

It should be stressed that, in most of the courses, such disciplines do not represent more than 3% of the total hour load of the course. Both the course Technology in Logistics – Campus Canoas and the Technical Course in Administration – Campus Osório are noteworthy, for the disciplines related to the theme under investigation in this study represent the highest percentages in relation to the total hour load when compared to the other courses (10,29% and 7,69%, respectively).

It is important to note that the Technical Course in Commerce in the modality Proeja – Campus Bento Gonçalves, despite not having any disciplines directly related to sustainability, emphasizes the themes Social Economy and Cooperation, offering three related disciplines. However, because these topics are not directly related to the scope of the present study, they were not considered.

The analysis of the PPCs of the courses revealed that only five of the campuses (Canoas, Osório, Porto Alegre, Restinga and Sertão) mention the issue of sustainability. The course Technology in Managerial Processes and the Technical Course in Administration Integrated to High School – Campus Osório mention the topic in one of the specific objectives: “Capacitate the development and critical analysis of business plans, aiming at ventures aligned with the principles of sustainability, feasible in the average and long run” (IFRS 2010a; 2010c).

Table 1: Courses related to the field of management and business and their modalities per IFRS campus; disciplines that deal with sustainability and the total hour load of the courses, of the disciplines and their corresponding percentage in relation to the total hour load of the courses

Campus	Course	Modality	Disciplineses that deal with the theme of sustainability or other related topics	Load (hours)		Percentage (Discipline|Course)
				Discipline	Course	
Bento Gonçalves	Technical Course in Commerce	Proeja	–	–	2760	–
Bento Gonçalves	Technology in Logistics	Undergraduate	Environmental Management and Social Responsibility	60	2205	2,72
Canoas	Technical Course in Administration	Integrated to High School	–	–	3439,9	–
Canoas	Technology in Logistics	Undergraduate	Environmental Management Society, Science and Technology Reverse Logistics	70 70 70	2040	10,29
Caxias do Sul	Technical Course in Administration	Proeja	Environmental Management	60	2650	2,26
Erechim	Technology in Marketing	Undergraduate	Ethics and Society	36	1736	2,07
Erechim	Technical Course in Sales	Subsequent	–	–	800	–
Farroupilha	Technology in Managerial Processes	Undergraduate	Corporate Ethics	63	1669,5	3,77
Feliz	Technology in Managerial Processes	Undergraduate	Environmental Management	40	1740	2,30
Osório	Technology in Managerial Processes	Undergraduate	Environmental Management and Social Responsibility	60	2180	2,75
Osório	Technical Course in Administration	Subsequent	Environmental Management and Quality Ethics and Human Relations	60 30	1170	7,69
Osório	Technical Course in Administration	Integrated to High School	–	–	3400	–
Porto Alegre	Technology in Managerial Processes	Undergraduate	Environmental Management and Social Responsibility	36	1870	1,93
Porto Alegre	Technical Course in Administration	Subsequent	Social and Environmental Responsibility	40	817	4,90
Porto Alegre	Technical Course in	Subsequent	–	–	900	–
Restinga	Technical Course in Administration	Subsequent	Environmental and Quality Management	36	1230	2,93
Sertão	Technical Course in Commerce	Proeja	-	–	2300	–

Source: Research data

The Technical Course in Administration in Porto Alegre refers to the theme when talking about the professional's profile – "Know the basics and principles of sustainability and social responsibility in organizations." – and required skills – "Follow the law that regulates social and environmental management to promote sustainability" and "Execute the principles of sustainability in organizations" (IFRS 2010d).

Similary, the course of the campus Sertão refers to sustaintability in the specific aspect of professional's profile – "are able to enter the world of work in a committed way to sustainable regional development" (IFRS 2011c).

The campuses of Osório and Restinga bring up the subject when presenting the profile of their courses, stating that "The course aims at forming professionals technically and ensuring they are politically prepared to meet the demands of society by stimulating entrepreneurship and cooperation in the field of Administration, thus respecting regional sustainability" (IFRS 2010a; 2010b; 2010c; 2010e).

Additionally, there is the PPC of the Technical Course in Administration Integrated to High School – Campus Canoas, which will be analyzed in more detail in the next section. The other PPCs do not mention sustainability in a direct manner, although it is important to say that some approach-correlated issues like ethics, socio-environmental responsibility, systemic thought, cooperation and social economy do.

Research findings show that the inclusion of sustainability in the curricula of management-related courses is already happening at the IFRS, though not yet in all campuses. This is seen as a positive fact, an important first step. However, it is our belief that sustainability-related issues must permeate all courses, occupying a central position in every PPC. Nevertheless, it was observed that, of all courses analyzed, only a few mention sustainability in their PPCs, generally in a superficial way. However, there are already some actions in the opposite direction. In this sense, a project carried out at the IFRS Canoas with a view of fostering integration and having sustainability as a core theme is presented as follows.

Interdisciplinary Project at the IFRS – Campus Canoas

One of the challenges faced by the IFRS is to form professionals able to deal with ever-changing scientific and technological knowledge and effectively apply it to society and to the working environment. Thus, the possibility of forming professionals able to deal with scientific and technological advances and to take a proactive attitude must fulfill three basic requirements: a scientific-technological and humanistic background, flexibility and continued education (IFRS 2011a).

In such a context, technical courses integrated in high schools represent an opportunity for youngsters to experience something different from what is usually offered by high schools, which in general tend to focus on preparing for university entrance examinations, and on regular technical courses, whose purpose is to form professionals able to perform activities related to a specific occupation. According to Ciavatta (2010), the interplay between general and professional education demands that the basis of both thought and production go beyond professional education practices and the theories of the propaedeutic education focused on preparing for entrance examinations only, because these are considered operational and mechanistic rather than humanistic.

It is believed that professional education must be viewed in a comprehensive way, so as to grant access to culture, science and work in an integrated manner. From this perspective, professional education and elementary education need to be understood as complementary instances, and overlapping or dominance should be avoided. This line of thought implies the concept of an integrated, holistic curriculum, which has been historically denied in the Brazilian professional education due to the positivist tradition and the mechanistic character of sciences, which resulted in a fragmented knowledge and a dichotomy between specific – with a professional character – and general – with a propaedeutic character – areas (IFRS 2011a).

According to Machado (2010: 81)

If reality is an integrated totality, so must be the resulting system of knowledge produced by individuals within its limits to act in it and transform it. Such understanding of totality is also expressed in the teaching and learning praxis. For didactic reasons, what is united is divided and separated. For didactic reasons, one can also aim at recomposing the whole.

One of the challenges in the elaboration of the PPC of the Technical Course in Administration Integrated to High School at IFRS – Campus Canoas was to find possible ways of developing a proposal that contemplated integration. How can an integrated curriculum be designed considering the specific characteristics of the context in which it will be developed and, at the same time, without ignoring that the proposition of an integrated curriculum is not the mere sum of its constituent disciplines of basic and specific education? For Ramos (2010: 116), "the integrated curriculum organizes knowledge and develops the teaching-learning process in a way that the concepts are learnt as a system of relationships pertaining to a concrete totality that is intended to be explained/understood".

The debate among professionals involved in this reality is necessary, and it has already happened at the IFRS – Campus Canoas. However, it should be noted that both basic and professional education faculties, despite being experienced professionals and having solid academic backgrounds, had never participated in the elaboration of an integrated curriculum before. Thus, the shared study and reflection became the strategy adopted to elaborate a proposal that opened research possibilities towards the construction of knowledge about integration.

According to Ciavatta (2010: 100), "it's necessary to discuss and attempt to elaborate on collectively, the academic-scientific strategies of integration". This movement would result in theoretical and practical considerations focused on developing and improving integration.

Therefore, with a view to getting group discussion started, a discipline entitled Interdisciplinary Project in the Technical Course in Administration Integrated to High School at IFRS – Campus Canoas was proposed, with the objective of working on themes related to different disciplines and having sustainability as a central theme, thus reinforcing the notion of integration. The discipline was thought of as a way to guarantee, in the curriculum structure, space for all the involved professionals to meet and act.

The Technical Course in Administration Integrated to High School at IFRS – Campus Canoas totals 3483 hours, distributed in over four years. The Interdisciplinary Project should be carried out during the three final years of the course, totaling 160 hours of instruction, with the objective of fostering research and curriculum integration (IFRS 2011b).

The project, to be carried out in small groups of students under the orientation of teachers from different areas, will have as its main objective the elaboration of a business plan, a preliminary study that evaluates the advantages and disadvantages of starting a business, presenting operational, logistic, marketing financial and human resources plans. In addition, students will also present the proposal for the creation of a new organization (a company, a cooperative, an association), with the objective of solving problems related to the reality they are inserted in. As a first step, students will have to brainstorm socio-environmental problems of their region and, afterwards, proceed to the creation of an organization that aims at reducing the problem(s) identified. The aim here is to guide the research towards eliciting local problems, as well as coming up with sustainable solutions.

The project will be formatted as problems and needs are listed and solutions are proposed. Both the professional in charge of coordinating the discipline (from the area of administration) and the teachers responsible for tutoring the different projects (who must necessarily act in other fields of knowledge) will closely monitor the groups, in order to identify what has already been accomplished in each group and, little by little, build new concepts and insights. There will also be discussion sessions and periodic seminars, whose objective is to allow students and teachers to share concepts and come up with sustainable solutions.

The idea is that the problems elicited by students, together with the challenges proposed by the teachers, provoke cognitive imbalance, leading students to search for solutions, thus stimulating the production of knowledge as a result of the interaction between teachers, students and subject matters. The main goal is to empower students to take a critical stand on reality and build new bodies of knowledge.

According to Piaget (1972),

> "Knowledge is not a copy of reality. Knowing an object or event does not mean simply looking at it and making a mental copy or image of it. To get to know an object it is necessary to act upon it. Knowing means changing, transforming the object, and understanding the process underlying this transformation and, consequently, understanding the way through which the object is built." (Piaget 1972)

The Interdisciplinary Project was built following the principles of constructivism; students will be in contact with the subject under study and will be invited to come up with new solutions to the problems found in their reality, having a critical look before them and the solutions presented so far. That means students will no longer be mere receptors and, when facing new challenging situations, will be demanded to act and try to solve the problems.

Considering that the implementation of the project is supposed to begin in 2012, teachers will be asked to propose actions for the disciplines listed in the curriculum, so that the issue of sustainability can be analyzed within different contexts. The goal is to identify how students view this matter and then, based on this information, elaborate problems from which new concepts can be built, in order to broaden student's understanding about the topic. This is considered a necessary step, for students will have the opportunity to make connections between theory and practice. For this reason, the support provided by the teachers from different areas is vital, because it will permit the evaluation of the difficulties faced and of what has been learnt by students throughout the project.

This way, we hope to integrate theory and practice and motivate students to act towards transforming their reality by applying in their professional practice the knowledge acquired in the course, always keeping in mind the concern with sustainable development.

Conclusions

This essay highlights the importance of an educational system oriented towards sustainability, particularly in management-related courses, bringing up the debate about two important issues: sustainability and integration. However, the implementation of this new proposal remains a challenge to be overcome.

The preoccupation with these themes is embedded in the proposal for creation of the IFs in Brazil, and they have shown to be a propitious place for the development of ideas and projects related to these concepts. Notwithstanding, the research carried out at the IFRS reveals a very scarce number of effective actions proposed by the PPCs of management-related courses. It is believed that this is due to a tendency to resist to changes and, at times, to the lack of previous successful experiences.

However, the Interdisciplinary Project proposal developed by the faculty at the IFRS – Campus Canoas stands as a way to start an integration process, having sustainability as its main theme, and can serve as a reference to other courses and teaching levels. It is hoped that it can instigate the debate about the theme and the implementation of actions towards this end in other campuses and teaching institutions as well.

It can be said that a fragmented view of the world and the lack of a critical and long-term view are some of the factors underlying the socio-environmental problems we face today. For this reason, it is believed that the insertion of a more critical and integrated preparation since high school will help form individuals able to understand the reality and the concept of sustainability in a broader and more systemic way.

The need to build a curriculum proposal for technical education integrated to high school is a challenge that professionals at IFRS – Campus Canoas have been facing as an investigative process to be carried out collectively, under constant evaluation. According to Machado (2010), didactic proposals are always a dynamic construction of pedagogical relationships, whose success depends on the participation and on the self-evaluation capacity of the subjects that shape them. In this context, the assumption that the issue of sustainability comprises relevant characteristics to promote curriculum integration, is guided by the possibilities of creating conditions that encourage students to become questioning, critical and reflexive individuals whose actions result from a multi-dimension analysis.

This interdisciplinary project is also a challenge for teachers, once it forces them to change paradigms, work collectively and teach in a different way. They need to understand how disciplines and knowledge are inter-related, and constantly rethink their roles in the process of building knowledge. By doing that, they will be more prepared to understand and promote both integration and sustainability. Perhaps this is one of the greatest challenges to be overcome throughout the implementation of the project.

It is important to remember that the Interdisciplinary Project presented here this is still in its first year of implementation, which does not allow a reliable evaluation of its outcomes. Nevertheless, a constant follow up is advisable, for it can provide interesting data to be presented in further studies.

In a further phase of the research, coordinators and teachers at the IFRS will be interviewed in order to access: a) their knowledge about the themes of sustainability and integration; b) whether or not they are aware these issues are highlighted in IFs documents, and c) what they are doing in their disciplines and courses concerning this topic.

Additionally, it should be stressed that the present research was limited to only one IF and its respective campuses, so results cannot be generalized to the entire Federal Network of Professional, Scientific and Technological Education. Therefore, a study involving other IFs is necessary for a more comprehensive analysis of the network as a whole.

References

Barbieri, J.C. (2007), *Gestão Ambiental Empresarial: conceitos, modelos e instrumentos*. Saraiva, São Paulo.

Barcelos, V. (2008), *Educação Ambiental: sobre princípios, metodologias e atitudes*. Vozes, Rio de Janeiro.

Blewitt, J., Cullingfor, C. (2004), *The Sustainability Curriculum: the challenge for higher education*. Cromwell, London.

Brasil (2004), *Decreto nº 5.154, de 23 de julho de 2004*. Available at: http://www.planalto.gov.br/ccivil_03/_ato2004-2006/2004/decreto/d5154.htm (accessed 16 January 2012).

– (2008), *Lei n.° 11.892, de 29 de dezembro de 2008*. Available at: http://www.jusbrasil.com.br/legislacao/92587/lei-11892-08 (accessed 16 January 2012).

Ciavatta, M. (2010), A formação integrada: a escola e o trabalho como lugares de memória e identidade. In Frigotto, G., Ciavatta, M., Ramos, M. (Ed.), *Ensino Médio Integrado: concepções e contradições*. Cortez, São Paulo.

Dias, R. (2006), *Gestão Ambiental Empresarial: responsabilidade social e sustentabilidade*. Atlas, São Paulo.

Gil, A.C. (1999), *Métodos e técnicas de pesquisa social*. Atlas, São Paulo.

Gladwin, T.N., Kennelly, J.J. and Krause, T.S. (1995), Shifting Paradigms for sustainable development: implications for management theory and research. *Academy of Management Review* 20 (4), pp. 874-907.

Graaf, H.J., Musters, C.J.M. and Leurs, W.J. (1996), Sustainable Development: looking for new strategies. *Ecological Economics* 16, pp. 205-216.

Hopwood, B., Mellor, M., O'Brien, G. (2005), Sustainable Development: Mapping Different Approaches. *Sustainable Development* 13 (1), pp. 38-52.

IFRS (2010a), *Projeto Pedagógico do Curso Superior de Tecnologia em Processos Gerenciais*. Osório.

– (2010b), *Projeto Pedagógico do Curso Técnico de Nível Médio Subsequente em Administração*. Restinga.

– (2010c), *Projeto Pedagógico do Curso Técnico em Administração Integrado ao Ensino Médio*. Osório.

– (2010d), *Projeto Pedagógico do Curso Técnico em Administração*. Porto Alegre. Available at: http://www.poa.ifrs.edu.br/?page_id=298 (accessed 28 November 2011).

– (2010e), *Projeto Pedagógico Técnico em Administração*. Osório.

– (2011a), *Instituto Federal de Educação, Ciência e Tecnologia do Rio Grande do Sul*. Available at: http://www.ifrs.edu.br/site/ (accessed 1 December 2011).

– (2011b), *Projeto Pedagógico do Curso Técnico em Administração Integrado ao Ensino Médio*. Canoas.

– (2011c), *Projeto Pedagógico do Curso Técnico em Comércio – Modalidade Concomitância Externa*. Sertão.

Jacobi, P. (2003), Educação Ambiental, Cidadania e Sustentabilidade. *Cadernos de Pesquisa* No. 118, pp. 189-205.

Kay, J.J., Regier, H.A., Boyle, M. and Francis, G. (1999), An ecosystem approach for sustainability: addressing the challenge of complexity. *Futures* 31 (7), pp. 721-742.

Leal Filho, W. (2000), Dealing with misconceptions on the concept of sustainability. *International Journal of Sustainability in Higher Education* 1 (1), pp. 9-19.

Lehtonen, M. (2004), The environment-social interface of sustainable development: capabilities, social capital, institutions. *Ecological Economics* 49 (2), pp. 199-214.

Machado, L. (2010), Ensino médio e técnico com currículos integrados. Propostas de ação didática para uma relação não fantasiosa. In Moll, J. (Ed.), *Educação profissional e tecnológica no Brasil contemporâneo. Desafios, tensões e possibilidades*. Artmed, Porto Alegre.

Mayumi, K. and Giampetro, M. (2006), The epistemological challenge of self-modifying systems: Governance and sustainability in the post-normal science era. *Ecological Economics* 57 (3), pp. 382-399.

MEC (2011), *Ministério da Educação – Catálogo Nacional dos Cursos Superiores em Tecnologia e Catálogo Nacional dos Cursos Técnicos*. Available at: http://portal. mec.gov.br/ (accessed 1 December 2011).

Munda, G. (2004), Social multi-criteria evaluation: methodological foundations and operational consequences. *European Journal of Operational Research* 158 (3), pp. 662-677.

Palma, L.C., Oliveira, L.M., Viacava, K.R. (2011), Sustainability in Brazilian Federal Universities. *International Journal of Sustainability in Higher Education* 12 (3), pp. 250-258.

Piaget, J. (1972), Developmente and Learning. In Lavatelly, C.S. and Stendler, F., *Reading in child behavior and development.* Hartcourt Brace Janovich, New York.

Ramos, M. (2010), Possibilidades e desafios na organização do currículo integrado. In Frigotto, G., Ciavatta, M., Ramos, M. (Ed.), *Ensino Médio Integrado: concepções e contradições.* Cortez, São Paulo.

Ravetz, J. (2004), The post-normal science of precaution. *Futures* 36 (3), pp. 347-357.

Reigota, M. (1994), *Meio Ambiente e representação social.* Cortez, São Paulo.

Roesch, S.M.A. (2005), *Projetos de estágio e de pesquisa em administração: guia para estágios, trabalhos de conclusão.* Dissertações e estudos de caso, Atlas, São Paulo.

Shrivastava, P. (1995a), Ecocentric management for a risk society. *Academy of Management Review* 20 (1), pp. 118-137.

– (1995b), The role of corporations in achieving ecological sustainability. *Academy of Management Review* 20 (4), pp. 936-960.

Souza, R.S. de (2000), *Entendendo a questão ambiental: temas de economia, política e gestão do meio ambiente.* Edunisc, Santa Cruz do Sul.

Starik, M. and Rands, G.P. (1995), Weaving an integrated web: multilevel and multisystem perspectives of ecologically sustainable organizations. *Academy of Management Review* 20 (4), pp. 908-935.

Steurer, R., Langer, M.E., Konrad, A., Martinuzzi, A. (2005), Corporations, Stakeholders and Sustainable Development I: A Theoretical Exploration of Business-Society Relations. *Journal of Business Ethics* 61 (3), pp. 263-281.

Tenório, F.G. (2006), *Responsabilidade Social Empresarial: teoria e prática.* FGV, Rio de Janeiro.

Toffler, A. (1995), *A empresa flexível.* Record, Rio de Janeiro.

Veiga, J.E. (2005), *Desenvolvimento sustentável – desafio do século XXI.* Garamond, Rio de Janeiro.

Chapter 72

Sustainable Coastal Management at the Public University in Uruguay: A Southern Cone Perspective

Daniel Conde, Daniel de Álava, Denise Gorfinkiel, Rodrigo Menafra and Ingrid Roche

Abstract

Since Rio 92, Integrated Coastal Management (ICM) has been considered the most appropriate strategy for planning coastal areas in the context of sustainable development. ICM seeks balance between social, environmental and economic aspects, addressing coastal conflicts in terms of ecosystem-based management, stakeholders' involvement and development of operational tools. In Uruguay and the Southern Cone, the consolidation of ICM with the support of Canadian partners at the University of the Republic Montevideo allowed the establishment of an ICM Regional Master's Program and the Interdisciplinary Center for Integrated Coastal Management of the Southern Cone, addressing research, education and outreach through the integration of university departments, faculty, disciplines and communities in six countries (Uruguay, Argentina, Brazil, Chile, Mexico, Canada). The Master's program, based on the phases of the management cycle, is an innovative program where students and professors interact in a teaching-learning environment, addressing complex coastal cases in cooperation with communities, to experience the complexity of coastal management processes in real situations. This paper addresses the evolution of ICM in Uruguay and its contribution to coastal sustainability through innovation in capacity building and research activities by the university and its coastal regional partners.

Integrated coastal management and sustainable development

On a night map of the Earth we may notice that approximately 50% of the human population and most anthropic activities are concentrated only less than 100 km from the coastline. Ensuring people's survival in this constricted area implies strong pressures and threats on coastal ecosystems and resources. Along the coastal zone of all continents, impacts from unplanned urban growth, land-based pollution, unsustainable aquaculture and fishing, eutrophication and acidification, widespread alien species or intensive tourism are major problems commonly included in coastal diagnosis (UNEP 2010a). This situation points to an urgent need to encourage environmentally sound approaches to manage coastal ecosystems and resources. Besides human-rooted reasons for coastal deterioration, climate change synergistically emerges as a new challenge, which will exacerbate erosion, increase saltwater intrusion into aquifers and diminish biodiversity, impacting on human settlements and activities (UNEP 2010b; IPCC 2007).

Since the 1972 UN Conference on Human Environment, governments increasingly recognize that environmental protection requires an integrated approach. A major paradigm shift occurred in 1992 Earth Summit in Rio, where a new vision on sustainable development was embraced. A decade later, at the World Summit on Sustainable Development in Johannesburg, states deepened their commitment by setting specific targets through the Johannesburg Plan of Implementation (UN 2011).

Agenda 21 (Chapter 17) deals with the protection of coastal areas and living resources and stresses the importance of coasts as life-supporting systems. From then on, Integrated Coastal Management (ICM) has been considered an appropriate strategy to cope with coastal issues and plans. ICM seeks balance between social, environmental and economic aspects, addressing conflicts on fundamental principles: ecosystem-based management (EBM), stakeholders' involvement and operational tools. Nowadays, countries and international organizations have committed to ICM and adopted EBM approaches like the convention on biological diversity.

The coast is recognized as a zone of emerging opportunities for sustainable development because many areas still exhibit good environmental conditions. This could be recognized by strategic planning instead of forcing new regulatory measures to protect already degraded environments, as it happens in highly developed watersheds (UN 2011). Economic and social benefits for coastal populations may then emerge from protecting ecosystems' integrity.

Despite their relevance, coastal and marine areas are still managed unsystematically and most governance structures have been deficient in tackling pressures on coastal ecosystems everywhere. Major obstacles to successful ICM include insufficient data and information on ecosystem processes, low capacity to develop more comprehensive EBM and ineffective decision frameworks to deal with coastal complexity (Cicin-Sain et al. 2011).

A major recommendation that commonly arisis is the need to develop capacity through national and regional efforts. There is an urgent need to build capacity for coastal management that provides long-term competence in ICM and EBM and enhances leadership for high-level decision makers, through the development of new university programs to educate the next generation of ICM leaders (IOC 2011).

According to Chircop (2000), ICM programs are challenging to develop and implement, both for students and instructors, and universities need to further facilitate these types of initiatives. Universities still struggle, however, with demands for interdisciplinary programs which are needed to deal with complex environmental impacts. Universities addressed this challenge through problem-oriented programs, balancing knowledge and skills, modifying classroom setting and exploiting the knowledge carried by postgraduate students. It is therefore concluded that diverse contexts will require dissimilar tactics and that simple solutions should not be expected to succeed.

Environmental literacy helps societies and communities to understand the relevance of incorporating the environmental dimension into daily issues, facilitating the practice of responsible decision-making for their own sustainable development. Education for sustainable development and curriculum greening (i.e. progressively orienting a curriculum towards sustainability) is currently a major objective at universities, and represents a worldwide ethical response to environmental challenges (Wemmenhove and de Groot 2001; AAU 2011). Progresses aside, in developing countries' universities this response is still not a widespread practice, where propensities to separate environment and development, as well as top-down teaching-styles to address environmental issues, continue to dominate.

Curriculum greening results from explicit decisions and therefore it is not a simple declaration of good purposes or defining budget priorities, but a process involving changes in fundamental conceptions, both institutionally and personally (UNESCO 2005; Ciurana and Leal Filho 2006). Furthermore, despite general consensus on the unavoidability of the interdisciplinary approach to address society's most complex issues, modern universities still face hard internal barriers to fully achieve this precondition. Sustainable development can also be pursued by promoting formal and informal partnerships beyond higher education institutions

(Clugston and Calder 1999), through community engagement, e.g. by developing projects with adult education centers, by redesigning education field practices, by promoting interdisciplinary through approaches open seminars or by instituting educational chairs for sustainability (Junyent 2007).

According to the ACES model (Ciurana and Leal Filho 2006), the implementation of sustainability principles in higher education includes the integration of the paradigm of complexity, the seeking of flexibility and permeability among disciplines, the contextualization of the curricular projects, working within a prospective orientation of alternative scenarios, adapting to new learning approaches, and creating reflection and participation spaces.

Threats to coastal sustainability in Uruguay

The Uruguayan coastal zone covers six provinces (Departments) along 700 km, from the inner part of Rio de la Plata estuary eastward to the border with Brazil (Fig. 1). It receives an outstanding water flow of 20,000 m^3 s^1 from a basin covering five countries. Along this narrow micro-mareal coast, freshwater with a diverse degree of pollution and fairly pristine seawater from the Brazil and Malvinas currents permanently blend together, creating a unique and complex environmental gradient, both aquatic and terrestrial. Along a short distance (ECOPLATA 2010) a wavering succession of modified urban areas and relatively natural beaches, wetland and rocky environments can be experienced.

The coast represents a major feature for Uruguay's economy, where tourism is growing into the nation's foremost economy in spite of it representing just a fifth of the country's area; the coast contains 70% of the population and generates almost 80% of the country's gross value added, from tourism, port operation, other industries and traditional cattle raising (Gorfinkiel 2006). Rapid coastal development has led to increasing conflicts between uses. Environmental crisis may synergistically surface as a consequence of severe land-use changes, intensive tourism, discharge of untreated effluents, beach erosion, informal settlements and unsustainable use of resources by impoverished populations, mostly derived from sector-oriented or absence of planning (Lemay 1998). Articles in Menafra et al. (2006) address specific major environmental issues on the Uruguayan coast.

Along with increasing anthropogenic pressure, the Uruguayan coast is suffering considerable climatic variability. Climate change is expected to accelerate sea level rise (11 cm has been recorded in 100 years), modify rainfall and freshwater discharges and change coastal winds (Marengo et al. 2010). The coastal capital at risk has been estimated at US$ 1.8 billion under a 1 meter sea level rise scenario due to infrastructure losses in the capital Montevideo and the touristic Punta del Este. This figure is considered much larger than expenses of protection options (Hareau et al. 1999).

Fig. 1: The coastal zone of Uruguay

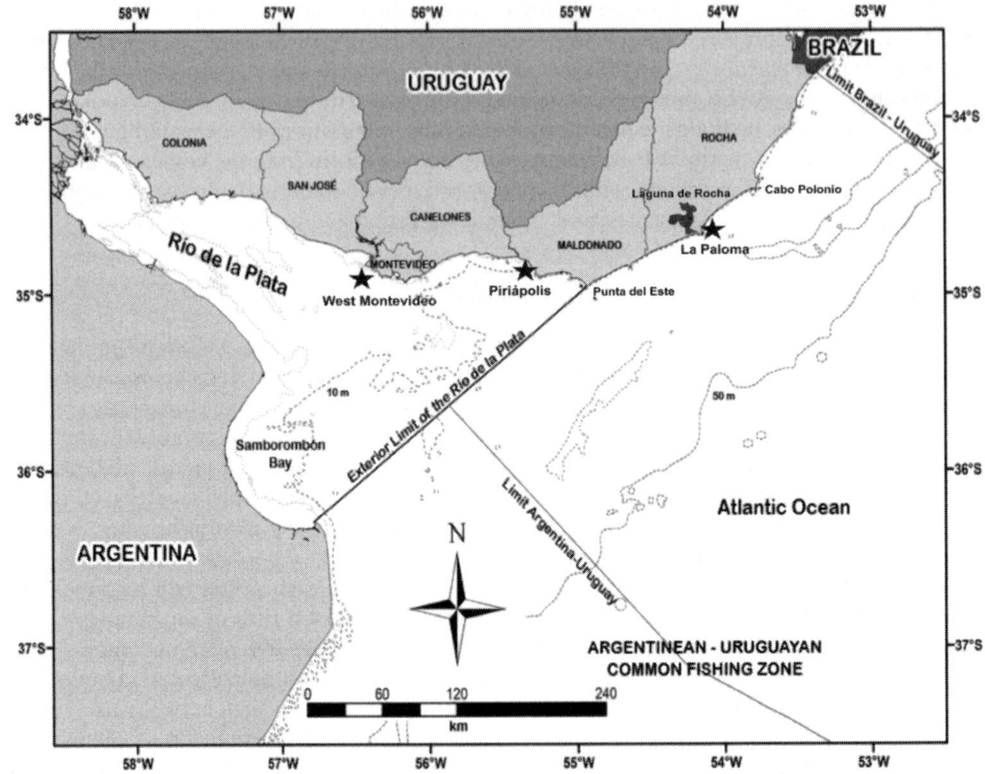

Modified from Menafra et al. (2006). Six coastal Departments are indicated as well as Rio de la Plata estuary and the Atlantic Ocean. Black stars identify three selected coastal cases addressed later in the text.

Local approaches to integrated coastal management

Although Uruguay and other Southern Cone countries have experienced advances over the last two decades towards more integrated approaches to solve coastal problems and issues, there is still a long way to go before achieving sustainable development at regional level (Barragán 2001; 2010). Five issues emerge when addressing the basis for coastal sustainability: the legal array, public participation, inter-institutional coordination, capacity-building and research competencies.

While national and local legislation for the management of the coastal zone exists, it still lacks an integrated approach in order to contribute significantly to sustainability. Nevertheless, some legislation is coastal-specific and is considered a legal framework for sustainability (e.g. the Natural Protected Areas Law, with contributes to sustainability through specific management plans; the Law of Territorial Planning and Sustainable Development, with several instruments for conflict resolution and the Law of Decentralization, promoting locally-orientated management through municipalities).

The Ordinance for Sustainable Development of the Coast of Rocha was the first attempt to unify national and local legislation for sustainable planning, defining ICM as a principle.

Stakeholder involvement in decision and planning processes has started, but a major deficiency is the lack of permanent institutional support required to guarantee participation.

An intricate institutional array of overlapping duties and jurisdictions still works against ICM success. From more than 40 institutions related with the coast, two thirds show disciplinary approaches to management and only a minority developed confident interdisciplinary practices (Menafra et al. 2009). Several initiatives did not evolve into long-lasting practices, with the exception of ECOPLATA, a two-decade program prompted by the Canadian Cooperation to promote ICM. ECOPLATA is a bridge between national policies and local management, encouraging decisions based on the principle of "connecting knowledge with action". The Coastal/Marine Management Unit at the National Directorate of the Environment represents the institutional recognition of the relevance of the coast, and is expected to play a key role in implementing management decentralization. The national parliament is presently addressing a draft coastal law which will induct Uruguay into the short list of countries with coastal legislation in Latin America (Barragán 2010).

Major challenges facing the implementation of ICM in Uruguay include: awareness on the wise use of coastal values as a vehicle of sustainability, institutionalization of ICM in the national agenda, promotion of decentralized decision-making systems based on local priorities, and generating an information baseline to evaluate ICM success (ECOPLATA 2010). A key challenge will be learning from the ICM international long-term experience and adjusting to specific local contexts for implementation. Recent efforts on specific legislation, new opportunities for institutional coordination, and increasing interest from local actors and municipal authorities are an indication of positive directional changes.

Addressing coastal sustainability at the university

Past efforts

Although coastal research has a long tradition at the University of the Republic (UdelaR) in Uruguay, it has been accused for decades of producing hardly any pertinent information for the purpose of management. The first research efforts indirectly related to sustainability in the coastal zone in Uruguay can be tracked back to the late 80s and early 90s, when Marine Sciences became an area of major interest at UdelaR (at the new faculty of Sciences) and for the Basic Sciences Development Program (PEDECIBA) (ECOPLATA, 2010). At that time, support from Argentinean and Brazilian research groups, as well from France (CRNS, IFREMER) and Canada (CIDA, Univ. Dalhousie), was critical to strengthen and consolidate local groups. The initiation of coastal research and management of large programs in the 90s and later (e.g. ECOPLATA, PROBIDES and FREPLATA), indicated the relevance and progresses of coastal topics in Uruguay.

In the last 20 years more than 250 projects and publications related to coastal management have been developed by UdelaR and other institutions, although they are mostly disciplinary. Currently, the most conspicuous technical knowledge of the coastal zone relates to natural processes, deriving from studies performed by university groups from disciplines like Biology, Ecology, Environmental Sciences, Engineering and Land Planning. On the other hand, Social Sciences or Law are less represented (PROBIDES 1999; ECOPLATA 1999; FREPLATA 2005; GEO Uruguay 2008; Menafra et al. 2009).

Furthermore, during the last decade a series of Master's programs addressing sustainability have emerged at UdelaR (e.g. Land Planning, Environmental Engineering, Environmental Sciences, and Local Development). Unlike in outreach and collaborative activities, in most cases these educational programs are multidisciplinary. More recently, an Environmental Thematic

Network (RETEMA) as well as a new academic initiative called Interdisciplinary Space, crossing horizontally through schools, programs and disciplines have evolved at UdelaR. RETEMA has organized interdisciplinary undergraduate courses on environmental issues, for example in cooperation with the School of Architecture, a situation that would have been unlikely to occur a few years ago (Instituto de diseño 2010). Nevertheless, in none of these programs was the coastal zone the focal study object.

Education in integrated coastal management today

An assessment by Ávila et al. (2004) recognized the implications of the lack of an education program in ICM in Uruguay and the surrounding area. The study established the demand for capacity building in ICM, specifically in topics like protected areas, territorial planning, local participation, and conflict resolution.

In 2002, with support from the Canadian Cooperation (through Dalhousie University) a process of creating a Master's program in ICM evolved from the School of Sciences at UdelaR and it rapidly expanded to several schools and beyond frontiers, including partners from regional universities (Argentina, Chile, Brazil and Mexico), as well as UNESCO. An interdisciplinary regional Master's program in ICM started in 2007 in Montevideo, with its third edition starting in 2012 in a decentralized location (Maldonado).

Fig. 2: Organizational diagram of the *Interdisciplinary Center for Integrated Coastal Management of the Southern Cone* (C-MCISur)

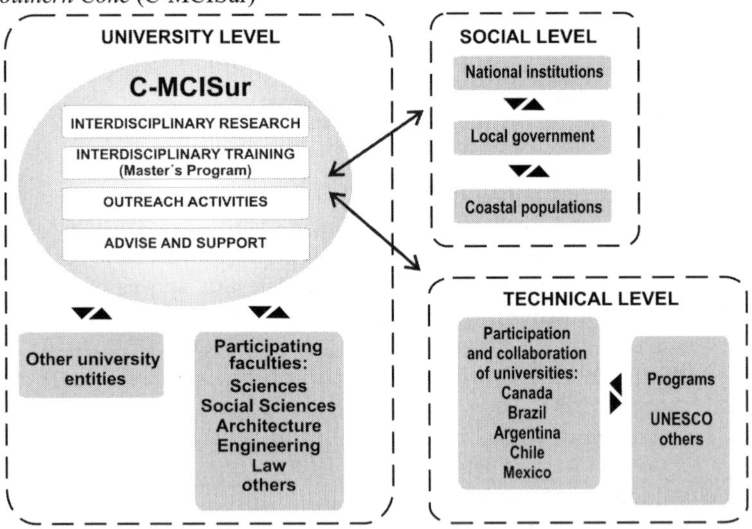

Main links and interactions between partners at different levels are represented (www.mcisur.edu.uy).

Since 2009, the Master's program is also part of the new "Interdisciplinary Center for Integrated Coastal Management of the Southern Cone" (C-MCISur), which constitutes one of the first programs of the new Eastern Regional University Center of UdelaR, hosting several new programs on the environment and development. The new center also adheres to an emergent policy of promoting interdisciplinary initiatives at UdelaR. The center and the Master's program were recently awarded an UNESCO Chair.

C-MCISur embraces professors and researchers in natural and social sciences from UdelaR and Latin-American and Canadian universities. Its main goal is to contribute to sustainable solutions for coastal issues and problems in the Southern Cone through several dimensions of integrated actions. Objectives include promoting interdisciplinary research, improving capacity-building in ICM (i.e. the Master's program), contributing to the conceptual development of ICM, and developing methodological approaches and tools for managers (Fig. 2).

Education: This dimension includes the Master`s program MCISur and other educational activities at various levels, such as an ICM course for mayors and local coastal technicians. The Master's program aims to train professionals beyond previous training in specific disciplines so that they are able to address coastal management through a critical, interdisciplinary and participatory perspective. The program constitutes an innovative program where students and teachers from different disciplines (e.g. lawyers, architects, biologists, oceanographers, sociologists and psychologists) interact with coastal populations and managers. The two-year program is based on large modules comprised of short interdisciplinary thematic units addressing ICM phases, with the objective of building theoretical knowledge and field performance (Fig. 3).

Research: The center focuses on seven research topics: adaptation of the coast to climate change, ecosystem fragility and processes of anthropic transformation of the coast, coastal/marine governance, EBM and co-management, sustainable economic development, adaptive planning and tools for coastal management. Ongoing research projects are regional in scope through collaborations with other research centers. A major effort is focused on the establishment of a Coastal Processes Observatory of the Western South Atlantic in the near future, with the support of the GEOS initiative and UNESCO.

Outreach: This area of activity involves the development of projects with participation of local stakeholders, mainly through the development of interdisciplinary case studies by students and professors of the Master's program (see details of three cases below). A major ambition is to attain coordination between scientific and traditional local knowledge, which in turn contributes to consensus decisions for approaching sustainable solutions.

Support to coastal populations and managers: Based on the conceptual approach of ICM, the center aims to contribute to local management through specific tools; e.g. the design and implementation of a multi-criteria decision-making system for the artificial opening of the sand bar of a protected coastal lagoon (Laguna de Rocha), in support of the National Committee on Global Change.

A major aim of the program is to provide strategies and tools to help experience the complexity of the management processes in real situations, two major challenges faced were: 1) to reach consensus on how to facilitate a holistic view to develop coastal development plans, and 2) attaining a working methodology involving the effective participation of coastal populations, as a means of securing sustainability. The first challenge was successfully addressed by basing the academic curriculum and the set of activities of the program on the phases of the management cycle: identification of key issues, planning, formal approval, implementation and evaluation.

Fig. 3: Structural diagram of the MCISur Master's program

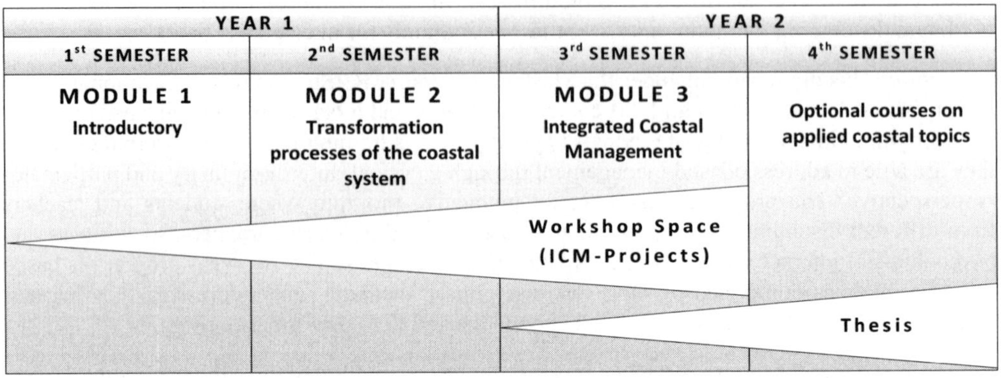

The sequence of the various interdisciplinary ensemble of activities is shown.

The second challenge was met through the inclusion of a specific set of activities in the curriculum, made up of interdisciplinary seminars and field work, called "Workshop space". It is a flexible space of analysis and development of proposals for specific coastal cases and aims to allow students to implement the application of concepts and methodologies and to experience the complexity of real management processes. The most important activities of this workshop are the "Projects in Integrated Coastal Management" (ICM-Projects).

ICM-Projects involve the integration of groups of students and professors from different disciplines over several months. They implement an interdisciplinary process by applying the principles and concepts of ICM to specific coastal areas, addressing the challenge of approaching a balance between socio-economic development and the vision of environmental sustainability. A final draft management proposal must derive from the process and must include documentation of work done with the participation of local stakeholders at each project site, with whom progress towards a shared vision to sustainable development must be attained.

Three coastal cases attempting for environmental sustainability

Here, a draft of main outputs from three selected ICM projects is presented, taken from the set of eight cases performed during the first two editions of the program recently published (C-MCISur 2011). Comprising of a variety of topics, these eight cases addressed the combination of industrial, port and tourist activities with the conservation of ecosystems and the preservation of the coastal heritage as core issues.

The cases summarized here were carried out on coastal sites on Río de la Plata middle estuary (Montevideo) and on the Atlantic Ocean (Maldonado and Rocha). The three cases selected focused on sites with new local jurisdictions since 2010 (see comments on the Law of Decentralization), thus representing examples of the new scenario where it is imperative to foster local level participation over the next few decades. The selected cases incorporated key topics of coastal stakeholders by addressing local municipal agendas.

Long-term social involvement in riverine West Montevideo: a suburban complex case for ICM

The objective was the elaboration of guidelines for ICM for a space of collective use along the west coast of Montevideo city, strengthening local capacities in existing areas of participation. It is a place of coexistence of numerous national and municipal programs as well as local groups with latent conflicts. The principles that guided the action in this case were defined as "to contribute more than to resolve, to participate more than to observe, and to search for processes rather than outcomes". For this, the group of students and professors was gradually integrated into the "Inter-social West" network and now this has evolved towards a formal co-management initiative to implement a coastal park based on ICM concepts.

ICM planning in estuarine Piriápolis (Maldonado): a traditional touristic coastal case

Here, the group initiated an ICM plan to ensure the preservation of the local socio-environmental values against the pressure of traditional tourism development. Participation by local residents allowed the main concerns, problems and expectations to be made explicit. The projected plan includes sub-projects to enhance archaeological, cultural and heritage coastal values, preserving relicts still remaining in a pristine state. They concluded that local natural and cultural values prompt this coastal zone to adopt a policy of conservation, providing an opportunity to test sustainable coastal development. Since the local community is well organized and has management capacities through the municipality, a pilot ICM plan is being pursued. This could be the first coastal area plan in Uruguay with high potential for success.

Governance in a touristic socio-ecological system: a marine case for ICM (La Paloma, Rocha)

The foundations of a governance system required to build an ICM plan was the goal of this group. The use of different approaches and tools, such as geo-historical and geographic analysis systems, allowed the characterization of major environmental values in the area and to survey key management issues relevant to the community and the local municipality. The analysis of development trends and scenarios allowed design proposals for new territorial integration to take place. Specific actions were recommended to develop an agenda, including joint actions through a network of governance sub-areas or "Meeting Platforms", in order to generate a sustainable process to preserve coastal heritage with tourism proposals in harmony with the existing protected areas of Rocha Lagoon and Cabo Polonio.

Major contributions to coastal sustainability

The consolidation of the Master's program in Integrated Coastal Management as well as the new research center focusing on coastal sustainability, both based on an interdisciplinary approach and an innovative educational-learning processes, suggests a formal recognition from UdelaR of a unique opportunity to contribute significantly to the socio-ecological integrity to the coastal zone of Uruguay.

UdelaR has addressed the recent economic and political improvement of Uruguay by pursuing a major advancement in superior education as a substantial aspect of integral development. According to UdelaR's Rector (unpub.) this progress must be encompassed by addressing critical threats and weaknesses like social inequity, poverty, and the innovation in production, as well as dealing with major educational and environmental issues. Improving quality of life whilst simultaneously seeking sustainability is a key challenge that requires a major contribution from the educational perspective. Given the relevance of the coastal zone in Uruguay, integral development must rely heavily on recognizing the necessity that sustainable practices and strategic planning in this zone are needed to prevent further degradation. In this context, coastal education for sustainability is a crucial requirement where UdelaR can play a major role.

The originality of the contribution of UdelaR's integrated coastal management initiative to environmental sustainability of Uruguay's coastal zone relies on eight basic pillars, explicitly pursued by the leading group (i.e. a regulated teleology sensu Vallega 1996):

- environmental sustainability is an explicit objective of the initiative;
- interdisciplinary, inter-institutional and integral facets of the program;
- an innovative and participative learning setting;
- the regional perspective pursued in most activities;
- a strong positive feedback between the educational program, research and outreach;
- applied research explicitly directed towards management;
- bidirectional interaction and learning with coastal communities and managers;
- an academic group with a ten-year life history and a long-term perspective.

Conclusions, lessons learned and perspectives

Coastal sustainable development and ICM concepts are closely associated with one another and it is fully accepted that ICM must be based on the best available scientific information, which highlights the need for new capacity-building programs and research that address coastal topics from a sustainability point of view (Leal Filho 2011). In Uruguay, diverse initiatives at UdelaR sustained by regional and Canadian partners are revealing an increasing understanding of the approach to sustainable coastal development through ICM, regionally accepted as the most important conceptual basis for long term coastal sustainability purposes.

Two major outcomes from a decade of fostering interdisciplinary initiatives are highlighted. One is the relevance of blending scientific academic research with traditional knowledge, which can only be done by working tightly with coastal stakeholders. The other one is related to the relevance of a regional perspective for dealing with coastal issues, in our case "the Southern Cone perspective".

Lessons learned from ICM projects indicate the benefit of local processes that merged academia, municipalities and local communities for real action in the planning and management of coastal sites towards sustainable goals. Integration was expressed as the formal participation in different areas and organizational networks existing in each locality, by members of the ICM projects and also by C-MCISur. For example, in the Piriapolis case depicted above, the process resulted in the present participation of the Center as a representative of UdelaR in the Council of the local network of groups. In the La Paloma case, the participation resulted in leading a series of workshops to develop the basis for good governance practices. C-MCISur is now formulating a co-management project for the development of a coastal cluster of Maldonado and Rocha, with the major aim of strengthening local participation processes and stakeholders in ICM.

ICM projects must be seen as participatory processes for developing proposals and identifying conflicts, as well as generating recommendations to be included in future ICM plans. Many cases addressed tourism as an alternative with the power of transforming itself into a sustainable and relevant source of income, especially given recent changes from conventional tourism into a tourism geared towards promoting regional cultural and natural elements.

A cross-cutting question of all ICM projects was how to reconcile economic development in a mainly privately-owned coast where 70% of the population and 60% of the productive activities occur, with the conservation of the natural and cultural assets. The cases agreed on the requisite of developing strategies that included identification of local scale development potentials, development of sustainable tourism, environmental education, improvement of the environmental and social conditions, and promotion of co-management as instruments for sustainable processes.

The goal of the whole educational program has been to deliver, on a coastal setting, core fundaments of sustainable development, with the aim of pursuing a new generation of environmentally aware students. These students will be more ethical and responsible and will soon act as professional decision makers. They are expected to be more proactive and have developed skills to cope with sustainable development challenges (Parker 2010).

Concerning the Southern Cone perspective of the capacity-building and research activities of C-MCISur, we highlight the benefits of an expanded knowledge of the coastal zone management issues in the region, the increased experience in participating in complex projects, and well-developed long-time relationships among partners as key factors for success. Improved knowledge of other countries' ICM experiences helps to broaden understanding of the topics and gives opportunities to contextualize local management approaches to apply ICM theory to new settings. In addition, regionally enhanced links led to potentially expanded networks and further collaborations (e.g. a new Latin American and Caribbean Network of Masters in ICM-FORMIC). Aside from deficiencies and drawbacks, major outcomes of initiatives with a regional perspective outweigh costs because of the synergies created, which result in increased development of global capacities to contribute to coastal sustainability.

A major lesson learned was to realize that engagement of faculty coming from different disciplines in an interdisciplinary dialogue is possible, although time for a natural evolution of the process is required, which in our case was almost a decade. This long-term engagement was not only for course teaching, but also for open academic work to discuss research topics, teaching challenges, and interdisciplinary issues relevant to coastal sustainability. In our case, as in other experiences elsewhere (Steiner and Posh 2006), links with students and communities for developing research in ICM, data collection and workshops, showed the benefits of encouraging faculty-student-stakeholders horizontal relationships.

Presently, the major challenge for the ICM global initiative at UdelaR is associated with the task of carrying on sound research useful for the management of major coastal issues and problems. The construction of a solid long-standing interdisciplinary research group, derived from the original team that carried out the educational program in ICM during the last decade, has shown to be demanding, unlike teaching, of other personal, disciplinary and institutional skills not previously foreseen. The oversimplification of naïvely transforming a teaching task force into a solid research group has proved to be a major challenge and it is expected to face several barriers before success is fully achieved. For this purpose, the complete group of teachers and researchers that builds up C-MCISur is presently working with occupational psychologists with the major task of developing an organizational internal process to build up a competent interdisciplinary academic research group before embarking into more complex coastal sustainability topics.

References

AAU (2011), *The Promotion of Sustainable Development by Higher Education Institutions in sub-Saharan Africa*. GUNi/IAU/AAU. Available at: http://www.aau.org/sites/default/files/esb/promotion-of-sd-by-HEIs-sub-saharan-africa.pdf (accessed 31 March 2012).

Ávila, S., Gorfinkiel, D. and Menafra, R. (2004), *Estudio sinóptico: Identificación de la demanda de capacitación en gestión para el desarrollo sustentable y gobernanza de los espacios costeros*. Informe de ECOPLATA, Montevideo. Available at: www.ecoplata.org (on demand) (accessed 31 March 2012).

Barragán Muñoz, J.M. (2001), The Coasts of Latin America at the End of the Century. *Journal of Coastal Research* 17 (4), pp. 885-899.

– (2010), *Manejo Costero Integrado y Política Pública en Iberoamérica: Un diagnóstico*. Necesidad de Cambio, Red IBERMAR (CYTED), Cádiz.

Chricop, A. (2000), Teaching Integrated Coastal Zone Management: Lessons from the Learning Arena. *Ocean and Coastal Management* 43, pp. 343-359.

Cicin-Sain, B., Balgos, M., Appiott, J., Wowk, K. and Hamon G. (2011), *Oceans at Rio+20: How Well Are We Doing in Meeting the Commitments from the 1992 Earth Summit and the 2002 World Summit on Sustainable Development? Summary for Decision Makers*. Global Ocean Forum, Newark. Available at: http://www.globaloceans.org/sites/udel.edu.globaloceans/files/Rio20SummaryReport.pdf (accessed 31 March 2012).

Ciurana A.G. and Leal Filho, W. (2006), Education for sustainability in university studies: Experiences from a project involving European and Latin American universities. *International Journal of Sustainability in Higher Education* 7 (1), pp. 81-93.

Clugston, R.M. and Calder, W. (1999), Critical Dimensions of Sustainability in Higher Education. In Leal Filho, W. (Ed.), *Sustainability and University Life*, pp. 31-46. Peter Lang Scientific Publishers, New York, NY.

C-MCISur (2011), *Manejo Costero Integrado en Uruguay: ocho ensayos interdisciplinarios*. Centro para el Manejo Costero Interdisciplinario del Cono Sur, UdelaR/CIDA, Montevideo.

ECOPLATA (1999), *Diagnóstico Ambiental y Socio-Demográfico de la Zona Costera Uruguaya del Río de la Plata: compendio de los principales resultados*. ECOPLATA, Montevideo. Available at: www.ecoplata.org.uy/site/docs/compendio.pdf (accessed 31 March 2012).

– (2010), *The sustainability of integrated management in the coastal zone of Uruguay*. ECOPLATA/IDRC, Montevideo.

FREPLATA (2005), *Análisis Diagnóstico Transfronterizo del Río de la Plata y su Frente Marítimo*. Documento Técnico, Proyecto Protección Ambiental del Río de la Plata y su Frente Marítimo, PNUD/GEF/RLA/99/G31. Available at: www.freplata.org/documentos/ ADT/default.asp (accessed 31 March 2012).

GEO Uruguay (2008), *Informe del Estado del Ambiente*. PNUMA/CLAES/DINAMA, Montevideo. Available at: www.ambiental.net/noticias/reportes/GeoUruguay2008.pdf (accessed 31 March 2012).

Gorfinkiel, D. (2006), The Economic Valuation of Coastal Areas: The Case of Uruguay. *Ocean Yearbook* 20, pp. 411-434.

Hareau, A., Hofstadter, R. and Saizar, A. (1999), Vulnerability to climate change in Uruguay: potential impacts on the agricultural and coastal resource sectors and response capabilities. *Climate Research* 12 (2-3), pp. 185-193.

Instituto de Diseño (2010), *La frontera del agua. El paisaje costero de Uruguay*. UdelaR/CSIC, Facultad de Arquitectura, Montevideo.

IOC (2011), *A Blueprint for Ocean and Coastal Sustainability*. IOC/UNESCO/IMO/FAO/UNDP, Paris. Available at: www.unesco.org/new/fileadmin/MULTIMEDIA/HQ/SC/pdf/interagency-blue-paper-ocean-rioPlus20.pdf (accessed 31 March 2012).

IPCC (2007), *Contribution of Working Groups I, II and III to the Fourth Assessment Report of the Intergovernmental Panel on Climate Change*. IPCC, Geneva. Available at: www.ipcc.ch/publications-and-data/publications-ipcc-fourth-assessment-report-synthesis-report.htm (accessed 31 March 2012).

Junyent, M. (2007), The ACES Network: Greening the Curriculum of Higher Education. In McKeown, R. (Ed.), *Good Practices in Education for Sustainable Development: Teacher Education Institutions*, pp. 29-34. UNESCO, Paris.

Leal Filho, W. (2011), About the role of universities and their contribution to sustainable development. *Higher Education Policy* 24, pp. 427-438.

Lemay, M. (1998), *Manejo de los recursos costeros y marinos en América Latina y el Caribe*. Informe técnico ENV-128, BID, Washington. Available at: http://ideas.repec. org/p/idb/brikps/30239.html (accessed 31 March 2012).

Marengo, J.A., Liebmann, B., Grimm, A.M., Misra, V., Silva Dias, P.L., Cavalcanti, I.F.A., Carvalho, L.M.V., Berbery, E. H, Ambrizzi, T., Vera, C.S., Saulo, A.C., Nogues-Paegle, J., Zipser, E., Seth A. and Alves, L.M. (2012), Recent developments on the South American monsoon system. *International Journal of Climatology* 32 (1), pp. 1-21.

Menafra, R., Conde, D., Roche, I., Gorfinkiel, D., Píriz, C., Baliero, W., Biasco, E., Fossati, M., Lorenzo, E., Cortazzo R. and Fournier, R. (2009), Challenges and perspectives for integrated coastal management in Uruguay. *Ocean Yearbook* 23, pp. 403-432.

Menafra, R., Rodríguez, L., Scarabino, F. and Conde, D. (Eds.) (2006), *Bases para la conservación y el manejo de la costa Uruguaya*. Vida Silvestre, Montevideo.

Parker, J. (2010), Competencies for interdisciplinarity in higher education". *International Journal of Sustainability in Higher Education* 11 (4), pp. 325-338.

PROBIDES (1999), *Plan Director de la Reserva de Biosfera Bañados del Este*. Mosca, Montevideo.

Steiner, G. and Posch, A. (2006), Higher education for sustainability by means of transdisciplinary case studies: an innovative approach for solving complex, real-world problems. *Journal of Cleaner Production* 14, pp. 877-890.

UN (2011), *Sustainable development in Latin America and the Caribbean 20 years on from the Earth Summit: progress, gaps and strategic guidelines. Preliminary version*. United Nations. Available at: http://www.eclac.cl/rio20/noticias/paginas/9/43799/2011-565-Rio+20-Summary-WEB.pdf (accessed 31 March 2012).

UNEP (2010a), *UNEP Emerging Issues: Environmental Consequences of Ocean Acidification: Threat to Food Security*. Available at: www.grid.unep.ch/product/publication/download/Environmental-Consequences-of-Ocean-Acidification.pdf (accessed 31 March 2012).

– (2010b), *Latin America and the Caribbean: Environment Outlook*. UNEP/GEO LAC 3, Panama. Available at: www.pnuma.org/geo/geoalc3/ing/index.php (accessed 31 March 2012).

UNESCO (2005), *United Nations Decade of Education for Sustainable Development (2005-2014)*. Available at: www.unesco.org/new/en/education/themes/leading-the-international-agenda/education-for-sustainable-development (accessed 31 March 2012).

Vallega, A. (1996), The coastal use structure within the coastal system. A sustainable development-consistent approach. *Journal of Marine Systems* 7, pp. 95-115.

Wemmenhove, R. and de Groot, W.T. (2001), Principles for university curriculum greening – An empirical case study from Tanzania. *International Journal of Sustainability in Higher Education* 2 (3), pp. 267-283.

Chapter 73

Environmental Management on University Campuses in Southern Brazil

Claudio Senna Venzke, Luis Felipe Nascimento, Luciana Paulo Gomes and Darci Barnech Campani

Abstract

The largest universities in the world have infrastructures and populations equivalent to or greater than many cities. However, they do not have efficient environmental management systems and so often fail to set a good example for the local communities.The environmental management systems of these universities are considered references. One of them is the Universidade do Vale do Rio dos Sinos – (University of the Sinos Valle – UNISINOS), which has a population of over 26,000 students and is located in São Leopoldo city. It was the first university in Latin America to obtain an ISO 14001 certificate for installing an efficient environmental management system (EMS). This system has produced excellent results by reducing environmental impacts especially in terms of water, energy consumption and waste management. The other university is the Universidade Federal do Rio Grande do Sul – (Federal University of Rio Grande do Sul – UFRGS), whose student population is around 46,000. It stands out because it has established a decentralized environmental management system which focuses on training and qualifying its collaborators. They are expected to take on the role of "environmental agents" in their workplaces and press for the adoption and control of solid residue and efficient water and energy consumption programs, besides taking part in other programs such as "green shopping". It is hoped that by reporting the experiences gained in these two projects other universities might follow the same path, since establishing an environmental management system in a university campus is also a way of promoting education for sustainability.

Introduction

Brazilian universities greatly influence the cities and regions where they are located. In some towns, it is not unusual for the local university to be the largest employer and have the largest budget in the municipality.

The campuses of major universities reproduce the structure of cities, because they have student housing, bars, restaurants, hospitals, shops, sports and leisure facilities, local authorities, etc. In large universities the university community is bigger than the population of most Brazilian municipalities. Therefore, an Environmental Management System – EMS implemented on a college campus can be used, with due adjustments, as the basis for the implementation of an EMS in a city with a similar population.

In Brazil, few universities have implemented EMS in their campuses, whilst papers and dissertations were written about these EMSs, for example Termignoni (2012) examined the environmental management systems (EMS) from four universities in Southern Brazil and found that there is a wide disparity among the four systems. The UNISINOS's EMS was analyzed by Venzke (2005) and UFRGS's EMS was analyzed by Rodrigues et al. (2010).

Besides the UNISINOS (University of Vale do Rio dos Sinos) and the UFRGS (Federal University of Rio Grande do Sul), the experiences of the State University of Campinas (UNICAMP), the Federal University of Rio Grande do Norte (UFRN), the Federal University Paraíba (UFPB) and the Pontifical Catholic University of Rio Grande do Sul (PUCRS) are of particular note.

UNICAMP began with a Hazardous Waste Management program involving the management of biological, chemical and radioactive waste produced at the university. In 2007 this program expanded to an EMS, involving the environment in which the university is located.

The UFRN created an Environment Directorate (*Diretoria de Meio Ambiente* – DMA) within the Infrastructure Inspectorate, which is responsible for environmental policy at the university. The DMA is also responsible for the training of human resources for the design, development and maintenance of environmental control systems.

In March 2011, the university prefecture of the UFPb, with support from the multimedia center, implemented the solid waste management system at the university.

In 2010, the PUCRS created the Environmental Management Committee, comprised of faculty and staff from various units, in order to help the university administration formulate policies and activities related to environmental management in their units.

These examples show that higher education institutions can play a leading role in sustainability in their external environment, by developing, cooperating in and generating programs, projects, courses and publications focused on sustainability, involving all possible areas of learning. To achieve this, these institutions must go beyond formal education, which occurs through teaching, research and extension courses, since, because they are recognized and legitimazed by much of the population, they should also take a role in disseminating knowledge by communicating with the external environment. Universities, besides having the potential to be leaders in research and technology, teaching and learning can and, moreover, should involve the community in their results.

This involvement is consistent with the proposals contained the Decade of Education for Sustainable Development (DESD), which was launched in 2005 by UNESCO. The action plan proposed for the decade is based on the vision of a world where everyone has the opportunity to benefit from education and learn the values, behavior and lifestyles required for a sustainable future and for positive social transformation. Thus, the DESD emphasizes the central role of education in the common quest for sustainable development, and its overall goal of integrating the values inherent in this development in all aspects of learning in order to promote behavioral changes that will create a sustainable and more equitable society for all (UNESCO 2005).

UNISINOS and UFRGS are regarded as references in the establishment of environmental management systems on university campuses. This article aims to analyze these two projects and identify the difficulties experienced in the implementation of the respective EMSs, as well as the results and impacts on the external environment of these two universities. Therefore, documents were analyzed and interviews held with key people in the process of implementing the EMSs. It should be pointed out that the authors of this article participated actively in the design and implementation of their respective systems. The theoretical framework used to understand the relationship between universities and their environments involved concepts borrowed from institutional theory (especially the concept of isomorphism), relating them to the role of universities in the pursuit of environmental sustainability. It is hoped that by publishing this article other universities might be encouraged to follow this path.

The Role of Universities in Promoting Sustainability

According to Moore et al. (2005), issues related to sustainability are closely linked to the functions of universities, because education is essential to bring about changes in the expected behavior of the individual. Moreover, social, scientific and technological research and innovation act as levers of sustainable development. In addition, universities are institutions of significant economic importance, helping to accelerate the dynamism of the region where they operate, thus representing an important element in the implementation of a sustainable development strategy. Therefore, education should be seen as a means of achieving sustainable development and not only as an objective of this development. It is worth noting that the role of universities should also go beyond an environmental education focusing only on a rational paradigm with training and information transfer, and should promote teaching methods centered on heightening awareness, changing behavior, developing skills, assessment capability and participation of teachers and students.

This more proactive position of universities is consistent with the proposal from Shrivastava (1995), who argues that corporations (this article focuses specifically on universities) should interact, mutually and reciprocally, with government policies, the community and other stakeholders. These elements can then move in together towards achieving environmental sustainability.

This interaction between actors can be understood from the perspective of the concepts of institutional theory, not as a structuralist approach, within the determinist viewpoint, which sees the environment as the major determinant of changes, but instead based on the interpretative character grounded on the social construction of reality as expounded by Berger and Luckmann (1985). This approach proposes that institutions should aim for and propose social meanings, which, when shared socially, become organizational and institutional rules and procedures that can be used as tools to justify actions. Thus, the Institutional Theory approach adopted as the basis of this article assumes that influence is exerted on organizations by groups or social norms, both internal and external, as proposed by Scott (2001), while the strategies that organizations adopt can be understood as the consequences of institutionalized patterns in the organizational environment.

These patterns can be influenced by coercive, normative or mimetic mechanisms, according to DiMaggio and Powell (1991), who use the concept of isomorphism to explain how organizational characteristics are modified to increase compatibility with environmental characteristics. For these authors, institutional isomorphism is largely responsible for the strategic homogeneity of organizations. Taking into account that the institutional isomorphism assumes that organizations respond similarly to other organizations that are, to some extent, adjusted to the environment, one can consider that the external influence on universities takes place in a coercive manner. This is because this arises from pressures from other organizations on which they are dependent, as well as being due to the cultural expectations of society. Examples of this type of pressure on Brazilian universities would be government mandates, primarily the Ministry of Education, which make universities increasingly homogeneous within the given fields, and organized around rituals of conformity with wider institutions.

Accordingly, in the mimetic and normative mechanisms proposed by DiMaggio and Powell (1991), there is the basis for understanding this influence. Mimetic isomorphism suggests that organizations model themselves on other organizations, and that uncertainty is a powerful force that encourages imitation. It occurs when organizations find themselves in uncertain environments, when their goals are ambiguous or when the technologies being used are challenged. Thus, to obtain a faster response they imitate organizational models that have proven successful in such environments.

This mimicry is defended by Leal Filho and Schwarz (2008), who report the experience of the Regional Centre of Expertise in Education for Sustainable Development – RCE) in Hamburg, Germany. The authors point out that, through projects and initiatives based on action, synergies can be achieved between regional centers of expertise in education for sustainable development and agencies based at the local level, influencing other organizations to adopt a more sustainable posture, by replicating successful projects.

Based on the above, this article examines the cases of UNISINOS and UFRGS that have sought to maintain and improve their credibility within the academic community and the region by focusing on environmental awareness, conforming to the specific legislation and promoting growth in harmony with nature.

Methodology

A qualitative research design was adopted in an attempt to further the understanding of a social phenomenon, in the case of this article, environmental management on university campuses. This study is exploratory since it procures data from the universities, recorded systematic data and non-systematic data obtained during interviews. The descriptive character refers to the reporting of the actions of the universities regarding the introduction of educating for sustainability with their populations and within their environments. As the study's objective was to investigate a current event, within a real context, it was decided to study two important cases.

The two cases were defined as the units of study based on the following criteria: leadership in establishing environmental management systems on university campuses in Brazil, universities whose academic high standing is recognized at a national level, a university community of more than 25,000 inhabitants, a private university – the *University of Vale do Rio dos Sinos* – UNISINOS and a public university – the *Federal University of do Rio Grande do Sul* – UFRGS.

UNISINOS, which is located in São Leopoldo, Rio Grande do Sul, has been in existence for over 40 years. It has over 900 teachers, of whom 86% hold postgraduate qualifications, approximately 900 employees and about 26,000 students. UFRGS is located in Porto Alegre, the capital of Rio Grande do Sul, and has campuses in other cities in the state. The first faculties that were to form UFRGS were founded in 1895 and the university was founded in 1934. In 2011 it had 46,000 students enrolled, of whom 18,500 are students in graduate courses. It has 2,541 teachers and 2,553 employees.

The data collection process consisted of documental searches, whereby the documents regarding the environmental management systems (EMS) of the two universities concerning courses related to environmental projects were reviewed. Between August and October 2011, semi-structured interviews were held with the managers of the EMSs at UNISINOS and UFRGS. The script of these interviews addressed the process of implementation; motivations and difficulties in the implementation process and maintenance of the EMS, achievements and prospects for those systems.

A qualitative approach was used in the data analysis. The data were selected and subsequently simplified. They were then organized so as to enable systematic analysis of the similarities and differences, as well as their relationship to the theme and later conclusion, during which the data were purified.

This approach aimed to present and to discuss the implementation of these EMS for futures references and studies. Our objective is not to provide formulas or a prescriptive way to implement an EMS, but to stimulate universities and other educational institutions to think and make decisions about sustainability.

Data Presentation and Analysis

Below, the EMS projects at UNISINOS and UFRGS are presented separately. In item 4.3 the relations between the two systems are examined.

The Environmental Management System at UNISINOS

As from 1997, project called "Green Campus" was developed with the purpose of directing the environmental actions at UNISINOS, so as to promote the preservation, improvement and restoration of environmental quality. In the beginning, the selective waste paper collection project was developed, which over time has been improved and continues to operate.

In 2002, the first planning meetings were held to organize the implementation of the Environmental Management System (EMS) with the object of formalizing and restructuring the work already being performed on the university campus. The goal was to provide UNISINOS with an effective management system that could achieve and demonstrate its environmental goals by controlling the impacts of its activities on the environment. To support this, it was decided to apply for the ISO 14001, the international standard dealing with Environmental Management Systems, which was reached in November 2004.

From then on, the challenge was to make the campus a huge laboratory or field of experiments for courses and research carried out in the institution. It is worth remembering that the initial project was founded on the strategic orientation of UNISINOS, which is committed to regional development, with transdisciplinarity and education for life. Thus, the expected result was the continuous improvement of environmental performance, pollution prevention and environmental conservation and transmission of the knowledge generated.

The certification process, the difficulties encountered in the implementation and the results achieved are described below.

The Certification Process

The description of the certification process refers to certain items contained in the ISO 14001, but it does not address each item separately and not in the same order the items appear in the standard, since it seeks to describe the steps in a chronological order.

The certification process officially began in April 2003 with the Chancellor signing the resolution certifying the commitment of senior members of the university administration to the certification process. This was essential for successful implantation and conforms with NBR ISO 14001 (ABNT 2004), regarding the important role of senior management in the certification process.

Table 1 shows the factors that influence the implementation of an EMS, showing the analysis of internal and external environments, especially the identification of the opportunity to create new courses in the environmental field.

Table 1: Factors influencing the implementation of the EMS

<table>
<tr><th colspan="2" rowspan="2">Strategic diagnosis for implementing the EMS</th><th colspan="2">Analysis of the internal environment</th></tr>
<tr><th>Predominance of vulnerabilities</th><th>Predominance of potentialities</th></tr>
<tr><td rowspan="2">Analysis of the external environment</td><td>Predominance of threats</td><td>SURVIVAL
Cost savings due to greater control of processes.</td><td>MAINTENANCE
Occupation of niches, being the first university in Brazil with environmental certification.</td></tr>
<tr><td>Predominance of opportunities</td><td>GROWTH
Innovation through the certification influencing and strengthening the entire institution in the search for new ways of working.</td><td>DEVELOPMENT
Development of new courses dealing with environmental issues</td></tr>
</table>

The Environmental Policy at UNISINOS

The environmental policy is made up of a statement of principles and intentions of the organization in relation to its environmental performance that should guide the planning of actions and establish goals and targets (ABNT 2004). Therefore, it is essential that it is formally presented and published, since it ensures that the public have access to the vision and commitment of the organization to the environment, which should be relevant to the nature, scale and environmental impacts of activities, products and services of the organization, as expressed by Almeida (2000).

The description of how UNISINOS aims to reduce the negative effects of its activities and its commitment to the continuous improvement of its environmental performance are expressed in its environmental policy, as follows:

> "The *Universidade do Vale do Rio dos Sinos,* by promoting and defending life, is committed to act in the prevention of pollution and conservation of the environment, in compliance with the current legislation and other requirements, providing continuous improvement to the Environmental Management System for the sustainable development of its campus and opportunity for the generation and transfer of knowledge and technology to the community" (UNISINOS 2011).

According to the university's EMS manual UNISINOS (UNISINOS 2011), the chancellery defined the environmental policy to reflect its commitment to continuous improvement, the prevention of pollution, and in compliance with the law, applicable environmental standards and other requirements subscribed by UNISINOS. This document sets out the objectives, targets and environmental management programs, and may be reviewed as a result of critical analysis, in view of changes in legislation, advances in science and technology, lessons learned from local, regional or global environmental accidents, market trends, strategic positioning, records of complaints, in addition to changes in expectations of stakeholders.

Through a website and its channels of communication, UNISINOS promotes internal communication among the various levels and functions of the institution in terms of its environmental and other requirements of its EMS, as well as receiving, documenting, and responding to communications from external interested parties.

Controlling and Assessing the Environmental Management System

The UNISINOS EMS has a procedure for managing the operational controls of the system, which includes mechanisms of informing suppliers of the requirements they have to meet.

The group that implemented the system identified operations and activities related to environmental aspects that required operational controls and developed operating instructions (OIs), as well as other documents to cover situations where their absence could lead to deviations from the goals and objectives of the university's environmental policy.

The OIs describe the activities step by step, including operating parameters and/or a flowchart of the activities, aimed at standardizing the execution of tasks, activities and services associated with the identified significant environmental aspects.

Auditing is an efficient way of evaluating the effectiveness of the EMS, by which it is possible to check that the EMS is being conducted in accordance with the requirements of the standard. Accordingly, UNISINOS has a procedure for the development of periodic environmental audits of their EMS, including the selection of internal auditors in accordance with specific criteria. These audits are intended to determine whether the EMS at UNISINOS:

- Conforms to provisions planned for environmental management, including the requirements of the ISO 14001
- Is being properly implemented and maintained.

The internal audits are planned and programmed according to the environmental importance of each activity involved, as well as the results of previous audits. This programming and the related documents include the areas and activities to be audited, the frequency and methodology of the audits, as well as the responsibilities and requirements for their implementation, and communication of results to the chancellery. This is, according to Moore (2005), the purpose of the integration of planning, decision-making and evaluation.

The Difficulties Encountered

The major challenge was to develop an internal communication plan which would be capable of attending to the basic beliefs of the people, so that they would make the effort to behave as required in relation to environmental issues, that is, they would adopt environmental attitudes. This point is reinforced by the search for communication guidelines used by organizations seeking to conform to the environmental questions present in the ISO 14063 (ABNT 2006), where there are two key points for a more recent view of institutional communications. One point is to consider communication as a process focused on interest groups in the same way as the quality process, for example, focused on customer satisfaction as suggested by Seiffert (2007). The other point to consider is the environmental communication as a process of sharing information on environmental issues between organizations and their stakeholders in order to build trust, credibility and partnerships, to educate those involved, and then use the information in decision making.

Achievements

This item presents the results obtained with the environmental certification of the university and is divided into two sub-items, one, related to internal issues, is focused primarily on the objectives of the EMS, and the other is related to projecting sustainability into the external environment.

The Internal Results

The results achieved internally as a result of the certification are expressed through the Environmental Management Programs (EMPs). UNISINOS has a procedure for creating and managing the EMPs. These are established via the implementation groups and conducted by their respective managers, in order to achieve objectives and targets, manage significant environmental aspects, as well as assist in facilitating and implementing the university's environmental policy. Each EMP assigns responsibilities at the relevant functions and levels, defining the means and timeframe for each objective and goal to be reached.

With respect to the development of academic programs, a point worth mentioning is the creation of courses in Environmental Management in 2005, and Environmental Engineering in 2010. The educational projects for these courses were strongly influenced by the knowledge generated during the implementation process together with a study of the market for these areas, as proposed by Miller et al. (2011) who argue that to produce knowledge aimed at sustainability, the process of knowledge production itself must be changed to be more adaptable and engaged with society.

Another result of certification, internally, has been a reduction in the consumption of water and energy and greater credibility and trust, which has been achieved in the course of attempting to increase the academic community's awareness of environmental issues. Certification facilitated conformity to the requirements of environmental legislation, encouraged continuous improvement in environmental performance, thus making the campus a huge laboratory or experimental area for the courses and research carried out at UNISINOS. At this point, one can expect induction by mimicry, as suggested by DiMaggio and Powell (1991), to all areas of the university.

The External Results

UNISINOS has the Social Technologies for Solidarity Enterprises Project. An example of the result of this project was the participation of teachers and students from different university courses in the Cleaner World training group, which is composed of women from communities in the vicinity of UNISINOS and housewives from low-income families. This group began making soap, detergent and fabric softener from used cooking oil, initially accompanied by the *Caritas* network (the Catholic Church's social activities network), then the Social Technology Group from UNISINOS joined the process in 2009, so that they could expand their production and market. This project expanded the frontiers of the university's activities, in accordance with Tachizawa's (2010) recommendation regarding organizations seeking new areas of influence. The activity involved students from Environmental Management, Environmental Engineering, Food Engineering, Design, Mechanical Engineering, Production, Physical Therapy and Psychology. Aside from the students going to the location where the Cleaner World group runs its activities, the group members also go to the campus, the laboratories and training meetings, and so are exposed to new experiences. The experiment was indicated for and received a prize of 40,000 Reais to expand the project from the Banco Real Solidarity University Project in 2010.

The project has stimulated considerable interest from other educational institutions through visits and interest in knowing the environmental management systems, which highlights it level of external influence. This represents another example of institutional isomorphism, through the mechanism of mimicry.

The Environmental Management System at UFRGS

Over the years, several activities aimed at increasing the environmental awareness of the university community have been developed at UFRGS, as well as external actions on the initiative of students or teachers. However, there was no link between the different units or between the researchers working on issues related to environmental management. The environmental awareness initiatives of the University Chancellery were not connected with the projects and research groups.

The proposal to establish a link between researchers and the formation of a group arose due to an external demand. The *Companhia Petroquímica do Sul* – Copesul, the producer of raw materials at the Triunfo Petrochemical Complex in Rio Grande do Sul, sponsored events and projects related to environmental issues in some units of UFRGS. Its managers asked the administration of UFRGS to develop integrated actions in relation to urban solid waste management and design proposals for a broader sustainable development. To this end, the University invited researchers in engineering, chemistry, biology, ecology, management, education, among others, to examine this request. Thus, in 1999 the Interdisciplinary Group for Environmental Management (*Grupo Interdisciplinar de Gestão Ambiental* – GIGA) emerged (Cela 2011).

Of particular note among the difficulties encountered in the structuring the GIGA are the different languages and paradigms of its participants. Although everyone worked with issues related to the environment, some terms were found to have different meanings for the engineer, the educator or the biologist. The understandings on how to solve the problems were different, as were the priorities for implementing the projects. IA long period of coexistence was necessary in order to reach some degree of homogenization of the vocabulary and knowledge of how each researcher worked in their respective areas.

The GIGA was them an informal articulation of researchers within UFRGS. In 2004, the Group produced a document and presented it to all the candidates for the elected post Chancellor of UFRGS, proposing a formal structure to promote and coordinate environment management at the University. The elected candidate fulfilled the campaign promise and, in 2005, created the Environmental Management Coordinating Committee (*Coordenadoria de Gestão Ambiental* – CGA), which now has the active participation of members of GIGA (Campani et al., 2005). With the activities being channeled to the CGA, the GIGA ceased to exist.

The Environmental Management Coordinating Committee – CGA

The CGA became the organ at UFRGS responsible for developing, implementing and monitoring the University's Environmental Management System. This Coordination Committee, directly linked to the Chancellors Office, has is responsible for managing all the programs and projects upon within the EMS, as well as keeping track of all processes within the University that include aspects related to environmental issues.

The CGA has a coordinator, appointed by the chancellor, and a permanent technical team. Faculty and staff from various units at the university also take part. This Coordination Committee held discussions regarding the development of Environmental Policy at UFRGS which was defined as follows:

> "The UFRGS, through its Centralized Administration and Directors of its Bodies, is committed to continually improving its environmental performance, preventing pollution, adopting procedures and practices aimed at preventing negative environmental impacts, in accordance with the legal requirements, generating alternatives for sustainability of the university community and society as a whole and developing a strategy for cultural change through a policy of environmental education." (UFRGS 2012)

The CGA used the "asset agents" model, in which each unit of the university is required to have a person who is responsible for controlling its assets, to create the figure of the "environmental agent". The environmental agent is an employee or teacher who takes responsibility for promoting and ensuring the environmental management in their unit. To this end, the CGA has organized extension courses to train the environmental agents, and each university building should have at least one environmental agent. As from the seventh group, it also began to train monitors and environmental inspectors.

At UFRGS the EMS is based on the environmental agents, who are responsible for reporting the proposals made by the CGA and overseeing the implementation of environmental management actions within their units. In addition to environmental agents, the following bodies at the UFRGS also formally integrate the EMS: The Center for Management and Treatment of Chemical Waste, the Radiation Protection Service and the Wildlife Refuge.

The ISO 14000 certification remains a goal, but the certification process should occur on the individual campuses and units at different times, depending on the conditions of the EMS on each campus.

The CGA became the university's link with society and government policies for the sector. For example, the Environmental Agenda in Public Administration – A3P of the Environment Ministry is being implemented at UFRGS through the actions and projects of the CGA. The A3P includes criteria ranging from the changing of investment plans, procurement and contracting of services for public institutions to proper management of waste generated and natural resources used.

The Implementation and Management of the EMS

The environmental management actions occur inside and outside the UFRGS. Internally, in addition to training program for the environmental agents, management tools were used to identify the environmental aspects and impacts of each unit. The "FMEA" (Failure Mode and Effect Analysis) tool was used to identify the environmental problems and the "5W2H" [What – what will be done (steps); Why – why is it done (justification) Where – where it will be done (local); When – when it will be done (time); Who – who will do it (responsibility) How – how it is done (method) How much – how much will it cost (cost)] was used to coordinate the action plan.

The problems encountered were classified using another spreadsheet, the Environmental Risk Index (ERI), which classified each problem. The total number of problems represents the ERI for each unit. The spreadsheet provides the unit manager and the system manager with a map of environmental issues and an indication of how critical each problem is.

The process of identifying measures to minimize potential impacts was facilitated by the use of the FMEA and 5W2H. With each round of the FMEA there was a reduction in the ERI score.

The Environmental Risk Index (ERI) varies from 1 to 10,000 points, and the sum of ERIs provides the full amount – the higher the ERI score, the greater the priority in solving the problem. The aspects and impacts are analyzed according to the following criteria (Rodriguez 2010):

- Severity of impact (S): Measures the severity of an environmental impact that may result from a potential failure.
- Occurrence of the cause (O): Deals with the probability of a cause or mechanism occurring.

- Degree of detection (D) Relationship between the detection and resolution of an occurrence.
- Ease of implementing the recommended action (E): Lists costs, the number of people involved and the time taken to implement the action plan.

The Environmental Risk Index is the result of the multiplication of different factors, where ERI = Index S x Index O index D x Index E.

The 5W2H tool is used to develop the action plan by facilitating the data interpretation. Table 2 demonstrates how, by means of the seven questions (What, Who, When, Where, Why, How and How much), the action plan is clearly defined.

Table 2: The action plan (5W2H)

N	Action	In Charge	Time	Place	Reason	Procedure	Budget
	What?	Who?	When?	Where?	Why?	How?	How much?

The Results

With the implementation of the EMS in a pilot unit of the university, after a year of monitoring, the consumption of water and electricity were reduced. There was a 16.25% reduction in water consumption, which equates to 283,000 liters of drinking water saved, which is significant for a building that maintained the same population and the same services.

The water and energy savings resulted from the installation of self-closing taps in the bathrooms and occupancy sensors in hallways and library and particularly, from the awareness of the teachers, staff and students who circulate daily through the building of the pilot unit. Simultaneously, other measures were adopted, such as the purchase of recycled A4 paper and the removal of disposable plastic cups in the administrative sector, which meant each employee was required to use their own mug to drink water and coffee.

Twelve months after implementing the EMS, there was a reduction of 124,224 points (56.13%) in the ERI. It should be noted that in this period there were no significant investments, the measures adopted were simple and easy to adopt, requiring changes of habits among the people working in the building.

Internally, the pilot unit trained three environmental agents and implemented an EMS that included a program of selective waste collection, reducing water and energy consumption and installing battery and cooking oil collectors. In relation to the academic programs, the content of all the subjects in the courses offered by the unit were analyzed and the inclusion of content related to environmental management was proposed. The teachers responsible for their respective disciplines were informed where they could find the literature and websites for their research. Therefore, despite having little or no background in the area, teachers were quickly able to assimilate the content suggested for their disciplines.

Externally, a partnership was established with a waste sorting cooperative which receives the dry waste from the university. Themed blogs were created, to which the students contributed throughout the semester with information on topics such as water, energy, waste, conscious consumption, climate change and green marketing.

Workshops were developed in order to spread the word regarding education for sustainability in elementary and high schools. The workshops covered topics like conscious consumption, water and energy conservation, climate change and green washing, among other topics. In these workshops the students participated in practical activities related to the topics discussed.

Within the UFRGS as a whole, over ten editions of the extension courses offered by the university professors to qualify the environmental agents have been held. These courses are currently open to the community. The CGA develops several projects internally and represents the UFRGS at events, committees and community organizations.

The actions of the CGA go beyond those required by the environmental control agencies. An example of this is the investments in green purchasing currently being carried out by the university.

A Comparison of the EMSs at UNISINOS and UFRGS

The EMSs in the two analyzed cases are similar in some respects and different in others. Table 3 provides an opportunity to compare some features of the EMSs at of the two universities.

The environmental policies of the two universities are similar to one another, as are the difficulties encountered in implementing their EMSs. Following the implementation of EMS, the internal results show that UNISINOS created two undergraduate courses in the area, while UFRGS invested more in extension courses aimed at training environmental agents. Both systems led to reductions in consumption, the prevention of accidents and compliance of the universities to Environmental Law, according to Moura's (2002) proposals.

As external results, the UFRGS and in UNISINOS both formed closer ties with the poorer sectors of society.

Table 3: Features of the EMSs at UNISINOS and UFRGS

EMSs	UNISINOS	UFRGS
Implementation of EMS	2002	2005
Implementation of ISO 14000	2004	Does not have
Environmental Policy	The Universidade do Vale do Rio dos Sinos, by promoting and defending life, is committed to act in the prevention of pollution and conservation of the environment, in accordance with the current legislation and applicable other requirements, continuously improving the Environmental Management System for the sustainable development of its campus and providing opportunities for the generation and transfer of knowledge and technology to the community.	UFRGS, through its Centralized Administration and the Directors of its Bodies, is committed to continually improving its environmental performance, preventing pollution, adopting procedures and practices aimed at preventing negative environmental impacts, in accordance with the legal requirements, generating alternatives for sustainability of the university community and society as a whole and developing a strategy for cultural change through a policy of environmental education
Control Mechanisms	– Operational Instructions – Internal Audits	– Support from environmental agents – Use of management tools (FMEA, 5W2H)
Difficulties encountered	– Making people adopt environmental attitudes – Stimulating awareness among the university community	– Differences in language and paradigm among the participants – Stimulating awareness among the university community

EMSs	UNISINOS	UFRGS
Internal Results	– the creation of courses: Environmental Management (2005) and Environmental Engineering (2010) – continuous improvement in environmental performance – awareness among the university community	– water and electricity savings – incentive for the inclusion of environmental content throughout tem course offered by the pilot unit – changes in the habits of the community
External Results	– Involvement of teachers and students from various courses in the university in training members of the Cleaner World Group – Prize from the Banco Real Solidarity University – interest from other universities in knowing the SEM at UNISINOS	– blogs with information on topics related to sustainability – Partnership with the waste sorters cooperative – Workshops on sustainability in primary and secondary schools

Conclusions

The results already obtained by the UFRGS and UNISINOS, together with the implementation of their EMSs, confirm the proposals made by Tachizawa (2010), in relation to new frontiers. According to the author, the management of modern organizations goes beyond their borders as it also influences those around them, Here, in the analyzed cases, this influence covers suppliers, students, the surrounding community and other educational institutions.

In this sense, it can be seen that both the universities are in an advanced stage of environmental responsibility, as they have already implemented EMSs and are anticipating the legal requirements and the demands of the public. Moreover, the experiences of UNISINOS and UFRGS are encouraging other institutions to follow their path, as shown by the number of visits and requests for information about the operation of their EMSs. This stimulus for the external environment is similar to the mimetic isomorphism suggested by DiMaggio and Powell (1991), and internally, the experiences serve as experiments and proof that the concepts discussed in class are real and achievable.

Thus, the experience of environmental management on university campuses has been shown to be a strong argument supporting awareness and behavior change for a significant portion of the university community. The repercussions of the training of students will be perceived in the medium and long term, strengthening the proposal of Leal Filho and Schwarz (2008), to engage interested parties in the context of sustainability. These initiatives also concur with the objectives of the Decade of Education for Sustainable Development, which recommends that everyone should have the opportunity to benefit from education and learn the values, behavior and lifestyles required for a sustainable future and the positive transformation of society.

The characteristics of social and environmental sustainability have a profound effect on how education institutions should be structured to include it in their objectives; one of the roles of academia is generation of interdisciplinary research and scientific knowledge based on solving problems in society. Therefore, this paper aims to stimulate this discussion in higher education.

References

ABNT. NBR ISO 14001 (2004), *Sistemas da gestão ambiental – Requisitos com orientações para uso*. ABNT, Rio de Janeiro, RJ.

ABNT. NBR ISO 14063 (2006) *Sistemas da gestão ambiental – Comunicação ambiental*. ABNT, Rio de Janeiro, RJ.

Almeida, J.R. (2000), *Gestão ambiental: planejamento, avaliação, implantação, operacionalização e verificação*. Ed Thex, Rio de Janeiro, RJ.

Backer, P. (2002), *Gestão Ambiental: administração verde*. Ed Qualitymark, São Paulo, SP.

Berger, P., Luckmann, T. (1985), *A construção social da realidade*. Vozes, Petrópolis, RJ.

Campani, D.B. et al. (2005), A Gestão Ambiental da Universidade Federal do Rio Grande do Sul. In: *V Congresso de AIDIS – Sección Uruguaya*. Montevideo, Uruguay.

Cela, S.R. (2011), *Barreiras à Implantação do Sistema de Gestão Ambiental na Escola de Administração da UFRGS*. Monograph. PPGA/EA/UFRGS, Porto Alegre, RS.

DiMaggio, P.J., Powell, W.W. (1991), The iron cage revisited: institutional isomorphism and collective rationality in organizational fields. In Powell, W.W., DiMaggio, P.J., *The new institutionalism in organizational analisys*, pp. 63-82. The University of Chicago Press, Chicago, IL.

Leal Filho, W., Schwarz, J. (2008), Engaging stakeholders in a sustainability context: The Regional Centre of Expertise on Education for Sustainable Development in Hamburg and region. *International Journal of Sustainability in Higher Education* 9 (4), pp. 498-508.

Miller, T.R., Muñoz-Erickson, T. and Redman, C. (2011), Transforming knowledge for sustainability: towards adaptive academic institutions. *International Journal of Sustainability in Higher Education* Vol. 12 (2), pp. 177-192.

Moore, J., Pagani, F., Quayle, M., Robinson, J., Sawada, B., Spiegelman, G., Wynsberghe, R.V. (2005), Recreating the university from within: collaborative reflections on the university of British Columbia's engagement with sustainability. *International Journal of Sustainability in Higher Education* 6 (1), pp. 65-80.

Moore, J. (2005), Seven recommendations for creating sustainability education at the university level: A guide for change agents. *International Journal of Sustainability in Higher Education* 6 (4), pp. 326-339.

Moura, L.A.A. (2002), *Qualidade e gestão ambiental: sugestões para implantação das Normas ISO 1400 nas empresas*. Ed. Oliveira Mendes, São Paulo, RS.

Rodrigues, A., Campani, D., Loguercio, A., Cela, S. e Winckler, M. (2010), *Sistema de gestão ambiental na Escola de Administração – UFRGS*. 2º Congresso Internacional de Tecnologias para o Meio Ambiente. CITMA, Bento Gonçalves, RS.

Scott, W.R. (2001), *Institutions and organizations* (2nd. ed.). Sage Publications. Thousand Oaks, CA.

Seiffert, M.E. (2007), *ISO Sistemas de gestão ambiental*. Atlas, São Paulo, SP.

Shrivastava, P. (1995), The role of corporations in achieving ecological sustainability. *Academy of Management Review* 20 (4), pp. 936-60.

Tachizawa, T. (2010), *Gestão ambiental e responsabilidade social corporativa*. Ed Atlas, São Paulo, SP.

Termignoni, L.D.F. (2012), *Framework de sustentabilidade para instituições de ensino superior comunitárias*. Master degree dissertation. PUC-RS, Porto Alegre, RS.

UNESCO (2005), *Década das Nações Unidas da Educação para o Desenvolvimento Sustentável, 2005-2014: documento final do plano internacional de implantação*. UNESCO-OREALC, Brasília, DF.

UNISINOS (2011), *Sistema de Gestão Ambiental da Universidade do Vale do Rio dos Sinos*. Available at www.unisinos.br/sga (accessed 15 December 2011).

UFRGS (2012), *Sistema de Gestão Ambiental da Universidade Federal do Rio Grande do Sul*. Available at http://paginas.ufrgs.br/sga/ (accessed 10 January 2012).

Venzke, C.S. (2005), A certificação ambiental do campus da Unisinos. In *VIII ENGEMA – Encontro Nacional sobre Gestão Empresarial e Meio Ambiente*. ENGEMA, Rio de Janeiro, RJ.

Chapter 74

A Development Model of Social Responsibility Competencies for Sustainable Development in the School of Economics and Business at the University of Chile

Oscar Jerez Yañez, Freddy Coronado Martínez and Gabriela Valenzuela Arcuch

Abstract

Professional social responsibility has gained importance in universities in recent years due to several key events that have occurred around the world. Clear examples of such events are the Enron case, the subprime crisis, the oil disaster in the Gulf of Mexico, and the case of the 33 miners trapped underground for two months in the north of Chile. Even though these events occurred in different industries, they have a clear pattern in common. They have occurred mainly due to the irresponsibility of professionals who have most probably been educated at the most prestigious higher education institutions in the world. Since its foundation in 1842, the University of Chile has declared as its main priority the education of students who in their professional career will best serve the nation. Consequently, the University of Chile should provide an academic program that forms both sustainability-conscious and environmentally responsible professionals. In line with this ethos, the School of Economics and Business of the University of Chile has created a model facilitating the development of social responsibility as a fundamental skill in the undergraduate profile. Processes and learning experiences need to foster the development of key competencies for sustainable development. For this reason, a model based on criteria and experiences that favor the development of these competencies has been designed. The model considers four key aspects, each with their own respective criteria, indicators and evidence. These aspects are: the link between the environment and its needs, teaching and learning processes, extra-curricular activities, and internships or final degree activities. This model was implemented in the School of Economics and Business at the University of Chile in order to guide student practice, and as a diagnosis based on the undergraduate experience and associated activities. The lessons learned after the initial application of the model and the challenges that arose from this process have been systematized and are now made available as key references to consider when implementing these types of mechanisms in higher education.

Introduction

During the Rio Declaration (1992), the concept of sustainable development was adopted internationally as a result of the 1987 Brundtland Report of the World Commission on Environment and Development of the United Nations. For its part, the ISO 26000 (2010) of Social Responsibility (SR), states that a firm has the objective of contributing to sustainable development and society's welfare through ethical behavior and being transparent about its impacts. The way in which firms are to be managed underwent great change, placing emphasis on assuming responsibility in terms of impact of the environment on society as a whole.

These considerations must also be present in the formation of future professionals. Since the United Nations Conference in Environment and Development (1992), higher education has been considered a key factor in providing solutions in sustainable development. At present, higher education institutions and business schools in particular, have been presented with rather special challenges (Kaivola and Rohweder 2007; Steinemann and Asce 2004). Firstly, they must be in possession of a curricular design that includes sustainability challenges. Secondly, teaching and learning methods promoting the development of responsible professional behavior must be implemented (Svanström et al. 2008). Thirdly, the development of research and innovation in this area must gain importance. Finally, universities must have a university management model that includes these guidelines to responsibly manage its impacts.

Concerning the teaching and learning process, research has concluded that interpersonal abilities must be considered as a basis for developing key competencies for sustainability. In addition, sub-competences such as systemic thinking, anticipation, understanding current regulations, strategic thinking, and communication skills must also be developed (Fry et al. 2009; Willard et al. 2010; Nijhuis et al. 2007; Puntambekar and Hubscher 2005). Higher education literature states that to achieve effectiveness in developing these competencies, it is important that the teaching/learning processes maintain a learning environment (Warburton 2003), consisting not only of formal classes, but also additional curricular or extracurricular activities. One example of these activities could be interdisciplinary cooperation activities, as well as also volunteer work (Barth et al. 2007).

It is within this context that the School of Economics and Business at the University of Chile, through its University Social Responsibility Unit (NexoRSU), has been developing a series of initiatives that take into consideration the formation of our undergraduate student and the great challenge our nation faces in terms of sustainability. Based on the implementation of these initiatives and the analysis of the results, a model was developed as a framework to articulate and align student experiences with the development of competencies for sustainable development. In addition, this model is also intended to provide guidelines to practical questions such as: What formative activities further promote the development of Social Responsibility (SR)? or What attributes or criteria are key when considering such initiatives?

The following chapter introduces the model. The third chapter presents the main lessons learned by implementing the model. The fourth chapter summarizes the contribution of the different initiatives to the dimensions identified in the model. Finally, the fifth chapter provides the main conclusions of this experience and the future challenges that remain.

Development Model of Social Responsibility Competence

This model was created through the systematization of lessons learned and practices made in the School of Economics and Business at the University of Chile during recent years. For this reason, based on the student's evaluations of our initiatives, extensive interviews with professors, directors and pedagogical experts, the key criteria in order to take action were identified, which were in turn organized into key training focus points. In summary, the model can be defined as a group of criteria linking the formation and competency of Professional Social Responsibility (PSR), grouped by the key dimensions of the teaching and learning process, thereby allowing for the implementation of initiatives and activities alongside the educational experience as a great step forward towards sustainable development. The main innovation of this model is that it facilitates the creation of a learning environment in which competence in social responsibility can be developed. This model is not only based on cognitive knowledge, but also on teaching and learning experience with replication possibilities. Thus, it suggests general conditions and specific activities that can be implemented in other institutions.

Four key elements can be considered: dimensions, criteria, indicators or conditions and the evidence or results of action initiatives. The dimensions mentioned are those parts of formation that interact dynamically with one another, and through which the required actions for professional development may be carried out (Brundiers et al. 2010). Criteria are orientations that define and link each one of the dimensions in direct relation to the PSR. Indicators or conditions are observable and evidential realizations of the fulfillment of the criteria. Finally, the product's or action's evidence matches those initiatives or effects where it is possible to determine how it was developed and in what way the competence of PSR is achieved. These actions or products are variable, and are indistinctively established by each respective institution, depending on context, resources, capabilities and educational projects, amongst others.

The dimensions of the teaching and learning process considered are:

a) **Build a link between the environment and its needs**: the purpose of this dimension is to create a guideline to connect the theoretical analysis of a social dilemma with realistic requirements based on stakeholders analysis (Brockbank and McGill 1998; Light et al. 2009; Michelsen and Rieckmann 2008; Brundiers 2010). This dimension has four criteria and it is transversal to the other three, giving new meaning to the question of "what for" in PSR. The first two are intended to allow third parties involved in each activity to play an active role in the definition and modeling of the problems, proposals and also the implementation of solutions. Below, the third criterion requires that each activity involves a reflective evaluation process by the stakeholders. Finally, the fourth criterion is intended to provide a monitoring mechanism to each activity and to develop certain basic abilities in the stakeholders that may facilitate the sustainability of the presented solutions.

b) **Curricular Activities**: refers to specific academic courses or requirements included by design on the students' program of study. This dimension has five criteria. The first two allow for the planning process of the courses, both in the original syllabi and in specific activities, to intentionally include the objectives related to the development of PSR, the required knowledge, and the necessary conditions to achieve these objectives at the student level (Clark 2010; Diamond 2008; Russell et al. 2010). The third and fourth criteria refer to the importance of having a professor monitor the activity in order to guarantee opportune feedback and profound analysis, and thereby allowing the student to meet their learning objectives. Finally, the fifth criterion intends to ensure that professors are adequately prepared to fulfill their role as facilitators of the teaching and learning process (Hobson et al. 2009; Nevins Stanulis and Floden 2009).

c) **Extra-curricular Activities**: These activities provide an opportunity for students to become engaged and involved in their education outside of their academic course. Participation is voluntary; the activities are organized by the students themselves and simultaneously monitored by the students and the university. These types of initiatives include freedom of choice and therefore favor the development of PSR competencies in a manner that feels natural, allowing the students to express themselves and to develop their individual interests (Schugurensky 2000). The three criteria defined in this dimension seek to make this freedom of choice available to be used under conditions that ensure the sustainability of the relationship between the faculty and third parties (first and second criteria), as well as the development of PSR competencies (third criterion).

d) **Internships and final degree activities**: this area refers to those actions or activities that intend to consolidate, by way of complete professional immersion, the real teaching and learning process (Bell and Kozlowski 2009; Gibbs 2009; Wood and Smith 2007). These activities are usually developed individually by the student in organizations outside of the university (professional internships) or guided by a professor (undergraduate thesis). In this context, it is possible to show the impact of the model in professional practice. The three criteria included intend to ensure an adequate support process that allows students to: link PSR and their profession (first criterion), to fuse the activities and PSR criteria (second criterion) and ultimately assess the experience in terms of achievements of PSR competence.

Overall, this model is characterized by

- The use of a variety of actions, procedures or initiatives, intending the development of PSR in students.
- The customization of these actions to the manner in which the student learns (metacognitive) and appreciates the importance (self-system) of the achievement of competency in PSR.
- The impact evaluation on the development of PSR competence in students.
- The guiding of the development of PSR competency, as an integral part of both the daily life of the community and the teaching and learning environment.
- The inclusion of criteria for each dimension based on both the literature and the experiences during the model implementation.
- Table 1 presents a summary of the model. The first column includes the four dimensions of the model. The second column presents the criteria developed for each dimension. Finally, the third column provides a proposal of indicators or conditions of achievement to facilitate the monitoring and management of each criterion.

Table 1: Development Model of Social Responsibility Competencies for Sustainability

Dimensions	Criteria	Indicators or conditions of achievement
a.- Relationship between the environment and its needs.	a.1.- The group or community involved is the one that must establish the real need to solve a problem or take the chance to improve with the help of the university.	a.1.1.- The community communicates explicitly and formally that it wants to solve or feels it needs in terms of an improvement opportunity. a.1.2.- Formal participation commitment on the part of the community, students and university through tasks, duties and resources. a.1.3.- Agreement on the objectives and real scope of the Project in an academically established time period.
	a.2.- The initiatives or actions must be made during the process "with" and "for" the community or group.	a.2.1.- Increasing expectations about the issues to be solved. a.2.2.- Gathering of relevant information on the problems. a.2.3.- The establishment of a joint diagnosis, from the information gathered . a.2.4.- Propose possible solutions to implement, according to the players' experience. a.2.5.- Agreement on a joint plan of action with established deadlines, results, resources and responsibilities. a.2.6.- Implementation of the agreed actions according to the different responsibilities, systematizing the related evidence .
	a.3.- The people involved must assess and systematize the impacts generated by the link.	a.3.1.- The students establish the impacts of their own learning and development in PSR. a.3.2.- The community values the actions and results of the agreed project. a.3.3.- The obtained results are aligned to the defined targets at the start of the project. a.3.4.- The students and the community record the activities and lessons learned. a.3.5.- The university makes orientations and suggestions from the evaluations and impacts on the different people and the community.
	a.4.- The development of abilities in the group or the community must be taken into consideration.	a.4.1.- Gather people's perceptions, about the ability to continue on their own. a.4.2.- Monitor the project and its sustainability in time.

Table 1: Development Model of Social Responsibility Competencies for Sustainability – *continued*

Dimensions	Criteria	Indicators or conditions of achievement
b.- Curricular Activities	b.1.- The design of the syllabus must include the development of PSR competencies.	b.1.1.- The objectives or learning outcomes must reflect PSR competence in some way. b.1.2.- The knowledge required for PSR competence must be evident or related to the discipline. b.1.3.- There must be coherence between the course's purposes intended with PSR through knowledge, learning and evaluation activities.
	b.2.- Learning activities must be characterized by considering the cognitive, experiential, metacognitive and self-system dimensions related to PSR.	b.2.1.- Identify the variables that have had influence on all the level's good practices (cognitive). b.2.2.- There must be field activities at least in every achievement or significant stage of the project (experiential). b.2.3.- Identify, monitor and adjust the resources and knowledge required in each of the stages or achievements, to effectively reach the proposed targets (metacognitive). b.2.4.- Give meaning and importance to the problem to solve in terms of own professional development (self-system).
	b.3.- During the implementation of activities, it is necessary to provide mentoring, monitoring and tracking processes during key milestones to reach the learning sets and the objectives to reach with the group or community.	b.3.1.- Mentoring, monitoring and tracking actions are already planned in the syllabus. b.3.2.- The tutorials allow the students to compare and answer questions and envision interventions with the professor. b.3.3.- The professor monitors and follows various actions and activities carried out by the students do. b.3.4.- Professors and students give feedback on the purposes, actions and obtained results.
	b.4.- The evaluation process of lessons learned must consider the implications and impacts on the groups and communities and systematization of the experience.	b.4.1.- Strategies to assess the possible impacts on the students, groups and communities' learning results are designed. b.4.2.- Evaluation strategies are adjusted and implemented during and after the actions. b.4.3.- The student's learning outcome and the effects on the community are seamlessly systematized.

Table 1: Development Model of Social Responsibility Competencies for Sustainability – *continued*

Dimensions	Criteria	Indicators or conditions of achievement
	b.5.- The professor must be able to carry out methodologies with the community that facilitate the achievement of PSR competence.	b.5.1.- The design and execution processes of the innovative course's first implementation carried out by the professor is accompanied by a couple who have experience in methodology. b.5.2.- The professor, after at least two continuous implementations of the same course, systematizes his experience, identifying lessons learned, challenges and improvements. b.5.3.- The professor communicates and comes with other professors to implement the methodology, to encourage continuous improvement in its own course and that of its pair.
c.- Extracurricular Activities. Relates to those activities or complementary actions or outside the curriculum that favor the development of PSR competence in a natural way and according to the students' interests.	c.1.- The design and implementation of the extracurricular activity must be consistent with relating criteria with the environment and sustainability over time.	c.1.2.- The Project must ensure continuity for at least three regular years. c.1.3.- Consider the link with the environment indicators in both the proposal and the implementation. c.1.4.- The project's results are assessed considering the initial diagnosis, the planned objectives, the implementation process of activities and the project's impact. c.1.5.- The project must have actions that allow it to be scalable.
	c.2.- The project's participation must consider students, professors and staff in an active way, if possible.	c.2.1.- Consider auto-selection regarding the area in which the activity will be selected. c.2.2.- The design and implementation of the project must consider at least five full-time faculty students. c.2.3.- The project must consider the participation of at least one faculty professor as an adviser. c.2.4.- The project must create the necessary conditions for staff participation. c.2.5.- The initiative must consider actions to manage and promote active and continuous participation of its members.
	c.3.- Consider monitoring process based on project milestones and the development of PSR competence.	c.3.1.- Define its own initial learning challenges concerning PSR that the project requires in its implementation. c.3.2.- Monitors and reflects about the development of its own PSR in each stage of the project (diagnosis, planning, implementation and evaluation). c.3.3.- Assesses the development of PSR according to its initial challenges and the project's development.

Table 1: Development Model of Social Responsibility Competencies for Sustainability – *continued*

Dimensions	Criteria	Indicators or conditions of achievement
d.- Internships and final degree activities.	d.1.- Accompanying spaces and processes must be encouraged during the final activities to favor the link between profession and PSR.	d.1.1.- During the moments prior to the start of the closing professional activities, the university presents the professional activity's general framework. d.1.2.- Spaces of reflection and experiential analysis are created alongside the process in terms of PSR. d.1.3.- Once the experiences are over, situations that allow for the synthesis and evaluation of PSR are created.
	d.2.- The students link these professional closing activities with PSR criteria.	d.2.1.- Identifies the relevant professional challenges that have been put at stake in terms of PSR. d.2.2.- Identifies the players and variables involved in the professional challenge. d.2.3.- Weighs up the variables and its implications related to PSR. d.2.4.- Proposes various solutions which are more consistent with PSR.
	d.3.- Development of PSR competence is evaluated as a result of the closing professional activities.	d.3.1.- Establishes the student' s own initial and final perceptions according to the achievement of PSR competence. d.3.2.- Defines the achievement level of PSR competence according to the firm's or institution's established parameters in terms of professional performance. d.3.3.- General integrated evaluations are established, by setting future professional challenges.

Implementing the Model in the School of Economics and Business at the University of Chile

The School of Economics and Business at the University of Chile has implemented this model since 2010 with the objective of articulating existing initiatives, systematizing the lessons learned, and monitoring its evolution. In this chapter this experience is summarized for each dimension.

a) **Link with the environment and its needs**: Main actions are related to creating the link between the university and the community. Thus, the university has created protocols in order to formalize this relationship and to obtain better results for all the parties involved. These protocols include monitoring milestones and intermediate evaluations over obtained results based on the stakeholders' opinion. The initiatives contributing to this dimension are intended to develop trust and a long-term relationship with the community partners by focusing on the needs and problems that they propose to the university, not vice versa. As a consequence, it is important to measure stakeholders' satisfaction with the initiative in order to make the adjustments to the initiative. Among the projects being carried out that contribute to this dimension, the university has different student projects that work with middle schools, municipalities, and micro-entrepreneurs. In the curricular field, consultancies for organizations are performed as learning-service activities and applied research theses. Finally, it is critical for a successful implementation of the model to have a university team in charge of implementing and monitoring the protocols, formal communication with students and stakeholders, and the fast and effective response to unexpected problems.
b) **Curricular Activities**: Correspond to courses with pedagogical methodologies like the Learning-Service and Learning based on Professional and Business Challenges which consist of both class discussion on professional topics and consultancy for organizations where student have the opportunity to practice the material covered in classes. These initiatives include courses such as Cost Accounting, Marketing III, and Evaluation of Social Project among others. Although the professors are responsible for their course, they must be accompanied by teaching and learning experts in order to provide an environment that is aligned with the model and thus facilitates the development of PSR. Finally, it is also important to have monitoring mechanisms that evaluate the accomplishment of learning goals and competency levels.
c) **Extracurricular Activities**: In general, these initiatives are volunteer projects, which are organized and managed by the students and are monitored by the university through the university social responsibility unit. This unit promotes the continuous development of these projects through competitive funds, methods offered and guidelines for managing projects, and provides consulting services in areas such as strategic planning, budget management, and project finance.
d) **Internships and final degree activities**: Considers compulsory professional internships and undergraduate theses. Internships are developed in firms, NGOs and the public sector. The search for internships and professional jobs is made by a university office specialized in this matter. Concerning the theses, the topics are collected by the university social responsibility unit, which after revising the requirements, establishes the contact between professors and students who want to address the topic. In order to facilitate the student reflection, these initiatives must be accompanied by formal meetings to discuss the development of PSR competency. On the same line, socializing early experiences and results among students and the community are also important for the development of the student self-system.

After identifying the most relevant activities for PSR development in our faculty, the model presented in the past chapter was used to assess if the conditions in which said activities are carried out promote PSR competencies. The following chapter shows the results of this evaluation.

Assessment of the developed experiences in FEN according to the presented model

To be able to make a descriptive analysis of how each activity contributed to the achievement of each dimension, the following procedure was carried out. Table 2 was taken as a base where the model was presented and a column for each developed activity in the School of Economics and Business was added. Subsequently, each activity was evaluated in terms of its contribution to each indicator, assigning a "1" if the activity contributed or "0" if it did not contribute to the corresponding indicator. Finally, to be able to assess the extent in which each activity contributes to each dimension, the contribution percentage that this activity adds to the total for each dimension was calculated.

Table 2: Relative Contribution of Activities to each Dimension

Dimension	**SR internship**	**Internships**	**Courses with Methodology within the Community**	**Student Projects**	**Ethics Course**	**Theses**	**SR Theses**	**Total**
Link with the environment and its needs	12%	2%	28%	28%	2%	2%	28%	100%
Curricular Activities	14%	14%	25%	NA	13%	12%	22%	100%
Extracurricular Activities	NA	NA	NA	100%	NA	NA	NA	100%
Internships and final degree activities	27%	27%	NA	NA	NA	20%	27%	100%

NA= Not Applicable

Table 2 shows the comparative results of each activity's relative contribution concerning each dimension. At dimension level we can observe that the performed activities can contribute to the four dimensions included in the model.

Regarding the **link with the environment and its needs** dimension, we highlight the almost negligible contribution of traditional activities such as internships, ethics courses and theses. The results show that these activities do not facilitate relating the environment and its needs, and as a result, they do not necessarily develop competencies for sustainability. In addition, the goal-directed activity that contributes to a lesser extent to the indicators is the "SR Internship" criterion. Since these internships depend on an organization outside the university, it doesn't necessarily allow the students to participate in the implementation and evaluation of a project with the community, to enable them to implement all the indicators contained in this criterion.

In the case of **curricular activities,** even though all activities contribute significantly, the courses with Learning with the Community methodologies create, in relative terms, a greater contribution to this dimension. This occurs mainly because this type of experience is intended to expose the student to a teaching-learning process that mixes theory and practice in a significant social problem.

In the dimension of **extra-curricular activities**, the only activity that is carried out is Management of Student Projects. Due to the course being intentionally designed, it allows not only to the contribution of every indicator or condition related to this dimension, but it also contributes to the development of more generic professional competencies of the SR like respect for diversity, team work, and the ability to jointly analyze the problems from different disciplines to give a solution for a complex problem.

Finally, in the case of activities that contribute to the dimension "**Internships and final degree activities**", the Faculty's evidence doesn't show significant differences.

Conclusions

The model presented in this paper is the result of a series of experiences implemented in the School of Business at the University of Chile intended to develop SRP competency. The implementation of this model has provided important lessons that we believe can help other higher education institutions throughout this journey.

First of all, the higher education institution must create a learning environment that facilitates the development of PSR competency through the whole educational experience. The model presented in this paper facilitates this environment by developing an integrated set of four criteria that provides guidelines and principles for different initiatives such as internships, theses, curricular, and extracurricular activities. In addition, the criteria also provide conditions to create long-term relationship with the different stakeholders of our community based on values such as respect, credibility and trust.

Secondly, although it is important to implement methodologies based on real cases that exposed the student to stakeholders outside the university, this methodology is not enough by itself. The university should have a group of experts that provide consulting services to the professors implementing these methodologies. Specifically, this team have to support the professor in preparing syllabi, managing the relationship with the stakeholders, designing classroom experiences aligned with PSR competency, promoting the reflection of the experience, and developing assessment tools in order to have evidence on the level of achievement of specific competencies. Finally, this team should also provide workshops and mentoring to develop the professors' skills that facilitate a teaching and learning experience consistent with the proposed model.

Finally, the implementation of this environment is a dynamic experience; as a result, the higher education institutions should have the mechanisms to provide enough flexibility to adapt to new challenges. For example, in the case of the University of Chile, the increasing number of courses demanding methodologies that involve stakeholders outside of the university, created the need for developing long-term relationships with stakeholders in order to have them as partners in multiple initiatives. This long-term relationship needs to be managed by a centralized unit that not only monitors the implementation of several courses, internships, theses or projects, but also should be open to the stakeholders' needs in order to provide fast and effective responses to unexpected problems.

Future research should consider measuring the impact of this model's implementation on students' competencies, university culture, and stakeholders' satisfaction. Other lines of research may investigate the impact of the different initiatives on the level of PSR competency, in order to identify the best practices for developing such a competency.

References

Barth, M., Godemann, J., Rieckmann, M. and Stoltenberg, U. (2007), Developing key competencies for sustainable development in higher education. *International Journal of Sustainability in Higher Education* 8 (4), pp. 416-430.

Bell, B.S. and Kozlowski, S.W. (2009), *Toward a Theory of Learner-Centered Training Design: An Integrative Framework of Active Learning*. Learning, Training, and Development in Organizations, 263 pp.

Brockbank, A. and McGill, I. (1998), *Facilitating Reflective Learning in Higher Education,* 297 pp. Taylor & Francis, Bristol.

Brundiers, K. (2010), *Educating students in real-world sustainability research: Vision and implementation*. Innovative Higher Education.

Brundiers, K., Wiek, A. and Redman, C.L. (2010), Real-world learning opportunities in sustainability: from classroom into the real world. *International Journal of Sustainability in Higher Education*.

Clark, P. (2010), Book Review: Review of Designing and Assessing Courses and Curricula by R.M Diamond. *Journal of Food Science Education* 9 (3), p. 80.

Diamond, R.M. (2008), *Designing and assessing courses and curricula: a practical guide*. 487 pp. Jossey-Bass.

Fry, H., Ketteridge, S. and Marshall, S. (2009), *A handbook for teaching and learning in higher education: enhancing academic practice*. Taylor & Francis.

Gibbs, P. (2009), Learning agreements and work-based higher education. *Research in Post-Compulsory Education* 14 (1), pp. 31-41.

Hobson, A.J., Ashby, P., Malderez, A. and Tomlinson, P.D. (2009), Mentoring beginning teachers: What we know and what we don't. *Teaching and Teacher Education* 25 (1), pp. 207-216.

Kaivola, T. and Rohweder, L. (2007), *Towards Sustainable Development in Higher Education – Reflections*. Education.

Light, G., Calkins, S. and Cox, R. (2009), *Learning and teaching in higher education: The reflective professional*. Sage Publications Ltd.

Michelsen, G. and Rieckmann, M. (2008), *Internationaler Masterstudiengang: Sustainable Development and Management*. Higher Education for Sustainability, Lüneburg.

Nevins Stanulis, R. and Floden, R.E. (2009), Intensive Mentoring as a Way to Help Beginning Teachers Develop Balanced Instruction. *Journal of Teacher Education* 60 (2), pp. 112-122.

Nijhuis, J., Segers, M. and Gijselaers, W. (2007), The interplay of perceptions of the learning environment, personality and learning strategies: a study amongst International Business Studies students. *Studies in Higher Education* 32 (1), pp. 59-77.

Puntambekar, S. and Hubscher, R. (2005), Tools for Scaffolding Students in a Complex Learning Environment: What Have We Gained and What Have We Missed? *Educational Psychologist* 40 (1).

Russell, W.D., Block, B., Kissock, S. and Kraft, J. (2010), *A Critical Thinking-Based Approach to Teaching University Wellness Courses*.

Schugurensky, D. (2000), *The forms of informal learning: Towards a conceptualization of the field*. Centre for the Study of Education and Work, OISE/UT. Available at http://tspace.library.utoronto.ca/ (accessed 1 February 2012).

Steinemann, A. and Asce, M. (2004), Implementing Sustainable Development through Problem-Based Learning. *Pedagogy and Practice* 129 (4) (October), pp. 216-224.

Svanström, M., Lozano-García, F.J. and Rowe, D. (2008). Learning outcomes for sustainable development in higher education. *International Journal of Sustainability in Higher Education* 9 (3), pp. 339-351.

United Nations (1992), *Report of the United Nations Conference on Environment and Development*, 3-14 June 1992, Rio de Janeiro, Brazil, Vol. 1, Resolutions Adopted by the Conference. UN, New York.

Warburton, K. (2003), Deep learning and education for sustainability. *International Journal of Sustainability in Higher Education* 4 (1), pp. 44-56.

Willard, M., Wiedmeyer, C., Warren Flint, R., Weedon, J.S., Woodward, R., Feldman, I. and Edwards, M. (2010), The sustainability professional: 2010 competency survey report. *Environmental Quality Management* 20 (1), pp. 49-83.

Wood, L.N. and Smith, N.F. (2007), Graduate attributes: teaching as learning. *International Journal of Mathematical Education in Science & Technology* 38 (6), pp. 715-727.

Chapter 75

Solar Power to the People: The Solar Energy Workshop at *Universidad de Chile*

Martín Jacques-Coper, Pablo Sánchez, Ernesto Frodden, Francisco del Campo, José Ignacio Sánchez, Diego Oyarzo, Juan Macías, Francisco Bucchi and Ismael Aguilera

Abstract

A key mainstay for a sustainable campus is the role of higher education institutions in promoting and guiding society towards sustainability. In this regard, universities are encouraged to develop technologies and provide sustainable solutions to meet community needs. This statement is of particular relevance in Chile, where a large part of the population does not have access to basic energy-related needs like hot water. Chile has high natural potential for solar, wind, and geothermal power. However, as things stand, solar energy is something that only middle and upper-class families can afford, mainly due to the considerable costs of the technology and installation. In this context, students at Universidad de Chile, motivated by their sustainable campus initiative and supported by a professor, have established the Solar Energy Workshop (*Taller Solar,* TS) and have taken on the challenge of revolutionizing Chile's energy system from the bottom up. The strategy comprises of teaching low-income communities how to make their own solar collectors to heat water and reduce expenditures on electricity and gas in their households. However, the aim is not just welfare – far from assistencialism, the goal is to empower individuals to harness and maintain their own energy sources instead of relying entirely on large energy corporations. This transfer of information and technology is a process which relies on receiving feedback from the focus communities. It makes renewable energy accessible to wide sectors of society, in a way that is adaptable to their respective needs. Besides, a wide-ranging political discussion about the national energy matrix has led to the development of *Energy Revolution*. In this paper, the authors review the experience gained by TS since 2007 and propose that students should have a key role as catalysts of the interaction between scientific/theoretical solutions developed by academics and researchers, and the needs of broad sectors of society. This article presents the strategy employed in the implementation of this successful initiative, which has the potential to be developed as part of university outreach in the framework of the sustainable campus. Finally, lessons learnt can be extrapolated for various projects of a similar nature.

Introduction

Today, economic, social and environmental evolution maintains a close relationship with the generation and consumption of energy, influencing nations' and communities' polices. This relationship, focused on the concept of sustainable development, has been a common discussion topic since the last half of the 20th century for politicians, academicians, and citizens, who seek greater harmony between economy, society and environment. A key challenge on the road towards a sustainable society is to develop energy sources that are environmentally friendly, clean, locally generated, cost efficient, and related to people's energy needs. Nowadays Chile is far from achieving this ideal situation: 71% of the primary energy supply of the country comes from fossil fuels – coal, oil and gas – (CADE 2011; AETS SA 2011), leading to a dependent and polluting energy matrix. It lacks a long-term plan and does not include a mature use of Non-Conventional Renewable Energies (NCRE). The future developments of the energy matrix are defined by investors who seek short-term profit maximization and orient it towards the inclusion of mega-power plants (Guzowski and Recalde 2010; Bórquez 2011). Hence, this policy is clearly unconnected to people's real energy needs, and leads to unequal access to energy.

The domestic fossil fuel coverage of national consumption is relatively low, and this is what partially caused the marginal cost of electric energy to increase five times in the 2007-2010 period (CCTP 2011) and drive up the cost of transport and basic necessities. Census data from 2002 shows that more than 40% of the Chilean population does not have access to basics like hot water (INE 2002). While a great part of this fraction does have gas water heaters in their homes, they do not have enough income to afford the fuel, which on average costs around CLP 25,000 (US$42) per month, close to 20% of the Chilean minimum wage. However – and contradictorily – Chile is one of the countries with the highest potential of solar, wind and geothermal power in Latin America (Gebremedhin et al. 2009; Hall et al. 2009; Pontt et al. 2008; DGF and CNE 2012). If Chile took advantage of solar power, it could save 70% of the energy consumption for hot water nationwide. Due to this huge solar potential, some authors indicate that it is adequate to start energy planning activities – although not yet for proper power plant design and dimensioning, in view of the current level of assessment of the resource (Ortega et al. 2010). But as things stand, solar energy is something only middle and upper class society can afford, mainly because of the substantial costs of the technology and installation. In this context, major and drastic changes concerning the energy production and distribution are to be carried in order to develop a sustainable energy matrix. The authors believe said system should be rearranged to include guiding principles like local autonomy, self-management, democracy, environmental awareness, and cost-efficiency, in order to respond to people's energy needs. Because these needed changes are radical and urgent, this process implies an *Energy Revolution*. An undefined social process is meant by this, arising from the critics of the actual system of production-distribution-consumption of energy. Up until now, work guided by similar principles has been led mainly by NGOs through local experiences, and has strived to attain widespread and sustainable use of renewable energies. Such is the case of EcoMaipo (www.ecomaipo.cl) and Intillapu (intillapu.jimdo.com), which develop their activities in cooperation with communities in northern and central Chile, funded by a variety of private and public sources.

In this context, higher education institutions are key entities for promoting the needed change of paradigm, and for collaborating in proposing a new sustainable development project for humankind. These institutions have the capability to create the knowledge needed to solve sustainability challenges, to train professionals who are aware of environmental issues, and to divulge solutions (Sharp 2002; Conde Hernández et al. 2006). As already pointed out, the energy issue in Chile is critical for sustainability, and therefore should also be addressed by universities from a technical and social point of view; particularly regarding development, implementation, and divulging of clean-energy technologies to solve the needs of communities.

Consequently, students from the Universidad de Chile (UCh) have taken on the challenge of revolutionizing Chile's energy system from the bottom up, by teaching low-income communities how to make their own low-cost solar collectors to heat their water and reduce what they spend on electricity and gas in their households. But this is not simply about welfare for those in need. The goal is to empower individuals to harness and maintain their own energy sources instead of relying exclusively on large energy companies. They are bringing the technology, and the information about how to use it, to poorer segments of society, in order to make renewable energy accessible at a low cost.

In this paper, the authors suggest that students have a key role in divulging the knowledge generated in universities, especially the knowledge that is applicable to environmental issues. Their key role is to act as catalysts of the interaction between scientific/theoretical solutions developed by academics, and the needs of broad sectors of society. This article presents the strategy employed in implementing a successful initiative which addresses communities' energy needs, and which has the potential to be developed as part of the outreach of universities in the framework of the sustainable campus.

The Solar Workshop at Universidad de Chile (UCh)

The Solar Energy Workshop (TS for its name in Spanish – *Taller Solar*) is a multidisciplinary initiative, composed mainly of students from UCh, which aims at encouraging low-cost solar energy use in society. Its strategy is to train people in the construction and installation of low-cost solar devices such as flat plate collectors. In addition, its participants discuss the necessary paradigm shift regarding the way society satisfies energy demand in Chile, which validates its project. In this section, the specific conditions which made TS possible and led to its creation are presented. The argument will be developed regarding three main groups of actors which will be characterized: academics, students, and the university.

At UCh, there is an enormous amount of accumulated experience and technical and theoretical knowledge. However, many possible social benefits which might be derived from this experience and knowledge usually remain unexploited. For instance, the traditional academic career paths followed by professors are usually constrained to fulfill teaching, research and administrative functions. Consequentially, they have little time left to participate in innovative social projects. Frequently, university extension is barely institutionalized as part of their activities. The birth of TS was due in part to the support and motivation provided by Prof. Roberto Román (Dept. of Mechanical Engineering and EcoMaipo NGO), who is not an exception in terms of having little free time. On his own initiative, Prof. Román created several formative instances, which were essential for the creation of TS. More generally, a professor's commitment to such a project will have a great impact, through the sharing of his valuable experience and knowledge with students, through extra-curricular activities, and also – and perhaps more importantly – by motivating the students, and by exemplifying a different way of understanding, the relationship between knowledge and society.

Students, in comparison, have much more freedom, as their schedule is normally not full, and their formal responsibilities are more or less limited to complying with course requirements. They however, are frequently eager for opportunities to learn more, and to go beyond the normal exigencies the university imposes on them. These characteristics are the reason why innovative projects and new points of view on academic knowledge can flourish naturally from this diverse group.

The idea behind TS was to implement an existing technical solution (low-cost flat plate collectors) in order to solve a widespread need in Chile: domestic hot water. Most of the students attracted by TS were already involved in social and/or environmental organizations. This engagement is partially motivated by viewing their own university experience as an opportunity to generate alternative life projects to help people and society.

In the particular view of TS members, a university is not only an institution where people get a technical or professional education, and where professors produce applied and theoretical knowledge. It is also a human group which is assigned the task of meeting society's challenges as profoundly as possible and to serve its needs. Applied knowledge, when used under this principle, can directly make people's lives better. For instance, the invention of a cheap and accurate medical test might both save lives and improve a patient's quality of life. Furthermore, the innovative use of an existing concept or technology, like flat-plate collectors, can create a practical solution to a social problem. Theoretical efforts, for example in basic sciences, are frequently far from any practical application. Nevertheless, they are essential for achieving a profound understanding of technologies and natural phenomena, and should not be neglected in this context. The university is also a strategic place for this kind of initiative to flourish: there are ways of obtaining internal competitive funding for initiatives and also to access the necessary infrastructure and interdisciplinary knowledge.

Therefore, higher education institutions (HEI) can directly support society through university outreach, making use of their theoretical approach to sustainability – which is conceptualized as the sustainable campus, which addresses transformations in research, outreach, and teaching (ULSF 1990). In this sense, a noteworthy experience in Chile is that of GEA (www.gea.usm.cl), a group of students and academics from Universidad Técnica Federico Santa María, who have been dedicated since 2000 to developing renewable energy solutions adapted to the needs of isolated sectors of the country. Another interesting case, concerning fresh water supply in northern Chile, is the contribution made by national universities to the understanding of the coastal stratocumulus deck (the so-called "Camanchaca") and to developing technologies aimed at collecting cloud water directly from the Camanchaca (Schemenauer et al. 1988; Román 1999). Therefore, universities are privileged places for developing new solutions to society's needs, and to provide a shortcut from theory to concrete applications. Last but not least, when working with communities, the name of the university as a well-known institution might induce a positive prejudice that might help gain people's confidence more quickly.

The present paper covers the case of the TS project, and discusses the outreach of Chilean HEI's regarding sustainability. As far as the authors know, no other experience of this kind has been documented in this way at national level. This successful initiative shows a way in which universities can contribute sustainable solutions to meeting the concrete needs of society. It also encourages the development of projects related to sustainability, and demonstrates a possible strategy for achieving their objectives.

History, Strategy and Achievements

The birth of the TS is linked with the initiatives undertaken by the Oikos student group at the Faculty of Physical and Mathematical Sciences (FCFM) of UCh for the purpose of establishing a sustainable campus (see Bucchi et al. 2012, in this volume). Although its initiative relies mostly in "learning-by-doing", the authors are convinced that this particular experience can contribute to achieving the goals of similar initiatives. The key lessons that may be learned are: the need to precisely outline objectives, the establishment of alliances with other university bodies (academics in particular), putting together an interdisciplinary team, and the ability to properly formulate proposals to be presented to university authorities and to competitive funding. These are all essential factors for developing projects in the outreach dimension of a sustainable campus initiative.

In 2007, following a strategy based on possible alliances with academics and researchers, Prof. Román offered the possibility to four Oikos members of attending a summer school devoted to renewable energies (EVER Chile 2007). It was organized by the Solar-Institut of Jülich (Germany) and Universidad de Chile, Universidad de Concepción, and Universidad Técnica Federico Santa María. A key aspect of this experience took place during the last week, when practical experiments were carried out. The Oikos participants recognized that the building of low-cost water heaters, also known as solar collectors (see Fig. 1), represented a real possibility for divulging this technical knowledge among the Chilean population. Thus, they were determined to create an instance that would materialize that great potential.

Fig. 1: Schematic view of a low-cost solar water heater, developed by EcoMaipo NGO and built at TS

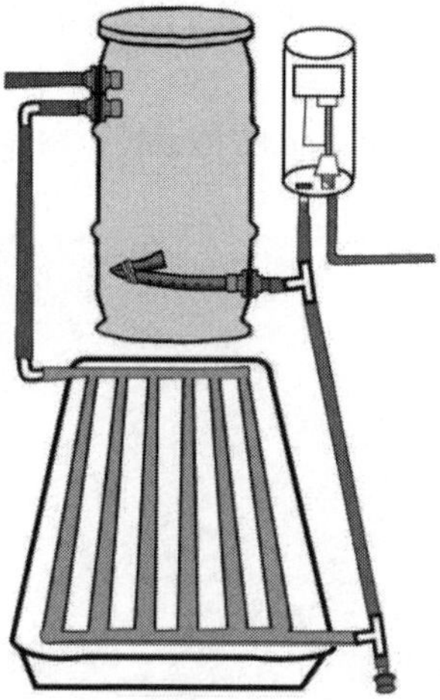

As some members of Oikos were also elected representatives in UCh's Student Union (FECh), there was substantial support for the idea of forming a Solar Energy Group during the 2008 Summer Volunteer Works Program. These volunteer works are carried out each year in poor and usually rather isolated regions of the country, and include projects covering a broad spectrum of disciplines (simple building construction, health services, legal assistance, etc.). Thus, the first TS experience took place in Mincha Norte, a rural village in the Coquimbo region, northern Chile. The aim of the Solar Energy Group consisted of building six flat plate collectors and five solar cookers (Fig. 2). The funding for this project was obtained through FCFM sponsorship, thanks to a decision taken by the FCFM faculty dean. Following the precept of bringing know-how, rather than assistance to the community, a prior assessment was carried out with staff from the local municipality, in order to identify families who needed such an artifact the most, and who were also willing to be involved in this initiative and to provide assurances about the future maintenance and proper use of the devices. After two weeks of joint work with the neighbors, technically assisted by Rodrigo Valdovinos (EcoMaipo NGO), the experience was deemed to be successful. Indeed, the original impetus of the organizers was such, that they considered it possible to consolidate Mincha as a kind of "solar capital" of Chile in a couple of years' time. In fact, this particular project was also viewed as a way of partially solving the local problem of lack of opportunities in Mincha for young people, which forces them to migrate to bigger urban areas. The idea was therefore to create a local factory of solar artifacts. Therefore, a proposal was sent to a tender for competitive project funding organized each year by the Chilean government through the Fund for Environmental Protection (FPA). Although the initial steps in the evaluation of this initiative were positively rated, the project failed in getting the necessary funds. A crucial factor which accounts for said failure was the fact that the local community never played a leading role in this initiative. In fact, it should be remembered that in Chile, after the harsh experience of oppression during the Pinochet dictatorship (1973-1990), the so-called "social capital" was greatly weakened or even destroyed. This fact thus began to be considered in the subsequent projects, which included the participation of specialists from the social disciplines at the university. As will be shown, the resulting interdisciplinarity of the group would then become an important strength. The group took this disappointment as an opportunity for learning new strategies and so, the following year, it prepared a fresh proposal for the 2009 Volunteer Works in Putaendo, a rural municipality in the Valparaíso region. This time, the project installed twelve artifacts (solar collectors, cookers and fruit dryers) and left behind two qualified neighbors for socializing the knowledge.

Fig. 2: Solar water heater built and installed at the locality of Mincha Norte (Coquimbo region) during FECh Volunteer Works 2008.

After having achieved this goal, the solar group started to discuss the creation of an *Energy Revolution,* when referring to the concept which should be embraced in future implementations of the initiative. It was clear that, as past activities lacked a systematic intervention and were limited to a few weeks of work per year in different localities, such experiences were condemned to remain as simple anecdotes. This is why a change of strategy was considered necessary. Therefore, the objective of the TS was now to establish a permanent work instance at the University of Chile, in order to work continuously on this issue and attract more members from inside and outside the university. It was also necessary to demonstrate that, contrary to popular opinion, low-cost solar devices were also an efficient solution for covering the energy needs in urban, not-necessarily poor, areas. In this way, a direct impact could also be developed in Chile's capital and mega city, Santiago.

So, the TS was born anew, formed mainly by participants in the preceding volunteer works. In order to get a physical space to manage materials and tools and to offer free theoretical and practical lessons concerning the basic aspects of energy (both technical and social), their members presented a proposal to the dean of the Faculty of Architecture and Urbanism (FAU). Finally, the first workshops were launched in March 2009. Up until the present day, this initiative has had five one-semester-long versions, free of charge for any interested participant (see Fig. 3). The workshops have been held every Saturday, and have been given by volunteer teams. The latest versions have been housed at the FECh building, where TS lately moved.

Fig. 3: A practical session of the FAU Solar Workshop, where instructors learn how to build solar devices

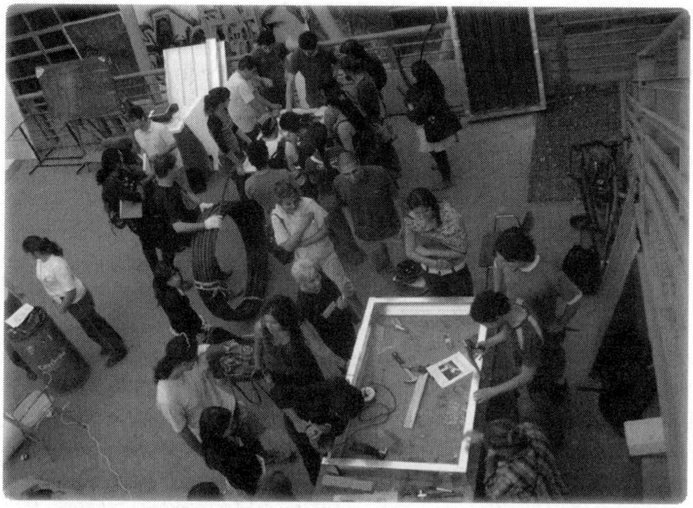

Since that first workshop, the TS has developed, after countless discussions, its own philosophy based on spreading the knowledge through free lessons, open to every interested person from inside and outside the university. As stated above, besides the theoretical background on solar energy and the technical operation of the devices, the main emphasis is on discussing the related environmental issues and, especially, the national energy policy. In Chile, national-level decisions are taken top-down in every order of things, without the possibility of any real direct participation from the population. The energy sector, in particular the electricity sector, is a prime example of this. In this field, the controversial project for five hydro dams in Patagonia – which would lead to irreversible environmental consequences – has recently had a tremendous impact on every level of society, unveiling a major crisis in Chile's pseudo-democratic political system. Hence, a fundamental aim of TS's efforts is to encourage communities to become involved and aware of the role of simple citizens in proposing, and working for, a better society. The group's interdisciplinarity is essential for accomplishing this important objective. For instance, during the workshop at FAU and FECh, the students from social disciplines promote special intragroup dynamics to clarify the role of social collaboration as a crucial factor for the achievement of objectives. To this end, they encourage environmental discussions. Also in line with these views on what a democratic organization should be, the internal organization of the Solar Workshop is horizontal, and tries to avoid a hierarchical structure. Nevertheless, specific tasks (accounting, etc.) are assigned to individuals, in order to ensure efficiency. TS's philosophy could be synthesized as "no one is indispensable, everyone is important".

The final goal of educating qualified people on low-cost solar devices is to revolutionize the country's energy sector through bottom-up initiatives. In other words, the members of TS want to show that with the natural renewable resources available in most of the country – and in this particular case, solar energy – one has the opportunity to satisfy, at least partially, everyday energy needs. TS is interested in training people who can replicate this initiative in their respective areas of influence: neighborhoods, educational institutions, etc. Therefore, a decisive strategic aspect of its efforts consists of establishing alliances and joint work initiatives with community-based organizations such as schools, cultural and social centers, neighborhood committees, etc. Concerning the teaching strategy, far from simple assistencialism, its members are interested in transferring theoretical and practical know-how, which nevertheless has to be adapted to every single reality. In line with this ideal, they expect every experience to have the active support of the local community, not only in terms of logistics aspects, but also in terms of finding ways to raise funds for the joint work.

Achievements

As the authors understand things nowadays, the activities that TS has conducted so far fall into three categories. Thus, a distinction is made between FECh Volunteer Works, the free courses offered at the University Workshop (FAU and FECh) and the Territorial Workshops. Table 1 shows a summary of all concrete experiences.

- *FECh Volunteer Works*: since 2008, TS has conducted five versions of the two-week long Volunteer Works in line with the other activities organized by the University Student Union (FECh) in the rural localities of Chile. One of these (2009) was the first group activity that led to the creation of the TS. Some of the major goals of these activities are: attracting participants among university students to experience the reality of teaching, manufacturing and installing devices; as well as working side by side with members of a community. In this respect, TS expects to recruit future members who can act as instructors during the permanent teaching activities at the University in Santiago.
- *FAU/FECh Workshop*: conceived as a permanent space devoted to teaching, discussing, building and improving the devices, this instance physically supports the existence of the group behind the TS. Strategically, it constitutes an essential qualitative step in order to project the impact of TS beyond transient interventions, such as the above. It broadens the spectrum of people accessible to TS's activities, spatially anchors them, and allows the spread of territorial workshops not only in the metropolitan region, but also across the whole country.
- *Territorial Workshops*: alliances are sought and created with community-based organizations, such as schools, cultural and social centers, neighborhood committees, etc., in order to train neighbors in the manufacture, installation and maintenance of solar devices. In parallel, societal issues linked with broad energy issues, such as national energy policies, society empowerment, citizens' participation, etc. are covered by means of group discussions and dynamics.

Table 1: Summary of activities led by TS at UCh and at territorial level, 2008-2012

Project name	Funding source	Assigned amount [USD]	Grant date	Execution place(s)	Time frame	Achievements	Observations
FECh Volunteer Works 2008	Deanship, FCFM, UCh	$1,650	jan-2008	Mincha Norte, Coquimbo Region	2 weeks, Feb 2008	6 solar collectors, 5 box cookers, 20 participants	
Development and divulgation of ways of using solar energy at low cost	"Blue Award", Office of Student Affairs, UCh	$3,080	dec-2008	FAU (UCh) and Territorial Workshops in La Legua, El Bosque, La Cisterna, Ñuñoa, P. Aguirre C., La Florida (Santiago) and Valparaíso	Jan 2009 – Mar 2010	5 Workshops: 11 solar collectors, more than 150 participants	Selected as the best project of the contest
FECh Volunteer Works 2009	Deanship, FCFM, UCh	$1,650	jan-2009	Rinconada de Guzmanes, Putaendo, Valparaíso Region	2 weeks, Feb 2009	8 solar collectors, 2 fruit dryers, 2 box cookers, 25 participants	
Divulgation of solar energy at low cost	FIPE, FCFM Students' Union (CEI)	$310	apr-2009	UCh	Apr 2009 – Jun 2009	divulgation material	
Solar energy workshop at low-cost	Fund for Institutional Development, Ministry of Education, Chile	$6,390	dec-2009	FECh Workshop (UCh) and La Granja, Santiago	Jan 2010 – Mar 2011	3 Workshops: 9 solar collectors, 100 participants	
FECh Volunteer Works 2010	Deanship, FCFM, UCh	$1,650	jan-2010	La Hacienda, Lolol, O'Higgins Region	2 weeks, Feb 2010	6 solar collectors, 25 participants	
Towards local Sustainable Development based on the use of solar cookers	Dutch Embassy in Chile	$3,030	oct-2011	Los Boldos, Colbún, Maule Region	Sep 2011	26 Parabolic solar cookers	Organized by Servicio País

Table 1: Summary of activities led by TS at UCh and at territorial level, 2008-2012 – *continued*

Project name	**Funding source**	**Assigned amount [USD]**	**Grant date**	**Execution place(s)**	**Time frame**	**Achievements**	**Observations**
FECh Volunteer Works 2011	Deanship, FCFM, UCh	$1,650	jan-2011	Paredones, O'Higgins Region	2 weeks, Feb 2011	7 solar collectors, 15 participants	
Manufacture of solar devices	Neighborhood Council of Botalcura	$266	aug-2011	Botalcura, Maule Region	1 weekend, Sep 2011	2 solar collectors and one built with reused materials.	Resources were raised by the community
Training Network for Socialization of Appropriate Technologies in Peñalolén	Fund for Environmental Protection, Ministry of Environment, Chile	$8,250	nov-2011	Peñalolén, Santiago	Jan 2012 – Dec 2012	In progress	
Manufacture of solar water heaters	GNL Quintero S.A.	$6,200	oct-2011	Quintero, Valparaíso Region	feb-2012	26 solar water heaters, a couple of solar fruit dryers; 50 people involved	Workshop offered to the Neighborhood of Las Brisas de Loncura, self-raised funds: $600

Although TS promotes the principle of volunteer work, monetary funding is needed for ensuring a set of necessary tools, supplementary devices (such as overhead projectors/data shows for the lessons), materials and logistic issues (e.g. transport). Fortunately, at UCh and particularly at FCFM, competitive funding is available for students' initiatives. In this manner, the first experiences were funded in this way. Nevertheless, since the scope of the projects has tended to increase, more ambitious proposals were presented to higher instances, such as the university's Office of Student Affairs and, later, to the government, which offers annual partial funding for diverse environment-related projects (see Table 1 for an overview of TS's funding sources). However, TS's members have always kept in mind the necessity of staying as self-managed as possible. Hence, territorial community-based initiatives are encouraged to raise their own funds through community activities, which also strengthen links within the local groups.

As supplementary high-impact achievements, a number of written and audiovisual materials have been developed over the last few years. Namely, some members developed their own "do it yourself" manual for building a low-cost flat plate collector. It explains in simple words the physical principles involved, the materials and tools needed, and sets out the steps to be followed in the construction and installation of the solar water heater. There are some examples of communities which have already obtained successful results with this guide, even without any direct connection with TS. In addition, several documents about alternative ways of self-constructing sustainable devices are available at the TS website, which also contains photographs of some of the territorial workshops (taller-solar.wikispaces.com/Difusion, www.taller solar.cl). There is also a documentary film by Metáfora Films about the experience of the Volunteer Works at Putaendo 2009 (*Energy Revolution,* vimeo.com/34204624), which has turned out to be an effective way of promoting TS's initiative.

Discussion

As is to be expected from an initial experience, the achievements of TS are mainly qualitative. They principally refer to the creation of a trained human team, able to spread the know-how related to solar energy devices. However, in addition to just technical skills, members of the Solar Workshop also have a well-grounded discursive framework, which constitutes an indication that the group is overcoming its formative period during which theoretical and practical maturities needed to be acquired. Notwithstanding the above, TS has not overlooked the quantitative aspect in terms of real impact towards an *Energy Revolution*. In this regard, the initiatives carried out have managed to install a large number of solar devices, aside from transferring know-how. This allows TS to concretely show its material results.

TS's experience shows that, for such an initiative to be effective it is important to put together an interdisciplinary team, inspired by the vision of developing practical and social solutions for sustainability challenges. Both aspects need to be considered when trying to achieve significant and ambitious advances towards real change in the energy paradigm for a sustainable future. This is why the active political discussion within the group has permitted its participants to mature their own concept of *Energy Revolution* and, furthermore, to link their activity with the national energy context. This concept pursues an ideal energy system that incorporates the principles of sustainability, real participation from the various sectors of society, diversification of the energy sources, distributed generation adapted to local needs, self-sufficiency and technological innovation, among other principal aspects.

The experience of the Solar Workshops has had a vital impact on its participants' personal education and training (both students and people from the communities). Participants in the workshops who later joined the group faced various organizational challenges, ranging from simple logistics to the formulation of long-term policies, including encouraging and inviting more participants, designing week-to-week programs for them, defining discussion topics and applying for funding, etc. In addition, thanks to the weekly discussions and to the contact with people from outside the university context, participants developed an awareness of the social and environmental issues of real communities. A side-effect of these experiences is the fact that some members of TS have decided to dedicate their future careers to the study and promotion of non-conventional renewable energies (NCRE) in the Chilean context. Furthermore, the TS experience has also proven to be a source of motivation and entrepreneurship for their participants, who have been organizing similar initiatives in their own communities and, in some cases, have even considered using this knowledge for an eventual business.

As the discussions have revealed, students organized in TS play a prominent role in the outreach of knowledge generated in universities, and in particular, the basic theory applicable to energy issues (Bucchi et al. 2012, this issue). In this context, the role of students is to apply and transfer this knowledge to real-world needs. A characteristic of this linkage is the horizontal point of view, naturally developed at student level. Apart from teaching, students are open to learning from the specific experiences of communities. In this context, communities usually establish a sincere and close relationship with students. Furthermore, communities are more receptive and at the same time increase their confidence to criticize the ideas exposed by students. Thus, the devices and the technology are not "sacred": the final truth and the solutions to environmental issues are developed jointly by the community and the Solar Workshop.

However, TS has lacked a stable structure, it is susceptible to generational changes (a natural phenomenon within students' groups). As members leave university for other horizons, they also weaken or sever their link with TS. More specifically, as a result, there is an absence of continuity in the accumulation of experience and specific knowledge. Consequently, many discussions and previously overcome mistakes are repeated. This could be solved by institutionalizing the TS, for instance by modifying the university's extension policy. However, this shouldn't be taken as the only alternative, as it might turn out to contradict the concept of *Energy Revolution* that TS supports, insofar as it should be a process involving the whole society, and not only students or academia.

Another, more subtle point that the authors view as a barrier to developing TS's project, has to do with the social fabric Chilean society is embedded in. Contrary to other historical moments (previous to the coup d'état in 1973), it is common nowadays to find poorly organized communities in Chile. Therefore, in many cases, people committed to organizing territorial Solar Workshops need to dedicate considerable time and effort to developing an essential local coordination.

Conclusions

Achievements, Motivations, Obstacles and Opportunities

The Solar Energy Workshop (TS) is a successful experience in the framework of Sustainable Development in Higher Education Institutions, because it shows how theoretical knowledge developed at UCh is put into practice through the use of low-cost solar energy devices. From an academic point of view, students in the TS play a prominent role in the outreach of scientific knowledge and ideas generated at UCh. Furthermore, and from a social perspective, the TS has contributed to bringing people together to discuss the current unsustainable Chilean energy system.

Participants in the TS have taken on the challenge of revolutionizing the Chilean energy system from the bottom-up, through active political discussion and spreading know-how regarding simple solar energy devices. The initiatives carried out have taught hundreds of participants how to make their own solar collectors to heat water, and have installed tens (perhaps up to a hundred) of solar devices in low-income communities. Achievements go beyond the material sphere: due to the methodology that was used, participants developed an awareness of the social and environmental issues affecting their communities.

Although motivations varied widely, participants usually joined TS workshops because of their interest in reducing the environmental impact of fossil fuels, and achieving savings in their electricity and gas bills. Both motivations are related to energy independence. Another attractive aspect of the TS, is the possibility to participate in a technology transfer process through a learning-by-doing methodology.

The main strength of the TS is its horizontal structure and the heterogeneous and multidisciplinary nature of its members. In part, its current success and positive past experiences are due to the support of Prof. Roberto Román (EcoMaipo), FECh and UCh. Finally, TS has benefitted from the current national debate on NCRE. On the other hand – as in many volunteer organizations –, the weaknesses of TS are the scarcity of human resources, the monitors' inability to dedicate themselves full-time to the TS, and the lack of sufficient financial resources.

Lessons learned

Some of the crucial lessons learnt through TS's experience, and which should be taken into account when dealing with projects of a similar nature, are as follows:

- A key role of undergraduate students towards creating sustainable society is to link the theoretical and practical knowledge developed at HEI with real world needs, and to apply it. In particular, student organizations could play a prominent role in the diffusion of knowledge generated in universities, which should entail institutional support of their activities.
- Although there is plenty of accumulated knowledge in universities for solving everyday problems, only the community itself knows its real needs. Therefore the community must actively participate in each project.
- Sustainability challenges, and especially those involving communities, need an interdisciplinary approach aimed at developing practical and social solutions.
- When working in communities, neighbors should feel part of the project and empowered within the work team. Therefore, specific efforts should be made to overcome asymmetries in the relationship. In that sense, students should be encouraged to learn from the community's particular experience.
- Consequently, devices and technology are not "sacred": the final truth and the solutions to environmental issues are developed jointly by the community and the Solar Workshop.
- In Chile, and in much of the developing world, most of the communities lack a basic infrastructure, so territorial Solar Workshops need to dedicate considerable time and effort to developing an essential local coordination. Unfortunately, when working with very precarious and unorganized communities, it is unlikely that the project will be successful.
- Up to now, the necessary financial support has been obtained from UCh and governmental grants. However, when working with communities, self-management and local financial sources should be sought to overcome "assistencialism".
- Active political discussion within the group has been a decisive boon for the maturation of an own concept of Energy Revolution and, furthermore, for linking TS's activity with the national energy context.

The Future

The next challenge for TS is to incorporate new technologies, which will be modified and adjusted jointly with each community to meet its energy needs. Strategies to achieve this goal are: inclusion of professionals and faculty members; proposing technical challenges for thesis/research projects, and appropriate courses in the context of the curricula of a sustainable campus. Nevertheless, the discussion concerning possible structural changes and the establishment of a permanent formal relationship with Universidad de Chile, is still open.

Acknowledgments. The authors would like to thank to Prof. Roberto Román and Rodrigo Valdovinos (EcoMaipo) for their inspiration, motivation and constant support. The various achievements of TS have only been possible through the profound conviction in cooperative group work ("*no one is indispensable, everyone is important*"). This paper is therefore dedicated to all participants who have contributed to this project. Last but not least, the authors are immensely grateful to Giancarlo Bucchi, who enthusiastically made possible the proper redaction of this article.

References

AETS SA (Application européenne de technologies et services) (2011), *Caracterización de los Actores Locales del Mercado Energético – Informe Final*. Available at: http://www.minenergia.cl/documentos/estudios/caracterizacion-de-los-actores-locales. html (accessed 25 January 2012).

Bórquez, R. (2011), *Actualización de Catastro de Proyectos de Generación Eléctrica en Chile*. Available at: http://www.terram.cl/images/app/app52_catastroproyectoselectricos_actualizacion_2011_final-sep.pdf (accessed 25 January 2012).

Bucchi, F., Jacques-Coper, M. and Sánchez, P. (2012), Towards a Sustainable Campus at Universidad de Chile: the Key Role of Students. In Leal Filho, W. (Ed.), *Sustainable Development at Universities: New Horizons*. Peter Lang, Frankfurt (this issue).

CADE (Comisión Asesora para el Desarrollo Eléctrico) (2011), *Informe de la Comisión Asesora para el Desarrollo Eléctrico*. Available at: http://www.minenergia.cl/documen tos/estudios/informe-de-la-comision-asesora-para-el.html (accessed 25 January 2012).

CCTP (Comisión Ciudadana Técnico Parlamentaria para la Política y la Matriz Eléctrica) (2011), *Chile necesita una gran reforma energética*. Available at: http://www.chile sustentable.net/?file_id=141 (accessed 25 January 2012).

Conde Hernández, R., González Castillo, O. and Mendieta Márquez, E. (2006), Hacia una gestión sustentable del campus universitario. *Casa del Tiempo* 8 Ep. III, No. 93, pp. 15-25.

DGF and CNE (2012), *Solar Energy Explorer*. Available at: http://ernc.dgf.uchile.cl/Explorador/Solar2/ (accessed 25 January 2012)

Gebremedhin, A., Karlsson, B. and Björnfot, K. (2009), Sustainable energy system – A case study from Chile. *Renewable Energy* 34 (5), pp. 1241-1244.

Guzowski, C. and Recalde, M. (2010), Latin American electricity markets and renewable energy sources: The Argentinean and Chilean cases. *International Journal of Hydrogen Energy* 35 (11), pp. 5813-5817.

Hall, S., Román, R., Cuevas, F. and Sánchez, P. (2009), *¿Se necesitan represas en la Patagonia? Un análisis del futuro energético chileno*. Ocho Libros Editores, Santiago de Chile.

INE (Instituto Nacional de Estadísticas) (2002), *Censo 2002 – Síntesis de Resultados*. Available at: http://www.ine.cl/cd2002/sintesiscensal.pdf (accessed 25 January 2012).

Ortega, A., Escobar, R., Colle, S., Luna de Abreu, S. (2010), The state of solar energy resource assessment in Chile. *Renewable Energy* 35 (11), pp. 2514-2524.

Pontt, J., Maldonado, P., Aedo, M.P. and Larraín, S. (2008), *Aporte Potencial de Energías Renovables No Convencionales y Eficiencia Energética a la Matriz Eléctrica, 2008-2025.* Available at http://www.prien.cl/documentos/UEEE-SIC/ENERGIA_web.pdf (accessed 25 January 2012).

Román, R. (1999), Obtención de agua potable por métodos no tradicionales. Obtención de agua a partir de las Camanchacas. *Ciencia al Día Internacional* 2 (2), pp. 1-13.

Schemenauer, R.S., Fuenzalida, H. and Cereceda, P. (1988), A Neglected Water Resource: The Camanchaca of South America. *Bulletin of the American Meteorological Society* 69 (2), pp. 138-147.

Sharp, L. (2002), Green campuses: the road from little victories to systemic transformation. *International Journal of Sustainability in Higher Education* 3 (2), pp. 128-45.

ULSF (University Leaders for a Sustainable Future) (1990), *Talloires Declaration.* Available at: http://www.ulsf.org/pdf/TD.pdf (accessed 25 January 2012).

Chapter 76

Sustainable Campus Building in Complex Conditions

Felipe Rodrigo Horta Jaques, Carolina Andrea Cares Inostroza and Guido Leonardo Olave Reyes

Abstract

Any university that wishes to take the current sustainability challenge seriously needs to be aware of and manage its environmental footprint. Managing the environmental impact of a university is recognized as a key element in promoting sustainable behavior in students and contributing to overall sustainability. However, Campus Sustainability is often thought of as an expensive and time-consuming endeavor, which is not highly regarded. This is why it is frequently seen as exclusive to rich universities, and considered an impossible goal for public universities or those in developing countries. It is also common for university leaders to appear keen on incorporating new courses on sustainability, as well as running other quick and visible initiatives, whilst really being reluctant to take serious steps towards reducing their campus' environmental footprint. This study proposes a model to achieve a more sustainable campus even in spite of a lack of institutional support and financing. It is inspired by the experiences of the School of Economics and Business of the University of Chile, along with relevant international literature on campus sustainability and its leadership. Besides this model, this paper will include conclusions drawn and lessons learnt that might be of use to other academic institutions, and raise questions that might inspire further research in this field. This model is meant as a guide for universities where there is no formal leadership or funding for Campus Sustainability, and provides a framework in the form of a practical guide to moving a campus towards sustainability.

International Context

Concern for the impact human activities may have upon the environment started to become widespread in the 70s. The scientific world was the first to react to the exploitation of nature and the sustained damaging of natural systems.

However, it was not until 1987 that several nations came together to produce a socio-economic report for the UN, whereby "sustainable development" was defined as *development that meets the needs of the present without compromising the ability of future generations to meet their own needs*[1]. This report is mainly based on the relationship between sustainability and ecology. The former has been associated since the 60s to an ability to remain over time. Thus, to exploit a mine for 50 years without considering the impact this would have on the environment was considered "sustainable" at the time[2].

When referring to development, the report states two restrictions: firstly, the ability to preserve the planet; secondly, giving up all levels of consumption which are not available for

1 World Comission on Environment and Development (1987) *Our Co*mon Future: Brundtland Report. Capítulo 2.

2 Boullón, Roberto (2006), *Espacio turístico y desarrollo sustentable.* Aportes y transferencias, Vol. 10, Núm. 2, pp. 17-24. Universidad Nacional Mar del Plata, Argentina.

everybody. The report, known as the Brundtland Report, establishes the following long-term aims: economic growth in poor countries; demographic control through the regulation of birth rates; the guarding of natural systems that sustain life; conservation of the ecosystems which are needed for human well-being; and the efficient use of non-renewable resources.

In addition to establishing the need for incorporating these aims into each country's growth strategies, the importance of periodically monitoring progress and new challenges was recognized, leading to conferences and international meetings during the 80s. These were soon translated into statements and letters. Correspondingly, in 1983, the United Nations Environment Program (UNEP) and the UN World Commission on Environment and Development were established.

In 1989, the UN started the Agenda 21 action plan, designed to promote sustainable development and ascribing a key role to local communities. In Chapter 28, the following aim is stated: "By 1996, most local authorities in each country should have undertaken a consultative process with their populations and achieved a consensus on 'a local Agenda 21' for the community"[3]. It was also established that "all local authorities in each country should be encouraged to implement and monitor programs which aim at ensuring that women and youth are represented in decision-making, planning and implementation processes"[4], thus conveying the importance of the participation of all social actors in attaining the proposed aims.

University Context

It was in this international scenario, where importance was ascribed to young people, that some universities, mainly in the United States, started to incorporate concern for academia's own environmental impact. The kick-start was given by UCLA, through its publication *In Our Backyard: Environmental Issues at UCLA, Proposals for Change, and the Institution's Potential as a Model*[5], and it was followed by other universities, such as the University of Wisconsin with its "Campus Ecology" program, and Brown University with "Brown is Green"[6].

Many universities started to join in, with new initiatives being launched at a slow but steady rate. Amongst them was the Association of University Leaders for a Sustainable Future, which set a milestone when it released its "Talloires Declaration". This statement was signed by 22 representatives from different universities around the world in 1990[7]. Currently, it has more than 400 affiliated schools and universities[8].

Aside from being a leader in the field, it established the following guidelines:

- Encourage all universities to engage in education, research, policy formation, and information exchange on population, environment, and development to move toward global sustainability.
- Establish programs to produce expertise in environmental management, sustainable economic development, population, and related fields to ensure that all university graduates are environmentally literate and have the awareness and understanding to be ecologically responsible citizens.

---

3 http://www.un.org/esa/dsd/agenda21_spanish/res_agenda21_28.shtml.
4 http://www.un.org/esa/dsd/agenda21_spanish/res_agenda21_28.shtml.
5 Brink et al. En Conde et al. Hacia una gestión sustentable del campus universitario.
6 Conde et al.
7 Conde et al.
8 http://www.ulsf.org.

- Set an example of environmental responsibility by establishing institutional ecology policies and practices of resource conservation, recycling, waste reduction, and environmentally sound operations. Thus, institutions are not only to participate but must also push the change forward.

Since then, there have been several international meetings aiming to promote Campus Sustainability, incorporating a growing number of universities and actors from the political and academic worlds. Whilst in 1990 only 31 university leaders signed the Talloires Declaration, in 1993 650 universities participated in the round-table meeting before the signing of the Kyoto Declaration.

However, why is it important to have universities actively participating? According to *Towards a Sustainable Management of the University Campus*[9], environmental management specialists are scarce. In addition, professionals from all other disciplines are unaware of the impact their daily activities may have on the environment, and the deleterious effects they could have on their health. Universities form professionals who frequently end up in positions managing society's institutions, which is why they have the responsibility as educational institutions to deliver awareness and knowledge, as well as the available technology and tools to promote a sustainable environment.

On the other hand, in the *Declaración Por la Sustentabilidad de y desde la Universidad* (Sustainability Declaration for and from the University)[10], it is established that "through education, research, knowledge spreading, access to information, and culture production, we can become leaders in social values, attitudes and behavior, pushing forward other economic, political, scientific and social agents to act accordingly as well".

Likewise, the international process supporting the transformation towards sustainability in higher education has been strengthened by the onset of the United Nations Decade of Education for Sustainable Development (2005-2014), which has promoted this process worldwide[11].

University and Sustainability in Chile

Although the first declarations around the world were signed during the 90s, it was not until April 29, 2010 that the first Protocol for a Sustainable Campus was signed voluntarily by a group of universities in Chile, aiming at achieving sustainability practices. These institutions are the following: *Universidad de Chile* (University of Chile), *Universidad Tecnológica Metropolitana* (Technological Metropolitan University – UTEM), *Universidad Andres Bello* (Andres Bello University – UNAB), *Universidad de Santiago* (University of Santiago – USACH), *Universidad Iberoamericana de Ciencias y Tecnologia* (Iberoamerican University of Science and Technology), *Universidad de Talca* (University of Talca), *Universidad Metropolitana de Ciencias de la Educacion* (Metropolitan University of Education Sciences – UMCE), and *Universidad Bolivariana* (Bolivarian University), plus the Middlebury College of the United States to support the incorporation and implementation processes.

---

9 Conde et al.

10 Arias, Raúl, Declaration of the Organización Universitaria Interamericana "Por la sustentabilidad de y desde la universidad". Website: http://www.udca.edu.co/es/noticias/382-rectoria/1642-decalaracion-sustentabilidad-universidad.html

11 Bravo, María Teresa (2008), La educación superior en México: Avances y desafíos en la reconversión ambiental de sus instituciones, in *Educación Ambiental para la sustentabilidad en México*.

The document produced in this meeting incorporates some of the tasks universities should aspire to, such as "the development of science for sustainability; the installation and permanent evaluation of sustainable management models, and clean production in each campus; systematic interaction with government agencies such as the CONAMA and the *Consejo Nacional de Produccion Limpia* (National Council for Clean Production); the development of transversal and interdisciplinary education methodologies for sustainability, which should be applicable to the wide variety of curricula in undergraduate, postgraduate and continuing education; the formation of professors and professionals with a solid background in the science of sustainability, ecology and good pedagogic practice."[12]

Ever since these protocols have come into effect, universities have had differing experiences during the implementation of the models on sustainable management. This is mainly due to their current organizational schemes and integrated planning. As of January 2012, USACH has five projects running under the Sustainable Campus program: the MAPUSACH (from the group of engineering students in agricultural business); the *Confederación de Estudiantes de Ciencias Ambientales de la Educación Superior* (Higher Education Student Confederation of Environmental Sciences); the CECADES (a group formed by students from different universities, which organizes training programs for different courses in the Metropolitan Region, helping schools obtain FNCAE certification); *Vocalía Ambiental Feusach* (which aims to provide a means through which student initiatives and questions related to the environment can be channeled); and Gama (a group of researchers dedicated to generating agroeconomic knowledge at the university, leading several in and off-campus sustainability activities)[13]. In spite of all this, it can be difficult at first to detect any integrated coordination beyond professors and students from related majors.

On the other hand, at the *Universidad Andres Bello* directors and academics of the Sustainability Center recently formed a "Committee of Sustainability", which will lead the project "UNAB Green Campus". The Center, launched in 2010, performs research in areas such as sustainable architecture, energy, ecological design, pollution, biodiversity, recycling, and climatic change, among others. The existence of a center exclusively dedicated to research in the field reflects the effort and commitment shown by this university.

However, what is it that makes Campus Sustainability management develop differently at different universities?

According to Conde, Gonzalez and Mendieta's research, the internal reality of each university prevents, or at least hinders, the development of committees dedicated exclusively to the implementation of Campus Sustainability plans. This can be due to the complexity of the organization, "combined with the complexity of the environmental requirements, which hinders most attempts to reach a consensus in goals, alternatives and solutions". Researchers thus explain that, as the focus is lost, so is the drive for organizational impact.

Following the USACH model, the School of Economics and Business (*Facultad de Economia y Negocios* – FEN) has started academic and student initiatives in spite of the lack of institutional support from the University of Chile, managing to have a positive impact on its surroundings. However, it is very hard to talk about a Sustainable Campus when only a few isolated initiatives exist.

---

12 http://www.conama.cl/educacionambiental/1142/article-47228.html.
13 http://www.usach.cl/index.php?id=12435.

Case Study: University of Chile

Sustainability initiatives in universities are always accompanied by typical higher education elements, which can help or hinder their development. Negative factors, as established in *Deterring Sustainability in Higher Education Institutions*, are the following:

- Lack of detection, interest and involvement;
- Particularities in the organizational structure;
- Resistance to change;
- Lack of human and material resources;
- Lack of communication and appropriately timed information;
- Lack of rigorous regulation;
- Lack of interdisciplinary research;
- Lack of operational indicators;
- Lack of institutional policies;
- Lack of an institutional consensus.

As a pioneer in the field within the university, the School of Economics and Business diagnosed the institution as suffering from many of these problems, and stated that several arose from the Chancellor's office. This became evident in the Project for Institutional Development of the University of Chile (which has been active since 2006), which lacks a corporate policy for sustainable development and sustainability. The latter term is only mentioned in terms of the University's ability to finance itself.

Likewise, sustainable development is mentioned within the mission of the University, but the mechanisms through which it should be achieved are not mentioned.

> "It is the University's responsibility to watch over the country's cultural heritage and national identity, promoting the improvement of the educational system. Therefore, the University should respond to the Nation's requirements as an intellectual reserve, which promotes social awareness, critical thinking and ethical responsibility, recognizing it as part of its mission to attend to the country's problems and needs. To this end, the complete knowledge of the nation's current state must be known, and its development through research and creation promoted. An integrated, balanced, and sustainable development of the country should be carried out, contributing to the solving of problems from the university's perspective. Common wellbeing and a civic formation inspired in democratic values should be achieved making sure cultural tradition is safeguarded and enriched at the national and universal level."[14]

Therefore, due to the lack of policies and institutional agreements the implementation of Sustainable Campus without the main support of the University becomes very complex. Furthermore, it has been detected that professors are neither familiarized with these initiatives nor interested in participating in them, except for a very limited group which, on isolated occasions, has tried to introduce the subject into classrooms.

In addition, FEN has tried to start dialogs to establish policies for Sustainable Campus at the University of Chile's Andres Bello Campus (which is shared by the faculties of Economics and Business, Architecture and Design, and the Public Affairs Institute), and the only reply received stated that no decision would be made as long as the Chancellor's office did not request it directly.

---

14 PDI http://www.uchile.cl/portal/presentacion/senado-universitario/documentos/30825/proyecto-de-desarrollo-institucional.

EcoFen, the Kick-Start

Interest in turning the School of Economics and Business into a sustainable faculty so that it would not only measure its environmental impact, but also work towards minimizing it, began with a group of students in 2007, who gathered together to generate projects for the Faculty's sustainable management and to make their classmates and professors aware of environmental issues.

The EcoFen project was born five years ago, aiming to study the possibility of recycling most of the waste generated at the Faculty. This is how the "FEN Recycles" initiative arose. It had to go through many stages to show that it was not only viable and sustainable in time, but also a necessary movement. During the first stage, students researched which materials would be recycled, where the containers would be placed, and the logistics of the project. This was followed by the characterization and measurement of the waste produced at the Faculty, where it was defined that paper, glass, cans, plastic (pet-1), tetra-pack and batteries would be recycled.

During the third semester of 2008, students performed a campaign to gather funds to implement the project in 2009. At the beginning of the latter, informative talks were given to the cleaning staff, making them aware of the campaign's aims, functioning, and why they played such an important part in the process. In March, all containers were placed, and a diffusion campaign to inform the whole community of the new recycling system at the Faculty was carried out. Finally, those in charge of the project took care of the logistics, establishing gathering points, the frequency with which recycling companies were to remove the waste, etc. This initiative worked under the responsibility of EcoFen during 2009, and in 2010 it was incorporated as an official project at the Faculty through the Unit of General Services, Infrastructure and Physical Plant.

FEN, Towards a Sustainable Campus

Before starting the Sustainable Campus campaign, the School of Economics and Business noticed the importance of actively involving existing autonomous and organized groups in the Faculty, such as EcoFen, student councils, social projects, and volunteering groups. Until 2010, these initiatives worked separately. However, they were integrated into the *Unidad Nexo Responsabilidad Social Universitaria* (Liaison of University Social Responsibility – NexoRSU), which gave institutional support to the students' work. In addition, this unit was in charge of the gradual incorporation of sustainability into the curriculum through courses and "Learning and Service" methodologies, as well as elective or free classes. Amongst them were Responsible Trekking, Sustainable Economic Development, Economy of Natural Resources and the Environment, Economy, Humans and Values, Ethics, and others. On the other hand, this unit has progressed to the cutting edge of the field, setting an example that is now starting to be followed by other university campuses.

Fig. 1: NexoRSU Methodological Framework

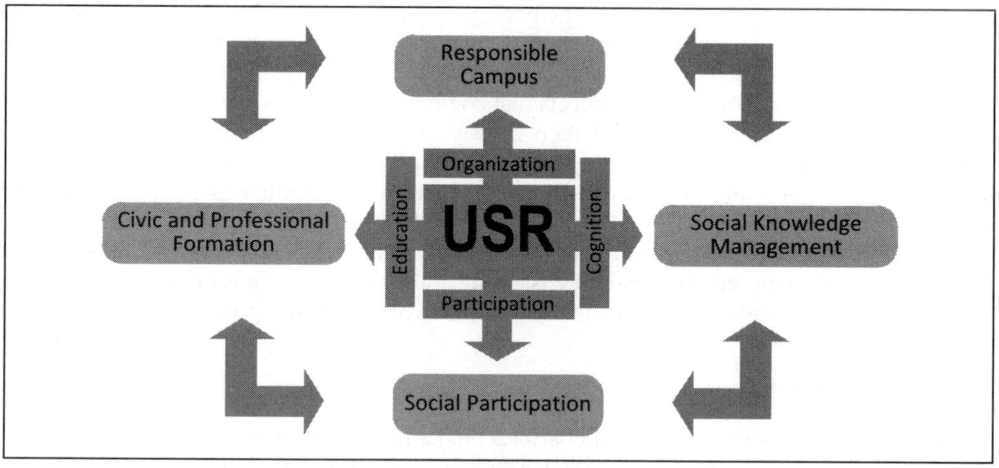

As shown in Fig. 1, University Social Responsibility can be defined through the impact the institution or Faculty has upon its environment. These can be classified into organization, education, cognition and participation, defined as follows:

1. **Organization:** impacts the life of the personnel (administrative, faculty and students) and the environment (waste, deforestation, transport, etc.).
2. **Education:** influences the formation of young people and professionals. Values, ethics, and how to interpret the world and behave in it – all of these influenced by the university.
3. **Cognition:** impact on knowledge production. This is the most important category, because it is through knowledge and paradigms that future leaders build society.
4. **Participation:** impact upon society's development.

The types of identified impacts allow the definition of four axes of University Social Responsibility:

1. **Responsible Campus:** implies encouraging an organizational behavior responsible for each and every member of the university community.
2. **Civic and Professional Formation:** the socially responsible management of academic formation; that is, curricula should be oriented towards society's real problems.
3. **Social Knowledge Management:** the socially responsible management of production and diffusion of knowledge, research and epistemological models promoted from the classroom.
4. **Social Participation:** carrying out projects together with other actors so that bonds may be built for mutual learning and social development.

These four axes are permanently providing feedback for each other, in order to have a dynamic continual improvement of responsibility.

Since the NexoRSU unit was created and became fully functional, the School of Economics and Business has established a Committee for Sustainable Campus, where student, staff, NexoRSU, and academic representatives participate to coordinate the activities taking place at FEN. These aim to improve environmental management, improve energy efficiency and waste management, and promote an environmentally friendly culture. The Committee is also in charge of diffusion campaigns and external communication of Campus Sustainability.

The School of Economics and Business requested external consultancy to establish the basis of the Sustainable Campus project. The consultancy allowed the identification of ideas and values related to the concept of sustainability; the scope and prospects for the management; the convergences and divergences; and the perception of an organizational culture. Good practice guidelines in internal FEN projects that could be replicated were addressed, as well as concluded and upcoming actions and their scopes. At this stage, interviews indicated once again that the main obstacles to the implementation of this project were the lack of initiative from the Chancellor's and Dean's offices within the Faculty. Further major obstacles were also identified, including: a wrong diagnosis of the FEN energetic consumption; the lack of importance ascribed to sustainability by the higher authorities; an excess of bureaucratic processes; management and transparency issues characteristic of complex organizations such as the University of Chile; and personal interests being placed above those of the Faculty.

Public-Private Alliances

We have therefore observed a series of obstacles hindering the correct implementation of a Sustainable Campus project, and financing is very likely one of the main issues. The implementation of different initiatives proposed in the sustainability project implies a large investment in human and financial resources, making it seem an unviable option for public institutions in developing countries. Thus, the Faculty has decided to involve public and private companies in this project, integrating companies that value and see the need for Social Responsibility.

The Faculty is currently working on several projects, such as green walls, clean spots open to the community (in one of the city's most important downtown neighborhoods), solar energy-based cellular phone chargers, etc.

Moreover, several companies have expressed interest to the Faculty in being involved in more sustainable productive processes – this will improve their image amongst consumers.. Thus, the diffusion platform that the Faculty offers (student profiles, strategic geographic sites) is becoming an interesting scene to motivate these organizations.

Lessons and Challenges

The process carried out by different Faculty areas for the Sustainable Campus project has yielded several lessons that will serve as the basis for the implementation of this project, as well as an impulse to face the new challenges better. First, the current condition of the University, whereby each faculty has to finance itself, leaves no option but to look for external financing. One of the main challenges is to motivate even more companies so that it is possible to start a project that not only utilizes resources, but is also sustainable over time. Furthermore, the high degree of commitment, creativity and motivation shown by students throughout the generation of innovative projects in the field makes it necessary to create new spaces for them to be involved in the Faculty processes that interest them.

On the other hand, it is key to know the interests and motivations of the staff in the field, in order to detect those who can exert a positive leadership effect and help achieve aims. To motivate professors whose connection with other areas is reduced or non-existent is also highly relevant, seeing as they are the ones who have direct contact with students.

Faculty areas related to the Sustainable Campus project have made endeavors in several processes, from which many lessons have been learned. These will be useful for the implementation of the project, and also as an impulse to face the new challenges in a better way. Amongst these was the Faculty's capability to generate future attractive projects that may allow the securing of external finance, connecting the public and private worlds through sustainable initiatives. The current Chilean university funding model requires that each school generate a significant amount of their own resources, for which it is key to generate alliances with other organizations. All of these lessons will be important to allow these initiatives to be replicated at a university, campus and faculty level.

Another lesson is the high degree of commitment, creativity, and motivation students have shown in generating innovative projects in the area. This shows the importance of promoting working spaces and meetings to allow their participation and involvement in transforming the Faculty into the place they expect it to become.

Given that the University is still not developing an integrated Sustainable Campus project, the biggest challenge is to encourage other units and faculties to start to adopt environmental good practices and initiatives. These should be aimed at unifying the efforts to achieve a university at the cutting edge of the field. Similarly, another great challenge is to establish a diffusion plan according to the scope of the project, so that professors, students and staff are completely aware of what the project implies and how they can actively participate.

The establishment of Sustainable Campus at the School of Economics and Business of the University of Chile has been very complex due to the characteristics defining this organization. However, the impulse provided by the students has been the main driver resulting in serious and responsible plans that have been made tangible, turning into attractive projects promoting the support of public and private companies. The main innovation in this area is that the Faculty has decided to lead the Campus Sustainability initiative of the University through its Committee and tangible actions, where students are the main axis, so as to promote the establishment of other equivalent movements at Faculty and central levels. Thus, the University of Chile has decided to lead the way towards sustainability within higher education institutions in this country.

The year 2012 will be crucial to establishing sustainability policies at the School of Economics and Business. After the maturing of its institutionalism, it will be possible to see the first results of the work that started in 2007. It is hoped that the first projects will be implemented in alliance with those companies that will have a direct impact in the university community and its closest surroundings. Another aim is to implement new lectures and other academic instances that will allow topics related to sustainability to be better integrated in the curriculum.

References

Association of University Leaders for a Sustainable Future. Available at http://www.ulsf.org (accessed 11 January 2012).

Boullón, Roberto (2006), Espacio turístico y desarrollo sustentable. *Aportes y transferencias* 10 (2), pp. 17-24.

Brink, T. et al. (1989), En nuestro patio trasero: la cuestión ambiental en la Universidad de California en Los Ángeles (ucla): propuestas para el cambio y el potencial institucional como modelo Comprehensive research proyect, Graduate School of Architecture and Urban Planning. In Conde, R., González, O. y Mendieta, E. *Hacia una gestión sustentable del campus universitario.* Tiempo Vol. 8 Issue 3, 10 pp.

Conde, R., González, O. y Mendieta, E. (s.a.), Hacia una gestión sustentable del campus universitario. *Casa del tiempo* VII Época 3 (93-94), pp. 15-25.

Corporación Nacional de Medio Ambiente *Universidades se comprometen con Campus Sustentable.* Available at http://www.conama.cl/educacionambiental/1142/article-47228.html (accessed 11 January 2012).

Programa de acción de las Nacionales Unidas de Río. *Programa 21.* Available at: http://www.un.org/esa/dsd/agenda21_spanish/res_agenda21_28.shtml (accessed 13 January 2012).

Universidad de Chile, *Proyecto de Desarrollo Institucional: El Compromiso de la Universidad de Chile con el País – Resumen Ejecutivo*. Available at http://www.uchile.cl/portal/presentacion/institucionalidad/58048/proyecto-de-desarrollo-institucional (accessed 15 January 2012).

Universidad de Santiago *Campus Sustentable. Available* at http://www.usach.cl/ index.php?id=12435 (accessed 10 January 2012).

World Commission on Environment and Development/WCED (1987), *Our Common Future*. United Nations. Oxford University Press, Oxford, 400 pp.

Chapter 77

Learn How to Manage Sustainability through Designing an Eco-Taxes Proposal

Gabriela Toledo Román and Oscar Jerez Yañez

Abstract

Ecological Economics is the science of sustainability management, which is fundamentally transdisciplinary. The present article is a systematization of teaching and learning methods about sustainability, developed in the course of Ecological Economics and taught to students in the final years of their Commercial Engineering course at the University of Chile. The course aims to develop sustainability criteria to help graduates face decision-making in their professional activities.

The methodology of the proposed learning model consists of five phases guiding the teaching and learning process, oriented to design an eco-tax proposal to manage sustainability in a particular economic sector. The stages are: a) awareness and context; b) diagnosis; c) literature and experience review; d) development of an eco-tax; and e) communicate and redesign. Each stage involves specific activities with appropriate pedagogical foundations.

This paper analyzes how the students are able to visualize the process of managing change and sustainability through an experience linked to future professional challenges. Finally, the paper reviews learning outcomes and summarizes the main lessons and challenges in implementing the methodology. The main value of this systematization is to propose a learning model linking deep learning in sustainability with the formation of professionals with key skills to manage sustainable development.

Introduction

The Ecological Economics Course (EEC) is an elective course for fifth-year students of Commercial Engineering at the University of Chile. The course emerges like a natural projection of an existing course of general formation. Its main objective is to engage students in creating solutions that improve the relationship with the environment, such as an eco-taxes proposal. This goal is guided by the need to train Economics students with critical and reflective elements that allow for dialogue with other disciplines in addressing major challenges such as sustainability management.

The society perceives a conflicting relationship between economics and the environmental and financial crisis lately experienced. Schools of economics have been partly blamed in this process, and the challenge of facing these criticisms still remains (Issa et al. 2011). Moreover, Springett (2005), posits that education for sustainability defies the rationality of a capitalist paradigm of production and consumption, which constitutes a major challenge to addressing sustainability in the curriculum, particularly in faculties of economics. We will work with this postulate, addressing this challenge using the proposed model.

The literature about teaching sustainability to economics students is encouraging. Stubbs and Cocklin (2007) analyze the advantages of employing active learning methodologies to develop critical thinking and capacities for discussion in economics students. Moreover, Cordano et al. (2003) show that the attitude of economics students towards environmental issues improves towards the end of environmental modules within the course. This paper systematizes a learning model innovation that was introduced in 2011 into the Ecological Economics course in the Faculty of Economy and Business at the University of Chile. Our working hypothesis is that the proposed learning model innovation allows meaningful learning and develops skills needed to manage sustainability. The model is based primarily on the literature on education for sustainable development, and was carried out with the following consideration in mind: "A learning process that responds to current challenges of learning in universities should have the following requirements: a) oriented skills, b) oriented to society and c) centered on the individual (in learning)" (Barth et al. 2007).

Section 3 reviews the main features of the model, the context, stages, activities and their foundations; Section 4 outlines the learning outcomes sought by the model, and finally Section 5 summarizes the discussion and future challenges.

The innovative experience

Context

Within the Faculty of Economics and Business at the University of Chile, there is a strong movement promoting debate on sustainability and pushing for changes. In 2011, the Faculty began the process of establishing a Sustainable Campus (SC). Related to the EEC, there is a growing demand for courses about protection of the environment and natural resources. Until the creation of the EEC, courses on environmental topics and natural resources were available, but with a strict economic perspective that did not tackle sustainability.

In the process of innovating in the teaching of sustainability, teachers of the EEC worked with a program from the same university called Teaching Excellence Network. The objectives of the Network are to give teachers tools to develop methods of learning according to their disciplines and objectives. A team working with the network developed a syllabus and goals for the course focusing on the development of competences.

Characteristics of the model

Transdisciplinary: Essential characteristic of the model

Ecological Economics is the science of sustainability management, and is rooted in a transdisciplinary perspective (Costanza 1991). Transdisciplinarity is a requirement in Education for Sustainable Development (ESD).

As in the vast majority of economics schools, students are only exposed to neoclassical economics (Stubbs and Cocklin 2007). In Chile, this is reinforced by institutions and the economic system, where neoliberal policies have been firmly entrenched since the fall of the military dictatorship (Olavarría 2003). Conflicts between the economy and the environment in Chile arise frequently due to the Chilean development strategy, which is based on a model of exporting natural resources in a deregulated and highly neoliberal scheme (Schurman 1996). Therefore, the environmental problem in Chile has a strong economic component. Transdisciplinarity in this context aims to: a) develop the capacity for dialogue, tolerance, respect for ideas and cooperation as key competencies for sustainable development (Barth et al. 2007); b) install the discussion of sustainability from a transdisciplinary and multi-criterial perspective (expanding the prevailing economic vision); and c) assess the tools of economics and other disciplines in terms of their ability to respond to environmental problems, replacing the view that market tools in themselves are the most efficient solution.

The model integrates a transdisciplinary perspective in different scales of intervention, classes in different disciplines in the curriculum, and the teaching and learning activities themselves (Davies and Devlin 2010).

There are two kinds of academic in this debate: an ecologist and an economist. Each is in charge of teaching and guiding students in different topics perspectives. For example, when ecology topics are seen from the point of view of an ecologist they have much more impact than learning takes place traditionally through texts. Finally, as we will see, learning activities are geared towards transdisciplinarity, both through the curriculum contents of the course and through the activities.

Competency-based training

Competency-based training is a way of responding to personal, social, professional and cultural problems in our present-day society, especially those related to the environment and human relationships. It represents a change in mindset from the traditional approach, i.e. it is a shift in focus from teachers teaching to students learning (Villa and Poblete 2007).

Training for skills development means integration of knowledge, resources and behaviors that move in a specific context of performance. The notions of mobilized integration and specific context are essential here. Competence is not simply the sum of knowledge and/or particular skills; competence articulates, composes, and constantly ponders these diverse resources and is the result of their integration (Le Boterf and Durand-Gasselin, P. Pechenart 1991). For this approach the educational value of experience must be assessed and highlighted (Irigoin and Vargas 2002). Learning by doing, creating conditions similar to actual or work contexts, becomes a necessary condition from the point of view of didactics. This makes sense, as this learning experience enables students to generate solutions to urgent everyday and complex problems.

The key competences to be developed in the EEC need to be defined. First of all, problem-solving. In terms of sustainability, there are many things that can be improved; more explicitly, there is something that is going wrong. We aspire that students will be able to construct knowledge and create solutions to this problem. Secondly, ethical commitment is necessary to really comprehend the relevance of sustainability. To foster this competency implies a dialogue between intellect and emotion, ultimately engaging with this issue because of its importance for our survival.

In summary, this experience is characterized by:

- Acquiring and developing the defined key competences related to sustainability;
- Developing skills through "learning by doing", i.e. from the design of an eco-tax;
- Favoring active student learning and including job performance issues related to sustainable development;
- Assessing learning; linking processes to students' own interests and the challenges of sustainability in the context professional responsibility..

Learning Model

The model is composed of articulators' milestones that facilitate the generation of learning using the context of the eco-tax. The milestone is a key moment of the learning process that gives meaning to the training process through the professional challenges. Thus, learning is generated through the experience, contrast, integration and linkage of professional work (Brockbank and McGill 1998; Fink 2003; Najjar and Simon 2009). There are five consecutive key milestones:

- Awareness and context;
- Diagnosis;
- Literature and experience review;
- Development of an eco-tax;
- Communicate and redesign.

Each milestone, in turn, has three essential elements: purpose, activities and rationale (basis). The purpose is intended to give focus to the milestone, i.e. a formative intention at that specific moment of the process. Each purpose is worked towards through activities. These activities or actions reveal how the milestone is executed according to a predefined intention. The activities are characterized as active, i.e. they put students in a leadership role, with the teacher acting as facilitator of the process. Each group of activities generates evidence of learning and reveals how to achieve the proposed purposes of the milestones. Each action is able to motivate the student, giving the process meaning. Finally, each of these articulated elements (milestones, goals and activities) is based on the linking of pedagogical foundations and sustainability. In the following, the model is reviewed in more detail.

Development of each milestone

Details of the purpose of each milestone, and its educational foundation in Higher ESD, are presented below.

a) Awareness and context

Purposes:

- Recognize the relationship between society and nature as an inter-disciplinary matter containing present different valuation criteria.
- Conceptualizing sustainability and its elements.
- Analyze the conflict between a mono-criterial perspective (economic) and the multidimensional nature of the situation.
- Assess biophysical criteria included in the analysis of sustainability.

Activities:

- Problematize and reflect on important elements in the conflict setting through a conflict assessment document (economic, social or environmental).
- Address the complexity and unpredictability of ecosystem functions through reading, allowing the mapping of relationships between causes and effects, seeking group reflection.
- Ask students to bring to class news items showing elements of an environmental conflict. Then, in groups, they compare news items from previously established categories.
- Ask students to rate the sustainability of different situations using a concept map of the energy cycle.

Assessment: Evidence of learning

- Students must submit an essay on the topic "Why include biophysical criteria in the analysis of sustainability?"
- In pairs, students are asked to choose an environmental issue, then think of what makes it a conflict and consider why it needs to be managed for sustainability.

Basis:

This milestone is fundamentally problematizing, i.e. it exposes the conflicts that arise from a mono-criterial analysis focused on economic growth. The conflict or problematization is a powerful internal motivational trigger for learning (Fink 2003; Marzano and Kendall 2007), and also an excellent way to promote integration and transdisciplinary vision on the issue of sustainability.

By questioning the current approach to tackling environmental issues, a wider debate about the economic model is enabled, incorporating elements from different disciplines in a functional way in order to solve the problem. Problem-solving orientation allows the students more freedom in committing to sustainability, since it does not impose a particular vision, leading to a commitment that is stronger and more durable (Norton and MacVaugh 2012).

This milestone is marked by activities in the classroom; as Barth et al. (2007) propose, "learning environments should be designed in such a way that also encourages informal learning processes." These informal settings are particularly favorable for the development of skills for sustainability (Barth et al. 2007). In this sense, group activities are simply an invitation for discussion and reflection, promoting a deeper exchange of ideas.

b) Diagnosis

Purposes:

- Identify patterns of sustainable consumption in different types of goods and services.
- Compare different methods for diagnosing sustainability.
- Incorporate the logic of sustainability into professional work situations by conducting a diagnostic analysis applied to the chosen sector.
- Establish a critique of the dominant development model from the analyzed cases.

Activities:

- Students are asked to discuss in groups the definition of sustainability and sustainable patterns in relation to renewable and non-renewable resources, waste and biodiversity.
- Students are asked to develop a sustainability diagnosis in a given economic sector (e.g. salmon fisheries). They must gather information on economic, social and environmental aspects and bring the information to class. Then, they are given a pattern of analysis to guide the diagnostic process.
- A comparison of the different elements present in the diagnosis of the sustainability of the economic sectors reviewed is carried out.
- Based on a text, students reflect on progress and developments.
- Social science professional talks to the group with the aim of instigating and informing conversation about environmental conflicts.

Assessment: Evidence of learning

A report must be submitted on the sustainability diagnosis in the economic sector previously chosen by the students. This must include a description of the sector and environmental conflicts and of course a diagnosis of sustainability including patterns of consumption/sustainable use of relevance to the sector.

Basis:

This milestone continues the problematizing role of the previous milestone, but in a more tangential way. The examination and diagnosis of sustainability in four critical sectors of the economy will reveal the nature of the Chilean economic model. Values and practices, and the issue of how to address the problem, are therefore examined. According to Warburton (2003), deep learning is influenced predominantly by the motivation of the student. In this sense, we think utilizing sectors of the economy illustrates elements of sustainability that are close to their realities. This therefore promotes the student's sense of responsibility for problems involving the society in which they are embedded.

However, the main focus of this milestone is placed on developing the skills to manage sustainability. At this stage, the students must comprehend the logic of sustainability and become familiar with different methods and tools for diagnosis. To do this, we review the four sectors of the economy, and in each one of them the student progresses towards taking a leading role in his or her learning and self-directing the activities of diagnosis. The learning environment must foster self-confidence and the self-direction of the learning process in a phased manner, and encourage the development of key competencies for sustainability (Barth et al. 2007). Bell and Morse (2003) use the approach of learning-by-doing with a particular focus on measuring sustainability, which promotes student leadership and engagement in the process. Finally, incorporating the social dimension of the sustainability diagnosis is the most difficult challenge, as this is more abstract and subjective (González et al. 2006). To this end, a professional from the social sciences is invited to speak to the students to expose socio-environmental conflicts with economic sectors. The professional discipline of the guest will differ according to the students. This is a starting point for understanding the importance of the social dimension and generating questions. The conversation itself is transdisciplinary and with ethical content, which allows a sensitive approach to the subject.

c) Review of the literature on selected issues

Purposes:

- Contrast various information sources for the generation of critical judgments.
- Generate relevant literature searches based on predefined criteria.
- Conduct a literature review of the current status of chosen economic sector with a view to generating a new proposal.
- Analyze instruments and tools for sustainability management in similar cases by establishing links between the literature review and a possible contribution.
- Apply screening criteria to review and synthesize the literature.

Activities:

- A professional librarian leads a workshop on article searches on bibliographical databases.
- Through case studies, students review the elements present in the eco-tax reforms. They have to identify elements present in successes and failures of eco-taxes.
- Extensive literature review and exercises on the comprehension and analysis of literature are completed.
- Students are asked to introduce a revised paper on eco-taxes and explain it to the class.

Assessment: Evidence of learning

A written report reviewing the national and international literature on both eco-taxes and the problems addressed.

Basis:

The changing conditions of sustainable solutions require the updating of information and the study of past experiences. It is still a challenge to build, implement and evaluate new solutions for sustainability. That is why this milestone aims to develop research capacity towards new proposals; this is encouraged by the activities involving exercise of inquiry, analysis and synthesis of the current state of the chosen sector.

Learning is reinforced with the proposed activities, which require the processing of information and immediate use of it. For example, generating summary sheets of relevant papers, a synthesis of the literature and an essay on a relevant paper all comes together to represent a more challenging activity that enables deeper learning.

This milestone fosters self-confidence and a self-directed learning model for example through endowing students with the responsibility to conduct a search of relevant papers and a free choice of how to approach the review.

d) Proposal development:

Purposes:

- Make decisions in the development of an eco-tax proposal, justifying it as a solution for managing sustainability in the area addressed.
- Learn that the challenge of managing a new sustainability proposal is affordable and achievable for themselves.
- Systematize the knowledge acquired in the process, developing the capacity for synthesis.

In many cases, proposals are original and not previously been addressed by research.

Activities:

- Presentation of eco-tax case studies is continued, with emphasis on experience related to the topic chosen.
- Steps towards adoption of an eco-tax described for discussion.

Assessment: Evidence of learning

Students submit a written eco-taxes proposal.

Basis:

The eco-tax development strongly promotes integration of knowledge and competences through the coordination of creative and implementable solutions, whereas diagnosis promotes integration based on the detection of problems and possible solutions.

Since Bloom's Taxonomy, the level of creation has constituted the most complex level of learning (Anderson et al. 2001). This means that confronting the student with the creation of an eco-tax proposal achieves a deep and meaningful learning environment, with the student able to define the scope and possible challenges (Steinemann and Asce 2004; Huntzinger et al. 2007). Creation is a complex process requiring the mobilization not only of disciplinary content, but also of the general competences described above. Finally, a flexible approach is required to implement creative solutions.

e) Redesign and communication

Purposes:

- Communicate progress to the group's proposal, setting out its scope and main criticisms.
- Participate in the discussion group formation on sustainability.
- Integrate new ideas and improvements to the project by feedback.

Activities:

- Proposal is presented to the course and teachers for feedback.
- Proposal is reworked according to observations made by the teacher and course students.

Assessment: Evidence of learning

- Reworked proposal incorporating observations of teachers and students.
- Results of observations of the proposal oral presentation.

Basis:

At this milestone, there are two elements involved in terms of learning: the redesign and communication. The redesign is a high impact process that allows students to monitor and assess the quality and accuracy of their own proposals. Cognitive processes and systems involved are metacognition and self-system (Marzano and Kendall 2007), i.e. how I generated the proposal and with what quality; both systems can create conditions for deep learning.

Communication is a key process in terms of the written and oral production of the proposal. Without a proper communication process, it is not possible for the student complete the training cycle. Learning to communicate effectively allows the student, on the one hand, to face a realistic task as in a professional environment and, secondly, to mobilize mental processes of synthesis, regulation of body attitude and above all, the handling of words and the student's own emphasis on oral discourse (Fink 2003). Many sustainability initiatives fall by the wayside simply because of a lack of adequate strategic communication.

Learning Outcomes

Learning Outcomes (LO) become relevant when we want to educate for sustainable development. They are the effects we want to see in our students after the teaching and learning process. There must be clear targets for learning that can be measured and evaluated, especially when the challenge is great and must be addressed promptly.

LO and the assessment criteria are communicated to students. This is in order that they become aware and are involved in their learning process.

1. *Establish the characteristics of sustainability through the analysis of the dialectical relationship between economic activity and nature, in order to identify their effects on the global crisis.*

Through this LO it is intended that the student becomes able to define sustainability and explain its characteristics. We understand this first step as essential in training future professional change agents. In Svanström et al. (2008), ACPA (College Students Educators International) delineates "Knowledge of the Environmental, Economic, and Social issues related to sustainability (Understanding)" as the main skills required to be an effective change agent.

To link the first result of learning with the practice, students must follow the criteria of learning. This will be evaluated as to whether students: 1.1) argue the logic of sustainability; 1.2) choose an appropriate study on sustainability issues.

Through the adequate satisfaction of the criteria for learning, the student shows that he or she knows and recognizes the elements of sustainability, relates to the real-world situation, and understands the multidimensionality of the analysis.

2. *Diagnose sustainability in economic activity through the application of patterns of consumption/sustainable use of different types of natural resources, in order to detect conflicts in the current development model.*

This LO is strongly related to the previous one, since both have to do with understanding the logic of sustainability and the elements of the discussion. To address this learning outcome, students must: 2.1) conduct an assessment of the sustainability situation in a given area.

The diagnosis makes it clear whether the student understands what sustainability is and is also able to use appropriate instruments to apply to a selected area. It also develops decision-making ability, teamwork and cooperation. Thus, all activities under this LO are directed towards the use and internalization of the definition of sustainability and its components.

3. *Build tools within the framework of ecological economics for application to economic, environmental and social problems, identifying their benefits to manage sustainability.*

The purpose of this LO is to achieve self-directed learning, decision-making, creative ability and the clear communication of ideas.

A student must: 3.1) describe the current regulatory framework, making a real contribution, since the proposal could be implemented; 3.2) evaluate the economic and non-economic tools available in the study area, recognizing what is already being done about a situation and assessing it in terms of the usefulness of tools for the management of sustainability; 3.3) assess the need for a new proposal in the area of study – this involves establishing the novelty and contribution of the proposal – as well as recognize its scopes and strengthen the case for future implementation.

Through properly satisfying the evaluation criteria the student visualizes in themselves the ability to materialize a proposal and the opportunity to make a substantive change.

Considerations in practice

All proposals were original, and most were well founded. 30% of the proposed eco-taxes were almost ready to be applied. Moreover, the students were motivated and committed to the course and to their work. The students put considerable effort into developing the proposal, to the detriment of time spent working through the reading material.

The participation of the students in the group activities works well and allows reflection and discussion. In this aspect the model works extremely well. It would possible to take more advantages from it by asking and encouraging students to read some specific information beforehand. Also, students may show some evidence of learning class by class.

Conclusions and challenges

It seems relevant to communicate and socialize the proposals, spreading their potential scope to society and potential users. In this sense, two essential practices as a university would be promoted: the link with the environment and the responsibility to transfer knowledge to real companies and institutions.

We found that the model is very successful in fostering competency in problem-solving related to managing sustainability, but still remains an inconclusive means of encouraging reviewing texts about content and transdisciplinary tools. There is a constant trade-off in assessing competences and contents jointly, which constitutes a great challenge to be tackled in a future version of the course. It is necessary to decide consciously where to place the emphasis, considering that students still respond with more commitment to evaluations.

It is necessary to evaluate the effect of methodology on actual learning outcomes of students, as well as on motivation and commitment. To the extent that the methodology causes significant changes in disciplinary knowledge integration based on the challenge of sustainability, we will be generating a significant contribution to their training. The substantial difference is given by the need for sustainability criteria, which were previously invisible and must now be considered.

Learning must not be restricted to the classroom. It is pertinent to assess the impact of the course in sustainability competences felt during the first years of students' employment. But, at this early stage, it is becoming evident that there is a need to establish links between the professional profile of the graduate, ESD and the expected learning outcomes of courses addressing sustainability.

The *Talloires Declaration* and the international movement for SC recognize the great importance of curriculum and research in achieving the purpose of producing change agents towards a better society. In this context, the contribution of this systematization is the process, is to give a first approach to including SC perspectives in the curriculum, and establishing both its major advantages and the challenges ahead.

References

Anderson, L., Krathwohl, D. and Bloom, B. (2001), *A taxonomy for learning, teaching, and assessing: A revision of Bloom's taxonomy of educational objectives*. Allyn and Bacon, Boston.

Barth, M., Godemann, J., Rieckmann, M. and Stoltenberg, U. (2007), Developing key competencies for sustainable development in higher education. *International Journal of Sustainability in Higher Education* 8 (4), pp. 416-430.

Bell, S. and Morse, S. (2003), *Measuring the sustainability*. Earthscan publications, London.

Le Boterf, G., Durand-Gasselin, P. and Pechenart, J.M. (1991), *Cómo invertir en formación*. Gestión 2000, Barcelona.

Brockbank, A. and McGill, I. (1998), Facilitating Reflective Learning in Higher Education. In Moon, J. (Ed.), *Higher Education* 39, pp. 489-491. Kluwer Academic Publishers, Netherlands.

Cordano, M., Ellis, K. and Scherer, R. (2003), Natural Capitalists: Increasing Business Students' Environmental Sensitivity. *Journal of Management Education* 27, pp. 144-157.

Costanza, R. (1991), Ecological economics: A research agenda. *Structural Change and Economic Dynamics* 2 (3), pp. 335-357.

Davies, M. and Devlin, M. (2010), Interdisciplinary higher education. In Davies, M., Devlin, M. and Tight, M. (Eds.), *Interdisciplinary higher education: perspectives and practicalities,* Parte 1, Emerald Group, pp. 3-29.

Fink, L.D. (2003), *Creating significant learning experiences: an integrated approach to designing college courses*. John Wiley and Sons.

González, C., Ríos, H., Brunett, L., Zamorano, S. and Villa, C. (2006), ¿Es posible evaluar la dimensión social de la sustentabilidad? Aplicación de una metodología en dos comunidades campesinas del valle de Toluca, México. *Convergencia. Revista de Ciencias Sociales [en línea]* 13 (40), pp. 107-139.

Huntzinger, D.N., Hutchins, M.J., Gierke, J.S. and Sutherland, J.W. (2007), Enabling Sustainable Thinking in Undergraduate Engineering Education*. *International Journal of Engineering* 23 (2), pp. 218-230.

Irigoin, M. and Vargas, F. (2002), *Competencia laboral: manual de conceptos, métodos y aplicaciones en el sector salud*, 252 p. Cinterfor, Montevideo.

Issa, Theodora, Issa, Tomayess and Chang, V. (2011), Would Teaching Sustainable Development Business Strategies Shift Students' Mindsets? An Australian Experience. *The International Journal of Environmental, Cultural, Economic and Social Sustainability* 7 (5), pp. 257-272.

Marzano, R. and Kendall, J. (2007), The new taxonomy of educational objectives. Corwin Press.

Najjar, J. and Simon, B. (2009), *Learning Outcome Based Higher Education: iCoper Use Cases, IEEE International Conference on Advanced Learning Technologies (ICALT)*. Workshop on European Workshop On E-Learning Standards Best Practice. Riga – Latvia.

Norton, M. and MacVaugh, J. (2012), Introducing Sustainability into Business Education Contexts Using Active Learning. *International Journal of Sustainability in Higher Education* 13 (1), pp. 73-87.

Olavarría, M. (2003), Protected Neoliberalism: Perverse Institutionalization and the Crisis of Representation in Postdictatorship Chile. *Latin American Perspectives* 30 (6), pp. 10-38.

Schurman, R.A. (1996), Snails, Southern Hake and Sustainability: Neoliberalism and Natural Resource Exports in Chile. *World Development* 24 (11), pp. 1695-1709.

Springett, D. (2005), 'Education for sustainability' in the business studies curriculum: a call for a critical agenda. *Business Strategy and the Environment* 14 (3), pp. 146-159. John Wiley & Sons, Ltd. Retrieved from http://dx.doi.org/10.1002/bse.447.

Steinemann, A. and Asce, M. (2004), Implementing Sustainable Development through Problem-Based Learning: Pedagogy and Practice. *Journal of Professional Issues in Engineering Education and Practice* 129 (4), pp. 216-224.

Stubbs, W. and Cocklin, C. (2007), Teaching sustainability to business students: shifting mindsets. *International Journal of Sustainability in Higher Education* 9 (3), pp. 206-221.

Svanström, M., Lozano-García, F.J. and Rowe, D. (2008), Learning outcomes for sustainable development in higher education. *International Journal of Sustainability in Higher Education* 9 (3), pp. 339-351. Retrieved from http://www.emeraldin sight.com/10.1108/14676370810885925.

Villa, A. and Poblete, M. (2007), Aprendizaje basado en competencias: una propuesta para la evaluación de las competencias genéricas, 196 pp. U. de Deusto, Bilbao.

Warburton, K. (2003), Deep learning and education for sustainability. *International Journal of Sustainability in Higher Education* 4 (1), pp. 44-56.

About the Authors

Dr. **Beatriz Acevedo** works as a senior lecturer in sustainable management at the Lord Ashcroft Business School, Anglia Ruskin University (Cambridge, UK). She combines her academic work with her artistic practice as a painter. She has considerable experience in research and academic consultancy on processes of social and sustainable development in Latin America. Dr. Acevedo's research concerns processes of organisational change; she is currently working on the intersections between aesthetics, visuality and sustainability in the context of higher education and business studies.

Lord Ashcroft Business School – Anglia Ruskin University – East Road – Cambridge CB1 8PT – United Kingdom – Beatriz.Acevedo@anglia.ac.uk *(corresponding author contact)*

Dr. **Libia Patricia Peralta Agudelo** is a designer and PhD in Landscape Ecology at the University of Nottingham in the UK. She currently teaches Product Design at the Unibrasil private faculty in Curitiba, and owns the EcoStudio consultancy company that develops sustainable architecture and design projects. Here main area of interest is Ecodesign applied to housing and furniture.

EcoStudio Soluções Sustentáveis – R. Leopoldo Teuber nº9 – CEP: 80520-660 – Curitiba (PR) – Brazil – pperalta.design@gmail.com

Ismael Aguilera is an Industrial Engineering graduate and researcher at Universidad de Chile. He presided over the Industrial Engineering Students' Union in 2009 and was a founding member of *Taller Solar*. His research is currently focused on promoting productivity in public hospitals.

Departamento de Ingeniería Industrial – Facultad de Ciencias Físicas y Matemáticas, Universidad de Chile – Chile

Daniel de Álava. Oceanographer and Master in Environmental Sciences. Professor of C-MCISur (UdelaR-CURE). His research field focuses on anthropic transformation processes in relation to future strategies for coastal zone management and climate change vulnerability. Coastal dynamics and enhancing local population knowledge and participation for an integrated coastal management are major research focus.

Centro Interdisciplinario para el Manejo Costero Integrado del Cono Sur – Universidad de la República – Montevideo (UdelaR) – Uruguay – daniel.dealava@gmail.com

Dr. **Azlan Amran** is an associate professor of the Graduate School of Business, Universiti Sains Malaysia (USM). Currently he holds a position as the Programme Manager of MBA Sustainable Development at the Graduate School of Business. His research interest is in corporate social responsibility, sustainability and corporate governance.

Graduate School of Business – Universiti Sains Malaysia – Malaysia – azlan_amran @usm.my

Emily Anderson is a Liu Scholar and PhD student at the Institute for Resources, Environment and Sustainability (IRES) at the University of British Columbia (UBC). Her research looks at the local level environmental and socioeconomic consequences of small farmer participation in global commodity and environmental service markets in developing countries. Emily's Master's thesis research at IRES looked at the generation of climate and development co-benefits from small farmer carbon projects in Uganda. She is a founding member of the International Development Research Network (IDRN) and a current steering committee member.

International Development Research Network – Liu Institute for Global Issues – University of British Columbia – – 6476 NW Marine Dr. – Vancouver BC V6T 1Z2 – idrn.ubc@gmail.com; – Institute for Resources – Environment and Sustainability – University of British Columbia – 2202 Main Mall – Vancouver BC V6T 1Z4 – Canada – emilka@interchange.ubc.ca *(corresponding author contact)*

Gabriela Valenzuela Arcuch. Business Administrator, University of Chile; Diploma in Management Control in Corporate Social Responsibility, University of Chile. Director of unit at the Social University Responsibility (NexoRSU) and teacher at the faculty of Economy and Business, University of Chile.

Director of Department of University Social Responsibility – School of Economics and Business – University of Chile – gvalenzu@fen.uchile.cl

Ulisses Miranda Azeiteiro was educated at Aveiro University, Portugal, where he received his Master Science Degree in 1994; Coimbra University, Portugal, where he received his PhD in Ecology in 1999; and the Universidade Aberta, Portugal, where he received his Aggregation in Biology in 2006. He holds an Assistant Professor teaching position at the Universidade Aberta (Assistant Professor with Aggregation and Tenure), Portugal, in the Department of Exact Sciences and Technology, where he teaches *Environmental Education, Education for Sustainable Development (EDS), Biology, Biodiversity* and *Research Methodologies*. He is Senior Researcher at the *Centre for Functional Ecology* of Coimbra University where he is the Coordinator of the group *Ecology and Society.*

Universidade Aberta – Departament of Science and Technology – Rua da Escola Politécnica 141 – 1269-001 Lisboa –; – Centre for Functional Ecology (CFE) – Universidade de Coimbra – 3001-401 Coimbra – Portugal

Susan L. Baker (BSc, MA, PhD) is the Coordinator for Outreach and Education for the National Centers for Coastal Ocean Science, a division of the National Oceanic and Atmospheric Administration in Silver Spring, Maryland.

Coordinator for Outreach and Education – National Centers for Coastal Ocean Science – National Oceanic and Atmospheric Administration – 1305 East West Highway – N/SCI – Silver Spring MD 20910 – USA – Susan.Baker@noaa.gov

Susan A. Baldwin is a chemical and biological engineering professor teaching a third-year laboratory course as well as senior undergraduate and graduate-level chemical and biological kinetics and reactor design courses. For many years she facilitated the final-year project design course for the biological engineering option students. Her area of expertise is remediation of industrial pollution using naturally based chemical and biological processes.

Chemical and Biological Engineering – University of British Columbia – 2360 East Mall – Vancouver V6T1Z3 – British Columbia – Canada – Sbaldwin@chbe.ubc.ca

Prof. **Dan Bălteanu**, the coordinator of the consortium, is the Director of the Institute of Geography, Member of the Romanian Academy, and Chair of the Romanian National Committee on Global Environmental Change. He is senior researcher at the Institute of Geography, PhD supervisor and professor at University of Bucharest. His academic competences are in the field of natural hazards, land use/land cover change, environmental and climate change. He is the author and editor of over 15 volumes and more than 150 scientific papers published in national and international journals, many of which deal with the impact of extreme climate events on the environment. Prof. Balteanu coordinated several national and international projects in the field of environmental sciences.

Romanian Academy – Institute of Geography – 12 Dimitrie Racovita St – Sector 2 – Bucharest 023993 – Romania

Dr. **Esther Barazzone** became President of Chatham University in 1992, after a career as a faculty member and administrator at Hamilton and Kirkland Colleges, the University of Pennsylvania, Swarthmore College and Philadelphia University. Dr. Barazzone holds a PhD and MA in European Intellectual History from Columbia University, where she was a Fellow of the Faculty. Dr. Barazzone was a Charter Scholar in the first graduating class of New College in Sarasota, Florida, where she earned her BA in Philosophy and History. She received a US Student Fulbright award to Spain and studied at the Wharton School of Business Administration and at Harvard University's Institute for Educational Management. Dr. Barazzone has led Chatham through a period of major institutional renewal and expansion leading to national recognition and elevation to university status in 2007. Since 1992, the institution has added graduate programs through the doctoral level, reinvigorated its undergraduate program, renewed and expanded its physical campus and infrastructure, engaged in significant fundraising, and assumed a new leadership role in the local and national community. An early advocate of campus sustainability efforts, Dr. Barazzone was a founding advisory board member and signatory of the American College and University Climate Commitment, and has given presentations on Chatham's sustainability efforts around the world.

Chatham University – 1 Woodland Road – Pittsburgh PA 15044 – USA

Paul A. Barresi (BSc, JD, MALD, PhD) is Professor of Political Science and Environmental Law and coordinator of the Environmental Management program at Southern New Hampshire University in Manchester, New Hampshire, USA.

School of Arts and Sciences – Southern New Hampshire University – 2500 North River Rd – Manchester NH 03106 – USA – p.barresi@snhu.edu

Svetlana Batarilo was born in Belgrade in 1970. Graduated from the Faculty of Architecture, University of Belgrade and began her PhD studies in 2005. As part of practical work, she has designed several structures in Serbia and abroad, independently or as a member of a team. She has taken part in architecture and urban design competitions, local and international, winning several awards, as well as exhibiting her projects. From 1996 to 1999, she worked at the Faculty of Architecture, University of Belgrade, teaching a course on Contemporary Architecture and Design. Since 2001, she has worked as a Teaching Assistant at the Transport and Traffic faculty, University of Belgrade, teaching the following subjects: Urban Planning, Fundamentals of Urban Planning, Sustainable Development and Environmental Protection, Fundamentals of Construction/architectural segment. Scope of research: urban morphology, urban reconstruction, urban design, sustainable development of cities.

University of Belgrade – Faculty of Transport and Traffic – 305 Vojvode Stepe St – 11000 Beograd – Serbia – s.batarilo@sf.bg.ac.rs

David Löw Beer is a PhD candidate at the Leuphana University Lüneburg in the field of economic education for sustainable development. He has studied economics and politics at the Universities of Tübingen and Niterói, Brasil. He also worked for two years in a comprehensive school as a fellow of Teach First Deutschland. In Tübingen, he led the student initiative Greening the University, e.V. from 2008 to 2009.

Leuphana University Lüneburg – Scharnhorststraße 1 – 21335 Lüneburg – Germany – loewbeer @leuphana.de

Leonardo Bezerra Pimentel, MSc (in progress) in Management System with emphasis on Total Quality. BSc in Production Engineering from Fluminense Federal University. Business Process Management (BPM) consultant, teacher and researcher at LATEC/UFF.

Fluminense Federal University – LATEC – Rua Passo da Pátria 156 – CEP 24210-240 – Niterói RJ – Brazil – phone +55 21 2629 5620 – fax +55 21 2717 6390 – leonardo@latec.uff.br

Dr. **Ružica Bogdanović** is a Professor of Urban Planning and Sustainable development at the Faculty of Transport Engineering, University of Belgrade and Faculty of Architecture, Union University – Nikola Tesla in Belgrade. Her main areas of interest in research and teaching include: urban planning, urban design and sustainable development. Bogdanovic has initiated, conducted and implemented a number of scientific research projects as team leader and principal investigator and has organized and moderated numerous panel discussions, round tables, national and international symposia. She is the author of numerous books and articles in professional journals and has presented papers at numerous national and international congresses and symposia. Dr. Ružica Bogdanović contributes to the profession by active engagement in different professional organizations. She is a member of IsoCaRp, a delegate of Serbia in the ECTP, member of the Association of Serbian Town Planners, Society of Belgrade Urban Planners (chairperson for eight years, in two terms), Belgrade Association of Architects.

She has received several awards between 1994 to 2009 for projects and books, as well as the Life Achievement "Emilijan Josimovic Grand Award" in 2003.

University of Belgrade – Faculty of Transport and Traffic – 305 Vojvode Stepe St – 11000 Beograd; – Union University Nikola Tesla – Faculty of Architecture – 62-64 Cara Dušana St – 11000 Beograd – Serbia – urbs@eunet.rs

Charlotte Bonner works in the Ethical and Environmental team at NUS. She leads the Green Impact programme, which works with over 100 organisations to improve environmental performance and engage people in the sustainability agenda. She is also studying for an MSc in Energy and Environmental Technology and Economics at City University, London. Her research interests encompass behaviour change and the psychology of personal engagement with sustainability. Charlotte is a keen advocate of sustainable transport.

National Union of Students – Drummond Street – London NW1 3HP – United Kingdom

Kelly Boulton is Sustainability Coordinator at Allegheny College. Boulton's research focuses on innovative green campus initiatives and climate action planning.

Allegheny College – 520 North Main Street – Meadville PA 16335 – USA – kelly.boulton @allegheny.edu

Francisco Bucchi is a senior-year student of Geology at the Faculty of Physical and Mathematical Sciences (FCFM), of Universidad de Chile. He has been a member of the Oikos group since 2000, and of the Sustainable Campus Commission (SCC) since 2011, in which he is responsible for organizing workshops to review international experiences related to sustainability and sustainable campuses. He acted as a general coordinator of the SCC during 2011. His past experiences include being a founding member of the *Taller Solar*, a group dedicated to the divulgation of low-cost uses of solar energy, and membership of the Common Program Student Union of the (CEPC).

Departamento de Geología – Facultad de Ciencias Físicas y Matemáticas – Universidad de Chile – francisco.bucchi@gmail.com *(corresponding author concact)*

Sue Bumpous is Program and Communications Manager for the North American Association for Environmental Education, where she manages the association's certification and accreditation programs.

North American Association for Environmental Education – 2000 P Street NW – Suite 540 – Washington DC 20036 – USA – sbumpous@naaee.org

Sandra Sofia Caeiro holds a PhD (2004) in Environmental Engineering from the Universidade Nova (New University), Lisbon, Portugal. She is currently Assistant Professor in the Department of Science and Technology at the Universidade Aberta and a researcher at the *Center for Environment and Sustainability Research (CENSE*) and *IMAR – The Marine and Environment Research Centre*. Her main research and teaching areas include environmental risk management, environmental management tools, Education for Sustainable Development, sustainable consumption and e-learning in science education. She is the institutional coordinator of the *European Virtual Seminar* in Sustainable Development and former coordinator of the *Master Program in Environmental Citizenship and Participation* as well as the BSc programme in Environmental Sciences at UAb. She is on the editorial board of the International Journal of Ocean and Coastal Management.

Universidade Aberta – Departament of Science and Technology – Rua da Escola Politécnica 141 – 1269-001 Lisboa; – CENSE-Center for Enviornmental and Sustainability Reseach – Faculdade de Ciências e Tecnologia – Universidade Nova de Lisboa – 2829-516 Caparica – Portugal

Darci Barnech Campani, undergraduate degree in Agricultural Engineering at the Federal University of Rio Grande do Sul – UFRGS (1978) and Physical Education at UFRGS (2008). Professor at the Department of Mechanical Engineering at UFRGS, currently heads the coordination of UFRGS'S Environmental Management. Has expertise in Energy and Environmental Planning and experience in Sanitary Engineering, with emphasis on Domestic Solid Waste and Environmental Management.

Universidade Federal do Rio Grande do Sul – UFRGS – Rua Luiz Englert s/nº – Sala 6 – Predio 12109 – Porto Alegre RS – Brazil – campani@ufrgs.br

Francisco del Campo is a graduate student at Universidad de Chile, where he first pursued an education in the Earth Sciences. Having always had the intention to pursue an academic career, he discovered his true passion a few years ago: physics. With the purpose of dedicating his future working life to this branch of science, he is currently undertaking a Master's degree in Physics at Universidad de Chile, performing research in non-linear optics.

Departamento de Física – Facultad de Ciencias Físicas y Matemáticas – Universidad de Chile – Chile – Colectivo Viento Sur – Chile

Dr. **Eloy F. Casagrande Jr.** is a designer and PhD in Mineral Resources Engineering and Environment at the University of Nottingham in the UK. He is currently the coordinator of the Green Office at the Federal University of Technology of Paraná (UTFPR) and teacher on the postgraduate programs in Technology and Civil Engineering. His main areas of interest are Sustainable Construction, Green Certification and Environmental Education.

PPGTE – Universidade Tecnológica Federal do Paraná (UTFPR) – Av. Sete de Setembro nº 3165 – CEP: 80230-901 – Curitiba (PR) – Brazil – eloy.casagrande@gmail.com

Valentina Castellani, PhD, graduated in Environmental Science in 2005 from the University of Milano Bicocca with a dissertation entitled "Definition of an index about economic and environmental efficiency concerning tourism management in the Italian Alps and optimization through Data Envelopment Analysis (DEA)". Since 2006 she has been studying for a PhD research project within the Research Unit on Sustainable Development (GRISS) at the University of Milano Bicocca, working on the development of methodologies and indicators to assess sustainability in tourism and agriculture sectors. She has taken part in several research projects about sustainable tourism, sustainable consumption, production patterns and sustainable planning.

GRISS – Gruppo di Ricerca sullo Sviluppo Sostenibile – Department of Environmental Science and Technology – University of Milano – Bicocca – Piazza della Scienza 1 – 20126 Milano – Italy – valentina.castellani1@unimib.it

Gisela Cebrián holds a Bachelor's degree in Environmental Sciences from the Universitat de Girona and a Master's degree in Mathematics and Science Education Research from the Universitat Autònoma de Barcelona. She is currently undertaking her PhD research at the Southampton Education School. Her research focuses on embedding Education for Sustainability within the undergraduate curriculum at the University of Southampton. Gisela is a member of the EDUSOST Research Network on Education for Sustainability and the Research Group Complex at the Universitat Autònoma de Barcelona. She has also been involved in research projects on Catalan universities and climate change, professional competences in sustainability in Higher Education and Problem Based Learning. Gisela's research interests encompass Education for Sustainability and innovative teaching and learning approaches within the Higher Education context.

Southampton Education School – University of Southampton – Building 32 – Southampton SO17 1BJ – United Kingdom – G.Cebrian-Bernat@soton.ac.uk

Prof. **Adriano Ciani** is Full Professor of Agricultural Economics and Rural Appraisal at the Department of Agricultural Economics, Farm Appraisal and Food Sciences (DSEEA) at Perugia University. He is also Professor of Environmental Economics and Farm Management. His current research is in: Sustainable Management of Territory, Common Goods, Water Resources, ICT and Sustainable Rural Development. He has written several books and 235 other publications. He has taken part in collaborative research, teaching, and technical and dissemination activities in over 50 countries. He founded the

BIOSPHERA Association and, in 2000, organized the First World Forum on Rural Tourism and Agro-tourism. He also founded the International Association of Experts in Rural Tourism and Agro-tourism (IAERT). From 2001 to 2009 he was commissioned by the MoFA as Scientific Attaché at the Italian Embassy Tirana, where he promoted the wide scientific programme "A Sustainable Development for Albania". He is Director of the International Summer School in "Sustainable Management and Promotion of Territory". In the years 1980-1990 he was and adviser to the Provincial Government of Perugia, to SASE Lt. Airport (Perugia), Vice-President of the Umbria-Tuscany Irrigation Authority, and Auditor of ERI New Editions. He is an honorary member of the Academic Senates of the seven universities of Romania. He has been appointed a "Symbol of the City of Durres" and "Honour of the City" in the Girocastro Region in Albania.

Department of Agricultural Economics – Farm Appraisal and Food Sciences (DSEEA) – Agricultural Faculty – Perugia University – 06121 Perugia – Italy – phone +39 (0)75 5857 141; fax: +39 (0)75 5857 146; mobile +39 (0)347 5109 486 – ciani@unipg.it – Skype: cianiunipg – www.adrianociani.it

Dr. **Amelia Clarke** has been working on environment and sustainability issues since 1989. She is now a faculty member in the School of Environment, Enterprise and Development (SEED) at the University of Waterloo, Canada, and is Director of their Master of Environment and Business (MEB) program. Her main research areas include: campus environmental management, community sustainable development strategies, collaborative strategic management, and youth-led social entrepreneurship. She holds a PhD in Management from McGill University. Amelia Clarke was also the President of Sierra Club Canada from 2003 to 2006 and founded the Sierra Youth Coalition in 1996. For more information or to contact Dr. Clarke, see: http://www.environment.uwaterloo.ca/seed/faculty-staff/clarke/

School of Environment – Enterprise and Development (SEED) – University of Waterloo – 200 University Avenue West – Waterloo N2L 3G1 – Ontario – Canada – phone: +1-519-888-4567 x 38910 – amelia.clarke@uwaterloo.ca

Daniel Conde, PhD in Aquatic Ecology. Professor of the Limnology Section (Faculty of Sciences, UdelaR), researcher Level II of the National System of Researchers and researcher First level of the Basic Sciences program PEDECIBA. He works in coastal lagoons and integrated costal management, and leads an interdisciplinary group in coastal management at UdelaR. He coordinates the new Interdisciplinary ICM Center of the Southern Cone and the Master's program in ICM. He has been responsible for several international and national research projects and has authored of more than 50 international publications and books.

Centro Interdisciplinario para el Manejo Costero Integrado del Cono Sur (Universidad de la República, Uruguay) – Centro Universitario Regional Este – Espacio Interdisciplinario – Facultad de Ciencias – vladddcc@gmail.com *(corresponding author contact)*

Dr. **Luis Eduardo Velázquez Contreras** has 20 years of experience as an Industrial Engineer and, since 1994, has served as Director of the Sustainable Development Group in the Engineering College. He has conducted several investigations in the Sustainability, Cleaner Production and Pollution Prevention fields, as well as in the study of sustainable universities. Dr. Velazquez holds a doctoral degree majoring in Cleaner Production and Pollution Prevention from the University of Massachusetts Lowell. Currently, he is a Professor and researcher at the University of Sonora in Mexico and Adjunct Professor at the University of Massachusetts Lowell in the USA. For almost two decades, he has focused on preventing, eliminating, and/or reducing environmental risks at source in Mexican enterprises.

Engineering College – University of Sonora – Mexico; – Work Environment – University of Massachusetts Lowell – USA – Luis_Velazquez@industrial.uson.mx

Diana C. Cooper, PhD, Lecturer at the University of Minnesota, Carlson School of Management, holds a doctorate from the University of Minnesota, an MBA from University of California, Irvine, and an AB from Stanford University. Her research focus is sustainable and cross-cultural organization development. From 2009-2011 she managed a major US Department of Labor grant for training dislocated workers for "green careers" in Minnesota public and private organizations. Prior to her academic career, Dr. Cooper was a systems consultant and led international leadership seminars for US executive MBA students. She travels extensively and has lived in the US, Costa Rica, the UK and Italy.

Department of Work and Organizations – University of Minnesota – 4005 Natchez Avenue South – Minneapolis MN 55416 – USA – phone +1-952 200-5903 – coope048@umn.edu

Martín Jacques-Coper is a PhD student of Climate Sciences at the University of Bern, Switzerland. He studied Geophysics (BSc) and holds an MSc degree in Meteorology and Climatology from the University of Chile (UCh). He has been involved in Wind Resource Assessments in Chile, and in 2010 he completed an internship at the ForWind Center for Wind Energy Research of the University of Oldenburg, Germany. At UCh, besides having been a representative at Student Unions (CEPC and FECh), he was actively involved in the Oikos student association and the *Taller Solar*, of both of which he is a founding member. These organizations are primarily devoted to the development of a sustainable campus proposal, and to promoting the use of solar thermal energy by implementing low-cost techniques in rural and urban areas of northern and central Chile, respectively. He is also a member of the organization Colectivo Viento Sur.

Institute of Geography and Oeschger Centre for Climate Change Research – University of Bern – Switzerland – jacques@giub.unibe.ch *(corresponding author contact)*

Freddy Coronado. Assistant Professor at the Department of Management Control and Information System at the School of Economics and Business at the University de Chile. He has been director of undergraduate and graduates programs and currently is the academic director of the Teaching and Learning Center at the School of Economics and Business. Professor Coronado holds a PhD in Business Administration from Michigan State University and a BA in Management Control Engineering and Accounting from the University of Chile. He works in strategic planning, management control, organizational behavior, corporate social responsibility and higher education. He is also a consultant for the government and private sector.

Academic Director – Teaching and Learning Center – School of Economics and Business – University of Chile – Chile – fcoronadom@fen.uchile.cl

Peter Blaze Corcoran is Professor of Environmental Studies and Environmental Education at Florida Gulf Coast University, where he serves as Director of the Center for Environmental and Sustainability Education. He has been a Faculty Member at College of the Atlantic, Swarthmore College, and Bates College, and a visiting professor in Australia, the Netherlands, and Fiji. Currently, he serves as a Visiting Professor at Universiti Sains Malaysia attached to the Centre for Global Sustainability Studies. He also serves as an Adjunct Professor at the University of the South Pacific, attached to the School of Islands and Oceans. Corcoran is a member of the editorial board of several scholarly journals, including the *Southern African Journal of Environmental Education, Ethics and Action*, the *Journal of Education for Sustainable Development*, and the *International Journal of Sustainability in Higher Education*. His recent books are *Young People, Education, and Sustainable Development: Exploring Principles, Perspectives, and Praxis* with Philip Molo Osano, Wageningen Academic Publishers (2009); *A Voice for Earth: American Writers Respond to the Earth Charter* with A. James Wohlpart, University of Georgia Press (2008); *The Earth Charter in Action: Toward a Sustainable World* published in the Netherlands by Royal Tropical Institute (KIT) Publishers (2005); and *Higher Education and the Challenge of Sustainability: Contestation, Critique, Practice, and Promise* with Arjen Wals published in the Netherlands by Kluwer Academic Press (2004).

Center for Environmental and Sustainability Education – Florida Gulf Coast University – 10501 FGCU Boulevard – Fort Myers – Florida 33965-6565 – USA

Debby Cotton has a doctorate in Environmental Education from St Anne's College, Oxford (Teaching Controversial Environmental Issues in Geography) and has been working in Higher Education for 15

years. Her current role involves responsibility for educational enhancement (particularly faculty development), and research in teaching and learning in higher education. She also has a key role in managing the newly-formed Pedagogic Research Institute and Observatory (PedRIO). Debby has a wide range of pedagogic research interests and has published widely in areas including ESD, e-learning, internationalisation and the student experience of fieldwork. Her recent research has focused largely on sustainability in higher education, transformative learning, informal learning and the hidden curriculum. Her full CV can be found at: http://www.plymouth.ac.uk/staff/dcotton. Some papers and a copy of Debby's thesis are available via Academia.edu: http://plymouth.academia.edu/DebbyCotton

Pedagogic Research Institute and Observatory (PedRIO) – Teaching and Learning Directorate – Plymouth University – PL4 8AA – UK – D.Cotton@plymouth.ac.uk

Dr. **Cheryl Desha** is a lecturer in sustainable development in the Science and Engineering Faculty (School of Earth, Environment and Biological Systems), Queensland University of Technology, Australia. She is also a principal researcher with The Natural Edge Project research group (TNEP), an academic partnership for research and capacity building for sustainable development. Cheryl and the TNEP team have developed a number of internationally renowned books on sustainable development, including 'Factor Five', 'Cents and Sustainability', 'Whole System Design', and 'The Natural Advantage of Nations'. Her PhD and post-doctoral research focuses on rapid curriculum renewal, grounded in capacity building for sustainable development and engineering education. This includes developing and trialling models and approaches to embedding new knowledge and skills, from trans-disciplinary inquiry into curriculum renewal theory and organisational change theory in the context of urgent and challenging times.

Queensland University of Technology – Brisbane – Australia – The Natural Edge Project – cheryl.desha@qut.edu.au

Manuel Diaz is Director of the Program for Environmental Economics and Management (PROGEA) at the University of Chile. He is a project consultant and researcher in Environmental Management and Energy, Mining, Climate Change, Risk Analysis, Environmental Economics and Information Systems. He received a Master of Science degree in Project Management for Environmental and Energy Engineering, Ecole des Mines de Nantes, in France in 2006; and a Master of Science degree in Industrial Engineering and Mining Engineering from the University of Chile, Santiago, in 1997.

Department of Electrical Engineering – University of Chile – 2007 Av. Tupper – Santiago-Centro – Santiago 837 0451 – Chile – phone: 56-2-978-4203 – mdiaz@ing.uchile.cl

Antje Disterheft is currently doing her PhD in *Social Sustainability and Development* at the Universidade Aberta, Lisbon, Portugal. She is a member of the *Centre for Functional Ecology* (*Ecology and Society*), at the Universidade de Coimbra, Portugal. She holds a degree in *Social Work* from the University of Applied Sciences, Hanover, Germany ('Diplom-Sozialarbeiterin/Sozialpädagogin') and a Master's degree in *Environmental Citizenship and Participation* from the Universidade Aberta. Since her first studies, she has been very much interested in the social questions related to Ecology, having participated in several campus projects. She has also worked for five years as an academic coordinator for international programmes at one of Lisbon's state universities.

Universidade Aberta – Departament of Science and Technology – Rua da Escola Politécnica 141 – 1269-001 Lisboa; – Centre for Functional Ecology (CFE) – Universidade de Coimbra – 3001-401 Coimbra – Portugal – phone +351 (0) 92 2125 358 – antje.disterheft@uc.pt *(corresponding author contact)*

Moritz A. Drupp studied economics, with a special focus on sustainability economics, at the Universities of Tübingen and Hagen (both Germany) as well as Tufts University (USA). He was a member of the advisory council for sustainable development of the University of Tübingen and has co-led the student initiative Greening the University, e.V.

University Tübingen – Friedrich-Dannenmann-Str. 6 – 72070 Tübingen – Germany – moritz.drupp@fulbrightmail.org

Lauren C. Edmonds is a graduate of The College of William and Mary, where she conducted this research for her honors thesis in Environmental Science and Policy. She was an original member of the William and Mary Committee on Sustainability, the administrative body dedicated to promoting sustainability on campus, and a leader in the William and Mary Student Environmental Action Coalition, the student group for environmental activism. For her service to The College, she was awarded the Sullivan Award and Lauren Burns Environmental Service Award upon graduation. Edmonds presented her research at the 2011 conference of the Association for the Advancement of Sustainability in Higher Education.

Environmental Science and Policy Program – The College of William and Mary – Williamsburg – Virginia – USA – Lcedmonds11@gmail.com *(corresponding author contact)*

Golda Arpudhalin Edwin received her MSc degree from the Department of Zoology at the Pondicherry University and is currently pursuing her PhD. Her area of research is sustainable water management practices with special reference to gray water management using constructed wetlands. She has authored several articles and presented in national and international conferences and workshops. Her honors include general proficiency for two consecutive years, as well as being a gold medalist in the Master's degree program. She is a member of the International Water Association.

Department of Ecology and Environmental Sciences – Pondicherry University – Puducherry 605014 – India – edwingolda@yahoo.com

Dr. **Tatjana Eremina** – PhD in physical and mathematical sciences, Associate Professor at the Russian State Hydrometeorological University; has been head of the Baltic Floating University cruises since 1996.

Russian State Hydrometeorological University – Malookhtinsky pr. 98 – St. Petersburg 195196 – Russia – tanya@rshu.ru

MSc **Alexandra Ershova** is a PhD student in her final year at Russian State Hydrometeorological University and co-editor of the Baltic Floating University Research Bulletin (since 2005). She has been involved in the Baltic Floating University project since 2002, starting as a BFU student, and has now participated in eight cruises. Ms. Ershova is currently a teacher with the Baltic University Programme (BUP) and teaches courses on the Baltic Sea Environment at RSHU.

Russian State Hydrometeorological University – Malookhtinsky pr. 98 – St. Petersburg 195196 – Russia – ershova@rshu.ru

Alejandro Esguerra is a PhD candidate at the Free University Berlin. Most recently he attended the Department of Science & Technology Studies at Cornell University, in order to deepen his research on the production of knowledge in transnational, hybrid arenas. In Amherst (USA) and Tübingen, he studied international relations and philosophy (MA), and in 2006 co-initiated the student initiative Greening the University, e.V.

Free University Berlin – Ihnestr. 26 – 14195 Berlin – Germany – a.esguerra@fu-berlin.de

Lukas Paul Fesenfeld is a third-year student at University College Maastricht. He is about to complete his Bachelor's degree in International and European Studies with a major in European Law and a minor in Development and Sustainability Studies. As a volunteer for the Maastricht University Green Office and co-president of the UCM Sustainability committee he actively works in the field of Sustainable Development.

University College Maastricht – Maastricht University – Zwingelput 4 – 6211 KH Maastricht – The Netherlands

Kieran M. Findlater is a Liu Scholar and PhD student at the Institute for Resources, Environment and Sustainability (IRES), UBC. His current research interests include energy access, technology diffusion, and rural development in non-industrialized countries, with ongoing work in India and South Africa, while his past academic and professional research experience spans climate, energy policy, agriculture, land use, and water governance. Along with a Bachelor of Science degree from the University of Alberta, he also holds a Master's degree from IRES for which he conducted research on biofuels and land use, with fieldwork in India. He is a founding member of IDRN and a current steering committee member.

International Development Research Network – Liu Institute for Global Issues – University of British Columbia – 6476 NW Marine Dr. – Vancouver BC V6T 1Z2; – Institute for Resources, Environment and Sustainability – University of British Columbia – 2202 Main Mall – Vancouver BC V6T 1Z4 – Canada – idrn.ubc@gmail.com

Eric Fitch (BSc, MEn, PhD) is Associate Professor of Environmental Science and Leadership and Director of the Environmental Science Program at Marietta College in Ohio, and the current Chair of the Advisory Board for the Interdisciplinary Environmental Association.

Environmental Science Program – Marietta College – 215 Fifth St – Marietta OH 45750 – USA – fitche@marietta.edu

William J. Focht (BSc, BE, MS, PhD) is Associate Professor of Political Science, Director of the Fire and Emergency Management Administration Program, and a past Director of the Environmental Institute and the Environmental Science Graduate Program at Oklahoma State University.

Department of Political Science – Oklahoma State University – 003 Life Sciences East – Stillwater OK 74078 – USA – will.focht@okstate.edu

Sérgio Luiz Braga França, PhD in Civil Engineering from UFF with emphasis on Management, Production, Quality and Sustainable Development. Professor at the School of Engineering – UFF. Coordinator of the Innovation and Technology Center for Sustainability (NITS/UFF). Engineer of Work Safety from UFF. Lead Auditor in Environmental Management System (BVQI). Since 2004 has worked in R&D projects in the areas of Sustainability, Environmental, Social Responsibility, Safety and Product Development.

Federal Fluminense University – LATEC – Rua Passo da Pátria 156 – CEP 24210-240 – Niterói RJ – Brazil – phone + 55 21 2629 5618 – fax +55 21 2717 6390 – sfranca@latec.uff.br

Olivia E. Freeman is a Master's student at the Institute for Resources, Environment and Sustainability, UBC. With a background in Natural Resource Conservation and International Development, she is currently completing her MSc thesis, examining the tradeoffs in realizing win-win benefits through carbon credit improved cookstove programs, and the implications of the carbon credit market as a new distribution model for improved cookstoves. Broadly, Olivia is interested in how to foster sustainable development through balancing resource management while improving livelihoods in the development context. She is a founding member of IDRN and a current steering committee member.

International Development Research Network – Liu Institute for Global Issues – University of British Columbia – 6476 NW Marine Dr. – Vancouver BC V6T 1Z2; – Institute for Resources – Environment and Sustainability – University of British Columbia – 2202 Main Mall – Vancouver BC V6T 1Z4 – Canada – idrn.ubc@gmail.com

Ernesto Frodden is a PhD candidate in Theoretical Physics at Universidad Católica de Chile and Aix-Marseille Université, France. His research interests are the foundations of space-time. To deal with this phenomenon, he is specializing in quantum gravity and specifically in black holes, for which the theoretical framework used is loop quantum gravity. He pursued a BSc in Physics in 2007 at Universidad de Chile. Besides academia, from 2008 he has been involved in *Taller Solar*, where he is an organizer and tutor in several activities aiming to develop and spread low-cost, environmentally friendly technologies in local communities. Some of these activities (summer volunteer works and formation courses) were undertaken in cooperation with the Students' Federation of Universidad de Chile (FECh).

Departamento de Física – Facultad de Física – Pontificia Universidad Católica de Chile – Chile

Ranka Gajić, MSc, is a Teaching Assistant (subjects: Urbanism, Sustainable Development and Environmental Protection and Constructions/Architectural segment) at the Faculty of Traffic and Transport

Engineering in Belgrade. She graduated from the Faculty of Architecture in Belgrade, as well as obtaining a Master's degree, and in 2010 she was approved as a PhD candidate at the faculty. She has been involved, both as an associate and team leader, in drafting town planning documentation for cities in Serbia (particularly Belgrade). As part of theoretical work, she participated in research projects and organizing symposia. She is the author of one monograph, supplementary textbook and over 35 published papers. She is a member of the Commission for Advancement of the Profession and ISOCARP. Since 2004, she has been Secretary of the Board of the Town Planners Association Belgrade. Scope of research: urban morphology, urban reconstruction, sustainable development of cities, sustainable urban land use.

University of Belgrade – Faculty of Transport and Traffic – 305 Vojvode Stepe St – 11000 Beograd – Serbia – igra@open.telekom.rs

Dr. **Salfarina Abdul Gapor** is currently a senior lecturer at the Universiti Sains Malaysia under the Development Planning and Management at the School of Social Science. She was trained as a town planner at the University of Sheffield and obtained her Master's degree from the University of Sussex, United Kingdom, in the field of Environment, Development and Policy. She then received her PhD in Physical Geography, focusing on indigenous knowledge and sustainable development from the University of Hull, United Kingdom. She teaches courses such as Environment, Natural Resource and Environment, and Theory and Issues in Development. At postgraduate level she is teaching Policies of the European Union and Natural Resource and Environmental Management. She has also been involved in a number of consultancies from UPEN, Penang and PERDA on Housing, Poverty and Livelihoods of the Bumiputera in Penang.

Center for Global Sustainability Studies – Universiti Sains Malaysia – 11800 Penang – Malaysia

Jairo Garcia is an Associate Research Assistant at the Office of Academic and Research Programs at The Earth Institute, Columbia University. His area of research is in higher education for sustainable development with emphasis in the interlinking of education and economic/environmental policies.

Office of Academic and Research Programs – The Earth Institute – Columbia University – 2910 Broadway – Hogan Hall – New York NY 10025 – USA – phone: +1-404-210-7575 – jhg2127@columbia.edu

Since March 2011, **Kira Giesecke** has been a PhD Student at the University of the West of Scotland and the University of Applied Sciences Hamburg. Her research focuses on the development of a Lifestyle Index of Sustainability, which measures the lifestyle areas of an individual, in relation to three dimensions of sustainability. She graduated from Edinburgh Napier University with a Bachelor of Honours Degree (First Class) in 2010. She lives and works in Hamburg, Germany.

Business School – University of West Scotland – Great Britain and University of Applied Science in Hamburg – Berliner Tor 5 – 20099 Hamburg – Germany –kira.giesecke@haw-hamburg.de or kiragiesecke@googlemail.com

Susan Gill (BA, MA, PhD) is the Director of Education for the Stroud Water Research Center in Pennsylvania.

Stroud Water Research Center – 970 Spencer Rd – Avondale PA 19311 – USA – sgill@stroudcenter.org

Dr. **Mark Giltrow** is the Program Head for Sustainable Business Leadership in the School of Business at the British Columbia Institute of Technology. Before BCIT, Dr. Giltrow was a Principal Consultant at The Canadian Centre for Local Government Research, and a Corporate Psychologist at Florida Power Corporation, BellSouth Corporation and United Parcel Service. Dr. Giltrow holds a Doctorate in Organizational Psychology from the Georgia Institute of Technology. Dr. Giltrow's research interests include sustainability, managerial leadership, and organizational creativity.

Program Head – School of Business – British Columbia Institute of Technology – British Columbia – Canada – mark_giltrow@bcit.ca

João Gois is a Master's degree student on the postgraduate program in Technology at the Federal University of Technology of Paraná (UTFPR). His main areas of interest are Energy Efficiency and Environmental Management.

PPGTE – Universidade Tecnológica Federal do Paraná (UTFPR) – Av. Sete de Setembro n° 3165 – CEP: 80230-901 – Curitiba (PR) – Brazil – joaogois@gmail.com

Luciana Paulo Gomes, PhD in Civil Engineering / Hydraulics and Sanitation at the School of Engineering of São Carlos / University of São Paulo. Currently she is a professor at the University of Vale do Rio dos Sinos – UNISINOS, a researcher at the Graduate Program in Civil Engineering and graduate courses in Civil Engineering and Environmental Management. She is a coordinator of the UNISINOS's Environmental Management System, the first university certified by ISO 14001 in Latin America. She works in sanitary engineering, with emphasis on Solid Waste, Domestic and Industrial Group Leader of CNPq, Environmental Sanitation, Researcher at CNPq.

Universidade do Vale do Rio dos Sinos – UNISINOS – Av. Unisinos 950 – São Leopoldo – Brazil – lugomes@unisinos.br

Louise Gonsalvez is a PhD student in educational policy and leadership at the University of Windsor. Her interests and publications focus on transformational educational leadership, social justice, sustainability, institutional governance, peace, and citizenship. As a representative of the British Columbia Teachers' Federation (BCTF) Committee for Action and Social Justice, she acted as a policy advisor, workshop facilitator and writer for the BCTF. She was also a co-author of the *British Columbia Ministry of Education's Social Justice 12 Integrated Resource Package* and the accompanying *Social Justice 12 Teacher Guide*. She is very interested in moving towards sustainability via education, governance, policy, social justice frameworks and leadership. Louise has travelled extensively and worked on a number of overseas projects (e.g. Australia, China, Ecuador, Ghana and South Africa). Louise is also a high school teacher at Sparwood Secondary School.

University of Windsor – Box 1232 – Fernie BC – Canada V0B 1M0 – lgonsalvez@hotmail.com

Melissa Goodall is the Assistant Director of the Yale Office of Sustainability. Her specific areas of focus are connecting global policy to local action, sustainability planning, and sustainable food. She is a member of the Executive Committee of the Yale Climate and Energy Institute and chairs the University's Paper Reduction Committee and Sustainability Microloan Committee. Prior to joining the Sustainability team, Melissa was the Associate Director of the Yale Center for Environmental Law and Policy. During her time with the Center, Melissa managed a range of projects related to environmental governance, environmental attitudes and behavior, and data-driven environmental decision-making. Melissa is currently a doctoral student at Antioch University New England and a UNEP GEO-5 research fellow. She earned her a Master of Science in Resource Management and Administration at Antioch, and her Bachelor of Fine Arts degree from New York University.

Yale University Office of Sustainability – 70 Whitney Avenue – PO Box 208275 – New Haven – Connecticut 06520-8275 – USA – sustainability.yale.edu – melissa.goodall @yale.edu

Denise Gorfinkiel. She joined the United Nations Education, Scientific and Cultural Organization as a Programme Officer for the Natural Sciences Sector in December 2005. Since 2001, she has been a professor and member of the Academic Committee of the Master's program in ICM at Universidad de la República (UDELAR, Uruguay). Her research, which dates back to 1995, focuses on international integration, environment and international relations and on economic and social management of the environment.

Centro Interdisciplinario para el Manejo Costero Integrado del Cono Sur – Universidad de la República – Montevideo (UdelaR) – Uruguay – denise.gorfinkiel@gmail.com

Marcus Grace taught science at comprehensive schools in London and Essex. He spent five years in Japan as a researcher at Yokohama National University and was a head teacher at a Berlitz school in

Tokyo. He is a Senior Lecturer in science education at the Southampton Education School at the University of Southampton, UK, and until recently he was Deputy Head. He is a member of the Southampton LifeLab project, an integrated school and hospital-based programme to improve scientific and health literacy among teenagers. Marcus reviews papers for the International Journal of Science Education, and Environmental Education Research, and was guest editor for the School Science Review 2010 issue on Education for Sustainable Development. His current research interests include: teaching and learning about socio-scientific issues; engaging students in health and environmental issues; and the science underpinning education for biodiversity, sustainability and citizenship.

Southampton Education School – University of Southampton – Building 32 – Southampton SO17 1BJ – United Kingdom

Viv Grant has 27 years' experience as a creative practitioner working in Community, Further and Higher Education contexts within the UK. This project utilises Viv's expertise in educational theory and practice to underpin and inform the diverse research currently emerging from sustainability discourses and practices. Viv also works as a part-time associate lecturer in Education Studies teaching on a newly validated module called Engaged Pedagogy, Love and the Practice of Care in Educational Settings.

Pedagogic Research Institute and Observatory (PedRIO) – Teaching and Learning Directorate – Plymouth University – Plymouth PL4 8AA – UK

Dr. **Ines Grigorescu** is a scientific researcher at the Institute of Geography, Romanian Academy, Environment and GIS Laboratory. Her research interests are related to the study of man-environment relationships, integrated environmental assessment and change and GIS. She has attended several national and international conferences and symposia, as well as having published over 50 papers on environmental issues in Romania.

Romanian Academy – Institute of Geography – 12 Dimitrie Racovita St. – Sector 2 – Bucharest 023993 – Romania

Thomas S. Grindsted is a PhD student in environmental management of universities, Department of Environmental, Social and Spatial Change, Roskilde University.

Roskilde University – Department of Environmental – Social and Spatial Change – P.O. Box 260 – Universitetsvej 1 – DK-4000 ROSKILDE – Denmark

José Baltazar Salgueirinho Osório de Andrade Guerra, PhD, Dean and Professor at Unisul Business School (Universidade do Sul de Santa Catarina – UNISUL, Brazil). Director, at UNISUL, of two research projects: JELARE – Joint European-Latin American Universities Renewable Energies Project; and REGSA – Promoting Renewable Electricity Generation in South America. Both projects financed by the European Union (through ALFA III and the thematic program for environment and sustainable management of natural resources, including energy). Author and organizer of several books in the fields of Economics, International Relations Environment, Sustainability and Renewable Energy.

Universidade do Sul de Santa Catarina – UNISUL – Rua Trajano 219 – 88010-010 – Florianópolis – Santa Catarina – Brazil – phone +55-48-3279.1927 – baltazar.guerra@unisul.br

Gowri Betrabet Gulwadi, (PhD in Architecture: Environment Behavior Studies, LEED AP) is an Associate Professor of Interior Design at the University of Northern Iowa. She facilitates the learning process of design students, emphasizing the importance of socio-cultural context and sustainability, and engaging pedagogical techniques that encourage reflection. Her research addresses health and restoration from stress, the ways in which people seek restorative experiences in their everyday environments, and design interventions that can help make challenging environments more manageable. She is affiliated with the Environmental Design Research Association (EDRA), the Interior Design Educators Council (IDEC), and the International Association of People-Environment Studies (IAPS).

University of Northern Iowa – 213 Latham Hall – School of Applied Human Sciences – Cedar Falls – Iowa 50614-0332 – USA – phone +1-319-273-2603 – betrabet@uni.edu

Zakri Abdul Hamid is Science Advisor to the Prime Minister of Malaysia. He is Director of the Centre for Global Sustainability Studies (CGSS@USM) and the holder of the Tuanku Chancellor Chair at Universiti Sains Malaysia. He was formerly Director of the Institute of Advanced Studies, United Nations University and Co-Chair of the Millennium Ecosystem Assessment. He is currently a member of the Arab Fund Fellowship Program, IAEA Senior Advisory Group on Technical Assistance and Cooperation, and the ICSU Executive Board. Zakri served as the Secretary General of the Society for the Advancement of Breeding Researches in Asia and Oceania (SABRAO) from 1981-89 and was Deputy Vice-Chancellor at Universiti Kebangsaan Malaysia from 1992-2000. He was the Founding President of the Genetics Society of Malaysia. Prof Zakri's interests include biodiplomacy, education for sustainable development and biotechnology and biodiversity policies for developing countries. He was the recipient of a Fulbright-Hays Fellowship (1981), and is a Fellow of the Academy of Sciences Malaysia, the Academy of Sciences of the Developing World, the World Academy of Art and Science, and the Islamic World Academy of Sciences. In 1998 he received the Langkawi Award, a national laureate for outstanding contribution in the field of environment in Malaysia. Three species known to science are named after him: a beetle (*Paleosepharia zakrii*), a cicada (*Pomponia zakrii*), and a pitcher plant (*Nepenthes zakriana*).

Centre for Global Sustainability Studies – Universiti Sains Malaysia – 11800 Penang – Malaysia

Suzanne Savanick Hansen is the Sustainability Manager at Macalester College. She is currently implementing the college's first comprehensive sustainability plan and is developing new ways to use the campus buildings and grounds as teaching tools. She earned her PhD in Conservation Biology from the University of Minnesota, a Master's degree in Environmental Management from Duke University and a Bachelor's degree in Geology from Carleton College.

Macalester College – 1600 Grand Ave – St. Paul MN 55105 – shansen2@macalester.edu

Charlie Hargroves is the Sustainable Development Fellow at the Entrepreneurship, Commercialization and Innovation Centre, University of Adelaide, Australia. He is also a co-founder and a principal researcher with The Natural Edge Project research group (TNEP), an academic partnership for research and capacity building for sustainable development. Charlie and the TNEP team have developed a number of internationally renowned books on sustainable development including 'Factor Five', 'Cents and Sustainability', 'Whole System Design', and 'The Natural Advantage of Nations'. His doctoral research with Professor Peter Newman focuses on adaptive governance and knowledge strategies for carbon structural adjustment, focusing on: implementation of new and innovative technology and processes; renewal of higher education programs and professional development, and efforts by local municipalities to encourage community behaviour change.

Karlson 'Charlie' Hargroves – University of Adelaide – Adelaide – Australia – The Natural Edge Project – karlson.hargroves@adelaide.edu.au

John Harraway is Senior Lecturer in the Department of Mathematics and Statistics at the University of Otago, New Zealand. John is currently President of the International Association for Statistical Education and has research interests in applied statistics and statistics education.

Department of Mathematics and Statistics – University of Otago – Dunedin – New Zealand

Leila Harris is an Assistant Professor at the Institute for Resources, Environment and Sustainability and the Centre for Women's and Gender Studies, UBC. She is trained as a socio-cultural and political geographer with interest in resource issues and inequalities, gender, and developing contexts. Her research to date has focused primarily on water politics in the Middle East/Turkey, water governance in the Global South, and gender dimensions of resource use and access. Given her focus on inequalities, work is also strongly connected to work in political ecology, environmental justice, and gender studies. Dr. Harris is one of two IDRN faculty mentors.

International Development Research Network – Liu Institute for Global Issues – University of British Columbia – 6476 NW Marine Dr. – Vancouver BC V6T 1Z2.; – Institute for Resources – Environment and Sustainability – University of British Columbia – 2202 Main Mall – Vancouver BC – Canada V6T 1Z4 – idrn.ubc@gmail.com

David M. Hassenzahl is founding Dean of the School of Sustainability and the Environment at Chatham University. He is an internationally recognized scholar of sustainability and risk analysis, and has spent more than two decades addressing subjects as diverse as climate change, energy, toxic chemicals, and public health. His research focuses on incorporating scientific information and expertise into public decisions, with particular emphasis on the management, interpretation, and communication of uncertainty. Author of numerous books and papers, Dr. Hassenzahl's efforts in climate change education have been supported by the National Science Foundation and the National Aeronautics and Space Administration, and recognition for his educational work includes the Society for Risk Analysis Outstanding Educator Award and the UNLV Foundation Distinguished Teaching Award. He is a Senior Fellow of the National Council for Science and the Environment and founding council member of the Association of Environmental Studies and Sciences.

Chatham University – 1 Woodland Road – Pittsburgh PA 15044 – USA – dhassenzahl@ chatham.edu *(correspondinz author contact)*

Adnan A. Hezri is a Visiting Researcher at the Centre for Global Sustainability Studies at Universiti Sains Malaysia (CGSS@USM), and a Research Fellow at the Institute for Environment and Development (LESTARI) at Universiti Kebangsaan Malaysia (UKM). Dr. Hezri also holds visiting appointments in the Fenner School of Environment and Society, Australian National University and in the think-tank The Institute of Strategic and International Studies (ISIS) Malaysia. He conducts research in areas of comparative environmental policy and the institutional aspects of sustainability. Hezri has been awarded research fellowships by various other institutions, including the Australia-Malaysia Institute in the Department of Foreign Affairs, Commonwealth of Australia, and the Nippon Foundation's Asian Public Intellectuals Fellowship.

Centre for Global Sustainability Studies – Universiti Sains Malaysia – 11800 Penang, Malaysia

Brandon P. Hollingshead is an Instructor in Interdisciplinary Studies and Civic Engagement at Florida Gulf Coast University, his alma mater, where he is also a faculty and editorial associate at the Center for Environmental and Sustainability Education. His research interests are in environmental communication, the rhetoric of sustainability, and humanities and sustainability. Hollingshead's Master's thesis at the University of Utah was on "Crafting Principles for Sustainable Development: Negotiations in the Drafting of the Earth Charter and Johannesburg Declaration on Sustainable Development." He has worked on a number of Earth Charter research projects and has published on youth participation in sustainable development.

Center for Environmental and Sustainability Education – Florida Gulf Coast University – 10501 FGCU Boulevard – Fort Myers 33965-6565 – Florida – USA – bphollin@gmail.com *(corresponding author contact)*

Tove Holm is responsible for enhancing sustainable development within all activities with an integrated management system for quality and environment which is certified according to ISO 9001 and ISO 14001 at Novia University of Applied Sciences where she holds the position of Environmental Coordinator and Quality Coordinator for the Unit of Research and Development. Her research interest is on the same theme.

Department of Biology – University of Turku – FI-20014 TURKU – Finland – tojoho@utu.fi *(corresponding author contact)*

Prof. **Debra Humphris** is Pro Vice Chancellor for Education and Student Experience at the University of Southampton and Professor of Health Care Development, she has extensive experience of educational leadership, organisational change, and a strong commitment to strategic innovation. Currently this includes leading the Curriculum Innovation Programme, designed to open up new interdisciplinary modules shaped by the University's research into major global challenges.

Southampton Education School – University of Southampton – Building 32 – Southampton SO17 1BJ – United Kingdom *(corresponding author contact)*

Carolina Andrea Cares Inostroza. Journalist, Universidad de Chile. Currently journalist in the undergraduate school at FEN, Universidad de Chile. Communication Coordinator at Observatorio de la Reconstrucción, Universidad de Chile and freelance journalist in human rights organizations.

Universidad de Chile – School of Economics and Business – Diagonal Paraguay 257 – Santiago – Chile – phone 56-2-9783999 – ccares@fen.uchile.cl

Felipe Rodrigo Horta Jaques. MBA in Administration, Universidad de Chile, Chile; Business Administration, Universidad de Chile, Chile. Currently Manager of Infrastructure Department at the School of Economic and Business (*FEN* in Spanish – *Facultad de Economía y Negocios*); member of Sustainable Campus Committee, FEN; and Leader of FEN Recycles and Energetic Efficiency program. Professor of Excel for Business course, FEN.

Universidad de Chile, School of Economics and Business – Diagonal Paraguay 257 – Santiago – Chile – phone 56-2-9783494 – fhorta@fen.uchile.cl

Govindran Jegatesen holds an Honors Degree in English Language and Literature Studies with a minor in Southeast Asian Studies and is currently pursuing his Master's in Science and Environmental Journalism at Universiti Sains Malaysia. He is currently also working as a Research Officer at the Centre for Global Sustainability Studies (CGSS) at Universiti Sains Malaysia (USM). His academic areas of interest are Disaster Risk Management, Women and Children's Development, Education for Sustainable Development, Biodiversity and Environmental Conservation. When not working on projects or assignments, he spends most of his time reading on Evolutionary Science and Wildlife Diversity – his personal areas of interest.

Center for Global Sustainability Studies – Universiti Sains Malaysia – 11800 Penang – Malaysia

Dr. **Guillermo Jimenez-Estevez** was born in Bogotá, Colombia. He received the BSc degree in Electrical Engineering from the Escuela Colombiana de Ingeniería, Bogotá, in 1998 and an MSc and a PhD from the University of Chile, Santiago, in 2003 and 2010 respectively. He is currently part of the research and development team of the Energy Center of Physical and Mathematical Sciencies Faculty of the University of Chile. His main research interests are Power Systems Planning and Operation, Renewable Energy, Smart Grids and Climate Change.

Department of Electrical Engineering – University of Chile – 2007 Av. Tupper – Santiago-Centro – Santiago 837 0451 – Chile – gjimenez@ing.uchile.cl

Sarah Johnson has a first-class Honours degree in Environmental Science from the University of Plymouth. She is working as the Environment and Utilities Officer at Anglia Ruskin University (UK), covering all aspects of environmental management ranging from working on travel plans, to biodiversity issues to waste and energy and fair-trade issues. Sarah has a passion for sustainability that extends into her personal life – she has volunteered in South America and enjoys working on her allotment.

Lord Ashcroft Business School – Anglia Ruskin University – East Road – Cambridge CB1 8PT – United Kingdom

Tim Jowett is research assistant to the Education for Sustainability Research Group at the University of Otago and currently completing a master's thesis in statistics.

Higher Education Development Centre, HEDC – Department of Mathematics and Statistics – University of Otago – Dunedin – New Zealand

Prof. **Lev Karlin** – doctor of physical and mathematical sciences, Rector of the Russian State Hydrometeorological University. He is a member of the Scientific Expert Council of the Marine Board under the government of the Russian Federation, Intergovernmental National Oceanographic Commission, Marine Council under the government of St.Petersburg, national coordinator of the IOC/UNESCO TEMA project, and has been a director of the BFU programme since 1993.

Russian State Hydrometeorological University – Malookhtinsky pr. 98 – St. Petersburg 195196 – Russia – rector@rshu.ru

Simon Kemp is Director of Employability and Employer Engagement in the Faculty of Engineering and the Environment, University of Southampton. A UK National Teaching Fellow, his main responsibilities include leading the University's Green Academy programme, and teaching a range of sustainability modules. His research interests include ESD and employability, carbon footprinting, waste management, and environmental management in industry. He is currently the HEA Academic Lead for ESD.

Faculty of Engineering and the Environment – University of Southampton – Southampton – Hampshire SO17 1BJ – United Kingdom – phone +44 (0)2380 595 868 – s.kemp@soton.ac.uk *(corresponding author contact)*

Lena Keul studied international economics at the Universities of Tübingen (Germany) and Recife (Brasil). After her studies, she worked on projects relating to sustainability, education for sustainable development and climate change at the University of Nürtingen-Geislingen as well as the City of Cologne. Since February 2011, she has worked in the sustainability research domain of Ökoworld LUX S.A. In Tübingen, she has co-initiated the student initiative Greening the University, e.V.

Ökoworld Lux S.A – Proactiv-Platz 3 – 40721 Hilden – Germany – lenakeul@hotmail.com

Hamoon Khelghat-Doost holds a Master's degree in Communication Studies and also a First Class Honors Bachelor of Arts in Humanities. He holds the position of Research Officer at the Centre for Global Sustainability Studies (CGSS) at Universiti Sains Malaysia (USM). He has completed courses on Comparative Religion, Jewish Studies and Soviet Studies at Dalhousie University, Halifax, Canada. He was an Archaeological Research Assistant for UNESCO and Laboratory of Geo-archaeology at Kazakhstan Ministry of Education and Science. He has been also a member of the editorial board for Southeast Asian Studies Conflict Network (SEACSN) magazine.

Center for Global Sustainability Studies – Universiti Sains Malaysia – 11800 Penang – Malaysia – hamoonk@hotmail.com *(corresponding author contact)*

Linas Kliucininkas has an extensive record of publications in different fields of environmental and sustainability sciences. The major part of these publications is of an applied character and has an interdisciplinary nature. During the last decade Kliucininkas has initiated and participated in a number of international programmes and projects supporting environmental and sustainability studies and research. He is a member of the Editorial Board of the International Journal of Environment and Sustainable Development. Kliucininkas is an expert on the FP6 programme Environment and Energy Infrastructures, and he has long-term experience of study quality evaluation at the universities in the Baltic States. Kliucininkas is director of the Baltic University Programme Centre for Lithuania.

Department of Environmental Engineering – Kaunas University of Technology – Radvilėnų pl. 19 – LT-50254 Kaunas – Lithuania – linas.kliucininkas@ktu.lt

Bruno Kobiski is a Master's degree student on the postgraduate program in Civil Engineering at the Federal University of Technology of Paraná (UTFPR). His main areas of interest are Carbon Sequestration in Building Materials and the Clean Development Mechanism (CDM) of the Kyoto Protocol.

PPGEC – Universidade Tecnológica Federal do Paraná (UTFPR) – Av. Sete de Setembro nº 3165 – CEP: 80230-901 – Curitiba (PR) – Brazil – kobiski@gmail.com

Ljubov Kohutnicka is a PhD student of Faculty of Geography of University of Warsaw, Poland. Having graduated from Maritime Academy in Estonia in 2006 and getting a MSc in Natural Sciences in 2009 made the author combine the two different fields of research – maritime transport and sustainable development and try to find ways to implement the idea of sustainability in the improvement of the development of maritime transport systems in the Baltic Sea Region.

Laboratory of Computer Education and GIS – Faculty of Geography and Regional Studies – University of Warsaw – Krakowskie Przedmieście 30 – 00-972 Warsaw – Poland – luka10710@gmail.com

Kanayathu Chacko Koshy is a faculty member and Professor of Sustainability at the Centre for Global Sustainability Studies at Universiti Sains Malaysia (CGSS@USM), Penang. He was the former Professor of Environmental Chemistry at The University of the South Pacific in Fiji and the Founding Director of the Pacific Centre for Environment and Sustainable Development. Dr. Koshy was Director of the Pacific Centres of the global change research and capacity building networks START, the System for the Analysis, Research and Training based in Washington, D.C. and the Asia Pacific Network in Kobe. Before joining The University of the South Pacific, he was a Faculty Member of the University of the West Indies, Jamaica, and the State University of New York at Buffalo. An integral part of his current work is on science, technology and innovation; education for sustainable development, sustainability policy, and climate change responses. He has served as Co-ordinator of the United Nations University Regional Centre of Expertise for Oceania, the UNESCO-ACCU Centre of Excellence for Education for Sustainable Development in the Pacific region, Principal Investigator of the Asia-Pacific Cultural Centre for UNESCO, European Union, Global Environmental Facility, and AusAID projects for the Pacific region, and the coordinator of the technical team for the development of the Pacific Regional Education for Sustainable Development Action Plan. He has lectured and published widely in his field of expertise.

Centre for Global Sustainability Studies – Universiti Sains Malaysia – 11800 Penang – Malaysia

Valdir de Jesus Lameira, PhD in Finance from PUC-Rio. MSc in Production Engineering from UFF. Graduated in Engineering from UFRJ. Author of books, chapters of books and articles in journals and conferences in Brazil and abroad. Reviewer of several magazines in Brazil and abroad. Worked on the Rio de Janeiro Stock Exchange, at Price Waterhouse and Bank of Brazil. Capital market analyst licenced by the Brazilian Securities and Exchange Commission. Teacher of IBMEC-Rio, Coppead and FGV. Member of the National Investor Relations Institute in Brazil and member of the Advisory Board of the Association of Investors in Portugal. Currently works as a researcher at INESC Coimbra, an institute linked to the University of Coimbra, having been approved under the international selection in a program called Science 2008.

INESC Coimbra – Portugal – Rua Antero de Quental 199 – 3000-033 Coimbra – Portugal – phone/fax +34-98-6 812-685 – vlameira@inescc.pt

Walter Leal Filho (BSc, PhD, DPhil, DSc, DL, DEd, DLitt) is a Senior Professor at London Metropolitan University (UK), Uppsala University (Sweden) and the Hamburg University of Applied Sciences in Germany, where he heads the Research and Transfer Centre "Applications of Life Sciences." He holds a doctorate in environmental education (PhD), a higher doctorate (DSc) in sustainable development and, commensurate with his scientific performance and outputs, the titles of Doctor of Letters (DL), Doctor of Literature (DLitt), Doctor of Philosophy (Phil) and Doctor of Education (DEd). He is an editor of the award-winning book series *Environmental Education, Communication and Sustainability*, with over 30 volumes produced to date, and founding editor of the *International Journal of Sustainability in Higher Education*. He has written, edited or co-edited over 70 books and has over 200 published papers and book chapters to his credit.

Research and Transfer Centre for Applied Sciences – University of Applied Sciences Hamburg – Lohbruegger Kirchstr. 65 – 21033 Hamburg – Germany – walter.leal@haw-hamburg.de

Ana Louise Tavares Leary is a graduate student at York University, Toronto, in the Master in Environmental Studies program. She completed her Bachelor of Arts degree at McGill University in political science and international development studies. At York, Ana is a graduate assistant for the CCAA project. Her research focuses on climate change policy, international development and waste diversion.

Prof. **Michael Lee** has spent 20 years teaching, researching and consulting in sustainability and related areas. He received his doctorate in geography from the London School of Economics in 1989, where he researched water harvesting systems, then consulted internationally on water conservation and protection. He first taught at the Pan-American School of Agriculture in Honduras, where he managed part of a USAID-funded global Sustainable Agriculture and Natural Resources Management program. He joined the Department of Geography and Environmental Studies at California State University East Bay (CSUEB) in 1996 and soon thereafter developed his course on Sustainable Development. He also teaches classes on Sustainable Resource Management, Sustainable Cities, and Global Change among others, and has consulted nationally and internationally on different aspects of sustainability, most recently on sustainable water resources and sustainable food systems. He was appointed CSUEB's Faculty in Residence for Sustainability & Pedagogy for 2011-12. He would like to thank all the students who have taken GEOG 4330 Sustainable Development over the years for their creative visioning and for providing the inspiration for this paper.

Department of Geography and Environmental Studies – California State University East Bay – 25800 Carlos Bee Blvd. – Hayward CA 94542 – USA – phone +1 510-885-3155 – fax +1 510-885-2353 – michael.lee@csueastbay.edu

Jordan Levine is a Liu Scholar and PhD student in the Interdisciplinary Studies Graduate Program, UBC. He is interested in a range of interconnected issues revolving around sustainability and justice. His current research focuses on adapting cognitive anthropological methods to improve multistakeholder resource management in UNESCO Biosphere Reserves. He is a founding member of IDRN and a current steering committee member.

International Development Research Network – Liu Institute for Global Issues – University of British Columbia – 6476 NW Marine Dr. – Vancouver BC V6T 1Z2; – Interdisciplinary Studies Graduate Program – University of British Columbia – 153A-6201 Cecil Green Park Rd. – Vancouver BC V6T 1Z1 – Canada – idrn.ubc@gmail.com

Paula Lindroos, PhD, is a director for the Baltic University Programme; she works at the BUP coordination secretariat at Uppsala University. From 2002 to 2011 Lindroos was director of the Centre for Lifelong learning at Åbo Akademi University in Turku, Finland. Her area of interest is education for sustainable development and she has authored numerous publications in this area. She has worked as an expert in many national and international working groups, projects and networks. Lindroos was adviser to the coordinator of the Baltic 21 education sector, and she coordinated the National Resource Centre of education for sustainable development in higher education for Finland.

Baltic University Programme – Uppsala Centre for Sustainable Development – Uppsala University – Villavägen 16 – SE-752 36 Uppsala – Sweden –paula.lindroos@csduppsala.uu.se

Brent Lovelock (PhD) is Associate Professor in the Department of Tourism, University of Otago, New Zealand and Co-Director of the University of Otago's Centre for Recreation Research. Brent researches the application of sustainable tourism within protected natural areas.

Department of Tourism – University of Otago – Dunedin – New Zealand

Heather Luna is a freelance Education for Sustainable Development (ESD) consultant. She worked with the Higher Education Academy as Project Coordinator for their ESD project, and was a lecturer in Economics in New York City. She is a tutor on the Ellen MacArthur Foundation/University of

Bradford Post-Graduate Certificate in the Circular Economy. Active in the Transition movement and the Dark Mountain Project, she is a student of long-term thinking.

92 Osprey Park – Thornbury – Bristol BS35 1LZ – United Kingdom – phone +44 (0)7725 034 010 – heather.a.luna @gmail.com.

Juan Macías is a graduate student of Physics at Universidad de Chile, currently undertaking a Master's degree. He works at the Laboratory of Matter out of Equilibrium of the Nonlinear Physics group at the Faculty of Physical and Mathematical Sciences (FCFM). He worked as a science teacher in a school for children at social risk and as a physics teacher, supporting students for the application test to enter Chile's universities system. He is currently an active member of *Taller Solar* and works as a teacher in workshops, as well as organizing the FECh Volunteer Works.

Departamento de Física – Facultad de Ciencias – Universidad de Chile – Chile

Prof. **Siarhei Malchenka** is Director at the Minsk Branch of MESI, and holds a PhD in Chemistry. His research areas are knowledge management, innovation management, environmental protection and life safety. As a director, he undertakes a large amount of administrative work.

Minsk Branch of MESI – 127 – Build.2 – Mayakovskogo Str. – Minsk 220028 – Belarus – office@mfmesi.ru

Romas Malevicius has a Master's degree in Thermal Engineering (Vilnius Gediminas Technical University, Lithuania) and a Bachelor's degree in International Business Strategy (Anglia Ruskin University, UK). Currently, he works as an associated lecturer at Lord Ashcroft Business School and is starting his PhD in pedagogy for sustainability and experiential learning, in higher education. Romas is an accomplished photographer and enthusiastic filmmaker.

Lord Ashcroft Business School – Anglia Ruskin University – East Road – Cambridge CB1 8PT – United Kingdom

Robert W. Marans (PhD) is Research Professor at the University of Michigan's Institute for Social Research and Professor Emeritus of Architecture and Urban Planning in the university's Taubman College of Architecture and Urban Planning. For more than three decades, he has conducted evaluative studies and research dealing with various aspects of communities, neighborhoods, housing, and parks and recreational facilities. His research has focused on user requirements and the manner in which attributes of the physical and sociocultural environments influence individual and group behavior and the quality of community life. Dr. Marans' most recent book, *Investigating Quality of Urban Life: Theory, Methods, and Empirical Research* was published by Springer (2011). His current research considers the impact of the built and natural environments on quality of life, the role of neighborhood in the health of Detroit residents, and issues of sustainability and energy conservation in buildings and institutional settings.

Institute for Social Research – University of Michigan – Ann Arbor MI. 48106 – USA – phone +1-734-763-9818 – marans@umich.edu

Amelia Mayahi is the Sustainability Coordinator at the University of Memphis. She earned her Bachelor of Arts degree at the University of Tennessee at Martin and her Master of Arts degree at the University of Memphis. Throughout high school and both her BA and MA degrees, Amelia has worked in facilities operations for higher educational institutions. Amelia is currently pursuing LEED accreditation. With a strong background and passion for sustainability, Amelia pursued making sustainability an integrated element in both facilities operations and student life. While earning her MA at the University of Memphis, Amelia was accepted for the position of Sustainability Coordinator. Through her past experience in Student Government and facilities operations, Amelia strives to change the culture of the University of Memphis and the surrounding communities to be more environmentally conscious. With the assistance of the Green Fee and supporting faculty, staff, and students, Amelia was able to build an outstanding recycling and bike-share program, create a volunteer network called the Green Team, implement an efficient sustainability committee, lead sustainability trainings classes, host numerous sustainability events and competitions, and implement an effective energy conservation team and initiatives. Amelia also currently serves as a board member for both the Tennessee Recycling Coalition (TRC) and Ecolaborative Memphis. While her priorities are grounded in the University of Memphis, Amelia is actively engaged in regional, state, and national sustainability initiatives for higher educational institutions.

Sustainability Coordinator – University of Memphis – Physical Plant – 325 Ray Herzog Building – Memphis 38152 – Tennessee – USA – phone +1.901.678.5543 – askosta@memphis.edu

David McInally is Executive Vice President and Treasurer at Allegheny College. McInally is interested in structural, financial and systemic changes required to enable sustainable campus development.

Allegheny College – 520 North Main Street – Meadville PA 16335 – USA – dave.mcinally @allegheny.edu

Dr. **Molly Mehling** is an Assistant Professor of Sustainability at Chatham University's School of Sustainability and the Environment. Her research and teaching interests include sustainability science, global water resources, community ecology, and women's and children's environmental health. She works to better understand the impact of human activities on mountain lake shorelines. She has an avid interest in the role of visual communication and information design in engaging stakeholders, transferring knowledge, and facilitating evidence-based decision-making. She holds a PhD in Zoology (2011) and an MEn in Environmental Science (2006) both from Miami University, and a BSc in Environmental Biology (2000) at Mount Union College (now University of Mount Union). Dr. Mehling has coordinated public service-oriented photography workshops for scientists at Ecological Society of America and scholarship awardees with the North American Nature Photography Association College Program. As a 2011 Research Ambassador Fellow, Dr. Mehling merged her science and photography to teach preschoolers about aquatic biota. She founded SustainableFocus.org – a collaborative endeavor to support the science-photography community as it shares and evaluates methods and projects to communicate issues of science and sustainability to diverse audiences.

Chatham University – 1 Woodland Road – Pittsburgh PA 15044 – USA

Marcelo Jasmim Meiriño, PhD (in progress) in Civil Engineering (UFF). MSc in Civil Engineering from UFF. Vice Coordinator of the Innovation and Technology Center for Sustainability (NITS/UFF). Engineer of Work Safety. Specialist in Energy Efficiency in Buildings, Commissioner of Social Responsibility of the Brazilian Petroleum Institute (IBP) and Gas. Architect and urban planner at UFRJ. Professor at the School of Engineering UFF, Researcher and Consultant LATEC / UFF.

Federal Fluminense University, LATEC – Rua Passo da Pátria 156 – CEP 24210-240 – Niterói RJ – Brazil – phone +55-21-2629-5618 – fax +55-21-2717-6390 – marcelo@latec.uff.br

Simon Meisch studied political science and German literature at the Universities of Tübingen and Edinburgh. He is a researcher at the International Centre for Ethics in the Sciences and Humanities and a lecturer at the University of Tübingen, working on issues of education for sustainable development, ethics and historical institutionalism, and has led the student initiative Greening the University, e.V.

University Tübingen – International Centre for Ethics in the Sciences and Humanities – Wilhelmstraße 19 – 72074 Tübingen – Germany – simon.meisch@uni-tuebingen.de

Rodrigo Menafra. He has over twelve years of experience in coastal and marine conservation and management, having participated in numerous interdisciplinary, multi-stakeholder projects and teaching activities in Uruguay, Iceland and Canada with emphasis on coastal policy, marine protected areas and integrated coastal management. He has recently acted as Marine Conservation Coordinator for the Canadian Parks and Wilderness Society, Nova Scotia Chapter, promoting the creation of marine protected areas in the region, with a particular focus on near-shore protection.

Centro Interdisciplinario para el Manejo Costero Integrado del Cono Sur – Universidad de la República – Montevideo (UdelaR) – Uruguay –

Charlotte Filiz Merten is about to complete her Bachelor's degree at University College Maastricht in International Relations, Economics and Development Studies. She is a volunteer in the Education & Research portfolio of the Maastricht University Green Office.

University College Maastricht – Maastricht University – Zwingelput 4 – 6211 KH Maastricht – The Netherlands

Arian Meyer graduated with the Bachelor of Arts in European Studies from the Faculty of Arts and Social Sciences in 2011. Currently, he is talking a the Master of Arts degree in Analyzing Europe at the University of Maastricht. His main research interests are to be found in the European Union in general and in organizational theory in particular. At the moment he is working on the issue of the principal-agent relationship between the EU and its agencies. During his Bachelor studies, he has spent one semester at the Institut d'Etudes Politiques (Sciences Po) Paris. Mr. Meyer has also been involved from early on in the construction of the Maastricht University Green Office in the position of Student Coordinator. His interest in sustainability studies stems from the theoretical-strategic angle.

Maastricht University Green Office – Sint Servaaskloster 32 – 6211TE Maastricht – The Netherlands

Andjelka Mihajlov: University professor, environmental consultant, promoter and scientist. By education she is a chemical engineer, with a PhD in Technical Sciences and more than 300 scientific references (in recent years mostly related to environmental governance and education, waste management, energy and environmental security). She is a full-time professor of environmental engineering at the Faculty of Technical Sciences at the University of Novi Sad. She is a member of the National Commission and is responsible for cooperation with UNESCO (she is also Chair of MAB and the Climate Change Sub-Committee). For 20 years she has been working for Environmentally Friendly and Sustainable Future, initiating the Environmental Cohesion Process of the Western Balkans as a tool for (regional sustainable) development. For five years (2005-2010) she was a member of EBRD ESAC *(Environmental and Social Advisory Council)*, and from 2008-2011 Dean of the Faculty of Environmental Governance and Corporate Responsibility. As Minister for the Protection of Natural Resources and Environment in the Serbian Government, she had a prominent role in leading the Reform Agenda of Serbia in the Environmental Sector from 2001 to 2004. In 2006 she was acknowledged by UNEP as an environmental leader of the Western Balkans. She is a member of the UN Network of Women Ministers and Leaders for the Environment.

Faculty of Technical Sciences – University of Novi Sad – Trg Dositeja Obradovica 6 – 21000 Novi Sad – Serbia – anmi@eunet.rs

Cynthia Morinville is a Master's student at the Institute for Resources, Environment and Sustainability, UBC. She holds a Bachelor of Arts with a double major in Environment Studies and International Studies from York University. Her research interests include questions of access to resources, participation in decision making and governance processes, as well as their intersection with, and implications for, identity making and notions of citizenship. Her Master's research more specifically looks into questions of water governance and access for residents of underserved informal settlements of the capital city of Accra, Ghana. Cynthia is a current IDRN steering committee member.

International Development Research Network – Liu Institute for Global Issues – University of British Columbia – 6476 NW Marine Dr. – Vancouver BC V6T 1Z2; – Institute for Resources – Environment and Sustainability – University of British Columbia – 2202 Main Mall – Vancouver BC V6T 1Z4 – Canada – idrn.ubc@gmail.com

Natalia Moudrak holds a Bachelor of Arts Degree in Economics (Honours) and a Master of Arts in Planning from the University of Waterloo. Natalia currently works as a Project Manager at Pollution Probe conducting research and developing educational initiatives to support policy development in the areas of energy, transportation and technology deployment. Examples of her work include analysis of barriers and opportunities for energy efficiency and conservation in the Ontario building sector, research on electric vehicles and smart grid integration, and analysis of greenhouse gas emissions reductions policies.

School of Environment – Enterprise and Development (SEED) – University of Waterloo – 200 University Avenue West – Waterloo – Ontario N2L 3G1 – Canada –nmoudrak@uwateloo.ca

Muthu Nandhivarman obtained his MSc degree in Zoology and has over ten years of work experience as an environmentalist, researcher and an activist for sustainable development. He is currently pursuing his doctorate at Pondicherry University, India. His research interests include an extensive study to evolve and implement the Green Campus Initiative at Pondicherry University. The main objective of this research is to assess and provide factual solutions for educational institutions such as schools and colleges, to make them sustainable with special reference to water, energy and waste.

Department of Ecology & Environmental Sciences – Pondicherry University – Puducherry 605014 – India – muthunandhi@yahoo.in

Naveed Nanjee is originally from Sugar Land, Texas and is currently studying on the Graduate Program in Islamic Studies and Humanities at the Institute of Ismaili Studies in London. Naveed completed his undergraduate degree with high honours from Vanderbilt University in Nashville, Tennessee. He double majored in Cognitive Studies and created his own interdisciplinary major called International Development and Sustainability. At Vanderbilt, in collaboration with various faculty members, he laid the foundation for an academic curriculum on sustainability, which was presented at the Cumberland Project Conference in 2010. Naveed spent a semester abroad in South Africa studying development at Stellenbosch University and interning at the Sustainability Institute.

Class of 2014 – Graduate Programme in Islamic Studies and Humanities – Institute of Ismaili Studies – 210 Euston Road – London NW1 2DA – United Kingdom – phone +44 (0)207 756 2700 – nnanjee@iis.ac.uk

Luis Felipe Machado do Nascimento, undergraduate degree in Electrical Engineering (1985) and Master's degree in Production Engineering (1989), both at the Federal University of Santa Maria; PhD in Economics and the Environment (1995) at the University Gesamthochschule Kassel, Germany. In 2002-2003 he held a post-docotal position at the University of Massachusetts, United States. He is an Associate Professor at the School of Management, Federal University of Rio Grande do Sul and a researcher of CNPq. Themes of research include: social and environmental management, sustainability and innovation, education for sustainability, sustainability of production chains, cleaner production, ecodesign, green marketing and clean development mechanisms.

Universidade Federal do Rio Grande do Sul – UFRGS – Av. Washington Luis – 855/414 Porto Alegre RS – Brazil – nascimentolf@gmail.com

Mostafa Nejati is a PhD Research Fellow at the School of Management, Universiti Sains Malaysia (USM). He is also the Editor-in-Chief of the Journal of the Global Business and Management Research (GBMR). His main research interests include sustainable development, sustainability in higher education, performance management, corporate governance, and healthcare management.

School of Management – Universiti Sains Malaysia – Malaysia – mostafa.nejati @gmail.com *(corresponding author contact)*

Associate Professor Dr. **Norizan Md. Nor** is the Director of the Centre for Global Sustainability Studies (CGSS) at Universiti Sains Malaysia (USM). His fields of expertise are Cross-Boundary Study, City and Community Sustainability Study and Tourism Geography. He was the chairperson for the Geography and Planning department at the School of Humanities, Universiti Sains Malaysia (USM) from 1994 to 1997. Later, from 1999 until 2000, he held the position of Deputy Dean at the School of Humanities. Dr. Norizan also served USM as the Dean of the School of Humanities from 2000 until 2007. In 2010, he coordinated the Master's program in Sustainable City and Community at USM. His main areas of interest include Human Geography, Town Geography, Area Development Planning and Social and Space Planning.

Center for Global Sustainability Studies – Universiti Sains Malaysia – 11800 Penang – Malaysia

David Nugent is the Director of the Master's in Development Practice Program at Emory University. He also holds an appointment as a professor in the Anthropology Department. His research focuses on poverty, development and underdevelopment among the world's most disadvantaged groups. He has done extensive theoretical and applied work on these issues in a variety of settings, from Latin America to East Africa to North America.

Emory University – Department of Anthropology – Emory University – 1557 Dickey Drive – Atlanta GA 30322 USA – phone +1-404-727-4164 – david.nugent@emory.edu

Diego Oyarzo studied for three academic years at the Science Faculty of Universidad de Chile, pursuing a Bachelor's degree in Physics. He has devoted his time to doing classes and developing didactic material for science education on elementary schools, and working as a volunteer at different instances of community-based self-construction workshops, as a member of Universidad de Chile's social and environmental organization *Taller Solar*.

Departamento de Física – Facultad de Ciencias – Universidad de Chile – Chile

Prof. **Paul Pace** is the Director of the Centre for Environmental Education & Research and Deputy Dean of the Faculty of Education. He has been active in environmental education as chairperson of Malta's National Environmental Education Strategy, founder of the Malta Association of Environmental Educators and as a researcher at the Faculty of Education, University of Malta, where he is a full-time associate professor in environmental education and science education. Appointed EkoSkola National Coordinator for Nature Trust (Malta) and the GLOBE (Global Learning and Observations to Benefit the Environment) Programme National Coordinator, he was also a member in the National Commission for Sustainable Development and the National Climate Change Committee. He is the vice-chairperson of the Interdiocesan Commission for the Environment of the Maltese Archdiocese and is currently on the National Curriculum Framework Review committee and the National Science Strategy committee.

Centre for Environmental Education & Research – Faculty of Education – University of Malta – Msida MSD 2080 – Malta – paul.j.pace@um.edu.mt

Arminda do Paço is a Professor at the University of Beira Interior, Portugal and researcher at NECE (Research Unit in Business Sciences). Her research interests include public, non-profit marketing and social marketing, environmental marketing and entrepreneurship education.

University of Beira Interior – Department of Business and Economics – Research unit NECE – Pólo IV – Estrada do Sineiro – 6200-209 Covilhã – Portugal – apaco@ubi.pt

Eric Pallant is Professor of Environmental Science at Allegheny College. Pallant's research incorporates international sustainable development in small communities including college campuses.

Allegheny College – 520 North Main Street – Meadville PA 16335 – USA – eric.pallant @allegheny.edu

Lisiane Celia Palma is a Professor and researcher at the Federal Institute of Education, Science and Technology of Rio Grande do Sul (IFRS) in the city of Canoas, Brazil; she is also a a PhD candidate on the postgraduate program in Management (PPGA) at the Federal University of Rio Grande do Sul (UFRGS), as well as a research member of the Technological Innovation and Education Group at the IFRS and the Organizations Study Group (GESTOR) at UFRGS, in the city of Porto Alegre, Brazil. Currently, she coordinates a research project entitled "Sustainability in management education" and the extension project "Metamorphosis – promoting change from environmental education" at the IFRS. Her research interests encompass sustainability, organizations, management and education. She has worked in projects related with these topics in Brazil, Italy and Colombia and has worked with secondary and higher education and also as a professor in MBA courses. She holds a Master's degree from the Center for Agribusiness Studies and Research (CEPAN) at UFRGS sponsored by the Coordination for Higher Education Personnel Improvement (CAPES).

Federal Institute of Education – Science and Technology of Rio Grande do Sul (IFRS) – Dr. Maria Zélia Carneiro de Figueiredo Street 870-A – Canoas/RS – Brazil – phone +55 51 34158200 – lisiane.palma@canoas.ifrs.edu.br

Jae Park, PhD, is an Assistant Professor at the Hong Kong Institute of Education. His research interests include sociology of education (technological, cultural and ecological discourses) and philosophy of education (East/West comparison particularly in ethics).

10 Lo Ping Road – Taipo – New Territories, Hong Kong – Department of International Education and Lifelong Learning – The Hong Kong Institute of Education – phone (852) 9184-8136 – jpark@graduate.hku.hk

Kelly A. Parker, PhD, is Professor of Philosophy, Environmental Studies, and Liberal Studies at Grand Valley State University in Allendale, Michigan, USA. His writings on American philosophy and environmental philosophy have appeared in *Environmental Ethics, Environmental Values,* and *Contemporary Pragmatism,* and in the volume *Environmental Pragmatism* (Routledge, 1996). He is the author of numerous articles and book chapters on the philosophy of John Dewey, William James, Josiah Royce, and Charles S. Peirce, and is the author of *The Continuity of Peirce's Thought* (Vanderbilt, 1998). He has served as a board member of the West Michigan Environmental Action Council, and was a founder of the Greater Grand Rapids Food Systems Council. He is presently at work on a critical and historical anthology on the philosophy of sustainability.

B3-105 Mackinac Hall – Grand Valley State University – Allendale 49401 – Michigan – USA – phone +1-616-331-3257 – fax +1-616-331-2601 – parkerk@gvsu.edu

Anupama Pasricha is an assistant professor of textiles and apparel in the School of Business and Leadership at St. Catherine University, St. Paul, Minnesota. She earned a PhD in Family and Consumer Sciences Education at Iowa State University. Her research interests include sustainable fashion and apparel; sustainability education within fashion and apparel; sustainable design, product development, and supply chains; service learning; and adaptive apparel. She has published her research and practice in national and international journals. She is a member of Educators for Socially Responsible Apparel Business (ESRAB) and currently serves on its Education Working Group. She also serves on the Curricular Development and Review Committee of the International Textiles and Apparel Association (ITAA). She is a member of the Ethics Committee of the American Association of Family and Consumer Sciences (AAFCS) and is an invited member of Sol Inspirations: EcoFashion in the Minneapolis-St. Paul area. She is a recipient of the 2011 AAFCS National New Achiever Award. Anupama Pasricha can be contacted via e-mail at apasricha@stkate.edu.

School of Business and Leadership – St. Catherine University – 2004 Randolph Avenue – St. Paul, Minnesota 55105 – USA

Megan Peloso is a Master's student at the Institute for Resources, Environment and Sustainability, UBC. Originally from Ottawa, Ontario, the focus of her thesis research is community water access and governance in urban informal settlements in the Accra Metropolitan Area (AMA) of Ghana, where she plans to conduct fieldwork in the summer of 2012. Megan holds a Bachelor of Social Sciences with Honors in International Development and Globalization from the University of Ottawa. She is a current IDRN steering committee member.

International Development Research Network – Liu Institute for Global Issues – University of British Columbia – 6476 NW Marine Dr. – Vancouver BC V6T 1Z2; – Institute for Resources – Environment and Sustainability – University of British Columbia – 2202 Main Mall – Vancouver BC V6T 1Z4 – Canada – idrn.ubc@gmail.com

Dr. **Javier Esquer Peralta** is an Industrial and Systems Engineer with a Doctor of Sciences degree in Cleaner Production from the University of Massachusetts-Lowell (UMass-Lowell) as well as a Sustainable Development Certificate from the University of Sonora (UNISON), Mexico. He has been an enthusiastic promoter of sustainable development in his community and has participated in several initiatives on environmental protection through UNISON. Dr. Esquer has also been a speaker at events of several kinds, locally, regionally and internationally.

Engineering College – University of Sonora – Mexico – jesquer@industrial.uson.mx

Patricia E. (Ellie) Perkins is an associate professor in the Faculty of Environmental Studies, York University, where she teaches and advises students in the areas of ecological economics, interdisciplinary research design, and community development. Her research focuses on feminist ecological economics, local economies, climate justice, and watershed governance. She currently directs a "Climate Change Adaptation in Africa" (CCAA) project funded through the International Development Research Centre and the UK Department for International Development.

Faculty of Environmental Studies – HNES 109 – York University – 4700 Keele Street – Toronto M3J 1P3 – Ontario – Canada – esperk@yorku.ca

Royann Jean Petrell is a chemical and biological engineering professor teaching fourth-year laboratory courses, senior-level hazardous waste course, and a graduate-level sustainability and environmental course in chemical and biological engineering. Her area of expertise is habitat restoration with a focus on chemical and biological remediation.

Chemical and Biological Engineering – University of British Columbia – 2360 East Mall – Vancouver V6T1Z3 – British Columbia – Canada – Rpetrell@chbe.ubc.ca

Julia Planko teaches Economics, Marketing and Business Ethics at the HU University of Applied Sciences Utrecht, and works for the research group Business and Innovation. Her research field is 'Sustainability and Innovation'.

HU University of Applied Sciences Utrecht – Padualaan 101 – 3584 CH Utrecht – Netherlands – julia.planko@hu.nl

Allison Potteiger serves as the 2011-2012 Yale University Office of Sustainability Fellow. Throughout the summer and this academic year, she has been assisting with the development of the Yale Medical Campus Sustainability Strategic Plan. She has also been working with the Yale Athletic Department as the project leader of Bulldog Sustainability. Over the summer she further established and implemented an overarching Communications Strategy for *Yale's Sustainability Strategic Plan 2010-2013*. Allison's focus spring semester will be Athletics and two events, Spring Salvage, Yale's move out program and Celebrate Sustainability, a weeklong series of events related to human health and well-being. A recent graduate of Ohio University, Allison earned a degree in public relations in addition to obtaining an Environmental Studies Certificate. While at school she also worked, interned and participated in various environmental organizations, including Ohio University's College of Osteopathic Medicine Student Affairs Office, Office of Sustainability, Residential Housing, and as a public relations intern for the Voinovich School of Leadership and Public Affairs.

Yale University Office of Sustainability – 70 Whitney Avenue – PO Box 208275 – New Haven – Connecticut 06520-8275 – USA – sustainability.yale.edu – allison.potteiger@yale.edu

Dr. **Gopalsamy Poyyamoli**, Associate Professor, Department of Ecology & Environmental Sciences Pondicherry University, Puducherry, India (gpoyya9@gmail.com). PhD in Ecology, Madurai Kamaraj University, Madurai; areas of specialization: social ecology and sustainable agro-ecology and ecosystem services, environmental education for sustainable development, green campus, industrial ecology, sustainable tourism, sustainable solid waste management and integrated coastal zone management. He is a recipient of the Indo-German cultural exchange program and the Canadian Studies faculty research fellowship; has completed three major and three minor research projects; is a member of the State Expert Appraisal Committee and the Commission on Ecosystem Management, IUCN, Switzerland; he is also a member of the Global Experts Directory on Ecosystem Services (IUCN, WRI, WBCSD & Earthwatch Institute), and a network member of the National Ecosystem Services Research Partnership, US EPA. He has published 25 international journal articles and contributed 14 invited articles to the SAGE series on Green Society, SAGE Publications, USA; he has written four book chapters and supervised eight PhD candidates. He is currently supervising seven PhD students.

Department of Ecology and Environmental Sciences – Pondicherry University – Puducherry 605014 – India – gpoyya9@gmail.com

Dr. **Arun Prasath Ramaswamy** has been working as Assistant Professor in the Center for Green Energy Technology, Pondicherry University since June 2010. As a faculty member in the Center for Green Energy Technology, he is involved in teaching, research, and outreach programs focused on renewable energy technologies and their application for sustainable development. He obtained his doctoral degree in Chemical Science/Materials from Anna University, India. He was a recipient of prestigious DAAD fellowship, (1999–2001) for his doctoral research work at the Max-Planck Institute for Polymer Research, Mainz, Germany. After his doctoral degree he worked as a material researcher in several prestigious institutes, as a research associate at the Indian Institute of Science, Bangalore, India (2002–2004), as a postdoctoral researcher at the University of Strathclyde, Glasgow, United Kingdom (2004–2006), and at the University of New South Wales, Sydney, Australia (2006–2008). He was also a senior researcher at Ghent University (2008–2010), with a special fellowship called BOF. He has published more than 20 peer-reviewed journal articles, and is the co-inventor of three international patents, as well as two European patent applications.

Laboratory for Energy Materials and Sustainability – Centre for Green Energy Technology – Pondicherry University – Puducherry 605014 – India – raprasath.get@pondiuni.edu.in, raprasath@gmail.com *(corresponding author contact)*

Rudi Pretorius is from the Department of Geography at the University of South Africa, where he has been lecturing in "Geography of Tourism", "Ecotourism" and "Climate Change" over a number of years. In addition he is the academic coordinator of the undergraduate programme in environmental management and currently vice-chair of the Department of Geography. He holds a Master's of Science in Geography and a Master's in Business Leadership. He is chair of the Division of Mathematical and Physical Sciences of the South African Academy of Science and Arts and is currently registered for a PhD in Geography, focusing on the potential and role of Geography in education for sustainable development within an ODL context.

Department of Geography – College of Agriculture and Environmental Sciences – University of South Africa – Florida Campus – Private Bag X6 – Florida 1710 – South Africa – pretorw@unisa.ac.za

Osvaldo Luiz Gonçalves Quelhas, PhD in Production Engineering from UFRJ. MSc in Civil Engineering from UFF. Graduated in Civil Engineering from UFF. Coordinator of the Technology, Business Management and Environmental Laboratory (LATEC/UFF). Professor of the Civil Engineering Master and Doctorate Programs in Civil production Management – UFF: Management System with emphasis in Sustainability (environmental, social responsibility and work safety). Worked as an Engineer and Process Manager in several national and international organizations such as the Brazilian Naval Arsenal in Rio de Janeiro and Shell Brasil S.A. President of Brazilian Association of Production Engineering (ABEPRO) (2006-2009). Permanent member of Social Responsibility Commission of Brazilian Oil Institute (Instituto Brasileiro de Petróleo-IBP).

Federal Fluminense University – LATEC – Rua Passo da Pátria 156 – CEP 24210-240 – Niterói RJ – Brazil – phone +55 21 2629 5618 – fax +55 21 2717 6390 – quelhas@latec.uff.br

Antonius Raghubansie is a senior lecturer at the University of Worcester, acting as team leader of the Green Academy programme. An expert in communications and branding, his main managerial experience is in the international FMCG beverages market, having held specialist senior marketing and general management roles working with brands such as Pepsi, Tropicana and El Dorado Rums.

Worcester Business School – University of Worcester – Worcester – United Kingdom – phone +44 (0)1905 542 297 – a.raghubansie@worc.ac.uk

Dr. **Mariano Ramirez Jr** is a Senior Lecturer in Industrial Design at the University of New South Wales (UNSW) in Sydney, Australia. Previously he was Academic Chair for Industrial Design at the De La Salle College of Saint Benilde, Associate Professor at the Mapúa Institute of Technology, and Senior Industrial Design Specialist at the Product Development and Design Centre of the Philippines. He trained in Sustainable Product Development at the Delft University of Technology and at the TNO

Institute of Industrial Technology, Netherlands. He is the Philippine Liaison for the O2 International Sustainable Designers Network, advisor to the Australian National Ecodesign Curriculum Development Committee, and convenor of the Oceania regional chapter of the LeNS Learning Network on Sustainability.

Faculty of Built Environment – University of New South Wales – Kensington – Sydney NSW 2052 – Australia – m.ramirez@unsw.edu.au

Dr. **Peter Rands** has previously worked in ICT in UK higher and further education sectors. Since 2009, he has been driving forward sustainability at CCCU and now leads the CCCU Green Academy team, which has resulted in the 'Futures Initiative'.

Sustainability Development Manager – Canterbury Christ Church University – North Holmes Road – Canterbury – Kent CT1 1QU – United Kingdom – phone +44 (0)1227 782648 – Peter.rands @canterbury.ac.uk

Dennis Redeker works as a project manager at the Maastricht University Green Office (Maastricht, The Netherlands). His responsibilities include the development of new educational activities on sustainable development, i.e. creating a 'green' curriculum, as well as the managing the inventory of ongoing courses and projects in Education for Sustainable Development. A student at University College Maastricht, he is about to complete his Bachelor's degree in the Social Sciences with a focus on International Relations and sustainable development.

Maastricht University Green Office – Sint Servaasklooster 32 – 6211 TE Maastricht – The Netherlands – greenoffice@maastrichtuniversity.nl *(corresponding author contact)*

Associate Professor **Kimberly D. Reiter** (BA, MA, PhD) is Interim Chair of the Department of History at Stetson University in Florida, a National Councilor for the Council on Undergraduate Research, and current President of the Interdisciplinary Environmental Association.

Department of History – Stetson University – 421 North Woodland Blvd – DeLand FL 32723 – USA – kreiter@stetson.edu

Associate Professor **Michael A. Reiter** (BSc, MS, PhD) is Graduate Chair and Director of the B.J. Moore Center for Integrated Environmental Science at Bethune-Cookman University, past President of the Interdisciplinary Environmental Association, and Associate Editor for the international journal Interdisciplinary Environmental Review.

Department of Integrated Environmental Science – Bethune-Cookman University – 640 Dr. Mary McLeod Bethune Blvd – Daytona Beach FL 32724 – USA – reiterm@cookman.edu

Guido Leonardo Olave Reyes Senior Graphic Designer, Universidad de Antofagasta Chile; Master's in Designing and Multimedia Communication, Instituto TRACOR Spain; digital means qualification (FAU), Universidad de Chile; Joomla Technology qualification (FACSO), Universidad de Chile; direction and management of web projects. Currently Professor in undergraduate digital imaging, FEN, Universidad de Chile.

Universidad de Chile – School of Economics and Business – Diagonal Paraguay 257 – Santiago 56-2-9783936 – Chile – guido@fen.uchile.cl

Dr. **Andrea Guadalupe Zavala Reyna** received her doctoral degree from the Autonomous University of Baja California. She is the project manager of the Institutional Sustainability Management System, which is certified on ISO 14001. She has been working in collaboration with small and medium-sized enterprises to create awareness of environmental risks. Dr. Zavala teaches environmental impact courses in the Civil Engineering Department.

Engineering College – University of Sonora – Mexico – azavala@dicym.uson.mx

Dr. **Alastair Robertson** is responsible for overseeing all Higher Education Academy work in Scotland. In his role as the HE Academy National Co-ordinator for Scotland, Alastair is the principal conduit for all aspects of HEA work in Scotland. He also has a UK-wide role for managing the HEA's work on policy development. His background is as a scientific researcher, and he joined QAA Scotland in 2003, co-managing support for the QAA Enhancement Themes. Alastair leads the HEA work and strategy on ESD.

Universities Scotland – 53 Hanover Street – Edinburgh EH2 2PJ – United Kingdom – phone +44 (0)131 225 0716 – Alastair.robertson@heacademy.ac.uk

Dr. **Zoe Robinson** is Director of ESD and Senior Lecturer in Environmental Science at Keele University. She leads the Keele Green Academy team in addition to acting as Programme Director for the BSc Environment and Sustainability programmes. Her research interests include ESD; climatic impacts on water resources in glacial environments; fuel poverty intervention strategies; and energy literacy and energy knowledge networks in communities.

School of Physical and Geographical Sciences – William Smith Building – Keele University – Staffordshire ST5 5BG – United Kingdom – phone +44 (0)1782 734 303 – z.p.robinson@esci.keele.ac.uk

Ingrid Roche. Researcher, Professor and Executive Director of the Urbanism Institute of the Faculty of Architecture (UdelaR). She is a lecturer and member of the Academic Committee of the ICM Master's program. She has participated in land planning programs related to environmental issues. She was advisor of Urban and Territorial Planning Program of Montevideo Municipality.

Centro Interdisciplinario para el Manejo Costero Integrado del Cono Sur – Universidad de la República – Montevideo (UdelaR) – Uruguay – ingridroc@gmail.com

Matthew Rockall teaches full-time in the British Columbia Institute of Technology Communications Department, in particular teaching Business Communications in the School of Business Marketing Management Program, but also teaching in the Fish, Wildlife and Recreation and Technical Writing Programs. Recent completion of a Master's in Environmental Education and Communication from Royal Roads University supports a long-term goal of taking part in the "greening" of BCIT, collaborating in widespread incorporation of sustainability in the curriculum.

British Columbia Institute of Technology – 3700 Willingdon Avenue – Burnaby BC VON1V4 – Canada – matthew_rockall@bcit.ca

Lucy Rodina is a Master's student at the Institute for Resources, Environment and Sustainability, UBC. She is also affiliated with the Program on Water Governance, and the Environment and Development: Gender, Equity and Sustainability research group (EDGES) at UBC. Her research focuses on issues of the Human Right to Water and water governance in Cape Town, South Africa. Lucy holds a Bachelor's degree in International Relations from UBC. She is a founding member of IDRN and a current steering committee member.

International Development Research Network – Liu Institute for Global Issues – University of British Columbia – 6476 NW Marine Dr. – Vancouver BC V6T 1Z2; – Institute for Resources – Environment and Sustainability – University of British Columbia – 2202 Main Mall – Vancouver BC V6T 1Z4 – Canada – idrn.ubc@ gmail.com

Lucia Rodríguez serves as the Director of the Global Master's in Development Practice (MDP) Secretariat, at The Earth Institute, Columbia University. Her research interests are in the development of sustainable development degree programs and the internationalization of academic institutions.

The Earth Institute – Columbia University – 475 Riverside Drive – Suite 239R – New York NY 10115 – USA – phone +1-212-280-2793 – lrodriguez@ei.columbia.edu

J. Terry Rolfe is Principal of Integrity Research and Communications in Surrey, British Columbia, Canada.

Integrity Research and Communications – 3145 Semiahmoo Trail – Surrey BC V4P1H4 – Canada – tsrolfe@aol.com

Gabriela Toledo Román, Master of Economics at the University of Chile. She conducts research in the economics of education, particularly with regard to teachers, preschool and and education funding for several research centers and universities. She is a teacher and researcher in sustainability issues at the University of Chile, and also part of the sustainable campus committee of the Faculty of Business and Economics, University of Chile.

Facultad de Economía y Negocios – Universidad de Chile – Colectivo Viento Sur – Chile – gtoledo@fen.uchile.cl

Felix Roosen-Runge studied physics, mathematics and philosophy at the University of Tübingen, Germany, and the ETH Zürich, Switzerland. At present, he is a PhD-candidate in biophysics in Tübingen. In Tübingen, he contributed to the development of an SD curriculum, conceptually as well as practically as a lecturer of seminars.

University Tübingen – Institute for Applied Physics – Auf der Morgenstelle 10 – 72076 Tübingen – Germany – f.roosen-runge@student.uni-tuebingen.de

Lars Rydén is a Professor at Uppsala University. In 1991 he initiated and was the first director of the Baltic University Programme (BUP) – a network of universities in the Baltic Sea region. The major concerns of L. Rydén are issues of sustainable development, environmental protection, and democracy in the Baltic Sea region. He has written and edited a series of books on environmental science, sustainability science and regional development. In cooperation with universities and municipalities he has initiated several projects on sustainable urban development of cities and towns in the Baltic Sea region. Prof. Rydén extensively cooperates with social partners and promotes environmentally friendly lifestyles.

Baltic University Programme – Uppsala Centre for Sustainable Development – Uppsala University – Villavägen 16 – SE-752 36 Uppsala – Sweden – lars.ryden@csduppsala.uu.se

Serenella Sala, PhD, graduated in Environmental Science and obtained a PhD in Ecology in 2005 from the University of Milano Bicocca. Since 2005, she has been the coordinator of the Research Unit on Sustainable Development (GRISS) headed by Prof. Demetrio Pitea at the University of Milano Bicocca. Her research activities are related to methods and tools for sustainable development and risk assessment, mainstreaming sustainability science into applied research through the development of sustainability indicators. She was the coordinator of several environmental projects both in the private and public sector and she has developed and implemented several interdisciplinary projects promoting public and private partnerships related to sustainable production and consumption patterns, with a specific focus on tourism activities.

GRISS – Gruppo di Ricerca sullo Sviluppo Sostenibile – Department of Environmental Science and Technology – University of Milano – Bicocca – Piazza della Scienza 1 – 20126 Milano – Italy – serenella.sala@unimib.it

Kaisu Sammalisto (PhD, MBA) is senior lecturer in industrial environmental economics and holds the position of Vice President for Sustainable Development at the University of Gävle in Sweden. Her research focus is on management systems both in industry, universities and in supply chains as a way to enhance sustainability.

University of Gävle – Faculty of Engineering and Sustainable Development – SE-80176 GÄVLE – Sweden

José Ignacio Sánchez is a Mechanical Engineering graduate student at Universidad de Chile, where he is developing one of the first functional designs of Stirling engines, a topic focused on renewable energies. In 2011, he conducted research on optical characterization of specular concentrating surfaces at the Spanish National Renewable Energies Center (CENER). He is one of the initiatiors of *Taller Solar*, where he has participated in several empowering interventions within communities, by building and using solar energy low-cost devices. As part of the university's organized group Oikos, he was part of the initiation of a campus recycling program, a minor elective on renewable energies and many other activities aiming to promote the diffusion of environmental issues.

Departamento de Ingeniería Mecánica – Facultad de Ciencias Físicas y Matemáticas – Universidad de Chile – Chile

Pablo Sánchez is a PhD candidate in Geology at the University of Chile (UCh) where he is working at the Andean Geothermal Center (CEGA). His research interests include geothermal systems in the Andes and the development of the geothermal energy as a renewable and sustainable source of energy for Chile. During his Geology studies (BSc) at the UCh, he participated in political and environmental organizations. He is a founding member of the Oikos student association and the Taller Solar, organizations devoted to achieving a sustainable campus and to encouraging the use of low-cost solar energy technologies, respectively. Currently he is part of the Sustainable Campus Commission at UCh and is a member of the organization Colectivo Viento Sur.

Departamento de Geología – Facultad de Ciencias Físicas y Matemáticas – Universidad de Chile – Colectivo Viento Sur – Chile

Tamara Savelyeva, PhD, is a sustainability scholar at the Faculty of Education, the University of Hong Kong, which she joined in 2010 after receiving her training in natural resources and sustainability education at Cornell and Virginia Tech universities, USA. Her current research focuses on issues of sustainability in higher education and curriculum development.

Faculty of Education, The University of Hong Kong – Hong Kong – China – Tamara SAVELYEVA – 110 Hui Oi Chow Science Building – Faculty of Education, The University of Hong Kong – Pok Fu Lam Rd – Hong Kong SAR – phone (852) 9337-9136 – tsavelyeva@gmail.com

Veronika Schulte. Dipl. Biol., Biologist at the Hamburg University of Applied Science in Germany in the Research and Transfer Centre “Applications of Life Sciences”. Coordinator of several EU-funded projects on technology transfer and development cooperation in the renewable energy sector in Latin America as well as African, Pacific and Caribbean small island developing states.

Hamburg University of Applied Sciences – Research and Transfer Centre “Applications of Life Sciences” – Lohbruegger Kirchstraße 65 – 21033 Hamburg – Germany – phone +49-40-42875-6362 – fax +49-40-42875-6079 – veronika.schulte@haw-hamburg.de

Christoph Michael Schulze is a third-year student at University College Maastricht, about to finish his Bachelor’s degree. His academic focus is International Relations and Political Science. Vice-chair of the student party NoVUM and member of the UCM sustainability committee, he is actively involved in matters of sustainable development.

University College Maastricht – Maastricht University – Zwingelput 4 – 6211 KH Maastricht – The Netherlands

Dr. **Stephen Scoffham** is a Principal Lecturer in the Faculty of Education, Canterbury Christ Church University (CCCU) and Honorary Publications Officer for the UK Geographical Association. His research interests centre on education in the 21st century and include creativity, the environment, intercultural understanding and the global dimension. An established author/consultant for school atlases, he currently has a leading academic role in developing sustainability and futures perspectives in the curriculum at CCCU.

Faculty of Education – Canterbury Christ Church University – North Holmes Road – Canterbury – Kent CT1 1QU – United Kingdom – phone +44 1227 782367 – stephen.scoffham @canterbury.ac.uk.

John Sedgwick is Professor of Film Economics and Chairman and founding Director of CIBS. Sedgwick researches into the economics and economic history of film and is particularly interested in risk analysis, having published articles in the journals Business History, Economic History Review, Explorations in Economic History, Historical Journal of Film, Radio and Television, Journal of Cultural Economics, and Journal of Economic History, as well as numerous book chapters. He has also published a monograph on the film industry in Britain during the 1930s, Popular Filmgoing in 1930s Britain: a Choice of Pleasures’ (Exeter, Exeter University Press, 2000), and with Pokorny edited an anthology of papers, ‘An Economic History of Film’, (London, Routledge 2005). Sedgwick was a Leverhulme Research fellow in 2000-01, a Sir Robert Menzies Research Fellow in 2006 and a RMIT/AFTRS Visiting Research Fellow in 2007.

Centre for International Business and Sustainability – London Met Business School – London Met University – 84 Moorgate – London EC2M 6SQ; – ABM Analytics Ltd – Suite 17 125 – 145-157 St. John Street – London EC1V 4PW – UK

Dr. **Amirul Shah Md Shahbudin** is a senior lecturer and the chairperson of the Accounting Department at the School of Management, Universiti Sains Malaysia (USM). His main research interests include management control, performance management and sustainability.

School of Management – Universiti Sains Malaysia – Malaysia – amirulshah@usm.my

Kerry Shephard (PhD) is Professor of Higher Education Development in the Higher Education Development Centre at the University of Otago, New Zealand. His research interests include eLearning, educational development and the public understanding of science. Kerry has been working with a multidisciplinary team from the University of Otago to benchmark the sustainability attributes of university students and to develop processes to monitor how they change.

Higher Education Development Centre, HEDC – University of Otago – Dunedin – New Zealand – kerry.shephard@otago.ac.nz *(corresponding author contact)*

Chris Shiel has championed global perspectives and education for sustainable development across higher education for over ten years. She is an Associate Professor at Bournemouth University and Director of the Centre for Global Perspectives. She has led the global perspectives agenda at Bournemouth since 2000, and in 2005 was awarded a Higher Education Leadership Foundation Fellowship for her work. Her approach is based on the development of global citizens who understand the need for sustainable development and are better prepared for global employability. She is a member of HEFCE's Sustainable Development Strategy group and a Board Member of Think Global. Her research interests include leadership behaviour for sustainable development, change management and behaviour change for SD.

Centre for Global Perspectives – Bournemouth University – Poole – Dorset BH12 5BB – UK – cshiel@bournemouth.ac.uk

Prof. **Mikhail Shilin** – doctor in biology and geography, Professor at the Russian State Hydrometeorological University (since 1990). Has been head of the Hydrobiology Lab of the Baltic Floating University cruises for the last 15 years and is editor of the Baltic Floating University Research Bulletin.

Russian State Hydrometeorological University – Malookhtinsky pr 98 – St. Petersburg 195196 – Russia shilin@rshu.ru

Michael Shriberg, PhD is Education Director at the Graham Environmental Sustainability Institute and Lecturer in the Program in the Environment at the University of Michigan. In addition to teaching courses such as Sustainability & the Campus and Sustainability Leadership Development, he leads the Graham Undergraduate Sustainability Scholars Program, the Graham Sustainability Doctoral Fellow's Program, and efforts to integrate sustainability across the curriculum at the University of Michigan. Dr. Shriberg is the North American editor of the International Journal of Sustainability in Higher Education. He recently returned to academia after serving as Director of Environment and Policy Director at Michigan Ecology Center, leading policy efforts on Great Lakes protection, climate and energy, as well as environmental health and toxics. Previously, he served as Program Director, Assistant Professor of Environmental Studies and Interim Director of the Rachel Carson Institute at Chatham University in Pittsburgh, PA, as well as an adjunct lecturer at Carnegie Mellon University. Dr. Shriberg earned a Master's and a PhD in Resource Policy & Behavior from the University of Michigan's School of Natural Resources and Environment and a BSc in Biology & Society from Cornell University. His research and practice focuses on organizational change and sustainability leadership, with an emphasis on higher education institutions.

Graham Environmental Sustainability Institute & Program in the Environment – University of Michigan – 625 E. Liberty Street – Suite 300 – Ann Arbor MI 48104 – USA – phone +1-734-647-6227 – mshriber@umich.edu

Dr. **Anne Sibbel**, PhD; M. Env. Sc.; Grad. Dip Ed (Applied Science); Bach Ed. (Physics); Dip. Frontline Management; Bach Appl. Sc. (Food Science & Technology) Currently at RMIT University, Anne has worked in sciences in the higher education sector over several decades, particularly in the areas of public health and nutrition education, food and environmental sciences. She has a strong interest in building the capacity of higher education to contribute to the challenge of achieving sustainability.

School of Applied Sciences – RMIT University – Building 14.5.9 GPO – Box 2476V – Melbourne – Victoria – Australia – 3001 – anne.sibbel@rmit.edu.au

Daniela Rodrigues da Silva is a Professor and researcher at the Federal Institute of Education, Science and Technology of Rio Grande do Sul (IFRS) in the city of Canoas, Brazil; she is also a PhD candidate on the postgraduate program in Science Education at UFRGS, in the city of Porto Alegre, Brazil, and a research member of the Technological Innovation and Education Group at the IFRS. Her research interests encompass integration and science education. Currently, she is coordinating a research project entitled "Problem solving as a strategy for learning chemistry: a case study at the IFRS – Campus Canoas" and is a collaborator in the research project "Sustainability in management education" as well as in the extension project "Metamorphosis – promoting change from environmental education" at the IFRS. She holds a Master's degree from the postgraduate program in Science Education at UFRGS and has experience in primary, secondary and higher education.

Federal Institute of Education – Science and Technology of Rio Grande do Sul (IFRS) – Dr. Maria Zélia Carneiro de Figueiredo Street 870-A – Canoas/RS – Brazil – phone +55 51 34158200 – daniela.silva@canoas.ifrs.edu.br

Dr. **Mihaela Sima** has been a researcher at the Institute of Geography since 2001. Her main interests are in the field of environmental sustainability and quality with a special emphasis on water quality, land degradation, natural and technological hazards and GIS. Dr. Sima has been involved in research projects dealing with the evaluation of the long-term impact of mining activities on rivers with an emphasis on heavy metal pollution and environmental quality. Currently, she coordinates the IGAR team in the FP7 project ECLISE.

Romanian Academy – Institute of Geography – 12 Dimitrie Racovita St. – Sector 2 – Bucharest 023993 – Romania – simamik@yahoo.com *(corresponding author contact)*

Gerald Singh is a PhD student at the Institute for Resources, Environment and Sustainability (IRES), UBC. He is affiliated with the Connecting Human and Natural Systems Lab at UBC. His research interests include conservation biology, community ecology of top predators, cumulative impacts, and zooarchaeology Gerald holds a Bachelor of Science in Conservation Biology from the University of Alberta and a Master of Science from IRES. He is an active IDRN member.

International Development Research Network – Liu Institute for Global Issues – University of British Columbia – 6476 NW Marine Dr. – Vancouver BC V6T 1Z2; – Institute for Resources – Environment and Sustainability – University of British Columbia – 2202 Main Mall – Vancouver BC – Canada V6T 1Z4 – idrn.ubc@gmail.com

Arja Sinkko is working at Kymenlaakso University of Applied Sciences as Development Engineer and lecturer in Technology and Transport Department. She has her MSc. in Industrial Engineering and Management from Lappeenranta University of Technology, Finland in 2006. Her main responsibilities consist of the university's sustainability activities and university – business co-operation development as well as teaching (quality, sustainability and occupational safety issues). She has been a member of research groups creating sustainable development indicators for secondary and tertiary education. She has written and published (in Finnish) several articles focused on sustainable development and quality assurance in higher education.

Kymenlaakso University of Applied Sciences (KyUAS) – POB 9 – FI-48401 Kotka – Finland – phone +358-(0)44-702 8219 – arja.sinkko@kyamk.fi

Sheila Skeaff (PhD) is Senior Lecturer in the Department of Human Nutrition, University of Otago, New Zealand. Sheila researches the iodine status of at-risk groups, childhood eating patterns and issues affecting women's health.

Department of Zoology – University of Otago – Dunedin – New Zealand

Liz Slooten (PhD) is Associate Professor in the Department of Zoology, University of Otago, New Zealand. Liz researches the assessment of risk and uncertainty using resource management models, with particular interests in marine mammals.

Department of Human Nutrition – University of Otago – Dunedin – New Zealand

Prof. **Richard Smardon** (BSc, MLA, PhD) is the Coordinator of the undergraduate Environmental Studies program at the State University of New York's College of Environmental Science and Forestry, and has reviewed some ten ES programs in the last four years as an external reviewer.

Department of Environmental Studies – SUNY College of Environmental Science and Forestry – 1 Forestry Drive – Syracuse NY 13210 – USA – rsmardon@esf.edu

Sarah Speight is Deputy Head of the School of Education, Chair of the University's Grand Challenge group on Education for Sustainable Development and leader of Nottingham's Green Academy team. Her remit is to support colleagues in integrating sustainability into curricula, teaching and learning across all disciplinary areas.

School of Education – The University of Nottingham – Jubilee Campus – Nottingham NG8 1BB – United Kingdom – Sarah.speight@nottingham.ac.uk

Jurgis Kazimieras Staniškis – Director of the Institute of Environmental Engineering (APINI) at Kaunas University of Technology. Since 1991 he has been the manager and/or the main expert on more than 80 international projects on sustainable industrial development, cleaner production, environmental management, product eco-design, integrated waste management and sustainable development education. Prof. Staniškis J.K. is the member of the UNEP/UNIDO global network "Resource Efficient and Cleaner Production (RECP-Net)" Executive Committee for Central and Eastern Europe and Middle Asia. He is Chairman of the Sustainable Development Committee for the Tripartite Council of the Lithuanian Republic.

Institute of Environmental Engineering (APINI) – Kaunas University of Technology – Kaunas – Lithuania – Jurgis.Staniskis@ktu.lt

William M. Stigliani received his PhD in Physical Chemistry from Princeton University. He is a Full Professor of Chemistry at the University of Northern Iowa (UNI), and was Director of UNI's Center for Energy & Environmental Education for 14 years. He is currently Senior Adviser to the Provost on sustainability curriculum, and develops and teaches new courses in sustainability. His interest in sustainability stems from tenures as Senior Research Scholar in the Environment Program at the International Institute for Applied Systems Analysis, near Vienna, Austria, and the Environmental Studies Board of the National Research Council, US National Academy of Sciences in Washington. His research interest is in the area of interactions of physical and chemical factors governing accumulation and mobilization of toxic materials in soils and sediments; he has applied his model to climate change/land use changes to help discern the potential impact of global change on mobilization of toxins in soils and waters. He was appointed to the Governor's *Iowa Climate Change Advisory Council* (2007-2010) that prepared a report outlining a sustainable energy pathway in Iowa for 2050. He has co-authored four university textbooks on environmental chemistry.

University of Northern Iowa – Center for Energy & Environmental Education – 111C – Cedar Falls 50614-0293 – Iowa – USA – phone +1-319) 273-7150 – William.stigliani@uni.edu

Alline Storni, PhD, graduated in Biology in 1999 and obtained a PhD in Ecology in 2004 at the Universidade do Estado do Rio de Janeiro, Brazil. From 2005 until 2007 she worked at Mamiraua Development Sustainable Reserve (Amazon, Brazil), measuring the impact on fauna of the presence of eco-tourists. In 2008 she came to Italy to do post-doctorate research in ecology and since 2010 she has worked at the University of Milano Bicocca at the Research Unit on Sustainable Development (GRISS), working on the development of methodologies and indicators to assess sustainability in tourism and on research projects for sustainable consumption.

GRISS – Gruppo di Ricerca sullo Sviluppo Sostenibile – Department of Environmental Science and Technology – University of Milano – Bicocca – Piazza della Scienza 1 – 20126 Milano – Italy – alline.storni@unimib.it

Mick Strack (PhD) is a lecturer in the New Zealand National School of Surveying. Mick's surveying interests are in land law, Māori land and aboriginal title, environmental ethics, urban design and planning, and our relationship with land, property and resources.

School of Surveying – University of Otago – Dunedin – New Zealand

Dr. **Alexey Suzyumov** – PhD in geology; 1969-1982: senior scientist at the Institute of Oceanology, Academy of Sciences of the USSR. 1982–2005: program specialist in the science sector of UNESCO and its Intergovernmental Oceanographic Commission (IOC), responsible for capacity-building projects including the "floating university – training-through-research" program with several branches in the Atlantic Ocean, Baltic and Caspian Seas, and in the Indo-Pacific Region. 2006–present: consultant for capacity-building program activities.

OC UNESCO – Paris – France – e-mail: suzyumov@gmail.com

John Swaddle is an evolutionary behavioral ecologist and professor of Biology at The College of William and Mary, and former Director of their Environmental Science and Policy program. Professor Swaddle was also a founding member of William and Mary's Committee on Sustainability, which addresses the environmental performance and stewardship of The College. Professor Swaddle has published more than 60 peer-reviewed articles and one book, for Oxford University Press in their prestigious *Ecology and Evolution* series. He has also been the recipient of a Royal Society of London University Research Fellowship, a National Science Foundation CAREER award, selected as the Most Outstanding New Investigator by the Association for the Study of Animal Behavior, and was awarded the Young Investigator Prize by the American Society of Naturalists. Professor Swaddle teaches courses related to human change of the environment and evolutionary biology, and he also coaches rugby at William and Mary.

Environmental Science and Policy Program – The College of William and Mary – Williamsburg – Virginia – USA – jpswad@wm.edu

Valentin Tappeser is a recent graduate of University College Maastricht with a Bachelor of Arts in Liberal Arts and Sciences in 2012. His research interest lies in interdisciplinary studies linked to humanities and social sciences. In particular Mr. Tappeser is interested in sustainability studies, sociology and political philosophy. Alongside his studies, Mr. Tappeser involves himself in a plethora of projects dealing with local student initiatives, art and music projects. He was one of the founding members of the Maastricht University Green Office before he was able to increase his knowledge of applied sustainability governance in Santa Cruz, where he spent a semester as part of his studies. In the future, Mr. Tappeser will follow the Global Studies Programme, a social science research Master's program at the Universities of Freiburg, Buenos Aires and Delhi, where he intends to do further research on grass roots sustainability initiatives around the globe.

Maastricht University Green Office – Sint Servaaskloster 32 – 6211TE Maastricht – The Netherlands – email: valentin@tappeser.com *(corresponding author contact)*

Dawit Tesfamichael is a PhD student at the Fisheries Centre, UBC. His research focuses on methods of assessment, evaluation and management of fishery resources in terms of ecological, social and economic sustainability attributes. He has done extensive research and published using ecosystem modeling, rapid evaluation, interview based and long time series analysis. He is an active IDRN member.

International Development Research Network – Liu Institute for Global Issues – University of British Columbia – 6476 NW Marine Dr. – Vancouver BC V6T 1Z2; – Fisheries Centre – University of British Columbia – 2202 Main Mall – Vancouver BC – Canada V6T 1Z4 – idrn.ubc@gmail.com

Dr. **Tatiana Teslenko** (Kandydat of Philological Sciences, 1989, PhD, 2000) lives in Vancouver, Canada and holds the position of Senior Instructor at the Centre for Professional Skills Development in the Faculty of Applied Science, the University of British Columbia. Her research interests include education for sustainable development, engineering communication, and community service learning. Her recent publications include articles and books on composition theory, genre studies, and a textbook for engineering students: *Rhetoric and Ideology of Genre* (Hampton, 2002), *Feminist Utopian Novels of the 70s* (Routledge, 2003), *Fundamental Competencies for Engineers* (Oxford University Press, 2006). She teaches professional communication courses and has designed and launched several programs for international students (such as ASSIST UBC and the Graduate Teaching Assistants training program). She is the founder and inaugural Director of the Centre for Professional Skills Development in the Faculty of Applied Science at the University of British Columbia.

The University of British Columbia – Vancouver – Canada – tteslenko@gmail.com

Jennifer "Jenna" Thompson is a LEED accredited Adjunct Professor of Architecture at the University of Memphis. She earned a Master of Architecture Degree from the University of Memphis and both a Bachelor of Architecture degree and a Bachelor of Interior Architecture degree from Auburn University. Jenna is currently pursuing a Doctor of Education in Higher and Adult Education at the University of Memphis.While at Auburn, Jenna encapsulated sustainable design in every project, the most notable of which is her extensive research of biodegradable plastics in interior design applications. This research yielded Jenna the distinguished Auburn University Undergraduate Research Fellowship Grant, which is awarded to only 20 students per year for the entire University, and the 2006 AIAS Student Research Honors Award. Her passion for creating awareness and action to social issues of environmental stewardship permeates her philosophy of design, adult education and teaching methodology. Her sustainable stewardship extends beyond her personal studies into other green initiatives. She currently sits on the Sustainable Shelby Advisory Committee for the Greater Memphis region, is the appointed Sustainable Coordinator for the Department of Architecture, University of Memphis, and represents various construction projects as a sustainable design consultant.

Adjunct Professor of Architecture – MArch – LEED BD+C – ID+C – University of Memphis – Department of Architecture – 404 Jones Hall – Memphis 38152 – Tennessee – USA – phone +1.901.678.2724 – Jlthmps5@memphis.edu

Natalia Tihomirova (Dr. and Professor in Economic Sciences) is Rector of Moscow State University of Economics, Statistics and Informatics (MESI). Her scientific interests lie in the areas of knowledge management, quality management and innovation management. She is an expert in e-learning. As a rector, Natalia Tihomirova conducts a large amount of administrative work.

MESI – 7 – Nezhinskaya Str. – Moscow 119501 – Russia – NTihomirova@rector.mesi.ru

Felix Tilmann Vahle is currently finishing his Bachelor's degree at University College Maastricht with a major in Sustainability Studies, Economics and International Relations. Being a volunteer in the operations portfolio steering group of the Maastricht University Green Office and president of the UCM sustainability committee, he is preparing for a career in the field of sustainable development.

University College Maastricht – Maastricht University – Zwingelput 4 – 6211 KH Maastricht – The Netherlands

Dr. **Luis Vargas** holds a Master's degree in Electrical Engineering from the University of Chile and a PhD Degree from the University of Waterloo. He is now a faculty member in the Electrical Engineering Department at the University of Chile, Santiago, and is a member of the Expert Panel for Chilean Electricity Law. His main research areas include: Climate Change Impact on Power Systems, Electrical Energy Market Analysis, and New and Renewable Energy. In these fields Dr. Vargas has published over 50 technical papers in international journals, chapters of books, and conferences. For more information or to contact Dr. Vargas, see: http://146.83.6.25/lvargas/index.htm.

Department of Electrical Engineering – University of Chile – 2007 Av. Tupper – Santiago-Centro – Santiago 837 0451 – Chile – phone 56-2-978-4203 – lvargasd@ing.uchile.cl

Dr. **Nora Elba Munguía Vega** holds a doctoral degree from the University of Massachusetts Lowell in the field of Cleaner Production and Pollution Prevention. She is a professor/researcher in sustainability issues at the University of Sonora, Mexico. She has conducted research studies on sustainability and education issues as well as on Occupational Health and Safety, Cleaner Production and Pollution Prevention fields. Her most recent work involves promoting toxic use reduction strategies in the Sonoran industry.

Engineering College – University of Sonora – Mexico – nmunguia@industrial.uson.mx

Dr. **Fayyaz Vellani** is Head of the Department of Graduate Studies at the Institute of Ismaili Studies in London. Prior to joining the Institute in 2010, Dr. Vellani served as Associate Director of the Critical Writing Program and Faculty-in-Residence at the University of Pennsylvania for four years. Dr. Vellani has a PhD in Geography from the University of London, an MA in Higher Education Administration from New York University and a Bachelor of Environmental Studies from the University of Waterloo. He has also served as Associate Registrar of Yale Law School and Director of Admissions and Financial Aid at the University of Toronto's Faculty of Law.

Department of Graduate Studies – Institute of Ismaili Studies – 210 Euston Road – London NW1 2DA – United Kingdom – phone. +44 (0)207 756 2700 – fvellani@iis.ac.uk

Cláudio Senna Venzke, undergraduate degree in Business Administration at Federal University of Rio Grande do Sul (1999) and a Master's degree in Business Administration at the Federal University of Rio Grande do Sul (2002). He is currently a PhD student at the PPGA Federal University of Rio Grande do Sul and a professor at the University of Vale do Rio dos Sinos – UNISINOS. Experience in administration, with emphasis on Social and Environmental Management, acting on the following themes: environmental management, corporate strategy, sustainable development, waste management and ecodesign.

Universidade do Vale do Rio dos Sinos – UNISINOS – Av. Unisinos 950 – São Leopoldo – Brazil – senna@unisinos.br

Dr. **Vlasios Voudouris** is the Director of the Centre for International Business and Sustainability (CIBS) at LondonMet Business School and the CEO of ABM Analytics Ltd – a high-level research consultancy in the energy/fuels markets. Voudouris has a wealth of experience in evidence-based policy, quantitative analysis and risk-adjusted investment processes. He was a Senior Advisor for the NEEG of Defra – Central UK Government. He also wrote a number of evidence-based reports for the COBRA meetings chaired by the UK Prime Minister. Voudouris also contributed to the House of Commons Energy and Climate Change Select Committee inquiry 'The UK's Energy Supply: Security or Independence?' Voudouris is on the Advisory Board of Strategic Asia where is building an effective portfolio of clients, policy papers and strategies. He also contributes to the Pacific Energy Summmit following an invitation by the US National Bureau of Asian Research.

Centre for International Business and Sustainability – London Met Business School – London Met University – 84 Moorgate – London EC2M 6SQ ; – ABM Analytics Ltd – Suite 17 125 – 145-157 St. John Street – London EC1V 4PW – UK – v.voudouris@londonmet.ac.uk *(corresponding author contacts)*

Timo Vuorisalo is senior lecturer of environmental science at the University of Turku, Finland, since 1992. He received his PhD in ecological zoology is from the University of Turku in 1989. In 1989-1990 he was a postdoctoral fellow at Indiana University, Bloomington, and before that a research assistant of the Academy in Finland. He has written several papers on evolutionary ecology, environmental history, and urban ecology.

University of Turku – Department of Biology – FI-20014 TURKU – Finland

Joseph P. Weakland is Editorial Associate at the Center for Environmental and Sustainability Education and a doctoral student in English at University of Florida. His various research and teaching interests include media representations of technology, exploring consumer culture in the composition classroom, posthumanism, environmentalism, critical theory, science fiction, and cyberpunk. Alongside his graduate studies, he assisted Peter Blaze Corcoran on the book project, *Young People, Education, and Sustainable Development: Exploring Principles, Perspectives, and Praxis* (Wageningen Academic Publishers, 2009). He is deeply interested in the intersection of cultural texts, technology, and environmental issues. In 2010, Joseph was named Florida Gulf Coast University's Graduate Student of the Year.

Center for Environmental and Sustainability Education – Florida Gulf Coast University – 10501 FGCU Boulevard – Fort Myers 33965-6565 – Florida – USA

Christopher W. Wells is an associate professor of Environmental History in the Department of Environmental Studies at Macalester College. His book, *Car Country: An Environmental History*, is forthcoming from the University of Washington Press. He holds a Bachelor's degree from Williams College as well as a Master's and PhD in US History from the University of Wisconsin-Madison.

Environmental Studies Department – Macalester College – 1600 Grand Ave – St. Paul MN 55105 – USA – wells@macalester.edu

Menno de Lind van Wijngaarden teaches Business Administration, Corporate Social Responsibility and Sustainable Management at the HU University of Applied Sciences Utrecht. He is project leader in curriculum development in the field of Business Ethics and Sustainability.

HU University of Applied Sciences Utrecht – Padualaan 101 – 3584 CH Utrecht – Netherlands – menno.lind@hu.nl

Jennie Winter is an Educational Developer at Plymouth University and her main research interest is Education for Sustainable Development (ESD). Jennie has been involved in a number of research projects exploring student and staff experiences of ESD, including how the university campus mediates informal learning for sustainability. She is currently working on a project investigating energy literacy amongst university students. Jennie also teaches about sustainability topics in both Geography and Education at Plymouth University. Find out more about Jennie at www.plymouth.ac.uk/staff/j1winter

Pedagogic Research Institute and Observatory (PedRIO) – Teaching and Learning Directorate – Plymouth University – PL4 8AA – UK – jennie.winter@plymouth.ac.uk

Oscar Jerez Yañez, PhD in Psychology and Education and Diploma of Advanced Studies in Psychology and Education, Universidad de Granada, Spain. Master of Education at the Universidad de Chile. Bachelor of Philosophy from Pontificia Università Salesiana di Roma and Bachelor of Education from Pontificia Universidad Católica de Chile. Expert in Teaching and Learning in Higher Education, Vocational and Job Training. Teacher, researcher and consultant to several universities, government and companies in Chile and Ibero-America. Currently Coordinator of Innovation and Development for Teaching Excellence Network, Universidad de Chile.

Executive Director – Teaching and Learning Center – School of Economics and Business and Coordinator of Innovation and Development for Teaching Excellence Network – University of Chile – Chile – ojerez@uchile.cl

Youssef Ahmad Youssef, PhD, Professor of the School of Business at Humber Institute of Technology and Advanced Learning in Canada. Active researcher on the subject of business, sustainability and renewable energies since 2004. Co-supervisor to two major EuropeAid- funded joint European and Latin American projects: JELARE and REGSA between 2008 and 2012. Founder and president of the Federation of Canadian Brazilian Businesses (FCBB), a not-for-profit organization that aims to foster sustainable innovation and entrepreneurial initiatives between Canada and Brazil. His research interests include informal entrepreneurship in emerging economies, renewable energies, entrepreneurship and sustainability and knowledge management, fields on which he has published several books and journal articles over his academic career.

Humber Institute of Technology and Advanced learning – School of Business – 3199 Lake Shore Boulevard West – Room H215 – Toronto M8V 1K8 – Ontario – Canada – phone +1 647390-8508 – youssef.youssef@humber.ca

Catherine Zeman received her PhD in Preventive Medicine from the University of Iowa. She is an Associate Professor in the Health Division, School of HPELS, at the University of Northern Iowa (UNI), and teaches classes in epidemiology, human diseases, environmental health, and environmental and occupational health regulations. Dr. Zeman is also Director of UNI's Recycling and Reuse Technology Transfer Center, an applied research, service and outreach center that focuses on issues of sustainability and solid waste. She is a member of the American Indian Science and Engineering Society and an Affiliate of the National Association of Native American Studies. Her research interests include: nitrates in the environment and their impact on human health with focus on children's health; industrial ecological principles of precision manufacturing, international health issues, understanding the health issues of underserved populations, environmental sustainability, workplace health and safety and wellness; and documenting pollution prevention practices in worker health and safety. Dr. Zeman was a Fulbright Scholar to Romania on environmental health issues during the spring/summer of 2004.

University of Northern Iowa – 207 Wellness/Recreation Center – Division of Health – School of Health, Physical Education & Leisure Services – Cedar Falls 50614-0184 – Iowa – USA – phone +1-319 273-7090 – Catherine.zeman@uni.edu

Siarhei Zenchanka (Prof.) is Head of the Department of International Relations and Marketing at the Minsk Branch of MESI, and has a PhD in Applied Science. His research focus is in the areas of quality management and environmental management, where he is an auditor.

Minsk Branch of MESI – 127, Build. 2 – Mayakovskogo Str. – Minsk 220028 – Belarus – szenchenko@mfmesi.ru

Hisham Zerriffi is an Assistant Professor and the Ivan Head South/North Research Chair at the Liu Institute for Global Issues, UBC. He conducts research at the intersection of technology, energy, and the environment, with a particular focus on rural areas of the developing world. Much of his research focuses on institutional factors impacting the diffusion of new technology, determinants and patterns of household energy choice and the welfare implications of rural energy use. Prior to joining UBC, he was a Postdoctoral Fellow with the Program on Energy and Sustainable Development at Stanford University. He holds a PhD from Carnegie Mellon University in Engineering and Public Policy. Dr. Zerriffi is one of two IDRN faculty mentors.

International Development Research Network – Liu Institute for Global Issues – University of British Columbia – – 6476 NW Marine Dr. – Vancouver BC V6T 1Z2; – Liu Institute for Global Issues – University of British Columbia – 6476 NW Marine Dr – Vancouver BC V6T 1Z2 – Canada – idrn.ubc@gmail.com

Thematic Index

Umweltbildung, Umweltkommunikation und Nachhaltigkeit
Environmental Education, Communication and Sustainability

Herausgegeben von / Edited by Walter Leal Filho

Band 1 Walter Leal Filho (Hrsg.): Umweltschutz und Nachhaltigkeit an Hochschulen. Konzepte-Umsetzung. 1998.

Band 2 Walter Leal Filho / Farrukh Tahir (eds.): Distance Education and Environmental Education. 1998.

Band 3 Walter Leal Filho (ed.): Environmental Engineering: International Perspectives. 1998.

Band 4 Walter Leal Filho / Mauri Ahlberg (eds.): Environmental Education for Sustainability: Good Environment, Good Life. 1998.

Band 5 Walter Leal Filho (ed.): Sustainability and University Life. 1999. 2nd, revised ed. 2000.

Band 6 Wout van den Bor / Peter Holen / Arjen Wals / Walter Leal Filho (eds.): Integration Concepts of Sustainability into Education for Agriculture and Rural Development. 2000.

Band 7 Manfred Oepen / Winfried Hamacher (Eds.): Communicating the Environment. Environmental Communication for Sustainable Development. 2000.

Band 8 Walter Leal Filho (ed.): Communicating Sustainability. 2000.

Band 9 Walter Leal Filho (ed.): Environmental Careers, Environmental Employment and Environmental Training. International Approaches and Contexts. 2001.

Band 10 Rolf Jucker: Our Common Illiteracy. Education as if the Earth and People Mattered. 2002.

Band 11 Walter Leal Filho (ed.): Teaching Sustainability at Universities. Towards curriculum greening. 2002.

Band 12 Walter Leal Filho (ed.): International Experiences on Sustainability. 2002.

Band 13 Lars Wohlers (Hrsg.): Methoden informeller Umweltbildung. 2003.

Band 14 Ulisses Azeiteiro / Fernando Conçalves / Walter Leal Filho / Fernando Morgado / Mário Pereira (eds.): World Trends in Environmental Education. 2004.

Band 15 Walter Leal Filho / Arnolds Ubelis (eds.): Integrative approaches towards sustainability in the Baltic Sea Region. 2004.

Band 16 Walter Leal Filho / Michael Littledyke (eds.): International Perspectives in Environmental Education. 2004.

Band 17 Daniela Krumland: Beitrag der Medien zum politischen Erfolg. Forstwirtschaft und Naturschutz im Politikfeld Wald. 2004.

Band 18 Walter Leal Filho / Bernd Delakowitz (Hrsg.): Umweltmanagement an Hochschulen: Nachhaltigkeitsperspektiven. 2005.

Band 19 Walter Leal Filho / Volker Lüderitz / Gunther Geller (Hrsg.): Perspektiven der Ingenieurökologie in Forschung, Lehre und Praxis. 2006.

Band 20 Walter Leal Filho (ed.): Handbook of Sustainability Research. 2005.

Band 21 Walter Leal Filho / Dieter Greif / Bernd Delakowitz (eds.): Sustainable Chemistry and Biotechnology – A Contribution to Rivers Management. 2006.

Band 22 Walter Leal Filho / David Carpenter (eds.): Sustainability in the Australasian University Context. 2006.

Band 23 Walter Leal Filho / Arnolds Ubelis / Dina Berzina (eds.): Sustainable Development in the Baltic and Beyond. 2006.

Band 24 Walter Leal Filho (ed.): Innovation, Education and Communication for Sustainable Development. 2006.

Band 25 Walter Leal Filho / Mario Salomone (eds.): Innovative Approaches to Education for Sustainable Development. 2006.

Band 26 Walter Leal Filho / Franziska Mannke / Philipp Schmidt-Thomé (eds.): Information, Communication and Education on Climate Change – European Perspectives. 2007.

Band 27 Ulisses Miranda Azeiteiro / Fernando Gonçalves / Ruth Pereira / Mário Jorge Pereira / Walter Leal Filho / Fernando Morgado (eds.): Science and Environmental Education. Towards the Integration of Science Education, Experimental Science Activities and Environmental Education. 2008.

Band 28 Walter Leal Filho / Nils Brandt / Dörte Krahn / Ronald Wennersten (eds.): Conflict Resolution in Coastal Zone Management. 2008.

Band 29 Walter Leal Filho / Franziska Mannke (eds.): Interdisciplinary Aspects of Climate Change. 2009.

Band 30 Mahshid Sotoudeh: Technical Education for Sustainability. An Analysis of Needs in the 21st Century. 2009.

Band 31 Walter Leal Filho (ed.): Sustainability at Universities – Opportunities, Challenges and Trends. 2009.

Band 32 Walter Leal Filho (ed.): World Trends in Education for Sustainable Development. 2011.

Band 33 Fernando Gonçalves / Ruth Pereira / Walter Leal Filho / Ulisses Miranda Azeiteiro (eds.): Contributions to the UN Decade of Education for Sustainable Development. 2012.

Band 34 Walter Leal Filho (ed.): Sustainable Development at Universities: New Horizons. 2012.

www.peterlang.de